The
Handbook
of Highway
Engineering

February 8, 2006

Dear Customer:

Thank you for your purchase of *The Handbook of Highway Engineering*, edited by T. F. Fwa.

Addendum:

Companion software referenced in Chapters 9 and 20 may be downloaded free of charge from the CRC Press website at:

http://www.crcpress.com/e_products/downloads

To locate the files from the downloads site, enter the title: *The Handbook of Highway Engineering* or ISBN: 0849319862.

We sincerely regret any inconvenience not having this information earlier may have caused you. Please let us know if we can be of any assistance regarding this title or any other titles that Taylor & Francis publishes.

Best regards,
Taylor & Francis Group

#1986/0-8493-1986-2

The Handbook of Highway Engineering

Edited by
T. F. Fwa

Taylor & Francis
Taylor & Francis Group
Boca Raton   London   New York

A CRC title, part of the Taylor & Francis imprint, a member of the
Taylor & Francis Group, the academic division of T&F Informa plc.

Published in 2006 by
CRC Press
Taylor & Francis Group
6000 Broken Sound Parkway NW, Suite 300
Boca Raton, FL 33487-2742

International Standard Book Number-10: 0-8493-1986-2 (Hardcover)
International Standard Book Number-13: 978-0-8493-1986-0 (Hardcover)
Library of Congress Card Number 2005051480

Library of Congress Cataloging-in-Publication Data

The handbook of highway engineering / edited by T.F. Fwa.
   p. cm.
  Includes bibliographical references and index.
  ISBN 0-8493-1986-2 (alk. paper)
  1. Highway engineering--Handbooks, manuals, etc. I. Fwa, T. F.

TE151.H344 2006
625.7--dc22
                                                2005051480

Taylor & Francis Group
is the Academic Division of T&F Informa plc.

Visit the Taylor & Francis Web site at
http://www.taylorandfrancis.com

and the CRC Press Web site at
http://www.crcpress.com

This Handbook is dedicated to all highway engineers
who have contributed significantly to
human mobility and interaction

Preface

A safe and efficient land transportation system is an essential element of sustainable regional or national economy. Roads have been and continue to be the backbone of the land transportation network that provides the accessibility for the required mobility to support economic growth and promote social activities. As more and more advanced and speedy modes of transportation are developed over time, and as the economic activities of the human society grow in pace and sophistication, the roles of roads have multiplied and their importance increased. At the same time, the potential adverse impacts of road development have also grown in magnitude, especially when proper planning, design, construction or management is not carried out.

To fully exploit the benefits of highway development and minimize possible adverse influences, the study of highway engineering must expand from merely meeting the basic needs of offering safe and speedy access from one point to another, to a field of study that not only covers the structural and functional requirements of highways and city streets, but also addresses the socio-economic and environmental impacts of road network development. Traditional engineering curriculum does not adequately cover these somewhat "softer" aspects of highway engineering and the societal roles of highway engineers. It is the intention of this Handbook to provide the deserved attention to these topics by devoting Part A with five chapters on issues related to highway planning and development. Few professionals will disagree that the highway engineer today must have sufficient knowledge in the areas of highway financing, access management, environmental impacts, road safety and noise. The five chapters should provide the necessary information on the social and environmental responsibilities of a highway engineer to the undergraduate student of civil engineering and the graduate research student in highway engineering. In addition, the highway engineer and the general reader would find an in-depth up-to-date account of the trend toward privatization of highway development and financing of highway projects.

Parts B and C of the Handbook cover the more traditional core aspects of highway engineering. Part B on the functional and structural design of highways is organized into 8 chapters. The chapters offer an extensive coverage on the technical issues of highway and pavement engineering. The chapter contributors have made special efforts to explain the latest developments and comment on the future trends in their respective chapters. These chapters adequately address the undergraduate and graduate curricular needs in understanding the principles and theories of highway and pavement engineering. They also update the professional highway engineer on new concepts and ideas in the field of study. The chapter on highway materials is especially timely in view of the experiences gathered since the mid 1990s from implementation of the Superpave technology in asphalt mix design and performance grading of asphalt cement. The chapters on structural design of pavements and pavement overlay design also present the concepts of the new mechanistic-empirical design approaches advanced by the 2002 Design Guide which has yet to be adopted by AASHTO. The chapter on design of concrete pavements introduces new closed-form solutions for deflection and stress computation of multi-slab systems, and easy-to-use software is provided with the Handbook. The software is available on the CRC website.

Part C deals with construction, maintenance and management of highways. While maintenance and management of highways are of primary concern in developed countries with an established highway network, it would be unwise for developing countries to ignore them in their highway network development programs where road construction activities are taking central stage. Experience in both developing and developed countries has shown that a sustainable highway infrastructure development program must adopt a total highway management approach that takes into consideration the entire life-cycle needs of the road network. This concept is well explained in the three chapters that address highway asset management, pavement management, and bridge management, respectively. Equipment, tools and analytical techniques for condition surveys, and structural and safety performance evaluation in support of the total highway management are found in other chapters in Part C. The chapter on pavement evaluation presents useful software for non-destructive pavement structural evaluation. It contains closed-form backcalculation computer programs for both rigid and flexible pavements. The software is found on the CRC website. It should be highlighted that highway agencies have in the past decades begun to apply the concept of asset management to the development, operation and improvement of highway assets in a systematic manner. The chapter on Highway Asset Management provides the reader with the background, concepts and principles involved.

Overall, this Handbook adopts a comprehensive and integrated approach, and offers a good international coverage. It contains 22 chapters, covering the entire spectrum of highway engineering, from planning feasibility study and environmental impact assessment, to design, construction, maintenance and management. The completion of this Handbook would not have been possible without the commitment by the chapter contributors, all experts in their respective fields. The editor is most grateful to them for their efforts towards producing this meaningful Handbook, to the great benefit of the professional transportation engineers, undergraduate civil engineering students, and graduate research students specializing in highway engineering.

T. F. Fwa
Editor
Professor and Head
Department of Civil Engineering
National University of Singapore
Republic of Singapore

Editor

 **T. F. Fwa** is the head of the Department of Civil Engineering where he is also a professor and the director of the Centre for Transportation Research, National University of Singapore. He received his BEng (First Class Honors) from the then University of Singapore (now known as the National University of Singapore), MEng from the University of Waterloo, Canada, and PhD from Purdue University, USA.

Dr. Fwa's research in the last 20 years covers all aspects of highway engineering, with special emphasis in the areas of pavement design, maintenance and management, and pavement performance evaluation and testing. He has published more than 200 technical papers in journals and conference proceedings, with nearly 130 of them in leading international journals. His work has led to three patents in nondestructive pavement testing and evaluation.

A widely respected researcher, Dr. Fwa has been invited to lecture and make technical presentations in more than 15 countries, including keynote lectures at a number of international conferences and symposia. He is currently the Asia Region Editor for the *ASCE Journal of Transportation Engineering*. He also serves on the editorial board of two other international journals: *International Journal of Pavement Engineering*, and *International Journal of Road Materials and Pavement Design*. He has received a number of awards for his academic and research contributions, including 1985 Eldon J. Yoder Memorial Award by Purdue University, USA, the 1992 Katahira Award by the Road Engineering Association of Asia and Australasia, the 1992 Arthur M. Wellington Prize by the American Society of Civil Engineers, the 1995 Katahira Award by the Road Engineering Association of Asia and Australasia, the 2000 Engineering Achievement Award by the Institution of Engineers, Singapore, Innovation Award 2003 by the Ministry of Transport, Singapore, and the 2005 Frank M. Masters Transportation Engineering Award by the American Society of Civil Engineers.

Dr. Fwa is also active professionally in the area of international highway engineering. He is currently vice president of the International Society for Maintenance and Rehabilitation of Transport Infrastructure, board member of the Eastern Asia Society for Transportation Studies, and special advisor to the International Association of Traffic and Safety Sciences. He is the founding president of the Pavement Engineering Society (Singapore) and the Intelligent Transportation Society (Singapore).

In the last 15 years, Dr. Fwa has devoted a great deal of time promoting highway engineering professional activities and developments in Asia Pacific. He has been responsible for bringing major international conferences and events in highway engineering to the region. This has been a benefit to the large number of professionals in developing countries. He heads the Executive Committee that manages the International Conference on Road and Airfield Pavement Technology and the Asia Pacific Conference on Transportation and the Environment. These are the two conference series that serve to raise professional awareness in the Asia Pacific region of the new knowledge and technologies in the area of highway engineering, and the importance of sustainable transportation development.

Contributors

M.A. Aziz
Department of Civil Engineering
National University of Singapore
Republic of Singapore
E-mail: creaziz@nus.edu.sg

John W. Bull
School of Civil Engineering and Geosciences
University of Newcastle upon Tyne
Newcastle upon Tyne, U.K.
E-mail: john.bull@ncl.ac.uk

R.L. Cheu
Department of Civil Engineering
National University of Singapore
Republic of Singapore
E-mail: cvecrl@nus.edu.sg

Anthony T.H. Chin
Department of Economics
National University of Singapore
Republic of Singapore
E-mail: anthonychin@nus.edu.sg

Kieran Feighan
PMS Pavement Management Services Ltd.
Dublin, Ireland
E-mail: kfeighan@iol.ie

T.F. Fwa
Department of Civil Engineering
National University of Singapore
Republic of Singapore
E-mail: cvefwatf@nus.edu.sg

K.N. Gunalan
Parsons Brinckerhoff Quade & Douglas, Inc.
Murray, UT, U.S.A.
E-mail: Gunalan@plsworld.com

Pannapa Herabat
School of Civil Engineering
Asian Institute of Technology
Pathumthani, Thailand
E-mail: pannapa@ait.ac.th

Zahidul Hoque
Pavement Specialist
Melbourne, Victoria, Australia
E-mail: zahidul_hoque@yahoo.com.au

Yi Jiang
Purdue University
West Lafayette, IN, U.S.A.
E-mail: jiang2@purdue.edu

Ian Johnston
Monash University Accident Research Centre
Monash University, Clayton Campus
Victoria, Australia
E-mail: Ian-johnston@general.monash.edu.au

Laycee L. Kolkman
Jacobs Engineering
Las Vegas, NV, U.S.A.
E-mail: laycee. kolkman@jacobs.com

Khaled Ksaibati
Wyoming Technical Transfer Center
University of Wyoming
Laramie, WY, U.S.A.
E-mail: khaled@uwyo.edu

Arun Kumar
RMIT University
Melbourne, Australia
E-mail: arun.kumar@rmit.edu.au;
        arunkumar@telstra.com.au

Michael S. Mamlouk
Department of Civil and Environmental Engineering
Arizona State University
Tempe, AZ, U.S.A.
E-mail: mamlouk@asu.edu

Sue McNeil
Urban Transportation Center
University of Illinois at Chicago
Chicago, IL, U.S.A.
E-mail: mcneil@uic.edu

K. Raguraman
Department of Geography
National University of Singapore
Republic of Singapore
E-mail: geokrk@nus.edu.sg

Stephen Samuels
TEF Consulting (Traffic, Environmental and
    Forensic Engineers)
Sydney, Australia
and
School of Civil and Environmental Engineering
University of New South Wales
Cronulla, NSW, Australia
E-mail: s.s@tefconsult.com. au

Kumares C. Sinha
School of Civil Engineering,
Purdue University
West Lafayette, IN, U.S.A.
E-mail: sinha@ecn.purdue.edu

Weng On Tam
Consultant
Austin, TX, U.S.A.
E-mail: wengtam@gmail.com

Mang Tia
Department of Civil & Coastal Engineering
University of Florida
Gainesville, FL, U.S.A.
E-mail: tia@ce.ufl.edu

Waheed Uddin
Department of Civil Engineering
Center for Advanced Infrastructure Technology
University of Mississippi
University, MS, U.S.A.
E-mail: cvuddin@olemiss.edu

Ian van Wijk
Africon Engineering International
Pretoria, South Africa
E-mail: ianvw@africon.co.za

Derek Walker
RMIT University, City Campus
Graduate School of Business
Melbourne, Victoria, Australia
E-mail: derek.walker@rmit.edu.au

Liu Wei
Department of Civil Engineering
National University of Singapore
Republic of Singapore
E-mail: cvelw@nus.edu.sg

Contents

Part B   Functional and Structural Design of Highways

Part C   Construction, Maintenance and Management of Highways

Part A

Highway Planning and Development Issues

1

Financing Highways

Anthony T.H. Chin
*National University of Singapore
Republic of Singapore*

1.1 Introduction

Countries in the developed world are faced with high maintenance cost of aging transportation highway s and replacing them in the face of budget deficits and cuts are no longer automatic options. Further, funds for transport highways have to compete with compelling priorities. In developing and growing economies, the demand made on highways with the growth in motor vehicles and accompanying economic growth have ensured the insatiable demand for better, safer, less congested highways and increase in personal travel. Growth in commerce also imposes growing demands on the highway network. This and many other public sector programs such as, educational institutions, housing and defense have largely drawn from budgetary surpluses in the past.[1]

---

[1]Most highway programs are based on "pay-as-you go", i.e., paying for construction, maintenance and administration as money became available from user fees, bonds and government grants. In many countries, even in good times revenues may not be enough. Increased public demand for transportation services continue to strain existing highway and drained financial resources.

However, several circumstances have come-up to influence change in the way public highways are financed and maintained. Firstly, the structural changes in the economy may not lead to large budget surpluses as happened in the past as the dynamics of globalization leads to shifts in foreign direct investments. Secondly, the shrinking tax base as experienced in matured economy. Thirdly, priorities in public sector expenditure might focus on the provision of services and other areas such as the aged.

It is also important to note that highways have "dual-nature". They have some characteristics of public goods and some of private goods. From the public good perspective, beyond the direct users (passengers and freight), highways provide benefits in many instances for by reducing the cost of transportation and distribution of goods. In the past the cost of collecting fees for highway use was high, politically unacceptable and not always practical. The government over the years has played a large role in the provision of highways.

Highways have some characteristics of private goods as well. Limited-access highways can exclude motorists who are not willing and able to pay tolls. When highways are congested, one motorist's use of the highway imposes costs in the form of delay on other users. Those mixed-goods characteristics of highways, changing economic environment and the advances in highway pricing technology should motivate and explain the need for a new approach to how highways are managed, financed and maintained (Estache, 1999; Estache et al., 2000). This chapter takes the perspective of looking at user charges as a means of achieving allocative efficiency in highway usage and examines the conditions for efficient private participation in the financing and provision of highway services.[2]

1.1.1   Towards the User Pay Principle

The five-user fee principles are: equity, simplicity, public good, value and flexibility. If one is concerned about efficiency and fairness, the user pay principle charged at levels corresponding to the economic benefits received from public goods or services (or the costs those are imposed on society) and this should be the guiding principle in charging for highway use.[3]

Highway pricing in theory serves three distinct purposes:

   (a) *Financial*: means to collect funds to pay for provision of highway services
   (b) *Efficiency*: to discourage over utilization of highway space
   (c) *Equity*: to reduce traffic congestion and increase mobility of efficient modes.

The basic guide lines involved in designing a highway pricing policy include marginal cost pricing, ability to pay principle, net-benefit principle and full cost pricing. These guidelines are put into practice through a variety of pricing/charging methods (such as distance traveled pricing, two-part fee, time-of-day pricing, level of congestion pricing etc.). The efficiency and equity arguments of highway pricing are well established and are potentially powerful means for generating revenues. The challenge lies in establishing and accommodating various users and beneficiaries to improve the effectiveness of pricing.

Should all users be charged? Which pricing methods are most realistic and effective in achieving efficient use of highway space? If other financial objectives are important such as covering operating costs, what are the implications for equity and efficiency of use of resources? What are the appropriate mechanisms to implement the effective pricing schemes? What practical methods are available for allocation of costs among user groups in a complex highway network? The overall objective must be *to ensure the use of highway space in an effective, efficient and equitable manner, consistent with the social, economic and environmental needs of present and future generations.* Highway pricing is an important element of the demand management strategy.

---

[2]Economic efficiency means allocating scarce resources to uses that are of the highest value to society as a whole.

[3]One other area where the principle can be applied is that of fuel excise tax. Petrol and other fossil fuels affect the environment. Restructuring the fuel excise tax based on user/polluter pay principle can raise the same revenue. The restructured tax would include other sources of pollutants and ensure that polluting activities bear a more appropriate charge for the use of air, water and land.

However, the philosophical distinction among allocative efficiency, cost recovery and revenue generation is not always easy to make. Nevertheless, a few principles are fundamental, and when applied to assess the purpose of the various proposals for user fees, can help to justify the practical solutions. The first is equity. Unless there is equitable access to highways, some users will not have freedom of mobility. User charges must not present the risk that freedom will be curtailed for those who are unable to pay.

There are two time-honored principles that should guide us. First, the *benefit principle* suggests that people should pay according to the benefits received from highway usage. Anyone who receives more benefits should pay more. For example, drivers who buy a lot of petrol pay more fuel-taxes. But they also drive a lot and reap much benefits from the system of highways. If some of the money is diverted to support mass transit, they may benefit through reduced congestion on the highways. Electronic Road Pricing (ERP) is an application of user charge. This is not a tax, but rather prices charged to motorists for the use of highway services. Drivers are charged directly for the services they consume, which again link payments to the benefits received.

The second time-honored principle is the *ability-to-pay principle*. People should pay charges in accordance with their ability to pay. People with a greater ability should pay more than those with lesser. Two concepts related to this principle are *horizontal equity* and *vertical equity*. Horizontal equity is said to exist when people with the same ability to pay charges pay the same amount of taxes. Problems arise when we try to apply this idea, because it is not always easy to compare people in terms of their abilities to pay. Income is often considered to be the best proxy as an indicator for assessing the ability to pay.[4]

The idea of vertical equity is even more difficult to apply. Vertical equity is said to exist when people with different abilities to pay in fact do pay appropriately different charges. The additional difficulty here (besides the already discussed problem of measuring the ability to pay) is to determine the degree to which taxes should differ when ability to pay differs.[5]

1.2   Financing Structure and Sources

Traditionally given the *public goods* nature of highways the financing structure and sources of finance are closely related to the role of government. Many of these attempt to minimize risk to the government (Heggie and Vikers, 1998; Irwin et al., 1998; Chin, 2003). These are discussed briefly below:

(i)   Toll financing
(ii)   Equity financing
(iii) Subordinated loans
(iv)   Senior commercial bank loans and debt securities
(v)   Institutional investors/highway investment funds
(vi)   Initial Public Offerings (IPO)
(vii) Asset securitization
(viii) The portfolio approach

---

[4]But consider two individuals, each with a SGD50,000 annual incomes. One of them has a portfolio of stocks and bonds worth SGD1 million, while the other has little or no wealth. Do equal incomes mean equal ability to pay? If not, how do we enter wealth into the calculation? Next consider two married couples, each with a SGD50,000 annual incomes. One couple has three dependent children, while the other has none. Do equal incomes mean equal ability to pay? If not, how do we factor the number of dependants into the calculation? We are far from exhausting the possible complexities, but finally consider two individuals, say both medical doctors, who have the *capacity* to earn SGD100,000 per year. One of them works hard and earns SGD100,000, while the other one is something of a slacker, doing half as much work, playing lots of golf, and earning only SGD50,000. Should we treat these two as having the same ability to pay, or should we regard the one as having twice as much ability to pay as the other? (Should we reward sloth?)

[5]If person A has twice the ability to pay of person B, should A pay twice the amount of charges that B pays? If you say yes, you are in favor of a proportional ERP rate structure. If you say A should pay more than twice the amount of charges, you are in favor of a progressive structure. If you say A should pay less than twice the amount of charges, you are in favor of a regressive structure. Implementation is impossible.

(ix) Pinpoint equity with indexed highway bond issues
 (x) The option concept
(xi) Value capture.

1.2.1   Advantages and Disadvantages of Toll Financing

The advantages and disadvantages of toll financing of highways as compared to financing from other sources such as fuel taxes or fiscal instruments depend on the nature of the economy. Toll highways are predominant in Northern Europe, North America, and Australia. The decision of whether to toll a particular highway or not is important where traffic levels are relatively low or generated traffic has yet to be realized in the immediate future.

The decision should be justified by economic and financial analyses since the perceived objectives (e.g., raising additional revenue, fairness in terms of the user-pays principle, optimal pricing and resource allocation) are seldom achieved. Also, the costs of establishing a toll system, the collection costs, and the diversion of tolls by collectors can be high. For example, the case studies and other evidence indicate that additional construction costs can range between 2 to 8% of initial costs and that operating expenses can range between 5 to 20% of toll revenue, depending on whether an open or closed tolling system is employed.

Toll highways in France has resulted in increased construction costs of about 10% and increased operating cost equal to 10 to 12% of revenues, which are considered comparable or lower than the collection costs and economic distortions of alternative revenue sources. Tolls have not resulted in misallocation of traffic between toll highways and parallel untolled highways. An estimated 6 to 7% of potential toll highway users diverted to parallel routes as a result of tolls. The economic costs of raising revenue by tolling in Vietnam suggest (including capital costs, collection costs, leakage and traffic diversion/suppression cost) should be lower than the cost of raising revenue by alternative means, and these economic costs should not be higher than 15 to 20% in the case of captive traffic. Leakage in one African country led to the introduction of a fuel tax because the collection rate through tolls was only 60%.

1.2.2   Equity Financing

It is relatively easy to attract domestic capital of both debt and equity for smaller projects, if the capital cost is less than $100 million. Moreover, it is very beneficial for a toll highway project to obtain domestic financing to avoid the exchange rate risk between local currency toll revenues and foreign currency debt. However, in many countries, local capital markets are not sufficiently developed to provide the long-term capital required for toll highway projects. The financial crisis in late 1990s in Asia followed by the recession has worsened the situation.

Malaysia has been successful in domestic financing of the North–South Expressway, a mega-project which cost a total capital cost of $ 3.192 billion. Out of this $755 million (25% of the total capital cost) comprised of shareholders' equity and convertible preference shares issued to the contractors, industrial groups, and institutional investors in Malaysia. The project was financed entirely on domestic markets; a generous government-support package and the capacity of domestic institutional investors, to take large preference share issues, played an important role in the successful equity financing of the project.

Thailand's Second Stage Expressway (SSE) and China's Guangzhou–Shenzhen Super Highway involved foreign equity and debt financing. Good project structure attracts foreign equity and foreign commercial bank loans. Although both of these projects faced serious problems, the SSE structure was backed up by the equity participation of the Asian Development Bank (ADB), major commercial banks in Thailand, and the Royal Property Bureau of Thailand, together with revenue sharing with the First Stage Expressway. The Guangzhou–Shenzhen Super Highway Project secured a firm repayment guarantee of bank loans by GITIC and obtained a government support provision including the acquisition of all necessary land at no cost and commercial development rights at interchanges.

Another important issue is the treatment of equity in a contractor-driven project in which the contractor tends to limit the amount of equity, and to sell down its equity as much and as early as possible after the completion of construction. Lenders generally limit such actions in the loan agreement with the concessionaire, but it is always an issue as to how much equity should be injected in the beginning and how long and to what extent the concessionaire should hold it.

1.2.3   Subordinated Loans

There are two important roles for subordinated loans:

(1) to fill the gap between the equity and the senior loans in the original finance structure, and
(2) to provide stand-by financial support in case of revenue shortfalls and cost overruns.

They may address to the difficulty of procuring equity (due to equity's slowness in recouping investment through dividend payments) by providing a stable cash payment stream with a higher interest rate than senior debts from the beginning years of the project. Because of the subordinate nature of repayment to senior debts, second to the equity injection, government's support and sponsors' support in the form of subordinated loans should be more acceptable to the senior debt providers than ordinary loans. This approach, while common, should be applied carefully since an excessive use of subordinated loans may considerably increase the capital costs and impair the sponsor's commitment to the project.

1.2.4   Senior Commercial Bank Loans and Debt Securities

Procurement of long-term bank loans for a privately financed toll highway projects is a critical issue in developing and transitioning economies. The longest tenure that a toll highway project company can obtain in a commercial bank loan in the East Asian countries is about 5 years, which is far too short to recoup the investment, whereas in many developed countries such as in the United States and United Kingdom, the tenure of commercial bank loans may extend to 15 to 30 years, i.e., matching the concession period.

To address this issue, various measures have been implemented in toll highway projects in developing countries. A straightforward, but difficult solution is to have a sufficiently sound contract structure with a hedge mechanism for foreign exchange risk to attract long-term off-shore debts. Alternative solutions that have been tried include:

(i) long-term loans of government controlled banks,
(ii) shareholders loans from an off-shore parent company that raises funds on off-shore capital markets,
(iii) domestic bond issues underwritten by government controlled institutional investors,
(iv) credit enhancement with respect to domestic fund raising and direct loans from donor agencies,
(v) securitization of existing toll highways, and
(vi) credit enhancement through revenue sharing with existing facilities.

There are many examples where the approaches set out above have been successfully applied. Indonesia used long-term loans of government controlled banks in many of their toll highway projects. China has used shareholder loans from an off-shore parent company that raises funds on off-shore capital market, as well as asset securitization of existing toll highways. The North–South Expressway in Malaysia and the M1/M15 in Hungary adopted domestic bond issues underwritten by government controlled institutional investors. The M1/M15 and M5, the Linha Amarela in Brazil, and the Cali–Candelaria–Florida toll highway in Colombia have adopted credit enhancement of domestic fund raising and direct loans from donor agencies, and credit enhancement through revenue sharing with existing facilities.

1.2.5   Institutional Investors/Highway Investment Funds

Institutional investors can be a good source of financing for toll highway projects since the long-term maturity of their funds matches the duration of a toll highway concession. The Employees Provident Fund has invested in Malaysia's North–South Expressway and insurance companies in Hungary have invested in M1/M15. However, since institutional investors in developing countries are not active in the highway sector in general, foreign institutional investors from developed countries can play an important role in filling the gap. Institutional investors, especially insurance companies and pension funds in the United States, have been actively pursuing investment opportunities in privately financed highway projects in Latin America and Asia. They have invested in toll highway projects directly, through various investment funds (the Asian Highway Fund [AIF], and the Asian Highway Development Company, Ltd [AIDEC]), and have purchased debt securities such as *144a* bonds in private placement (this is a kind of global bond that is regulated under [the United States] Securities and Exchange Commission Rule 144a). The procedure for issuance and underwriting was simplified in 1990 and limited only to investors termed "Qualified Institutional Buyers (QIB)", who are professional institutional investors such as insurance companies and pension funds.

1.2.6 Initial Public Offerings

An IPO of a single asset company with a Build Operate Transfer (BOT) arrangement can be difficult as the duration of future cash flow is limited by the fixed concession period and the enterprise is affected to a great extent by general stock market sentiment at the time of IPO. On the other hand, an IPO based on multiple assets with a portfolio of stable cash-generating toll highway projects may become an appropriate solution to fund raising issues in developing countries. China has gone into IPO, often for multi-asset companies; in many instances both expressway and holding companies listed their shares on the Hong Kong and Shenzhen Stock Exchanges. Examples include: Anhui Expressway (Hong Kong, 11/96), Guandong A Share (Shenzhen, 1/96), Guangdong B Share (Shenzhen, 8/96), Jiangsu Expressway (Hong Kong, 6/97), Sichuan Expressway (Hong Kong, 10/97), Zhejiang Expressway (Hong Kong, 5/97), Shenzhen Expressway (Hong Kong, 3/97); Holding Companies: Cheung Kong Highway (Hong Kong, 7/96), New World Highway (Hong Kong, 10/95), and Highway King Highway (Hong Kong, 6/96). In Indonesia Jasa Marga, the public toll highway company, planned an IPO but it was postponed due to the Asian financial crisis.

1.2.7 Asset Securitization

One innovative approach is the leveraging of existing highway assets to raise new funds in capital markets. This approach can be attractive to private investors, since they need to take only limited construction/completion risks and the transactions offer the prospect of high returns. The approach is also attractive to governments, since it permits them to obtain additional financing with relative ease, including for financially less attractive but economically viable projects.

China, the pioneer of this approach, by securitization of existing highway assets, including highways financed with World Bank assistance, has been able to raise large sums of additional capital from foreign investors. Major developments with respect to asset-based capital markets toll highway financing in China have included the following:

(i) the raising of $100 million in 1994 by Sichuan Province through the private placement of equity shares in off-shore markets to finance the development of the 90-km Chengdu–Mianyiang Expressway;

(ii) an equity offering of B shares on the Shenzhen Stock Exchange by the Guangdong Provincial Expressway Company in 1996;

(iii) completion of a $200 million Eurobond issue by Zhuhai Highway Company Ltd of Guangdong Province in 1996 and

(iv) the listing of at least nine China-related highway stocks on the Hong Kong Stock Exchange by 1997, including Chueng Kong Highway and Highway King.

However, there are some concerns with the asset securitization approach such as the possibility of over-leveraging the asset at the expense of the obligation to repay the original loan. Another problem using the capital markets is that the timing and the volume of fund raising is inherently affected to a great extent by prevailing market sentiment. Investors are very sensitive to the political risk (seen recently in East Asia and the Russian Federation) as well as the features of alternative investment opportunities such as the coupon rate for various bonds, the profitability of stock markets, and ups and downs of real estate markets. Therefore, financing capital–market–financing should not be relied upon as a perpetual, stable source of funds.

1.2.8   Portfolio Approach

The portfolio approach adopted by Road King Infrastructure Limited (RKI) of Hong Kong may provide a solution for developing countries where the procurement of foreign debt is difficult and both domestic debt and equity for the financing of the country's toll highway development are scarce. RKI leverages the creditworthiness and the track record of their projects in China, thereby diversifying its investment portfolio into various regions (high growth centers) and risk profiles, and attaining favorable income distribution and a minimum income undertaking from local partners for most projects.

RKI's portfolio approach contains very specific elements such as: (i) its location in the Hong Kong SAR, a quality international financial center; (ii) its parent company's credibility, engineering know-how, and long involvement in the Chinese market; and (iii) its ability to diversify geographically into its existing assets throughout China. RKI's toll highway projects in China are financed through a combination of IPO, the cash flow from the existing portfolio, a note issue-2 and a transferable loan certificate (TLC) issue-3. At its IPO, the company had interests in ten highways in China, of which eight were in operation and six at different stages of negotiation. The company's interests in these toll highway projects were held through wholly owned subsidiaries, which, together with the relevant Chinese joint venture partners, established various cooperative joint venture (CJV) companies for investment in different projects. There are no green-field toll highway projects in the company's portfolio. As of June 1998, its toll highway assets were diversified in eight provinces in China, totaling 974.6 km of highways.

Highway Management Group Limited (HMG) in the United Kingdom has adopted another form of the portfolio approach to toll highway development. Two concession companies, Highway Management Services (Peterborough) Limited for the A1(M) and Highway Management Services (Gloucester) Limited for the A419/A417 — entered into Design, Build, Finance & Operate (DBFO) contracts with the Secretary of State for Transport to widen and improve the highways and to operate and maintain them for 30 years. The financing for the two highway projects was raised through Highway Management ConsolidatesPlc, a newly created special purpose financing entity, 100% controlled by HMG, which provides funds to the individual RMS companies through back-to-back onloans. In March 1996, Lehman Brothers and SBC Warburg underwrote a £165 million, 25-year, fixed rate bond issue to partially fund the two projects, and arrange a £111million, 25-year European Investment Bank (EIB) loan facility to provide the remainder of the required senior debt financing.

This structure allowed cross-application of dividends so each project could support the other. It enabled projected interest coverage levels to be tighter than they would have been otherwise, thus lowering the cost of financing. Combining two different highways diversified the lenders' risks. It eliminated the need for two separate financing, minimized the duplication of documentation and negotiation with financing parties, and created a public bond offering large enough to be liquid and to meet demand at the long-end of the sterling bond-market. Although the issue here was not the difficulty of fund raising as in the previous example, the approach gave HMG a better risk-profile for their business as a whole by having two different DBFO highway projects and formed a base for evolving into a large toll-highway operating company.

Distinctive features of the HMG case are as follows:

- first bond offering under the United Kingdom's Private Finance Initiative;
- large magnitude compared to other sterling Eurobond offerings;
- first monoline-guaranteed project–finance–bond in the United Kingdom;
- bond financing prior to construction;

- acceptance of deferred annuity structure by United Kingdom investors;
- dual-project structure with limited cross-collateralization;
- shadow-tolls as a basis for project revenue; and
- joint and several guarantees by sponsors with varying credit capacity.

Financing the two HMG projects together made sense because it created portfolio diversification and economies of scale. This kind of approach may provide a foundation for a company to grow into a large operating company to perform more pooled financing, and eventually raise funds on its own based on the financial strength of its underlying projects.

1.2.9 Pinpoint Equity with an Indexed Highway Bond Issue

Pinpoint equity (high debt-to-equity ratio) coupled with inflation-indexed bond issues may be used to relieve investors of the problem of slow returns on their investment through dividends. The issues here are: (i) the need for the investor to flexibly recoup its investment without hindering the opportunity of private financing; (ii) lowering the capital cost to achieve a more affordable toll rate; (iii) the need to provide the investor with incentives that lead to higher returns; and (iv) the need of the investor for liquid investments.

This concept was created in United Kingdom. The project company of the Second Severn Crossing, Severn Plc, was formed with the minimum allowable equity capital for a private limited company, £100,000, of which the ordinary stock amounted to £50,000 and the preferred shares amounted to £50,000. The debt component comprised £131 million in indexed-linked debt, £150 million in EIB loans, with £150 million in a standby letter of credit, and a £190 million of floating rate bank loan. Although the project company has succeeded in the operation of the existing Severn Crossing, mitigating the start-up risk of the new project, the debt–equity ratio of the newly constructed portion of the project is only about 0.02%. This pinpoint equity approach frees investors from the problem of slow recoupment through dividends and also may lower the cost of capital. This in turn lowers the debt–equity ratio to 0.02% = £0.1million/(£131 m + £150 m + £190 m) required toll rate to facilitate an earlier transfer of the bridge back to the government. In the case of Second Severn Crossing project, this approach was coupled with an RPI-indexed bond-issue. The bond was listed so that investors were able to access immediate liquidity in order to address to the drawback of the project financing approach of tying up investor's capital over the long-term. Although this is a good practice for addressing these issues, it requires a very mature financial market to succeed.

1.2.10 Option Concept

The option concept is applied in many aspects of toll highway financing, e.g., convertible preference shares and bond issues. It has also been used to provide liquidity for shares held by the contractors and the sponsor companies. It gives flexibility to the financing structure and may broaden the horizon of fund providers for privately financed toll highway projects. The North–South Expressway (Malaysia) and the M2 motorway (Australia) projects involved application of the concept to provide liquidity for shares held by the contractors and the sponsor companies.

1.2.11 Value Capture

The value-capture concept is an approach by which the increase in real estate values created by a transport project, such as a toll highway, are "captured" to help pay for the transport highway. The approach is risky because it relies on favorable trends in the real estate market. The concept has been planned or actually applied in a number of contexts, for example, the Guangzhou–Shenzhen Superhighway in China, the Hopewell Bangkok Elevated Highway and Track System in Thailand, and the Malaysia–Singapore Second Crossing in Malaysia, with the experience indicating that over-dependence on real estate investment earnings to structure a transport concession is risky given the volatility of real estate markets.

1.3 Role of Donor Agencies

Aid agencies can address to a variety of the financing, institutional, and regulatory issues of toll highway development through: (i) long-term loans, (ii) interim financing, (iii) credit enhancement with respect to political risks, (iv) the arrangement of finance, (v) the provision of guarantees, and (vi) technical assistance and human resource development.

1.3.1 Long-Term Loans

One way in which donor agencies can provide assistance for toll highway development is long-term financing. In many developing and transitioning economies, even ones that have large capital markets, it is difficult for domestic capital to cover the costs of needed highway projects mainly because of the lack of long-term financing tools.

The Inter-American Development Bank (IDB) provided long-term loans to Brazil and Colombia. The National Economic and Social Development Bank (BNDES) is the sole provider of long-term financing in Brazil, with financing typically for 10 years at the Brazilian long-term Interest rate plus a spread of 3 to 4%. Access to long-term international project finance debt may be enhanced through the involvement of a multilateral institution in a project. This is because loans under the umbrella of institutions such as the International Finance Corporation (IFC), regional development banks, or the Export–Import (EXIM) Bank of Japan are exempt from Bank of International Settlements provisioning requirements and carry less sovereign risk. IDB's participation encouraged the international commercial banks to participate in the first green-field toll highway project in Brazil, Linha Amarela (the Yellow Line Highway), which reached financial closure in 1996; IDB contributed a USD14 million, 10-year loan at LIBOR plus 375 basis points to this project. IDB also played a critical financial role in the Cali–Candelaria–Florida toll highway project in Colombia, where long-term financing is very limited and expensive; IDB provided a 10-year loan of $10 million, at LIBOR plus 325 basis points and helped mobilize local long-term financing at reasonable cost.

1.3.2 Interim Financing

Donor agencies can also provide *interim* financing. The European Bank for Reconstruction and Development (EBRD) supported the Budapest Orbital Motorway Project (MO) with 2 billion Hungarian forint ($27 million at the time) in *interim* financing for the country's Highway Fund. This freed up monies for the M1/M15 motorway when preparatory work on the M1/M15 project, an important pilot project in EBRD's view, was stalled due to a lack of reserves in the fund.

1.3.3 Credit Enhancement with Respect to Political Risks

Donor agencies provide various measures of credit enhancement with respect to political risks, including political risk insurance, Export Credit Agency (ECA) cofinancing and guarantees, and Partial Credit and Risk Guarantees. In Colombia, Partial Risk Guarantees (PRG) provided by the World Bank was an option given to bidders. This would protect project lenders or bond holders against debt-service default due to the government's inability to meet its payment obligations as a result of the Ministry of Transport not authorizing toll adjustments agreed in the concession contract. Political *forcemajeure*, or changes in the law adversely affecting the project's ability to service its debt and leading to default. Although the successful bidder declined the Bank's PRG in its bid, the offering of PRG assured increased competition with its attendant benefits for the government.

1.3.4 Arranger of Finance

A donor agency can play the role of financial arranger. The financing package for the M1/M15 motor-way project in Hungary was co-arranged by EBRD and *Banque Nationale de Paris* in December 1993. The syndication loan for the M5 toll motor-way project was also arranged by EBRD, *Commerzbank,* and ING Bank. In the case of the M1/M15, the EBRD issued 5- and 12-year bonds and provided a full guarantee of the bonds launched by the project sponsor.

1.3.5   Provision of Guarantees

Another role played by donor agencies is to guarantee loans. For example, for the M5 motor-way in Hungary, the EBRD offered guarantees to participating banks for the final repayment of their loans.

1.3.6   Technical Assistance and Human Resource Development

Donor agencies, particularly the World Bank, can add value by providing expertise and contributing to training in various areas related to toll highway development. For example, the Bank is now actively assisting China in the development of a policy framework conducive to attracting private capital for new toll highways. The new BOT framework is to be put forward as a provisional decree on highway projects with private investment with approval from the State Council pending. The State Development Planning Commission has designated five pilot highway projects to implement the new BOT policy framework, including the Wuhan Junshan Yangtze River Bridge in Hubei Province and a toll expressway in Guangdong Province. The World Bank provided grant funds to conduct a commercial feasibility study of the Yangtze River Bridge, with the aim of enhancing the "bankability" of the project by the private sector. The World Bank is also funding studies to assess the development of public toll highway authorities in Hunan and Hubei Provinces, which would give the provinces the capacity to manage large expressway systems, supplementing the planning, financing, and construction of highways by provincial communications departments.

Support for master planning, feasibility studies, and concession contract have been provided as part of technical assistance. Japan International Cooperation Agency (JICA) supported the preparation of highway network master plans in Malaysia, Thailand, and Philippines, Chile, utilized World Bank support for feasibility studies and the preparation of basic engineering for projects identified by the Ministry of Public Works as candidates for concessions as well as consultant services for the review of the regulatory framework. EBRD was instrumental in improving the drafting of certain parts of the concession contract in Hungary (e.g., relating to *force majeure* and consequences of government action affecting operation). The United States Agency for International Development (USAID) assisted in the establishment of a BOT Center in the Philippines, with the responsibilities of information dissemination, training, and provision of assistance for the drafting of bid documents and conducting the bidding.. The center provides technical assistance to various government agencies involved with private sector participation, but has no regulatory role or real power outside offering suggestions.

1.4   An Analysis of Highway Project Financing

1.4.1   Developed Countries

1.4.1.1   Financing Structure

Highway financing in developed countries financed their projects primarily through equity financing, issuing of various kinds of bonds, securities and debentures. In addition, loans from both international banks and local banks were made for debt financing. All necessary calculations and analysis on cost and benefits, economic and financial internal rate of return were obtained to evaluate the financial viability and benefits such as time and operating savings from the project. Sensitivity Analysis was also conducted to test the robustness of the project against changes in prices and interest rates. There was no government or federal funding involved in these projects except for projects in Canada where most projects were subsidized or funded by the government. However, no complacency was allowed as the private developers were obliged to payback the government from toll revenue collected. Contingency funds were allocated for any possible unanticipated costs escalation. Insurance coverage for loans undertaken was secured from commercial banks and other trading and financial institutes. All the above sources of finance were obtained by the relevant private developers' consortium team.

1.4.1.2   Institutional Structure

All projects were announced by the respective government or ministerial department to private developers for competitive bidding. Legislation set up relevant departments and institutes to monitor and cooperate with the private developer independently of the government. Guidelines for prequalification of these proposals were also implemented by government bodies for efficient and better short listing of proposals obtained. Strict prerequisites such as the financial performance and relevant domain knowledge and expertise for the construction of the project were also required to be submitted as reports to the respective government departments. The successful bidder for the project will be the one that has the required resources and accumulated past experience with related projects and a proposal that best fulfils the government's priorities and objectives at a reasonable cost and appropriate risks. The consortium thus will consist of several different companies or corporations that specialize in all required aspects such as construction, engineering, government regulations etc in the project. Both the government department and private entity would have to be engaged in obtaining public support in favor of the project. Appropriate measures were undertaken to preserve or sometimes rebuild natural habitats such as the Dulles Greenway project. Due care was undertaken to ensure minimum disruption to the existing residents in the community.

1.4.1.3   Why Did Some Projects Fail?

Despite all the necessary planning and implementation measures taken by both the government departments and the private entities, to ensure successful construction and eventual sustainability of these projects, some projects encountered problems. Others managed to overcome the challenges without incurring additional costs. We shall look at some critical factors that caused some projects to fail.

1.4.1.3.1   Public Relations

Proposed highway projects such as State Road 522 Corridor Improvements (Arizona, U.S.A.) and State Road 18 (Washington, U.S.A.) failed to be implemented because of strong opposition from the public to these toll highways. Sixteen thousand citizens signed petitions against State Road 522 and a majority of the city council members rejected the project. Polls indicate that residents will not travel through the highway if toll were imposed. Both the state officials and private developers were accused of attempting to extract "extra dollars" from their pockets from the toll. In Arizona, the highway project VUE 2000 was also defeated due to poor public relations. In addition both the ADOT and developers also did not make any concerted effort to obtain the necessary support. Other states realizing the situation in Washington and Arizona, made developers submit proposals to local communities after submitting to the government before entering into any form of negotiations with the government.

1.4.1.3.2   Realistic Budgets and Appropriate Design

The Channel Tunnel although received good response in terms of its toll revenue collection. The initial appraisal budget was unrealistic resulting in a massive increase in actual cost of financing. However, it was not enough to cover the Channel Tunnel's debt service, resulting in massive losses and needed to arrange for debt–equity swap although it was at the brink of bankruptcy. The tunnel could not be closed or sold, thus such arrangements had to be made.

1.4.1.3.3   Lack of Cooperation

There was lack of cooperation between British and French governments for the Channel Tunnel project and the reluctance to intercede with the activities for formulation of safety regulations. Had these factors been absent, millions of dollars would have been saved.

1.4.1.3.4   Inclusion of Qualified Expertise

The exclusion of transportation experts in the development of the Channel Tunnel project resulted in delays and costs escalation.

1.4.1.3.5 Trust and Initiative

Providers of finance distrusted contractors of the Channel Tunnel due to repeated cost overruns in the construction. Contractors in Britain therefore were considered as inefficient by financial institutes and this caused disputes between the two parties. Lack of initiative from the Channel Tunnel Group Limited/France-Manche S.A. (CTG/FM) in bridging the gap of distrust and misunderstanding was apparent during the early stages of the Channel Tunnel. In contrast, the State Road 91- highway project in United States had a very enthusiastic developer, Peter Kiewit, who obtained alternative sources of finance for his project on his own initiative without which, the project would not have been implemented.

1.4.1.3.6 Other Factors

Sensitive environmental and social issues, such as in the Dulles Greenway, rebuilding of the wetlands were undertaken by the developer. Also, responding to antinoise and air pollution measures quickly enhanced the trustworthiness of the developer. Conservative and prudent forecasts of traffic volume in these highways should be exercised to prevent over-optimistic estimates in accordance with forecasts in economic outlook of the country or region.

1.4.2   Developing Economies

1.4.2.1   Financing Structure

Highway projects were financed mainly from development banks like ADB, WORLD BANK in the form of loans. Finances also came from domestic development banks like in China, the China Development Bank (CDB) and the government themselves. Development banks financed all foreign exchange funds and part of the costs in local currency, except in China where both the government and ADB shared the costs between foreign exchange and local currency. Similarly, interest payments were financed mostly by these banks (ADB) in terms of foreign exchange. Contingency funds and other forms of insurance financing were also mostly provided by the banks; however respective government had to provide counterpart funds on their part. No or little private entity financing was raised for these projects. All the necessary financial and economic risks assessment was done mostly by the government agencies, in addition to evaluation reports that had to be submitted to ADB or other related development banks financing the project.

1.4.2.2   Planning and Implementation

All the projects undertaken were executed by existing government bodies or departments with a higher ministry as the overall in-charge. Government officials were given relevant training in all aspects of management and construction overseas and in China before starting on the project. Mostly international consultants were hired to join the project supervision and advisory team, except in China. The government also announced the projects for international competitive bidding, there were generally less bidders as compared to developed countries. Also, the successful bidder did not have the required resources or related past experience in any projects before. The private entity did not have a consortium of specialized institutes or corporations related to different aspects of the construction of the project such as cost benefits analysis and quality control. The respective government department or ministry did not have any prequalification procedure for bidders; in terms of contractor experience or expertise, helping to short list them for bidding. Similarly, no guidelines and criteria were given from the government, stating clearly the roles and responsibilities of the private developer during construction, after construction and the repayment of the loan incurred from tolls collected. All these requirements were prepared by development banks, and were only general guidelines to follow, and not specific to any highway projects.

1.4.2.3   Why Projects Fail?

Highways and other highway projects in developing countries have been plagued by huge delays resulting in cost overruns. Even after the highway was completed and opened, there has been insufficient toll collection from over-optimistic forecasts of future traffic volume and economic environment.

Such adversities seemed to shadow highway projects in developing countries, although even with these conditions many projects were successful as well. We shall look into some crucial factors that resulted in some projects to fail.

1.4.2.3.1 Exchange Rate Risks
Since most project funds have a large share of foreign exchange component in it, such risks are inevitably high. Projects like the Gujarat Rural Roads and State Roads Programme planned for construction in India were abandoned or reduced in scope due to devaluation of the domestic currency. The situation in Indonesia was similar during the 1997 financial crisis when depreciation of the Rupiah caused some contractors to go bankrupt and pulling out of projects (like the Eastern Islands Roads project) and other remaining contractors had to bear the higher costs. However, the project was not abandoned but completed with lower than expected traffic volume flow. No tolls was introduced in this project.

1.4.2.3.2 Exclusivity
The exclusivity of a highway project is ultimately its elasticity of demand. The introduction of another highway with the same purpose of transportation to the same destination will dampen the toll collection on both highways. The development of the Suzhou Industrial Park (a joint venture between Suzhou of China and Singapore) involved the construction of a highway as part of the industrial park. However no exclusivity or second facility was granted by the Beijing government. This resulted in the construction of another highway feeding into Suzhou resulting in reduced toll fare collection.

1.4.2.3.3 Legal Framework
Property rights law must be well developed and enforced effectively by the regulatory body in the affected area in order to prevent informal transport provider from playing the role as an effective substitute against the formal highway. In Jamaica, such laws were ineffectively enforced and highway projects had encountered losses as a result. The political will to enforce traffic rules in toll collection and penalties or fines in Jamaica became a problem in ensuring the success of the Urban Transport Project. Brazil achieved greater success because of its ability to enforce the rule of law.

1.4.2.3.4 Economic Environment
After the 1997 financial crisis, many completed highway projects in Asia suffered massive losses from inadequate toll collection. Tolls from the Malaysian North–South Expressway were not able to cover operating expenses and had to resort to the unpopular measure of raising toll rates. Revenue collection did not improve.

1.4.2.3.5 Role of Government
The government has to switch its role from a provider of highway to a regulator. In other words, more private entity sector involvement is needed especially in financing of the project. Highway projects in Malaysia and China were and are still executed by stated-owned enterprises. The long-run efficiency is questionable.

1.4.3 General Issues in Highway Financing

1.4.3.1 Planning

All highway projects regardless of whether they are provided by the government or the private sector go through formal planning. We shall discuss some of the factors that support the need for formal planning in the public interests.

(1) *Economic argument*: Highways are classified as public goods with *externality* properties which are not *internalized* by the free market. It spurs on economic growth and is an important asset to the host country. These projects require huge funding from either the government or the private sector and therefore have massive economic implications be it a success or failure.

(2) *Political argument*: As mentioned above, highway projects involve huge money and respective interests groups such as industries, landowners, environmentalists etc. These interests groups can sway government decisions on highway projects against the majority of individuals who have little power or influence over decisions. Having a good management system and plan acts as a check against corruption especially if this is extensively publicized. Being transparent at various stages of implementation will make corruption practices more controllable.

(3) *Social argument*: Basic amenities such as clean drinking water and roads are fundamental in developing countries. The construction of basic highway infrastructure has to be distributed in accordance with the measurable social needs of the region or country. It is economically efficient to make the user pay for the usage of highway such as roads. In this way, the highway should be built only if users can afford to pay for it. However, issues of fairness and injustice will result in protests if a proper reconciliation plan is not available.

1.4.3.1.1 Issues in Planning

Demographic changes such as future trends in population and employment play a vital role in assessing the criteria for the demand for future highway. Demand and cost risks are the two most important risk factors that render a highway project financially not viable. Trends in transportation data such as the ownership of cars, distance traveled are assessed as well together with the capacity of present highway assets are often misread or miscalculated. The fulfillment of highway projects objectives such as urban access enhancement, rural access enhancement, environmental impact etc must be closely assessed. All members and interests groups should have the lists of objectives and what are achieved against that through meetings or workshops. Trend identification and forecasting of elements of cost and demand are art as much as a technical skill. Thus strategic planning, good information and data, and a good grasp and macro understanding of the wider contribution of the highway to the local and wider economy is essential. Alternative plans are to be tested against scaled checklists with the scores weighted accordingly in order to fulfill forecast requirement parameters. These alternative plans will contain physical, operational and regulatory constraints as well that need to be rated against forecast requirements. In addition to Net Present Values (NPV) and Internal Rate of Return (IRR) drawn from Cost–Benefit Analysis (CBA) and financial analysis evaluation of highway projects should also include Multi Criteria Analysis which incorporates nonmonetized elements.

1.4.3.1.2 Normative Aspects of Planning

Planning strategic decisions involves many people and groups interacting and making "smaller" decisions. Judgment forms the backbone of communication and arguments reconciliation. Judgment involves manipulating complex situations, leadership that accepts ambiguity, conflict and confusion would be the possible key to effective decision making on highway project financing.

1.4.3.2 Issues in Build Operate and Transfer

The role of government is central to BOT arrangements. Traditionally, the BOT arrangements transfer administration and financial burden relief to the private developer. The government focuses on social issues and provides the necessary funding such as health care. The revenue stream of the project will be of particular interest to the sponsor. The government guarantees against credit risks of the project and granting of development property rights to the sponsor to safeguard the ownership of the project. The pay-back period during operation of the highway for the funds provided should be stated and any delays or cost overruns to be specified. Development phase of the project begins from support for the project within the government departments. Diplomatic arrangements are essential for cross-border highway projects. Proposals from different developers are assessed and finally approved by the government.

Design consultations are needed to minimize design iterations and costs overruns. The final demand or need for the highway as well as absolute exclusivity of the highway asset will determine the revenue level. During the construction phase apart from design iterations and adequate financing, management of labor and other related personnel such as project engineers, project consultants, field consultants are essential. In the operational phase maintenance and collection costs are to be monitored and minimized. Foreign exchange fluctuations are especially crucial for third world nations. Volatility of interest rate determines the repayment cost of loans. Price changes as a result of demand and supply changes in the cost of production of the project. Accuracy in the traffic forecasts and toll rates charged pave the way for the success of the project. Autonomy in flexibility of the toll rate charged in order to be indexed to inflation and other cost indicator increases. Sovereignty risks in the highway asset from the private developer against changes in government policies and composition or legislation. War and hostility between nations especially are important for cross border projects. Social stability such as terrorism and strikes also play a role.

1.4.3.3 Issues in Privatization

Privatization has been the "rage" for the past two decades. However this takes on many forms and arrangements. The rationale for this including the understanding that increased competition and profit motive would minimize costs and enhance revenue. Reduced labor costs, economies of scale are some examples of these costs savings. Future costs incurred by the government agency and the contractor have to be borne by the government. The former would mean compensation costs incurred to lay off previously government employed workers. The latter would include costs such as contract changes and potential bankruptcy or service interruption by the contractor. Current costs which include salaries, fringe benefits and other cost of services and supplies or equipment, other indirect costs include a decrease in department or overhead charges may be offset by increases to other departments or services. In the event of bankruptcy or service interruption alternative arrangements will have to be made. Although private managers may exercise greater flexibility in labor management and be more susceptible to innovations or extra training, care must be taken to prevent "cutting of corners" to increase profit and compromise quality service. Private employees are more concerned about their performance as the chances of being fired due to poor performance are higher than the public sector. Sensitivity to citizen complaints is thus better emphasized.

Local government may oppose the decision to privatize as it is a loss of prestige and power. Government unions face the threat of their officials being voted out of office. In the process of bidding, the transfer would not take place until qualified bidders have been assessed with their proposals with the price offered by the bidder. Therefore, the price offered for the transfer, the qualifications and their proposals are essential for successful bidding. There must be careful monitoring and management of the project such that objectives and standards of the government are met by the vendor, contract amendments or agreements for acceptable changes are negotiated. A criterion for systems analysis and evaluation must be established.

1.4.3.4 Summary of Issues

There are common issues involving three parties in a highway project in developed or developing countries. They are the users, the government and the private entity. The issues will include before implementation, during implementation (construction) and after implementation.

1.4.3.4.1 Users

(1) *Tolls*: The amount of the toll charged or any form of price discrimination for different classes of road users and taxes imposed in place of tolls will cause negative feelings towards the project, thereby affecting its public opinion and support. Tolls charged should minimize the effect of transferring congestion to other nontoll roads or highways.

(2) *Social and natural living environment*: Construction of the project will result in uprooting and resettlement of residents in the area. Damage to the surrounding natural habitats will be inevitable. Care must be taken in handling these sensitive issues especially when compensation of resettlement or uprooting is concerned.

(3) *Pollution*: During construction, noise and air pollution are a major hazard and should be reduced to the minimum whenever possible.

(4) *Safety*: Road safety precautions and speed-limits effectiveness will determine whether the highway is both safe and efficient to road users.

(5) *Time saving costs*: Ultimately, the effectiveness of the highway will rest upon how much time is really saved for motorists when they pay the toll to use it as compared to other nontoll roads.

1.4.3.4.2 Government

(1) *Economic growth*: Highway development has always been playing an important role in economic growth and it will continue to do so. Employment opportunities are created and livelihood is secured for the people.

(2) *Spillover effects*: Engagement of foreign or international consultants and developers together with local government and private entities would help the local group to acquire management knowledge and expertise in handling similar highway projects.

(3) *Public support*: Obtaining enough support from the public is essential especially in developed countries. Participation from the community in certain areas like by creating a mascot to represent the project will aid positively towards ensuring support for the project.

(4) *Political support*: Opponents of the highway projects will reject any tax or toll imposed to finance it. Tax rebates which pay back taxpayers a certain percentage from their transport tax would help alleviate some pressure from opposition.

(5) *Vision*: The government's vision of the country's transport network will shape the outcome of future highway projects.

(6) *Risk sharing*: Guarantees in the form of loan security help allocate risks and enable such projects to be financially viable.

1.4.3.4.3 Private Sector

(1) *Profitability*: Profitability of the project will be of primary concern.

(2) *Demand*: Traffic volume forecasts and other parameters such as rate of return, financial rate of return and sensitivity analysis should be analyzed prudently to ensure financial viability of the project.

(3) *Benefits*: Relevant costs and benefits in terms of vehicle operation and time savings to be evaluated.

(4) *Transparency*: Clear government policies and objectives submitted to both the public and the private entity will enhance the credibility of the government to these parties.

(5) *Toll*: Toll charges are to be flexible or fixed when current demand does not cover for operating expenses although fully indexed for inflation. Issues of concern will be possible protests and the elasticity of demand for the highway.

(6) *Risks sharing*: This is the apportioning of risk to be borne by the government and the necessary safety nets to be set in place to mitigate against risk.

(7) *Specific risks*: External shocks pose a potential threat to these projects as they affect economies of nations. Internal management risks and commercial risks are to be reviewed and preventive measures are to be undertaken. Foreign exchange risks can be minimized through hedging. Risks should never be reduced to zero, but to a reasonable level.

(8) *International agencies*: Help in the form of technical and financial assistance may be drawn from development banks such as the Asian Development Bank (ADB), the World Bank (WB), and the Multilateral Guarantee Agency (MIGA). These institutes are able to provide counter guarantees for the host government.

1.5 "Template" for Successful Highway Financing

On comparing the above review of two alternative models of highway financing in developed and developing countries, there exist some common factors and situations which can aid substantially in ensuring a successful implementation of projects as well as their future financial viability and sustainability (Fisher and Babbar, 1996; Chin, 2002a). These fall primarily into three areas.

1.5.1   The Government

(1) Government should play a role of regulator as much as possible AND NOT A PROVIDER for the development of the highway project in terms of initial planning and construction with its objectives and priories clearly stated to allow private entities to take over the role of developing the highway.

(2) Government should be flexible and responsive to create necessary departments or agencies to cooperate with and facilitate the private developer in obtaining all essential approvals and agreements with other related government bodies quickly and efficiently.

(3) Government should not incorporate other priories that are "inappropriate" to the development of the project other than the project's primary aim. This would mean that the aim of creating massive employment opportunities from a highway project may compel the developer to employ higher costs labor intensive methods of construction. Instead, an appropriate technological approach should be adopted in accordance with the cost structure of the economy. For example, China has an economy with low labor costs and labor intensive methods of construction is more feasible there than in Singapore.

(4) Government should as far as possible not take part in any form of financing on the project including providing financial security in terms of contingency funds or insurance coverage for the various loans taken.

(5) Government should ensure political stability in order to safeguard the overall business and consumer confidence. Also, exchange rate stability is definitely a crucial factor that requires government's assurance.

(6) Government should enforce prerequisites in terms of financial performance and expertise for private developers and they should qualify on these counts before bidding for any project. In addition, there should also be proper guidelines and criteria to be adhered to according to government's objectives for the short listed proposals.

(7) Government should provide some form of incentives such as tax exemption for the private developers.

1.5.2   The Private Sector

(1) Ideally the project team should comprise of a consortium team of qualified specialists in the construction and other project related areas with a good past record. The consortium should have adequate liquidity and financial assets in the corporation to draw confidence from both the government and public.

(2) Regular consultations and communication for working closely with the government especially to provide updates on construction phases, implementation of quality control and reports of any potential problems in development that might surface are important. This enhances productivity of the project.

(3) Initiative must be exercised to acquire all sources of finance in event of deficiency in funds and solve other possible management problems within the entity or with the government. A concerted effort must be made with the related government agency to gain support from the general public against any opponents of the project. Proposals should be submitted to the public as well before further negotiations are made with the government. Sensitivity and responsiveness to social and environmental issues as a result of the project are important to gain positive support for the project.

(4) Efficient contractors and subcontractors should be employed to mobilize resources effectively and timely with other logistical supporting elements in the construction of the project. Emphasis must

be on the quality of the project in terms of its design criteria to suit the anticipated traffic volume and subsequently to keep maintenance and operation costs at the minimum.

1.5.3   Scope of the Project and its Environment

(1) Intracity highway projects should be integrated into the existing urban road network which is an integral part of a development vision for the city. This holistic approach is important as it will affect the collection of tolls.

(2) Intracity projects within urban landscape (not industrial estates or parks) are significantly affected by population growth, automobile ownership, tourism and cost of public transport system, fuel and vehicles taxes as well. Therefore, higher weightage should be given to the estimation of these parameters for more accurate traffic volume forecasts.

(3) Intercity highway projects should have absolute exclusivity in the route that they serve. Respective cities should enter into agreement guaranteeing exclusivity involving the federal or central government.

(4) Intercity projects should be encouraged when there is already an existing inefficient transport system in place, therefore implementation will merely be an improvement to the existing transport system.

(5) Intercity projects in urban landscapes are more significantly affected by economic growth, tourism and industrial growth between the affected cities. The presence or the potential of any industrial estate or park joint venture in future between the cities would largely increase the success of the project.

1.5.4   Economic Growth and External Shocks

(1) The presence and potential of strong growth, either in the country or the region where it is, would encourage the implementation of the highway projects.

(2) A sound budget and balance of payments surplus would inject confidence and encourage the implementation of the project.

(3) External shocks such as financial crisis, war, oil price shock and health epidemics would cripple any potential plans for highway projects.

1.5.5   Sustainability

(1) Preservation of the surroundings should be seriously considered not only during construction, but also after construction of the project.

(2) Room and potential for expansion of the same project should be taken into consideration during construction. Tolls should be fully indexed for inflation.

(3) Sufficient safety measures and fire precautions should be applied to keep public opinion positive.

1.5.6   Management of Highway

(1) Small road networks should be centralized to a single management.

(2) Management and ownership of the undesignated roads should be transferred among road or highway authorities.

(3) Lease period should be made flexible to ensure private entities do not incur any losses. For example, in U.K. the lease period will be reduced if traffic volume is greater than expected and *vice versa*.

(4) Tolls should not be charged only on some lanes of the highway or expressway but on all the lanes.

(5) Toll collection should be efficient and smooth without compromising the flow of the traffic, for example, Automatic Vehicle Identification (AVI) is used in the United States.

(6) Toll funds should be kept separate from government funds in order to finance future maintenance and even upgradation of the highway.

1.5.7 Money, Capital Markets, and Development Banks

(1) Developing countries should develop their domestic bond and equity markets to alleviate their reliance on development banks for finance in future.
(2) Development banks should disburse funds for financing responsively to keep the line of credit "alive"
(3) Development banks should exercise close supervision and consulting in the construction of the project.
(4) Development banks should exercise initiative and in the event of inadequacy from the host government prepare all necessary technical assistance such as economic, traffic forecasts and proposal guidelines for them.

1.5.8 Training and Education

(1) Qualified training and education should be provided to the younger generation and government officials to aid them in managing current and future projects.
(2) Highway financing and engineering should be incorporated in college courses curriculum.
(3) Involvement of local technical expertise in all aspects of the implementation and management of the project in developing countries should be encouraged without diluting the role of the foreign developer.

1.5.9 Public Opinion

There are a number of issues for road pricing to gain maximum public support. We shall discuss some of the main concerns here.

(1) *The objectives of the scheme must meet the public concerns*: A road pricing scheme would most likely succeed if it addresses to the traffic-related issue plaguing most of the countries.
(2) Implications on the type of charging scheme such as congestion reducing should be equated with a reduction in vehicle-miles traveled at peak periods.
(3) *Demonstrate that there are no alternative solutions to the related traffic problems*: Charging for road use is more acceptable to public when viewed as a last resort solution. It should also display the role of providing the means of improving the road condition and limiting environmental deterioration.
(4) *Revenues should be as far as possible hypothecated*: The revenue earned from tolls should flow back to road users in other forms such as fuel or vehicle tax cuts or subsidies and even for the construction and improvement of better roads.
(5) *Scheme should be simple and straight forward*: Drivers should understand the system fairly easily and allow tem to be able to calculate costs of their journey in advance. However, trade-offs from simplicity results in less efficient or equitable solutions, whereby costs are not borne fairly by the road users.
(6) *Kind of technology considered*: The main concern is the technology of charging the road users individually must work accurately. Apart from this, monitoring of vehicles may intrude privacy and having a meter in the vehicle may also spur negative feelings.
(7) *Equity issues must be addressed*: Charges should be related to vehicle and engine size, so as to provide cheaper travel for smaller vehicles. Rebates such as free units should be allocated to all drivers allowing them to trade the units for income (redistribution of income). Exemption of road charges should be monitored closely to prevent dilution of revenues especially from low income travelers.

1.5.10 Pricing Roads

(1) *Charges should be closely related to amount of road use*: This requirement is best met by some kind of metering device attached to the vehicle.(Price to charge: Marginal Costs + Congestion + Marginal Costs from accidents and environment damage).

(2) *Prices vary across different times of the day, and for different classes of vehicles as well*: Prices should b e higher during peak periods and when pollution and congestion costs are high.

(3) *Prices should be stable*: This allows drivers to plan their route cost effectively in advance before undertaking them.

(4) *"Fair" system*: Price levels are ensured to be defensible and no room for discrimination haphazardly amongst road users.

(5) *Reliable technology and equipment*: Failure of the metering device in terms of over charging would be fatal to the survival especially if itis a new pricing system.

(6) *Durability of the system*: Vehicle populations are numbered in millions and require the system to be "smart" to recognize them individually.

(7) *Small payments as far as possible*: Large payments such as area licenses are not appropriate when it comes to shoppers and tourists who only utilize the roads occasionally.

(8) *The system should be able to accommodate users from other states or towns*: Just like our individual phone bills, we only pay one monthly bill for our phone charges. Similarly, for road charges users should be able to pay their bills from just one single account. Indication of future demand of road space: The system should be able to provide information on the extent of travelers who are prepared to pay for specified routes for future investment.

Traditionally highway financing drawn from government grants, international aid agencies, bond issues, various forms of fiscal instruments (such as vehicle import tax, registration fees, fuel tax and road license tax), employing financial instruments (such as bonds and debentures) and highway tolls. These are administered either through a public transport agency or authority or a road fund. Be it traditional or innovative approaches to highway financing should result in increasing incomes, wealth, and well being of society. It should not result in additional costs and like other investments highways should be judiciously selected with the potential to yield benefits over many years (Fisher and Babbar, 1996). If the benefits exceed the costs of a project, after factoring-in the cost of capital that could have been used for alternative income-generating ventures, the project is a worthwhile investment for the community. Borrowing to finance the construction of a highway enables people to enjoy its benefits sooner but passes on the cost of debt servicing to future generations. As such there is no and cannot be "one mode" to successful highway financing (Chin, 2002a,b). This is usually influenced by potential benefits over costs, circumstances surrounding the project and the trade off between the needs of present versus future generations. This chapter has reviewed several approaches and concluded with a suggested template only to be used as a guide.

References

Chin, A.T.H. 2002a. *Study on Toll Rates on the Hefei–Anqing Expressway*, Research Report prepared for the World Bank, Washington, DC and the Anhui Province Communications Department, People's Republic of China, 250 pp.

Chin, A.T.H. 2002b. *Study on the Planning, Finance and Operation of Toll Highways Network*, Research Report prepared for the World Bank, Washington DC and the Anhui Province Communications Department, People's Republic of China, 180 pp.

Chin, A.T.H. 2003. *Optimal Timing of Projects and Dealing with Risk and Uncertainty*, Research Report, Land Transport Authority, Singapore, 25 pp.

Estache, A. 1999. *Privatization and Regulation of Transport Infrastructure in the 1990s: Successes ... and Bugs to Fix for the Next Millennium*, Unpublished paper, 35 pp.

Estache, A., Romero, M.,, and Strong, J. 2000. *The Long and Winding Path to Private Financing and Regulation of Toll Roads*. The World Bank, Washington, DC, 49 pp.

Fisher, G. and Babbar, S. 1996. *Private Financing of Toll Roads*, RMC Discussion Paper Series 117. The World Bank, Washington, DC, 47 pp.

Heggie, I.G. and Vikers, P. 1998. *Commercial Management and Financing of Roads*, World Bank Technical Paper No. 409. The World Bank, Washington, DC, 171 pp.

Irwin, T., Klein, M., Perry, G.E., and M.Thobani, eds., 1998. *Dealing with Public Risk in Private Infrastructure*, The World Bank, Washington, DC, 173 pp.

Further Reading

Asian Development Bank, 2000. *Developing Best Practices for Promoting Private Sector Investment in Infrastructure — Roads*, Manila, Philippines.

Davis, H.A. 2003. *Project Finance: Practical Case Studies, 2nd Ed.*, Resources and Infrastructure, Vol. 2. Euromoney Books, London, UK.

European Conference of Ministers of Transport, 2002. *Tolls on Interurban Road Infrastructure: An Economic Evaluation*. ECMT Publications, Paris, France.

http://rru.worldbank.org/Documents/Toolkits/Highways/. An excellent and step by step guide to highway privatization and financing.

http://www.fhwa.dot.gov/innovativefinance/index.htm. United States Department of Transportation — Federal Highway Administration.

Gomez-Ibanez, J.A. 1993. *Going Private: The International Experience with Transport Privatization*. The Brookings Institution, Washington, DC, 310 pp.

Levinson, D.A. 2002. *Financing Transportation Networks*. Edward Elgar Publishing, 232 pp.

The Congress of the United States Congressional Budget Office, 1998. *Innovative Financing of Highways: An Analysis of Proposals*, Washington, DC, 82 pp.

2

Access Management of Highways

K. Raguraman
National University of Singapore
Republic of Singapore

Kumares C. Sinha
Purdue University
West Lafayette, IN, U.S.A.

2.1   Introduction

A region's social and economic fabric is weaved through intricate movements of people, goods and services. Roadways are the capillaries, veins, and arteries of life, carrying the pedestrians, bicycles, cars, buses, and cargo trucks to their intended destinations. In the conventional approach to the planning of road systems, the growth of travel demand is largely addressed by adding capacity, including widening existing roads and building new roads. However, every time capacity is added, additional traffic is generated, especially in situations where traffic demand controls are minimal or absent. The extent of the induced traffic generated by road infrastructure improvements varies from place to place but the fact that the potential of higher speeds offered by new capacity results in more trips, longer trips, and redistributed trips has been fairly well documented (Goodwin, 1996; Hills, 1996; Hansen and Huang, 1997; Noland, 2001; Cervero, 2002). The additional traffic generated is also, in part, attributable to new activities that locate on these new and upgraded roads (Ewing and Lichtenstein, 2002). In a matter of time, therefore, the road returns to being congested and unsafe, demanding capacity expansion.

In many cities and regions, the approach of incrementally adding supply to meet increasing demand is close to reaching its physical limits. As an area becomes more crowded with activities, space may simply not be available for expansion of roads. While some authorities resort to cumbersome land acquisition procedures to procure more space for new road development, this measure is generally not preferred as it involves sensitive issues of displacement, resettlement, and high compensation costs (ADB, 1998). In some areas, roads have expanded upwards (viaducts) and downwards (tunnels) so as to not take land away from existing uses but the cost of such projects are prohibitive. Where opportunities are more readily available, road capacity expansion may perhaps be useful and necessary. However, it is critical that efforts are taken to put in place an effective system of managing existing and new roads to extend their life and reduce the need for costly upgrading, to keep the traffic flowing smoothly and improve levels of safety.

A safe and efficient transport system drives the urban and regional economy allowing for the quick movement of vehicles within the urban communities and between cities in a region. Every property owner and community wants access to the road system so as to benefit from this participation in the space-economy. On the other hand, every motorist on the road wants smooth and efficient mobility to get to their destinations. More access connections to a major road result in more traffic conflicts, contributing to a deterioration of the functional integrity of the entire road system. With the development of new access points to a highway or arterial and with the new traffic signals erected, the traffic speed and capacity of the roadway are reduced increasing the potential for congestion and accidents.

The planning process striving to balance the provision of direct links to communities simultaneously ensuring the smooth and safe flow of traffic in the road system is broadly termed Access Management. It involves aspects of planning of road design in terms of the number and form of links to major roads, as well as the management of the flow of traffic between these roads. Various techniques are used to manage entrances and exits and related turning movements onto and off the major roadways, as well as maintain design criteria and standards necessary to preserve the operational capacity, speed and safety of the roadways. Every access point can contribute to the deterioration of traffic level of service and these impacts increase geometrically over time as both the traffic volumes and number of access points increase. Access management must therefore accommodate not only the current condition but also the future growth. Systematic application of planning, design, and operation of access management features includes approaches (driveways and street connections to a roadway), medians, median openings, signals, auxiliary lanes, and interchanges (Oregon DOT, 2001). The targets of access management efforts have traditionally been the highways and major arterial roads. However, the experience of road planning in many places currently shows that access management is increasingly extended to minor roads as well. Planners now seek to develop a comprehensive and integrated land use-access management policy that builds on a systematic classification of all roadways according to function and traffic characteristics in relation to the nature of activities of the adjacent lands.

2.2   Principles of Access Management

Access management is not a modern concept; it is simply an application of a concept that was first identified in the early 1900s. However, access management has been undervalued or virtually ignored as an engineering and safety element in roadway design and decision making. There are number of established guidelines for access management, which have been developed over the years arising from studies on traffic flow and safety levels on different classes of roads. Much of the published work originates from the United States, where extensive research on access management has been actively carried out by the Federal Highway Administration (FHWA) as well as the state transport agencies. Although access management as a congestion management tool has been around for a long time, interest in it has only picked up in the mid 1990s, when it was recognized that it is important to effectively manage existing road spaces instead of adding new capacity in response to growth in traffic demand.

One of the first states to have a system wide comprehensive access management program was Colorado. The state legislature declared in 1979 that all state highways were controlled-access highways. What made the Colorado new approval process different from the earlier permit systems in Colorado and other states, was the application of the principles of access management to all state routes, including principal arterials, secondary roads, local frontage roads, freeways, and expressways (Demosthenes, 1999). A systematic access management policy addresses the questions of why, when, where, and how access should be provided or denied, and what legal or institutional changes are needed to enforce these decisions (AASHTO, 2001: 89). Depending on the function of the road, access may be fully controlled where priority is given to through-traffic or partially controlled to preserve a desirable balance between access and mobility. Building on traffic engineering concepts and ideas, access management seeks to reduce and separate conflict points in a roadway system and minimize interference to traffic flow. A carefully conceived and well implemented access management plan can help to preserve the capacity, speed, and safety of traffic on a road system extending its life span and reducing the need to build new roadways.

The access management principles highlighted in this section represent a synthesis of ideas drawn from the wide collection of papers and reports produced by academics and transport professionals from government (State Departments of Transportation, Transportation Research Board (TRB), AASHTO, and FHWA), and private consultancy agencies. They can be broadly summarized into the following:

1. Develop a functional system of classification of roadways, and define the roadways in the system in terms of these classes: the classification scheme will show where each roadway in an area is placed in its functional range between access and mobility, and provide a framework for the assessment of the need for different types of access management and control measures.
2. Provide a balanced road circulation system: the functional integrity of a highway system at the city and regional scales rests on the development of a graduated system of roadways that facilitates a balanced and well distributed circulation of traffic. This establishes a sound framework for the effective functioning of roadways in accordance with their defined categories in the classification system.
3. Establish access control to roadways with higher functional classifications: freeways and expressways are reserved for the mobility function. Direct property access should be restricted and the entrances to these higher class roadways need to be carefully planned in terms of location, spacing, and other design features.
4. Establish standards on driveways that feed into collector and arterial roads: junctions between driveways and the main roads represent conflict points that can affect smooth traffic flow and increase the potential for accidents. Regulations and guidelines on driveway spacing, frequency, design, and location will help to reduce the severity of these conflicts.
5. Plan the location and design of major intersections: the flow of traffic through intersections needs to be carefully coordinated to allow for smooth progression. Driveways need to be located at a reasonable distance away from intersections to reduce traffic interferences.
6. Promote the development of dedicated turning lanes and other special turning treatments: if the lanes on a roadway carry both through-flow and turning traffic, disruptions and conflicts arise, and these can result in bottle-necks that pose congestion and potential accident problems. By separating turning traffic from through-flow traffic, these conflicts can be reduced, resulting in the safe and effective functioning of the roadway.
7. Consider the application of median treatments including raised medians and two-way left-turn lanes (TWLTL): these are among the most controversial measures of access management but are highly effective in reducing conflicts along roadway stretches brought about by turning and crossing traffic.
8. Assess economic impacts and involve public participation in access management policy initiatives: this helps to assess level of acceptance, create awareness of the long-term benefits of access management, and evaluate the effectiveness of measures.

The following sections of the Chapter examines each of these guidelines in some detail, highlighting the central theoretical ideas and planning considerations of managing and controlling access within each principle. The main tools and methods that can be employed in access management are discussed and the impacts evaluated, drawing on actual practices in different cities or states.

2.2.1 Classification of Roadways

The first question that is asked when a road is built relates to the function that it is going to serve. At the lowest level, there is a local road that provides access to individual housing units in small residential neighborhood. The local roads feed into collector roads that link up multiple neighborhoods. The collector roads feed into the minor arterial roads, which in turn link to the major arterials that provide connections between large scale community networks. At the highest level, there are the freeways and expressways that carry regional and intercity traffic. The size of roads in each road class depends on the density of the activities, be it housing, industrial, office, or other activities, as this will have a bearing on the volume of traffic generated.

While the names given to different categories of roads may vary from country to country, in most cases there is an attempt to have a classification system according to function of the roadways. In the United States, the functional classification system of roadways developed by FHWA distinguishes between four classes — local, collector, minor arterial, and principal arterials. The last category is further subdivided into interstate highways, freeways, and expressways.

The ultimate goal of road planning is to ensure safe and smooth flowing traffic conditions through the entire system of roads. While it is important that the roadways must have sufficient capacity to accommodate the traffic demand, it is critical to recognize that road design and management should be undertaken to ensure that the available capacity is not unduly affected by creation of unnecessary bottlenecks and conflicts.

Freeways and expressways are designed for maximum mobility with high speeds and capacities. If there are limited or no controls on the access points to a major freeway or expressway, the capacity, efficiency of travel, and safety levels of the roadway may be compromised. This is why access to the large regional highways needs to be strictly controlled to interchanges spaced many kilometers apart. On the other hand, a local street, which has individual links to adjacent properties cannot be expected to serve a long distance mobility function. With a significant reduced volume of traffic flowing at low speeds, the function of these local roadways is primarily one of access. The remaining roadways, the collectors and arterials serve a mix of access and mobility functions, and the land use and traffic characteristics would largely determine where they are placed between the access–mobility continuum (Figure 2.1). However, the decision regarding what level of access control to be implemented can be difficult. The AASHTO geometric design policy (AASHTO, 1994) provides some general guidelines regarding access control, but more detailed and comprehensive guidelines are lacking. The decision to implement access control often involves engineering judgments as there are no uniform national guidelines regarding access control (Brown et al., 1998).

Tables 2.1 and 2.2 present the road classification system and roadway length statistics for the United States and Singapore. It can be discerned that the broad framework for the road classification scheme is similar in the two countries, and this observation generally applies to most of the other countries as well.

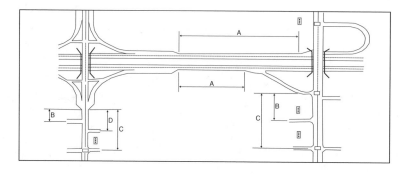

FIGURE 2.1   Access-mobility function (*Source*: TRB, 2003).

TABLE 2.1 Public Road Length in the United States by Functional System

	Urban	Rural	Total
Principal arterial			
Interstate	13,640	33,107	46,747
Other freeways and expressways	9,377	98,945	108,322
Other	53,679	132,052	185,731
Subtotal	76,696	137,857	214,553
Minor arterial	90,922	431,754	522,676
Collector	89,845	271,371	361,216
Local	644,449	840,982	1,485,431
Total	901,912	2,106,725	3,008,637

Source: Compiled from FHWA, 2002.

Functions of roads evolve over time. As a region becomes more intensively built-up, traffic volumes on a major arterial may increase and speeds may be reduced. The building of a new freeway that by-passes these built-up areas to serve through-flow traffic easing congestion may lead to the arterial taking on more of an access function. The development of new major roadways requires a reexamination of the classification system of the region's entire road network and some displacement of roads from their original classes may be necessary. The changes in the categorization of roads provide useful opportunities for the redistribution of traffic so that the functional integrity of the entire road system is structurally sustained.

A long term land-use transport plan should be developed to determine not only the functional class of a particular roadway but also to establish the estimated life of that roadway within the class. An area may avoid expensive modifications of existing roadways to sustain its functional position within a particular class if it is clear from the long-term plan that the roadway may be displaced to an access function when a proposed new roadway is built later on. In the process, planners also avoid dealing with the fallout negative impacts of road modifications, perceived or real, on the adjacent communities.

TABLE 2.2 Classification of Roads and Spacing Standards in Singapore

Road System	Road Type	Road Reserve (m)	Number of Lanes	Design Speed (km/h)	Road Capacity (Vehicle/h)	Road Spacing (Km)	Buffer Standards (m)
		Function: movement					
Expressway system	Expressway	52.9	8	80–90	14,400	2.5–9.0	30
	—	45.5	6	—	10,800	—	—
	Semi-expressway	45.4	8	60–70	9,600	1.5–4.0	12–15
	—	38.6	6	—	7,200	—	—
Arterial system	Major arterial	45.4	8	60–70	6,000	1.2–1.5	12–15
	—	38.6	6	—	4,500	—	—
	Arterial	38.6	6	50–60	4,500	0.9–1.2	7.6–10
	—	31.8	4	—	3,000	—	—
		Function: access					
Collector system	Primary access	31.8	4	50–60	—	0.6–0.9	7.6
	—	26.2	4	—	—	—	—
	—	21.4	2	—	—	—	—
Local road system	Local access	18.0	2	40–50	—	0.1–0.6	2.0
	—	15.4	2	—	—	—	—
	—	14.2	2	—	—	—	—
	—	12.2	2	—	—	—	—
	—	11.6	2	—	—	—	—
	—	11.1	2	—	—	—	—

Source: Urban Redevelopment Authority (1992), cited in Chan, 1999.

2.2.2 Promote a Balanced Road Circulation System

Once a long-term transport plan is mapped out with a clear functional hierarchy of existing and proposed roadways, planners can develop standards and controls on the number of access points and adopt various techniques to improve the form of the access at each point. In his landmark study, Marks (1974) proposed a graduated system of roadways in which each functional class of roads was served by the next class in the hierarchy. This concept of graduated access represents one useful foundation for guiding the practice of access management. Such a system of roadways allows a driver to make a smooth transition from a lower speed roadway to a higher speed one in a way that promotes safety and efficient traffic flow. If, for instance, there is a collector road with a design speed of 40 km/h feeding directly into a freeway with a speed of 90 km/h, it would be difficult and dangerous for the driver to increase the vehicle speed as he enters the freeway or reduce the speed as he leaves the freeway. Unless special provisions are made, this large speed differential will create bottleneck situations in the ramp area and potentially impede smooth traffic flow. In California, for example, the established standards on highway design specify that where the local facility connects to a freeway or expressway (such as ramp terminal intersections), the design speed of the local facility shall be a minimum of 55 km/h. However, the design speed should be 75 km/h when feasible (Caltrans, 2001). If it becomes necessary to connect two roads with a big difference in design speeds, the slip roads need to be designed in such a way as to enable a vehicle entering (acceleration-lane) or leaving (deceleration-lane) a highway to increase or decrease its speed, respectively, to a value at which it can safely merge with through traffic.

A graduated system of roadways will provide good circulation of traffic only if the supply of roadways at each level is adequate and if the roadways of different regions are well-connected. The inadequate provision of secondary roads, for example, will result in the arterial roads accommodating both local and through traffic, when they are functionally designed primarily for the latter. If a large percentage of drivers intending to travel from one traffic zone to an adjacent zone, go into a freeway and then take the next exit to enter the adjacent zone, the function of the freeway to provide a through-link for travel between spatially distant zones becomes compromised.

As communities grow and land is subdivided for development, it is essential to ensure continuation and extension of the local street system. Dead end streets and gated communities force more traffic onto higher level roadways and increase demand for direct roadway access to them. Levinson (1999) highlights the examples of road systems in Las Vegas and Asian cities such as Bangkok where there is an over concentration of traffic on relatively few major multi-lane roadways which are served by narrow discontinuous streets. The traffic volume in these arterials builds up quickly resulting in congestion and safety problems.

"Joint and cross-access strategies help to relieve the demand on major roadways for short trips, thereby helping preserve roadway capacity" (CUTR, 1998: 4). A graduated system of roadways is desirable but it is important to be well balanced and integrated. Completeness and connectivity in the road systems mean that efficient alternate routes are available for a driver to get between two points those are relatively close together.

Poor access management design leads to increased delays and also causes unnecessary air pollution and fuel consumption (time spent idling at red lights or in stop-and-go traffic). A study by Ohio-Kentucky-Indiana Regional Council of Governments concluded that 40% of all fuel consumption in highway transportation was attributable to vehicles stopped and idling at traffic signals (OKI, 1999). Frequent and poorly spaced traffic signals can reduce roadway capacity to over 50%. Furthermore, access management is also a safety issue as 50 to 60% of crashes are access related.

2.2.3 Establish Standards on Access Controls to Freeways, Expressways, and Major Arterials

The availability of good internal circulation between neighboring districts may still not deter drivers from using the high-speed roadways for short trips. Access controls to the major roadways is therefore necessary to encourage drivers to use the system of minor arterials and other lower order roadways for short distance commutes, and take the freeway only for long distance rides. The access standards to be established depend on the functional classification of streets used in a particular city or region and they need to be enforced through legislation. For example, the access standards established by the Iowa Department of Transportation specifies eight levels of roadway classes and four levels of access classes, and each has their own standards and regulations concerning design and access control (Table 2.3).

While access to a freeway is always through grade-separated interchanges, irrespective of its location being urban or rural, an expressway may occasionally have signalized at-grade intersections. As a principle, there should be full access control on freeways and expressways where direct property access is prohibited. As far as possible, full access rights should be purchased along the length of the freeway and expressway. Existing private accesses and public road connections that do not meet the spacing standards should be removed during project development. The key considerations for access management into high-speed freeways relate to the design or form of access points and the spacing between them. Freeways serve continuous traffic flow and, as such, should not encounter any signalized intersections. Opposing traffic movements must be separated by physical medians or concrete rail barriers, and cross traffic should be separated by grade separation structures. Directional ramps, in the form of interchanges, limit the access to freeways.

For expressways and principal arterials, private access is generally not permitted. In some special instances, parcels fronting only on the expressway may be given access to another public road or street by constructing suitable connections if such access can be provided at reasonable cost (Caltrans, 2001: 100–3). The criteria for ascertaining reasonable access from a local street will include factors such as function, capacity, and safety concerns, as well as whether this access will cause operational problems within the local street system (Nevada DOT, 1999: 15). Often, these special access provisions are only temporary, and they will be removed once alternative reasonable access is subsequently developed. Effective interchange management requires participation at the local level, where land development

TABLE 2.3 Roadway Category and Classification

Category	Roadway Classification	Function	General Design Features
1	Freeway	Interstate and interregional traffic movements	Multi-lane with medians, interchange access
2	Expressway	Interstate, intrastate, interregional, intraregional, intercity and intracity traffic movements	Multi-lane with median and widely spaced public access points
3	Regional highway	Primary: interregional, intraregional and intercommunity traffic movements secondary: land access	May be two or multi-lane facilities
4	Rural highway	Balances rural travel needs with land access	Generally two lanes
5	Principal arterial	Primary: inter and intra-city and inter and intra-regional traffic movements secondary: land access	Multi-lane with median
6	Minor arterial	Primary: intercommunity and intracity traffic movements secondary: land access	May be two or four lanes, may have median
7	Collector	Balances traffic movements with land access	Generally two lanes may be four lanes
8	Frontage or service road	Land access	Two lanes

Source: CTRE, 2000a, 2000b.

decisions are made. This separation of jurisdiction between state and the local planning authority may make it difficult to sustain the efficiency and safety of interchange areas. It is necessary to review state policies to identify changes that may be needed to facilitate local participation in managing interchange development (Land, 2001).

Guidelines for the spacing of interchanges take into account two key goals:

1. preservation of smooth traffic flow on the high-speed freeway
2. safe and efficient entrance or departure from the freeways from or into the connecting arterials.

These, in turn, depend on a number of factors including the perception–reaction time of drivers, the deceleration and braking requirements, traffic volume, composition of traffic, design of the interchanges, gradient and angle of roadway, and visual intrusions at the roadsides. In general, most state agencies within the United States specify a minimum distance of 1 to 2 miles in the spacing between freeway interchanges and major arterial intersections in urban areas. The spacing standards are usually higher in rural areas because vehicle speeds are higher.

Another aspect of access management on freeways relates to the functional areas of interchanges where merging and diverging of traffic take place. The functional area must have sufficient clearance to allow for vehicles to weave and merge along the exit ramps, and to change lanes. Vehicles will then be able to make a smooth and safe transition from a freeway into a road with at-grade access points. The potential problem posed by closely spaced accesses adjacent to the ramp terminal can cause traffic back onto the freeway, interrupt the flow of traffic, and impact the smooth operation of the adjoining facility. In the U.S., the spacing standard for urban areas is usually about 1320 ft from the end of the off-ramp to the first private driveway on the left side, median opening, or intersection with a public road (see Tables 2.4 and 2.5, and corresponding Figures 2.2 and 2.3). Deviations from this standard may be permitted in "fully developed urban areas" where urban densities are achieved in more than 85% of the developable frontage. When only right-turns into driveways or public roads are involved, a shorter clearance area of 750 ft may be used (Oregon DOT, 2002: 6–11; Missouri DOT, 2003: 20). A systematic evaluation of the alternatives for improving traffic flow and safety at a freeway interchange is presented by Gluck and King (2003), drawing from their analysis of the complex urban Highbridge Interchange of the Cross Bronx Expressway (Interstate 95) and the Major Deegan Expressway (Interstate 87) in New York City.

The significance of principal arterials and the need for access controls in these large roads cannot be underestimated. Roads belonging to these classes usually account for a small percentage of a nation's roads but carry significantly more than proportionate volume of trips. In the United States, for example, the 160,000-mile National Highway System, which includes the interstate highways and principal state highways, accounts for about 5% of the total road network but carries 40% of the nation's traffic. The interstate highways alone comprise a little more than 1% of the country's roads but carry more than 24% of travel, including 41% of total truck-traffic. The remaining traffic is largely carried on the 3-million mile network of arterials, collectors, and local roads. It is interesting to note that 78% of the nation's road network is in rural areas, but they account for only 39% of the traffic. Urban areas account for 22% of total roads and 61% of the traffic (AASHTO, 2002: 27–32).

2.2.4   Standards on Number, Location and Design of Driveways, and Signalized Intersections

While access to freeways and principal arterials is generally restricted, access to smaller arterials and collector streets also needs to be regulated and properly designed. In the local development guide plans or ordinances, special conditions could be stipulated to limit the number of driveways to one per parcel, and to encourage the use of side roads or shared driveways. Developers can be given incentives such as density-bonuses or reduced frontage requirements to utilize access from existing side roads or to construct side roads rather than directly access an arterial or a collector road (Nashua Regional Planning Commission, 2002: 4). This approach reduces the number of conflict points along an arterial road reducing the potential risk for accidents. There are no fixed standards on the number of driveways per mile (access frequency) but many states in the U.S. recommend 20 to 30 driveways per mile as a maximum standard for a multi-lane arterial with a posted speed limit of 35 mi/h. The accident rate becomes unacceptably high above this access frequency (CTRE, 2003a, 2003b).

The number of driveways per mile of collector or arterial road may be well within the recommended density standards, but it is also important that they are adequately spaced from each other to reduce the bunching of conflict situations that may lead to deterioration in traffic flow and accidents. While there are no technical standards on driveway spacing, some guidelines duly considering the posted speed of the roadway, the traffic volume and composition, and land use factors (plot size, type of activities and frontage requirements) may be established. The TRB, for example, has suggested that unsignalized access spacing on major arterials should be between 300 to 500 ft, and on minor arterials between 100 to 300 ft, depending on operating speeds (Koepke and Levinson, 1992).

Apart from considering the number and spacing of driveways, it is important also to plan carefully their location and design (driveway corner clearance and geometrics). The location should be such that drivers leaving the driveway and those on the main thoroughfare should be within adequate sight of each other. This allows for better judgment and safe responses when such conflict situations are encountered. The turns into and out of the driveways should not be sharp and shallow. Driveway widths and turning radii are also important considerations. Vehicles turning into the driveway may encounter difficulties and have to slow down drastically, and if this is not executed properly, it may lead to unsafe conditions (broadside and rear end collisions) and holding up traffic behind the turning vehicle.

TABLE 2.4 Minimum Spacing Standards Applicable to Freeway Interchanges with Two-Lane Cross-Roads for State of Oregon, USA

Category Of Mainline	Type of Area	Spacing Dimension			
		A (mi)	B (ft)	C (ft)	D (ft)
Freeway	Fully developed urban	1	750	1,320	750
	Urban	1	1,320	1,320	990
	Rural	2	1,320	1,320	1,320

Source: Oregon DOT (2002).

Notes: (1) If the cross-road is a state highway, these distances may be superseded by the Access Management Spacing Standards, providing the distances are greater than the distances listed in the above table.

(2) No four-legged intersections may be placed between ramp terminals and the first major intersection.

A = Distance between the start and end of tapers of adjacent interchanges.

B = Distance to the first approach on the right (right-in or right-out only).

C = Distance to first major intersection (no left turns allowed in this roadway section).

D = Distance between the last right-in or right-out approach road and the start of the taper for the on-ramp.

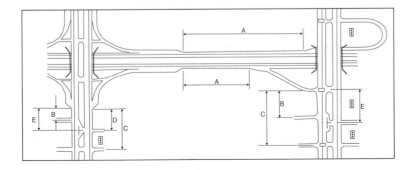

FIGURE 2.2 Measurement of spacing standards for Table 2.4.

Geometric design has a direct impact on vehicle operating speed, traffic engineering, and the eventual safety of low speed urban streets. Geometric design must balance access, pedestrian or bicycle use, and right-of-way issues with speed and safety. If vehicles operate above the intended speed, speed differentials can result between vehicles and increase the risk and severity of potential accidents. The potential for speed differential between vehicles also increases with the number of conflict points created when vehicles enter streets from driveways and intersections (Poe et al., 1996).

Efficiently designed, well-located and adequately spaced driveways help to reduce congestion and accidents on the main streets. The typical reduction in free-flow speed (for one direction) is approximately 0.15 mi/h per access point and 0.005 mi/h per right-turning movement per hour per mile of road. This translates into a speed reduction of 2.5 mi/h for every ten access points per mile, up to a maximum of 10 mi/h reduction at 40 access points per mile (Reilly, 1990). In terms of accident rates, a TRB study done correlating crash rates with access density found that an increase from 10 to 20 access points per mile would increase crash rates by roughly 30% (Gluck et al., 1999). Possible guidelines on standards for driveway spacing will depend on the posted speed of the particular main road, the density of use of the adjacent land, and the classification of the road.

TABLE 2.5   Minimum Spacing Standards Applicable to Freeway Interchanges with Multi-Lane Cross-Roads for State of Oregon, USA

Category of Mainline	Type of Area	Spacing Dimension				
		A	B	C	D	E
Freeway	Fully developed urban	1 mile	750 ft	1,320 ft	990 ft	1,320 ft
	Urban	1 mile	1,320 ft	1,320 ft	1,320 ft	1,320 ft
	Rural	2 miles	1,320 ft	1,320 ft	1,320 ft	1,320 ft

Source: Oregon DOT (2002).

Notes: (1) If the cross-road is a state highway, these distances may be superseded by the Access Management Spacing standards, provided the distances are greater than the distances listed in the above table.

(2) No four-legged intersections may be placed between ramp terminals and the first major intersection.

   A = Distance between the start and end of tapers of adjacent interchanges.

   B = Distance to the first approach on the right.

   C = Distance to first major intersection (no left turns allowed in this roadway section).

   D = Distance between the last approach road and the start of the taper for the on-ramp.

   E = Distance to the first directional median opening. (No full median openings are allowed in non-traversable medians to the first major intersection).

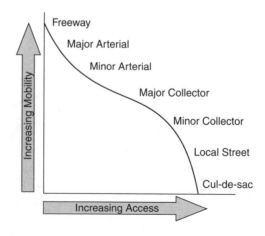

FIGURE 2.3   Measurement of spacing standards for Table 2.5.

2.2.5 Promoting Through-Flow at Signalized Intersections

For major at-grade signalized intersections such as between an arterial and a collector street, minimum distance standards need to be maintained to preserve traffic flow, improve safety and sustain the functional hierarchy of the road system. A corollary effect of increasing the distance between signals is the improvement of air quality in heavily traveled high-density corridors. Traffic signals serve to assign right-of-way to conflicting movements of traffic at intersections. While they reduce the number of right angle collisions at an intersection, they can cause an increase in rear-end collisions. A balance therefore needs to be achieved with due safety considerations to maintain a smooth traffic flow. It is important that signalized access points on major roadways must be considered in a total city or region wide traffic coordination scheme.

As with the standards for driveway spacing, the appropriate spacing between signalized intersections on a particular arterial stretch depends on the speed of the road and density of traffic, but it is noted that "anything greater than two per mile has a significant impact on congestion and safety" (FHWA, 2003). However, depending on the density of land use, this spacing standard may be reduced but a minimum spacing of at least 1/4 mile should be maintained, as any further reduction will lead to unacceptable disruptions in traffic flow in the form of delays and queues at intersections.

The Institute of Transportation Engineers (ITE) and FHWA have presented interesting summaries of the effects of signal spacing on traffic speeds, environmental effects, and accident rates. Aggregate of studies on access management highlights the fact that "if the number of access points are held constant at less than 20 unsignalized access points per mile, and the number of signals per mile are categorized as less than two, as compared to two to four signals per mile, there is a 50% increase in the crash rate (from 2.6 to 3.9 million vehicle miles of travel [VMT])" (ITE, 2004: 1–2). In terms of travel time, a study of a newly signalized intersection in Cincinnati found a 20% increase in peak travel times, while a "demonstration project in Colorado revealed that half-mile signal spacing and raised medians on a 5-mile roadway segment reduced total hours of vehicle travel by 42% and total hours of delay by 59%, compared to a quarter mile signal spacing" (FHWA, 2003).

Apart from the spacing of signalized intersections, the quality of flow along a street is also affected by the speed of traffic on the street, and length of the traffic signal cycle. Synchronization of adjacent traffic signals (green waves) helps to reduce stops and delays. This can be best achieved using a system of traffic-actuated signals, which adjusts the green time according to the volume of traffic, optimizing flow at the intersections. It is more difficult to achieve in two-way streets because of the need to coordinate opposing traffic movements. Several cities have implemented some variant of this system on their arterial road network. Access and mobility management efforts can be optimized when they are complemented with such road management measures that use intelligent transportation system (ITS) tools.

Where there are parallel two-way high-density arterials which are just a few blocks apart, transport authorities may want to consider converting them into two one-way streets, since progressed movement on one-way streets are much more readily achieved. This conversion has a significant bearing on access as one direction of traffic flow has been removed. Distance traveled will be increased but this can be compensated by time savings from smoother traffic flow. Pedestrian safety can be improved as crossings become simplified, but the increase in speeds afforded in one-way streets may present a new set of problems, especially on wider arterials. A comprehensive impact analysis of proposed conversions, including an evaluation of the traffic circulation in the surrounding road system, needs to be carried out before the modification to one-way streets is carried out. In fact, many cities across the United States are switching back from one-way to two-way streets for traffic calming, more convenient public transport access, and easier navigation.

2.2.6 Treatments on Turning Lanes

Access controls to major arterials in the form of driveway consolidation and other management measures facilitate the speedier movement of through-flow traffic. However, when if a lane on the roadway serves both turning and through-flow traffic, disruptions referred to as "side friction" may be encountered because of the sudden deceleration of speeds of the traffic intending to turn at intersections. This creates conflicts or bottlenecks under heavy traffic flow bringing about long queues of traffic waiting to turn left and right, and constrictions in the passageway of through-flow traffic.

Where an arterial passes through a developed area having numerous cross streets and driveways, and where it is impractical to limit left turns, the TWLTL is the only practical solution. As left-turning vehicles are provided a separate space to slow down and wait for gaps in traffic, the interference to through-traffic in the remaining lanes is minimized (AASHTO, 2001: 479). The construction of dedicated lanes for traffic turning at intersections can therefore help to separate turning traffic from through-flow traffic, reduce conflicts and promote safety and improved traffic flow. Major arterials with high traffic volumes and relatively high posted speeds can benefit most from such lanes, including right-turn acceleration lanes, auxiliary left-turn lanes, and auxiliary right-turn lanes.

Exclusive turning lanes are best suited for roadways that have moderate to high volumes of traffic. The most effective way to construct turning lanes is to develop them in combination with medians. By building an additional lane on an existing two or four lane road, the middle lane can then be converted to a median with gaps for dedicated left-turn and right-turn lanes. At intersections, a further lane on each side of the road is required as it is necessary to cater simultaneously to right- and left-turning traffic. The medians can be developed just at the intersections (limited raised median) or along the entire roadway. A traffic impact study will have to be carried out to determine the need for turning lanes based on the existing traffic volumes, speed and the projected impact of the proposed use, and make recommendations on the design of the turning lane. In general, the length of the turning lane should be a minimum two-car-lengths and based on the number of vehicles likely to arrive in a 2-minute period at peak hour (unsignalized) or the signal length and timing (Nashua Regional Planning Commission, 2002: 17).

Studies have shown that the addition of the exclusive turn lanes contributes significantly to improved safety, reduced congestion and delay. Left turns into driveways cause conflicts when they stop in a lane also serving through traffic, and they degrade the arterial function by affecting the average travel speed of those through-vehicles. They also expose themselves and other vehicles to risks because of speed differences (McShane et al., 1996). A major synthesis of research on left-turn lanes demonstrated that exclusive turn lanes reduce crashes by between 18 and 77% (average 50%) and reduce rear-end collisions between 60 and 88% (Gluck et al., 1999). Left turn lanes also substantially increase the capacity of many roadways. A shared left-turn and through lane has about 40 to 60% of the capacity of a standard through lane (Gluck et al., 1999). Another compilation of research in this topic found a 25% increase in capacity, on average, for roadways that added a left-turn lane (S/K Transportation Consultants, 2000).

Some States and cities have adopted indirect turns to reduce these conflicts. In New Jersey, the jug-handle left turn requires a right turn onto a feeder street, followed by a left turn onto a cross street. Detroit has extensively used an indirect U-turn that requires a U-turn past an intersection, followed by a right turn instead of a regular U-turn. Like dedicated left turn lanes, indirect turns reduce crashes, improve congestion and add capacity. Crashes decline by 20% on average and 35% if the indirect turn intersection is signalized. Capacity typically shows a 15 to 20% gain (Gluck et al., 1999).

2.2.7   Median Treatments

Traffic flow on an arterial is often impeded by the intermittent stoppage of vehicles intending to turn left into small driveways, but this does not generally pose severe problems when the traffic volume is low. When the flow becomes heavy, for example, during peak hours serious bottleneck situations may arise, and such turning traffic may contribute to massive congestion and safety problems. Median treatments for roadways represent one of the most effective means to control access to small driveways. They serve a dual role of separating opposing traffic movements and managing left turns and crossing movements between driveways and arterials. By controlling left turning movements, conflicts between through and turning traffic can be reduced, resulting in improved safety. Medians are most effective in roadways with high volumes and four or more lanes of traffic.

TABLE 2.6   Accident Rates in Roadways with Different Median Types

Total Access Points Per Mile[a]	Representative Accident Rates (Crashes per Million VMT) by Type of Median Urban and Suburban Areas		
	Median Type		
	Undivided	Two-Way Left-Turn Lane	Non-Traversable Median
<20	3.8	3.4	2.9
20.01–40	7.3	5.9	5.1
40.01–60	9.4	7.9	6.8
>60	10.6	9.2	8.2
Average rate	9.0	6.9	5.6

Source: Gluck et al., 1998.

[a] Includes both signalized and unsignalized access points.

The two major median treatments include TWLTL and raised medians. The TWLTLs are open and traversable for vehicle turns at designated zones, which are marked. Raised medians fully separate opposing traffic and are non-traversable for turns and crossings. There are breaks in the median to provide access to properties on either side of the road or for U-turns. It is desirable to limit the number of median breaks as they contribute to conflict points and safety hazards. They are usually provided only at shared driveways and intersections. Access to a driveway with only a single business or residence from the opposite side of a major road can only be made through a U-turn further down the road or a more circuitous route involving a driveway where a break is available.

Because median treatments reduce access along an arterial, they are often resisted by property and business owners who feel their customers, sales, and property values may become negatively affected. The need for a median should therefore be assessed carefully through a traffic engineering study to evaluate the safety, traffic flow, and economic impacts of the increased access control. Such a study would take into account considerations such as through traffic volumes, speed, number and configuration of lanes, left turn volumes, access point density, right-of-way width, and land uses along the roadway. The results of the study would not only ascertain the need for a median, but also provide directions on the appropriate type of median and its design features.

Generally, the TWLTL design works well where the speed on the arterial highway is relatively low (40 to 70 km/h or 25 to 45 mi/h) and there is no heavy concentration of left turn traffic (AASHTO, 2001: 479). At higher traffic volumes and speeds and higher driveway densities, the TWLTL becomes less effective. Some useful insights on the situational requirements of a TWLTL and raised medians have been undertaken. One study suggests that where the average daily traffic (ADT) volume exceeds 20,000 vehicles per day and the demand for mid-block turns is high, a raised median should be considered (Texas DOT, 2004: 1–5). However, if driveway densities are low, TWLTLs may still be used at higher levels of ADT. However, if the ADT is projected to exceed 28,000 a raised median is recommended (CTRE, 2003a, 2003b).

In terms of safety, a study of corridors in several cities in Iowa found that TWLTLs reduced crashes by as much as 70%, improved level of service by one full grade in some areas, and increased lane capacity by as much as 36% (Iowa DOT, 1997). However, several other studies have shown that roadways with non-traversable medians are always safer than those with continuous TWLTL. The average crash rate is about 30% lower than roadways with TWLTL. Table 2.6 summarizes the representative crash rates by median type for urbanized areas. A well-designed median can also provide a pedestrian refuge at crossings. In Georgia, it was found that raised medians reduced pedestrian-involved crashes by 45% and fatalities by 78%, compared to TWLTLs (Parsonson et al., 2000). Another study looking at pedestrian exposure risk along a four lane suburban roadway in New Jersey following the construction of a raised median found that the risk reduced by 28% (King et al., 2003). Raised medians can also add to the overall aesthetics of a roadway corridor by incorporating landscaping or other items of visual interest (Texas DOT, 2004).

2.3 Economic Impacts of Access Management and Public Participation

The discussion above on different types of access management measures has highlighted the significant positive impacts on capacity, speed, and safety of roadways that have undergone some form of partial or full access control. There are also benefits in the form of improved pedestrian safety, better aesthetic quality of the community, and significant environmental benefits. The comparative assessment of access management practices in various places is useful for promoting a better understanding of the limitations and potentials of different tools (Frawley and Eisele, 2001). Some of the studies have yielded useful models quantifying the impact of access management techniques on safety and traffic flow (Lomax et al., 1997; Gluck et al., 1999). These models may be applied to new contexts where proposed access management effects are evaluated, provided there are site specific adjustments that take into account land use, road, and traffic configurations (Miller et al., 2001). Recent developments in simulation tools and analyses have also made it easier to construct site specific models since critical "parameters and variables can be controlled efficiently and the underlying effects be studied more systematically" (Prassas and Chang, 2000: 17; Yang and Zhou, 2004).

While the extent of benefits of access management measures relating to the effective functioning of a roadway system may vary from place to place, it is clear that the benefits are there and are significant. However, the performance of various traffic indicators is only one dimension of the overall impact of access management. There are also corollary economic and community welfare impacts brought about by the implementation of access management schemes, and in this area, there is much less agreement on the value of such schemes.

From the point of view of a business enterprise located along an arterial, direct access of the property to the roadway is often desired as it is perceived to contribute to higher levels of customer flow, increased sales and higher property value. Businesses become naturally concerned about the negative impact of the reduced accessibility resulting from the need to make circuitous trips following various access management measures such as driveway consolidation, median treatments, and the conversion of streets from two-way to one-way driveways. Such perceived negative impacts can sometimes lead to a stalemate in proposed access management projects.

The resistance to such measures is more pronounced when existing roadways where the businesses are already located become earmarked for new access management treatments. It is likely that when the roadway was first built, the high growth rates in the traffic volumes was perhaps not expected, and the land use and access standards were not that tightly regulated. However, the surge in traffic and, consequently, the serious congestion and safety problems may leave city and state authorities with little choice and they have to take these measures to improve the efficiency of movement of traffic along major roads. In newly developed land, the problem of acceptance is less serious but still prevails. It is easier to review and modify the local ordinances and building plans to ensure that access management standards and guidelines are followed in the development of new residential and business districts. However, "considerable political pressure is often exerted to reduce the amount of access control because of fears that developers may be turned off and decide to build elsewhere or that voters may object to perceived inconvenience and risks" (Giguere, 2003).

Interestingly, the experiences from many corridors where various access management projects have been implemented actually show that there are few adverse economic and community impacts. Since the mid 1990s, several studies on the economic and social impacts of access management projects have been carried out. A large majority of these studies focus on the effects of median treatments, which restrict left-turning vehicles on business activity as these constitute one of the most controversial schemes in access management. Most of these studies are not based on actual sales data because of the proprietary nature of this information. Rather they largely focus on the opinions and perceptions of owners of commercial businesses about the impacts of access management measures, before and after case examples, or generalized comparisons of business activity across corridors (Williams, 2000). Some of the findings of these studies are presented below.

In 1995, a survey was conducted both on drivers and business owners in Florida on the impact of raised medians on specific corridors and the results indicated that while the general perception was an additional inconvenience would caused by the median, it was more than compensated by smoother traffic flow and improved safety. 78% of the drivers surveyed felt safety had improved while 84% felt traffic moved better. However, about 30% of the business owners noted that they had experienced some negative impacts such as delivery difficulties and reduction in business volume (Ivey et al., 1995). Another study looking at impacts of access management projects in southeast Florida found that 26% of the business owners reported a loss in profits after the measures were implemented. About 5% reported a gain while the remaining large majority of businesses saw little or no negative impact of their profitability, number of customers, or property values (Vargas and Gautam, 1989).

A synthesis of more recent studies on the economic impacts of median projects has been undertaken by Williams (2000). She examined the results of work done in Kansas, Iowa, Texas, and Florida, and reported similar conclusions to the two studies mentioned above. The adverse economic impacts were small, and were mostly restricted to particular businesses that relied on pass-by traffic, such as gas stations and convenience stores. While a few businesses reported increases in sales, the majority experienced no significant change in business activity. Specifically, the Texas and Florida studies highlighted that business expectations of negative impacts before project implementation were higher than what was actually the case after. The general conclusion that can be made from a broad survey of the impact studies is, access managed corridors remain attractive for business, and well-managed access does not contribute, for the most part, a declining customer base. The Texas study furthermore indicated that corridors with access control improvements experienced an 18% increase in property values after construction (Eisele and Frawley, 1999, cited in FHWA, 2003).

It is apparent that a reduction in the level of direct access is a real cause for concern for owners of businesses and other developments. However, it is often not appreciated that the conflicts resulting from poor or excessive access points contribute to congestion and delays in the community's roadway system, making the very region where the businesses are located inaccessible and unattractive. In the long term, the region will lose its economic vitality and businesses will move out to higher quality locations. Clearly, there is a need for transportation authorities to engage with the community in its access management policy, to explain the rationale, demonstrate the community-scale and long-term benefits by drawing from similar projects elsewhere, and ensure that the project is planned and designed in a way that incorporates the more critical concerns of the stakeholders. With a more direct and meaningful involvement of affected businesses and landowners, it is likely that proposed projects may find greater acceptance and success in achieving the access management objectives (Williams, 2000: 4).

Opportunities must be presented for greater public engagement in access management policy making, and the entire community, not just the affected businesses, should be encouraged to participate in the process. Without public participation, it will be likely that there will be strong reactions to the project but these will be from those who are directly affected by the project. This group will tend to focus on how the project will affect them individually and will not look at the broader and longer term benefits to the community. Without the wider community involvement, therefore, "it may appear to decision makers as if a project has little public support, even if it does. The result is pressure for major changes in the project as affected parties appeal their case to elected and agency officials" (Iowa CTRE, 2000a, 2000b: 6.2).

Depending on the scope of a project, public involvement has to be incorporated from the conception or study phase right through to the design and implementation phase of the project. This will ensure that key issues on the ground are not overlooked, and that the public is well informed about what the project entails. The forms of public involvement can range from public forums, information newsletters, presentations by planners, public plan displays, special local citizens' task forces, and feedback groups. The goal is to build a bridge between planners and the people, to nurture a working partnership that strengthens the credibility and acceptance of transportation agencies and their policies. A study on median projects in Florida found that the state's DOT offices with a public involvement for its projects had greater success in achieving their projects' objectives and fewer appeals for administrative hearings than those who relied on public hearings (Williams, 2000).

2.4   Conclusion

The growth of direct access links to arterials and highways and the poor management of roads in relation to turning and through traffic contribute to the deterioration of the functional integrity of the roadway system. The development of new land uses and access points coupled with the traffic volumes reduces the speed and capacity of the adjacent roadways and increases congestion and accidents. Access management can ameliorate these problems by reducing and separating the conflict points through various techniques such as creating dedicated turn lanes, limiting the number of left-turn access points by installing medians, and promoting driveway consolidation. The Chapter has outlined the key principles of access management and provided some insights on the established methods, with illustrations on the benefits and impacts using specific case studies.

The road and highway system is an important asset for the circulation of people and goods, for sustaining the economic and social life of a city or region. Roads are also expensive assets, and it is not prudent to simply add on new capacity in the roadway system when it is not able to cope with the growing traffic volume. A more efficient and cost-effective solution would be to examine closely the structure and flows in the existing roads and highways, and to evaluate the possibility of easing or removing the bottlenecks to smooth traffic flow. Access management is all about better managing existing capacity, and ensuring that new roads are built to last in the face of future land developments.

It is the responsibility of transportation authorities to ensure that reasonable access is provided to properties, and it is important to get property owners to understand that reasonable access does not equate to direct access to a main arterial or highway. Clear rules and guidelines have to be in place to specify when direct access may or may not be provided, and these rules should be consistently applied on a wider scale, preferably across a state jurisdiction. Many access management projects focus on particular corridors and the affected community may not accept the measures if they do not understand why the same rules do not apply to other corridors. This points to the need for a comprehensive access management policy at the state jurisdiction level that addresses clearly the questions of why, when, where, and how direct access should be provided or denied with a good explanation for exceptional circumstances. The policy has to be placed within a sound legal and regulatory framework to provide the authorities the necessary power to enforce the decisions.

While the traffic impacts of access management may yield visible positive impacts, the perceived non-traffic impacts deserve considerable attention as they bring social, political and institutional issues to the surface, and may lead to cumbersome legal battles and protestations that may derail the projects. Public involvement at the wider community level is crucial for the acceptance and successful implementation of well-conceived projects, but this can only be effectively executed if transportation planners and officials have a broad-based understanding of the implications of proposed measures.

Land use planning issues are tightly interwoven with highway planning, especially in terms of location, and design of roadways. Access management strategies should therefore be considered alongside a long-range comprehensive regional land use plan, where the transportation demands of future proposed land uses are factored into a systematic highway development and access management plan. Access management is a multidisciplinary field that involves not only traffic engineering and land use planning, but also transportation management, real estate, land law, right-of-way issues, and it is "important that all the individuals involved with each of these functions be on the 'same page' whether at the program or project level" (Giguere, 2003).

In many countries across the world, it can be observed that traffic volumes, both in terms of number of vehicles and miles driven, have outstripped the increase in road capacity. Increasing the capacity through building new highways or expanding current ones will be necessary but this option is difficult and expensive, given the competing demands for land in a growing region. The need to better manage existing road capacity has naturally become more critical under these circumstances, and the application of access management techniques will undoubtedly gather much momentum in the coming years. Even though the field is new, many of the tools and methods that come under its umbrella have been around for much longer. The challenge is to consolidate the existing knowledge and experiences, to explore and experiment with new methods, and to develop and market comprehensive and systematic programs that will receive community support.

Acknowledgment

The assistance of Gabriel Tine in the preparation of the Chapter is thankfully acknowledged.

References

AASHTO (American Association of State Highway and Transportation Officials), 2001. *A Policy of Geometric Design of Highways and Streets*, Washington, DC.

AASHTO (American Association of State Highway and Transportation Officials), 2002. *Bottomline*, Washington, DC.

ADB (Asian Development Bank), 1998. *Handbook on Resettlement: A Guide to Good Practice*, Manila, http://www.adb.org/Documents/Handbooks/Resettlement/default.asp

Brown, H.C., Labi, S., Tarko, A.P., and Fricker, J.D. 1998. *A Tool for Evaluating Access Control on High-Speed Urban Arterials*, FHWA/IN/JTRP-98/7, Indiana Department of Transportation, Final Report. Purdue University. West Lafayette, Indiana.

Caltrans (California Department of Transportation), 2001. *Highway Design Manual*, November.

Cervero, R., Induced travel demand: research design, empirical evidence and normative policies, *J. Plan. Lit.*, 17, 1, 3–20, 2002.

Chan, C.S. 1999. Measuring Physical Density: Implications on the Use of Different Measures on Land Use Policy in Singapore, Thesis (M.C.P.), Massachusetts Institute of Technology, Department of Urban Studies and Planning, Cambridge, MA.

CTRE (Center for Transportation Research and Education), 2000. *Access Management Handbook*. Iowa State University, Ames, IA.

CTRE (Center for Transportation Research and Education), 2000. *Access Management Research and Awareness Program Phase 4 Final Report*, CTRE Management Project 97-1. Iowa State University, Ames, IA.

CTRE (Center for Transportation Research and Education), 2003. *Access Management Frequently Asked Questions 2: Driveway spacing*. Iowa State University, Ames, IA, http://www.ctre.iastate.edu/research/access/toolkit/2.pdf

CTRE (Center for Transportation Research and Education), 2003. *Access Management Frequently Asked Questions 3: Driveway Density and Driveway Consolidation*. Iowa State University, Ames, IA, http://www.ctre.iastate.edu/research/access/toolkit/3.pdf

CUTR (Center for Urban Transportation Research), 1998. *Ten Ways to Manage Roadway Access in Your Community*. University of South Florida, http://www.cutr.eng.usf.edu

Demosthenes, P. 1999. Access management policies: an historical perspective. *The International Right-of-Way Association Conference*, Albuquerque, NM, July 23, http://www.cutr.usf.edu/research/access_m/ada70/History%20of%20AM.pdf

Eisele, W. and Frawley, W. 1999. A methodology for determining economic impacts of raised medians: data analysis on additional case studies, *Research Report 3904-3*, Texas Transportation Institute, College Station, TX, October.

Ewing, R. and Lichtenstein, A. 2002. *Induced Traffic and Induced Development*. Chester County Planning Commission, West Chester, PA, Unpublished Report, October, http://www.chesco.org/planning/pdf/InTrInDv.pdf

FHWA (Federal Highway Administration, US), 2002. *Highway Statistics*. FHWA, Washington, DC.

FHWA (Federal Highway Administration, US), 2003. *Benefits of Access Management*, Information Brochure, http://safety.fhwa.dot.gov/geometricdsgn/accessmgmtbrochure/accessmgmtbrochure.pdf

Frawley, W.E. and Eisele, W.L. 2001. Lessons learned: access management programs in selected states. In *Eighth TRB Conference on the Application of Transportation Planning Methods*, pp. 305–313. Corpus Christi, Texas, April 22–26.

Giguere, R.K. 2003. *Access Management, A1D07: Committee on Access Management.* Transportation Research Board, Washington, DC.

Gluck, J. and King, P., Access management at a high-volume interchanges, *Transport. Res. Rec. — J. Transport.*, 133–141, 2003.

Gluck, J., Levinson, H.S., and Stover, V. 1998. Overview of NCHRP project 3–52: impacts of access management techniques. In *Proceedings of the Third National Conference on Access Management*, Ft. Lauderdale, FL.

Gluck, J., Levinson, H.S., and Stover, V. 1999. *Impact of Access Management Techniques*, NCHRP Report 420. National Cooperative Highway Research Program, Transportation Research Board, National Academy Press, Washington DC.

Goodwin, P.B., Empirical evidence on induced traffic: a review and synthesis, *Transportation*, 23, 1, 35–54, 1996.

Hansen, M. and Huang, Y., Road supply and traffic in California urban areas, *Transport. Res. Part A*, 31A, 3, 205–218, 1997.

Hills, P.J., What is induced traffic?, *Transportation*, 23, 1, 5–16, 1996.

Iowa DOT (Department of Transportation), 1997. *Access Management Awareness Program: Phase II Report.* Iowa DOT, Ames, IO.

ITE (Institute of Transportation Engineers, USA), 2004. *Access Management: A Key to Mobility and Safety*, Issue Briefs No. 13, April.

Ivey, Harris, and Walls, Inc. 1995. Corridor land use, development and driver/business survey analysis. *District Wide Median Evaluation Technical Memorandum*, Final report prepared for the Florida Department of Transportation, Winter Park, Florida, November 1995.

King, M.R., Carnegie, J.A., and Ewing, R., Pedestrian safety through a raised median and redesigned intersections, *Transport. Res. Rec.*, 1828, 56–66, 2003.

Koepke, F.J. and Levinson, H.S. 1992. *Access Management Guidelines for Activity Centers*, National Cooperative Highway Research Program, Report 348, Transportation Research Board, National Research Council, 1992.

Land, L.A. 2001. Land development and access management strategies for interchange areas. In *Eighth TRB Conference on the Application of Transportation Planning Methods*, pp. 3–6. Corpus Christi, Texas, April 22–26.

Levinson, H.S. 1999. Street spacing and scale. In *Proceedings of the Urban Street Symposium*, pp. 28–30. Dallas, June.

Lomax, T., Turner, S., Levinson, H.S., Pratt, R., Bay, P., and Douglas, T. 1997. *Quantifying Congestion*, NCHRP Report No. 398. Transportation Research Board, Washington, DC.

Marks, H. 1974. *Traffic Circulation Planning for Communities*, Gruen Associates under Commission with Motor Vehicle Manufacturers Association Inc., Los Angeles, CA.

McShane, W.R., Choi, D.S., Eichin, K., and Sokolow, G. 1996. Insights into access management details using TRAF–NETSIM. In *Proceedings of the Second National Conference on Access Management*, pp. 471. Vail, Colorado.

Miller, J.S., Hoel, L.A., Kim, S., and Drummond, K.P., Transferability of models that estimate crashes as a function of access management, *Transport. Res. Rec.*, 1746, 14–21, 2001.

Missouri DOT (Department of Transportation), 2003. *Access Management Guidelines*, 12 September 2003, Jefferson City, MO, http://www.modot.state.mo.us/newsandinfo/documents/Access MgmtGuidelines_1003.pdf

Nashua Regional Planning Commission, 2002. *Access Management Guidelines.* New Hampshire Department of Transportation, Nashua, NH.

Nevada DOT (Department of Transportation), 1999. *Access Management System and Standards*, Carson City, NV, July.

Noland, R.B., Relationships between highway capacity and induced vehicle travel, *Transport. Res. Part A*, 35A, 1, 47–72, 2001.

OKI (Ohio-Kentucky-Indiana) Regional Council of Governments, 1999. *The Case For Access Management*, http://www.oki.org/transportation/caseforaccess.html

Oregon DOT (Department of Transportation), 2001. *Access Management Manual: Approach Application and Permit Process*, Planning Section, Access Management Program Unit.

Oregon DOT (Department of Transportation), 2002. *Highway Design Manual*, Salem, OR, http://www.odot.state.or.us/techserv/engineer/pdu/PDFs/Highway%20Design%20Manual%202002/2002Chp6-Freeway.pdf

Parsonson, Peter S., Waters III Marion G., and Fincher, James S. 2000. Georgia study confirms the continuing advantage of raised medians over two-way left-turn lanes. In *Fourth National Conference on Access Management*, Transportation Research Board and FHWA's Office of Technology Applications, Oregon DOT, Portland, August 14.

Poe, M.P., Tarris, J.P., and Masin, J.M. 1996. Influence of access and land use on vehicle operating speeds along low-speed urban streets, In *Proceedings of the Second National Conference on Access Management*, p. 339. Vail, Colorado.

Prassas, E.S. and Chang, J.I., Effects of access features and interaction among driveways as investigated by simulation, *Transport. Res. Rec.*, 1706, 17–28, 2000.

Reilly, W. 1990. *Capacity and Service Procedures for Multi-lane Rural and Suburban Highways*, Final Report NCHRP Project 3–33. JHK and Associates and Midwest Research Institute.

S/K Transportation Consultants Inc., 2000. *Access Management*, National Highway Institute Course Number 133078, Location and Design.

Texas DOT (Department of Transportation), 2004. *Access Management Manual*, Design Division, Austin, TX, June.

TRB (Transportation Research Board), 2003. *Access Management Manual*. TRB Committee on Access Management, Washington, DC.

Vargas, F. and Gautam, Y. 1989. *Problem: Roadway Safety vs. Commercial Development Access*, ITE 1989 Compendium of Technical Papers, pp. 46–50. ITE (Institute of Transportation Engineers, USA), Washington, DC.

Williams, K.M. 2000. *Economic Impacts of Access Management*, Center for Urban Transportation Research. University of South Florida, Tampa, January 28.

Yang, X.K. and Zhou, H.G., CORSIM-based simulation approach to evaluation of direct left turn versus right turn plus U-turn from driveways, *J. Transport. Eng. -ASCE*, 130, 1, 68–75, 2004.

3

Environmental Impact Assessment of Highway Development

M.A. Aziz
National University of Singapore
Republic of Singapore

3.1 Introduction

Highway development enhances mobility and is critical to the economic growth of a community and a country as a whole. Unfortunately, inappropriately planned, designed, and constructed highways can aggravate the conditions of the poor, and harm the natural and socio-economic environment. The common adverse impacts of highway development include damage of natural landscape, habitat and bio-diversity, destruction of cultural and social structure of affected communities, creation of air and water pollution, and generation of noise and vibration.

To minimize adverse environmental and socio-economic impacts, highway infrastructure must be built to a high quality and maintained to a high standard. This can be achieved by integrating environmental considerations into highway development planning, design, and construction. The process consists of three key elements:

(a) Identification of the full range of possible impacts on the natural and socio-economic environment;
(b) Evaluation and quantification of these impacts and
(c) Formulation of measures to avoid, mitigate and compensate for the anticipated impacts.

The above process which systematically deals with these elements is called Environmental Impact Assessment (EIA). The end results are presented in an EIA report including an Environmental Management Plan (EMP). It must be emphasized that an EIA process is not solely aimed at identifying, quantifying, and mitigating the negative impacts of a highway development project. It should also be used to optimize positive impacts of the project (World Highways, 2001).

3.1.1   Environmental Impacts of Highway Development

Possible adverse impacts of highway development projects are varied. Some of the adverse impacts that can be caused by highway development activities are summarized in Table 3.1.

3.1.1.1   Types of Environmental Impacts

Environmental impacts of highway development can be grouped under the following categories: (a) direct and indirect, (b) cumulative, (c) local and widespread, (d) short- and long-term, (e) temporary and permanent and (f) random and predictable (WB, 1997).

Direct impacts. These are caused by various highway development processes such as land acquisition, removal of vegetal cover, and erosion of productive soils. For example, taking of soils from borrows pits; and sands, gravels, and stones from quarries for use in highway construction are obvious direct impacts. In these cases, the land area in which the pit and quarry sites are located has been directly affected by activities associated with the highway development project. Direct impacts generally are relatively easy to inventorize, asses, and control than indirect ones.

Indirect impacts. Indirect impacts of highway development, also known as secondary, tertiary, or chain impacts, may have more profound consequences on the environment than direct impacts. Indirect impacts are more difficult to measure. Over time, they can affect larger geographical areas of the environment than anticipated. Examples include degradation of surface water quality by the erosion of land cleared and excavated for the construction of new highways. Another common indirect impact associated with the construction of new highways is increased deforestation of an area or the influx

TABLE 3.1   Possible Adverse Environmental Impacts of Highway Development Activities

Activities	Possible Adverse Environmental Impacts
Land acquisition for roadway and right-of-way	Displacement of residences and businesses
	Divided community with restriction to mobility and reduced social activities
	Loss of cultural and historical heritage
Site preparation for road construction	Loss of vegetative cover and top soil
	Degradation of habitat and possible loss of valued ecosystem components
	Accelerated erosion
	Reduced retention of surface water
	Deterioration of water quality
	Damage to flora and fauna
Earthwork of road construction	Change of landscape
	Interference with surface and groundwater hydrology
	Disruption of natural drainage system
	Unstable slopes and landslides
	Damage to flora and fauna
Road construction operations	Generation of dust, debris and waste
	Air quality degradation
	Noise, vibration and odors
	Water and soil contamination
	Traffic congestion and travel delay
	Disturbances to local community by construction workers
	Spread of bacterial and viral diseases

of settlers. In areas where wild life is plentiful, new highways often lead to the rapid depletion of animals and vegetation due to poaching. The acquisition of land to build a new highway may displace farmers and business communities directly or it may interfere with their cropping, livestock raising, fisheries, and business activities.

Cumulative impacts. Cumulative environmental impacts may arise from any of the following types of highway events: (a) single large events, i.e., large highway projects (b) multiple interrelated events, such as highway network projects within a region (c) catastrophic sudden events, such as major landslides (d) incremental, widespread, slow change, such as poorly designed culvert or drainage system along a highway extending through a watershed. These can generate additive, multiplicative, or synergetic impacts, which can cause disruption to the functions of one or more ecosystems. Cumulative impacts might occur due to the destruction of vegetation and eventual erosion. The vegetation usually does not have enough time to recover because of traffic on the highway and the problem is exacerbated over time.

Local and widespread impacts. Local impacts include adverse effects in the immediate vicinity of a highway development project such as the destruction of buildings or restricted access to farms. Widespread impacts can occur relatively far from the highway development site. These impacts are often linked to indirect adverse effects that arise due to ribbon development along the highway right-of-way which include the influx of settlers, unauthorized construction of new houses, shops, and small industries.

Short- and long-term impacts. Short-term impacts are those, which appear during or shortly after the highway construction. Long-term impacts may arise during the operation and maintenance phases of the highway and may last for decades.

Temporary and permanent impacts. Temporary impacts are those whose occurrence is not long lasting and which eventually reverses. Permanent impacts are irreversible. The affected ecosystem does not return to its previous state on a human timescale.

Random and predictable impacts. In highway development projects, it is useful to distinguish between highly probable impacts from random or unpredictable impacts with a low probability of occurring but may have serious consequences. For example, the construction of a highway through a large, densely populated area will result in population displacement and business loss. On the other hand, incidents such as accidental pollution, fire, or spillage of toxic products are, by nature, unpredictable.

3.1.2   Need for EIA

The EIA evaluates the environmental impacts expected at various stages of highway development and offers a useful basis for planners, consultants, contractors, design, and construction engineers to formulate and select the most desirable strategies for different phases of the highway development process. To achieve the best results, EIA should be performed as a part of the highway development process.

The EIA process consists of the following basic stages (ESCAP, 1990):

1. Identification and quantification of possible positive and negative environmental impacts resulting from the proposed highway development project.
2. Recommendation of measures to reduce or offset the negative impacts of the proposed highway development project with the aim of achieving the minimum level of environmental degradation. The measures can be corrective in nature or as modifications to the proposed plan.
3. Monitoring of the level of enforcement and degree of effectiveness of the recommended environmental protection measures during the project implementation phase, and performing postconstruction auditing and evaluation of the degree of effectiveness of the implemented environmental protection measures.

For a sustainable highway development, it is a must to conduct an EIA study. An EIA team requires experienced environmental management professionals supporting the engineering team. To be effective, the EIA process should be initiated from an early stage of a highway development project.

3.1.3   EIA Guidelines for Highway Development Projects

EIA guidelines for general types of infrastructure development projects are varied. EIA guidelines for highway development have been developed by international organizations, bilateral and multilateral development agencies such as the United Nations Environmental Programme (UNEP), United Nations Economic and Social Commission for Asia and the Pacific (UNESCAP), World Bank (WB), United States Agency for International Development (USAID), Asian Development Bank (ADB), Organization for Economic Cooperation and Development (OECD), and Inter-American Development Bank (IADB). Some references for examples of highway specific EIA guidelines are listed in Table 3.2.

Countries such as Australia, Britain, Canada, China, India, Japan, Malaysia, Philippines, the Republic of Korea, Thailand, and U.S.A. have developed various forms of EIA guidelines specifically on highway development projects. A brief review of the highway specific EIA guidelines adopted by some international organizations (ADB, 1997; WB, 1997) reveals that there are some basic differences among these guidelines. In some countries, EIA is mandatory as a statutory requirement. In most developing countries, EIA is required only for projects financed by international funding agencies. Some countries deal only with the procedural aspects rather than technical aspects of EIA, while others deal with both. Some guidelines concentrate on the procedure for EIA study and preparation of the EIA report, leaving approval and implementation of EIA to other agencies.

3.1.4   Procedural and Substantive Issues of EIA Process

The major issues of the EIA process can be classified into two main groups: procedural and substantive issues. Tables 3.3 and 3.4 list the key procedural and substantive issues, respectively. Institutional barrier is a major procedural issue in effective implementation of EIA. Very often EIA for highway development projects has been conducted loosely and taken as a supplementary requirement secondary to the overall economic and engineering issues. The first step is therefore to reorient the conventional planning process of highway development projects. Parallel to the consideration of economic benefits for highway development projects, equal emphasis should be given to environmental and socio-economic issues in all phases of the highway development process. A strong coordination, cooperation, and interrelationship between various environmental professionals, particularly the EIA-undertaking agencies and EIA-approving authorities, need to be maintained.

TABLE 3.2   References for Highway Specific EIA Guidelines

S/no.	EIA Guidelines
1	Environmental impact assessment of road transport development: the state-of-the-art in selected countries of the ESCAP region: Background paper submitted to the Seminar on the EIA of Road Transport Development, UNESCAP, Bangkok, 3–8 November 1986
2	Report of the ESCAP/UNEP Seminar on Environmental Impact Assessment of Road Transport Development, UNESCAP, Bangkok, 3–8 November 1987
3	Environmental Impact Assessment of Road Transport Development: The State-of- the-Art in Selected Countries of the UNESCAP Region. ESCAP Bangkok, 1988
4	Environmental Impact Assessment: Guidelines for Transport Development. Publication No. ST/ES-CAP/785, UNESCAP, Bangkok, 1990
5	Environmental Impact Assessment of Roads, Organisation for Economic Cooperation and Development (OECD). Paris, 1994
6	Roads and the Environment: A Handbook Report TWU 13. Transport Division, World Bank, Washington DC, 1994
7	Environmental Assessment in the Transportation Sector — Guidelines for Managers. Inter-American Development Bank IADB. Washington DC, 1996
8	Roads and the Environment: A Handbook Transport Division, World Bank Paper No. 376, Washington DC, 1997
9	Multistage Environmental and Social Impact Assessment: Guidelines for a Comprehensive Process. UNESCAP. Publication No. ST/ESCAP/2177, 2001

TABLE 3.3   Issues of Procedural Consideration in EIA Process

Procedural Consideration	Issues
1. Institutional infrastructure and legal frameworks	Key aspects of environmental protection and conservation for the specific project
	Integration of EIA into the road project approval process
	Policy framework for integration of EIA into the road project cycle
	Measures of monitoring and evaluation for compliance
2. EIA requirements	Identification of the extent of scope of EIA
	Financial resources available
	Institutional and technical capabilities
	Competency of consultants and contractors
3. EIA process	Stages of the EIA process to be adopted
	Topics in the EIA study
	Main environmental issues to be focused on
	Information to be provided to planners and decision makers
4. Collection of information and data	Availability of planning data
	Different agencies to be involved and responsibilities
	Adequacy of available data
	Reliability and compatibility of available data
	Effectiveness of available standards and norms for EIA
	Institutional framework for coordinated data collection
	Guidelines on methods of data collection
	Accessibility of EIA data and information by the public and nongovernmental organizations
5. EIA implementation	Roles of politicians, senior government officials and decision makers
	Political and management interference in EIA implementation
	Availability of EIA guidelines
	Requirements of EIA reports
	Institutional and administrative framework for implementation of EIA
6. Enforcement of EIA	Adequacy of institutional infrastructure
	Commitment of relevant agencies
	Environmental awareness of relevant agency officials
	Budget and manpower availability
	Availability of guidelines and standards
	Site inspection and monitoring capability
7. Public participation	Guidelines for public hearing and involvement of stakeholders
	Institutional arrangement for conducting public hearings
	Guidelines for acquiring information, data and feedback from the public
	Provision of public involvement at various stages of the road EIA process

Public participation is not generally given due consideration in the present highway specific EIA process. Proper public participation helps to mitigate environmental and socio-economic impacts. Some impacts will only surface and be felt at the end of the highway construction and at the beginning of operation. While the number of people affected and the magnitude of direct impacts during the highway construction phase are usually limited and easier to deal with, the postconstruction consequences tend to last for a long period of time and cover a much wider area geographically. In addition, the adversely affected population is usually dispersed along a long stretch of the highway and their voice is often not heard in the implementation of the highway development project.

Most EIA process focuses on broad environmental considerations such as national forests and wildlife, historical and cultural sites, critical land features and water-bodies, guided by the general environmental protection laws. For better protection of the natural and human environment, the substantive level

TABLE 3.4 Specific Substantive Environmental Concerns in EIA Process

Parameter	Issues of Environmental Concern
1. Soil	Earth movements and slope failures
	Soil erosion
	Changes of surface relief
	Sedimentation of roadside waterbodies and drains
	Loss of productive topsoil in borrow areas
	Soil contamination
2. Water	Changes in flow of surface water and groundwater
	Water quality degradation by waste materials, equipment lubricants, fuels and detergents
	Sedimentation of surface waterbodies
3. Ecosystem	Damage of vegetation, habitat and biodiversity
	Destruction of reproduction and food zones for fish, aquatic and migratory birds
	Contamination of biota
4. Air	Generation of dust and toxic gases
	Vehicle and equipment emissions from construction activities
	Degradation of air quality caused by construction wastes
5. Landscape	Destruction of natural landscape
	Changes in surface drainage network
	Soil erosion and movements
	Damage to vegetation and trees
6. Social and cultural activities	Disintegration of social activities caused by physically divided community
	Disruption of traditional nonmotorized modes of transport
	Poor living environment of roadside ribbon development
	Displacement and resettlement of inhabitants, businesses, private and public institutions
7. Human health and safety	Degradation of living environment due to pollution of air and water, noise and vibration
	Road accidents

of environmental considerations needs to be addressed closely. Appropriate mitigation measures could be directly transformed into standard construction contracts automatically complying with the requirements.

3.2 EIA in Phases of Highway Development

The highway development process has five phases: (a) conception (b) planning (c) design (d) construction (e) operation. To ensure sustainable highway development, it is essential that EIA be carried out in all the five phases.

The EIA process must be fully integrated with different phases of the highway development cycle, including the postconstruction environmental auditing and evaluation. This means the EIA process should be activated with the initiation of environmental screening as and when the highway development project conception phase commences. The EIA process must not be terminated prematurely with the conclusion of the highway construction. It must be continued through the postconstruction environmental auditing and evaluation stage, covering the entire life cycle of the highway concerned. Table 3.5 shows the integration of the various stages of the EIA process into the highway development cycle.

3.2.1 Highway Project Conception

The development of new highways, in most cases, is conceived at the policy level along with a master plan for the city or the region. The highway development strategies adopted and the constraints imposed at

TABLE 3.5 Integration of Different Stages of EIA Process into Highway Development Program

Stage of EIA Process	Phase of Highway Development
Environmental screening	Conception phase — project conception and initial feasibility assessment of highway project
Initial environmental examination	Planning — identifying goals and developing broad development alternatives
Environmental and social impact analysis	Design — evaluation of the viability of alternatives to arrive at the most desired alternative, and produce the detailed design
Monitoring of environmental measures	Construction — construction of highway
Postconstruction environmental auditing and social evaluation	Operation — highway operations management and road network management

the conception stage have wide ranging effects on various environmental issues in the subsequent stages of the highway development process. Performing a comprehensive, system wise, macro-analysis of the master plan must be considered at this stage with due considerations to the environmental impacts.

The overall environmental impacts of the proposed highway development projects need to be evaluated at this conception phase because remedial measures can be incorporated to achieve the best results. Hence, the environmental screening stage of the EIA process should be initiated during this phase of the highway development. It serves the important function of identifying highway development projects that would potentially create adverse environmental impacts and short listing these projects for further examination.

3.2.2 Highway Project Planning

The highway project planning process consists of identifying goals and developing alternatives. It includes taking inventories and forecasting economic and transport needs. Inventories are taken to determine current conditions and existing facilities as a basis for predicting the future situation. Environmental screening and the ensuing IEE are the EIA activities required for this phase of highway development project to prepare the terms of reference (TOR) of the EIA. The environmental screening identifies the probable adverse environmental impacts of the proposed highway development project, while the IEE (scoping) provides estimates of the impacts to ascertain if a fullscale EIA is needed.

A highway location analysis as part of the planning process will include the EIA analysis to compare the relative merits of possible corridors for the proposed highway. The highway authority must work closely with environmental specialists on various procedural issues and substantive parameters to provide reasonable assessments of the various environmental impacts and conclude with appropriate recommendations concerning the various development alternatives.

3.2.3 Highway Design

In the design phase of the highway development process, the viable alternatives identified during the planning phase are simultaneously considered and compared with one another to determine the best option. The effectiveness of various environmental mitigation measures is assessed and the best measures are recommended to meet prescribed criteria and requirements. This falls within the scope of the environmental impact analysis stage of the EIA process. Based on the findings of the environmental screening and IEE, the environmental analysis examines in detail all the major environmental impacts of the highway development projects. The environmental analysis also makes recommendations on suitable remedial measures for each of the adverse environmental impacts analyzed.

3.2.4 Highway Construction

The implementation phase of the highway development process encompasses detailed design, right-of-way acquisition and construction. The required EIA activities in this phase consist of incorporating the recommendations of the EIA report into the final design, preconstruction mobilization, mitigation schemes, and execution of the highway. These activities cover three stages of the EIA process, namely (a) the environmental impact analysis which produces the EIA report with the EMP (b) the monitoring of environmental measures which ensures that the procedures recommended in the approved EIA report are adhered to by various agencies involved in construction supervision (c) the postconstruction auditing and evaluation during which monitoring and evaluation are performed on those mitigation measures implemented during various stages of highway construction.

For effective implementation of the recommended environmental measures, a combined team of highway engineers and environmental specialists is recommended for successful completion of a highway project.

3.2.5   Highway Operation and Maintenance

The postconstruction auditing and environmental evaluation phase extends beyond the end of the construction of a highway. It examines the performance of the highway during its service life. The implementation of the EIA recommendations during this phase ensures that the adverse environmental impacts are properly controlled, and helps to identify good practices and technologies for highway construction. This covers both the operation and maintenance period of the highway. As postconstruction highway operation and maintenance constitute an ongoing phase of the entire life cycle of the highway development and management, it is important that the postconstruction environmental evaluation must also monitor and evaluate the impacts of these activities. The engineers and others responsible in this phase should be familiar with various environmental requirements. Regular consultation with environmental specialists is required in this postconstruction auditing and evaluation stage.

3.3   EIA Activities in Highway Development

3.3.1   Major Steps of EIA Process

The following are the major steps in an EIA process:

Step 1   Perform ES (Environmental Screening)
Step 2   Perform environmental scoping and IEEs
Step 3   Prepare TOR for EIA
Step 4   Perform EIA study and prepare EIA Report
Step 5   Review and approve EIA Report
Step 6   Formulate EMP and monitor implementation of recommended environmental protection measures
Step 7   Conduct postconstruction environmental audit and evaluation

Activities involved in various steps of the EIA process are shown in Table 3.6. Table 3.7 lists the roles and responsibilities of stakeholders against the major steps of an EIA process.

3.3.1.1   Environmental Screening

Environmental screening is the process undertaken to decide the extent of environmental review that a highway project requires. In some countries, it is simply a decision as to whether an EIA is required or not using a prescribed list of criteria. For the most part, the screening criteria for determining the level of review required are relatively well defined. In some cases, there is considerable discretion in determining whether or not an EIA should be carried out. There is generally a threefold categorization as follows (ADB, 1997; WB, 1997).

Project Category A. Highway development projects in this category require a fullscale EIA. The potentially significant environmental issues for these projects may lead to changes in land use, as well as

TABLE 3.6 Activities in Major Steps of EIA Process for Highway Development

Major EIA Steps	Activities Involved
1. Environmental screening	Identification of possible major adverse environmental impacts of the road project
	Decision on the level of EIA study that the project must undergo
2. Environmental scoping or initial environmental examination	Estimation of magnitudes of adverse environmental impacts
	Identification of areas of major environmental concerns
	Development of terms of reference (TOR) for EIA study
3. Preparation of EIA report	Analysis and assessment of environmental impacts
	Proposal of mitigation measures for adverse environmental impacts
	Formulation of monitoring plan
	Development of environmental management plan
	Arrangements for public consultation and public participation
4. Review of EIA report	Forming of reviewing panel
	Establishment of EIA review criteria
	Development of quality control measures
	Conduct of formal review
	Preparation of review report
5. Approval or rejection of proposed project	Rejection or approval of the project by authority
	Preparation of terms and conditions for approved project
	Specifying of specific environmental protection measures
	Specifying monitoring and evaluation techniques
6. Environmental management plan and monitoring	Implementation of environmental management plan for mitigation measures
	Enforcement of monitoring programs
7. Postconstruction audit and evaluation	Evaluation of the degree of implementation of the environmental management plan
	Assessment of effectiveness of monitoring programs
	Evaluation of the degree of success of mitigation measures

changes to the social, physical, and biological environment. An environmental specialist's advice needs to be sought to determine the scope of the EIA necessary for compliance with environmental policies.

Project Category B. This projects under this category require an IEE, not a fullscale EIA. The scale is often the only difference between projects in this category and those in category A. The projects under this category are not located in environmentally sensitive areas and the environmental impacts are less severe than those of projects in category A. Mitigation measures for these projects are more easily prescribed. An environmental specialist will be required to assist in formulating the TOR for the IEE so that the IEE report will comply with the authority's policies.

Project Category C. Projects under this category typically do not require an EIA. These projects are unlikely to have adverse environmental and socio-economic impacts.

3.3.1.2 Environmental Scoping

Environmental scoping is the process of determining the issues to be addressed, the information to be collected and the analysis required to assess the environmental impacts of highway development projects. The primary output of environmental scoping is the TOR required to conduct an EIA and to prepare the EIA report.

3.3.1.3 IEE

In some countries, scoping is conducted for an EIA process in the context of an IEE. After a highway project has been screened and found to have potentially significant environmental impacts, an IEE is

TABLE 3.7 Roles and Responsibilities of Stakeholders in EIA

EIA Activity	Stakeholders				
	Government EIA Agency	Government Road Agency	Other Government Agencies	EIA Consultants	Public and Interest groups
EIA screening	Screen project	Gather project related information	Identify relevant issues and concerns, and offer review comments	Assist by providing technical information and advice	—
Scoping or initial environmental examination (IEE)	Approve terms of reference (TOR) and review IEE	Issue TOR and provide initial environmental examination	Raise issues and concerns, and offer review comments	Assist in preparing and reviewing TOR; preparing and reviewing IEE	Provide feedback and comments directly or through public hearing
Environmental management plan	Evaluate management plan	Implement environmental protection measures and monitoring	Provide technical support in monitoring	Conduct monitoring	—
Postconstruction audit and evaluation	Evaluate performance and impacts of project	Gather project related information	Provide technical support in evaluation	Assist in evaluation of project performance and impacts	Provide feedback and comments

undertaken to determine the probable environmental impacts associated with the project and ascertain whether a fullscale EIA is required. The IEE is usually conducted with a limited budget and is based on existing information and the professional judgment of people, knowledgeable about the impacts of similar highway projects. The following are the primary objectives of the IEE:

- Identification of the nature and severity of specific, significant environmental issues associated with the project
- Identification of easily implementable mitigating or offsetting measures for the significant environmental issues. The IEE serves as the final impact assessment and no further EIA is required if the IEE shows that there are no significant environmental issues which need further study
- Development of the TOR for the fullscale EIA study in case more detailed impact assessment, or any special information and data are needed.

Conducting an IEE ensures a focused TOR for a fullscale EIA because it identifies and provides background information on the issues requiring resolutions. The objectives of the IEE may be accomplished without extensive financial and human resources, thereby increasing efficiency. Since evaluations and decisions are based on limited information, sound judgment, and appropriate experience are the most crucial requirements for IEE execution. Competent EIA practitioners need to be involved in the IEE-phase because the decisions made at this stage affect the composition and scope of the EIA performed on a highway project. A poor IEE report could result in failure to recognize significant environmental impacts, where as a good report can result in efficient resolution of significant environmental issues.

3.3.1.4 Fullscale EIA

A highway project must undergo a fullscale EIA if it is explicitly prescribed by law, regulation or if the IEE results indicate that an EIA is required. A fullscale EIA normally involves a rigorous study whereby new environmental information and data are collected. Number of environmental experts are generally required. A fullscale EIA may involve elaborate review procedures and require public consultation. A detailed EIA report is required as part of a fullscale EIA. EIA reports are generally prepared by EIA practitioners. In most cases, EIA consultants follow the guidelines developed by the reviewing agency. These guidelines specify what are to be included in the EIA report.

3.3.1.5 TOR

OECD (1994) has developed procedural guidelines that include a framework TOR for EIA of various development projects. The guidelines were prepared for use by environmental specialists of bilateral aid and implementing agencies of the recipient country. The framework TOR outlines the requirements for two qualitatively different types of information: (a) detailed project justification (b) detailed EIA information.

Detailed project justification includes information on (i) the problem or development goal (ii) the proposed solution (iii) the cooperation amongst donors, lenders, and the developing country (iv) the objectives of the assessment (v) the legal and policy considerations (vi) the institutional capacity (vii) the alternatives to the project and within the project (viii) the institutional cooperation; and (ix) the public involvement.

Detailed EIA information includes: (i) description of the highway development project (ii) description of the environment (iii) information quality (iv) positive impacts and negative impacts on natural and human environment (v) resettlement and compensation (vi) mitigation measures (vii) EMP (viii) environmental monitoring program or plan.

3.3.1.6 Preparation of EIA Report

The final EIA report is to be prepared by the project proponent or the EIA practitioners who are responsible for the EIA. An EIA report typically contains the following items:

- executive summary of the EIA findings
- description of the proposed highway development project
- major environmental, socio-economic, and natural resource issues that need clarification and elaboration
- adverse impacts on the natural and socio-economic environment — their identification and prediction
- discussion of options for mitigating adverse impacts and shaping the project to suit its environment and an analysis of the trade-off involved in choosing between alternative actions
- overview of gaps or uncertainties in the information and data
- summary of the EIA for the general public, especially for the people affected by the project
- conclusions and recommendations
- list of references
- annexures and appendices

3.3.1.7 Review and Approval of EIA Report

Different agencies use different methods for the review of the EIA report on highway development projects. The EIA report is reviewed by a reviewing agency or by a special standing committee or commission established to review projects in a given sector. In most cases, a technical evaluation of the EIA report is made by specialists. This technical evaluation provides the basis for the review. The output of the review report is either a rejection or an approval of the project. In case of approval the report outlines the terms and conditions for the project to proceed with. These terms and conditions are attached to any license, permit, or certificate issued by the approving authority.

In most cases, the results of an EIA review are provided to the agency responsible for ultimately approving the proposed project. In many jurisdictions, project approval also depends on approval from the EIA agency. One output of the EIA review process is the terms and conditions that are attached to the approval. These terms and conditions define the environmental protection measures that must be integrated into the project. The terms and conditions may also specify environmental monitoring that must be undertaken in conjunction with the project construction and postconstruction auditing for project evaluation.

3.3.1.8   EMP and Monitoring

Environmental management is that part of the project management which is responsible for the implementation of the mitigation measures and the environmental monitoring. The EMP contains the following information:

- Details of the mitigation measures that will be undertaken to ensure compliance with environmental laws and regulations.
- Objectives and management of the monitoring program, and the specific information and data to be collected.
- Institutional responsibility, reporting requirements, enforcing capability, and resources required in terms of funds, skilled staff, equipment, and training needs.

3.3.1.9   Postconstruction Environmental Auditing and Evaluation

Postconstruction follow-up is required to determine whether the conditions of project approval such as the environmental protection measures and monitoring program have been undertaken as required. Further follow-up is required to determine if the environmental protection measures are successful and the monitoring data have been analyzed and acted upon.

The postconstruction evaluation reports and project performance audit reports recommended by ADB (1997) and WB (1996) include a final assessment on the following:

- the degree by which the projects satisfied the EIA recommended environmental requirements
- the effectiveness of mitigating measures and institutional development
- whether any unanticipated impacts occurred as a result of the project implementation activities.

For each highway development project, the EIA report reviews and assesses the following: (a) beneficial and detrimental environmental impacts of the project (b) location and design and operational alternatives considered with reasons for final choice (c) environmental protection measures adopted and the effect of such measures upon project costs and economic evaluation of the project (d) environmental aspects of the project in relation to overall cost-benefit analysis.

3.3.2   Tasks of EIA Study Team

To ensure that the study addresses all the important issues to the decision makers and includes all the issues and concerns raised by various groups, an EIA study team shall hold discussions with the highway project developer, decision makers, regulatory agencies, scientific and engineering institutions, local community leaders, and others. The study team should select primary environmental impacts to focus on and determine for each the magnitude, geographical extent, local sensitivity, and significance to decision makers.

In performing their tasks, the EIA study attempts to answer the following questions:

- (a) For identification of impacts — What will happen as a result of the project?
- (b) For prediction of impacts — What will be the extent of the changes?
- (c) For evaluation of impacts — Do the changes matter?
- (d) For mitigation of impacts — What can be done about them?
- (e) For documentation and preparation of EIA report — How can decision makers be informed of what needs to be done?

In practice, EIA is a repetitive process of asking the first four questions repeatedly, til a workable solutions can be offered to the decision makers.

3.3.2.1 Identification of Impacts

The identification of impacts in an EIA study generally uses the following methods:

- Compilation of a comprehensive list of key environmental impacts such as changes in air and water quality, noise levels, wildlife habitats, bio-diversity, landscape, social and economic systems, cultural heritage, settlement patterns, and employment levels.
- Identifying all the sources of impacts such as dust, spoils, vehicles emissions, water pollution, construction camps, etc using checklists or questionnaires. This is followed by listing possible receptors in the environment (e.g., crops, communities, and migrant labors) through surveying the existing environmental and socio-economic conditions and consultation with concerned parties.
- Identifying and quantifying various environmental and socio-economic impacts through the use of checklists, matrices, network, overlays, models, and simulations.

3.3.2.2 Prediction of Impacts

Prediction of impacts technically characterises the causes and effects of impacts, and their secondary and synergistic consequences for the environment and the local community. It examines each impact within a single environmental parameter into its subsequent effects in many disciplines (e.g., deterioration of water quality, destruction of fisheries, adverse economic effects on fishing villages, and resulting socio-cultural changes). It draws on physical, biological, socio-economic, and anthropological data and techniques. In quantifying impacts, it employs mathematical models, physical models, socio-cultural models, economic models, experiments, or expert judgments.

To prevent unnecessary expenses, the sophistication of the prediction methods used should be kept in proportion to the scope of the EIA. For instance, a complex mathematical model of atmospheric dispersion should not be used if only a small amount of relatively harmless pollutants is emitted. Simpler models are available and are sufficient for the purpose. Also, it is unnecessary to undertake expensive analyses if not required by the target decision-makers. All prediction techniques of environmental impacts, by their nature, involve some degree of uncertainty. The study team should quantify the uncertainty of prediction in terms of probabilities or margins of error.

3.3.2.3 Evaluation of Impacts

Each predicted adverse impact is evaluated to determine whether it is significant enough to warrant mitigation. This judgment of significance can be based on one or more of the following: (a) comparison with laws, regulations or accepted standards; (b) consultation with the relevant decision makers; (c) reference to preset criteria such as protected sites, or endangered species (d) consistency with government policy objectives (e) acceptability to the local community or the general public.

3.3.2.4 Mitigation of Impacts

If the previous step has found the impacts to be significant, then the EIA study team will proceed to analyze mitigation measures. The possible mitigation measures include: (a) changing project sites, routes, processes, raw materials, construction methods, operating methods, and disposal locations of wastes and spoils (b) introducing pollution controls, waste treatments, monitoring, phased implementation, landscaping, personnel training, special social services, or public education (c) offering (as compensation) restoration of damaged resources, money to affected persons, concessions on other issues, or off-site programs to enhance some other aspects of the environment or quality of life for the community. Cost of all mitigation measures must be quantified.

Various mitigation measures are then compared and trade-offs between alternative measures are weighed. The EIA study team will propose one or more action-plans, which may include technical control measures, an integrated management scheme (for a major highway development project), monitoring, contingency plans, operating practices, and project scheduling. The study team should explicitly analyze the implications of adopting different alternatives, to help make the choices clearer for decision makers. Several analytical techniques are available for this purpose:

(a) Cost-benefit analysis in which all quantifiable factors are converted to monetary values, and actions are assessed for their impacts on the highway project costs and benefits. However, the unquantifiable factors can be equally important, and need to be taken into account in the decision making process;

(b) Explaining what course of actions would follow from various broad value judgments (e.g., those socio-economic impacts are more important than resources);

(c) A simple matrix of environmental and socio-economic parameters vs. mitigation measures, containing brief descriptions of the effects of each measure;

(d) Pair wise comparisons, whereby the effects of an action are briefly compared with the effects of each of the alternative actions, one pair at a time.

3.3.2.5   Documentation and Report

The results of EIA must be documented in the form of a well organized report to be communicated to decision makers. This document of the fullscale EIA report is sometime called an Environmental Impact Statement (EIS) of the project, especially when it is submitted as part of a permit application.

As mentioned earlier in Section 3.3.1, the contents of an EIA report for highway development projects should, at a minimum, contain: (a) an introduction (b) a project description (c) a detailed description of the environment (d) an assessment of environmental impacts and mitigation measures (e) an EMP (f) an environmental monitoring plan.

In many jurisdictions, the highway specific EIA report also contains an evaluation of alternatives, environmental, social and economic analyses including a cost-benefit analysis, and a description of the public participation program.

3.4   EIA Methodologies

3.4.1   Impact Assessment

A large number of impact assessment methods are available, varying widely in their levels of technical complexity, the amounts of data required, and the levels of precision and certainty. All these factors should be considered when selecting a method. Improved practices of EIA and advances in information technology have greatly expanded the range of tools available to the EIA practitioner. For example, geographical information systems (GIS), global position system (GPS) and expert systems have been developed to help in environmental screening, environmental scoping, developing TOR, and conducting IEE. The following are some of the more techniques and methods that have evolved to cover factors not normally considered in the traditional cost-benefit analysis.

3.4.1.1   *Ad hoc* Methods

This is a useful method when time constraints and lack of information require that the EIA must rely exclusively on expert opinion. This method is very easy to use but has a few drawbacks. It may not encompass all the relevant impacts and hence they are not comparable. Even the relative weightages of various impacts cannot be compared. It is inefficient as it requires considerable effort to identify and assemble an appropriate panel of experts for each impact assessment. It provides minimal guidance for impact analysis while suggesting broad areas of possible impact.

3.4.1.2   Checklists

Checklists list the types of impacts associated with particular types of highway development project. This method primarily organizes information and ensures that no potential impact is overlooked. This is a more formalized process compared to the *ad hoc* approach in that specific areas of impacts are listed and instructions are supplied for impact identification and evaluation. Common checklists include: (a) scaling checklists in which the listed impacts are ranked in order of magnitude or severity,

(b) scaling–weighting checklist in which numerous environmental parameters are weighted (using expert judgment), and an index is then calculated to serve as a measure for comparing project alternatives.

There are four general types of checklist:

- *Simple Checklist*. This is a list of environmental parameters with no guidelines on how they are to be measured and interpreted.
- *Descriptive Checklist*. This includes an identification of environmental parameters and guidelines on how to measure data on particular parameters.
- *Scaling Checklist*. It is similar to a descriptive checklist, but with additional information on subjective scaling of the parameters.
- *Scaling–Weighting Checklist*. This is similar to scaling checklist, with additional information for the subjective evaluation of each parameter with respect to all the other parameters.

3.4.1.3 Matrices

Matrix methods identify interactions between various project actions and environmental parameters and components. They incorporate a list of project activities with a checklist of environmental components that might be affected by these activities. A matrix of potential interactions is produced by combining these two lists (placing one on the vertical axis and the other on the horizontal axis). One of the earliest matrix methods was developed by Leopold et al. (1971). In a Leopold matrix and its variants, the columns of the matrix correspond to project actions while the rows represent environmental conditions. The impact associated with the action columns and the environmental condition row is described in terms of its magnitude and significance.

Most matrices were built for specific applications, although the Leopold Matrix itself is quite general. Matrices can be tailor-made to suit the needs of any highway development project that is to be evaluated. They should preferably cover both the construction and the operation phases of the highway project, because sometimes, the former causes greater impacts than the latter. Simple matrices are useful for (a) screening (b) scoping or IEE (c) identifying areas that require further research (d) identifying interactions between project activities and specific environmental components.

However, matrices also have their disadvantages:

- they tend to overly simplify impact pathways
- they do not explicitly represent spatial or temporal considerations
- they do not adequately address synergistic impacts.

Matrices require information about both the environmental components and project activities. The cells of the matrix are filled in using subjective (expert) judgment or by using extensive databases. There are two general types of matrices: (a) simple interaction matrix (b) significance or important-rated matrix. Simple matrix methods simply identify the potential for interaction. Significance or importance-rated methods require either databases or more experience to prepare. Values assigned to each cell in the matrix are based on scores or assigned ratings, not on measurement and experimentation. For example, the significance or importance of impact may be categorized as no-impact, insignificant impact, significant impact, or uncertain. Alternatively, it may be assigned a numerical score for example, zero is no impact and ten is maximum impact.

3.4.2 Sectoral Guidelines

Sectoral guidelines help bringing collective experience with environmental impacts of specific highway development project types to bear during initial assessments. They normally contain a comprehensive listing of (a) project types covered by the guidelines (b) activities that fall within each project type (c) environmental components that may possibly be affected by the project activities (d) significant issues that must be addressed in project planning (e) suggested mitigation measures that might be incorporated into the project (f) recommended monitoring requirements.

In practice, the sectoral guidelines are most useful in early stages of an EIA when TOR are unavailable or are in process of preparation. They help in impact identification and in the development of detailed TOR for conducting an EIA study. They also provide guidance on how to present information in proper format to aid in the review and evaluation of EIA report.

3.4.3   Impact Prediction

The credibility of an EIA of highway development projects relies on the ability of the EIA practitioners to estimate the nature, extent, and magnitude of change in environmental components that may result from the project activities. Information about predicted changes is needed for assigning impact significance, prescribing mitigation measures, and designing and developing EMP and monitoring programs. The more accurate the predictions, the more confident the EIA practitioner will be in prescribing specific measures to eliminate or minimize the adverse impacts of highway development projects.

Canter (1996) and Canter and Sadler (1997) provide excellent reviews, based on American experience, of many of these prediction techniques. In many EIA applications, combinations of the basic prediction techniques are commonly used. This is particularly true when using computerized modeling for specific applications, as the application of a computer model usually requires collection of environmental information to set baseline values for the model's variables and to determine the values for model's parameters. The three most common types of models used in EIA are physical, experimental, and mathematical models.

3.4.3.1   Physical Models

Physical models are small scale models of the environmental system under investigation in which experiments can be carried out to predict future changes. There are two types of physical models: illustrative or visual models, and working physical models. Illustrative or visual models depict changes to an environmental system caused by a proposed development activity using pictorial images. These images are developed from sketches, photographs, films, three dimensional scale models, and by digital terrain models or digital image processing systems. Working physical models, on the other hand, simulate the processes occurring in the environment and uses reduced scale models so that the resulting changes can be observed and measured. Such models, however, cannot satisfactorily model all real life situations. Error may occasionally creep-in as a result of the scaling process. Technical expertise and large quantities of data are required to construct working physical models that adequately simulate the behavior of the real environment. Validation and interpretation of the results of modeling may require time and technical expertise.

3.4.3.2   Experimental Models

Scientific data from laboratory or field experiments provide basic information on the relationships between environmental components and human activities. Research results are used to construct empirical models that can simulate the likely effects of an activity on an environmental component. Examples of experiments in which tests are carried out in the actual environment include *in situ* tracer experiments to monitor the movement of releases into the environment; controlled experiments in small parts of potentially affected ecosystems; noise tests to determine levels of disturbance and pumping tests on groundwater. Experimental modeling requires substantial amounts of money, effort, time, and expertise in specialized fields.

3.4.3.3   Mathematical Models

Mathematical models use mathematical equations to represent the functional relationships between variables. In general, sets of equations are combined to simulate the behavior of environmental systems. The number of variables in a model and the nature of the relationships between them are determined by the complexity of the environmental system being modeled. Mathematical modeling aims to limit, as much as possible, the number of variables keeping the relationships amongst variables as simple as possible without compromising the accuracy of representation of the environmental system.

Mathematical models are described according to the following: (i) empirical or internally descriptive (ii) generalized or site-specific (iii) stationary or dynamic (iv) homogenous or non-homogenous (v) deterministic or stochastic.

Mathematical models require varying amounts of resource inputs. A simple model, such as the river water quality model used in the above example, may require minimal input data and simple manual calculation while a complex Gaussian plume model may require sophisticated computer techniques and demand considerable resources of input data, time, and expertise.

3.4.4   Economic Analysis

Traditionally, the EIA is meant to be an independent report related to the environmental impacts of infrastructure development projects. The assessment has minimal links with economic analysis. There have been considerable discussions among various stakeholders involved in planning, financing, designing and implementing infrastructure development projects on how to measure the economic importance of expected environmental impacts. Due consideration is therefore given to the emerging role of economic evaluation of environmental impacts, specifically on how to use such information in impact assessment.

The economic analysis of highway development projects has had a relatively long history. Initially, environmental impacts were deemed external to development projects, and were excluded from economic analysis. Subsequently, it became the practice to describe environmental impacts quantitatively. Since the mid 1980s, there has been a growing interest in placing monetary values on environmental impacts combining these values into the overall project analysis. In this regard, ADB's "Economic Analysis of the Environmental Impacts of Development Projects" is an excellent example (ADB, 1997). In 1996, the ADB published "Economic Evaluation of Environmental Impacts: A Workbook" which provides the practitioner with a step-by-step procedure as to how the economic values of environmental impacts can be integrated into project analysis. The roles of environmental economics in an EIA usually fall under three categories: (a) use of economics for "cost-benefit analysis" as an integral part of project selection (b) use of economics in the assessment of activities suggested by the EIA (c) economic assessment of the environmental impacts of the development projects.

The use of economic analysis in EIA can aid in assessing the proposed project more objectively. If the EIA exercise is used as planning tool in an iterative manner, it is possible to reduce the negative environmental impacts and capture more positive environmental components if the economic analysis of such impacts is possible with every interaction. The result of integrating economic analysis of environmental impacts can be very useful in enhancing the quality of a project.

3.4.5   Social Impact Assessment

Evaluation of the social implication of a highway development project is tightly linked to the scrutiny of its social and economic objectives. Social impact assessments must go well beyond determining a project's adverse impacts. As a methodology, social impact assessment refers to a broad range of processes and procedures for incorporating social dimensions into highway development projects. In some jurisdictions and agencies, the social impact assessment is conducted in conjunction with the EIA while in others, it is conducted separately. In both cases, the social impact assessment influences project design and the overall approval of the project.

Many of the methods and techniques discussed so far have evolved in the context of international development assistance provided by bilateral agencies and the multilateral development banks. As such, they focus on people as beneficiaries. They also consider people as vulnerable groups that may be adversely affected by the project. The social impact assessment aims to determine the social costs of the project and the degree to which the benefits of a project will be distributed in an equitable manner. Social impact assessments are necessary to help ensure that the project will accomplish its development goals such as poverty reduction, enhancement of the role of women in development, human resources development, population planning, and avoiding or mitigating negative effects on vulnerable group, and protecting these groups.

By addressing the specific development goals in the social impact assessment of highway development projects, the developers, the lenders and the government agencies can help ensure that project benefits are realized and negative social impacts are minimized. Various methods and approaches have been developed to consider the following social dimensions: (a) social analysis (b) gender analysis (c) indigenous people's plans (d) involuntary resettlement plans (e) cooperation with non-governmental organizations (f) use of participatory development processes and (g) benefits monitoring and evaluation.

3.4.5.1   Social Impact Assessment and Project Cycle

As with environmental considerations, the need to analyze the social factors which influence and are influenced by a highway development project, continues throughout the entire life of a project. The major activities involved in incorporating social dimensions into the project are summarized in Table 3.8. The project preparation stage, in particular the preparation of the feasibility study, is the focus of many social impact assessment activities. To ensure that social dimensions are adequately addressed, it is imperative to provide clear, focused TOR and specific guidance on how to carry out the necessary analyses to those who are responsible to prepare the EIA study.

3.4.5.2   Conducting Social Assessment

The assessment method for incorporating social dimensions into a project covers the following components: (a) social analysis, i.e., identification of the project affected people (PAP); assessment of needs; assessment of demands; (b) assessment of absorptive capacity (c) conducting gender analysis (d) assessment of adverse impacts such as poverty impact assessment, indigenous people and involuntary resettlement on vulnerable groups (e) targeting; (f) designing participatory development processes (g) formulating delivery mechanisms (h) benefit of monitoring and evaluation.

Social analysis involves three principal steps: initial issue identification; preliminary assessment of all issues; and detailed social analysis of the potential for the major impacts. Initial issue identification may be carried out in an *ad hoc* or informal way, by seeking expert opinion, and by public involvement. The key to success is to incorporate a range of perspectives in the process. Since the widest range of social, economic, cultural, resource use and infrastructure effects occur at the local level, local people generally identify most potential impacts and are key to the identification of issues. The success of a social analysis

TABLE 3.8   Activities of Social Dimensions in Highway Development Cycle

Project Stage	Activities Undertaken
Screening and initial environmental examination	Identification of social dimensions and associated processes that may be important in the project
	Selection of key elements of social analysis
	Identification of initial potential social issues and impacts
	Initial social assessment
EIA study	Social analysis
	Involuntary resettlement planning
	Indigenous peoples planning
	Gender analysis
	Poverty impact analysis
	Benefit monitoring and evaluation planning
Implementation	Arrangements for resettlement
	Information dissemination on role of beneficiaries
	Ongoing stakeholder consultation
	Strengthening beneficiary organizations
	Improving absorptive capacity of target groups
	Mitigating adverse effects on vulnerable groups
Monitoring	Monitoring of social indicators developed during the project design
	Reviewing missions to assess social dimensions and associated processes
	Progress reporting by the executing agency on beneficiary participation by number, gender, income group; participation by adversely affected groups; formation of beneficiary groups numbers by gender and income class

can be enhanced by taking the following measures (i) involving a qualified social impact specialist with a solid background in social sciences (ii) incorporating some form of participatory development process (iii) hiring local experts (iv) using local knowledge as well as scientific data and using realistic assumptions for development practices such as construction practices.

An important requirement in any social analysis is to gather baseline information on client groups. As not all project beneficiaries will have the same needs and demands, the population is divided into subgroups. A basic profile for each subgroup should be developed. This profile should include: (i) number in each subgroup (ii) differentiation by gender (iii) number of single-headed households (iv) household size (v) occupations (vi) income and asset levels (vii) levels of education and access to education (viii) health problems and access to health services (ix) social organization and group formation and (x) ethnic or cultural distinctions.

In conducting social assessment, a resettlement action plan (RAP) may have to be examine to assess the social and economic impacts on the displaced due to a highway development project. Various salient features of a well-designed RAP are given in Table 3.9. RAPs are often focused on the issues associated with land, ownership, and identification of who is entitled to compensation as a result of the highway development project.

3.4.6 Public Participation

Most highway development projects affect a wide range of people with varied interests. Public participation is required to allow the affected people to identify significant environmental and socio-economic issues. An effective EIA process takes issues raised by the public into account in the project planning and design, or addresses the issues through appropriate environmental protection measures. Many highway development projects have failed because their planning and design did not address the local needs. Although most developing countries worldwide have no formal requirements for public participation in highway development projects, communities are sometimes consulted by the EIA Team during its preparation of the EIA report. While this practice of community consultation is relatively new. It is assuming increasing importance and becoming more prevalent.

3.4.6.1 Functions of Public Participation Program

The EIA process for highway development projects is incomplete without the effective public participation. Public participation is an integral part of the EIA process because it serves the following important functions:

(a) Informing project-affected people in advance enabling them to make necessary arrangement to minimize adverse impacts.
(b) Disseminating information to all interested parties, professionals and the general public to gather opinions and useful inputs.
(c) Presenting useful channels for collection of specific environmental and socio-economic information through the local people.
(d) Making effective communication with project- affected people to gain support for the project.
(e) Making effective consultation and dialogue with local inhabitants for cost-effective environmental remedial measures by means of indigenous techniques using locally available materials and manpower.

A successful public participation program requires the following three elements to be effectively executed: (i) dissemination of information to the PAP; (ii) getting feedback from the PAP and other influential parties; and (iii) consultation with local people to collect data for baseline conditions.

3.4.6.2 Main Tasks in Public Participation Program

The main tasks of the public participation program are: (i) selection of the team for executing the public participation programme; (ii) identification of stakeholders (PAPs, NGOs, and others interest groups;

TABLE 3.9   Salient Features of Resettlement Action Plan

---

1. Objectives of resettlement and policy framework
    Purpose and objectives of resettlement
    National and local and compensation laws
    Description of donor policies and how these will be achieved under the project
    Principles and legal/policy commitments from borrower/executing agency
2. Project design and scope of the resettlement
    Scope of resettlement and how resettlement relates to the main investment project
    Alternative options, if any, considered to minimize resettlement
    Special consideration given to how the project will impact indigenous people and other vulnerable groups, including
        women
    Responsibility for resettlement planning and implement
3. Socio-economic information and entitlements
    Impact of land acquisition on potential affected peoples
    Identify all losses to resettlers and host communities
    Details of common property resources
    Cut-off dates of eligibility
    New eligibility of policy and entitlement matrix
4. Resettlement site development and income restoration
    Location, quality of site, and development needs
    Layout, design and social infrastructure
    Safeguarding income and livelihoods
    Income restoration programs
    Gender issues and other vulnerable groups
    Integration with host communities
5. Institution framework for resettlement implementation
    Mandate of resettlement agency
    Establishing a resettlement unit and staffing
    Technical assistance for capacity building
    Role of NGOs and PAPs organizations in resettlement
    Grievance redress committees
6. Consultation and community participation
    Identification of project stakeholders
    Mechanisms for participation
    Participatory settlement management
    Institutions in participation
    NGOs as a vehicle for participation
7. Resettlement budget and financing
    Land acquisition and resettlement costs
    Budgetary allocation and timing
    Sources of funding and approval process
8. Monitoring and evaluation
    Establishing a monitoring and evaluation system
    Monitoring and reporting
    NGOs/PAPs participation in monitoring and evaluation
    Resettlement impact evaluation

(iii) mechanisms by which stakeholders can obtain relevant information of the project; (iv) procedures by which stakeholders are to channel their feedback and opinions to relevant authority; (v) telephone numbers and addresses of the contact persons; (vi) listing of topics for which public opinions and feedback are required.

3.4.6.3   Formats of Public Participation Programs

To achieve an effective public participation in the highway specific EIA process, it is necessary to study the profile of the communities and inhabitants to plan in advance how the relevant information will be disseminated and solicited to conduct consultation to obtain the needed feedback.

Depending on the profile of the stakeholders, their cultural and educational background, different formats of executing the public consultation and participation program can be adopted. The following are the usual formats of public participation: (i) public displays; (ii) interview surveys; (iii) questionnaire surveys; (iv) meetings with local people; (v) public hearings; (vi) consultation with interest groups; and (vii) on-site discussions and interviews.

3.4.7 EMP

The primary goal of EIA is to develop procedures to ensure that all mitigation measures and monitoring requirements specified in the approved EIA will actually be carried out in subsequent stages of a highway development project. These mitigation measures and monitoring requirements are normally set out in an EMP.

A well structured EMP usually covers all phases of the highway development project, from preconstruction right through to decommissioning. The EMP outlines mitigation and other measures that will be undertaken to ensure compliance with environmental laws and regulations and to reduce or eliminate adverse impacts. It specifically outlines the following:

(a) technical work program including details of the required tasks and reports, and necessary skilled staff supplies, and equipment;
(b) detailed accounting of estimated costs to Implement the EMP
(c) planned operation or implementation of the EMP including a manpower deployment schedule indicating activities and inputs from various government agencies.

The main mechanism for the implementation of the EMP for a highway development project is the establishment of an Environmental Management Office (EMO). The EMO is established with sufficient staff and budget, as a part of the project management office (PMO). Environmental staff in the PMO works alongside the construction and operation personnel to ensure that the measures and requirements outlined in the EMP are properly carried out. The establishment and funding of an EMO is essential insurance for environmentally sound and ecologically friendly highway projects.

3.4.7.1 Implementation of EMP

The implementation of an EMP for a highway development project requires the following:

- Detailed final design (plans and specifications) for the project must incorporate all mitigation measures specified in the approved EIA.
- Contract for construction of the project must include all mitigation measures to be implemented.
- The mitigation measures should be sufficiently detailed so that the construction contractor, in preparing his bid, will be clearly aware that he is required to comply with the mitigation measures.
- Construction contractors' performance should be duly monitored for compliance with the EMP by competent environmental construction inspectors provided by the EMO.
- Upon completion of construction of the project, inspection should take place to check that the works, as built, meet all significant environmental requirements before the project is officially accepted.
- Operation stage monitoring program must be implemented as specified in the EMP.

3.5 Summary

Highway development is essential to economic growth but if not properly planned, its adverse environmental and social impacts can outweigh the benefits it creates. The adverse impacts of highway development include air and water pollution, noise disturbance and visual impairment, destruction of natural habitat and bio-diversity, and lowering of living quality of displaced inhabitants. The positive and negative impacts of highway development need to be examined together in a logical manner to ensure that sustainable social and economic growth can be achieved through the highway development program. An effective strategy to achieve this goal is to apply EIA to the highway development process. By integrating environmental and social considerations into the complete highway development process, the negative impacts could be minimized or eliminated, while the positive benefits are enhanced.

The EIA process is a systematical procedure that aims at identifying, quantifying, characterizing and mitigating negative impacts of highway development projects striving to optimize the positive benefits. It is crucial that the EIA process be initiated as early as possible together with the highway development project. This Chapter has outlined the framework in which different stages of the EIA process could be integrated with the phases of the complete highway development cycle, including environmental screening during the highway project conception phase, and the postconstruction evaluation and monitoring during the operation and maintenance phase of the completed highway.

References

ADB, 1997. *Environmental Impact Assessment for Developing Countries. Vol. 1 — Overview, and Vol. 2 — Case Studies.* Asian development Bank, Manila, Philippines.

Canter, L. 1996. *Environmental Impact Assessment*, 2nd ed. McGraw–Hill, New York, USA.

Canter, L. and Sadler, B. 1997. A Tool Kit of Effective EIA Practice — A Review of Methods and Perspectives on Their Application: A Supplementary Report of the International Study of the Effectiveness of Environmental Assessment. IAIA, Environmental and Groundwater University of Oklahoma, OK, USA.

ESCAP, 1990. *Environmental Impact Assessment: Guidelines for Transport Development (ST/ESCAP/785).* United Nations, New York, USA.

Leopold, L.B., Clarke, F.E., Hanshaw, B.B., and Balsley, J.R. 1971. A procedure for evaluating environmental impact. United States Geological Survey, Geological Survey Circular 645, Government Printing Office, Washington, DC.

OECD, 1994. *Environmental Impact Assessment of Roads.* Organization for Economic Cooperation and Development, Paris, France.

UNESCAP, 2001. *Multistage Environmental and Social Impact Assessment: Guidelines for a Comprehensive Process*, Publication No. ST/ESCAP/2177. United Nations, New York, USA.

WB, 1996. *Strategic Urban Road Infrastructure Projects*, Appraisal Report. WB Transport Division, Washington, DC, USA.

WB, 1997. *Roads and the Environment: A Handbook*, WB Technical Paper No.376, J.H. Christopher and K. Tsunokawa, eds., Washington, DC, USA.

World Highways. 2001. Environmental Impact. http://www.worldhighways.com

4

Highway Safety

Ian Johnston
Monash University
Victoria, Australia

Part A — Fundamental Concepts in Road Traffic Safety

4.1   Safe System Design Is Paramount

Of all the man-made systems created to enrich our daily lives, the road transport system exacts, by far, the highest price in human injury and death. Globally, as many as 50 million persons are injured in road traffic crashes each year and about 1.2 million of them die (World Health Organization, 2004). Road traffic injury is the single largest cause of (unintentional) death in the first five decades of human life.

Tragically, many governments appear to accept this high human price as inevitable. Evans (2002), for example, contrasts the equanimity with which we (as societies) accept the road toll with the clamour for "protection" with regard to the numerically far smaller death toll from terrorism.

There is a widespread assumption that accidents, injuries and deaths are neither predictable (in the particular) nor preventable (in the aggregate). This assumption arises from the persistent belief that careless, negligent, or, at times, reprehensible behavior of road users is the primary cause and that, since the injured are "at fault", there is little governments can do. Indeed, the road user has been assigned, by decree, the ultimate responsibility for his/her own safety. In most countries, there are general rules (in traffic law) to the effect that one should always behave in such a way as to avoid a crash. When a crash occurs then by definition at least one party must have broken the general rule (Tingvall, 1998).

The purpose of this chapter is to ensure that this inappropriate line of thinking is not perpetuated by the next generation of highway engineers. Roads, traffic management systems and vehicles are all man-made products. The road traffic injury problem is thus a man-made problem and can be remedied by man-made solutions. Highways can, literally, be as safe as we are prepared to make them.

The challenge we face is that the road traffic system is an unstable man–machine system. In contrast, quality assured man–machine systems — in industry for example — are stable systems tolerant of human error. What we need to do is design, build and operate the road transport system to be far more tolerant of the normal range of human performance.

FIGURE 4.1   A Swedish "2 + 1" rural arterial, divided by flexible barrier (*Source*: Larsson et al., 2003).

Surprisingly, the primacy of systems design in the thinking of road safety professionals is a recent paradigm shift not yet widely embraced at institutional and government levels. This chapter explores the current state of road traffic safety thinking and knowledge and builds a framework for its application by highway engineers. It is less about providing a menu of engineering solutions to various road traffic safety problems and more about providing a mental map for developing the most effective solutions for given circumstances.

An example will serve to demonstrate why it is so important to focus on principles. Sweden has a substantial road traffic injury problem on its arterial, high speed, rural roads involving head-on crashes between vehicles in opposing flows. The problem appears to have originated in the relative width (13 m) of the sealed pavement, which was such as to encourage operation as a pseudo three-lane road. The design solution was to make it a formal three-lane road by using flexible barrier in the "center" of the road to create a divided road, with two lanes of traffic in one direction and one in the other, with a reversal every few kilometers to provide equality of overtaking opportunity (see Figure 4.1). Deaths on the treated road lengths have fallen by 90% (Larsson et al., 2003).

This is an excellent example of how a design innovation overcame a "historical" design problem, with a safe system achieved at a fraction of the cost of upgrading the roads to freeway standard.

4.2 The Dimensions of the Road Traffic Injury Problem

4.2.1 The Absolute Numbers

It is difficult to be precise about the global, annual toll of road traffic injury and death as there is substantial under-reporting, particularly in the motorizing countries and particularly (in all countries) for the less serious injuries. Even in the case of death, not all countries comply with the international convention of counting, as a road traffic death, those deaths occurring up to 30 days after the fatal-injury-producing road traffic event.

The best available estimates of the global road traffic injury problem put the annual number of deaths at 1.2 million and the number of injured persons at up to 50 million (World Health Organization, 2004).

As a rough rule of thumb, for every road traffic death, there are of the order of 15 injuries severe enough to require hospitalization and more than a further 50 injuries requiring medical treatment, but not hospitalization (see Figure 4.2). While governments, communities and the media tend to focus on road traffic deaths, deaths represent only the tip of the road toll iceberg.

Tragically, the high human price we currently pay for our road transport system is forecast to increase dramatically over the next two decades. The global, annual number of road traffic deaths is forecast to more than double to well over 2,000,000 deaths by 2020 (Kopits and Cropper, 2003; World Health Organization, 2004). The reason is simple; much of the world is in the early, rapid stages of motorization; including several countries with some of the largest populations (for example, China, India and Indonesia).

Table 4.1 illustrates that, already, almost 90% of global road traffic deaths occur in motorizing countries, particularly the rapidly motorizing countries of the Asia Pacific Region.

It is forecast (see Table 4.2) that, by 2020, the fall of some 30% in deaths from road traffic crashes in the highly motorized nations will be more than offset by dramatic increases in those nations already rapidly motorizing, and by those very early in the motorization process.

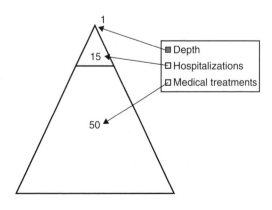

FIGURE 4.2 The "road toll" iceberg.

TABLE 4.1 The Proportion of Global Road Traffic Deaths × Stage of Motorization

		Global Road Traffic Deaths (%)
Motorized countries		14
Motorizing countries		86
Asia/Pacific	44	
Latin America/Caribbean	13	
Central and Eastern Europe	12	
Africa	11	
Middle East	6	
	86	

Is this dramatic global increase in death and injury an inevitable price of progress? The forecasts are based on the assumption that the road traffic safety interventions put in place in motorizing nations will tend to follow the historical patterns of the now motorized nations. This need not be a self-fulling prophecy if the rapidly motorizing nations avoid making the same mistakes.

4.2.2 Road Traffic Injury Relative to Other Public Health Problems

While the absolute numbers are large they are only part of the story and it is important to set road traffic injury in the context of other public health issues, because governments make judgments based (in part) on relative priorities.

The World Health Organization (WHO) uses a standardized measure of "disability adjusted life years" (DALYS) to compare different diseases and injuries. Table 4.3 shows that, in 1990, road traffic injury ranked ninth in the list of ten most prevalent global diseases. The table also shows that, as a result of the rapid motorization of much of the world, road traffic injury is expected to rank third by 2020. As the developing nations bring the diseases related to poverty under control the disease related to mobility threatens to get out of control!

Moreover, what are obscured by the data in Table 4.3 are the differential demographics of the diseases listed. As already pointed out, road traffic injury is a disease of the young and is, by far, the most common single cause of death from postinfancy through to the fifth decade of life.

When one examines only injury-related deaths, road traffic injury forms the largest single proportion, close to one quarter. The next most common is suicide. Road traffic injury causes seven times more deaths than war and more than twice as many deaths as result from other forms of violence (World Health Organization, 2004).

TABLE 4.2 Projected Proportionate Increases (Decreases) in the Absolute Number of Road Traffic Deaths, in Selected Regions, in 2020 against the Base Year 2000

	Increase in 2020 vs. 2000 (%)
South Asia	145
East Asia and Pacific	80
Sub-Saharan Africa	80
Middle East and North Africa	70
Latin America/Caribbean	50
Central and Eastern Europe and Central Asia	20
Motorized countries	(30)

Source: derived from Kopits and Cropper (2003).

TABLE 4.3 Top 10 Leading Contributors to the Global Burden of Disease

Disease or Injury	
1990	2020
Lower respiratory infections	Ischaemic heart disease
Diarrhoeal diseases	Unipolar major depression
Perinatal conditions	Road traffic injuries
Unipolar major depression	Cerebrovascular disease
Ischaemic heart disease	Chronic obstructive pulmonary disease
Cerebrovascular disease	Lower respiratory infections
Tuberculosis ↑	Tuberculosis
Measles	War
Road traffic injuries	Diarrhoeal diseases
Congenital abnormalities	HIV/AIDS

Source: Murray, CLJ, Lopez, AD, eds. *The Global Burden of Disease: A Comprehensive Assessment of Mortality and Disability from Diseases, Injuries, and Risk Factors in 1990 and Projected to 2020,* Harvard University Press, Boston, 1996. Epidemiologists use estimated disability-adjusted life years (DALYs) lost as the measure of the burden of disease. Reproduced with kind permission of the World Health Organization from World Health Organization (2004).

4.2.3 Personal, Socio-Economic and Economic Impacts

The burden of road traffic injury falls disproportionately on the socio-economically disadvantaged. There are several explanations. First, in road traffic crashes the least protected road users are the most vulnerable to injury. These are pedestrians, cyclists and motorcyclists, and the socio-economically disadvantaged are much more likely to meet the bulk of their personal transport needs from these cheaper modes of transport. Secondly, given the preponderance of younger persons among road traffic injury victims, the impacts upon a household of the removal of the most economically active members are greatest. Serious injury, particularly long-term disability, has an even greater impact, as these family units are in the worst position to bear the costs of meeting the long-term needs of the injured person.

The economic impact of road traffic injury upon national economies is considerable. The estimates range from 1 to 2% of gross national product, with the higher estimate being for the high-income countries.

Finally, at a personal level, the impacts of serious injury are often immense. Loss of employment, marital difficulties and depression are among the less obvious forms of disability that often accompany the more obvious physical effects.

In summary, no matter how the road traffic injury burden is measured there can be no doubt that it ranks among the major public health problems facing 21st century man; a problem rapidly worsening as the bulk of the world's population seeks to reap the benefits of motorized, personal mobility.

4.3 Measuring Road Traffic Safety Performance

4.3.1 Transport Safety, Personal Safety and Motorization

Having clearly established the high price in human injury and life exacted by our road transport system we need to answer the separate question of how dangerous road use really is. If most of the world's populations are road users at some time — and many, if not most, are road users not only on a daily basis but several times each day — then the sheer volume of exposure to risk will result in a large absolute number of injury events. The proper question then becomes: what is the minimum level of risk we can reasonably aspire to attain and how far are we from achieving that level of risk?

This question is not as easy to answer as it might at first appear. We know that the risk of severe injury, given an injury producing event, moves down the scale from pedestrians through bicyclists and motorcyclists to the occupants of passenger vehicles. However, we do not know very much about the probability of an injury producing event occurring for each of these forms of road use under the variety of conditions within which each takes place. What we typically estimate is a composite of both probabilities — crash event and injury severity. For example, in countries of the European Union, the death risk for motorcyclists is said to be 20 times that of car users, while pedestrians have a ninefold risk and bicyclists an eightfold risk (European Transport Safety Council, 2001).

One of the reasons for our lack of understanding of the crash event risk of different forms of road use is the absence of exposure data. Many countries use surrogate measures for road use, such as the number of registered motor vehicles or estimates of vehicle kilometers driven derived either from household surveys of vehicle use or from fuel consumption figures. But these are crude aggregate figures which take no account of nonmotorized movement (walking, cycling, etc.) or the proportion of movement in urban and rural areas, by night and day, and so on. While a small amount of information is available from travel surveys of the proportion of travel performed as pedestrians, bicyclists or motorized vehicle occupants; recreational versus commuting driving; driving on high speed rural roads and low speed urban roads, etc., most such data are collected for transport planning rather than road traffic safety purposes.

To illustrate the importance of understanding the nature of risk exposure, consider the finding that motorcycling in India is safer than motorcycling in the U.S.A despite the fact that, overall, the U.S.A. operates a far safer road transport system (Mohan and Tiwari, 1998). India has only two thirds of the U.S. rate of motorcyclist deaths per million motorcycles because motorcycling in India is the dominant form of personal transport and is undertaken on small, low-powered machines in congested environments while, in the U.S.A., motorcycling is predominantly a recreational activity, frequently undertaken on powerful machines and often in high speed environments.

Nevertheless, the traditional performance measure for the level of road transport safety has been death and serious injury per 10,000 registered motor vehicles or per 1,000,000 motorized vehicle kilometers of travel. This measure has been termed *Transport Safety* (Trinca et al., 1988) as it reflects the risk of serious injury per unit transport use. Road, traffic and transportation agencies have long favored this measure.

In more recent times, many governments have come to place their primary focus on the public health measure of *Personal Safety*. This measure is the rate of death or serious injury per 100,000 persons. It is a rate rather than a risk measure and, as such, its use facilitates direct comparison with other forms of disease.

Both *Transport Safety* and *Personal Safety* are intimately, but differently, related to the level of motorization in any given nation. Figure 4.3 is a schematic illustration of the general form of these relationships. When motorization is low the risk of death per unit road use is high. As motorization increases the risk of death per unit road use falls — as better roads are built, traffic management systems improve, safer vehicles are used and other programs of road traffic safety measures are implemented — and at high, but still increasing, levels of motorization, it approaches asymptote. In contrast, when motorization is low the rate of death per head of population is also low, because total population exposure to risk is low. It rises as more and more of the population become exposed to motorized road use and only commences to fall again once motorization has reached a level at which road infrastructure and traffic management systems reach maturity and government interventions to control road traffic safety become both systematic and intense (Trinca et al., 1988).

Figure 4.4 presents a case study of Australian data for the period 1950 through 2000 to particularize the general relationship. It demonstrates

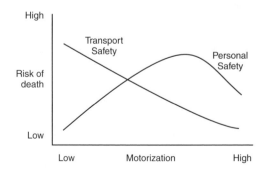

FIGURE 4.3 The general form of the relationship between motorization and both Transport Safety and Personal Safety. *Source*: Derived from Trinca et al. (1988).

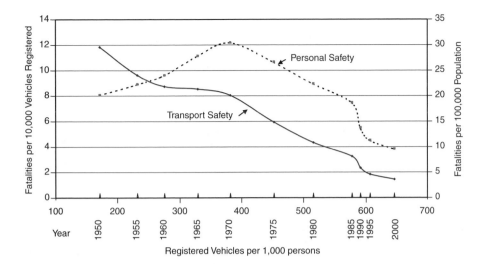

FIGURE 4.4 Trends in Transport Safety and Personal Safety in Australia as motorization increased (for the period 1950–2000).

that the rate of road traffic death rose while the risk of death per unit road use fell, because the gains in safety from improved roads and vehicles were not sufficient to offset the growth in population exposure to risk. It was in the late 1960s and early 1970s that the absolute number of deaths reached a level which communities, and therefore governments, were no longer prepared to tolerate. Special road traffic safety bodies were created and systematic, evidence-based intervention programs commenced (Vulcan, 1988). Not only did the population rate of road traffic death then begin to fall but the rate of decrease in the risk of road traffic death per unit road use accelerated.

4.3.2 Benchmarking Performance

Most national governments in motorized countries have formal road traffic safety policies, strategic plans and countermeasure implementation programs. In setting safety targets to strive for, it is common to benchmark safety performance against that achieved in other nations and to seek to achieve "best practice".

There are many traps in benchmarking. From the forgoing discussions, the most obvious need is to ensure that the nations chosen for comparison are at a reasonably similar level of motorization. As a nation commences to motorize, the first generation of vehicles are motorized bicycles and low powered motorcycles. Frequently, these are adapted to carry more passengers and become unique three-wheeled vehicles (Mohan and Tiwari, 1998). In the 1990s in Australia, less than 5% of registered motor vehicles were motorcycles while in Malaysia the proportion was closer to 60% and in Vietnam of the order of 90%. As already noted, the probability of injury, given an injury producing event, is highest for pedestrians, cyclists and motorcyclists.

Motorized countries, of course, only benchmark against other motorized countries. However, there are substantial differences in motorization levels among the nations classified as motorized and even greater differences in their patterns of road use. Table 4.4 compares *Transport Safety* and *Personal Safety* in motorized countries classified by their (relative) degree of motorization. The general relationship between level of motorization and traffic safety performance remains — the worst record is in the U.S.A., far and away the most highly motorized country, and the best record is shared by the three countries with relatively low levels of motorization (among motorized nations).

What is less obvious is that the differences in levels of motorization (among nations considered fully motorized) largely reflect geographic area and land use patterns which, in turn, shape the patterns of road

TABLE 4.4   *Transport Safety* and *Personal Safety* for a Range of Motorized Countries × Relative Level of Motorization (2001)

Relative Level of Motorization	Country and Specific Motorization Rate (Mot. Vehs./1,000 pop.)	Personal Safety (Deaths per 100,000 pop.)	Transport Safety (Deaths per 10,000 Reg. Motor Vehs.)
Very high	U.S.A. (777)	14.8	1.9
High	New Zealand (667)	11.8	1.7
	Australia (644)	8.9	1.4
	Japan (625)	7.9	1.3
	Spain (606)	13.8	2.3
Moderate	France (590)	13.8	2.3
	Canada (582)	9.5[a]	1.6[a]
	Sweden (551)	6.2	1.1
	U.K. (517)	6.1	1.2
	Netherlands (513)	6.2	1.2

Source: compiled from Australian Transport Safety Bureau (2003).
[a] 2000.

transport. For example, the U.S.A., Canada and Australia are all large countries of relatively recent urban development (all part of the "New World") and there are large distances between major population centers. In all three countries, well over two thirds of the road traffic injuries and deaths occur to the occupants of conventional passenger vehicles. In the U.K., a small country in area, far more of the personal transport task is able to be performed by public transport and by foot and, correspondingly, the proportion of road traffic injuries and deaths among conventional passenger vehicle occupants is far lower, and the proportion accounted for by pedestrians far higher.

The level at which motorization approaches its asymptote in a given country depends strongly upon the size of the country, its patterns of regional and urban development and its economic patters of land use. At one extreme is the geographically small nation of Singapore where car use is tightly constrained by government regulation and where some two thirds of the passenger travel task is achieved by public, rather than private transport.

Meaningful road traffic safety performance benchmarking can only result from comparing like with like.

4.3.3   What Levels of Safety Have Been Achieved?

Most motorized nations with mature road and traffic systems and systematic, (largely) evidence-driven road traffic safety programs have achieved transport safety levels in the range of one to two deaths per 10,000 registered vehicles and less than one death per 100,000,000 vehicle kilometers driven. However, while the current *Transport Safety* levels are at historic lows, there is no reason to suppose that the reduction in risk per unit of road use has yet reached asymptote. Table 4.5 illustrates the continuation of improvement in *Transport Safety* — in each of a range of countries — over the most recent 25 year period.

TABLE 4.5   Selected International Comparisons of *Transport Safety* over 25 Years

	U.S.A.	Australia	Sweden	U.K.	France
1975	3.2	5.9	3.8	3.9	8.1
1980	3.2	4.3	2.5	3.4	6.3
1985	2.6	3.2	2.2	2.5	4.6
1990	2.4	2.3	1.8	2.2	4.2
1995	2.1	1.8	1.3	1.5	3.1
2000	1.9	1.4	1.2	1.2	2.4

Source: compiled from Australian Transport Safety Bureau (2003). Road traffic deaths per 10,000 registered vehicles.

TABLE 4.6 Selected International Comparisons of *Personal Safety* over 25 Years

	U.S.A.	Australia	Sweden	U.K.	France
1975	20.7	26.6	14.3	11.9	27.3
1980	22.5	22.3	10.2	11.1	25.4
1985	18.4	18.6	9.7	9.4	20.6
1990	17.9	13.7	9.1	9.4	19.8
1995	15.9	11.2	6.5	6.4	15.3
2000	15.2	9.5	6.7	6.0	13.6

Source: compiled from Australian Transport Safety Bureau (2003). Road traffic deaths per 100,000 population.

The gap between these five nations has narrowed. In 1975, France (the worst performer in Table 4.5) had 2.5 times the rate of the U.S.A. (the best at that point). By 2000 France (still the worst) had only twice the rate of the best performers (Sweden and the U.K.). Note also that the U.S.A. — the most highly motorized nation throughout the 25 years — has made the least gains in *Transport Safety* and has moved from a leader to one of the worst performers among the motorized nations (see also Table 4.4).

The population-based rate of road death (*Personal Safety*) is closely dependent upon the (relative) level of motorization and, most importantly, the nature of the patterns of road use. While the geographically larger countries, which are more heavily dependent upon the private passenger vehicle and with large distances between major population centers, have experienced larger proportionate reductions in population based rates they have been unable to match the rates achieved in the less highly motorized (of the motorized) nations. Current "best practice" is around six deaths per 100,000 persons, again with no suggestion that a minimum has been reached (see Table 4.6). Unlike *Transport Safety* the ratio of best to worst performers has not decreased. Again, the U.S.A. has made the smallest proportionate gains.

We can conclude from this analysis that there is much scope to further reduce the road toll. While excellent gains have been made, death and serious injury remain a crucial public health problem, even for motorized nations, and despite the inexorable pressures of ever-increasing road use, we have not approached the limits of our ability to reduce the burden of mobility.

4.4   Is Safety Just a "By-Product" of the Road Transport System?

4.4.1   The Place of Transport in Economic and Social Development

All nations strive continuously to improve the quality of life of their citizens. While there is no single definition, the indicators of gains in the quality of life include rises in per capita literacy and numeracy rates, rises in average life span (indicating improvements in hygiene, diet, and access to medical treatment), and rises in average family income and increases in recreational time and opportunity. The generic indicator of "progress" is increases in gross national product per capita, that is, in continuous economic growth.

Although the relationship is not simple, transport development is a vital cog in the economic development engine (ECMT, 2002). A nation's ability to get its goods efficiently to market and to improve the effectiveness of its production systems requires efficient freight transport. The freight task is shared by all transport modes; however, in most countries, road transport is the dominant mode for other than bulk, long-distance freight movement.

Road transport is also a vital cog in a nation's social development engine. Access to schools, to new forms of workplace, to medical services and to recreational facilities requires effective personal mobility. There is even some evidence of psychological benefits (for individuals) from private motor vehicle transport (Ellaway et al., 2003). There can be no doubt that personal motorized transport is a highly cherished value in most developed nations.

In short, continuous improvement in transport is a fundamental underpinning for continuous economic and social development. Transport is, clearly, a substantial public good and road transport will long remain the dominant transport mode for both passenger and freight movements, except in small densely populated nations and in a relatively small number of very large metropolises.

The issue then is not how to replace personal motorized transport but how to manage the downsides. This chapter focuses on reducing the price paid in human injury and death. The other major externalities — to use the term favored by economists to describe the unintended outcomes of economic systems — are:

- the adverse human health impacts of pollutants from internal combustion engines
- the environmental impacts of pollutants and the longer-term implications of the consumption of nonrenewable fossil fuels
- the loss of urban amenity as the road transport paraphernalia intrude upon the urban landscape and as traffic congestion restricts the opportunities of citizens in large metropolises.

4.4.2 Defining Sustainable Transport

The term "sustainable transport" evolved in recognition of the need explicitly to manage the impacts of all the externalities arising from transport growth, not just for the current generation of citizens but for future generations as well.

Having committed to the concept of sustainable transport, governments still face the complex task of determining what the acceptable limits are for each of the adverse by-products, and then for managing the system to ensure that these limits are not exceeded. For example, the imposition upon motor vehicle manufacturers of a progressive regulatory time frame for, on the one hand, the reduction of various noxious emissions and, on the other, the increase in the fuel efficiency of internal combustion engines, are steps towards reducing adverse outcomes. Other examples include the imposition of road pricing in central London to decrease the volume of traffic and the introduction, in most motorized nations, of systematic road traffic safety intervention programs. Nevertheless, each is an isolated step. We have no vision of the desired integrated end point and no satisfactory way of knowing how close we are to having a sustainable road transport system (Johnston, 2002b).

This matters because decisions have to be made every day at the micro (project) level. Governments frequently use the cost/benefit ratio to justify individual project investment decisions. The common metric is, of course, the dollar and the "economic" benefits of the proposed action are weighed up against the "economic" costs, not only of taking the action but of the (incidental) negative outcomes. The problem with this approach is that our ability to assign dollar values to the range of benefits and costs is very primitive.

For example, one of the principal benefits of transport project proposals is the anticipated reduction in travel time. Throughout transportation history, journey time has been a primary driver of transport development. Quantum leaps were made when the railways replaced the horse-drawn carriages, when air transport replaced sea transport for international journeys, and when the motorized vehicle replaced the pedestrian and the bicycle. But is speed of journey the right goal if sustainable transport is the end objective? It is at least arguable that the continual push for faster journeys is the major barrier to achieving a sustainable system — speed is inimical to safety, increases emissions and increases fuel use (Haworth and Symmons, 2002). Since the key factor in personal quality of life is accessibility, not minimum journey time, ease of access may be a better objective, yet we have no current metric for accessibility.

As a further example of the paucity of our decision support tools, there are at least two ways of placing a dollar value on human injury and death and they result in substantially different estimates. The first (the human capital method) involves a direct calculation of costs incurred — the lost productivity from the congestion the crash caused, the costs of the medical recovery, treatment and rehabilitation facilities and processes, the vehicle repair costs, the lost productivity of the injured parties, and so on. The second method, which results in an approximate doubling of the estimated dollar value of a human life in a high income country, uses the "willingness to pay" method in which people make complex judgments, as citizens, about the levels of investment they are prepared to see governments make in order to save a human life.

Yet we continue to apply the benefit/cost ratio as though it were a magic ruler. Given our inadequate tools, it is hardly surprising that, at the macro level, governments have begged the question of setting formal limits to the intertwined adverse outcomes of road transport. However, there is another, perhaps more basic, reason. The broad road transport interests — vehicle manufacturers and all the attendant industries, the road asset construction and maintenance industry, and the plethora of personal mobility interests — are so strong and so economically and politically important in western, motorized economies that their collective vested interest in promoting motorization provides substantial resistance to any move away from continuous road transport growth.

What is certain is that we are a long way from defining a sustainable road transport system. Most governments continue to take the pragmatic approach of seeking to reduce — at an aggregate level — each of the identified adverse impacts without disturbing the fundamental thrust of road transport system development. They then use the benefit/cost ratio to justify — at the project level — their continued tolerance of adverse effects using an argument of net economic worth.

Even on those occasions when there are no other externalities to consider — when the decision to be taken is to do nothing or to take one of a limited number of options to reduce serious injury crashes — the use of the benefit-cost ratio is problematic. Ivey and Scott (2004a) examine the case for leaving a roadside utility pole unguarded in its current location against the cases for re-locating it, or guarding it using one of several available technologies. No option involves a change in travel time, fuel usage or pollution, so the only benefit is reduced injury and the only cost the cost of the treatment.

They examine several scenarios, ranging from a (specific) pole being struck once a year to eight poles struck over 5 years from a contiguous set of 20 utility poles. The benefit-cost ratio is a very useful tool for choosing among all options *except* the "do-nothing" option. Only for the "do-nothing" option is there a threat of potential liability in the event of a subsequent crash and the potential cost is several orders of magnitude greater than treatment option costs. As Ivey and Zegeer (2004) point out, if the site does not have an atypical number of collisions (historically) and it complies with generally applied standards, the risk of a successful liability claim can be effectively ignored and the "do-nothing" option becomes economically attractive. In reality, a very small number of specific objects are struck more than once in a 5 year period.

4.4.3 Sustainable Road Traffic Safety

At least one government has taken a radical position with regard to road safety by, in effect, removing human life from the road transport, trade-off decision-making process. In 1997, the Swedish parliament endorsed a road traffic safety strategy entitled Vision Zero (Tingvall, 1998). In essence, the philosophical position is that no road user, while acting lawfully, should be killed or seriously injured in a road crash in Sweden. It requires the Swedish Road Administration to design, build and operate the road and road traffic system to be tolerant of normal human error, and the normal range of behaviors, to a degree that ensures that crashes do not result in death or serious injury. But it does not require the road agency to provide a road and traffic system to cater for grossly illegal behavior, such as driving under the influence of alcohol or drugs or driving at excessive speed. For the first time, anywhere in the world, it places the *leadership* role for road transport system safety squarely in the hands of the road transport system designers and operators, albeit in partnership with others. A (slightly) less radical model of "sustainable safety" has been adopted in Netherlands. The implications of both models for the accountability of road and traffic agencies, and the professions that serve them, are detailed in later sections.

Vision Zero is at one extreme of the conceptual spectrum. At the other is the economic rationalist position that crashes (and their outcomes) are but one of many externalities in road transport, that all externalities can be assigned an economic value — along with all mainstream effects — and that the optimum decisions are those which maximize the economic value (of the project or issue at hand) to the nation.

4.5   Modern Thinking about Road Traffic Safety

4.5.1   A Historical Background

In most countries the police are responsible for the collection of basic information on road traffic crashes and injury. This role was accepted in the early days of motorization as a natural extension of the police function of "keeping order" in most facets of community activity — and because traffic law enshrined the (legal) principle of behaving in a "safe manner". Traffic control was also historically a police function but was quickly replaced, in most now motorized countries, by professional traffic engineers. The police investigation of crashes focused, naturally, on allocating blame to one or more of the involved road users and prosecuting breaches of traffic law. As the motor vehicle insurance industry grew, so too its focus was on apportioning blame to the various involved parties.

Traffic law, and the early institutional framework for safety management, fostered a widespread community belief that the primary cause of crashes was the errors made by, or the misbehaviors perpetrated by, road users. Police and insurance investigators lacked the skill, the inclination or the mandate, to look much beyond human failure.

Beginning in the 1960s scientists undertook broader, more systematic investigations of the causes of road traffic crashes but, inadvertently, their results served to reinforce the already widespread view of road traffic safety. While emphasizing that crashes are multicausal and that each crash was the outcome of a chain of events and circumstances — the removal of any one link of which might have prevented the crash — the take-away message for the general public and decision-makers alike was that human error could be identified in the causal chain in something in excess of 95% of crashes (for example, Shinar, 1978; Johnston, 1994 — see also Figure 4.5).

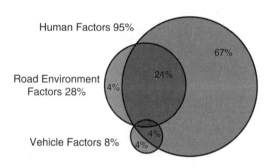

FIGURE 4.5   Typical findings from "At Scene" crash investigation studies — expressed as categories of contributing factors.

The early road traffic safety interventions, outside of traffic law enforcement, were typically the responsibility of "National Safety Councils" which were staffed by lay persons whose primary response to the growing injury problem was to conduct public education campaigns exhorting road users to behave more carefully and more responsibly. As there was no evaluation of the effectiveness of the various educational and enforcement efforts they remained at the very general level rather than being targeted at specific types of crash problem. (For a satirical — and powerfully insightful — analysis of this era, see Haight, 1973.)

The official road safety mantra was "the three Es" — Engineering, Education and Enforcement — but there was typically no co-ordination or integration of effort. The "Engineering" meant little more than that the civil engineers were responsible for road building and maintenance, that the traffic engineers were responsible for controlling traffic flow and that the vehicle engineers designed and built vehicles. Neither road nor traffic authorities were given any substantive accountability for the level of safety in the road systems they built and operated; their primary role was to facilitate traffic flow. Nor were vehicle engineers held accountable for the safety of the vehicles they put into the traffic stream. There were certainly no incentives for dialogue among the three professions.

As the now-motorized nations went through their period of rapid motorization in the 1950s and 1960s so the absolute number of road traffic injuries and deaths ballooned and community concern grew.

While a small number of pioneers were making major contributions to safety knowledge, their efforts were not having much influence on government or institutional policy and practice. From injury epidemiology, came the concept of reducing the transfer to the human body, of the kinetic energy "released" in a crash, to a level which the body could tolerate. The early seat belt systems, and other injury attenuation devices, were practical results from this work. However, it was not until the late 1960s that automobile manufacturers began to be compelled to improve the levels of protection offered to vehicle occupants. It took the advocacy of people like Ralph Nader to force such government action. (See Robertson, 1992, especially Chapter 10.)

As long ago as 1966 the United States federal government enacted the Highway Safety Act. It required both vehicle and road designers to comply with safety standards, particularly, with regard to crash-worthiness. As Michie points out, it caused infrastructure providers to stop designing "only for drivers who remained on the roadway" (Michie, 2004, p. 37). Unfortunately, what happened in practice was disappointing, particularly outside the U.S.A. A few public health professionals also began to transfer to the fledgling road traffic safety field the host-agent-environment construct so influential in tackling disease (Haddon et al., 1964). This enabled road safety strategists to look beyond the simplistic notion that human error could only be reduced through education and enforcement and, in effect, gave birth to the science of road safety.

However, it was not until the absolute toll of injury and death exceeded community tolerance that governments began to take systematic remedial action. For most of the now motorized countries this watershed occurred somewhere between the middle 1960s and the middle 1970s. Separate road traffic safety bodies were created and, in most places, they quickly assumed what had been the traditional responsibility of the National Safety Councils and they rapidly diversified the range of road traffic safety interventions.

4.5.2   The Development of Strategies

Once a scientific, evidence-driven approach began to be taken to the management of road traffic safety, progress was rapid. The first breakthrough was the acceptance that crashes have complex, multi-dimensional causal chains and that the probability of crashes can be reduced by interventions at one or more points in the common types of chain. The second breakthrough was the understanding that injury is the result of the imposition of (mostly mechanical) energy upon the human body at a level beyond the body's ability to tolerate; that "energy (is) the necessary and specific cause of injury" (Robertson, 1992, p. 212). In short, the conceptual watershed in thinking was that no road traffic crash has a simple, single cause and, even if it did, it does not follow that the most effective remedial action is to focus only, or even primarily, upon that direct cause.

Robertson (1992) expresses this very clearly. Think of a road user in motion as energy. The amount of energy is, of course, defined by mass and speed ($E = \frac{1}{2}mv^2$). Serious injury occurs when the energy is transformed — as in a crash — and is beyond the body's tolerance levels. Injury control is thus the management of energy transfer to the body. It follows, then, that the cause of the crash is less the focus of attention than the modifiability of the energy transfer process.

A popular "mental map" that resulted was the "Haddon Matrix". William Haddon Jr. was a public health professional who transferred his expertize to road traffic safety as the inaugural Director of the United States National Highway Traffic Safety Administration. Table 4.7 outlines the conceptual matrix and Table 4.8 provides an example of its application to a specific road crash problem.

With the concepts of primary safety (reducing the probability of a crash) and secondary safety (reducing the probability of injury, given that a crash occurs) firmly established, a range of basic road traffic safety strategies were developed (for a more detailed summary see Trinca et al., 1988). The five

TABLE 4.7   The Haddon Matrix

	Human Factors	Vehicle Factors	Physical Environment
Pre-event			
Event			
Post-event			

Pre-event phase: all factors that determine whether the event leading to injury will take place. *Event phase*: all factors that determine whether an injury will occur once the incident has been initiated. *Post-event phase*: all factors that determine the final damage and the permanent handicap stemming from injury.

major categories of strategy are:

- exposure control
- crash prevention
- injury control
- behavior modification
- post-injury management

Each will be briefly described.

4.5.2.1   Strategy 1 — Exposure Control

Shaping public policy in a way that reduces the amount of travel, or that substitutes safer forms of travel for less safe forms, will lead to decreases in road traffic injury and death. Car pooling, for example, reduces exposure to risk during commuting journeys by substituting one journey for two, or more. Encouraging commuters to travel to work by bus reduces the exposure to risk even further, and commuting by either light or heavy rail reduces exposure by an even greater margin. (Unfortunately, the

TABLE 4.8   Application of the Haddon Matrix in Order to Identify Potential Countermeasures to the Problem of "Single Vehicle, Run-Off-Road, Struck Fixed Object" Crashes

Human Factors	Vehicle Factors	Physical Environment
	Pre-event	
Training in skid recovery	Fatigue monitoring/warning device	Sealed shoulders
Fatigue management education	Traction control system	"Tactile" edge lines
Speed enforcement	Antilock brakes	Skid resistant pavement surface
	Alcohol ignition-interlock	
	Event	
Seat belt wearing law	Seat belt pretensioners	Clearing roadsides of rigid objects
Reduced speed limits	Airbags	Guarding rigid objects with energy absorbing systems
	Collapsible steering columns	
	Padded interiors	Frangible poles
	Post-event	
First aid training for motorists	Automatic May-day system linked to crash sensor	Location markers to facilitate emergency system access
	Reduced probability of postcrash fire through fuel tank design	Strategic location of trauma treatment centers
		Air ambulance for remote access

Note: The table entries are illustrative only.

commuting journey has a lower risk of serious injury or death than journeys for many other purposes so the overall effect on road trauma is far less than most advocates of a wholesale shift to public transport imagine.)

Similarly, schemes to discourage motorcycling in favor of conventional passenger vehicle use transfer exposure from a higher to a lower risk form. (Much of the early gains in road traffic safety in rapidly motorizing countries occur as the population shifts its primary transport vehicle from two to four wheel form.) The introduction of zero blood alcohol laws for probationary drivers — which has been implemented in many motorized countries — is a form of exposure reduction, because it seeks to remove one of the highest risk forms of behavior from one of the highest risk subpopulations of road users.

4.5.2.2 Strategy 2 — Crash Prevention

Many aspects of road design influence the probability of a crash. Geometric road design standards limit both horizontal and vertical curvature, require minimum sight distances, and prescribe minimum lane widths, pavement shoulders, and so on. Divided roads prevent the possibility of head-on crashes (unless the median can be traversed by errant vehicles); access control limits the number of potential conflict points; and intersection design is fundamental to controlling potential conflict opportunities. Skid resistant pavements decrease the risk of loss of control.

Similarly, in vehicle design, there are many opportunities to decrease the probability of a crash occurring. Improvements to vehicle suspensions, braking systems, tires, and so on, fit this category. Many recent developments such as anti-lock braking systems, traction control and electronic stability systems show promise although there is, as yet, little evidence of their effectiveness in crash and, more importantly, injury prevention.

4.5.2.3 Strategy 3 — Injury Control

Seat belts, airbags, burst-proof door latches, collapsible steering columns, body panel "crumple zones", frangible utility poles, roadside barrier systems and bicycle/motorcycle helmets are all examples of interventions designed to limit the amount of kinetic energy that is transferred to the more vulnerable parts of the human body during a crash.

4.5.2.4 Strategy 4 — Behavior Modification

Traffic law and its enforcement and road user education and training are the common tools for seeking to constrain road user behavior to its lowest risk forms. This strategy is frequently used in conjunction with one of the other strategies.

Regulation and its enforcement apply to institutions as well as individual road users. For example, bicycle helmet wearing has been shown to be maximally effective when it is required by legislation (O'Hare et al., 2002). Regulation is also required to ensure that vehicle manufacturers, infrastructure owners and operators provide minimum levels of primary and secondary safety in the vehicles and roads they produce.

4.5.2.5 Strategy 5 — Postinjury Management

Improvements in emergency service response times, in both on-scene and hospital-based trauma management, and in rehabilitation facilities and processes, have decreased the probability that a seriously injured person will die and that a serious injury will result in long-term disability.

4.5.3 From Strategies to Effective Implementation

A key element in the evolution of a systematic science of road traffic safety was the gradual creation, from a zero base, of a body of knowledge of the effectiveness of the plethora of interventions attempted and the relative effectiveness of different strategies, alone and in combination. Creating this body of knowledge involved a commitment to the scientific evaluation of countermeasure efforts and programs which, in turn, depended upon quantum improvements in the quality and coverage of routine databases about crashes and their outcomes.

Despite the very considerable advances in data systems, in the underpinning knowledge set and in scientifically based conceptualizations and strategic thinking, serious barriers to the implementation of effective safety countermeasure programs have persisted.

One of the defining features of the field of road traffic safety is its institutional complexity. Roads are built and maintained, and traffic is managed, by, or through, road and traffic agencies. Vehicles are designed, built, marketed, sold and maintained through a (mostly) independent set of industries and agencies. Traffic law policy, the enforcement of traffic law and the punishment of offenders is the responsibility of yet another set of agencies. Educating road users is split between the public education system, specific traffic safety education agencies and separate driver training organizations. Emergency response and trauma management systems add another layer of institutional complexity, as do public and private insurance schemes which deal with the compensation processes in the aftermath of crashes.

While institutional complexity is common in the systems that control or influence many aspects of everyday human activity, the natural inclination of institutions to limit themselves to the boundaries of their core businesses was, for a long while, a barrier to truly effective road traffic safety program implementation. For a long time, the institutional separation of the "three Es" constrained progress. Progressive road safety policy makers now talk about the "three Cs"[1]:

- *Commitment*: unless a government exercises meaningful leadership in road traffic safety, progress will be suboptimal. Yet, many governments continue to take the line of least resistance, giving in to the irresistible forces promoting continuous road transport expansion.
- *Co-operation*: given that the actions of institutions in one part of the road transport system can run directly counter — in safety outcomes — to actions in another part of the system, co-operation is essential. For example, the reduction of speed limits in urban areas, or the introduction of traffic management techniques to slow vehicle speeds, are directly at odds with the marketing and advertising of vehicles on the basis of their power and performance and the promotion of driving as exhilarating.
- *Co-ordination*: even where co-operation is secured, it is vital that the efforts of the different institutions are effectively coordinated. For example, in one State in Australia, the co-ordination of urban speed management activities among the road traffic agency, the enforcement process and the public education process resulted in demonstrable benefits (Cameron et al., 2003).

The current "best practice" model for road traffic safety policy making, intervention programming and effective implementation involves a whole-of-government approach to the coordinated implementation of integrated countermeasure programs. Successful programs are based upon:

- routine surveillance of safety progress, using comprehensive, high-quality data systems, covering the gamut of road traffic safety problems
- strategic targeting of the key problems using evidence-based strategies and program options
- the provision of adequate resource for meaningful implementation
- rigorous evaluation of the effectiveness of the interventions
- continuous improvement in implementation based upon the evaluation results and maximum co-ordination among all relevant institutions.

4.5.4   The Defining Visions

While the evolution of the science of road traffic safety has been international, there are marked national and cultural differences in the strategies selected and the countermeasure programs implemented.

Australia, for example, has behavioral control as its predominant strategy (Johnston, 2002a, 2003). Australia has led the way with legislation to mandate self-protection — it was the first country to require helmet wearing by both motorcyclists (in the 1960s) and bicyclists (in the 1990s); it was the first

---

[1] I am indebted to Charles Melhuish of the Asian Development Bank for introducing me to this concept.

with compulsory seat belt wearing (in the 1970s); it has one of the lowest permissible blood alcohol concentrations; and it operates one of the most intense enforcement programs in the world to curtail both drink driving and speeding. While Australia does not ignore the other strategies, behavioral control predominates.

In contrast, the U.S. is far less willing to mandate and control behavior and, as Lave and Lave (1990) point out, is libertarian to the point of respecting a citizen's right to make mistakes, even fatal ones, unless the actions are harmful to others. Thus, the U.S. has legislated to prohibit "drunk" (as opposed to drink) driving — that is, most States have legal limits of 0.10% BAC or higher — yet seat belt wearing (in all seat positions) and helmet wearing by motorcyclists are not mandatory in all States. The United States ranks 14th out of 18 motorized nations in the proportion of vehicle occupants wearing seat belts and only 20 States require all motorcyclists to wear helmets (Insurance Institute for Highway Safety, 2002). There is a philosophical preference for passive, secondary safety measures, that is, measures which do not require road users to take specific actions. Consequently, the design of airbags in the United States, does not assume a seat belted occupant whereas the design of seat belts in Australia and in Europe does and, as a result, the bags in the latter countries are less "aggressive".

Adams (1995) is, perhaps, the most outspoken advocate against interventions to control behaviors that are risky only (or primarily) for those who engage in them. Hauer (1990) goes so far as to characterize the U.S. as still being at the "leaches, spells and exorcism" stage of road traffic safety thinking. Certainly, the U.S. has among the worst road traffic safety rates of the highly motorized nations. In part this is because it is by far most highly motorized but it is also in part because it has chosen not to implement many of the behavior control strategies and measures proven effective elsewhere.

These brief observations of the cultural and socio-political influences upon road traffic safety policies and intervention programs underline again the dominance of personal motorized transport in western culture and the critical role played by the plethora of industries and institutions dependent upon, in one way or another, continuous growth in road traffic.

Nevertheless, the motorized world appears to be on the brink of the next quantum leap in road traffic safety thinking. A small number of countries, most notably Sweden and The Netherlands, have formalized radical visions for the future of road traffic safety. The most important thing to note about these visions is that their primary purposes seem to be to change the way communities, and the institutions which serve them, conceptualize and prioritize the road traffic safety problem and to commit governments to a radically new way of approaching the road transport system decision making process.

The Swedish Vision Zero philosophy has already been outlined. The philosophy starts with the explicit acknowledgement that humans are fallible and will make mistakes. Since crashes cannot always be avoided the focus is on "… designing roads, vehicles and transport services in a way that someone can tolerate the violence of an accident without being killed or seriously injured." (Koornstra et al., 2002). For the first time, this places a direct accountability squarely on the shoulders of the road and traffic agency. The agency's professionals have to ensure that the road and traffic system is designed and operated in such a way that it is tolerant of the most frequent forms of human error and that, when a crash occurs, the kinetic energy transferred to the human body is within tolerance limits.

It is important to note that the strategy of behavioral control is neither ignored nor underplayed. Vision Zero requires vehicle, road and traffic engineers explicitly to design and operate the road transport system, but only within the bounds of reasonably expected behavior. That is, the system must tolerate normal human performance but not grossly illegal behaviors. It explicitly bounds the design prerogatives within a framework of rules for behavior. Thus, vehicle designers may assume that car occupants are belted and that motorcyclists are wearing helmets and road and traffic engineers may assume that road users (mostly) comply with traffic rules (such as intersection priority rules and speed limits). They are required, however, to satisfy themselves that the rules themselves are in accord with the bounds of normal human performance. Intersection priority rules, for example, must not demand unusually complex judgments on the part of road users.

Perhaps the most radical element of Vision Zero is that the Swedish government has made human life paramount. As Tingvall (2004) states: "Life and health are not allowed in the long run to be traded off against benefits of the road transport system, such as mobility. Mobility and accessibility are, therefore, functions of the inherent safety of the system." In effect, this removes serious injury and death from cost/benefit analyses. Before a potential transport project can be considered it needs to be demonstrated that the design and operation principles are in accord with the Vision Zero philosophy.

This stands in stark contrast with the traditional economic rationalist approach, typified in the following quote from a recent report of the U.S. Transportation Research Board arising from a review of utility poles as a roadside safety issue;

> A primary objective... is to develop a strategy of minimizing utility pole crashes that maximizes the benefit to society for every action and expenditure. A secondary, but nonetheless important, objective is to provide a good defensive position relative to litigation.   (Ivey and Scott, 2004b)

It seems that, strictly applied, the Vision Zero decision-making process would not achieve conventionally acceptable benefit-cost ratios. Consider the following extract from Johnston (2002a) which is based upon the work of Elvik and Amundsen (2000):

> The Swedish government asked the Institute for Transport Economics in Norway to evaluate overall benefits and costs of three possible strategic packages that could be implemented over the decade to 2010 (Elvik and Amundsen, 2000):
>
> • The first was based strictly upon the benefit/cost ratio. What happens if we put in place all known measures where the evidence in the literature is that the aggregate marginal benefits (safety, mobility and environment) over the 10-year period are predicted to be greater than the aggregate marginal costs?
> • Secondly, what happens if you implement the Vision Zero principle of minimizing energy transfer at impact, particularly through speed control and road design modifications?
> • The third package was the same as the first, but where every measure was to be implemented with the sort of intensity and resource that would be needed to get the maximum potential effect implied in the evaluation literature. Table 4.9 contains the results.

The strict benefit/cost program was estimated to lead to an overall benefit/cost ratio of 1.25, with major benefits in accident cost reductions, some added cost in increased travel time, some savings in vehicle operating costs, some environmental benefits and some reduction in induced traffic. However, the Vision Zero program, on a conventional benefit/cost basis was estimated to be negative. Nearly twice the savings in accident costs compared with the strict/benefit process of countermeasure selection, but with 50 times the added travel time costs, resulting from the constraining of traffic speeds. There was estimated also to be some increase in vehicle operating

TABLE 4.9   Evaluation of Safety Strategy Packages (Million SEK, 10 Years 2002–2011)

	Strict c:b	Vision Zero	Maximum Potentials
Benefits			
Accident cost savings	81,890	140,542	190,011
Travel time savings	(2,925)	(142,284)	(109,237)
Veh. op. cost savings	2,028	(2,493)	(5,961)
Env. cost savings	4,137	4,129	1,182
Induced traffic savings	366	(7,854)	(6,025)
	85,496	**(7,960)**	**69,970**
Costs	68,436	384,514	667,992
Benefit/cost ratio	1.25	(0.02)	0.10

Source: Reproduced from Johnston (2002a) (based upon Elvik and Amundsen, 2000).

costs, positive impacts on the environment and negative impacts on induced traffic, resulting, overall, in a negative cost/benefit ratio. Finally, the "maximum potentials" program was estimated to provide larger accident reduction benefits than even the Vision Zero program but also with large increases in travel time, substantial increases in vehicle operating costs, smaller environmental benefits and quite substantial disbenefits in induced traffic. The benefit/cost ratio was estimated at 0.10. This is compelling evidence that further major gains in safety are available if nations are prepared to pay the mobility cost. The Swedish vision asserts that the price must be paid, that is the reduction of death/catastrophic injury over-rides other considerations.

In The Netherlands an overarching vision similar in application to the Swedish model — "Sustainable Safety" — has been formulated. However, it stops short of explicitly removing serious injury and death from the trade-off decision-making process.

The following passage from Koornstra et al. (Koornstra et al., 2002, pp. 13,14) summarizes the Dutch vision:

> The starting point of the concept of "Sustainable Safety" is to drastically reduce the probability of accidents in advance, by means of infrastructural design. In addition, where accidents still occur, the process that determines the severity of these accidents should be influenced in such a way that serious injury is virtually excluded… A sustainable safe traffic system has an infrastructure that is adapted to the limitations of human capacity, through proper road design, vehicles equipped with tools to simplify the tasks of man and constructed to protect the vulnerable human being as effectively as possible, and a road user who is adequately educated, informed and, where necessary, controlled. The key to arrive at a sustainable safe traffic system lies in the systematic and consistent application of three safety principles:
>
> – functional use of the road network by preventing unintended use of roads;
> – homogenous use by preventing large differences in vehicle speed, mass and direction;
> – predictable use, thus preventing uncertainties among road users, by enhancing the predictability of the course of the road and the behavior of other road users.

The similarities between the Dutch "Sustainable Safety" and the Swedish "Vision Zero" are obvious. Both place the primary accountability for road traffic safety with the designers of vehicles, roads and traffic management systems. This is a radical departure from tradition. Safety has been, and in many places still is, a secondary purpose of vehicle, road and traffic system design, with its secondary status firmly cemented in the belief that road users carry almost total responsibility for their own crashes.

In neither country has the vision yet been fully realized; both governments acknowledge that their respective visions are long-term and that the concepts will only progressively come to influence decision making.

In the past — and still in most nations — road and traffic agencies had a primary mission of meeting the community's demand for increasing levels of personal mobility and accessibility. Now, at least some motorized countries are explicitly determining that serious injury and death can no longer be traded for improvements in personal mobility. Proposals to improve the efficiency and effectiveness of the road transport system in its primary task of moving people and goods must have a zero tolerance to serious injury and death. It is not that personal motorized transport improvements have become any less important, simply that death and serious injury cannot be part of the price. It must also be emphasized that it is not a "no crash" philosophy, it is a "no death" philosophy.

While the governments of many motorized nations have national road traffic safety policies, no others have yet embraced such an explicit vision. Nevertheless, many reveal at least the beginnings of a paradigm shift in thinking away from a primary focus on direct, conventional measures of behavioral control and more towards an emphasis on improving road, vehicle and traffic system design and performance (see, for example, Johnston, 2002a,b).

This paradigm shift is coming to be known as a model of social responsibility. In such a model, vehicle designers have an obligation to provide vehicles matched to human performance capability and with a level of protection that will minimize serious injury in the event of the most common forms of crash. Road and traffic engineers have a responsibility for designing a road and traffic system which is tolerant of normal human error and which protects against death and serious injury to the greatest extent practicable — under the assumption that the vehicle manufacturers have met their social responsibility, that is, that the vehicles they are providing infrastructure for are safely designed. Finally, road users have a responsibility to conform to the rules and norms of road use required by the design of the safe road transport system.

It would be naïve to suggest that highway engineers, vehicle engineers or road users have fully embraced the concept of social responsibility. For example, the trend, in some countries, to the rapid growth of large 4WDs (SUVs, RUVs, etc.) clearly places personal safety ahead of social responsibility. Ratings of vehicle crashworthiness (protection to occupants in the rated vehicle) are now being accompanied by ratings of vehicle "aggressivity" (level of harm to occupants in vehicles struck by the rated vehicle) in an endeavor to inform purchasers of the net social harm likely to flow from the increased disparity in mass among the passenger vehicle fleet (Newstead et al., 2003).

Part B — The "Tool Box" for Road and Traffic Engineers

4.6   A Safety Engineering Mindset

Continuous improvement to road infrastructure and advances in the management of traffic have played a major part in the improvements in *Transport Safety* as nations have motorized. Much of the safety gain, however, has been incidental in that the primary purpose of infrastructure and traffic management improvements has been to cope with the rapidly increasing traffic volumes and the (apparently) insatiable desire for ever shorter travel times. High speed rural motorways and freeways and urban ring roads, for example, physically separate opposing traffic flows and strictly control access — both characteristics are fundamental to efficiency and extremely valuable for safety.

While both *Transport Safety* and *Personal Safety* continue to improve, the sheer volume of exposure to risk ensures that the absolute numbers of road traffic deaths and serious injuries remains at, or close to, the top of the public health problems confronting the "developed world". To take the next quantum leap in road traffic safety requires not only a new mindset but a new form of leadership on the part of road and traffic engineers and the institutions for which they work. No longer can safety be a secondary component in their thinking.

In part, this is because the safety gains from the application of other road traffic safety strategies are slowing. Consider, for example, vehicle crashworthiness. The continual growth of urban traffic is leading to an increase in the frequency of side-impact crashes at intersections. Protecting a vehicle occupant (in the struck vehicle) from injury in a side-impact is particularly difficult given the lack of space between the human body, the host vehicle structure and the intruding vehicle structure (see Figure 4.6). In side-impacts, serious injury is highly probable at impact speeds as low as 45–50 km/h. In head-on crashes, in contrast, a raft of features — collapsible steering columns, frontal structure crumple zones, seat

Source: www.civil.ubc.ca/transportation.htm

FIGURE 4.6   The difficulty of protecting occupants in side impacts.

belts and airbags — combine to attenuate the kinetic energy "released" in a crash, resulting in a minimum reaching the human body. Even here, serious injury is highly probable at impact speeds (for either of the vehicles) beyond 65–70 km/h.

Compounding the problem for vehicle designers is the increasing diversity in the masses of registered vehicles. As a result of changes in methods of production and distribution the demand for road freight transport continues to increase. In many motorized nations, the rate of growth in the number of registered commercial vehicles is outstripping the rate of growth in the number of registered private passenger vehicles quite substantially (see for example, Department of Transport and Regional Services, 2000). In addition, some motorized nations are seeing an explosion in the number of larger four-wheel-drive or sports and recreational utility vehicles at one end of the private passenger vehicle market and a similar rapid growth in very small (and light) commuting style passenger vehicles at the other end of the market (Johnston, 2002a). The consequence is an increase in the probability of crashes between vehicles of unequal masses with a consequent imbalance in the kinetic energies to be attenuated if serious injury is to be avoided.

In motorizing countries, much of the road traffic death and serious injury problem stems from the extremely high proportions of the so-called vulnerable road users — pedestrians, cyclists and motorcyclists — who fare badly in collisions with the rapidly growing number of trucks, buses and passenger vehicles. In such countries, design attention to the exterior of motorized vehicles and, most importantly, the control of travel speeds in order to constrain impact speeds, is more fundamental.

In short, in motorized countries, vehicle crashworthiness gains are being offset (to a degree) by an increasing disparity in the distribution of masses among the vehicle populations and an increase in side-impact crashes.

Similarly, many motorized nations have put in place the most obvious behavioral controls to improve safety and, for several of these, the returns appear to be static or diminishing.

Mandating self-protection is a common, though by no means universal, strategy. The wearing of helmets by bicyclists and motorcyclists and the wearing of seat belts by passenger vehicle occupants are the most obvious examples. In the case of Australia, which mandated seat belt wearing in the early 1970s, wearing rates are among the highest in the world. In excess of 95% of front seat occupants in passenger vehicles wear their seat belts and yet more than 20% of fatally injured passenger vehicle occupants have been found to be not wearing their belts at the time of their crash (Johnston, 2003). Thus, a very small minority of the vehicle occupant population accounts for a substantial proportion of the vehicle occupant fatal injuries. Further public education and enforcement is unlikely to reach this "deviant" group of dedicated nonwearers.

Most motorized nations place a legal limit on a driver's blood alcohol level in an attempt to reduce the number of deaths and serious injuries arising out of crashes in which the consumption of excessive alcohol played a substantial, causative role. In Australia, the proportion of fatally injured drivers with blood alcohol levels above the legal limit (of 0.05 mg%) was approximately 50% before very intensive random breath testing and supporting public education was introduced in the mid to late 1980s. Over the next 5 years this proportion more than halved but, despite continued intensive enforcement and supporting public education, it has not been further reduced in the last decade (Johnston, 2003). Just as with the minority of habitual nonwearers of seat belts there appears to be a minority of road users resistant to conventional behavioral control measures.

Following detailed analyses of this type, many road traffic safety strategic plans have concluded that the largest gains in future safety — at least for those nations which have systematically applied both vehicle crashworthiness technology and known, effective behavioral controls — lie in realizing the potential of road and traffic system design and management (arrive alive!, 2001). To repeat, the most important paradigm shift needed for substantial future road traffic safety improvements is the acceptance by road and traffic engineers that safety must no longer be a secondary consideration. The basic tenet of the Swedish Vision Zero is that traffic safety shall not be a function of mobility, rather mobility must be a function of traffic safety.

If road and traffic engineers, and the institutions for which they work, are to accept that they must lead the next major stage in the evolution of road traffic safety then they will need also to:

1. Abandon the outmoded belief that behavioral control should be the frontline road traffic safety countermeasure strategy simply because human behavior or error can be identified as a cause in the vast majority of crashes.

2. Accept that crashes have multiple causes and that successful interventions can occur at many points in the causal chains.

3. Embrace the philosophy of loss reduction, rather than crash prevention — accepting that in such a universal, daily human activity, crashes are inevitable and safe road use must be defined as road use with minimum probability of death or serious injury.

4. Temper the use of economic rationalist decision-making, applying it only to choose the most cost-efficient treatments and not to justify the "no action" option.

5. Base all program decisions upon systematic analysis of reliable data and research evidence of likely safety effectiveness.

6. Evaluate all interventions and seek continuously to build the stock of road and traffic safety engineering knowledge.

7. Integrate road and traffic engineering interventions with the interventions of vehicle engineers, legislators, traffic police, and others addressing common crash problems.

4.7 Matching Systems to Users

4.7.1 The Fundamental Relationship between Design and Safety Outcomes

Hauer (1990) describes three fundamental ways in which road and traffic engineers influence the inherent level of safety in the road transport system. These are:

1. *Influencing the opportunity for a crash to occur*. The potential for a two (or more) vehicle collision is related both to the number of possible conflict points for vehicles on intersecting paths, and the frequency with which those conflict points are experienced. (Scott and Ivey (2004b) would add that the greater the number of decisions required of road users at a given conflict point the greater the frequency of crashes at that point.) Hauer identified 32 possible conflict points for two vehicles on intersecting paths at a conventional cross intersection and only nine conflict points at a T-intersection.

One can extend the analysis beyond intersections. On a divided carriageway, the only potential for a collision between vehicles in opposing flows is if one vehicle crosses the median and enters the opposing flow. In contrast, on an undivided road there is an omnipresent potential for collision between vehicles in opposing flows. Similar analyses can be made regarding the potential for single vehicle crashes; for example, the opportunity for a single vehicle run-off-road crash is greatest on a narrow, undivided road with adverse camber on a small radius horizontal curve.

2. *Influencing the probability of a crash occurring per given crash opportunity*. While there may be 32 potential conflict points at a cross intersection the engineer has considerable influence over the probability of a crash occurring at any one of these conflict points. For example, a signal-controlled intersection decreases the probability of several forms of potential intersection crash compared with an uncontrolled intersection. Similarly, the addition of slip lanes or the introduction of signalized turn phases reduce the probability of collisions involving turning maneuvers. Stop signs reduce collision probabilities to a greater extent than do give way signs for two reasons — first, they reduce complexity in the driver's decision process as judgments of clearance times and distances to potentially conflicting traffic are largely taken out of play; and secondly, the approach speeds of potentially conflicting vehicles are reduced to a minimum.

3. *Controlling the process of energy exchange when a crash does occur*. Of all intersection designs a roundabout ensures shallow angle crashes which are less harmful. Similarly, the use of guard rail

dissipates the kinetic energy, in contrast with the focused energy exchange when a vehicle hits a utility pole or a tree.

What follows from these three safe engineering enablers are a small number of basic (often overlapping) principles for safe design:

1. Match the road design and traffic management control systems to the fundamental purpose of each part of the road network. For roads that primarily serve the need for access to residences — or where volumes of pedestrians and bicyclists are high — a low speed environment must be created and potential conflict points with higher speed traffic flows must be minimized. Ideally this is done at the time of original design but can be achieved by remedial engineering treatments, generally encompassed within the term local area traffic management (see van Schagen, 2003). At the other end of the road function spectrum are roads designed for high speed travel between urban centers, or from one sector to another within a large urban area. On such roads, all vulnerable road users should be denied access or provided with protected, segregated "lanes", other vehicle access control should be strictly managed, opposing traffic flows must be separated, and various other steps taken to facilitate safe, high-speed movement.

2. Make separate provision for essentially incompatible forms of road use. The provision of a network of bicycle paths in a large city, for example, facilitates bicycling — both for commuting and for recreation — while separating cyclists from motorized traffic. Pedestrian underpasses and overpasses are similar examples of separating one class of vulnerable road user from the predominant traffic stream.

3. Minimizing conflict points, and reducing the probability of collision at those points are also critical. Minimizing the number of access points, for example from a residential subdivision to the arterial road network, is fundamentally more important than the traffic management controls at each access point. The higher the design speed of the surrounding collector roads the more important the control of access points becomes.

4. Managing for crash severity is crucial. A common failing in the design of high-speed roads is to assume that safe design is limited to the traveled way. Highway engineers seem to assume that drivers will not leave the paved surface because the design has paid careful attention to lane width and shoulder width, to vertical and horizontal curvature, to skid resistance, and to pavement markings and other forms of delineation. While these measures undoubtedly reduce the probability of road departures they cannot guarantee it. In Australia, some one third of all fatalities — and 50% of fatalities on high-speed rural roads, including those of high geometric design standard — are the result of a vehicle running off the road and colliding with a tree, pole, embankment, or otherwise overturning on untraversable terrain (Corben et al., 2003). In the United States some 30% of all road crash fatalities in 2000 were the result of striking a fixed object (Scott and Ivey, 2004a,b). We will return to the importance of managing roadside safety a little later.

5. Perhaps the most fundamental of the basic design principles is that of speed management — matching traveling speed to the level of safety inherent in the road and roadside environment. We will shortly return to this issue and explore it in great detail.

4.7.2 Understanding Human Behavior and Its Limitations

Road use, in all its forms, involves complex information processing and decision making. For a text on human factors for highway engineers see Fuller and Santos (2002). For a comprehensive review of current knowledge regarding driving see Groeger (2000). For a simpler overview, see Chapter 3 in Ogden (1996).

Unfortunately, there is no such thing as the "design standard road user". Novice drivers perceive and respond to the road and traffic environment in ways quite different from those of experienced drivers. Older road users similarly react and respond in different ways. Young children, as pedestrians, exhibit quite unpredictable behavior. Yet engineers need to make assumptions about human abilities when they design signs, delineation systems, signal phasings and so on. They do so reasonably well in specific areas — for example the conspicuity and legibility of road signs and pavement markings is based upon knowledge of the human visual system. However, generally speaking, highway engineers understand little about human behavior.

The engineering substitute for understanding complex human behavior is to design according to prescribed standards and manuals, in the belief that those standards and manuals are founded upon a sound knowledge of human performance. Unfortunately, that is far from the case. Let us consider four examples.

The first relates to the common practice of placing guardrail along stretches of rural roadside where the land falls away steeply from the edge of the road reserve. This has nothing to do with the probability of a vehicle departing the road on one of these stretches and is only superficially related to the likely severity of a crash should such a departure occur. While running off the road into a ravine will certainly lead to a high severity crash, running off a section of similar road into an immediately adjacent row of trees will lead to a crash of at least equal severity. This common engineering preference of protecting vehicles from "falls" is related to the way humans perceive danger. We perceive a clear risk of falling from heights, but we are unable to see kinetic energy. Figures 4.7 and 4.8 illustrate the point dramatically. No engineer would design the road shown in Figure 4.7, that is, without protection from the possibility of a road departure resulting in a "fall" into the ravine. However, Figure 4.8 has similar potential for a devastating transfer of kinetic energy but is typical of our undivided rural road systems with unprotected roadsides.

The second example is a case history demonstrating how our traffic law and intersection traffic management protocols only slowly began to take account of normal human performance and limitations.

In Australia, the basic traffic law governing behavior at unsignalized intersections was that, in the event of a potential conflict between two vehicles on intersecting paths, the vehicle on the right had right of way (Australians drive on the left). This was a simple rule, for both police and insurance companies, as in every crash between vehicles on intersecting paths the (legally) "guilty party" was obvious. Unfortunately, the judgments required of drivers under this regime were very complex and mistakes were extremely common. As a driver (A) approached an intersection (see Figure 4.9) he checked to see if there was a vehicle (B) approaching from his right to which he might have to give way. If there was he had then to decide his likelihood of clearing the intersection before the conflicting vehicle arrived, which required a complex estimation of distance and closing speed, both of his and the potentially conflicting vehicle. However, if a third vehicle (C) was approaching the intersection from the opposing flow to (A) then the vehicle (B) on (A)'s right would be required to give way to (C) (because it was on (B)'s right)! This required (A) to make an even more complex judgment, since now there are time, distance and closing speed judgments to be made with respect to three vehicles.

In addition to the complexity of the decision making, this law encouraged aggressive driving in that, if there was uncertainty, you could speed up and try and clear the intersection in advance or simply "bully" the other driver into slowing down.

As traffic volumes increased so, too, did the number of intersection collisions and, eventually, the law was modified. However, as is so common in the history of road and traffic engineering, the initial change was made from the stand point of maximizing traffic flow. A priority road system was introduced in which the general give-way-to-the-right rule still applied unless an intersection had a sign indicating that it did not. The major roads with the higher traffic volumes then had their intersections with minor roads signed to indicate that vehicles on the major roads had priority. This did not decrease crashes, even at these modified intersections, as the minor road drivers had to take greater risks because of their limited opportunities to enter the high volume roads, and drivers on the high volume roads increased their travel speeds in the belief that they would not experience conflict at intersections. Finally, the give-way-to-the-right rule was scrapped — along with its priority road exception scheme — and all intersections were controlled either by stop or give way signs, or by signals.

FIGURE 4.7 Our fear of falling makes the danger in kinetic energy obvious. *Source*: Published with the kind permission of Professor Claes Tingvall, Swedish National Road Administration.

The third example is drawn from Carsten (2002) who relates the history of the development of the "mini roundabout" in the United Kingdom over 30 years ago. Traditionally, priority at conventional roundabouts was accorded to those entering a roundabout, over those circulating. Trials, driven by a desire to increase capacity at roundabouts, reversed the priority in favor of those circulating. The desired results of substantial reductions in delay and increases in saturation flow capacity were achieved. This, in turn, lead to the "invention" of the mini-roundabout which began to replace sign and signal controlled cross intersections with substantial capacity gains. Serendipitously, casualty crashes also decreased markedly. Carsten (2002) argues that this was a result of the simplification of the intersection negotiation task.

The key conclusion Carsten draws is that intuitive techniques have been the norm for advances in traffic engineering and that a systematic approach integrating engineering and behavioral knowledge would facilitate more rapid gains. This conclusion is supported by the case study of the evolution of priority rules at uncontrolled intersections in Australia.

The fourth, and final, example is drawn from Hauer (1990). Hauer describes the history of the method for computing sight distance in the standards for geometric road design. It was recognized that in order to anticipate danger and respond appropriately a driver needed to see a potential collision at a distance

FIGURE 4.8 Our inability to sense danger in kinetic energy. *Source*: Published with the kind permission of Professor Claes Tingvall, Swedish National Road Administration.

sufficient to facilitate stopping from the road's design speed. Those responsible for framing the standard knew the average height of the average driver's eye in the average passenger vehicle (at that point in time) and then selected as the "design danger object" an object with a height of four inches (a "dead dog" Hauer wondered!). This made the mathematical calculation of maximum vertical curvature for a given desired design speed a simple process. However, vehicle design is not static and, over time, the average driver's eye height in the average passenger vehicle decreased. The implications for geometric road design were of obvious concern so the "design danger object" was simply amended from a four to a six inch high object — a very pragmatic decision but hardly one anchored in a knowledge of human performance (perhaps dogs were getting bigger!). Over time, with further changes, the six inch object went from mandatory to desirable in the standard, and a fifteen inch high "danger design object" became allowable. Hauer's examination of standards outside the United States revealed a wide range of object heights upon which the sight distance calculations depend, underlying further the arbitrary nature of the standard setting process.

Hauer went on to challenge the underlying assumption that a vertical curve was unsafe if the stopping distance for a vehicle traveling at the road's design speed exceeded the design sight distance. There is evidence to support his challenge. For example, McLean (1977), using Australian rural road crash data, established that while the crash rate was higher on short rather than long radius (horizontal) curves it was highest for "unexpected" short radius curves on otherwise high standard road. In other words, drivers recognize, albeit in a crude way, the relative safety of different road environments and manage (to a degree) their speed accordingly. On low standard roads in constrained environments the crash rate on short radius curves is not exceptional.

The moral of these examples is that road and traffic engineers cannot assume that they are designing, building and operating safe road systems matched to normal levels of human performance simply because they are complying with the raft of road design standards and manuals of uniform traffic control devices.

FIGURE 4.9 "A" has to make continuous judgments of the speed, distance and time to conflict for three vehicles under the "give way to the right" rule of intersection priority.

The shift in mindset that is gathering pace is well expressed in the principles underpinning the Dutch vision of "Sustainable Safety". The concept is to have a small number of functional categories of road, with each road type distinctively and easily recognizable to all road users. This is likely to engender appropriate behavior for that road type, given that the function and its design are properly grounded in valid models of behavior (SWOV, 2003)

The concept of using road design and traffic management as the primary means of promoting appropriate road user behavior and decreasing the probability of the most common errors is well captured in the term "self explaining road" (Theeuwes and Godthelf, cited in Saad, 2002). Driving is a complex, unstructured task requiring adaptive behavior. The road rules leave much latitude to road users and information acquisition is often informal, thus the more homogeneous and consistent the road the more likely the user is to behave in predictable ways (Saad, 2002).

4.8 Moving beyond Standards

The above discussion should not be interpreted as implying that we should discard our road design and traffic management standards. Far from it; they provide a consistent base for design and a platform from which we can improve. In addition, they are a vital anchor in civil litigation defenses by infrastructure owners and operators against alleged breaches of duty of care. The standard of care sought by courts is often proof of compliance with established standards (Scott, 2004). Unfortunately, the "...tidal wave of liability suits..." (Scott, 2004, p. 34) works against safety innovations unless they can be clearly established to be safer than extant standards.

Ivey and Zegeer describe strategies (for making decisions on improving roadside safety) as "best offense", "best bet" and "best defense" to guide infrastructure owners from the potential liability perspective. While this may be pragmatic it is equally important that we continue to question everything we do, evaluate everything we implement, and gradually increase our stock of knowledge so that we genuinely seek to minimize death and serious injury.

As Carsten (Carsten 2002, p. 18) puts it:

> The development of better solutions will be an iterative procedure, in which new solutions are compared with old ones... An eclectic, empirical approach will produce greater dividends than an arbitrary selection of a single "right" approach. Human behavior is not so predictable that we can

identify *a priori* how road users will behave in response to a particular implementation. Behavioral adaptation — how road users adjust their behavior when faced by a change in the traffic environment or vehicle engineering — can be hypothesized, but its form and extent can only be confirmed through actual experience.

In so doing, it is vital that road and traffic engineers have regular dialogue with vehicle designers and engineers. For example, the rake angle of a windscreen, and the amount of tinting permitted, both have a major influence on what drivers can see. Eye heights, and other aspects of driving position, impact substantially on the way the driver views his road environment. In a similar way, understanding the limits of vehicle crashworthiness has major implications for the energy attenuation innovations of road and traffic engineers. Traditionally, there has been very little dialogue between the two professions.

Most importantly, standards should not be regarded as providing a sufficient base for safe design. Standards change very slowly, in part because they are the first line of defense against claims of liability so there is resistance to moving away from time-honored minima. Innovations, no matter how effective, lead only slowly to design change yet, as we have already seen, further progress in road traffic safety requires innovation on the part of road and traffic engineers.

In the case of vehicle design, the globalization of the industry has added a further constraint to safety rule making. Vehicle safety design rules are harmonized internationally to ensure ready access to all markets but this has driven rulemaking to the lowest common denominator, that is, to the minimum requirements of any one nation or group of nations. This is not to suggest that vehicle designers and manufacturers do not seek continuously to improve the primary and secondary safety of their vehicles. But those safety innovations become optional extras, which need to be specified, and paid for, by individual purchasers and, as yet, there is no widespread market for vehicle safety features. Indeed, in many motorized nations, the marketing and selling of automobiles is anchored in the promotion of power, performance and speed, which runs directly counter to developing a safe motoring culture (Ferguson et al., 2003)

However there is hope. In some nations — or groups of nations such as the European Union — governments, consumer groups (particularly "automobile clubs") and manufacturers are working together to provide "market standards" for safety. In Europe, there is a star rating system where points are awarded for a variety of primary and secondary safety features and manufacturers strive to reach the "five star" safety rating. This is known as EuroNCAP (New Car Assessment Program). These market-driven standards, while having no regulatory force, are having a major impact upon the level of safety in vehicles but essentially only on vehicles sold, at present, within the European "common market". When such vehicles are exported to other markets many of these "standard" (but not regulated) safety features are omitted, typically to allow these vehicles to compete in the new markets on price.

A more recent development is the commencement of an analogous system for rating the level of safety of different classes of road. Known as the European Road Assessment Program (or EuroRAP) it has, according to its web site, the objectives of:

- systematically assessing risk and identifying safety shortcomings that can be addressed with practical road-improvement measures;
- putting the assessment of risk at the heart of strategic decisions on route improvements, crash protection and standards of route management.
- providing road planners and engineers with vital benchmarking information to show them how well — or badly — their roads are performing compared with others, both in their own country and in other countries.

One of the most innovative, and promising, features is the linkage between EuroNCAP and EuroRAP. It encourages road and traffic engineers to take into account the energy-absorbing capabilities of the vehicle and the road as an integrated whole. For example, the web site states that a EuroNCAP five-star rated car cannot protect its occupants in a crash above 70 km/h on a EuroRAP one-star rated road.

EuroRAP has developed two protocols — risk rate mapping and star rating of roads through a road protection score. The risk rate mapping produces geographic plots (within a defined area) of crash risks and crash densities by road class for different types of road user. The star rating protocol involves visual inspections by trained assessors of each road's major safety features and hazards. On a rural road, for example, the inspection focuses on measures to prevent head-on collisions, measures to prevent collisions with unfenced roadside objects and side impacts at rural junctions.

While still at the development and trial stage, EuroRAP is an excellent example of how road and traffic engineers can exhibit leadership in the push for a quantum advance in road traffic safety and move well beyond regulatory standards to seek genuine best practice.

Most importantly, it is a demonstration of how to manage the institutional interfaces that have plagued progress in road traffic injury reduction. The approach holds promise for achieving a true balance in the desire for continual gains in travel speed with the need to minimize death and serious injury. In effect, the primary responsibility lies with the road and traffic engineer to establish a travel speed that achieves the safety objective. If a travel speed of, say, 110 km/h is desired on a particular route, the highway engineer needs to satisfy himself that a restrained occupant in a three or more-star rated vehicle can be protected on that route in the key situations of run-off-road, head-on and intersection crash scenarios, either by removing the potential for such conflict or by attenuating the energy adequately in the event of a crash. In so doing, the highway engineer is entitled to assume that the road user will comply with the legislated behavioral norms relating, for example, to speed limits, drink-driving, etc.

It is important to stress that approaches like those embraced in the Vision Zero concept do not seek to reduce travel speed *per se* but rather to insist that a high-speed road must provide an appropriate level of protection from death and serious injury.

One final issue warrants discussion in this section. In many nations, the delivery of road asset design, construction and maintenance is moving from its traditional government agency base to the private sector. There are many degrees of "outsourcing" and many forms of "privatization" of road infrastructure ownership and management. The lines of accountability for providing safe roads need special attention in this process. It is, as yet, rare to find in road construction and maintenance outsourcing contracts — or even in typical BOOT (build, own, operate, transfer) schemes — any criteria for safety management. While the contractor will almost always be held accountable for keeping pavement roughness levels within specified bounds, or ensuring that average travel speeds are maintained at nominated levels, there are almost never any criteria for ensuring that death and serious injury is kept to nominated levels.

4.9 Travel Speed and Journey Time — the Horns of the Mobility and Safety Dilemma

The history of transport is dominated by man's desire continuously to reduce travel time — steam ships replacing sail, railways replacing horse-drawn wagons and the automobile replacing the horse and the bicycle. The "horsepower" of the average passenger vehicle is at an historic high, the top speed capability of the automobile has steadily increased and most modern automobiles have speedometers calibrated to well over 200 km/h!

The search for ever shorter journey times has placed a premium on speed of travel. But with increasing speed goes increasing danger. The laws of physics are immutable. For every increase in travel speed there is a disproportionately larger increase in stopping distance in an emergency, a disproportionately larger increase in impact speed in the event of a crash, an even larger increase in impact energy and a very large increase in the probability of death and serious injury. This series of steep nonlinear relationships multiplies together.

In reaction, we seek to constrain travel speed through the imposition of speed limits and through the enforcement of these limits and, particularly in urban areas where roads serve a multiplicity of functions, we augment speed limits with special road and traffic measures such as chicanes, speed humps, and similar devices.

There is, not surprisingly, a constant tension between the desire for continual reductions in road transport journey times and the need to provide for man's information processing and decision-making limitations and to limit the amount of kinetic energy to be managed in the event of a crash.

There is no dispute about the nature of the relationships between travel speed and stopping distance, between travel speed and impact speed and between impact speed and severity of injury. There is, however, controversy about the relationship between travel speed and the probability of a crash occurring.

Many vehicle design innovations are intended to reduce crash probability at higher travel speeds, for example, anti-lock braking systems, traction control systems and electronic stability systems. Unfortunately, the extent to which they do so is, as yet, unknown. Similarly, there is much the highway engineer can do to facilitate safe, higher travel speeds. Of roads with speed limits of 100 or 110 km/h those with the lowest crash rate per million vehicle kilometers are those with a median to divide opposing traffic flows and full control of access for entering and departing traffic streams. Even on two-lane, two-way rural roads, crash rates can be reduced by using sealed rather than gravel shoulders, by using high standard curve delineation, by using "rumble strip" pavement edge marking, and so on (Oxley et al., 2003).

The question is whether we have yet found the right balance between the twin objectives of minimum journey time and minimum death and serious injury. The literature suggests that we have not yet learned effectively to match speed limits with crash probabilities and outcomes.

There is no satisfactory definition of speeding. Illegal speed behavior, that is travel speeds above the legally posted limit, is commonly divided into moderate and excessive speeding, where excessive speeding is typically defined as more than 20 km/h above the posted limit. There is also "inappropriate speeding" in which the speed, while below the posted limit, is inappropriate for the level of traffic flow, the frequency of vulnerable road users, or the existence of adverse weather conditions such as heavy rain or fog. Safety advocates place the proportion of crashes caused by speeding — of all three types — at between 30 and 40% (arrive alive!, 2001) while mobility advocates insist it is less than 10% (Buckingham, 2003). Unfortunately, the argument generates heat but no light and is a red herring in the speed management debate. Earlier, we saw how the fixation on determining the proportion of crashes involving human error as a causative factor engendered a false belief in the efficacy of behavioral control strategies. So too in this debate, by definition, speed plays some role in every crash and its outcome. But, as we have seen, causation and effective preventive strategies do not always go hand in hand.

Of far more significance, are the literature reporting evaluations of the changes in death and serious injuries following changes in speed limits on particular roads or within particular areas. Let us examine, first, the literature on speed limit changes on high speed rural roads. The most quoted case study comes from the United States when the oil crisis of the early 1970s led the U.S. Federal Government to reduce the speed limit on the interstate highway network to 55 mph (89 km/h). An evaluation in the early 1980s indicated that some 2000 to 4000 lives had been saved per year, along with some US$2 billion in fuel costs, but that U.S. motorists had incurred an additional one billion hours of driving (Committee for the Study of the Benefits and Costs of the 55 mph National Maximum Speed Limit, 1984). Note, again, the implied trade-off — lives and fuel saved but time lost.

In 1987, with the fuel crisis long over and the mobility lobby active, the speed limit on the U.S. interstate network was raised to 65 mph (105 km/h). By 1988, 40 States had raised their limits and an evaluation estimated fatalities to have risen 30% above that expected had the limit not changed (NHTSA, 1998).

In 1995, all federal controls on the ability of States to set their own speed limits were abolished. Some States reduced their limits on the interstate network to 60 mph, some increased their speed limits to 70 mph and some to 75 mph. A recent evaluation estimated an increase of 35% in fatalities per mile for those States raising the limit to 70 mph and of 35% in those States raising the limit to 75 mph (Patterson et al., 2002).

The U.S. story is by no means unique. Nilsson (1984), Anderson and Nilssen (1997) reviewed a large number of similar studies across a range of countries and reported very similar outcomes. They concluded that, for a given type of road, fatal crash rates increase with speed change to the fourth power with less severe crash rates increasing to lesser powers. Finch et al. (1994) concluded that a 1 km/h increase in mean speed increases injury crashes by 3%. While the precise relationships are still debated their broad form is now well accepted.

Strong supporting evidence that injury crash frequency increases with increases in travel speed comes from the U.K. (Taylor et al., 2002). They focused on single carriageway rural roads, all with a 60 mph speed limit. They studied 174 road sections classified into four groups — low quality, lower than average quality, higher than average quality and high quality. Quality was operationally defined by the density of bends, junctions and accesses and relative mean free speeds. Thus low quality roads were hilly, with a high bend density and low traffic speed, while high quality roads had a low density of bends and junctions and a high traffic speed.

Taylor et al. (2002) found that injury crash frequency was greatest on low quality roads and least on high quality roads thus confirming the vital role of road standard variations even among roads serving the same basic function. Moreover crash frequency increased by nearly one third for each additional cross road per kilometer, underlining the vital importance of controlling conflict points.

But the most important finding — for this discussion — is that all types of crash increased rapidly with increases in mean speed such that a 10% increase in mean speed resulted in a 26% increase in crash frequency. Road designers need to understand that "safety margins" above the "design speed" almost always lead to increases in mean and 85th percentile speeds.

Speed management is vital to safety. High speed roads must provide a commensurately high level of protection from harm.

The evidence that we have not yet struck the appropriate balance between the journey time and safety objectives is compelling. This is perhaps not surprising as speed limit setting has been an essentially arbitrary affair. Historically, speed limits were chosen based on prevailing speed distributions on different classes of road, with the 85th percentile free travel speed being chosen as the best guide to the appropriate speed limit for that type of road. The apparent logic was that drivers make reasonable judges of safe travel speed. There is no scientific evidence to support this assumption: on the contrary, there is a reasonable body of experimental work to suggest that drivers are poor judges of speed and distance and that the higher travel speed the greater the error in judgment (Groeger, 2000).

Speed limits cannot be viewed in isolation. The extent to which road users comply with limits is strongly related to the nature and intensity of enforcement. Most drivers exceed posted limits regularly for at least part of their journey.

Where there are enforcement tolerances — that is, where offenders are not ticketed for speeds not dramatically greater than the posted limited — then average free speeds exceed posted limits. Thus, in many places, illegal (moderate) speeding is (unofficially) sanctioned.

Silcock et al. (2000) surveyed 1000 U.K. motorists, undertook focus group discussions and had over 200 drivers make ratings of a "video-drive" in one of the most comprehensive studies of attitudes to, and perceptions of, speeding. They found that:

- the community sees only "excessive" speed as dangerous;
- and considers an appropriate speed to be just above the posted limit (with over 85% reporting regularly driving over the limit by small margins);
- the above beliefs are anchored in an understanding of police enforcement tolerances and the positive reinforcement of speeding by the media, popular culture and motor vehicle advertising;
- drivers are highly critical of what they judge to be inconsistent speed zoning practices and, possibly of greatest importance, respond to their "reading" of the prevailing road and traffic environment.

It is not surprising, then, that speed limits are viewed by many road users as maximum speeds only in adverse conditions and helps explain why public controversy frequently accompanies enforcement blitzes. In Victoria, Australia, a (successful) attempt to reduce urban fatalities by reducing the enforcement tolerance, vastly increasing enforcement intensity and keeping the enforcement covert became a substantial political issue (Johnston, 2003).

In short, much current speed management practice reinforces prevailing beliefs that (moderate) speeding is not unsafe. The little (case/control) research that exists on the relationship between travel speed relative to posted limits and the probability of occurrence of casualty crashes suggest an exponential relationship, but with moderate increases in risk at levels just above the posted limits (Kloeden et al., 2001, 2002). Given the high frequency of moderate speeding — and the low frequency of excessive speeding — it may well be that a focus on reducing the average speed of travel (for roads without the highest level of protection from serious injury in the event of a crash) will bring the greater safety return, particularly in urban areas where there are the highest proportions of vulnerable road users. This research underpinned the successful "experiment" in Victoria, Australia.

4.10   Will "Information Technology" Be the Springboard to a Safer System?

Intelligent Transport Systems (ITS) is the generic term for the application of developments in communication and computing technologies to transport systems and services. Myriad applications have appeared over the last two decades and large scale research programs are current in (at least) the European Union, North America and Japan. The field of ITS is, arguably, now a professional specialty. Many claims are made for the potential of ITS to provide quantum leaps in the capacity of road infrastructure, the efficiency of traffic flow, the environmental sustainability of road transport and the level of road system safety. A picture is often painted of a future in which private vehicles will only be under the active control of the driver for the local street origin and destination components of a journey with linked platoons of automatically guided vehicles completing the high speed intra-metropolis or high speed inter-city components of trips — intelligent vehicles operating on intelligent highways.

At present, however, and in the immediate future the emphasis is on improving the information available to road users. As Lucas and Montoro (2004) put it:

> Eventually, improvement in the road transport system is to be achieved by presenting the right information at the right place in the right moment and in a short, clear, understandable way to road users.

The premise is that the problems of inefficient and unsafe traffic flow are problems of inadequate information. The link between this concept and the "self explaining road" that underpins the Dutch approach to sustainable safety is obvious. Unfortunately, there is something of a rush to develop, patent and market specific applications and insufficient attention is being given to an integrated, system-wide "design revolution". Particularly with regard to the potential of ITS to reduce death and serious injury a recent review concluded:

"There is a need for a move away from technology-driven ITS research development to user-oriented design…" and "… Human factors principles and knowledge must be incorporated into design of these systems and they need to cater for the special needs of various road user groups. Failure to do so could seriously compromise the safety of the entire road transport system." (Regan et al., 2001)

A simple classification of ITS technologies includes:

- *Vehicle-based ITS technologies*: sensors on the vehicle receive and process data and display the processed data to the driver in the form of instructions, warnings, and so on. Examples include advising the driver of the speed limit applicable to the current road section, alerting the driver when he is in excess of that speed limit by a predetermined margin for a predetermined time, forward-looking radar to warn the driver when his following distance to the vehicle immediately ahead approaches a predetermined minimum, rearward-looking radar to alert a reversing driver to an object in his path, fatigue monitoring systems, and so on.

- *Infrastructure-based technologies*: these comprise road-based sensors gathering data and conveying it to drivers via such traffic management tools as variable message signs. Examples include warning signs of unusual environmental conditions such as the possibility of "black ice" on the pavement or of unusual traffic conditions such as the existence of blocked lanes due to a crash ahead, sensing a vehicle's speed and "instructing" the driver to slow down for example on the approach to a substandard horizontal curve, to a "blind" intersection, and so on.
- *Co-operative technologies*: in these, information collected by road-based sensors might be transmitted to in-vehicle displays, data collected by in-vehicle sensors might be conveyed to road-based information displays or information might be collected from both types of source and transmitted each to the other. For example, a vehicle approaching an intersection might be sensed by in-road detectors and that information may be transmitted to other legs of the intersection where further transmitters might broadcast that vehicle's arrival to other vehicles equipped to receive the transmission.

Of special interest is the potential of ITS to radically improve our capacity to manage speed behavior to match the inherent safety of the infrastructure and the prevailing ambient traffic and environmental conditions. We have already seen that the conventional static posting of a speed limit is interpreted by most drivers as a limit only for adverse conditions with the majority of drivers exceeding the posted limits frequently in normal conditions. Motorists are critical of what they see as inconsistent setting of speed limits.

The most prolific safety focused ITS research area is in-vehicle "intelligent speed adaptation" (ISA) systems (Regan et al., 2003). In its simplest form ISA provides a continuous indication of the prevailing posted speed limit on the section of road being traversed. The vehicle is fitted with a GPS and a "look up table" of the posted limits on all roads in a given geographical area. In addition to advising the driver of the current posted limit, variants of ISA:

- advise the driver when he has been exceeding the limit by a preset amount for a predetermined time
- provide audible "encouragement" to slow down
- or, through engine or breaking management systems, directly govern the speed of the vehicle

All such ISA systems are, of course, dependent upon the static posted speed limits. If these limits are changed either permanently as the road function changes over time or temporarily as in the case of road works then the "look up table" will be inaccurate. There are ISA systems under development that are dynamic, in that they communicate with road based transmitters which signal temporary changes in speed limit or adverse conditions such as a high probability of black ice on the pavement and seek to influence the driver's speed behavior accordingly. For a thorough recent review of ISA developments see Regan (2004).

The environment-based equivalent of the in-vehicle ISA systems has not received commensurate attention. This is somewhat surprising as the broader application of variable message signs is under rapid development (for example, Lucas and Montoro (2004)). Recall that the underlying principle of ITS is to raise the consistency and predictability of road user behavior through the provision of critical information in the most timely, unambiguous manner. Traffic signs and pavement markings are the most widespread means of communicating information to drivers. However, they are static and cannot respond to changes in traffic conditions or other ambient environmental conditions. Variable message signs overcome these barriers. There are some attempts to modify speed behavior via variable message signing. These tend to be confined to specific situations. For example, on the approach to a substandard horizontal curve, pavement sensors have been used to measure the speed of individual vehicles and, if these are in excess of a safe curve negotiating speed, a variable message sign is triggered which advises the driver to slow down as he/she approaches the curve (McDonald and Winnett, 2003).

The system-wide application of variable message signs would enable posted limits to vary more or less continuously as the traffic and other environmental conditions varied. This is likely to overcome the concerns most drivers have with the status of static speed limits. While system-wide application would be cost prohibitive at the current stage of technology of variable message signing it warrants serious consideration for the high-speed roads where it is most difficult to protect road users from serious injury in the event of a crash.

In summary, the field of ITS offers enormous potential to achieve the "self explaining roads" that underpin the Dutch concept of sustainable safety. What must be carefully addressed, however, are the human factors aspects of ITS design. There are many traps in designing the human–machine interface and, unfortunately, there has been a tendency in the ITS field to rush products to market before their influence on human behavior has been properly tried and the unintended side effects designed out. For an excellent overview of the issues to be considered see Fuller and Santos (2004).

4.11 The Decision Making Process

4.11.1 Reactive or Proactive?

The optimum time to ensure mobility with safety is at the time of original design and construction of a road project. The process begins with a clear statement of road function. Functionality determines design speed and volumes which, in turn, determine the fundamentals of road geometry, access control and roadside management.

However, road function commonly changes as a metropolis, or a geographic region, develops. What began as a rural collector linking farm access roads with a rural arterial, may itself become an inter-city route. In many nations, as development progresses, funds are not available to "immediately" upgrade the collector to rural arterial design standard and interim measures are taken to cope with the increase in traffic volumes. This is the point at which safety is often compromised. It is here that road and traffic engineers must be most vigilant. Measures to ensure minimal risk of serious injury crashes — and minimal risk of serious injury when crashes occur — must be a formal step in the planning of how to manage a road with a changed primary function.

One way of achieving this is to adopt a systematic road safety audit process. There are well-developed safety audit tools available (see, for example, Ogden, 1996). Audits should always be conducted as part of the planning process for new roads and for road improvement programs. The audit tool can also be used systematically to assess the "design safety" of the entire network, that is as a pro-active measure to identify roads, or road sections, that have inadequate safety design — whether or not that road section is currently on a road improvement program. With a baseline audit of the network, regular monitoring can occur as volumes increase, as traffic mixes change and as functionality beings to alter. This network-wide proactive audit is the conceptual underpinning of the Euro-RAP program already discussed.

While road safety auditing is becoming widespread as part of the road project planning process its use as a network-wide safety status assessment and ongoing monitoring tool remains in its infancy.

Much of the infrastructural road safety effort continues to be reactive. Crash and injury data sets are used to identify "black spots" or "black lengths" — locations where disproportionately high frequencies of crashes have occurred. These black spots are then examined at an individual level to determine the reasons behind the frequency of crashes at each location, appropriate site specific remedial measures are selected and applied. (For a good outline of how to apply a black spot program see Ogden, 1996).

These reactive, remedial safety programs have been outstandingly successful in reducing crashes at the treated locations and in returning high cost/benefit ratios (Newstead and Corben, 2001). Their overall impact on the level of network safety depends, however, upon the total level of investment set aside for such programs. While a vital part of the road and traffic engineer's safety tool box, black spot programs are not a panacea. They successfully resolve problems of the past, they cannot foresee these problems in advance. Network wide auditing, and specific auditing of road project plans, are essential elements of a prospective approach to ensuring that new black spots are not created.

4.11.2 Frequency or Risk?

We have already seen the nature of the relationship between exposure to risk and crash and injury frequency. While it is important to seek to reduce both the probability of crash and the probability of injury given a crash they must not become the primary drivers.

A divided, limited access, high geometric standard rural highway will have a markedly lower serious injury crash rate than a two-lane, two-way rural arterial yet it will, typically, account for a far larger proportion of the road toll. While the number of casualties per million vehicle kilometers traveled is low the sheer volume of travel will result in a high absolute number of causalities.

Nowhere is the frequency/risk quandary more apparent than in the slavish adherence to arbitrary standards such as the "nine meter clear zone". An historical *assumption* was that 80% of vehicles leaving a rural road at highway speed could recover within nine meters and, hence, eliminating hazards within that distance of the pavement edge was assumed to provide adequate safety (Michie, 2004). Not only is the assumption false (see Carsten, 2002) but on very high volume roads a large absolute number of potential harm-producing events remain.

Recall that most nations now use the population based rate of death and serious injury as their primary tool to monitor road safety progress and that the media, the community and the politicians focus on absolute numbers. Hence, the road and traffic engineer must seek to reduce the absolute numbers of death and serious injury on each class of road and not be fooled into focusing only on those classes of road with the highest casualty rates per vehicle kilometers of travel.

The risk measure is a valuable benchmarking tool but the absolute number of deaths and serious injuries on a given road length is the best indicator of whether or not the engineer needs to take further action.

Part C — Summary and Conclusions

4.12   Bringing It All Together

It is difficult to think of a technology more pervasive in its impacts upon everyday living than the personal, motorized vehicle. It shapes our cities, it helps determine where we live and work and where and how we play; it threatens the "livability" of our current and future environment and it threatens our very lives.

Yet, despite a few qualms, we embrace it and support its continuous growth. No "motorized" nation has yet ceased to increase its number of motor vehicles per capita and the "developing" nations strive to motorize, and succeed, at a rapid pace.

More people die, are permanently disabled, or experience serious injury as a result of road crashes than from any other cause of trauma — war, terrorism, or violence; or injury in workplace, domestic or community settings. As global motorization continues road trauma will increase dramatically and will soon rank third among human "diseases". The disease of personal mobility is unquestionably one of the most critical global public health problems.

Tragically, for too long we have tended to regard a quantum of road trauma as the inevitable price of motorization. But we have not sought to define that quantum. Most motorized countries have decreased the population-based rate of road death to around one third of what it was three decades or so ago. Many nations have achieved a road transport safety rate of at or less then one death per 100,000,000 vehicle kilometers of travel. Is this the irreducible minimum? It would appear not as most motorized nations continue to achieve reductions in both their rates of *Transport Safety* and *Personal Safety*.

In part, the recent gains are the result of a shift from diverse, un-coordinated countermeasure efforts from a plethora of public and private sector agencies to an integrated effort by key (mostly public sector) organizations which have accepted an accountability for designing and operating a safe system — tolerant of normal human error and capable of mitigating the trauma of kinetic energy to the human body in the event of a crash — and decreasing the traditional emphasis on facilitating mobility and avoiding responsibility by allocating the blame of crashes to the injured parties.

We are, finally, coming to realize that highways can, literally, be as safe as we are prepared to make them. There are man-made solutions to this man-made problem. We are questioning the traditional economic rationalist approach to decision-making — the greatest economic good for the greatest number — and considering removing death and serious injury from the economic equation.

In so doing, we are coming to realize that the greatest future gains lie in the traditionally most neglected area — road and traffic engineering. The minimization of death and serious injury must continue to become the primary driver, with the growth of personal mobility optimized as the secondary objective.

References

Adams, J. 1995. *Risk*. UCL Press, London.

Anderson, G. and Nilssen, G. 1997. *Speed management in Sweden*, VTI: Linköping arrive alive! Victoria's Road Safety Strategy 2002–2007, www.arrivealive.vic.gov.au

Australian Transport Safety Bureau (ASTB), 2003. *International Road Safety Comparisons: The 2001 Report*. Australian Government, Canberra.

Buckingham, A. 2003. *Speed Traps: Saving Lives or Raising Revenue*, Policy, Vol. 19(3). Centre for Independent Studies, Sydney.

Cameron, M., Newstead, S., Diamantopoulou, K., and Oxley, P. 2003. *The Interaction Between Speed Camera Enforcement and Speed-Related Mass Media Publicity in Victoria*, Report No. 201. Monash University Accident Research Centre, Melbourne, Australia.

Carsten, O. 2002. *Multiple perspectives. Human Factors for Highway Engineers*, R. Fuller and J.A. Santos, eds., Pergamon (Elsevier Science), New York.

Committee for the Study of the Benefits and Costs of the 55 mph National Maximum Speed Limit, 1984. *55: A Decade of Experience*, Transport Research Board, Special Report 204. National Research Council, Washington, DC.

Corben, B., Tingvall, A., Fitzharris, M., Newstead, S., and Johnston, I.R. 2003. A review of guidelines for the use of median barriers, *Proceedings 21st ARRB Transport Research and 11th Road Engineering Association of Asia and Australasia Conference*, May, Cairns.

Department of Transport and Regional Services, 2000. *The Commonwealth's Transport Directions: Task and Outlook*. DoTRS, Canberra.

European Conference of Ministers of Transport (ECMT), 2002. *Transport and Economic Development: Report of the Round Table on Transport Economics*, 119.5. OECD, Paris.

Ellaway, A., Macintyre, S., Hiscock, R., and Kearns, A., In the driving seat: psychosocial benefits from private motor vehicle transport compared to public transport, *Transportation Res. Part F*, 6, 217–231, 2003.

Elvik, R. and Amundsen, A.H. 2000. *Improving Road Safety in Sweden — Summary Report*, Report 489/2000. Institute of Transport Economics, Norwegian Centre for Transport Research, Oslo.

EuroRAP http://195.167.162.57/inside.htm

European Transport Safety Council, 2001. *Transport Safety Performance Indicators*. ETSC, Brussels.

Evans, L. 2002. A plan for traffic safety. An opinion piece published in San Francisco's *The Examiner* newspaper, accessed on 1701/2003 from www.scienceservingsociety.com/pubs/OpEd/plan.html

Ferguson, S.A., Hardy, A.P., and Williams, A.F., Content analysis of television advertising for cars and minivans: 1983–1998, *Accid. Anal. Prev.*, 35, 6, 825–831, 2003.

Finch, D.J., Kompfner, P., Lockwood, C.R., and Maycock, G. 1994. *Speed, speed limits and accidents*, Report 58. Transport Research Laboratory, Crowthorne.

Fuller, R. and Santos, J.A., eds., 2002. *Human Factors for Highway Engineers*, Pergamon (Elsevier Science), New York.

Groeger, J.A. 2000. *Understanding Driving*. Psychology Press (Taylor and Francis), London.

Haddon, W. Jr., Suchman, E.A., and Klein, D. 1964. *Accident Research: Methods and Approaches*, Factors that Determine Injury. Harper and Row, New York, chap. 9.

Haight, F.A. A traffic safety fable, *J. Saf. Res.*, 5, 4, 226–228, 1973.

Hauer, E. 1990. *The engineering of safety and the safety of engineering. Challenging the Old Order — Towards New Directions in Traffic Safety Theory*, J.P. Rothe, ed., Transaction, New Brunswick.

Haworth, N. and Symmons, M. 2002. *The Relationship Between Fuel Economy and Safety Outcomes*, Report No. 1881. Monash University Accident Research Centre, Melbourne, Australia.

Insurance Institute for Highway Safety, 2002. *Status Report*, Special Issue: Low Priority Assigned to Highway Safety, Vol. 37(10).

Ivey, D.L. and Scott, C.P. 2004. *Solutions in Transportation Research Board*, State of the Art Report 9: Utilities and Roadside Safety.

Ivey, D.L. and Scott, C.P. 2004. *Summary in Transportation Research Board*, State of the Art Report 9: Utilities and Roadside Safety.

Ivey, D.L. and Zegeer, C.V. 2004. *Strategies in Transportation Research Board*, State of the Art Report 9: Utilities and Roadside Safety.

Johnston, I.R. 1994. *Human Factors in Road Accidents*, Queensland Division Technical Paper, Institution of Engineers, Australia.

Johnston, I.R., Will a 4WD strategy work in the shifting sands of policy?, *Road Transport Res.*, 11, 1, 66–72, 2002a.

Johnston, I.R. 2002b. Sustainable transport — separating the meaning from the mantra, *Keynote Address to the Third International Conference on Seamless and Sustainable Transport*, 25–27 November, Nanyang Technological University, Singapore.

Johnston, I.R. 2003. Improving road safety in the longer term — finding the right buttons to push, *Keynote Address to 2003 Road Safety Research, Policing and Education Conference*, 24–26 September, Roads and Traffic Authority, Sydney, NSW.

Kloeden, C.N., McLean, A.J., and Glonek, G. 2002. *Reanalysis of Travelling Speed and the Risk of Crash Involvement in Adelaide, South Australia*, Report CR207, Australian Transport Safety Bureau, Canberra.

Kloeden, C.N., Ponte, G., and McLean, A.J. 2001. *Travelling Speed and the Risk of Crash Involvement on Rural Roads*, Report CR204. Australian Transport Safety Bureau, Canberra.

Koornstra, M., Lynam, D., Nilsson, G., Nordzij, P., Petterson, H.E., Wegman, F., and Wouters, P. 2002. *SUNflower: A Comparative Study of the Development of Road Safety in Sweden, the United Kingdom and the Netherlands.* SWOV, Leidschendam.

Kopits, E. and Cropper, M. 2003. *Global Road Traffic Fatality Projections 2000–2020.* Working Paper for World Health Organization, Geneva.

Larsson, M., Candappa, N., and Corben, B.F. 2003. *Flexible Barriers Along High-Speed Roads — A Lifesaving Opportunity*, Report No. 210. Monash University Accident Research Centre.

Lave, L.B. and Lave, C.A. 1990. *Barriers to increasing highway safety. Challenging the Old Order — Towards New Directions in Traffic Safety Theory*, J.P. Rothe, ed., Transaction, New Brunswick.

Lucas, A. and Montoro, L. 2004. *Some critical remarks on a new traffic system: VMSs part II. The Human Factors of Traffic Signs*, C. Castro and T. Horberry, eds., CRC Press, Boca Raton, FL, USA.

McDonald, N. and Winnett, M. 2003. Persuasion or Stick? Latest advances in speed management in the UK, *Proceedings 2003 Road Safety Research, Policing and Education Conference*, 24–26 September, Roads and Traffic Authority, Sydney, NSW.

McLean, J.R. 1977. *The Inter-Relationship Between Accidents and Road Alignment*, Report AIR 000-68, ARRB Transport Research, Melbourne, Australia.

Michie, J. 2004. Professionalism *in* Transportation Research Board, State of the Art Report 9: Utilities and Roadside Safety.

Mohan, D. and Tiwari, G. 1998. *Traffic Safety in Low Income Countries: Issues and Concerns Regarding Technology Transfer from High Income Countries, Reflections on the Transfer of Traffic Safety Knowledge to Motorising Nations*. Global Traffic Safety Trust, Melbourne, Australia.

Newstead, S.V. and Corben, B.F. 2001. *Evaluation of the 1992–1996 Transport Accident Commission Funded Accident Black Spot Treatment Program in Victoria*, Report No.182. Monash University Accident Research Centre, Melbourne, Australia.

Newstead, S.V., Delaney, A., and Watson, L. 2003. *Vehicle safety ratings estimated from combined Australian and New Zealand real crash data. Pilot Study: Stage 5*, Report No. 203. Monash University Accident Research Centre, Melbourne, Australia.

NHTSA, 1998. *Report to Congress: The Effect of Increased Speed Limits in the Post-NMSL Era*. US Department of Transportation, Washington, DC.

Nilsson, G. 1984. *Speeds, Accident Rates and Personal Injury Consequences for Different Road Types*, Report No. 277. Swedish National Road and Transport Institute, Linköping.

Ogden, K.W. 1996. *Safer Roads: a Guide to Road Safety Engineering*. Avebury Technical, UK.

O'Hare, M., Langford, J., Johnston, I.R., and Vulcan, P. 2002. *Bicycle Helmet Use and Effectiveness*. Monash University Accident Research Centre, Sweden, Contract Report to SNRA.

Oxley, J., Corben, B., Koppel, S., Fildes, B., and Jacques, N. 2003. *Cost Effective Infrastructure Measures on Rural Roads*. Monash University Accident Research Centre, Sweden, Contract Report to SNRA.

Patterson, T.L., Frith, W.J., Povey, L.J., and Keall, M.D. 2002. The effect of increasing rural interstate speed limits in the U.S.A. *Traffic Injury Prev.*, 3, 4, 316–320.

Regan, M.A. 2004. *A sign of the future 1: intelligent transport systems. The Human Factors of Traffic Signs*, C. Castro and T. Horberry, eds., CRC Press, Boca Raton, FL, USA.

Regan, M., Oxley, J.A., Godley, S.T., and Tingvall, C. 2001. *Intelligent Transport Systems: Safety and Human Factors Issues*, Report No. 01/01. Royal Automobile Club of Victoria (RACV), Melbourne, Australia.

Regan, M., Young, K., and Haworth, N. 2003. *A review of literature and trials of Intelligent Speed Adaptation devices for light and heavy vehicles*, Austroads, AP-0R237/03.

Robertson, L.S. 1992. *Injury Epidemiology*. OUP, London.

Saad, F. 2002. *Ergonomics of the driver's interface with the road environment: the contribution of psychological research. Human Factors for Highway Engineers*, R. Fuller and J.A. Santos, eds., Pergamon (Elsevier Science), New York.

Scott, C.P. 2004. *Legal Issues in Transportation Research Board*, State of the Art Report 9: Utilities and Roadside Safety.

Scott, C.P. and Ivey, D.L. 2004. *Utility Pole Collisions in Transportation Research Board*, State of the Art Report 9: Utilities and Roadside Safety.

Scott, C.P. and Ivey, D.L. 2004. *Initiatives in Transportation Research Board*, State of the Art Report 9: Utilities and Roadside Safety.

Shinar, D. 1978. *Psychology on the Road — the Human Factor in Traffic Safety*. Wiley, New York.

Silcock, D., Smith, K., Knox, D., and Beuret, K. 2000. *What Limits Speed? Factors that Affect How Fast We Drive*. AA Foundation for Road Safety Research, London.

SWOV. 2003. *SWOV Programme 2003–2006*, Report R-2003-18, available at www.swov.nl

Taylor, M.C., Baruya, A., and Kennedy, J.V. 2002. *The Relationship Between Speed and Accidents on Rural Single-Carriageway Roads*, Report TRL 511. Transport Research Laboratory, Berkshire, UK.

Tingvall, C. 1998. The Swedish Vision zero and how Parliamentary approval was obtained, *Proceedings Road Safety Research, Policing and Education Conference*, 16–17 November, Wellington, New Zealand, pp. 6–8.

Tingvall, C. 2004. *The Swedish Vision Zero*, World Report on Road Traffic Injury Prevention. World Health Organization, Geneva.

Trinca, G.W., Johnston, I.R., Campbell, B.J., Haight, F.A., Knight, P.R., Mackay, G.M., McLean, A.J., and Petrucelli, E. 1988. *Reducing Traffic Injury — A Global Challenge.* Royal Australasian College of Surgeons, Melbourne.

van Schagen, I., ed., 2003. *Traffic calming schemes: opportunities and implementation strategies. Report R-2003-22*, SWOV Institute for Road Safety Research, Leidschendam, The Netherlands.

Vulcan, A.P. 1988. Strategies for implementation of road safety measures, *Keynote Address to Road Traffic Safety Research Council*, Wellington, New Zealand, September, pp. 1–17.

World Health Organization, 2004. *World Report on Road Traffic Injury Prevention.* World Health Organization, Geneva.

5

Road Traffic Noise

Stephen Samuels
*TEF Consulting (Traffic, Environmental
and Forensic Engineers)
Sydney, Australia
University of New South Wales
Sydney, Australia*

5.1 Introduction and Fundamentals

5.1.1 Sound and Noise

It is important to have some understanding of the nature of sound, because every noise is a type of sound. Noise is usually defined as unwanted sound. Thus it can present with variations in magnitude that may be temporal or spatial in nature or a combination of both. A particular type of noise is known as environmental noise which is noise that is generated by a variety of outdoor sources and then propagates from outdoors to receivers that may be situated either indoors or outdoors. In turn, road traffic noise is one type of environmental noise.

There are many textbooks available which provide detailed treatments of the nature and physics of sound, but some of the key features of relevance to the present chapter are summarized below.

- Sound is an elastic wave disturbance.
- It occurs in any medium that has the properties of both mass and stiffness.
- The most common medium in which environmental sound (noise) occurs is air, where it travels as pressure fluctuations above and below atmospheric pressure.
- The magnitude of these pressure fluctuations is termed the "Sound Pressure." It is generally measured in micro Pascals (μPa).
- Sound pressures of typical sounds might range from 20 μPa to around 200,000,000 μPa.

5.1.2 Sound Pressure Level

Because the range of sound pressures involved with typical sounds is very large, as set out above, it has been found convenient to use a logarithmic function to quantify sound and noise. This function is termed the Sound Pressure Level (SPL) and it is defined in the equation below. The SPL is a function of the ratio of the square of the sound pressure of a given sound to the square of a reference sound pressure. In fact, the reference sound pressure is the smallest sound that young, good ears can just hear, i.e., $20 \, \mu$Pa. As also shown above, the unit of the SPL is deciBels, abbreviated to dB. (A dB is a relative scale, in this case relative to P_0, the reference sound pressure.)

$$SPL = 10 \log_{10}\left(\frac{P^2}{P_0^2}\right)$$

where SPL is the sound pressure level (dB re P_0), P is magnitude of pressure fluctuations (Pa) and P_0 is the magnitude of a reference pressure is $20 \, \mu$Pa.

Typical values of Sound Pressures and their accompanying SPLs are given in Table 5.1. Another useful feature of the SPL is that it is a logarithmic function, since human auditory response is also logarithmic. It is generally accepted that an increase in SPL of 3 dB is about the smallest change that can be detected by the human auditory system. Furthermore, an increase of 10 dB is generally assessed subjectively as being a "doubling of sound volume."

To summarize, sound, and environmental noise in particular, is always quantified by means of the logarithmic SPL. This means that when comparing different sounds, the process involved is a logarithmic one, not a linear one. Most people tend to compare things in a linear fashion, so the use of a logarithmic context requires a shift in approach. For example, one might assess one bag of apples as being twice as heavy as another, or the visibility through a haze as being about one third as that without any haze present. This linear approach is not possible when noise levels are being compared. Such comparisons must be made in terms of how many dB one differs from another. What follows on the next two pages are two examples of logarithmic mathematics showing how SPLs are combined. While not essential, it is recommended that some understanding of these examples is gained.

Example 5.1

A sound source originally generates a sound pressure P_1 which produces a SPL_1. The source is altered so that it subsequently generates a sound pressure P_2 which is double the original P_1. For most sound sources this is a very difficult thing to do because of the increased amount of energy required. Nevertheless, the calculations show that the increase in SPL is 6 dB. Note that the following calculations

TABLE 5.1 Typical Sound Pressures and Sound Pressure Levels

Apparent loudness	Sound Pressure (μPa)	Sound Pressure Level (dB)	Source
Deafening	200,000,000	140	Jet aircraft
		130	Pain threshold
Very loud	20,000,000	120	Loud horn at 1 m
	2,000,000	100	Noisy factory
	200,000	80	Noisy office
Loud		70	Average street noise
	20,000	60	Average office
Moderate	2200	40	Private office
Faint	200	20	Rustling leaves
		10	Normal breathing
Very faint	20	0	Hearing threshold

have been undertaken in accord with the mathematical laws of logarithms. Recall that $10\log 4 = 6$. As will be shown subsequently in this unit, a change of 6 dB is considerable. It would, as indicated earlier, be subjectively assessed as quite noticeable.

$$\text{Original sound pressure} = P_1$$

$$\text{New sound pressure } P_2 = 2P_1$$

$$\text{SPL}_1 = 10 \log_{10}\left(\frac{P_1^2}{P_0^2}\right)$$

$$\text{SPL}_2 = 10 \log_{10}\left(\frac{P_2^2}{P_0^2}\right) = 10 \log_{10}\left(\frac{(2P_1)^2}{P_0^2}\right) = 10 \log_{10}\left(\frac{4P_1^2}{P_0^2}\right) = 10 \log_{10}\left(\frac{P_1^2}{P_0^2}\right) + 6$$

i.e.,

$$\text{SPL}_2 = \text{SPL}_1 + 6 \text{ dB}$$

Example 5.2

A sound source originally generates a sound pressure P_1 which produces a SPL_1. Another source which generates a sound pressure P_2 and a SPL_2 is set up beside the first. With both sources operating the total sound pressure level will be SPL. But when both sources produce the same sound pressure as one another, the calculations show that the increase in SPL in this case is 3 dB, which be subjectively assessed as only just detectable. The converse of this particular example is that if one of two identical, adjacent noise sources is turned off, the net effect will be only just able to be detected by the human auditory system. Such an outcome has profound implications for controlling environmental noise, particularly road traffic noise, which will be considered subsequently in the present chapter.

Source 1: sound pressure P_1
Source 2: sound pressure P_2
Total sound pressure $= P^2$

$$P^2 = P_1^2 + P_2^2$$

If

$$P_1 = P_2 : P^2 = 2P_1^2$$

$$\text{Total SPL} = 10 \log_{10}\left(\frac{P^2}{P_0^2}\right) = 10 \log_{10}\left(\frac{2P_1^2}{P_0^2}\right) = 10 \log_{10}\left(\frac{P_1^2}{P_0^2}\right) + 3$$

i.e.,

$$\text{Total SPL} = \text{SPL}_1 + 3$$

The two examples above illustrate how SPLs must be manipulated logarithmically. One further illustration of this is shown in Figure 5.1, which demonstrates how the SPLs from four independent, equal noise sources combine to produce the total SPL. The logarithmic additions involved here are in accord with the chart that follows the diagram. If the four sources in the diagram were the jet engines of an aircraft, a powerful insight

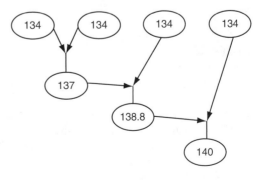

FIGURE 5.1 Adding four independent noise.

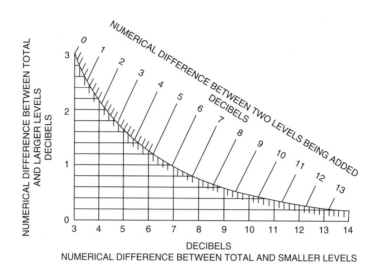

FIGURE 5.2 Adding sound pressure levels.

into the difficulties in reducing environmental noise is gained by inspecting the diagram in the opposite direction of the arrows. The total noise from the four engines is 140 dB. If one engine is shut down, the noise reduction obtained is 1.2 dB, which is completely undetectable. Even with two engines shut down, the reduction is just noticeable at 3 dB. Shutting down two of four engines on an aircraft is definitely not a practical noise control solution. But even if it were possible, the noise reductions would be small. A chart to assist with the addition of SPLs is given in Figure 5.2.

5.1.3 Human Auditory Response

As mentioned previously, the human auditory system responds to sound, and therefore noise, in a logarithmic manner. The limit of human hearing is termed the audibility threshold and is shown in Figure 5.3. The curve in this diagram plots the minimum SPLs that young good ears can hear. This graph actually plots SPL against sound frequency, which is measured in cycles/sec and which is also called Hertz (abbreviated as Hz). Frequency is another characteristic of sound and it quantifies the rate at which the pressure fluctuations in the sound wave repeat over time. It is mostly the case that, sound and noise sources emit a number of sound waves, each of which occurs at a different frequency.

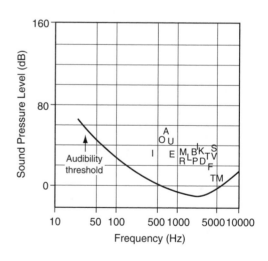

FIGURE 5.3 Human auditory response.

Superimposed on the SPL/frequency graph of Figure 5.3 are some of the sounds of the English language. For example, by saying "A" in a normal conversational speech level, one would generate a sound of about 60 dB at a frequency of approximately 700 Hz. One characteristic of the English language is that the majority of meaning in words is carried by the consonants, which clearly occur at higher frequencies than the vowels.

In quantifying environmental noise, experiences over many years have shown that it is useful to adjust the value of a SPL to allow for human auditory response. In this way it has been found that assessments of the noise more closely emulate the way in which people might respond to or be affected by the noise.

To make this adjustment the audibility threshold curve in the diagram has been adopted. Instruments known as Sound Level Meters are used to measure noise levels (SPLs). These instruments are equipped with electronic filters that, when activated, adjust or weight the measured noise in accord with the audibility threshold response. When this is done, the resulting measured noise level is said to be "A-weighted", where the "A" refers to audibility. These noise levels have the unit of A-weighted deciBels, which is commonly abbreviated as dB(A). Environmental noise is most commonly quantified in terms of A-weighted noise levels.

5.1.4 Environmental Noise

Environmental noise sources may include one or more of the following attributes.

- Continuous, where the SPL of the noise is constant or varies slowly
- Single events, such as the passby of a motorcycle on an isolated road or the discharge of a firearm
- Repetitive single events, which are generally regarded as reoccurrences of single events.

Propagation paths can vary widely:

- Along roads
- Over parklands and outdoor spaces
- Over barriers, walls or hills
- Through building elements such as windows.

Similarly there are many receivers:

- Residents at home undertaking various activities such as conversing, reading, watching TV or sleeping
- Students working in educational institutions
- Office workers
- Patients in hospitals.

Note that none of the above lists is exhaustive but together they serve to illustrate the wide range of situations and conditions associated with environmental noise.

In order to allow for the characteristics of environmental noise such as those set out above, a range of descriptors has evolved over the years. The choice of which descriptor to adopt depends primarily on the nature of the environmental noise being investigated. However, an important and often critical consideration involves what is specified in the policies and procedures of the relevant Regulatory Authority. In Australia, for example, these tend to be the Environment Protection Agencies in each State. Such Authorities usually specify how environmental noise should be quantified and assessed. Some descriptors that have commonly been adopted in many countries include the following.

- L_{eq}. The energy equivalent level of the noise being investigated.
- $L_{10}(1\ h)$. The noise level exceeded for 10% of a 1-hour period. This is usually regarded as quantifying the average maximum level of the noise being investigated.
- $L_{90}(1\ h)$. The noise level exceeded for 90% of a 1-hour period. This is usually regarded as quantifying the background or ambient noise level and is measured in the absence of the noise being investigated.

5.1.5 Noise Generation and Propagation

5.1.5.1 Generation

Environmental noise is generated by a range of sources, some of which have been considered already in this chapter. It is common that environmental noise sources comprise substantial infrastructure such as a petrochemical complex. In this type of situation there are many sources within the infrastructure that contribute to the overall noise generation process. In the case of road traffic noise, the sources comprise the individual vehicles in the traffic stream. That is, the total noise generated by road traffic is the aggregation of the noise generated by each vehicle in the traffic. Further details of this process are addressed subsequently in the present chapter.

5.1.5.2 Propagation

Sound propagation outdoors is a topic on which much has been researched and written. There are three primary factors that govern the manner in which sound propagates outdoors, in the absence of obstructions:

- Atmospheric or air absorption of sound
- Ground effects
- Meteorological effects.

Propagation is also an important factor involved in the prediction of environmental noise and this is particularly so in the prediction of road traffic noise.

5.1.6 Noise Exposure and Impacts

5.1.6.1 Exposure

Once environmental noise is generated it propagates outdoors and ultimately may impact, to various degrees, on receivers. From a broader community perspective both the range and extent of any such impacts may be investigated in terms of exposure. For example, Table 5.2 shows the nature of adverse affectation to environmental noise typically found in surveys conducted in Sydney. These figures illustrate the significance of transportation as an environmental noise source. For example, 73% of people in the Sydney area are adversely affected by environmental noise as a result of exposure to road traffic noise.

Surveys in Sydney over time have consistently shown that around 11% of the Sydney population are usually exposed to what is termed an "unacceptable" level of road traffic noise. Furthermore, about 39% of the population are generally exposed to "undesirable" levels of road traffic noise. It is not surprising, therefore, to find that about 85% of the respondents to these surveys indicated that road traffic noise is a source of common annoyance. As indicated in the above table, rail transport and industrial noise are typically responsible for similar, but lower levels of adverse affectation.

Air transport noise, on the other hand, is typically responsible for a considerable 17% of adverse affectation. Indeed, for those residents close to an airport or within designated take-off or landing flight paths, aircraft noise is usually their most annoying source of environmental noise. Reported affectation by aircraft noise generally increases when plans to expand an existing airport or construct a new airport appear in the public domain. There were many examples of this in and around Sydney during the 1980s and 1990s when debate and discussion about Sydney's future airports were taking place. Complaints to the Aviation Authorities increased substantially during the period when expansion of Sydney's Kingsford Smith Airport was being debated (Kinhill Engineers, 1990).

TABLE 5.2   Sydney People Adversely Affected by Environmental Noise

Noise Source	Proportion of Population Adversely Affected (%)
Road traffic	73
Aircraft traffic	17
Rail traffic	6
Industrial sites	4

After NSW EPA (1990).

5.1.6.2   Impacts

The primary impacts of environmental noise such as that from road traffic lie in the arena of annoyance rather than in others such as hearing damage. Generally this annoyance is of a type that interferes with activities like conversing and reading. However, sleep disturbance is one important potential impact that has been receiving much attention in recent years (Environment Council of Alberta, 1980). The concern here is that regular sleep disturbance can possibly lead to other illnesses, although the emerging evidence for any such relationship between these two is generally regarded as not yet being strong enough to make robust conclusions.

Environmental noise impacts of the annoyance/interference type are commonly quantified by means of empirically determined dose/response relationships. Again, the psychological literature abounds with background material concerning the nature, form, and determination of dose/response relationships. Many studies have also been conducted on such relationships for environmental noise in recent years. Typical of these studies are those reported by Butcha and Vos (1998), Fields (1998), and Miedema and Vos (1998). These papers deal with a variety of common environmental noise sources and reveal different ways of determining both the doses and the responses. A common outcome from many studies concerning the impacts of environmental noise is that, the extent of reported annoyance tends to increase with increased doses or levels and durations of environmental noise. It would be fair to say that this is a most complex issue, particularly when one considers the potential impacts that very low levels of noise might have on human response in certain circumstances. A good example of the latter is the very issue of sleep disturbance mentioned above. In this situation the disturbance can possibly arise from very low levels of noise such as a dripping tap or a ticking clock. On the other hand, it can also arise from higher noise levels such as road traffic or music from nearby residences.

5.2   Road Traffic Noise

5.2.1   Introduction

As set out in Section 5.1.6, road traffic is a major source of noise to which the community is exposed. This is the significant part road traffic noise plays in the environmental impact assessment (EIA) process associated with new road proposals or the redevelopment of existing road infrastructure. In addition, it is often the case that, during the conduct of this assessment process resident action or other groups will form in opposition to the road proposal. Such groups usually have particular reasons for their opposition and common examples of these are the following.

- Maintenance of an existing urban situation, without the imposition of additional traffic and road infrastructure
- Safety concerns, particularly for children and the elderly
- The general "Not In My Back Yard" syndrome

While all these type of concerns are both perfectly legitimate and reasonable, many of them are not amenable to either scientific analysis or precise quantification of their potential impacts on the community. For instance, how would one go about quantifying the potential nuisance associated with parts of a community becoming physically separated from one another as a consequence of the alignment of a new road? Consequently, what often happens in such situations is that the action groups seek out some other factor that can be analyzed and quantified and use this factor prominently in mounting their case against the road proposal. Experience has shown that the issue of road traffic noise is ideal here, since established techniques are available for measuring, analyzing, predicting, and assessing it. Furthermore, noise is a physical phenomenon with which the majority of the community can identify. Consequently, it is very important indeed for professional practitioners to understand the nature of traffic noise so that their measurements, analyses, predictions, and assessments are well founded on a scientific, technical basis.

TABLE 5.3   Major Factors Associated With Road Traffic Noise

Technical Aspect of Traffic Noise	Major Influencing Factors
Generation	Traffic volume
	Traffic speed
	Traffic composition
	Traffic operating conditions
	(free or interrupted flow)
	Road type
	Road surface
Propagation	Ground cover
	Source–receiver distance
	Screening
	Weather and atmospheric conditions
Reception	Receiver location (in or outside)
	Ambient noise conditions

5.2.2   Sources of Road Traffic Noise

From a technical perspective, road traffic noise may be regarded as the aggregation of the noise produced by individual vehicles in the traffic stream. The major factors that influence the generation, propagation, and reception of traffic noise are listed in Table 5.3. Further details about these factors and how each behaves will be given subsequently in Sections 5.2.5 (Prediction) and 5.2.7 (Control) of this chapter. Another important aspect of road traffic noise is that it exhibits temporal variations that may be considered as microscopic and macroscopic in nature, both of which occur simultaneously.

5.2.2.1   Microscopic Variations in Traffic Noise

These occur over relatively small time periods such as a few minutes and are associated with short-term fluctuations in the traffic conditions. An example might be what one would observe whilst standing at the roadside in the vicinity of a signalized intersection of two arterial roads. The flow of traffic here would vary as the signal phases changed and one would observe the short-term fluctuations in noise as platoons of vehicles passed by the observation location.

Two examples of the microscopic variations of traffic noise are given in Figures 5.4 and 5.5. Both figures depict the noise/time history that would be observed at the roadside as an isolated platoon of vehicles drives by the observation location. In the

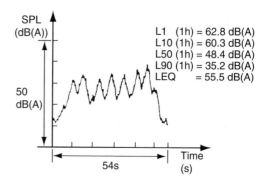

FIGURE 5.4   Noise/time history of a constant speed driveby of a platoon of seven cars.

first, a platoon comprising 7 vehicles drives past at constant speed. As the platoon approaches the noise level increases above the existing ambient level. It then fluctuates in response to the passby of the individual vehicles in the platoon and returns to the ambient level as the platoon departs. The second arises from a similar platoon accelerating away from a set of traffic signals as the light turns green. Within this noise/time history the first part ensues from the vehicles at the stationary, idle condition. As they accelerate away the noise increases and fluctuates again, but in a very different way from the constant speed passby case. Note that the individual vehicle components are also evident in Figure 5.5.

These two noise/time histories are typical of what regularly occurs in practice. It is partly because they are so different in form that they would be observed as sounding distinctly different. The various noise indices used to quantify traffic noise are also shown in Figures 5.4 and 5.5 for each case, although these

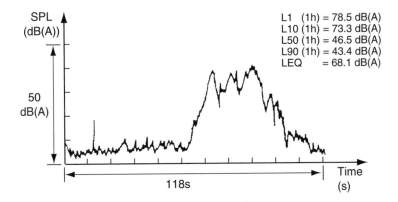

FIGURE 5.5 Noise/time history of a platoon of vehicles moving off from an intersection.

should be interpreted cautiously because of the small sample times involved. Nevertheless, they give some indication of the relativities between these indices in typical traffic noise situations.

5.2.2.2 Macroscopic Variations in Traffic Noise

Such variations take place over longer time periods of hours. Typically these variations are associated with the gradual change in traffic noise that occurs as the total traffic flow on a road varies during the course of a day. Along side a freeway with freely flowing traffic, for example, the traffic noise levels would be higher in the morning and evening peak hours compared to the mid day business time periods when the traffic volumes are less. An example of the macroscopic variations in traffic noise appears in Figure 5.6. These data were monitored on a reasonably busy arterial road in the suburbs of Sydney where the noise climate is dominated by road traffic which itself is fairly uniform throughout the day. The drop-off in noise during the early morning hours reflects the much reduced traffic at that time. Throughout the rest of the day the noise is reasonably uniform although slight increases are just apparent during the morning and evening peak hour periods. This reflects the general uniformity of the traffic conditions existing along that particular road.

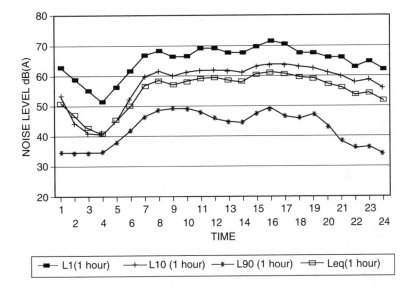

FIGURE 5.6 Typical macroscopic variations in traffic noise throughout a day.

Again a variety of traffic noise indices are shown in Figure 5.6, but the most commonly adopted are the $L_{10}(1\ h)$ and the $L_{eq}(1\ h)$ indices, both of which will feature prominently in what appear subsequently in Sections 5.2.4 and 5.2.5. For road traffic noise alone these two indices are related according to the empirically determined relationship of Equation 5.1. Note how this relationship is evident in Figure 5.6 data over the period from approximately 0600 to 2400 h. The observation that it does not hold from around 0100 to 0600 may be interpreted as indicating that during this period the noise climate at that location was caused primarily by sources other than road traffic. Application of Equation 5.1 in this manner can be a very useful technique in examining data that are purported to be traffic noise data. When Equation 5.1 is observed to hold, it may be regarded as a strong indication that the data are indeed traffic noise.

$$L_{10}(1\ h) = L_{eq}(1\ h) + 3 \tag{5.1}$$

5.2.2.3 Major Sources of Traffic Noise

As mentioned above, traffic noise is generated by the vehicles in the traffic stream. This generation mechanism may be conveniently analyzed by considering the sources of traffic noise as being vehicle related and the non-vehicle related. The major sources of traffic noise are summarized in Table 5.4, which should be considered in conjunction with Table 5.3, since the two are closely related. In effect, the non-vehicle sources of Table 5.4 represent the factors that influence or control the generation and propagation of the noise from individual vehicles and the means by which the individual vehicle noise components are aggregated to produce the total traffic noise. Both these vehicle and non-vehicle sources will now be considered in further detail.

5.2.2.4 The Vehicle Noise Sources

All of these sources operate simultaneously to produce the total noise emitted by a vehicle. Much research and development effort has been spent on understanding how these sources behave and the outcomes of these efforts are seen in gradually reducing vehicle noise outputs with each model year, particularly in the case of passenger cars. The mechanisms of vehicle noise generation are largely outside the scope of the present chapter, but interested readers could find a wealth of publications on this topic in the open literature such as international acoustics conferences and the various relevant technical journals.

TABLE 5.4 Summary of the Major Sources of Traffic Noise

Vehicle Sources	Non-Vehicle Sources
Engine	Traffic conditions
Exhaust	Road type and conditions
Tire/road interaction	Site conditions
Aerodynamic effects	Other infrastructure
Air intake and cooling fan	Weather and climate

TABLE 5.5 Vehicle Noise Source Levels

Noise Source	Roadside Noise Level at 15 m From Passenger Car Traveling at 60 km/h (dB(A))
Engine/mechanical	77
Exhaust	76
Air intake	74
Cooling fan	75
Tire/road interaction	80
Total	83

An important feature of the generation of vehicle noise is that each of the component noise sources of Table 5.4 operates in a speed dependent manner. In the case of a typical passenger car, for example, the contributions of the various sources to the roadside noise produced by the car during a constant speed drive-by would be as given in Table 5.5. In this particular example the dominant sources are the engine and the interaction of the tires with the road. Both these sources vary with speed but they do so somewhat differently from one another, as depicted in Figure 5.7. Again much is known and has been published about the behavior of these noise sources. What can be said

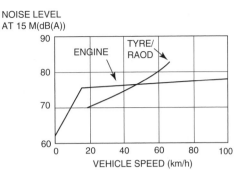

FIGURE 5.7   Vehicle noise source variations with speed.

here is that the general observations are that, with the exception of motorcycles and the like, for just about all vehicles in a state of reasonable maintenance, tire/road interaction represents the primary source of noise for all constant speeds in excess of around 40 to 50 km/h (Sandberg and Ejsmont, 2002). This is a most significant issue, the implications and applications of which will become apparent subsequently when the control of traffic noise is considered.

5.2.2.5   The Non-Vehicle Noise Sources

As mentioned previously, the non-vehicle sources listed in Table 5.4 represent those factors that control and influence the means by which individual vehicle noise is generated, aggregated with the noise from other vehicles and then propagated to roadside receivers. Yet, these are the factors that are the subject of most attention when traffic noise is being measured, predicted, assessed or ameliorated. Functional relationships, most of which are empirically based, are available to describe and quantify how each of these factors affects traffic noise. These relationships are embodied in noise prediction methods and will be considered subsequently in Section 5.2.5. However, consideration of some aspects of these factors now follows.

Road traffic noise varies with the volume, composition (mix of vehicle types in the traffic) and speed of the traffic at any particular location. Considering just the total volume of traffic, it has been well established that the effect of this factor may be described according to the relationship of Equation 5.2 (UK DoT, 1988).

$$L_{FLOW} = K + 10 \log_{10}(q) \tag{5.2}$$

where L_{FLOW} is the traffic flow component of L_{10} (1 h) (dB(A)), K is a constant, and q is traffic volume (Vehicles/h).

There are profound implications for the assessment and control of traffic noise in this rather simple relationship. One of the suggestions frequently made by members of the community for the control of traffic noise is to reduce the traffic volume. Equation 5.2 reveals that by halving the traffic volume a reduction of only 3 dB(A) in the traffic noise level would ensue. Since a change of this magnitude would only just be detected by good, unimpaired ears, it might almost be regarded as being of little benefit. Yet halving any traffic volume is a most difficult, generally impossible task, particularly along freeways or major arterial roads. Consequently, it may be concluded that there is little to be gained by way of traffic noise control by just reducing traffic volume alone. As will be outlined subsequently, control of traffic noise requires the simultaneous application of several strategies.

5.2.3   Pavement Effects on Road Traffic Noise

Road type is another important factor in the generation of traffic noise, especially given the significance of tire/road interaction amongst the sources of vehicle noise, as described previously. Again the effect of pavement type on traffic noise is an issue about which much is known throughout Australia and internationally (Samuels, 1982; Samuels and Dash, 1996; Samuels and Parnell, 2003; Sandberg and Ejsmont, 2002). Of particular relevance to the present discussion are the acoustic performances of various pavement types that have been scientifically investigated. Results from a variety of such studies undertaken in Australia by the present author have been summarized in Table 5.6. It is clear that there is a considerable range of noise levels produced by these pavements. Both this range and the relative performance of the pavement types included are comparable and similar to those determined internationally. The pavement types included in Table 5.6 are common throughout Australia and are briefly described below.

A chip seal is a thin pavement comprising a layer of bitumen onto which crushed rock has been placed and rolled.

Portland cement concretes are reinforced cement concrete pavements that have various surface textures applied by tyning and similar processes.

Asphaltic concretes are comprised of fine crushed rock in a bituminous binder and are commonly applied in arterial roads and residential streets throughout Australia. The so-called dense graded types are smooth, uniform pavements that typically vary in depth from around 25 mm to 250 mm, depending on the traffic loading conditions. Open graded types incorporate an upper, porous layer that provides a water drainage path. Such pavements are known to produce low noise levels, as evidenced in Table 5.6.

The results of Table 5.6 were produced in empirical studies, such as those of Samuels and Dash (1996) and Samuels and Parnell (2003) that involved collecting and analyzing various types of data. Included were basic traffic noise, controlled test pass by noise and statistical pass by noise data. Controlled test pass by data are collected by measuring the pass by noise of test vehicles as they repeatedly drive over a pavement of interest at specified speeds. Statistical pass by data are obtained by measuring the pass by noise of individual vehicles in the normal traffic stream operating on a given pavement. Both of these techniques are well established and refined and are specified in relevant Australian and International Standards such as ISO (1997).

It is particularly important to note that the figures in Table 5.6 are averages and that the traffic noise produced on any given pavement type (for a given set of relevant factors such as traffic and propagation conditions) will exhibit some degree of variability from site to site. Typically this variability is in the order of 2 to 3 dB(A) and is considered to arise from factors such as variations in materials and construction procedures.

Finally, the other non-vehicle sources listed in Table 5.4 relate to how traffic noise is propagated and attenuated. Since they form a significant part of the noise prediction and control processes, they will be dealt with in detail in Sections 5.2.5 and 5.2.7.

TABLE 5.6   Average Variations in Pavement Noise Found in Australian Studies Undertaken by the Present Author

Pavement Type	Average Variation in Traffic Noise (dB(A))
14 mm Chip seal	+4
Portland cement concrete with lateral tynes and dragged surface finish	+2.5
Cold overlay slurry seal	+2
Dense graded asphaltic concrete	0
Stone mastic asphaltic concrete	−2
Portland cement concrete with exposed aggregate surface	−3
Open graded asphaltic concrete	−4

5.2.4   Measurement of Road Traffic Noise

Typically, road traffic noise is measured outdoors in either a free field situation or at the façade of a potentially affected building. These are very different conditions from those of the laboratory where it is usually possible to control or minimize the effects of externalities that may influence or interfere with the signal being monitored or measured. Consequently, when measuring road traffic noise it is always important to ensure that the conditions under which the measurements are conducted are as good as possible. Here the term "conditions" refers to factors such as the weather (e.g., wind, temperature, rain, humidity, etc.) and the operation of other sources of noise that are not related to the road traffic noise being investigated. The latter can include a multitude of sources like rustling foliage, animals, insects, plant, and equipment. Further, it is critical that all relevant conditions and factors that could potentially affect the measurements are both monitored and recorded. While this may at first appear to be an intuitively obvious requirement, it is often both difficult and time consuming to achieve. One key reason for this is that it is often not clear just what conditions and factors are relevant and could affect the measurements.

Well-tried and documented techniques for the measurement of road traffic noise are set out in comprehensive detail in various documents such as Standards Australia (1984). Given what appears there, the overall process of measuring traffic noise requires attending to the following six issues.

1. *Objectives and data applications.* Decide and set out clearly the objectives of undertaking the measurements and how the data collected are going to be applied subsequently. For example, are the data required for an EIA of a new road proposal, or will they be tendered and debated during legal proceedings as a consequence of traffic noise complaints from residents in one house on a main road? These are potentially two very different situations that would require rather different approaches and very different amounts of noise data.

2. *Sites and measurement locations.* Where and when the measurements are to be undertaken are important issues that can only be addressed once the matters in point 1 above have been resolved. According to the present author's extensive experience in measuring road traffic noise over many years, selection of suitable sites and of the measurement locations within each site usually turn out to be about the most taxing elements of the entire data collection process. This is generally because of the externalities referred to previously that have the potential to affect the measurements. For instance, traffic noise data may be required at four or five residential facades that front on to a 5 km section of an urban freeway midway between, say, two interchanges. An initial inspection of the section revealed that there are ten such residences, but five of them are near to some light industrial sites. Since the industrial noise was clearly audible and was also detected by your instrumentation at these five residences, it would interfere with the traffic noise measurements, so these five potential sites were eliminated. Of the other five sites, one is almost inaccessible and a viscous, huge dog inhabits another. This leaves only three that turn out to be suitable.

3. *Record all site details.* This is a critically important procedure that, in essence, involves measuring and recording all the relevant site data such as those necessary to conduct a traffic noise prediction. It is a good practice to prepare site drawings and to take several photographs of each site. Only in this way can the noise data be confidently applied both after it has been collected and possibly again in the future. These are common situations where before/after studies are involved.

4. *Monitor noise levels.* Here the required noise data are collected using appropriate instrumentation systems according to procedures such as those of Standards Australia (1984). It is important to ensure that all instrumentation systems are suitably calibrated and will measure the noise indices of interest, such as the L_{eq} or the L_{10} (1 h).

5. *Monitor traffic conditions.* In the vast majority of cases it is mandatory to monitor the traffic conditions simultaneously with the noise measurements. This requires measuring flows, speeds, and compositions. Generally the nature and format of these traffic data are aligned with the traffic noise prediction model adopted by the organization undertaking the measurements. Without such data it would not be possible, for example, to apply the measured noise data in any form of before/after study. An example to illustrate this point follows subsequently in the case study of Section 5.3.

6. *Document all results fully.* All site, traffic and noise data, along with any other relevant information, should be recorded in a document such as a report.

While the above six issues may appear straightforward and somewhat routine, they are frequently not attended to with sufficient rigor. Further, it is often also the case that one or more of them are neglected all together. A consequence of either of these situations is usually that the noise data cannot be applied for the purpose for which they were collected. This is, of course, usually a most unsatisfactory outcome that can only be rectified by repeating the entire data collection exercise. The consequent time, resource, and financial implications of this situation are obvious. The case study of Section 5.3 exemplifies how misleading outcomes can ensue from road traffic noise measurements that were not collected in accord with the above six principles.

5.2.5   Prediction of Road Traffic Noise

Any given road traffic noise prediction model usually involves a series of algorithms that describe and quantify the manner in which noise is generated, propagated, and attenuated. That is, noise prediction models incorporate mathematical descriptions of what have been set out previously in this chapter concerning the generation, propagation, and attenuation of road traffic noise. Generally road traffic noise is measured and predicted utilizing the $L_{10}(1\text{ h})$ and the $L_{eq}(1\text{ h})$ indices, which are related according to Equation 5.1 of Section 5.2.2. Some derivatives of these have also been adopted such as $L_{10}(18\text{ h})$. This particular index is the arithmetic average of 18 $L_{10}(1\text{ h})$ values over the period 0600 to 2400 on a given day. It originated in the U.K. and has been widely applied there (UK DoT, 1988) and throughout Australia. Use of this index is based on the assumption that there are negligible levels of traffic noise during the 6-hour period from 2400 to 0600. Until recently this assumption was generally true across countries such as Australia and U.K. so that the $L_{10}(18\text{ h})$ index was the accepted measure of traffic noise. Over the last few years, however, there have been increasing instances where the assumption has not been valid and this has prompted the use of some alternative indices. These indices have either covered the entire day or parts of the day (Houghton, 1994; NSW RTA, 2001). In addition, there has been a move towards increasing use of L_{eq} indices.

Some typical examples of these indices are as follows.

L_{10} (16 h)   A daytime index, being the arithmetic average of the $L_{10}(1\text{ h})$ indices from 0600 to 2200 h
L_{10} (15 h)   As per L_{10} (16 h), but from 0700 to 2200 h
L_{10} (8 h)   A night time index, being the arithmetic average of the $L_{10}(1\text{ h})$ from 2200 to 0600 h
L_{10} (9 h)   As per $L_{10}(18\text{ h})$, but from 2200 to 0700 h

There are also L_{eq} indices corresponding to each of the above four indices.

It has regularly been observed in Australian and international studies over several years that there are readily determined correlations between the various traffic noise indices. These typically take the form of empirically determined relationships such as those of Equation 5.1. A recent Australian study by Huybregts and Samuels (1998) produced the relationships reproduced in Equation 5.3. Of these the first four are novel whilst the fifth is consistent with previous relationships of this type published in the open literature over a number of years (Brown, 1989, for example).

$$L_{10}(18\text{ h}) = L_{eq}(16\text{ h}) + 2.2, \qquad L_{10}(18\text{ h}) = L_{eq}(8\text{ h}) + 6.7, \qquad L_{10}(16\text{ h}) = L_{eq}(16\text{ h}) + 2.5,$$
$$\text{(5.3)}$$
$$L_{10}(8\text{ h}) = L_{eq}(8\text{ h}) + 2.6, \qquad L_{10}(18\text{ h}) = L_{eq}(24\text{ h}) + 3.2$$

Along with Equation 5.1, these relationships may be applied to transform the output of a traffic noise prediction model from one index to another as may be required.

5.2.5.1 Calculation of Road Traffic Noise

There are many traffic noise prediction models and associated computer software packages available around the world. In this author's experience there are many technical similarities between most of these models. Moreover, most of the models perform to a generally similar degree of accuracy. It is usually the case that if one becomes experienced in using one particular model it is a relatively easy task to become familiar with another model. One popular and widely known model originated in the U.K. and is known as CoRTN — the Calculation of Road Traffic Noise (UK DoT, 1988). This particular model was first released in 1975 and was subsequently updated in 1988. It has been comprehensively evaluated under Australian conditions (Saunders et al., 1983). The CoRTN model is relatively simple in construct, is well tried and tested and is rather typical of many traffic noise prediction models available around the world. CoRTN predicts either $L_{10}(18\,\text{h})$ or $L_{10}(1\,\text{h})$.

The following explanatory points assist in understanding the CoRTN model.

1. To undertake predictions at a site with the CoRTN model, the site is firstly divided into segments. A prediction is conducted at the same receiver location for each segment within the site and the predicted levels for all segments added (logarithmically) to produce the overall predicted level at the receiver location. A segment is a sector, within which each of the relevant (noise generating) parameters remains constant. For example, say, the site is a four lane divided freeway where the pavement type on the northbound lanes is very different from that on the southbound lanes. In this case the pavement effect on traffic noise is not constant because, all else being constant, the noise from the northbound lanes would differ from that from the southbound lanes. This site would therefore be divided into two segments, one for each pair of lanes. Predictions would then be made for both segments and added to give the total noise level at the receiver location. Segment selection such as this example can be made on the basis of any of the relevant CoRTN input parameters. It should be pointed out that this is often a difficult task that requires some experience.

2. CoRTN predicts either $L_{10}(18\,\text{h})$ or $L_{10}(1\,\text{h})$ noise indices and thus the prediction process commences with either of the following two relationships in Equation 5.4 and Equation 5.5. From there the process proceeds in exactly the same manner for either noise index.

$$\text{Basic Noise Level (1 h)} = -42.2 + 10\log_{10}q \tag{5.4}$$

$$\text{Basic Noise Level (18 h)} = -29.1 + 10\log_{10}Q \tag{5.5}$$

where q is the hourly traffic flow (Vehicles/h), Q = Daily traffic flow (Vehicles/18 h from 0600 to 2400).

3. The relationships between noise and traffic flow in Equation 5.4 and Equation 5.5 are for a constant speed of 75 km/h and a traffic stream comprised of cars only on a flat road. That is, the percentage of heavy vehicles in the traffic (p) is zero as is the gradient (G).

4. Having determined the Basic Noise Level (BNL) as above, a subsequent relationship is then applied to generate a correction factor that allows for the actual speed and heavy vehicle content of the traffic. This correction is added arithmetically to BNL.

5. From there, further relationships are then utilized to provide two corrections that are required if a gradient is involved. These particular relationships also allow for varying heavy vehicle content in the traffic. The gradient correction factors thus obtained are applied to BNL, in the same manner as in point 4 above. Note that, in those cases when a site is comprised of segments based on traffic lanes (as in the example given above), the gradient corrections are only applied to the traffic proceeding uphill. Both corrections are applied where the traffic is proceeding in both directions within a segment, such as in the case of a two lane, two way road.

6. The next step is to determine the attenuation of the traffic noise that occurs as the noise propagates to the receiver. This is done in two parts, the first of which involves determining the direct distance from the traffic noise source to the receiver. The traffic noise line source is deemed to be situated 3.5 m in from the edge of the carriageway and 0.5 m above the pavement surface. For a two lane two-way road the source line is in from the nearside carriageway. If a divided road is involved where there are segments for the two directions of traffic flow, there are two source lines. One is in from the edge of the nearside

carriageway while the other is in from the far edge of the far carriageway. The edge of the carriageway means, in effect, the edge of that part of the road where the traffic flows. This is frequently taken to be the road curb. However where parked vehicles are present, the edge line must be located just inside the parking "lane." The receiver is situated a distance d from the edge of the nearside carriageway, which is equivalent to $d + 3.5$ from the (nearside) source line. Further the receiver is positioned at a height h above the source line, which is equivalent to $h + 0.5$ above the pavement surface. The values of d and h are determined from site inspections or from scale drawings or other suitable means. From there, the second part involves determining the attenuation over what is known as hard ground is determined from a relationship presented in CoRTN. A typical example of hard, or reflective, ground would be a paved car parking area. For a given combination of h and d the correction is determined as just explained and added to the figure obtained thus far in point 5 above.

7. In the situation where the intervening ground is not all hard, allowance must be made for the additional attenuation that occurs due to the effects of the soft, absorbent ground present. This is done via a subsequent CoRTN relationship and the resulting correction factor is added to the figure obtained thus far in point 6 above.

8. The next factor to be considered is the presence of any intervening infrastructure that might have an attenuating effect on the noise as it propagates to the receiver. The CoRTN relationship used for this purpose involves the attenuation provided by a noise barrier. Other intervening structures are modeled as though they were barriers. This particular relationship gives the barrier attenuation as a function of the path length difference $(a + b - c)$ between the direct noise path from source to receiver (c) and the path diffracted over the top of the barrier $(a + b)$ from the source to the receiver. Values of a, b, and c are determined from the site geometry and usually require some relatively simple Pythagorean calculations. Two further observations can be made about this process and the relationship involved, which can also serve as a barrier design tool. Firstly, it is generally difficult in practice to achieve path length differences greater than around 0.5 m. This is typically a result of dimensional and similar constraints that occur at any given site. Secondly, the situations of the path length difference being equal to zero (known as the grazing incidence condition) or less than zero can provide some attenuation. These situations must not be overlooked when determining the relevant site parameters to incorporate in a particular prediction exercise. The barrier correction factor thus determined is added to the value arrived at so far in point 7.

9. If the receiver has an uninterrupted view of the road in both directions, this view is deemed to subtend an angle of 180°. Frequently this is not the case, so an appropriate correction factor is determined and added to the total noise level calculated thus far. The angle of view correction is usually invoked for sites where there is more than one segment, and is an important means by which the contribution of each segment to the total noise is determined. For example, say, a receiver was situated behind one side of a long building that provides noise attenuation (and, therefore, is modeled as a barrier). Such a situation would require two segments — one for the obscured part and one for the unobscured part. When looking to the left the receiver's view of the road is obscured by the building, while to the right the view is unobstructed. From a scale drawing of this situation the angle of view of the obscured segment was found to be, say, 50°. Then the angle of view of the unobscured segment was 130°. Then the angle of view correction applied to the obscured segment noise prediction would be, according to the relevant CoRTN relationship, about -5.6 dB(A). The corresponding correction for the other segment would be -1.4 dB(A). These corrections allow for the greater contribution overall of the unobscured segment for which the angle of view is greater.

10. Once all the above corrections have been determined and added to BNL in each segment, the resulting noise components from each segment are combined by logarithmic addition as per Figure 5.2. Examples of this well known process are as follows.

$$76 + 70 = 77.0$$

$$76 + 73 = 77.8$$

$$76 + 76 = 79.0$$

TABLE 5.7   Calibration Adjustments of CoRTN for Australian conditions

Location of Receiver	Calibration Adjustment (dB(A))	Accuracy of Calibrated Prediction (dB(A))
Free field	− 0.7	± 3.6
1 m in front of a facade	− 1.7	± 5.0

11. Then a correction is made for the effects of pavement type. These corrections for Australian pavements have been presented previously in Table 5.6. Note that, should segments made on the basis of pavement type be involved, these pavement noise corrections would be made within each segment. It must also be pointed out that these pavement type corrections would serve as a guide to the effects of pavement type on road traffic noise in other countries. However it is recommended that locally derived pavement noise factors should always be applied. The procedure presented here for allowing for the effects of pavement type on the predicted noise level is one which has been developed in Australia and which has been adopted in Australia. It differs slightly from the original CoRTN procedures adopted in U.K. as set out in UK DoT (1988).

12. A further correction is then made if required for the reflective effects of building facades or the like. It is often the case that noise predictions are conducted 1 m in front of the front facade of a residence or building. At such a position the noise reaching the receiver would be the aggregation of the directly propagated signal with that reflected from the building façade. The correction to be applied in this case is also given in CoRTN. Sometimes there might be continuous, reflective walls or the like on the side of the road opposite the receiver. An additional correction for this relatively rare situation (in Australia) is also given in CoRTN.

13. The final correction to be made is that for the accuracy of CoRTN. In Australia this accuracy was determined by Saunders et al. (1983) and is summarized in Table 5.7. The calibration adjustments shown there are added to the final noise level determined above. The accuracy figures represent the 95% confidence limits around the calibrated predictions. Again it is emphasized that this procedure has been developed and applied in Australia, so it is recommended that a locally developed procedure be applied.

14. When using CoRTN it is possible to proceed through the process described above by visually reading values off the various charts and tables — a so called manual method of prediction. Surprisingly, this can provide reasonably good predictions and it is a procedure that is often undertaken at the beginning of an extensive prediction exercise in order to obtain a "ballpark" estimate of the likely predicted noise levels. A rather common approach is, however, to employ the algorithms which are given throughout CoRTN to calculate BNL and the subsequent correction factors. These algorithms form the basis of the several commercially available computer versions of CoRTN and in practice most predictions are conducted using such software. Irrespective of whether a manual or software-based prediction technique is adopted, the most critical aspect of the entire prediction process is the determination of the input parameters required and the values of each parameter. This aspect is, in fact, independent of the prediction model adopted and is critical in any traffic noise prediction procedure.

5.2.5.2   Traffic Noise Model

There are many traffic noise prediction models available around the world and detailed studies of these are outside the scope of the present chapter. The models have many features in common so that once one such as CoRTN has been understood and mastered it is a reasonably straightforward process to pick up another model and use it for predicting traffic noise. One model that is currently receiving some attention throughout U.S.A. and Australia is the American Federal Highway Administration's Traffic Noise Model — TNM. After several years of well-publicized development, TNM was finally released in 1998 (Menge et al., 1998). At the time of writing the present chapter TNM was being evaluated in several states across Australia (e.g., Samuels and Huybregts, 2001). The outcomes of these evaluations are expected to provide calibration factors for TNM under Australian conditions similar to those presented previously in Table 5.7 for CoRTN.

The primary reasons for the interest in TNM are that, it incorporates some of the latest technological developments in areas such as vehicle noise source types and outputs in relation to noise propagation theory. In addition, TNM has been configured to predict the noise indices L_{eq} which, as mentioned previously, are becoming more commonly adopted for traffic noise studies. Furthermore TNM is a computerized model that includes some powerful interactive data input and output routines. Note that use of TNM requires running a sophisticated computer package. However the care to detail in translating any given situation into values of model input parameters necessary to conduct predictions with CoRTN is also required when using TNM. Indeed, such care is necessary when using any traffic noise prediction model.

5.2.6   Assessment of Road Traffic Noise Impacts

The process of EIA and the procedures involved are issues that cover a wide range of parameters and factors (OECD, 1994, 1995). When new infrastructure projects, such as roads, airports or open cut coal mines are proposed, arrangements are in place throughout countries like Australia that require an EIA of the proposal to be conducted. One outcome of such an assessment is a document commonly known as an Environmental Impact Statement (EIS). While dealing in detail with both the EIA process and the resulting EIS are outside the scope of the present chapter, the subsequent treatment of the assessment of traffic noise impacts might be better understood if the following objectives of the EIA process and the typical aspects addressed in an EIS are appreciated.

5.2.6.1   Primary Objectives of EIA

- To ensure that decisions are taken following timely and sound environmental advice.
- To involve the community.
- To ensure that project proponents assume their environmental responsibility.
- To facilitate environmentally sound proposals that minimize the adverse environmental aspects and maximize the benefits to the environment.
- To allow for ongoing environmental management.
- To promote education and awareness.

5.2.6.2   Typical Environmental Aspects Addressed in an EIS

- Site location and conditions
- Visual environment
- Biology (Flora and Fauna)
- Noise
- Land use
- Education
- Heritage
- Physical environment
- Climate (Ozone and Greenhouse effects)
- Meteorology (Air quality)
- Water quality
- Social impacts
- Economic aspects
- Aboriginal and/or archaeological aspects

While neither of the above listings purports to be exhaustive or complete, they do serve to illustrate the wide-ranging nature of the EIA process and of the resulting EIS document. Environmental noise, of which road traffic noise is one type, is usually an important issue for infrastructure proposals such as those mentioned above. Obviously noise is one of many environmental factors that are considered in the EIA and addressed in the EIS. In this context it is desirable for the noise impact assessment process to be comparable to the processes adopted for the other environmental factors. In this way the impacts of each of the factors may be assessed in a consistent and fair manner and subsequent comparisons between these impacts facilitated. However, it is relevant to recall what is mentioned earlier in Section 5.2.1 regarding how the issue of noise may assume a high profile within communities opposed to a particular development proposal.

Assessment of environmental noise impacts does not, of course, occur only within the ambit of the EIA process. There are many other situations where such assessments are conducted. Typical examples include the assessment might be required in response to complaints about noise produced by existing plant and equipment or by an established industrial complex. Common sources of complaints within the community involve noise from entertainment venues, from swimming pool filtration systems and from domestic air conditioning units. However, as mentioned previously in Section 5.1.6, road traffic noise is a major source of adverse affectation within the community.

5.2.6.3   Selection of an Impact Assessment Procedure

There are general procedures available for the assessment of environmental noise impacts which have evolved over recent years and which form the basis of both regulatory authorities' requirements and various regulations and standards (e.g., International Standards Organisation (1996), EPA NSW (1999)). An outline of these procedures is presented below, but it must be remembered that the procedure adopted in any particular case must be tailored to suit the conditions and features of that case.

The usual aims of environmental noise impact assessment are to describe and assess the noise exposures and impacts of residents in a community. In this context exposures are taken to be the values of the predicted or measured noise indices at the most exposed façades of residential buildings. The noise impacts are then determined by comparing these exposures to criteria, which are often set by the relevant regulatory authority. While this procedure may appear simple and straightforward, there are several issues that must be addressed in order to make the procedure effective.

The noise index adopted must be agreed to by the various parties involved in the assessment process. Commonly adopted indices are L_{eq} and $L_{10,}$ but this is not always the case. In any event there must be technology available for the measurement and/or the prediction of the chosen noise index. The accuracy and precision of the chosen technology should also be known and considered when the measured or predicted indices are subsequently applied. In addition, the chosen technology should be well tried and tested and should lend itself to independent verification or, where appropriate, to scientific scrutiny. The techniques set out in Sections 5.2.4 and 5.2.5 of the present chapter for the prediction and measurement of road traffic noise fall into this category.

The criteria also come in a variety of forms and again are usually set by the relevant regulatory authority. It should be noted here that the nature of and the values associated with environmental noise criteria are continuously evolving issues about which there is often considerable discussion and debate. The following list presents some typical environmental noise criteria that are either currently in use or are under consideration for use in the immediate future.

5.2.6.4   The Maximum Noise Level Criteria

These are the levels of a particular noise index which should not be exceeded and may be regarded as the traditional types of environmental noise criteria. Usually they are well documented and accompanied by a considerable amount of explanatory material which sets out how the criterion is determined and the assessment made. For example, EPA NSW (1998) sets a criterion for stationary noise sources. Simply put, and without involving all the explanatory details, the criterion is that the L_{10} level of the noise source, measured over a 15 min period, should not exceed the background level by more than 5 dB(A). Here the background level is set as the L_{90} level measured in the absence of the noise source.

For road traffic noise, EPA NSW (1999) specifies day time (7AM to 10PM) and night time criteria (10PM to 7AM) for a variety of road types located within a range of land use types. For example, in the case of a new freeway in a residential area, the day time criterion is an L_{eq} level of 55 dB(A) while the night time criterion is an L_{eq} level of 50 dB(A). It is strongly emphasized again that criteria such as these must be considered within the context of the considerable amount of explanatory material included in the source document referenced above in which the criteria are set out.

5.2.6.5   The Population Annoyance Criteria

These criteria have been under development and discussion amongst the international acoustics professional community over the last few years. Typically they set limits on the number or proportion of the community that are either seriously or highly annoyed by the noise from the particular source under assessment. ISO (1996), for example, sets out this process. These criteria are based on empirically derived curves that show annoyance increasing with noise level.

There is not as yet any significant degree of uniformity in the manner in which criteria of this type are set or applied. Within the Environmental Protection Authorities across Australia there is support for the view that no more than around 10% of the population should be adversely affected by any particular environmental variable, including noise. However, in applying such an approach the need for compromise arises rapidly because of the trade offs between achieving the criteria and the costs of doing so (e.g., RTA NSW, 1992, 2001).

One benefit in working with population annoyance criteria is that it facilitates determination of areas or groupings of the community that experience positive and adverse environmental noise impacts from the source being assessed. This approach has been evident in some recent EISs for road and air transport development proposals. For example, RTA NSW (1994) presents the EIS for an urban freeway that passes through established residential areas of Sydney. Amongst the noise assessment component of that particular EIS were results showing clusters of residences where the noise climate would improve and where it would deteriorate after the freeway became operational.

5.2.6.6   A General Assessment Procedure Framework

As might be expected from what has preceded, there are many possible methods by which an environmental noise impact assessment may be undertaken. In any given situation the particular technique adopted must be tailored to the situation itself so that all relevant factors are included as appropriate in the assessment. What now follows, therefore, is a generalized framework within which a specific assessment procedure may be developed. Assessments of road traffic noise fall well within this framework.

5.2.6.6.1   Set the Objectives

Specify clearly what is to be assessed and why. Generally this involves identifying the noise sources and the potentially affected receivers. Environmental noise impact assessments generally deal with the determination and assessment of the exposure of residents in the community to noise from a specified source or combination of sources.

5.2.6.6.2   Specify the Context and Aims

Usually the aim is to determine the levels of a specified noise index at locations within an area potentially affected by the noise sources under assessment. From there the noise impacts are assessed by comparing the noise index levels with relevant criteria.

5.2.6.6.3   Determine the Assessment Area

The assessment is conducted over an area that surrounds or adjoins the noise sources. This area must be carefully specified, yet doing so can often become quite complicated. In a new road proposal, for example, the initial approach might be to constrain the assessment area to the road corridor plus, say, two or three rows of houses on either side of that corridor. But if the new road will have the effect of altering traffic patterns on nearby existing roads, then it might be appropriate to extend the assessment area to include residences on or around these existing roads. Consideration must also be given to other relevant conditions in the possible area and the existence of other noise sources that might interact with the noise sources being assessed.

5.2.6.6.4   *Locate the Receiver Locations*

All potentially affected or benefited receiver locations within the assessment area must be located and specified. It may not be necessary, however, to conduct noise measurements or predictions at each and every receiver location. What is required is to ensure that the noise exposure of all receivers is determined fairly, accurately, and representatively. This may involve a consideration of the number of residents within each residence and the duration and frequency of the residents' time in the residence.

5.2.6.6.5   *Determine the Noise Exposure at Nominated Receiver Locations*

This is done either by measurement or by prediction or by various combinations of both. It is important to specify how the exposures were determined and to justify the techniques adopted. If measurements are conducted, principles such as those described in Section 5.2.4 of the present chapter should be adopted and documented. In the case of predictions, the efficacy of the prediction model adopted must be set out, along with the accuracy of the model. The accuracy figures should be presented and used in a manner similar to that demonstrated for CoRTN in Section 5.2.5 of the present chapter. The noise exposures can be specified in terms of the noise indices, such as those mentioned in Section 5.2.5 for road traffic noise, at the various receiver locations. This does not necessarily mean that a noise index will be provided for each receiver location. It might be possible to use particular individual locations that are representative of a given area or cluster of residences. Alternatively, noise contours might be produced that cover very wide and extensive areas at or near to the noise sources. Another possibility is for the exposures to be presented as areas where specified noise levels are exceeded or within which noise levels fall within a certain range. Again they might be expressed in terms of the numbers of residents exposed to certain noise levels or ranges of levels. Which format to adopt will depend on a number of factors such as, the nature and operation of the noise sources, and the type of assessment criteria adopted.

5.2.6.6.6   *Assess Against Criteria*

At this stage the noise exposures are compared with the relevant criteria to effect the assessment.

5.2.6.6.7   *Consider Remedial Treatments*

Should the assessment reveal that the noise exposures are excessive, it is usually the next step to consider the introduction of remedial treatments by way of noise control. Doing so can be an extensive operation and is covered in some detail subsequently in Section 5.2.7 for road traffic noise. Once some treatments have been selected, their effectiveness must be estimated and the ensuing noise exposures determined. From there the assessment is repeated and this process continues until the exposures are deemed to comply with the criteria. Sometimes this outcome cannot be achieved for technical or cost or some other reasons. Such a situation generally leads to some compromises that are usually negotiated with the relevant regulatory authority. It should also be pointed out at this stage that, what is referred to here as remedial treatments is a generic term that covers a wide range of possibilities, again as will be detailed in Section 5.2.7 in the case of road traffic noise

5.2.6.6.8   *Document the Procedure and Outcomes*

Documentation of the assessment is obviously an important component of the entire process. The document should cover all of the matters addressed in the present section.

5.2.7   Control of Road Traffic Noise

This section deals with how road traffic noise may be controlled or attenuated. In so doing it brings together much of what has preceded in the chapter. Noise control is, of course, an extensive area of technology. Therefore, it is not the intention here to present much detailed material or to cover aspects such as traffic noise control design techniques. Rather the approach is one of exploring the types of control options currently available for road traffic noise and from there to consider how the options might be selected and their effectiveness evaluated.

Throughout this section it is most important to remember that the control of road traffic noise is based on the following two fundamental concepts.

- The issue of traffic noise should always be included as a fundamental parameter in the planning, design, construction, and operational phases of a road infrastructure or similar project that has the potential to impact upon or otherwise affect the community.
- When considering low noise designs or noise control treatments for road traffic noise, do not limit the consideration to one treatment. Rather, opt for solutions that include as many treatments or noise control strategies as possible.

5.2.7.1   The General Approach to Road Traffic Noise Control

In concert with the classic theories of engineering noise control, road traffic noise may be controlled in the following three areas.

- At the source
- Along the transmission path
- At the receiver

Examples of each abound and many have been explored directly or indirectly in previous sections of the present chapter. Some typical examples are presented below, which is by no means an exhaustive or exclusive listing.

5.2.7.2   At Source

- Design of quieter motor vehicle engines and exhaust systems
- Traffic management schemes
- Restricted speed limits and limiting the access to a road by heavy vehicles
- Road pavement surface treatments

5.2.7.3   Along the Transmission Path

- Land use planning to separate noise sensitive areas away from major highways and freeways
- Roadside noise barriers

5.2.7.4   At the Receiver

- New building design
- Retrofitting existing buildings with components such as windows with greater noise attenuation performance

Determination of the estimated noise attenuation provided by each technique such as those listed above can and do become rather complex matters. However, they are also very important matters, particularly when road traffic noise is, as recommended above, included as a baseline parameter in the design of some new road infrastructure such as a freeway. In general terms, a good way of approaching these matters is to apply noise predictive models, such as those introduced in Section 5.2.5, for both the design and attenuation estimation processes.

One good example of a design-based approach towards road traffic noise control at the receiver is what is known in Australia as the Quiet House. It is a demonstration project that was jointly sponsored by the state road and environmental agencies in the State of New South Wales, Australia. The project took the form of an architectural competition to design a residential house that could be built on a busy road with high levels of traffic noise. That is, the design had to be such that with the house exposed to high levels of traffic noise, the acoustic environment of the interior spaces and rooms of the house were suitable for the activities undertaken in these areas. Furthermore, the construction of the house was required to be typical of those in modern, suburban residential developments and to involve the use of conventional building materials. Several entries were received and a winning design selected. This particular house was constructed on a busy arterial road in suburban Sydney where it remains on public exhibition. Despite the high levels of traffic noise to which the Quiet House is exposed, its inside acoustical climate is very good and all the relevant acoustical criteria for activities such as conversing, reading, and sleeping have been met. A diagram, freely available in the public domain, of the Quiet House has been reproduced below in Figure 5.8.

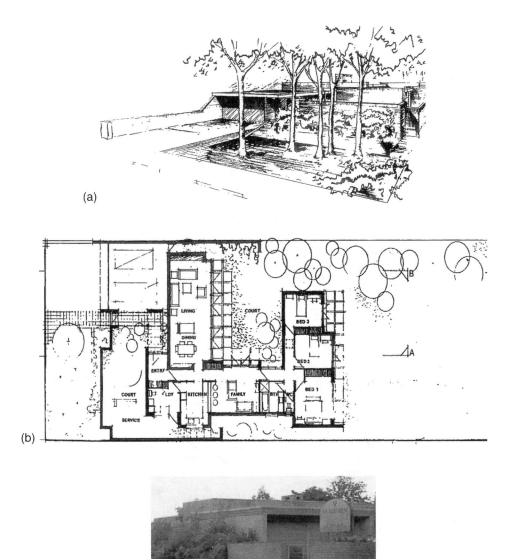

FIGURE 5.8 The quiet house (a) street view (b) plan view (c) photograph.

5.2.7.5   Selection and Evaluation of Road Traffic Noise Control Treatments and Strategies

Given what has preceded in the present section, it is apparent that there are likely to be several potential treatments that could be considered in any given road traffic noise situation. Again it is emphasized that the finally adopted solution to each road traffic noise problem should include as many of the potential treatments as possible. In concert with what has been presented above, the control of road traffic noise tends to involve combinations of general approaches such as the following.

- Original design of new roads or redesign of existing roads
- Placing limits or conditions on traffic operations
- Retrofitting noise control treatments
- Planning the introduction and/or operation of roads
- Management of roads and their environs

The two issues to be addressed now are: firstly, how to approach the process of selecting the appropriate set of road traffic noise control treatments or strategies, and secondly, how to evaluate these treatments or strategies. In this context, treatments tend to refer to physical or engineering solutions such as installation of a noise attenuation barrier, while strategies involve measures such as land use planning to align areas of similar noise tolerance. While treatments and strategies differ in both their construct and application, what now follows will not differentiate between them nor treat them separately.

5.2.7.6   A Framework for the Selection and Evaluation of Treatments or Strategies

The approach here is similar to that of Section 5.2.6, where a generalized framework was presented, which could be used subsequently as the basis for developing a detailed procedure for a specific situation. The mooted framework comprises four steps.

1. Identify the road traffic noise problem and, where possible, divide it into categories that describe or quantify the various forms of the problem and the relevant attributes of each category.
2. Establish a set of noise control treatments and/or strategies that is as comprehensive as possible.
3. Formulate a set of noise control attributes by which the effectiveness of each treatment or strategy may be evaluated.
4. In the context of the specific situation being considered, apply point 3 above to point 2 above, for each of the categories of point 1 above.

The outcome of this process should be a comprehensive listing of treatments and/or strategies with an accompanying set of recommendations for the application of each treatment and/or strategy.

5.2.7.7   An Example of the Selection and Evaluation of a Set of Road Traffic Noise Control Treatments/Strategies

The following example presents one way of conducting the selection and evaluation of a set of road traffic noise control treatments/strategies. Much of what has been presented in this chapter concerning road traffic noise has been primarily concerned with major road infrastructure. However, it is not always the case and road traffic noise issues arise from many situations. Indeed, there are many other situations and these will become apparent in the following example. This particular example ensued from the deliberations of an Australian Environment Council (as it was then known) Working Group of which the present chapter's author was a member. The Working Group addressed the process set out in the above framework in Australia. AEC (1988) documents the Group's output, which is a selection and evaluation of road traffic noise control treatments and strategies.

5.2.7.7.1   Problem Identification

Road traffic noise ensues from a wide range of road types and it impacts on a similarly wide range of receivers. It can be conveniently divided into the following four categories. While these categories were selected on the basis of the above Working Group members' experiences and collective wisdom, it is recognized that other categories could be assembled to replace those below. Nevertheless the following categories do relate well to the current situation in Australia. What should also be noted is that any consideration of road traffic noise must incorporate a clear distinction between new (or proposed) and existing roads.

Category #1. Large volume traffic flows along major arterials, highways and freeways, impacting on residential, and other noise sensitive properties. (The examples contained in the present chapter have been mainly concerned with this category. It is usually this category that comes to mind first when the issue of road traffic noise is raised.)

Category #2. Night time movement of heavy vehicles along major and minor arterials, impacting principally on residential areas.

Category #3. Day and night time intrusion of light and heavy vehicles in local residential roads and streets.

Category #4. Aggregate increases in ambient noise across residential and other areas, primarily as a result of road traffic operations, impacting on the community at large.

5.2.7.7.2 Noise Control Treatments and Strategies

As already evident in the present chapter, there are many approaches to the control of road traffic noise. A summary of the primary techniques and strategies is set out below, and some further details of each will appear subsequently.

- Road design
- Traffic management
- Planning
- Incentives
- Community awareness
- Roadside barriers

5.2.7.7.3 Noise Control Attributes

There were eight attributes selected with which to evaluate the above treatments and strategies.

- *Noise control effectiveness.* That is, what level of noise reduction is provided by the treatment or strategy
- *Function.* What is involved in implementing the treatment or strategy
- Indirect benefits and disbenefits
- *Variability of effectiveness.* Will the noise reduction be sustained over time or will some direct or external effect cause the reduction to vary over time?
- Implementation feasibility
- Dependency on other strategies
- Cost
- Implementation time and scheduling.

5.2.7.7.4 Selection and Evaluation Process

By applying point 3 to point 2 above for each of the categories mentioned in point 1, the outcome was a substantial matrix, part of which is reproduced in Table 5.8. The effectiveness ratings included in the matrix were again determined by the Working Group on the basis of the members' collective expertise and wisdom. The following techniques and strategies were consistently rated as generally being highly effective.

Road design
  ◦ Road pavement surface treatments
  ◦ Source-receiver distance

TABLE 5.8 A sample of the effectiveness matrix (After AEC (1988)): A road in a New Area For The Four Traffic Noise Categories

Noise Control Treatment	Effectiveness Of Treatment			
	Category #1	Category #2	Category #3	Category #4
Distance	High	High	Not applicable	Low
Gradient	Medium	Medium	Low	Low
Road surface	High	High	Low	Low
Road elevation	Low	Low	Not applicable	Not applicable
Road in cut	Medium	Medium	Low	Not applicable
Traffic controls	Medium	High	Medium	Medium

Traffic management
- Traffic channelization
- Local Area Traffic Management (LATM)
- Restricted use routes

Roadside noise barriers
- Along freeways
- Along some major arterials

Planning
- Land zoning and/or classification
- Noise tolerant land use substitution
- Residential cluster developments
- Separation of noise sensitive land use.

5.3 A Case Study

A case study is now presented to illustrate many of the concepts and principles set out in the present chapter. It is loosely based on a matter in which the present author provided some advice.

5.3.1 The Story

A reasonably sized surgical hospital is situated on an important collector road in an urban area of a large city such as Sydney. The road is important since it forms a key link in the heavy transport operations that take place in the area. That is, the road carries a substantial volume of heavy trucks, particularly during the 7AM to 7PM time period each working day of the week. Both patients and staff have been complaining of the traffic noise to which the hospital is exposed from the collector road and the medical staff in particular began to become quite concerned at the effects of this noise on the welfare of convalescing patients.

A traffic noise committee was thus established by the hospital and comprised both medical and general staff of the hospital. The committee was charged with the tasks of identifying the causes of the hospital's traffic noise problem and suggesting one or more solutions. One of the first things the committee did was to meet together in the hospital grounds between the façade of the hospital building and the collector road. After watching and listening to the traffic for a while the committee formed the view that the major cause of the noise problem was the volume of heavy trucks on the road. After some further discussion they decided that if there were no trucks the traffic noise problem would be greatly improved. Although no member of the committee had any prior experience in either traffic engineering or acoustics they were convinced that the problem was quite straightforward and that they had come across the perfect solution.

The committee did realize, however, that it was likely to be very difficult to implement this solution. They could foresee objections from the road transport industry and further thought that convincing the Local Government Authority (LGA) responsible for the road to put in place a heavy truck ban on the road would not be easy. In presenting their recommendations to the hospital board the traffic noise committee also recommended that they monitor the traffic noise from the road before and after the ban was implemented. They thought that such a move would be appreciated by the LGA who would then be more likely to effect the truck ban. By collecting the traffic noise data the committee felt that they would be readily able to demonstrate the acoustic benefits of the truck ban. The committee's recommendation was accepted by the board, who then set about having it implemented.

What then followed was a well publicized series of meetings and political maneuverings that involved the hospital board, the LGA, the road transport industry, local residents and sundry members of State Parliament. All these took several weeks, after which the LGA finally decided to put in place a trial ban on heavy trucks in the collector for a period of 12 weeks. This trial was subject to the before/after traffic noise monitoring program being undertaken by the hospital's traffic noise committee as they had originally suggested. A decision as to whether the ban would be extended or made permanent by the LGA would depend on the outcomes of the monitoring program.

On hearing this decision the traffic noise committee immediately hired a sound level meter from a local supplier. They positioned the meter in the hospital grounds at about the location of their original meeting, turned it on and set it up to measure the L_{10} (1 h) and as instructed by the supplier. They then left the meter there to collect the "before" noise data. The results they obtained indicated that the L_{10} (1 h) turned out to be remarkably constant throughout the day and could be recorded as 71.8 dB(A). After the truck ban had been in place for about 10 weeks the committee felt it was then the time to conduct the "after" noise measurements. Again they hired the same instrument and set in up again at exactly the same location in the grounds as for the "before" measurements. This time the L_{10} (1 h) was also constant and they recorded a value of 70.0 dB(A). At this stage the committee became very concerned with such a small 1.8 dB(A) difference. By now they had done a little homework and realized that such a change would be regarded as negligible. On this basis they knew that the LGA would suspend the truck ban which they had worked so hard to achieve. Reconvening a meeting at the measurement location in the grounds to consider the results and their next steps, they listened again to the traffic noise. They agreed that, while it sounded a little different than before the truck ban, they could not say the noise sounded "quieter".

5.3.2   Review, Analysis and Commentary

This particular story is typical of many that have occurred in practice and which have resulted in outcomes that are generally seen to be unsatisfactory by those involved, such as the hospital's traffic noise committee. Firstly, the details of the traffic and ensuing acoustical situation along that collector road will now be examined. This will be done on the basis of the algorithms of CoRTN and you should refer to these and verify for yourself what follows. Then consideration will be given to what the committee did, in relation to the six issues set out above that must be addressed when conducting traffic noise measurements. From there, recommendations for rectifying the situation will be made.

Along the collector road the "before" traffic flows comprised 20% heavy vehicles. This is indeed high by Australian standards where busy arterial roads would typically experience heavy vehicle compositions in the range 5% to 10%. It would partly explain the committee's focus on heavy vehicles. Table 5.9 sets out the "before" and "after" traffic conditions and their associated noise levels. Also shown are how both the traffic and noise changed over time from the "before" to the "after" conditions. Note here that the "before" and "after" conditions refer to those times when the traffic noise committee undertook their measurements.

Examination and interpretation of Table 5.9 will explain what actually occurred along the collector road at the hospital site. What should be noted now is that this is a very typical series of events as far as changes in traffic and associated noise levels are concerned. Firstly, note the traffic conditions of speed, flow and heavy vehicle content operating during the "before" condition. The immediate effects of the truck ban were to eliminate heavy vehicles in the traffic composition and also therefore to reduce the traffic volume. Together these two changes would bring about a noise reduction of 4.6 dB(A) which may

TABLE 5.9 Traffic Conditions and Noise Levels at The Hospital

Traffic Conditions			L_{10} (1 h) (dB(A))	Change in L_{10} (1 h) (dB(A))	Comments
Speed (km/h)	Flow (Vehicle/h)	Heavy Vehicles (%)			
75	500	20	71.8	0	"Before" condition
75	400	0	67.2	− 4.6	Effect of truck ban
75	500	0	68.2	− 3.6	Increased traffic
85	500	0	69.2	− 2.6	Increased traffic and speed
85	600	0	70.0	− 1.8	"After" condition

be regarded as the real effect of banning the trucks in the collector road. It is a useful reduction that would be noticed by observers within the hospital site. While you might have expected a greater reduction, a key observation to be made in this example is the overall level of effort and difficulty experienced by the hospital to achieve such a reduction. As a general rule it is difficult to achieve substantial reductions in traffic noise by implementing one strategy alone.

However, not long after the truck ban was implemented and the volumes reduced, motorists realized that the collector road now provided an easier route through the area. Consequently the traffic volumes soon increased as shown in the third row of data in Table 5.9. This had the effect of slightly increasing the noise level so that the original noise reduction was reduced to 3.6 dB(A). Not long after the traffic speeds also increased, primarily because of the improved conditions when there were no heavy vehicles in the traffic. Again a small increase in noise occurred and the original reduction now became 2.6 dB(A). Finally the traffic volumes increased gradually to the conditions prevailing when the committee took their second, "after" measurements. It is clear in Table 5.9 how the committee found their outcome of a 1.8 dB(A) reduction.

The committee took only two noise measurements at the same location in the hospital grounds. It may be given that the weather conditions were the same during both measurements and that the hospital grounds were also in the same state on both occasions. Consequently it may be concluded in this particular instance that the noise propagation and attenuation factors were constant for both sets of measurements. So the variations in traffic noise can be attributed solely to the traffic conditions shown in Table 5.9. As the committee did not measure all the relevant traffic parameters, they were not able to explain their observed reduction of 1.8 dB(A). This is a key point of the present example. It illustrates the importance of taking great care when assessing environmental noise and proposing noise control strategies.

The example may be reviewed in terms of six issues related to the activities of the committee in measuring noise and assessing the noise control outcomes.

1. *Objectives and data applications.* The committee had a clear view of what and why they wanted to measure and how the measurements would be applied. They mistakenly thought that all they had to do was to measure noise alone and nothing else. This is a very common misconception in traffic noise situations and indeed in many environmental noise studies.

2. *Site and measurement locations.* The committee was fortunate that they decided to take both measurements at the same location. In this case the site was readily determined. Often this is not possible in practice because a variety of things might change at the chosen site between the two sets of measurements.

3. *Site details.* The committee did record relevant site details to the extent that they were able to conduct the second measurements at the same location as the first. However, if they had to take the second set of measurements at a slightly different location, further variability would have been introduced into their data because of the change in propagation conditions.

4. *Monitor noise.* This was done readily with a modern, user-friendly instrument capable of many functions and applications. Such instruments are capable of producing copious volumes of data which may appear impressive in the tables and diagrams of a report. However, they are of little value if

the six issues currently under review are neglected. It is a very good practice as far as traffic noise measurements are concerned to measure more than one noise index. In the present example it would be prudent to measure both the L_{10} (1 h) and the L_{eq} (1 h) simultaneously, which is well within the capability of modern instrumentation systems used for environmental noise studies. Having done this, the data could be assessed against Equation 5.1 given previously in the present chapter. Since this particular relationship holds only for road traffic noise, it is further evidence that the measured data ensue from traffic if the measured indices conform to the relationship. This can become an important matter where it is suspected that some traffic noise data might have been affected by some other sources of noise.

5. *Monitor traffic conditions.* A major mistake made by the committee was their failure to monitor the traffic conditions during both measurements. This has been explored in some detail already via the discussion surrounding Table 5.9. Again it is emphasized that such mistakes are common, not only in traffic noise cases but in the general areas of environmental noise measurement and assessment.

6. *Document results.* Presumably some form of documentation was finally produced by the committee and no doubt it would have made interesting reading, given their outcomes. The point here is that if all of the preceding five issues are addressed completely, the ensuing report will be technically sound and any recommendations will be supported by the robust evidence contained in the report.

If the six issues had been adequately addressed by the committee and complete, good quality data thus been obtained, the situation revealed in Table 5.9 would have emerged. Perhaps only the first and last lines of data would have been produced, but the intervening lines could have been readily predicted. Note that in some cases these predictions can be a little difficult, because values of some parameters have to be estimated and the accuracy of such estimates can vary. Nevertheless, it is usually found that these predictions allow reasonable interpretations of what has been occurring. Indeed, it is often unnecessary to determine exactly what took place between the two sets of measurements.

5.3.3 A Recommended Strategy

In any event, given what appears in Table 5.9, it would be reasonable to recommend the following strategy in order to effect the real benefits following the heavy vehicle ban.

1. Introduce a traffic management plan in the local area around the hospital, so that the collector road is no longer attractive to motorists. A key objective of such a plan would be to reduce the traffic volumes in the road back to, and hopefully less than, 400 vehicles/h.
2. Possibly as part of the traffic management plan, introduce a speed restriction and a speed limit in the road. In concert with current practices throughout many local government areas in Australia, a limit of 50 km/h has merit. Implementing such a condition may be impeded by a number of other factors, so a viable alternative might be 60 km/h.
3. Aim the strategies to achieve a noise reduction exceeding 6 dB(A).

You should verify for yourself that if the traffic volumes were to drop to 300 vehicles/h and that if a speed limit of 50 km/h was introduced and enforced, the resulting noise benefit would become 63.1 dB(A), a reduction of 8.7 dB(A). This probably represents the maximum possible reduction from traffic measures alone. If the traffic were maintained at 400 vehicles/h and a speed limit of 60 km/h achieved, the noise reduction would become 6.3 dB(A), and again you should verify this. Note how the various traffic parameters influence the overall noise level and how they must all be considered when determining the effects of any traffic-based noise control strategy.

5.4 Summary

This chapter commenced with an introduction to sound and noise, citing the common definition of noise as unwanted sound. It then introduced the concept of SPL and explained the logarithmic basis to this parameter, the unit of which is the deciBel (dB). A detailed and important treatment of manipulating

SPLs followed, which is fundamental to what would follow later concerning road traffic noise levels and noise control principles. Some information on human auditory response was then addressed, which demonstrated the reason behind quantifying noise in A-weighted dB. From there a generalized explanation of environmental noise, again of which road traffic noise in one type, was given. This was followed by a brief explanation of the generation and propagation of environmental noise along with a mention of noise exposure and impacts on the community.

Road traffic noise was then introduced and some of the sources of this particular noise discussed. How road traffic noise varies both microscopically and macroscopically throughout the course of a day was considered. A more detailed treatment of the sources and generation of road traffic noise then followed. These sources were categorized as being either Vehicle sources (such as the engine and tire/road interactions) or Non-vehicle sources (such as traffic conditions and road type). Particular emphasis was given to the important effects of pavement surface type on the generation of road traffic noise. An outline was then given of the principles involved in the measurement of road traffic noise and this was followed by a rather more detailed treatment of how road traffic noise might be estimated or predicted.

At this stage the chapter turned to how road traffic noise impacts might be assessed, especially within the broader context of EIA. The road traffic noise level criteria adopted in this process were also discussed in a general manner. This led on to a consideration of the control of road traffic noise and involved both noise control treatments and strategies and how these might be selected and evaluated. A case study completed the chapter and was configured to illustrate many of the concepts and principles covered in the chapter.

References

Australian Environmental Council, 1998. *Strategies For The Control of Traffic Noise Using Non-Vehicle Based Methods*. Australian Environmental Council, Canberra, ACT, Australia.

Australian Standards Association, 1984. *Acoustics — Methods for the measurement of road traffic noise*, AS 2072. Standards Australia, Homebush, NSW, Australia.

Australian Standards Association, 1989. *Acoustics — Description and measurement of environmental noise. Part 1: General procedures. Part 2: application to specific situations. Part 3: acquisition of data pertinent to land use*, AS 1055. Standards Australia, Homebush, NSW, Australia.

Australian Standards Association, 1989. *Acoustics — Road traffic noise intrusion — Building siting and construction*, AS 3671. Standards Australia, Homebush, NSW, Australia.

Beranek, L.L. 1988. *Acoustic Measurements*. Acoustical Society of America, Cambridge, MA.

Bies, D.A. and Hansen, C.H. 1988. *Engineering Noise Control — Theory And Practice*. Unwin Hyman Ltd, London, UK.

Brown, A.L. 1989. *Some simple transformations for road traffic noise scales*, Australian Road Research, Vol. 19, No. 4. Australian Road Research Board, Vermont South, Vic., pp. 309–312.

Buchta, E. and Vos, J., A field survey on the annoyance caused by sounds from large firearms and road traffic, *J. Acoust. Soc. Am.*, 104, 5, 2890–2902, 1998.

Environment Council of Alberta, 1980. Noise In The Human Environment, Vol. 1 & 2. Environment Council of Alberta, Edmonton, Alta., Canada.

Environment Protection Authority of NSW, 1990. *The State of The Environment Report*. EPA, Chatswood, NSW.

Environment Protection Authority of NSW, 1998. *Stationary Noise Source Policy*. EPA, Chatswood, NSW, Australia.

Environment Protection Authority of NSW, 1999. *Environmental Criteria For Road Traffic Noise*. EPA, Chatswood, NSW, Australia.

Fields, J.M. Reactions to environmental noise in an ambient noise context in residential areas, *J. Acoust. Soc. Am.*, 104, 4, 2245–2260, 1998.

Houghton, J. 1994. *Transport and the environment*, Eighteenth Report of The UK Royal Commission on Environmental Pollution. HMSO, London, England.

Huybregts, C.P. and Samuels, SE. 1998. New relationships between L_{10} and L_{eq} for road traffic noise. *Proceedings of InterNoise 98*, Christchurch, New Zealand.

International Standards Organisation, 1996. *Acoustics — description, assessment and measurement of environmental noise. Part 1 — Basic quantities and assessment procedures*, ISO/DIS 1996-1-WD 4 N-88. ISO, Geneva, Switzerland.

International Standards Organisation, 1997. *Acoustics — Method for measuring the influence of road surfaces on traffic noise — Part 1: the statistical passby method*, ISO 11819-1. ISO, Geneva, Switzerland.

Jones, H.W. 1979. Noise in The Human Environment, Vol. 2 and 3. Environment Council of Alberta, Edmonton, Alta., Canada.

Kinhill Engineers, 1990. *Environment Impact Statement — Proposed third runway for Sydney (Kingsford Smith) Airport*. Kinhill Engineers Pvt. Ltd, Chatswood, NSW.

Mason, K.D. 1993. *The future road transport noise agenda in the UK*, Volume 3 of the UK Environmental Foresight Project. HMSO, London, England.

Menge, C.W., Rossano, C.F., Anderson, G.S., and Bajdek, C.F. 1998. *FHWA traffic noise model — version 1 technical manual*, Report FHWA-PD-96-010. Federal Highways Administration, Department of Transportation, Washington, DC.

Miedema, H.M.E. and Vos, H., Exposure–response relationships for transportation noise, *J. Acoust. Soc. Am.*, 104, 6, 1998, 1998.

Ministry of Transport Denmark, 1991. *Noise barriers — a catalogue of ideas*, Road Data Laboratory Report 81. Road Directorate, Ministry of Transport, Herlev, Denmark.

Organisation for Economic Cooperation and Development, 1994. *Environmental Impact Assessment Of Roads*. OECD, Paris, France.

Organisation for Economic Cooperation and Development, 1995. *Roadside Noise Abatement*. OECD, Paris, France.

Roads and Traffic Authority NSW, 1991. *Noise Barriers and Catalogue of Selection Possibilities*. Roads and Traffic Authority, Sydney, NSW, Australia.

Roads and Traffic Authority of NSW, 1992. *Environment Manual Volume 2 — Interim Traffic Noise Policy*. RTA, Sydney, NSW, Australia.

Roads and Traffic Authority of NSW, 1994. *M5 East Motorway — Fairford Road to General Holmes Drive — Environmental Impact Statement*. RTA/NSW and Manidis Roberts Consultants, Sydney, NSW.

Roads and Traffic Authority of NSW, 2001. *Environment Noise Management Manual*. RTA, Sydney, NSW, Australia.

Samuels, S.E. 1982. *The generation of tyre/road noise*, ARRB Report ARR 121. ARRB-Transport Research, Vermont South, Vic..

Samuels, S.E. and Dash, D. 1996. Development of low noise pavements in Australia. *Proceedings of PIARC Road Surfacing Symposium*, Christchurch, New Zealand.

Samuels, S.E. and Huybregts, C.P. 2001. *Experiences with implementing the American FHWA traffic noise model in Australia*, Proceedings of Internoise 2001. International Institute of Noise Control Engineering, The Hague, The Netherlands, August, pp. 295–299.

Samuels, S.E. and Nichols, J. 1994. *Reducing traffic noise on cement concrete pavements*, Proceedings of IMEA Conference. Institute of Municipal Engineers, Sydney, NSW, Australia, pp. 65–68.

Samuels, S.E. and Parnell, J. 2003. A novel technique for allowing for the effects of pavement type in the prediction of road traffic noise. *Proceedings of 21st ARRB and 11th REAAA Conference*, Cairns, Queensland, Australia, May.

Saunders, R.E., Samuels, S.E., Leach, R., and Hall, A. 1983. *An evaluation of the UK DoE traffic noise prediction method*, Australian Road Research Board Research Report ARR 121. ARRB, Vermont South, Vic.

Sandberg, U. and Ejsmont, J.A. 2002. *Tyre/road Noise Reference book*. Informex, Kisa, Sweden.
Transportation Research Board, 1997. *Environmental issues in transportation*, Transportation Research Record No 1601. TRB, Washington, DC.
UK Department of Transport, 1988. *The Calculation of Road Traffic Noise*. HMSO, Welsh Office, UK.
Verband Der Automobilindustrie, 1978. *Urban Traffic And Noise*. VDA, Frankfurt am Main, Germany.

Further Reading

Australian Bureau of Statistics, 1997. *Australian Transport and The Environment*. ABS, Canberra, ACT, Australia.

Part B

Functional and
Structural Design
of Highways

6

Highway Geometric Design

R.L. Cheu
National University of Singapore
Republic of Singapore

6.1 Design Concept and Philosophy

6.1.1 Design Objectives

Highway geometric design refers to the calculations and analyses made by transportation engineers (or designers) to fit the highway to the topography of the site while meeting the safety, service and performance standards. It mainly concerns with the elements of the highways that are visible to the drivers and users. However, the engineer must also take into consideration the social and environmental impacts of the highway geometry on the surrounding facilities.

Usually, highway geometric design has the following objectives:

1. Determine, within the allowance permitted by the design standard and right-of-way, the routing of proposed highway.
2. Incorporate, within the design standard, various physical features of the road alignment to ensure that drivers have sufficient view of the road (and obstacles) ahead for them to adjust their speed of travel to maintain safety and ride quality.
3. Provide a basis for the highway engineers to evaluate and plan for the construction of a section of the proposed highway.

In the United States, the American Association of State Highway and Transportation Officials (AASHTO) has published a series of standards and guidelines for highway geometric design. The content of this chapter is mainly based on the latest design policy (AASHTO, 2001).

6.1.2   Design Considerations

To meet the objective of fitting the highway to site topography and yet satisfy the safety, service and performance standards, the following considerations have to be properly addressed in the design process.

- Design speed
- Design traffic volume
- Number of lanes
- Level of service (LOS)
- Sight distance
- Alignment, super-elevation and grades
- Cross section
- Lane width
- Horizontal and vertical clearance

The above factors are interactive as drivers react to the combination of them (and among themselves) to produce the observable operational performance of the highway.

6.1.3   Access Control and Management

Access control refers to the regulation of road users' right to use the properties and facilities adjacent to the highway. Access management involves providing and managing the access to lane development while minimizing the interruption of traffic flow on the highway and ensure safety of all users.

Vehicles (and users) on a highway wish to access the adjacent properties needs to turn off from the highway. Full control of access means that preference is given to through traffic by providing access connections by means of ramps at only selected points or with selected local roads, and by prohibiting crossing at grade. Full control access is used at freeways. In partial control access, connections between through traffic and local properties are provided at at-grade intersections or grade-separated interchanges, and private driveways. The access points, although relatively more frequent and more direct, are still limited to designated locations. A highway agency can control and manage access by means of statutory control, land-use regulations, geometric designs and driveway regulations. Geometric design features, such as curbs, medians, frontage roads and channelization at intersections are ways to facilitate or limit control access. For details, refer to Koepke and Levinson (1992).

6.1.4   Design Process

A highway designer is concerned with at least four major areas of design at different stages of project planning and design phases: (1) location design; (2) alignment design; (3) cross sectional design; and (4) access design.

Location design takes place at the earlier stage of project planning. It refers to the macro-level routing of a planned highway connecting two points through the existing highways, communities, natural terrain. Normally, information such as lane-use master plan; existing and projected population distribution; survey maps; maps of existing infrastructure; geology, ecological, biological, and environmental information; and aerial photographs are among the essential inputs. Inputs are also sought from civil engineers, planners, economists, ecologists, sociologists, environmental experts, and lawyers. With all the necessary inputs, several potential routes are drawn up by the designer on a contour map (e.g., on a scale of 1:10,000). The designer then goes through the various iterative and consultative steps with the stake holders to modify and select the most feasible layout. The consultative process is perhaps the most time consuming, which may take several months. A more detailed site survey is then carried out to locate the key control points of the alignments, in terms of geo-coordinates and elevations. The designer then proceeds with the detailed alignment, cross sectional and assess design.

6.2 Highway Alignment Design

6.2.1 Overall Alignment

The alignment of a highway is a three-dimensional problem because the highway itself negotiates through the terrain in connecting two points. The highway may be visualized as segments of connected horizontal and vertical curves (or their combination). The alignment of a highway is best represented by its centre line in a three-dimensional coordinate system (e.g., longitude, latitude, and elevation).. However, for the ease of interpretation of the construction drawing, the convention of plan and profile views has been adopted.

The plan view gives the horizontal alignment of a highway. The length of the highway is measured along the plan view, on a horizontal plane. The length is expressed in terms of distance from a reference station, in terms of stations. Each station is 100 m. A highway normally starts from a fixed reference station. The distance from the reference station, together with the direction from the reference station or subsequent stations, spells out the horizontal alignment.

The vertical alignment (including the gradients and vertical curves) are represented in a profile view. The profile view is the view along the length (including the true length of horizontal curve) of the highway. The elevations of all the points at regular intervals or when necessary are specified in the profile view.

6.2.2 Horizontal Alignment

A horizontal curve provides the directional transition on the horizontal plane, between two straight sections of the highway running in different directions. Horizontal curves are expressed as circular curves with constant radii, or successive curves with different radii. Spiral curves with constant rate of change of radius can also be found. A curve can be described by its radius or by its degree of curvature. This chapter covers horizontal curves with constant radii.

Figure 6.1 shows the properties of a curve with a constant radius (R) connecting two straight sections of a highway. The curve starts at point of curvature (PC), ends at point of tangent (PT). The point of intersection (PI) is the intersecting point if the two straight lines are extended. Δ is the central angle of the curve, expressed in degrees.

The length of tangent (T) is

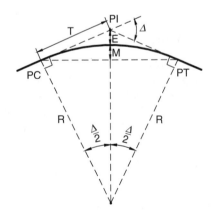

FIGURE 6.1 Horizontal curve.

$$T = R \tan\left(\frac{\Delta}{2}\right) \tag{6.1}$$

The middle ordinate M is

$$M = R\left[1 - \cos\left(\frac{\Delta}{2}\right)\right] \tag{6.2}$$

The external distance E is

$$E = R\left[\frac{1}{\cos\left(\dfrac{\Delta}{2}\right)} - 1\right] \tag{6.3}$$

The degree of curvature (D) is the central angle subtended by a 100 m arc of the curve.

$$D = \frac{5729.6}{R} \tag{6.4}$$

The length of the curve (L) is

$$L = 100\frac{\Delta}{D} \tag{6.5}$$

Example 6.1

A horizontal curve is to be designed to connect two straight sections of a highway. The PI is originally determined to be stn 180 + 00, and Δ is 30°. If the radius of the curve is fixed at 403.15 m, what are the station numbers of PC and PT, and the length of the curve?

$$T = R\tan\left(\frac{\Delta}{2}\right) = 403.15x\tan\left(\frac{30°}{2}\right) = 108.02 \text{ m}$$

$$PC = \text{stn}(180 + 00) - \text{stn}(1 + 08.02) = \text{stn}(178 + 91.98)$$

$$L = 100\frac{\Delta}{D} = 100\Delta\left(\frac{R}{5729.6}\right) = 100(30)\left(\frac{403.15}{5729.6}\right) = 211.09 \text{ m}$$

$$PT = \text{stn}(180 + 00) + \text{stn}(2 + 11.09) = \text{stn}(182 + 11.09)$$

6.2.3 Vertical Alignment

A vertical curve provides a smooth transition between two tangent grades. There are two types of vertical curves: crest vertical curves and sag vertical curves. Example profiles of crest and sag vertical curves are shown in Figure 6.2, with the initial grade G_1, final grade G_2, and their signs.

As a departure from the horizontal curve, the points of curvature, intersection and tangent of a vertical curve are denoted by *PVC*, *PVI* and *PVT*, respectively. The length of curve L is the distance between *PVC*

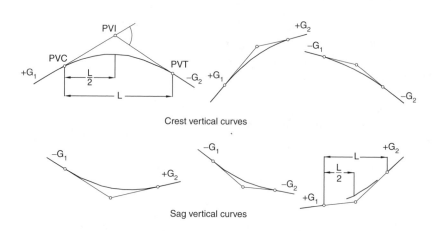

Crest vertical curves

Sag vertical curves

FIGURE 6.2   Crest vertical curves and sag vertical curves.

and *PVT* measured along the horizontal plane. The *PVI* is at the midpoint between *PVC* and *PVT* along the horizontal plane.

In the specification of a highway's vertical alignment, the elevations of points along the highway's center line are required. The vertical curves are parabolic in form. The parabolic curve has been used for the purpose of calculation because it has a constant rate of change of slope and equal curve tangents on both ends of the curve.

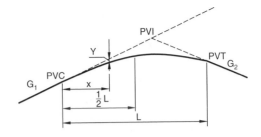

FIGURE 6.3 Profile of a crest vertical curve.

A crest vertical curve is shown in Figure 6.3. Note that an uphill is expressed in a positive gradient while a downhill is expressed in a negative gradient.

A vertical curve starts at the point of vertical curvature (*PVC*) and ends at the point of vertical tangent (*PVT*). The length of highway between *PVC* and *PVT* is *L*. The initial and final grades are denoted by G_1 and G_2, respectively, expressed in %. Based on the equation of a parabolic curve, the vertical offset y at any distance x from the projected initial gradient is

$$y = \frac{G_2 - G_1}{100L} x^2 \tag{6.7}$$

Negative values of y mean a downward offset from the projected tangent from *PVC* (as in the case of crest vertical curves) while positive values of y mean a upward offset from the projected tangent from *PVC* (in sag vertical curves).

The highest or lowest point on the curve is given by

$$x' = \frac{LG_1}{G_1 - G_2} \tag{6.8}$$

$$y' = \frac{LG_1^2}{200(G_2 - G_1)} \tag{6.9}$$

Example 6.2

A highway must traverse a 6% followed by a -2% grade. The length of the crest vertical curve is 2040 m. Calculate the elevation for the first 600 m of the vertical curve at 100 m intervals, and the highest point of the curve.

$$\text{Given that } L = 2040 \text{ m}, \ y = \frac{G_2 - G_1}{100L} x^2 = \frac{-2 - (6)}{100(2040)} x^2 = -3.9216 \times 10^{-5} x^2$$

x (m)	100	200	300	400	500
y (m)	-0.392	-1.568	-3.529	-6.275	-9.804

The negative value means that the elevation is measured downward from the projected original gradient, based on G_1.

The highest point on the vertical curve is

$$y' = \frac{LG_1^2}{200(G_2 - G_1)} = \frac{2040(6)^2}{200(-2 - 6)} = -45.900 \text{ m}$$

below the projected gradient based on G_1, which happens at

$$x' = \frac{LG_1}{G_1 - G_2} = \frac{2040(6)}{6 - (-2)} = 1530 \text{ m from PC}$$

6.3   Design Control and Criteria

6.3.1   Road Types

The basic functional types of roads are locals, collectors, arterials and freeways. Two major considerations in the classification of highway functional types are access to land use and mobility. On the two extremes, the design of local streets emphasizes access with little consideration for mobility, while the design of freeways emphasizes mobility with limited access. The design of collectors and arterials falls in between, with collectors emphasizes more for access and arterials favors mobility.

6.3.2   Design Vehicles

Key controls in highway geometric design are the physical and dynamical characteristics of vehicles using the highway. Considering the many types of vehicles in the traffic stream, it is necessary to establish several classes of vehicles, and select a representative vehicle within each class for design use. These selected vehicles are termed design vehicles, and their dimensions, weights and operating characteristics are essential in the establishment of design control criteria. For the purpose of highway geometric design, each design vehicle has larger physical characteristics than most of the vehicle in its class.

There are generally four classes of design vehicles: (1) passenger cars, (2) buses, (3) trucks, and (4) recreational vehicles. The passenger car category includes sport utility vehicles, minivans, vans, and pickup trucks. The bus and truck categories include buses and trucks of all sizes, respectively. The highway designer should exercise his judgment in selecting the appropriate design vehicle for design control, based on the intended use of the facility. For example, the design vehicle from the passenger car category is adequate for the design of parking lots and their access roads. On the other hand, a city transit bus should be used for the design of a street in the city along bus route, with little or no truck traffic.

Turning radius limits the design of horizontal curves. Important vehicle characteristics that affect the minimum turning radius are: minimum center line turning radius, wheelbase, track width, and out-of-track width. AASHTO has provided the templates for turning paths of 17 design vehicles traveling at 15 km/h. The minimum design turning, center line turning and minimum inside radius are listed in Table 6.1.

6.3.3   Driver Characteristics

Geometric design of a highway should consider users, especially drivers' performance limits. There are limits to a driver's vision, perception, reaction, concentration, and comfort that could impact the highway safety and operating efficiency.

When driving, most drivers receive information visually from their views of the roadway alignment, markings and signs. They do receive other information through vehicle feedback from the suspension system and steering control, and roadway noise.

The information received by a driver needs time to be processed before a response action takes place. A well-known study on the brake-reaction time has been made by Johansson and Rumar (1971). They reported that when an event is expected, the driver's reaction time has an average value of 0.6 sec. For an unexpected event, the average reaction time is 0.8 sec. The average brake-reaction time of a driver (including decision time), is 2.5 sec. This is dependent on the driver's alertness. Brake-reaction time is important in determining sight distance in highway geometric design. Koppa (2000) has summarized the results obtained from recent studies on brake-reaction time. These findings are consistent with those obtained by Johansson and Rumar. Readers may refer to Koppa (2000) for more details.

Driver expectancies are built up over time, with consistent road design. Unusual or unexpected geometric design or event always leads to longer reaction and response time. The geometric design of highway should be in accordance with the driver's expectation.

TABLE 6.1 Minimum Turning Radii of Design Vehicles

Design Vehicle Type	Passenger Car	Single Unit Truck	Inter-City Bus (Motor Coach)		City Transit Bus	Conventional School Bus (65 Passenger)	Large[a] School Bus (84 Passenger)	Articulated Bus	Intermediate Semi-Trailer	Intermediate Semi-Trailer
Symbol	P	SU	BUS-12	BUS-14	CITY-BUS	S-BUS11	S-BUS12	A-BUS	WB-12	WB-15
Minimum Design Turning Radius (m)	7.3	12.8	13.7	13.7	12.8	11.9	12.0	12.1	12.2	13.7
Center-line[b] Turning Radius (CTR) (m)	6.4	11.6	12.4	12.4	11.5	10.6	10.8	10.8	11.0	12.5
Minimum Inside Radius (m)	4.4	8.6	8.4	7.8	7.5	7.3	7.7	6.5	5.9	5.2

Design Vehicle Type	Interstate Semi-Trailer	Interstate Semi-Trailer	"Double Bottom" Combination	Triple Semi-Trailer/Trailers	Turnpike Double Semi-Trailer/Trailer	Motor Home	Car and Camper Trailer	Car and Boat Trailer	Motor Home and Boat Trailer	Farm Tractor w/ One Wagon[c]
Symbol	WB-19[d]	WB-20[e]	WB-20D	WB-30T	WB-33D[d]	MH	P/T	P/B	MH/B	TR/W
Minimum Design Turning Radius (m)	13.7	13.7	13.7	13.7	18.3	12.2	10.1	7.3	15.2	5.5
Center-line[b] Turning Radius (CTR) (m)	12.5	12.5	12.5	12.5	17.1	11.0	9.1	6.4	14.0	4.3
Minimum Inside Radius (m)	2.4	1.3	5.9	3.0	4.5	7.9	5.3	2.8	10.7	3.2

Note: Numbers in table have been rounded to the nearest tenth of a meter.

[a] School buses are manufactured from 42 to 84 passenger sizes. This corresponds to wheelbase lengths of 3.350 to 6,020 mm, respectively. For these different sizes, the minimum design turning radii vary from 8.78 to 12.01 m and the minimum inside radii vary from 4.27 to 7.74 m.

[b] The turning radius assumed by a designer when investigating possible turning paths and is set at the centerline of the front axle of a vehicle. If the minimum turning path is assumed, the CTR approximately equals the minimum design turning radius minus one-half the front width of the vehicle.

[c] Turning radius is for 150 to 200 hp tractor with one 5.64 m long wagon attached to hitch point. Front wheel drives is disengaged and without brakes being applied.

[d] Design vehicle with 14.63 m trailer as adopted in 1982 Surface Transportation Assistance Act (STAA).

[e] Design vehicle with 16.16 m trailer as grandfathered in with 1982 Surface Transportation Assistance Act (STAA).

In recent years, there has been increased concern for older drivers. The percentage of older drivers among the driving population has increased over the years. Older drivers tend to have longer reaction time, and this should be reflected in the design.

6.3.4   Design Volume

A highway should be designed for a traffic volume and characteristics of the traffic stream it serves. The volume and vehicle composition directly affects the number of lanes, lane width, alignments and grades. The design volume and vehicle composition are normally given. These figures are normally the projected numbers given by a transportation planning tool.

The volume may come in the form of average daily traffic or peak-hour traffic. In any case, it is recommended that the design volume should be taken as the 30th highest hourly volume that a highway is expected to handled in a year. AASHTO (2001) has provided several factors for converting average daily traffic into 30th highest hourly volume.

Vehicle of different sizes and weights have different performance characteristics. A truck is equivalent to several passenger cars. In geometric design, the passenger car equivalent factor is dependent on the grade and other factors. For geometric design, all vehicles are classified into passenger cars and trucks. Trucks are defined as vehicles having more than 4000 kg of gross vehicle weight and having dual tires on at least one rear axle. Truck traffic is expressed as the percentage of total traffic during the design hour.

6.3.5   Design Speed

The design speed is the speed selected to be used in the design calculations, in determining the geometric dimensions of the highway. The selected design speed should be consistent with the functional classification of the highway, its surrounding land use, topography and driver's anticipated operating speed. The designer should adopt a highest practical speed that will result in a conservative geometric design. A design speed of 110 km/h should be used for freeways, expressways, and other rural highways. Urban arterials should have design speeds of between 30 to 70 km/h. The lower range of design speed should be used for residential streets, collectors, and downtown crowded areas, while speeds in higher range are applied to suburban arterials. The actual operating speed of an arterial depends on the spacing of the intersections and types of traffic control. The selected design speed should be in increment of 10 km/h.

6.4   Design Elements

6.4.1   Sight Distance

Sight distance is the roadway ahead that is visible to the driver. Various sight distance criteria exist in highway geometric design to provide drivers with sufficient warning of potential obstacle or conflict ahead.

6.4.2   Stopping Sight Distance

Stopping sight distance is the distance traveled during a driver's brake reaction time plus the braking distance for the vehicle to come to a complete stop. The equation to compute stopping sight distance without vehicle skidding is

$$d = Vt + \frac{V^2}{2(a + Gg)} \tag{6.10}$$

where V is the design speed, a the constant deceleration rate, G the grade (in decimal, positive value for upgrade, and negative value for downgrade), and $g = 9.81$ m/sec^2. AASHTO recommends that $t = 2.5$ sec and $a = 3.4$ m/sec^2 be used in determining the minimum stopping sight distance.

6.4.3   Decision Sight Distance

Decision sight distance is the distance needed for a driver to detect and perceive an obstacle or information, and select an appropriate maneuver. This is important when a driver is approaching a traffic control device, or posted information signs. Because decision sight distance is for drivers to a maneuver or evasive action rather than just to stop, it is greater than stopping sight distance. The decision sight distance for change in speed, path or direction on rural, suburban, and urban road may be calculated from

$$d = Vt \tag{6.11}$$

AASHTO (2001) recommends a range of $10.2 \le t \le 14.5$ sec.

6.4.4   Passing Sight Distance on Two-Lane Road

In a two-lane road, the sight distance required when pulling out to the opposing lane to pass a slow moving vehicle is critical in determining where no-passing zone should exist. The passing sight distance is the sum (Figure 6.4):

$$d = d_1 + d_2 + d_3 + d_4 \tag{6.12}$$

where d_1 is the initial maneuver distance, which is the sum of distances traveled during perception and reaction time plus the initial period of acceleration until the vehicle encroaches the passing lane. The corresponding time for this initial maneuver is t_1. d_1 is given by the expression:

$$d_1 = t_1\left(v - m + \frac{at_1}{2}\right) \tag{6.13}$$

in which v is the average speed of the passing vehicle, and m is the relative speed of the passed and passing vehicles, and a is the average acceleration. d_2 is the distance traveled while the vehicle is occupying the passing lane. The corresponding time is t_2.

$$d_2 = vt_2 \tag{6.14}$$

d_3 is the clearance length, for margin of safety between the passing and opposing vehicles. AASHTO recommends that $30 \le d_3 \le 90$ m for $56 \le v \le 100$ km/h. d_4 is the distance traveled by the opposing

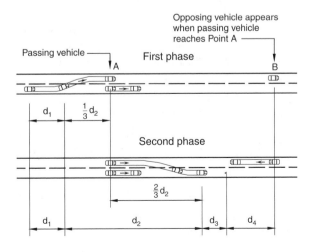

FIGURE 6.4   Passing sight distance for two-lane highway.

vehicle during the passing maneuver during time $t_1 + t_2$. AASHTO recommends that the opposing vehicle may be assumed to travel with speed v, and d_4 may be taken as

$$d_4 = \frac{2}{3} d_2 \tag{6.15}$$

6.4.5   Criteria for Measuring Sight Distance

Other than stopping sight distance, decision sight distance and passing sight distance it is assumed that an object of specific size or height is continuously visible to the driver. The distance is dependent on the height of the driver's eye above the road surface, the object height (or size) above the road surface, and the height and lateral position of sight obstructions within the driver's line of sight. These have effect on the design of horizontal and vertical curve, which will be covered later.

For sight distance calculations, the following values usually apply:

- Height of the driver's eye above road surface = 1.080 m for passenger cars, 2.330 m for trucks
- Height of object: 600 mm for stopping sight distance, 1.080 m for passing sight distance

The considerations of sight distance for horizontal and vertical curves will be discussed later in this chapter.

6.4.6   Horizontal Alignment

The objectives of horizontal alignment design is to provide a smooth transition between two straight road sections that also provide proper cross sectional drainage, ensure vehicle/driver safety while traveling at a design speed. Vehicle safety is safeguarded by providing adequate stopping sight distance and incorporating provision to prevent vehicle overturning and skidding.

The simplest curve that connects the tangents of two straight road sections is a curve with a single, constant radius as shown in Figure 6.1. The above discussion has so far assumed that the radius of the curve has been determined. The radius of a horizontal curve (R) depends on the design speed, super-elevation, friction factor, and sight distance as discussed below.

6.4.7   Radius of Horizontal Curve

The design of horizontal curves should be based on a combination of design speed, curvature, and super-elevation, subject to the laws of physics and limitations of human comfort and tolerance.

When a vehicle with weight W is traveling at speed v in a horizontal curve with radius R, the vehicle, driver and passengers experience a centripetal force of magnitude $Wv^2/(gR)$ towards the center of the curvature. This acceleration is to some extent countered by the force exerted by the weight onto itself (N) and the frictional force between the tire and pavement (F). Figure 6.5 shows the cross sectional view on the forces experienced by a vehicle. The governing equation, based on the law of mechanics, is

$$\frac{e+f}{1-ef} = \frac{v^2}{gR} \tag{6.16}$$

where:

- e = rate of super-elevation, in decimal;
- f = side friction factor between the tire and the pavement surface;
- v = vehicle speed.

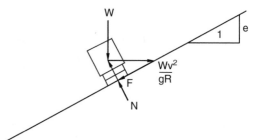

FIGURE 6.5   Super-elevation.

R = radius of horizontal curve.

g = gravitational constant = 9.81 m/sec^2.

The designer should use the above governing equation to select a suitable combination of e and R that satisfies v. There are, however, practical limitations of e and f. Firstly, $e \leq 0.12$ (almost 1 in 8). High e value should only be used on low volume road, and when no snow or ice exists. Most common values are $e \leq 0.04$ to $e \leq 0.06$. Studies have shown that f depends on tire and pavement conditions and v, with higher speeds corresponding to lower f. AASHTO recommends the use of $f = 0.10$ for $v = 110$ km/h to $f = 0.15$ for $v = 60$ km/h. When the product of e and f becomes negligible, the above equation may be re-written into conservative form of

$$R \geq \frac{v^2}{(e+f)g} \tag{6.17}$$

which governs the minimum horizontal turning radius.

Example 6.3

What would be shortest radius for a horizontal curve with a design speed of 110 km/h with super-elevation of 1/20?

Given that $e = 1/20 = 0.05$, and $v = 100$ km/h = 30.50 m/s, and assume that $f = 0.10$

$$R \geq \frac{v^2}{(e+f)g} = \frac{30.50^2}{(0.05+0.10)9.81} = 632.18 \text{ m}$$

6.4.8 Sight Distance on Horizontal Curve

When a driver travels along a horizontal curve, his sight distance is limited by a physical obstruction, such as sidewall, slope or building, at the inside of the curve. Figure 6.6 illustrates this point, where M is the offset from the center line of the inside lane to the obstruction. For design purpose, the stopping sight distance is the length of curve along the center line of the inside lane, where the vehicle will be traveling or braking (with the vehicle in full control).

For stopping sight distance (S) shorter than the length of the curve (L), the following equation applies

$$M \geq R\left[1 - \cos\left(\frac{90S}{\pi R}\right)\right] \tag{6.18}$$

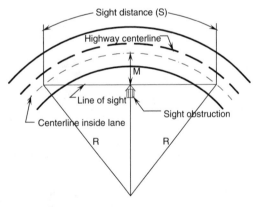

FIGURE 6.6 Sight distance on horizontal curve.

This equation may be used to calculate minimum offset M, when locating any infrastructure that may pose as an obstruction to the drivers. The following equation may be used to check the adequacy of stopping sight distance, if M is fixed:

$$S \geq \frac{\pi R}{90}\left[\cos^{-1}\left(\frac{R-M}{R}\right)\right] \tag{6.19}$$

Example 6.4

A sound wall is to be constructed at the edge of shoulder, along the inside of a horizontal curve of an urban freeway. The inside lane is 3.8 m wide, with a shoulder of 1.20 m. The radius of the curve, measured up to the outer edge of the shoulder is 45 m. Determine the sight distance of this section of the curve with the sound wall.

$$\text{Given } M = 1.20 + 3.80/2 = 3.1 \text{ m}$$

$$R = 45 + 1.20 + 3.80/2 = 48.10 \text{ m}$$

Assume $L > S$

$$S \geq \frac{\pi R}{90}\left[\cos^{-1}\left(\frac{R - M}{R}\right)\right] = \frac{\pi(48.10)}{90}\left[\cos^{-1}\left(\frac{48.10 - 3.1}{48.10}\right)\right] = 34.73 \text{ m}$$

6.4.9 Vertical Alignment

The purpose of vertical alignment design is to determine the elevation of selected points along the roadway, to ensure proper drainage, safety, and ride comfort.

6.4.9.1 Crest Vertical Curve

The stopping sight distance is the controlling factor in determining the length of a crest vertical curve. Figure 6.7 shows the design concept and geometry. The minimum length of L is such that when the driver of a vehicle climbs over the crest he/she has enough stopping distance if there is a 150 mm object on the road. In this case, height of the driver's eye (h_1) and height of the object (h_2) are two important inputs. Other controlling factors are driver's stopping sight distance (S), absolute change in gradient before and after the crest curve ($A = |G_1 - G_2|$, in %). The design equations are

$$L = \frac{AS^2}{100(\sqrt{2h_1} + \sqrt{2h_2})^2} \qquad \text{for } L \geq S$$

$$L = 2\left[S - \frac{100(\sqrt{h_1} + \sqrt{h_2})^2}{A}\right] \qquad \text{for } L \leq S \qquad (6.20)$$

The 100 in the above equations are to convert A from % into decimals. Typical design height for h_1 and h_2 are 1.080 m and 0.600 m, respectively.

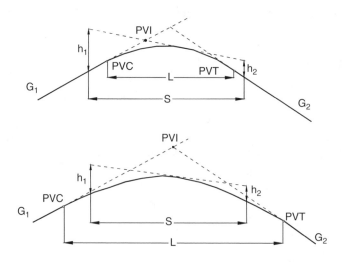

FIGURE 6.7 Sight distance for crest vertical curve.

For night driving on a highway without street lighting, a driver's vision is limited by the area illuminated by the vehicle's headlight. At higher speed, the distance covered by the headlight is smaller than the stopping sight distance. In this case, $h_1 = 0.600$ m (equivalent to the height of the headlights) should be used.

For passing sight distance (in the section of the vertical curve is designed as a passing zone), the object to be detected ahead is the opposing vehicle. In this case $h_2 = 1.080$ m should be used.

Example 6.5

A vertical curve is to be constructed between a 3.5% grade and a -4% grade. The required sight distance is 300 m. The dangerous object is considered to be on the pavement surface, and the driver's eye level is at 1.05 m above the pavement surface. Determine the length of the vertical curve that will satisfy the sight distance requirement.

Given $h_1 = 1.05$ m, $h_2 = 0.00$ m, $S = 300$ m, $G_1 = 3.5\%$, $G_2 = -4^a\%$

We have $A = |G_1 - G_2| = 7.5\%$

First assume $L > S$, we may use

$$L = \frac{AS^2}{100(\sqrt{2h_1} + \sqrt{2h_2})^2} = \frac{7.5(300)^2}{100(\sqrt{2(1.05)} + \sqrt{2(0)})^2} = 3214.29 \text{ m}$$

which is $>S$.

If we assume $L < S$, then

$$L = 2\left[S - \frac{100(\sqrt{h_1} + \sqrt{h_2})^2}{A}\right] = 2\left[300 - \frac{100(\sqrt{1.05} + \sqrt{0})^2}{7.5}\right] = 572.00 \text{ m}$$

which violates the assumption of $L < S$. Therefore we should use $L = 3214.29$ m.

6.4.9.2 Sag Vertical Curve

Figure 6.8 shows the driver's sight limitation when approaching a sag vertical curve. The problem is more obvious during the night time when the sight of the driver is restricted by the area projected by the headlight beams of his/her vehicle. Hence, the angle of the beam from the horizontal plane is also important. This design control criteria is known as headlight sight distance. The headlight height of $h = 0.600$ m and upward angle for the headlight projection cone of $\beta = 1^0$ is normally assumed. The governing equations are

$$L = \frac{AS^2}{200(h + S \tan\beta)} \qquad \text{for } L \geq S$$

$$L = 2S - \frac{200(h + S \tan\beta)}{A} \qquad \text{for } L \leq S \qquad (6.21)$$

A driver may experience discomfort when passing a vertical curve. The effect of discomfort is more obvious on a sag vertical curve than a crest vertical curve with the same radius, because the gravitational and centripetal forces are in the opposite directions. Some of the ride discomfort may be compensated by combination of vehicle weight, suspension system and tire flexibility. The following equation has been

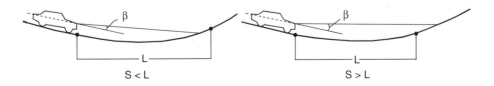

FIGURE 6.8 Sight distance for sag vertical curve.

recommended by AASHTO as the minimum length of a vertical curve that will provide satisfactory level of ride comfort

$$L = \frac{Av^2}{395} \tag{6.22}$$

The curve length to satisfy this criterion is usually about half the length that is needed to meet the stopping sight distance criteria.

6.4.9.3   Sight Distance at Undercrossings

Sight distance on the roadway that is the lower tier of a grade separated intersection should also consider obstruction of the sight pose by the structure. Most of the roadways at the lower tier of an interchange are designed as sag vertical curves. Figure 6.9 illustrates a typical design situation.

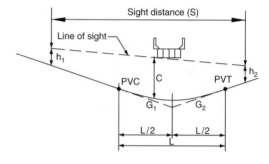

FIGURE 6.9   Sight distance at undercrossing.

The general equation for a sag vertical curve length at an undercrossing is

$$L = 2S - \frac{800\left[C - \dfrac{h_1 + h_2}{2} \right]}{A} \qquad \text{for } L \leq S$$

$$L = \frac{AS^2}{800\left[C - \left(\dfrac{h_1 + h_2}{2} \right) \right]} \qquad \text{for } L \geq S \tag{6.23}$$

where C is the vertical clearance from the road surface to the bottom of the obstructing structure.

The above equations have several applications. In deciding the length of the vertical curve for a given A, the design may select the sight distance S to cater for, with appropriate h_1 and h_2 to calculate L. The designed curve length should be at least L. On other hand, given a designed curve length L, A and C, the designer may check whether S satisfies the sight distance criteria (with corresponding h_1 and h_2). Also, given L, A, S, h_1 and h_2, the designer may use these equations to decide the minimum clearance C.

Example 6.6

Determine the minimum length of curve required to connect a descending 4% grade to an ascending 3% grade. The vertical clearance should be 5.1 m and the required sight distance is 300 m. The height of eye for a commercial vehicle is 1.83 m and the hazardous object is 0.46 m above the pavement surface.

Given $h_1 = 1.83$ m, $h_2 = 0.46$ m, $S = 300$ m, $C = 5.1$ m, $G_1 = -4\%$, $G_2 = +3^a\%$

We have $A = |G_1 - G_2| = 7\%$.

First assume $L \leq S$, we may use

$$L = 2S - \frac{800\left[C - \dfrac{h_1 + h_2}{2} \right]}{A} = 2(300) - \frac{800\left[5.1 - \dfrac{1.83 + 0.46}{2} \right]}{7} = 148.00 \text{ m}$$

which is $< S$. Therefore our assumption is correct.

On the contrary, if we assume $L \geq S$ and use

$$L = \frac{AS^2}{800\left[C - \left(\dfrac{h_1 + h_2}{2} \right) \right]} = \frac{7(300)^2}{800\left[5.1 - \left(\dfrac{1.83 + 0.46}{2} \right) \right]} = 199.12 \text{ m}$$

which violates our assumption of $L \geq S$.

6.4.9.4   Other Design Considerations for Vertical Curves

AASHTO has recommended a few design considerations for vertical curves based on driver comfort, expectation, and aesthetics:

- A smooth and gradual change in grade along the vertical curve is preferred over several short sections at constant grades.
- The "roller-coaster" type of curve, i.e., series of alternate crest and sag vertical curves should be avoided. This is because drivers will experience an even greater degree of discomfort when there is a sudden change in the direction of the centripetal force.
- The effect of downgrade will lead to increasing vehicle speed. The effect of this on the design speed and the impact on safety should be considered.
- Drainage at the lowest point of a sag vertical curve should be designed properly.

6.4.10   Combination Curves

When a section of the highway requires both horizontal and vertical curves in its alignment, the curves should not be designed independently. Poor combinations of horizontal and vertical may have negative effect on driving comfort, reduce the operational efficiency and in the worst case, affect safety.

The following points should be noted when combining horizontal and vertical curves:

- Curvature (or radius) and grade should be in proper balance at the chosen design speed, to counter the centripetal force act on the vehicles and drivers.
- Gradual change in grades and radius are preferred than successive short sections each with a constant radius and grades, to avoid humps which are visible to drivers, and may impede the operational efficiency.
- Sharp horizontal curve should not be placed at the crest of a vertical curve. This is because drivers are not able to see a change in horizontal alignment when climbing a vertical curve. This can be avoided if the length of the horizontal curve is greater than that of the vertical curve.
- Sharp horizontal curve should not be placed at the low point of a sag vertical curve. This is because a driver approaching the lowest point on the downgrade at night has his/her vision limited by the highlight coverage. Even in daylight, the view ahead is often distorted.
- In a two-lane rural highway, it is necessary to provide passing zones at regular intervals. In this case, passing sight distance, instead of stopping sight distance should be use, and the curves are normally of greater lengths than required.
- The radius of horizontal curves should be as big as possible, and grades of vertical curves should be as flat as possible, when the highway is approaching an intersection.

6.4.11   Cross Section

A typical cross section of a highway consists of the following components: traveled way (traffic lanes), shoulders (on both edges, paved or unpaved). Important elements in the geometric design are: cross slope of travel lane, lane width, width and slope of shoulder, and curb (if it is used).

6.4.11.1   Type of Pavement Surface

For the purpose of cross section in highway geometric design, AASHTO classifies pavements into high-type and low-type. High-type pavements retain their shape and do not ravel at the edge if placed on a stable subgrade. Their surfaces are usually smooth. Low-type pavements have a tendency toward raveling and are usually used for low volume road. This type of pavement is normally constructed from gravel, earth, or crushed stone.

6.4.11.2   Cross Slope

Undivided highways on tangents, or on flat curves, have two types of cross slope designs: (1) a crown or high point in the middle and a cross slope downward toward both edges; (2) a high point at one edge, and a cross slope across slope across the entire width.

On divided highway, each one-way roadway may be treated as an undivided highway as mentioned above. Figure 6.10 shows a few possible designs. A cross section with each roadway crowned separately provides a more rapid drainage of storm water, but at the expense of more drainage inlets and underground lines. This type of cross sectional design should be limited to area with heavy rainfall. Roadways with unidirectional cross slopes tend to provide more comfort for drivers when they change lanes. Downward slopes toward the median have the advantage of requiring only one drainage conduit at the median.

The rate of cross slope is an important element in cross section design. The slope should be steep enough to direct the rainfall runoff to flow towards the edge of the roadway (instead of along the slope of a vertical curve), provide sufficient drainage capacity for the design storm, but at the same time, avoid the tendency for vehicles traveling straight to drift towards the low point of the cross section (unless corrected by steering control). Cross slope of up to 2% are normally not noticeable by drivers in terms of steering control, more than 2% will normally require steering correction. Excessive cross slope also presents danger for lateral skidding when vehicle apply emergency brake even on dry pavement. The design of cross road may be used to counter consistent crosswind. If a roadway has a crown at the centre, the cross slopes on both sides of the crown line should range from 1.5% to 2%. In any case, the slopes on both sides of the crown line should not be more than 2%. Using slopes of more than 2% on both sides of the crown will subject drivers to more than 4% differential slope when moving across it. Subsequent lanes toward the edge may have the same slope, or have the slope increased by 0.5% to 1%. The above guideline applies for high-type pavements.

Low-type pavements are relatively porous. To prevent water from absorbing into the materials that make up the surface layer, high slopes of 2% to 6% are applicable.

6.4.11.3   Lane Width

The width of a travel lane has significant effect on the driving speeds of the vehicles on that lane, and the passing vehicles on adjacent lanes. This also affects the highway capacity and level of service. The design of lane width should not only consider widths of the vehicles, but also the driving psychology. The width of a lane is measured from the pavement marking on both sides. In general, lane widths vary from of 2.7

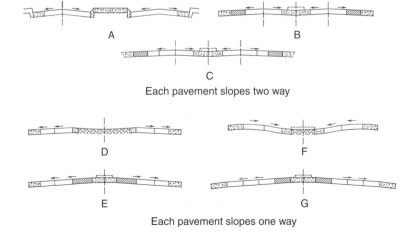

FIGURE 6.10   Cross slope designs for divided highways.

to 3.75 m. The narrower lanes are mainly used in residential areas and in urban roads with low travel speeds (e.g., turning storage lanes, at intersections) where majority of the vehicles are passenger cars. In residential areas, lane width should not be less than 2.7 m. In urban streets, lane width of less than 3.0 m should be avoided. Larger lane widths are used when commercial vehicles are present. On a two-lane undivided highway, ideal lane width should be 3.6 m or more to provide sufficient clearance for opposing vehicles. When there is restriction on the right-of-way and a narrower-than-ideal lane width is necessary, the designer should strike a balance between the widths of different lanes for different vehicle types. On a unidirectional highway with multiple lanes where it is not possible to use the same width for all the lanes, the outer lanes should be wider than the middle lane because commercial vehicles and trucks are expected to use that lane. This may also provide space for cyclists. The lane next to the curb should also be wider than the middle lane.

6.4.11.4 Shoulder

A shoulder is the part of the highway near the edges of the paved surface that is designed to provide structural lateral support for the pavement. A shoulder also provides additional space for drivers to make corrective actions, for stopped vehicles and for bicycle use, to increase sight distance on horizontal curves, and to provide clearance for placement of road signs and guardrails. The width of a shoulder is measured from the pavement marking of the outermost lane to the intersection of the shoulder slope and foreslope planes. Not the entire width of the shoulder (according to the definition above) may be paved.

The ideal width of a shoulder is 3.0 m, so that a vehicle that is stopping on the shoulder will not interfere with the movement of vehicles using the outermost lane. Shoulder wider than 3.0 m may encourage driver to use it as a normal travel lane in case of congestion. Frequently used shoulder width of 1.80 to 2.40 m do provide sufficient clearance for vehicles on the outermost lane, but large vehicles may have to take evasive action. Regardless of the width, a shoulder should be continuous along the roadway.

The slope of the shoulder should be designed to be steeper than the traveled way to encourage rapid drainage of the collected runoff from the traveled way to drainage conduit. On the other hand, the slope should not be too steep to pose difficulty for vehicles to use. Typical range of the slope is from 2 to 6%.

6.4.11.5 Curb

A curb is a steep raised element of a roadway that provides the following functions: drainage control, roadway edge delineation, right-of-way control and delineation of pedestrian walkways. Curbs are used extensively in low-speed urban streets, but not on high-speed rural highways and freeways. This is because a vehicle may overturn when hitting a curb at high speed.

There are generally two types of curbs: vertical curbs and sloping curbs. Vertical curbs are either vertical or nearly vertical, with a height of 150 to 200 mm. They may prevent or discourage vehicles from leaving the roadway. Sloping curbs have slopes that range from $1V : 2H$ to $1V : 1H$. The height is between 100 to 150 mm. They are designed such that a vehicle in emergency may go over the curb.

6.4.12 Intersection Alignment

An intersection is an area where two or more highways join or cross. There are generally two types of intersections: at-grade intersections and grade separated interchanges. This section deals with at-grade intersections. Grade separated interchanges will be covered in Section 6.4.13.

The main objective of at-grade intersection design is to facilitate the convenience, comfort, and safe movement of vehicles and pedestrians that use the intersection.

An intersection is defined by both its functional and physical areas. Physical area refers to the region where the paths of vehicles in different movements cross. This is where traffic conflicts occur. The functional area extends upstream and downstream of the physical intersection area, in all the approaches.

The functional area consists of three basic elements: perception–reaction distance, maneuver distance, and queue storage area.

The type of at-grade intersection is primarily determined by the number of approaches. Three-leg intersections are generally in "T" configuration, with the major highway carrying the through traffic. The three-leg intersections are either unchannelized or channelized. Unchannelized intersection is for junctions of minor or local roads with relatively more important highways. In area where speed and turning movements are high, additional area of surfacing or flaring may be provided. Additional turning lanes to serve turning vehicles (as auxiliary lanes or waiting lanes) may improve the efficiency of the intersections.

The basic types of four-leg intersections: unchannelized and channelized, are shown in Figure 6.11.

Intersections are points of conflicts between vehicles and pedestrians. The alignment design of an intersection should permit users to recognize the presence of the intersection, traffic control devices and vehicles in the vicinity, and take appropriate actions. To achieve this, the alignment should be as straight and grade as flat as possible, as all the approaches. The intersecting approaches should meet each other at right angle, or as close to a right angle as possible. In no cases should the angle of intersection be more then 30° from the right angle. Approach gradient should not be more than 3%. Some adjustment to an approach's upstream horizontal and vertical alignments may be necessary to bring the approach at the desired angle and grade.

The width of roadways for vehicle to make turning maneuvers may be wider than the lane width for highways. The width of a turning roadway is governed by the volume of the turning traffic and the types of vehicle it serves. The minimum turning radii should be based on the minimum turning path of the selected designed vehicles, at a particular design speed. AASHTO has provided a list of the minimum turning radii and turning paths of different types of design vehicles. As a guide, radii of 4.5 to 7.5 m are adequate for passenger vehicles, while radii of 9 m should be designed for small volume of trucks, while radii of 12 m should be given for large trucks.

Channalized islands may be provided in an intersection to separate conflicting traffic streams, control the angle of conflict, regulate traffic, show indication of the proper use of lanes, offer protection of pedestrians and waiting vehicles, and provide space for placement of traffic control devices. The islands should be obvious to the approaching vehicles.

The drivers approaching an intersection should have an unobstructed view of the entire intersection, including traffic control devices and other approaching vehicles that have potential to cause a collision. The view of the driver ahead should be at least equal to the stopping sight distance. For the purpose of intersection sight distance calculations, the driver's height is taken as 1.080 m, the conflicting vehicle's height is 1.330 m. The height of a truck's driver's eye is taken as 2.330 m.

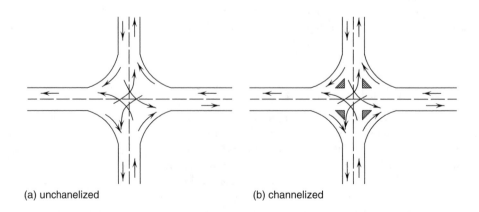

(a) unchanelized   (b) channelized

FIGURE 6.11 Typical 4-leg unchannelized and channelized intersections.

6.4.13 Interchanges

AASHTO (2001) defines an interchange as a system of connecting roadways in conjunction with one or more grade separations that provides for the movement of traffic between two or more roadways or highways at different levels. Interchanges are more efficient in handling higher volume than intersections, by separating conflicting traffic movements and ensure their uninterrupted flow.

The configurations of interchanges vary from a single ramp connecting a highway to a local street, to complex "spaghetti" layout connecting three freeways. The configuration depends on many factors, among which are the right-of-way, site topography, overall capacity, ramp capacity and storage, volumes of through and turning movements, truck traffic, design controls, and construction cost.

Whenever possible, the design on interchanges should be uniform and consistent. Ramps or connectors should be provided to serve all the major traffic movements. There should be only one exit and one entry point that serve a particular movement.

6.4.13.1 Three-Leg Interchanges

Most of the three-leg interchanges are designed to connect two highways in "T" or "Y" form. An important factor in influencing the choice of the layout is the turning volumes. The "T" layout is suitable for heavy through traffic and smaller turning volume and the "Y" layout favors a more balanced turning percentages. Figure 6.12 shows several configurations of three-leg interchanges. Figures 6.12a and b are the so-called trumpet interchange. The design depends on the approach angles of the intersecting highways, and the volumes of the movements that are involved in the loop ramp. The loop ramp should be designed for the smaller turning volume. Figures 6.12c–e show three typical designs of the "Y" configurations. All the movements are grade separated. Obviously, the "Y" configurations are more costly to construct, require more right-of-way, but also have higher capacity than the "T" configurations.

6.4.13.2 Four-Leg Interchanges

Four-leg interchanges may be classified into four designs: (1) diamond interchanges; (2) single-point urban interchanges (SPUI); (3) cloverleaf interchanges; and (4) interchanges with direct and semi-direct connections.

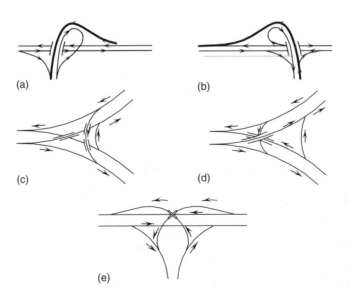

FIGURE 6.12 Three-leg interchanges.

The diamond interchange (see Figure 6.13) is the simplest and most commonly found four-leg configuration. It is commonly used to connect traffic from a major highway (or freeway) with an urban arterial. Its characteristics are: (1) one-way diagonal ramp at each quadrant; (2) the ramps are aligned in approximately the same direction as the major highway, allowing vehicles to enter and exit the major highway at almost the same speed; (3) the ramps require relatively less right-of-way

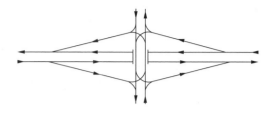

FIGURE 6.13   Diamond interchange.

compared to other configurations. Diamond interchanges, however, require all left turning movements to cross at least one conflicting traffic stream. In urban area where the arterial cross street has moderate or high volume, it is always necessary to have a pair of signalized intersections at the intersections of ramps and the arterial cross street, and the signal timing plans of the two pair of intersections have to be properly coordinated. The capacity of the diamond interchange is limited by capacity of these two intersections. The ramps should provide sufficient queue capacity such that vehicles will not be split back onto the major highway. A diamond interchange with the major highway crossing below the arterial is preferred from traffic operations point of view. This is because the upgrade at the exit ramps will help to decelerate the vehicle, and the downgrade at the on-ramp will assist in vehicle acceleration. Drivers on the on-ramp will have a better view of the traffic ahead when merging into the major highway.

6.4.13.3   Single-Point Urban Interchanges

The SPUI is also known as single point diamond interchange. As its name suggest, this type of interchange is used in urban area, connecting a major highway with an arterial. The main characteristics of SPUI is, all the left turning movements are handled by a signalized intersection (hence, the word single-point). Figure 6.14 shows a typical layout of a SPUI. Like the diamond interchange design, the capacity of the SPUI is limited by the capacity of the signalized intersection. The advantage of SPUI over diamond

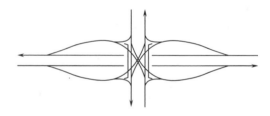

FIGURE 6.14   Single point urban interchange.

interchange is that, it is easier to maximize the capacity of one intersection, than a pair of intersections. The SPUI requires the same right-of-way but may involve higher construction cost than the diamond interchange. This is because the turning radius of the left-turning traffic may require a larger bridge span. Use of a smaller turning radius will reduce the capacity of the left-turning movements.

6.4.13.4   Cloverleaf Interchanges

Cloverleaf interchanges are generally defined as interchanges that use loop ramps in at least two quadrants to handle left-turning movements.

An interchanges with loop ramps in all four quadrants is termed "full cloverleaf" while others are referred to as "partial cloverleaf". Full cloverleaf is used for two intersecting major highways or freeways. A partial cloverleaf, with two loop ramps at two quadrants, may be used at an intersection of a major highway (freeway) and an arterial.

The major advantage of the cloverleaf interchanges is that all the turning movements are separated with minimal structural cost (only a two-tier structure is required at highway to highway crossing). All the vehicles are able to move continuously through the interchange area, irrespective of the directions of travel. However, the main disadvantage is that, along a major highway the off-ramps are downstream of the on-ramps. Thus, entering and exiting vehicles have to undergo weaving maneuvers. The length of the weaving section is limited by the space between two ramps. The capacity of the weaving section may pose some operational difficulty for drivers when the turning volumes are high. Therefore, cloverleaf interchanges are only recommended for highways where the turning volumes are moderate or low. Because of the design of the loop ramps, left-turning vehicles have to make a clockwise motion of 270°, which can be counter-intuitive to drivers' expectation. To enable drivers to travel on the loop ramps without significant reduction in speeds, the radius of the loop ramps have to be large, and this requires more right-of-way. Cloverleaf inter-changes require relatively more right-of-way than most of other interchange configurations, and therefore is seldom used in urban areas.

The weaving problem associated with the fully cloverleaf interchanges may be overcome by partial cloverleaf interchanges with loop ramps at diag-onally opposite quadrants (Figure 6.15).

6.4.13.5 Ramp and Connector Designs

Once the configuration of an interchange has been decided, the design problem may be decomposed into the design of individual ramp and connector. The design speeds of ramps are always lower than the running speed of a highway, while those of direct connectors may only be slightly lower than the normal running speed. Compound or spiral horizontal curves are desirable to provide a smooth transition between the speed of the

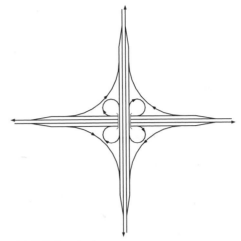

(a) Full cloverleaf

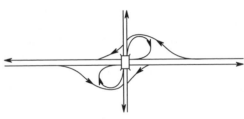

(b) Partial cloverleaf

FIGURE 6.15   Some cloverleaf interchanges.

through highway and the ramps and connectors. Abrupt and unexpected change in direction and speed should be avoided. Sight distance along a ramp should be greater than the stopping sight distance on horizontal and vertical curves. Sight distance for passing is not required, but the driver should have a clear view of the termination point (intersection, merging or weaving area) ahead. Ramp grade may be higher than a through highway, but should be as flat as practical. A check should be made on the effect of ramps on trucks that may slow down the following traffic.

6.5   Level of Service Consideration

Given a fixed volume of traffic flow and vehicle composition, the geometric design of highways, such as the horizontal and vertical curves, grades, lane width, number of lanes, etc., affects the travel speed of vehicles and the interaction between vehicles. The outcome of such effect is the observable operational performance of the highway, qualitatively represented by the highway level of service. The concept of LOS and methods of evaluation for different highway types are covered in great detail in the Highway Capacity Manual (TRB 2000).

The highway designer should strive to provide a LOS that is as high as possible. AASHTO (2001) has published a list of acceptable LOS for different road types (Table 6.2).

It is recommended that, after the designer has come out with the geometric elements of the highway (horizontal and vertical curves, grades, super-elevation, etc.) after checking through all the design elements, he should estimate the LOS based on the road type, designed volume, vehicle composition, number of lanes, and lane width. The geometric design should be revised if the LOS is not meeting

TABLE 6.2   Acceptable LOS for Different Road Types

Functional Class	Appropriate Level of Service for Specified Combinations of Area and Terrain Type			
	Rural Level	Rural Rolling	Rural Mountainous	Urban and Suburban
Freeway	B	B	C	C
Arterial	B	B	C	C
Collector	C	C	D	D
Local	D	D	D	D

the expectation (for examples, by modifying the grades, radius of curvature). He should also ensure that continuous segments of the same highway do not have sudden change in LOS that may cause surprise to the drivers.

6.6   Summary

This chapter covers the basic elements of highway geometric design. The alignment of a highway is defined by a series of horizontal or vertical curves (or a combination of both). The geometry of the curves must accommodate the dimensions and operating characteristics of the design vehicles, the ride comfort of drivers (and passengers), the corresponding road type, design volume, and speed. These are the features that either decide the input to the design or form the basis for checking the design output. The design of horizontal and/or vertical curves must provide sufficient sight distances for different types of vehicle maneuvers for the purpose of safety. After deciding the horizontal and vertical alignments, the cross section of a highway is considered. The impact of the geometry of the cross section on vehicle operations and driving comfort must be checked in combination with the horizontal and vertical curves. Highway geometric design also includes the design of intersection and grade separated interchanges. Typical intersection and interchange layouts are covered in this chapter. Finally, after completing the alignment design and checking through and adjusting the alignment to meet all the design controls and criteria, a transportation engineer must, based on the designed geometry, evaluate the traffic operating conditions and level of service. The highway alignment or design elements may need to be revised based on the feedback provided by the level-of-service analysis.

There are other factors a transportation engineer may wish to consider in highway geometric design, including cost: social and institutional issues regarding right of way, construction cost, vehicle operating cost, provision for other vehicles or mode of transportation, and availability of computer aided software that complies with the design standard. They cannot be covered exhaustively in this chapter. Readers should refer to the other chapters in this Handbook or relevant references for details. Regardless of the factors considered, a transportation engineer must always have the primary purpose of the highway functions, and the design objectives in mind. That is, to provide sufficient safety standard and yet maintain acceptable level of service and performance.

References

AASHTO, 2001. *A Policy on Geometric Design of Highway and Streets, 4th Ed.*, American Association of State Highway and Transportation Officials.

Johannson, C. and Rumar, K., Driver's brake reaction time, *Hum. Factors*, 13, 1, 22–27, 1971.

Koepke, F.J. and Levinson, H.S. 1992. *Access Management Guidelines for Activity Centers*, NCHRP Report 348. Transportation Research Board.

Koppa, R.J. 2000. *Human Factors*, Traffic Flow Theory: A State of the Art Report. Transportation Research Board.

TRB, 2000. *Highway Capacity Manual 2000*. Transportation Research Board.

7

Highway Materials

Weng On Tam
Consultant
Austin, TX, U.S.A.

7.1   Introduction

From the early days of the Roman Empire to the interstate highway system in the United States, roadway networks have been developed to support military operations. Over the years, however, the materials used for roadway construction have progressed with time. This advancement in materials has been accompanied with corresponding advancements in methods with which these materials are characterized and applied to pavement structural design.

Currently, there are two primary types of pavement surfaces — Portland cement concrete (PCC) and hot-mix asphalt concrete (HMAC). Below this wearing course are material layers that provide structural support for the pavement system. These may include either (a) the aggregate base and subbase layers, or (b) treated base and subbase layers, and the underlying natural or treated subgrade. The treated layers may be cement-treated, asphalt-treated or lime-treated for additional structural support.

There are various methods by which pavement layers are designed. For example, HMAC may be designed using the Marshall, Hveem, or Superpave mix design systems. PCC may be designed using the American Concrete Institute (ACI) or the Portland Cement Association (PCA) method.

While this chapter is titled "Highway Materials," the focus of this chapter will be on HMAC with particular attention on Superpave technology. Other materials, such as PCC, aggregates, and soils, will only be briefly discussed.

7.2   Hot-Mix Asphalt Concrete

HMAC consists primarily of mineral aggregates, asphalt cement (or binder), and air. It is important to have suitable proportions of asphalt cement and aggregates in HMAC so as to develop mixtures that have desirable properties associated with good performance. These performance measures include the resistance to the three primary HMAC distresses: permanent deformation, fatigue cracking, and low-temperature cracking.

Permanent deformation refers to the plastic deformation of HMAC under repeated loads. This permanent deformation can be in the form of rutting (lateral plastic flow in the wheelpaths) or consolidation (further compaction of the HMAC after construction). Aggregate interlock is the primary component that resists permanent deformation with the asphalt cement playing only a minor role. Angular, rough-textured aggregates will help reduce permanent deformation. To a significantly lesser extent, a stiffer asphalt cement may also provide some minor benefit.

Cracking can be subdivided into two broad categories: load associated cracking and non-load associated cracking. Load associated cracking has traditionally been called fatigue cracking. In this scenario, repeated stress applications below the maximum tensile strength of the material eventually lead to cracking. Factors associated with the development of fatigue cracking include the *in-situ* properties of the structural section, asphalt cement, temperature, and traffic. Non-load associated cracking has traditionally been called low-temperature cracking. During times of rapid cooling and low temperatures, the stress experienced by the HMAC may exceed its fracture strength. This leads to immediate cracking.

7.2.1 Background to Mineral Aggregates

Mineral aggregates make up 90 to 95% of a HMA mix by weight or approximately 75 to 85% by volume. Their physical characteristics are responsible for providing a strong aggregate structure to resist deformation due to repeated load applications. Aggregate mineralogical and chemical makeup are important in evaluating characteristics such as hardness (toughness), soundness (durability), shape, and stripping potential.

In ASTM D8 (ASTM, 2003), aggregate is defined as "a granular material of mineral composition such as sand, gravel, shell, slag, or crushed stone, used with cementing medium to form mortars or concrete or alone as in base courses, railroad ballasts, etc." These aggregates can be divided into three main categories — natural, processed, and synthetic (artificial) aggregates. Natural aggregates are mined from river or glacial deposits. They are frequently referred to as pit- or bank-run materials. Gravel and sand are examples of natural aggregates. Gravel is normally defined as aggregates passing the 3 in. (75 mm) sieve and retained on the No. 4 (4.75 mm) sieve. Sand is usually defined as aggregate passing the No. 4 sieve with the silt and clay fraction passing the No. 200 (0.075 mm) sieve. These aggregates in their natural form tend to be smooth and round.

Processed materials include gravel or stones that have been crushed, washed, screened, or otherwise treated to enhance the performance of HMAC. Processed materials tend to be more angular and better graded. Synthetic aggregates are not mined or quarried. Rather, they are manufactured through the application of physical and/or chemical processes as either a principal product or a by-product. They are often used to improve the skid resistance of HMAC. Blast furnace slag, lightweight expanded clay, shale or slate are examples of synthetic aggregates.

In additon to the use of traditional aggregates mentioned above, there has been an increase in the use of waste products in HMAC. Scrapped tires and glass are the two most commonly used waste products that have been "disposed of" in HMAC. Some of these waste products enhance the performance of HMA concrete while others neither benefit nor adversely affect its performance. Waste products that are detrimental to HMAC performance should not be arbitrarily applied in preparing the mix.

7.2.2 Aggregate Specifications and Tests

Traditional aggregate specifications for HMA include the American Association of State Highway and Transportation Officials (AASHTO) M29 (ASTM D1073) "*Standard Method of Test for Fine Aggregate for Bituminous Paving Mixtures,*" ASTM D692 "*Standard Specification for Coarse Aggregate for Bituminous Paving Mixtures,*" and ASTM D242 "*Standard Specification for Mineral Filler for Bituminous Paving Mixtures.*" During the Strategic Highway Research Program (SHRP, 1994), no new research was conducted on aggregate properties. However, experts were polled (using a modified Delphi method) on the important aggregate tests and corresponding appropriate test values to ensure adequate performance in HMAC. This served as the basis for determining the consensus properties and source properties used in the Superpave mix design system. The consensus properties are critical properties required to develop a desirable HMA mix. These properties are:

- coarse aggregate angularity
- fine aggregate angularity
- flat, elongated particles, and
- clay content

During the SHRP program, criteria were developed for consensus properties and have been modified to reflect later developments in Superpave research, as will be discussed under the Superpave mix design section. In addition to the consensus properties, properties specific to each aggregate source (source properties) are also important. Since the critical values are source specific, a range of universally acceptable values could not be established. Instead these values should be established by individual agencies for sources specific to their locality based on knowledge and experience in the area. The source properties are:

- toughness
- soundness, and
- deleterious materials

In addition to the Superpave consensus and source properties, there are other properties that may influence the performance of aggregates in HMA — particle index, plasticity index (PI), affinity for asphalt, and absorption. This section discusses additional aggregate characteristics identified as playing an important role in HMA performance.

7.2.2.1 Coarse and Fine Aggregate Angularity

Coarse aggregate angularity is defined as the percent by weight of aggregates retained on the No. 4 (4.75 mm) sieve with one or more fractured face. This property is determined using ASTM D5821 "*Standard Test Method for Determining the Percentage of Fractured Particles in Coarse Aggregate*" (ASTM, 2003). In this procedure, individual aggregates are manually examined for the presence of fractured faces. Once the aggregates are sorted by the number of fractured faces, their percentages of aggregate with at least one or two fractured faces are calculated. A fractured face is defined as any angular, rough, or broken surface of an aggregate particle that occupies more than 25% of the outline of the aggregate particle visible in that orientation. The percent of fractured particles in coarse aggregate can provide an indication of inter-particle shear friction or stability.

Fine aggregate angularity is defined as the percent of air voids present in a loose compacted aggregate sample that passes the No. 8 (2.36 mm) sieve. This property is determined using AASHTO T304 "*Standard Method of Test for Uncompacted Void Content of Fine Aggregate — Method A*" (AASHTO, 2003) . In this procedure, a nominal 100 cm^3 calibrated cylinder measure is filled with fine aggregate through a funnel placed at a fixed height above the measure. When the measure is overfilled, it is struck off and the mass of aggregate inside the measure is determined by weighing. The uncompacted void content is the difference between the volume of the cylindrical measure and the absolute volume of the fine aggregate (calculated from its mass and bulk dry specific gravity). For fine aggregate with a given gradation, the void content determined from this test provides an indication of the aggregate's angularity, sphericity, and surface texture relative to other fine aggregates with the same gradation.

Let us consider a stockpile with crushed (angular) aggregate and one with rounded aggregates. If both stockpiles are of equivalent mass, the crushed aggregate would form a steeper (greater angle of repose), more stable stockpile compared to the one with rounded aggregates. This is due to the greater degree of aggregate-interlock in the crushed aggregate stockpile.

From a theoretical standpoint, this can be explained using the Mohr–Coulomb theory. This theory states that the shear strength of an aggregate mix (τ) is dependent on cohesion of the aggregates (c), the normal stress experienced by the aggregates (σ), and the internal friction of the aggregates (ϕ) as shown in the equation below:

$$\tau = c + \sigma \cdot \tan \phi \tag{7.1}$$

By themselves, aggregates have relatively little cohesion. Therefore the source of shear strength in an aggregate mix comes primarily from the angle of internal friction. The greater the angle of internal friction, the greater the increase in shear strength with increasing normal stress. This angle of internal friction is higher for aggregates with good aggregate-interlock.

Rounded aggregates, such as natural gravel and sands, are more workable and require less effort to compact. However, HMA mixtures comprised of rounded aggregates may continue to densify under traffic loading and lead to rutting. Angular aggregates, on the other hand, are harder to compact due to the aggregate-interlock, which gives the mix greater shear strength. These mixes tend to be more stable and resistant to rutting. Realizing this, many owner agencies have a minimum coarse and fine aggregate angularity requirement.

7.2.2.2 Flat and Elongated Particles

Flat or elongated particles are defined as aggregate having a ratio of width to thickness or length to width greater than a specified value. This property is determined using ASTM D4791 "*Standard Practice for Flat Particles, Elongated Particles, or Flat and Elongated Particles in Coarse Aggregate*" (ASTM, 2003). This test is conducted on aggregates retained on the No. 4 (4.75 mm) sieve. In this procedure, a proportional caliper is used to measure the dimensional ratio of a representative sample of coarse aggregate. Aggregates exceeding the 5 to 1 ratio are considered flat and elongated in the Superpave mix design system. The percent of flat or elongated aggregates is reported as a percentage of total aggregates tested. Some people believe that this requirement should be more stringent, possibly at 3 to 1 or 2 to 1.

Aggregates used in HMA mixes should be cubicle rather than disproportionate in their dimensions. Aggregates particles that are significantly longer in one dimension than in the other one or two dimensions have a propensity to break during the construction process or under traffic loading.

7.2.2.3 Clay Content

Clay content is defined as the percentage of clay material contained in the aggregate fraction that passes the No. 4 (4.75 mm) sieve. This property is determined using AASHTO T176 "*Standard Method of Test for Plastic Fines in Graded Aggregates and Soils by Use of the Sand Equivalent Test*" (AASHTO, 2003). In this procedure, a sample of fine aggregate is placed in a graduated cylinder with a flocculating solution. The cylinder is then agitated to loosen the clayey fines within and surrounding the aggregate particles. The presence of a flocculating solution and the agitation to the container cause the clayey material to go into suspension above the aggregate. After allowing the constituents to settle for a specific length of time, the height of suspended clay and sedimented aggregate is measured. The sand equivalent value is the ratio of the sand reading to the clay reading as a percentage.

A low sand equivalent value, or high clay content, means that there is "dirt" on the surface of the aggregates. This "dirt" can reduce the bond between the aggregate and asphalt cement. Consequently, the mix would have a greater tendency for stripping. Cleaner aggregate with higher sand equivalent values will enhance the performance of HMA.

7.2.2.4 Toughness

Toughness is the percent loss of material from an aggregate blend during the Los Angeles Abrasion test. This property is determined using AASHTO T96 (ASTM C131) "*Standard Method of Test for Resistance to Degradation of Small-Size Coarse Aggregate by Abrasion and Impact in the Los Angeles Machine*" (AASHTO, 2003). This test covers a procedure for testing aggregates up to 1.5 in. (37.5 mm) in size. ASTM C535 should be used for aggregates with a larger maximum size up to 3 in. (75 mm). In this procedure, aggregate is degraded through abrasion, impact, and grinding in a rotating steel drum containing steel spheres. The Los Angeles abrasion loss is the difference between the original and final mass of the sample is expressed as a percentage of the original mass after washing off the No. 12 (1.70 mm) screen.

This test provides some indication of the aggregate's ability to resist degradation from processes that it would encounter through its life as an aggregate in HMA. HMA aggregates can be degraded during stockpiling, processing (through a HMA plant), placing, and compacting. It may also degrade due to traffic loads. Typical test values range from 10% for extremely hard rocks (e.g. basalt) to 60% for soft rocks (e.g. limestone). ASTM D692 "*Standard Specification for Coarse Aggregate for Bituminous Paving Mixtures*" (ASTM, 2003) specifies a loss not greater than 40% for surface courses or 50% for base courses. Highway agencies typically limit the percent loss to between 40 and 60%. It should be noted that the Los Angeles abrasion test primarily evaluates an aggregate's resistance to degradation by abrasion and impact. However, results from this test do not correlate satisfactorily with HMA performance in the field. For example, slag and soft limestones often produce good performing mixes even though their Los Angeles abrasion loss is high.

7.2.2.5 Soundness

Soundness is the percent loss of material from an aggregate blend during the sodium or magnesium sulfate soundness test. This property is determined using AASHTO T104 "*Standard Method of Test for Soundness of Aggregate by Use of Sodium Sulfate or Magnesium Sulfate*" (AASHTO, 2003). In this procedure, aggregate samples are put through repeated cycles of immersion in saturated solutions of sodium or magnesium sulfate followed by oven drying. The percent loss of material is determined by taking the difference between the original and final (after the specified number of cycles) masses expressed as a percentage of the original mass.

This test evaluates the aggregate's ability to resist breaking down or disintegrating due to weathering (i.e., wetting and drying and/or freezing and thawing). ASTM D692 "*Standard Specification for Coarse Aggregate for Bituminous Paving Mixtures*" (ASTM, 2003) specifies a five-cycle weighted loss of not more than 12% when sodium sulfate is used or 18% when magnesium sulfate is used. The limit specified for magnesium sulfate is greater than that for sodium sulfate because magnesium sulfate creates a more severe test. While this test was initially developed to evaluate the effect of freeze–thaw conditions on aggregates, it has been adopted rather universally to screen aggregates.

7.2.2.6 Deleterious Materials

The percentage of deleterious materials in blended aggregate is determined using AASHTO T112 (ASTM C142) "*Standard Method of Test for Clay Lumps and Friable Particles in Aggregate*" (AASHTO, 2003). In this procedure, aggregates are individually subjected to finger pressure (while soaking) to determine materials that are friable or clay lumps. The percent of clay lumps and friable particles is determined by taking the difference between the original and final mass retained on a No. 200 (0.075 mm) sieve, after wet sieving, expressed as a percentage of the original mass. The percent of deleterious materials can range from 0.2 to 10%.

7.2.2.7 Particle Index (Shape and Texture)

Particle index is an overall measure of aggregate particle shape and texture. This property is determined using ASTM D3398. "*Standard Test Method for Index of Aggregate Particle Shape and Texture*" (ASTM, 2003). In this test method, the percent voids in the aggregate compacted in two stages according to a specified procedure is used to calculate the particle index (I_a) value based on the equation below:

$$I_a = 1.25V_{10} - 0.25V_{50} - 32.0 \qquad (7.2)$$

where V_{10} and V_{50} are percent voids in aggregate compacted respectively with 10 and 50 drops of standard tamping rod per layer.

Typically, rounded particles with smooth surface textures may have a particle index of 6 or 7 while a highly-angular crushed particle with rough surface textures can have particle indices of 15 to 20 or higher.

The role of particle shape was discussed earlier in the section on "Coarse and Fine Aggregate Angularity" and "Flat, Elongated Particles." Therefore, the discussion in this section will focus on surface texture. Surface texture, like particle shape, influences the workability and strength of HMA. Aggregates with rough textures, such as crushed limestone or gravel, tend to form stronger bonds with asphalt cement and increases the strength and asphalt cement demand of a mix. On the other hand, aggregates with smooth textures, such as river gravels and sands, tend to form weaker bonds with asphalt cement which leads to reduced strength and decreased asphalt cement demand. However, smooth aggregate surface textures may provide more workability.

7.2.2.8 Plasticity Index

The Plasticity Index (PI) is a measure of the degree of plasticity of fines (material passing the No. 200 sieve). It can provide an indication of the amount and type of fines. This property is determined using ASTM D4318 "*Standard Test Method for Liquid Limit, Plastic Limit, and Plasticity Index of Soils*" (ASTM, 2003). The PI is defined as the difference between the liquid limit (LL) and plastic limit (PL) as shown in the equation below:

$$PI = LL - PL \tag{7.3}$$

7.2.2.9 Affinity for Asphalt

An aggregate's affinity for asphalt cement is its propensity to attract and remain attached to asphalt cement. Asphalt cement must coat the aggregate, stick to the aggregate, and resist stripping of the asphalt film in the presence of water. Though there are several theories surrounding the chemical interactions that take place in stripping, none of them has completely explained this phenomenon. It is important to recognize that some aggregates appear to have a greater affinity for water than for asphalt cement. These hydrophilic (water-loving) aggregates have a tendency to get stripped (asphalt film gets detached from the aggregate) with exposure to water. Siliceous aggregates such as quartzite and some granites are examples of hydrophilic aggregates. On the other hand, hydrophobic (water-hating) aggregates have a greater affinity for asphalt cement. Limestone and dolomite are examples of hydrophobic aggregates. There are various tests that evaluate a mixture's propensity for stripping. They include AASHTO T283, ASTM D3625, and AASHTO T165 (ASTM D1075). Unfortunately, none of these tests could accurately predict the stripping potential of HMA mixes in different environments.

   The test selected for use in the Superpave mix design procedure, AASHTO T283 "*Standard Method of Test for Resistance of Compacted Asphalt Mixtures to Moisture-Induced Damage*" (AASHTO, 2003), is commonly used to evaluate stripping. In this procedure, two sets of replicate samples are compacted and tested. One set is put through a conditioning regiment while the other is not. The tensile strength ratio (TSR) reported is the ratio of the indirect tensile strength of the conditioned set to that of the unconditioned set.

7.2.2.10 Absorption

Absorption is a measure of an aggregate's porosity. While porosity is generally associated with the absorption of water, a porous aggregate also tends to absorb asphalt cement. Porous aggregates have a greater asphalt cement demand and require additional asphalt cement for a comparable mix. Therefore, highly porous aggregates are generally not used for HMA unless the aggregates possess certain desirable qualities that outweigh the cost of additional asphalt cement. Blast furnace slag and other synthetic aggregates are examples of lightweight, wear resistant aggregates that are used in spite of their high porosity.

7.2.3   Aggregate Gradation

Aggregate gradation is the distribution of particle size expressed as a percentage of the total sample weight. This property is determined using AASHTO T27 (ASTM C136) "*Standard Method of Test for Sieve Analysis of Fine and Coarse Aggregate*" (AASHTO, 2003). In this test, aggregate passed through

TABLE 7.1 Standard Sieve Sizes Used in the Superpave System

Sieve Designation	Sieve Opening (in.)	Sieve Opening (mm)
1-in.	1.00	25.0
3/4-in.	0.75	19.0
1/2-in.	0.500	12.5
3/8-in.	0.375	9.5
No. 4	0.187	4.75
No. 8	0.0937	2.36
No. 16	0.0469	1.18
No. 30	0.0234	0.600
No. 50	0.0117	0.300
No. 100	0.0059	0.150
No. 200	0.0029	0.075

sieves with progressively smaller openings. The mass of aggregate retained is then used to determine the percent of aggregate retained on and/or passing each sieve.

Gradations can be represented graphically using "percent passing" (by weight) as the ordinate and particle size as the abscissa. Fuller and Thompson (1907) developed one of the best-known grading charts in the early 20th century. The equation for Fuller's maximum density curve is as follows:

$$P = 100(d/D)^n \tag{7.4}$$

where d is the diameter of the sieve in question, P total percent pasing or finer than the sieve, and D is the maximum size of the aggregate.

Studies by Fuller and Thompson showed that an aggregate's maximum density line was achieved when "n" was equal to 0.5. In the early 1960s, the Federal Highway Administration (FHWA) introduced a grading chart similar to Equation 7.4 but with a "0.45" exponent instead of "0.5". Using this gradation chart, the maximum density line can be obtained by drawing a straight line from the origin on the lower left of the chart to the point where the nominal aggregate size intersects the ordinate value of 100%. A gradation hugging the maximum density line should be avoided because there is insufficient space between aggregates for the asphalt cement to adequately coat the aggregates. As the deviation between the aggregate gradation and the maximum density line increases, the allowable space for the asphalt cement to coat the aggregates increases. This allows for higher voids in mineral aggregate (VMA) in the mix.

Standard sieve sizes used in the Superpave system are shown in Table 7.1. Figure 7.1 provides examples of general descriptions of four different types of gradations.

HMA mixes are often categorized based on aggregate size. For example, the Superpave mix design system uses the nominal maximum aggregate size to represent the size of aggregates in a mix. The nominal sieve size is defined as one sieve size larger than the first sieve to retain more than 10% of combined aggregate. The maximum sieve size is defined as one sieve size larger than the nominal maximum sieve (McGennis et al., 1995).

7.2.4 Background to Asphalt Cement

Asphalt cement is a dark brown- or black-colored bituminous material used in HMA paving. It occurs naturally in geologic strata and was used in the late 19th century for paving roads. The Trinidad Lake deposit is a well-known source of naturally occurring asphalt. Asphalt cement can also be derived synthetically from the petroleum refining process. Since asphalt cement consists primarily of the highest boiling fraction of petroleum, it is captured as the residue from the vacuum tower. Tar is a distinctly different product that is often mistaken for asphalt cement. It was also used in late 19th century paving. However, it is a product derived from the destructive distillation of bituminous coal. Due to environmental and health concerns as well as inherent undesirable physical characteristics, tar is seldom used for paving purposes today.

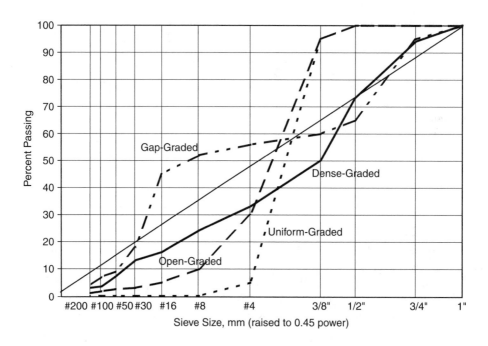

FIGURE 7.1   Types of aggregate gradations.

7.2.4.1   Asphalt Cement Behavior

Asphalt cement is a viscoelastic material. This means that asphalt cement exhibits both viscous and elastic behavior. However, the extent to which asphalt cement behaves as a viscous material or an elastic material is dependent on the temperature and rate of loading. For example, asphalt cement at 60°C (140°F) may flow the same amount in 1 h as it would at 25°C (77°F) in 5 h. This behavioral relationship is sometimes called time–temperature superposition. At high temperatures under slow loading conditions (e.g. slow-moving traffic), asphalt cements behave like a viscous fluid and flow. This is often called plastic behavior because after the asphalt cement has flowed, it does not return to its original position. This may lead to permanent deformation in the asphalt mix. It should be noted that aggregates play a significant role in resisting permanent deformation. At low temperatures under rapid loads (e.g. fast-moving traffic), asphalt cements behave like elastic solids. They will deform under loads and return to their original position after the load is removed. If the load applied causes stresses greater than the strength of the mix, then low temperature cracking will occur. At intermediate temperatures, asphalt cements exhibit both viscous and plastic behavior.

Asphalt cements are primarily composed of organic molecules. These molecules react with oxygen in the environment and the asphalt cement becomes more brittle. This process is called oxidation or age hardening. In practice, age hardening occurs during the mixing and compacting phase of construction (short-term aging) as well as in the long-term in-service phase of the pavement (long-term aging). In the mixing stage, there are thin films of asphalt cement coating the aggregates. With a significant surface area in contact with oxygen in a high temperature environment, oxidation occurs fairly rapidly. Oxidation continues to occur in the storage, transportation, and paving stages of HMA construction, but at a much slower rate. Researchers and practitioners have strived to simulate this hardening of asphalt cement in the laboratory. The thin-film oven (TFO) and rolling thin-film oven (RTFO) tests are used to simulate short-term aging in the laboratory. The pressure aging vessel (PAV) is used to simulate long-term aging in the laboratory. These tests will be described in greater detail later in this chapter.

7.2.5 Asphalt Cement Specifications and Tests

Over the years, there have been a number of methods for classifying asphalt cement. The penetration, viscosity, and aged-viscosity grading systems were commonly used up till the late 1980s. With the 1990s came the advent of the Superpave binder grading system. This system is a Superpave product developed during SHRP and enhanced over the past 10 years. Currently, almost every state in the United States uses the Superpave binder grading system as the backbone of their HMA binders. Some states have incorporated minor deviations from the AASHTO specification. For example, Texas uses a "PG plus" specification that adds some traditional requirements on top of the Superpave binder requirements. The following sections will discuss some of the different asphalt cement specifications and the tests associated with them. The Superpave binder specifications will be discussed using AASHTO M320 (AASHTO, 2003).

7.2.5.1 Penetration Grading System

In the early 20th century, the penetrometer was the principal means of measuring and controlling consistency of asphalts. In 1918, the Bureau of Public Roads (BPR), now the FHWA, introduced the penetration grading system. The BPR and the ASTM were instrumental in making the penetration test a standard in asphalt cement testing. Table 7.2 shows ASTM D946 "*Standard Specification for Penetration-Graded Asphalt Cement for Use in Pavement Construction*" (ASTM, 2002). This specification includes five penetration grades ranging from a hard asphalt graded at "40-50" to a soft asphalt cement graded at "200-300." The sections below discuss the tests used to classify penetration grades.

7.2.5.1.1 Penetration Test

Penetration is the number of units of 0.1 mm penetration depth achieved during the penetration test. It is an empirical measure of the asphalt cement's hardness. This property is determined using AASHTO T49 (ASTM D5) "*Standard Method of Test for Penetration of Bituminous Mixtures*" (AASHTO, 2003). In this procedure, a needle is typically loaded with a 100-g weight and allowed to penetrate into an asphalt cement sample for 5 sec. Prior to conducting the test, the asphalt cement sample is brought to the testing temperature, typically 25°C (77°F). A harder asphalt cement will have a lower penetration while a softer asphalt cement with have a higher penetration. This test may also be conducted at 0°C (32°F) with a 200-g load for 60 sec or at 46.1°C (115°F) with a 50-g weight for 5 sec.

7.2.5.1.2 Flash Point Test

Flash point is the temperature to which asphalt cement may be heated without the danger of causing an instantaneous flash in the presence of an open flame. This property is determined using AASHTO T48

TABLE 7.2 Standard Penetration Grades — ASTM

Test	Penetration Grade									
	40-50		60-70		85-100		120-150		200-300	
	Min.	Max.	Min.	Max.	Min.	Max.	Min.	Max.	Min.	Max.
Penetration at 25°C (77°F) 100 g, 5 sec	40	50	60	70	85	100	120	150	200	300
Flash Point, °F (Cleveland open cup)	450	—	450	—	450	—	425	—	350	—
Ductility at 77°C (25°C) 5 cm/min, cm	100	—	100	—	100	—	100	—	100[a]	—
Solubility in trichloroethylene, %	99.0	—	99.0	—	99.0	—	99.0	—	99.0	—
Retained penetration after thin-film oven test, %	55+	—	52+	—	47+	—	42+	—	37+	—
Ductility at 77°F (25°C) 5 cm/min, cm after thin-film oven test	—	—	50	—	75	—	100	—	100[a]	—

Source: From ASTM, 2002. ASTM D946 Standard Specification for Penetration-Graded Asphalt Cement for Use in Pavement Construction, *Annual Book of ASTM Standards*, West Conshohocken, Pennsylvania. With permission.

[a] If ductility at 77°F (25°C) is less than 100 cm, material will be accepted if ductility at 60°F (15.5°C) is 100 cm minimum at the pull rate of 5 cm/min.

(ASTM D92) "*Standard Method of Test for Flash and Fire Points by Cleveland Open Cup*" (AASHTO, 2003). In this procedure, a brass cup partially filled with asphalt cement is heated at a given rate. A flame is passed over the surface of this cup periodically and the temperature at which this flame causes an instantaneous flash is reported as the flash point. Minimum flash point requirements are typically incorporated into asphalt cement specifications for safety reasons. However, a change in flash point may indicate the presence of contaminants.

7.2.5.1.3   Ductility Test

Ductility is the number of centimeters a standard briquette of asphalt cement will stretch before breaking. This property is determined using AASHTO T51 (ASTM D113) "*Standard Method of Test for Ductility of Bituminous Mixtures*" (AASHTO, 2003). In this procedure, a standard briquette is molded such that its smallest cross-section is 1 cm^2. This sample is brought to test temperature in a water bath maintained at 25°C. This briquette is then stretched at a rate of 5 cm/min until it breaks. This test is sometimes run at 4°C with the sample pulled at 1 cm/min.

7.2.5.1.4   Solubility Test

Solubility is the percentage of an asphalt cement sample that will dissolve in trichloroethylene. This property is determined using AASHTO T44 (ASTM D2042) "*Standard Method of Test for Solubility of Bituminous Materials*" (AASHTO, 2003). In this procedure, an asphalt cement sample is dissolved in trichloroethylene and then filtered through a glass-fiber pad where the weight of the insoluble material is measured. The solubility is calculated by dividing the weight of the dissolved portion by the total weight of the asphalt cement sample. This test is used to check for contamination in asphalt cement. Most specifications require a minimum of 99% solubility in trichloroethylene. Technicians conducting this test need to take the necessary precautions while handling trichloroethylene, as it is a carcinogen. Studies have been conducted to replace trichloroethylene with a safer solvent.

7.2.5.1.5   Thin-Film Oven Test

The TFO test is used to approximate the effect of short-term aging during the mixing process. This test is conducted using AASHTO T179 (ASTM D1754) "*Standard Method of Test for Effect of Heat and Air on Asphalt Materials (Thin-Film Oven Test)*" (AASHTO, 2003). In this procedure, a 50-g asphalt cement sample is placed on a cylindrical flat-bottom pan to a depth of about 3.2 mm (0.125 in.). The pan is then placed on a shelf that rotates at 5 to 6 rev/min per minute in a ventilated oven maintained at 163°C for 5 h. The sample, considered short-term aged after it is removed from the oven, is then tested in accordance with the specification requirements.

7.2.5.2   Viscosity (AC) and Aged Residue (AR) Viscosity Grading Systems

In the 1960s, the FHWA, ASTM, AASHTO, industry, and a number of state highway agencies wanted asphalts to be graded by viscosity at 60°C (140°F). The main reasons for this shift was to replace an empirical measure with a more fundamental material property and to measure a property at a temperature which approximates the average pavement surface temperature on a hot summer day. Table 7.3 shows the viscosity grading requirements from ASTM D3381 "*Standard Specification for Viscosity-Graded Asphalt Cement for Use in Pavement Construction*" (ASTM, 2003). This specification includes five viscosity grades ranging from a hard asphalt graded at "AC-40" to a soft asphalt cement graded at "AC-2.5." At approximately the same time, the California Department of Highway (now Caltrans) was developing a parallel AR viscosity grading specification with cooperation from the Pacific Coast User Producer Group. Table 7.3 also shows the aged-viscosity grading requirements from ASTM D3381 "*Standard Specification for Viscosity-Graded Asphalt Cement for Use in Pavement Construction*" (ASTM, 2003). This specification includes five AR viscosity grades ranging from a hard asphalt graded at "AR-160" to a soft asphalt cement graded at "AR-10." The following sections discuss asphalt cement tests used in the Viscosity and Aged Residue Viscosity grading systems not previously discussed.

TABLE 7.3(A)   Standard Specification for Viscosity-Graded Asphalt — ASTM. (a) Requirements for Asphalt Cement, Viscosity Graded at 140°F (60°C)

Test	Viscosity Grade				
	AC-2.5	AC-5	AC-10	AC-20	AC-40
Viscosity, 140°F (60°C) P	250 ± 50	500 ± 100	1000 ± 200	2000 ± 400	4000 ± 800
Viscosity, 275°F (135°C), min, cSt	80	110	150	210	300
Penetration, 77°F (25°C), 100 g, 5 sec, min	200	120	70	40	20
Flash Point, Cleveland open cup, min. °F (°C)	325 (163)	350 (177)	425 (219)	450 (232)	450 (232)
Solubility in trichloroethylene, min, %	99.0	99.0	99.0	99.0	99.0
Tests on residue from thin-film oven test:					
Viscosity, 140°F (60°C), max, P	1250	2500	5000	10,000	20,000
Ductility, 77°F, (25°C), 5 cm/min, min, cm	100[a]	100	50	20	10

Note. Grading based on original asphalt. *Source*: From ASTM, 2003. ASTM D3381 standard specification for viscosity-graded asphalt cement for use in pavement construction, *Annual Book of ASTM Standards*, West Conshohocken, Pennsylvania. With permission.

[a] If ductility is less than 100, material will be accepted if ductility at 60°F (15.5°C) is 100 minimum at a pull rate of 5 cm/min.

7.2.5.2.1   *Absolute and Kinematic Viscosity Tests*

Viscosity can be defined as a fluid's resistance to flow. In the asphalt paving industry, two tests are used to measure viscosity — absolute and kinematic viscosity tests. The relationship between absolute and kinematic viscosity is shown below:

$$\text{Kinematic Viscosity} = \frac{\text{Absolute Viscosity}}{\text{Density}} \qquad (7.5)$$

Absolute viscosity is determined using AASHTO T202 (ASTM D2171) "*Standard Method of Test for Viscosity of Asphalt by Vacuum Capillary Viscometer*" (AASHTO, 2003). In this procedure, a partial vacuum pulls an asphalt cement sample through the viscometer maintained at a temperature

TABLE 7.3(B)   Requirements for Asphalt Cement, Viscosity Graded at 140°F (60°C)

Test	Viscosity Grade					
	AC-2.5	AC-5	AC-10	AC-20	AC-30	AC-40
Viscosity, 140°F (60°C), P	250 ± 50	500 ± 100	1000 ± 200	2000 ± 400	3000 ± 600	4000 ± 800
Viscosity, 275°F (135°C), min, cSt	125	175	250	300	350	400
Penetration, 77°F (25°C), 100 g, 5 s, min	220	140	80	60	50	40
Flash Point, Cleveland open cup, min. °F (°C)	325 (163)	350 (177)	425 (219)	450 (232)	450 (232)	450 (232)
Solubility in trichloroethylene, min, %	99.0	99.0	99.0	99.0	99.0	99.0
Tests on residue from thin-film oven test:						
Viscosity, 140°F (60°C), max, P	1250	2500	5000	10,000	15,000	20,000
Ductility, 77°F, (25°C), 5 cm/min, min, cm	100[a]	100	75	50	40	25

Note. Grading based on original asphalt. *Source*: From ASTM, 2003. ASTM D3381 standard specification for viscosity-graded asphalt cement for use in pavement construction, *Annual Book of ASTM Standards*, West Conshohocken, Pennsylvania. With permission.

[a] If ductility is less than 100, material will be accepted if ductility at 60°F (15.5°C) is 100 minimum at a pull rate of 5 cm/min.

TABLE 7.3(C)   Requirements for Asphalt Cement, Viscosity Graded at 140°F (60°C)

Test on Residue from Rolling Thin-Film Oven Test:[a]	Viscosity Grade				
	AR-1000	AR-2000	AR-4000	AR-8000	AR-16000
Viscosity, 140°F (60°C),P	1000 ± 250	2000 ± 500	4000 ± 1000	8000 ± 2000	16000 ± 4000
Viscosity, 275°F (135°C), min, cSt	140	200	275	400	550
Penetration, 77°F (25°C), 100 g, 5 s, min	65	40	25	20	20
% of original penetration, 77°F (25°C), min	—	40	45	50	52
Ductility, 77°F, (25°C), 5 cm/min, min, cm	100[b]	100[b]	75	75	75
Tests on original asphalt:					
Flash Point, Cleveland open cup, min. °F (°C)	400 (205)	425 (219)	440 (227)	450 (232)	460 (238)
Solubility in trichloroethylene, min, %	99.0	99.0	99.0	99.0	99.0

Note. Grading based on residue from rolling thin-film oven test. *Source*: From ASTM, 2003. ASTM D3381 standard specification for viscosity-graded asphalt cement for use in pavement construction, *Annual Book of ASTM Standards*, West Conshohocken, Pennsylvania. With permission.

[a] Thin-film oven test may be used but the rolling thin-film oven test shall be the referee method.

[b] If ductility is less than 100, material will be accepted if ductility at 60°F (15.5°C) is 100 minimum at a pull rate of 5 cm/min.

of 60°C (140°F). Two marks on the viscometer indicate the starting and ending time for the test. The time taken to pull the asphalt cement from the beginning point to the end point is recorded and compared against the time for a fluid of known viscosity. The ratio in these two times is used to calculate the viscosity of the asphalt.

Kinematic viscosity is determined using AASHTO T201 (ASTM D2170) "*Standard Method of Test for Kinematic Viscosity of Asphalts (Bitumens)*" (AASHTO, 2003). In this procedure, an asphalt cement sample is allowed to flow between two timing marks within a Zeitfuchs Cross-Arm viscometer under the force of gravity while the viscometer is maintained at a temperature of 135°C (275°F). Two marks on the viscometer indicate the starting and ending time for the test. The time taken to pull the asphalt cement from the beginning point to the end point is recorded. The kinematic viscosity is calculated by multiplying this time with the calibration factor supplied by the viscometer tube's manufacturer.

7.2.5.2.2   Rolling Thin-Film Test
The RTFO test is specified in the performance-graded (PG) binder system and is discussed in that section.

7.2.5.3   Performance-Graded System
The PG system for grading asphalt cements emerged from the SHRP in the early 1990s. Unlike the aggregate portion of the program, new tests were developed to test asphalt cements. To begin, let us discuss the concepts behind the PG system.

When the Penetration grading system was initially developed, it consisted of testing asphalt cements at one temperature (i.e. 25°C or 77°F). However, the performance of HMA cannot be adequately captured with one test at one temperature. There needed to be a way of evaluating asphalts at both high and low temperatures. From this emerged requirements within the viscosity (AC) and aged-residue viscosity (AR) grading system for tests to be conducted at 25°C (penetration), 60°C (absolute viscosity), and 135°C (kinematic viscosity). Requirements at three temperatures allow for better definition of the temperature dependency (difference in asphalt cement properties at different temperatures) of asphalt cements specified. This concept is illustrated in Figure 7.2. In this example, Binders A and B have the same temperature dependency over the range of temperature shown with Binder A being stiffer. Binder C has similar characteristics with Binder A at low temperatures and Binder B at high temperatures. Binder C also has significantly greater temperature susceptibility than Binders A and B. This graph indicates that Binders A, B, and C will have different characteristics and performance in hot and cold weather conditions. These control points limit the extent to which asphalt cement properties can vary. However, it does not tie down the performance of the asphalt cement at specific temperatures.

In the PG binder system, threshold values were developed for binder properties to address distresses at different temperatures. For example, the minimum values for "$G^*/\sin\delta$" was established at 1.00 and 2.20 kPa for unaged and RTFO-aged

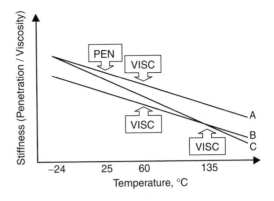

FIGURE 7.2 Characteristics of three asphalt cements of the same grade.

binders, respectively, to address permanent deformation. Therefore, the asphalt cement should be tested at the anticipated in-service temperature to see if the value for "$G^*/\sin\delta$" falls below these values. The PG binder tests are not tied to a fixed temperature. Instead they are tied to the anticipated in-service temperatures at which a particular distress occurs. For example, "$G^*/\sin\delta$" is determined using the dynamic shear rheometer (DSR) at test temperatures corresponding to hot summer days. Similarly, threshold values for other properties were established to address low-temperature cracking and fatigue cracking at the respective temperatures. Figure 7.3 shows the chart from ASTM D6373 "*Standard Specification for Performance Graded Asphalt Binder*" (ASTM, 2003).

A PG binder is specified using the high and low temperature grades. For example, a PG 64-22 is specified for an average 7-day maximum design temperature of 64°C and a minimum pavement design temperature of −22°C. The first step in using this chart is to find your path down it. For a PG 64-22, match the "64" in the "performance grade" row with the "22" below it. From this point forward, just follow that column down to the bottom of the chart. The first set of three tests are on the original (or unaged) binder. The first test is the flash point test. As discussed earlier in this section, this is a safety test. The second test is the rotational viscosity test run at 135°C. This test is to evaluate the high-temperature workability of the asphalt cement. The third test is the DSR test on the unaged binder at 64°C. The DSR test is used to evaluate the asphalt cement's ability to resist permanent deformation.

The second set of tests involves the short-term (RTFO) aging of the unaged binder and subsequent tests on the RTFO-aged binder. Other than aging the binder, the RTFO test is also used to evaluate the loss of volatiles in the short-term aging process. Next, the RTFO-aged sample is tested using the DSR for a second evaluation of the asphalt cement's ability to resist permanent deformation. The third set of tests involves the long-term aging (pressure aging vessel or PAV) of the RTFO-aged binder and subsequent tests on the PAV-aged binder. The DSR test is run on the PAV-aged binder to evaluate the binder's susceptibility to fatigue cracking. The bending beam rheometer (BBR) test, and possibly the direct tension test are also run on the PAV-aged binder to evaluate the binder's ability to resist low-temperature cracking.

The following sections will discuss tests developed during the SHRP.

7.2.5.3.1 *Rotational Viscosity Test*

The rotational viscosity is used to evaluate high-temperature workability of asphalt cement. This property is determined using AASHTO T316 (ASTM D4402) "*Standard Method of Test for Viscosity Determination of Asphalt Binder Using Rotational Viscometer*" (AASHTO, 2003). In this procedure, asphalt cement is poured into a sample chamber, which is subsequently placed into a thermo-container. This container keeps that asphalt cement at a temperature of 135°C (275°F). The torque required to maintain a constant rotational speed of a cylindrical spindle submerged in the asphalt cement sample is used to calculate the rotational viscosity. This test is used to ensure that asphalt cement is sufficiently fluid

Performance Grade	PG 46			PG 52							PG 58					PG 64						PG 70						PG 76					PG 82				
	-34	-40	-46	-10	-16	-22	-28	-34	-40	-46	-16	-22	-28	-34	-40	-10	-16	-22	-28	-34	-40	-10	-16	-22	-28	-34	-40	-10	-16	-22	-28	-34	-10	-16	-22	-28	-34
Average 7-day maximum Pavement Design Temperature, °C	< 46			< 52							< 58					< 64						< 70						< 76					< 82				
Minimum Pavement Design Temperature, °C [A]	> -34	> -40	> -46	> -10	> -16	> -22	> -28	> -34	> -40	> -46	> -16	> -22	> -28	> -34	> -40	> -10	> -16	> -22	> -28	> -34	> -40	> -10	> -16	> -22	> -28	> -34	> -40	> -10	> -16	> -22	> -28	> -34	> -10	> -16	> -22	> -28	> -34
Original Binder																																					
Flash Point Temp., D92: min °C	230																																				
Viscosity, D4402: max. 3 Pa·s Test Temp., °C	135																																				
Dynamic Shear, P246: [C] G*/sinδ, min. 1.00 kPa 25 mm Plate, 1 mm Gap Test Temp. at 10 rad/s, °C	46			52							58					64						70						76					82				
Rolling Thin Film Oven Test (Test Method D2872)																																					
Mass Loss, max. percent	1.00																																				
Dynamic Shear, P246: [C] G*/sinδ, min. 2.20 kPa 25 mm Plate, 1 mm Gap Test Temp. at 10 rad/s, °C	46			52							58					64						70						76					82				
Pressure Aging Vessel Residue (AASHTO PP1)																																					
PAV Aging Temperature, °C [D]	90			90							100					100						100 (110)						100 (110)					100 (110)				
Dynamic Shear, P246: G*sinδ, max. 5000 kPa 8 mm Plate, 2 mm Gap Test Temp. at 10 rad/s, °C	10	7	4	25	22	19	16	13	10	7	25	22	19	16	13	31	28	25	22	19	16	34	31	28	25	22	19	37	34	31	28	25	40	37	34	31	28
Creep Stiffness, P245: [E] S, max. 300 Mpa, m-value, min. 0.300 Test Temp. at 60s, °C	-24	-30	-36	0	-6	-12	-18	-24	-30	-36	-6	-12	-18	-24	-30	0	-6	-12	-18	-24	-30	0	-6	-12	-18	-24	-30	0	-6	-12	-18	-24	0	-6	-12	-18	-24
Direct Tension, P252: [E] Failure strain, min. 1.0% Test Temperature at 1.00 mm/min., °C	-24	-30	-36	0	-6	-12	-18	-24	-30	-36	-6	-12	-18	-24	-30	0	-6	-12	-18	-24	-30	0	-6	-12	-18	-24	-30	0	-6	-12	-18	-24	0	-6	-12	-18	-24

[A] Pavement temperatures are estimated from air temperatures using an algorithm contained in the SUPERPAVE software program, or are provided by the specifying agency.

[B] The referee method shall be D4402 using a #21 spindle at 20 RPM, however alternate methods may be used for routine testing and quality assurance. This requirement may be waved at the discretion of the specifying agency if the supplier warrants that the asphalt binder can be adequately pumped and mixed at temperatures that meet all applicable safety standards.

[C] For quality control of unmodified asphalt cement production, measurement of the viscosity of the orginal asphalt cement may be substituted for dynamic shear measurements of G*/sinδ at test temperatures where the asphalt is a Newtonian fluid. Any suitable standard means of the viscosity measurement may be used, includingcapillary or rotational viscometry (Test Methods D2170 or D2171).

[D] The PAV aging temperature is based on simulated climatic conditions and is one of three temperatures 90°C, 100°C, or 110°C. The PAV aging temperature is 100 °C for PG 64- and above, except in desert climates, where it is 110°C.

[E] If the creep stiffness is below 300 MPA, the direct tension test is not required. If the creep stiffness is between 300 and 600 MPA, the direct tension failure strain requirement can be used in lieu of the creep stiffness requirement. The m-value requirement must be satisfied in both cases.

FIGURE 7.3 Performance graded binder specifications. [*Source:* From ASTM, 2003. ASTM D6373 Standard Specification for Performance Graded Asphalt Binder, *Annual Book of ASTM Standards*, West Conshohocken, Pennsylvania. With permission.]

when pumping and mixing. This test is often run at a higher temperature (e.g., 165°C) in conjunction with determining the laboratory mixing and compaction temperatures in the Superpave mix design system. This process is discussed in Section 7.2.8.4.2.

7.2.5.3.2 Dynamic Shear Rheometer Test

DSR is used to characterize the viscous and elastic behavior of asphalt cements. This characterization is accomplished using AASHTO T315 *"Standard Method of Test for Determining Rheological Properties of Asphalt Binder Using a Dynamic Shear Rheometer (DSR)"* (AASHTO, 2003). In this procedure, a disk of asphalt cement is placed between an oscillating spindle and the base plate. Since asphalt cement is a viscoelastic material, there is a time lag (δ) between the applied shear stress and the resulting shear strain. This is shown in Figure 7.4. The maximum shear stress (T_{max}) and the maximum shear strain (γ_{max}) are calculated using the following equations:

$$\tau_{max} = 2T/\pi f^3 \qquad (7.6)$$

$$\gamma_{max} = \Theta r/h \qquad (7.7)$$

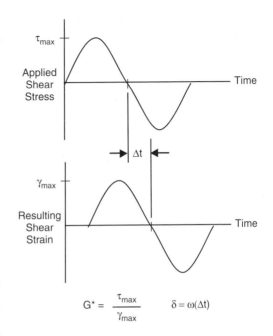

$$G^* = \frac{\tau_{max}}{\gamma_{max}} \qquad \delta = \omega(\Delta t)$$

FIGURE 7.4 Stress–strain response of a viscoelastic material.

where T is the maximum applied torque (Nm), r is the radius of specimen or plate (4 or 12.5 mm), Θ is the deflection (rotation) angle, h is the specimen height (1 or 2 mm).

This test is used to evaluate an asphalt cement's ability to resist permanent deformation and fatigue cracking. The test is conducted at the average 7-day maximum pavement design temperature to evaluate resistance to permanent deformation and at the PG intermediate temperature to evaluate resistance to fatigue temperature. The fatigue test temperature can be read off the specification chart. A minimum $G^*/\sin\delta$ of 1.00 and 2.20 kPa at the average 7-day maximum pavement design temperature for the un-aged and RTFO-aged samples, respectively, have been established to minimize permanent deformation. A maximum $G^*/\sin\delta$ of 5000 kPa has been established to minimize fatigue cracking. Post-SHRP research has indicated that the fatigue cracking criterion had not been successful in preventing fatigue cracking. Further research and development continues on this issue.

7.2.5.3.3 Rolling Thin-Film Oven Test

The general purpose of the RTFO test is similar to that of the TFO test. It is intended to simulate short-term aging of asphalt cements in the laboratory. The advantage of the RTFO test over the TFO test is that it is quicker and a larger number of samples can be tested at the same time. This test is conducted in accordance with AASHTO T240 (ASTM D2872) *"Standard Method of Test for Effect of Heat and Air on a Moving Film of Asphalt (Rolling Thin-Film Oven Test)"* (AASHTO, 2003). In this procedure, approximately 35-g samples are poured into cylindrical glass bottles and placed in a rack within an oven maintained at 163°C (325°C). This rack rotates at approximately 15 rev/min with air being circulated into the bottles at a rate of approximately 4000 ml/min at one point in the rotation. After 85 min, the test is completed and the sample is considered short-term aged after it is removed from the oven. In this test, the mass loss during the aging process is calculated. This mass loss, due to the loss of volatiles, provides an indication of potential aging that may take place with this asphalt cement during the mixing and compaction. The PG system allows for a maximum mass loss of 1%.

7.2.5.3.4   *Pressure Aging Vessel Test*

The PAV is used to simulate long-term field aging of asphalt cements in the laboratory. This test is conducted in accordance with AASHTO R28 "*Standard Recommended Practice for Accelerated Aging of Asphalt Binder Using a Pressure Aging Vessel (PAV)*" (AASHTO, 2003). In this procedure, approximately 50-g samples are poured into PAV pans and placed in the sample rack within a pressurized (2070 kPa) oven maintained at approximately 100°C (212°C). The oven temperature may also be set to 90°C (194°F) or 110°C (230°F) based on the in-service climatic conditions. After approximately 20 h, the test is completed and the sample is considered long-term aged and is ready for the tests.

7.2.5.3.5   *Bending Beam Rheometer Test*

The BBR is used to measure the stiffness of asphalt cements at low temperature. The test provides two parameters. Creep stiffness is a measure of how the asphalt resists loading and is measured after 60 sec. The m-value is a measure of how the asphalt stiffness changes as a load is applied and is represented by the slope of the log stiffness vs. log time curve at 60 sec. These properties are measured using AASHTO T313 "*Standard Method of Test for Determining the Flexural Creep Stiffness of Asphalt Binder Using the Bending Beam Rheometer (BBR)*" (AASHTO, 2003). In this procedure, a beam of asphalt cement that has been RTFO- and PAV-aged is maintained at a constant temperature inside a fluid bath. The middle of the beam is subjected to a constant loaded by a blunt-nosed shaft while being supported at two locations near its ends. The deflection of the beam is measured and plotted against time as shown in Figure 7.5.

Equation 7.8 shows classic beam theory used to calculate the flexural beam stiffness:

$$S(t) = \frac{PL^3}{4bh^3 \delta(t)} \tag{7.8}$$

where $S(t)$ is the creep stiffness (at time 60 sec), P is the applied constant load (980 mN), L is the distance between beam supports (102 mm), b is the beam width (12.5 mm), h is the beam thickness (6.25 mm), and $\delta(t)$ is the deflection (at time 60 sec).

The desired critical stiffness and m-value at the minimum pavement design temperature is after 2 h of loading. However, the SHRP researchers discovered that these values could be obtained after 60 sec of loading if the test temperature was raised 10°C. This is based on the theory of time–temperature superposition.

A maximum stiffness of 300 kPa and minimum m-value of 0.300 are required to minimize low-temperature cracking. If both these criteria are met, there is no need for additional testing. If the maximum stiffness is between 300 and 600 kPa and the m-value is greater than 0.300, then the direct tension test may be conducted to check for compliance with the PG specifications for that grade of asphalt cement. Once again, these critical values are measured at the anticipated critical temperature, i.e. the minimum pavement design temperature.

7.2.5.3.6   *Direct Tension Test*

DT measures the low-temperature ultimate tensile strain of asphalt cement. The test provides the strain at which the maximum load occurs. This property is measured using AASHTO T314 "*Standard Method of Test for Determining the Fracture Properties of Asphalt Binder in Direct Tension*" (AASHTO, 2003). In this

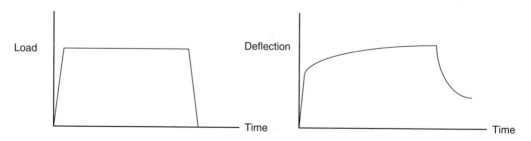

FIGURE 7.5   Plot of load and deflection vs. time.

procedure, a small dog-bone shaped asphalt cement sample is maintained at a constant temperature inside a fluid bath. This specimen is then loaded in tension at a constant strain rate of approximately 1.00 mm/min. The failure strain (ε_f) is the strain corresponding to the maximum applied stress (σ_f) and is calculated using Equation 7.9:

$$\varepsilon_f = \Delta L/L \qquad (7.9)$$

where ε_f is the failure strain, ΔL change in length, and L is effective gauge length. An example of the stress–strain plot for a test is shown in Figure 7.6.

The desired critical strain at the minimum pavement design temperature is 1%. This test is also conducted at 10°C above minimum pavement design temperature.

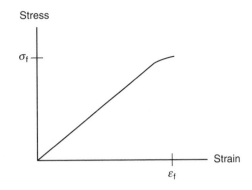

FIGURE 7.6 Example stress–strain relationship for a direct tension test.

7.2.6 Polymer Modified Asphalts

Polymer modified asphalts have been used in the asphalt paving industry for over 60 years. However, the birth of the PG system for grading asphalts provided new impetus for increased product development and usage of polymer modified asphalts. Adding polymers to asphalt generally increases the high temperature grade of the asphalt binder. However, depending on the type of polymer, the low temperature grade may increase or decrease. Polymer modified asphalts may improve resistance to permanent deformation, thermal cracking, fatigue cracking, and moisture damage. Many specifying agencies are now requiring asphalts with performance grades where polymer modification is required. As a rule of thumb, performance grades with a numeric difference of greater than 92 require polymer modified asphalts. For example, the numeric difference for a PG 72-28 is 100. The asphalt cements meeting this grade are likely to be polymer modified.

Polymers used in asphalt may be classified by their physical properties. Based on their response to stretching with some force, they may be classified as plastomers or elastomers. Plastomers yield and remain stretched after the force is released. Elastomers return to their original shape when released. Most polyolefins, such as polyethylene, polypropylene, and ethylene vinyl acetate (EVA), are considered plastomers. Styrene–butadiene rubber (SBR) and styrene–butadiene–styrene (SBS) are categorized as elastomers.

7.2.7 Volumetric Properties of Asphalt Mixtures

A compacted asphalt concrete mix consists primarily of aggregate, asphalt, and air. The volumetric properties associated with the combination of these three components are widely used for mix design and production control. Since it is impractical to measure the volume of constituent components within a HMA mix in the laboratory or in the field, mass–volume relationships are used to convert the measurable masses into their corresponding volumes. The mass of a constituent is directly proportional to its volume as shown in the following equation:

$$M = VG\rho_w \qquad (7.10)$$

where M is the mass of constituent, V is the volume of constituent, G is the specific gravity of constituent, and ρ_w is the density of water (1.0 g/cm^3).

The relationship between mass and volume is illustrated in Figure 7.7. This phase diagram is similar to that used in soil mechanics, except that the liquid in HMA is asphalt cement instead of water. It shows the air component at the top followed by the effective (free) asphalt component, the absorbed asphalt

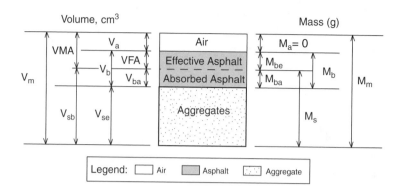

FIGURE 7.7 Mass–volume relationship diagram.

(i.e. absorbed by the mineral aggregate) component, and the mineral aggregate component. The total asphalt in HMA is the sum of the effective and absorbed asphalt. Some volumetric properties are of particular interest for HMA mix design. They can be found in "*Background to Superpave Asphalt Mixture Design and Analysis*" published by the FHWA (McGennis et al., 1995). This reference is listed in the references at the end of the chapter.

Example 7.1

This example involves the computation of (a) percent air voids, (b) VMA, (c) voids filled with ashphalt (VFA), (d) effective asphalt content, (e) absorbed asphalt content, and (f) maximum theoretical specific gravity based on the following information:

Mixture Bulk Specific Gravity, G_{mb} = 2.331
Asphalt Binder Specific Gravity, G_b = 1.013
percent binder, P_b = 5.0% by mix
Aggregate Bulk Specific Gravity, G_{sb} = 2.707
Aggregate Effective Specific Gravity, G_{se} = 2.733

Figure 7.8 illustrates the information available in Example 7.1 in a mass–volume diagram.

Solution

Figure 7.8 shows the phase diagram for Example 7.1 with the assumption that calculations are based on a unit volume (i.e. 1 cm^3). As air has no mass, we can determine that the mass of air is 0 g. Based on the mass–volume relationships, the following calculations can be made.

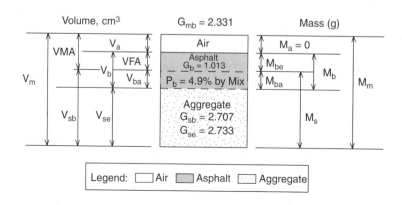

FIGURE 7.8 Initial mass–volume diagram for Example 7.1.

Mass of Mix:

$$M_m = V_m G_m \rho_w = 1 \text{ cm}^3 \times 2.331 \times 1.0 \text{ g/cm}^3 = 2.331 \text{ g} \tag{7.11}$$

Mass of Asphalt:

$$M_b = P_b M_m = 5/100 \times 2.331 \text{ g} = 0.117 \text{ g} \tag{7.12}$$

Mass of Aggregate:

$$M_a = M_m - M_b = 2.331 \text{ g} - 0.117 \text{ g} = 2.214 \text{ g} \tag{7.13}$$

Bulk Vol. Agg.:

$$V_{sb} = \frac{M_s}{G_{sb}\rho_w} = \frac{2.214 \text{ g}}{2.707 \times 1.0 \text{ g/cm}^3} = 0.817 \text{ cm}^3 \tag{7.14}$$

Effective Vol. of Agg.:

$$V_{se} = \frac{M_s}{G_{se}\rho_w} = \frac{2.214 \text{ g}}{2.733 \times 1.0 \text{ g/cm}^3} = 0.810 \text{ cm}^3 \tag{7.15}$$

Vol. of Total Asphalt:

$$V_b = \frac{M_b}{G_b\rho_w} = \frac{0.117 \text{ g}}{1.013 \times 1.0 \text{ g/cm}^3} = 0.115 \text{ cm}^3 \tag{7.16}$$

Vol. of Absorbed Asphalt:

$$V_{ba} = V_{sb} - V_{se} = 0.817 \text{ cm}^3 - 0.810 \text{ cm}^3 = 0.007 \text{ cm}^3 \tag{7.17}$$

Vol. Effective Asphalt:

$$V_{be} = V_b - V_{ba} = 0.115 \text{ cm}^3 - 0.007 \text{ cm}^3 = 0.108 \text{ cm}^3 \tag{7.18}$$

Vol. Air Voids:

$$V_a = V_m - V_{se} - V_b = 0.115 \text{ cm}^3 - 0.007 \text{ cm}^3 = 0.108 \text{ cm}^3 \tag{7.19}$$

Mass of Absorbed Asphalt:

$$M_{ba} = V_{ba} G_b \rho_w = 0.007 \text{ cm}^3 \times 1.013 \times 1.0 \text{ g/cm}^3 = 0.007 \text{ g} \tag{7.20}$$

Mass of Effective Asphalt:

$$M_{be} = V_{be} G_b \rho_w = 0.108 \text{ cm}^3 \times 1.013 \times 1.0 \text{ g/cm}^3 = 0.109 \text{ g} \tag{7.21}$$

% of Voids:

$$P_a = \frac{V_a}{V_m} 100\% = \frac{0.073 \text{ cm}^3}{1.0 \text{ cm}^3} \times 100\% = 7.3\% \tag{7.22}$$

Voids in Mineral Aggregate:

$$\text{VMA} = \frac{V_a - V_{be}}{V_m} 100\% = \frac{0.075 \text{ cm}^3 + 0.108 \text{ cm}^3}{1.0 \text{ cm}^3} \times 100\% = 18.3\% \tag{7.23}$$

Voids Filled With Asphalt:

$$\text{VFA} = \frac{V_{be}}{\text{VMA}} 100\% = \frac{0.108 \text{ cm}^3}{0.183 \text{ cm}^3} \times 100\% = 59.0\% \tag{7.24}$$

Effective Asphalt Content (% Mass of Mix):

$$P_{be} = \frac{M_{be}}{M_m} 100\% = \frac{0.109 \text{ g}}{2.331 \text{ g}} \times 100\% = 4.7\% \tag{7.25}$$

Absorbed Asphalt Content (percent Mass of Aggregate):

$$P_{ba} = \frac{M_{ba}}{M_s}\,100\% = \frac{0.007\text{ g}}{2.314\text{ g}} \times 100\% = 0.3\% \tag{7.26}$$

Maximum Theoretical Specific Gravity:

$$G_{mm} = \frac{G_{mb}}{(V_m - V_a)/V_m} = \frac{2.331}{(1.0\text{ cm}^3 - 0.075\text{ cm}^3)/1.0\text{ cm}^3} = 2.520 \tag{7.27}$$

The solution to Example 7.1 above made use of fundamental relationships to solve mass–volume calculations. The calculated volumes and masses are shown in Figure 7.9. However, mix designers often use formulas that can be easily used in spreadsheets to automate these calculations. These formulas are provided below.

Bulk Specific Gravity of Combined Aggregate:

$$G_{sb} = \frac{P_1 + P_2 + \cdots + P_n}{\dfrac{P_1}{G_1} + \dfrac{P_2}{G_2} + \cdots + \dfrac{P_n}{G_n}} \tag{7.28}$$

where P_1, P_2, P_n are individual percentages by mass of aggregate and G_1, G_2, G_n are individual bulk specific gravities of aggregate.

Effective Specific Gravity of Aggregate:

$$G_{se} = \frac{P_{mm} - P_b}{\dfrac{P_{mm}}{G_{mm}} - +\dfrac{P_b}{G_b}} \tag{7.29}$$

where G_{mm} is the theoretical maximum specific gravity of mix, P_{mm} is the percent by mass of total loose mix (100%), P_b is the percent asphalt by total mass of mix, and G_b is the percent asphalt binder specific gravity.

Theoretical Maximum Specific Gravity of Mix:

$$G_{mm} = \frac{P_{mm}}{\dfrac{P_s}{G_{se}} - +\dfrac{P_b}{G_b}} \tag{7.30}$$

where P_{mm} is the percent by mass of total loose mix (100%), P_s is the percent of combined aggregate by total mass of mix, P_b is the percent asphalt by total mass of mix, G_{se} is the effective specific gravity of combined aggregate, and G_b is the asphalt binder specific gravity.

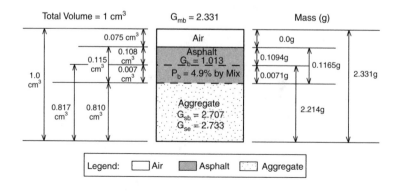

FIGURE 7.9 Final mass–volume diagram for Example 7.1.

Asphalt Absorption:

$$P_{ba} = 100 \frac{G_{se} - G_{sb}}{G_{se} G_{sb}} G_b \qquad (7.31)$$

where G_{se} is the effective specific gravity of combined aggregate, G_{sb} is the bulk specific gravity of combined aggregate, and G_b is the asphalt binder specific gravity.

Percent Effective Asphalt Content:

$$P_{be} = P_b - \frac{P_{ba}}{100\%} P_s \qquad (7.32)$$

where P_a is the percent asphalt by total mass of mix, P_{ba} is the percent absorbed asphalt by mass of combined aggregate, and P_s is the percent aggregate by total mass of mix.

Percent Voids in Mineral Aggregate:

$$VMA = 100\% - \frac{G_{mb} P_s}{G_{sb}} \qquad (7.33)$$

where G_{sb} is the bulk specific gravity of combined aggregate, G_{mb} is the bulk specific gravity of compacted mix, and P_s is the percent of combined aggregate by total mass of mix.

Percent Air Voids (by Volume):

$$P_a = 100\% - \frac{G_{mm} - G_{mb}}{G_{mm}} \qquad (7.34)$$

where G_{mm} is the theoretical maximum specific gravity of mix and G_{mb} is the bulk specific gravity of compacted mix.

Percent Voids Filled with Asphalt:

$$VFA = 100\% - \frac{VMA - V_a}{VMA} \qquad (7.35)$$

where VMA is the voids in mineral aggregate and V_a is the air voids in compacted mix by percent of total volume.

7.2.8 Design of Hot-Mix Asphalt Concrete

The objective of mix designs is to produce HMA that performs well both structurally and functionally. From a performance standpoint, the HMAC should be durable and be able to resist pavement distresses (such as permanent deformation, load-induced fatigue, thermal fatigue, low temperature cracking, and moisture-induced damage). From a construction standpoint, the mix should be workable enough to place and compact with reasonable effort. Additionally, surface courses should provide sufficient skid resistance for safety considerations.

7.2.8.1 A Brief History

Crawford (1989) traced asphalt mix design back to the 1860s with the first bituminous pavements placed mostly in Washington, DC from about 1868 to 1873. In those days, tar and naturally occurring Trinidad Lake asphalt were used as binders. The tar mixes did not perform well as surface mixes but were satisfactory as bases. Richardson (1912) later developed what could be considered the first mix design test, the "Pat Test," to determine optimum binder content in a HMAC. This was a visual test that determined if the optimum asphalt content was achieved by evaluating the stain created by a sheet sample of asphalt against a brown manila paper.

Several others also contributed to the advancement of asphalt technology. However, it was the work of Francis Hveem (mid 1920s) and Bruce Marshall (late 1930s) that led to mix design procedures still widely used around the world today. Vallerga and Lovering (1985) provided a detailed account of the evolution of the Hveem procedure. Foster (1982), and White (1985) gave a detailed account of the development of the Marshall method.

From October 1987 through March 1993, the SHRP conducted a $50 million research effort to develop new ways to specify, test, and design asphalt materials. From this research effort and continued development over the last 10 years, the Superpave mix design methodology was developed and refined. The Superpave mix design method has caught on in the United States with many states agencies specifying some form of it. The following sections will take a closer look at these three mix design procedures. It should be noted that these mix design procedures are primarily intended for regular dense HMAC. Additional information for designing HMAC with recycled asphalt pavement (RAP) can be found in National Center for Asphalt Technology (NCAT) Report No. 96-5 (Kandhal and Foo, 1997). There are other procedures available for mixes such as the stone mastic asphalt (SMA) (Brown et al., 1997) and open-graded friction course (OGFC) (Mallick et al., 2000).

7.2.8.2 Marshall Mixture Design Method

Bruce Marshall, while serving as Bituminous Engineer with the Mississippi State Highway Department, put together the concepts behind the Marshall mix design procedure. The U.S. Corps of Engineers conducted extensive research based on Marshall's concepts and eventually established the mix design method named after Marshall. AASHTO adopted this mix design procedure as AASHTO R-12 "*Standard Recommended Practice for Bituminous Mixture Design Using the Marshall and Hveem Procedures*" (AASHTO, 2003). Prior to the development of the Superpave mix design procedure, the Marshall mix design was the most commonly used mix design procedure in the United States. Even now, it is still the most commonly used procedure in many parts of the world.

This procedure, contained in the Asphalt Institute's publication entitled "*Mix Design Methods for Asphalt Concrete and Other Hot Mix Type*" (Asphalt Institute, 1997), can be broken into seven major steps. A summary of these steps is found below. More details on these steps can be found in the Asphalt Institute's "*Mix Design Methods for Asphalt Concrete and Other Hot-Mix Types*" (MS-2) (Asphalt Institute, 1997).

7.2.8.2.1 *Step 1. Aggregate Evaluation*

The aggregates proposed for use in the mix design should be evaluated against the aggregate requirements set forth by the specifying agency. Descriptions of the commonly used aggregate tests are described in Section 7.2.2. However, individual agencies may have requirements different from those used in the Superpave mix design procedure. If the aggregates meet the applicable requirements above, the aggregate specific gravity, absorption, and gradation are determined. These will be used for determining the volumetric properties of the mix. Last, but not least, the individual aggregate gradations are combined in varying proportions to develop trial blends. The gradations of these trial blends should be evaluated based on past agency experience and other general principles (e.g. those described in Section 7.2.3). With the theoretical calculations completed, batching sheets should be completed showing the batch weights for the relevant sieve sizes for each individual aggregate source used.

7.2.8.2.2 *Step 2. Asphalt Cement Evaluation*

The asphalt cement used in the mix design should be suitable for the geographical location where the mix will be placed. Run the required tests to ensure that the asphalt cement meets all the agency specifications. Finally, determine the mixing and compaction temperatures based on viscosities of 170 ± 20 and 280 ± 30 centistokes (cSt), respectively.

7.2.8.2.3 *Step 3. Preparation of Marshall Specimens*

Prepare the Marshall specimens in accordance to the requirements set in AASHTO R-12. Compact three replicate specimens at five asphalt contents. The asphalt contents should be selected at 0.5% asphalt increments with two asphalt contents falling above and below the "optimum" asphalt content. These specimens are compacted using the Marshall compactor. Three loose specimens should also be prepared for determining the maximum theoretical specific gravity near the "optimum" asphalt content.

Determine the bulk specific gravity of the compacted specimens and maximum theoretical specific gravity of the loose mix using AASHTO T166 (ASTM D2726) "*Standard Method of Test for Bulk Specific Gravity of Compacted Asphalt Mixtures Using Saturated Surface-Dry Specimens*" and AASHTO T209 "*Standard Method of Test for Theoretical Maximum Specific Gravity and Density of Bituminous Paving Mixtures*," respectively. ASTM D6752 "*Standard Test Method for Bulk Specific Gravity and Density of Compacted Bituminous Mixtures Using Automatic Vacuum Sealing Method*" is a more recently approved method for determining bulk specific gravity of compacted bituminous mixtures.

7.2.8.2.4 *Step 4. Marshall Stability and Flow*
Determine the Marshall stability and flow using ASTM D1559. Stability is defined as the maximum load carried by a compacted specimen tested at 140°F at a loading rate of 2 in./min. The two primary factors in determining the stability are the angle of internal friction of the aggregate and the viscosity of the asphalt cement. Therefore mixes with angular aggregates will have a higher stability than mixes with rounded aggregates. Similarly, mixes with more viscous asphalt cements will have a higher stability than mixes with less viscous asphalt cements. The primary use of the Marshall stability is to evaluate the effect of asphalt cement in the Marshall mix design procedure. It is not specifically correlated to the stability of mixes in the field. Increasing the Marshall stability in the laboratory does not automatically translate to increased stability of mixes in the field. Flow is the vertical deformation of the sample at failure. High flow values typically indicate a plastic mix that could be susceptible to permanent deformation. Low flow values may indicate low air voids that may lead to premature cracking.

7.2.8.2.5 *Step 5. Density and Void Analysis*
Using the bulk specific gravity and maximum theoretical specific gravity test results and the equations in Section 7.2.7, the volumetric properties of the mix can be determined. This information is used in Step 6.

7.2.8.2.6 *Step 6. Tabulating and Plotting Test Results*
With the completion of Steps 4 and 5, the average (from three replicates) results can be tabulated and plotted. The following plots can then be made to evaluate the mix:

- Density (or Unit Weight) vs. Asphalt Content
- Marshall Stability vs. Asphalt Content
- Flow vs. Asphalt Content
- Air Voids vs. Asphalt Content
- VMA vs. Asphalt Content
- VFA vs. Asphalt Content

The density plot typically shows a trend of increasing density until the peak is reached. After this peak, the density begins to decrease. The Marshall stability has a similar trend but its peak is typically at a lower asphalt content than density. Some recycled mixes may show a decreasing stability with increasing asphalt content with no peak. Flow typically increases with increasing asphalt content. The percent air voids should decrease and the VFA increase with increasing asphalt content. VMA is another property that increases with asphalt content until it reaches its peak and then decreases with additional increase in asphalt content.

7.2.8.2.7 *Step 7. Optimum Asphalt Content Determination*
The criteria used to select the optimum asphalt content can vary considerably between agencies. The Asphalt Institute method (Asphalt Institute, 1997) and the National Asphalt Pavement Association (NAPA) procedure (NAPA, 1982) are described below.

In the Asphalt Institute method, the following asphalt contents are determined:

- Asphalt content at maximum stability
- Asphalt content at maximum density
- Asphalt content at the mid-point of specified air void content (typically at 4%)

TABLE 7.4   Marshall Mix Design Requirements

Marshall Method Mix Criteria	Light Traffic Surface and Base		Light Traffic Surface and Base		Light Traffic Surface and Base	
	Min	Max	Min	Max	Min	Max
Compaction, number of blows each end of specimen	35		50		75	
Stability, N (lb)	3336 (750)	—	5338 (1200)	—	8006 (1800)	—
Flow, 0.25 mm (0.01 in.)	8	18	8	16	8	14
Percent Air Voids	3	5	3	5	3	5
Percent VMA			See Table 10.3			
Percent VFA	70	80	65	78	65	75

Notes. (1) All criteria, not just stability value along, must be considered in designing an asphalt paving mix. Hot mix asphalt bases that do not meet these criteria when tested at 60°C (140°F) are satisfactory if they meet the criteria when tested at 38°C (100°F) and are placed 100 mm (4 in.) or more below the surface. This recommendation applies only to regions having a range of climatic conditions similar to those prevailing throughout most of the United States. A different lower test temperature may be considered in regions having more extreme climatic conditions. (2) Traffic classifications: Light, Traffic conditions resulting in a Design EAL < 104; Medium, Traffic conditions resulting in a Design EAL between 10^4 and 10^6; Heavy, Traffic conditions resulting in a Design EAL > 10^6. (3) Laboratory compaction efforts should closely approach the maximum density obtained in the pavement under traffic. (4) The flow value refers to the point where the load begins to decrease. (5) The portion of the asphalt cement lost by absorption into the aggregate particles must be allowed for when calculating percent air voids. (6) Percent voids in mineral aggregate is to be calculated on the basis of the ASTM bulk specific gravity for the aggregate. *Source*: From Asphalt Institute, 1997. *Mix Design Methods for Asphalt Concrete and Other Hot-Mix Types*, Manual Series No. 2, MS-2, 6th ed., Lexington, Kentucky. With permission.

Determine the proposed optimum asphalt content by averaging the results from the three criteria shown above. Compare the proposed optimum asphalt content against the criteria in Tables 7.4 and 7.5. The proposed optimum asphalt content is selected if it meets the criteria in Tables 7.4 and 7.5. If not, the mix should be redesigned.

In the NAPA approach, the proposed optimum asphalt content is selected at the mid-point of the air void content criterion. Again, this is typically at 4% air voids. This Marshall stability, flow, VMA, and VFA at the proposed optimum asphalt content should be checked against the applicable specifications. If the specifications are met, the optimum asphalt content is confirmed. If not, the mix should be redesigned.

TABLE 7.5   Marshall Mix Design Minimum Voids in Mineral Aggregate — MS2

Nominal Maximum Particle Size[a,b]		Minimum VMA, percent		
mm	in.	Design Air Voids, percent[c]		
		3.0	4.0	5.0
1.18	No. 16	21.5	22.5	23.5
2.36	No. 8	19	20	21
4.75	No. 4	16	17	18
9.5	3/8	14	15	16
12.5	1/2	13	14	15
19.0	3/4	12	13	14
25.0	1.0	11	12	13
37.5	1.5	10	11	12
50.0	2.0	9.5	10.5	11.5
63.0	2.5	9	10	11

Source: Asphalt Institute, 1997. *Mix Design Methods for Asphalt Concrete and Other Hot-Mix Types*, Manual Series No. 2, MS-2, Sixth Edition, Lexington, Kentucky. With permission.

[a] Standard Specification for Wire Cloth Sieves for Testing Purposes, ASTM E11 (AASHTO M92).
[b] The nominal maximum particle size is one size larger than the first sieve to retain more than 10%.
[c] Interpolate minimum voids in the mineral aggregate (VMA) for design air void values between those listed.

7.2.8.3   Hveem Mixture Design Method

Francis Hveem, former Materials and Research Engineer for the California Department of Transportation, developed and advanced the concepts in the Hveem mix design procedure. This procedure was developed through extensive research on hot-mix asphalt pavements. AASHTO has adopted this mix design procedure as AASHTO R-12 "*Standard Recommended Practice for Bituminous Mixture Design Using the Marshall and Hveem Procedures*" (AASHTO, 2003). This procedure, contained in the Asphalt Institute's publication entitled "*Mix Design Methods for Asphalt Concrete and Other Hot Mix Types*," (Asphalt Institute, 1997) can be broken into ten major steps. A summary of these steps is found below. More details on these steps can be found in the Asphalt Institute's "*Mix Design Methods for Asphalt Concrete and Other Hot-Mix Types*" (MS-2) (Asphalt Institute, 1997).

7.2.8.3.1   Step 1. Aggregate Evaluation
The aggregate evaluation step for the Hveem mix design procedure is the same as that for the Marshall mix design procedure and will not be repeated here.

7.2.8.3.2   Step 2. Asphalt Cement Evaluation
The asphalt cement evaluation step for the Hveem mix design procedure is the same as that for the Marshall mix design procedure and will not be repeated here.

7.2.8.3.3   Step 3. Estimated Optimum Asphalt Content
Use ASTM D5148 "*Standard Test Method for Centrifuge Kerosene Equivalent*" to determine the centrifuge kerosene equivalent (CKE) for aggregate passing the No. 4 sieve. Determine the percent oil retained on aggregate retained on the No. 4 sieve. Using the two results above, use the five charts provided in the Hveem method of design to estimate the approximate bitumen ratio (ABR). Additional information on this step can be found in the reference above.

7.2.8.3.4   Step 4. Preparation of Stabilometer Specimens
Prepare the Hveem specimens in accordance to the requirements set forth in AASHTO R-12. Compact specimens at seven asphalt contents. The asphalt contents should be selected at 0.5% asphalt increments with two asphalt contents falling below the "optimum" asphalt content and four specimens falling above the "optimum" asphalt content. These specimens are compacted using a kneading compactor. Loose specimens should also be prepared for determining the maximum theoretical specific gravity near the "optimum" asphalt content.

7.2.8.3.5   Step 5. Hveem Stabilometer Test
The stabilometer test is performed in accordance with ASTM D1560 "*Standard Test Method for Resistance to Deformation and Cohesion of Bituminous Mixtures by Means of Hveem Apparatus*." This test provides information used to calculate the stabilometer value:

$$S = \frac{22.2}{\dfrac{P_h D_2}{(P_v - P_h)} + 0.222} \tag{7.36}$$

where P_v is the vertical pressure, P_h is the horizontal pressure gauge reading at 5000 lb compressive load, and D_2 is the specimen displacement.

The Hveem stability (S) is expressed as a number between 0 and 100. A higher S-value indicates higher stability.

7.2.8.3.6   Step 6. Density and Void Analysis
The bulk specific gravities and maximum theoretical specific gravities of the compacted mix can be determined using the same test methods described for the Marshall mix design procedure. With this information, the air void content and density can be determined using the volumetric relationships in Section 7.2.7.

7.2.8.3.7 *Step 7. Analysis of Test Results*
The following information should be plotted to check for reasonable trends:

- Hveem stability vs. asphalt content
- Density vs. asphalt content
- Percent air voids vs. asphalt content

The Hveem stability should decrease with increasing asphalt content. Density will generally increase but may not hit a peak. Air voids should decrease with increasing asphalt content. The first step is to select the four highest asphalt contents that do not exhibit surface flushing or bleeding. Then select the highest asphalt content that does not have an air void content or stability below the minimum threshold of 4% air and the minimum allowable stability based on traffic. The minimum stability values are shown in Table 7.6.

7.2.8.4 Superpave Mixture Design Method
The Superpave mix design method is a product of the SHRP conducted between October 1987 and March 1993. The original product included three levels of mix design. The Level I mix design approach, based on mixture volumetrics, was intended for use on roadways with less than one million equivalent single axle loads (ESALs). The Level II approach added performance prediction tests to the mixture volumetrics and was intended for roadways with between 1 and 10 million ESALs. Level III included enhanced performance prediction tests and was intended for roadways with greater than 10 million ESALs. However, the performance prediction tests did not predict performance accurately enough. As a result, the Level I mix design approach is currently being employed for all traffic levels. The National Cooperative Highway Research Program (NCHRP) and other research funding agencies are continuing efforts to develop more accurate performance prediction tests and models. Over the last 10 years, there has also been a significant amount of research to enhance the Level 1 mix design. These findings have been implemented into the current version of the mix design procedure, AASHTO MP2 "*Superpave™ Volumetric Mix Design*" (AASHTO, 2003).

7.2.8.4.1 *Step 1. Selection of Materials*
The selection of materials can be broken out into two parts. The first part addresses the selection of an appropriate asphalt binder and the second part addresses the selection of suitable aggregates.

7.2.8.4.2 *Selection of Asphalt Cement*
The performance grade of the asphalt cement is selected based on the climatic conditions experience at the proposed location of the asphalt concrete pavement. The minimum and/or maximum criteria set for asphalt binder properties tested are the same. The only change between different binder grades is the temperature at which these properties are measured. Within the Superpave software, there is weather

TABLE 7.6   Hveem Stability Criteria — MS2

Traffic Category Test Property	Heavy		Medium		Low	
	Min.	Max.	Min.	Max.	Min.	Max.
Stabilometer Value	37	—	35	—	30	—
Swell			less than 0.762 mm (0.030 in.)			

Notes. (1) Although not a routine part of this design method, an effort is made to provide a minimum percent of air voids of approximately 4%. (2) All criteria, and not stability value alone, must be considered in designing an asphalt paving mix. (3) Hot-mix asphalt bases that do not meet these criteria when tested at 60°C (140°F) are satisfactory if they meet the criteria when tested at 38°C (100°F) and are placed at 100 mm (4 in.) or more below the surface. This recommendation applies only to regions having a range of climatic conditions similar to those prevailing throughout most of the United States. A different lower test temperature may be considered in regions having more extreme climatic conditions. (4) Traffic Classifications: Light, Traffic conditions resulting in a Design EAL $< 10^4$; Medium, Traffic conditions resulting in a Design EAL between 10^4 and 10^6; Heavy, Traffic conditions resulting in a Design EAL $> 10^6$. *Source*: From Asphalt Institute, 1997. *Mix Design Methods for Asphalt Concrete and Other Hot-Mix Types*, Manual Series No. 2, MS-2, 6th ed., Lexington, Kentucky. With permission.

information for 6092 reporting weather stations in the United States and Canada (AASHTO, 1993). The SHRP researchers concluded that the 7-day average maximum air temperature and the 1-day minimum air temperature are required for determining the high and low temperature conditions that should be used for design. The 7-day average maximum air temperature is defined as the average highest air temperature for a period of 7 consecutive days within a given year. The 1-day minimum temperature is defined as the lowest air temperature recorded in a given year. The mean and standard deviation of these temperatures were calculated for each year within the Superpave software. Note that the Superpave software only uses weather stations with a minimum of 20 years of data. For designers outside of the United States and Canada, the appropriate local weather information will have to be analyzed to calculate the 7-day average maximum air temperature and the 1-day minimum air temperature.

However, the PG binder selection is based on pavement temperatures, not air temperatures. Superpave defines the high pavement design temperature as the temperature at a depth of 20 mm below the pavement surface and the low design pavement temperature at the surface of the pavement surface. Based on theoretical models, the following equation was developed for determining the temperature 20 mm below the pavement surface:

$$T_{20\ mm} = (T_{air} - 0.00618Lat^2 + 0.2289Lat + 42.2)(0.9545) - 17.78 \qquad (7.37)$$

where $T_{20\ mm}$ is the high pavement design temperature, T_{air} is the 7-day average maximum air temperature, and Lat is the geographical latitude of the project in degrees.

The low pavement design temperature at the pavement surface is the same as the 1-day minimum temperature since the air temperature is the same as the pavement surface temperature. Unlike other asphalt cement specifications, the PG binder system can accommodate reliability concepts in the selection of the binder grade.

Consider an example in City A. The Superpave software is used to determine the climatic conditions in City A. The 7-day average maximum air temperature is 38°C (degrees Celsius) with a standard deviation of 2.5°C and the 1-day minimum air temperature is − 21°C with a standard deviation of 3°C. Using Equation 7.37, the high pavement design temperature is 57°C. The low pavement design temperature is the 1-day minimum air temperature. Figure 7.10 illustrates the pavement design temperatures for City A and the associated reliabilities. The high and low pavement design temperatures would be 57 and − 21°C, respectively, for a minimum 50% reliability. At this minimum level of reliability, a PG 58-22 asphalt cement would be sufficient. The minimum 98% reliability level would be at two standard deviations above the high pavement design temperature and below the low pavement design temperature. The high and low pavement design temperatures would be 62 and − 27°C, respectively, for a minimum 98% reliability. At this minimum level of reliability, a PG 64-28 asphalt cement would have to be used. If a higher level of reliability (98%) is desired at the high temperature end due to

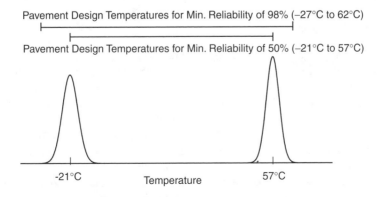

FIGURE 7.10 Pavement design temperatures for City A.

concerns with permanent deformation but a lower level of reliability is acceptable at the low temperature end because low-temperature cracking is not a concern, then a PG 64-22 may be an economical choice without having to compromise conservatism.

While reliability is a major advantage of the PG system, other factors may also impact the final selection of the asphalt cement. For example, the PG system assumes highway speed traffic with a specific level of traffic. If the proposed roadway will carry slower moving vehicles, then the high-temperature performance grade should be "bumped up" to accommodate this condition. For example, the designer may select a PG 70-XX

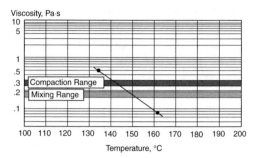

FIGURE 7.11 Selection of laboratory mixing and compaction temperatures. [*Source*: From Asphalt Institute, 2001. *Superpave Mix Design*, Superpave Series No. 2 (SP-2), Lexington, Kentucky. With permission.]

binder instead of a PG 64-XX binder. If the anticipated traffic loading is between 10 and 30 million ESALs, the designer may consider a similar bump in the high-temperature performance grade. If the anticipated traffic loading is greater than 30 million ESALs, the designer should "bump up" the high temperature performance grade.

Once the asphalt cement has been selected for a project, the temperature–viscosity relationship is determined. The rotational viscosity is determined at two temperatures (possibly at 135 and 165°C) and a linear relationship is plotted through the two points in a viscosity (log scale) vs. temperature (arithmetic scale). From here, the temperatures corresponding to viscosities of (0.17 ± 0.02) and (0.28 ± 0.03) Pa s are used as the mixing and compaction temperature, respectively. This process, illustrated in Figure 7.11, is suitable for asphalt cements that are not polymer modified. The mixing and compaction temperatures for polymer-modified asphalts should be based on the manufacturer's recommendations.

7.2.8.4.3 Selection of Aggregates
No new aggregate tests were developed as part of SHRP. The Superpave mix design approach incorporated existing aggregate tests that are used across the United States. These tests have already been covered in Section 7.2.2 and will not be discussed here. Instead, Table 7.7 is provided to show the aggregate consensus properties requirements for the combined gradation.

It should be noted that the consensus aggregate requirements in Table 7.7 are not for the individual aggregate sources. The requirements are for the combined gradation. Also, local agencies may specify other aggregate tests such as the source properties. While the Superpave requirements are applied to the combined gradation, some designers may elect to conduct the consensus and source property tests on all the individual aggregate sources to help narrow down the acceptable ranges of blend percentages for the aggregates. While this information is useful to the designer, it requires additional time and has a corresponding cost associated with it. In addition to these tests, aggregate specific gravities must also be determined for the volumetric analysis.

7.2.8.4.4 Step 2. Selection of Design Aggregate Structure
The Superpave mix design system includes aggregate gradation control points through which the aggregate gradation must pass. There are control points at the maximum aggregate size sieve, nominal maximum aggregate size sieve, the No. 8 (2.36 mm) sieve, and the No. 200 (75 μm) sieve. The control point requirements are summarized in Table 7.8. In addition to these control points, there is a restricted zone through which it is generally recommended that gradation bands do not pass. This zone covers an area on both sides of the maximum density line. Gradations that pass through the restricted zone from below the maximum density line have often been labeled "humped" gradations. In many cases, this characteristic hump in the gradation is caused by a mixture with excessive sand or one with too much fine sand relative to the total sand. This type of mix has been linked to the tender mixes in the field, which have been challenging to compact. For mixes with a nominal maximum aggregate size of 19 mm or less,

TABLE 7.7 Superpave Aggregate Consensus Properties Requirements

Design ESALs[a]	Coarse Aggregate Angularity (Percent), Minimum		Uncompacted Void Content of Fine Aggregate (Percent), minimum		Sand Equivalent (Percent), minimum	Flat and Elongated[c] (Percent), maximum
	≤ 100 mm	> 100 mm	≤ 100 mm	> 100 mm		
< 0.3	55/—	—/—	—	—	40	—
0.3 to < 3	75/—	50/—	40	40	40	10
3 to < 10	85/80[b]	60/—	45	40	45	10
10 to < 30	95/90	80/75	45	40	45	10
≥ 30	100/100	100/100	45	45	50	10

Source: From Asphalt Institute, 2001. Superpave Mix Design, Superpave Series No. 2 (SP-2), Lexington, Kentucky. With permission.

[a] Design ESALs are the anticipated project traffic level expected on the design lane over a 20-year period. Regardless of the actual design life of the roadway determine the design ESALs for 20 years and choose the appropriate Ndesign level.

[c] Criterion based upon a 5:1 maximum-to-minimum ratio. (If less than 25% of a layer is within 100 mm of the surface, the layer may be considered to be below 100 mm for mixture design purposes.)

[b] 85/80 denotes that 85% of the coarse aggregate has one fractured face and 80% has two or more fractured faces.

TABLE 7.8 Aggregate Gradation Control Points for Superpave Mixes

	Aggregate Gradation Control Points (Percent Passing)									
Sieve, mm	9.5 mm Mix		12.5 mm Mix		19 mm Mix		25 mm Mix		37.5 mm Mix	
	Min.	Max.	Min.	Max.	Min.	Max.	Min.	Max.	Min.	Max.
50.0	—	—	—	—	—	—	—	—	100	—
37.5	—	—	—	—	—	—	100	—	90	100
25.0	—	—	—	—	100	—	90	100	—	90
19.0	—	—	100	—	90	100	—	90	—	—
12.5	100	—	90	100	—	90	—	—	—	—
9.5	90	100	—	90	—	—	—	—	—	—
4.75	—	90	—	—	—	—	—	—	—	—
2.36	32	67	28	58	23	49	19	45	15	41
1.18	—	—	—	—	—	—	—	—	—	—
0.600	—	—	—	—	—	—	—	—	—	—
0.300	—	—	—	—	—	—	—	—	—	—
0.075	2	10	2	10	2	8	1	7	0	6

TABLE 7.9 Restricted Zone Boundaries for Superpave Mixes

	Restricted Zone Percent Passing Requirements (Percent Passing)									
Sieve, mm	9.5 mm Mix		12.5 mm Mix		19 mm Mix		25 mm Mix		37.5 mm Mix	
	Min.	Max.	Min.	Max.	Min.	Max.	Min.	Max.	Min.	Max.
4.75	—	—	—	—	—	—	39.5	39.5	34.7	34.7
2.36	47.2	47.2	39.1	39.1	34.6	34.6	26.8	30.8	23.3	27.3
1.18	31.6	37.6	25.6	31.6	22.3	28.3	18.1	24.1	15.5	21.5
0.600	23.5	27.5	19.1	23.1	16.7	20.7	13.6	17.6	11.7	15.7
0.300	18.7	18.7	15.5	15.5	13.7	13.7	11.4	11.4	10.0	10.0

the zone extends from the No. 8 (2.36 mm) sieve through to the No. 50 (300 μm) sieve. For mixes with a nominal maximum aggregate size greater than 25 or 37.5 mm, the zone extends from the No. 4 (4.75 mm) sieve through to the No. 50 (300 μm) sieve. The gradation control points and restricted zone boundaries are shown in Tables 7.8 and 7.9, respectively. A graphical representation of these requirements for a 19.0 mm Superpave mix is shown in Figure 7.12.

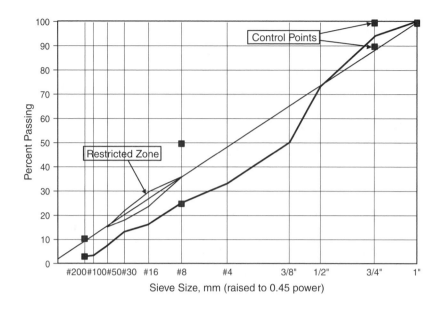

FIGURE 7.12   Gradation control points for a 19.0 mm Superpave mix.

When the Superpave mix design procedure was first introduced, the recommendation was to use gradations that passed below the restricted zone. However, this is not a requirement. In fact, several highway agencies in the United States have found success using gradations that pass above the restricted zone. Experience has also shown that some gradations passing through the restricted zone perform satisfactorily. These gradations should be carefully scrutinized to ensure adequate VMA and address any concerns over the possibility of having a tender mix.

In the mix design process, individual aggregate sources are combined to develop trial blends that meet the gradation requirements as well as consensus and agency-specific requirements. In some states, testing the source properties for aggregate source may not be necessary. For example, frequently used aggregate sources in the state of Texas are regularly tested by the State for source approval. If an approved source maintains its "approved" status in the "Quality Monitoring Program," mix designs using this aggregate source do not require additional specific source property testing.

In a typical mix design, gradations from the available aggregate sources are combined mathematically to develop three trial blends that meet the gradation requirements and aggregate property requirements. Following this, the trial asphalt binder content for each trial batch has to be determined. The designer may use experience with similar mixes to approximate the trial asphalt binder content or use a theoretical approximation. In this approximation, the first step is to determine the effective specific gravity of the blend:

$$G_{se} = G_{sb} + 0.8(G_{sa} - G_{sb}) \tag{7.38}$$

where G_{se} is the aggregate effective specific gravity, G_{sa} is the aggregate apparent specific gravity, and G_{sb} is the aggregate bulk specific gravity.

The "0.8" factor is based on regular aggregates. If absorptive aggregates are used, the designer should adjust the value closer to 0.5 or 0.6. Next the volume of asphalt binder can be determined using the equation below:

$$V_{ba} = \frac{P_s(1 - V_a)}{\left(\dfrac{P_b}{G_b} + \dfrac{P_s}{G_{se}}\right)} \times \left(\frac{1}{G_{sb}} - \frac{1}{G_{se}}\right) \tag{7.39}$$

where V_{ba} is the volume of absorbed binder (cm^3/cm^3 of mix), P_b is the percent of binder (assume 0.05), P_s is the percent of aggregate (assume 0.95), G_b is the specific gravity of binder (assume 1.02), and V_a is the volume of air voids (assume 0.04 cm^3/cm^3 of mix).

The volume of effective binder is then given by Equation 7.40:

$$V_{be} = 0.081 - 0.02931[Ln\ (S_n)] \tag{7.40}$$

where V_{be} is the effective binder content by volume and S_n is the nominal maximum sieve size of aggregate blend (in.).

Next, the weight of aggregate is determined using Equation 7.41:

$$W_s = \frac{P_s(1 - V_a)}{\left(\dfrac{P_b}{G_b} + \dfrac{P_s}{G_{se}}\right)} \tag{7.41}$$

where W_s is the weight of aggregate (in g), P_s is the percent of aggregate (assume 0.95), V_a is the volume of air voids (assume 0.04 cm^3/cm^3 of mix), P_b is the percent of binder (assume 0.05), G_b is the specific gravity of binder (assume 1.02), and G_{se} is the aggregate effective specific gravity.

The trial asphalt binder content can be calculated using Equation 7.42:

$$P_{bi} = \frac{G_b(V_{be} + V_{ba})}{[G_b(V_{be} + V_{ba})] + W_s} \times 100 \tag{7.42}$$

where P_{bi} is the trial asphalt binder content (percent by weight), G_b is the specific gravity of binder (assume 1.02), V_{be} is the effective binder content by volume, V_{ba} is the volume of absorbed binder (cm^3/ cm^3 of mix), and W_s is the weight of aggregate (in g).

If the actual (measured) values are available for any of the equations above, they should be used in lieu of the assumed values. With the batch percentages determined, it is time to calculate the batch weights and prepare the samples. Note that two replicates of each of the trial blends should be compacted using the Superpave gyratory compactor (SGC) in accordance with AASHTO T312 *"Standard Method of Test for Preparing and Determining the Density of the Hot-Mix Asphalt (HMA) Specimens by Means of the Superpave Gyratory Compactor"* (AASHTO, 2003). Two specimens from each trial blend should also be prepared for determining the trial blend's maximum theoretical specific gravity. Batch weights of 4500 and 2000 g are typically sufficient for the gyratory and maximum theoretical specific gravity specimens, respectively. The individual components from the trial blend are brought to the mixing temperature prior to mixing. Similarly, the trial blend is brought to the compaction temperature prior to compaction. Between these two processes, the mix is maintained in the oven to simulate short-term aging. The number of gyrations used for compaction is based on the traffic level as shown in Table 7.10 below.

The trial blends are compacted to N_{design} using the SGC with the specimen height continually monitored by the compactor. After the compaction process is completed, the specimen is extruded from the mold and allowed to cool. The gyratory specimens are then tested in accordance with AASHTO T166 *"Standard Method of Test for Bulk Specific Gravity of Compacted Bituminous Mixtures Using Standard Surface-Dry Specimens"* (AASHTO, 2003). The maximum theoretical specific gravity samples are tested in accordance with AASHTO T209 *"Standard Method of Test for Theoretical Maximum Specific Gravity and Density of Bituminous Paving Mixtures"* (AASHTO, 2003). Using the measured bulk specific gravity from the specimens at N_{design} and the maximum theoretical specific gravity, the percent compaction can be determined after each gyration based on the ratio of specimen heights after each gyration and the height of the specimen compacted to N_{design}. The relationships obtained can be used to determine the percent compaction ($\%G_{mm}$) at different numbers of gyrations. Figure 7.13 shows the compaction

TABLE 7.10 Superpave Gyratory Design Compactive Effort — SP2

Design ESALs (millions)	Compaction Parameters			Typical Roadway Applications
	$N_{initial}$	N_{design}	N_{max}	
<0.3	6	50	75	Very light traffic (local/county roads; city streets where truck traffic is prohibited)
0.3 to <3	7	75	115	Medium traffic (collector roads; most county roadways)
3 to <30	8	100	160	Medium to high traffic (city streets; state routes; U.S. highways; some rural interstates)
≥30	9	125	205	High traffic (most of the interstate system; climbing lanes; truck weighing stations)

When specified by the agency and the top of the design layer is ≥100 mm from the pavement surface and the estimate design traffic level ≥0.3 million ESALs, decrease the estimated design traffic level by one, unless the mixture will be exposed to significant main line and construction traffic prior to being overlaid. If less than 25% of the layer is within 100 mm of the surface, the layer may be considered to be below 100 mm for mixture design purposes. When the design ESALs are between 3 to <10 million ESALs the agency may, at their discretion, specify $N_{initial}$ at 7, N_{design} at 75, and N_{max} at 115, based on local experience. *Source*: From Asphalt Institute, 2001. *Superpave Mix Design*, Superpave Series No. 2 (SP-2), Lexington, Kentucky. With permission.

information for an example gyratory specimen. Example 7.2 provides an example of properties that need to be calculated in the Superpave mix design process.

Example 7.2

Below is some information from the bulk specific gravity, maximum theoretical specific gravity and gyratory compaction test results from one specimen. Normally these properties need to be calculated for each specimen used in the trial blends. However, the calculations for only one sample will be shown in this example.

Specimen Height at $N_{initial}$ = 128.6 mm
Specimen Height at N_{design} = 115.0 mm
Bulk Specific Gravity at N_{design} = 2.452
Maximum Theoretical Specific Gravity (G_{mm}) = 2.567
Trial Blend Percent Aggregate = 95.5% and Percent Binder = 4.5%
Aggregate Bulk Specific Gravity, G_{sb} = 2.700
Aggregate Effective Specific Gravity G_{se} = 2.756
Binder Specific Gravity, G_b = 1.02
Percent Passing No. 200 Sieve = 3.5%

Based on the information provided, determine the following properties: (a) percent compaction at N_{design}, (b) percent compaction at $N_{initial}$, (c) percent air voids at N_{design}, (d) %VMA at N_{design}, and (e) estimated asphalt content to achieve 4% air voids. Based on this estimated asphalt content to achieve 4% air voids, calculate the (f) estimated %VMA, (g) estimated %Gmm at $N_{initial}$, (h) estimated effective binder content, and (i) estimated dust proportion.

Solution

(a) Percent Compaction at N_{design}:

$$\%G_{mm}(N_{design}) = \frac{BSG}{G_{mm}} \times 100$$

$$= \frac{2.452}{2.567} \times 100 = 95.5\% \quad (7.43)$$

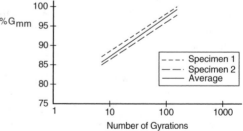

FIGURE 7.13 Example gyratory compaction data and curve.

(b) Percent Compaction at $N_{initial}$:

$$\%G_{mm}(N_{initial}) = \frac{(\text{SpecimenHeight}@N_{design})}{(\text{SpecimenHeight}@N_{initial})} \times \%G_{mm}@N_{design} = \frac{115.0\,mm}{128.6\,mm} \times 95.5\% = 85.4\% \quad (7.44)$$

(c) Percent air voids at N_{design}:

$$\%\text{AirVoids} = 100 - \%G_{mm}@N_{design} = 100 - 95.5 = 4.5\% \quad (7.45)$$

(d) % VMA at N_{design},

$$\%\text{VMA} = 100 - \left\{ \frac{(\%G_{mm}@N_{design})G_{mm}P_s}{G_{sb}} \right\} = 100 - \left\{ \frac{95.5 \times 2.567 \times 0.955}{2.700} \right\} = 13.3\% \quad (7.46)$$

(e) Estimated asphalt content to achieve 4% air voids:

$$P_{b,estimated} = P_{bi} - 0.4(4 - V_a) = 4.5 - 0.4(4 - 4.5) = 4.7\% \quad (7.47)$$

(f) Estimated VMA:

$$\%\text{VMA}_{estimated} = \%\text{VMA}_{initial} + C(4 - V_a) \quad (7.48)$$

where $C = 0.1$ if %Air Voids < 4 and 0.2 if %Air Voids > 4

$$\%\text{VMA}_{estimated} = 13.3 + 0.2(4 - 4.5) = 13.2\%$$

(g) Estimated $\%G_{mm}$ at $N_{initial}$:

$$\%G_{mm,estimated}@N_{initial} = \%G_{mm,trial}@N_{initial} - (4.0 - V_a) = 85.4 - (4.0 - 4.5) = 85.9\% \quad (7.49)$$

(h) Estimated effective binder content:

$$P_{be,estimated} = -(P_s G_b) \times \left(\frac{G_{se} - G_{sb}}{G_{se}G_{sb}} \right) + P_{b,estimated} = -(95.5 \times 1.02) \times \left(\frac{2.756 - 2.700}{2.756 \times 2.700} \right) + 4.7$$

$$= 4.0$$

$$(7.50)$$

(i) Estimated dust proportion:

$$DP_{estimated} = \frac{P_{0.075}}{P_{be,estimated}} = \frac{3.5}{4.0} = 0.89 \quad (7.51)$$

The estimated properties are compared against the mix design criteria, the specific traffic, and nominal maximum aggregate size. The Superpave volumetric mixture requirements are shown in Table 7.11 below. The designer may now select the trial blend that meets these requirements. If none of the trial blends meets the criteria, the designer should develop new trial blends with different percentages from each source or consider using new sources that may allow greater flexibility in achieving the volumetric requirements.

7.2.8.4.5 *Step 3. Selection of Design Asphalt Binder Content*

Once the design aggregate structure is determined, the next step is to select the design asphalt binder content. Compact two replicate specimens each of the design aggregate structure at the estimated asphalt content, $\pm 0.5\%$ of the estimated asphalt content and $+1.0\%$ of the estimated asphalt content. These four asphalt contents are the minimum required in the Superpave system. Two specimens should also be prepared at the estimated optimum asphalt for determining the maximum theoretical specific gravity. These test specimens should be prepared and tested in the same manner as the trial blends in Step 2. Plot three of the properties (percent air voids, VMA, and VFA) against asphalt content to determine the design asphalt content. Figures 7.14 through 7.16 show an example of these three graphs.

The criteria shown in Figures 7.14 through 7.16 are for 19.0 mm Superpave mix with a design traffic loading of 20 million ESALs. Based on Figure 7.16, a design air void content of 4% gives an asphalt content of 4.8%. This 4.8% asphalt corresponds to a 13.2% VMA and 69% VFA as shown in Figures 7.14

TABLE 7.11 Superpave Volumetric Mixture Design Requirements — SP2

Design ESALs (million)	Required Density (Percent of Theoretical Maximum Specific Gravity)			Voids-in-the-Mineral Aggregate (Percent), minimum					Voids Filled With Asphalt (Percent)	Dust-to- Binder Ratio
	$N_{initial}$	N_{design}	N_{max}	Nominal Maximum Aggregate Size (mm)						
				37.5	25.0	19.0	12.5	9.5		
<0.3	≤91.5								70–80	0.6–1.2
0.3 to <3	≤90.5								65–78	
3 to <10	≤89.0	96.0	< = 98.0	11.0	12.0	13.0	14.0	15.0	65–75	
10 to <30										
≥30										

Design ESALs are the anticipated project traffic level expected on the design lane over a 20-year period. Regardless of the actual design life of the roadway, determine the design ESALs for 20 years, and choose the appropriate N_{design} level. For 9.5-mm nominal maximum size mixtures, the specified VFA range shall be 73 to 76% for design traffic levels ≥ 3 million ESALs. For 25.0-mm nominal maximum size mixtures, the specified lower limit of the VFA shall be 67% for design traffic levels <0.3 million ESALs. For 37.5-mm nominal maximum size mixtures, the specified lower limit of the VFA shall be 64% for all design traffic levels. If the aggregate gradation passes beneath the boundaries of the aggregate restricted zone, consideration should be given to increasing the dust-to-binder ratio criteria from 0.6–1.2 to 0.8–1.6. *Source*: From Asphalt Institute, 2001. *Superpave Mix Design*, Superpave Series No. 2 (SP-2), Lexington, Kentucky. With permission.

and 7.16, respectively. Assuming that this mix met the $N_{initial}$, N_{design}, and $N_{maximum}$ requirements set forth in Table 7.11, this mix meets the Superpave volumetric requirements.

7.2.8.4.6   *Step 4. Evaluate Moisture Sensitivity*

The last step in the Superpave mix design process is a check for moisture susceptibility. This test is discussed in Section 7.2.2.9 and will not be repeated here.

7.3   Emulsified and Cutback Asphalts

HMA concrete uses asphalt cement that has been heated for construction purposes. However, asphalt cement can be emulsified with an emulsifying agent and water to form asphalt emulsions or dissolved in suitable petroleum solvents to form cutback asphalts. In asphalt emulsions, the water evaporates leaving the asphalt cement residue to perform its function. For cutback asphalts, the solvent evaporates leaving the asphalt cement to perform its function.

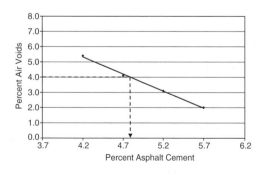

FIGURE 7.14 Percent air voids vs. percent asphalt content.

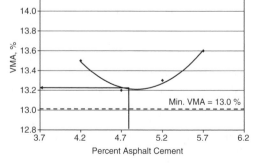

FIGURE 7.15 Percent VMA vs. percent asphalt content.

7.3.1 Cutback Asphalts

Cutback asphalts consist primarily of asphalt cement and a solvent. The speed at which they cure is related to the volatility of the solvent (diluent) used. Cutbacks made with highly volatile solvents will cure faster as the solvent will evaporate more quickly. Conversely, cutbacks made with less volatile solvents will cure slower as the solvent will evaporate slower.

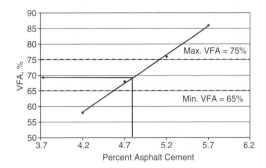

FIGURE 7.16   Percent VFA vs. percent asphalt content.

There are three types (RC, MC, and SC) cutback asphalts, which indicate the rate at which the solvent evaporates. Rapid-curing (RC) cutback asphalts are composed of asphalt cement and a light diluent of high volatility, generally in the gasoline or naphta boiling point range. Medium curing (MC) cutback asphalts are composed of asphalt cement and a medium diluent of intermediate volatility, generally in the kerosene boiling point range. RC and MC cutback asphalts are used for a variety of roadway, airfield, and industrial applications. Slow-curing (SC) cutback asphalts, often called road oils, contain asphalt cement and oils of low volatility (The Asphalt Institute, 1989). SC cutback asphalts are primarily used in prime coats and dust control. In addition to categorization by type, cutbacks asphalts are also divided into grades (e.g. 70, 250, 800, or 3000), which indicate the allowable kinematic viscosity.

The standard practice for selecting cutback asphalts is covered in ASTM D2399 *"Standard Practice for Selection of Cutback Asphalts"* (ASTM, 2003). Table 7.12 shows the specification requirements for RC cutback asphalts as described in ASTM 2028 *"Standard Specification for Cutback Asphalt (Rapid-Curing Type)"* (ASTM, 2003). Other cutback asphalt specifications can be found in the respective AASHTO and ASTM manuals.

TABLE 7.12   Requirements for Rapid-Curing Cutback Asphalts — ASTM

Designation	RC-70		RC-250		RC-800		RC-3000	
	Min	Max	Min	Max	Min	Max	Min	Max
Kinematic viscosity at 60°C (140°F), mm² s	70	140	250	500	800	1600	3000	6000
Flash point (Tag open-cup), °C (°F)	—	—	27+ (80+)	—	27+ (80+)	—	27+ (80+)	—
Distillation test:								
Distillate, volume percent to total distillate to 360°C (680°F):								
to 190°C (374°F)	10	—	—	—	—	—	—	—
to 225°C (437°F)	50	—	35	—	15	—	—	—
to 260°C (500°F)	70	—	60	—	45	—	25	—
to 316°C (600°F)	85	—	80	—	75	—	70	—
Residue from distillation to 360°C (680°F), percent volume by difference	55	—	65	—	75	—	80	—
Test on residue from distillation:								
Viscosity at 60°C (140°F), Pa s[a]	60	240	60	240	60	240	60	240
Ductility at 25°C (77°F), cm	100	—	100	—	100	—	100	—
Solubility in trichloroethylene, %	99.0	—	99.0	—	99.0	—	99.0	—
Water, %	—	0.2	—	0.2	—	0.2	—	0.2

Source: From ASTM, 2003. ASTM D2028 Standard Specification for Cutback Asphalt (RC Type), *Annual Book of ASTM Standards*, West Conshohocken, Pennsylvania. With permission.

[a] Instead of viscosity of the residue, the specifying agency, at its option, can specify penetration at 100 g: 5 sec at 25°C (77°F) of 80 to 120 for Grades RC-70, RC-250, RC-800, and RC-3000. However, in no case will both be required.

Due to increasing environmental regulations and concerns, the use of cutback asphalts has been significantly restricted or even prohibited in many parts of the United States. Economic (use of two petroleum products and the need for additional heat) and safety (low flash point) factors have also contributed to the decline of cutback asphalts. With availability of new formulations and improved laboratory procedures, asphalt emulsions have increasingly been used to substitute cutback asphalts.

7.3.2 Asphalt Emulsions

Asphalt emulsions consist primarily of asphalt cement, water, and an emulsifying agent. They should be stable enough for pumping, mixing, and prolonged storage. While the emulsifying agent keeps the asphalt cement globules apart prior to its application, the emulsion should "break" (i.e. the process of the asphalt cement separating from the water) rapidly upon contact with aggregate in a mixer or after spraying on the roadbed. When the water has evaporated, the asphalt cement residue performs its function with the original adhesion, durability, and water-resistance of the original product.

There are three types of asphalt emulsions. Anionic emulsions have electro-negatively charged asphalt globules, cationic emulsions have electro-positively charged asphalt globules, and nonionic emulsions have neutral asphalt globules. In practice, the first two types of emulsions are most commonly used in roadway applications. Emulsions are further classified by grades based on the rate at which they set. They are termed rapid-setting (RS), medium-setting (MS), slow-setting (SS), and quick-setting (QS) based on their relative setting times.

According to the Asphalt Institute (1993), RS emulsions have little or no ability to mix with aggregates, MS emulsions are expected to mix with coarse but not fine aggregate, and SS and QS emulsions are designed to mix with fine aggregate, with the QS emulsions expected to break more quickly than the SS emulsions.

Asphalt emulsions are also identified by alphabetical and numerical characters. The "C" preceding certain grades denotes a cationic asphalt emulsion and its absence denotes an anionic emulsion. For example, a CSS-1 is a cationic emulsion and an SS-1 is an anionic emulsion. Grades with a "2" are more viscous than grades with a "1." For example, an RS-2 is more viscous that an RS-1. The "HF" preceding some of the anionic grades indicates "high float" from the float test (AASHTO T50, ASTM D139). High float emulsions contain chemicals that allow a thicker film of asphalt to coat aggregate particles and minimize drain off of asphalt cement from the aggregate. For polymer-modified asphalt emulsions, an additional letter (usually P, S, or L) is added at the end of the emulsion grades to denote this property even though these grades are not currently used in the AASHTO and ASTM specifications. For example, an HFRS-2P normally designates a modified asphalt emulsion. Table 7.13 shows standard AASHTO (M140 and 208) and ASTM (D977 and 2397) grades for asphalt emulsions (Asphalt Institute, 1993).

TABLE 7.13 Standard AASHTO and ASTM Grades for Asphalt Emulsions

Anionic Asphalt Emulsion	Cationic Asphalt Emulsion
RS-1	CRS-1
RS-2	CRS-2
HFRS-2	—
MS-1	—
MS-2	CMS-2
MS-2h	CMS-2h
HFMS-1	—
HFMS-2	—
HFMS-2h	—
HFMS-2s	—
SS-1	CSS-1
SS-1h	CSS-1h
QS-1h	CQS-1h

7.4   Portland Cement Concrete

PCC is one of the most widely used construction materials. It is used in used in all types of structural elements (e.g. bridges, tunnels, subways) as well as in pavement applications (e.g. highways). This section will provide a brief discussion on the constituents of PCC and mix design procedures.

7.4.1   Constituents of Portland Cement Concrete

PCC is composed of cement paste and aggregates. The cement paste consists of Portland cement mixed with water while the aggregates are composed of fine and coarse fractions. Within the PCC lies entrained and entrapped air. The following section provides a brief discussion on the primary constituents in PCC. Additional information can be found in the PCA's "*Design and Control of Concrete Mixtures*" (PCA, 1988).

7.4.1.1   Portland Cement

Portland cement was first patented by Joseph Aspdin in 1824. This cement was prepared by calcining some finely-ground limestone, mixing this with finely divided clay, and calcining the mixture in a kiln until the carbon dioxide was driven off. The resulting mixture was finely ground and used as Portland cement. The name "Portland" was coined by Aspdin because the hardened cement appeared similar to a naturally occurring building stone quarried in Portland, England.

In the modern plant, Portland cement is manufactured by pulverizing clinker consisting primarily of hydraulic calcium silicates along with some calcium aluminates and calcium minoferrites with one or more forms of calcium sulfate (gypsum). Materials used in the manufacturing process must contain the appropriate proportions of calcium oxide, silica, alumina, and iron oxide. The ingredients may be blended using the dry process or the wet process. In the dry process, grinding and blending of the ingredients is done with the dry materials. In the wet process, grinding and blending is conducted with the materials in a slurry form. After blending is complete, the raw mix passes through a kiln where temperatures of 1430°C (2606°F) to 1650°C (3002°F) cause a chemical reaction resulting in grayish-black pellets of cement clinker. These approximately 12.5 mm (0.5 in.) pellets are allowed to cool before being pulverized into Portland cement. Gypsum may be added at this juncture to regulate the setting time. Analyses of the ingredient components are essential to maintain uniformity in the manufacturing process and ensure quality of the resulting product.

Portland cement is composed of approximately 60 to 65% lime (CaO), 20 to 25% silica (SiO_2), and 7 to 12% iron oxide (Fe_2O_3) and alumina (Al_2O_3). The percentages of the different components may be varied to meet different physical and chemical requirements based on its intended use. Table 7.14 describes eight types of Portland cement.

Type I cement is general-purpose cement suitable for regular use where special properties are not required. Type II cement is used when sulfate concentrations in groundwater are higher than normal but not unusually severe. It will usually generate less heat at a slower rate than Type I cement. Type III cement has the same composition as Type I cement but it is ground finer. The increased surface area causes a quicker chemical reaction and higher early strengths. Type IV cement is used when the heat of hydration

TABLE 7.14   Type of Portland Cement

Type of Portland Cement	Description
Type I	Normal
Type IA	Normal, air entraining
Type II	Moderate sulfate resistance
Type IIA	Moderate sulfate resistance, air entraining
Type III	High early strength
Type IIIA	High early strength, air entraining
Type IV	Low heat of hydration
Type V	High sulfate resistance

must be minimized. While it develops strength slower than the other types of cement, it may be used in massive structures where excessive heat is detrimental. Type V cement is used when soils or groundwater have a high sulfate content. The low tricalcium aluminate (C_3A) content increases the cement's sulfate resistance.

7.4.1.2   Aggregates

Aggregates form approximately 60 to 75% of the concrete volume, which is equivalent to approximately 70 to 85% by weight. Fine aggregates generally consist of natural sands or crushed stone with particles predominantly smaller than the No. 4 sieve. Coarse aggregates consist of one or a combination of gravels and crushed aggregates with particles primarily greater than the No. 4 sieve.

Aggregate characteristics and their contributions to the mix design as well as properties of fresh and hardened concrete shall only be covered briefly in this chapter. Aggregates are typically characterized according to its particle size distribution, particle shape and surface texture, specific gravities, absorption and surface moisture. Particle size distribution can be further divided into coarse-aggregate and fine-aggregate grading. The particle size and distribution has an impact on the workability of the mix. In general, the larger the maximum particle size, the less Portland cement is necessary. Particle shape and surface texture have a greater impact on the properties of fresh concrete than they do on the properties of hardened concrete. The rougher and more angular the particles are, the more water is required to produce workable concrete. However, rougher and more angular particles tend to have a stronger bond with the cement and water mixture. Voids between aggregates also increase with increased aggregate angularity. Specific gravity is not generally considered a measure of aggregate quality, but is required in the mix design process. Absorption and surface moisture are used to determine the aggregate's appetite for moisture and its current moisture condition. The results of these tests are used to control concrete batch weights.

7.4.1.3   Water

Almost all naturally occurring water that is safe for drinking should be suitable for making PCC. Water of questionable quality may be used if the 7-day strengths of mortar cubes made using ASTM C109 are at least 90% of companion samples made with drinkable or distilled water. Additionally, ASTM C191 *"Standard Test Method for Time of Setting of Hydraulic Cement by Vicat Needle"* (ASTM, 2003) should be conducted to ensure that impurities in the mixing water do not adversely shorten or extend the setting time. The criteria for acceptable water can be found in ASTM C94 *"Standard Specification for Ready-Mix Concrete"* (ASTM, 2003). Water containing excessive impurities may affect setting time and concrete strength as well as reduce durability and/or resistance to corrosion of reinforcement.

7.4.1.4   Chemical Admixtures

Chemical admixtures may be used to enhance Portland cement properties based on the requirements for a specific application. The primary reasons for using admixtures are to reduce the cost of concrete construction, to enhance certain concrete properties, and to ensure the quality of concrete during the different stages of construction.

Air-entraining admixtures are used to introduce microscopic air bubbles in concrete. Air-entrainment dramatically enhances the durability of concrete exposed to moisture during cycles of freezing and thawing. Entrained air also increases the workability of fresh concrete while reducing segregation and bleeding. Specifications and test methods for air-entraining mixtures can be found in ASTM C260 and C233. Water-reducing admixtures are used to reduce the quantity of mixing water required to produce concrete of a specific slump, reduce the water–cement ratio, or increase slump. Regular water reducers may decrease the water content by 5 to 10%. High-range water reducers (also called Superplasticizers) may decrease the water content by 12 to 30%. While high-range water reducers are typically more effective than regular water reducers, they are more expensive. Water reducers typically produce an increase in strength because of the reduction in the water–cement ratio. The effectiveness of water reducers is dependent on its chemical composition, concrete temperature, cement composition, cement fineness, cement content, and the presence of other admixtures. The effectiveness of water reducers diminishes with time after it is introduced into the batch. Specifications and test methods for water reducers can be found in ASTM C494 and ASTM 1017.

Retarding admixtures are used to slow down the rate at which concrete sets. Retarders may be used to compensate for accelerated setting due to hot weather or delay initial set for prolonged concrete placements. The presence of retarders may reduce early (first few days) strength gain. Accelerating admixtures have the opposite effect from retarding admixtures in that they increase early strength gain. However, the use of accelerating admixtures may lead to increase in drying shrinkage, potential reinforcement corrosion, discoloration, and scaling. Specifications and test methods for retarding and accelerating admixtures can be found in ASTM C494.

7.4.1.5   Finely Divided Mineral Admixtures

Finely divided mineral admixtures are powdered or pulverized materials added to PCC to enhance the properties of fresh and/or hardened concrete. They may be broadly put into three categories: cementitious materials, pozzolanic materials, pozzolanic and cementitious materials, and inert materials. Fly ash, ground granulated blast-furnace slag, and condensed silica fume are commonly used mineral admixtures. Specific effects of different mineral admixtures on fresh and hardened concrete will not be discussed here.

7.4.1.6   Mix Design

The goal of designing a PCC mix is to determine the appropriate proportions of each constituent to economically produce a mix that has sufficient strength, workability, and durability for the intended purpose. Strength is often specified based on compressive strength at 28 days. Workability is a measure of how easily the mix can be transported, placed, and consolidated. Durability is a measure of how well the mix withstands the *in-situ* service conditions over time.

American Concrete Institute (ACI) Recommended Practice 211 contains both a weight and volume method for proportioning (or designing) PCC mixtures. Another approach to designing a mix is the PCA water–cement ratio method. Both these procedures can be found in the PCA's "*Design and Control of Concrete Mixtures*" (PCA, 1988) and will not be described here.

7.5   Base and Subbase Materials

While HMAC and PCC are primarily used on the surface of pavements, there are several underlying layers that play a critical role in the performance of a pavement. This includes bases, subbases, and subgrades. These layers consist primarily of treated and untreated aggregates. AASHTO and ASTM tests are currently available to test these materials to determine if they are suitable for use under HMAC or PCC pavements. The materials characterization and how their properties impact the performance of these layers and the overall pavement structure are covered in Chapter 18 of this handbook.

References

AASHTO, 2003. *Standard Specifications for Transportation Materials and Methods of Sampling and Testing. Part I and II*. American Association of State Highway and Transportation Officials, Washington, DC.

AASHTO, 1993. *Superpave for the Generalist Engineer and Project Staff*, Instructor Manual. American Association of State Highway and Transportation Officials, Washington, DC.

Asphalt Institute, 1989. *The Asphalt Handbook, 1989th Ed.*, Manual Series No. 4 (MS-4). The Asphalt Institute, Lexington, KY.

Asphalt Institute, 1993. *A Basic Asphalt Emulsion Manual, 3rd Ed.*, Manual Series No. 19 (MS-19). The Asphalt Institute, Lexington, KY.

Asphalt Institute, 1997. *Mix Design Methods for Asphalt Concrete and Other Hot Mix Types, 6th Ed.*, Manual Series No. 2 (MS-2). The Asphalt Institute, Lexington, KY.

ASTM, 2003. *Annual book of ASTM standards. Volume 04.03*, Road and Paving Materials; Vehicle-Pavement Systems. ASTM International, West Conshohocken, Pennsylvania.

Brown, E.R., Haddock, J.E., Mallick, R.B., and Lynn, T.A. 1997. *Development of a mixture design procedure for stone matrix asphalt (SMA)*, NCAT Report No. 97-3, Auburn, Alabama.

Crawford, C. 1989. *The Rocky Road of Mix Design*. National Asphalt Pavement Association, Hot Mix Asphalt Technology.

Foster, C.R. 1982. *Development of Marshall Procedures for Designing Asphalt Paving Mixtures*, National Asphalt Pavement Association, NAPA, IS84.

Fuller, W.B. and Thompson, S.E., The laws of proportioning concrete, *Journal of Transportation Divsion, American Society of Civil Engineers, Volume 59*, 1907.

Kandhal, P.S. and Foo, K.Y. 1997. *Designing recycled hot mix asphalt mixtures using superpave technology*, NCAT Report No. 96-5, Auburn, Alabama.

Mallick, R.B., Kandhal, P.S., Cooley, L.A., and Watson, D.E. 2000. *Design, construction, and performance of new-generation open-graded friction courses*, NCAT Report No. 2000-01, Auburn, Alabama.

McGennis, R.B., Anderson, R.M., Kennedy, T.W., and Solaimanian, M. 1995. *Background of superpave asphalt mixture design and analysis*, Publication No. FHWA-SA-95-003. U.S. Department of Transportation, Federal Highway Administration, Springfield, Virgina.

NAPA, 1982. *Mix Design Techniques — Part I*, NAPA TAS-14, National Asphalt Paving Association, Instructors Manual.

PCA, 1988. *Design and Control of Concrete Mixtures, 13th Ed.* Portland Cement Association, Skokie, Illinois.

Richardson, C. 1912. *The Modern Asphalt Pavement, 2nd Ed.* Wiley, New York.

SHRP, 1994. *Level one mix design: materials selection, compaction, and conditioning*, Report SHRP-A-408, Strategic Highway Research Program. National Research Council, Washington, DC.

Vallerga, B.A. and Lovering, W.R., Evolution of the hveem stabilometer method of designing asphalt paving mixtures, *Proceedings, Association of Asphalt Paving Technologists, Volume 54*, 1985.

White, T.D., Marshall procedures for design and quality control of asphalt mixtures, *Proceedings, Association of Asphalt Paving Technologists, Volume 54*, 1985.

Further Information

Additional information on highway materials can be found from many sources. There are hot-mix related publications from the AI and NAPA as well as PCC related publications from the ACI and PCA. Additionally, the latest research and developments in highway materials can be found from conference proceedings and technical journals. Examples of these materials include the Transportation Research Records published by the Transportation Research Board and the Journal of the Association of Asphalt Paving Technologists.

Below is a list of websites that contain useful information and publications that can be purchased:

1. American Concrete Institute (www.aci-int.org)
2. Association of Asphalt Paving Technologists (www.asphalttechnology.org)
3. International Center for Aggregate Research (www.ce.utexas.edu/org/icar)
4. National Asphalt Pavement Association (www.hotmix.org)
5. National Center Asphalt Technology (www.ncat.us)
6. Portland Cement Association (www.portcement.org)
7. The Asphalt Institute (www.asphaltinstitute.org)
8. Transportation Research Board (www.trb.org)
9. Turner-Fairbanks Highway Research Center (www.tfhrc.gov) and Federal Highway Administration (www.fhwa.dot.gov)

8

Design of Flexible Pavements

Michael S. Mamlouk
Arizona State University
Tempe, AZ, U.S.A.

8.1 Introduction

8.1.1 Pavement Layers and Materials

A typical flexible (or asphalt) pavement consists of surface, base course, and subbase built over compacted subgrade (natural soil) as shown in Figure 8.1. In some cases, the subbase layer is not used, whereas in a small number of cases both base and subbase are omitted.

     The surface layer is made of hot-mix asphalt (HMA) (also called asphalt concrete). The material for the base course is typically unstabilized aggregates. The aggregate base could also be stabilized with asphalt, portland cement, or another stabilizing agent. The subbase is mostly a local aggregate material. Also, the top of the subgrade is sometimes stabilized with either cement or lime.

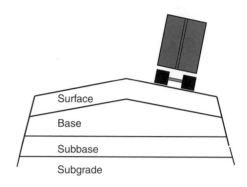

Unlike rigid pavement, when the traffic load is applied on top of the surface layer a localized deformation occurs under the load, while the load is distributed in a manner as shown in Figure 8.2. It can be seen that the load is distributed on a small area at the surface. As the depth increases, the same load is distributed over a larger area. Therefore, the highest stress occurs at the surface and the stress decreases as the depth increases. Thus, the highest quality material needs to be at the surface and as the depth increases lower quality materials can be used. When the load is

FIGURE 8.1   Typical layers of flexible pavement.

removed the pavement layers rebound. A very small amount of deformation, however, could stay permanently which could accumulate over many load repetitions causing rutting in the wheel path. The name, flexible pavement, is used because of the localized deformation and the rebounce that happens every time the traffic load is applied and removed.

The required thickness of each layer of the flexible pavement varies widely depending on the materials used, magnitude and number of repetitions traffic load, environmental conditions, and the desired service life of the pavement. These factors are generally considered in the design process so that the pavement would last for the required designed life without excessive distresses. In most cases, the surface layer varies from 1 to 10 in., which could include a number of overlays. The base layer typically varies from 4 to 12 in. and the subbase varies from 6 to 20 in.

8.1.2   Unique Properties of Flexible Pavements

Pavement is unique when compared to other civil engineering structures. Some of the unique properties of flexible pavement are discussed below.

8.1.2.1   Fast Deterioration with Time

Each traffic load application contributes to some extent to pavement distresses. Different types of distress could happen and accumulate over the years such as rutting, fatigue cracking, material disintegration,

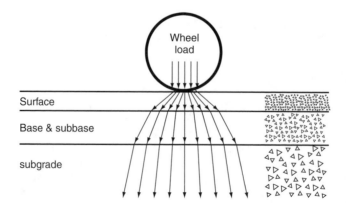

FIGURE 8.2   Load distribution in flexible pavement.

roughness and bleeding. When one or more of these distresses reach a certain unacceptable level, the pavement is considered as failed. The typical life of a flexible pavement varies from case to case, with an average value of 10 to 15 years. A good method of pavement design should include the designed life, or how long the pavement is expected to last before failure. The incorporation of the designed life in the design process is one of the hardest tasks faced by the pavement designer. In many cases, the expected designed life does not match with the actual service life. Unlike pavements, the design of other civil engineering structures does not consider the factor of designed life. In such cases the designer assumes that if the structure is safe under the maximum possible load, it will be safe for an extended period of time. This concept does not work with pavements because of their fast deterioration and short service lives.

8.1.2.2 Repeated Loads

When a traffic wheel moves on the pavement surface it creates a stress pulse. This stress pulse creates a dynamic pavement response, which is harder to analyze as compared to static response. Dynamic waves propagate throughout the pavement layers and subgrade, and involve reflections and refractions at the layer interfaces.

8.1.2.3 Variable Load Configuration

Different vehicular axle configurations are available with a different number of wheels at the end of each axle. Axles can be single, tandem, tridem, or multiple, while wheels can be either single or dual. Passenger cars have single axles and single wheels. However, trucks can take different combinations of axle and wheel configurations as shown in Figures 8.3 and 8.4. Different axle and wheel configurations result in stress interactions within the pavement structure, which in turn influence pavement performance.

FIGURE 8.3 Trucks with different axle and wheel configurations.

The maximum legal load limit varies from one location to other, but are typically in the order of 20,000 to 22,000 Lb for a single axle and 32,000 to 36,000 Lb for a tandem axle.

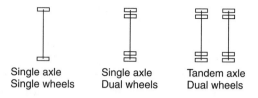

Single axle   Single axle   Tandem axle
Single wheels   Dual wheels   Dual wheels

FIGURE 8.4 Schematic of common axle and wheel configurations.

8.1.2.4 Variable Load Magnitude

Traffic loads vary from light to heavy for passenger cars and loaded trucks respectively. Since pavement materials have non-linear response, doubling the load magnitude does not result in doubling the stress or strain. More importantly, doubling the load magnitude does not result in doubling the rate of pavement deterioration. In fact, increasing the load magnitude exponentially increases the rate of pavement deterioration.

8.1.2.5 Variable Tyre Pressure

Trucks have much higher tyre pressures than passenger cars. Typical tyre pressures of passenger cars are in the order of 30–35 psi, while trucks have tyre pressures of 100–115 psi. Higher tyre pressures result in higher contact pressures at the surface of the pavement and, in turn, faster deterioration of the surface layer. Truck tyre pressures have been increasing over the years, challenging pavement engineers to improve the quality of the HMA material in order to reduce premature pavement failure.

8.1.2.6 Traffic Growth

Pavement is designed to carry future traffic, which usually increases over the years. Predicting future traffic growth is not always accurate. This inaccuracy in predicting future traffic affects the accuracy of predicting pavement performance and consequently pavement designed life.

8.1.2.7 Change of Material Properties with Environmental Conditions

Environmental conditions have large effect on the properties of pavement materials. For example, HMA gets softer at high temperatures resulting in rutting, and harder at low temperatures resulting in thermal cracking. Also, rain and freeze–thaw cycles weaken the HMA materials and reduce the load carrying capacity of base, subbase and subgrade. In addition, HMA ages with time resulting in increasing its stiffness and its susceptibility to cracking.

8.1.2.8 Change of Subgrade Properties with Distance

Since pavement is built to cover a large distance, the same road might be built over different types of subgrade materials with different properties. Moreover, the road could be built over cut or fill subgrade sections having different material properties. The change of subgrade properties requires different thicknesses of pavement layers in order to support the same traffic load and produce the same performance.

8.1.2.9 Channelized Traffic Load

Traffic load is applied in the wheel path. This channelization of traffic load results in faster deterioration in the wheel path as compared to the area between wheel paths. The design process should consider the proper stress and strain distributions within the pavement structure to determine critical locations and possible deteriorations.

8.1.2.10 Multi-Layer System

The pavement structure consists of several layers built over the subgrade. These layers have different materials with different properties. The distribution of stresses and strains within the multi-layer pavement system depends on the thickness and material properties of these layers.

8.1.2.11   Unconventional Failure Definition

Failure of typical civil engineering structures is defined as break or fracture. This usually happens when the applied stress exceeds the maximum allowable value, or the strength of the material. Unlike other civil engineering structures, the applied stresses in pavement are usually much smaller than the strength of the material. Therefore, one load application does not fail the pavement, but causes an infinitesimal amount of deterioration. This deterioration gradually increases until it reaches an unacceptable level, or failure. Because of this failure mechanism, different types of distresses can occur in asphalt pavements as discussed in the next section. Thus, pavement failure does not happen because of a collapse of the pavement structure, but when one or more of the distresses reach an unacceptable level.

These unique properties of pavement call for a design concept that is different than that used for other structures. The designer has to incorporate these factors to ensure that failure is not reached until the end of the intended designed life.

8.1.3   Pavement Distresses and Performance

8.1.3.1   Common Pavement Distresses

Different types of distress can occur in asphalt pavement. These distresses could be developed due to traffic load repetitions, temperature, moisture, aging, construction practice, or combinations. The common types of distresses in flexible pavements are described in detail by NRC (1993), and are discussed below.

8.1.3.1.1   Fatigue Cracking

Fatigue cracks are a series of longitudinal and interconnected cracks caused by the repeated applications of wheel loads. This type of cracking generally starts as short longitudinal cracks in the wheel path and progress to an alligator-cracking pattern (interconnected cracks) as shown in Figure 8.5. This type of cracking happens because of the repeated bending action of the HMA layer when the load is applied. This generates tensile stresses that eventually create cracks at the bottom of the asphalt layer. Cracks gradually propagate to the top of the layer and later progress and interconnect. This type of distress will eventually lead to a loss of the structural integrity of pavement system.

FIGURE 8.5   Advanced stage of fatigue cracking.

Recent studies have demonstrated that a less common fatigue cracking may initiate from the top of the pavement surface and propagates downward (top-down cracking). This type of fatigue is not as well defined from a mechanical viewpoint as the more classical "bottom-up" fatigue. It is hypothesized that critical tensile and shear stresses develop at the surface and cause extremely large contact pressures at the tyre edges-pavement interface, coupled with highly aged (stiff) surface layer and these are responsible factors for this kind of cracking. Top-down cracking first shows up as relatively long longitudinal cracks adjacent to the tyres within the wheel paths.

8.1.3.1.2   Rutting

Rutting is defined as permanent deformation in the wheel path as shown in Figure 8.6. Rutting can occur due to: (a) unstable HMA, (b) densification of HMA, (c) deep settlement in the subgrade as demonstrated in Figure 8.7.

Unstable HMA can occur because of one or more of different reasons such as too much asphalt binder, too soft asphalt binder, rounded aggregate particles, smooth aggregate texture, or too many fines in the HMA mix. Densification of HMA can occur because of the poor compaction during construction. Deep settlement can happen because of poor drainage or weak subgrade.

8.1.3.1.3 Roughness
Roughness is defined as the irregularities in the pavement profile which causes uncomfortable, unsafe, and uneconomical riding. Roughness affects the dynamics of moving vehicles, increasing the wear on vehicle parts and the handling of vehicles. Thus, road roughness has an appreciable impact on vehicle operating costs and the safety, comfort, and speed of travel. It also increases the dynamic loading imposed by vehicles on the surface, accelerating the deterioration of the pavement structure.

8.1.3.1.4 Thermal Cracking
As the temperature decreases the HMA material contracts. Since the material is restrained from movement due to the friction with the underlying material, tensile stresses develop within the HAM material. If the tensile stress exceeds the tensile strength of the material, thermal cracks develop as shown in Figure 8.8. Thermal cracks typically occur in the transverse direction perpendicular to the direction of traffic. This type of cracking is usually equally spaced. This is a non-load associated type of cracking and it starts during the winter season. The width of the thermal cracks usually changes from summer to winter. In some cases, small cracks heal during the summer season. In other cases, the width of the crack increases from one year to another.

8.1.3.1.5 Shoving
Shoving is a form of plastic movement resulting in a localized bulging of the pavement surface. Shoving can take a number of different forms such as upheaval (Figure 8.9), "wash-boarding" or ripples across the pavement surface, or crescent-shaped bulging. Shoving occurs in the asphalt layers that lack stability because of too much asphalt binder in the HMA mix, too soft asphalt binder, rounded aggregate particles, smooth aggregate texture, or too many fines in the mix.

8.1.3.1.6 Bleeding or Flushing
Bleeding or flushing is the upward movement of the asphalt binder resulting in the formation of a film of asphalt on the surface as shown in Figure 8.10. Bleeding occurs when the HMA mix is too rich with asphalt binder that is forced to the surface when traffic load is applied especially in hot weather. Bleeding could be hazardous because it makes the pavement slippery when wet.

FIGURE 8.6   Rutting.

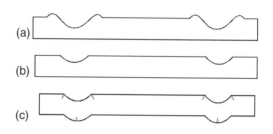
FIGURE 8.7   Rutting due to: (a) Unstable asphalt concrete, (b) Densification of asphalt concrete, and (c) Deep settlement.

FIGURE 8.8   Thermal cracking.

8.1.3.1.7 *Raveling*

Raveling is the progressive separation of aggregate particles in a pavement from the surface downward or from the edge inward as shown in Figure 8.11. Usually fine aggregate particles are separated first followed by coarse aggregates. Raveled surfaces are aged and typically look dry and weathered. Raveling is caused by one or more of several reasons such as lack of compaction, dirty or disintegrating aggregate, too little asphalt in the mix, or overheating of the mix.

8.1.3.1.8 *Polished Aggregate*

Aggregate particles in the HMA may get polished smooth and create a slippery pavement surface when wet (Figure 8.12). Some aggregates, particularly some types of limestone, become polished rather quickly under traffic.

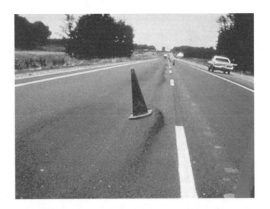

FIGURE 8.9 Shoving.

8.1.3.1.9 *Reflection Cracking*

Cracks in the underneath layer might reflect in the overlay as shown in Figure 8.13. Reflection cracking occurs frequently in asphalt overlays on concrete pavement and cement treated basis. They also occur when cracks in the old asphalt layer are not properly repaired before overlay. Reflect cracks may take several forms depending on the pattern of the crack in the underneath layer.

8.1.3.2 Pavement Performance

Figure 8.14 shows how the pavement condition varies with time or traffic applications. When the road is first built it typically has a good condition. With time and with the continuous applications of traffic loads the pavement gradually deteriorates and the condition gets worse. The change of pavement condition with time or traffic is defined as performance. Performance is affected by several factors as discussed in the next section and cannot easily be predicted.

When pavement condition reaches a certain unacceptable level the pavement reaches the end of its serviceable life. Performance prediction is important in order to ensure that the pavement reaches the unacceptable condition at the end of its

FIGURE 8.10 Bleeding or flushing.

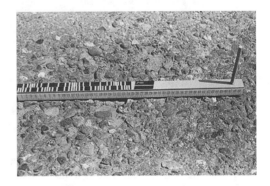

FIGURE 8.11 Raveling.

designed life. Currently, there is no completely mechanistic (or theoretical) method to predict pavement performance. Empirical or mechanistic-empirical methods are currently being used to predict performance.

8.1.3.3 Serviceability

User's opinion of how well they are being served by the road is largely subjective. The serviceability of a given road may be expressed as the average evaluation given by all users of the road. Performance, therefore, is the overall appraisal of the service-ability history of a pavement. Present serviceability rating (PSR) is the average of user assessment or rating of the quality of the pavement. Present serviceability index, (PSI) is an estimate of the PSR of a pavement based on objective measures of the pavement quality and a correlation equation for relating these measures to the PSR. PSR and PSI follow a scale from zero to five, where zero is for an impassable road condition and five for an excellent condition. A well-constructed new pavement has a PSI of 4.5–4.6. Agencies typically define failure as a terminal serviceability index, p_t, of 2.0, 2.5, or 3.0 as shown in Figure 8.14.

FIGURE 8.12   Polished aggregate.

8.1.4 Factors Affecting Performance of Flexible Pavements

Pavement performance is affected by several factors, which are traffic, soil and pavement materials, environment, and construction and maintenance practice.

8.1.4.1 Traffic

Traffic has a major effect on pavement perform-ance. Traffic characteristics that affect perform-ance are traffic load, traffic volume, tyre pressure, and vehicle speed. Traffic load produces stresses and strains within the pavement structure and the subgrade, which gradually contribute to the development of pavement distresses. For example, heavier loads result in higher potential for fatigue cracking and rutting (Figures 8.5 and 8.6). Traffic volume affects pavement performance since larger number of load repetitions increases the chance for fatigue cracking. Also, higher tyre pressure produces higher stress concentrations at the pavement surface that could result in rutting and shoving in the HMS layer (Figure 8.7a). Finally, vehicle speed affects the rate of applying the load. Since asphalt concrete is a visco-elastic plastic material, its response is affected by the rate of load application. Slow or stationary vehicles have more chances of developing rutting and

FIGURE 8.13   Reflection cracking.

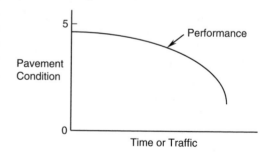

FIGURE 8.14   Change of pavement condition versus time.

shoving than high-speed vehicles. On the other hand, high travel speeds cause more severe bouncing of vehicles, and result in larger dynamic loading and increased roughness.

8.1.4.2 Soil and Pavement Materials

Soil and pavement materials significantly affect pavement performance. Of course, high quality materials are needed to provide good support to traffic loads under various environmental conditions. Important material properties include mechanical properties such as elasticity, visco-elasticity, plasticity, temperature susceptibility, durability and aging characteristics. These properties affect how the material responds to traffic loads and environmental conditions such as temperature, freeze-thaw effect, and rain.

8.1.4.3 Environment

Environmental conditions that affect pavement performance include moisture, temperature, and their interaction. For example, moisture may reduce subgrade support and weakens various pavement layers. High temperatures soften asphalt concrete and could create rutting within the surface layer. Temperatures below freezing have a bad effect on pavement performance, especially cycles of freeze and thaw. As demonstrated in Figure 8.15, if the subgrade is wet and the temperature drops below freezing, ice lenses form. Since these ice lenses have larger volume than water, frost heave will develop and may create bulges in the pavement surface. If the temperature fluctuates above and below freezing points in the same season combined with poor drainage, subgrade support will be significantly reduced and could result in excessive deterioration of the pavement structure within a short period of time as shown in Figure 8.16.

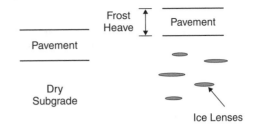

FIGURE 8.15 Effect of freeze and thaw on pavement response.

FIGURE 8.16 Alligator cracking developed due to cycles of freeze and thaw combined with poor drainage.

8.1.4.4 Construction and Maintenance Practice

In many cases, defects in pavement start during construction and propagate during service. In fact, poor construction procedure will almost always ensure poor pavement performance. For example, poor compaction of subgrade or any pavement layer allows excessive further compaction by traffic, which appears in the form of rutting and surface cracking. Poor placement of HMA during construction may result in weak transverse or longitudinal construction joints that are susceptible to early cracking and deterioration. Excessive air voids in the HMA layer due to poor compaction will result in fast aging followed by cracking. In contrast, too much compaction of HMA will result in too small amount of air voids that could create rutting or bleeding. Lack of smoothness of the pavement during construction increases the dynamic impact of traffic, and consequently, speeds up the rate of developing roughness during service.

8.2 Design Objectives, Constraints, and Evolution

It is important to define the basic objectives and constraints of the design process so that the procedure would be efficient and successful.

8.2.1 Design Objectives and Constraints

The objectives of pavement design can be listed as follow (Haas et al., 1994).

1. Maximum economy, safety, and serviceability over the design period
2. Maximum or adequate load-carrying capacity in terms of load magnitude and repetitions

3. Minimum or limited deteriorations over the design period
4. Minimum or limited noise or air pollution during construction
5. Minimum or limited disruption of adjoining land use
6. Maximum or good aesthetics

Some of these objectives conflict with each other. For example, maximizing load-carrying capacity competes with the objective of maximizing economy. Therefore, a trade-off among some of these objectives is needed.

The pavement designer typically faces several economic, physical, and technical design constraints such as,

1. Availability of time and fund for design and construction
2. Minimum allowable level of serviceability before rehabilitation
3. Availability of materials
4. Minimum and maximum layer thickness
5. Capabilities of construction and maintenance personnel and equipment
6. Testing capabilities
7. Capabilities of structural and economic models available
8. Quality and extent of the design data available

8.2.2 Design Approaches

Many methods of designing flexible pavement have been developed by various transportation agencies and evolved throughout the years. These methods range from very simple in concept to highly sophisticated. Although different agencies have been using design procedures that satisfy their local conditions, pavement design methods can be grouped into four distinct approaches:

8.2.2.1 Methods Based on Experience

Many agencies have been adopting standard pavement sections for different ranges of traffic levels and environmental conditions. These standard sections are mostly based on previous experience and are applicable to local materials and budget practice. Although these methods are old, they are still being used by relatively small agencies because of their simplicity, low design cost, and reliability under certain conditions. These methods, however, do not allow for comparison between alternatives. They also do not recognize the varying serviceability with age. These methods also assume average material properties, traffic levels, and environmental conditions. If any of these variables change, this approach looses its validity.

8.2.2.2 Methods Based on Soil Formula or Simple Strength Tests

These methods are based on empirical correlations between the required pavement thickness and soil classification or simple strength tests of subgrade materials such as California Bearing Ratio (CBR). This approach is also old and assumes that traffic load is mostly carried by subgrade, whereas pavement layers are mainly used for smoothness and dust control. Similar to the previous approach, these methods are simple, have low design cost, and could be reliable under certain conditions. The disadvantages are, these methods do not recognize the varying serviceability with age. These methods also assume average pavement material properties, traffic levels, and environmental conditions.

8.2.2.3 Methods Based on Statistical Evaluation of Pavement Performance

These methods are based on extensive field observation of pavement performance under different conditions and developing empirical relations between pavement thickness and material properties, traffic, and environmental conditions. Once these empirical relations are defined, the designer can input various input parameters and determine the required thicknesses of different layer. A typical example of this approach is the 1993 AASHTO (American Association of State Highway and Transportation Officials) design method (AASHTO, 1993). The main advantage of this approach over previous approaches is that the method considers the change of serviceability with pavement age. Thus, the designer can design a pavement section to last for a certain designed life with a predetermined serviceability level. This approach also considers in-service conditions and is not based of simple theoretical assumptions. It also allows for economic comparison between design alternatives.

This approach, however, still suffers from the dependency on empirical relations that are limited to the conditions under which they were developed. If changes occur in any input parameters such as increasing axle loads and tyre pressure or if a new pavement material is used such as modified asphalt binders, the method would not be valid.

8.2.2.4 Methods Based on Structural Analysis of Layered Systems

This approach is more fundamental than all other approaches since it considers basic material responses such a stresses, strains, and deformations. In such cases, the traffic load is applied on a simulated multilayered-pavement system and the critical material responses are calculated. These critical response parameters are then correlated to performance using transfer functions, typically based on empirical relations. The designer, therefore, has the capability to determine the required layer thicknesses so that the pavement would last for the required designed life without exceeding predetermined distress levels. This approach represents a major improvement over others due to its accuracy and reliability. However, this approach requires extensive testing and computations. Methods based on this approach also incorporate empirical correlations, although the degree of empiricism is small. In addition, theoretical models require extensive calibration and verification since the incorporated assumptions may not exactly match field conditions.

The proposed AASHTO mechanistic-empirical pavement design method (NCHRP, 2003) follows the last approach. Future approaches are expected to be more rational with less, or even no, dependency on empirical relations.

8.3 Stresses and Strains in Flexible Pavements

When the traffic wheel moves on the pavement surface a load pulse is applied. This load pulse creates stresses and strains throughout the multi-layer pavement system, including subgrade. As indicated earlier, stresses and strains resulting from a single load application are much less than those required to cause failure. However, when these stresses and strains are repeated many times throughout the years, pavement gradually deteriorates until it reaches to an unacceptable condition, or failure. Therefore, the magnitudes and the number of repetitions of stresses and strains affect pavement performance and determine its service life. Thus, it is important to accurately estimate stresses and strains in the multi-layer pavement system so that the pavement can be accurately designed. As discussed above, earlier pavement design approaches were empirical in nature and did not make use of stresses and strains developed by traffic loads. As the dependency on empirical relations has been decreasing and the degree of rationality has been increasing in the design process, estimating stresses and strains is becoming more vital. Before we discuss specific methods used to estimate stresses and strains in the pavement structure, basic concepts need to be defined such as the relation between traffic load, tyre pressure and contact area (tyre imprint) and how they affect stresses and strains in the pavement system.

8.3.1 Tyre Pressure, Contact Pressure and Contact Area

The interaction between vehicle and pavement is complex since pavement roughness excites the dynamic forces generated by vehicles, while these dynamic forces simultaneously increase the pavement roughness (Mamlouk, 1997). Research has been conducted to have better understanding of how the vehicle dynamic loads affect contact pressure within the tyre imprint for different pavement roughness levels, vehicle types and speeds, tyre types, and tyre pressures (Gillespie et al., 1993; De Beer, 1994; Cebon and Winkler, 1991).

Current design methods have used simplified procedures, which assume a static tyre load applied on the pavement surface (Yoder and Witczak, 1975; Huang, 2003). Figure 8.17 shows the tyre pressure (p_t), contact pressure (p_c), and contact area (A). The contact pressure between the tyre and pavement is slightly different from the tyre pressure because of the tension in the tyre walls. Also, the contact pressure is not exactly uniform throughout the contact area depending on the amount of tyre pressure. For example, for high-pressure tyres the contact pressure at the center of the tyre may be greater than that under

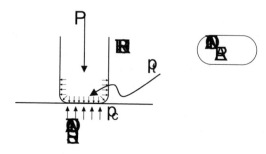

FIGURE 8.17 Tyre pressure, contact pressure, and contact area.

the tyre walls. For low-pressure tyres the reverse is true. However, in most cases we make two assumptions: (1) the contact pressure is uniformly distributed throughout the contact area, and (2) the contact pressure is equal to the tyre pressure. Thus,

$$p_c = p_t = p \qquad (8.1)$$

The contact area can be related to the load and tyre pressure as:

$$A = \frac{P}{p} \qquad (8.2)$$

where A is the contact area, P is the load, and p is the tyre pressure (assumed to be equal to the contact pressure). Assuming that the contact area is circular, the radius, a, can be obtained as:

$$a = \sqrt{\frac{A}{\pi}} = \sqrt{\frac{P}{p\pi}} \qquad (8.3)$$

Other shapes of tyre imprint have been assumed by various researchers such as rectangles and ellipses with an area following Equation 8.2.

8.3.2 Effect of Changing Traffic Load and Tyre Pressure

When a static load is applied on the pavement surface, the vertical stress at the surface will be equal to the tyre pressure as assumed earlier. As the depth within the pavement layers and into the subgrade

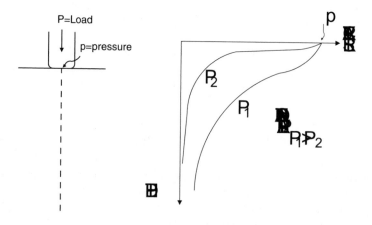

FIGURE 8.18 Effect of changing load.

increases, the vertical stress decreases. If the load increases while the tyre pressure is kept constant, the vertical stress at the surface remains the same but the vertical stresses increase in deeper layers as shown in Figure 8.18. On the other hand, if the load is kept constant while the tyre pressure is increased, the vertical stress at the surface increases to match the new tyre pressure while the stresses in deeper layers do not significantly change as demonstrated in Figure 8.19. Therefore, it can be concluded that the load affects deeper layers, while tyre pressure affects surface layers. This concept is very important in pavement design since the required thicknesses of the different pavement layers are mostly a function of vehicle load, while the required quality of the surface is a function of tyre pressure.

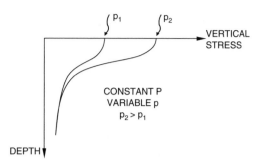

FIGURE 8.19 Effect of changing pressure.

8.3.3 Stresses and Strains in Multi-Layered Systems

Many methods have been developed throughout the years to estimate basic engineering parameters such as stresses and strains in multi-layered pavement systems. To simplify the problem, Boussinesq in 1885 was able to compute the vertical stress at any depth due to a static point load applied on a single-layer system, or a half-space (Figure 8.20). He also assumed a homogenous, isotropic, and linear elastic material. Based on the relationship, stresses and surface deformations can be computed assuming a uniformly distributed circular load applied on a homogeneous single-layer system.

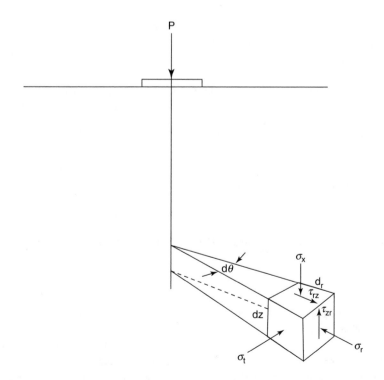

FIGURE 8.20 Stresses developed in a one-layer system due to the application of a surface load.

In 1943, Burmister provided analytical expressions for stresses and displacement in two- and three-layer elastic systems due to static loads. Other researchers developed tables and graphs to simplify the estimation process. Details of different solutions by various researchers are given by Yoder and Witczak (1975) and Huang (2003).

Numerical solutions to a multi-layered system have been developed where stresses, strains, and displacements could be estimated with a good degree of accuracy. Different researchers have used different assumptions related to the type of load, material properties, and boundary conditions. Load types include static, constant-magnitude moving, and dynamic. Material properties include linear and non-linear with elastic, visco-elastic or plastic behavior. Boundary conditions include finite or infinite depth of the bottom layer as well as different levels of friction at the interfaces between layers. Since the solution involves extensive computations, computer programs have been developed such as Chevron (Warren and Diekmann, 1963 ELSYM5 [Kopperman et al., 1985]), BISAR (Michelow, 1963), Kenpave (Huang, 2003), and JULEA (Peutz et al., 1968).

Stresses and strains in the multi-layered pavement systems have been recently incorporated into pavement design. Critical response parameters are estimated and then correlated to pavement performance. For example, the tensile strain at the bottom of the asphalt layer has been correlated to the fatigue life of the pavement section. Also, the compressive strain on top of the subgrade has been correlated to the rutting behavior. Therefore, the pavement can be designed to allow a certain level of fatigue and rutting at the end of the designed life. The procedure proposed for the recently developed Mechanistic-Empirical Design Guide (NCHRP, 2003) uses stresses, strains and displacements at various locations in the pavement system to estimate the designed life of the pavement structure.

8.4   Major Road Experiments

A number of experimental road tests have been conducted by various agencies to develop better understanding of pavement performance for different material types, load levels, and environmental conditions. In these road tests, pavement sections are built under typical field conditions and subjected to accelerated traffic loading until failure. New materials that have not been used before can also be used in these road tests to evaluate their applicability and performance. The condition of the pavement is periodically evaluated using condition surveys, roughness measurements, field load testing, and coring and sampling followed by laboratory testing.

The results obtained from experimental road tests provide useful information for pavement design and performance. Relations can be developed to predict performance and validate theoretical models. The use of different materials and their ability to sustain traffic loading and environmental conditions can also be obtained from the test results.

In spite of their use and importance in pavement studies, the relations obtained from these experimental road tests are mostly empirical. Thus, these relations are applicable to conditions used in the tests. For example, the results cannot accurately be used for materials, traffic loading or environmental conditions that were not available during the test.

8.4.1   AASHO (American Association of State Highway Officials) Road Test

Pavement design up to the late 1950 s was primarily based on empirical knowledge and engineering judgment. The enormous investment in the Interstate Highway system in the United States required improved design methods. The AASHO Road Test was designed to provide empirical design methods for the selection of the thickness of highway pavements (HRB, 1962).

The test was one of the early major experimental road tests that had a significant effect of modern pavement design. It was conducted at Ottawa, Illinois, in the U.S. from 1958 to 1960. It consisted of six pavement loops, each of which had a variety of designs of both flexible and rigid pavement. By having only one type of truck operated on each of the loops, data were gathered to allow the development of specific relationships for the damage caused by different truck loads. The pavement condition and ride quality were surveyed every 2 weeks and the present serviceability rating, PSR, was recorded.

The results of the AASHO Road Test were used to develop the serviceability-performance concept (Figure 8.14) and to define pavement failure. Empirical equations for the selection of pavement thickness were developed. The concept of equivalent axle load that will be discussed in Section 8.5 was also developed. Statistics was extensively used for the design of the experiment, construction quality control, and the analysis of the performance data. The 1993 AASHTO Pavement Design Guide and previous versions were based on the results of the AASHO Road Test. These AASHTO design guides have been extensively used for many years to design pavements worldwide.

The AASHO Road Test, however, has some limitations including:

- One environment that may not apply to other environments
- One set of material properties
- Trucks with small tyre pressures that are not typical of current vehicles
- Accelerated test with limited environmental exposure.

8.4.2   Long-Term Pavement Performance Program (LTPP)

In 1987, the LTPP program, a comprehensive 20-year study of in-service pavements, began a series of rigorous long-term field experiments to monitor more than 2400 asphalt and portland cement concrete pavement test sections across the U.S. and Canada (FHWA, 1993). These pavement sections covered different material types, traffic volumes, and environmental conditions. Established as part of the Strategic Highway Research Program (SHRP) and now managed by the Federal Highway Administration (FHWA), LTPP was designed as a partnership with the U.S. States and Canadian Provinces. The LTPP's goal was to help the States and Provinces make decisions that will lead to better performing and more cost-effective pavements.

The data that have been collected through the LTTP program and are available in the DataPave CD (FHWA, 2002) include:

- Condition surveys
- Annual roughness measurements
- Annual nondestructive structural testing using the Falling Weight Deflectometer
- Coring, sampling, and laboratory testing
- Annual skid measurements

Some of the LTPP data have been used in the development of the proposed AASHTO Mechanistic-Empirical Pavement Design Guide, which is discussed in Section 8.6. The LTTP data have been used for the validation and calibration of the design models in the proposed AASHTO M-E Pavement Design Guide.

8.4.3   Other Road Experiments

Other major road experiments that have been conducted by various transportation agencies throughout the world include WesTrack in Nevada (NCHRP, 2000), MnRoad in Minnesota (MDOT, 1994), the National Center for Asphalt Technology (NCAT) Pavement Test Track (Metcalf, 1996), New Zealand's CAPTIF (Metcalf, 1996), TRL loading facility in the U.K. (Metcalf, 1996). These road experts have been used by various agencies to evaluate the performance of different types of materials under various traffic levels and environmental conditions.

8.5   AASHTO, 1993 Design Method

8.5.1   Background

The AASHTO, 1993 design method is based the statistical evaluation of the performance data obtained from the AASHO Road Test presented in Section 8.4. The 1993 version (AASHTO, 1993) was preceded by several versions and followed by the proposed AASHTO mechanistic-empirical design method presented in Section 8.6. Although the 1993 method is not the latest, it is currently the most commonly used method of design of asphalt pavement and is expected to be used for several more years until agencies adopt the proposed AASHTO M-E method.

8.5.2 Equivalent Single Axle Load (ESAL)

Traffic loads applied on the pavement surface range from light passenger cars to heavy trucks. Heavy traffic loads have more harmful effect to pavement than light loads. Also, as the number of repetitions of the same load increases the effect on the pavement increases. To design a pavement section the damage caused by all axle loads that will be applied on the pavement during its designed life has to be considered.

According to the AASHTO (1993) design method axles with different magnitudes and different numbers of repetitions are converted to an equivalent number of repetitions of a standard axle load that causes the same damage to the pavement. A standard axle load was selected as 18000 Lb (80 kN) applied on a single axle with a dual wheel at each end. The ESAL is the equivalent number of repetitions of the 18-kip (80 kN) standard axle load that causes the same damage to the pavement caused by one-pass of the axle load in question. The 1993 AASHTO guide developed Equivalent Axle Load Factors (*EALF*) to relate the damage caused by different load magnitudes and axle configurations to the standard axle load as shown in Equation 8.4.

$$EALF = \frac{W_{t18}}{W_{tx}} \tag{8.4}$$

where W_{t18} is the number of 18-kip (80-kN) single-axle load applications to time t (failure) and W_{tx} is the number of x-axle load applications to time t (failure). Based on the data obtained at the AASTO Road Test, the following regression equation was developed.

$$\log\left(\frac{W_{tx}}{W_{t18}}\right) = 4.79 \log(18+1) - 4.79 \log(L_x + L_2) + 4.33 \log(L_2) + \frac{G_t}{\beta_x} - \frac{G_t}{\beta_{18}} \tag{8.5a}$$

$$G_t = \log\left(\frac{4.2 - p_t}{4.2 - 1.5}\right) \tag{8.5b}$$

$$\beta_x = 0.40 + \frac{0.081(L_x + L_2)^{3.23}}{(SN + 1)^{5.19} L_2^{3.23}} \tag{8.5c}$$

where L_x is the load in kips on one single axle, one set of tandem axles, or one set of triple axles; L_2 is the axle code (1 for single, 2 for tandem axles, and 3 for triple axles); p_t is the terminal serviceability index (See Section 8.1); β_{18} is the value of β_x when L_x is equal to 18 and L_2 is equal to one. SN is the structural numbers, which is an index that combines the effect of material properties, layer thicknesses and drainage quality according to Equation 8.6. Tables 8.1–8.3 show numerical values of *EALF* for single, tandem, and triple axles, respectively. In order to get the *EALFs* the structural number has to be known or assumed.

$$SN = a_1 D_1 + a_2 D_2 m_2 + a_3 D_3 m_3 \tag{8.6}$$

where:

    a_1, a_2 and a_3 = Structural layer coefficients (Defined in the next section, Step 8).
    D_1, D_2 and D_3 = Thicknesses of surface, base and subbase, respectively.
      m_2, m_3 = Drainage coefficients (Table 8.4).

Since the *EALFs* are not very sensitive to SN, a SN value of 5 may be assumed in most cases. Unless the design thickness is significantly different, no iterations will be needed.

As shown in Table 8.1, the *EALF* corresponding to a single axle of 18000 Lb is 1, since it is the standard axle load. For tandem and triple axles (Tables 8.2 and 8.3), an *EALF* of 1 corresponds to an axle load of

TABLE 8.1 Axle Load Equivalency Factors for Flexible Pavements, Single Axles

Axle Load (kips)	$p_t = 2.0$				$p_t = 2.5$				$p_t = 3.0$			
	SN				SN				SN			
	3	4	5	6	3	4	5	6	3	4	5	6
2	0.0002	0.0002	0.0002	0.0002	0.0003	0.0002	0.0002	0.0002	0.0006	0.0003	0.0002	0.0002
4	0.003	0.002	0.002	0.002	0.004	0.003	0.002	0.002	0.006	0.004	0.002	0.002
6	0.011	0.010	0.009	0.009	0.017	0.013	0.010	0.009	0.028	0.018	0.012	0.010
8	0.036	0.033	0.031	0.029	0.051	0.041	0.034	0.031	0.080	0.055	0.040	0.034
10	0.090	0.085	0.079	0.076	0.118	0.102	0.088	0.080	0.168	0.132	0.101	0.086
12	0.189	0.183	0.174	0.168	0.229	0.213	0.189	0.176	0.296	0.260	0.212	0.187
14	0.354	0.350	0.338	0.331	0.399	0.388	0.360	0.342	0.468	0.447	0.391	0.058
16	0.613	0.612	0.603	0.596	0.646	0.645	0.623	0.606	0.695	0.693	0.651	0.622
18	1.00	1.00	1.00	1.00	1.00	1.00	1.00	1.00	1.00	1.00	1.00	1.00
20	1.56	1.55	1.57	1.59	1.49	1.47	1.51	1.55	1.41	1.38	1.44	1.51
22	2.35	2.31	2.35	2.41	2.17	2.09	2.18	2.30	1.96	1.83	1.97	2.16
24	3.43	3.33	3.40	3.51	3.09	2.89	3.03	3.27	2.69	2.39	2.60	2.96
26	4.88	4.68	4.77	4.96	4.31	3.91	4.09	4.48	3.65	3.08	3.33	3.91
28	6.78	6.42	6.52	6.83	5.90	5.21	5.69	5.98	4.88	3.93	4.17	5.00
30	9.2	8.6	8.7	9.2	7.9	6.8	7.0	7.8	6.5	5.0	5.1	6.3
32	12.4	11.5	11.5	12.1	10.5	8.8	8.9	10.0	8.4	6.2	6.3	7.7
34	16.3	15.0	14.9	15.6	13.7	11.3	11.2	12.5	10.9	7.8	7.6	9.3
36	21.2	19.3	19.0	19.9	17.7	14.4	13.9	15.5	14.0	9.7	9.1	11.0
38	27.1	24.6	24.0	25.1	22.6	18.1	17.2	19.0	17.7	1.9	11.0	13.0
40	34.3	30.9	30.0	31.2	28.5	22.5	21.1	23.0	22.2	14.6	13.1	15.3

TABLE 8.2 Axle Load Equivalency Factors for Flexible Pavements, Tandem Axles

Axle Load (kips)	$p_t = 2.0$				$p_t = 2.5$				$p_t = 3.0$			
	SN				SN				SN			
	3	4	5	6	3	4	5	6	3	4	5	6
10	0.008	0.007	0.006	0.006	0.011	0.009	0.007	0.006	0.020	0.012	0.008	0.007
12	0.016	0.014	0.013	0.012	0.023	0.018	0.014	0.013	0.039	0.024	0.017	0.014
14	0.029	0.026	0.024	0.023	0.042	0.033	0.027	0.024	0.068	0.045	0.032	0.026
16	0.050	0.046	0.042	0.040	0.070	0.057	0.047	0.043	0.109	0.076	0.055	0.046
18	0.081	0.075	0.069	0.066	0.109	0.092	0.077	0.070	0.164	0.121	0.090	0.076
20	0.124	0.117	0.109	0.105	0.162	0.141	0.121	0.110	0.232	0.182	0.139	0.119
22	0.183	0.174	0.164	0.158	0.229	0.207	0.180	0.166	0.313	0.260	0.205	0.178
24	0.260	0.252	0.239	0.231	0.315	0.292	0.260	0.242	0.407	0.358	0.292	0.257
26	0.360	0.353	0.338	0.329	0.420	0.401	0.364	0.342	0.517	0.476	0.402	0.360
28	0.487	0.481	0.466	0.455	0.548	0.534	0.495	0.470	0.643	0.614	0.538	0.492
30	0.646	0.643	0.627	0.617	0.703	0.695	0.658	0.633	0.788	0.773	0.702	0.656
32	0.843	0.842	0.829	0.819	0.889	0.887	0.857	0.834	0.956	0.953	0.896	0.855
34	1.08	1.08	1.08	1.07	1.11	1.11	1.09	1.08	1.15	1.15	1.12	1.09
36	1.38	1.38	1.38	1.38	1.38	1.38	1.38	1.38	1.38	1.38	1.38	1.38
38	1.73	1.72	1.73	1.74	1.69	1.68	1.70	1.73	1.64	1.62	1.66	1.70
40	2.15	2.13	2.16	2.18	2.06	2.03	2.08	2.14	1.94	1.89	1.98	2.08
42	2.64	2.62	2.66	2.70	2.49	2.43	2.51	2.61	2.29	2.19	2.33	2.50
44	3.23	3.18	3.24	2.31	2.99	2.88	3.00	3.16	2.70	2.52	2.71	2.97
46	3.92	3.83	3.91	4.02	3.58	3.40	3.55	3.79	3.16	2.89	3.13	3.50
48	4.72	4.58	4.68	4.83	4.25	3.98	4.17	4.49	3.70	3.29	3.57	4.07

TABLE 8.3 Axle Load Equivalency Factors for Flexible Pavements, Triple Axles

Axle Load (kips)	$p_t = 2.0$				$p_t = 2.5$				$p_t = 3.0$			
	SN				SN				SN			
	3	4	5	6	3	4	5	6	3	4	5	6
20	0.029	0.026	0.024	0.023	0.042	0.032	0.027	0.024	0.069	0.044	0.031	0.026
22	0.042	0.038	0.035	0.034	0.060	0.048	0.040	0.036	0.097	0.065	0.046	0.039
24	0.060	0.055	0.051	0.048	0.084	0.068	0.057	0.051	0.132	0.092	0.066	0.056
26	0.083	0.077	0.071	0.068	0.114	0.095	0.080	0.072	0.174	0.126	0.092	0.078
28	0.113	0.105	0.098	0.094	0.151	0.128	0.109	0.099	0.223	0.168	0.126	0.107
30	0.149	0.140	0.131	0.126	0.195	0.170	0.145	0.133	0.279	0.219	0.167	0.143
32	0.194	0.184	0.173	0.167	0.247	0.220	0.191	0.175	0.342	0.279	0.218	0.188
34	0.248	0.238	0.225	0.217	0.308	0.281	0.246	0.228	0.413	0.350	0.279	0.243
36	0.313	0.303	0.288	0.279	0.379	0.352	0.313	0.292	0.491	0.432	0.352	0.310
38	0.390	0.381	0.364	0.353	0.461	0.436	0.393	0.368	0.577	0.524	0.437	0.389
40	0.481	0.473	0.454	0.443	0.554	0.533	0.487	0.459	0.671	0.626	0.536	0.483
42	0.57	0.580	0.561	0.548	0.661	0.644	0.597	0.567	0.775	0.740	0.649	0.593
44	0.710	0.705	0.686	0.673	0.781	0.769	0.723	0.692	0.889	0.865	0.777	0.720
46	0.852	0.849	0.831	0.818	0.918	0.911	0.868	0.838	1.01	1.00	0.920	0.865
48	1.02	1.01	0.999	0.987	1.07	1.07	1.03	1.01	1.13	1.15	1.08	1.03
50	1.20	1.20	1.19	1.18	1.24	1.25	1.22	1.20	1.30	1.31	1.26	1.22
52	1.42	1.42	1.41	1.40	1.44	1.44	1.43	1.41	1.47	1.48	1.45	1.43
54	1.66	1.66	1.66	1.66	1.66	1.66	1.66	1.66	1.66	1.66	1.66	1.66
56	1.93	1.93	1.94	1.94	1.90	1.90	1.91	1.93	1.86	1.85	1.88	1.91
58	2.42	2.23	2.25	2.27	2.17	2.16	2.20	2.24	2.09	2.06	2.13	2.20

TABLE 8.4   Recommended Drainage Coefficients for Untreated Bases and Subbases in Flexible Pavements (AASHTO, 1993)

Quality of Drainage		% of Time Pavement Structure is Exposed to Moisture Levels Approaching Saturation			
Rating	Water Removed Within	<1%	1–5%	5–25%	>25%
Excellent	2 h	1.40–1.35	1.35–1.30	1.30–1.20	1.20
Good	1 day	1.35–1.25	1.25–1.15	1.15–1.00	1.00
Fair	1 week	1.25–1.15	1.15–1.05	1.00–0.80	0.80
Poor	1 month	1.15–1.05	1.05–0.80	0.80–0.60	0.60
Very Poor	Never drain	1.05–0.95	0.95–0.75	0.75–0.40	0.40

about 34 kips and 48 kips, respectively. Note that the *EALF* is very sensitive to the magnitude of the axle load regardless of the axle configuration. In fact, the *EALF* exponentially increases as the load increases indicating a significant increase in pavement damage.

Knowing the *EALF* and the estimated number of repetitions of axle loads for different axle groups in the first day of opening the road to traffic, the initial daily ESAL ($ESAL_o$) is computed as:

$$ESAL_o = \sum_{i=1}^{m} (N_i \, EALF_i) \tag{8.7}$$

where N_i is the number of repetitions of axle group i, $EALF_i$ is the equivalency factor for axle group i, and m is number of axle groups. The cumulative ESAL during the designed life of the pavement (W_{18}) is then calculated as:

$$W_{18} = \sum_{i=1}^{n} ESAL = \frac{ESAL_o(365)}{\log_e(1+i)}[(1+i)^n - 1] \tag{8.8}$$

where n is the designed life of the pavement in years and i is the expected annual traffic growth rate.

8.5.3   Design Procedure

The main requirement is to determine the thicknesses of various pavement layers to satisfy the design objectives stated in Section 8.2. Assuming that the pavement section consists of surface, base and subbase, three thicknesses: D_1, D_2 and D_3 are required for the three layers, respectively. The design procedure can be divided into 12 steps as presented below. These steps have been incorporated in several computer programs to facilitate the design procedure such as DARWin (AASHTO, 2001).

8.5.3.1   Step 1 — Reliability

A reliability level (R) is selected depending on the functional classification of the road and whether the road is in urban or rural area. The reliability is the chance that pavement will last for the design period without failure. A larger reliability value will ensure better performance, but it will require larger layer thicknesses. Table 8.5 shows reliability levels suggested by the 1993 AASHTO design guide. The reliability

TABLE 8.5   Suggested Levels of Reliability for Various Functional Classifications (AASHTO, 1993)

Functional Classification	Recommended Level of Reliability	
	Urban	Rural
Interstate and other freeways	85–99.9	80–99.9
Principal arterials	80–99	75–95
Collectors	80–95	75–95
Local	50–80	50–80

levels shown in Table 8.5 have a wide range to accommodate different field conditions. Different agencies typically select reliability values from the table that match their local conditions.

8.5.3.2   Step 2 — Overall Standard Deviation

The overall standard deviation (S_o) takes into consideration the variability of all input data. The 1993 design guide recommends an approximate range of 0.4 to 0.5 for flexile pavements. An overall standard deviation value (S_o) is selected by the designer within this range.

8.5.3.3   Step 3 — Cumulative Equivalent Single Axle Load

In this step, the designer assumes a designed life, typically in the range of 10 to 20 years. The cumulative expected 18-kip (80-kN) ESAL (W_{18}) during the designed life in the design lane is then determined as discussed earlier. If the cumulative two-directional 18-kip ESAL is known, the designer must factor the design traffic by directions by multiplying by the directional distribution factor (D) to get the ESAL in the predominate direction. For example, if the traffic split during the peak hour is 70–30%, D is taken as 0.7. To get the ESAL in the design (right) lane, the design traffic in the predominant direction is multiplied by the lane distribution factor (L) shown in Table 8.6.

8.5.3.4   Step 4 — Effective Roadbed Soil Resilient Modulus

Determine the resilient modulus (M_R) of the roadbed soil in the laboratory according to AASHTO T307 method (AASHTO, 2004). Since the resilient modulus of the soil depends on the moisture content, different resilient moduli will be obtained in different seasons depending on the amount of rain or snow in each season. Thus, an effective roadbed soil resilient modulus is needed to represent a weighted average value for the whole year.

   Figure 8.21 can be used to estimate the effective roadbed soil resilient modulus. In this method, the year is divided into a number of distinct seasons where the resilient modulus is significantly different. The relative damage (u_f) corresponding to each M_R value is determined using the scale in Figure 8.21 and recorded in the table. The u_f values are averaged and the corresponding M_R value is obtained from the same scale and reported as the effective roadbed soil resilient modulus.

8.5.3.5   Step 5 — Resilient Moduli of Pavement Layers

The resilient moduli (M_R) of the surface, base, and subbase layers are either determined using laboratory testing or estimated using previously developed correlations.

8.5.3.6   Step 6 — Serviceability Loss

The serviceability loss is the difference between the initial serviceability index (p_o) and the terminal serviceability index (p_t) (See Figure 8.14).

$$\Delta PSI = p_o - p_t \tag{8.9}$$

The typical P_o value for a new pavement is 4.6 or 4.5. The recommended values of p_t are 3.0, 2.5 or 2.0 for major roads, intermediate roads and secondary roads, respectively.

8.5.3.7   Step 7 — Structural Numbers

As shown in Equation 8.6, the structural number (SN) is an index value that combines layer thicknesses, structural layer coefficients, and drainage coefficients. In this step the structural numbers required above

TABLE 8.6   Lane Distribution Factor (AASHTO, 1993)

No. of Lanes in Each Direction	% of 18-kip ESAL in the Design Lane
1	100
2	80–100
3	60–80
4	50–70

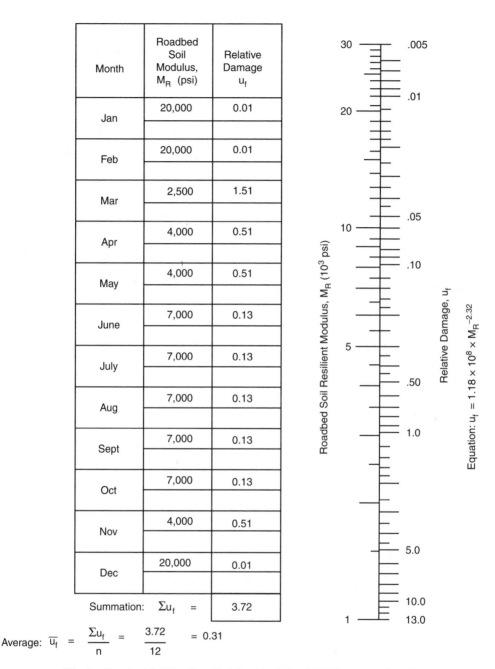

Month	Roadbed Soil Modulus, M_R (psi)	Relative Damage u_f
Jan	20,000	0.01
Feb	20,000	0.01
Mar	2,500	1.51
Apr	4,000	0.51
May	4,000	0.51
June	7,000	0.13
July	7,000	0.13
Aug	7,000	0.13
Sept	7,000	0.13
Oct	7,000	0.13
Nov	4,000	0.51
Dec	20,000	0.01
Summation: Σu_f =		3.72

Average: $\overline{u_f} = \dfrac{\Sigma u_f}{n} = \dfrac{3.72}{12} = 0.31$

Effective Roadbed Soil Resilient Modulus, M_R (psi) = 5,000 (corresponds to $\overline{u_f}$)

FIGURE 8.21 Worksheet for estimating effective roadbed soil resilient modulus (AASHTO, 1993).

the subgrade, subbase, and base layers are determined. The required structural number above the subgrade (SN_3) is determined first using either Equation 8.10 or Figure 8.22.

$$\log W_{18} = Z_R S_o + 9.36 \log(SN+1) - 0.2 + \frac{\log[\Delta PSI/(4.2-1.5)]}{0.4 + 1094/(SN+1)^{5.19}} + 2.32 \log M_R - 8.07 \quad (8.10)$$

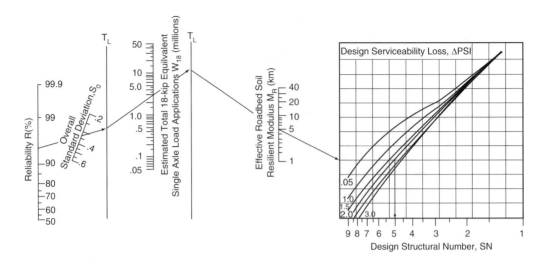

FIGURE 8.22 Design chart for flexible pavements based on using mean values for each input (AASHTO, 1993).

where:

W_{18} = Cumulative expected 18-kip ESAL during the designed life in the design lane
Z_R = Normal deviate for a given reliability R (3)
S_o = Standard deviation
M_R = Effective roadbed soil resilient modulus (step 4)

Note that SN can be obtained from Equation 8.10 either by trial and error or by iteration using a computer program.

This process is repeated two more times to obtain the required structural number above the subbase (SN_2) and the required structural number above the base (SN_1). To obtain SN_2, M_R value of the subbase should be used in Equation 8.10 (or Figure 8.22). Similarly, to obtain SN_1, M_R of the base should be used.

8.5.3.8   Step 8 — Structural Layer Coefficients

The structural layer coefficient is a measure of the relative ability of a unit thickness of a given material to function as a structural component of the pavement. Three structural layer coefficients (a_1, a_2 and a_3) are required for the surface, base and subbase, respectively. These coefficients can be determined from road tests, as was done in the AASHO Road Test, or from correlations with material properties as shown in Figures 8.23, 8.24 and 8.25 (Van Til et al., 1972). It is recommended that the structural layer coefficients be based on the resilient modulus, which a more fundamental material property. A typical a_1 value for the dense-graded HMA is 0.44, which corresponds to a resilient modulus of 450,000 psi as shown in Figure 8.23.

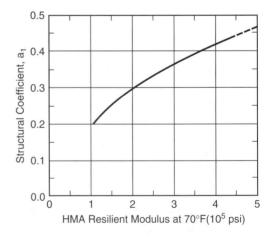

8.5.3.9   Step 9 — Drainage Coefficients

Drainage coefficients are measures of the quality of drainage and the availability of moistures in

FIGURE 8.23 Chart for estimating structural layer coefficient of dense-graded asphalt concrete based on the elastic (resilient) modulus (Van Til et al., 1972).

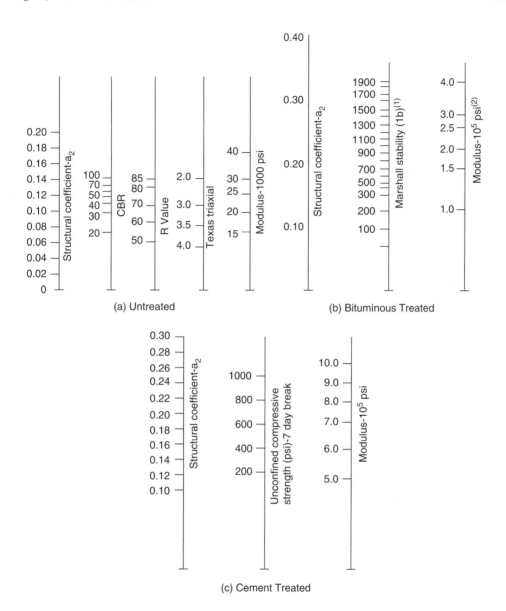

FIGURE 8.24 Correlation charts for estimating resilient modulus of bases (Van Til et al., 1972).

the granular base and subbase. Two equal drainage coefficients (m_2 and m_3) are needed for the base and subbase, respectively. The drainage coefficient values for the untreated base and subbase recommended by the AASHTO 1993 design guide are shown in Table 8.4.

8.5.3.10 Step 10 — Layer Thicknesses

Using the structural numbers required above the base, subbase and the subgrade (SN_1, SN_2 and SN_3) obtained in Step 7, the layer thicknesses of the surface, base and subbase (D_1, D_2 and D_3) can be obtained from Equations 8.11, 8.12 and 8.13, respectively. First, Equation 8.11 is used to solve for D_1 and the value is round up to the next 1/2 in. increment. The rounded value of D_1 is used in Equation 8.12 to solve for D_2 and the value is rounded up to the next 1 in. increment. Finally, the rounded values of D_1 and D_2 are

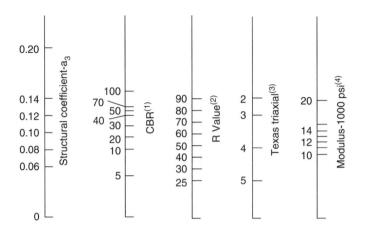

FIGURE 8.25   Correlation chart for estimating resilient modulus of subbases (Van Til et al., 1972).

used in Equation 8.13 to solve for D_3 and the value is rounded up to the next 1 in. increment.

$$SN_1 \leq a_1 D_1 \tag{8.11}$$

$$SN_2 \leq a_1 D_1 + a_2 D_2 m_2 \tag{8.12}$$

$$SN_3 \leq a_1 D_1 + a_2 D_2 m_2 + a_3 D_3 m_3 \tag{8.13}$$

The values of D_1, D_2 and D_3 have to meet certain minimum practical thicknesses as shown in Table 8.7.

Note that Equations 8.11 to 8.13 may allow for thickness compensations among layers. For example, a larger value for D_1 may be used that would allow for a smaller value of D_2. Since the costs of materials at different locations are different, the designer can make use of the thickness compensation concept to obtain the most economic pavement section.

8.5.3.11   Step 11 — Freeze or Thaw and Swelling

If the pavement is located in an area where freeze or thaw and soil swelling exists, the AASHTO (1993) design guide recommends additional procedure to estimate the reduction in the service life due to this environmental effect (AASHTO, 1993).

8.5.3.12   Step 12 — Life-Cycle Cost

In step 3, a pavement design period was assumed, which may not produce the least life-cycle cost. In this step, the designer assumes a few other design periods and repeats the design process for each design period. A life cycle cost analysis is then performed, as discussed in Chapter 18, to obtain the most economic design strategy. In this analysis all the costs included in the analysis period are considered such

TABLE 8.7   Minimum Thickness (in.) (AASHTO, 1993)

Traffic, ESAL's	Asphalt Concrete	Aggregate Base
Less than 50,000	1.0 (or surface treatment)	4
50,001–150,000	2.0	4
150,001–500,000	2.5	4
500,001–2,000,000	3.0	6
2,000,001–7,000,000	3.5	6
Greater than 7,000,000	4.0	6

as the costs of initial construction, maintenance, rehabilitation, and the salvage value of the pavement section at the end of the analysis period. The users cost may also be considered.

8.5.4 Sample Problem

Design the pavement for an expressway consisting of an asphalt concrete surface, a crushed-stone base, and a granular subbase using the 1993 AASHTO design chart (Figure 8.22). The cumulative ESAL in the design lane for a design period of 15 years is 7×10^6. The area has good quality drainage with 10% of the time the moisture level is approaching saturation. The effective roadbed soil resilient modulus is 7 ksi, the subbase has a CBR value of 80, the resilient modulus of the base is 40 Lb, and the resilient modulus of asphalt concrete is 4.5×10^5 psi. Assume a reliability level of 95% and S_o of 0.45.

Solution

Step 1
   Reliability $(R) = 95\%$ (Given)
Step 2
   overall standard deviation $(S_o) = 0.45$ (Given)
Step 3
   $W_{18} = 7 \times 10^6$ (Given)
Step 4
   Effective road-bed soil resilient modulus = 7 ksi (Given)
Step 5
   Resilient modulus of subbase = 20 ksi (Figure 8.25)
   Resilient modulus of base = 40 ksi (Given)
   Resilient modulus of asphalt concrete surface = 450 ksi (Given)
Step 6
   Assume initial serviceability index $(p_o) = 4.6$
   Assume terminal serviceability index $(p_t) = 3.0$
   $\Delta \text{PSI} = 4.6 - 3.0 = 1.6$
Step 7
   $SN_3 = 5.2$ (Using Figure 8.22 and subgrade M_R of 7 ksi)
   $SN_2 = 3.5$ (Using Figure 8.22 and subbase M_R of 20 ksi)
   $SN_1 = 2.7$ (Using Figure 8.22 and base M_R of 40 ksi)
Step 8
   $a_3 = 0.14$ (Figure 8.25)
   $a_2 = 0.17$ (Figure 8.24)
   $a_1 = 0.44$ (Figure 8.23)
Step 9
   Drainage coefficients = $m_2 = m_3 = 1.1$ (Table 8.4)
Step 10
   Equation 8.11: $2.7 \leq 0.44\, D_1$
   $D_1 = 6.1$ in. (Round to 6.5 in.)
   Equation 8.12: $3.5 \leq 0.44 \times 6.5 + 0.17 \times D_2 \times 1.1$
   $D_2 = 3.4$ in. (Use a minimum value of 6 in.) (Table 8.7)
   Equation 8.13: $5.2 \leq 0.44 \times 6.5 + 0.17 \times 6 \times 1.1 + 0.14 \times D_3 \times 1.1$
   $D_3 = 7.9$ in. (Round to 8 in.)
Step 11
   No information given on Freeze–thaw or swelling
Step 12
   No information given on costs

Therefore,

$D_1 = 6.5$ in.
$D_2 = 6$ in.
$D_3 = 8$ in.

8.6   Proposed AASHTO Mechanistic-Empirical Design Method

At the time of preparing this section a trial version of the AASHTO Mechanistic-Empirical Pavement Design Guide (formerly known as 2002 Pavement Design Guide) was available for verification (NCHRP, 2003). Therefore, this section covers basic approaches used in the proposed AASHTO M-E Pavement Design Guide and is not intended to be used for actual design. For more detailed information the reader needs to consult with the final version when it is published.

The proposed AASHTO M-E Pavement Design Guide (NCHRP, 2003) was developed to avoid the limitations of the 1993 guide and to provide the highway community with a state-of-the-practice tool for the design of new and rehabilitated pavement structures based on mechanistic-empirical principles. The proposed M-E Pavement Design Guide represents a major change in the way pavement design is performed. The designer first considers site conditions (traffic, climate, subgrade) and construction conditions in proposing a trial design. The trial design is then evaluated for adequacy through the prediction of key distresses and smoothness. If the design does not meet desired performance criteria, it is revised and the evaluation process is repeated as necessary.

Unlike empirical procedure, the mechanistic-empirical format of the proposed M-E Pavement Design Guide provides a framework for continuous improvement to keep up with changes in trucking, materials, construction, design concepts, computers, and others. It is expected to reduce early failures and increase pavement longevity. The "mechanistic" part of the procedure refers to the application of the principles of engineering mechanics, which leads to a rational design process. The mechanistic portion includes the use of stresses, strains (linear or nonlinear), and time dependency. The "empirical" part includes relating the theoretical computation of damage (which is, in turn, a function of pavement deflection, strain, or stress responses) at some critical locations to measured field distress.

8.6.1   Design Approach

The design approach provided in the proposed AASHTO M-E Pavement Design Guide consists of three major stages: (1) Evaluation, (2) Analysis, and (3) Strategy selection, as summarized in Figure 8.26. The design process for new pavement structures includes consideration of the following:

- Foundation or subgrade,
- Paving materials,
- Construction factors,
- Environmental factors (temperature and moisture),
- Traffic loadings,
- Sub-drainage,
- Pavement performance (key distresses and smoothness),
- Design reliability, and
- Life cycle costs.

8.6.2   Hierarchical Design Inputs

The hierarchical approach is employed with regard to traffic, materials, and environmental inputs. In general, three levels of inputs are provided.

- Level 1 inputs provide for the highest level of accuracy and, thus, would have the lowest level of uncertainty or error. Level 1 inputs would typically be used for designing heavily trafficked

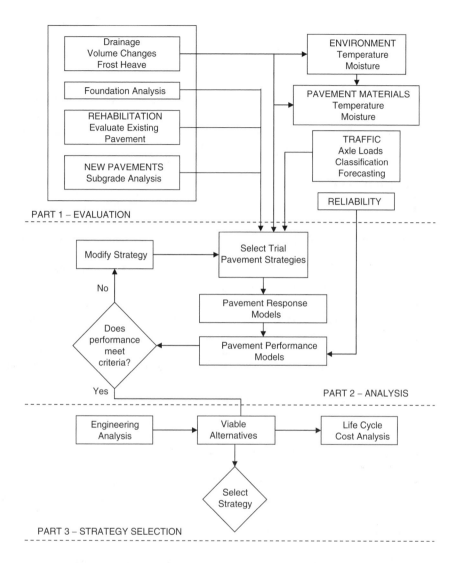

FIGURE 8.26 Conceptual schematic of the three-stage design process (NCHRP, 2003).

pavements or wherever there are dire safety or economic consequences of early failure. Level 1 material inputs require laboratory or field testing, such as the dynamic modulus testing of hot-mix asphalt concrete or site-specific axle load spectra data collections. Obtaining Level 1 inputs requires more resources and time than other levels.

• Level 2 inputs provide an intermediate level of accuracy and would be closest to the typical procedures used with earlier editions of the AASHTO Guide. This level could be used when resources or testing equipment are not available for tests required for Level 1. Level 2 inputs typically would be user-selected, possibly from an agency database, could be derived from a limited testing program, or could be estimated through correlations. Examples would be estimating asphalt concrete dynamic modulus from binder, aggregate, and mix properties or using site-specific traffic volume and traffic classification data in conjunction with agency-specific axle load spectra.

• Level 3 inputs provide the lowest level of accuracy. This level might be used for design where there are minimal consequences of early failure (e.g., low volume roads). Inputs typically would be user-selected values or typical averages for the region. An example would be default unbound materials resilient modulus values used by an agency.

8.6.3 Traffic Characterization

The proposed AASHTO M-E Pavement Design Guide considers truck traffic loadings in terms of axle load spectra. The full axle load spectra for single, tandem, tridem, and quad axles are considered. The ESAL approach used in the 1993 AASHTO Guide is no longer used in the M-E method as a direct design input. In a few cases, axle load spectra are converted internally in the software to ESALs as a means of making use of earlier mathematical models that have not been converted to an axle load spectra basis. The software uses the number of heavy trucks as an overall indicator of the magnitude of truck traffic loadings (FHWA class 4 and above).

8.6.4 Environmental Effects

Changing temperature and moisture profiles in the pavement structure and subgrade over the designed life of a pavement are considered through a sophisticated climatic modeling tool called the Enhanced Integrated Climatic Model (EICM). The EICM is a one-dimensional coupled heat and moisture flow program that simulates changes in the behavior and characteristics of pavement and subgrade materials in conjunction with climatic conditions over several years of operation. The EICM is a modified version of previous work by others (Larson and Dempsey, 1997).

The EICM software is linked to the software accompanying the proposed AASHTO M-E Pavement Design Guide and internally performs all the necessary computations. The user inputs to the EICM are entered through interfaces provided as part of the Design Guide software. The EICM processes these inputs and feeds the processed outputs to the three major components of the Design Guide's mechanistic-empirical design framework — materials, structural responses, and performance prediction.

8.6.5 Structural Modeling of the Pavement Structure

8.6.5.1 Structural Response Models

Structural response models are used to compute critical stresses, strains, and displacements in flexible pavement due to traffic loads. These responses are then utilized in damage models to accumulate damage, month by month, over the design period. The accumulated damage at any time is related to specific distresses such as fatigue cracking or rutting, which is then predicted using a field-calibrated cracking model.

The structural models for flexible pavements include the multi-layer elastic program JULEA for linear elastic analysis. If the user chooses to use the Level 1 hierarchical inputs to characterize the non-linear moduli response of any unbound layer materials (bases, subbases and subgrades), then the two dimensional finite element program DSC-2D is utilized (Desai, 2000) to conduct finite element analysis.

The outputs from the pavement response model are the stresses, strains, and displacements within the pavement layers. Of particular interest are the critical response variables required as inputs to the pavement distress models in the mechanistic-empirical design procedure. Examples of critical pavement response variables include:

- Tensile horizontal strain at the bottom or top of the HMA layer (for HMA fatigue cracking)
- Compressive vertical stresses and strains within the HMA layer (for HMA rutting)
- Compressive vertical stresses and strains within the base and subbase layers (for rutting of unbound base layers)
- Compressive vertical stresses and strains at the top of the subgrade (for subgrade rutting)

8.6.5.2 Analysis of Trial Design

The approach is iterative and begins with the selection of initial trial designs. A trial design is selected based on past agency experience or on general design catalogs (Darter et al., 1997). Each design strategy analyzed includes all details, such as initial estimates of layer thickness, required repairs to the existing pavement, and pavement materials characteristics. The trial sections are analyzed by accumulating incremental damage over time using the pavement structural response and performance models. The outputs of the analysis (the expected amounts of damage over time) are then used to estimate distress over time and traffic through calibrated distress models. Modifications are made to the trial strategies and further iterations performed until a satisfactory design that meets the performance criteria and design reliability is obtained.

8.6.6 Performance Prediction

The concept of pavement performance used in the proposed AASHTO M-E Pavement Design Guide includes consideration of functional performance, structural performance, and safety. The proposed Guide is primarily concerned with structural and functional performance. The structural performance of a pavement relates to its physical condition such as fatigue cracking and rutting or other conditions that would adversely affect the load-carrying capability of the pavement structure or would require maintenance. The functional performance of concerns how well the pavement serves the highway user. Ride quality, or smoothness, is the dominant characteristic of functional performance.

The design and analysis of a given pavement structure is based upon the accumulation of damage as function of time and traffic. The primary distresses considered in the design guide for flexible pavements are:

- Permanent deformation (rutting).
- Fatigue cracking (bottom-up and top-down).
- Thermal cracking.

In addition, pavement smoothness (IRI) is predicted based on these primary distresses and other factors.

8.6.6.1 Permanent Deformation in Asphalt Mixtures

The overall permanent deformation for a given season is the sum of permanent deformation for each individual layer. The permanent deformation is calculated for each load level, sub-season, and month of the analysis period. In the proposed AASHTO 2002 Design Guide the permanent deformation is only estimated for the asphalt bound and unbound layers. The estimation of permanent deformation for asphalt bound and unbound layers is discussed in the following paragraphs.

The constitutive relationship to predict rutting in the asphalt mixtures is based upon a field calibrated statistical analysis of laboratory repeated load permanent deformation tests. The model is shown below:

$$\frac{\varepsilon_p}{\varepsilon_r} = k_1 \times 10^{-3.4488} T^{1.5606} N^{0.479244} \tag{8.14}$$

where:

ε_p = Accumulated plastic strain at N repetitions of load (in./in.)
ε_r = Resilient strain of the asphalt material as a function of mix properties, temperature and time rate of loading (in./in.)
N = Number of load repetitions
T = Temperature (deg F)

$$k_1 = (C_1 + C_2 \times depth) \times 0.328196^{depth}$$

$$C_1 = -0.1039 \times H_{ac}^2 + 2.4868 \times H_{ac} - 17.342$$

$$C_2 = 0.0172 \times H_{ac}^2 - 1.7331 \times H_{ac} + 27.428$$

The permanent deformation model for the unbound granular base is as follows:

$$\delta_a(N) = 1.673 \left(\frac{\varepsilon_0}{\varepsilon_r} \right) e^{-\left(\frac{\rho}{N} \right)^\beta} \varepsilon_v h \tag{8.15}$$

The calibrated model for all subgrade soils is as follows:

$$\delta_a(N) = 1.35\left(\frac{\varepsilon_0}{\varepsilon_r}\right)e^{-\left(\frac{\rho}{N}\right)^\beta}\varepsilon_v h \qquad (8.16)$$

The total rutting in the pavement structure is equal to the summation of the individual layer permanent deformation for each season. The total rutting can be expressed by the following equation:

$$\text{PD}_{\text{Total}} = \text{PD}_{\text{AC}} + \text{PD}_{\text{GB}} + \text{PD}_{\text{SG}} \qquad (8.17)$$

where PD_{AC}, PD_{GB}, and PD_{SG} are the permanent deformations in the asphalt layer, granular base, and subgrade, respectively.

8.6.6.2 Fatigue Cracking

Estimation of fatigue damage is based upon Miner's Law, which states that damage is given by the following relationship.

$$D = \sum_{i=1}^{T} \frac{n_i}{N_i} \qquad (8.18)$$

where:

    D = damage.
    T = total number of periods.
    n_i = actual traffic for period i.
    N_i = traffic allowed under conditions prevailing in i.

The relationship used for the prediction of the number of repetitions to fatigue cracking is shown in Equation 8.19, which is based on the Asphalt Institute model (Asphalt Institute, 1991).

$$N_f = 0.00432 \times k_1' \times C\left(\frac{1}{\varepsilon_t}\right)^{3.9492}\left(\frac{1}{E}\right)^{1.281} \qquad (8.19)$$

where:

$$k_1' = \cfrac{1}{0.000398 + \cfrac{0.003602}{1 + e^{(11.02 - 3.49 \times h_{ac})}}} \quad \text{For the bottom-up cracking}$$

$$k_1' = \cfrac{1}{0.001 + \cfrac{29.844}{1 + e^{(30.544 - 5.7357 \times h_{ac})}}} \quad \text{For the top-down cracking}$$

where h_{ac} is the total thickness of the asphalt layers.
    The transfer function to calculate the bottom-up and top-down fatigue cracking in asphalt mixtures is expressed as:

a. For bottom-up cracking (% of total lane area)

$$\text{FC}_{\text{bottom}} = \left(\frac{6000}{1 + e^{(C_1 \times C_1' + C_2 \times C_2' \times \log 10(D \times 100))}}\right) \times \left(\frac{1}{60}\right) \qquad (8.20)$$

where:

    $\text{FC}_{\text{bottom}}$ = bottom-up fatigue cracking, % lane area
    D = bottom-up fatigue damage

$$C_1 = 1.0$$
$$C_1' = -2 \times C_2'$$
$$C_2' = 1.0$$
$$C_2' = -2.40874 - 39.748 \times (1 + h_{ac})^{-2.856}$$

b. For top-down cracking (feet/mi)

$$FC_{top} = \left(\frac{1000}{1 + e^{(7 - 3.5 \times \log 10(D \times 100))}} \right) \times (10.56) \qquad (8.21)$$

where:

FC_{top} = top-down fatigue cracking, ft/mi
D = top-down fatigue damage, decimal

8.6.6.3 Thermal Cracking Model

The amount of transverse cracking expected in the pavement system is predicted by relating the crack depth to an amount of cracking (crack frequency) by the following expression:

$$C_f = \beta_1 \times N\left(\frac{\log C/h_{ac}}{\sigma} \right) \qquad (8.22)$$

where:

C_f = Observed amount of thermal cracking.
β_1 = Regression coefficient determined through field calibration.
$N(z)$ = Standard normal distribution evaluated at (z).
σ = Standard deviation of the log of the depth of cracks in the pavement.
C = Crack depth.
h_{ac} = Thickness of asphalt layer.

The amount of crack propagation induced by a given thermal cooling cycle is predicted using the Paris law of crack propagation:

$$\Delta C = A\Delta K^n \qquad (8.23)$$

where:

ΔC = Change in the crack depth due to a cooling cycle.
ΔK = Change in the stress intensity factor due to a cooling cycle.
A, n = Fracture parameters for the asphalt mixture.

8.6.6.4 Smoothness Model

The proposed AASHTO 2002 Guide uses pavement smoothness, as indicated by the IRI, as the functional performance indicator. The approach employed is to predict changes in IRI over time as a function of pavement distress, site conditions, and maintenance. Typical initial values of as-constructed IRI (IRI_i) range between 50 and 100 in./mi, whereas typical terminal IRI values (IRI_t) are between 150 and 200 in./mi.

The major factors influencing loss of smoothness of a pavement are distresses such as cracking and rutting, which are in turn influenced by design, materials, subgrade, traffic, age, and environment. Smoothness is accounted for by considering the initial smoothness at construction and subsequent changes over the pavement life by summing the damaging effect of distress, as shown in smoothness model in Equation 8.24. The model is based on an additive combination of initial smoothness, change in smoothness due to the increase in individual distress, change in smoothness due to site conditions, and maintenance activities.

$$S(t) = S_0 + (a_1 S_{D(t)1} + a_2 S_{D(t)2} + \cdots + a_n S_{D(t)n}) + b_j S_j + c_j M_j \qquad (8.24)$$

where:

$S(t)$ = pavement smoothness at a specific time, t (IRI, in./mi)
S_0 = initial smoothness immediately after construction (IRI, in./mi)
$S_{D(t)(i=1 \text{ to } n)}$ = change of smoothness due to ith distress at a given time t in the analysis period
$a_{(i=1\ldots n)}, b_j, c_j$ = regression constants
S_j = change in smoothness due to site factors (subgrade parameters and age)
M_j = change in smoothness due to maintenance activities

8.6.7 Design Reliability

An analytical solution that allows the designer to design for a desired level of reliability for each distress and smoothness is available. Design reliability is defined as the probability that each of the key distress types and smoothness will be less than a selected critical level over the design period. For examaple, the design reliability for smoothness (IRI) is defined as shown in Equation 8.25.

$$R = P[\text{IRI over Design Period} < \text{Critical IRI Level}] \qquad (8.25)$$

Note that this definition varies from that in the 1993 AASHTO Design Guide. Table 8.8 provides recommended reliability values.

8.6.8 Overview of Flexible Pavement Design Process

The overall iterative design process for asphalt pavements is illustrated in Figure 8.27. The main steps in the design process include the following:

1. Assemble a trial design for specific site conditions — define subgrade support, asphalt concrete and other paving material properties, traffic loads, climate, pavement type and design and construction features.
2. Establish criteria for acceptable pavement performance at the end of the design period (i.e., acceptable levels of rutting, fatigue cracking, thermal cracking, and IRI).
3. Select the desired level of reliability for each of the applicable performance indicators (e.g., select reliability levels for rutting, cracking, and IRI).
4. Process input to obtain monthly values of traffic inputs and seasonal variations of material and climatic inputs needed in the design evaluations for the entyre design period.
5. Compute structural responses (stresses and strains) using multilayer elastic theory or finite element based pavement response models for each axle type and load and for each damage-calculation increment throughout the design period.
6. Calculate accumulated distress and damage at the end of each analysis period for the entyre design period.

TABLE 8.8 Recommended Levels of Reliability (NCHRP, 2003)

Functional Classification	Recommended Level of Reliability	
	Urban	Rural
Interstate and Freeways	85–97	880–95
Principal Arterials	880–95	75–90
Collectors	75–85	70–80
Local	50–75	50–75

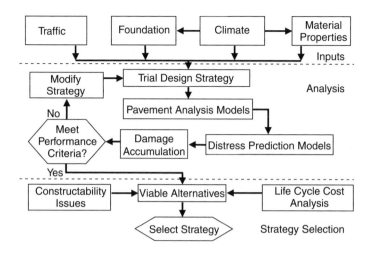

FIGURE 8.27 Overall design process for flexible pavements (NCHRP, 2003).

7. Predict key distresses (rutting, bottom-up/top-down fatigue cracking, thermal cracking) at the end of each analysis period throughout the designed life using the calibrated mechanistic-empirical performance models provided in the Guide.

8. Predict smoothness (IRI) as a function of initial IRI, distresses that accumulate over time, and site factors at the end of each analysis increment.

9. Evaluate the expected performance of the trial design at the given reliability level.

10. If the trial design does not meet the performance criteria, modify the design and repeat the steps 4 through 9 above until the design does meet the criteria.

The designs that satisfy performance criteria are considered feasible from a structural and functional viewpoint and can be further considered for other evaluations, such as life cycle cost analysis.

8.6.9 Design Software

A user-oriented computational software and documentation based on the proposed AASHTO M-E Pavement Design Guide procedure was developed. The designer begins the design process by configuring a trial design. The software then provides a prediction of key distress types and smoothness over the designed life of the pavement. This prediction is based on mean or average values for all inputs. The distresses and smoothness predicted therefore represent mean values that can be thought of as being at a 50% reliability estimate. Typically, the designer will require a higher probability that the design will meet the performance criteria over the designed life. The more important the project in terms of consequences of failure is, the higher is the desired design reliability.

8.6.10 Calibration to Local Conditions

Any agency interested in adopting the design procedure described in this Guide should prepare a practical implementation plan. The plan should include training of staff, acquiring of needed equipment, acquiring of needed computer hardware, and calibration and validation to local conditions.

8.7 Summary

A typical flexible pavement consists of surface, base course, and subbase built over compacted subgrade. Pavement is subjected to repeated loads with variable magnitudes, tyre pressures, and configurations.

Distresses are developed due to traffic load repetitions, temperature, moisture, aging, construction practice, or combinations. Common distresses include fatigue cracking, rutting, roughness, and thermal cracking. Pavement fails when one or more of the distresses reach an unacceptable level.

The design of flexible pavement has evolved throughout the years from pure empirical to mechanistic-empirical. No pure mechanistic method of design has been developed. Currently, the most common method of design is the 1993 AASHTO method, which is based the statistical evaluation of the performance data obtained from the AASHO Road Test. The proposed AASHTO mechanistic-empirical method was developed to provide the highway community with a state-of-the-practice tool for the design of new and rehabilitated pavement structures. The implementation of the proposed AASHTO M-E Pavement Design Guide is expected to grow in the next several years. The pavement design practice is also expected to gradually depend on more "mechanistic" approaches and reduce the dependency on imperial procedure.

References

AASHTO, 1993. *AASHTO Guide for Design of Pavement Structures.* AASHTO, Washington, DC.

AASHTO, 2001. *AASHTOWare Software for Design of Pavement Structures*, DARWin Version 3.1, User's Guide. AASHTO, Washington, DC.

AASHTO, 2004. *Standard Method of Test T307 for Determination of Resilient modulus of Soils and Aggregate Materials, 24th Ed.*, Standard Specifications for Transportation Materials and Methods of Sampling and Testing. AASTHO, Washington, DC.

Asphalt Institute, 1991. *Thickness Design — Asphalt Pavements for Highways and Streets*, Manual Series No.1 (MS-1). Asphalt Institute, Lexington, KY.

Cebon, D. and Winkler, C.B. 1991. *A Study of Road Damage Due to Dynamic Wheel Loads Using a Load Measuring Mat*, Report SHRP-ID/UFR-91-518. Strategic Highway Research Program, Washington, DC.

Darter, M.I., Von Quintus, H.L., Jiang, Y.J., Owusu-Antwi, E.B., and Killingsworth, B.M. 1997. *Catalog of Recommended Design Features*, CD-ROM, NCHRP Project 1-32. TRB. National Research Council, Washington, DC.

De Beer, M., 1994. *Measurement of Tyre/Pavement Interface Stresses Under Moving Wheel Loads*, Presented at the Third Engineering Foundation Conference on Vehicle–Road and Vehicle–Bridge Interaction, Holland, June.

Desai, C.S., 2000. User's Manual for the DSC-2D Code for the 2002 Design Guide (in preparation).

FHWA, 1993. *Data Collection Guide for LTPP Studies*, Operational Guide No. SHRP-LTPP-OG-001. Federal Highway Administration, LTPP Division, Washington, DC.

FHWA, 2002. DataPave, CD-Rom, Version 3, Federal Highway Administration, Washington, DC.

Gillespie, T.D., Karamihas, S.M., Sayers, M.W., Nasim, M.A., Hansen, W., Ehsan, N., 1993. *Effects of heavy-vehicle characteristics on pavement response and performance*, Report No. 353, NCHRP, Transportation Research Board.

Haas, R., Hudson, W.R., Zaniewski, J. 1994. *Modern Pavement Management.* Krieger Publishing Company, Melbourne, FL.

HRB, 1962. *The AASHO Road Test, Report 5 — Pavement Research*, Special Report 61E. Highway Research Board, Washington, DC.

Huang, Y.H. 2003. *Pavement Analysis and Design, 2nd Ed.* Prentice Hall, Englewood Cliffs.

Kopperman, S., Tiller, G., Tseng, M. 1985. *ELSYM5 Interactive Version*, User's Manual. SRA Technologies, Inc.

Larson, G., Dempsey, B.J. 1997. *Enhanced Integrated Climatic Model*, Version 2.0, Report No. DTFA MN/DOT 72114. University of Illinois at Urbana-Champaign, Urbana, IL.

Mamlouk, M.S., General outlook of pavement and vehicle dynamics, *ASCE, J. Trans. Eng.*, 123, 6, 515–551, 1997, Tech. Note.

MDOT, 1994. Minnesota Road Research Project, Load Response Instrumentation Installation and Testing Procedure, Report No. MN/PR-94/01, Minnesota Department of Transportation, Maplewood, MN.

Metcalf, J.B. 1996. *Application of Full-Scale Accelerated Pavement Testing*, NCHRP synthesis 235. Transportation Research Board, Washington, DC.

Michelow, J. 1963. *Analysis of Stresses and Displacements in an N-Layered Elastic System Under a Load Uniformly Distributed on a Circular Area*. California Research Corporation, Richmond, CA.

NCHRP, 2000. Performance-Related Specification — Part I, Final Draft Report, WesTrack Team, NCHRP, Washington, DC.

NCHRP, 2003. 2002 Design Guide, National Cooperative Highway Research Program Project 1-37A, Draft Final Report, NCHRP, Washington, DC.

NRC, 1993. Distress Identification Manual for the LTPP Project, Report SHRP-P-338, National Research Council, Washington, DC. (website: http://www.trb.org/mepdg).

Peutz, M.G.F., Van Kempen, H.P., and Jones, A. 1968. *Layered Systems Under Normal Surface Loads*, Highway Research Record 228. Highway Research Board, Washington, DC.

Van Til, C.J., McCullough, B.F., Vallerga, B.A., and Hicks, R.G., 1972. Evaluation of AASHO Interim Guides for Design of Pavement Structures, NCHRP Report 128.

Warren, T. and Diekmann, W.L. 1963. *Numerical Commutation of Stresses and Strains in Multiple-Layer Asphalt Pavement System*, Internal Report. Chevron Research and Technology Company, Richmond, CA.

Yoder, E.J. and Witczak, M.W. 1975. *Principles of Pavement Design, 2nd Ed.* John Wiley, New York.

9

Design of Rigid Pavements

T.F. Fwa
National University of Singapore
Republic of Singapore

Liu Wei
National University of Singapore
Republic of Singapore

9.1   Characteristics of Rigid Pavements

Rigid pavements are mostly found in major highways and airports. They also serve as heavy-duty industrial floor slabs, port and harbor yard pavements, and heavy-vehicle park or terminal pavements. Rigid highway pavements, like flexible pavements, are designed as all-weather, long-lasting structures to serve modern day high-speed traffic. They offer high quality riding surfaces for safe vehicular travel, and function as structural layers to distribute vehicular wheel loads in such a manner that the induced stresses transmitted to the subgrade soil are of acceptable magnitudes. The load transmission mechanisms by which the two forms of pavement achieve the load distribution requirement, however, are very different. Figure 9.1 shows that while the flexible pavement is designed to provide sufficient thickness to distribute the applied load with depth, the rigid pavement relies on rigid slab action to spread the load over a large area.

The most common type of material used for rigid pavement slab construction is Portland cement concrete, mainly because of economic reasons and its easy availability. The concrete slab must be designed to withstand repeated traffic loadings. Fatigue failure of pavement due to repeated loadings caused by daily traffic is a major design consideration of rigid pavements. Fatigue failure occurs when a load, though smaller than the failure load of the concrete slab, is repeatedly applied on the pavement a sufficiently number of times. This form of failure is common for highway pavements because a typical highway will receive millions of wheel passes during its service life. While the design life of a flexible pavement may be in the range of 15 to 20 years, it is common for a concrete pavement to be designed with a service life of 30 to 40 years.

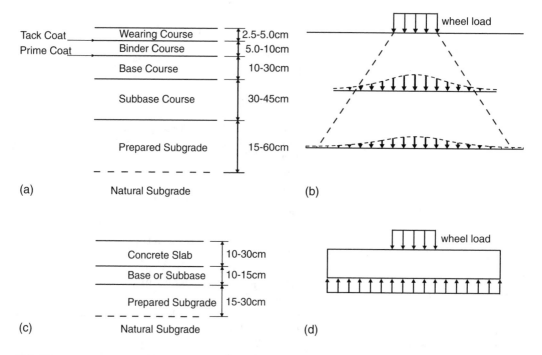

FIGURE 9.1 Geometric and load transmission characteristics of flexible and rigid pavements: (a) typical cross section of flexible road pavement; (b) load transmission in flexible pavement; (c) typical cross section of rigid road pavement; (d) load transmission in rigid pavement.

Besides traffic loadings, thermally induced tensile stresses must also be considered in the design. It has long been recognized by highway engineers that stresses in concrete pavements resulted from temperature changes can be of equal magnitude to stresses induced by wheel loadings. Thermal stresses are thus of major design concern because of the relatively low tensile strength of concrete materials. This leads to the following additional design issues for concrete pavements: (i) choice of plan dimensions of slab panels, which has a direct influence on the magnitude of thermal stresses induced within the slabs; (ii) design

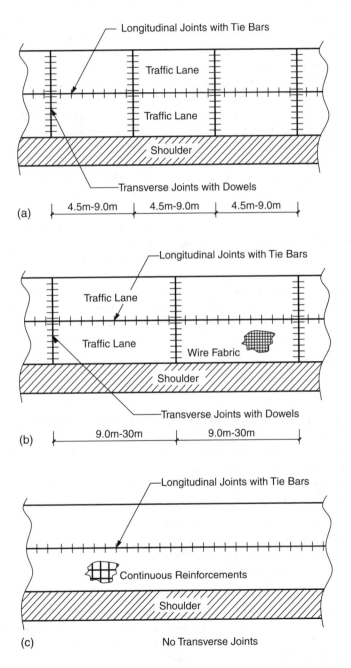

FIGURE 9.2 Types of concrete pavement: (a) jointed plain concrete pavement (JPCP); (b) jointed reinforced concrete pavement (JRCP); (c) continuous reinforced concrete pavement (CRCP).

of temperature reinforcements to control crack widths and crack spacings; and (iii) joint and joint reinforcement design to effect load transfer between adjacent slabs.

Being supported continuously by the underlying subbase layer (if provided) or subgrade, typical deflections of a structurally sound rigid slab under wheel loads are much less than 1.0 mm. Steel reinforcements are not necessary in rigid pavement slabs to resist traffic loadings. Even when temperature reinforcements are provided, their contributions to resisting traffic loadings are nominal and are usually ignored in rigid pavement design.

The choice of joint spacing (which is dictated by the plan dimensions of slab panels chosen), and temperature steel design are interrelated design decisions. Depending on the design preference of the designer with regard to joint spacing and percent steel reinforcement used, three main types of rigid pavement (see Figure 9.2) can be differentiated:

Jointed plain concrete pavement (JPCP). JPCP has been successfully used worldwide without temperature steel reinforcements. This is possible by limiting the joint spacing to about 6 m (20 ft), although joint spacing of up to 9 m (30 ft) has been used. U.K. practice as spelled out in Road Note 29 (Road Research Laboratory, 1970) adopts 5 m (16.5 ft) joint spacing for JPCP. In U.S.A., a performance survey of in-service pavements by Nussbaum and Lokken (1978) suggested a maximum joint spacing of 6 m (20 ft) for doweled joints and 4.5 m (15 ft) for undoweled joints.

Jointed reinforced concrete pavement (JRCP). The use of temperature steel reinforcements in JRCP allows joint spacings of up to 30 m (100 ft) to be constructed. Doweled joints are usually required because of the longer joint spacings. The required steel reinforcement, expressed as percent of the cross-sectional area of the concrete slab, rarely exceeds 0.75%.

Continuously reinforced concrete pavements (CRCP). Joint related distresses are major maintenance concerns for concrete pavements. The main idea of using CRCP is to eliminate joints altogether in concrete pavement slab construction. The percent steel reinforcement required usually ranges from 0.4 to 1.0.

Prestressed concrete pavement is another type of concrete pavement that has been used in practice, though to a much lesser extent than the other three types. By applying prestressing to partly or wholly neutralize either load or thermally induced tensile stresses, thinner slab thickness can be used. However, since the practical minimum slab thickness is about 150 mm (6 in.), the savings on thickness for highway pavements are not likely to be attractive. The design of prestressed concrete pavement is not covered in this chapter.

9.2   Considerations for Structural Design of Rigid Pavement

A pavement must be designed and constructed to serve its function to provide a durable all-weather traveled surface for safe and efficient movements of people and goods with an acceptable level of comfort to the motorists. In meeting these functional requirements, due considerations must be given to the following aspects during the design and construction stages:

 (a) Determination of soil properties, design traffic loadings and environmental parameters
 (b) Selection of materials for various pavement layers
 (c) Structural thickness design of pavement layers
 (d) Drainage design for the pavement system
 (e) Safety and geometric design

Topics (a) and (c) for rigid pavement are discussed in the subsequent sections of this chapter. The other topics are covered in other chapters of this Handbook. Discussions of pavement materials are found in Chapters 7 and 14. Pavement drainage design is addressed in Chapter 13. Safety issues and road geometric design are dealt with in Chapters 4, 6, and 21.

Besides the main considerations mentioned above, there are several other design considerations specific to concrete pavement that need to be highlighted. The key concern in the structural design of concrete pavement is cracking failure caused by one or more of the following actions: (i) load-induced tension, (ii) thermally induced tension, and (iii) tension generated by subgrade soil movements or moisture changes. The common design practice for concrete pavement has been to consider load-induced stresses separately from the other forms of stresses when steel reinforcements are used. The thickness of the pavement slab is designed for the design traffic loadings, while the steel reinforcements are designed to resist tensile stresses caused by temperature changes or subgrade soil movements. It is thus common to position the steel reinforcements at mid depth of the pavement slab, although other positions (such as lower or upper 1/3 depth) have also been used in practice. In the case of plain concrete pavement where no reinforcements are provided, the total tension due to the various appropriate factors combined must be considered in the pavement slab thickness design.

Another important consideration in concrete pavement design is joint failure caused by either pumping or erosion of the support materials. Pumping refers to the ejection of water and fine-grained materials from the supporting layers due to downward slab deflection at the affected joint under moving traffic loads. This type of problem has occurred in jointed rigid pavements under the repeated applications of heavy truck loads. Erosion of support materials can occur due to inadequate drainage provisions caused by poor sub-drainage design or weak support materials. To prevent pumping and to offer a stable support to the pavement slab, a subbase layer (sometimes also called base layer for the case of rigid pavement construction) is provided. Subbase may be constructed of gravel, crushed stones, cement-modified soil, asphalt mixture, or low-strength concrete.

To allow for thermal expansion, contraction, warping, or breaks in the construction of concrete slabs, joints are provided in concrete pavements. Performance records of in-service jointed concrete pavements in different parts of the world have shown that joint related distresses are a major maintenance concern to highway agencies. Joint and joint reinforcement design, therefore, is a critical element that deserves special attention in the design and construction phases of jointed concrete pavement. It is covered in Section 9.11 of this chapter.

9.3   Computation of Design Traffic Loading

Although it is convenient to describe the design life of a pavement in years, the design parameter that ultimately governs the total length of service life of a pavement is the total traffic loading. It is thus more appropriate to describe the design life of a pavement in terms of the total design traffic loading. For example, a rigid pavement designed for 40 years with an assumed traffic growth of 3% will reach the end of its design life sooner than 40 years if the actual traffic growth is higher than 3%.

For highway pavements, the computation of design traffic loading involves the following steps:

  (a) Estimation of the initial year traffic volume and composition
  (b) Estimation of the annual traffic growth rate by vehicle type
  (c) Estimation of directional split of design traffic
  (d) Estimation of design lane traffic
  (e) Estimation of the magnitudes of wheel loads by vehicle type
  (f) Computation of the number of applications of wheel loads in the design lane

Information concerning (a) and (b) can be obtained from traffic survey and forecast based on historical trends or prediction using transportation models. The analysis required for steps (c), (d), (e) and (f) are explained in the following sub-sections.

9.3.1   Directional Split and Design Lane Traffic Loading

It is common practice to report traffic volume of a highway to include flows for all lanes. To determine the design traffic loading in the design lane, one must split the traffic by direction and distribute the directional traffic by lane. In circumstances where uneven directional split occurs, pavements are designed based on the heavier directional traffic loading.

TABLE 9.1 Lane Distribution of Traffic for Pavement Design

Number of Lanes per Direction	Recommendations by AASHTO (1993) Percentage of ESAL in design lane[a]	Recommendations by Asphalt Institute (1991) Percentage of trucks in design lane	Recommendations by PCA (1984) Percentage of trucks in design lane[b]
1	100	100	100
2	80–100	90 (70–96)	$1 - (\log_{10}\text{ADT} - 3)/5.34$
3	60–80	80 (50–96)	$0.875 - (\log_{10}\text{ADT} - 3)/5.23$
4	50–75	80 (50–96)	—

[a] See Section 9.3.3 for definition of ESAL which stands for equivalent single axle loads.
[b] Equations are derived from a chart by PCA (1984).

The design lane for pavement structural design is usually the slow lane (lane next to the shoulder in most cases) in which a large proportion of the heavy vehicle traffic is expected. Table 9.1 shows examples of recommended lane-use distributions for design lane traffic loading computation. It is noted that the recommendations of Portland Cement Association (PCA) and Asphalt Institute (AI) deal only with truck traffic, while those of AASHTO are based on axle loadings, which are predominantly contributed by heavy truck traffic. This is because, as will be explained in Example 9.1 of Section 9.3.3, the damaging effects of light vehicles such as passenger cars are thousands of times less than heavy trucks.

9.3.2 Traffic Loading Computation

Structural analysis and design of pavement requires the knowledge of (a) the magnitudes of axle loads in the design traffic, and (b) the number of times each of these loads will be applied on the design lane during the design life of the pavement. Two forms of field survey are required to obtain the required information from similar highway type within the same region. First, traffic count surveys must be conducted to determine the number of vehicle types in the design traffic. For pavement design, it is necessary to classify vehicles by size and axle configuration, such as cars, buses, single-unit trucks, and different types of multiple-unit trucks. Table 9.2 shows the common vehicle classification system used. The other form of survey is to measure the axle or wheel loads of each vehicle type. Such axle or wheel load survey can be performed at weighing stations or using weigh-in-motion devices. Data collected from the two forms of survey enable one to compute the number of repetitions by axle type (i.e., by single axle, tandem axle, and tridem axle, etc.), as illustrated in Table 9.3.

9.3.3 Characterizing Effects of Applied Loads of Mixed-Traffic Stream

The combined loading effects of different axle types in the design traffic must be considered when designing a pavement. There are two common approaches by which the combined effects are evaluated: one based on the hypothesis of cumulative damage, and the other by the method of equivalent load.

Under the hypothesis of cumulative damage, the allowable number of repetitions by each load type can be established for a given form of pavement damage. A damage ratio D_i for load type i is defined as

$$D_i = (n_i/N_i) \tag{9.1}$$

where n_i is the design repetitions of load type i and N_i is the allowable repetitions. The total amount of damage caused by the mixed traffic is computed as the sum of damage ratios of all load types.

The method of equivalent load expresses the effects of mixed loads of different configurations and magnitudes in terms of an equivalent number of a preselected standard load. However, it must be noted that there exist different basis of damage upon which a load can be converted into the preselected standard load, and the computed answers of equivalent number vary with the basis adopted. The most common form of equivalent load used in pavement design is the "equivalent single axle load of 18 kips (80 kN),"

TABLE 9.2 Vehicle Classification for Pavement Design

Description	Axle Configuration	Total Number of Axles	Number of Single Axles, Double Axles, and Tridem Axles
Small passenger car		2	2S
Large passenger car		2	2S
Two-axle single unit truck		2	2S
Two-axle single unit bus		2	2S
Passenger car with single-axle trailer		3	3S
Three-Axle single unit truck		3	1S-1D
Three-axle single unit container truck		3	3S
Passenger car with two-axle trailer		4	4S
Four-axle single unit truck		4	2S-1D
Four-axle single unit container truck		4	2S-1D
Five-axle single unit container truck		5	1S-2D
Five-axle double unit truck		5	5S
Six-axle double unit truck		6	4S-1D
Seven-axle double unit truck		7	3S-2D
Eight-axle B-train double unit truck		8	1S-2D-1T
Nine-axle B-train double unit truck		9	1S-1D-2T
Eleven-axle B-train double unit truck		11	1S-5D
Twelve-axle B-train triple unit truck		12	1S-1D-3T
Twelve-axle B-train triple unit truck		12	1S-1D-3T

Note: S, D, and T stand for single, double, and tridem axles, respectively.

abbreviated as ESAL. It was introduced at the well-known AASHO Road Test where equivalency factors to convert one pass of any given axle load to equivalent passes of an 18-kip (80-kN) single axle load were determined from the road test data (Highway Research Board, 1962). The equivalency factors, known as the ESAL factor, were derived based on the relative damaging effects of various axle loads, with the damage caused by the standard 18-kip (80-kN) single axle taken as unity. Pavement damage in this case is computed as the loss of PSI (present serviceability index). PSI is a measure of pavement serviceability on a scale of 0 to 5 (Carey and Irick, 1960), as shown in Figure 9.3. Table 9.4 presents the ESAL factors of axle loads for different thicknesses of rigid pavements with a terminal serviceability index of 2.5.

Example 9.1

Computation of ESAL of vehicles (Table 9.5). The axle loads of three vehicles are as follows (1 kip = 4.448 kN):

Car — Front single axle = 2 kips; Rear single axle = 2 kips
Bus — Front single axle = 10 kips; Rear single axle = 10 kips
Truck — Front single axle = 10 kips; Middle single axle = 20 kips; Rear tandem axle = 32 kips

TABLE 9.3   Example of Axle Load Data for Pavement Design

Single Axle		Double Axle		Tridem Axle	
Axle load (kips)	Number of repetitions/day	Axle load (kips)	Number of repetitions/day	Axle load (kips)	Number of repetitions/day
Less than 3	1438	9–11	2093	25–27	588
3–5	3391	11–13	1867	27–29	515
5–7	3432	13–15	1298	29–31	496
7–9	6649	15–17	1465	31–33	448
9–11	9821	17–19	1743	33–35	225
11–13	2083	19–21	1870	35–37	372
13–15	946	21–23	2674	37–39	474
15–17	886	23–25	2879	39–41	529
17–19	472	25–27	2359	41–43	684
19–21	299	27–29	2104	43–45	769
21–23	98	29–31	1994	45–47	653
		31–33	1779	47–49	527
		33–35	862	49–51	421
		35–37	659	51–53	363
		37–39	395	53–55	298
		39–41	46	55–57	125
				57–59	84
				59–61	67
				61–63	46
				63–65	42
				65–67	31
				67–69	28
				69–71	21
				71–73	16
				73–75	12

Note: 1 kip = 1000 lbs = 4.448 kN.

Consider a terminal serviceability index of 2.5, and a rigid pavement of slab thickness equal to 9 in. (225 mm). Table 9.4 is used to obtain the appropriate ESAL factors. The ESAL contributions are $(0.0002 + 0.0002) = 0.0004$ for the car, $(0.082 + 0.032) = 0.114$ for the bus, and $(0.082 + 1.57 + 1.49) = 3.142$ for the truck. The ratios of ESAL contributions are (car):(bus):(truck) = 1:285:7,855. It can be seen from this example that the damaging effects of a truck and a bus are respectively, 7,855 and 285 times that of a passenger car. This explains why passenger car volumes are often ignored in traffic loading computation for pavement design.

Example 9.2

ESAL computation using axle load data (Table 9.6). Calculate the total daily ESAL of the traffic data of Table 9.3 for a rigid pavement with slab thickness of 10 in. The design terminal serviceability index for pavements is 2.5. The data in Table 9.3 are repeated in columns (1) and (2) of the table below. The ESAL factors in column (3) are obtained from Table 9.4 for slab thickness of 10 in. (250 mm). The ESAL contribution by each axle group is computed by multiplying its ESAL factor by the number of axles per day. The total ESAL of the traffic loading is 33742 for the rigid pavement.

9.3.4   Formula for Computing Total Design Loading

Depending on the information available, the steps in the computation of the total design load may differ slightly. Assuming the initial year total number of applications $(N_i)_1$ of load i is known,

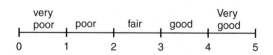

FIGURE 9.3   Five-point scale of present serviceability index.

TABLE 9.4 AASHTO Load Equivalency Factors for Rigid Pavements Based on Terminal Serviceability Index of 2.5

Axle Load (kips)	Slab Thickness, D (inches)								
	6	7	8	9	10	11	12	13	14
(a) Single Axles and p_t of 2.5									
2	0.0002	0.0002	0.0002	0.0002	0.0002	0.0002	0.0002	0.0002	0.0002
4	0.003	0.002	0.002	0.002	0.002	0.002	0.002	0.002	0.002
6	0.012	0.011	0.010	0.010	0.010	0.010	0.010	0.010	0.010
8	0.039	0.035	0.033	0.032	0.032	0.032	0.032	0.032	0.032
10	0.097	0.089	0.084	0.082	0.081	0.080	0.080	0.080	0.080
12	0.203	0.189	0.181	0.176	0.175	0.174	0.174	0.173	0.173
14	0.376	0.360	0.347	0.341	0.338	0.337	0.336	0.336	0.336
16	0.634	0.623	0.610	0.604	0.601	0.599	0.599	0.599	0.598
18	1.00	1.00	1.00	1.00	1.00	1.00	1.00	1.00	1.00
20	1.51	1.52	1.55	1.57	1.58	1.58	1.59	1.59	1.59
22	2.21	2.20	2.28	2.34	2.38	2.40	2.41	2.41	2.41
24	3.16	3.10	3.22	3.36	3.45	3.50	3.53	3.54	3.55
26	4.41	4.26	4.42	4.67	4.85	4.95	5.01	5.04	5.05
28	6.05	5.76	5.92	6.29	6.61	6.81	6.92	6.98	7.01
30	8.16	7.67	7.79	8.28	8.79	9.14	9.35	9.46	9.52
32	10.8	10.1	10.1	10.7	11.4	12.0	12.3	12.6	12.7
34	14.1	13.0	12.9	13.6	14.6	15.4	16.0	16.4	16.5
36	18.2	16.7	16.4	17.1	18.3	19.5	20.4	21.0	21.3
38	23.1	21.1	20.6	21.3	22.7	24.3	25.6	26.4	27.0
40	29.1	26.5	25.7	26.3	27.9	29.9	31.6	32.9	33.7
42	36.2	32.9	31.7	32.2	34.0	36.3	38.7	40.4	41.6
44	44.6	40.4	38.8	39.2	41.0	43.8	46.7	49.1	50.8
46	54.5	49.3	47.1	47.3	49.2	52.3	55.9	59.0	61.4
48	66.1	59.7	56.9	56.8	58.7	62.1	66.3	70.3	73.4
50	79.4	71.7	68.2	67.8	69.6	73.3	78.1	83.0	87.1
(b) Tandem Axles and p_t of 2.5									
2	0.0001	0.0001	0.0001	0.0001	0.0001	0.0001	0.0001	0.0001	0.0001
4	0.0006	0.0006	0.0005	0.0005	0.0005	0.0005	0.0005	0.0005	0.0005
6	0.002	0.002	0.002	0.002	0.002	0.002	0.002	0.002	0.002
8	0.007	0.006	0.006	0.005	0.005	0.005	0.005	0.005	0.005
10	0.015	0.014	0.013	0.013	0.012	0.012	0.012	0.012	0.012
12	0.031	0.28	0.026	0.026	0.025	0.025	0.025	0.025	0.025
14	0.057	0.052	0.049	0.048	0.047	0.047	0.047	0.047	0.047
16	0.097	0.089	0.084	0.082	0.081	0.081	0.080	0.080	0.080
18	0.155	0.143	0.136	0.133	0.132	0.131	0.131	0.131	0.131
20	0.234	0.220	0.211	0.206	0.204	0.203	0.203	0.203	0.203
22	0.340	0.325	0.313	0.308	0.305	0.304	0.303	0.303	0.303
24	0.475	0.462	0.450	0.444	0.441	0.440	0.439	0.439	0.439
26	0.644	0.637	0.627	0.622	0.620	0.619	0.618	0.618	0.618
28	0.855	0.854	0.852	0.850	0.850	0.850	0.849	0.849	0.849
30	1.11	1.12	1.13	1.14	1.14	1.14	1.14	1.14	1.14
32	1.43	1.44	1.47	1.49	1.50	1.51	1.51	1.51	1.51
34	1.82	1.82	1.87	1.92	1.95	1.96	1.97	1.97	1.97
36	2.29	2.27	2.35	2.43	2.48	2.51	2.52	2.52	2.53
38	2.85	2.80	2.91	3.03	3.12	3.16	3.18	3.20	3.20
40	3.52	3.42	3.55	3.74	3.87	3.94	3.98	4.00	4.01
42	4.32	4.16	4.30	4.55	4.74	4.86	4.91	4.95	4.96
44	5.26	5.01	5.16	5.48	5.75	5.92	6.01	6.06	6.09
46	6.36	6.01	6.14	6.53	6.90	7.14	7.28	7.36	7.40
48	7.64	7.16	7.27	7.73	8.21	8.55	8.75	8.86	8.92
50	9.11	8.50	8.55	9.07	9.68	10.14	10.42	10.58	10.66
52	10.8	10.0	10.0	10.6	11.3	11.9	12.3	12.5	12.7
54	12.8	11.8	11.7	12.3	13.2	13.9	14.5	14.8	14.9

(Continued)

TABLE 9.4 Continued

Axle Load (kips)	Slab Thickness, D (inches)								
	6	7	8	9	10	11	12	13	14
56	15.0	13.8	13.6	14.2	15.2	16.2	16.8	17.3	17.5
58	17.5	16.0	15.7	16.3	17.5	18.6	19.5	20.1	20.4
60	20.3	18.5	18.1	18.7	20.0	21.4	22.5	23.2	23.6
62	23.5	21.4	20.8	21.4	22.8	24.4	25.7	26.7	27.3
64	27.0	24.6	23.8	24.4	25.8	27.7	29.3	30.5	31.3
66	31.0	28.1	27.1	27.6	29.2	31.3	33.2	34.7	35.7
68	35.4	32.1	30.9	31.3	32.9	35.2	37.5	39.3	40.5
70	40.3	36.5	35.0	35.3	37.0	39.5	42.1	44.3	45.9
72	45.7	41.4	39.6	39.8	41.5	44.2	47.2	49.8	51.7
74	51.7	46.7	44.6	44.7	46.4	49.3	52.7	55.7	58.0
76	58.3	52.6	50.2	50.1	51.8	54.9	58.6	62.1	64.8
78	65.5	59.1	56.3	56.1	57.7	60.9	65.0	69.0	72.3
80	73.4	66.2	62.9	62.5	64.2	67.5	71.9	76.4	80.2
82	82.0	73.9	70.2	69.6	71.2	74.7	79.4	84.4	88.8
84	91.4	82.4	78.1	77.3	78.9	82.4	87.4	93.0	98.1
86	102.0	92.0	87.0	86.0	87.0	91.0	96.0	102.0	108.0
88	113.0	102.0	96.0	95.0	96.0	100.0	105.0	112.0	119.0
90	125.0	112.0	106.0	105.0	106.0	110.0	115.0	123.0	130.0
(c) Triple Axles (i.e., Tridem Axles) and p_t of 2.5									
2	0.0001	0.0001	0.0001	0.0001	0.0001	0.0001	0.0001	0.0001	0.0001
4	0.0003	0.0003	0.0003	0.0003	0.0003	0.0003	0.0003	0.0003	0.0003
6	0.001	0.001	0.001	0.001	0.001	0.001	0.001	0.001	0.001
8	0.003	0.002	0.002	0.002	0.002	0.002	0.002	0.002	0.002
10	0.006	0.005	0.005	0.005	0.005	0.005	0.005	0.005	0.005
12	0.011	0.010	0.010	0.009	0.009	0.009	0.009	0.009	0.009
14	0.020	0.018	0.017	0.017	0.016	0.016	0.016	0.016	0.016
16	0.033	0.030	0.029	0.028	0.027	0.027	0.027	0.027	0.027
18	0.053	0.048	0.045	0.044	0.044	0.043	0.043	0.043	0.043
20	0.080	0.073	0.069	0.067	0.066	0.066	0.066	0.066	0.066
22	0.116	0.107	0.101	0.099	0.098	0.097	0.097	0.097	0.097
24	0.163	0.151	0.144	0.141	0.139	0.139	0.138	0.138	0.138
26	0.222	0.209	0.200	0.195	0.194	0.193	0.192	0.192	0.192
28	0.295	0.281	0.271	0.265	0.263	0.262	0.262	0.262	0.262
30	0.384	0.371	0.359	0.354	0.351	0.350	0.349	0.349	0.349
32	0.490	0.480	0.468	0.463	0.460	0.459	0.458	0.458	0.458
34	0.616	0.609	0.601	0.596	0.594	0.593	0.592	0.592	0.592
36	0.765	0.762	0.759	0.757	0.756	0.755	0.755	0.755	0.755
38	0.939	0.941	0.946	0.948	0.950	0.951	0.951	0.951	0.951
40	1.14	2.15	1.16	1.17	1.18	1.18	1.18	1.18	1.18
42	1.38	1.38	1.41	1.44	1.45	1.46	1.46	1.46	1.46
44	1.65	1.65	1.70	1.74	1.77	1.78	1.78	1.78	1.79
46	1.97	1.96	2.03	2.09	2.13	2.15	2.16	2.16	2.16
48	2.34	2.31	2.40	2.49	2.55	2.58	2.59	2.60	2.60
50	2.76	2.71	2.81	2.94	3.02	3.07	3.09	3.10	3.11
52	3.24	3.15	3.27	3.44	3.56	3.62	3.66	3.68	6.68
54	3.79	3.66	3.79	4.00	4.16	4.26	4.30	4.33	4.34
56	4.41	4.23	4.37	4.63	4.84	4.97	5.03	5.07	5.09
58	5.12	4.87	5.00	5.32	5.59	5.76	5.85	5.90	5.93
60	5.91	5.59	5.71	6.08	6.42	6.64	6.77	6.84	6.87
62	6.80	6.39	6.50	6.91	7.33	7.62	7.79	7.88	7.93
64	7.79	7.29	7.37	7.82	8.33	8.70	8.92	9.04	9.11
66	8.90	8.28	8.33	8.83	9.42	9.88	10.17	10.33	10.42
68	10.1	9.4	9.4	9.9	10.6	11.2	11.5	11.7	11.9
70	11.5	10.6	10.6	11.1	11.9	12.6	13.0	13.3	13.5

(Continued)

TABLE 9.4 Continued

Axle Load (kips)	Slab Thickness, D (inches)								
	6	7	8	9	10	11	12	13	14
72	13.0	12.0	11.8	12.4	13.3	14.1	14.7	15.0	15.2
74	14.6	13.5	13.2	13.8	14.8	15.8	16.5	16.9	17.1
76	16.5	15.1	14.8	15.4	16.5	17.6	18.4	18.9	19.2
78	18.5	16.9	16.5	17.1	18.2	19.5	20.5	21.1	21.5
80	20.6	18.8	18.3	18.9	20.2	21.6	22.7	23.5	24.0
82	23.0	21.0	20.3	20.9	22.2	23.8	25.2	26.1	26.7
84	25.6	23.3	22.5	23.1	24.5	26.2	27.8	28.9	29.6
86	28.4	25.8	24.9	25.4	26.9	28.8	30.5	31.9	32.8
88	31.5	28.6	27.5	27.9	29.4	31.5	33.5	35.1	36.1
90	34.8	31.5	30.3	30.7	32.5	34.4	36.7	38.5	39.8

Source: AASTO. 1993 AASHTO Guides for Design of Pavement Structures. Copyright 1993 by the American Association of State Highway and Transportation Officials, Washington, DC. Used by permission.

and a constant growth of N_i at a rate of $r\%$ per annum is predicted, the design lane loading for an analysis period of n years can be computed by the following equation:

$$(N_i)_{\mathrm{T}} = (N_i)_1 \cdot \frac{(1 + 0.01r)^n}{0.01r} \cdot f_{\mathrm{D}} \cdot f_{\mathrm{L}} \qquad (9.2)$$

where:

$(N_i)_{\mathrm{T}}$ = total design application number of load i for n years,
$(N_i)_1$ = initial year design application of load i,
r = annual growth rate of load i in percent,
f_{D} = directional split factor, and
f_{L} = lane-use distribution factor.

The lane distribution factors recommended by various agencies are given in Table 9.1. For design methods that rely on ESAL, instead of computing cumulative applications for individual load types, the total ESAL is calculated by replacing $(N_i)_1$ in the equation with the initial year ESAL.

Example 9.3

The daily ESAL computed in Example 9.2 is based on the initial traffic estimate for both directions of travel in an expressway with three lanes in each direction. The annual growth of traffic is estimated at 2.5%. The directional split is assumed to be 40 and 60%. The expressway is to be constructed of rigid pavements. Calculate the design lane ESAL for a service life of 15 years. Adopting the AASHTO lane distribution factor shown in Table 9.1, we have $f_{\mathrm{L}} = 0.7$. Given $f_{\mathrm{D}} = 0.6$, $L = 15$, $i = 2.5$, and from Example 2 $(\mathrm{ESAL})_1 = (33{,}742 \times 365)$, the total design lane $(\mathrm{ESAL})_{\mathrm{T}}$ is

$$(\mathrm{ESAL})_{\mathrm{T}} = (33{,}742 \times 365) \frac{[1 + 0.01(2.5)]^{15} - 1}{0.01(2.5)} (0.7)(0.6) = 92.76 \times 10^6$$

9.4 Material Properties for Design of Rigid Pavement

The structural properties of pavement slab and foundation materials are required as input to structural analysis for pavement design to check that the proposed design is adequate and provides sufficient safety margins against possible modes of failure. Table 9.7 lists the material properties that have to be determined for pavement design.

TABLE 9.5 Compuation of Total Daily ESAL for Example 9.2

Axle Load (kips)	Repetitions per Day	ESAL Factor	ESAL
(1)	(2)	(3)	(2) × (3)
S2	1438	0.0002	0.2876
S4	3391	0.002	6.782
S6	3432	0.01	34.32
S8	6649	0.032	212.768
S10	9821	0.081	795.501
S12	2083	0.175	364.525
S14	946	0.338	319.748
S16	886	0.601	532.486
S18	472	1	472
S20	299	1.58	472.42
S22	98	2.38	233.24
T2-10	2093	0.012	25.116
T2-12	1867	0.025	46.675
T2-14	1298	0.047	61.006
T2-16	1465	0.081	118.665
T2-18	1743	0.132	230.076
T2-20	1870	0.204	381.48
T2-22	2674	0.305	815.57
T2-24	2879	0.441	1269.639
T2-26	2359	0.62	1462.58
T2-28	2104	0.85	1788.4
T2-30	1994	1.14	2273.16
T2-32	1779	1.5	2668.5
T2-34	862	1.95	1680.9
T2-36	659	2.48	1634.32
T2-38	395	3.12	1232.4
T2-40	46	3.87	178.02
T3-26	588	0.194	114.072
T3-28	515	0.263	135.445
T3-30	496	0.351	174.096
T3-32	448	0.46	206.08
T3-34	225	0.594	133.65
T3-36	372	0.756	281.232
T3-38	474	0.95	450.3
T3-40	529	1.18	624.22
T3-42	684	1.45	991.8
T3-44	769	1.77	1361.13
T3-46	653	2.13	1390.89
T3-48	527	2.55	1343.85
T3-50	421	3.02	1271.42
T3-52	363	3.56	1292.28
T3-54	298	4.16	1239.68
T3-56	125	4.84	605
T3-58	84	5.59	469.56
T3-60	67	6.42	430.14
T3-62	46	7.33	337.18
T3-64	42	8.33	349.86
T3-66	31	9.42	292.02
T3-68	28	10.6	296.8
T3-70	21	11.9	249.9
T3-72	16	13.3	212.8
T3-74	12	14.8	177.6
		Total:	33741.59

The prefix S stands for single axle, T2 for tandem axle, and T3 for tridem axle.

TABLE 9.6 Fatigue and Erosion Analysis for Example 9.3

Axle Load (kips)	Design Load (kips)	Design n	Fatigue		Erosion	
			N_1	(n/N_1)	N_2	(n/N_2)
52T	62.4T	3,100	800,000	0.004	800,000	0.004
50T	60.0T	32,000	2,000,000	0.016	1,000,000	0.030
48T	57.6T	32,000	10,000,000	0.0032	1,200,000	0.027
46T	55.2T	48,000	unlimited	0	1,700,000	0.028
44T	52.8T	158,000	unlimited	0	2,000,000	0.079
42T	50.4T	172,000	unlimited	0	2,800,000	0.061
40T	48.0T	250,000	unlimited	0	3,500,000	0.071
30S	36.0T	3,100	25,000	0.124	1,700,000	0.002
28S	33.6T	3,100	70,000	0.044	2,200,000	0.001
26S	31.2T	9,300	200,000	0.045	3,000,000	0.002
24S	28.8T	545,000	800,000	0.682	5,000,000	0.033
22S	26.4T	640,000	10,000,000	0.064	9,000,000	0.071
			Total:	0.982		0.41

9.4.1 Properties of Concrete

To design against fatigue cracking failure under repeated traffic loadings, the elastic modulus and Poisson's ratio, and the elastic properties of the foundation are needed to compute the stresses and strains under each type of axle load. Next, based on the modulus of rupture (or flexural strength) of concrete and the load-induced tensile stresses, the allowable number of repetitions for each axle load can be estimated. modulus of rupture is commonly determined by means of the third-point load test in accordance with the standard test procedure of ASTM C78 (ASTM, 2003a). Alternatively, according to the American Concrete Institute (ACI, 2001), modulus of rupture, of normal-weight, MR, normal-strength concrete can be estimated from its compressive strength f_c as follows:

$$MR = 0.6\sqrt{f_c}(MR \text{ and } f_c \text{ are in MPa}) \tag{9.3a}$$

$$MR = 7.5\sqrt{f_c}(MR \text{ and } f_c \text{ are in lb/in.}^2) \tag{9.3b}$$

where f_c is determined using cylindrical specimens by test method ASTM C39 (ASTM, 2003b). The compressive strength f_c of commercially produced concrete with normal aggregate usually falls in the range of 3000 to 10,000 lb/in.^{2} (20 to 70 MPa).

TABLE 9.7 Material Properties for Design of Concrete Pavement

Failure Mode	Material Properties Required	
	Concrete Slab	Foundation[a]
Fatigue cracking	Elastic modulus Poisson's ratio Modulus of rupture	Modulus of subgrade reaction Elastic modulus and Poisson's ratio
Temperature-induced cracking	Elastic modulus Poisson's ratio Tensile strength Coefficient of thermal expansion Coefficient of drying shrinkage	Coefficient of friction at interface between slab and foundation

[a] Foundation refers to subgrade, or where subbase is provided, the combined layer of subgrade plus subbase.

Similarly, the elastic modulus E_c of concrete can also be estimated from the compressive strength by the following expressions:

$$E_c = 4730\sqrt{f_c}\,(E_c \text{ and } f_c \text{ are in MPa}) \tag{9.4a}$$

$$E_c = 57,000\sqrt{f_c}\,(E_c \text{ and } f_c \text{ are in lb/in.}^2) \tag{9.4b}$$

The tensile strength of concrete, often measured in terms of the indirect tensile strength or the tensile splitting strength, can be determined using ASTM C496 Standard Test Method for Splitting Tensile Strength of Cylindrical Concrete Specimens (ASTM, 2003c). The indirect tensile strength has a lower value than the modulus of rupture. Its value is usually in the range 0.1 to 0.2 of the compressive strength f_c. It may also be estimated from the following expression (Mindess et al., 2003):

$$f_t = 0.305 f_c^{0.55} \,(E_c \text{ and } f_c \text{ are in MPa}) \tag{9.5a}$$

$$f_t = 0.305 f_c^{0.55} \,(E_c \text{ and } f_c \text{ are in lb/in.}^2) \tag{9.5b}$$

Poisson's ratio can be determined by direct strain measurements in the uni-axial compression test by test method ASTM C469 (ASTM, 2003d). The typical values of the Poisson's ratio of concrete are within the range of 0.15 to 0.20. Since the structural response of concrete slab is rather insensitive to the value of Poisson's ratio, a value within the typical range is usually assumed for pavement design without any practical loss of accuracy in the analysis.

The coefficient of thermal expansion α_c of concrete is dependent on the thermal properties of its two constituents, namely cement paste and aggregate. The following are typical values of α_c reported in the literature (Neville, 1996):

Concrete with limestone aggregate: $\alpha_c \approx 6 \times 10^{-6}$ to 8×10^{-6} per °C
Concrete with granite aggregate: $\alpha_c \approx 8 \times 10^{-6}$ to 10×10^{-6} per °C
Concrete with sandstone aggregate $\alpha_c \approx 10 \times 10^{-6}$ to 12×10^{-6} per °C
Concrete with blast-furnace slag aggregate $\alpha_c \approx 9 \times 10^{-6}$ to 12×10^{-6} per °C

The coefficient of drying shrinkage of concrete is influenced by its water and cement ratio and aggregate content. Its value typically varies within the range of 2×10^{-4} to 8×10^{-4} mm/mm or in./in. (Neville, 1996).

9.4.2 Properties of Subgrade and Subbase

Depending on the theoretical model adopted to represent the foundation, different parameters are employed to characterize the structural response of the foundation materials. For concrete pavement design, the most common practice is to consider the foundation as a dense liquid, as was assumed by Westergaard (1926a, 1939) and many subsequent researchers in computing the structural response of rigid pavements. Under this assumption, the surface deflection w of a pavement is related to the vertical applied pressure p by the following equation,

$$p = k \cdot w \tag{9.6}$$

where k is termed as the modulus of subgrade deflection. k can be determined using Equation 9.6 by means of a plate loading test. The standard plate size of 30 in. (762 mm) diameter is used. Yoder and Witczak (1975) recommended that an applied load intensity of 10 lb/in² (69 kPa) would give a satisfactory k value for rigid pavement design. Typical values of k are from 50 to 200 pci (1.3×10^4 to 5.4×10^4 kN/m³) for low and high plasticity clayey soils, from 200 pci (5.4×10^4 kN/m³) for sandy soils and gravels, and up to 300 pci (8.1×10^4 kN/m³) or more for stabilized soils.

When a subbase layer is provided above the subgrade, the effective k value will increase. Plate loading test can be conducted to determine the improved k directly. If the subgrade k value is known, the PCA

(1984) and AASHTO (1993) provide easy-to-use design aids for designers to read off the increased k for various thickness of subbase added, as presented in Section 9.9.1 of this chapter.

Instead of using the dense liquid model, pavement foundation can also be modeled as an elastic solid. The required structural properties for the foundation materials are the resilient modulus and Poisson's ratio. The Poisson ratio, though necessary, has very little effect on the computed surface deflections of the pavement. Poisson ratios of 0.35 and 0.40 may be used for untreated granular subgrade and fine-grained soils, respectively, and a value of 0.20 may be used for cement or lime treated subgrade soils. Procedures for testing untreated subgrade soils and untreated base or subbase materials for determination of resilient modulus M_r are found in AASHTO test method T 307 (AASHTO, 2004). However, many highway laboratories are not equipped to perform the resilient modulus test for soils and base and subbase materials, it is common practice to estimate M_r through empirical correlation with other soil properties. Equation 9.7 is one such correlation suggested by the 1993 AASHTO Guide for fine-grained soils with soaked California Bearing Ratio (CBR) of 10 or less:

$$M_r \text{ (lb/in.}^2) = 1500 \times CBR \tag{9.7}$$

Other correlations are also found in the literature such as the work by Van Til et al. (1972). For unbound base and subbase materials, M_r may be estimated from the following correlations:

$$M_r \text{ (lb/in.}^2) = 740 \times CBR \qquad \text{for} \theta = 100 \text{ lb/in.}^2 \tag{9.8}$$

$$M_r \text{(lb/in.}^2) = 440 \times CBR \qquad \text{for} \theta = 30 \text{ lb/in.}^2 \tag{9.9}$$

$$M_r \text{(lb/in.}^2) = 340 \times CBR \qquad \text{for} \theta = 20 \text{ lb/in.}^2 \tag{9.10}$$

$$M_r \text{(lb/in.}^2) = 250 \times CBR \qquad \text{for} \theta = 10 \text{ lb/in.}^2 \tag{9.11}$$

where θ is the sum of principal stresses. CBR of subgrade soil, base or subbase material is determined by means of a standard laboratory test as specified by ASTM test method D1883 (ASTM, 2003e). CBR of these materials can also be measured using the in-situ procedure outlined in ASTM test method D4429 (ASTM, 2003f).

9.5 Computation of Load-Induced Stresses: Conventional Methods

9.5.1 Westergaard's Closed-Form Solutions

Westergaard (1926a, 1926b) was the first to propose a complete theory of structural behavior of rigid pavements. This theory is still the basis of computing load-induced stresses in many of the design procedures in use today. Westergaard modeled the pavement structure as a homogenous, isotropic, elastic, thin slab resting on a Winkler (dense liquid) foundation. From existing test data and experience, he identified the three most critical loading positions, the interior (also called center), edge, and corner, as illustrated in Figure 9.4 and derived equations for computing the critical stresses and deflections for loads placed at the edge, corner and center, respectively.

Westergaard made the following simplifying assumptions in his analysis: (1) The concrete pavement of known thickness acts as an infinitely large, homogenous, isotropic elastic slab; (2) The foundation acts like a bed of springs (dense liquid foundation model) under the slab; (3) There is full contact between the slab and foundation; (4) All forces act normal to the surface where shear and frictional forces are negligible; (5) The semi-infinite foundation has no rigid bottom; and (6) The slab is of uniform thickness, and the neutral axis is at its mid depth. Westergaard's original equations have been modified by different researchers, mainly to bring them into better agreement with field measurements.

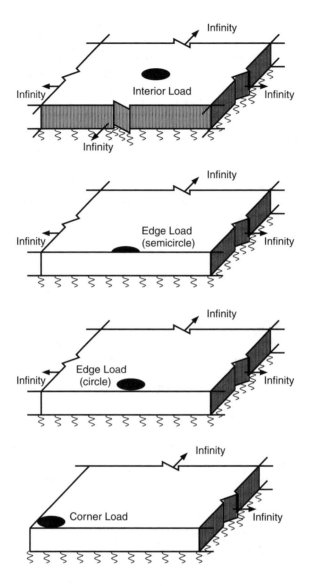

FIGURE 9.4 Problem definition and loading positions considered by Westergaard.

9.5.1.1 Interior Loading

Westergaard considered interior loading as a case when the load is applied at a "considerable distance from the edge." In the derivation, Westergaard defined the radius of relative stiffness, ℓ as follows:

$$\ell = \left[\frac{Eh^3}{12(1 - \mu^2)k} \right]^{\frac{1}{4}} \tag{9.12}$$

where E is the Young's modulus and μ is the Poisson's ratio of concrete, h is the thickness of the concrete slab, and k is the modulus of subgrade reaction. He derived the following expressions for the maximum stress σ and maximum deflection w under a circular distributed wheel load P acting at the interior loading position (Westergaard 1926b, 1939):

$$\sigma = \frac{3(1+\mu)P}{2\pi h^2}\left(\ln\frac{\ell}{b} + 0.6159\right) \tag{9.13}$$

$$w = \frac{P}{8k\ell^2}\left\{1 + \frac{1}{2\pi}\left[\ln\left(\frac{a}{2\ell}\right) - 0.673\right]\left(\frac{a}{\ell}\right)^2\right\} \tag{9.14}$$

where $b = a$ when $a \geq 1.724h$; and $b = \sqrt{1.6a^2 + h^2} - 0.675h$ when $a < 1.724h$.

9.5.1.2 Corner Loading

Westergaard considered a slab that extends to infinite length and width from the loaded corner. Using a method of successive approximation, Westergaard derived the following formulas for computing the maximum bending stress and deflection w when the slab is subjected to a corner load P:

$$\sigma = \frac{3P}{h^2}\left[1 - \left(\frac{a\sqrt{2}}{\ell}\right)^{0.6}\right] \tag{9.15}$$

$$w = \frac{P}{k\ell^2}\left[1.1 - 0.88\frac{a\sqrt{2}}{\ell}\right] \tag{9.16}$$

9.5.1.3 Edge Loading

Westergaard defined edge loading as the case when "the wheel is at the edge of the slab, but at a considerable distance from any corner." Two possible scenarios were considered: (1) a circular load with its center placed a radius length from the edge, and (2) a semicircular load with its straight edge in line with the slab. For the case of circular loading, the maximum bending stress σ and the maximum deflection w are computed as

$$\sigma = \frac{0.803P}{h^2}\left[4\ln\left(\frac{\ell}{a}\right) + 0.666\left(\frac{a}{\ell}\right) - 0.034\right] \tag{9.17}$$

$$w = \frac{0.431P}{k\ell^2}\left[1 - 0.82\left(\frac{a}{\ell}\right)\right] \tag{9.18}$$

The maximum bending stress and deflection for a semicircular loading at the edge are

$$\sigma = \frac{0.803P}{h^2}\left[4\ln\left(\frac{\ell}{a}\right) + 0.282\left(\frac{a}{\ell}\right) + 0.650\right] \tag{9.19}$$

$$w = \frac{0.431P}{k\ell^2}\left[1 - 0.349\left(\frac{a}{\ell}\right)\right] \tag{9.20}$$

The above expressions that include modifications made to the original Westergaard equations were recommended by Ioannides et al. (1985).

9.5.1.4 Limitations of Westergaard Solutions

Westergaard's solutions suffer from several limitations due to the assumptions made in their derivation. The main limitations are: (a) Stresses and deflections can be calculated only for the specific interior, edge and corner loading conditions; (b) Effects of finite dimensions of actual pavement slabs are not considered; (c) Load transfer across joints or cracks is not considered; (d) The solutions were derived based on thin-slab theory, ignoring the transverse shear deformation that exists in actual pavement slabs; (e) The Winkler foundation adopted does not take into account the additional support provided by the surrounding subbase and subgrade; (f) Multiple wheel loads cannot be considered.

9.5.2   Finite Element Solutions

9.5.2.1   2-D Finite Element Solutions

The finite element method provides a modeling alternative that is well suited for applications involving systems with irregular geometry, unusual boundary conditions or nonhomogenous composition. The most widely used finite element computer codes are based on the slab model that idealizes concrete pavement slabs as classical thin plates supported by a subgrade idealized either as Winkler foundation or semiinfinite elastic solid foundation. Examples of such two-dimensional (2D) computer software include KENSLABS developed by Huang (1974, 1985, 1993), ILLI-SLAB (Tabatabaie and Barenberg, 1978), WESLIQUID and WESLAYER (Chou and Huang, 1979, 1981), FEACONS (Tia et al., 1987), and ILSL2 (Khazanovich and Yu 1998) and ISLAB2000 (Khazanovich et al., 1999), which are enhanced versions of ILLI-SLAB. These computer programs are able to study the response of a multi-slab system under the action of multiple loads at any locations, represent the exact finite dimensions of each slab, include joint details and load transfer across joints, and simulate interface contact conditions between the pavement slab and the underlying layer.

9.5.2.2   3-D Finite Element Solutions

With the rapid development of computer technology in recent years, there has been marked improvements in the computational efficiency and capability of computing devices. The use of three-dimensional (3-D) finite element programs for pavement analysis is no longer a forbidden task. The general-purpose 3-D finite element code ABAQUS (Hibbitt, Karlsson & Sorensen Inc., 2002) offers a powerful analytical tool not available previously for solving complex pavement problems. Examples of applications include analysis of pavement joints by Uddin et al. (1995), evaluation of pavement sub-drainage systems by Hassan and White (2001), and study of skid resistance of pavement materials by Liu et al. (2003).

EverFE, a special-purpose code developed at the University of Washington (Davids et al., 1998), is the latest addition to the family of 3-D finite element programs for the analysis of rigid pavements. It is equipped with an efficient computational model to operate on personal computers. It can solve multi-slab systems consisting of up to nine slabs, with up to three base layers. Load transfer across joint through either dowel action or aggregate interlock can be considered. EverFE is able to model wheel loading and temperature curling effects. It also allows loss of contact between slab and base to be studied.

9.5.2.3   Limitations of Finite Element Solutions

Finite element methods are a numerical technique with the correctness and accuracy of their solutions dependent critically on the choice of appropriate mesh representation. Hence, in performing finite element analysis, the design of finite element mesh as well as convergence verification and accuracy assessment require skills and professional knowledge of the technical nature of the problem analyzed. The right solutions and sufficiently accurate answers are usually obtained through a trial-and-error process. Each computer run of finite element analysis, and 3-D finite element analysis in particular, calls for time consuming input preparation effort, and the analysis itself is highly demanding computationally even with today's high-speed computer technology. A change of loading locations requires a fresh input preparation and a separate computer run. These shortcomings are obstacles to finite element methods being adopted for routine structural analysis and design of pavements.

9.6   Computation of Load-Induced Stresses: Improved Methods

9.6.1   Closed-Form Solutions for Slabs with Finite Dimensions

A number of researchers (Kelley 1939, Teller and Sutherland 1943, Ioannides et al., 1985) have highlighted the following deficiencies of Westergaard's solutions for slab responses under applied vertical loads: (a) a range of slab sizes exist within which the accuracy of the solutions for deflections and flexural stresses are questionable; and (b) the solutions agreed well with experimental observations for the interior loading condition, but less so for the edge and corner loading conditions. Its simplified thin-slab representation and the assumption of infinite plan dimensions of the slab are considered to be the other limitations.

A theoretical solution has been developed at the National University of Singapore (Shi et al., 1994) for the structural response of a rectangular slab under vertical applied load by considering the actual finite dimensions of the slab, and the thick-plate action. Using this model and assuming a Winkler foundation, Fwa et al. (1996) showed that the Westergaard's infinite slab assumption is valid if the slab length is more than four times the radius of relative stiffness of the pavement system. However, because of thick-plate action, not all solutions for slab responses converge to Westergaard's values when slab sizes are large. For large slabs where the effect of slab dimensions is negligible, the difference of thick-plate solutions and Westergaard solutions increases with the radius of relative stiffness of the slab system.

For interior loading, Table 9.8 indicates the required slab dimensions for the infinite slab assumption to be acceptable for practical applications, and the differences between Westergaard solutions and the thick-plate solutions when the infinite slab assumption is valid. Data of the maximum bending stresses and the maximum deflections are presented for two slab configurations: (a) square slabs and (b) 3.66 m (12 ft) wide rectangular slabs. Tables 9.9 and 9.10 provide the corresponding information for edge and corner loading, respectively. Square slabs are found in heavy industry pavement systems, and airport and seaport pavement systems, while rectangular slabs are commonly used for highways.

For a slab with dimensions less than those that satisfy the infinite slab assumption, considering the pavement responses reported in the three aforementioned tables (i.e. maximum bending stress and maximum deflection) under the respective loading conditions, the differences between the Westergaard solutions and the corresponding thick-plate solutions increase as the slab dimensions become smaller. The closed-form solutions provided by the computer code for multiple-slab systems, to be described in the next section, can be applied to compute the pavement responses for single-slab systems as well.

9.6.2 Closed-Form Solutions for Multiple-Slab System

A jointed rigid pavement system does not behave like a homogeneous solid slab because most joints represent discontinuities that do not offer complete transfer of all forms of forces and moments. For instance, as explained later in Section 9.11, while contraction joints, warping joints or expansion joints are designed to transfer shear forces from one slab to the next, the amount of moment transfer, if any, is typically not considered in their design. One would therefore expect the pavement responses of an individual slab in a jointed pavement system to be different from the responses of a single slab with free edges.

Liu and Fwa (2004) developed a nine-slab theoretical model that provides closed-form solutions for a jointed slab system supported on a Winkler foundation and subjected to vertical loads. The loaded slab of interest is represented by the central slab, with eight surrounding slabs to take into account the effects of jointed pavement system. Only shear force transfer is assumed across a joint. Shear force transfer is specified in terms of the efficiency of joint shear transfer (see Section 9.11.3) which is defined as the ratio of vertical deflections of the two edges along the joint between two slabs. The theoretical closed-form solution to the nine-slab model was derived by superposing solutions of specially selected elemental slabs. It provides a useful analytical tool to compute the stresses and deflections of a multiple-slab system under an interior load.

Liu and Fwa (2005) have also developed a six-slab thick plate solution to compute the structural responses of edge slab of a multiple-slab jointed concrete pavement system. The jointed concrete pavement system is idealized as a six-slab system resting on a Winkler foundation. The six slabs are arranged in two rows with three slabs in each row. The loaded slab of interest is represented by a middle slab, with the five surrounding slabs to take into consideration the effects of jointed pavement system.

It should be mentioned that, as presented in Section 9.5.2, 2-D and 3-D finite element programs are available to solve multiple-slab systems. However, a closed-form theoretical solution is always

TABLE 9.8 Comparison of Thick Plate and Westergaard Solutions for Interior Loading (Radius of Loaded Area = 150 mm)

	Maximum Bending Stress σ_{max}				Maximum Deflection σ_{max}			
	Minimum slab size required for infinite slab assumption		Does thick plate solution converge to Westergaard solution as slab size approaches infinity?		Minimum slab size required for infinite slab assumption		Does thick plate solution converge to Westergaard solution?	
	By other studies	Thick plate solution	Square slab	Rectangular slab	By other studies	Thick plate solution	Square slab	Rectangular slab
	$A \geq 20$ ft (6 m) for 9 in. (225 m) thick slab by Bergstrom et al. (1949)	(square & 3.66 m wide rectangular slabs)	Yes for $L = 0.5$ m No for $L \geq 1.0$ m	No for all L values studied	$(A/L) \geq 8.0$ by Ioannides et al. (1985)	(square & 3.66 m wide rectangular slabs)	Yes for all values of L	No for all values of L studied
	$A \geq 16.5$ ft (7.8 m) by Yang (1972)	$A \geq 4$ m for $L = 0.5$ m	Westergaard solution underestimates σ_{max} by 5% at $L = 1.0$ m, 10% at $L = 1.5$ m, 15% at $L = 2.0$ m	Westergaard solution overestimates σ_{max} at $L \leq 1.0$ m; underestimates σ_{max} at $L > 1.0$ m. At $A = 10$ m & $L = 2$ m, Westergaard solution underestimates σ_{max} by about 25%		$A \geq 2$ m for $L = 0.5$ m		Westergaard solution under-estimates σ_{max}. For $A = 10$ m, σ_{max} is underestimated by 10% at $L = 0.5$ m 70% at $L = 2.0$ m
	$(A/L) \geq 3.5$ ft by Ioannides et al. (1985)	$A \geq 4$ m for $L = 1.0$ m $A \geq 6$ m for $L = 1.5$ m $A \geq 7$ m for $L = 2.0$ m				$A \geq 4$ m for $L = 1.0$ m $A \geq 6$ m for $L = 1.5$ m $A \geq 8$ m for $L = 2.0$ m		

Note: A = slab length, L = radius of relative stiffness.

TABLE 9.9 Comparison of Thick Plate and Westergaard Solutions for Edge Loading (Radius of Loaded Area = 150 mm)

	Maximum Bending Stress σ_{max}				Maximum Deflection σ_{max}			
	Minimum slab size required for infinite slab assumption		Does thick plate solution converge to Westergaard solution as slab size approaches infinity?		Minimum slab size required for infinite slab assumption		Does thick plate solution converge to Westergaard solution?	
	By other studies	Thick plate solution	Square slab	Rectangular slab	By other studies	Thick plate solution	Square slab	Rectangular slab
	$(A/L) \geq 5.0$ ft by Ioannides et al. (1985) (square & 3.66 m wide rectangular slabs)	$A \geq 2$ m for $L = 0.5$ m; $A \geq 4$ m for $L = 1.0$ m; $A \geq 6$ m for $L = 1.5$ m; $A \geq 8$ m for $L = 2.0$ m	No for all L values studied. Westergaard solution over-estimates σ_{max} by 7% at $L = 5.0$ m, 10% at $L = 1.0$ m, 12% at $L = 2.0$ m values studied	No for all L values studied. Westergaard solution overestimates σ_{max}, at $A = 16$ m, σ_{max} overestimated by about 14% for all L values studied	$(A/L) \geq 8.0$ by Ioannides et al. (1985)	(square & 3.66 m wide rectangular slabs) $A \geq 2$ m for $L = 0.5$ m; $A \geq 4$ m for $L = 1.0$ m; $A \geq 7$ m for $L = 1.5$ m; $A \geq 9$ m for $L = 2.0$ m	Yes for all values of L	No for all values of L studied. Westergaard solution under-estimates σ_{max}. For $A = 10$ m, σ_{max} is underestimated by 10% at $L = 0.5$ m 30% at $L = 2.0$ m

Note: A = slab length, L = radius of relative stiffness.

TABLE 9.10 Comparison of Thick Plate and Westergaard Solutions for Corner Loading (Radius of Loaded Area = 150 mm)

	Maximum Bending Stress σ_{max}				Maximum Deflection σ_{max}			
	Minimum slab size required for infinite slab assumption		Does thick plate solution converge to Westergaard solution as slab size approaches infinity?		Minimum slab size required for infinite slab assumption		Does thick plate solution converge to Westergaard solution?	
	Thick plate solution	By other studies	Square slab	Rectangular slab	By other studies	Thick plate solution	Square slab	Rectangular slab
	(square & 3.66 m wide rectangular slabs)	$(A/L) \geq 4.0$ ft by Ioannides et al. (1985)	No for all L values studied	No for all L values studied	$(A/L) \geq 5.0$ by Ioannides et al. (1985)	(square & 3.66 m wide rectangular slabs)	Thick plate solutions for all values of L converge to within 5% of Westergaard solutions	No for all values of L studied
	$A \geq 2$ m for $L = 0.5$ m		Westergaard solution underestimates σ_{max} by 15% for all values of L	Westergaard solution underestimates σ_{max}. At $A = 16$ m, σ_{max} overestimated by about 15% for all L values studied		$A \geq 2$ m for $L = 0.5$ m		Westergaard solution underestimates σ_{max}. For $A = 10$ m, σ_{max} is underestimated by 12% at $L = 0.5$ m 40% at $L = 0.5$ m 40% at $L = 2.0$ m
	$A \geq 4$ m for $L = 1.0$ m	(see Notes 2 & 3 for Bradbury and Pickett solutions, respectively)				$A \geq 4$ m for $L = 1.0$ m		
	$A \geq 6$ m for $L = 1.5$ m					$A \geq 6$ m for $L = 1.5$ m		
	$A \geq 7$ m for $L = 2.0$ m					$A \geq 7$ m for $L = 2.0$ m		

Note

1: A = slab length, L = radius of relative stiffness.

2: Thick plate solutions for σ_{max} at $L \geq 1.0$ m converge to Bradbury solution, but at $L = 0.5$ m converges to a value equal to 88% of Bradbury (1938) solution.

3: Thick plates solutions for σ_{max} at $L \geq 1.0$ m converge to about 82% of Pickett solution, but that at $L = 0.5$ m converges to a value equal to about 72% of Pickett (1946) solution.

preferred from the stand-point of ease of application, computational efficiency and accuracy. The computer code MSLAB-LOAD is accessible on the CRC Press website. The users' manual and input instructions are found in Appendix 9A. Each computer analysis of either a nine-slab or a six-slab problem takes less than 2 seconds on a Pentium IV personal computer of 2.4 MHz CPU speed with a RAM of 512 MB.

9.7 Computation of Thermal Stresses: Conventional Methods

9.7.1 Warping Stresses Due to Temperature Differential

Figure 9.5 shows a typical γ-shape pattern of extreme-temperature profile registered within a concrete pavement slab in a daily cycle. The large extent of the soil mass beneath the pavement structure produces a reservoir effect that maintains a stable temperature regime. As a result, the temperature at the bottom face of the pavement slab remains more or less constant. In the early hours of a day when the air temperature is the lowest, the temperature at the top pavement surface will be lower than that of at the bottom pavement surface. The situation is reversed when the air temperature rises to the day's highest point in the afternoon.

The temperature differential between the top and bottom faces of the pavement slab gives rise to warping stresses. In the hot afternoon, the top face at higher temperature will expand more than the

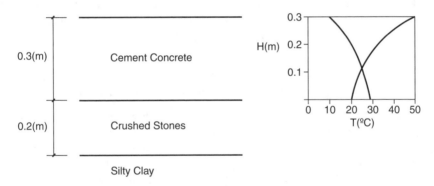

(a) A concrete pavement section and its temperature profile

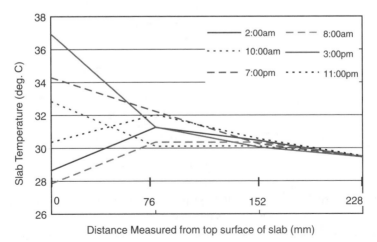

Distance Measured from top surface of slab (mm)

(b) Temperature data recorded in April in Illinois

FIGURE 9.5 Temperature distribution profile in concrete pavement.

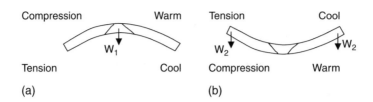

FIGURE 9.6   Warping stresses in concrete pavement: (a) daytime warping; (b) nighttime warping.

bottom face if both are free to move, and curling would form a shape convex upward as shown in Figure 9.6(a). In reality, however, the weight of the slab prevents curling from taking place, resulting in compression in the top face, and tension in the lower face of the slab. In the early morning, with lower temperature in the upper face and higher temperature in the lower face, tensile stresses are found in the upper face and compressive stresses in the lower face because the weight of the concrete prevents the slab from forming an upward concave shape, as depicted in Figure 9.6(b).

9.7.2   Solutions for Warping Stresses

Traditionally, in the design for and analysis of concrete pavements, the temperature distribution across the slab thickness has been assumed to be linear. Westergaard (1927) was the first to solve for the warping stresses induced by linear temperature profiles. He developed solutions of warping stresses in slabs supported on Winkler foundation for the following three cases:

(a) A slab infinite in the horizontal x and y direction
(b) A slab infinite in the plus y and the plus or minus x direction
(c) A slab with a finite width in y direction but infinite in the plus and minus x direction

Based on the solution for case (c) by Westergaard, Bradbury (1938) superposed the solution for an infinite strip with width x and that for another infinite strip with width y, and obtained a solution for a slab with finite x and y dimensions. This solution is widely used today in rigid pavement design for warping stress analysis. Of special interest to pavement designers are the stresses at the following locations: center of the slab, mid-point of the longitudinal edge of the slab, and mid-point of the transverse edge of the slab. These stresses are computed by the following formulas:

$$(\sigma_x)_{\text{center}} = \frac{E\alpha_t\Delta_t}{2(1-\mu^2)}(C_x + \mu C_y) \tag{9.21a}$$

$$(\sigma_y)_{\text{center}} = \frac{E\alpha_t\Delta_t}{2(1-\mu^2)}(C_y + \mu C_x) \tag{9.21b}$$

$$(\sigma_x)_{\text{edge}} = \frac{E\alpha_t\Delta_t}{2}C_x \tag{9.21c}$$

$$(\sigma_y)_{\text{edge}} = \frac{E\alpha_t\Delta_t}{2}C_y \tag{9.21d}$$

where $(\sigma_x)_{\text{center}}$ and $(\sigma_y)_{\text{center}}$ are slab-center warping stresses in x and y directions, respectively, $(\sigma_x)_{\text{edge}}$ and $(\sigma_x)_{\text{edge}}$ are slab-edge warping stresses in x and y directions, respectively, E the elastic modulus of concrete, α_t the coefficient of thermal expansion of concrete, μ the Poisson's ratio of concrete, Δ_t the difference between the temperatures at top and bottom faces of the slab, and C_x and C_y are warping stress coefficients. The values of C_x and C_y, given in Figure 9.7, are dependent on the length L_y and width L_x of the slab, and its radius of relative stiffness ℓ.

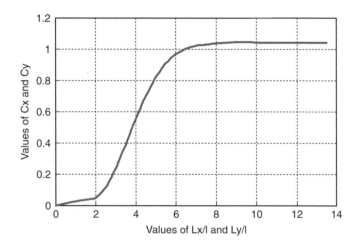

FIGURE 9.7   Warping stress coefficients for concrete pavement.

It should be mentioned that finite element codes, such as KENSLABS, WESLIQUID, WESLAYER, ILLI-SALB, ILSL2, ISLAB2000, and EverFE described earlier in Section 9.5.2 can also be applied to analyze thermal warping stresses in concrete pavements.

9.7.3   Stresses Due to Subgrade Restraint

When the ambient temperature falls or rises, besides the tendency to warp vertically, the pavement slab also undergoes horizontal contraction or expansion which leads to some relative movements between subgrade and the bottom face of the slab. Friction between the pavement slab and subgrade provides restraint to these relative movements, thereby creating stresses within the slab. The pavement designer is more interested in the case where temperature falls, because tensile stresses will develop as the subgrade friction restrains the slab from contracting.

Figure 9.8 shows schematically the frictional resistance mobilized at the interface of pavement slab and subgrade to resist slab movements. It is common practice to consider an average coefficient of friction, μ_r, to estimate the magnitude of tensile stress f_t generated within the slab as follows:

$$f_t = \frac{\mu_r \left(\frac{1}{2} \gamma_c BLh \right)}{Bh} = \frac{\gamma_c \mu_r L}{2} \tag{9.22}$$

where B, L and h are respectively, the width, length and thickness of the slab, and γ_c the density of the concrete slab. Depending on the subgrade or subbase material in contact with the concrete slab, the value of μ_r ranges from about 1.0 to more than 2.0 as given in Table 9.11. For normal concrete, γ_c can be taken as 2,500 kg/m³. Based on Equation 9.22, a theoretical "crack-free" length of concrete pavement can be defined by equating f_t to the tensile strength of the concrete slab as given by Equation 9.23. This can offer as a guide of the maximum panel length permitted of a JPCP:

$$\text{Crack-free length } L = \frac{2f_T}{\gamma_c \mu_r} \tag{9.23}$$

where f_T is the tensile strength of concrete.

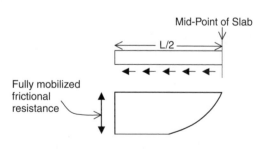

FIGURE 9.8   Frictional resistance against slab movements.

TABLE 9.11   Values of Friction Factor Between Concrete Slab and Support Materials

Friction Factor	Type of Material beneath Slab
Surface treatment	2.2
Lime stabilization	1.8
Asphalt stabilization	1.8
Cement stabilization	1.8
River gravel	1.5
Crushed stone	1.5
Sandstone	1.2
Natural subgrade	0.9

Source: AASHTO. 1993 AASHTO Guides for Design of Pavement Structures.
Copyright 1993 by the American Association of State Highway and Transportation
Officials, Washington, DC. Used by permission.

For normal concrete materials, warping stresses tend to be more critical for pavements with slab length in the range of about 8 to 15 m (25 to 48 ft), while subgrade restrained stresses become important only for very long slabs exceeding about 30 m (100 ft) in length.

9.8   Computation of Thermal Stresses: Improved Methods

9.8.1   Effect of Nonlinear Temperature Distribution

As early as the 1930s, Teller and Sutherland (1935) had reported based on tests conducted on concrete pavements that temperature distributions across concrete pavement thickness were highly nonlinear, and that stresses arising from restrained temperature warping are equal in importance to those produced by the heaviest wheel loads. As depicted earlier in Figure 9.5, a typical nonlinear temperature profile is featured with a relatively rapid change of temperature within the top quarter of the slab thickness, followed by a more gradual change towards the bottom face. Recent studies showed that for the temperature ranges encountered in a temperate region, this assumption of linear temperature profile could lead to errors of 30% or more in the computed peak warping stresses (Choubane and Tia, 1992).

Advanced finite element programs such as ISLAB2000 and EverFE (see Section 9.5.2) are able to compute warping stresses caused by nonlinear temperature distributions. A number of researchers (Choubane and Tia, 1995; Wu and Larsen, 1993) have applied finite element programs to analyze the effects of nonlinear temperature distribution on rigid-pavement behavior. While this numerical analysis tool is adequate for analyzing the problem, it is largely restricted to specialized users as it requires coding skills and careful check for convergence of solutions. Zhang et al. (2003) developed a closed-form theoretical solution (coded as computer program WARP-NONLINEAR) that provides an answer in less than a second on a Pentium IV personal computer of 2.4 MHz CPU speed. The solution was developed for a single rectangular slab with four free edges resting on a foundation. It is derived based on the Reissner Thick-plate theory (Reissner, 1945) for a concrete pavement slab supported on either a Pasternak foundation (Pasternak, 1954) or a Winkler foundation (Westergaard, 1926c). The nonlinear temperature distribution across the slab thickness was divided into three components, and the final solutions were obtained by superposition of the stresses caused by the three components as shown in Figure 9.9. The model takes into consideration the exact slab dimensions, the effect of transverse shear deformation by considering thick-slab action, and the effect of subgrade interlocking if the Pasternak foundation model is selected.

Zhang et al. (2003) analyzed temperature profile data in Florida and Illinois using WARP-NONLINEAR. It was found that the assumption of linear temperature distribution under estimated the peak slab-top tensile warping stress by 15.9% in Florida, and 55.4% in Illinois, and over estimated the peak slab-bottom tensile warping stress by 25.1% in Florida and 74.9% in Illinois. The computer code

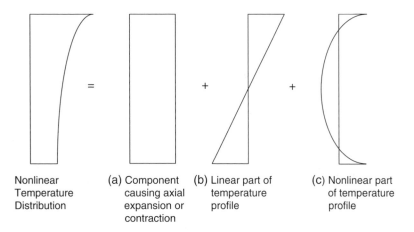

Nonlinear
Temperature
Distribution

(a) Component
causing axial
expansion or
contraction

(b) Linear part of
temperature
profile

(c) Nonlinear part
of temperature
profile

FIGURE 9.9 Decomposition of nonlinear temperature distribution into three components.

WARP-NONLINEAR is accessible on the CRC Press website. The users' manual and input instructions are found in Appendix 9B.

9.8.2 Effect of Pasternak Foundation

One of the main drawbacks of the Westergaard–Bradbury solutions for warping stresses (see Section 9.7.2) lies with the foundation model. Winkler model is a poor representation of the actual subbase or subgrade materials. For instance, it is known that transverse connection exists in actual foundations. Full-size experimental results by Yao and Huan (1984) have shown that warping stresses predicted by the Bradbury model are usually higher than those developed in actual pavement slabs. Better prediction of warping stresses can be achieved by replacing the Winkler foundation by a Pasternak foundation with an appropriate representation of the transverse connection of actual foundation materials.

Shi et al. (1993) derived a theoretical solution of warping stresses in concrete pavement slabs resting on a Pasternak foundation. The traditional procedure of computing warping stresses by the Westergaard–Bradbury approach with Winkler foundation is a special case of this more general solution. The transverse connection in the foundation is represented by a subgrade shear modulus, G. When G is set to zero, the foundation model reduces to a Winkler foundation. Figure 9.10 shows the

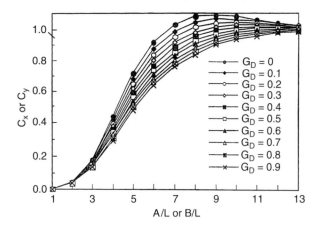

FIGURE 9.10 Coefficients of warping stresses for slab supported on Pasternak foundation.

computed values of warping stress coefficient for different G_D values. G_D is related to G by the following expressions:

$$G_D = \frac{G \cdot L^2}{2D} \quad \text{with} \quad L = \left(\frac{D}{k}\right)^{0.25} \quad \text{and} \quad D = \frac{Eh^3}{12(1-\mu^2)} \tag{9.24}$$

where L is the radius of relative stiffness of the slab system, D the flexural rigidity of the slab, k the modulus of subgrade reaction, and E and μ, respectively, the elastic modulus and Poisson's ratio of concrete. It is seen that the warping stress coefficient decreases with increasing G value. In other words, the Bradbury solution provides a conservative estimate of the warping stresses. Shi et al. (1993) suggested that G (in MN/m) can be computed as $(0.35k)$ where k is given in MN/m^3.

9.9 Thickness Design of Concrete Pavement

Two major approaches of thickness design are being used today for concrete pavements. The first approach relies on empirical relationships between thickness design and pavement serviceability performance based on data obtained from full-scale loading tests and in-service pavements. The thickness design procedure of AASHTO (1993) is an example. The second approach, which may be termed mechanistic-empirical approach, develops mechanism-based empirical relationships of failure developments in terms of the properties of pavement materials, and load-induced and thermal stresses, and calibrates these relationships with actual pavement performance data. The PCA (1984) method of design adopts this approach. Thickness design procedure by AASHTO and PCA are discussed in this section.

9.9.1 AASHTO Procedure for Thickness Design

The design procedure described in the AASHTO Guide (AASHTO, 1993) was developed based on the findings of the AASHO Road Test (Highway Research Board, 1962). The central idea was to provide sufficient slab thickness to ensure adequate level of pavement performance throughout the design life of the pavement. AASHTO defines pavement performance in terms of the present serviceability index (PSI) which varies from 0 to 5 (Carey and Irick, 1960). The PSI of newly constructed rigid pavements is about 4.5. For pavements of major highways, the end of service life is considered to be reached when PSI = 2.5. A terminal value of PSI = 2.0 may be used for secondary roads. Serviceability loss given by the difference of the initial and terminal serviceability indices, is required as input to the design procedure. The design traffic loading in ESAL is computed according to the procedure outlined in Section 9.3. Other input parameters are discussed in this section.

9.9.1.1 Reliability

The AASHTO Guide incorporates in the design a reliability factor R% to account for uncertainties in traffic prediction and pavement performance. R% indicates the probability that the pavement designed will have a performance level higher than the terminal serviceability level at the end of the design period. The ranges of R% suggested by AASHTO are 85 to 99.9%, 80 to 99%, 80 to 95% and 50 to 80% respectively, for urban interstates, principal arterials, collectors and local roads. The corresponding ranges for rural roads are 80 to 99.9%, 75 to 95%, 75 to 95% and 50 to 80%. The overall standard deviation, S_o, for rigid pavement performance developed at the AASHO Road Test 15.

9.9.1.2 Pavement Material Properties

The elastic modulus E_c and the 28-day modulus of rupture S_c of concrete are required input parameters. E_c and S_c are determined according to the procedures described in Section 9.4.1. Although the subgrade soil property required for thickness design is the modulus of subgrade reaction k, the AASHTO Guide specifies the resilient modulus M_r as the input. To account for time variations of soil properties within a year, an elaborate procedure for determining the correct design input value of k is described by the

AASGTO guide as follows:

(1) Divide the year into equal-length time intervals each equal to the smallest season, and determine the soil resilient modulus M_r in each interval. AASHTO Guide suggests that the smallest season should not be less than one-half month.
(2) Each M_r value is corrected for the presence of subbase (see Figure 9.11) and bedrock within a depth of 3 m (10 ft) (see Figure 9.12) to obtain the composite modulus of subgrade reaction.
(3) Estimate the relative damage u corresponding to each seasonal modulus of subgrade reaction by the following equation:

$$u = (D^{0.75} - 0.39k^{0.25})^{3.42} \tag{9.25}$$

where u is dimensionless, D is the projected slab thickness in inches, and k is modulus of subgrade reaction in lb/in.^{2}/in. or pci. Instead of Equation 9.25, Figure 9.13 can be used to obtain u.

(4) Sum up the u values of all seasons and divide by the number of seasons to obtain the average seasonal damage.

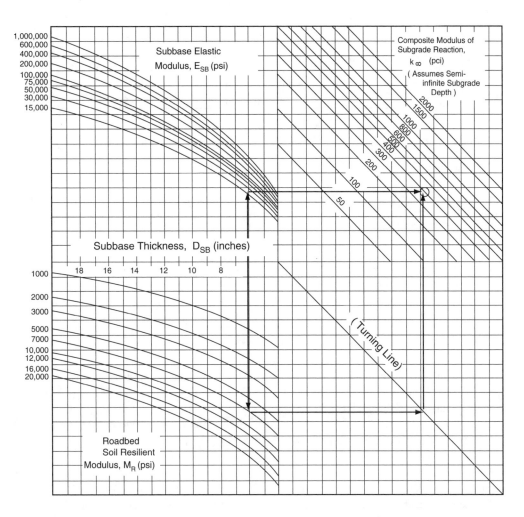

FIGURE 9.11 Chart for estimating composite k (*Source*: AASHTO, 1993.) AASHTO Guides for Design of Pavement Structures. Copyright 1993 by the American Association of State Highway and Transportation Officials, Washington, DC. Used by permission.

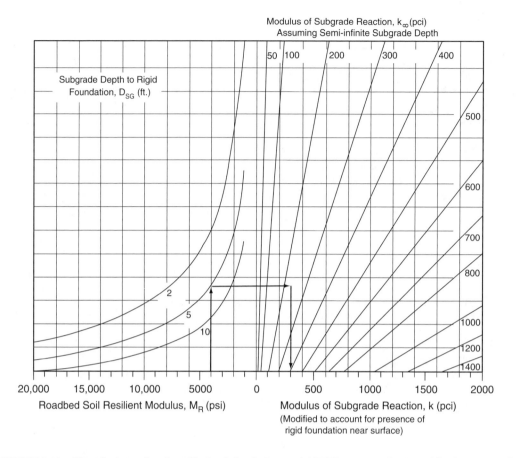

FIGURE 9.12 Chart for *k* as a function of bedrock depth. *Source*: AASHTO, 1993. AASHTO Guides for Design of Pavement Structures. Copyright 1993 by the American Association of State Highway and Transportation Officials, Washington, DC. Used by permission.

(5) Substitute the average seasonal damage into Equation 9.25 or Figure 9.13, and obtain the effective *k*.

(6) To take into account the effect of likely loss of support, apply Figure 9.14 To arrive at the design effective *k*. The loss of support value represented by LS in Figure 9.14 are given in Table 9.12.

Example 9.4

Computation of composite subgrade reaction. A concrete pavement is constructed on a 6 in. thick subbase with elastic modulus of 20,000 lb/in.^{2} The resilient modulus of the subgrade soil is 7000 lb/in.^{2} The depth of subgrade to bedrock is 5 ft.

Entering Figure 9.11 with $D_{SB} = 6$ in., $E_{SB} = 20,000$ lb/in.^{2} and $M_r = 7000$ lb/in.^{2}, obtain $k_\infty = 400$ pci. With bedrock depth of 5 ft., composite $k = 500$ pci from Figure 9.12.

Example 9.5

The value of composite *k* values determined at 1 month intervals are: 400, 400, 450, 450, 500, 500, 450, 450, 450, 450, 450, and 450 pci. Projected slab thickness is 10 in. and LS = 1.0. Determine effective *k*.

By means of Equation (9.25) or Figure 9.13, the relative damage for each *k* can be determined. Hence total $u = 100 + 100 + 97 + 97 + 93 + 93 + 94 + 94 + 94 + 94 + 94 + 94 + 94 = 1{,}144$. Average $u = 95.3$, and average $k = 470$ pci. Entering Figure 9.14 with LS = 1.0, read effective $k = 150$ pci.

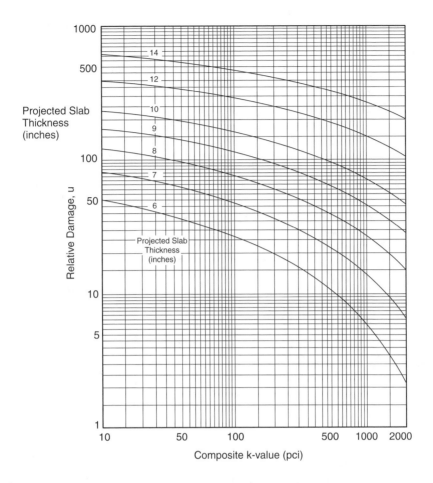

FIGURE 9.13 Chart for estimating relative damage to rigid pavements. *Source*: AASHTO, 1993. AASHTO Guides for Design of Pavement Structures. Copyright 1993 by the American Association of State Highway and Transportation Officials, Washington, DC. Used by permission.

9.9.1.3 Load Transfer Coefficient

Load transfer coefficient J is a numerical index introduced to account for the load transfer efficiencies of different joint designs. Table 9.13 presents the J values for the joint conditions at the AASHO Road Test. Lower J values are associated with pavements with load transfer devices (such as dowel bars) and those with tied shoulders. For cases where a range of J values applies, higher values should be used with low k values, high thermal coefficients, and large variations of temperature.

9.9.1.4 Drainage Coefficient

To allow for changes in thickness requirement due to differences in drainage properties of pavement layers and subgrade, a drainage coefficient C_d was included in the AASHTO thickness design. Setting $C_d = 1$ for the conditions at the AASHO Road Test, Table 9.14 shows the C_d values for other conditions. The percent of time during the year the pavement structure would be exposed to moisture levels approaching saturation can be estimated from the annual rainfall and the prevailing drainage condition.

9.9.1.5 Slab Thickness Requirement

The required slab thickness is obtained using the nomograph in Figure 9.15. It is noted that the moisture effect on subgrade strength, and the effect of drainage condition on pavement performance have

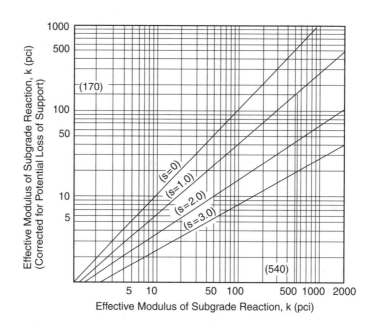

FIGURE 9.14 Correction of effective modulus of subgrade reaction for potential loss of subbase support. *Source*: AASHTO, 1993. AASHTO Guides for Design of Pavement Structures. Copyright 1993 by the American Association of State Highway and Transportation Officials, Washington, DC. Used by permission.

been considered in the design computation. Other environmental impacts such as roadbed swelling, frost heave, and deterioration due to weathering could result in some serviceability loss. This loss in serviceability can be added to that caused by traffic loading for design purposes.

Example 9.6

Apply AASHTO procedure to design a concrete pavement slab thickness for ESAL $= 11 \times 10^6$. The design reliability is 95% with a standard deviation of 0.3. The initial and terminal serviceability levels are 4.5 and 2.5, respectively. Other design parameters are $E_c = 5 \times 10^6$, $S'_c = 650$ psi, $J = 3.2$, and $C_d = 1.0$.
 Design PSI loss $= 4.5 - 2.5 = 2.0$. From Figure 9.15, $D = 10$ in.

9.9.2 PCA Procedure for Thickness Design

The thickness design procedure published by PCA (1984) adopts a mechanistic-empirical approach that addresses two basic forms of failure, namely fatigue cracking failure and erosion failure of the supporting

TABLE 9.12 Values for Loss of Support Factor LS

Type of Material	Loss of Support (LS)
Cement treatment granular base ($E = 1,000,000$ to $2,000,000$ lb/in.^{2})	0.0 to 1.0
Cement aggregate mixtures ($E = 500,000$ to $1,000,000$ lb/in.^{2})	0.0 to 1.0
Asphalt treated base ($E = 350,000$ to $1,000,000$ lb/in.^{2})	0.0 to 1.0
Bituminous stabilized mixtures ($E = 40,000$ to $300,000$ lb/in.^{2})	0.0 to 1.0
Lime stabilized ($E = 20,000$ to $70,000$ lb/in.^{2})	1.0 to 3.0
Unbound granular materials ($E = 15,000$ to $45,000$ lb/in.^{2})	1.0 to 3.0
Fine-grained or natural subgrade materials ($E = 3000$ to $40,000$ lb/in.^{2})	2.0 to 3.0

 Source: AASHTO. (1993) AASHTO Guides for Design of Pavement Structures.
 Copyright 1993 by the American Association of State Highway and Transportation Officials, Washington, DC. Used by permission.

TABLE 9.13 Values of Load Transfer Coefficient *J*

Shoulder	Asphalt		Tied PCC	
Load transfer device	Yes	No	Yes	No
Pavement type				
Plain jointed and jointed reinforced	3.2	3.8–4.4	2.5–3.1	3.6–4.2
CRCP	2.9–3.2	NA	2.3–2.9	NA

Source: AASTO. 1993 AASHTO Guides for Design of Pavement Structures.
Copyright 1993 by the American Association of State Highway and Transportation Officials, Washington, DC. Used by permission.

materials. It was developed by identifying the forces and material properties responsible for the failure mechanisms, and relating theoretically computed values of relevant stresses, deflections and applied forces to appropriate pavement performance criteria. The relationships adopted for design were derived from data of (i) major road test programs, (ii) model and full-scale tests, and (iii) performance of normally constructed pavements subject to normal mixed traffic.

Traffic loading data are gathered and compiled in terms of axle-load distribution as described in Section 9.3. Each axle load is further multiplied by a load safety factor (LSF) according to the following recommendations: (i) LSF = 1.2 for interstate highways and other multilane projects with uninterrupted traffic flow and high volumes of truck traffic; (ii) LSF = 1.1 for highways and arterial streets with moderate volumes of truck traffic; and (iii) LSF = 1.0 for roads, residential streets and other streets with small volumes of truck traffic. The flexural strength of concrete is determined by the 28-day modulus of rupture from third-point loading according to ASTM test method C78 (ASTM, 2003a). Subgrade and subbase support is defined in terms of the modulus of subgrade reaction *k*.

9.9.2.1 Fatigue Design

Fatigue design for slab thickness is performed with the aim to control fatigue cracking. The design is based on the most critical edge stress, with the applied load positioned at the mid-length of the outer edge as shown in Figure 9.16. The computed slab thickness is the same for JRCP, JPCP with either doweled or undoweled joints, and also for CRCP. The presence of tied concrete shoulder, however, must be considered since it significantly reduces the critical edge stress. The design analysis is based on the concept of cumulative damage given by

$$D = \sum_{i=1}^{m} \frac{n_i}{N_i} \qquad (9.26)$$

where *m* is the total number of axle load groups, n_i is the predicted number of repetitions for the *i*th load group, and N_i is the allowable number of repetitions for the ith load group.

TABLE 9.14 Values of Drainage Coefficient C_d

Quality of Drainage	Percent of Time Pavement Structure Is Exposed to Moisture Levels Approaching Saturation			
	Less than 1%	1–5%	5–25%	Greater than 25%
Excellent	1.25–1.20	1.20–1.15	1.15–1.10	1.10
Good	1.20–1.15	1.15–1.10	1.10–1.00	1.00
Fair	1.15–1.10	1.10–1.00	1.00–0.90	0.90
Poor	1.10–1.00	1.00–0.90	0.90–0.80	0.80
Very poor	1.00–0.90	0.90–0.80	0.80–0.70	0.70

Source: AASHTO. 1993 AASHTO Guides for Design of Pavement Structures.
Copyright 1993 by the American Association of State Highway and Transportation Officials, Washington, DC. Used by permission.

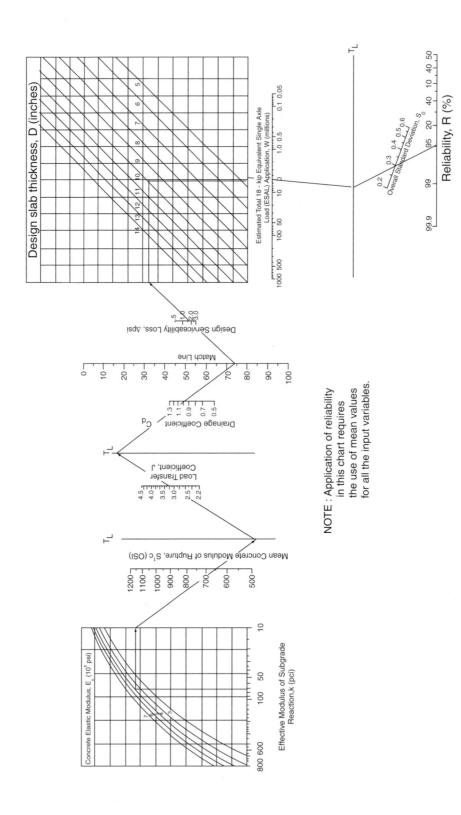

FIGURE 9.15 Rigid pavement thickness design chart. *Source:* AASHTO, 1993. AASHTO Guides for Design of Pavement Structures. Copyright 1993 by the American Association of State Highway and Transportation Officials, Washington, DC. Used by permission.

The steps in the fatigue design procedure are:

1. Multiply the load of each design axle load group by the appropriate LSF.
2. Assume a trial slab thickness.
3. Knowing the modulus of subgrade reaction k, obtain from Table 9.15 the equivalent stress for the projected slab

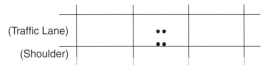

(Traffic Lane)

(Shoulder)

FIGURE 9.16 Critical loading position for fatigue analysis.

TABLE 9.15 Equivalent Stress for Fatigue Analysis

Slab Thickness (in.)	k of Subgrade–Subbase (pci)						
	50	100	150	200	300	500	700
(a) Equivalent Stress — no concrete shoulder (single axle and tandem axle)							
4	825/679	726/585	671/542	634/516	584/486	523/457	484/443
4.5	699/586	616/500	571/460	540/435	498/406	448/378	417/363
5	602/516	531/436	493/399	467/374	432/349	390/321	363/307
5.5	526/461	464/387	431/353	409/331	379/305	343/278	320/264
6	465/416	411/348	382/316	362/296	336/271	304/246	285/232
6.5	417/380	367/317	341/286	324/267	300/244	273/220	256/207
7	375/349	331/290	307/262	292/244	271/222	246/199	231/186
7.5	340/323	300/268	279/241	265/224	246/203	224/181	210/169
8	311/300	274/249	255/223	242/208	225/188	205/167	192/155
8.5	285/281	252/232	234/208	222/193	206/174	188/154	177/143
9	264/264	232/218	216/195	205/181	190/163	174/144	163/133
9.5	245/248	215/205	200/183	190/170	176/153	161/134	151/124
10	228/235	200/193	186/173	177/160	164/144	150/126	141/117
10.5	213/222	187/183	174/164	165/151	153/136	140/119	132/110
11	200/211	175/174	163/155	154/143	144/129	131/113	123/104
11.5	188/201	165/165	153/148	145/136	135/122	123/107	116/98
12	177/192	155/158	144/141	137/130	127/116	116/102	109/93
12.5	168/183	147/151	136/135	129/124	120/111	109/97	103/89
13	159/176	139/144	129/129	122/119	113/106	103/93	97/85
13.5	152/168	132/138	122/123	116/114	107/102	98/89	92/81
14	144/162	125/133	116/118	110/109	102/98	93/85	88/78
(b) Equivalent stress — concrete shoulder (single axle and tandem axle)							
4	640/534	559/468	517/439	489/422	452/403	409/388	383/384
4.5	547/461	479/400	444/372	421/356	390/338	355/322	333/316
5	475/404	417/349	387/323	367/308	341/290	311/274	294/267
5.5	418/360	368/309	342/285	324/271	302/254	276/238	261/231
6	372/325	327/277	304/255	289/241	270/225	247/210	234/203
6.5	334/295	294/251	274/230	260/218	243/203	223/188	212/180
7	302/270	266/230	248/210	236/198	220/184	203/170	192/162
7.5	275/250	243/211	226/193	215/182	201/168	185/155	176/148
8	252/232	222/196	207/179	197/168	185/155	170/142	162/135
8.5	232/216	205/182	191/166	182/156	170/144	157/131	150/125
9	215/202	190/171	177/155	169/146	158/134	146/122	139/116
9.5	200/190	176/160	164/146	157/137	147/126	136/114	129/108
10	186/179	164/151	153/137	146/129	137/118	127/107	121/101
10.5	174/170	154/143	144/130	137/121	128/111	119/101	113/95
11	164/161	144/135	135/123	129/115	120/105	112/95	106/90
11.5	154/153	136/128	127/117	121/109	113/100	105/90	100/85
12	145/146	128/122	120/111	114/104	107/95	99/86	95/81
12.5	137/139	121/117	113/106	108/99	101/91	94/82	90/77
13	130/133	115/112	107/101	102/95	96/86	89/78	85/73
13.5	124/127	109/107	102/97	97/91	91/83	85/74	81/70
14	118/122	104/103	97/93	93/87	87/79	81/71	77/67

Source: Portland Cement Association. 1984. Thickness Design for Concrete Highway and Street Pavements. With permission.

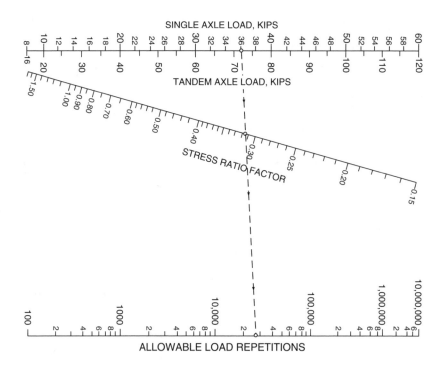

FIGURE 9.17 Allowable repetitions for fatigue analysis. *Source*: Portland Cement Association. 1984. Thickness Design for Concrete Highway and Street Pavements. p. 15. With permission.

thickness, and calculate the stress ratio factor as

$$\text{Stress Ratio Factor} = \frac{\text{Equivalent Stress}}{\text{Modulus of Rupture}} \qquad (9.27)$$

4. For each axle load i, obtain from Figure 9.17 the allowable load repetitions N_i.
5. Compute the cumulative damage D from Equation 9.26. If D exceeds one, select a larger trial thickness and repeat steps 3 through 5. The trial thickness is adequate against fatigue failure if D is less than or equal to one.

9.9.2.2 Erosion Design

The PCA erosion design for pavement thickness design is to guard against foundation and shoulder erosion, pumping and faulting. The critical deflection considered is at the corner, as shown in Figure 9.18. The presence of shoulder and the type of joint construction will affect the thickness design. The concept of cumulative damage as defined by Equation 9.26 is again applied. The steps are:

1. Multiply the load of each design axle load group by LSF.
2. Assume a trial slab thickness.
3. Obtain from Table 9.16 the erosion factor for the projected slab thickness and the modulus of subgrade reaction.

(Traffic Lane)

(Shoulder)

FIGURE 9.18 Critical loading position for erosion analysis.

TABLE 9.16 Erosion Factor for Erosion Analysis

Slab Thickness (in.)	k of Subgrade–Subbase (pci)					
	50	100	200	300	500	700
(a) Erosion factors — doweled joints, no concrete shoulder (single axle/tandem axle)						
4	3.74/3.83	3.73/3.79	3.72/3.75	3.71/3.73	3.70/3.70	3.68/3.67
4.5	3.59/3.70	3.57/3.65	3.56/3.61	3.55/3.58	3.54/3.55	3.52/3.53
5	3.45/3.58	3.43/3.52	3.42/3.48	3.41/3.45	3.40/3.42	3.38/3.40
5.5	3.33/3.47	3.31/3.41	3.29/3.36	3.28/3.33	3.27/3.30	3.26/3.28
6	3.22/3.38	3.19/3.31	3.18/3.26	3.17/3.23	3.15/3.20	3.14/3.17
6.5	3.11/3.29	3.09/3.22	3.07/3.16	3.06/3.13	3.05/3.10	3.03/3.07
7	3.02/3.21	2.99/3.14	2.97/3.08	2.96/3.05	2.95/3.01	2.94/2.98
7.5	2.93/3.14	2.91/3.06	2.88/3.00	2.87/2.97	2.86/2.93	2.84/2.90
8	2.85/3.07	2.82/2.99	2.80/2.93	2.79/2.89	2.77/2.85	2.76/2.82
8.5	2.77/3.01	2.74/2.93	2.72/2.86	2.71/2.82	2.69/2.78	2.68/2.75
9	2.70/2.96	2.67/2.87	2.65/2.80	2.63/2.76	2.62/2.71	2.61/2.68
9.5	2.63/2.90	2.60/2.81	2.58/2.74	2.56/2.70	2.55/2.65	2.54/2.62
10	2.56/2.85	2.54/2.76	2.51/2.68	2.50/2.64	2.48/2.59	2.47/2.56
10.5	2.50/2.81	2.47/2.71	2.45/2.63	2.44/2.59	2.42/2.54	2.41/2.51
11	2.44/2.76	2.42/2.67	2.39/2.58	2.38/2.54	2.36/2.49	2.35/2.45
11.5	2.38/2.72	2.36/2.62	2.33/2.54	2.32/2.49	2.30/2.44	2.29/2.40
12	2.33/2.68	2.30/2.58	2.28/2.49	2.26/2.44	2.25/2.39	2.23/2.36
12.5	2.28/2.64	2.25/2.54	2.23/2.45	2.21/2.40	2.19/2.35	2.18/2.31
13	2.23/2.61	2.20/2.50	2.18/2.41	2.16/2.36	2.14/2.30	2.13/2.27
13.5	2.18/2.57	2.15/2.47	2.13/2.37	2.11/2.32	2.09/2.26	2.08/2.23
14	2.13/2.54	2.11/2.43	2.08/2.34	2.07/2.29	2.05/2.23	2.03/2.19
(b) Erosion factors — aggregate-interlock joints, no concrete shoulder (single axle and tandem axle)						
4	3.94/4.03	3.91/3.95	3.88/3.89	3.86/3.86	3.82/3.83	3.77/3.80
4.5	3.79/3.91	3.76/3.82	3.73/3.75	3.71/3.72	3.68/3.68	3.64/3.65
5	3.66/3.81	3.63/3.72	3.60/3.64	3.58/3.60	3.55/3.55	3.52/3.52
5.5	3.54/3.72	3.51/3.62	3.48/3.53	3.46/3.49	3.43/3.44	3.41/3.40
6	3.44/3.64	3.40/3.53	3.37/3.44	3.35/3.40	3.32/3.34	3.30/3.30
6.5	3.34/3.56	3.30/3.46	3.26/3.36	3.25/3.31	3.22/3.25	3.20/3.21
7	3.26/3.49	3.21/3.39	3.17/3.29	3.15/3.24	3.13/3.17	3.11/3.13
7.5	3.18/3.43	3.13/3.32	3.09/3.22	3.07/3.17	3.04/3.10	3.02/3.06
8	3.11/3.37	3.05/3.26	3.01/3.16	2.99/3.10	2.96/3.03	2.94/2.99
8.5	3.04/3.32	2.98/3.21	2.93/3.10	2.91/3.04	2.88/2.97	2.87/2.93
9	2.98/3.27	2.91/3.16	2.86/3.05	2.84/2.99	2.81/2.92	2.79/2.87
9.5	2.92/3.22	2.85/3.11	2.80/3.00	2.77/2.94	2.75/2.86	2.73/2.81
10	2.86/3.18	2.79/3.06	2.74/2.95	2.71/2.89	2.68/2.81	2.66/2.76
10.5	2.81/3.14	2.74/3.02	2.68/2.91	2.65/2.84	2.62/2.76	2.60/2.72
11	2.77/3.10	2.69/2.98	2.63/2.86	2.60/2.80	2.57/2.72	2.54/2.67
11.5	2.72/3.06	2.64/2.94	2.58/2.82	2.55/2.76	2.51/2.68	2.49/2.63
12	2.68/3.03	2.60/2.90	2.53/2.78	2.50/2.72	2.46/2.64	2.44/2.59
12.5	2.64/2.99	2.55/2.87	2.48/2.75	2.45/2.68	2.41/2.60	2.39/2.55
13	2.60/2.96	2.51/2.83	2.44/2.71	2.40/2.65	2.36/2.56	2.34/2.51
13.5	2.56/2.93	2.47/2.80	2.40/2.68	2.36/2.61	2.32/2.53	2.30/2.48
14	2.53/2.90	2.44/2.77	2.36/2.65	2.32/2.58	2.28/2.50	2.25/2.44
(c) Erosion factors — doweled joints, concrete shoulder (single axle and tandem axle)						
4	3.28/3.30	3.24/3.20	3.21/3.13	3.19/3.10	3.15/3.09	3.12/3.08
4.5	3.13/3.19	3.09/3.08	3.06/3.00	3.04/2.96	3.01/2.93	2.98/2.91
5	3.01/3.09	2.97/2.98	2.93/2.89	2.90/2.84	2.87/2.79	2.85/2.77
5.5	2.90/3.01	2.85/2.89	2.81/2.79	2.79/2.74	2.76/2.68	2.73/2.65
6	2.79/2.93	2.75/2.82	2.70/2.71	2.68/2.65	2.65/2.58	2.62/2.54
6.5	2.70/2.86	2.65/2.75	2.61/2.63	2.58/2.57	2.55/2.50	2.52/2.45
7	2.61/2.79	2.56/2.68	2.52/2.56	2.49/2.50	2.46/2.42	2.43/2.38
7.5	2.53/2.73	2.48/2.62	2.44/2.50	2.41/2.44	2.38/2.36	2.35/2.31
8	2.46/2.68	2.41/2.56	2.36/2.44	2.33/2.38	2.30/2.30	2.27/2.24
8.5	2.39/2.62	2.34/2.51	2.29/2.39	2.26/2.32	2.22/2.24	2.20/2.18
9	2.32/2.57	2.27/2.46	2.22/2.34	2.19/2.27	2.16/2.19	2.13/2.13

(Continued)

TABLE 9.16   Continued

Slab Thickness (in.)	k of Subgrade–Subbase (pci)					
	50	100	200	300	500	700
9.5	2.26/2.52	2.21/2.41	2.16/2.29	2.13/2.22	2.09/2.14	2.07/2.08
10	2.20/2.47	2.15/2.36	2.10/2.25	2.07/2.18	2.03/2.09	2.01/2.03
10.5	2.15/2.43	2.09/2.32	2.04/2.20	2.01/2.14	1.97/2.05	1.95/1.99
11	2.10/2.39	2.04/2.28	1.99/2.16	1.95/2.09	1.92/2.01	1.89/1.95
11.5	2.05/2.35	1.99/2.24	1.93/2.12	1.90/2.05	1.87/1.97	1.84/1.91
12	2.00/2.31	1.94/2.20	1.88/2.09	1.85/2.02	1.82/1.93	1.79/1.87
12.5	1.95/2.27	1.89/2.16	1.84/2.05	1.81/1.98	1.77/1.89	1.74/1.84
13	1.91/2.23	1.85/2.13	1.79/2.01	1.76/1.95	1.72/1.86	1.70/1.80
13.5	1.86/2.20	1.81/2.09	1.75/1.98	1.72/1.91	1.68/1.83	1.65/1.77
14	1.82/2.17	1.76/2.06	1.71/1.95	1.67/1.88	1.64/1.80	1.61/1.74
(d) Erosion factors — aggregate-interlock joints, concrete shoulder (single axle and tandem axle)						
4	3.46/3.49	3.42/3.39	3.38/3.32	3.36/3.29	3.32/3.26	3.28/3.24
4.5	3.32/3.39	3.28/3.28	3.24/3.19	3.22/3.16	3.19/3.12	3.15/3.09
5	3.20/3.30	3.16/3.18	3.12/3.09	3.10/3.05	3.07/3.00	3.04/2.97
5.5	3.10/3.22	3.05/3.10	3.01/3.00	2.99/2.95	2.96/2.90	2.93/2.86
6	3.00/3.15	2.95/3.02	2.90/2.92	2.88/2.87	2.86/2.81	2.83/2.77
6.5	2.91/3.08	2.86/2.96	2.81/2.85	2.79/2.79	2.76/2.73	2.74/2.68
7	2.83/3.02	2.77/2.90	2.73/2.78	2.70/2.72	2.68/2.66	2.65/2.61
7.5	2.76/2.97	2.70/2.84	2.65/2.72	2.62/2.66	2.60/2.59	2.57/2.54
8	2.69/2.92	2.63/2.79	2.57/2.67	2.55/2.61	2.52/2.53	2.50/2.48
8.5	2.63/2.88	2.56/2.74	2.51/2.62	2.48/2.55	2.45/2.48	2.43/2.43
9	2.57/2.83	2.50/2.70	2.44/2.57	2.42/2.51	2.39/2.43	2.36/2.38
9.5	2.51/2.79	2.44/2.65	2.38/2.53	2.36/2.46	2.33/2.38	2.30/2.33
10	2.46/2.75	2.39/2.61	2.33/2.49	2.30/2.42	2.27/2.34	2.24/2.28
10.5	2.41/2.72	2.33/2.58	2.27/2.45	2.24/2.38	2.21/2.30	2.19/2.24
11	2.36/2.68	2.28/2.54	2.22/2.41	2.19/2.34	2.16/2.26	2.14/2.20
11.5	2.32/2.65	2.24/2.51	2.17/2.38	2.14/2.31	2.11/2.22	2.09/2.16
12	2.28/2.62	2.19/2.48	2.13/2.34	2.10/2.27	2.06/2.19	2.04/2.13
12.5	2.24/2.59	2.15/2.45	2.09/2.31	2.05/2.24	2.02/2.15	1.99/2.10
13	2.20/2.56	2.11/2.42	2.04/2.28	2.01/2.21	1.98/2.12	1.95/2.06
13.5	2.16/2.53	2.08/2.39	2.00/2.25	1.97/2.18	1.93/2.09	1.91/2.03
14	2.13/2.51	2.04/2.36	1.97/2.23	1.93/2.15	1.89/2.06	1.87/2.00

Source: Portland Cement Association. 1984. Thickness Design for Concrete Highway and Street Pavements. With permission.

4. For each axle load i, obtain from Figure 9.19 the allowable load repetitions N_i.
5. Compute D from Equation 9.26. If D exceeds one, select a greater trial thickness and repeat steps 3 through 5. The trial thickness is adequate if D is less than or equal to one.

The final design thickness of the pavement slab is equal to the larger of the thickness obtained from the fatigue and erosion analysis, respectively.

Example 9.7

Determine the required slab thickness for an expressway with the design traffic shown in Table 9.6. The pavement is to be constructed with doweled joint, but without concrete shoulder. Concrete modulus of rupture is 650 lb/in.^{2} The modulus of subgrade reaction k is 130 pci (Table 9.16).

Trial and error approach is needed by assuming slab thickness. Solution is shown only for slab thickness $h = 9.5$ in. For expressway LSF $= 1.2$. The design load is equal to (1.2 × axle load). From Table 9.15, equivalent stress for single axle is 206 and for tandem axle 192. The corresponding stress ratios are 0.317 and 0.295. N_1 for fatigue analysis is obtained from Figure 9.17. From Table 9.16, erosion factor is 2.6 for single axle and 2.8 for tandem axle. N_2 for erosion analysis is obtained from Figure 9.19. The results show that the design is satisfactory.

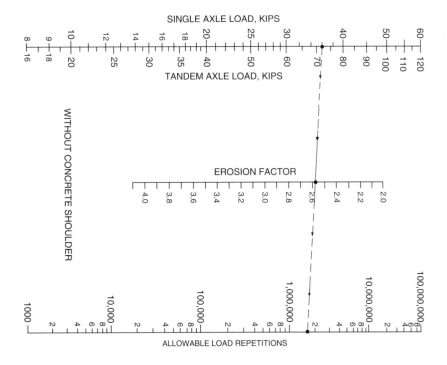

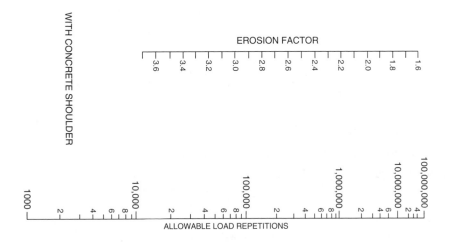

FIGURE 9.19 Allowable repetitions for erosion analysis. *Source*: Portland Cement Association. 1984. Thickness Design for Concrete Highway and Street Pavements. With permission.

9.10   Reinforcement Design of Rigid Pavement

The main function of steel reinforcements in concrete pavement is crack-width control. They are designed to hold cracks tightly closed so that the pavement would remain as an integral structural unit. The amount of reinforcement required is a function of slab length (or joint spacing) and thermal properties of the pavement material. Reinforcements are not required in jointed plain concrete pavements (JPCP) whose lengths are relatively short. The crack-free length defined in Section 9.7.3 serves as a useful guide of the maximum panel length permitted. The AASHTO Guide suggested that the joint spacing (in feet) for JPCP should not greatly exceed twice the slab thickness (in inches), and the ratio of slab width to length should not exceed 1.25. The AASHTO procedure of reinforcement design for JRCP (jointed reinforced concrete pavement) and CRCP (continuously reinforced concrete pavement) are presented in this section.

9.10.1   Reinforcement Design for JRCP

The amount of steel reinforcement required in JRCP can be computed as that needed to resist the tensile stress given by Equation 9.22. Hence, the percent steel reinforcement (either longitudinal or transverse reinforcement) can be calculated as

$$P_s = \frac{\mu_r \gamma_c L}{2 f_s} \times 100\% \tag{9.28}$$

where L = slab length, μ_r = friction factor between the bottom of slab and the top of underlying subbase or subgrade, γ_c = density of concrete slab, and f_s = allowable working stress of steel reinforcement. The AASHTO recommended values for μ_r are given in Table 9.11. The allowable steel working stress is equal to 75% of the steel yield strength. For Grade 40 and Grade 60 steel, f_s is equal to 30,000 lb/in.^{2} (210 MPa) and 45,000 lb/in.^{2} (315 MPa) respectively. For Welded Wire Fabric, f_s is 48,750 lb/in.^{2} (341 MPa).

Example 9.8

Determine the longitudinal steel reinforcement requirement for a 30 ft. (9.14 m) long JRCP constructed on crushed stone subbase.

Assume the density of the concrete slab γ_c = 0.09 pci (2,500 kg/m^3) and the allowable working stress of steel reinforcement f_s = 30,000 lb/in.^{2} (210 MPa). From Table 9.11, μ_r = 1.5. Percent steel reinforcement

$$P_s = \frac{30 \times 12 \times 1.5 \times 0.09}{2 \times 30,000} = 0.081\%$$

9.10.2   Reinforcement Design for CRCP

9.10.2.1   Longitudinal Reinforcement Design for CRCP

The AASHTO procedure of longitudinal reinforcement design for CRCP is an elaborate process. The steel reinforcements are designed to satisfy limiting criteria in the following three aspects: (a) crack spacing, (b) crack width, and (c) steel stress.

9.10.2.1.1   *Limiting percent reinforcements based on crack spacing control —* $(P_{max})_1$ *and* $(P_{min})_1$

The design requires that crack spacing be kept between 3.5 ft (1.1 m) and 8 ft (2.4 m). The lower limit minimizes punch-out and the upper limit minimizes edge spalling of cracks. Figure 9.20 is used to determine the percent reinforcement P required for the specified crack spacings, resulting in two values that define the range of acceptable percent reinforcement: $(P_{max})_1$ and $(P_{min})_1$.

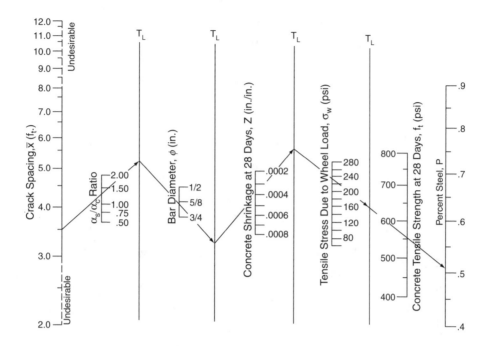

FIGURE 9.20 Minimum percent reinforcement to satisfy crack spacing criteria. *Source*: AASHTO, 1993. AASHTO Guides for Design of Pavement Structures. Copyright 1993 by the American Association of State Highway and Transportation Officials, Washington, DC. Used by permission.

The input variables for determining the required steel reinforcement include the coefficient of thermal expansion of concrete α_c, the thermal coefficient of steel α_s, diameter of reinforcing bar, concrete shrinkage Z at 28 days, tensile stress σ_w due to wheel load, and concrete tensile strength f_t at 28 day. The common values of α_c and Z are given in Table 9.17. A value of $\alpha_s = 5.0 \times 10^{-6}$ in./in./°F $(9.0 \times 10^{-6}$ mm/mm/°C) may be used. Steel bars of 5/8 and 3/4 in. (16 and 20 mm) diameter are

TABLE 9.17 Common Values of Coefficient of Thermal Expansion and Coefficient of Drying Shrinkage for Concrete

(a) *Approximate Relations between Shrinkage and Indirect Tensile Strength of Portland Cement Concrete*	
Indirect Tensile Strength (lb/in.^{2})	Shrinkage (in./in.)
300 (or less)	0.0008
400	0.0006
500	0.00045
600	0.0003
700 (or greater)	0.0002

(b) *Recommended value of the thermal coefficient of concrete as a function of aggregate types*	
Type of Coarse Aggregate	Concrete Thermal Coefficient $(10^{-6}/°F)$
Quartz	6.6
Sandstone	6.5
Gravel	6.0
Granite	5.3
Basalt	4.8
Limestone	3.8

Source: AASTO. 1993 AASHTO Guides for Design of Pavement Structures.
Copyright 1993 by the American Association of State Highway and Transportation Officials, Washington, DC. Used by permission.

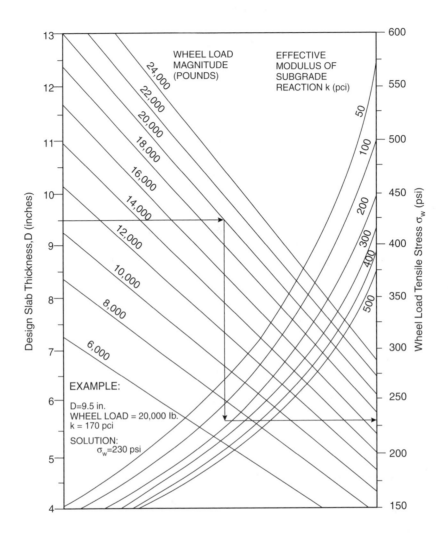

FIGURE 9.21   Chart for estimating wheel load tensile stress σw.

typically used and the 3/4 in. (20 mm) bar is the largest practical size for crack-width control and bond requirements. σ_w is determined using Figure 9.21 based on the design slab thickness, the magnitude of the wheel load, and the effective modulus of subgrade reaction.

9.10.2.1.2   *Limiting percent reinforcements based on crack width — (P_{min})2*

Steel reinforcements are required in CRCP to control crack width to within 0.04 in. (1.0 mm) to prevent spalling and water infiltration. The minimum percent steel needed, $(P_{min})_1$, to meet this requirement is obtained from Figure 9.22 with a selected bar size and input variables σ_w and f_t.

9.10.2.1.3   *Limiting percent reinforcements based on steel stress — (P_{min})3*

There is also the need to provide sufficient amount of steel to guard against steel fracture and excessive permanent deformation. AASHTO recommends a limiting value equal to 75% of the ultimate tensile strength. The minimum percent steel requirement, $(P_{min})_3$ is obtained from Figure 9.23. The determination of $(P_{min})_3$ requires the choice of allowable steel working stress (see Table 9.18 for Grade 60 steel) and the computation of a design temperature drop given by

$$\Delta T = T_H - T_L \tag{9.29}$$

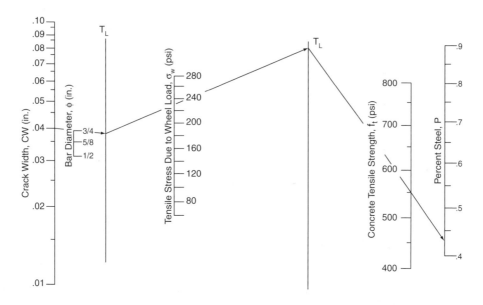

FIGURE 9.22 Minimum percent steel reinforcement to satisfy crack width criterion. *Source*: AASHTO, 1993. AASHTO Guides for Design of Pavement Structures. Copyright 1993 by the American Association of State Highway and Transportation Officials, Washington, DC. Used by permission.

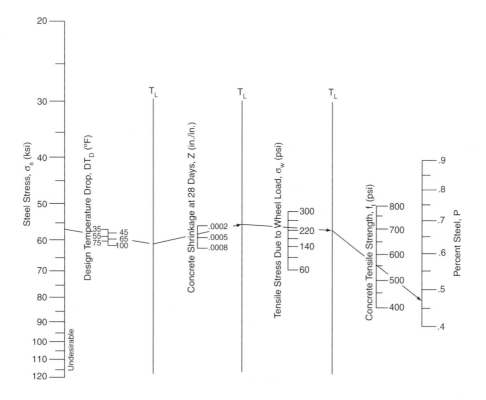

FIGURE 9.23 Minimum percent reinforcement to satisfy steel stress criteria. *Source*: AASHTO, 1993. AASHTO Guides for Design of Pavement Structures. Copyright 1993 by the American Association of State Highway and Transportation Officials, Washington, DC. Used by permission.

TABLE 9.18 Allowable Working Stress for Grade 60 Steel

Indirect Tensile Strength of Concrete at 28 days, lb/in.^{2}	Reinforcing Bar Size		
	No. 4	No. 5	No. 6
300 (or less)	65	57	54
400	67	60	55
500	67	61	56
600	67	63	58
700	67	65	59
800 (or greater)	67	67	60

Source: AASTO. 1993 AASHTO Guides for Design of Pavement Structures.
Copyright 1993 by the American Association of State Highway and Transportation Officials, Washington, DC. Used by permission.

where T_H is the average daily temperature during the month the pavement is constructed, and T_L is the average daily low temperature during the coldest month of the year.

The design percent longitudinal steel should fall within $(P_{max})_1$ and (P_{min}), the largest value of $(P_{min})_1$, $(P_{min})_2$ and $(P_{min})_3$. If $(P_{max})_1$ is less than (P_{min}), a revised design by changing some of the input parameters is required. If $(P_{max})_1$ is larger than (P_{min}), the number of reinforcing bars or wires required, N, is a value between N_{min} and N_{max} where

$$N_{min} = \frac{P_{min}\cdot(\text{Slab Thickness})\cdot(\text{Slab Width})}{100\cdot(\pi/4)\phi^2} \tag{9.30a}$$

$$N_{max} = \frac{P_{max}\cdot(\text{Slab Thickness})\cdot(\text{Slab Width})}{100\cdot(\pi/4)\phi^2} \tag{9.30b}$$

where ϕ is the diameter of reinforcing bar.

9.10.2.2 Transverse Reinforcement Design for CRCP

The procedure of reinforcement design for JRCP described under Section 9.10.1 is applicable for the design of transverse steel reinforcement for CRCP. The slab length L in Equation 9.28 now refers to the distance between free longitudinal edges. A longitudinal joint with tie bar is not considered a free edge.

Example 9.9

Longitudinal steel reinforcement design for CRCP. Design data: $D = 9$ in., coefficient of thermal expansion $\alpha_c = 5.3 \times 10^{-6}/F^0$ for concrete with granite coarse aggregate (see Table 9.17), indirect tensile strength of concrete, $f_t = 600$ lb/in.^{2} at 28 days, $Z = 0.0003$ (see Table 9.17), maximum construction wheel load = 18,000 lb, effective $k = 100$ pci, design temperature drop = 55°F.

Try steel bar diameter 0.5 in., from Figure 9.21, wheel load tensile stress, $\sigma_w = 240$. With $(\alpha_s/\alpha_c = 5.0/5.3 = 0.94)$, obtain $(P_{min})_1 = 0.4\%$ for $X = 8$ ft, and $(P_{max})_1 = 0.49\%$ for $X = 3.5$ ft from Figure 9.20. (b) Crack width control. Obtain $(P_{min})_2 = 0.4\%$ from Figure 9.22. (c) Steel stress control, $\sigma_s = 67$ psi from Table 9.18, obtain $(P_{min})_3 = 0.45\%$. Hence, overall $(P_{min}) = 0.4\%$ and $(P_{max}) = 0.49\%$ for a pavement width of 12 ft, apply Equations (9.30a) and (9.30b), $N_{min} = 29.7$ and $N_{max} = 32.3$. Use 30 numbers of 0.5 in. bars.

9.11 Joints and Load Transfer Design

9.11.1 Joint and Joint Spacing Design

The main kinds of joints commonly found in rigid pavements are contraction, warping, expansion, isolation, and construction joints. These different types of joints, with their typical forms of construction

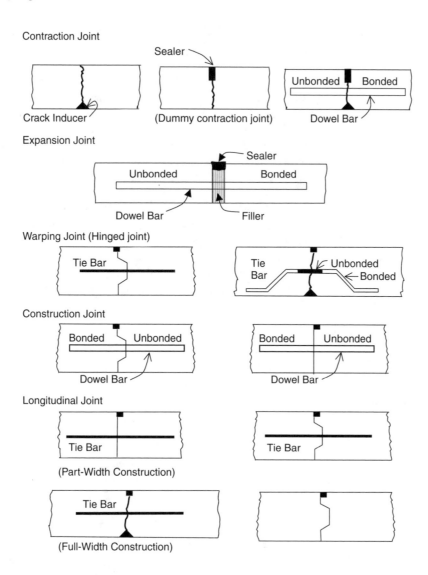

FIGURE 9.24 Types of joints in concrete pavements.

details shown in Figure 9.24, are provided in a concrete pavement system to serve various functions while maintaining certain degree of structural continuity (with the exception of isolation joints) through provisions for some form of load transfer. In addition, it is important to ensure that all joints must be sealed against water seepage and to exclude grit and debris. They must also not adversely affect the riding quality of the pavement surface, and not cause much interference in the process of pavement construction.

9.11.1.1 Contraction Joints

Contraction joints are transverse or longitudinal joints constructed to reduce slab length or width as a means to decrease temperature induced stresses (warping stresses as well as tensile stress caused by slab movements), control spacing and width of cracks caused by thermal forces or moisture changes, and to allow for thermal contraction of the concrete slabs. The maximum acceptable spacing of contraction joints is a function of slab thickness, amount of steel reinforcement provided, tensile strength of the concrete, thermal coefficient of thermal expansion of the concrete, magnitude of temperature change in the area, and frictional resistance at the bottom face of the pavement. Higher values of the first two factors will allow the joint spacing to increase, while any increase in the latter three factors will call for smaller joint spacing.

As a rough guide, AASHTO (1993) recommended that the spacing (in feet) of contraction joint for JPCP should not greatly exceed twice the slab thickness (in inches), and that the ratio of slab width to length should not exceed 1.25. For example, this means a joint spacing of not more than 20 ft (6.1 m) for a 10 in. (250 mm) thick slab. The FHWA (1990) recommends that the L/ℓ ratio (slab length divided by radius of relative stiffness) must not exceed 5.0 when determining the maximum slab length. The U.K. practice (Road Research Laboratory, 1970) adopted 5 m joint spacing for JPCP, and 10 m or more for JRCP following a linear relationship between joint spacing and the amount of steel reinforcement provided. According to this linear relationship, the required amounts of steel reinforcement for joint spacings of 35 m, 21 m and 10 m are 5.55, 3.41 and 1.73 kg per m^3 of slab volume, respectively.

A contraction joint is typically constructed by installing a sawed, formed or tooled groove of sufficient depth such that a full depth crack will form upon contraction of the concrete slab, and hence the name "dummy" contraction joint is sometimes used. AASHTO (1993) suggests that the depth of the groove should be 1/4 of the slab thickness for transverse contraction joints, and 1/3 of the slab thickness for longitudinal contraction joints. The width of the groove is dependent on the properties of the sealant used as described in Section 9.11.2.

9.11.1.2 Warping Joints

Warping joints, also known as hinge joints, are transverse or longitudinal breaks formed in the pavement slab to allow for a small amount of angular movement. The purpose is to relieve high tensile stresses due to restrained warping of the slab. Warping joints are also constructed by forming surface grooves. However, unlike contraction joints, opening of the joints due to contraction of concrete is prevented by tie bars or reinforcements.

9.11.1.3 Expansion Joints

Expansion joints are full-depth transverse or longitudinal joints designed to allow for concrete slab expansion, thereby relieving compressive stresses due to temperature rise, and avoiding blowups on hot summer days. However, experience in the United States and the United Kingdom (U.K.) had shown that expansion joints progressively closed during service. This has caused the nearby contraction joints to open up, resulting in loss of aggregate interlock and hence reduction of load-transfer efficiency. This in turn led to a host of joint related distresses, including faulting, corner breaks, edge spalling, sealant failures, and pumping. AASHTO (1993) states that the use of expansion joints is minimized due to cost, complexity and performance problem. Huang (1993) indicated that blowups of concrete pavements are related to a certain source and type of coarse aggregate, and their occurrences can be minimized if precaution is exercised in selecting the aggregates. He also reasoned that the plastic flow of concrete can gradually relieve the compressive stress, and concluded that it is not necessary to install the expansion joint except at bridge ends. ACPA (1992) recommended that expansion joints should not be used in concrete pavements built with normal aggregates under normal temperatures with contraction joints spaced less than 60 ft (18 m), and that they are only needed in (a) pavements with panel length of 60 ft (18 m) or more without contraction joints in-between, (b) pavements constructed when the ambient temperature is below 4°C (40°F), (c) pavements where contraction joints are allowed to be infiltrated by large incompressible materials, and (d) pavements constructed of materials with high expansion characteristics.

In the U.K, although the Road Note 29 (Road Research Laboratory, 1970) permits the omission of expansion joints in pavements constructed during the summer months, researchers and practitioners have cautioned against such practice in view of compression failures in abnormally hot weather (MOT, 1978; Croney and Croney, 1991). Road note 29 recommends that for JRCP, every third joint should be an expansion joint, and the rest contraction joints. For JPCP, expansion joints should be spaced at 60 m intervals for slabs thicker than 200 mm, and 40 m for thinner slabs. The gap width of the expansion joint is usually of the order of 20 to 25 mm (3/4 to 1 in.). The width required depends on the difference between the temperature at which the concrete is laid and the maximum service temperature, the coefficient of thermal expansion of the concrete, the frictional restraint provided by the subgrade or subbase, and the compressibility of the joint filling material.

9.11.1.4   Isolation Joints

Isolation joints are full depth joints provided to allow relative horizontal and vertical movements between a pavement and the adjacent structures with the purpose of avoiding damages to the pavement and the structures. This applies to structures such as bridge abutments, drainage inlets, manholes, footings, or other types of paved structures. Typical widths of isolation joints are similar to expansion joints.

9.11.1.5   Construction Joints

Construction joints are installed at the end of a day's placement or when casting of concrete slab is interrupted due to equipment breakdown or other reasons such as unfavorable weather conditions. Placement of concrete can be planned to have transverse construction joints coincide with either contraction or expansion joints. Longitudinal construction joints are usually positioned at slab edge to minimize their effect on riding quality.

9.11.2   Joint Sealant Design

The main function of joint sealant is to prevent water, grit and other incompressible objects from entering the joint. Sealant must be effective in sealing the joint, durable against weathering, and able to withstand repeated extension and compression. There are two types of sealant, namely (a) field-molded sealants which are applied in liquid or semi liquid form, and (b) premolded (or preformed) sealants which are manufactured products of specified shapes.

The groove formed in a joint, which serves as sealant reservoir, must be so shaped to achieve the optimum joint sealing effect. The sealant surface should be placed 1/8 to 1/2 in. (3 to 12 mm) below the surface of the pavement. For premolded sealant, appropriate dimensions of the sealant reservoir are specified by the manufacturers. For field-molded sealants, AASHTO (1993) recommended that the depth to width ratio of sealant should be within a range of 1 to 1.5, with a minimum depth of 3/4 and 1/2 in. (20 to 12 mm) for longitudinal and transverse joints, respectively. The required reservoir width W may be computed based on the estimated joint opening and the allowable sealant strain as follows:

$$W = \frac{C \cdot L \cdot (\alpha_c \cdot \Delta T + Z)}{S} \qquad (9.31)$$

where L is the joint spacing, α_c the coefficient of thermal expansion of concrete, ΔT the temperature change, Z the coefficient of drying shrinkage of concrete, S the allowable strain of joint sealant material, and C the adjustment factor for subbase and slab friction restraint. AASHTO (1993) recommended C values of 0.65 for stabilized subbase, and 0.80 for granular subbase.

9.11.3   Load-Transfer Design

The ability to effectively transfer vehicular loads across a joint from one slab to the next has an important impact on the long-term performance of a rigid pavement. Most joint related distresses and performance problems, such as faulting, pumping, and corner cracks can be attributed partly to inadequate load-transfer efficiency. These are caused by high slab stresses in the loaded edge of a joint when the next slab across the joint does not take up sufficient share of the load due to poor load transfer.

Shear and moment transfer can take place across a joint. Shear transfer across a joint can be achieved by one of the following means: (a) aggregate interlocking, (b) dowel bars, (c) keyed joints, and (d) tie bars. Shear transfer in contraction joints are realized through either aggregate interlock or dowel bars for load transfer. Warping joints rely on aggregate interlock, and expansion joints on dowel bars. Tie bars are commonly used in warping joints and longitudinal joints to promote shear transfer through aggregate interlock. Moment transfer may occur across a joint if it is closed. If a joint is not tightly held together, the transfer of moment across the joint is negligible.

Moment transfer across a joint is normally neglected, and the term load transfer commonly refers to shear transfer in joint design. There are several different basis upon which load-transfer efficiency can be defined, and the resulted numerical values may not be the same. A common form of definition adopted in practice is (Ioannides and Korovesis, 1992; Huang, 1993; Chou, 1995):

$$\text{Load-Transfer Effiiency (\%)} = \frac{\Delta_U}{\Delta_L} \times 100\% \qquad (9.32)$$

where Δ_U is the deflection of the unloaded slab, and Δ_L is that of the loaded slab. The load-transfer efficiency is 0% if Δ_U is zero, and 100% if $\Delta_U = \Delta_L$. At efficiency of 100%, each of the two slabs carries half of the applied load.

9.11.3.1 Aggregate Interlock

Aggregate interlock refers to the mechanical locking between the two surfaces of a crack. Load transfer through aggregate interlock is usually acceptable only for secondary roads or residential streets with low traffic volume. Aggregate interlock is progressively weakened through wear and tear as more and heavier traffic passes over. It is thus inadequate for high volume roads or roads that carry truck traffic.

Aggregate interlock is lost once the crack opens up. To guard against this, connectors such as tie bars can be used to ensure that the joint will stay tightly closed. FHWA (1990) has indicated that aggregate interlock is ineffective in cracks wider than about 0.9 mm (0.035 in.).

9.11.3.2 Dowel Bars

Dowel bars are smooth steel bars installed across a transverse joint to provide a mechanical connection between slabs without restricting horizontal joint movements. Dowelled joints are required for major roads, main urban streets, industrial roads, airport and harbor pavements. The design of dowel bars is mostly based on experience. AASHTO (1993) and PCA (1991) recommended the use of dowel bars with diameter equal to 1/8 of the slab thickness. Most road agencies adopt the standard center-to-center spacing of 300 mm (12 in.) regardless of the size of dowel bars. The typical length of dowel bars is 460 mm (18 in.). Dowel bars are usually coated with a bond-breaking substance to prevent bonding to the concrete. Corrosion of dowel bars may cause joints to lock up. Epoxy coated and stainless steel dowels have been used to prevent corrosion.

9.11.3.3 Tie Bars

Tie bars are deformed steel bars installed in contraction or warping joints to hold the faces of two adjacent slabs tightly together, thereby promoting aggregate interlock. They are also used in keyed joints to ensure load transfer. They are commonly used in longitudinal joints to provide load transfer through aggregate interlock, and to prevent lanes from separating. Tie bars are typically 12.5 mm (1/2 in.) in diameter with lengths between 0.6 and 1.0 m (24 and 39 in.), and spaced at intervals of 0.75 to 1.1 m (30 to 43 in.). The amount of tie bars required is determined based on the same consideration as that of the reinforcement design for JRCP (see Section 9.10.1), by means of the following equation:

$$P_s = \frac{\mu_r \gamma_c L}{f_s} \times 100\% \qquad (9.33)$$

where all variables are as defined in Equation 9.28, except that L is now the distance from the joint to the free edge. For longitudinal joints, L is equal to the lane width of 2- or 3-lane pavements. In the case of 4-lane pavements, L equals to the lane width for the two outer joints, and two times the lane width for the inner joint.

The length of tie bars T_L is given by the required bond length plus 75 mm (3 in.) to allow for misalignment:

$$T_L = 2\left(\frac{A_s f_s}{f_b \Sigma}\right) + \delta = \left(\frac{f_s D}{f_b}\right) + \delta \qquad (9.34)$$

where A_s is the cross-sectional area of one tie bar, f_s the allowable tensile stress of the steel bar, f_b is the allowable bond stress, Σ the perimeter of the steel bar, D the diameter of the steel bar, and δ the allowance for misalignment.

9.11.3.4 Keyed Joint

A keyed joint, also known as a tongue-and-groove joint, is a joint constructed by forming a protruding rib on the edge of one slab to fit into a groove in the edge of the adjacent slab. Keyed joints have been used for longitudinal construction joints and warping joints. The key, or keyway, should be formed at mid depth with a height of about 0.2 to 1/3 of the slab thickness. Keyed joints are not suitable for slab thinner than 250 mm (10 in.) due to possible joint failures caused by reduced shear strength at the joint. FHWA (1990) stated that tie bars are necessary when using keyways.

9.12 Future Developments

An ideal design method for concrete pavements would be one that is mechanistically based to address various failure modes, as well as the overall pavement performance with respect to riding quality. Unfortunately, the current knowledge in the field of pavement engineering does not permit one to develop a fully mechanistic procedure for pavement design. A more realistic and practical near- and medium-term goal would be to adopt the so called mechanistic-empirical principles to improve the current design methodology incrementally. In other words, mechanistic-empirical design methods that make use of mechanistic concepts to explain specific pavement distress developments or failure modes can be adopted with the support of appropriate empirical relationships. The thickness design procedure of PCA described in Section 9.9.2 is a good example of mechanistic-empirical approach that addresses fatigue cracking failure and erosion failure of the supporting materials. Possible improvements to this design method include checking against other forms of failure, and prediction of riding quality performance. The AASHTO procedure of reinforcement design for CRCP is another good example in which the design is arrived at by controlling crack width, crack spacing and steel stress.

Shortly after publishing the 1993 AASHTO Guide, AASHTO launched an ambitious project with the aim to develop a new 2002 Guide for the design of new and rehabilitated pavement structures based on mechanistic-empirical principles. The development work was undertaken as NCHRP Project 1-37A entitled Development of the 2002 Guide for the Design of New and Rehabilitated Pavement Structures (Pierce, 2003). The project report was published in mid-2004 (ARA Inc., 2004). It is noted that the serviceability concept adopted in the AASHTO 1993 Guide is an empirical approach that cannot be related directly to the structural condition of the pavement, nor any specific distress type and its severity or extent. It is noted that the concept of Equivalent Single Axle Load (ESAL) does not lend itself to mechanistic analysis of the effect of new axle configurations and heavier axle loads of modern day traffic. The concepts of serviceability performance and ESAL together make extrapolation of the design procedure to thicker pavements and their application to pavements of new materials questionable.

The proposed design procedures recommended by the NCHRP Project 1-37A contain two major deviations from the 1993 AASHTO Design Guide: (i) the abandonment of the ESAL concept and replacing with analysis that addresses the direct effects of axle loads, and (ii) the replacement of the panel-rating based serviceability concept with road smoothness prediction in the form of IRI (International Roughness Index) and prediction models for individual distress types. The intended design process of the proposed 2002 Guide, as shown in Figure 9.25, consists of the following five modules: (a) integrated climatic model, (b) material properties module, (c) axle loadings module, (d) pavement structure model, and (e) analysis module to make distress predictions and smoothness prediction. The perceived benefits

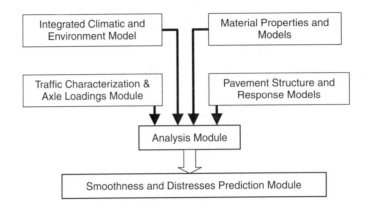

FIGURE 9.25   Design process of proposed AASHTO 2002 Guide.

of the proposed approach include the following:

- More efficient and cost-effective designs
- Improved design reliability
- Reduced life cycle costs
- Ability to predict various modes of failure
- Reduced premature failures
- Better evaluation of load effects
- More efficient use of available materials
- Better characterization of seasonal effects
- Improved rehabilitation design
- Ability to consider daily, seasonal and yearly changes in materials, climate and traffic into design process

The original schedule for the development of the proposed AASHTO 2002 Guide comprises four stages: Stage A began in February 1998 with the development of the entire framework for the design of new and rehabilitated pavement systems; Stage B would see the completion a working draft of the proposed Guide; Stage C was to cover the development of the final text materials and software, calibration, and fine tuning of the design procedure by the end of 2001; and Stage D was to address modifications and revisions by April 2002. Although the Final Report of NCHRP Project 1-37A has been published, the proposed procedure has not been adopted by AASHTO as the new design guide. It forms part of AASHTO's evaluation and education process to assist in preparing for a future edition of provisional mechanistic-empirical design guide.

References

AASHTO, 1993. *AASHTO Guide for Design of Pavement Structures 1993*. American Association of State Highway and Transportation Officials, Washington, DC.

AASHTO, 2004. Standard Method of Test T307 for Determination of Resilient modulus of Soils and Aggregate Materials, *Standard Specifications for Transportation Materials and Methods of Sampling and Testing*, 24th ed., American Association of State Highway and Transportation Officials, Washington, DC.

ACI, 2001. *ACI Manual of Concrete Practice*. American Concrete Institute, Farmington Hills, MI.

ACPA, 1992. *Proper Use of Isolation and Expansion Joints in Concrete Pavements, Concrete Information*. American Concrete Pavement Association, Stokie, IL.

ARA Inc., 2004. *Guide for Mechanistic-Empirical Design of New and Rehabilitated Pavement Structures*, Final Report, NCHRP Project 1-37A. National Cooperative Highway Research Program, Transportation Research Board, Washington, DC.

ASTM, 2003a. *C39 Standard Test Method for Compressive Strength of Cylindrical Concrete Specimens*, Annual Book of ASTM Standards, Vol. 04.02. American Society for Testing and Materials, West Conshohocken, PA.

ASTM, 2003b. *C78 Standard Test Method for Flexural Strength of Concrete Using Simple Beam with Third-Point Loading*, Annual Book of ASTM Standards, Vol. 04.02. American Society for Testing and Materials, West Conshohocken, PA.

ASTM, 2003c. *C496 Standard Test Method for Splitting Tensile Strength of Cylindrical Concrete Specimens*, Annual Book of ASTM Standards, Vol. 04.02. American Society for Testing and Materials, West Conshohocken, PA.

ASTM, 2003d. *C469 Standard Test Method for Static Modulus of Elasticity and Poisson's Ratio of Concrete in Compression*, Annual Book of ASTM Standards, Vol. 04.02. American Society for Testing and Materials, West Conshohocken, PA.

ASTM, 2003e. *D1883 Standard Test Method for CBR (California Bearing Ratio) of Laboratory-Compacted Soils*, Annual Book of ASTM Standards, Vol. 04.02. American Society for Testing and Materials, West Conshohocken, PA.

ASTM, 2003f. *D4429 Standard Test Method for CBR (California Bearing Ratio) of Soils in Place*, Annual Book of ASTM Standards, Vol. 04.02. American Society for Testing and Materials, West Conshohocken, PA.

Bergstrom, S.G., Fromen, E., and Linderholm, S. 1949. Investigation of Wheel Load Stresses in Concrete Pavements, Vol. 13, *Proceedings of Swedish Cement Concrete Research Institute*.

Bradbury, R.D. 1938. *Reinforced Concrete Pavements*. Wire Reinforcement Institute, Washington, DC.

Carey, W.N. and Irick, P.E. 1960. *The Pavement Serviceability Performance Concept*, Bulletin 250. Highway Research Board, Washington, DC.

Chou, Y.T. 1995. *Estimating Load Transfer from Measured Joint Efficiency in Concrete Pavements*, Transportation Research Record, No. 1482. Transportation Research Board, Washington, DC.

Chou, Y.T. and Huang, Y.H., 1979. A computer program for slabs with discontinuities, In *Proceedings. International Air Transportation Conference*, Vol. 1, pp. 121–136.

Chou, Y.T. and Huang, Y.H. 1981. A computer program for slab with discontinuities on layered elastic solids, In *Proceedings. 2nd International Conference on Concrete Pavement Design*. Purdue University, U.S.A, pp. 78–85.

Choubane, B. and Tia, M. 1992. *Nonlinear Temperature Gradient Effect on Maximum Warping Stresses in Rigid Pavements*, Transportation Research Record 1370. Transportation Research Board, Washington, DC, pp. 11–19.

Choubane, B. and Tia, M., Analysis and Verification of Thermal Gradient Effects on Concrete Pavement, *J. Transport. Eng. ASCE*, 121, 1, 75–81, 1995.

Croney, D. and Croney, P. (1991) The Design and Peformance of Road Pavements, 2nd Ed. McGraw-Hill Book Company, London.

Davids, W.G., Turkiyyah, G.M., and Mahoney, J. 1998. *EverFE — Rigid Pavement Finite Element Analysis Tool*, Transportation Research Record 1629. National Research Council, Washington, DC, pp. 41–49.

FHWA, 1990. *Concrete Pavement Joints*, Technical Advisory T 5040.30. Highway Administration, Washington, DC.

Fwa, T.F., Shi, X.P., and Tan, S.A., Analysis of concrete pavements by rectangular thick plate model, *J. Transport. Eng. ASCE*, 122, 2, 146–154, 1996.

Hassan, H.F. and White, T.D., 2001. Modeling Pavement Sub-drainage Systems, presented at the 80th Meeting of the Transportation Research Board, Washington, DC.

Hibbitt, Karlsson & Sorensen Inc., 2002. *ABAQUS, Explicit Version 6.3*. Pawtucket, Rhode Island, U.S.A.

Highway Research Board, 1962. The AASHO Road Test, Report 5 Pavement Research, HRB Special Report 61E, Washington, DC.

Huang, Y.H., Finite Element Analysis of Slabs on Elastic Solids, *J. Transport. Eng. ASCE*, 100, 2, 403–416, 1974.

Huang, Y.H. 1985. A Computer Package for Structural Analysis of Concrete Pavements, pp. 295–307. In *Proceedings 3rd International Conference on Concrete Pavement Design and Rehabilitation*. Purdue University, U.S.A.

Huang, Y.H. 1993. *Pavement Analysis and Design*. Prentice-Hall, Englewood Cliffs, New Jersey, U.S.A.

Ioannides, A.M. and Korovesis, G.T., Analysis and design of doweled slab-on-grade pavement systems, *J. Transport. Eng., ASCE*, 118, 5, 745–768, 1992.

Ioannides, A.M., Thomson, M.R., and Barenberg, E.J. 1985. *Westergaard Solutions Revisited*, Transportation Research Record, No. 1043. Transportation Research Board, Washington, DC.

Kelly, E.F. (1939) Applications of the results of research to the structural design of concrete pavements, *Public Roads*, 20, 6.

Khazanovich, L. and Yu, T. 1998. *ILSL2* — A Finite Element Program for Analysis of Concrete Pavements and Overlays, In *Proceedings of Fifth International Conference on Bearing Capacity of Roads and Airfields*, Trondheim, Norway.

Khazanovich, L., Yu, T., Rao, S., Galasova, K., and Shats, E., ISLAB2000 Version 1.1, ERES Consultants, 1999.

Liu, W. and Fwa, T.F., *Analysis of Concrete Pavement by Nine-slab Thick Plate Model*, CTR Technical Report 2004-1, Center for Transportation Research, National University of Singapore, Singapore.

Liu, W. and Fwa, T.F., Closed-form Six-slab Thick Plate Solution for Analysis of Edge Slab of Concrete Pavement, presented at 84th Annual Meeting of Transportation Research Board, Washington, DC. Transportation Research Record, Transportation Research Board, in press.

Liu, Y., Fwa, T.F., and Choo, Y.S., Finite element modeling of skid resistance test. *J. Transport. Eng.*, 129 (3), 316–321, 2003.

Mindess, S., Young, F.J., and Darwin, D. 2003. *Concrete, 2nd Ed.* Pearson Education, Inc., Upper Saddle River, NJ.

MOT (1978) Road Pavement Design, Technical Memorandum No. H6/1978, Her Majesty Stationery Office, London.Neville, A.M. 1996. *Properties of Concrete*, 4th ed., Wiley, New York.

Nussbaum, P.J. and Lokken, E.C. 1978. *Portland Cement Concrete Pavements, Performance Related to Design-Construction-Maintenance*, Report No. FHWA-TS-78-202. Federal Highway Administration, Washington, DC.

Pasternak, P.L. 1954. On a new method of analysis of an elastic foundation by means of two foundation constants, Gps. Izd. Lit. po Strait. I Arkh (in Russian).

PCA, 1984. *Thickness Design for Concrete Highway and Street Pavements*. Portland Cement Association, Skokie, IL.

PCA (1991) Design and Construction of Joints for Concrete Highways, Concrete Paving Technology, Portland Cement Association, Skokie, IL.

Pickett, G. 1946. *Appendix III: A Study of Stresses in the Corner Region of Concrete Pavement Slabs under Large Corner Loads*, Concrete Pavement Design. Portland Cement Association, Skokie, IL.

Pierce, L.M., 2003. 1-37A Design Guide — Project Status and State Viewpoint, presented at International Conference on highway Pavement Data, Analysis and Mechanistic Design Applications, 7th to 10th September, Columbus, OH.

Reissner, E., Effect of Transverse Shear Deformation on Elastic Plates, *J. Appl. Mech.*, 12, 69–77, 1945.

Road Research Laboratory, 1970. *A Guide to the Structural Design of Pavements for New Roads*, 3rd Ed., Department of the Environment, Road Note 29. Her Majesty Stationery Office, London.

Shi, X.P., Fwa, T.F., and Tan, S.A., Warping stresses in concrete pavements on Pasternak foundation, *J. Transport. Eng. ASCE*, 119, 6, 905–913, 1993.

Shi, X.P., Tan, S.A., and Fwa, T.F., Rectangular thick plate with free edges on Pasternak foundation, *J. Eng. Mech. ASCE*, 120, 5, 971–988, 1994.

Tabatabaie, A.M. and Barenberg, E.J., 1978. Finite Element Analysis of Jointed or Cracked Concrete Pavements, Transportation Research Record, 671, pp. 11–20.

Teller, L.W. and Sutherland, E.C., The structural design of concrete pavements, part 2: observed effects of variations in temperature and moisture on the size, shape, and stress resistance of concrete pavement slabs, *Public Roads*, 16, 9, 169–197, 1935.

Teller, L.W. and Sutherland, E.C. (1943) The structural design of concrete pavements, Public Roads, 23, 5, 167–212.

Tia, M., Armaghani, J.M., Wu, C.L., Lei, S., and Toye, K.L. 1987. *FEACONS III Computer Program for an Analysis of Jointed Concrete Pavements*, Transportation Research Record No. 1136, pp. 12–22.

Uddin, W., Hackett, R.M., Joseph, A., and Pan, Z. 1995. *Three Dimensional Finite Element Analysis of Jointed Concrete Pavement*, Transportation Research Record No. 1482.

Van Til, C.J., McCullough, B.F., Vallerga, B.A., and Hicks, R.G. (1972) *Evaluation of AASHTO Interim Guides for Design Pavement Structures, NCHRP 128*. Highway Research Board, Washington D.C.

Westergaard, H.M., 1926a. Computation of stresses in concrete roads, Vol. 5, pp. 90–112. In *Proceedings, Highway Research Board*.

Westergaard, H.M., 1926bb. Stresses in Concrete Pavements Computed by Theoretical Analysis, *Public Roads*, 7, 25–33.

Westergaard, H.M., 1926c. In Analysis of stresses in concrete pavement due to variations in temperature, Vol. 6, pp. 201–215. In *Proceedings, Highway Research Board*, Washington, DC.

Westergaard, H.M., Analysis of stresses in concrete pavements caused by variations of temperature, *Public Roads*, 8, 54–60, 1927.

Westergaard, H.M., 1939. Stresses in concrete runways of airports. Vol. 19, pp. 197–202. In *Proceedings. Highway Research Board*.

Wu, C.H. and Larsen, T.J. 1993. Analysis of Structural Response of Concrete Pavements under Critical Thermal Loading Conditions, pp. 317–340. In *Proceedings of the 5th International Conference on Concrete Pavement Design and Rehabilitation*, April 20th to 22nd, Purdue University, Lafayette, IN.

Yang, N.C. 1972. *Design of Functional Pavements*. McGraw-Hill Inc., New York.

Yao, Z.K. and Huan, L.Y. (1984) Experimental Analysis of Warping Stresses in Concrete Pavements, *J. East China Highway*, 3, 1–7.

Yoder, E.J. and Witczak, M.W. 1975. *Principles of Pavement Design*. Wiley, New York.

Zhang, J., Fwa, T.F., Tan, K.H., and Shi, X.P., Model for nonlinear thermal effect on pavement warping stresses, *J. Transport. Eng. ASCE*, 129, 6, 695–702, 2003.

Appendix 9A
User's Manual for MSLAB_LOAD

9A.1 Overview

This program presents the closed-form solutions for a nine-slab thick-plate model and a six-slab thick-plate model respectively for concrete pavements under vertical loading. The pavement slab in either model is supported on a Winkler foundation. For the nine-slab model, the loaded slab of interest is represented by the central slab, with eight surrounding slabs. For the six-slab model, the loaded slab of interest is a middle edge slab, with the five surrounding slabs to take into consideration the effects of jointed pavement system. Only shear force transfer is assumed across a joint. Shear force transfer is specified in terms of the efficiency of joint shear transfer, which is defined as the ratio of vertical deflections of the two edges along the joint between the unloaded and loaded slabs.

9A.2 System Requirements

The MSLAB_LOAD program runs on IBM compatible computer with Windows 2000 or Windows XP professional Version Operating System and MATLAB 6.0 or later versions. The personal computer must be installed with the MATLAB software before running the MSLAB_LOAD program.

9A.3   Installation

1. Access the drive that the MATLAB folder has been installed (default drive is c:), and double click to open the folder "work".
2. Download the folder MSLAB_LOAD from the CRC website to the desktop, open the folder and copy all its files (MSLAB_LOAD.m, MSLAB_LOAD.fig, nineslabload.m, sixslabload.m) to the folder "work" opened in step 1.
3. Close the folder "work"
4. To perform the computer run, double click the MATLAB icon on the desktop screen
5. Run the MSLAB_LOAD program by entering the following command at the prompt: >MSLAB_LOAD.

9A.4   Input

Input Parameter	Unit
Indicator for choice of analysis	9-slab or 6-slab analysis
Slab dimensions: slab length	m
slab width	m
slab thickness	m
Slab properties: Elastic modulus	MPa
Poisson's ratio	Numerical value
Subgrade property: Modulus of subgrade reaction	MN/m^3
Applied load: Magnitude of applied pressure	MPa
Length and width of loading area	M
Location of center of loading area	x and y coordinates in m (see Figure 9A.1 for location of the origin)
Load transfer coefficient of joints	Numerical value between 0 and 1, where 0 represents no shear transfer across the joint, and 1 represents 100 percent load transfer across the joint
Coordinates (x, y) where deflection and bending stresses are to be computed	x and y coordinates in m (see Figure 9A.1 for location of the origin)

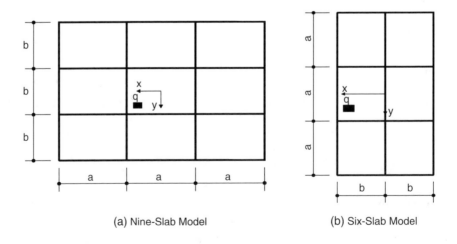

(a) Nine-Slab Model   (b) Six-Slab Model

FIGURE 9A.1   Coordinate systems for nine-slab model and six-slab model.

Screening display before keying in input

Screening display after keying in input

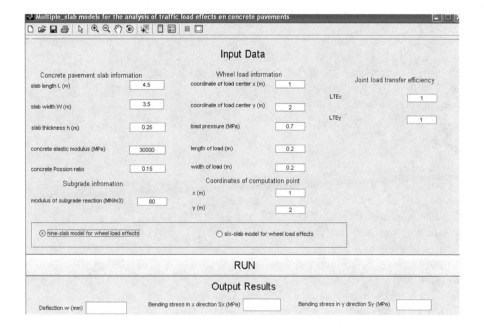

Screening display after running the MSLAB_LOAD program

9A.5   Output

Deflection, unit (mm)
Bending Stress at slab bottom along x direction, unit (MPa)
Bending Stress at slab bottom along y direction, unit (MPa)

9A.6   Example 1 — Nine-Slab Analysis

Input

Slab length: 4.5 m
Slab width: 3.5 m
Slab thickness: 0.25 m
Elastic modulus of concrete: 30,000 MPa
Possion's Ratio of concrete: 0.15
Modulus of subgrade reaction: 80 MN/m^3
Coordinate of load center (x_0, y_0): (1.0,2.0)
Load Pressure: 0.7 MPa
Length of load area: 0.2 m
Width of load area: 0.2 m
Coordinate of computational point (x,y): (1.0, 2.0)
Transverse joint load transfer efficiency (LTEx): 1.0
Longitudinal joint load transfer efficiency (LTEx): 1.0
Model selection: nine-slab model

Output

Deflection: 0.2656 mm
Bending Stress at slab bottom along *x* direction: 0.96599 MPa
Bending Stress at slab bottom along *y* direction, 0.75133 MPa

9A.7   Example 2 — Six-Slab Analysis

Input

Slab length: 4.5 m
Slab width: 3.5 m
Slab thickness: 0.25 m
Elastic modulus of concrete: 30,000 MPa
Possion's Ratio of concrete: 0.15
Modulus of subgrade reaction: 80 MN/m^3
Coordinate of load center (x_0, y_0): (1.0,2.0)
Load Pressure: 0.7 MPa
Length of load area: 0.2 m
Width of load area: 0.2 m
Coordinate of computational point (x,y): (1.0, 2.0)
Transverse joint load transfer efficiency (LTEx): 1.0
Longitudinal joint load transfer efficiency (LTEx): 1.0
Model selection: six-slab model

Output

Deflection: 0.29407 mm
Bending Stress at slab bottom along x direction: 1.3411 MPa
Bending Stress at slab bottom along y direction, 1.0431 MPa

Appendix 9B
User's Manual for WARP-NONLINEAR

9B.1   Overview

This computer program presents a closed-form solution for the analysis of warping stresses in a concrete pavement caused by non-linear temperature distribution across the pavement slab thickness. The analysis is based on a thick-plate model and takes into consideration the exact slab dimensions, and the effect of transverse shear deformation.

9B.2   System Requirement

The WARP_NONLINEAR program runs on IBM compatible computer with Windows 2000 or Windows XP professional Version Operating System.

9B.3   Execution

1. Download the folder WARP_NONLINEAR from CRC website to desktop
2. Double click the folder WARP_NONLINEAR
3. Double click the file WARP_NONLINEAR to run the program

9B.4   Input

Input Parameter	Unit
Slab dimensions: slab length	m
slab width	m
slab thickness	m

(Continued)

Input Parameter	Unit
Slab properties: Elastic modulus	MPa
Poisson's ratio	Numerical value
Coefficient of thermal expansion	1/(°C)
Subgrade properties: Modulus of subgrade reaction	MN/m^3
Shear modulus of subgrade	MN/m (zero value for Winkler foundation, positive value for Pasternak foundation)
Temperature data: 2 to 4 temperature measurements at different depths (Note: An analysis based on linear temperature distribution will be performed when only 2 temperature data points are entered. A nonlinear analysis will be performed when 3 or 4 temperature data points are entered	Temperature in °C Depth in m measured from top surface

Screening display before keying in input

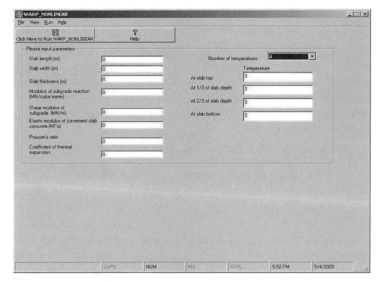

Screening display after keying in input

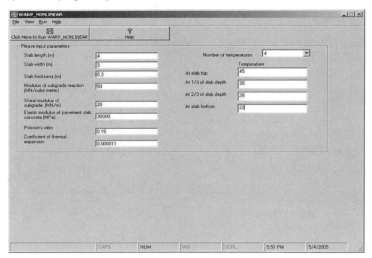

9B.5   Output

- Slab deflection unit (mm)
- Bending stress at slab top along *x* direction (MPa)
- Bending stress at slab bottom along *x* direction (MPa)
- Bending stress at slab top along *y* direction (MPa)
- Bending stress at slab bottom along *y* direction (MPa)

Screening display after running the WARP_NONLINEAR program

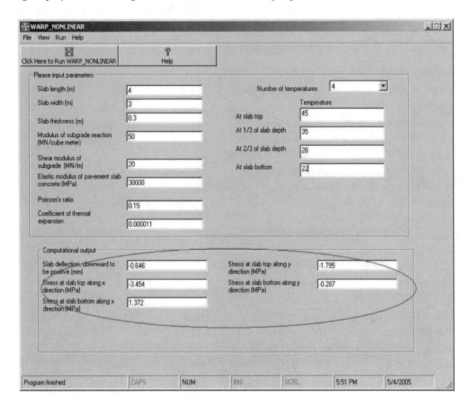

9B.6   Example 1 — Linear Temperature Profile Defined by Two-Point Temperature Data

Input

Slab length = 4 m
Slab width = 3 m
Slab thickness = 0.3 m
Modulus of subgrade reaction = 50 MN/m^3
Shear modulus of subgrade = 20 MN/m
Elastic modulus of pavement slab concrete = 30,000 MPa
Poisson's ratio of pavement slab concrete = 0.15
Coefficient of thermal expansion of concrete = 0.000011/°C
Number of Temperature data = 2
Temperature at slab top = 55°C
Temperature at slab bottom = 40°C

Output

> Slab deflection = − 0.438 mm
> Bending stress at slab top along x direction = − 1.849 MPa
> Bending stress at slab bottom along x direction = 1.849 MPa
> Bending stress at slab top along y direction = − 0.724 MPa
> Bending stress at slab bottom along y direction = 0.724 MPa

9B.7   Example 2 — Nonlinear Temperature Profile Defined by Three-Point Temperature Data

Input

> Slab length = 4 m
> Slab width = 3 m
> Slab thickness = 0.3 m
> Modulus of subgrade reaction = 50 MN/m^3
> Shear modulus of subgrade = 20 MN/m
> Elastic modulus of pavement slab concrete = 30,000 MPa
> Poisson's ratio of pavement slab concrete = 0.15
> Coefficient of thermal expansion of concrete = 0.000011/°C
> Number of Temperature data = 3
> Temperature at slab top = 45°C
> Temperature at slab mid-depth = 31°C
> Temperature at slab bottom = 22°C

Output

> Slab deflection = − 0.6724 mm
> Bending stress at slab top along x direction = − 3.992 MPa
> Bending stress at slab bottom along x direction = 1.679 MPa
> Bending stress at slab top along y direction = − 2.266 MPa
> Bending stress at slab bottom along y direction = − 0.047 MPa

9B.8   Example 3 — Nonlinear Temperature Profile Defined by Four-Point Temperature Data

Input

> Slab length = 4 m
> Slab width = 3 m
> Slab thickness = 0.3 m
> Modulus of subgrade reaction = 50 MN/m^3
> Shear modulus of subgrade = 20 MN/m
> Elastic modulus of pavement slab concrete = 30,000 MPa
> Poisson's ratio of pavement slab concrete = 0.15
> Coefficient of thermal expansion of concrete = 0.000011/°C
> Number of Temperature data = 4
> Temperature at slab top = 45°C
> Temperature at slab one-third depth = 35°C
> Temperature at slab two-third depth = 28°C
> Temperature at slab bottom = 22°C

Output

Slab deflection $= -0.6461$ mm
Bending stress at slab top along x direction $= -3.454$ MPa
Bending stress at slab bottom along x direction $= 1.372$ MPa
Bending stress at slab top along y direction $= -1.795$ MPa
Bending stress at slab bottom along y direction $= -0.287$ MPa

10

Design of Segmental Pavements

John W. Bull
University of Newcastle upon Tyne
Newcastle upon Tyne, U.K.

10.1   Introduction to Segmental Pavements

The definition of a segmental pavement used in this chapter is that of a preformed unit of material with a plan area ranging from 100 by 200 mm to 5000 by 10,000 mm and having a thickness between 50 and 600 mm. The segmental pavement is laid as individual units which may or may not be joined together to form a linked flexible or rigid pavement. The material may be natural stone, timber, precast concrete or baked clay.

10.2   History of Segmental Pavements

The earliest segmental pavements are dated around 3000BC in Crete, with similar segmental pavements being common in ancient Babylon, ancient Egypt and Greece (Lilley, 1991; Pritchard and Dowson, 1999). The segmental pavements in Crete were flat-topped boulders with wide joints, while other segmental pavements used limestone slabs. The growth of the Roman Empire required the building of adequate roads. These roads were built using the subgrade, a drainage layer, a sub-base, a road-base and a surface layer of stone setts with edge restraints (Lilley, 1991) (Figure 10.1).

In Europe, in medieval times, segmental pavements comprising cobbles or stone setts of typical size 75 by 200 by 225 mm deep were used in towns and cities (Croney and Croney, 1997). The setts were originally laid on a granular foundation, but later a lime–cement or cement–concrete foundation up to 300 mm thick was used (Croney and Croney, 1997). Stone sett pavements were preferred in Europe, but the United States preferred hard-burnt bricks. In Europe brick pavements were used, almost exclusively in Holland, until the 1960s when concrete block paving became the main segmental pavement material. Concrete block paving was introduced into the UK in the 1970s.

The industrial revolution required roads to

FIGURE 10.1   Roman segmental pavement in the United Kingdom.

transport goods and these roads were usually surfaced with stone setts, stone flagstones, or clay bricks. The uses of stone setts predominated due to their higher resistance to loads and to abrasion. With the development of railways and the higher concentration of goods at railway stations, better segmental pavements were developed and stone setts up to 200 by 300 by 225 mm thick were used.

Wood block pavements were introduced into European cities after the 1850s and were also used in Australia. The wooden blocks were of the same plan area as the stone setts. They were laid on a lime or cement mortar bed and were much cheaper than stone setts. The use of wooden blocks continued until the 1950s.

Other forms of segmental pavements have been used including asphalt blocks 250 mm square and 125 mm thick used for industrial floors, bridge approaches and viaduct floors, cast iron blocks used for heavy loadings, rubber blocks used for shock-absorbing, skid resistance and as a noiseless surface, vitrified bricks for their durability, impermeability, low tractive resistance and clay pavers 100 by 216 mm by between 65 to 100 mm thick.

The use of segmental pavements declined following the introduction of bituminous and cementitious binders in the early 1900s, but with the introduction of 100 by 200 mm plan area mass produced concrete blocks, in the 1950s, the use of concrete segmental pavements increased and the use of brick pavers reduced.

Concrete paving flags of plan area 600 by 900 mm were widely used for footways for the first eight decades of the twentieth century, but because of their large size, the difficulty in handling them and the increases in loads applied to them, smaller flags with plan areas of 300 by 300 mm, 400 by 400 mm and 450 by 450 mm were introduced.

10.3   Types of Segmental Pavements

Concrete block paving and clay pavers normally have a rectangular plan area of 100 by 200 mm although after the 1980s there was a multiplicity of block sizes and shapes. A range of thickness is available. Concrete block paving is available in a variety of textures, colors and multi-colors (Figure 10.2).

Precast concrete paving flags of plan area 300 by 300 mm [type G], 400 by 400 mm [type F], 450 by 450 mm [type E], 450 by 600 mm [type A], 600 by 600 mm [type B], 600 by 750 mm [type C] and 600 by 900 mm [type D] are available. Thickness ranges between 50 and 70 mm.

Large precast concrete paving units, also called raft units, are in the main rectangular in plan area with sizes ranging from 2000 by 2000 mm to 5000 by 10,000 mm; the smaller size predominate (Bull, 1992; Rollings and Chau, 1981). Raft unit thickness ranges between 140 and 600 mm. The size of the raft unit is limited only by the practical considerations of transportation and lifting capacity.

In general segmental pavements have an impermeable surface where up to 60% of the rainfall and stormwater runs off the surface. To assist down stream flood relief concrete block paving can be designed as a permeable surface to intercept the rainfall and the stormwater (Interpave, 2003).

FIGURE 10.2   Precast concrete pavers and paving flags.

There are three types of permeable concrete block paving.

- Total infiltration allows the rain water to infiltrate through the joints or voids between the concrete blocks, pass through the construction layers and eventually dissipate into the subgrade.
- Partial infiltration is similar to total infiltration in that it allows some of the rain water to infiltrate through the pavement. However a series of perforated pipes is added at the formation level to

allow water to be drained off into sewers, watercourses etc. The percentage of rainfall to be drained off is decided upon at the design stage of the segmental pavement.

• No infiltration stops the rainwater permeating beyond the impermeable membrane laid on top of the formation level. The rainwater is carried away by means of perforated pipes laid on top of the impermeable membrane. No infiltration is used where the subgrade has a low permeability, a low strength or where water needs to be collected and stored.

10.4   Where Segmental Pavements are Used

Segmental pavements are resistant to aviation fuel, hydraulic oils, de-icing chemicals and anti-icing chemicals and petroleum spillages. Segmental pavements (Figure 10.3):

• can accommodate major ground movements
• can be lifted out and relevelled
• can be opened up to underground services and then re-laid
• can be used to improve the appearance of pedestrian areas such as, town centers and other public spaces
• have different surface textures to increase driver awareness or pedestrian awareness of hazards and to increase vehicle skid resistance (BS 7932, 2003).

FIGURE 10.3 Segmental paving used to resist petroleum spillages.

10.4.1   Concrete Block Paving

Concrete block paving is suitable for aircraft hard standing, car parks, cycle paths, domestic drives, factory floors, industrial pavements, paving for exceptional loads, pedestrian areas, roads for low speed traffic, medium speed traffic and service areas.

Concrete block paving provides a durable surface that is comfortable to walk on, pleasant to look at, easy to maintain and ready for immediate use. Concrete block paving is available in a variety of colors, coatings, shapes and textures enabling different structural functions to be delineated. For example domestic driveways, footpaths, access roads, parking bays, pedestrian schemes in towns and other road and footway functions. Textured paving blocks are used in historic towns to give a "weathered" appearance.

Concrete block paving is used in heavily trafficked and loaded industrial areas, container ports and areas subject to large point loads, e.g., 15 tonne wheel loads or 40 tonne on a corner casting of a container. These heavy loads are resisted by the interlocking quality of the pavement where the blocks move together as a pavement, rather than as isolated blocks.

Concrete block paving is extremely durable and can resist harsh or very exposed environments, temperatures below −30°C and repeated freeze–thaw cycles.

Concrete block paving is used extensively for traffic calming where the intention is to improve safety by reducing traffic speeds. This is achieved by road and speed restrictions and in the pattern, coloring and texture of pavement materials. Concrete block paving is used to produce the various shapes and changes of level employed to control road speeds and to separate vehicles and pedestrians. These features include road humps and speed cushions which require vehicles to reduce speed to go over them. Build outs, pinch points and chicanes change the road line requiring the traffic to slow down and change direction.

10.4.2   Permeable Concrete Block Paving

Permeable concrete block paving is used where a structural pavement must allow rainwater to pass through the paving surface for dispersal into the ground. Permeable pavements are created using precast concrete blocks shaped either to produce voids between the blocks or to have enlarged joints. Both types provide an interlocking surface (Interpave, 2003). In towns, impermeable landscaping can produce rainwater runoff of up to 95% of the rainfall. Permeable pavements reduce the percentage of runoff and reduce the distance rainwater must travel before encountering a permeable opening. Any pollutants in the rainwater are trapped, filtered and will degrade over time. Following the laying of the blocks, the voids or the enlarged joints are filled with single sized aggregates to facilitate rainwater runoff between the blocks (Figure 10.4).

10.4.3   Concrete Paving Flags

Precast concrete paving flags have been used for footways since the end of the ninetieth century. Paving flags have a slip resistant surface and are available in a variety of colors and textures enabling different functions of the pavement to be delineated. For example tactile paving flags have raised parts and bars to assist visually impaired persons. Paving flags can be colored to match traditional stone paving in historic towns and conservation areas. Smaller paving flags are designed to accept overrunning by commercial vehicles (Figure 10.5).

Precast concrete paving flags can be divided into three main categories: standard, small element and decorative. Paving flags are used for approaches and surrounds to residential and industrial buildings, entrances to buildings, footways where vehicular overrun may occur, forecourts, paved areas in public and private gardens, paved areas surrounding shops and offices, pedestrian footways, pedestrian precincts, school playgrounds, swimming pool surrounds and vehicular driveways.

To warn visually impaired persons of specific hazards, eight different types of tactile concrete paving flags are manufactured. Their surface features are described below:

FIGURE 10.4   Permeable segmental paving.

FIGURE 10.5   Footpath surfaced with concrete paving flags.

- Blister paving, with flattened domes on its surface, is used at controlled crossings where pedestrians have priority over vehicles.
- Hazard warning paving with 6 mm high, 20 mm wide, rounded bars on its surface, spaced at 50 mm intervals, warn of specific hazards such as, steps, level crossings and the approach to on-street light rapid transport platforms.

- Platform edge off-street warning paving has 5 mm high, 25 mm wide, flat-topped domes on its surface, spaced in alternate rows at 66.5 mm intervals to warn of the edge of an off-street railway platform.
- Platform edge on-street warning paving has 6 mm high, 150 mm long by 83 mm wide flat-topped lozenge shapes on its surface to warn of the edge of an on-street light rapid transit platform.
- Segregated shared cycle track/footway paving is laid with bars parallel to the direction of travel of the cyclists and at 90 degrees to the direction of travel of the pedestrians. The paving has 5 mm high, 30 mm wide, flat-topped bars at 100 mm intervals on its surface to warn of possible conflicts with cyclists and to guide pedestrians away from the cyclists.
- Paving with a central delineating strip is used to separate the two areas of the segregated but shared cycle track/footway paving to assist pedestrians and cyclists in keeping to their respective areas of paving.
- Guidance path paving has 5 mm high, 35 mm wide, flat-topped bars at 80 mm intervals on its surface and is used where normal guides such as, buildings and kerbs are not present.
- Information paving has a detectable soft feel of an elastomeric surface to help locate amenities such as, telephone boxes etc.

10.4.4   Precast Concrete Raft Units

Precast concrete raft units are used where the quality of the subgrade is low, the applied loading is concentrated, there are high levels of traffic movement and substantial amounts of settlement (Bull, 1986, 1990a, 1990b; Rollings and Chau, 1981). Typical raft unit uses include airfields, airfield hard standing, bus bays, channeled traffic, crawler lanes for slow moving traffic, heavily trafficked road intersections, lorry parks, parking areas, pavements in industrial areas, port container areas, roundabouts and storage areas for bulk materials such as, aggregates, coal and salt. They are also used for temporary hard standing and for the rapid repair of highways and runways. In areas of substantial subgrade settlement, raft units can be lifted out, the sub-base relevelled and the raft units reinserted. Raft units can be quickly laid to make emergency temporary roads and then taken up and reused.

10.5   Traffic Loading

Concrete block paving is available in thickness of 50, 60, 80 and 100 mm. The 50 mm thickness is used for domestic drives, paths and patios. The 60 mm thickness is used for car parks, domestic drives, footways, lightly trafficked areas and residential roads with less than five commercial vehicles per day. The 80 mm thickness is used for aircraft pavements, factory floors, footways, industrial pavements, medium speed roads and residential roads with over five commercial vehicles per day. The 100 mm thickness is used in areas subjected to exceptionally high axle or point loads.

Concrete paving flags are available in a number of sizes, types A, B, C, D, E, F and G as described previously. Types A, B, C and D are available in thickness of 50 and 63 mm. Type E is available in thickness of 50 or 70 mm. Type F is available in thickness of 50 or 65 mm. Type G is available in thickness of 50 and 60 mm. Types A, B, C and D are considered as standard paving flags, while types E, F and G are considered as small element paving flags.

All concrete paving flag sizes and thickness can be used where there are no vehicles such as, footways protected by street furniture, pedestrian precincts, areas where there is very occasional use by cars or light mechanical sweepers, unprotected footways and in no parking areas where vehicular overrun can not happen. For footways where vehicles cross to reach a domestic driveway concrete paving flags of type A 63 mm thick, type B 63 mm thick, type E 70 mm thick, type F 65 mm thick and type G 60 mm thick are used. For footways overrun by cars and occasional commercial vehicles, for unprotected pedestrian precincts with up to 50 commercial vehicle crossings per day and for service vehicles or fire appliance access type E 70 mm thick, type F 65 mm thick and type G 60 mm thick are required.

10.6   Design Procedures for Segmental Pavements

10.6.1   Introduction

Initially the plan area of concrete paving flags was modeled on the plan area of natural stone paving flags. These initial concrete paving flags were large, either 600 by 600 by 63 mm thick or 900 by 600 by 63 mm thick, weighed between 50 and 75 kg and were difficult to handle. There was no design procedure as they carried only pedestrians. With the increase in both the numbers and the weight of vehicles, it became common for the flag to be overrun by vehicles. This resulted in the concrete paving flags breaking, rocking or settling. Because of these problems research and laboratory tests were carried out to develop a design procedure for segmental pavements.

To design any type of segmental pavement the designer needs to know the strength of the subgrade, the traffic loading, the volume of the traffic and the required design life. From this data the designer can calculate the required thickness and strength of the sub-base and the road base together with the plan area, thickness, strength and reinforcement requirements of the segmental pavement.

A site investigation is required to determine the classification and the bearing strength of the subgrade. The susceptibility of the subgrade to frost heave and the position of the water table are also required.

The required design life is specified either by the number of standard axles, or by the length of time in service. In specifying the number of standard axles, the required design life is stated clearly. If length of time in service is used and subsequently traffic volume and axle loads increase, the segmental pavement will fail well before its specified design life.

The traffic loading on a segmental pavement is related to the standard axle loading of 8200 kg. The damage done by traffic is determined from the actual axle load divide by the standard axle load increased by the fourth power (Croney and Croney, 1991). Further the volume of and the change in traffic with time needs to be estimated with care. For example a segmental pavement designed for use by vehicles of less than 1500 kg to gain access to a small night time only parking area, will soon fail if it is used by 40,000 kg heavy good vehicles to gain access to a steel works that operates 24 h per day.

A capping layer of material is used to cover a weak subgrade. Above the capping layer, or the subgrade, if there is no capping layer is the sub-base. The sub-base is a significant structural layer and is usually constructed using granular material or cement bound material. Above the sub-base is the road base. The road base is generally the main structural layer and may be constructed using dense bound macadam, hot rolled asphalt or lean concrete. Above the road base is a laying course usually of unbound sand that forms a regulating and continuous support layer to the segmental pavement. Not all the layers may be required, for example a segmental pavement used solely for pedestrians may not require a capping layer or a road-base layer.

10.6.2   Design Procedure for Concrete Block Paving

The design procedure for concrete block paving, which draws upon Road Note 29, LR1132 and BS 7533, is dependent upon whether the paving is to be used for heavy duty industrial areas, highways, lightly trafficked areas, surfaces used by aircraft or as an overlay of existing paving (BS 7533-1, 2001, BS 7533-2, 2001; Powell et al., 1984; Pritchard and Dowson, 1999; Transport and Road Research Laboratory, 1970).

10.6.2.1   Design Procedure for Lightly Trafficked Areas

The design of lightly trafficked areas is based on Road Note 29 and BS 7533 applies to cycle paths, domestic drives, estate roads, industrial vehicle parks, pedestrian routes and private vehicle parks (BS 7533-2, 2001; Transport and Road Research Laboratory, 1970). The following is a design example (Pritchard and Dowson, 1999) (Figure 10.6).

- A road to carry up to ten commercial vehicles per day is to be constructed into a small non industrial estate and is to have a design life of 20 years.
- The CBR of the subgrade is 3%; this value takes into account the subgrade condition and water table level.

- The total amount of the traffic, expressed as standard axles to be carried during the 20 year design life, has to be determined. This is achieved by calculating the number of commercial vehicles using the pavement during its lifetime and converting them to the number of standard axles. Ten commercial vehicles per day for six working days per week multiplied by 48 working weeks per year multiplied by the 20 year design life gives 57,600 commercial vehicles.
- Allowing for traffic growth and using a conversion factor for public roads carrying less than 250 commercial vehicles per day gives the number of standard axles over the lifetime of the road as 27,000.
- The sub-base thickness is determined using the 3% subgrade CBR and the 27,000 standard axles and gives a sub-base thickness of 280 mm.
- The design summary is a 280 mm thickness of Type 1 granular sub-base (GSB), a laying course of sand 50 thick and 80 mm thick concrete block paving (Department of Transport, 1992).

FIGURE 10.6   Lightly trafficked pedestrian area.

- If the subgrade is frost susceptible, the sub-base thickness must be increased to 320 mm.

10.6.2.2   Design Procedure for Highways

The procedure for designing concrete block paving for highways is based on BS 7533 (BS 7533-1, 2001, BS 7533-2, 2001). The following is a design example (Pritchard and Dowson, 1999).

- A highway is to be constructed with concrete block paving as its running surface.
- The highway has a speed limit of 48 kph and will carry 250 commercial vehicles per day. The forecast traffic growth is 2% and a 20-year design life is required.
- By converting the 250 commercial vehicles per day to the number of standard axles over the 20 year design life and together with a 2% growth in traffic the cumulative traffic is found to be 2.2 million standard axles.
- If, at a later date, the traffic speed is expected to be raised above 48 kph the number of standard axles must be multiplied by 2.
- The subgrade is heavy clay. To establish the subgrade CBR, the plasticity index, the degree of drainage and the construction conditions of the subgrade are required. From this information the subgrade CBR is found to be 3%.
- The thickness of the subgrade improvement layer for a 3% CBR subgrade and for traffic levels in excess of half a million standard axles is 350 mm.
- The thickness of the type 1 GSB is determined. As the sub-base is to be used by 500 standard axles for site access and there is a subgrade improvement layer the sub-base thickness will be 225 mm.
- The thickness of the road-base depends upon the number of standard axles over the design life, in this case 2.2 million, giving a road base thickness of 130 mm of cement bound material or dense bitumen macadam.
- The sand laying or bedding course is 30 mm thick with a concrete block paving thickness of 80 mm.

- The design summary is a 350 mm thick subgrade improvement layer overlaying the subgrade, 225 mm thick type 1 GSB, 130 mm thickness of road base, a sand laying course 30 mm thick and concrete block paving 80 mm thick.
- If the subgrade is susceptible to frost, the total pavement thickness of the sub-base, sand laying course and concrete paving blocks must be greater than 450 mm.

10.6.2.3 Design Procedure for Pavements for Heavy Duty Industrial Areas

Concrete block paving subjected to heavier than highway loading and to container corner casting loads is designed in accordance with BS 7533 or The Structural Design of Heavy Duty Pavements for Ports and Other Industries (BS 7533-1, 2001; Knapton and Meletiou, 1996). Knapton and Meletiou's design manual considers surface deformation in the order of 50 to 75 mm as serviceability failure and recommends the use of 80 mm thick rectangular blocks laid to a herringbone pattern as they exhibit a high level of stability and strength. For initial design purposes the paving surface is assumed to be 80 mm thick concrete paving blocks laid on 30 mm of bedding sand overlaying a base of a grade C10 (Young's modulus of 35 GPa) concrete bound material with an assumed flexural strength of 2 MPa. Beneath the base is the sub-base, which overlays the capping layer which itself overlays the subgrade (Figure 10.7).

FIGURE 10.7   Heavy industrial sheet steel loading.

The design procedure is as follows (Knapton and Meletiou, 1996):

- The loading regime is considered. The types of vehicle, their weight, both empty and fully loaded, their wheel track, their wheel spacing, and the number of passes they will make over the lifetime of the pavement and the CBR of the subgrade are determined.
- The tire pressures, the dynamic effects of the vehicles when cornering, accelerating and braking are determined, as is the possibility of channelized travel of the vehicles.
- The number of load repetitions is converted to an equivalent number of repetitions of the heaviest wheel so that the equivalent single load used in the design is derived from the heaviest wheel load values.
- Static loads due to the corner castings of containers, containers stacked on top of each other and trailer dolly wheels which all cause high bearing stresses on the concrete block paving are taken into account and recorded as container stacking loads related to the equivalent single load.
- The base material is assumed to be a C10 concrete bound material, but it can be exchanged for an equivalent amount of an alternative material such as, crushed rock, wet lean concrete, fiber reinforced concrete, or other cement bound materials.
- From the value of the equivalent single wheel load which usually ranges between 100 kN to 1500 kN and the number of passes, usually between 250,000 and 25,000,000 the thickness of the base can be determined. The thickness usually ranges between 200 to 700 mm.
- Although the design assumes concrete block paving is to be the surface layer, calculations for the use of concrete, precast concrete raft units and asphalt are also included in the design manual.
- The compacted sub-base thickness is usually between 150 to 225 mm depending upon the CBR of the subgrade. The sub-base should be constructed using a granular material comprising crushed rock or slag.
- A capping layer of thickness between 250 and 600 mm is used to cover subgrade with a CBR of less than 5%.

10.6.2.4   Other Design Procedures for Concrete Block Paving

Other areas where concrete block paving is used include airfields and overlaying existing paving.

- Where concrete blocks are used by aircraft alternative procedures such as those given by the Ministry of Defense are used (Ministry of Defense, 1996).
- Worn roads and paved areas may be strengthened and reconditioned by an overlay of concrete block paving (BS 7533-1, 2001, BS 7533-2, 2001). The existing paving must be assessed and where a sand laying course is to be laid over an existing impermeable surface, there must be adequate drainage to prevent the sand becoming saturated before the joints have sealed (Pritchard and Dowson, 1999).

10.6.3   Design Procedure for Structural Flag Paving

Many pedestrianisation schemes are now designed to support not only pedestrians but also commercial delivery vehicles. To accommodate these overrunning vehicles, three small element paving flags were developed; type E 450 by 450 by 70 mm thick, type F 400 by 400 by 65 mm thick and type G 300 by 300 by 60 mm thick. Paving flags of type E, F and G are designed to accept frequent car, occasional commercial vehicle overrun, use in pedestrian precincts with up to 50 commercial vehicle overruns per day and to provide fire service access paths. There are two modes of failure for type E, F and G paving flags. Types E and F fail due to increased stress in the concrete, while type G fails due to the subgrade becoming overstressed.

The design process ensures the concrete flag pavement remains serviceable without major reconstruction in its design life, by evaluating the required type and thickness of the sub-base or road-base to ensure the allowable stresses in the paving flag and in the subgrade are not exceeded (BS 7533-8, 2003; Pritchard and Dowson, 1999).

The design process is as follows (Pritchard and Dowson, 1999).

10.6.3.1   Design Procedure for Pedestrian Precincts

- A pedestrian precinct around a large shopping center is to be constructed using type F 400 by 400 by 65 mm paving flags and a Type 1 sub-base.
- Commercial vehicles will deliver goods to the shops on an average of 350 days per year. The commercial vehicles are expected to be 80% two axles, 10% three axles and 10% four axles.
- The pavement is to have a 20 year design life.
- The subgrade is silty clay. The CBR of the subgrade is determined as being between 3 and 5%. The lower value of 3% is used.
- The total amount of traffic is assessed as ten commercial vehicles per day. Converting the mix of two, three and four axle vehicles to standard axles gives a load of 0.7 standard axles per day. Over the 20 year design life some 49,000 standard axles will overrun the paving flags.
- The sub-base or road-base thickness has to be determined. A type 1 sub-base is to be used.
- To prevent over stressing of the paving flags a minimum sub-base thickness of 165 mm is required.
- To prevent over stressing of the subgrade a minimum sub-base thickness of 240 mm is required.
- The required minimum sub-base thickness of 240 mm is to be used.

10.6.3.2   Design Procedure for Overlaying Existing Pavements

- When using paving flags to overlay an existing pavement it is necessary to estimate the remaining life of the existing pavement to determine the required depth of the overlay.
- The existing pavement is inspected to determine the amount of cracking and the depth of the rutting. For example if the existing pavement is as new or is fully cracked the condition factor CF1 would be 1.0 or 0.2 respectively. If the depth of rutting is 10 mm or 50 mm the condition factor CF2 would be 1.0 or 0.4, respectively.
- The construction thickness of the existing pavement is multiplied by the appropriate CF1 and CF2 values to give the equivalent thickness of new construction.

- The design procedure for pedestrian precincts described in Section 10.6.3.1 is then used to determine the required depth of construction of the new pavement.
- If the equivalent thickness of new construction is greater than the required depth of construction of the new pavement then a sand bedding layer is laid over the existing construction and the new paving flags are laid on top.
- If the equivalent thickness of new construction is less than the required depth of construction of the new pavement then additional sub-base must be laid to achieve the required design life before the sand bedding layer and the new paving flags are laid.

10.6.4 Design Procedure for Precast Concrete Raft Units

The design procedure for precast concrete raft units is based on a three dimensional multi-layer linear elastic analysis using the finite element method. The procedure takes into account the applied loading, the number of load applications, the strength of the subgrade, the stresses and the deflections in the pavement layers and the stresses in the raft units.

The raft unit serviceability limit state is defined as the point where the raft unit requires some form of maintenance, usually relevelling. For the raft unit, the serviceability limit state is related to the load induced maximum principal concrete tensile stress σ, the concrete's modulus of rupture MR and the number of load application N the raft unit can sustain using Equation 10.1 (Bull, 1986, 1990a).

$$N = 225,000[MR/\sigma]^4 \tag{10.1}$$

The subgrade serviceability limit state is defined as the number of load applications N_s, the subgrade will accept before raft unit relevelling is required, using the maximum downward vertical compressive stress in the subgrade B (kPa) and the subgrade CBR in percent using Equation 10.2 (Bull, 1990a; Heukelom and Foster, 1960):

$$N_s = [[280 \times CBR]/B]^4 \tag{10.2}$$

For the raft units, Equation 10.1 is the serviceability limit state, however if the subgrade CBR is less than 5%, Equation 10.2 becomes the serviceability limit state.

As raft units, well past their calculated serviceability limit state, continue to carry their design loading an ultimate limit state for raft unit replacement was determined. The ultimate limit state load cycles N_u, the serviceability limit state load cycles N, the tire contact pressure P and the reference tire pressure of 0.875 MPa are related through Equation 10.3.

$$N_u = N[P/0.875]^C \tag{10.3}$$

C has a maximum value and reduces when the value of P is greater than 2 MPa (Bull, 1990a).

The raft unit design method is interactive and computer based, with the pavement designer being able to enter data at any point in the program. For the initial design the program assumes standard pavement parameters as shown in Table 10.1.

The program requests the required pavement fatigue life in terms of the heaviest wheel loading, the wheel configuration and the subgrade CBR. From this data, the program determines the maximum

TABLE 10.1 Standard Pavement Parameters

Layer	Plan size (mm × mm)	Thickness mm	Young's modulus	Poisson's ratio
Raft unit	2000 × 2000	150	34 GPa	0.15
Bedding sand	—	50	75 MPa 7.5% CBR	0.25
Sub-base	—	300	200 MPa 20% CBR	0.25
Subgrade	—	600	3 MPa 0.3% CBR	0.30

allowable concrete stress in the form of the design pavement stress (DPS) number and the maximum allowable subgrade stress in the form of the design subgrade stress (DSS) number. The designer then inputs the raft unit concrete strength. The program determines the applied pavement stress (APS) number and the applied subgrade stress (ASS) number. If the DPS and the DSS numbers are greater than the APS and ASS numbers respectively, the pavement design is satisfactory. The program requires the subgrade depth, the degree of saturation of the subgrade and details of the GSB. A lean concrete sub-base may also be used. From the difference between the design numbers and the applied numbers, the designer can determine the raft unit thickness and reinforcement requirements. The program will output the pavement life in terms of the raft unit serviceability limit state, the subgrade serviceability limit state and the raft unit ultimate limit state.

10.6.4.1   Design Example for a Precast Concrete Raft Unit Pavement

An airfield pavement is required to have a subgrade serviceability limit state and a raft unit serviceability limit state of 10,000 movements of an aircraft with a fully laden take off main gear load of 1000 kN spread equally between four wheels spaced at 2000 by 1500 mm. The subgrade has a CBR of 15% and a depth greater than 5 m. The GSB will be 600 mm thick and have a CBR of 30%. The raft unit is 2000 by 2000 by 175 mm thick. The maximum tire pressure is 3 MPa. The design procedure is shown in the following six stages with Table 10.2 showing the data input and the data output.

- Stage 1 shows the required DPS and DSS values as 218 and 420 respectively.
- Stage 2 considers the standard raft unit data of Table 10.1 but uses 60 MPa concrete and the loading of 1000 kN on the four wheels spaced at 2000 by 1500 mm to generate the APS and the ASS values of 460 and 115 respectively. Subtracting the DPS number of 218 from the APS number of 460 shows that the APS is 242 in excess of the DPS. The value of 242 must be reduced to a maximum of zero otherwise the raft units will be overstressed. Subtracting the DSS number of 420 from the ASS number of 115 shows that the ASS is 305 below the DSS and that the subgrade is under stressed.
- Stage 3 uses the standard subgrade data of Table 10.1 modified using the real subgrade CBR of 15% and the real subgrade depth which is in excess of 5 m. The results show that the APS is too high by 161 and must be reduced. The ASS is 82 less than the DSS number and the subgrade is still under stressed.
- Stage 4 uses the standard sub-base data of Table 10.1 modified using the 600 mm thick GSB compacted to a CBR of 30%. The results show that the APS is still too high by 94 and must be reduced. The ASS is now 133 less than the DSS number and the subgrade is still under stressed.

TABLE 10.2   Data Input and Output for the Six Stages in the Design of a Raft Unit Pavement

Stage	Data input	Data output
1	Load movements = 10,000 Subgrade CBR = 15%	DPS = 218 DSS = 420
2	Concrete strength = 60 MPa Total load = 1000 kN Four wheels 2 m by 1.5 m	APS − DPS = 460 − 218 = 242 ASS − DSS = 115 − 420 = − 305
3	Subgrade CBR = 15% Subgrade depth > 5 m	APS − DPS = 379 − 218 = 161 ASS − DSS = 338 − 420 = − 82
4	CBR of GSB = 30% Thickness of GSB = 600 mm	APS − DPS = 312 − 218 = 94 ASS − DSS = 287 − 420 = − 133
5	Raft thickness = 175 mm Normal raft unit reinforcement	APS − DPS = 218 − 218 = 0 ASS − DSS = 283 − 420 = − 137
6	Serviceability limit state Relevelling after Ultimate limit state	10,000 movements 48,500 movements 210,850 movements

- Stage 5 modifies the standard raft unit data of Table 10.1 by increasing the raft unit thickness to 175 mm. This change reduces the APS − DPS to zero and gives the raft unit the required design life of 10,000 movements. The ASS number is 137 less than the DSS number and gives the subgrade a design life of 48,500 movements.
- Stage 6 shows that the raft unit pavement has achieved its required service life of 10,000 movements. After 48,500 movements the raft units will require relevelling due to subgrade settlement. The raft units have an ultimate limit state life of 210,850 movements before they must be replaced.

The life of 48,500 movements before the raft units will require relevelling due to subgrade settlement can be reduced to 10,000 movements if the CBR value and the thickness of the GSB are reduced below 30% and 600 mm respectively. This would require additional iterations of the design process and reduce the cost of construction.

10.6.5   Design Procedure for Permeable Concrete Block Paving

The design of a permeable concrete block pavement has to satisfy both hydraulic and vehicular loading requirements. The design which requires the thicker sub-base is then used.

10.6.5.1   Design Example for Permeable Concrete Block Paving

A vehicular parking area is to be constructed using permeable concrete block paving. The paved area must retain the rainfall over a 24 h period. There is no requirement for runoff from surrounding impermeably paved areas to be accommodated by the permeable pavement.

- *Stage 1 Hydraulic requirements.* The subgrade has a CBR of 3% and a coefficient of permeability of 10^{-8} m/sec. The vehicles using the parking area have a maximum axle load of 2000 kg. The sub-base material is to be open graded crushed rock and as the CBR of the subgrade is low and unsuitable for the direct infiltration of rainwater a no infiltration system is to be used. For the hydraulic design, the ratio r of the 1 h storm rainfall depth to the two day maximum rainfall depth for a 5 year storm return period for the location of the car park is 0.39. For an r value of 0.39 and a 24 h rainfall retention period, a 155 mm thickness of permeable pavement sub-base is required.
- *Stage 2 Vehicular loading requirements.* The design load must be determined. Although the maximum axle load is given as 2000 kg this load must be multiplied by a partial load factor which depends on the certainty of the load. In this case the certainty of the load is well informed and the partial load factor of 1.6 gives a design load of 3200 kg. For the 3200 kg design load the sub-base thickness of open graded crushed rock for a 5% CBR is 300 mm. As the CBR of the subgrade is 3% the 300 mm thickness is increased to 550 mm. A partial safety factor of 1.1 is applied to the 550 mm to take account of the nature of the sub-base material, giving a sub-base design thickness of 605 mm.

The sub-base design thickness for water storage is 155 mm but for vehicular loading 605 mm is required. The higher value of 605 mm is used.

As a no infiltration system is to be used, the impermeable membrane must be durable, robust and able to withstand loads both during construction and throughout the full design life of the pavement. The membrane must be unaffected by potential pollutants and must be installed with fully watertight joints and discharge outlets.

10.7   Maintenance and Reinstatement of Segmental Pavements

10.7.1   Introduction

Guidance for the maintenance of concrete block paving and concrete flag paving is given by Pritchard (Pritchard, 2001). A significant feature of segmental paving is that the paving can be lifted out and then re-laid in such a way that the reinstatement is invisible. For the best results and to minimize subsequent settlement the disturbed pavement layers must be compacted adequately (BS 7533-11, 2003).

10.7.2   Maintenance and Reinstatement of Concrete Block Paving

Concrete paving blocks which have been in use for some time will be tightly locked together. Consequently, extracting the first block is carried out either by breaking it out or by using a bricklayer's small trowel to lever it out. Once a few blocks have been extracted, further blocks can be removed by hand, initially by using a vibration tool to loosen them. The area of extracted blocks must be sufficient large to ensure that when the blocks are replaced a level surface will result. Before relaying, the blocks will need cleaning and those around the area of extracted blocks will need checking to ensure they have not moved and reduced the area to be filled.

The material used for backfilling the hole, must be compacted in 135 mm thick layers to minimize subsequent settlement. Type 1 GSB material is recommended as backfill (Department of Transport, 1992). Foamed concrete may be used as a backfill because it is self levelling, flows to fill the void, does not compact after setting and allows the sand bedding layer and the concrete paving blocks to be laid 18 h after back filling. The sand bedding layer is screeded to a level up to 12 mm higher than the existing sand layer of the surrounding concrete block paving. After laying, joint filling and vibrating the blocks, vehicular compaction will ensure that the final running surface is at the same level as that for the undisturbed blocks.

10.7.3   Maintenance and Reinstatement of Concrete Flag Paving

Concrete paving flags that have been laid flexibly and have been in use for some time will be firmly locked together. The first flag is removed by using a bricklayer's small trowel to remove the sand between the flags and to lever out the flag. Once the first flag has been removed the remaining flags should be easy to lift out. Normally the area of flag paving to be removed is one complete row beyond the area to be excavated. If the flags are joined using mortar, the mortar must be broken out with a hammer and chisel. If the mortar is too strong the flag may have to be broken out. Once an open edge has been obtained the adjacent flags can be levered out. Any mortar still adhering to the flag can be removed using a hammer, a chisel and a wire brush.

The material used for backfilling the hole, must be compacted in 135 mm thick layers to minimize subsequent settlement. Type 1 GSB material is recommended for the backfilling (Department of Transport, 1992). Foamed concrete may be used as a backfill because it is self levelling, flows to fill the void, does not compact after setting and can be overlaid 18 h after backfilling. The sand bedding layer is screeded to a level usually 6 mm higher than the existing sand layer of the surrounding concrete flag paving. After laying and joint filling, the flags are vibrated to ensure that the final surface is level. If the laying course is mortar and not sand, the flag paving must not be used until the mortar has achieved a working strength.

10.7.4   Maintenance and Reinstatement of Concrete Raft Units

Maintenance of precast concrete raft units is usually related to either the very high localized loading of container corner castings or to excessive settlement caused by a weak subgrade. If the raft unit has failed due to high localized loading, only the raft unit need be replaced. If the pavement has failed due to excessive settlement, the subgrade will need strengthening with both the sub-base and the bedding sand being replaced. The original raft unit can still be used.

Maintenance is carried out by the joints, which are usually sand or sand slurry, being removed using a circular concrete saw. The raft units, which have precast lifting eyes, are lifted out using a fork lift truck or a small crane. The sub-base is dug out, replaced and compacted into place. A 50 mm thick sand bedding layer is compacted on top of the sub-base. The raft is then relayed.

10.8   Manufacture of Segmental Pavements

10.8.1   Introduction

Concrete segmental paving is produced in one of the following three ways, wet casting, wet pressing or semi-dry pressing.

10.8.1.1   Wet Casting

For wet casting, wet concrete is poured into timber, plastic or steel moulds. The upper surface of the concrete is leveled and the mould vibrated. The concrete remains in the mould until it is strong enough to be removed and stored. The mould is then cleaned and reused. The concrete will use the minimum amount of water to increase resistance to abrasion and to frost damage and be air-entrained to resist the action of de-icing salts. Pigments may be added to the mix to change the color of the concrete.

10.8.1.2   Wet Pressing

Wet pressing was first used in the early 1900s for precast concrete paving flags as the method produces well compacted products with good abrasion, frost and sulphate resistance. The concrete mix is very wet and is self levelling when poured into the mould. A high pressure is applied to the upper surface of the concrete long enough to squeeze out most of the water. The resulting concrete flag is immediately strong enough to be vacuum lifted out of the mould. The edges may be chamfered and pigments are used to provide a range of colors.

10.8.1.3   Semi-Dry Pressing

Semi-dry pressing is used for the production of concrete block paving. Many blocks are manufactured in individual moulds within one large mould during each pressing operation. The large mould is filled with concrete; the concrete is vibrated before and during the pressing operation. The top surface of the concrete is pressed down and the blocks taken out and stored. The semi-dry process requires higher cement content than the wet press process.

10.8.2   Manufacture of Concrete Block Paving

Concrete block paving in the UK is manufactured to BS 6717 and BS EN 1338 which specifies the strength, sampling and testing methods, sizes, tolerances and surface finish. (BS EN 1338, 2003; BS 6717, 2001).

10.8.3   Manufacture of Clay Pavers

Clay pavers are manufactured to BS EN 1344 in the same way as clay bricks (BS EN 1344 ,2002). Shale, mixed with water, is shaped by passing it through a die, cut to length, dried and baked at high temperatures until vitrified. BS EN 1344 specifies the requirements and test methods (BS EN 1344, 2002).

10.8.4   Manufacture of Concrete Paving Flags

Concrete paving flags are manufactured in accordance with BS 7263 (BS 7263-1, 2001). BS 7263 specifies the marking, sampling, sizes, surface finish, testing, tolerances and traverse bending strength. Flags can be finished to produce a variety of textured, profiled, ground or polished surfaces. BS EN 1339 specifies the requirements and test methods (BS EN 1339, 2003).

10.8.5   Manufacture of Stone Paving Flags

Stone paving flags are produced from blocks of stone. The flags are formed by splitting the block of stone along its natural planes. The thickness of the flags usually varies between 25 and 100 mm, with the plan area being cut to size depending upon the thickness. Flags may also be formed by sawing the blocks into a number of slices. These slices are then cut into a rectangular plan shape. BS EN 1341 specifies the requirements and test methods for slabs of natural stone while BS EN 1342 specifies the requirements and test methods for stone setts (BS EN 1341, 2001; BS EN 1342, 2000).

10.8.6 Manufacture of Concrete Raft Units

Concrete raft units are manufactured, to precise specifications of loading and support conditions, in steel or glass fiber moulds using the wet cast process. These conditions affect the choice of raft size, shape, thickness, concrete strength and the amount and type of the reinforcement (Bull, 1990a). The rafts may be unreinforced or reinforced with steel bars, fibers, or prestressing wires. The steel reinforcement is usually in the form of two layers of precisely placed steel bar or wire mesh. The reinforcement is symmetrical within each layer, but with its spacing reduced near to the edges of the raft. Fiber reinforcement may be added to improve the raft unit's resistance to impact loading and to spalling (Bull, 1990a). Two steel tubes are welded to the reinforcement to facilitate lifting. Steel edge surrounds may be added to the upper edges of the raft to improve impact resistance. To increase skid resistance, crushed rock is used as an aggregate, with natural sharp sand or crushed rock fines. Concrete voids are minimized by aggregate grading, the addition of microsilica or fly ash and by vibration. The required minimum characteristic strength is 50 MPa with the addition of a plasticizer to increase workability. The upper concrete surface is textured during curing. When the concrete has gained sufficient strength, the raft unit is removed from the mould.

10.9 Laying of Segmental Pavements

10.9.1 Introduction

The laying of concrete block paving is covered in BS 7533 (BS 7533-3, 1997). Further guidance for concrete block paving and concrete flag paving for cleaning, laying, maintenance and sealing is given in Pritchard (2001).

Segmental pavements are laid manually although machine laying of blocks is used if it is economic, if there is minimal cutting of the blocks and the area to be covered is large and unobstructed. Cutting segmental paving to size is carried out using a guillotine or a power driven abrasive or diamond tipped wheel. The power driven wheels give a cleaner and straighter cut.

10.9.2 Laying of Concrete Block Paving and Clay Pavers

The laying of concrete block paving and clay pavers must begin against a fixed restraint, with the direction of laying and the laying pattern, or bond, being specified. A minimum cross fall of 2.5% is specified.

Paving blocks and pavers are laid in one of three patterns, stretcher bond, herringbone and parquet sometimes called basket weave. The bond being used depends upon whether vehicles will use the paving.

- Stretcher bond is used where traffic loading is light and in areas where blocks are not going to be laid around curves. Stretcher bond requires skilled laying to maintain the straight joint lines, to ensure that the blocks do not creep and that the width of the joints do not widen.
- Parquet is usually restricted to pedestrian areas and again requires skilled laying.
- Herringbone is the most used and the easiest to lay pattern, but it requires careful planning of the layout. In vehicular areas, rectangular blocks in herringbone pattern either at 45° or 90° to the edge restraints are used.

The width of the joints between blocks or pavers is 2 to 5 mm. This thickness allows sand to be brushed into the joint and for the blocks to develop interlock and load transfer. If the joint width is too small the units will touch each other and cause spalling, if the joint width is too large, no interlock will occur.

Following lying, the blocks or pavers are compacted into place. This is carried out by passing a plate vibrator across the surface of the blocks. The number of passes will depend upon the power of the vibrator. The vibrator must not be used near an unrestrained edge as the blocks will move apart. After compaction, dry sand is brushed into the joints and a plate vibrator passed across the surface of the units. Once this has been done the segmental paving can be opened to traffic.

10.9.3 Laying of Concrete Paving Flags

Due to their size some cutting of concrete paving flags to fit around obstacles is inevitable, however good detailing and the use of blocks reduces the need for cutting. Edge restraints are required to prevent the movement of the flags but the cut edge of a flag should not be placed against an edge restraint. A minimum crossfall of 2.5% is specified (BS 7533-4, 1998).

The range of sizes of the flags allows a number of patterns to be created especially if the flags and the block are laid together. The two most common flag patterns are given below.

- Stack bond where the longitudinal and horizontal gaps between the flags are continuous straight lines.
- Broken bond where only the longitudinal gaps or the horizontal gaps between the flags are continuous straight lines. Broken bond is preferred where cars and commercial vehicles overrun the flags.

A sand laying course is used in conjunction with the small element concrete paving flags. For the larger flags a mortar bedding layer is used.

10.9.4 Laying of Precast Concrete Raft Units

Precast concrete raft units are laid on a 50 mm thick sand bedding layer, which is itself supported by a GSB layer (Bull, 1992). Rafts are quickly placed using either fork lift trucks or small cranes and are immediately available for trafficking. The minimum raft sub-base CBR and thickness is 20% and 300 mm respectively. If the sub-base is unbound, it must be permeable and well graded to allow both vertical and horizontal water flow. For very heavy loading, the sub-base should be bound, preferably with Portland cement. The bound sub-base stops all vertical draining, forcing water run off to flow either over the top of the raft if the joints are sealed or through the bedding sand if the joints are unsealed. To increase raft service life the sand bedding layer may be replaced by an asphalt layer. Load transfer between rafts is optional. Doweling the joints improves ride quality, reduced joint movement and increases pavement life. If the joint is undowled, some load transfer will take place depending upon joint width and the type of joint filler. The advantage of having undowled joints is that subgrade settlement is easily identified and rectified. The raft being lifted out, the sub-base relevelled and the raft re-laid.

10.10 Materials and Definitions Used for Segmental Pavements

The materials used for segmental pavement construction are described in this section.

- Bedding course is a layer of either sand, sand and cement or lime mortar used for the bedding of the segmental pavement onto the sub-base.
- Block paving is a paving surfaced with concrete blocks, stone setts, clay pavers or wood blocks.
- Brick paving is a pavement surfaced with clay or calcium silicate pavers or bricks. Calcium silicate pavers are not used where the pavement is subject to heavy loading or de-icing salts.
- California Bearing Ration (CBR) is a measure of the bearing strength of a pavement layer.
- Capping layer is a layer of selected material, with a CBR of at least 15%, used to provide a working platform and to protect sub-grades with CBRs of less than 5%.

- Channelization of wheel tracks occurs when vehicles follow the same path. This occurs if the lane markings are the same width as the vehicles and causes the pavement to rut or deflect to an excessive amount.
- Cobbles are rounded or angular stone, sized between 60 and 200 mm (BS 7533-7, 2002).
- Concrete block paving is a pavement construction with concrete paving blocks overlaying a laying course forming both the wearing surface and a structural element of the pavement.
- Concrete paving blocks are concrete elements fitting within a 295 by 295 mm plan area with a thickness of at least 60 mm. The normal plan area is 200 by 200 mm.
- Edge restraints are placed around all the boundaries of the segmental paving and are essential to prevent lateral movement of the segmental paving and loss of the sand laying course (BS 7533-6, 1999).
- Flags or flagstones are flat slabs of natural stone, artificial stone, or precast concrete with a plan area greater than 300 mm by 300 mm.
- Flexible pavement is a pavement built with little or no tensile strength, which deforms progressively as load is applied.
- Formation is the surface of the subgrade in its final shape after completion of the earthworks.
- Freeze–thaw resistance is the ability to resist damage resulting from cycles of freezing and thawing both with and without de-icing salts.
- GSB material Type 1 should be crushed rock, crushed slag, crushed concrete or well burnt non plastic shale.
- GSB material Type 2 should be natural sands, gravels, crushed rock, crushed slag, crushed concrete or well burnt non plastic shale.
- Hard standing is a paved area provided for outdoor parking, servicing or storage.
- Herringbone pattern is where blocks are laid such that the end of one block is against the side of the next block.
- Industrial pavement is a paved area used for parking, loading or unloading commercial vehicles and for storage in such places as factories, dockyards etc.
- Interlock is the effect of frictional force between paving units which prevents them moving vertically in relation to each other.
- Joint filler is a strip of compressible material that fills a joint space.
- Jointing materials are materials such as, sand, sand cement, mortar, mortar and sand and cement.
- Laying course is a layer of sand, cement or lime and sand mortars, on which the segmental paving units are laid to provide continuous support for the units.
- Lean concrete is a cement-stabilized material.
- Parquet laying patterns are ones where blocks are laid in pairs. The adjacent blocks are also laid in pairs but with the joints perpendicular to the first pair.
- Pavement is that part of the pavement structure above the subgrade.
- Pavers are clay-paving units with a plan area of approximately 100 by 200 mm.
- Paving blocks are concrete paving units with a rectangular plan area of 100 by 200 mm. Shapes other than rectangles are used.
- Raft units are large precast concrete paving units with sizes between 2000 by 2000 mm and 5000 and 10,000 mm.
- Rigid segmental paving is a paving where clay pavers, concrete flags, concrete blocks or stone setts are bedded onto a concrete slab.
- Road base is one or more layers between the sub-base and the laying course constituting the main structural element of a flexible pavement. The road-base may be dense tar macadam, dense bitumen macadam, rolled asphalt, aggregate cement, concrete, lean concrete or bituminous materials.
- Sealant is used to seal the surface of the segmental paving to prevent staining and to ease cleaning of the paved surface.
- Segmental paving is a form of paving where the surface is made from natural stone, precast concrete, baked or burnt clay, or timber blocks which are small enough to be handled manually.

- Semi dry mix is a concrete mix or a mortar mix with low moisture content.
- Serviceability failure in a heavy-duty segmental pavement occurs by either excessive vertical compressive strain in the subgrade or excessive horizontal strain in the base. For pavements with stabilized bases such as granular materials the tensile strain is the active design constraint.
- Setts are segmental paving units commonly of stone or concrete which are rectangular in plan 100, 150, or 200 mm long, 75, or 100 mm wide and 100, 125 or 150 mm deep.
- Skid resistance of segmental paving is determined by the degree of polishing that a surface will undergo as a result of trafficking (BS 7932, 2003).
- Small element paving are square precast concrete paving flags with a plan area of 300 by 300 mm, 400 by 400 mm and 450 by 450 mm.
- Stone setts are rectangular in plan commonly 100, 150, or 200 mm long, 75 or 100 mm wide and 100, 125 and 150 mm deep.
- Stretcher bond is a laying pattern with continuous joint lines in one direction and discontinuous joint lines in the perpendicular direction.
- Sub-base is one or more layers situated between the subgrade and the road-base or the laying course if no road-base is required. Sub-bases are made using unbound materials such as, crushed rock or natural sands and gravels. Wet mix materials made from a combination of sand and crushed rock are also used. Cement bound materials such as soil cement and aggregate cement is also used. Further, normal concrete with a low cement content, called wet-lean concrete may be used.
- Subgrade is the upper part of the soil, natural or constructed, which supports the loads transmitted by the overlaying segmental pavement.
- Surface course is a layer of segmental paving units that provides a wearing course for the applied loads.
- Traffic loading is related to the 8200 kg standard axle loading.
- Type 1 sub-base material is unbound, free draining and restricted to crushed rock, crushed concrete or well burnt non-plastic colliery shale. Compaction is usually carried out using vibrating rollers.
- Wearing course is the surface upon which the vehicles move, e.g., blocks, flags etc.
- Wet casting is a method of manufacturing precast concrete products by placing wet concrete in a mould and then compacting and vibrating it.

10.11   International Segmental Paving

It has not been possible to describe the full range of segmental paving available world-wide due to the differences in materials, climate, traffic and pavement construction. Many countries, including the following, have standards for the manufacture, design and laying of segmental paving: Australia, Austria, Belgium, Bulgaria, Canada, Columbia, Czech Republic, Denmark, Finland, France, Germany, Hungary, India, Ireland, Israel, Italy, Japan, Morocco, Netherlands, New Zealand, Norway, Poland, Romania, Russia, South Africa, Spain, Sweden, Switzerland, Taiwan, Turkey, U.S.A. and Yugoslavia (Lilley, 1991). The CEN committee CEN 178 have produced European Standards for concrete, clay and natural stone segmental paving (BS EN 1338, 2003; BS EN 1339, 2003; BS EN 1340, 2003; BS EN 1341, 2001; BS EN 1342, 2000; BS EN 1343, 2000; BS EN 1344, 2002).

These standards have been written to cover the following areas:

- For paving blocks: abrasion testing, dimensional tolerances, frost resistance, shapes, skid resistance, splitting tests, textures, visual appearance and water absorption. Plan areas standardize around 200 by 100 mm although lengths up to 280 mm are used with thickness' being between 60 and 140 mm. The definition of a paving block is given as the overall length divided by the thickness must be less than or equal to four.
- *For concrete paving flags.* Air-entertainment requirements, flatness tests, flexural tests, freeze – thaw tests and splitting tests. The overall length does not exceed 1000 mm and the overall length divided by the thickness is greater than four.

- *For cobbles, setts, stone paving blocks and stone paving flags.* Abrasion resistance, absorption, dimensions, freeze thaw, materials, e.g., granite and other selected rocks, skid/slip, strength and water absorption.
- *For clay pavers.* Abrasion, dimensional tolerances, size, slip/skip, thickness and transverse breaking load.
- *For wooden blocks.* Both with and without preservatives.
- *For concrete there are requirements regarding.* Admixtures, aggregates, cement, cement content, material requirements, mix proportions, water and water–cement ratio.

10.12   Model Specification Clauses for Segmental Paving

When a segmental pavement is to be laid it is necessary for the client, the contractor, those who will pay for the pavement, use the pavement and maintain the pavement to be clear on the requirements of the pavement. These requirements are achieved by using a contract with information in it similar to that described below.

10.12.1   The Pavement Surface

The color, shape, size, surface sealant and type of pavement surface must be specified. For example precast block paving or concrete paving flags should be specified by the relevant country's standards.

10.12.2   Bedding, Laying and Jointing of Concrete Block Paving

The sand used for the laying course and for the joints must be specified regarding grading, moisture content and the type of load it is going to sustain. For example, aircraft pavements, footways, industrial pavements, pedestrian areas and roads require different categories of material for the laying course. The laying course thickness, when laid on a sub-base will be 50 mm thick when compacted and when laid on a road base will be 30 mm thick when compacted.

The preparation of the laying course may be by one of three methods.

- The laying course may be precompacted by being spread out and compacted with a vibrating plate compactor and screeded to the required level.
- The laying course may be partially precompacted by the use of a vibrating plate compactor with a further 15 mm layer of loose material being laid over the top.
- The laying course may be compacted after the blocks have been laid. The laying course and the concrete blocks are compacted using a vibrating plate compactor.

The joints between the blocks must have a gap of between 2 and 5 mm. Sand must be brushed into the joints and a vibrating plate compactor applied to the blocks. No concrete blocks should be laid within 1000 mm of an unrestrained edge.

10.12.3   Bedding, Laying and Jointing of Concrete Paving Flags on Sand

For the standard flags of type A, B, C and D or small element flags of type E, F and G, the compacted laying course of sand shall have a nominal thickness of 25 mm. The type of sand must be specified and the top of the compacted sand must be loosened. The flags must be bedded down onto the sand using a paviors maul, or a vibrating plate compactor for flags of type E, F or G. No flags may be laid within 500 mm of an unrestrained edge. Typical restrained edging includes kerbs (BS EN 1340, 2003; BS EN 1343, 2000; BS 7263-2, 2001).

10.12.4   Bedding, Laying and Jointing of Concrete Paving Flags on Mortar

For the standard flags of type A, B, C or D the mix of the mortar laying course and any additives must be specified. The usual compacted thickness of the mortar is 25 mm. Where the joint thickness between the flags is 5 mm or less, jointing sand or a lime-sand mortar shall be used to fill the joint. If the joint thickness is between 5 and 10 mm, the joints are to be filled with compacted mortar. Joints less than 2 mm and greater than 10 mm are not acceptable.

10.12.5   Preparation of the Subgrade

The subgrade must be excavated to formation level and any weak or unsuitable subgrade must be removed and replaced with a subgrade improvement layer. The requirements of the subgrade improvement layer must be specified. All sub-soil drainage must be completed before the sub-base is laid. All trenches must be back filled and all back fill compacted. The surface level of the subgrade must be specified as must the allowable deviations from the surface level.

10.12.6   The Pavement Sub-Layers

The type and material used for each of the pavement sub-layers must be specified. The sub-layer materials are usually any of the following:

- type 1 GSB,
- dense bitumen macadam,
- hot rolled asphalt,
- pavement quality concrete,
- wet lean concrete,
- cement bound material or
- concrete.

Drainage must be completed in conjunction with the construction of the sub-layer. The sub-layer material must be placed in layers not exceeding a specified thickness and compacted in accordance with the specifications. Concrete and wet lean concrete must be compacted by tamping or vibration. Other materials must be compacted using an engine driven vibro tamper, a vibrating plate compactor or a vibrating roller. The surface requirements of the finished sub-layer must be specified in relation to lack of cracks, lack of movement when being compacted, texture, surface level and permitted deviations from this level. Concrete sub-layers and sub-layers containing cement must be protected with a curing membrane if there are time delays before they are covered with the next layer. There must be specified minimum times and minimum temperatures before these concrete sub-layers and sub-layers containing cement can be overlain with the segmental paving. Any areas that become defective must be removed, re-laid with new material and recompacted.

10.12.7   Preparation of Existing Bases as the Sub-Layer

Where segmental paving is to be laid over existing roads or similar types of construction, corrections to the levels of the existing construction may be needed. It is also necessary to ensure that any existing drainage can continue to function after the adjustment of the levels. Excess material should be removed by a planning process and limits in the deviations of the levels of the sub-base and the road-base must be specified. The required level of the sub-layer must be built up using material as specified and laid in layers and compacted as specified.

10.12.8   Edge Restraints and Surface Level

Edge restraints must surround all areas of segmental paving to prevent the loss of the sand laying course and the sideways movement of the segmental pavement. The maximum difference in surface level between adjacent segmental paving units must be specified as must the flatness tolerance of the pavement surface. The usual maximum values are between 3 and 10 mm depending upon the type of segmental paving and the applied loading.

10.13   The Future of Segmental Paving

The future of segmental paving depends upon the segmental paving industry's ability to satisfy and to predict consumer needs at economic prices. These requirements relate to the design, manufacture, laying, use and maintenance of segmental paving.

10.13.1   The Design of Segmental Paving

The design of segmental paving has changed from the late nineteenth century's empirically based procedures to today's analytical and computer based procedures which use laboratory and on site testing to ensure the efficient use of materials in a sustainable way.

Segmental paving can now be designed for any combination of loading and subgrade, with computer methods being able to assimilate the most diverse requirements into the design; for example, the use of permeable segmental paving. Initially segmental paving design required only knowledge of the applied loading and the strength of the subgrade. The added hydraulic design requirements that the paving surface must be permeable and that the pavement layers and the subgrade must be able to drain away and filtrate the rainfall at a rapid rate, have now been incorporated.

10.13.2   The Manufacture and Laying of Segmental Paving

The manufacture of segmental paving is an area that has long been under scrutiny by the manufactures. Manufacturing segmental paving is a highly efficient operation with little room for improving existing techniques. What is needed is increased versatility in the use of recycled materials and alternatives to existing materials used in the manufacture of segmental paving. For example can the setting time of the cement be reduced to allow increased usage of the manufacturing plant? Are there ways in which the laying of the sand laying course and the paving surface could be automated to a higher extent to reduce the labor cost of installation?

10.13.3   The Use of Segmental Paving

New uses and the development of existing uses for segmental paving will increase. This can be seen in the development of permeable segmental paving. This is a market which has only recently been recognized, but with the continued urbanization of land and the need to continue to refill aquifers rather than run rainwater into drains and into rivers, the increase in this market is assured (Figure 10.8).

The importance of maintaining the ambiance of historic town centers has meant that many towns wish to replace asphalt road surfacing with segmental paving. This will become an expanding and a lucrative market, providing the manufactures of segmental paving can produce paving that harmonizes with the natural materials of the townscapes.

One area of massive potential expansion is the domestic home market. Families want attractive

FIGURE 10.8   Segmental paving used to enhance ambiance.

drives, landscape areas, pathways and patios around their houses and they are prepared to pay for this requirement. This requires manufacturers to offer many new attractive designs, colors, textures and a personalized design service. Specialist design offices will be needed that offer a complete service of sales, design, construction and maintenance.

10.13.4   The Maintenance of Segmental Paving

There is little maintenance that needs to be carried out to segmental paving. Maintenance usually comprises the refilling of the joints with sand and the cleaning of the surface of the paving

FIGURE 10.9   Cleaning and sealing concrete paving flags.

(Pritchard, 2001). The development of new surface sealants and treatments to resist algae, bitumen stains, chewing gum, dirt, graffiti, lichens, moss, oil stains, paint, rust stains and tire marks is continuing BS 7533 (BS 7533-11, 2003) (Figure 10.9).

Acknowledgments

Acknowledgement of all the figures used in this chapter is given to Interpave, The Precast Concrete Paving and Kerb Association of 60 Charles Street Leicester, LE1 1FB, UK. The use of Interpave's publications is duly acknowledged.

References

BS EN 1338, 2003. *Concrete Paving Blocks, Requirements and Test Methods.* British Standards Institution, London.

BS EN 1339, 2003. *Concrete paving flags, Requirements and Test Methods.* British Standards Institution, London.

BS EN 1340, 2003. *Concrete Kerb Units, Requirements and Test Methods.* British Standards Institution, London.

BS EN 1341, 2001. *Slabs of Natural Stone for External Paving. Requirements and Test Methods.* British Standards Institution, London.

BS EN 1342, 2000. *Setts of Natural Stone for External Paving. Requirements and Test Methods.* British Standards Institution, London.

BS EN 1343, 2000. *Kerbs of Natural Stone for External Paving. Requirements and Test Methods.* British Standards Institution, London.

BS EN 1344, 2002. *Clay Pavers, Requirements and Test Methods.* British Standards Institution, London.

BS 6717, 2001. *Precast, Unreinforced Concrete Paving Blocks — Requirements and Test Methods.* British Standards Institution, London.

BS 7263, 2001. *Precast Concrete Flags, Kerbs, Channels, Edgings and Quadrants.* British Standard Institution, London, *Part 1* Precast unreinforced concrete paving flags and complementary fittings requirements and test methods. *Part 2* Precast unreinforced concrete kerbs, channels edgings and quadrants requirements and test methods.

BS 7533, *Pavements Constructed with Clay, Natural Stone or Concrete Pavers.* British Standards Institution, London. *Part 1* Guide for the structural design of heavy duty pavements constructed of clay pavers or precast concrete paving blocks, 2001. *Part 2* Guide for the structural design of lightly trafficked pavements constructed of clay pavers or precast concrete paving blocks, 2001. *Part 3* Code of practice for laying precast concrete paving blocks and clay pavers for flexible pavements, 1997. *Part 4* Code of practice for the construction of pavements of precast concrete flags or natural stone slabs, 1998. *Part 6* Code of practice for laying natural stone, precast concrete and clay kerb units, 1999. *Part 7* Code of practice for the construction of pavements of natural stone setts and cobbles, 2002. *Part 8* Guide for the structural design of lightly trafficked pavements of precast concrete flags and natural stone flags, 2003. *Part 11* Code of practice for the opening, maintenance and reinstatement of pavements of concrete, clay and natural stone, 2003.

BS 7932, 2003. *Determination of the Unpolished and Polished Pendulum Test Value of Surface Units*. British Standards Institution, London.

Bull, J.W., An analytical solution to the design of precast concrete pavements, *J. Num. Anal. Methods Geomech.*, 10, 115–123, 1986.

Bull, J.W. 1990a. *Precast Concrete Raft Units*. Blackie and Sons, Glasgow, chap. 2.

Bull, J.W. 1990b. The research and development of a design method for precast concrete pavement units used for highway loading and used for heavy industrial loading, pp. 338–347. In *Second International Conference on Structural Engineering Analysis and Modeling*, Kumasi, Ghana.

Bull, J.W. 1992. The analysis, design, manufacture and laying of precast concrete for highways and airfields. *Third International Symposium on Noteworthy Applications in Concrete Prefabrication*, Singapore, 250–258.

Croney, P. and Croney, D. 1991. *The Design and Performance of Road Pavements*, 2nd Ed. McGraw-Hill, New York, chap. 8.

Croney, P. and Croney, D. 1997. *The Design and Performance of Road Pavements*, 3rd Ed. McGraw-Hill, New York, chap. 2.

Department of Transport, 1992. *Manual of Contract Documents for Highway Works, Volume 1: Specification for Highway Works*. HMSO, London.

Heukelom, W. and Foster, C.R., Dynamic testing of pavements, *J. Soil. Mech. Found. Div. ASCE*, 86, 1960, 5MI, Proc Paper 2368.

Interpave, 2003. *Permeable Pavements, Guide to the Design, Construction and Maintenance of Concrete Block Permeable Pavements*. Interpave, Leicester.

Knapton, J. and Meletiou, M. 1996. *The Structural Design of Heavy Duty Pavements for Ports and other Industries*. Interpave, The British Precast Concrete Federation, Ltd, Leicester.

Lilley, A.A. 1991. *A Handbook of Segmental Paving*. E&FN SPON, London.

Ministry of Defense, 1996. *Concrete Block Paving for Airfields*, Defense Works Functional Standard 035. Defense Estates Organization/HMSO, London.

Powell, W.D., Potter, H.C., Mayhew, H.C., and Nunn, M.E. 1984. *The Structural Design of Bituminous Roads*. Transport and Road Research Laboratory, Report LR1132, TRRL, Crowthorne.

Pritchard, C. 2001. *Precast Concrete Paving: Installation and Maintenance*. Interpave, Leicester.

Pritchard, C. and Dowson, A. 1999. *Precast Concrete Paving: A Design Handbook*. Interpave, Leicester.

Rollings, R.S. and Chau, Y.T. 1981. *Precast Concrete Pavements*. US Army miscellaneous paper, GL-81-10, Vicksburgh, Mississippi, U.S.A.

Transport and Road Research Laboratory, 1970. *A Guide to the Structural Design of Pavements for New Roads, Department of the Environment Road Note 29*, 3rd edn. HMSO, London.

11

Overlay Design for Flexible Pavements

Mang Tia
University of Florida
Gainesville, FL, U.S.A.

11.1   Introduction

11.1.1   Need for Overlay

A flexible pavement, with accumulated traffic loads and time in service, may suffer one or more of the following deficiencies:

1. Excessive rutting
2. Excessive cracking
3. Inadequate ride quality
4. Inadequate skid resistance of surface

In some cases, a pavement may be adequately maintained and may not have the above-listed deficiencies, but the pavement may have the following problems:

1. Excessive maintenance costs
2. Inadequate structural capacity for the expected future traffic loads.

In all of the above cases, treating the pavement with an overlay is the most commonly used method for restoring or upgrading the pavement to its desired condition and level of serviceability.

11.1.2   Types of Overlay for Flexible Pavements

The types of overlay on flexible pavements include the following:

1. Asphalt concrete overlay on flexible pavement
2. Conventional Portland cement concrete overlay on flexible pavement, in which the concrete layer is placed unbonded over the asphalt surface.
3. Ultra-thin Portland cement concrete overlay on flexible pavement, in which a thin concrete layer of 10 cm (4 in.) or less is placed bonded over the asphalt surface.

11.1.3   Overlay Design Procedure

Overlay design procedures that have been used can be grouped into three main categories as follows (Finn and Monismith 1984; Monismith and Brown 1999):

1. Component analysis design
2. Deflection-based design
3. Analytically-based design

A component analysis overlay design procedure basically involves evaluating the condition of the components (pavement layers) in the existing pavement to be overlaid, and comparing them to equivalent thicknesses of new pavement materials to be placed. The required overlay thickness is equal to the difference between the required total thickness and equivalent thickness of the existing layer. The procedure usually requires making an engineering judgment based on visual inspection and laboratory testing of the existing pavement materials. Examples of component analysis overlay design procedures include (1) AASHTO, 1993 component analysis overlay design procedure (AASHTO, 1993), (2) Asphalt Institute effective thickness method (Asphalt Institute, 2000), and (3) U.S. Corps of Engineers component analysis method (Corps of Engineers, 1958).

A deflection-based overlay design procedure basically uses the surface deflection caused by a nondestructive test (NDT) to estimate the structural capacity of an existing pavement. The required overlay thickness is the additional pavement thickness that will be required to bring the NDT deflection to the desired level. This type of design procedure is usually based on the empirical correlations between certain NDT deflections and field performances. Because of the empirical nature of this procedure, the applicability of a particular procedure is usually limited to regions of similar conditions (such as, climate, soil type and pavement materials used), and the same NDT test procedure used.

Examples of deflection-based overlay design procedures include (1) Asphalt Institute deflection-based method (Asphalt Institute, 2000), (2) California Department of Transportation method (Caltrans, 1995), (3) AASHTO, 1993 deflection-based overlay design procedure (AASHTO, 1993), (4) Transport and Road Research Laboratory (U.K.) method (Lister et al., 1982), (5) Roads and Transport Association (Canada) method (CGRA, 1965), and (6) U.S. Army Corps of Engineers deflection-based method (Hall, 1978).

An analytically based (or mechanistic) overlay design procedure is based on the analysis of stresses and strains in a pavement due to the expected traffic loads, and the correlations of the analytical stresses and strains to performance. The required overlay thickness is the additional pavement thickness that will be required to bring the expected stresses and strains to the acceptable levels to prevent failure. This procedure requires extensive evaluation of the *in situ* properties of all the materials to be used in the pavement structure, including their damage characteristics (such as, creep and fatigue behavior.) This procedure also requires more extensive analysis as compared with the other two methods.

Examples of analytically based AC overlay design procedures include (1) Federal Highway Administration (FHWA) procedure developed by Austin Research Engineers (ARE) (ARE, 1975), (2) FHWA procedure developed by Resource International Incorporated (RII) (Majidzadeh and Ilves, 1980), (3) Shell Research Procedure (Claessen and Ditmarsch, 1977), (4) Washington DOT Everpave method (WSDOT, 1995), (5) University of Nottingham (U.K.) method (Brown et al., 1987), (6) Austroads method (Austroads, 1994), and (7) Florida REDAPS procedure (Ruth et al., 1990).

11.2   Evaluation of Pavement Performance for Overlay

In designing an overlay for an existing pavement, it is imperative to perform a thorough evaluation of the existing pavement to determine its areas of deficiency and deterioration. The causes of the deficiencies and/or deterioration need to be determined so that the proper pretreatments and materials for use in overlay can be selected.

This section describes the fundamentals in the evaluation of pavement's functional, structural and safety performance. The selection of proper pretreatments and materials for use in overlay for pavements of various deterioration conditions is also described in this section.

11.2.1   Functional Performance

The functional performance of a pavement is the ability of the pavement to serve its users in its primary function, which is to provide a safe and smooth driving surface. The most commonly used measure of functional performance of a pavement are its ride quality, which is commonly quantified in terms of Present Serviceability Rating (PSR), Present Serviceability Index (PSI), Riding Comfort Index (RCI), and Ride Number (RN). Roughness is commonly quantified in terms of International Roughness Index (IRI).

The concept of PSR and the standards for PSR were developed during the AASHO Road Test (Carey and Irick, 1960). PSR is a rating of pavement serviceability on a scale of zero (worst condition) to five (best condition), and is based on the average rating from a panel of 12 raters. At the time of the AASHO Road Test, a PSR of less than 2.5 was considered to be unacceptable for primary highways, and a PSR of less than 2.0 was considered to be unacceptable for secondary highways.

The concept of PSI was also developed during the AASHO Road Test. Since it was not practical to have a panel of 12 raters, who needed to be calibrated to the original AASHO standards, to rate pavements at all times, a relationship between PSR and some measured physical characteristics for flexible pavements was developed as follows:

$$PSR = 5.03 - 1.91 \log_{10}(1 + \overline{SV}) - 1.38\overline{RD}^2 - 0.01\sqrt{C + P} \qquad (11.1)$$

where:

$\overline{SV}$ = average slope variance in units of 10^{-6}.
$\overline{RD}$ = average rut depth in inches.
$C + P$ = cracking and patching in $\text{ft}^2/1000 \text{ ft}^2$.

The estimated PSR obtained from the measured physical characteristics is termed PSI.

Of the three physical characteristics used in the prediction of PSR, the slope variance is the most dominant factor, and can be used to estimate PSR alone. Slope variance is a measure of longitudinal surface roughness. Various equipment for measuring pavement roughness have been used for determination of PSI. The most commonly used equipment for measurement of pavement surface roughness, which is in turn used to determine PSI, include the Mays meter, PCA road meter and the road laser profiler, which are usually operated at a speed of 50 mi/h (80 km/h). It is to be noted that the calibration of these various equipment for determination of PSI is supposedly to be set to the standards as established during the AASHO Road Test. Ride Comfort Index (RCI) and Ride Number (RN) are similar to PSI, except that it is on a scale of zero (worst condition) to ten (best condition).

11.2.2   Structural Performance

Structural performance of a pavement is its ability to sustain the applied traffic loads without showing distress. Deficiencies in a pavement's structural performance can be manifested through its observed distresses. Evaluation of pavement distresses is usually done through a condition survey, which is conducted by a trained personnel making visual observation of the rated pavement and noting down the type, frequency and severity of the observed distresses. The three main types of distresses are (1) cracking, (2) rutting, and (3) raveling. The observed distresses can be used to determine the extent and the causes of deterioration in the pavement, and to determine the most appropriate pretreatments and materials to use for an overlay.

The structural capacity of an existing pavement can be evaluated by performing a component analysis, or by means of nondestructive testing. From a component analysis, the structural number (SN) as used in the AASHTO, 1993 Pavement Design Guide (AASHTO, 1993), or the TI as used in the Caltrans Pavement Design Guide (Caltrans, 1995) can be estimated. The measured deflections from NDT tests can be used to estimate the structural capacity of the pavement, based on the empirical relationship between certain NDT deflections and performance. They can also be used to back-calculate elastic moduli of the various layers in the existing pavement, which in turn can be used as inputs to analytical models to predict the behavior and performance of the pavement under various loading conditions.

11.2.3   Safety Performance

Two main common concerns in the evaluation of pavement safety are (1) the skid resistance of pavement surface under wet condition, and (2) the potential for hydroplaning. The pavement surface's skid resistance under wet condition is commonly measured by the Locked Wheel Skid Trailer Test, which is standardized by ASTM as E274 Standard Test Method for Skid Resistance of Paved Surfaces Using a Full-Size Tire (ASTM, 2003). This test measures the skid resistance as experience by a locked tire traveling at 40 mi/h (64.4 km/h) on a wet pavement surface. Measurements of skid resistance are made in terms of skid number (SN_{40}), which is defined as:

$$SN = 100 f \tag{11.2}$$

where:

f = friction factor, or coefficient of friction as measured by this test.

A skid number of over 35 on a wet surface is generally considered to be acceptable, while a skid number of less than 25 is considered unacceptable.

Hydroplaning is the condition when the vehicle tires loss contact with the pavement surface due to the accumulation of water on the pavement surface. This can be caused by excessive rutting and/or insufficient cross slope in the pavement, which can cause accumulation of water on the pavement surface. A rut depth of 0.5 in. (10 mm) or more can create a potential for hydroplaning. The absence of an open-graded friction course on the pavement surface also increase the potential for accumulation of water and thus the potential for hydroplaning.

11.2.4   Selection of Pretreatments and Materials for Overlay

In the design and placement of an overlay for an existing pavement, it is very important that the causes of problems or deficiencies in the existing pavement should be determined and sound engineering judgment made in order to achieve effective benefits from the overlay. This section describes the recommended pretreatments and overlay materials for various conditions of the existing asphalt pavement to be rehabilitated.

11.2.4.1   Pavement with Extensive Cracking

If an existing pavement is severely cracked, it is important that the cracked layer should be milled off, or a crack relief layer such as an asphalt rubber membrane interlayer (ARMI) be placed before the overlay is placed, to prevent reflective cracking through the overlay. While it is possible to design a thicker overlay to postpone the time for reflective cracking to propagate to the surface, this approach is usually not cost effective unless an increase in the elevation of the pavement surface is needed.

11.2.4.2   Pavement with Top-Down Cracking

If an existing pavement suffers from top-down cracking, simply milling off the cracked layer and overlaying it with the same asphalt mixture will lead to the same problem again. Unlike bottom-up cracking, top-down cracking is not related to the thickness design of the pavement, but rather, to the material characteristics of the asphalt mixture used and the high surface stresses induced by the high-pressure radial tires. These high surface stresses are not governed by the thickness of the asphalt layer. Asphalt mixtures with higher fracture energy (i.e., the amount of energy required to fracture the material) have been found to resist top-down cracking better. To reduce the potential for top-down cracking in this pavement in the future under the same traffic condition, the pavement needs to be overlaid with an asphalt mixture which has demonstrated to be resistant to top-down cracking in service. This may require the use of a polymer modified asphalt mixture for the overlay, or placing a polymer modified interlayer between the friction course and the structural layer.

11.2.4.3   Pavement with Low-Temperature Cracking

Evenly spaced transverse cracks in asphalt pavements are low-temperature cracks caused by rapid cooling of the pavement surface coupled with the use of an asphalt mixture that is too hard at the low temperature. These cracks initiate from the surface and usually propagate to the bottom of the asphalt layer. If the asphalt layer is not to be removed, a crack relief layer may be placed on top of this pavement before the overlay is placed to prevent the cracks to reflect through the overlay. The same mixture with low-temperature cracking problem should not be used as the overlay material.

11.2.4.4   Pavement with Severe Rutting

If an existing pavement suffers from severe rutting, the pavement needs to be milled off to sufficient depth to remove the rut before the overlay is placed so that overlay will be of uniform thickness. If rutting occurred because of a rut susceptible mixture, the entire layer containing this unstable mixture needs to be milled off. The rut susceptible mixture should not be used as the overlay material.

11.2.4.5   Pavement with Insufficient Cross-Slope

If an existing pavement has insufficient cross-slope, resulting in inadequate drainage, sufficient pavement surface material must be milled off to correct the cross-slope before the overlay is placed.

11.2.4.6   Pavement with Raveling Problem

If an existing pavement has a raveling problem due to moisture damage on the surface, the deteriorated material needs to be milled off, and the existing drainage problem needs to be corrected before an overlay is placed. A stripping-resistant mixture needs to be used as the overlay material.

11.2.4.7   Pavement with Unstable Base or Subbase

If an existing pavement has an unstable granular base or subbase layer due to water saturation, the unstable layer needs to be removed and replaced with a stable material. If the problem is due to high water table, the water table needs to be brought down by installation of a properly designed sub-drainage system.

11.2.4.8   Pavement with Poor Ride Quality

If an existing pavement has a poor ride quality, but has sufficient structural capacity from its thickness design, the pavement should be milled to sufficient levelness, and an overlay of minimum required thickness may be placed to restore the pavement's ride quality.

11.2.4.9   Pavement with Poor Skid Resistance

If an existing pavement has inadequate skid resistance at the surface, the existing friction course, if any, needs to be milled off, and a new friction course may be placed as an overlay to restore the skid resistance of the pavement surface.

11.2.4.10   Proper Selection of Overlay Materials and Friction Courses

A lot of pavement distresses are related to quality of the materials rather than the pavement's thickness design. Thus, it is very important that proper asphalt mixtures with adequate cracking, rutting and stripping resistance be used as the overlay material. The asphalt binders used for the asphalt mixtures should be of a grade that is suitable for the expected environmental condition. If a structural overlay is placed on a high-volume high-speed highway, an appropriate friction course should be placed on the pavement surface to provide sufficient skid resistance.

11.3   Procedures for Design of AC Overlay on Flexible Pavement

11.3.1   AASHTO, 1993 Pavement Design Guide

The procedure for design of AC overlay on flexible pavement in the AASHTO, 1993 Pavement Design Guide (AASHTO, 1993) consists of a component analysis method and a deflection-based analysis method. It is based on the concept that the structural capacity of a flexible pavement can be quantified by a SN. The required overlay thickness is the amount that will increase the effective SN of the existing pavement (after the necessary milling and repair before the overlay) to the required SN to meet the future traffic demand. The relationship between the required overlay thickness and the other parameters can be expressed by the following equation:

$$SN_{ol} = a_{ol}D_{ol} = SN_f - SN_{eff} \qquad (11.3)$$

where:

SN_{ol} = required overlay structural number.
a_{ol} = structural coefficient for the AC overlay.
D_{ol} = required overlay thickness in inches.
SN_f = required structural number for future traffic demand.
SN_{eff} = effective structural number of the existing pavement to be overlaid.

The procedures for determination of SN_{eff}, SN_f, a_{ol} and D_{ol} are described in the following sections.

11.3.1.1   Determination of Effective Structural Number of Existing Pavement (SN_{eff})

The effective structural number of an existing pavement to be overlaid (SN_{eff}) may be determined from (a) results of Non-Destructive Tests (NDT) (using a deflection-based procedure), (b) results of condition survey (using a component analysis), or (c) remaining life analysis.

11.3.1.1.1   Determination of SN_{eff} from NDT
The determination of SN_{eff} from results of NDT is based on the assumption that the structural capacity of a pavement is a function of its total thickness and overall stiffness. The relationship between SN_{eff}, thickness and stiffness as given in the AASHTO, 1993 Pavement Design Guide is as follows:

$$SN_{eff} = 0.0045D\sqrt[3]{E_p} \qquad (11.4)$$

where:

D = total thickness of all pavement layers above the subgrade in inches.
E_p = effective modulus of pavement layers above the subgrade in psi.

E_p may be determined through the following steps:

(1) Perform a Falling Weight Deflectometer (FWD) test on the existing pavement using a load magnitude of approximately 9000 pounds (40 kn). Measure deflections at the center of the load and at least one other location at a sufficient distance from the load.

(2) Determine the subgrade modulus (M_R) using the following equation:

$$M_R(\text{psi}) = \frac{0.24P}{d_r r}$$

(11.5)

where:

P = applied FWD load in pounds.

d_r = FWD deflection at a distance r from the center of the load in inches.

r = distance from center of FWD load in inches.

The distance r should be far enough such that the measured FWD deflection would be independent of the effects of the pavement layers above the subgrade, but close enough such that the deflection would not be too small to be measured accurately. The minimum distance is given as follows:

$$r \geq 0.7a_e$$

(11.6)

where:

$$a_e = \sqrt{\left[a^2 + \left(D\sqrt[3]{\frac{E_p}{M_R}} \right)^2 \right]}$$

(11.7)

a_e = radius of the stress bulb at the subgrade-pavement interface in inches.

a = FWD load plate radius in inches.

D = total thickness of all pavement layers above the subgrade in inches.

E_p = effective modulus of pavement layers above the subgrade in psi.

(3) E_p may be determined from the FWD deflection at the center of the load plate using the following equation:

$$d_0 = 1.5\,\text{pa}\left\{ \frac{1}{M_R\sqrt{1 + \left(\frac{D}{a}\sqrt[3]{\frac{E_p}{M_R}} \right)^2}} + \frac{\left[1 - \frac{1}{\sqrt{1 + \left(\frac{D}{a} \right)^2}} \right]}{E_p} \right\}$$

(11.8)

where:

d_0 = FWD deflection at the center of the load plate (and adjusted to a standard temperature of 68°F) in inches.

p = FWD load plate pressure in psi.

Given a known or assumed value of M_R, the value of E_p can be determined from the above equation. It is to be noted that d_0 is to be adjusted to a standard temperature of 68°F. The AASHTO, 1993 Pavement Design Guide provides a figure (Figure 5.6 in the AASHTO Guide) for use for adjusting d_0 for AC mix temperature for AC pavements with granular and asphalt-stabilized bases, and a figure (Figure 5.7 in the AASHTO Guide) for adjusting d_0 for AC pavements with cement- and pozzolanic-stabilized bases. These two figures are shown here as Figures 11.1 and 11.2, respectively.

11.3.1.1.2 *Determination of* SN$_{\text{eff}}$ *from Condition Survey*

The method of determination of SN$_{\text{eff}}$ from condition survey involves making an engineering judgment in assigning layer coefficients and drainage coefficients to the various layers of the existing pavement, and

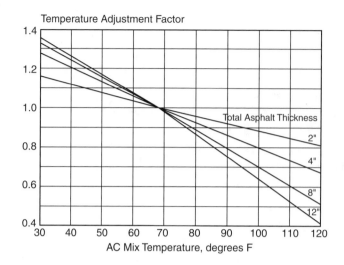

FIGURE 11.1   Adjustment to d_o for AC mix temperature for pavement with granular or asphalt-treated base. (*Source*: From Guide for Design of Pavement Structures, Volume I, 1993 by AASHTO, Washington, DC. With permission.)

calculating the SN_{eff} using the structural number equation as follows:

$$SN_{eff} = a_1 D_1 + a_2 D_2 m_2 + a_3 D_3 m_3 \qquad (11.9)$$

where:

$D_1, D_2, D_3 =$ thickness of existing pavement surface, base and subbase layers.
$a_1, a_2, a_3 =$ corresponding structural layer coefficients.
$m_2, m_3 =$ drainage coefficients for granular base and subbase.

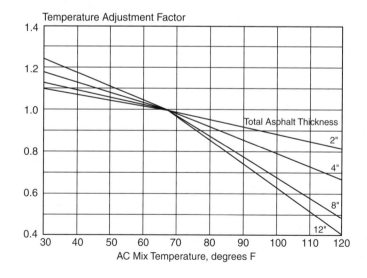

FIGURE 11.2   Adjustment to d_o for AC mix temperature for pavement with cement- or pozzolanic-treated base. (*Source*: From Guide for Design of Pavement Structures, Volume I, 1993 by AASHTO, Washington, DC. With permission.)

The assigning of drainage coefficients is similar to the case for new construction, which has already been described in Chapter 8 of this Handbook. Generally, the assigned layer coefficients of the in-service pavement materials should in most cases be less than the values that would be assigned to the same materials for new construction. However, limited guidance is available for the determination of layer coefficients for in-service pavement materials. Each highway agency usually adopts its own set of values to use based on local experience. The AASHTO, 1993 Pavement Design Guide provides a table (Table 5.2 in the AASHTO Guide) with suggested layer coefficients for various pavement materials with various levels of deterioration. It is shown here as Table 11.1.

11.3.1.1.3 *Determination of* SN_{eff} *from Remaining Life Analysis*

The determination of SN_{eff} from remaining life analysis is based on the fatigue damage concept that the structural capacity of a pavement diminishes gradually as the pavement is subjected to increasing number of traffic loads.

The remaining life of a pavement, as a percentage of its design life can be represented by the following equation:

$$R_L = 100\left[1 - \left(\frac{N_p}{N_{1.5}}\right)\right] \qquad (11.10)$$

where:

R_L = remaining life in percent.
N_p = total traffic to date in 18-kip (80 kn) ESAL.
$N_{1.5}$ = total traffic to pavement failure (PSI = 1.5) in 18-kip ESAL.

TABLE 11.1 Suggested Layer Coefficients for Existing AC Pavement Layer Materials

Material	Surface Condition	Coefficient
AC surface	Little or no alligator cracking and/or only low-severity transverse cracking	0.35–0.40
	<10% low-severity alligator cracking and/or <5% medium- and high-severity transverse cracking	0.25–0.35
	>10% low-severity alligator cracking and/or <10% medium-severity alligator cracking and/or >5–10% medium- and high-severity transverse cracking	0.20–0.30
	>10% medium-severity alligator cracking and/or <10% high-severity alligator cracking and/or >10% medium- and high-severity transverse cracking	0.14–0.20
	>10% high-severity alligator cracking and/or >10% high-severity transverse cracking	0.08–0.15
Stabilized base	Little or no alligator cracking and/or only low-severity transverse cracking	0.20–0.35
	<10% low-severity alligator cracking and/or <5% medium- and high-severity transverse cracking	0.15–0.25
	>10% low-severity alligator cracking and/or <10% medium-severity alligator cracking and/or >5-10% medium- and high-severity transverse cracking	0.15–0.20
	>10% medium-severity alligator cracking and/or <10% high-severity alligator cracking and/or >10% medium- and high-severity transverse cracking	0.10–0.20
	>10% high-severity alligator cracking and/or >10% high-severity transverse cracking	0.08–0.15
Granular base or subbase	No evidence of pumping, degradation, or contamination by fines	0.10–0.14
	Some evidence of pumping, degradation, or contamination by fines	0.00–0.10

Note: From Guide for Design of Pavement Structures, Volume I, 1993 by AASHTO, Washington, DC. With permission. Documents may be purchased from the AASHTO bookstore at 1-800-231-3475 or online at http://bookstore.transportation.org

To calculate R_L, the total amount of traffic the pavement has carried to date (N_p) and the total amount of traffic the pavement could be expected to carry to a terminal serviceability index of 1.5 ($N_{1.5}$) need to be determined. $N_{1.5}$ can be estimated using the AASHTO pavement design equations or nomographs and using a terminal PSI of 1.5 and a reliability of 50 percent. Using the AASHTO pavement design equation with a terminal PSI of 1.5 and a reliability of 50 percent, the relationship between $N_{1.5}$ and the original structural number, SN_o, can be expressed as follows:

$$Log_{10}N_{1.5} = 9.36 \log_{10}(SN_o + 1) + 2.32 \log_{10}M_R - 8.27 \tag{11.11}$$

The determined $N_{1.5}$ can be used to calculate the remaining design traffic, N_{eff}, as follows:

$$N_{eff} = N_{1.5} - N_p \tag{11.12}$$

where:

N_{eff} = remaining design traffic to pavement failure (PSI = 1.5) in 18-kip ESAL.

SN_{eff} can be determined from the following equation:

$$SN_{eff} = CF \times SN_o \tag{11.13}$$

where:

CF = condition factor, which is a function of R_L.

The AASHTO, 1993 Pavement Design Guide provides a figure (Figure 5.2 in the AASHTO Guide) for determination of CF as a function of RL. It is shown here as Figure 11.3.

It is to be pointed out that the AASHTO remaining life concept has been found to produce inconsistent overlay design thickness. Well performing pavements that have served beyond its design life and are still in good condition would have a negative Remaining Life according this method of analysis. Conversely, a poorly performing pavement that has failed prematurely would still have a high Remaining Life according to this method. Thus, this method has been recommended for exclusion from the AASHTO overlay design approach, as presented in Appendix M of the AASHTO, 1993 Pavement Design Guide (AASHTO, 1993).

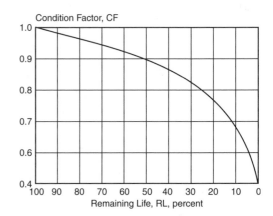

FIGURE 11.3 Relationship between condition factor and remaining life. (*Source*: From Guide for Design of Pavement Structures, Volume I, 1993 by AASHTO, Washington, DC. With permission.)

11.3.1.2 Determination of Required Structural Number for Future Traffic (SN_f)

The determination of the required SN for future traffic for an overlaid asphalt pavement (SN_f) is similar for a new construction. It requires the determination of (1) the future traffic (N_f), (2) the effective subgrade resilient modulus (M_R), (3) the design PSI loss (ΔPSI), (4) design reliability (R), and (5) overall standard deviation (S_o). The procedures for determination of these parameters are described in Chapter 8 of this Handbook.

The AASHTO, 1993 Pavement Design Guide recommended that, when M_R is back-calculated from deflections caused by a 9000 pound (40 kn) FWD load, the M_R value must be adjusted by a correction factor C of 0.33 for use in the determination of SN_f. The following equation for estimation of M_R is to be used:

$$M_R = C\left(\frac{0.24P}{d_r r}\right) \tag{11.14}$$

where:

recommended $C = 0.33$.
P = applied FWD load in pounds.
d_r = FWD deflection at a distance r from the center of the load in inches.
r = distance from center of FWD load in inches.

With the determination of the needed design parameters, the required SN for the overlaid pavement can be calculated using the AASHTO pavement equation for flexible pavement, which is given below:

$$\log_{10} N_f = Z_R S_o + 9.36 \log_{10}(SN_f + 1) + \frac{\log_{10}\left(\dfrac{\Delta PSI}{4.2 - 1.5}\right)}{0.40 + \dfrac{1094}{(SN + 1)^{5.19}}} + 2.32 \log_{10} M_R - 8.27 \quad (11.15)$$

11.3.1.3 Determination of Structural Coefficient for the AC Overlay (a_{ol}) and Required Overlay Thickness (D_{ol})

The required overlay thickness is computed as follows:

$$D_{ol} = \frac{(SN_f - SN_{eff})}{a_{ol}} \quad (11.16)$$

The determination of structural coefficient of the AC overlay is similar to that for new construction, and is described in Chapter 8 of this Handbook.

11.3.2 Asphalt Institute Deflection-Based Procedure

The Asphalt Institute's Deflection-Based AC Overlay Design Procedure is based on the concept that the structural capacity of a flexible pavement is related to the rebound deflections measured by a Benkelman Beam, or the equivalent Benkelman Beam rebound deflections as correlated from other NDT devices (Asphalt Institute, 2000). A pavement of higher structural capacity would have smaller Benkelman Beam rebound deflections, while one with a lower capacity would have higher Benkelman Beam rebound deflections. The required overlay thickness is the amount of overlay that will reduce the rebound deflections to an acceptable level for the design traffic. The design procedure consists of the following main steps:

(1) Run Benkelman Beam tests on the pavement to be overlaid using a truck weight of 18,000 pounds (80 kn) on a single axle with dual tires to determine the rebound deflections. It is recommended that 10 tests should be run per section, or a minimum of 20 tests per mile in the outer wheel path. If other NDTs instead of the Benkelman Beam tests are run, the NDT results must be converted to equivalent Benkelman Beam rebound deflections using appropriate correlation equations.

(2) Determine the 97.5 percentile rebound deflection, which is termed the Representative Rebound Deflection (RRD), as follows:

$$RRD = (\bar{x} + 2s)c \quad (11.17)$$

where:

$\bar{x}$ = mean of rebound deflections that have been adjusted for temperature.
s = standard deviation of rebound deflections.
c = critical period adjustment factor.

This means that only 2.5% of the time would the rebound deflection value be higher than the RRD.

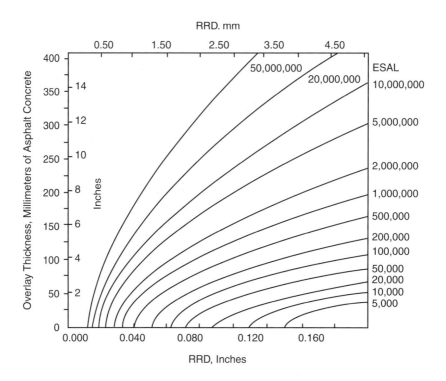

FIGURE 11.4 Asphalt concrete overlay thickness required to reduce pavement deflections from a measured to a design deflection value (rebound test). (*Source*: From Asphalt Overlays for Highway and Street Rehabilitation, 2000 by Asphalt Institute. With permission.)

(3) Estimate the design ESAL that the pavement is expected to support in the future after the overlay. The AASHTO load equivalency factors for flexible pavement are used to calculate the total design ESAL.

(4) Using an overlay thickness chart (Figure 8.2 in the Asphalt Institute's Manual Series No. 17) (Asphalt Institute, 2000), determine the required overlay thickness according to the RRD and the design ESAL. This is shown here as Figure 11.4.

It is to be pointed out that using the Benkelman rebound deflection as the sole indicator of structural capacity of a pavement (as in the case of the Asphalt Institute's deflection-based procedure) may sometimes lead to erroneous estimations. A pavement with a low subgrade modulus will give a high rebound deflection though it may be structurally adequate. Conversely, a pavement with a high subgrade modulus will give a low rebound deflection though it may be structurally inadequate.

11.3.3   Caltrans Deflection-Based Procedure

California Department of Transportation's AC Overlay Design Procedure is based on the concept that the structural capacity of a flexible pavement is related to the deflections as measured by the California Deflectometer, or equivalent California Deflectometer deflections. The required overlay thickness is the amount of overlay which will reduce the California Deflectometer deflections to an acceptable level for the design traffic. Structural capacity is quantified in terms of TI (Caltrans, 1995). The design procedure consists of the following main steps:

(1) Run Dynaflect tests on the pavement to be overlaid in the outside wheel path at 0.01-mile intervals if the project is less than 1 mile in length, or at 0.20-mile intervals if the project is over 1 mile in length. Convert the Dynaflect deflection values into equivalent California Deflectometer values.

TABLE 11.2   Tolerable California Deflectometer Values

	Tolerable Deflections (X 0.001 in.)											
DGAC Depth (foot)	Traffic Indexes (TI's)											
	5	6	7	8	9	10	11	12	13	14	15	16
0.00	66	51	41	34	29	25	22	19	17	15	14	13
0.05	61	47	38	31	27	23	20	18	16	14	13	12
0.10	57	44	35	29	25	21	19	16	15	13	12	11
0.15	53	41	33	27	23	20	17	15	14	12	11	10
0.20	49	38	31	25	21	18	16	14	13	12	10	10
0.25	46	35	28	24	20	17	15	13	12	11	10	9
0.30	43	33	27	22	19	16	14	12	11	10	9	8
0.35	40	31	25	20	17	15	13	12	10	9	8	8
0.40	37	29	23	19	16	14	12	11	10	9	8	7
0.45	35	27	21	18	15	13	11	10	9	8	7	7
0.50[a]	32	25	20	17	14	12	11	9	8	8	7	6
CTB[b]	27	21	17	14	12	10	9	8	7	6	6	5

Note: From Flexible Pavement Rehabilitation Manual, June 2001 by CALTRANS, California. With permission.
[a] For an AC thickness greater than 0.50 ft, use the 0.50 ft depth.
[b] Use the CTB line to represent treated base materials that are equal to or greater than 0.35 ft (105 mm) thick or if the base is a PCC pavement, regardless of the thickness of AC cover. If the underlying treated base thickness is less than 0.35 ft (105 mm), consider it an untreated base.

(2) Calculate the 80 percentile California Deflectometer value as follows:

$$D_{80} = \bar{x} + 0.83s \qquad (11.18)$$

where:

D_{80} = 80 percentile of California Deflectometer value.
$\bar{x}$ = mean California Deflectometer value.
s = standard deviation of all California Deflectometer values for a test section.

(3) Determine the Tolerable Deflection at the Surface (TDS) for the design TI from the Tolerable Deflection Chart (Table 1 in Caltrans' Flexible Pavement Rehabilitation Manual) (Caltrans, 2001). This is shown here as Table 11.2.

(4) Calculate the Percent Reduction in Deflection at the surface:

$$\text{PRD} = \frac{D_{80} - \text{TDS}}{D_{80}}(100) \qquad (11.19)$$

where:

PRD = percent reduction in deflection at the surface.
TDS = tolerable deflection at the surface.

(5) Determine the Gravel Equivalence (GE) required to reduce D_{80} to the TDS.

(6) Determine the overlay thickness as follows:

$$\text{Overlay thickness} = \frac{\text{GE}}{G_f} \qquad (11.20)$$

where:

G_f = Gravel factor.

For a dense graded asphalt concrete, use a G_f of 1.9.

The California Department of Transportation has compared the performance of overlays designed by engineering judgment, component analysis and deflection-based analysis. The result of the comparison indicated that the deflection-based method was judged to be the best among the three methods compared.

11.3.4 Analytically Based Procedures

An analytically based design procedure for AC overlay on flexible pavement basically involves the following main steps:

(1) Characterize the existing pavement materials through condition survey, NDT and laboratory testing to determine their pertinent properties that are related to pavement behavior. These properties may include (a) elastic modulus and Poisson's ratio, (b) tensile strength, (c) fatigue behavior, (d) shear strength, (e) creep properties, (f) fracture energy (or modulus of toughness), (g) thermal properties, and (h) aging characteristics. The effects of temperature on these properties through the expected range of pavement temperature also need to be determined.

(2) Select an appropriate pavement analysis model to be used for calculating the responses (such as, stress and strains) in the pavement subjected to load and environmental effects. Establish the failure criteria to be used to determine if a pavement is structurally adequate.

(3) Using the selected pavement analysis model and the determined pavement material properties as inputs, determine the responses (such as, stresses and strains) of the pavement under the design traffic loads.

(4) Compare the computed responses to the failure criteria. If the computed responses are acceptable according to the failure criteria, the pavement does not need a structural overlay. (It may still need a functional overlay to correct for poor skid resistance or ride quality.)

(5) If the computed responses are not acceptable according to the failure criteria, the pavement needs an overlay. Using the same pavement analysis model, calculate the responses of the pavement with different overlay thicknesses.

(6) The design overlay thickness is the minimum overlay thickness such that the computed responses are acceptable according to the failure criteria.

This following sections present the basic features of a few examples of analytically based overlay design procedures. The features to be presented include (1) the analytical model used, (2) method for determination of pavement material properties, (3) failure criteria used, and (4) method for determination of overlay thickness.

11.3.4.1 FHWA-ARE Procedure

The FHWA procedure developed by Austin Research Engineers (ARE) (ARE, 1975) uses the computer program ELSYM for analysis of pavement response. ELSYM models a flexible pavement as a multi-layer elastic system. Laboratory tests are performed on representative specimens taken from the pavement to determine the elastic modulus and Poisson's ratio of each layer material, which are needed as inputs to the ELSYM program. Results of NDTs on the pavement along with results of condition survey are to be used to establish analysis sections, as each analysis section should have similar conditions including NDT deflections. Any NDT equipment which can provide reliable deflection results can be used for this purpose.

This overlay design method uses a cracking failure criterion and a rutting failure criterion. The cracking criterion is given as:

$$N = 9.73 \times 10^{-15} \left(\frac{1}{\varepsilon_t} \right)^{5.16} \tag{11.21}$$

where:

N = allowable number of 18-kip single-axle load repetitions.

ε_t = maximum horizontal tensile strain on the bottom of the asphalt layer caused by an 18-kip (80 kn) single-axle load.

The rutting criterion is given as an equation of the following form:

$$\text{Log } N = f(R, \varepsilon_{z1}/\sigma_{z1}, \sigma_{z2}, \sigma_{x2}, \sigma_{z3}, \varepsilon_{z4}, \sigma_{z5}, \varepsilon_{z5}, d_T) \tag{11.22}$$

where:

R = allowable rut depth.
$\varepsilon_{z1}, \varepsilon_{z4}$ = vertical strain at bottom of layer 1 and 4, respectively.
$\sigma_{z1}, \sigma_{z2}, \sigma_{z3}$ = vertical stress at bottom of layer 1, 2 and 3, respectively.
$\sigma_{z5}, \varepsilon_{z5}$ = vertical stress and strain at top of layer 5.
σ_{x2} = horizontal stress parallel to load axle in bottom of second layer.
d_T = number of days per year when average daily temperature is equal to or greater than 64°F (based on 5-year average).

The overlay thickness is selected to limit fatigue cracking and to limit rutting for the design traffic.

11.3.4.2 FHWA-RII Procedure

The FHWA procedure developed by Resource International Incorporated (RII) (Majidzadeh et al., 1980) is based on the FHWA-ARE procedure. It also uses the computer program ELSYM for analysis of pavement response. The elastic moduli of the layer materials may be obtained by laboratory tests on representative specimens taken from the pavement, and/or by backcalculations from NDT deflections. The NDTs that may be run include the Dynaflect, Road Rater or Falling Weight Deflectometer. The ELSYM program is used for (1) calculating the analytical deflections, which are used to match the measured NDT deflections for estimation of layer moduli, and (2) calculating the stresses and strains in the pavement with and without the overlay.

This overlay design method uses a cracking failure criterion, which is given as follows:

$$N_f = 7.56 \times 10^{-12} \left(\frac{1}{\varepsilon_t} \right)^{4.68} \tag{11.23}$$

where:

N_f = allowable number of 18-kip single-axle load repetitions to failure.
ε_t = maximum tensile strain parallel to direction of traffic.

The overlay thickness is selected to limit fatigue cracking for the design traffic.

11.3.4.3 Shell Research Procedure

The Shell Research Procedure (Claessen and Ditmarsch, 1977) uses the computer program BISAR for analysis of pavement response. Similar to ELSYM, BISAR models a flexible pavement as a multi-layer elastic system. The elastic moduli of the layer materials are determined by backcalculations from NDT deflections using a falling weight deflectometer (FWD).

This overlay design method uses a cracking failure criterion and a rutting failure criterion. The cracking failure criterion is given as:

$$N = A \left(\frac{1}{\varepsilon_t} \right)^a \left(\frac{1}{E_1} \right)^b \tag{11.24}$$

where:

N = allowable number of 18-kip single axle loads.
ε_t = maximum tensile strain at the bottom of the asphalt layer.
E_1 = elastic modulus of the asphalt layer.
A, a, b = coefficients which are dependent on mixture characteristics.

The rutting failure criterion is given as:

$$(\varepsilon_v)_3 = 2.8 \times 10^{-2} \times N^{-0.25} \tag{11.25}$$

where:

N = allowable number of 18-kip (80 kn) single axle loads.

$(\varepsilon_v)_3$ = maximum vertical strain at the top of the subgrade.

The overlay thickness is selected to limit fatigue cracking and to limit rutting for the design traffic.

11.3.4.4 Florida REDAPS Procedure

The REDAPS (Rehabilitation Design of Asphalt Pavement Systems) procedure for evaluation of existing pavement and rehabilitation design was developed by the University of Florida for possible application by the Florida Department of Transportation (FDOT) (Ruth et al., 1990). This procedure has been shown to be effective in evaluating the structural performance of existing pavements and selecting effective rehabilitation designs for flexible pavements in Florida (Lu et al., 1998). However, this system has not yet been implemented by FDOT.

The REDAPS procedure uses the linear elastic computer program BISAR to compute the state of stress and deformation in the pavement to be evaluated. Either the Dynaflect or the Falling Weight Deflectomer (FWD) tests can be used to evaluate the in-service pavement conditions.

Because properties of asphalt materials are very sensitive to temperature changes, the elastic modulus of the asphalt concrete (E_1) is estimated by using the asphalt viscosity-temperature relationship. The viscosity of the asphalt at the specified temperature is first estimated by the following equation:

$$\log(\eta_{100}) = B_0 + B_1\log(°K) \tag{11.26}$$

where:

η_{100} = the viscosity of the asphalt concrete layer at a constant power of 100 W/m^3, and mean pavement temperature in degree Kelvin, Pa-s.

B_0, B_1 = coefficients, determined by linear regression analysis (Tia and Ruth, 1987).

When the viscosity-temperature relationship is known, the asphalt concrete layer modulus (E_1) is estimated by regression equations relating the η_{100} of the asphalt to the E_1 of the asphalt concrete.

When the viscosity-temperature relationship is unknown, the asphalt concrete layer modulus (E_1) is estimated as follows:

(1) If the pavement has no cracks:

$$\log(E_1) = 6.4147 - 0.148T$$

where E_1 is in psi and T is temperature in °C.

(2) If there exist considerable cracking (i.e., medium frequency class 2 and 3):

$$\log(E_1) = 6.4167 - 0.01106T$$

The elastic moduli of the base, subbase and subgrade (E_2, E_3, E_4) are initially estimated from the Dynaflect or FWD deflections by prediction equations. Subsequently, these moduli are tuned to obtain the best fit within the measured deflections (Ruth et al., 1989).

Different from the current overlay design procedures, which are generally based on fatigue failure criteria, the REDAPS model was formulated according to the Critical Condition Concept proposed by Ruth et al., (1989). This concept states that an asphalt pavement fails at a critical condition rather than long-term fatigue. This critical condition is reached when the combination of thermal, age hardening and the vehicular loading effects produce stress, strain, or energy sufficient to fracture the asphalt concrete. A pavement may fail due to rapid cooling to low temperatures where the age hardened binder is excessively brittle and stress relaxation due to creep is minimal. Also, failure can be the result of heavy wheel loads without rapid cooling of the pavement when high stresses are produced by the combined effects of load, binder hardness, and layer moduli, particularly that of the base course. The weakening of the foundation support due to moisture fluctuations relating to climate variation or change in other environmental conditions can also contribute to pavement failure. However, a pavement may never fail under the same loading condition, no matter how many load repetitions, if all other factors are favorable, e.g., ideal temperature (moderate and nearly constant), softer or aging resistant asphalts and good mix design which prevents rutting.

There are two options by which a pavement evaluation can be performed. They are: (1) the simplified evaluation, and (2) the detailed evaluation.

In the simplified evaluation, the load-induced stresses are calculated using the BISAR program to establish the approximate structural condition of the pavement based on the stress ratio criteria (SR = applied tensile stress/maximum tensile strength). According to the research conducted over the past 15 years for the FDOT, two levels of maximum stress ratio were chosen as the criteria for evaluating the structural performance of the asphalt concrete pavements in the Florida State (Ruth et al., 1990). When SR < 0.30, the pavement is considered to be in good condition. However, immediate rehabilitation may be necessary when SR > 0.55. A more comprehensive analysis is required to evaluate the combined effects of load and thermal induced stresses if $0.30 \leq SR \leq 0.55$.

The detailed evaluation is performed when $0.30 \leq SR \leq 0.55$. The combined vehicular load and thermal stresses under the specified cooling curves will be determined.

There are two overlay design approaches. One is the simplified design approach based on the BISAR analysis under heavy truck loads (24-kip (106.8 kn) single-axle as standard loads) at the typical regional minimum temperature using the stress ratio criterion. The other is the detailed design approach based on two independent analyses to assess the combined effect of rapid cooling and heavy vehicular loads using the fracture energy criterion. The simplified design procedure is developed using an empirical stress ratio criterion (SR = 0.25) observed in the field in the state of Florida incorporating indirectly the thermal and asphalt age hardening effects. It is therefore applicable only to the climatic conditions in Florida. The detailed design procedure, on the other hand, considers the actual cooling condition, binder age hardening and vehicular loads for any type of material and climatic conditions. Therefore, the detailed design procedure is generally recommended for use in practice.

11.4 Procedures for Design of PCC Overlay on Flexible Pavement

11.4.1 Conventional PCC Overlay on Flexible Pavement

11.4.1.1 Feasibility of PCC Overlay

A conventional PCC overlay is a feasible rehabilitation alternative for flexible pavement. A PCC overlay offers an advantage over an AC overlay in that it can provide a harder pavement surface, which will be more resistant to rutting caused by high tire pressure. In certain situations, it can be more cost-effective than an AC overlay. It is most cost-effective when the existing flexible pavement is badly deteriorated. In the case of a badly deteriorated asphalt pavement, the asphalt surface would have to be removed if an AC overlay was to be placed. However, a PCC overlay may be placed over the deteriorated asphalt surface without removing the asphalt layer. The PCC overlay is usually not bonded to the cracked asphalt surface to prevent propagation of cracks through the concrete.

Conditions under which a PCC overlay would not be feasible include following:

(1) There is not enough vertical clearance to accommodate the added thickness due to the thicker PCC overlay.
(2) If the existing pavement suffers from severe heaving and settlement, which require removal of the asphalt surface layer and reworking the base layer.
(3) If the amount of deterioration on the AC pavement is not large and other alternatives would be more cost-effective.

11.4.1.2   Design of PCC Overlay

When an existing asphalt pavement is overlaid with a conventional PCC overlay, the asphalt pavement will essentially act as a stiff foundation for concrete overlay. Thus, the design of a concrete overlay over an existing asphalt pavement is essentially the same as the design of a new concrete pavement over a stiff foundation. The effective modulus of subgrade reaction (k) of the stiff foundation as provided by the existing asphalt pavement needs to be properly determined. The effective k-value of the foundation can then be used in the design of the concrete pavement. Please refer to Chapter 9 of this Handbook on the Design of Rigid Pavements for procedures for determination of effective k-value and design of rigid pavements.

11.4.2   Ultra-Thin White Topping on Flexible Pavement

11.4.2.1   Features of Ultra-Thin Whitetopping

Ultra-Thin Whitetopping (UTW) is a thin concrete overlay, 2 to 4 in. (5 to 10 cm) thick and with a short joint spacing between 2 to 6 feet (61 to 183 cm), placed bonded to an existing asphalt pavement surface. UTW differs from a conventional concrete overlay in two main aspects. First, the concrete overlay in an UTW pavement is bonded to asphalt layer, while a conventional concrete overlay is not. Secondly, the joint spacing in an UTW pavement is much lower than that in a conventional concrete overlay. Because of the short joint spacing used, there is usually no dowel bar or tie bar used at the joints of the UTW.

Figure 11.5 illustrates the difference in load-induced stresses between an UTW and a conventional concrete overlay. Due to the bonding between the concrete and asphalt layers in an UTW pavement, the concrete and asphalt layers would act as one composite slab with a high combined thickness and thus a high bending rigidity. This special feature not only reduces the magnitude of the stresses in the concrete, it also allows the concrete layer to take the compressive stresses. This works to the advantage of the concrete since concrete is much stronger in compression than it is in tension.

Figure 11.6 illustrates the advantage of using a shorter joint spacing in an UTW as compared with a conventional concrete overlay. The shorter joint spacing allows the UTW slabs to deflect downward more to spread the load to the asphalt layer rather than to take the load more by bending action. The shorter joint spacing also reduces the possible curling of the concrete slab due to the effects of thermal gradient in the concrete, and thus reduces the maximum temperature-load induced stresses in the UTW pavement.

As compared with a conventional concrete overlay, an UTW offers the possible advantage of reduction of cost due to the use of a thinner overlay thickness and the absence of dowel bars and tie bars. As compared with an asphalt overlay, an UTW offers the possible advantage of increased rutting resistance due to the use of a hard concrete surface.

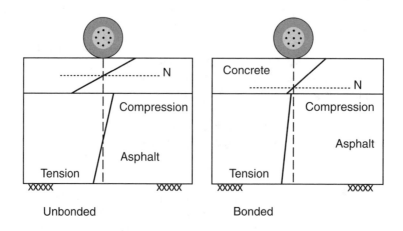

FIGURE 11.5   Effects of interface bonding.

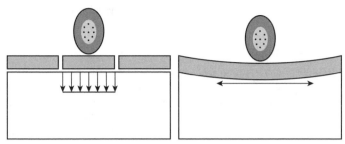

Short joint spacing allows the slabs to deflect instead of bend.
This reduces slab stresses to reasonable values.

FIGURE 11.6   Effects of short joint spacing

11.4.2.2   Feasibility of UTW

UTW is a recently developed technology for resurfacing of asphalt pavements. The first reported UTW experimental project in the U.S. was constructed on a landfill disposal facility near Louisville, Kentucky in 1991 (Risser et al., 1993). The concrete mixture was designed to provide relatively high early compressive strength, 3,500 psi (24 MPa) at 24 h. A low water-cement ratio of 0.33 was selected to achieve higher strength and reduce drying shrinkage. Two concrete slab thicknesses, 2 in. (5 cm) and 3.5 in. (9 cm), and two joint spacings, 2 ft (61 cm) and 6 ft (1.83 m) were used. The Louisville UTW pavement has performed well, carrying many more traffic loads than predicted by design procedures available at that time. Following the success of the Louisville UTW project, many other states, including Tennessee, Georgia, North Carolina, Kansas, Iowa, Pennsylvania, New Jersey, Colorado, Missouri, Mississippi, Virginia and Florida, have constructed and evaluated UTW projects. Results from these experimental studies have indicated that UTW is a viable rehabilitation technique for asphalt pavements.

The intended use of UTW has been primarily on low-truck volume roads, automobile parking areas and parking aprons in general aviation airports. The asphalt pavement to be overlaid with UTW must not be cracked and must have sufficient thickness (after the necessary milling to remove the surface unevenness and cracks, and to produce a rough surface texture for good bonding with the concrete overlay). Badly cracked asphalt pavements, where there is insufficient asphalt thickness after all the cracked materials are milled off, are not suitable candidate for UTW (Tia et al., 2002).

11.4.2.3   Construction Procedures for UTW

The construction procedures for UTW consist of the following steps: preparing asphalt surface, placing the concrete, finishing, surface texturing, curing, and sawing the joints.

The asphalt pavement surface must be milled prior to concrete placement in order to produce a rough asphalt surface to achieve a bond with the concrete overlay. Milling of the asphalt surface is to be followed by cleaning with compressed air to remove all laitance, dust, grit and all foreign materials. There must be an adequate asphalt thickness of a minimum of 3 in. (75 mm) after milling.

Concrete used in UTW can be produced at a ready-mix plant and delivered to the site by ready-mix trucks. Normal slipform pavers can be used to spread, screed, and consolidate the concrete in an efficient manner. After the concrete surface is finished and textured, a curing compound is usually immediately sprayed on the entire surface to provide adequate curing. The curing compound is generally applied at a rate twice the normal application rate for thicker concrete pavements because thin concrete slabs can lose water rapidly (ACPA, 1998). Joint sawing must be performed as soon as surface conditions permit, or when the concrete is able to support the equipment and the operator.

Usually joints are not sealed because joint openings are generally narrow due to the short joint spacing.

11.4.2.4   Factors Affecting Performance of UTW

Two conditions which can cause failure of a UTW pavement are: (1) when the stresses in the concrete layer is excessive causing fracture of the concrete, and (2) when the shear stresses at the concrete-asphalt interface is excessive causing failure of the bonding between the concrete and the asphalt. These two stresses must be kept low in order to ensure good performance of the UTW.

The results of a parametric analysis on the effects of the various material properties and pavement parameters on the maximum load and temperature-induced stresses in UTW pavements gave the following findings (Tia and Kumara, 2003; Kumara et al., 2003):

(1) The maximum stresses in the concrete layer in the UTW pavement increase significantly as (a) asphalt thickness decreases, (b) asphalt stiffness decreases, (c) concrete thickness decreases, (d) concrete stiffness increases, (e) base stiffness decreases, (f) concrete panel size (or concrete joint spacing) increases, and (g) temperature differential (which is the difference between the temperature at the top and the temperature at the bottom) in the concrete layer becomes more negative. The maximum stresses in the concrete are not significantly affected by changes in subgrade modulus.

(2) The maximum shear stresses at the concrete-asphalt interface increase significantly as (a) asphalt thickness decreases, (b) concrete thickness decreases, (c) base stiffness decreases, and (d) concrete panel size (or joint spacing) increases. The following parameters were observed to have little effects on the maximum interface shear stresses: (a) asphalt stiffness, (b) concrete stiffness, (c) temperature differential in the concrete layer, and (d) subgrade modulus.

11.4.2.5   PCA Design Procedure for UTW

In 1994 the Portland Cement Association sponsored a comprehensive research effort aimed at developing a mechanistic based design procedure for UTW (Wu et al., 1997; Mack et al., 1997). This research effort resulted in the development of the PCA design procedure for UTW. The following are the main steps in the developed design procedure.

11.4.2.5.1   *Step 1 — Determine Effective Radius of Relative Stiffness (l_e)*

For a chosen UTW design, calculate the effective radius of relative stiffness for a fully bonded composite pavement (l_e) as follows:

$$l_e = ((Eh^3/12)_e/(1 - \mu^2)/K)^{0.25} \tag{11.29}$$

where:

$$(Eh^3/12)_e = (E1 \times h1^3)/12 + E1 \times h1 \times (NA - h1/2)^2 + (E2 \times h2^3)/12$$
$$+ E2 \times h2 \times (h1 - NA + h2/2)^2 \tag{11.30}$$

   K = modulus of subgrade reaction, pci.
   μ = concrete Poisson's ratio = 0.15.
   $E1$ = concrete modulus of elasticity in psi.
   $h1$ = concrete slab thickness, in inches.
   $E2$ = asphalt modulus of elasticity in psi.
   $h2$ = asphalt layer thickness in inches.
 NA = neutral axis from top of concrete slab in inches.

$$NA = (E1 \times h1 \times (h1/2) + E2 \times h2 \times (h1 + h2/2))/(E1 \times h1 + E2 \times h2) \tag{11.31}$$

11.4.2.5.2 Step 2 — Compute the critical asphalt strains due to all anticipated single axle loads at the joint and critical concrete stress due to all anticipated single axle loads at the slab corner

Calculate the critical asphalt strain due to an 18-kip (80 kn) single axle load at the joint (JT_{18}), in microstrains, as follows:

$$Log_{10}(JT_{18}) = 5.194 - 0.927 \times Log_{10}(K) + 0.299 \times Log_{10}(L/le) - 0.037 \times le \quad (11.32)$$

where:

L = concrete slab joint spacing in inches.

Calculate the critical asphalt strains due to other single axle loads by multiplying JT_{18} by weight proportions. For example, the critical asphalt strain due to a 36-kip (160 kn) single axle load will be 2.0 times that of JT_{18}, and the critical asphalt strain due to a 9-kip (40 kn) single axle load will be 0.5 times that of JT_{18}.

Calculate the critical concrete stress due to an 18-kip (80 kn) single axle load at the corner of the slab (COR_{18}), in psi, as follows:

$$Log_{10}(COR_{18}) = 4.952 - 0.465 \times Log_{10}(K) + 0.686 \times Log_{10}(L/le) - 1.291 \times Log_{10}(le) \quad (11.33)$$

Similarly, calculate the critical concrete stresses due to other single axle loads by using weight proportions.

11.4.2.5.3 Step 3 — Compute the critical asphalt strains due to all anticipated tandem axle loads at the joint and critical concrete stress due to all anticipated tandem axle loads at the slab corner

Calculate the critical asphalt strain due to a 36-kip (160 kn) tandem axle load at the joint (JT_{36}), in microstrains, as follows:

$$Log_{10}(JT_{36}) = 5.997 - 0.891 \times Log_{10}(K) - 0.786 \times Log_{10}(le) - 0.028 \times le \quad (11.34)$$

Calculate the critical concrete stress due to a 36-kip (160 kn) tandem axle load at the corner of the slab (COR_{36}), in psi, as follows:

$$Log_{10}(COR_{36}) = 4.826 - 0.559 \times Log_{10}(K) - 0.963 \times Log_{10}(le)$$
$$+ 1.395 \times Log_{10}(L/le) - 0.088 \times (L/le) \quad (11.35)$$

11.4.2.5.4 Step 4 — Compute the critical temperature induced asphalt strains and concrete stresses

Calculate the critical temperature induced asphalt strains due to temperature differentials at the joint (JT_T), in microstrain, as follows:

$$(JT_T) = (-24.255) + 1.802 \times (\alpha 1 \times \Delta T) + 14.969 \times L/le \quad (11.36)$$

where:

ΔT = temperature differential in concrete slabs, °F.
$\alpha 1$ = concrete coefficient of thermal expansion, /°F.

Calculate the critical temperature induced concrete stresses due to temperature differentials at the slab corner (COR_T), in psi, as follows:

$$(COR_T) = 23.805 - 2.948 \times (a1 \times \Delta T) - 15.451 \times L/le \quad (11.37)$$

11.4.2.5.5 Step 5 — Compute total asphalt strains and total concrete stresses

Add the critical load-induced asphalt strains to the critical temperature-induced asphalt strains to obtain the total asphalt strains. Similarly, add the critical load-induced concrete stresses to the critical temperature-induced concrete stresses to obtain the total concrete stresses.

11.4.2.5.6 Step 6 — Compute stress ratios

For each level of axle load, compute the stress ratio (SR) by dividing the total concrete stress by the design concrete modulus of rupture.

11.4.2.5.7 Step 7 — Determine the allowable repetitions for the concrete layer

For each load, compute the allowable number of load repetitions for the concrete layer (N_C) as follows
For SR > 0.55

$$\text{Log}10(N_C) = (0.97187 - \text{SR})/0.0828 \tag{11.38}$$

For 0.45 < SR < 0.55

$$N_C = (4.2577/(\text{SR} - 0.43248))^{3.268} \tag{11.39}$$

For SR < 0.45

$$N_C = \text{unlimited} \tag{11.40}$$

11.4.2.5.8 Step 8 — Compute the total percent fatigue damage for the concrete layer

For each load, compute the percent fatigue by dividing the anticipated load repetitions by the allowable load repetitions and multiplying by 100. Sum up all the percents fatigue for both single and tandem axle loadings. If the total is less than 100 percent, the design is adequate in terms of resistance to cracking in the concrete.

11.4.2.5.9 Step 9 — Determine the allowable repetitions for the asphalt layer

For each load, compute the allowable load repetitions for the asphalt layer (N_A) as follows:

$$N_A = C \times 18.4 \times (4.32 \times 10^{-3}) \times (1/\varepsilon)^{3.29}(1/Ea)^{0.854} \tag{11.41}$$

where:

N_A = number of load repetitions for 20% or greater AC fatigue cracking.
ε = maximum tensile strain in the asphalt layer.
Ea = asphalt mixture dynamic modulus in psi.
C = a correction factor = 10^M.
$M = 4.84 \times (Vb/(Vv + Vb) - 0.69)$.
Vb = volume of asphalt, percent.
Vv = volume of air voids, percent.

11.4.2.5.10 Step 10 — Calculate the total percent fatigue for the asphalt layer

For each load, compute the percent fatigue by dividing the anticipated load repetitions by the allowable load repetitions and multiplying by 100. Sum up all the percents fatigue for both single and tandem axle loadings. If the total is less than 100 percent, the design is adequate in terms of resistance to fatigue cracking in the asphalt layer.

It is to be pointed out that UTW is still a developing technology. The PCA procedure as presented above represents the first published work in the mechanistic design of UTW pavements. There are still many areas of uncertainties in the understanding of the behavior and performance of UTW. It is expected that more development and refinement in this technology will be seen in the near future.

11.5 Summary

Treating a flexible pavement with an overlay is the most commonly used method for restoring or upgrading the pavement to its desired level of serviceability. The types of overlay on flexible pavements include (1) asphalt concrete overlay, (2) unbonded conventional Portland cement concrete overlay, and (3) bonded ultra-thin Portland cement concrete overlay. In designing an overlay for an existing pavement, it is imperative to perform a thorough evaluation of the existing pavement to determine its areas of deficiency. The causes of the deficiencies need to be determined so that the proper pretreatments and materials for use in overlay can be selected.

This chapter presents the fundamentals in the evaluation of a pavement's functional, structural and safety performance as related to its need for an overlay. The selection of proper treatments and materials for use in overlay for flexible pavements of various deterioration conditions is also described. The commonly used procedures for design of asphalt concrete overlay, conventional Portland cement concrete overlay and ultra-thin Portland cement concrete overlay over flexible pavements are presented.

References

AASHTO, 1993. *AASHTO Guide for Design of Pavement Structures*. American Association of State Highway and Transportation Officials, Washington, DC.

ACPA, 1998. *Whitetopping — State of the Practice*, ACPA Engineering Bulletin EB210P. American Concrete Pavement Association, Skokie, Illinois.

ARE Inc, 1975. *Asphalt Concrete Overlays of Flexible Pavements*, Report No. FHWA-RD-75-75 (Vol. I) & FHWA-RD-75-76 (Vol. II). Federal Highway Administration, Washington, DC.

Asphalt Institute, 2000. *Asphalt Overlays for Highway and Street Rehabilitation*. Asphalt Institute, Lexington, Kentucky.

ASTM, 2003. *Annual Book of ASTM Standards*. ASTM, West Conshohocken, Pennsylvania.

Austroads, 1994. *Pavement Design — A Guide to the Structural Design of Road Pavements*. Austroads Pavement Research Groups, Sydney, Australia.

Brown, S.F., Tam, W.S., and Brunton, J.M. 1987. Structural evaluation and overlay design: analysis and implementation, *Proceedings, 6th International Conference on Structural Design of Asphalt Pavements*, University of Michigan, Ann Arbor, Michigan.

Caltrans, 1995. *Highway Design Manual*. California Department of Transportation, California.

Caltrans, 2001. *Flexible Pavement Rehabilitation Manual*. California Department of Transportation, California.

Carey, W.N. and Irick, P.E. 1960. *The Pavement Serviceability-Performance Concept*, Highway Research Board Bulletin 250. National Research Council, Washington, DC.

CGRA, 1965. *A Guide to the Structural Design of Flexible and Rigid Pavement in Canada*. Canadian Good Roads Association, Canada.

Claessen, A.I.M. and Ditmarsch, R. 1977. Pavement evaluation and overlay design — The shell method, *Proceedings, Fourth International Conference on the Structural Design of Asphalt Pavements*, University of Michigan, Ann Arbor, Michigan.

Cole, L.W., Sherwood, J., and Qi, X. 1999. Accelerated pavement testing of ultra-thin whitetopping, Accelerated Pavement Testing International Conference, Reno, Nevada.

Corps of Engineers, 1958. *Engineering and Design - Flexible Pavements*, EM-1110-45-302. US Army Corps of Engineers, USA.

FDOT, 2003. *Flexible Pavement Condition Survey Handbook*. State Materials Office, Florida Department of Transportation, Gainesville, Florida.

Finn, F.N. and Monismith, C.L. 1984. *Asphalt Overlay Design Procedures*. Transportation Research Board, National Research Council, Washington, DC.

Hall, J.W., Jr. 1978. Nondestructive evaluation procedure for military airfields, Miscellaneous Paper S-78-7, US Army Engineers Waterways Experiment Station, Vicksburg, Mississippi.

Highway Research Board, 1962. The AASHO road test — Research Report 5, Pavement Research, *HRB Special Report 61E*, National Research Council, Washington, DC.

Kumara, M.W., Tia, M., and Wu, C.L. 2003. *Evaluation of Applicability of Ultra-thin Whitetopping in Florida*, Transportation Research Record 1823. Transportation Research Board, Washington, DC.

Lister, N.W., Kennedy, C.K., and Ferne, B.W. 1982. The TRRL method for planning and design of structural maintenance, *Proceedings, Vol. 1, Fifth International Conference on the Structural Design of Asphalt Pavements*, University of Michigan, Ann Arbor, Michigan.

Lu, D., Tia, M., and Ruth, B.E. 1998. Development of an enhanced mechanistic analysis and design system for evaluation and rehabilitation of flexible pavements, *Proceedings of 3rd International Conference on Road and Airfield Pavement Technology*, Beijing, China.

Mack, J.W., Wu, C.L., Tarr, S.M., and Refai, T. 1997. Model development and interim design procedure guidelines for ultra-thin whitetopping pavements, *Proceedings, Sixth International Conference on Concrete Pavement Design and Materials for High Performance*, Volume I, Purdue University, West Lafayette, Indiana.

Majidzadeh, K. and Ilves, G. 1980. Flexible pavement overlay design procedures, Report No. FHWA-RD-79-99, Federal Highway Administration, Washington, DC.

Monismith, C.L. and Brown, S.F. 1999. Developments in the Structural Design and Rehabilitation of Asphalt Pavements over Three Quarters of a Century, *Journal of the Asphalt Paving Technologists*, Volume 68A, The Association of Asphalt Paving Technologists, St. Paul, Minnesota.

Risser, R.J., LaHue, S.P., Voigt, G.F., and Mack, J. 1993. Ultra-thin concrete overlays on existing asphalt pavement, *5th International Conference on Concrete Pavement Design and Rehabilitation*, Vol. 2, Purdue University, West Lafayette, Indiana.

Ruth, B.E., Roque, R., Tia, M., Bloomquist, D.G., and Guan, L. 1989. Structural characterization and stress analysis of flexible pavement systems: computer adaptation, *Final Report*, Department of Civil Engineering, University of Florida, Gainesville, Florida.

Ruth, B.E., Tia, M., and Guan Liqiu 1990. Structural characterization and stress analysis of flexible pavement systems — rehabilitation design, University of Florida, Gainesville, Florida.

Tia, M. and Kumara, W. 2003. Evaluation of performance of ultra-thin whitetopping by means of heavy vehicle simulator (analysis, planning & design phase), *Research Report*, University of Florida, Gainesville, Florida.

Tia, M. and Ruth, B.E. 1987. *Basic Rheology and Rheological Concepts Established*, Asphalt Rheology: Relationships to Mixture, ASTM STP 941, O.E. Briscoe, ed., American Society for Testing and Materials, Pennsylvania.

Tia, M., Wu, C.L., and Kumara, W. 2002. Forensic investigation of the Ellaville Weigh Station UTW pavements, Research Report, University of Florida, Gainesville, Florida.

WSDOT, 1995. WSDOT pavement guide, Vol. 2, Washington State Department of Transportation, Washington.

Wu, C.L., Tarr, S.M., Refai, T.M., Nagi, M.N., and Sheehan, M.J. 1997. Development of ultra-thin whitetopping design procedure, Report prepared for Portland Cement Association, PCA Serial No. 2124, Skokie, Illinois.

12

Overlay Design for Rigid Pavements

T.F. Fwa
National University of Singapore
Republic of Singapore

12.1 Introduction

12.1.1 Functions of Overlays

When a rigid pavement reaches the end of its service life, the pavement engineer has two choices if it is required to provide the pavement with another service life span. The engineer could go for a reconstruction by removing or crushing the damaged pavement structure and designing an all-new pavement structure on the same location. Alternatively, if the pavement is still sound (fully or partially) from a structural point of view, an overlay could be constructed on top of the existing pavement slab. This latter option of pavement rehabilitation is usually more desirable in terms of cost and construction time.

Overlay can also be constructed on rigid pavement that has not yet reached the end of its service life. This could prove to be economically more attractive than rehabilitating the pavement at the end of its service life because a smaller overlay thickness is required on pavement that is still structurally sound than when the pavement is much older and affected by distresses such as cracking and surface deformation. The best timing for pavement rehabilitation is an important issue in pavement management, which is covered in Chapter 18. There may also be other nonstructural reasons for applying overlay on rigid pavement, such as increasing the safety of the pavement surface or improving the appearance of the riding surface. Such applications are more appropriately referred to as "resurfacing." This chapter deals with the structural design of overlays on rigid pavements.

12.1.2 Types of Overlays for Rigid Pavements

Depending on the type of material used, overlays constructed to rehabilitate rigid pavement can be classified as either Portland cement concrete (PCC) overlay or hot-mix asphalt (HMA) overlay, also known as rigid overlay and flexible overlay, respectively. Both rigid and flexible overlays can be used for rehabilitation of different types of rigid pavements, including JPCP, JRCP and CRCP.

Rigid overlays are further divided into three subgroups depending on the bonding condition provided at the interface between the overlay and the parent (i.e., existing) pavement slab. These are bonded overlays, unbonded overlays, and partially bonded overlays. In the construction of bonded overlays, a bonding agent is applied to fully bond the overlay to the parent pavement slab, thereby producing a monolithic or composite pavement structure. On the other hand, unbonded overlays are constructed by intentionally providing a bond-breaking interlayer to ensure that the overlay and the parent pavement slab act as two separate structural slabs. The third group are partially bonded overlays that are laid directly on the parent pavement slab without placing an interface bonding agent or a bond-breaking interlayer.

12.2 Concrete Overlays of Concrete Pavement

12.2.1 Design and Construction Considerations

The type of concrete overlay to be used is very much dependent on the distress state and structural condition of the parent pavement slab. On pavement slabs without any structural defects or surface distresses, any of the three forms of concrete overlay may be applied. All three forms of overlay are also applicable when minor structural defects or limited surface distresses are present, provided that the defects or distresses are repaired. However, when there are serious structural defects, such as severe cracking, bonded and partially bonded overlays are not suitable. This is because the cracks in the parent slabs are likely to induce stress concentrations in the overlay, thereby causing cracks (known as reflection cracks) to develop in the overlay. If the pavement slab is structurally sound, but suffers from extensive surface distresses, either fully bonded or unbonded overlays may be used, but not partially bonded overlays. The reason is that the distresses present may cause debonding to spread and affect the structural capacity of the partially bonded overlay pavement.

Unbonded overlays, being independent slabs themselves, can be used regardless of the condition of the parent pavement slab. However, it is likely to be the most expensive because the overlay thickness required is larger than either the bonded or partially bonded construction. A distress survey and a structural evaluation of the parent pavement must be conducted to provide the necessary information to select the most cost-effective form of overlay construction. Methods of distress survey are covered in Chapter 19, while pavement structural evaluation is dealt with in Chapter 20.

In the construction of bonded or partially bonded overlays, the surface of the parent pavement must be thoroughly cleaned to ensure that proper bonding can be achieved. Sandblasting, water-blasting or milling to remove deteriorated surface materials may be performed to prepare the surface for casting of the overlay. Surface distresses should be repaired so that they do not become the "seeds" for a debonding process. Cement grout or liquid epoxy may be used as the bonding agent for bonded overlay construction. The typical range of thickness of bonded overlays is from 50 to 150 mm (2 to 6 in.).

For unbonded overlay construction, a bond-breaking interface layer of 50 mm (2 in.) or less of HMA or sand-asphalt is applied. An unbonded overlay is itself a structural slab, which resists traffic loading together with a separated underlying parent pavement. As such, the thickness of an unbonded slab must not be smaller than 125 mm (5 in.). The common thickness of unbonded concrete overlays ranges from 125 to 300 mm (5 to 12 in.). The unbonded overlay can be JPCP, JRCP or CRCP, regardless of the type of concrete of the parent pavement. This also means that it is not necessary to match the joint type or joint location between the overlay and the parent pavement. The American Association of State Highway and Transportation Officials (AASHTO, 1993) recommends that the placement of joints in unbonded overlays should be mismatched from existing joints and cracks by at least 3 ft (0.9 m). It is believed that mismatching of joints will enhance load transfer between the overlay and the parent slab, with the understanding that the thick, unbonded overlay is sufficiently strong to prevent reflection cracking.

On the other hand, a bonded overlay or a partially bonded overlay must be designed having matching joint locations with those of the parent pavement. Plain concrete is most commonly used for bonded or partially bonded overlay, although steel reinforcements may be used in thicker overlays. The use of dowel bars in bonded or partially bonded overlays is not recommended (Huang, 1993). Owing to the uncertainty of the degree of bonding that can be achieved in the field, partially bonded overlay construction is not adopted in practice by many agencies.

12.2.2 U.S. Corps of Engineers' Method

The overlay thickness equations based on the effective thickness concept developed by the Corps of Engineers (1958) are probably the most well known and widely adopted relationships for rigid pavement overlay design. The equations take the following forms:

$$(D_{OL}) = (D_T) - C(D_{eff}) \tag{12.1}$$

$$(D_{OL})^{1.4} = (D_T)^{1.4} - C(D_{eff})^{1.4} \tag{12.2}$$

$$(D_{OL})^2 = (D_T)^2 - C(D_{eff})^2 \tag{12.3}$$

where:

D_{OL} = overlay thickness.

D_T = slab thickness required if a new pavement were to be constructed on the existing subgrade.

D_{eff} = effective thickness of the parent pavement.

C = condition factor of the parent pavement. C is equal to 1.0 for parent pavements in good structural condition with little or no structural cracking; 0.75 for parent pavements with little progressive distress such as spalling, multiple cracks, etc.; and 0.35 for parent pavements which are badly cracked and may show multiple cracking, shattered slabs, spalling and faulting.

Equation 12.1 to Equation 12.3 are empirical equations developed from full-scale accelerated tests.

Although the Corps of Engineers' overlay design equations were originally derived for airfield pavements, these have been widely adopted worldwide for rehabilitation design of both flexible and rigid highway pavements. The AASHTO design procedure described in the next section is an example.

12.2.3 AASHTO Design Procedure

12.2.3.1 AASHTO Design Procedure for Bonded Overlay

The 1993 AASHTO Design Guide (AASHTO, 1993) adopts the Corps of Engineers' concept of effective thickness for overlay design. The bonded overlay thickness is given by

$$D_{OL} = D_T - D_{eff} \tag{12.4}$$

in which all the variables are as defined in Equation 12.1.

12.2.3.1.1 Determination of D_T

Table 12.1 summarizes the design parameters required for the determination of D_T. The nondestructive method of back-calculating the structural properties of the existing rigid pavement based on deflection testing is the preferred procedure strongly recommended by AASHTO (1993). The AASHTO Design Guide employs the ILLI-BACK backcalculation analysis developed by Ioannides et al. (1989) to estimate the following properties of the existing pavement slab: the dynamic modulus of subgrade reaction, k, and the elastic modulus of the existing pavement slab, E_c. The following steps are involved in the determination of k and E_c:

1. Obtain from falling-weight deflectometer test the surface deflections D_0, D_{12}, D_{24} and D_{36} at 0, 12, 24 and 36 in. (0, 305, 610 and 915 mm) from the center of load, respectively.

TABLE 12.1 Parameters for Determination of D_T in Bonded Overlay Design

Parameters	Method of Determination
(a) Effective modulus of subgrade reaction k	Back-calculate effective dynamic k value from falling-weight deflectometer measured deflections with sensors located at 0, 12, 24 and 36 in. (0, 305, 610 and 915 mm) from the center of load, and divide the dynamic k value by 2 to obtain the effective static k value; or Remove slab at the test location and perform plate loading test in accordance with ASTM test method D 1196 (ASTM, 2003b)
(b) Elastic modulus E_c of existing concrete slab	Back-calculate from falling-weight deflectometer measured deflections with sensors located at 0, 12, 24, and 36 in. (0, 305, 610, and 915 mm) from the center of load
(c) Modulus of rupture S_c of concrete of existing slab	Estimate from back-calculated E_c by the following equation: $$S_c = 43.5 \times 10^{-6} E_c + 488.5$$ in which S_c and E_c are measured in psi; or Estimate from indirect tensile strength f_t by $$S_c = 210 + 1.02 f_t$$ in which S_c and f_t are measured in psi. The indirect tensile strength f_t is obtained by conducting ASTM C496 split cylinder test (ASTM, 2003a) on 6-in. (150 mm) diameter cores
(d) Design PSI (present serviceability index) loss	Estimate PSI after overlay construction, and subtract the design terminal PSI at the end of the selected design life
(e) Load transfer factor J	For JPCP and JRCP, conduct loading test at representative transverse joints in the outer wheelpath, and compute percent load transfer LT from $$LT(\%) = 100(\Delta_{UL}/\Delta_L)B$$ where Δ_{UL} and Δ_L are measured deflections at the loaded and loaded side of the joint, and B is the slab bending correction factor. B is given by ratio of deflections d_0 and d_{12} at 0 and 12 in. (0 and 305 mm), respectively, form the center of load; i.e., $B = (d_0/d_{12})$. $J = 3.2$ for LT > 70%, 3.5 for LT between 50 and 70%, and 4.0 for LT < 50% For CRCP, use $J = 2.2$ to 2.6 for overlay design
(f) Loss of support of existing slab, LS	Loss of support should be improved with slab stabilization, and assume a fully supported slab with LS $= 0$
(g) Design reliability R; overall standard deviation S_0; subdrainage capability of existing pavement after improvement	See Chapter 9 on design of rigid pavement

2. Calculate the parameter AREA in U.S. customary units as follows:

$$\text{AREA} = 6\left(1 + 2\frac{D_{12}}{D_0} + 2\frac{D_{24}}{D_0} + \frac{D_{36}}{D_0}\right) \qquad \text{(in.)} \qquad (12.5)$$

3. Enter Figure 12.1 with D_0 and AREA to obtain the effective dynamic k value. The effective static k value is equal to half the effective dynamic k value.
4. Enter Figure 12.2 with AREA and effective dynamic k to obtain a value for E_cD^3. Since the slab thickness D is known, the elastic modulus E of the existing pavement slab can be calculated.

Using the values of the parameters listed in Table 12.1 as input, the thickness of a new pavement D_T for the design traffic loading is computed by means of the procedure presented in Section 8.9.1. It should be

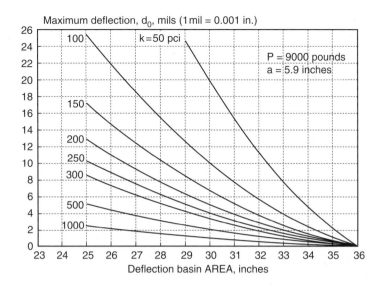

FIGURE 12.1 Effective dynamic k-value determination from d_0 and AREA. *Source*: AASHTO, 1993. *AASHTO Guides for Design of Pavement Structures*. Copyright 1993 by the American Association of State Highway and Transportation Officials, Washington, DC. Used by permission.

noted that the properties of the existing pavement, and not those of the overlay materials, are used in computing D_T.

The main disadvantages of the backcalculation algorithm based on AREA (Li et al., 1996; FHWA, 1997) are: (i) it relies on the index AREA which is computed from deflections at fixed locations, (ii) it is sensitive to the normalizing deflection D_0, (iii) the rigid formulation and solution scheme of ILII-BACK makes it unsuitable for assessing or comparing the relative performance of different deflection measuring devices for backcalculation applications, and (iv) the algorithm cannot be applied when there are missing

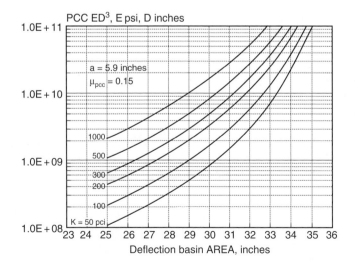

FIGURE 12.2 Elastic modulus determination from k-value, AREA, and slab thickness. *Source*: AASHTO, 1993. *AASHTO Guides for Design of Pavement Structures*. Copyright 1993 by the American Association of State Highway and Transportation Officials, Washington, DC. Used by permission.

or faulty deflection measurements at any of the specified sensor locations. A more general and robust closed-form backcalculation algorithm, known as NUS-BACK, for rigid pavements is presented in Chapter 20. Numerical examples of backcalculation analysis are also presented in Chapter 20.

12.2.3.1.2 Determination of D_{eff}

Two methods are proposed by the AASHTO Design Guide for the determination of the effective thickness D_{eff}, namely, the condition survey method and the remaining life method.

12.2.3.1.2.1 Condition Survey Method

Based on the existing condition of the parent (i.e., existing) slab, its effective thickness is computed as

$$D_{eff} = F_{jc}F_{dur}F_{fat}D \tag{12.6}$$

where:

D = thickness of the parent slab.
F_{jc} = joints and cracks adjustment factors.
F_{dur} = durability adjustment factor.
F_{fat} = fatigue adjustment factor.

When there are no deteriorated transverse joints or cracks, or if all such defects are effectively repaired, F_{jc} can be taken as 1.00. Otherwise, F_{jc} can be assigned according to a practically linear equation joining the points of $F_{jc} = 1.00$ for zero deteriorated transverse joints and cracks, and $F_{jc} = 0.56$ for 200 such joints and cracks per mile (1 mile = 1.609 km). F_{dur} has the value of 1.00 if there are no signs of durability problems, 0.96 to 0.99 if there is some durability cracking but no spalling, and 0.88 to 0.95 if both cracking and spalling exist. F_{fat} has a value of 0.97 to 1.00 if very few transverse cracks and punch-outs exist, 0.94 to 0.96 if a significant number of transverse cracks and punch-outs exist, and 0.90 to 0.93 if a large number of transverse cracks and punch-outs exist.

Example 12.1

Design a bonded concrete overlay for a 9-in. -thick old concrete pavement with less than 10 deteriorated transverse joints and cracks per mile. Spalling is not found along the cracks, and no durability cracks or punch-outs are observed. A thickness of 11 in. is required if a new pavement is to be constructed on the existing subgrade.

$F_{jc} = 0.98$, $F_{dur} = 1.0$, $F_{fat} = 0.97$, $D_{eff} = 0.98 \times 1.0 \times 0.97 \times 9 = 8.56$, $D_{OL} = 11 - 8.56 = 2.44$ in.
Use 2.50-in. overlay.

12.2.3.1.2.2 Remaining Life Method

Based on the percent remaining life of the existing pavement, its effective thickness can be estimated by the following equation:

$$D_{eff} = C_F D \tag{12.7}$$

where D is the thickness of the parent slab, and C_F the condition factor determined from Figure 12.3. To determine the condition factor C_F, the remaining life of the parent pavement must first be computed by the following equation:

$$RL = 100\left[1 - \frac{N_p}{N_{1.5}}\right] \tag{12.8}$$

where:

RL = the percent remaining life.
N_p = the total ESALs (equivalent single-axle loads) to date.
$N_{1.5}$ = the total ESALs to pavement "failure" at PSI = 1.5.

$N_{1.5}$ can be determined using the AASHTO thickness design nomograph presented in Figure 9.15 of Chapter 9. An implicit assumption of this approach is that the pavement has been constructed and maintained in accordance with the original design without durability or distress problems.

Example 12.2

Design a bonded concrete overlay for a 9.5-in.-thick old concrete pavement with $N_{1.5} = 1.2 \times 10^7$ ESAL. The total ESAL to date is 2,000,000. For a new additional design traffic volume, a thickness of 11 in. is required if a new pavement is to be constructed on the existing subgrade.

$RL = 100\{1 - (2,000,000/12,000,000)\} = 83.3\%$. From Figure 12.3, $C_F = 0.97$, $D_{eff} = 0.97 \times 9.5 = 9.2$ in. $D_{OL} = 11 - 9.2 = 1.8$ in. Use 2.0-in. overlay.

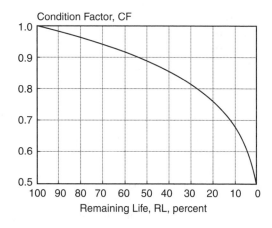

FIGURE 12.3 Relationship between condition factor and remaining life. *Source*: AASHTO, 1993. *AASHTO Guides for Design of Pavement Structures*. Copyright 1993 by the American Association of State Highway and Transportation Officials, Washington, DC. Used by permission.

12.2.3.2   AASHTO Design Procedure for Unbonded Overlay

According to the Corps of Engineers' concept of effective thickness for overlay design, the unbonded overlay thickness is given by

$$(D_{OL})^2 = (D_T)^2 - (D_{eff})^2 \tag{12.9}$$

in which all the variables are as defined in Equation 12.3.

12.2.3.2.1   *Determination of D_T*

The following design parameters are determined in the same manner as described in the preceding section for bonded overlay design: effective static k value, design PSI loss, load transfer factor, loss of support, overlay design reliability, overall standard deviation for rigid pavement, and subdrainage capability of existing pavement. The major difference is that the modulus of rupture and the elastic modulus of the concrete material of the new unbonded overlay are used in the computation of D_T.

12.2.3.2.2   *Determination of D_{eff}*

D_{eff} of the existing pavement may be estimated by either the condition survey method or the remaining life method. The computation of D_{eff} by the remaining life method is identical to the procedure described in the preceding section under bonded overlay design. For the condition survey method, only the joints and cracks adjustment factor F_{jcu} is applicable as shown by the following equation:

$$D_{eff} = F_{jcu}D \tag{12.10}$$

F_{jcu} is obtained by determining the number of transverse joints and cracks per mile, N_{jc}, and reading from a graph. The graph can be approximated by two straight lines, one joining the points ($N_{jc} = 0$, $F_{jcu} = 1.00$) and ($N_{jc} = 30$, $F_{jcu} = 0.97$), and the other joining the points ($N_{jc} = 30$, $F_{jcu} = 0.97$) and ($N_{jc} = 200$, $F_{jcu} = 0.90$).

Example 12.3

Design an unbonded concrete overlay for the old concrete pavement described in Example 12.1.

$F_{jcu} = 0.99$. $D_{eff} = 0.99 \times 9 = 8.91$ in. $(D_{OL})^2 = 11^2 - 8.91^2 = 41.61$, $D_{OL} = 6.45$ in. Use 6.50-in. overlay.

Example 12.4

Design an unbonded concrete overlay for the old concrete pavement described in Example 12.2.

$RL = 100\{1 - (2,000,000/12,000,000)\} = 83.3\%$. From Figure 12.3, $C_F = 0.97$, $D_{eff} = 0.97 \times 9.5 = 9.2$ in. $(D_{OL})^2 = 11^2 - 9.2^2 = 36.36$ in. $D_{OL} = 6.0$ in.

12.2.3.3 Remaining Life Concept of 1986 AASHTO Design Procedure

It is worthwhile noting that a remaining life concept was introduced in the 1986 AASHTO Guide (AASHTO, 1986), but was abandoned in the subsequent 1993 AASHTO Guide procedure that has been described in the preceding section. The remaining life concept essentially highlights that although a young pavement and an old pavement may have identical effective thickness at the time of evaluation, they have different remaining lives because the older pavement would have a higher rate of deterioration. This is depicted in Figure 12.4. It follows that for the same design traffic loading the thickness of overlay required on the younger and older pavements, respectively (see Figure 12.4), would not be the same. The corresponding equation for overlay thickness requirement is given as

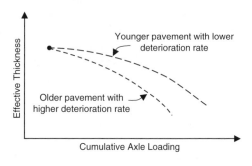

FIGURE 12.4 Remaining life concept for overlay design.

$$(D_{OL})^n = (D_T)^n - F_{RL}(D_{eff})^n \tag{12.11}$$

where F_{RL} is the remaining life factor to take into consideration the effect of remaining life (or more specifically, the effect of deterioration of the parent pavement slab); all other symbols are as defined earlier. The remaining life factor F_{RL} is a function of the remaining life of the existing parent pavement slab and the length of design overlay life selected. The value of F_{RL} varies from 0 to 1.

It is unfortunate that there was an error in the derivation of the F_{RL} in the 1986 AASHTO Design Guide (Elliott, 1989; Fwa, 1991), which leads to erratic overlay design thickness. Fwa (1991) confirmed the rationale of the remaining life concept and derived a new expression for the remaining life factor that effectively removes the inconsistencies of the 1986 AASHTO overlay design equation.

Equation 12.9 of the 1993 AASHTO Design Guide is a special case of Equation 12.11 with $F_{RL} = 1.0$. By taking F_{RL} as unity, it is assumed that the existing parent pavement slab with effective thickness D_{eff} will deteriorate at the same rate as a new pavement slab. This is apparently an overestimation of the F_{RL} value. As a result, Equation 12.9 provides a design overlay thickness that is thinner than the design based on the correct value of F_{RL}.

12.2.4 PCA Design Procedure

12.2.4.1 Design of Unbonded Overlay

The PCA design procedure is based on an equal-stress concept. It selects an overlay thickness that would have an edge stress equal to or less than the corresponding edge stress in an adequately designed new pavement under the action of an 18-kip (80-kN) single-axle load. Design charts in Figure 12.5 are provided for the following three cases:

Case 1. Existing pavement exhibiting a large amount of mid-slab and corner cracking; poor load transfer at cracks and joints.

Case 2. Existing pavement exhibiting a small amount of mid-slab and corner cracking; reasonably good load transfer across cracks and joints; localized repair performed to correct distressed slabs.

Case 3. Existing pavement exhibiting a small amount of mid-slab cracking; good load transfer across cracks and joints; loss of support corrected by undersealing.

The design charts are valid for modulus of subgrade reaction values between 100 and 300 pci (27 and 81 MN/m³), and were derived from finite element analysis with the following input properties: modulus of elasticity of 5×10 psi (35 GPa) for overlays and 3×10 and 4×10 psi (21 and 28 GPa) for the parent pavements. If a tied shoulder is provided, the thickness of the overlay may be reduced by 1 in. (25 mm), subject to the minimum thickness requirement of 6 in. (150 mm).

Example 12.5

Design a concrete overlay for an existing 9.5-in.-thick concrete pavement if the required new single slab thickness is 10 in.

For Case 1 condition, $T_{OL} = 9$ in. from Figure 12.5(a). For Case 2 condition, $T_{OL} = 6.5$ in. from Figure 12.5(b). For Case 3 condition, $T_{OL} < 6$ in. from Figure 12.5(c), use minimum 6 in.

12.2.4.2 Design of Bonded Overlay

A similar concept of stress equivalency is also adopted in the design of bonded overlay, except that the comparison is now made between the stress-to-strength ratios of the new and the overlaid pavements. Design charts (see Figure 12.6) are prepared for three different ranges of moduli of rupture S_c of the parent concrete slab. S_c may be estimated from the average splitting tensile strength f_t as follows:

$$S_c = 0.9A(f_t - 1.65s) \qquad (12.12)$$

where f_t is determined according to ASTM test method C496 (ASTM, 2003b), s is the standard deviation of splitting tensile strength, and A is a regression constant ranging from 1.35 to 1.55. An A-value of 1.45 is suggested in the absence of local information. The 0.9 factor in Equation 12.12 is to relate the strength of the concrete specimen to that near or at the edge. To determine the splitting

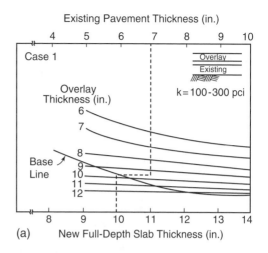

(a)

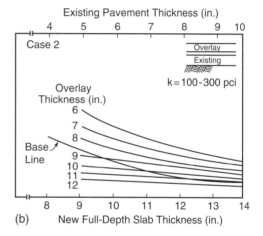

(b)

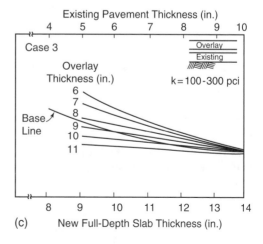

(c)

FIGURE 12.5 PCA design charts for unbonded overlays. *Source*: Proceedings of Third International Conference on Concrete Pavement Design and Rehabilitation. April 23 to 25, 1985, Purdue University, pp. 378–379. With permission.

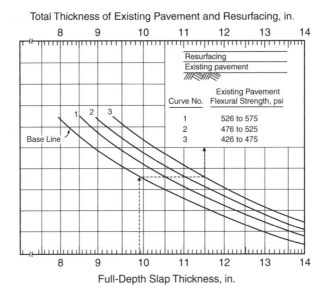

FIGURE 12.6   PCA design for bonded overlay. *Source*: Proceedings of Third International Conference on Concrete Pavement Design and Rehabilitation. April 23 to 25, 1985, Purdue University, pp. 379. With permission.

tensile strength, one core should be taken every 300 to 500 ft (91 to 152 m) at mid-slab and about 2 ft (0.6 m) from the edge of the outside lane.

Example 12.6

A 9.5-in. thick concrete pavement is to be strengthened to match the capacity of a new 10-in. concrete pavement. What is the required thickness of bonded overlay if the existing concrete has a flexural strength of 450 psi.

Use curve 3 of Figure 12.6, the thickness of existing pavement plus overlay = 11.5 in. Overlay thickness = 11.5 − 9.5 = 2.0 in.

12.3   Asphalt Concrete Overlays of Concrete Pavement

12.3.1   Design and Construction Considerations

Asphalt concrete overlay is commonly used on rigid pavements because of the convenience and the relatively short time the overlaid pavement can be open to traffic. In the design of flexible overlays on concrete pavements, besides determining the thickness required to provide adequate structural capacity of the total overlaid pavement to receive the design traffic loading, a major design concern is to ensure that the overlay design is sufficient to prevent reflection cracking, and to resist temperature-induced stresses.

The design objective is thus to provide an overlay design that meets the thickness requirement for the design traffic loading, with a sufficiently low risk of developing reflection cracking. Reflection cracking leads to a reduced service life span of the overlaid pavement due to premature cracking failures, loss of serviceability, and durability problems such as debonding of overlay and stripping in the asphalt concrete. Reflection cracking is caused by stress or strain concentrations in the overlay due to either load-induced or temperature-induced movements associated with the joints and cracks in the underlying pavement. Traffic loading induces differential vertical movements at existing joints and cracks in the underlying parent pavement slab. Daily and seasonal temperature changes induce horizontal movements such as widening or closing of the cracks or joints.

Besides isolated surface distresses, the major forms of rigid pavement distresses to be corrected include excessive slab deflections, excessive horizontal movements, differential deflections at joints and cracks, and moisture-related distresses at cracks and joints. Depending on the distress conditions of the existing rigid pavement, the critical design temperature differential in the region and the characteristics of design traffic, one of the following options may be adopted for the design of an asphalt concrete overlay on a given rigid pavement:

1. Provide a sufficiently thick layer of asphalt concrete layer on the rigid pavement.
2. Break the rigid pavement into smaller pieces to serve as a base course for the asphalt concrete overlay.
3. Install a crack-relief layer on the rigid pavement before placing an asphalt concrete overlay.
4. Lay a stress-absorbing interlay membrane before placing an asphalt concrete overlay.
5. Apply a fabric membrane interlayer placing an asphalt concrete overlay.
6. Create joints in the asphalt concrete overlay to match the transverse joints in the underlying parent rigid pavement, and seal the joints against water infiltration and ingression of foreign particles.

AASHTO (1993) classifies the methods of breaking concrete slabs into four categories based on the sizes of the broken pieces, and assigns layer coefficients accordingly. The following three methods may be employed to break the existing concrete slabs:

1. Breaking and seating refers to breaking of JRCP into pieces of about 1 ft (305 mm), rupturing the reinforcement or breaking its bond with the concrete, and seating the pieces firmly into the foundation by several passes of a 35- to 50-ton (32- to 45-tonne) rubber-tired roller.
2. Cracking and seating consists of cracking a JPCP into pieces of 1 to 3 ft (305 to 415 mm) in size, and seating by a 35- to 50-ton (32- to 45-tonne) rubber-tired roller.
3. Rubblizing and compacting is the process that fractures JRCP, JPCP or CRCP into pieces smaller than 1 ft (305 mm) and then compacts the layer with two or more passes of a 10-ton (9-tonne) vibratory compactor.

It is noted that all three processes, when properly executed, provide a uniform and firm support and no further crack control treatment is needed.

12.3.2 Asphalt Institute Method

The Asphalt Institute (1983) makes available two different approaches for overlay design, namely, the *effective thickness method* and the *deflection method*.

12.3.2.1 Effective Thickness Method

This method evaluates the effective thickness T_{eff} of the existing rigid pavement in terms of thickness of asphalt concrete, and determines the required overlay thickness T_{OL} as

$$T_{OL} = T - T_{eff} \tag{12.13}$$

in which T is the required thickness of a new full-depth pavement if constructed on the existing subgrade, to be determined from Figure 12.7.

The effective pavement thickness T_{eff} is obtained by multiplying the existing slab thickness by a conversion factor. The conversion factor is assigned a value of 0.9 to 1.0 if the concrete pavement is stable, undersealed, and uncracked; a value of 0.7 to 0.9 if it is stable and undersealed, has some cracking but contains no pieces smaller than about 10 ft^2 (1 m^2); a value of 0.5 to 0.7 if it is appreciably cracked and faulted, cannot be effectively undersealed, with slab fragments ranging in size from approximately 10 to 40 ft^2 (1 to 4 m^2), and has been well seated on the subgrade by heavy pneumatic-tired rolling; and a value of 0.3 to 0.5 if it has been broken into small pieces of 2 ft (0.6 m) or less in maximum dimension.

Example 12.7

An old 10 in. concrete pavement that is stable and undersealed, suffers from some degree of cracking but contains no pieces smaller than 10 ft^2 (1 m^2). Design an asphalt overlay for an expected traffic of 4×10^6 ESAL. The design CBR of the subgrade was found to be 4%. The mean annual air temperature is 60°F (16°C)

Design $M_r = 1500$ CBR $= 6000$ psi. From Figure 12.7, the required thickness of new full-depth asphalt pavement $T = 12.0$ in. Based on the condition of the concrete pavement, the conversion factor is taken as 0.8. $T_{eff} = 0.8 \times 10 = 8.0$ in. $T_{OL} = 12.0 - 8.0 = 4.0$ in.

12.3.2.2   Deflection-Based Method

Deflections at pavement edges and both the total and differential deflections at joints or cracks are of major concern in the Asphalt Institute overlay design method. Deflection measurements can be made using the Benkelman Beam (see Chapter 20) or other device at the following locations: (a) the outside edge on both sides of two-lane highways; (b) the outermost edge of divided highways; and (c) corners, joints, cracks, and deteriorated pavement areas. For JPCP and JRCP, the differential vertical deflection at joints should be less than 0.05 mm (0.002 in.), and the mean deflection should be less than 0.36 mm (0.014 in.). For CRCP, Dynaflect deflections (see Chapter 20) of 15 to 23 μm (0.0006 to 0.0009 in.) or greater lead to excessive cracking and deterioration. Undersealing or stabilization is required when the deflection exceeds 15 μm (0.0006 in.).

The required thickness of asphalt concrete overlay is determined by the amount of deflection reduction to be achieved. The Asphalt Institute considers that dense-graded asphalt concrete overlay can reduce deflections by 0.2% per mm (5% per in.) of its thickness. However, depending on the mix type and environmental conditions, it may be as high as 0.4 to 0.5% per mm (10 to 12% per in.). If a reduction in deflection of 50% or more is required, it will be more economical to apply undersealing so that thinner overlay can be used.

For a given slab length and mean annual temperature differential, the required overlay thickness is selected from Figure 12.8. The thicknesses are provided to minimize reflection cracking by taking into account the effects of horizontal tensile strains and vertical shear stresses. The design chart has three sections: A, B and C. In section A, a minimum thickness of 100 mm (4 in.) is recommended. According to the estimate of the Asphalt Institute, this thickness should reduce the deflection by 20%. In sections B and C, the thickness may be reduced if the pavement slabs are shortened by breaking and seating (denoted as alternative 2 in Figure 12.8) to reduce temperature effects. This is recommended when an overlay thickness approaches the 200 to 225 mm (8 to 9 in.) range. Another alternative is the use of a crack-relief layer (denoted as alternative 3 in Figure 12.8). The design of the crack-relief structure recommended by the Asphalt Institute consists of a 3.5-in. (90-mm)-thick layer of coarse open-graded HMA, containing 25 to 35% interconnecting voids and made up of 100% crushed material. This relief layer should be overlain by a dense-graded asphalt concrete surface course at least 1.5 in. (38 mm) thick and a dense-graded asphalt concrete leveling course of at least 2 in. (50 mm) thick.

Example 12.8

A Benkelman Beam test at a joint of a PCC pavement measured vertical deflections of 0.052 and 0.033 in. on the two edges. The pavement has a slab length of 40 ft. Design an asphalt concrete overlay on the concrete pavement. The design temperature differential is 70°F.

Mean vertical deflection = 0.0425 in., differential deflection = 0.019 in.

(Alternative 1 — thick overlay). From Figure 12.8, more than 9 in. of overlay is required. Use either alternative 2 or 3.

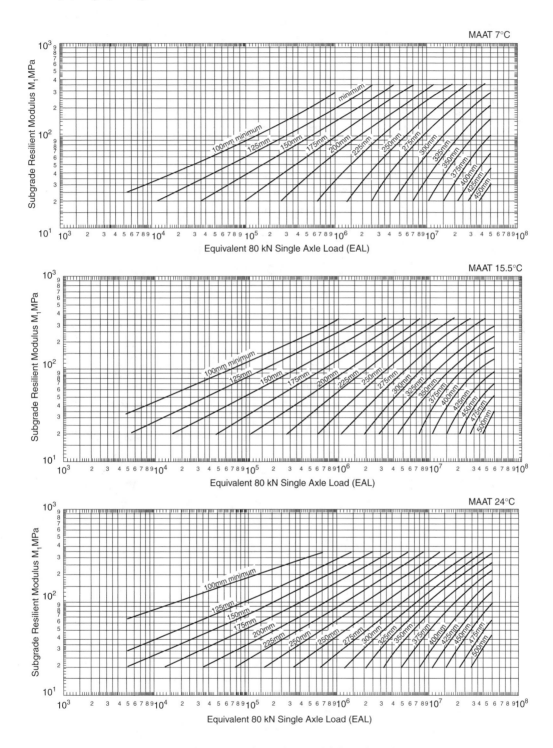

FIGURE 12.7 Asphalt Institute method for thickness design of full-depth asphalt pavement. *Source*: Asphalt Institute, 1991. Asphalt Overlays for Highway and Street Rehabilitation. Manual Series MS-1, p. 53, 59, and 65. With permission.

TEMPERATURE DIFFERENTIAL (°F)

Slab Length (ft)	30	40	50	60	70	80	Slab Length (m)
10 or less	100 mm (4 in.)	100 mm (4 in.)	100 mm (4 in.)	100 mm (4 in.)	100 mm (4 in.)	100 mm (4 in.)	3
15	100 mm (4 in.)	100 mm (4 in.)	100 mm (4 in.)	100 mm (4 in.)	100 mm (4 in.)	100 mm (4in)	4.5
20	100 mm (4 in.)	100 mm (4 in.)	100 mm (4 in.)	100 mm (4 in.)	125 mm (5 in.)	140 mm (5.5in)	6
25	100 mm (4 in.)	100 mm (4 in.)	100 mm (4 in.)	125 mm (5 in.)	150 mm (6 in.)	175 mm (7in)	7.5
30	100 mm (4 in.)	100 mm (4 in.)	125 mm (5 in.)	150 mm (6 in.)	175 mm (7 in.)	200 mm (8in)	9
35	100 mm (4 in)	115 mm (4.5 in.)	150 mm (6 in.)	175 mm (7 in.)	215 mm (8.5 in.)	Use Alternative 2 or 3	10.5
40	100 mm (4 in.)	140 mm (5.5 in.)	175 mm (7 in.)	200 mm (8 in.)	Use Alternative 2 or 3	Use Alternative 2 or 3	12
45	115 mm (4.5 in.)	150 mm (6 in.)	190 mm (7.5 in.)	225 mm (9 in.)	Use Alternative 2 or 3	Use Alternative 2 or 3	13.5
50	125 mm (5 in.)	175 mm (7 in.)	215 mm (8.5 in.)	Use Alternative 2 or 3	Use Alternative 2 or 3	Use Alternative 2 or 3	15
60	150 mm (6 in.)	200 mm (8 in.)	Use Alternative 2 or 3	Use Alternative 2 or 3	Use Alternative 2 or 3	Use Alternative 2 or 3	18
	17	22	28	33	39	44	

TEMPERATURE DIFFERENTIAL (°F)

FIGURE 12.8 Asphalt Institute method for thickness design of asphalt overlay on concrete pavement. *Source*: Asphalt Institute, 1983. Asphalt Overlays for Highway and Street Rehabilitation. Manual Series MS-17, p. 79. With permission.

(Alternative 2 — break and seat to reduce slab length). Break slab into 20-ft sections. From Figure 12.8, 5.0 in. of overlay is required. For the overlaid pavement, mean vertical deflection = $0.0425 - \{(5.0 \times 5\%) \times 0.0425\} = 0.0319 > 0.014$ in., vertical differential deflection = $0.019 - \{(5.0 \times 5\%) \times 0.019\} = 0.0142 > 0.002$ in. Undersealing is needed.
(Alternative 3 — crack relief layer). Use 3.5 in. crack relief course with 1.5 in. surface course and 2 in. leveling course, giving a total of 7 in. asphalt concrete courses. Similar procedure of deflection checks to those for Alternative 2 indicates that undersealing is required.

12.3.3   AASHTO Design Procedure

There are two different AASHTO procedures for flexible overlay design depending on whether fracturing of the existing concrete slab is required.

12.3.3.1 Asphalt Concrete Overlay of Fractured Concrete Pavement

For cases where fracturing of the existing concrete pavement has been performed, the required flexible overlay thickness is given by

$$SN_{OL} = SN_T - SN_{eff} \tag{12.14}$$

where SN_T is the structural number required if a new flexible pavement were to be constructed on the subgrade, and SN_{eff} is the effective structural number of the existing pavement after fracturing. The thickness of overlay is given by SN_{OL} divided by the layer coefficient of the overlay material (see Chapter 8 for common values of layer coefficient for asphalt concrete).

The method of determining SN_T is described in Chapter 8. SN_{eff} is determined by component analysis as follows:

$$SN_{eff} = a_2 D_2 m_2 + a_3 D_3 m_3 \tag{12.15}$$

where a_2, D_2, and m_2 are layer coefficient, thickness, and drainage coefficient, respectively, of the fractured slab, and a_3, D_3, and m_3 are the corresponding properties of the base layer. The value of a_2 suggested by AASHTO (1993) is 0.20 to 0.35 for JRCP after breaking and seating treatment, or JPCP after cracking and seating treatment; and 0.14 to 0.30 for rubblized CRCP, JRCP, or JPCP pavements. The value of a_3 is 0.00 to 0.10 for the normal granular base materials used for concrete pavement construction. Due to lack of information on drainage characteristics of fractured concrete pavements, AASHTO (1993) recommends a default m_2 value of 1.0. Suitable values of m_3 can be found in Chapter 8 dealing with flexible pavement design.

Example 12.9

Design an asphalt overlay for an old 10-in. concrete pavement with a 6-in.-thick granular base that has been rubblized. The structural number of 4.0 is required if a new asphalt pavement is to be constructed on the subgrade. m_3 of the base material is 1.20.

Take $a_2 = 0.25$ and $a_3 = 0.05$, $SN_{eff} = (0.25 \times 10.0 \times 1.0) + (0.05 \times 6.0 \times 1.20) = 2.50 + 0.36 = 2.86$. $SN_{OL} = 4.0 - 2.86 = 1.14$. Choose asphalt material $a_1 = 0.44$, thickness of asphalt overlay $= (1.14/0.44) = 2.59$ in. Use 2.75-in. asphalt concrete overlay.

12.3.3.2 Asphalt Concrete Overlay of Concrete Pavement

The required thickness of asphalt concrete overlay D_{OL} to increase the structural capacity for a specified design for future traffic is given by

$$D_{OL} = A(D_T - D_{eff}) \tag{12.16}$$

where D_T and D_{eff} are as defined in Equation 12.1 and the procedure for their determination is the same as that described in Section 12.2.3 for concrete bonded overlay. The A factor is for converting PCC thickness to asphalt concrete, and is computed by the following equation:

$$A = 2.2233 + 0.0099(D_T - D_{eff})^2 - 0.1534(D_T - D_{eff}) \tag{12.17}$$

The thickness of asphalt concrete overlay of concrete pavement may vary from 2 to 10 in. (50 to 250 mm).

Example 12.10

An old 10-in. concrete pavement has an effective thickness of 9.6 in. Determine the thickness of asphalt concrete overlay if the required thickness of a new concrete pavement is 11 in.

$A = 2.2233 + 0.0099(11 - 9.5)^2 - 0.1534(11 - 9.5) = 2.02$. $D_{OL} = 2.015(11 - 9.5) = 3.02$ in. Use 3.0-in. asphalt overlay.

12.4   Future Development

The overlay design methods currently in use are basically empirical in nature, much like the present state of development of the design methodologies for new pavements. A mechanistic design procedure based on analytical determination of critical stresses, strains or deflections under different possible forms of applied loading, and the assessment of safety margins against various types of failure mechanisms would be the ultimate approach to adopt. This is currently not achievable in view of the gaps existing in:

1. The ability to simulate analytically the pavement responses to actions and effects of applied loadings such as repeated dynamic traffic loading, cyclic thermal loading, and variations in moisture
2. The theoretical characterization of the behavior and deterioration of pavement materials under various forms of applied loadings

It is thus practical to first develop some form of mechanistic-empirical design procedure to replace the current empirical design methods that have been used in largely similar forms since the 1970s. This is the common understanding of the various efforts to improve pavement design methodologies in recent years, including the proposed 2002 Design Guide initiated by AASHTO (Pierce, 2003; ARA Inc., 2004). Unfortunately even this transitional step toward the ultimate mechanistic approach is not straightforward. More research and field verification efforts are needed, and a lengthy development and implementation process is expected before a satisfactory mechanistic-empirical design procedure can become a reality and adopted by most highway agencies.

Ideally, it is not necessary to consider new pavement design and overlay design separately once a complete mechanistic pavement design procedure is put in place. The general steps in the design would be similar since an overlaid pavement is just another form of multi-layer structure, and the overall service and performance requirements of an overlaid pavement are not different from those of a new pavement. This is, however, not likely the case for a mechanistic-empirical design method. Before a comprehensive mechanistic design approach becomes a reality, a mechanistic-empirical design procedure will still be the approach for practical design. Until then the design of new pavement and that of pavement overlay has to be treated separately.

References

AASHTO, 1986. *Guide for Design of Pavement Structures*. American Association of State Highway and Transportation Officials, Washington, DC.

AASHTO, 1993. *Guide for Design of Pavement Structures*. American Association of State Highway and Transportation Officials, Washington, DC.

ARA Inc., 2004. *Guide for Mechanistic-Empirical Design of New and Rehabilitated Pavement Structures*, Final Report, NCHRP Project 1-37A, National Cooperative Highway Research Program. Transportation Research Board, Washington, DC.

Asphalt Institute, 1983. Asphalt Overlay for Highway and Street Rehabilitation. Manual Series No. 17. College Park, MD

Asphalt Institute, 1991. Thickness Design — Asphalt Pavements for Highways & Streets. Manual Series No. 1. College Park, MD

ASTM, 2003a. *ASTM C496 Test Method for Splitting Tensile Strength of Cylindrical Specimens*, Annual Books of ASTM Standards. American Society for Testing and Materials, West Conshohochen, PA.

ASTM, 2003b. *ASTM D 1196 Standard Test Method for Non-repetitive Static Plate Load Tests of Soils and Flexible Pavement Components, for Use in Evaluation and Design of Airport and Highway Pavements*, Annual Books of ASTM Standards. American Society for Testing and Materials, West Conshohochen, PA.

Corps of Engineers, 1958. *Rigid Airfield Pavement*, Engineering and Design Manual EM-1110-45-303. Office of the Chief Engineer, US Department of the Army, Vicksburg, MI.

Elliott, R.P. 1989. An Examination of the AASHTO Remaining Life Factor, Transportation Research Record No. 1215, pp. 53–59.

FHWA, 1997. *LTTP Data Analysis Phase I: Validation of Guidelines for k-Value Selection and Concrete Pavement Performance Prediction*. Federal Highway Administration, US Department of Transportation, Publication No. FHWA-RD-96-198.

Fwa, T.F. 1991. Remaining-Life Consideration in Pavement Overlay Design, *J. Transport. Res., ASCE*, 117, 6, 585–601.

Huang, Y.H. 1993. *Pavement Analysis and Design*. Prentice-Hall, Engelwood Cliffs, NJ.

Ioannides, A.M., Barenberg, E.J., and Lary, J.A. 1989. *Interpretation of Falling Weight Deflectometer Results Using Principles of Dimensional Analysis*, Proceedings of the Fourth International Conference on Concrete Pavement Design and Rehabilitation, Purdue University, West Lafayette, IN, pp. 231–247.

Li, S., Fwa, T.F., and Tan, K.H. 1996. Closed-Form Back Calculation of Rigid Pavement Parameters, *J. Transport. Eng.*, 122, 1.

Pierce, L.M. 2003. 1-37A Design guide — project status and state viewpoint, presented at International Conference on Highway Pavement Data, Analysis and Mechanistic Design Applications, 7–10 September, Columbus, OH.

13

Highway Drainage Systems and Design

Khaled Ksaibati
University of Wyoming
Laramie, WY, U.S.A.

Laycee L. Kolkman
Jacobs Engineering
Las Vegas, NV, U.S.A.

13.1 Introduction

One of the most important factors in building a road is drainage. If every other aspect of highway design and construction is done perfectly but the drainage does not work well, the road will fail quickly. Figure 13.1 is a picture of a road less than a year old that did not have adequate drainage in the fill upon which the pavement structure was built. Even when the failure is not catastrophic, water in the highway structure will inevitably lead to premature failure. They are called "high ways" because early road builders knew it was necessary to construct roads above the ground and away from water.

Highway drainage and subdrainage systems are complex and location specific. There are many factors to consider when designing drainage systems. Most of the design guides and equations available come from empirical studies applicable only to limited geographic regions. Yet processes derived from empirical studies are often used indiscriminately. It is important to be aware of the situations for which a particular design process is applicable. In all but the simplest applications, a combination of appropriate methods and suitable materials is used to move water away from the pavement structure.

Erosion control is an integral part of highway drainage. Excessive erosion may plug culverts, leading to saturation of the pavement structure and failure of drainage systems, and finally resulting to pavement failure. Environmental considerations that the dictate erosion must be controlled.

Surface drainage must be provided to drain precipitation away from the pavement structure. In a simple example, cross slope directs water to the shoulder where it flows into a ditch, then down the ditch to a culvert and finally into an existing natural drainage. Water on the road surface can lead to hydroplaning and skidding which may result in accidents and fatalities. Surface drainage design includes the prediction of run-off and infiltration, as well as open channel analysis and culvert design for the movement of surface water to convenient locations or naturally occurring water paths.

FIGURE 13.1   Catastrophic slope failure.

During the design process, culverts must be analyzed to determine capacity, control, and performance. Linings must be added to open channels to minimize erosion and, in some cases, to decrease the velocity.

Highway design uses a combination of surface and subsurface drainage solutions. Subsurface drainage problems often lead directly to pavement failure. They may also lead to slope failures, such as the one in Figure 13.1, particularly when fills are constructed on existing slopes or when roads are constructed through steep cuts. There are many sources of subsurface water and just as many solutions. For example, longitudinal and transverse drains used in conjunction with drainage blankets can remove groundwater and water melting from ice lenses. Subsurface drainage design has progressed rapidly in recent years, leading to increased availability of products and improved pavement performance. Subsurface drainage that removes excess water from the pavement structure increases the performance and life of a pavement section, but they are not economically feasible in all situations. Proper cost–benefit analysis should be performed before drainage systems are installed.

Selecting a design approach for water flow requires sound engineering judgment. Consequences and likelihood of failure and economic constraints are the main considerations. Temporary construction drainage facilities are designed to a lower flow rate since the probability of, for example, a 50-year flood is slim during the duration of construction. Permanent facilities, on the other hand, are much more likely to suffer a 50-year flood. One must use a more conservative design approach when putting together plans for a critical roadway than for a road that can easily be bypassed. The *only road* to residences is more critical than the *only road* to a gravel pit. Selecting a design flow is a critical choice that must be made early in the design process. These decisions do not easily lend themselves to rigorous engineering analysis but they must be made with considerable care and forethought.

Cost and risk must be balanced during the design of drainage structures. In many cases, drainage structures and storm water control are mandated by regulations. One must consider the required drainage facilities and the economic benefits derivable from more durable roads when designing drainage systems.

This chapter covers the design methods that are currently considered good engineering practice, and are generally recommended by the Federal Highway Administration. However, each of the design methods should be used as a guideline and modifications will usually be necessary for specific locations.

13.2   Surface Drainage

13.2.1   Cross Slopes

The main objective of pavement cross slopes is to insure the rapid removal of rain, melting snow and ice from pavement sections. Cross slopes should not be so steep that they make driving difficult or uncomfortable. Drivers may experience difficulty when trying to steer their cars along steeper cross slopes. Removal of precipitation from the roadway is extremely important. Water on the roadways can cause hydroplaning when tires lose contact with the pavement surface. Estimation of how much water a certain pavement will retain should take into account pavement texture and roadway geometry, as well as the amount and duration of precipitation. By increasing the cross slope of roadways, drainage time and the risk of hydroplaning are reduced. The cross slopes as recommended for different surface types are in Table 13.1. Cross slopes should be steeper for surfaces that tend to retain water, such as soil and gravel surfaces; but flatter for surfaces that will retain less moisture, such as asphalt and concrete.

Shoulders should be sloped to encourage drainage away from the pavement surface. If shoulders are sloped toward the pavement surface, sediment and debris may accumulate causing traffic hazards. Drainage should be directed toward the swale or the median. The range of shoulder cross slopes can vary from 1.5 to 6.0%. The cross slope should be determined by the type of surfacing material, as well as the use of a curb. Swales are generally less hazardous to traffic than curbs, and they facilitate the removal of precipitation without erosion of fill sections.

Longitudinal slopes are of particular importance on sections utilizing curbs. These sections should have at least a 0.3% grade. This should be increased to 0.5% in areas where there is not a suitable crown . In areas where the topography remains very flat, a minimum of 0.2% grade can be used on curbed sections.

13.2.2   Roadside Drainage

The two most common types of ditches are flat-bottom and V-section ditches. A flat-bottom ditch is recommended where design constraints allow its use. Flat or rounded ditches reduce the shy distance, the distance from a steep grade or drop drivers are comfortable with; they also increase the probability that a driver can safely recover after leaving the roadway. Ditches should be designed so that the water in them does not saturate the subgrade. Roadside drainage must be designed to satisfy hydraulic specifications, as will be discussed later.

Drainage in cut sections is of particular importance due to the increased likelihood that runoff will accumulate on the roadway surface. By intercepting the runoff, the chance of flooding during high intensity precipitation is reduced, as is erosion in the cut. Runoff should be intercepted and redirected before it has a chance to accumulate on or near the road. This can be accomplished with the use of intercepting channels alongside the roadway. The need for intercepting channels increases in arid

TABLE 13.1   Recommended Cross Slope by Surface Type

Roadway Element	Range in Rate of Cross Slope (%)
High-type surface	1.5–2.0
Two-Lanes	1.5 min
Three or more lanes in each direction	Increase 0.5–1.0 per lane; 4.0 max
Intermediate surface	1.5–3.0
Low-type surface	2.0–6.0
Urban arterials	1.5–3.0 Increase 1.0 per lane
Shoulders	
Bituminous or concrete	2.0–6.0
With curbs	≥ 4.0

Source: From *Drainage of Highway Pavements*, Highway Engineering Circular No. 12, Federal Highway Administration, Washington, DC. (March 1984).

and semiarid climates where there is less vegetation. Interception ditches should be placed so that they redirect runoff before it reaches the roadway.

13.3   Erosion Prevention

Erosion must be prevented in drainage channels, along the right-of-way, and along the slopes surrounding the roadway. Erosion causes both soil loss and damages associated with sediment accumulation downstream, and is generally restricted by environmental regulations.

Curbs and gutters prevent erosion along shoulders and side slopes. This is also an acceptable method of erosion control in areas without paved shoulders. In this case the curb should be placed along the edge of the pavement. Water is then channeled to roadside drainage channels.

A common form of erosion prevention is growing vegetation. This is usually accomplished by seeding. Seeds may be manually spread, hydroseeded, or incorporated in a turf reinforcement mat or erosion blanket. Following grading, grasses or other small vegetation are established along the slope. The time taken to get established is the drawback to using vegetation as an erosion prevention approach. On steeper slopes, some form of soil protection should be used to reduce erosion as vegetation matures. Though it is not effective in areas with high water volumes or velocities, vegetation can economically reduce erosion.

13.3.1   Erosion Control Blankets

Erosion control blankets can be used by themselves or in combination with vegetation to reduce slope erosion. Erosion control blankets may be natural or synthetic. Natural blankets are the most economical choice and are generally 100% biodegradable. Vegetation growth is often protected with blankets made of natural fibers that will degrade once plants are established. One synthetic material is a turf reinforcement mat. Synthetic mats are designed to prevent erosion where high velocity water is expected. Figure 13.2 shows a synthetic blanket with vegetation becoming reestablished. These mats are permanent and are not intended to degrade. Synthetic mats can be used on slopes as steep as 1:1 with very little damage to the mat or underlying soil. Erosion is prevented by reducing the velocity of surface water and by protecting soil and vegetation.

13.3.2   Channel Erosion

Flexible and rigid linings can also be used to prevent channel erosion. Factors determining the use of a flexible lining may include the need for the lining to conform to a changing grade in the slope of the channel. Flexible linings are often more economical than rigid linings and look more natural. However, flexible linings are often not able to resist large forces caused by erosion. Some types of permanent flexible linings include riprap, gravel, and rock-lined ditches which combine a geotextile and riprap.

Channels may be lined with a number of materials. They may be either flexible or rigid. The choice of flexible or rigid channel lining

FIGURE 13.2   Synthetic blanket in ditch.

FIGURE 13.3   Rock-lined ditch.

depends on the circumstances of each application and cost. Flexible linings are cheaper, while rigid linings are better in preventing seepage and are more resistant to shear forces.

Rigid linings may be selected where space is limited or where water needs to be moved quickly. They also are better in withstanding freeze-thaw action, though in most applications some movement of a flexible lining is not detrimental.

There are several types of flexible ditch linings. Grass lining is acceptable for ditch sections those are relatively flat and wide in areas with adequate precipitation. Erosion blankets may be used to stabilize ditches while vegetation grows. Rock-lined ditches consist of notching a V-ditch, placing a geotextile separation fabric, then placing rip rap on top of the fabric, as shown in Figure 13.3.

13.4   Hydraulic Design

The way in which water is collected from the ocean, moved across land, and eventually deposited back to the sea is called the hydrologic cycle. The sun ultimately fuels this cycle by providing the energy needed for evaporation. The amount of water available globally is virtually constant, meaning water is neither created nor destroyed, it can only change forms through evaporation and precipitation. Precipitation can either be viewed as point precipitation or as aerial precipitation. Generally, precipitation or rainfall amounts are measured at a single point. It is necessary to convert the point values to an aerial value to find precipitation over large areas. There are several methods to accomplish this calculation.

Two methods for determining rainfall over a large area are the Isohyetal and Thiessen methods. The Isohyetal Method interpolates values between gauges. Isohyets or lines of equal rainfall, are drawn connecting points having identical rainfall. The result of this method is a sort of topographical map showing rainfall as contour lines over the entire watershed. The Isohyetal method has greater accuracy when consideration is given to the topographical features of the watershed that may affect precipitation. The Thiessen Method involves connecting points where rainfall data are known into triangles. Lines are then used to divide the triangles into polygons, ultimately leaving one rain gauge in the center of each polygon. These polygons are then weighted by area to find the average depth of precipitation over the entire watershed. The Thiessen Method should not be used in mountainous areas. It is best suited for relatively flat watersheds. Either method can be used to predict the rainfall over a watershed, and thus the amount of water entering a drainage structure.

Precipitation events are categorized by three factors, intensity, duration and frequency. Intensity is the rate at which precipitation falls. Duration is time over which the rain falls. Frequency is the probability that a certain intensity and duration will occur in any given year. It must be understood that this frequency is a probability which means that a 100-year storm has a 1% chance of occurring in any given year, not that it will occur only once every 100 years. Intensity, duration, and frequency data are collected at a particular point and are then used to construct an intensity-duration-frequency (IDF) curve. Numerous software packages can be used to calculate particular IDF curves. These IDF curves are used to determine the intensity of rainfall for a given duration and frequency.

13.4.1   Surface Runoff

Surface runoff is directly determined by the amount of excess precipitation. Excess precipitation is the precipitation that is left once all losses have been accounted for. Some of the losses include evaporation during the event, infiltration, and storage in depressions.

Evaporation is the process of transferring water vapor from land and water sources to the atmosphere. Another form of evaporation is transpiration, or the amount of water expelled from vegetation. Evaporation generally accounts for the largest portion of water losses from a storm event. The rate of evaporation for any given area depends on temperature, surface type, relative humidity, and wind.

Infiltrated precipitation is the amount of water filtered through the earth's surface. This water returns moisture to the soil and replenishes ground water supplies. Infiltration rates are affected by the type of soil and vegetation. Once the precipitation has fallen to the ground, it may be detained in depression storage areas from where the precipitation cannot runoff. Depression areas accumulate water that may infiltrate the surface if it is permeable. However, if the surface is not permeable, the water is left to evaporate. Depression storage does not add to runoff and must be deducted as a loss. Surface runoff is estimated for the drainage area in question. Topographic maps can be helpful in determining the drainage area since they accurately display the surrounding terrain. The drainage area is the entire area that will contribute to the runoff at the point where the amount of runoff is calculated.

13.4.2   Curve Number

A widely accepted method for determining the runoff for a given drainage area has been developed by the U.S. Soil Conservation Service (1986). It is called the Curve Number (CN) procedure.

The following procedure is used to establish a CN for the drainage in question:

(1) The drainage is divided into land use descriptions from Table 13.2 with the percentage of each land use type.
(2) The hydrologic soil group for each land use description as defined by the Natural Resources Conservation Service (NRCS) (U.S. Soil Conservation Service, 1986), is determined using the following four types:

   Type A   Low Runoff Potential, High Infiltration Rates: Sands and gravels, deep, well to very well drained.

   Type B   Moderate Runoff Potential, Moderate Infiltration Rates: Moderately coarse-textured soils, moderately deep to deep, moderately well to well drained soils.

   Type C   High Runoff Potential, Slow Infiltration Rates: Moderate to fine-grained soils or soils with a layer that impedes downward movement of water.

   Type D   Very High Runoff Potential, Very Low Infiltration Rates: Clay soils with high swelling potential, soils with a claypan or clay layer at or near the surface, and shallow soils over nearly impervious material.

TABLE 13.2   Runoff Curve Numbers for AMC II

Land-Use Description/Treatment/ Hydrologic Condition		Hydrologic Soil Group			
		A	B	C	D
Residential					
Average lot size:	Average percent impervious				
1/8 acre or less	65	77	85	90	92
1/4 acre	38	61	75	83	87
1/3 acre	30	57	72	81	86
1/2 acre	25	54	70	80	85
1 acre	20	51	68	79	84
Paved parking lots, roofs, driveways, etc.		98	98	98	98
Streets and roads					
Paved with curbs and storm sewers		98	98	98	98
Gravel		76	85	89	91
Dirt		72	82	87	89
Commercial and business areas (85% impervious)		89	92	94	95
Industrial districts (72% impervious)		81	88	91	93
Open spaces, lawns, parks, golf courses, cemeteries, etc.					
Good condition: grass cover on 75% or more of the area		39	61	74	80
Fair condition: grass cover on 50 to 75% of the area		49	69	79	84
Fallow					
Straight row	—	77	86	91	94
Row crops					
Straight row	Poor	72	81	88	91
Straight row	Good	67	78	85	89
Contoured	Poor	70	79	84	88
Contoured	Good	65	75	82	86
Contoured and terraced	Poor	66	74	80	82
Contoured and terraced	Good	62	71	78	81
Small grain					
Straight row	Poor	65	76	84	88
Straight row	Good	63	75	83	87
Contoured	Poor	63	74	82	85
Contoured	Good	61	73	81	84
Contoured and terraced	Poor	61	72	79	82
Contoured and terraced	Good	59	70	78	81
Close-seeded legumes or rotation meadow					
Straight row	Poor	66	77	85	89
Straight row	Good	58	72	81	85
Contoured	Poor	64	75	83	85
Contoured	Good	55	69	79	83
Contoured and terraced	Poor	63	73	80	83
Contoured and terraced	Good	51	67	76	80
Pasture or range	Poor	68	79	86	89
	Fair	49	69	79	84
	Good	39	61	74	80
Contoured	Poor	47	67	81	88
Contoured	Fair	25	59	75	83
Contoured	Good	6	35	70	79
Meadow	Good	30	58	71	78
Woods or forest land	Poor	45	66	77	83
	Fair	36	60	73	79
	Good	25	55	70	77
Farmstead	—	59	74	82	86

Source: After *"Hydrology" Supplement A to Section 4*, Engineering Handbook, US Department of Agriculture, Soil Conservation Service, 1968.

TABLE 13.3 Runoff Curve Numbers for AMC II

Antecedent Condition	Description	Dormant Season 5-Day Rainfall (in.)	Growing Season 5-Day Rainfall (in.)
AMC I	A condition of watershed soils where the soild are dry but not to the wilting point, and when satisfactory plowing or cultivatrion takes place	≤ 0.5	<1.4
AMC II	The average case for annual floods, that is, an average of the conditions that have preceded the occurrence of the maximum annual flood on numerous watersheds	0.5 to 1.1	1.4 to 2.1
AMC III	If heavy rainfall or light rainfall and low temperatures have occurred during the 5 days previous to the given storm and the soil is nearly saturated	>1.1	>2.1

Source: From *Natural Resources Conservations Service*.

(3) Table 13.2 is used to determine the CN based on the land type and the hydrologic soil group for each land use description. Each of the agricultural descriptions is categorized as having good, fair, or poor hydrologic conditions. The CN from Table 13.2 is for antecedent moisture condition (AMC) II.

(4) The AMC for the entire drainage is determined from Table 13.3. The AMC is determined by the rainfall in inches during the previous 5 days. For the purposes of design AMC II is usually used. If the AMC is Condition II, skip to Step (6).

(5) The CN for AMC II from Table 13.2 is carried to Table 13.4.

(6) Once individual curve numbers have been determined for each section of the drainage area, a composite curve number can be determined. The composite curve number is a weighted average of each of the individual curve numbers according to the percentage of the drainage area they represent.

13.4.3 Peak Discharge Using the SCS TR-55 Method

Once the curve number has been determined, the peak discharge can be determined. First, the maximum potential storage is calculated. The runoff is determined by removing the storage from the rainfall, then the time of concentration and the peak discharge per square mile per inch are determined. Finally the discharge is calculated.

Using the weighted curve number, the potential storage can be determined using Equation 13.1.

$$S = (1000/CN) - 10 \tag{13.1}$$

where:

CN is the weighted average curve number, S is the equivalent potential storage in inches.

The first step in this method is to determine the runoff (R) in inches. This can be accomplished by removing the losses from the precipitation over the drainage area. Equation 13.2 and Equation 13.3 are the SCS Equations for determining runoff amounts.

$$R = \frac{(P - I_a)^2}{P - I_a + S} \tag{13.2}$$

$$R = \frac{(P - 0.2S)^2}{P + 0.8S} \tag{13.3}$$

where:

R = runoff (in.).
P = accumulated rainfall (in.).

TABLE 13.4 Corresponding Runoff Curve Numbers for Conditions I and III

CN for Condition II	Corresponding CN for Condition	
	I	III
100	100	100
95	87	98
90	78	96
85	70	94
80	63	91
75	57	88
70	51	85
65	45	82
60	40	78
55	35	74
50	31	70
45	26	65
40	22	60
35	18	55
30	15	50
25	12	43
15	6	30
5	2	13

Source: Adapted from USDA Soil Conservation Service TP-149 (SCS-TP-149), *A Method for Estimating Volume and Rate of Runoff in Small Watersheds*, revised April 1973.

S = watershed storage (in.).

I_a = initial abstractions including surface storage, interception by vegetation and infiltration prior to runoff (in.).

$I_a = 0.2S$, (approximation from empirical studies). Note that if I_a is greater than P, the runoff will be zero.

To calculate the peak discharge, the time of concentration must be calculated. There are number of methods available. A method that accounts for differences in cover types was developed by the SCS:

$$t_c = \frac{1.67L^{0.8}[(1000/CN) - 9]^{0.7}}{1900S^{0.5}}$$

where:

t_c = time of concentration (h).

L = length of flow path (ft).

CN = SCS runoff curve number.

S = average watershed slope (%).

The value of the peak discharge for a 24-h storm at time of concentration can be determined graphically using Figure 13.4. This graph is only valid for type II storms and is not applicable to coastal regions or the lower Mississippi valley. These areas are shown in Figure 13.5. Graphs for type I, IA, and III storms, which generally occur along the coastal regions, are available from the U.S. NRCS, formerly known as the U.S. Soil Conservation Service. Once the drainage area, runoff, and peak discharge per square mile per inch of precipitation have been determined, the peak discharge at the outlet point is

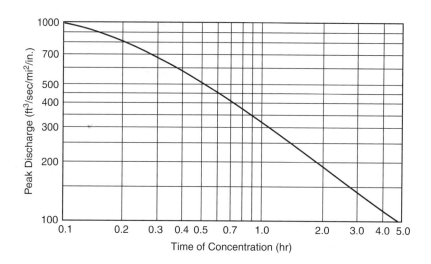

FIGURE 13.4 Peak discharge vs. time of concentration for 24 h, Type II storm distribution (*Source*: From Soil Conservation Service, 1973. *A Method for Estimating Volume and Rate of Runoff in Small Watersheads*).

calculated using Equation 13.4:

$$q_P = q_U AR \qquad (13.4)$$

where:

q_P = peak discharge (ft^3/sec).
q_U = peak discharge for 24-h storm at time of concentration (ft^3/sec/mi^2/in.).
A = drainage area (mi^2).
R = runoff (in.).

Example 13.1

Calculating runoff and peak discharge using the TR-55 method

A 5-mi^2 watershed is comprised of soil with a moderate infiltration rate, and an AMC I. Thirty percent of the area is residential with $\frac{1}{4}$ acre lot sizes, 20% is good condition grass cover, and 50% is good quality, straight row, small grain. If the time of concentration is 1.5 h and the 24-h precipitation is 8 in., calculate the runoff and peak discharge using the curve number method.
Answer:
Determine the soil type.

Since it has moderate infiltration, it is soil type B.

Determine the weighted curve number.

Residential $\frac{1}{4}$ acre lots CN for AMC II soil type B = 75 (from Table 13.2)
Use Table 13.4 to convert to AMC I = 57

$$CN = 0.30 \times 51 + 0.20 \times 41 + 0.50 \times 57 = 52.0$$

Solve for S using Equation 13.1.

$$S = \frac{1000}{CN} - 10 = \frac{1000}{52.0} - 10 = 9.23$$

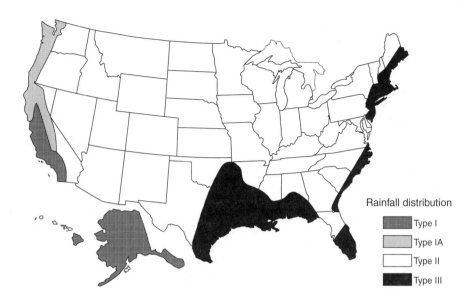

FIGURE 13.5 Storm distribution regions (*Source*: From *Engineering Field Manual for Conservation Practices*, US Soil Conservation Service, Washington, DC, November 1986).

Solve for runoff using Equation 13.3.

$$R = \frac{(P - 0.2S)^2}{P + 0.8S} = \frac{(8 - 0.2 \times 9.23)^2}{8 + 0.8 \times 9.23} = \frac{37.87}{15.38} = 2.46 \text{ in.}$$

Find q_U using Figure 13.1.

$$\text{For } t_c = 1.5 \text{ h}, \qquad q_U = 250 \text{ ft}^3/\text{sec/mi}^2/\text{in.}$$

Solve for peak discharge using Equation 13.4.

$$q_P = q_U AR = 250 \times 5 \times 2.46 = 3075 \text{ ft}^3/\text{sec}$$

13.4.4 Rational Method

The Rational Method is most often used for small urban watershed areas. The Rational Method assumes that the runoff from uniform rainfall intensity will reach a maximum at the point of design when all parts of the watershed are contributing. Runoff will be equal to the excess precipitation once the time of concentration has been met. This time of concentration is affected by the watershed size as well as the soil and vegetation within the watershed. The Federal Aviation Administration has adopted an equation to determine the time of concentration for airfield drainages. It has been used frequently for overland flow in urban areas. Time of concentration is shown in the FAA equation, Equation 13.5.

$$t_c = \frac{1.8(1.1 - C)L^{0.5}}{S^{0.333}} \qquad (13.5)$$

where:

t_c = time of concentration (min).
L = overland flow length (ft).

C = rational method runoff coefficient.
S = surface slope (%).

Runoff coefficients can be found in Table 13.5. These are typical runoff coefficient values for 5 to 10 year storm frequencies. The coefficients must be multiplied by a frequency factor if they are to be used for less frequent events. These multipliers are also found in Table 13.5.

With the determined frequency period an IDF curve should be produced for the area using a duration value equal to the time of concentration. The next step is to solve Equation 13.6 for the peak flow.

$$Q_P = CIA \qquad (13.6)$$

where:

Q_P = peak runoff rate (ft^3/sec).
C = runoff coefficient, use multiplier if less than 10-year frequency.
I = average rainfall intensity, from IDF curve where duration is t_c (in./h).
A = drainage area (acres).

If the duration of the design storm is greater than or equal to the time of concentration, Equation 13.6 is valid. For storm intensities where the duration is less than the time of concentration, the area must be adjusted in Equation 13.6 to accommodate the shorter duration.

TABLE 13.5 Typical C Coefficients for 5 to 10 year Frequency Design

Description of Area	Runoff Coefficients
Business	
Downtown areas	0.70–0.95
Neighborhood areas	0.50–0.70
Residential	
Single-family areas	0.30–0.50
Multiunits, detached	0.40–0.60
Multiunits, attached	0.60–0.75
Suburban	0.25–0.40
Residential (1.2 acre lots or more)	0.30–0.45
Apartment dwelling areas	0.50–0.70
Industrial	
Light areas	0.50–0.80
Heavy areas	0.60–0.90
Parks, cemeteries	0.10–0.25
Playgrounds	0.20–0.40
Railroad yard areas	0.20–0.40
Unimproved areas	0.10–0.30
Streets	
Asphaltic	0.70–0.95
Concrete	0.80–0.95
Drives and walks	0.75–0.85
Roofs	0.70–0.95
Return Period	Multiplier
2–10	1.0
25	1.1
50	1.2
100	1.25

Source: From *Hydraulic Design Series No. 19*, US Department of Transportation, Washington, DC, 1965.

Example 13.2

Calculating runoff using the Rational method

Calculate the runoff for a 75-acre urban area in Colorado. The area comprises 50% residential (suburban), 30% steep; heavy soil, lawns, and 20% asphalted streets. Use a 50-year frequency storm and a rainfall intensity of 1.5 in./h.

Answer:

First determine the Type of storm. Figure 13.5 shows Colorado in Type II rainfall area.

Second determine the runoff coefficients and the multiplier (Table 13.5, midpoint values):

Residential (suburban) = 0.325 × 1.2 × 0.50 = 0.195
Lawns; heavy soil, steep = 0.30 × 1.2 × 0.30 = 0.108
Asphalt streets = 0.825 × 1.2 × 0.20 = 0.198
Weighted runoff coefficient = C_W = 0.195 + 0.108 + 0.198 = 0.501
Using Equation 13.6 solve for peak runoff
$Q_P = CIA = 0.501 \times 1.5 \times 75 = 56$ cfs(ft^3/sec)

13.4.5 Unit Hydrograph

Unit hydrographs present storm data in graphical form. Rainfall and runoff are plotted as a function of time. A unit hydrograph is used for a precipitation event that produces precisely 1 in. of excess rainfall. Hydrographs can be for any duration storm, as the "unit" term refers to the one unit of runoff and not the time duration. Commonly, 1-, 2- or 24-h duration times are used. Hydrographs also show the effect of the drainage basin on the runoff. This accounts for slope, soil type, and vegetation. Hydrographs are constructed using storm or rainfall data for the drainage basin. It is desirable to have the most complete data for any given basin. This means that data should be collected from multiple reliable sources, for numerous different storm events. One source of rainfall data is the National Oceanic and Atmospheric Administration (NOAA).

13.4.6 Open Channel Design

Flow velocity is the main factor when designing an open channel. Flow velocity is important because a high velocity tends to erode channel linings while a low velocity will cause the water to deposit sediments in the bottom of the channel. Figure 13.6 shows a fairly flat open channel ditch with a small accumulation of sediment. Channel flow velocity usually depends on the shape and size of the channel, channel lining type, the quantity of water transported, and the type of suspended material. The most appropriate range of channel gradient to produce the required velocity is between 1 and 5%. For most types of lining, sedimentation is usually a problem when slopes are less than 1%, and excessive erosion of the lining will occur when slopes are higher than 5% (Garber and Hoel, 2002). Tables 13.6 and 13.7 give the maximum velocities recommended by the Federal Highway Administration.

FIGURE 13.6 Concrete linked ditch.

13.4.7 Basic Hydraulics

There are two types of flow in open channels, steady flow and unsteady flow. For steady flow, the flow depth at a given cross section does not change; it can be assumed to remain constant during the time interval under consideration. With unsteady flow, the depth of flow at a given cross section changes with time. Steady flow is uniform if the depth of flow is the same at every section along the length of the channel. Steady, uniform flow is an ideal condition which seldom occurs in natural channels. However, for many practical applications, changes in width, depth or direction are sufficiently small so that uniform flow can be assumed. Assuming uniform, steady flow in a channel, the mean velocity may be computed using the Manning equation shown here as Equation 13.7.

$$v = \frac{1.49}{n} R^{2/3} S^{1/2} \qquad (13.7)$$

where:

v = average velocity (ft/sec).
$R = A/P$ = Hydraulic radius (ft).
A = area of cross-sectional flow (ft^2).
P = wetted perimeter (ft).
S = slope of total head line (ft/ft).
n = manning's roughness coefficient.

The wetted perimeter of the channel is the perimeter of the channel in contact with water. Discharge from the channel is given as Equation 13.8.

$$Q = Av \qquad (13.8)$$

TABLE 13.6 Maximum Permissible Velocities in Erodible Channels, Based on Uniform Flow in Continuously Wet, Aged Channels[a]

Soil Type or Lining (Earth; No Vegetation)	Maximum Permissible Velocities For:		
	Clear Water	Water Carrying Fine Silts	Water Carrying Sand and Gravel
Fine Sand (noncolloidal)	1.5	2.5	1.5
Sandy loam (noncolloidal)	1.7	2.5	2.0
Silt loam (noncolloidal)	2	3.0	2.0
Ordinary firm loam	2.5	3.5	2.2
Volcanic ash	2.5	3.5	2.0
Fine gravel	2.5	5.0	3.7
Stiff clay (very colloidal)	3.7	5.0	3.0
Graded, loam to cobbles (noncolloidal)	3.7	5.0	5.0
Graded, silt to cobble (colloidal)	4.0	5.5	5.0
Alluvial silts (noncolloidal)	2.0	3.5	2.0
Alluvial silts (colloidal)	3.7	5.0	3.0
Coarse gravel (noncolloidal)	4.0	6.0	6.5
Cobbles and shingles	5.0	5.5	6.5
Shales and hard pans	6.0	6.0	5.0

Source: From *Design Charts for Open-Channel Flow*, Hydraulic Design Series No. 3, US Department of Transportation, Washington, DC, 1961.

[a] As recommended by Special Committee on Irrigation Research, American Society of Civil Engineers, 1926, for channels with straight alignment. For sinuous channels multiply allowable velocity by 0.95 for slightly sinuous, by 0.9 for moderately sinuous channels, and by 0.8 for highly sinuous channels.

TABLE 13.7 Maximum Permissible Velocities in Channels Lined with Uniform Stands of Grass Covers, Well Maintained[a,b]

Cover	Permissible Velocity on[a]:		
	Slope Range (%)	Erosion Resistant Soils (ft/sec)	Easily Eroded Soils (ft/sec)
Bermuda grass	0–5	8	6
	5–10	7	5
	Over 10	6	4
Buffalo Grass			
Kentucky bluegrass	0–5[b]	7	5
Smooth brome	5–10[b]	6	4
Blue grama	Over 10	5	3
Grass mixture	0–5	5	4
	5–10	4	3
Lespedeza sericea	0–5[c]	3.5	2.5
Weeping lovegrass			
Yellow bluestem			
Kudzu			
Alfalfa			
Crabgrass			
Common lespedeza[d]	0–5[c]	3.5	2.5
Sudangrass[d]			

Source: From *Design Charts for Open-Channel Flow*, Hydraulic Design Series No. 3, US Department of Transportation, Washington, DC, 1961.

[a] Use velocities over 5 (ft/sec) only where good covers and proper maintenance can be obtained.

[b] Do not use on slopes steeper than 10%.

[c] Use on slopes steeper than 10%.

[d] Annuals, used on mild slopes or as temporary protection until permanent covers are established.

where:

Q = discharge (ft^3/sec).
A = area of cross-sectional flow (ft^2).
v = average velocity (ft/sec).

By substituting Equation 13.8 into Equation 13.7, the Manning's formula will give the discharge as shown in Equation 13.9.

$$Q = \frac{1.49}{n} A R^{2/3} S^{1/2} \tag{13.9}$$

where all variables are as previously defined.

Recommended values of the Manning's roughness coefficients can be found in Table 13.8. Graphic solutions of the discharge for rectangular and trapezoidal cross sections as defined by the Federal Highway Administration are given in Figures 13.7 and 13.8.

Example 13.3

Open Channel Design

Find the depth and velocity of flow in a trapezoidal channel with 2:1 side slopes and a 5 ft bottom width. The slope of the channel is 2% and the discharge is 55 ft^3/sec.

Answer:

From Figure 13.8

TABLE 13.8   Manning's Roughness Coefficients

II. Open Channels, Lined (Straight Alignment):	Manning's, n, Range[a]
A. Concrete, with surfaces as indicated:	
1. Formed, no finish	0.013–0.017
2. Trowel finish	0.012–0.014
3. Float finish	0.013–0.015
4. Float finish, some gravel on bottom	0.015–0.017
5. Gunite, good section	0.016–0.019
6. Gunite, wavy section	0.018–0.022
B. Concrete, bottom float finished, sides as indicated:	
1. Dressed stone in mortar	0.015–0.017
2. Random stone in mortar	0.017–0.020
3. Cement rubble masonry	0.020–0.025
4. Cement rubble masonry, plastered	0.016–0.020
5. Dry rubble (riprap)	0.020–0.030
C. Gravel bottom, sides as indicated:	
1. Formed concrete	0.017–0.020
2. Random stone in mortar	0.020–0.023
3. Dry rubble (riprap)	0.023–0.033
D. Brick	0.014–0.017
E. Asphalt:	
1. Smooth	0.013
2. Rough	0.016
F. Wood, planed, clean	0.011–0.013
G. Concrete-lined excavated rock:	
1. Good section	0.017–0.020
2. Irregular section	0.022–0.027

Source: From *Design Charts for Open-Channel Flow*, Hydraulic Design Series No. 3, US Department of Transportation, Washington, DC, 1961.

[a] Ranges indicated for open channels, lined, are for good to fair construction (unless otherwise stated). For poor quality construction, use larger values of n.

$Q = 55$ ft^3/sec
Slope $= 0.02$ ft/ft
Reading from the graph velocity $= 6.2$ ft/sec
Depth $= 1.2$ ft

Flow in an open channel is governed by three basic equations, the continuity, momentum and energy equations. The continuity equation, Equation 13.10, simply means that discharges at various points along the channel are equal.

$$Q_1 = Q_2 = V_1 A_1 = V_2 A_2 \qquad (13.10)$$

where:

Subscripts 1 and 2 are points along the channel

$V =$ average velocity (ft/sec).
$A =$ cross-sectional area of flow (ft^2).
$Q =$ cischarge (ft^3/sec).

The momentum equation is derived from Newton's second law which states that the summation of all external forces on a system is equal to the change in momentum. The equation is a vector relationship and the form shown is for the x-direction, with similar forms in the y- and z-directions. The momentum

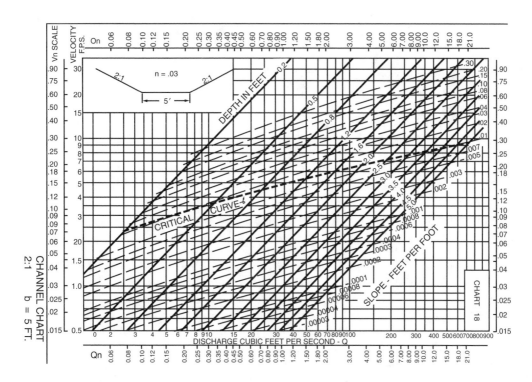

FIGURE 13.7 Graphical solution of Manning's equation for a 2:1 side trapezoidal channel (*Source*: From *Design Charts for Open Channels*, US Department of Transportation, Washington, DC, 1979).

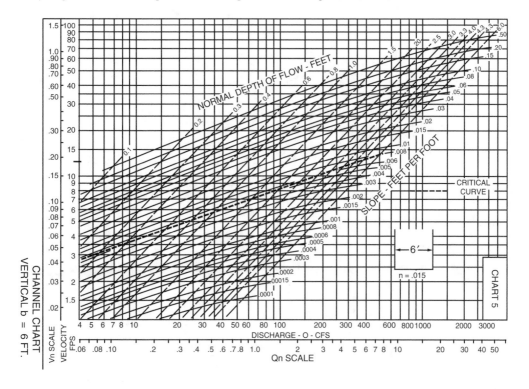

FIGURE 13.8 Graphical solution of Manning's equation for a rectangular channel, b-6 ft (*Source*: From *Design Charts for Open Channels*, US Department of Transportation, Washington, DC, 1979).

concept is important in grate design, hydraulic jump analysis, analysis of paved channels intersecting at angles and in design of energy dissipaters (Ayres, 2001). Equation 13.11 is the momentum equation.

$$\sum F_x = \rho Q(\beta_2 V_{x_2} - \beta_1 V_{x_1})$$ (13.11)

where:

F_x = forces in the x-direction.
ρ = density of water.
Q = discharge.
$\beta_{1,2}$ = momentum coefficient, usually assumed to be 1.
V_{x_1}, x_2 = velocities in the x-direction.

   Subcritical and supercritical flows are important in some situations. Subcritical flow is a condition where the water is relatively deep, the velocity remains low and the slope of the channel is mild. Subcritical flow occurs when the depth of flow, velocity, and orientation of flow are dependent upon downstream factors. The flow usually has a low velocity and is described as tranquil (Ayres, 2001). Supercritical flow occurs when there is shallow water running along steep slopes at relatively high velocities. Supercritical flow occurs when the depth of flow, velocity, and orientation of flow are dependent upon upstream factors. The flow usually has a high velocity and may be described as rapid, shooting, and torrential (Ayres, 2001). Critical depth is the point at which the flow transits from supercritical to subcritical. Transition from supercritical to subcritical flow will result in the formation of a hydraulic jump. Hydraulic jumps essentially absorb energy in the channel. When analyzing a hydraulic jump situation, as shown in Figure 13.9, Equation 13.12 can be utilized. The Froude number is equal to one at critical depth; however, if the depth is not critical, the Froude number can be determined using Equation 13.13.

$$y_2 = 0.5y_1(\sqrt{1 + 8\mathrm{Fr}^2} - 1)$$ (13.12)

where:

y_1 = depth of water upstream, supercritical flow (ft).
y_2 = depth of water downstream, subcritical flow (ft).
Fr = froude number, unitless.

$$\mathrm{Fr} = \frac{V}{\sqrt{gy}}$$ (13.13)

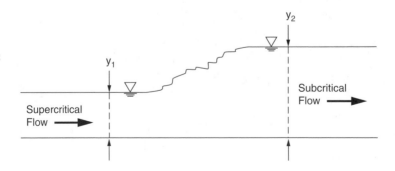

FIGURE 13.9   Hydraulic jump situation.

    Fr > 1, Supercritical flow
    Fr = 1, Critical flow
    Fr < 1, Subcritical flow
where:

    V = average flow velocity (ft/sec).
    g = acceleration of gravity, 32.2 ft/sec^2.
    y = flow depth (ft).

The amount of energy loss due to a hydraulic jump in a rectangular channel is shown in Equation 13.14.

$$\Delta E = \frac{(y_2 - y_1)^3}{4y_1 y_2} \tag{13.14}$$

where:

    ΔE = energy loss due to hydraulic jump (ft).
    y_1 = depth of water upstream, supercritical flow (ft).
    y_2 = depth of water downstream, subcritical flow (ft).

    Energy within an open channel is one of the most influential factors governing design. Energy within the system can be viewed in Figure 13.10. The Energy Grade Line (EGL) represents the total energy at any given section, defined as the sum of the three components of energy represented on each side of the energy equation. These components of energy are referred to as:

1. Velocity head
2. Pressure head
3. Elevation head

The Hydraulic Grade Line (HGL) is below the EGL by the amount of the velocity head, or the sum of the pressure head and elevation head (Ayres, 2001). Equation 13.15 shows this relationship mathematically.

$$Z_1 + y_1 + \frac{V_1^2}{2g} = Z_2 + y_2 + \frac{V_2^2}{2g} + h_L \tag{13.15}$$

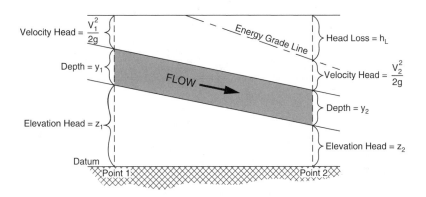

FIGURE 13.10   Energy within a system.

where:

Z_1 = elevation of the streambed at the upstream section (ft).
Z_2 = elevation of the streambed at the downstream section (ft).
y_1 = flow depth at the upstream section (ft).
y_2 = flow depth at the downstream section (ft).
V_1 = average flow velocity at the upstream section (ft/sec).
V_2 = average flow velocity at the downstream section (ft/sec).
h_L = head loss between upstream and downstream sections (ft).
g = gravitational acceleration, 32.2 ft/sec².

Example 13.4

Determination of head loss

The velocity of the upstream end of a rectangular channel 3.0 ft wide is 10.0 ft/sec, and the flow depth is 6.5 ft. The depth at the downstream end is 6.0 ft. The elevation at the upstream end is 1640.3 ft and at the downstream end is 1640.0 ft. Determine the head loss due to friction.

Answer:

Use the continuity equation, Equation 13.10 to find the velocity at the downstream end.

$$Q = VA$$

$$V_2 = \frac{Q}{A_2} = \frac{(10.0)(6.5)(3.0)}{(6.0)(3.0)} = 10.83 \text{ ft/sec}$$

Use the energy equation to find the head loss, h_L

$$\frac{V_1^2}{2g} + Y_1 + Z_1 = \frac{V_2^2}{2g} + Y_2 + Z_2 + h_L$$

$$\frac{(10.0)^2}{2(32.2)} + 6.5 + 1640.3 = \frac{(10.83)^2}{2(32.2)} + 6.0 + 1640.0 + h_L$$

$$h_L = (1.55 + 6.5 + 1640.3)$$
$$- (1.82 + 6.0 + 1640.0) = 0.53 \text{ ft}$$

The specific energy equation, Equation 13.16, does not take into account the elevation head and is therefore the energy relative to the bed of the channel. The relationship between the specific energy and the depth of flow is shown graphically in Figure 13.11.

$$E = y + \frac{V^2}{2g} \qquad (13.16)$$

where:

E = specific energy (ft).
y = flow depth (ft).
V = average velocity (ft/sec).
g = gravitational acceleration, 32.2 ft/sec².

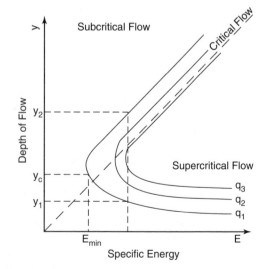

FIGURE 13.11   Specific energy vs. depth of flow (*Source*: Adapted from *FHWA Highways in the River Environment*, Training and Design Manual, 1990).

13.4.8 Channel Lining Design

Once the channel's hydraulic factors are known and it has been determined that a channel lining is needed, the lining type must be selected. Flexible linings are subject to forces that may cause erosion of the lining itself. Rigid linings are able to withstand more forces than flexible linings; however they tend to be more expensive. If space is restricted or the velocities in the channel do not allow for flexible linings, a more expensive rigid lining may be used.

The first step in determining a suitable lining is finding the channel's cross-sectional flow area. There are numerous graphs and tables for determining this area provided in the FHWA publication *Design of Roadside Channels with Flexible Linings HEC No. 15* (FHWA, 1988). Charts and graphs are given for most typical cross sections. However, if a chart is not available or computations need to be conducted manually, the Equation 13.17 to Equation 13.20may be used. Critical depths for rectangular sections may be calculated using Equation 13.17.

$$Y_c = \left(\frac{q^2}{g} \right)^{1/3} \qquad (13.17)$$

where:

Y_c = critical depth (ft).
q = flow per foot of width (ft^3/sec/ft).
g = gravitational acceleration, 32.2 ft/sec^2.

Example 13.5

Determining critical depth
    Find the cross-sectional area required for a rectangular channel to carry 200 ft^3/sec of runoff. Use 6 ft for the bottom width of the channel, 3% for the slope of the channel, and a Manning's n of 0.015. Determine the channel's critical depth.
Answer:

Depth of flow $= d$
Area of cross section $= 6d$
Wetted perimeter $= 6 + 2d$
Hydraulic radius $= 6d/(6 + 2d)$

Solve for the area using Equation 13.9

$$Q = \frac{1.49}{n} A R^{2/3} S^{1/2} \text{ then } 200 = \frac{1.49}{0.015} 6d \left(\frac{6d}{6 + 2d} \right)^{2/3} (0.03)^{1/2}$$

$$6d \left(\frac{6d}{6 + 2d} \right)^{2/3} = \frac{200 \times 0.015}{1.49(0.03)^{1/2}} = 11.63$$

Solving by trial and error, the depth of flow is approximately 1.79. A freeboard of 1 ft must be added bringing the required depth to approximately 3 ft.

Solving graphically, enter Figure 13.7 at $Q = 200$ ft^3/sec and follow up vertically until reaching the slope of 0.03. This yields a depth of about 1.8 ft. Reading across the chart, the critical velocity is approximately 18 ft/sec. This depth of flow is located above the critical curve on the graph, meaning the flow is supercritical. To determine the critical depth, Equation 13.17 is used

$$Y_c = \left(\frac{q^2}{g} \right)^{1/3} \text{ then } Y_c = \left(\frac{(200/6)^2}{32.2} \right)^{1/3} = 3.3 \text{ ft}$$

The critical depth can be read off from Figure 13.7, beginning at $Q = 200$ ft³/sec and following the graph vertically to the intersection with the critical curve, a value of 3.3 ft can be found.

The hydraulic depth, found by using Equation 13.18, is often referred to as the mean or average depth of an open channel.

$$Y_h = \frac{A}{T} \qquad (13.18)$$

where:

Y_h = hydraulic depth (ft).
A = cross-sectional flow area (ft²).
T = top width of the water surface (ft).

Where an energy dissipater is present, Equation 13.19 is often used to determine the equivalent depth of flow.

$$Y_e = \sqrt{\frac{A}{2}} \qquad (13.19)$$

where:

Y_e = equivalent depth (ft).
A = cross-sectional flow area (ft²).

Under critical flow conditions, the critical depth is found by using Equation 13.20.

$$\frac{Q_c^2}{g} = \frac{A_c^3}{T} \qquad (13.20)$$

where:

Q_c = critical discharge (ft³/sec).
g = gravitational acceleration, 32.2 ft/sec².
A_c = cross-sectional flow area (ft²).
T = top width of water surface(ft).

This equation can then be utilized to find the critical depth as shown in Example 13.6.

Example 13.6

Determining critical depth

Calculate the critical depth of a trapezoidal channel with 2:1 side slopes and a base width of 5 ft, a flow in the channel of 350 ft³/sec, and Manning's $n = 0.03$.

Answer:

The cross-section of the channel is shown in
   Figure 13.12, where the depth of flow is y.
The top width is
   $T = 5 + 4y$
The area is
   $A = 5y + 2y^2$

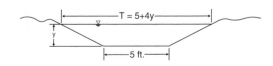

FIGURE 13.12   Channel cross-section for Example 17.6.

To find the critical depth Equation 13.20 is used

$$\frac{Q^2}{g} = \frac{A^3}{T} \text{ so } \frac{350^2}{32.2} = \frac{(5y + 2y^2)^3}{(5 + 4y)}$$

Solve for y using trial and error

$$y \approx 3.48 \text{ ft}$$

Solving graphically, begin with Figure 13.8 at $Q = 350 \text{ ft}^3/\text{sec}$ and follow up vertically until reaching the critical curve. This gives a depth approximately equal to 3.5 ft. Reading across the chart, critical velocity is approximately 8.3 ft/sec.

Freeboard is the vertical distance from the water surface to the top of the channel under design condition. It must be considered in the design. The importance of freeboard depends on the consequences of overflowing the channel bank. At a minimum, the freeboard should be sufficient to prevent waves or fluctuations in water surface from overflowing the sides. In a relatively flat, permanent roadway channel about 6 in. of freeboard should be adequate and for temporary channels no freeboard is necessary. Steep gradient channels should have a freeboard height equal to the flow depth. This allows for large variations to occur in the flow depth caused by waves, splashing, and surging. Lining materials should extend to the freeboard elevation (FHWA, 1988).

13.4.9 Channel Lining

The steps necessary to determine a suitable channel lining and the protection needed to prevent erosion is described by the Federal Highway Administration in *Hydraulic Engineering Circular No. 15* (FHWA, 1988). The first step in this process is to select a lining and determine the permissible shear stress in lb/ft^2. The flow can then be estimated by choosing an initial Manning's roughness coefficient n from Table 13.9 or from Figure 13.13. When calculating the normal flow depth, compare the normal depth to the estimated depth. If the two numbers do not agree, these first two steps must be repeated. Once the two

TABLE 13.9 Manning's Roughness Coefficients

Lining Category	Lining Type	Depth Ranges (n-value)		
		0–0.5 ft	0.5–2.0 ft	>2.0 ft
Rigid	Concrete	0.015	0.013	0.013
	Grouted riprap	0.040	0.030	0.028
	Stone masonry	0.042	0.032	0.030
	Soil cement	0.025	0.022	0.020
	Asphalt	0.018	0.016	0.016
Unlined	Bare soil	0.023	0.020	0.020
	Rock cut	0.045	0.035	0.025
Temporary[a]	Woven paper net	0.016	0.015	0.015
	Jute net	0.028	0.022	0.019
	Fiberglass roving	0.028	0.022	0.019
	Straw with net	0.065	0.033	0.025
	Curled wood mat	0.066	0.035	0.028
	Synthetic mat	0.036	0.025	0.021
Gravel riprap	1 in. D50	0.044	0.033	0.030
	2 in. D50	0.066	0.041	0.034
Rock riprap	6 in. D50	0.104	0.069	0.035
	12 in. D50	—	0.078	0.040

Note: Values listed are representative values for the respective depth ranges. Manning's roughness coefficients, n, vary with flow depth. *Source*: From *Design of Stable Channels with Flexible Linings*, Hydraulic Engineering, Circular No. 15, US Department of Transportation, Washington, DC, October 1975.

[a] Some temrorary linings become permanent when buried.

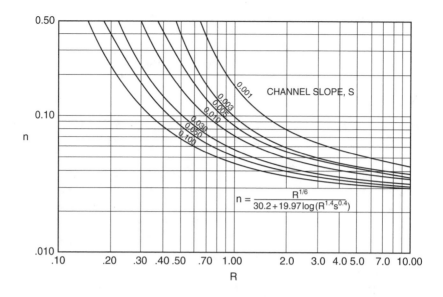

FIGURE 13.13 Manning's *n* vs. Hydraulic Radius, *R*, for Class C vegetation (*Source*: From *Design of Stable Channels with Flexible Linings*, Hydraulic Engineering Circular No. 15, US Department of Transportation, Washington, DC, October 1975).

values agree, the maximum shear stress at normal depth is calculated. The maximum shear stress must remain lower than the permissible shear stress.

A suitable lining may also be found using the maximum top width of the channel. The maximum depth d_{max}, should be appropriate and can be selected from the charts for each of the different lining types. Some examples of these charts are shown in Figures 13.14, 13.15, and 13.16, and depend on the type of material chosen to line the channel. Utilizing this maximum depth, the hydraulic radius and cross-sectional area can be computed mathematically or graphically from Figure 13.17. The flow is determined using this cross-sectional area and the velocity found in Figures 13.18 and 13.19, depending on the type of lining. This calculated flow must be similar to the designed flow for the material to be suitable. If it is lower than the design flow, the channel lining is inadequate. While if the flow is higher, the channel may need redesigning using a more cost effective lining. If it is determined that an alternate lining type is necessary, the process must be repeated and the new material should be checked for suitability.

Example 13.7

Determining channel linings

A median ditch is lined with a good stand of Kentucky bluegrass (approximately 8 in. high). The ditch is trapezoidal with a bottom width of 4 ft and side slopes of 4:1. The ditch slope is 0.010 ft/ft. Compute the maximum discharge for which this lining will be stable and the corresponding flow depth.

Answer:

From Table 13.10, Kentucky bluegrass has a retardance class of C and from Table 13.11, the permissible shear stress is $\tau_p = 1.0$ lb/ft².

Then the allowable depth can be determined by assuming $\tau_p = \tau_d$

$$d = \tau_p/(62.4S) = 1.0/(62.4 \times 0.01) = 1.6 \text{ ft}$$

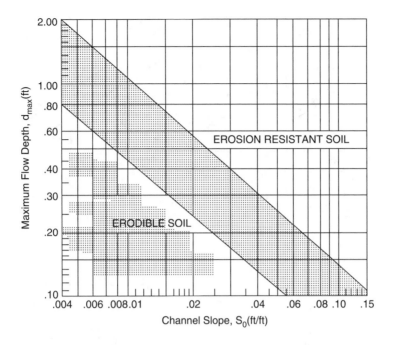

FIGURE 13.14 Maximum permissible depth of flow, for channels lined with jute mesh (*Source*: From *Design of Stable Channels with Flexible Linings*, Hydraulic Engineering Circular No. 15, US Department of Transportation, Washington, DC, October 1975).

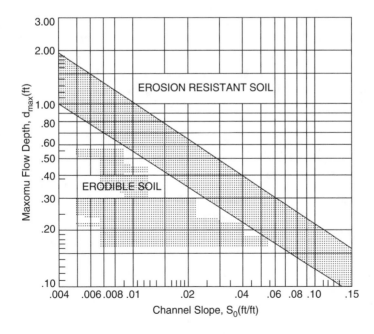

FIGURE 13.15 Maximum permissible depth of flow, for channels lined with 3/8 in. fiberglass mat (*Source*: From *Design of Stable Channels with Flexible Linings*, Hydraulic Engineering Circular No. 15, US Department of Transportation, Washington, DC, October 1975).

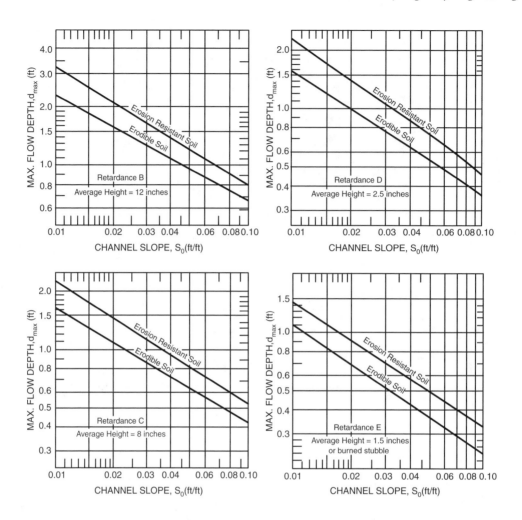

FIGURE 13.16 Maximum permissible depth of flow for channels lined with Bermuda grass, good stand, cut to various lengths (*Note*: Use on slopes steeper than 10% is not recommended) (*Source*: From *Design of Stable Channels with Flexible Linings*, Hydraulic Engineer Circular No. 15, US Department of Transportation, Washington, DC, October 1975).

Now determine the flow area A and hydraulic radius R

$$A = d(b + zd) = 1.6(4 + 4 \times 1.6) = 16.6 \text{ ft}^2$$

$$P = b + 2d(1 + z^2)^{1/2} = 4 + 2 \times 1.6(1 + 16)^{1/2} = 17.2 \text{ ft}$$

$$R = A/P = 16.6/17.2 = 0.97 \text{ ft}$$

Finally determine the Manning's n value from Figure 13.13 and solve for Q from Manning's formula. From figure 13.13, $n = 0.072$ and

$$Q = (1.49/n) AR^{2/3} S^{1/2}$$

$$Q = (1.49/0.072)(16.6)(0.97)^{2/3}(0.01)^{1/2} = 33.7 \text{ ft}^3/\text{sec}$$

(This method is called the maximum discharge method and is useful for determining the stable channel capacity for a variety of different linings for purposes of comparison.)

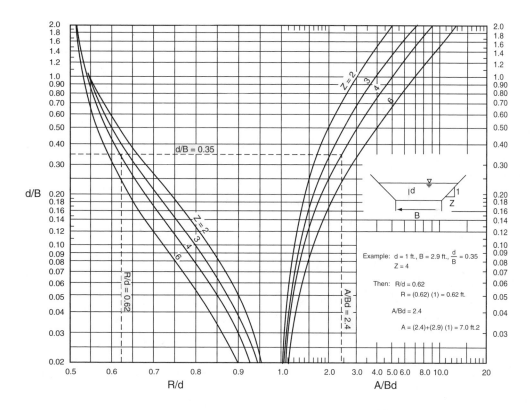

FIGURE 13.17 Trapezoidal channel geometry (*Source*: From *Design of Stable Channels with Flexible Linings*, Hydraulic Engineering Circular No. 15, US Department of Transportation, Washington, DC, October 1975).

Example 13.8

Determining channel linings

A rectangular channel has a 2.5% slope, and a bottom width of 5 ft, and is comprised of erodible soil. The flow through the channel is 200 ft^3/sec. Determine if a channel lining of 3/8 in. fiber–glass mat is suitable for this channel.

Answer:

Determine d_{max} from Figure 13.14

$d_{max} = 0.32$ ft

Wetted perimeter $= 2 \times 0.32 + 5 = 5.64$ ft

Cross-sectional area $0.32 \times 5 = 1.6$ ft^2

$$R = \frac{1.6}{5.64} = 0.28$$

Using the Figure 13.19 determine the flow velocity for the lining

$$V = 73.53(0.28)^{1.330}(0.025)^{0.512} = 2.05 \text{ ft/sec}$$

or reading off the graph $V \approx 2.0$ ft/sec

So the maximum allowable flow is

$$Q = 2.05 \times 1.6 = 3.28 \text{ ft}^3$$

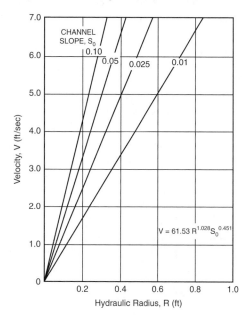

FIGURE 13.18 Flow velocity for channels lined with jute mesh (*Source*: From *Design of Stable Channels with Flexible Linings*, Hydraulic Engineering Circular No. 15, US Department of Transportation).

This means that fiber–glass mat is not acceptable for this situation. Another type of lining should be chosen and the analysis repeated.

13.4.10   Pipe Culvert Design

Pipe culverts are made from several types of materials including concrete, plastic, aluminum, and corrugated steel. The selection of culvert material for a particular location depends on topography, soil properties, material cost, climate and ease of installation. Some culverts are composed of different types of materials; for example in a highly corrosive or abrasive area a metal culvert may be coated with asphalt or a plastic material. Culverts come in several commonly used shapes including circular, box, elliptical, pipe arch, and arch. Shape selection is based on construction costs, limitations on upstream water surface elevation, roadway embankment height, and hydraulic performance (FHWA, 1985). Where culvert height is a concern, several pipes may be placed side-by-side. Culvert size is determined by the designed peak flow through the pipe. Peak flow rates can be determined from a hydrograph based on the rainfall or stream flow in the area.

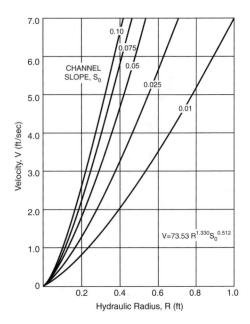

FIGURE 13.19   Flow velocity for channels lined with 3/8 in. fiberglass mat (*Source*: From *Design of Stable Channels with Flexible Linings*, Hydraulic Engineering Circular No. 15, US Department of Transportation, Washington, DC, October 1975).

Pipe culverts are usually located in existing channel beds. This is generally the cheapest placement since it involves the least earthwork and re-routing of the water. Exceptions may include mountainous areas where the channel may require an unusually long culvert, or winding channels that need a straight culvert. If channel relocation is the most cost effective solution, care must be taken to avoid abrupt stream transitions at either end of the culvert and erosion at the inlet or outlet.

Economic factors governing pipe culvert design and installation must account for costs incurred and damage done in the event of failure. This is referred to as risk analysis. The storm frequency is determined by finding a balance between the cost of construction, maintenance, and the risk of flooding, washout, or failure of the culvert. Under-design of the culvert results in a high risk of unacceptable damage, while over-design will usually be a waste of money.

13.4.11   Culvert Control

When designing culverts, it should be determined whether the flow through the culvert is controlled at the inlet or the outlet. With inlet control, the amount of water passing through is controlled as water enters the pipe. With outlet control, the amount of water is controlled downstream of the inlet.

13.4.11.1   Flow Types

Flow in pipe is generally described as either super-critical or subcritical. River rapids are supercritical, as water flows relatively quickly and unimpeded through a fairly steep pipe. A meandering stream is an example of subcritical flow. Hydraulic jumps may occur when water enters a pipe in super-critical flow, then the flow in the pipe becomes subcritical as water begins to back up due to downstream pressure as shown in Figure 13.20(B) and (D). Water may occasionally pass through a pipe under pressure flow

TABLE 13.10   Classification of Vegetal Covers as to Degrees of Retardancy

Retardance	Cover	Condition
A	Weeping lovegrass	Excellent stand, tall (average 30 in.)
	Yellow bluestem	
	Ischaemum	Excellent stand, tall (average 36 in.)
B	Kudzu	Very dense growth, uncut
	Bermuda grass	Good stand, tall (average 12 in.)
	Native grass mixture: little bluestem, bluestem, blue gamma other short and long stem midwest grasses	Good stand, unmowed
	Weeping lovegrass	Good stand, tall (average 24 in.)
	Laspedeza sericea	Good stand, not woody, tall (average 19 in.)
	Alfalfa	Good stand, uncut (average 11 in.)
	Weeping lovegrass	Good stand, unmowed (average 13 in.)
	Kudzu	Dense growth, uncut
	Blue gamma	Good stand, uncut (average 13 in.)
C	Crabgrass	Fair stand, uncut (10–48 in.)
	Bermuda grass	Good stand, mowed (average 6 in.)
	Common lespedeza	Good stand, uncut (average 11 in.)
	Grass-legume mixture: summer (orchard grass, redtop, Italian ryegrass, and common lespedeza)	Good stand, uncut (6–8 in.)
	Centipede grass	Very dense cover (average 6 in.)
	Kentucky bluegrass	Good stand, headed (6–12 in.)
D	Bermuda grass	Good stand, cut to 2.5 in.
	Common lespedeza	Excellent stand, uncut (average 4.5 in.)
	Buffalo grass	Good stand, uncut (3–6 in.)
	Grass-legume mixture: fall, spring (orchard grass, redtop, Italian reygrass, and common lespedeza)	Good stand, uncut (4–5 in.)
	Lespedeza serices	After cutting to 2 in. (very good before cutting)
E	Bermuda grass	Good stand, cut to 1.5 in.
	Bermuda grass	Burned stubble

Note: Covers classified have been tested in experimental channel. Covers were green and generally uniform.

Source: From *Design of Stable Channels with Flexible Linings*, Hydraulic Engineering Circular No. 15, US Department of Transportation, Washington, DC, October 1975.

where both the inlet and outlet are submerged, the pipe is full, and water flow is driven by the difference in head between the head-water and the tail-water.

13.4.11.2   Determining Culvert Control

Head-water and the tail-water depths must be known to determine the culvert control. Head-water is measured at the entrance of the culvert and this depth of water at the entrance is called *head-water depth*. Head-water should be low enough not to overtop the roadway, cause damaging flooding upstream, or divert the water from the culvert to an undesirable location. Head-water is measured from the culvert inlet invert, i.e., the culvert bottom at the inlet to the highest water surface at the inlet. Tail-water is defined as the depth of water downstream of the culvert measured from the outlet invert, the culvert bottom at the outlet, as water flows from the culvert exit. The tail-water depth is especially important when the culvert is outlet-controlled. High tail-water is usually due to obstructions downstream in the channel, but it can also occur as a result of resistance within the channel. Backwater calculations from the downstream control point are required to precisely define tail-water. When appropriate, normal depth approximations may be used instead of backwater calculations (FHWA, 1985).

Sometimes there is adequate area to retain large amount of water above a culvert inlet. In such cases, the upstream storage should be taken into account when designing a culvert's capacity. Upstream storage

TABLE 13.11   Summary of Shear Stress for Various Protection Measures

Protective Cover	Underlying Soil		τ_p (lb/ft^2)
Class A vegetation	Erosion	Resistant	3.70
		Erodible	3.70
Class B vegetation	Erosion	Resistant	2.10
		Erodible	2.10
Class C vegetation	Erosion	Resistant	1.00
		Erodible	1.00
Class D vegetation	Erosion	Resistant	0.60
		Erodible	0.60
Class E vegetation	Erosion	Resistant	0.35
		Erodible	0.35
Woven paper			0.15
Jute net			0.45
Single fiberglass			0.60
Double fiberglass			0.85
Straw with net			1.45
Curled wood mat			1.55
Synthetic mat			2.00
Plain grass, good cover		Clay	N/A
Plain grass, average cover		Clay	N/A
Plain grass, poor cover		Clay	N/A
Grass, reinforced with nylon		Clay	N/A
Dycel with grass		Clay	N/A
Petraflex with grass		Clay	N/A
Armorflex with grass		Clay	N/A
Dymex with grass		Clay	N/A
Grasscrete		Clay	N/A
Gravel			
$D_{50} = 1$ in.			0.40
$D_{50} = 2$ in.			0.80
Rock			
$D_{50} = 6$ in.			2.50
$D_{50} = 12$ in.			5.00
6 in. gabions		Type I	35.00
4 in. geoweb		Type I	10.00
Soil cement (8% cement)		Type I	>45
Dycel with out grass		Type I	>7.0
Petraflex with out grass		Type I	>32
Armorflec with out grass		Type I	$12-20$
Erikamat with 3 in. asphalt		Type I	$13-16$
Erikamat with 1 in. asphalt		Type I	<5
Armorflex class 30 with longitudinal and lateral cables, no grass		Type I	>34
Dycell 100, longitudinal cables, cells filled with mortar		Type I	<12
Concrete construction blocks, granular filter underlayer		Type I	>20
Wedge-shaped blocks with drainage slot		Type I	>25

ft/sec $\times$ 0.03048 = m/sec

lb/ft^2 $\times$ 47.87 = N/m^2

Source: From *Design of Stable Channels with Flexible Linings*, Hydraulic Entineering Circular No. 15, US Department of Transportation, Washington, DC, October 1975.

is dependent on topography and is generally high in areas with large fills and flat slopes. Storage is determined from contour maps or cross-sections.

Inlet control occurs when the culvert barrel has the capacity to carry more water than the inlet can accept. This means that the size of the inlet controls the amount of water in a culvert, not the barrel size. The inlet control point is just beyond the entrance of the culvert, resulting in super-critical flow

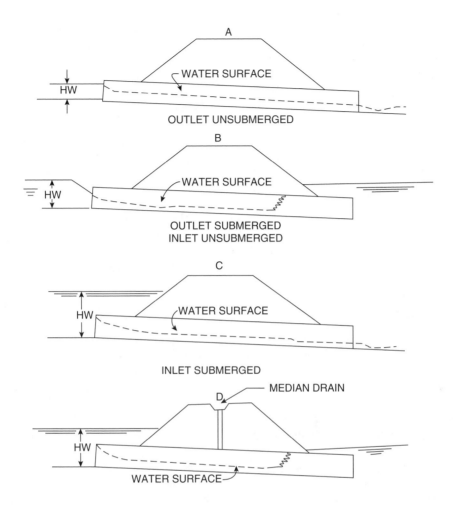

FIGURE 13.20 Types of inlet control (*Source*: From Norman, J.M., Houghtalen, R.J., and Johnston, W.J., in *Hydraulic Design of Highway Culverts*, Report No. FHWA IP-85-15, US Department of Transportation, Office of Implementation, McLean, VA, September 1985).

immediately past the control point. Conversely, in an outlet controlled system, the barrel of the culvert will not carry the amount of water introduced by the inlet. The outlet control point may lie directly at the culvert exit or it may be produced by downstream conditions. If a culvert is expected to perform under inlet control conditions some of the time and outlet control conditions for the remainder of the time, the design must be such that the culvert never performs at an unacceptable level under either control condition. This results in a culvert design that will sometimes perform better than necessary, but will be barely adequate under other flow conditions.

13.4.12   Inlet Control

Different types of inlet controls are shown in Figure 13.20. Each one of these examples is affected by the submergence of the inlet and outlet of the culvert.

Figure 13.20(A) depicts a condition where neither the inlet nor outlet end of the culvert is submerged. The flow passes through critical depth just downstream of the culvert entrance and the barrel flow is super-critical. The barrel flows partly full over its length, and the flow approaches normal depth at the outlet end.

Figure 13.20(B) shows that submergence of the outlet end of the culvert does not assure outlet control. In this case, the flow just downstream of the inlet is super-critical and a hydraulic jump forms in the culvert barrel.

Figure 13.20(C) is a more typical design situation. The inlet end is submerged and the outlet end flows freely. Again, the flow is super-critical and the barrel flows partly full over its length. Critical depth is located just downstream of the culvert entrance, and the flow is approaching normal depth at the downstream end of the culvert.

Figure 13.20(D) is an unusual condition illustrating that even submergence of both the inlet and the outlet ends of the culvert does not assure full flow. In this case, a hydraulic jump will form in the barrel. The median inlet provides ventilation of the culvert barrel. If the barrel were not ventilated, subatmospheric pressures could develop which might create an unstable condition causing the barrel to alternate between full flow and partly full flow (FHWA, 1985).

Factors that affect inlet control include head-water elevation, inlet area, inlet edge configuration and inlet shape. Several charts are available in the *Hydraulic Design for Highway Culverts* (FHWA, 1985). These charts can be used to determine the head-water depth that is required for adequate flow through the culvert.

13.4.13   Outlet Control

Outlet control is influenced by tail-water depth, fall, length, and roughness, as well as the factors that influence inlet control. Subcritical or pressure flow exists in the culvert under outlet control conditions. Some examples of outlet controlled flow are shown in Figure 13.21. In each of these examples, the control is located at the outlet or a downstream section. This is the section where the barrel is only partially full and the flow is subcritical.

Figure 13.21(A) represents the classic full flow condition, with both inlet and outlet submerged. The barrel is in pressure flow throughout its length. This condition is often assumed in calculations, but seldom exists.

Figure 13.21(B) depicts the outlet submerged and the inlet unsubmerged. For this case, the head-water is shallow so that the inlet crown is exposed as the flow contracts into the culvert.

Figure 13.17(C) shows the entrance submerged to such a degree that the culvert flows full throughout its entire length while the exit is unsubmerged. This is a rare condition which requires extremely high head-water to maintain full barrel flow with no tail-water. The outlet velocities are usually high under this condition.

Figure 13.21(D) is more typical. The culvert entrance is submerged by the head-water and the outlet end flows freely with a low tail-water. For this condition, the barrel flows partly full over at least part of its length, and the flow passes through critical depth just upstream of the outlet.

Figure 13.21(E) is also typical, with neither the inlet nor the outlet end of the culvert submerged. The barrel flows partly full over its entire length, and the flow profile is subcritical (FHWA HEC 5, 1985).

The analysis of outlet control conditions are based on *balance of energy*. The energy loss through the culvert can be found using Equation 13.21.

$$H_\mathrm{L} = H_e + H_\mathrm{f} + H_o + H_\mathrm{b} + H_\mathrm{j} + H_\mathrm{g} \qquad (13.21)$$

where:

H_L = total energy required (ft).
H_e = energy loss at entrance (ft).
H_f = friction loss (ft).
H_o = energy loss at exit (ft).
H_b = bend loss (ft).

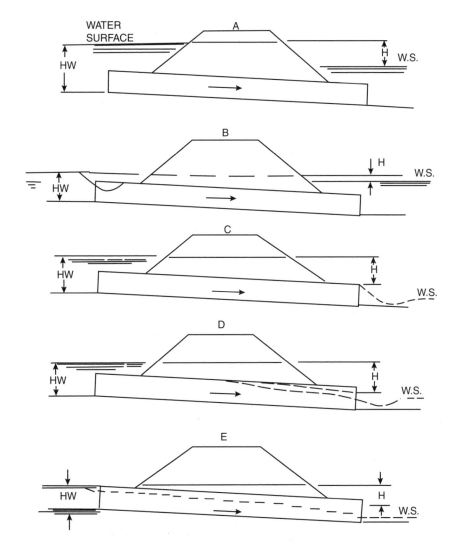

FIGURE 13.21 Types of outlet control (*Source*: From Normann, J.M., Houghtalon, R.J., and Johnston, W.J., in *Hydraulic Design of Highway Culverts*, Report No. FHWA-IP-85-15, US Department of Transportation, Office of Implementation, McLean, VA, September 1985).

H_j = energy loss at junction (ft).
H_g = energy loss at safety gates (ft).

If there are no bends, junctions, or grates included in the culvert design, the head loss equation can be simplified as shown in Equation 13.22.

$$H_L = \left[1 + k_e + \frac{29n^2 L}{R^{1.33}} \right] \frac{V^2}{2g} \qquad (13.22)$$

where:

H_L = total energy required (ft).
k_e = factor based on inlet configuration as shown in Table 13.12.

TABLE 13.12   Entrance Loss Coefficients

Type of Structure and Design of Entrance	Coefficient, k_e
Pipe, concrete	
Projecting from fill, socket end (groove-end)	0.2
Projecting from fill, sq. cut end	0.5
Headwall or headwall and wingwalls	
Socket end of pipe (groove-end)	0.2
Square-edge	0.5
Rounded (radius 1/12D)	0.2
Mitered to conform to fill slope	0.7
End-section conforming to fill slope	0.5
Beveled edges, 33.7° or 45° bevels	0.2
Side- or slope-tapered inlet	0.2
Pipe or pipe-arch, corrugated metal	
Projecting from fill (no headwall)	0.9
Headwall or headwall and wingwalls square-edge	0.5
Mitered to conform to fill slope, paved or unpaved slope	0.7
End-section conforming to fill slope	0.5
Beveled edges, 33.7° or 45° bevels	0.2
Side- or slope-tapered inlet	0.2
Box, reinforced concrete	
Headwall parallel to embankment (no wingwalls)	
Square-edged on 3 edges	0.5
Rounded on 3 edges to radius of 1/12 barrel dimension, or beveled edges on 3 sides	0.2
Wingwalls at 30° to 75° to barrel	
Square-edged at crown	0.4
Crown edge rounded to radius of 1/12 barrel dimension, or beveled top edge	0.2
Wingwall at 10° to 25° to barrel	
Square-edge at crown	0.5
Wingwalls parallel (extension of sides)	
Square-edged at crown	0.7
Side- or slope-tapered inlet	0.2

Source: Adapted from Norman, J.M., Houghtalen, R.J., and Johnston, W.J., *Hydraulic Design of Highway Culverts*, Report No. FHWA-IP-85-15, US Department of Transportation, Office of Implementation, McLean, VA, September 1985.

n = manning's coefficient for culverts as shown in Table 13.13.
R = hydraulic radius of the full culvert barrel = Area/Perimeter (ft).
L = length of the culvert barrel (ft).
V = velocity in the culvert barrel (ft/sec).
g = 32.2 ft/sec^2.

The following equations can be utilized to determine the losses due to a bend, a junction, or the addition of a grate. Equation 13.23 is used when a bend is added to the culvert, while Equation 13.24 and Equation 13.25 are for the addition of a junction. Equation 13.26 can be used when a bar grate is present, however an approximate estimate may be found using Equation 13.27. Since these equations are empirical, they must be used with great caution.

$$H_b = K_b \left(\frac{V^2}{2g} \right)$$

(13.23)

TABLE 13.13 Manning's Roughness Coefficients

Type of Conduit	Wall and Joint Description	Manning n
Concrete pipe	Good joints, smooth walls	0.011–0.013
	Good joints, rough walls	0.014–0.016
	Poor joints, rough walls	0.016–0.017
Concrete box	Good joints, smooth finished walls	0.012–0.015
	Poor joints, rough, unfinished walls	0.014–0.018
Corrugated metal pipes and boxes, annular	2–2/3 by 1/2 in. corrugations	0.027–0.022
corrugations (Manning n varies with barrel size)	6 by 1 in. corrugations	0.025–0.022
	5 by 1 in. corrugations	0.026–0.025
	3 by 1 in. corrugations	0.028–0.027
	6 by 2 in. structural plate corrugations	0.035–0.033
	9 by 2–1/2 in. structural plate corrugations	0.037–0.033
Corrugated metal pipes, helical corrugations,		0.012–0.024
full circular flow	2–2/3 by 1/2 in. corrugations 24 in. plate width	
Spiral rib metal pipe	3/4 by 3/4 in. recess at 12 in. spacing, good joints	0.012–0.013

Source: Adapted from Norman, J.M., Houghtalen, R.J., and Johnston, W.J., in *Hydraulic Design of Highway Culverts*, Report No. FHWA-IP-85-15, US Department of Transportation, Office of Implementation, McLean, VA, September 1985.

where:

H_b = head loss due to bar grate (ft).
K_b = bend loss coefficient as shown in Table 13.14.
V = velocity in the culvert barrel (ft/sec).
g = 32.2 ft/sec^2.

$$H_j = y' + H_{V_1} - H_{V_2} \tag{13.24}$$

where:

H_j = head loss through the junction in the main conduit (ft).
y' = change in HGL through the junction (ft) Equation 13.25.
H_{V_1} = velocity head in the upstream conduit (ft).
H_{V_2} = velocity head in the downstream conduit (ft).

$$y' = \frac{Q_2 V_2 - Q_1 V_1 - Q_3 V_3 \cos \theta_j}{0.5 [A_1 + A_2] g} \tag{13.25}$$

TABLE 13.14 Loss Coefficients for Bends

Radius of Bend Equivalent Diameter	Angle of Bend, Degrees		
	90°	45°	22.5°
1	0.50	0.37	0.25
2	0.30	0.22	0.15
4	0.25	0.19	0.12
6	0.15	0.11	0.08
8	0.15	0.11	0.08

Source: Adapted from Norman, J.M., Houghtalen, R.J., and Johnston, W.J., in *Hydraulic Design of Highway Culverts*, Report No. FHWA-IP-85-15, US Department of Transportation, Office of Implementation, McLean, VA, September 1985.

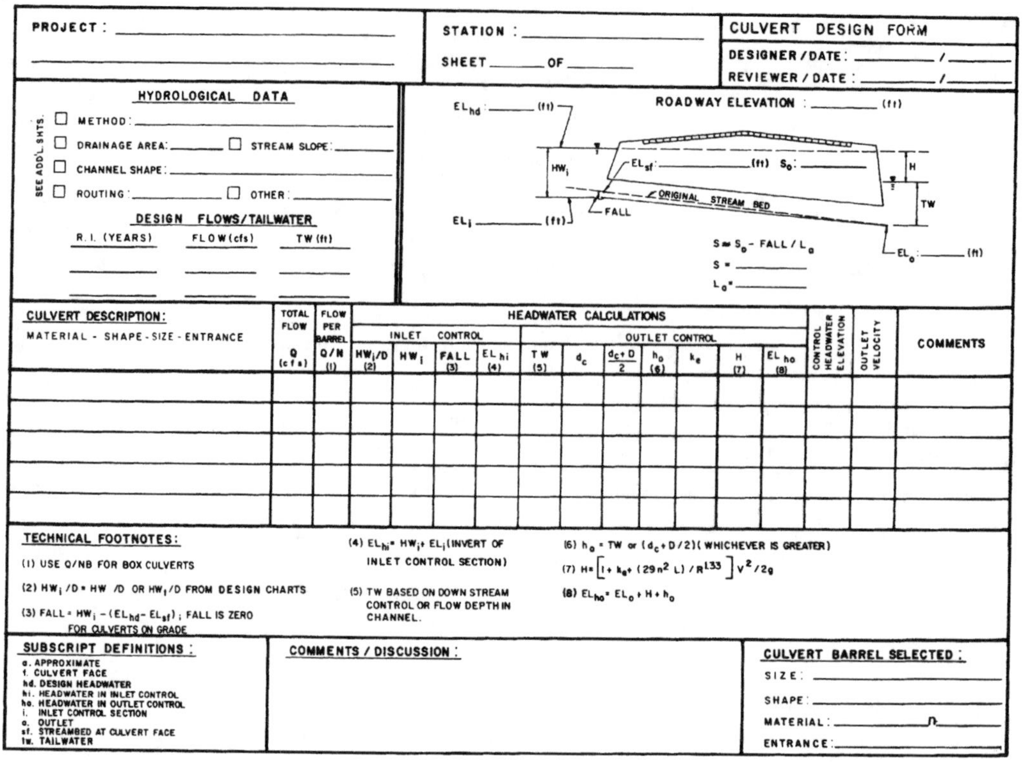
PROJECT :
STATION :
SHEET ___ OF ___
CULVERT DESIGN FORM
DESIGNER / DATE : ___ / ___
REVIEWER / DATE : ___ / ___
HYDROLOGICAL DATA
SEE ADD'L SHTS.
☐ METHOD :
☐ DRAINAGE AREA : ☐ STREAM SLOPE :
☐ CHANNEL SHAPE :
☐ ROUTING : ☐ OTHER :
DESIGN FLOWS/TAILWATER
R. I. (YEARS) FLOW (cfs) TW (ft)
EL_{hd} : ___ (ft)
ROADWAY ELEVATION : ___ (ft)
HW_i
EL_{sf} : ___ (ft) S_o : ___
H
TW
ORIGINAL STREAM BED
FALL
EL_i ___ (ft)
S = S_o - FALL / L_o
S =
L_o =
EL_o : ___ (ft)
CULVERT DESCRIPTION:
MATERIAL - SHAPE - SIZE - ENTRANCE
TOTAL FLOW Q (cfs)
FLOW PER BARREL Q/N (1)
HEADWATER CALCULATIONS
INLET CONTROL
HW_i/D (2) HW_i FALL (3) EL hi (4)
OUTLET CONTROL
TW (5) d_c (d_c+D)/2 h_o (6) k_e H (7) EL ho (8)
CONTROL HEADWATER ELEVATION
OUTLET VELOCITY
COMMENTS
TECHNICAL FOOTNOTES:
(1) USE Q/NB FOR BOX CULVERTS
(2) HW_i / D = HW / D OR HW_i / D FROM DESIGN CHARTS
(3) FALL = HW_i - (EL_{hd} - EL_{sf}) ; FALL IS ZERO FOR CULVERTS ON GRADE
(4) EL_{hi} = HW_i + EL_i (INVERT OF INLET CONTROL SECTION)
(5) TW BASED ON DOWN STREAM CONTROL OR FLOW DEPTH IN CHANNEL.
(6) h_o = TW or (d_c + D/2) (WHICHEVER IS GREATER)
(7) H = [1 + k_e + (29n² L) / R^{1.33}] V² / 2g
(8) EL_{ho} = EL_o + H + h_o
SUBSCRIPT DEFINITIONS :
a. APPROXIMATE
f. CULVERT FACE
hd. DESIGN HEADWATER
hi. HEADWATER IN INLET CONTROL
ho. HEADWATER IN OUTLET CONTROL
i. INLET CONTROL SECTION
o. OUTLET
sf. STREAMBED AT CULVERT FACE
tw. TAILWATER
COMMENTS / DISCUSSION :
CULVERT BARREL SELECTED :
SIZE :
SHAPE :
MATERIAL : ___ n ___
ENTRANCE :

FIGURE 13.22 Work sheet for culvert design (*Source*: From Normann, J.M., Houghtalen, R.J., and Johnston, W.J., in *Hydraulic Design of Highway Culverts*, Report No. FHWA-IP-85-15, US Department of Transportation, Office of Implementation, McLean, VA, September 1985).

where:

subscripts 1, 2, and 3 refer to outlet pipe, upstream pipe, and lateral pipe (Figure 13.22)

y' = change in HGL through the junction (ft).
Q = flow rate (ft^3/sec).
V = velocity (ft/sec).
A = area of the culvert barrel (ft^2).
θ_j = angle of lateral with respect to the outlet conduit (degrees).
H_{V_1} = velocity head in the upstream conduit (ft).
H_{V_2} = velocity head in the downstream conduit (ft).

$$H_g = k_g\left(\frac{W}{\chi}\right)\left(\frac{V_u^2}{2g}\right)\sin\theta_g \tag{13.26}$$

where:

H_g = head loss due to the bar grate (ft).
k_g = dimensionless bar shape factor, which is
- 2.42 for sharp-edged rectangular bars
- 1.83 for rectangular bars with semi-circular face upstream
- 1.79 for circular bars
- 1.67 for rectangular bars with semi-circular upstream and downstream faces.

V_u = approach velocity (ft/sec).
W = maximum cross-sectional width of the bars facing the flow (ft).
χ = maximum clear spacing between bars (ft).
θ_g = angle of the grate with respect to the horizontal (degrees).

$$H_g = 1.5\left(\frac{V_g^2 - V_u^2}{2g}\right)$$ (13.27)

where:

H_g = head loss due to the bar grate (ft).
V_g = velocity between the bars (ft/sec).
V_u = approach velocity (ft/sec).

When the culvert is flowing full, the tail-water, entrance conditions, and exit losses can be viewed as displayed in Figure 13.23. By setting the energy of section one equal to the energy at section two, the following Equation is created:

$$HW_o + \frac{V_u^2}{2g} = TW + \frac{V_d^2}{2g} + H_L$$ (13.28)

where:

HW_o = head-water depth above the outlet invert (ft).
V_u = approach velocity (ft/sec).
TW = tail-water depth above the outlet invert (ft).
V_d = downstream velocity (ft/sec).
H_L = sum of all head losses including entrance, friction, exit, and other losses (ft).
g = 32.2 (ft/sec^2).

In the majority of cases, the approach velocity will be relatively low and the approach velocity will be neglected, as will the downstream velocity. This means the equation can be simplified as follows:

$$HW_o = TW + H_L$$ (13.29)

where all values are as previously defined.

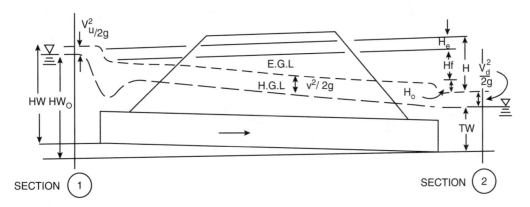

FIGURE 13.23 Full flow energy grade line (EGL) and hydraulic grade line (HGL) (*Source*: From Normann, J.M., Houghtalen, R.J., and Johnston, W.J., in *Hydraulic Design of Highway Culverts*, Report No. FHWA-IP-85-15, US Department of Transportation, Office of Implementation, McLean, VA, September 1985).

These Equations are applicable when the pipe is flowing full, or where the conditions pictured in Figure 13.21(A, B, C) exist. If conditions D or E exist (partially full flow conditions), then additional calculations must be done. The equations for these calculations can be found in the *Hydraulic Design of Highway Culverts* handbook. When the pipe is flowing full, it may be simpler to use nomographs for outlet control situations. These nomographs are available in the same handbook and take into account the entrance friction and exit losses, and will yield the necessary head-water depth needed for the design.

As mentioned earlier, once the head-water elevation needed for either the inlet or outlet control has been determined, the higher value will govern the design. If the inlet value is highest, the design will be inlet-controlled. Conversely if the outlet value is higher, the design will be outlet-controlled. The outlet velocity is then calculated by using this controlling head-water depth. If the inlet control value governs, the normal depth and velocity in the culvert barrel are used. This velocity at the normal depth can be assumed to be the outlet velocity. If the outlet control value governs the design, the outlet velocity is determined by using the barrel geometry and the following:

- Critical depth, if the depth of the tail-water is less than critical depth.
- Tail-water depth, if the depth of the tail-water lies between the critical depth and the top of the barrel.
- Barrel height, if the tail-water is higher than the top of the barrel.

This design process may take numerous reiterations in order to find an acceptable culvert configuration. To facilitate this process, the Federal Highway Administration has designed the worksheet shown in Figure 13.22.

Example 13.9

Design of pipe culverts (From FHWA, 1985. *Hydraulic Design of Highway Culverts*)

A new culvert at a roadway crossing is required to pass a 50-year flow rate of 300 ft^3/sec. Use the following site conditions:

EL$_{hd}$ = 110 ft based on adjacent structures
Shoulder elevation = 113.5 ft
Elevation of stream bed at culvert face = 100 ft
Natural stream slope = 2%
Tail-water depth = 4.0 ft
Approximate culvert length = 250 ft

Design a reinforced concrete box culvert for this installation. Try both square edges and 45° beveled edges in a headwall. Do not depress the inlet (no fall).

Answer:

Shown in Figure 13.24, the culvert design worksheet.

Example 13.10

Design of pipe culverts (From *Hydraulic Design of Highway Culverts*)

Design method for culverts without standard design charts

Use a long span culvert to pass the 25-year flood of 5500 ft^3/sec under a high roadway fill. The design flow should be below the crown of the conduit at the inlet, but the check flow (100-year flow) of 7500 ft^3/sec may exceed the crown by not more than 5 ft. Use the following site conditions:

EL$_{hd}$ = 240 ft
Elevation of stream bed at culvert face (EL$_{sf}$) = 220 ft
Shoulder elevation = 260 ft
Stream slope (S_o) = 1.0%

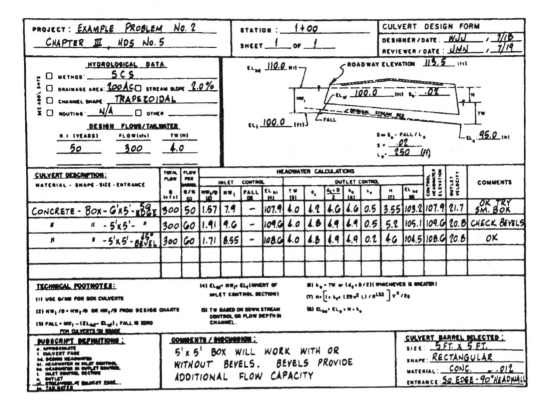

FIGURE 13.24 Solution to Example 13.9.

Approximate culvert length = 200 ft

Tail-water depth = 16 ft for $Q = 550$ ft^3/sec, 19 ft for $Q = 7500$ ft^3/sec

Design an elliptical structural plate corrugated metal conduit for this site. Use a head-wall to provide a square edge condition. Corrugations are 6 in. by 2 in. Try a 30 ft span by a 20 ft rise elliptical structural plate conduit for this site. From the manufacturer's specifications, $A = 487.5$ ft^2 and $D = 20$ ft. Neglect approach velocity.

Answer:

Inlet Control:

$AD^{0.5} = (487.5)(20)^{0.5} = 2180$

$Q/AD^{0.5} = 2.52$

Based on HW/$D = 0.90$ so

HW = HW$_f$ = (0.90)(20) = 18 ft

EL$_{hi}$ = 220 + 18 = 238.0 ft

For the check flow:

$Q/AD^{0.5} = 3.44$

Based on HW/$D = 1.13$ so

HW = HW$_f$ = (1.13)(20) = 22.6 ft

EL$_{hi}$ = 220 + 22.6 = 242.6 ft

Outlet Control:

Backwater calculations are necessary to check outlet control.

Backwater calculations (from hydraulic tables for elliptical conduits):

TABLE 13.15   Example 13.10

d/D	D	A/BD	A	R/D	R
0.65	13.00	0.5537	332.2	0.3642	7.28
0.70	14.00	0.6013	360.8	0.3781	7.56
0.75	15.00	0.6472	388.3	0.3886	7.77
0.80	16.00	0.6908	414.5	0.3950	7.90
0.85	17.00	0.7313	438.8	0.3959	7.92
0.90	18.00	0.7671	460.3	0.3870	7.74
0.95	19.00	0.7953	477.2	0.3649	7.30
1.00	20.00	0.8108	486.5	0.3060	6.12

$Q = 5500$ ft^3/sec, $d_c = 12.4$ ft
$Q = 7500$ ft^3/sec, $d_c = 14.6$ ft

Since TW $> d_c$, start backwater calculations at TW depth.
Determine normal depths (d_n) using hydraulic tables.

for $Q = 5500$ ft^3/sec, $n = 0.034$;
$d_n = 13.1$ ft
for $Q = 7500$ ft^3/sec, $n = 0.034$;
$d_n = 16.7$ ft

Since $d_n > d_c$ flow is subcritical
Since TW $> d_n$ water surface has an M-1 profile
Plot the area and hydraulic radius vs. depth from data obtained from tables, as shown in Table 13.15 and Figure 13.25.
Complete the water surface computation as shown in Figure 13.26.

HW = specific head(H) + $k_e(V^2/2g)$
neglecting approach velocity head:
for $Q = 5500$ ft^3/sec
HW = $18.004 + (0.5)(3.208) = 19.6$ ft
EL$_{ho}$ = $220 + 19.6 = 239.6$ ft
for $Q = 7500$ ft^3/sec
HW = $22.627 + (0.5)(3.89) = 24.6$ ft
EL$_{ho}$ = $220 + 24.6 = 244.6$ ft

therefore:

Design Q:
EL$_{hd}$ = 240.0   EL$_{hi}$ = 238.0   EL$_{ho}$ = 239.6
Check Q:
EL$_{hd}$ = 245.0   EL$_{hi}$ = 242.6   EL$_{ho}$ = 244.6

This culvert design meets the requirements stated in the problem.

13.4.14   Culvert Inlet Design

Inlet configuration is the shape of the entrance to the culvert barrel. The four standard types of inlets are shown in Figure 13.27. An inlet is generally the same shape as the barrel of the culvert. The following shapes are most commonly used: rectangular, circular, and elliptical. Cases where similar shapes are not used for both the barrel and the inlet may result in the addition of control sections within the barrel. Inlet edge configuration becomes increasingly important in inlet controlled situations and can be used

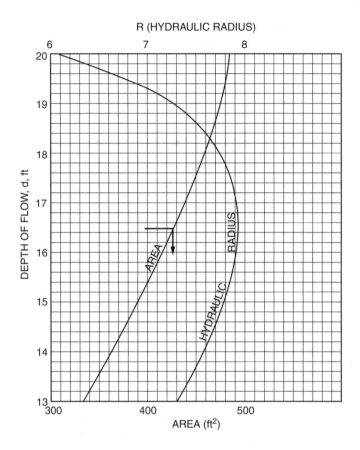

FIGURE 13.25 Plot of area and hydraulic radius vs. depth from Example 13.10 (*Source*: From Normann, J.M., Houghtalen, R.J., and Johnston, W.J., in *Hydraulic Design of Highway Culverts*, Report No. FHWA-IP-85-15, US Department of Transportation, Office of Implementation, McLean, VA, September 1985).

to improve the performance of the culvert. It is desirable to reach full or nearly full flow with the addition of the inlet configuration, thus increasing the capacity of the culvert. For culverts with outlet control, the addition of an inlet configuration is an unnecessary expense. This is due to the fact that outlet-controlled culverts generally have a designed discharge that produces the full flow condition. The type of inlet selected for a particular culvert should take into account each of the following factors:

- structural stability
- aesthetics
- erosion control and fill retention
- hydraulic performance
- economics and safety

Each of the different types of inlets has properties that increase the performance of the culvert.

13.4.15 Beveled Edge Inlets

The first type of inlet is a beveled-edge inlet and can be viewed in Figure 13.28. Beveled edges perform better than square edges causing minimal contraction of flow which leads to an increase in the effective face of the culvert. It is recommended that all culverts with both inlet and outlet control conditions have bevels. As shown in Figure 13.28, there are different angles to which the bevel can be carried. The larger

WATER SURFACE PROFILE COMPUTATIONS

Identification: EXAMPLE PROBLEM No.5, HDS No.5   By: JMN   Date: 3/14

Channel Shape: ELLIPTICAL C.M.P., B=30 ft, D=20 ft, d_n=13.1 ft, TW=16 ft, S_o=0.01
M. PROFILE, BARREL LENGTH=20 ft, START AT TW=16 ft

Manning n=0.034
Q= 5500 C.F.S.

d	A	R	V=Q/A	$\frac{V^2}{2g}$	H=d+$\frac{V^2}{2g}$	(1) ΔH	$R^{2/3}$	$AR^{2/3}$	(2) $S=\left(\frac{K_n}{AR^{2/3}}\right)^2$	$\bar{S}_i=(S_{11}\cdot S_{12})^{1/2}$	$\bar{S}_o-\bar{S}_i$	(3) $\frac{\Delta H}{\bar{S}_o-\bar{S}_i}$	L=ΣΔL
16	414.5	7.90	13.27	2.73	18.739	0.322	3.969	1645.3	0.00682	0.00601	0.00399	80.7	0.0
15.5	401.6	7.85	13.70	2.91	18.412	0.297	3.953	1587.39	.00620	.00618	.00352	84.4	80.7
15.0	388.3	7.77	14.16	3.115	18.115	0.111	3.926	1524.33	.00678	.00714	.00286	38.8	165.1
14.8	382.9	7.34	14.36	3.208	18.004		3.779	1447.12	.00752				203.9 OK

(1) SUBTRACT SECOND H FROM FIRST H VALUE
(2) $K_n=\frac{Qn}{1.49}=\frac{(5500)(.034)}{1.49}=125.50$
(3) IF ΔL IS +, PROFILE IS PROGRESSING UPSTREAM.
    -ΔL DENOTES DOWNSTREAM PROGRESSION.

WATER SURFACE PROFILE COMPUTATIONS

Identification: EXAMPLE PROBLEM No.5, HDS No.5   By: WJJ   Date: 3/14

Channel Shape: ELLIPTICAL CMP., B=30 ft, D=20 ft, d_n=16.7 ft, TW=19.0 ft, S_o=0.01
M. PROFILE, BARREL LENGTH=200 ft, START AT TW=19.0 ft

Manning n=0.034
Q= 7500 ft³/s

d	A	R	V=Q/A	$\frac{V^2}{2g}$	H=d+$\frac{V^2}{2g}$	(1) ΔH	$R^{2/3}$	$AR^{2/3}$	(2) $S=\left(\frac{K_n}{AR^{2/3}}\right)^2$	$\bar{S}_i=(S_{11}\cdot S_{12})^{1/2}$	$\bar{S}_o-\bar{S}_i$	(3) $\frac{\Delta H}{\bar{S}_o-\bar{S}_i}$	L=ΣΔL
19.0	477.20	7.30	15.72	3.836	22.835	0.151	3.765	1796.6	0.00907	0.00904	0.00096	157.6	0.0
18.8	474.39	7.408	15.86	3.833	22.684	.074	3.803	1803.4	.00901	.00900	.00100	74.1	157.6
18.7	472.65	7.457	15.88	3.9098	22.610	.057	3.820	1805.3	.00899	.00899	.00101	56.96	231.7 Too Much!
18.75	473.80	7.440	15.83	3.8911	22.627		3.814	1806.9	.00897				214.1 OK

(1) SUBTRACT SECOND H FROM FIRST H VALUE
(2) $K_n=\frac{Qn}{1.49}=\frac{(7500)(0.034)}{1.49}=171.14$
(3) IF ΔL IS +, PROFILE IS PROGRESSING UPSTREAM.
    -ΔL DENOTES DOWNSTREAM PROGRESSION.

FIGURE 13.26   Water surface computations for Example 13.10 (*Source*: From Normann, J.M., Houghtalen, R.J., and Johnston, W.J., in *Hydraulic Design of Highway Culverts*, Report No. FHWA-IP-85-15, US Department of Transportation, Office of Implementation, McLean, VA, September 1985).

angle of 33.7° will require modification to the head-wall; however it will produce better performance than the smaller 45° angle.

13.4.16   Tapered Inlets

When a culvert is operating under inlet control conditions, it is common to increase its performance by adding a tapered inlet. By adding a control section, tapered inlets increase the capacity of the culvert by reducing contractions in the throat section. If the flow is contracted, there is a reduction of the effective cross-sectional area of the barrel. This reduction may cause up to a 50% loss of available barrel area. It is therefore quite important to add a taper in inlet-controlled situations. There is not a significant increase in performance when adding tapers vs. bevels to outlet controlled situation, and it is therefore not recommended. Usually performance is improved with tapers and sidewalls before using the additional enhancement of a depressed inlet control. Using tapers and depressions will increase the performance of the culvert beyond that of a simple bevel. There are currently design charts and methods available for rectangular box culverts and circular pipe culverts utilizing side-taper and slope-taper inlets. The same general design may be used on other barrel shapes; however, there are no design charts currently available to simplify calculations. Side-taper inlets are used when depressions on the upstream

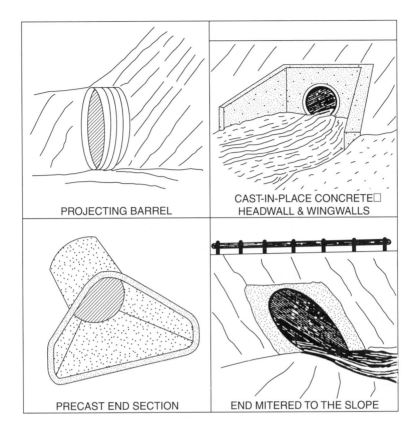

PROJECTING BARREL

CAST-IN-PLACE CONCRETE☐
HEADWALL & WINGWALLS

PRECAST END SECTION

END MITERED TO THE SLOPE

FIGURE 13.27 Four standard inlet types (*Source*: From Normann, J.M., Houghtalen, R.J., and Johnston, W.J., in *Hydraulic Design of Highway Culverts*, Report No. FHWA-IP-85-15, US Department of Transportation, Office of Implementation, McLean, VA, September 1985).

face are present as well as when they are absent. Slope-taper inlets may have a vertical face or a mitered face corresponding to the fill slope.

13.4.17 Side Tapered Inlets

Figure 13.29 (a) shows a side-tapered inlet. The side-tapered inlet has a distinctly enlarged face section and a tapered side wall transition to the culvert barrel. The barrel is extended through the inlet floor and the face section is generally as high as the barrel. As long as the barrel height does not exceed the face height by more than 10%, the top of the inlet may slope upward. The throat is the section of the tapered sidewall that intersects the barrel. The control is provided by either the face or the throat sections. As shown in the figure, HW_t is the point of measurement for the head-water depth when the throat is the control point, and HW_f, when the face is the control point. Since flow contraction can generally be eliminated at the throat, it is desirable to use this point as the control. Due to the lower throat in a side-tapered inlet, contractions are decreased and head is increased. These benefits can be further increased with the addition of an upstream depression. Two examples of the construction of such depressions are in Figures 13.29 (b and c).

The first example places the depression between the wing-walls. This type of depression requires that the base of the barrel be extended beyond the face by at least $D/2$ before increasing to a steep upward slope. Figure 13.29(c) is the second example and it shows the dimensions needed for the construction of a sump upstream from the face. When either of these designs is used, the length of the crest must be

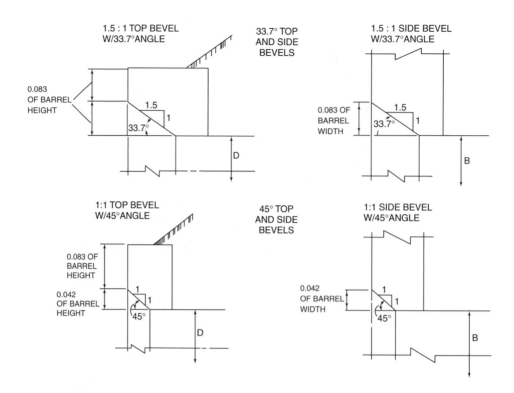

FIGURE 13.28 Beveled edges (*Source*: From Normann, J.M., Houghtalen, R.J., and Johnston, W.J., in *Hydraulic Design of Highway Culverts*, Report No. FHWA-IP-85-15, US Department of Transportation, Office of Implementation, McLean, VA, September 1985).

checked since the crest will become a weir if the length is too short. The crest must not control the flow for either the design flow or head-water. Inlets that are depressed or side tapered require more head at both the face and the throat sections. The increase in head will lower the required throat section and the size of the face.

Figure 13.29(d) shows the slope-tapered inlet design. Slope-tapered inlets generally have the same construction as the side-tapered inlets including an enlarged face section and sloping sidewalls which meet the barrel at the throat section. A vertical drop or fall is included to increase the head between the throat and the face. The junction of the steep slope of the inlet and the flat slope of the barrel forms a bend. This type of configuration will allow for three control sections at the face, bend, and throat. The design of the bend is only limited by the minimum distance it must be placed from the throat; procedures for determining the throat and face sections can be found in the *Hydraulic Design of Highway Culverts* manual. The highest level of efficiency for this particular design is achieved when the throat is the control.

13.5   Subdrainage Design

Pavement performance and integrity often depends on the removal of water below the pavement surface. Water often seeps down through cracks in the pavement into the base or subgrade. Some water gets under the pavement by way of groundwater through capillary action. Water and melting snow may seep down through the shoulders into the base or subgrade. All water introduced to the pavement structure should be accounted for.

Surface and subsurface drainage systems must be designed so they do not work against each other. It is important to take into account the surface drainage systems when designing the subdrainage. When designing for a particular area, it is important to take into account past experiences and failures.

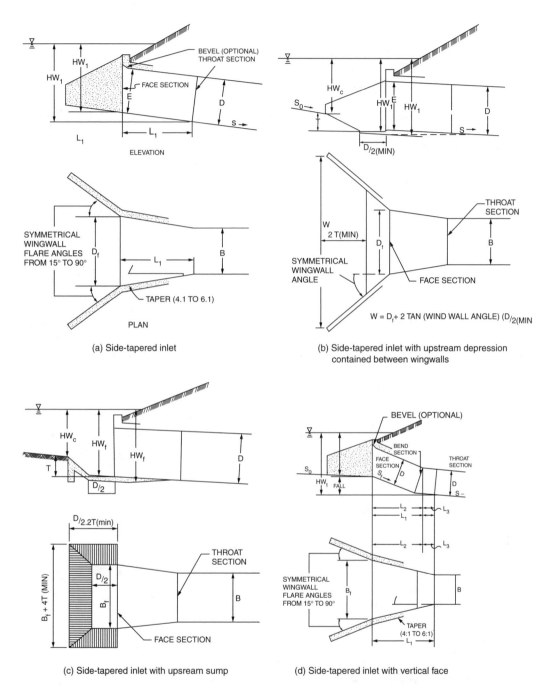

(a) Side-tapered inlet

(b) Side-tapered inlet with upstream depression contained between wingwalls

(c) Side-tapered inlet with upsream sump

(d) Side-tapered inlet with vertical face

FIGURE 13.29 Side-tapered inlets (*Source*: From Normann, J.M., Houghtalen, R.J., and Johnston, W.J., in *Hydraulic Design of Highway Culverts*, Report No. FHWA IP-85-15, US Department of Transportation, Office of Implementation, McLean, VA, September 1985).

Subsurface water sources may not be readily apparent during preliminary investigations; unidentified water sources may not be accounted for in the project design. When soft spots become apparent during construction, these water sources must be dealt with appropriately using thicker pavement sections, subdrainage, or both.

13.5.1   Surface Water Problems

The most common type of problems associated with the presence of subsurface water are slope instability and accelerated pavement deterioration. Slope instability is a direct result of increasing stresses along a plane, leading to shear failure of the soil. The introduction of water increases stresses and decreases strength speeding up the failure process. This failure may only be minor sloughing of the soil along the face of the slope; however, it may also contribute to a catastrophic slope failure removing a large portion of the same.

Pavement deterioration may be accelerated if the surface, base, or subgrade is infiltrated with excessive amount of water. When the pavement structure or subgrade becomes saturated with water, the ability of the pavement to carry traffic loads efficiently is decreased. Cracked asphalt pavements may have enough pressure in the free water to move suspended fines to the pavement surface. This action, called pumping, happens in concrete pavement as well but it occurs in a different manner as slabs move under the weight of traffic.

Concrete pavement slabs of Portland cement often curl upward resulting from temperature gradient within the slab causing small spaces between the slabs to increase. Increased spacing between the slabs allows the flow of moisture to penetrate reaching fines located directly beneath the slab. Once these fines are suspended in the water, a load moving across the slabs will cause the solution containing the fines to be ejected upwards to the pavement surface. Frequent repetitions of this process will increase the void spaces under the edges of the slabs by eroding away a portion of the material. Increasing void spaces will result in cracking of the pavement and permanent deformation of the slab. Joint seals and subsurface drainage designs will facilitate the removal of water from such joints, reducing the likelihood of pumping at joints in the concrete pavement.

Frost action will also damage the integrity of the pavement structure. Frost action occurs when moisture in the frost susceptible soil layers of the pavement structure are subjected to long periods of subfreezing temperatures. This leads to capillary migration of water and eventually to the formation of ice lenses beneath the pavement surface. Heaving of the pavement structure will result as the additional water from capillary action increases the size of the ice lenses. The pavement surface is left unsupported as the ice lenses melt, usually in the spring every year. Melting ice lenses cause the soil around them to become saturated and in some cases even super-saturated, greatly reducing the strength of the soil, which may lead to pavement failure.

13.5.2   Subdrainage Systems

There are four main types of drainage systems used in highway construction. They are

1. Longitudinal drains or edge drains
2. Transverse and horizontal drains
3. Drainage blankets
4. Well systems

Each of these drainage systems may intercept water above an impervious layer, lower the water table, or for transfer water from another drainage system.

13.5.3   Longitudinal Drains

Longitudinal drains, sometimes referred to as edge drains, consist of collector pipes placed in trenches with filters for protection. They run parallel to the centerline of the roadway. Some examples of longitudinal drains are shown in Figures 13.30–13.32.

The drain shown in Figure 13.30 is installed to draw down the water table so that it remains beneath the pavement structure. Since it is in a cut, the water table could easily be high enough to enter the pavement

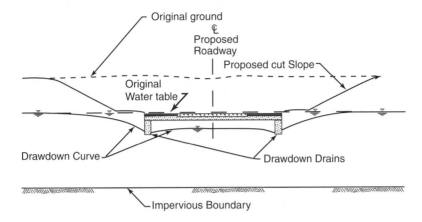

FIGURE 13.30 Symmetrical longitudinal drains used to lower the water table (*Source*: From *Highway Sub-Drainage Design*, Report No. FHWA-TS-80-224, US Department of Transportation, Washington, DC, August 1980).

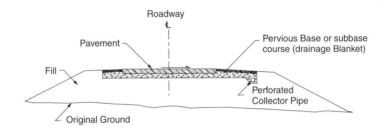

FIGURE 13.31 Longitudinal collector drain used to remove water seeping into pavement structural section (*Source*: From *Highway Sub-Drainage Design*, Report No. FHWA-TS-80-224, US Department of Transportation, Washington, DC, August 1980).

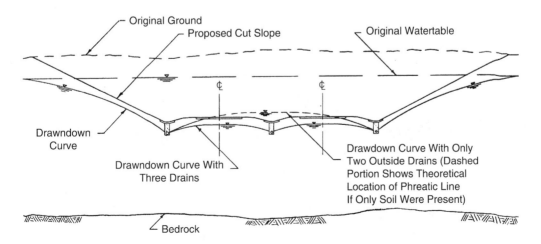

FIGURE 13.32 Multiple longitudinal drawdown drain installation (*Source*: From *Highway Sub-Drainage Design*, Report No. FHWA-TS-80-224, US Department of Transportation, Washington, DC, August 1980).

structure without edge drains. Figure 13.31 represents a situation where the water infiltrating the road-bed may be collected and transported to a more convenient location. In both of these cases, the roadway was narrow enough to allow for the placement of only one or two longitudinal drains. Figure 13.32 represents a situation requiring more than two longitudinal drains. In such cases, it is often desirable to locate the third drain beneath the median or in the center of a wide road-bed. This is often necessary in situation where the water table is high or at an interchange.

13.5.4 Transverse and Longitudinal Drains

The transverse drain is similar in construction to a lateral drain, but these drains generally run perpendicular to the centerline of the roadway or slightly skewed. The most common use of a transverse drain is to remove the water that may seep into the roadbed at joints as shown in Figure 13.33. Draining water at joints is a necessary activity; however, these types of drains should be used with great caution in areas prone to frost heave. Frost action may damage the roadway except above the drains, causing a wave to appear on the pavement surface. Horizontal drains are used in cut or fill slopes, and often empty directly into the side ditches. The pipes may enter directly into these side-ditches, or it may be necessary to use a treatment to prevent erosion, such as a paving the drainage ditches or placing riprap or splash-blocks at the drain outlets.

13.5.5 Drainage Blankets

A drainage blanket is usually comprised of materials that have high permeability coefficients, namely they drain quickly and easily. Generally base and subbase layers are not designed to act as drainage

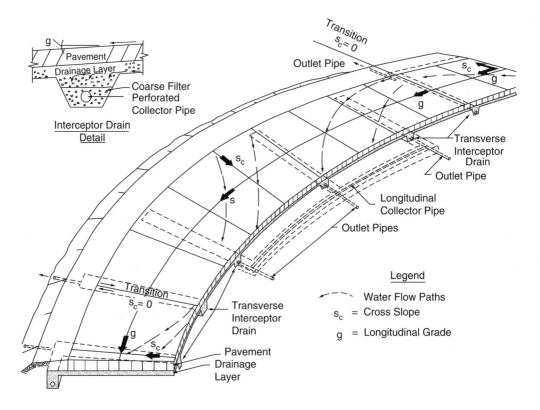

FIGURE 13.33 Transverse drains superelevated curves (*Source*: From *Highway Sub-Drainage Design*, Report No. FHWA-TS-80-224, US Department of Transportation, Washington, DC, August 1980).

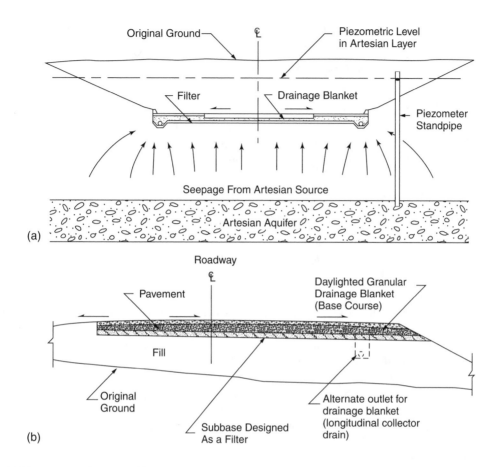

FIGURE 13.34 Applications of horizontal drainage blankets (*Source*: From *Highway Sub-Drainage Design*, Report No. FHWA-TS-80-224, US Department of Transportation, Washington, DC, August 1980).

blankets, therefore a drainage blanket should also be included in the design. The flow should be directed in the drainage blanket such that its width and length are large compared to its thickness. Drainage blankets designed properly beneath the roadway will eliminate most problems caused by infiltration and groundwater from gravity and artesian sources. Figure 13.34(a) and (b) shows some examples of drainage blankets. These examples use both drainage blankets and longitudinal drains for the removal of subsurface water. Care must be taken when designing a drainage layer so that the subgrade materials will not run into the drainage layer and clog the system. This is accomplished by the addition of a filter layer. In Figure 13.34(b) this is shown with the addition of a longitudinal drain, rather than simply allowing the water to daylight through the drainage blanket. One common use of drainage blankets in conjunction with longitudinal drains is to control groundwater in cut and fill slopes along the roadway itself.

13.5.6 Well Systems

Well systems are a collection of vertical wells drilled for the purpose of relieving pore pressures from slope areas. Well systems may be used as short term drainage systems or permanent ones. They are often used in conjunction with pump systems for short term applications, and with another drainage system for permanent applications. This can be accomplished with the addition of a drainage layer at the well outlet for the redirection of water.

13.5.7 Subdrainage Design Procedure

Design procedures for subdrainage systems as defined by the Federal Highway Administration in the *Highway Sub-Drainage Design Guide* are outlined as follows:

1. Collect data available for the drainage area.
2. Determine the net in-flow, or the amount of water to be removed.
3. Determine which system is necessary to remove that amount of water.
4. Analyze the capacity of the drainage system selected and change the design if necessary.
5. Evaluate the results, taking into account economic and performance factors.

13.5.7.1 Collecting Data

The data needed for designing a drainage system includes the following:

- Highway and subdrainage geometries
- Soil properties, including classification and permeability
- Climatological data
- Other information

The geometry of both the highway and the subsurface are needed to find potential problems that must be dealt with in the design process. Designers need to accumulate as much information as possible before beginning the design process. This information should include grades, elevations, pavement widths, thicknesses, and suggested slope configurations.

The primary concern, when considering soil characteristics, is the permeability of the soil surrounding the proposed or existing roadway. The ability to move water through the soil and the speed at which it flows are important design parameters. The coefficient of permeability should be determined by laboratory testing or *in situ* measurements, however some typical values can be found in Table 13.16. Permeability values, based on the Unified Soil Classification system, can be found in the FHWA publication *Highway Sub-Drainage Design* (FHWA, 1980).

Precipitation and climate data are used to determine the amount and source of water that may be present in the subsurface, and the frost action that may occur. Precipitation has been discussed at length earlier in this chapter. Maps, available from the National Weather Service, provide rainfall intensity, frequency, and duration information that is particularly useful. Frost penetration may be determined based on the geographic location of the roadway. Maps are also available from the Federal Highway Administration for determining the maximum depth of frost penetration. An example of this is shown in Figure 13.35.

Other pertinent information may include impacts resulting from the design of a subdrainage system in the area surrounding it, impacts on the current construction and on future construction. As with the construction of any system, all economic factors should be taken into account.

TABLE 13.16 Common Factors of Permeability

Soil Description	Coefficient of Permeability k (ft/day)[a]	Degree of Permeability
Medium and coarse gravel	>30.0	High
Fine gravel; coarse, medium and fine sand; dune sand	$30.0-3.0$	Medium
Very fine sand; silty sand; loose silt; loess; rock flour	$3.0-0.03$	Low
Dense silt; dense loess; clayey silt; silty clay	$0.03-0.0003$	Very low
Homogeneous clays	<0.0003	Impervious

Source: From *Highway Sub-Drainage Design*, Report No. FHWA-TS-80-224, US Department of Transportation, Washington, DC, August 1980.

[a] Note that 1 ft/day is equivalent to 3.529×10^{-4} cm/sec.

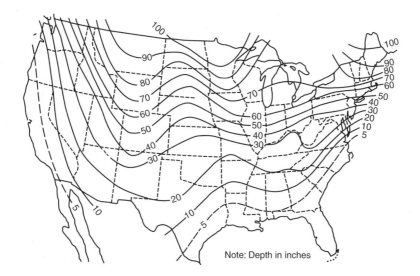

FIGURE 13.35 Maximum depth of frost penetration in the United States (*Source*: From *Highway Sub-Drainage Design*, Report No. FHWA-TS-80-224, US Department of Transportation, Washington, DC, August 1980).

13.5.7.2 Determine Net In-flow

The second step in this design process is to determine the net in-flow, or the amount of water that needs to be removed from the area in question. The net in-flow is the sum of all in-flow parameters including:

- Surface Infiltration
- Ground water
- Water resulting from the melting action of ice lenses (Melt water)
- Vertical flow of water from the pavement

13.5.7.3 Infiltration

Calculations for the amount of water infiltrating the pavement surface are complicated and depend on a number of parameters. The type and size of cracks, precipitation rate, and humidity in the air can all affect the infiltration rate, as well as the permeability of subsurface layers. It is therefore important to use a conservative method to determine the surface infiltration rate. Equation 13.30 is one conservative method provided by the Federal Highway Administration.

$$q_i = I_c \left[\frac{N_c}{W} + \frac{W_c}{WC_s} \right] + k_p \tag{13.30}$$

where:

q_i = design infiltration rate (ft^3/day/ft^2 of drainage layer).
I_c = crack infiltration rate (ft^3/day/ft of cracks) = 2.4 ft^3/day/ft for most applications.
N_c = number of contributing longitudinal cracks or joints = $N + 1$ for normal cracking
       on new pavements where N is the number of lanes.
W_c = length of contributing longitudinal cracks or joints (ft).
W = width of granular base or subbase subjected to infiltration (ft).
C_s = spacing of the transverse cracks or joints (ft).
k_p = rate of infiltration (ft^3/day/ft^2) = Coefficient of permeability through the
       un-cracked pavement surface.

Local experience should prevail when determining all values, and they should be increased or decreased as necessary. This is true for I_c, N_c, and C_s, although it is generally recommended that a value of 40 ft be used for C_s when the pavement is new. When past experiences in the area have shown high infiltration rates, it is necessary to develop values of C_s based on local conditions.

Example 13.11

Infiltration Into a Flexible Pavement (From FHWA, 1980. *Highway Sub-Drainage Design*)

Consider a new bituminous concrete pavement for two lanes of a four-lane divided expressway. The traffic lanes are 12 ft wide, with a 4 ft inside shoulder and a 10 ft outside shoulder, as shown in Figure 13.36. For normal cracking; $N_c = 3$; $C_s = 40$; $W_c = 38$; and $W = 24$.

Therefore, with $I_c = 2.4$ and $k_p = 0$ then Equation 13.30 is

$$q_i = 2.4 \left[\frac{3}{24} + \frac{38}{24(40)} \right] = 0.395 \text{ or } 0.4 \, \frac{\text{ft}^3/\text{day}}{\text{ft}^2}$$

13.5.8   Ground Water

Ideally ground water should be lowered or intercepted using the drainage techniques discussed earlier in this chapter, though it is not always possible to do so. In some cases this seepage must be accounted for in design. This ground water may be introduced in one of the two ways, the first is gravity drainage and the second is artesian flow. Gravity drainage computations begin by determining the radius of influence. This radius is determined by using Equation 13.31.

$$L_i = 3.8(H - H_0) \tag{13.31}$$

where:

L_i = radius of influence (ft).
H = thickness of the subgrade below the water table (ft).
H_0 = thickness of the subgrade below the pipe (ft).
$H - H_0$ = amount of draw-down (ft).

Once the value of L_i has been determined, Figure 13.37 can be used to find the total upward flow quantity, q_2. The average in-flow is then calculated using Equation 13.32.

$$q_g = \frac{q_2}{0.5W} \tag{13.32}$$

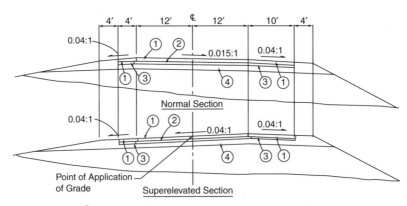

①- 6″ Hot mixed Asphaltic Concrete (3″ Surf., 3″ Binder)
②- 6″ Cement Stablized Aggregate Base
③- 6″ Aggregate Base
④- 12″ Dense Graded Aggregate Subbase

FIGURE 13.36   Flexible pavement section for Example 17.11 (*Source*: From *Highway Sub-Drainage Design*, Report No. FHWA-TS-80-224, US Department of Transportation, Washington, DC, August 1980).

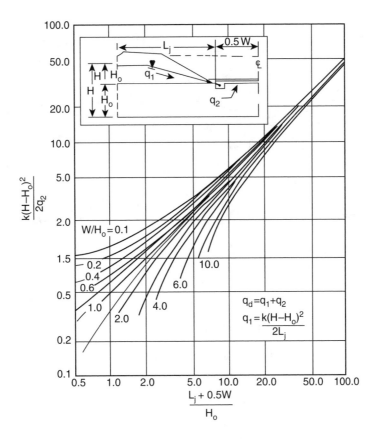

FIGURE 13.37 Chart for determining flow rate in horizontal drainage blanket (*Source*: From *Highway Sub-Drainage Design*, Report No. FHWA-TS-80-224, US Department of Transportation, Washington, DC, August 1980).

where:

q_g = in-flow rate for gravity drainage (ft^3/day/ft^2 of drainage layer).
q_2 = total upward flow into one-half of the drainage blanket (ft^3/day/linear ft of roadway).
W = width of drainage layer (ft).

Darcy's law or flow nets can be used to estimate the in-flow rate for artesian flow. This is done using Equation 13.33.

$$q_a = k \frac{\Delta H}{H_0} \tag{13.33}$$

where:

q_a = design in-flow rate from artesian flow (ft^3/day/ft^2 of drainage area).
ΔH = excess hydraulic head (ft).
H_0 = thickness of the subgrade soil between the drainage layer and the artesian aquifer (ft).
k = coefficient of permeability (ft/day).

It is important to notice that this quantity of water is small compared to the infiltration, yet it is not reasonable to neglect this additional amount of water; the accumulation of small amounts may affect the final design.

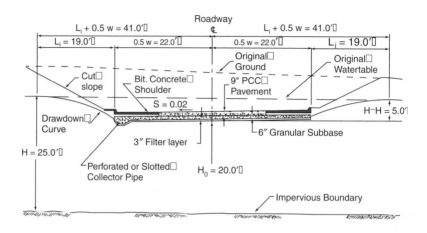

FIGURE 13.38 Rigid pavement section for Example 17.12 (*Source*: From *Highway Sub-Drainage Design*, Report No. FHWA-TS-80-224, US Department of Transportation, Washington, DC, August 1980).

Example 13.12

Gravity Flow of Groundwater into Pavement Drainage Layer (From FHWA)

Consider the flow situation shown in Figure 13.38. The native soil is a silty sand with a measured coefficient of permeability $k = 0.34$ ft/day. The average draw-down is

$$(H - H_0) = (25.0 - 20.0) = 5 \text{ ft}$$

Using Equation 13.31 the influence distance is

$$L_i = 3.8(5) = 19 \text{ ft}$$

$$\frac{(L_i + 0.5W)}{H_0} = \frac{(19 + 22)}{20}$$

$$\frac{W}{H_0} = \frac{44}{20} = 2.2$$

Figure 13.37 shows that

$$\frac{k(H - H_0)}{2q_2} = 0.74$$

Using Equation 13.32

$$q_g = \frac{q_2}{0.5W} = \frac{1.15}{22} = 0.052 \frac{\text{ft}^3/\text{day}}{\text{ft}^2}$$

Example 13.13

Artesian Flow of Groundwater into Pavement Drainage Layer (From FHWA)

Consider the flow situation shown in Figure 13.39. The subgrade soil above the artesian aquifer is a clayey silt with a coefficient of permeability, k, of 0.07 ft/day. A piezometer installed during the course of the subsurface exploration program at this site showed that the piezometric head of the water in the artesian layer was about 8 ft above the bottom of the proposed pavement drainage layer, as shown in Figure 13.39. Answer:

Using Equation 13.33 and

$$\Delta H = 8.0 \text{ ft}$$
$$H_0 = 15.0 \text{ ft}$$
$$k = 0.07 \text{ ft/day}$$

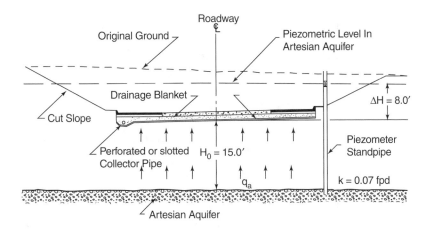

FIGURE 13.39 Artesian flow of groundwater into a pavement drainage layer for Example 17.13 (*Source*: From *Highway Sub-Drainage Design*, Report No. FHWA-TS-80-224, US Department of Transportation, Washington, DC, August 1980).

gives

$$q_a = \frac{0.07(8.0)}{15.0} = 0.037 \, \frac{ft^3/day}{ft^2}$$

13.5.9 Ice Lens Water Melt

The frost susceptibility of the soil is a major factor in determining the amount of frost action, and therefore the amount of melt water, from the formation of ice lenses. Groundwater may also increase the amount of melt water from these ice lenses when temperatures are low enough to permit freezing. When the air temperature rises, the ice lenses melt. This melt water must be removed efficiently. The rate at which this water flows is dependent on how fast the lenses melt, the permeability of the soil, and the pavement drainage system itself. There is no exact method for determining the rate at which the water will flow, however a good approximation can be found using Figure 13.40. Design flow rate from melt water, q_m, can be determined using this chart. This chart is a function of the frost susceptibility classification, which can usually be determined based on past experience in the area or sound engineering judgment. The chart utilizes the stress on the subgrade soil by the pavement, or σ_p. This stress is calculated using a one-square foot section of the pavement lying above the subgrade. The design flow rate, provided in the chart, accounts for the first day of full thaw from the ice lens, and is therefore conservative. This is because the rate of flow is a function of time and will continue to decrease from the value given in the chart. This conservative value may result in soil saturation and in cases where saturation is not acceptable, additional drainage devices will be necessary.

Example 13.14

Flow from a Thawing Subgrade Soil (*Highway Sub-Drainage Design*, Report No. FHWA-TS-80-224, U.S. Department of Transportation, Federal Highway Administration, 1980.)

Consider the case of a 9 in. thick concrete pavement with a 6 in. thick granular subbase, designed as a drainage layer, overlaying a silty subgrade soil. The soil has 39% of its particles finer than 0.02 mm and classifies as an ML soil under the Unified Soil Classification system (Corps of Engineers, 1952). The groundwater and temperature conditions at the pavement site are both conductive to frost action. It is assumed that the coefficient of permeability, k, of the thawed subgrade soil is 0.05 ft/day.

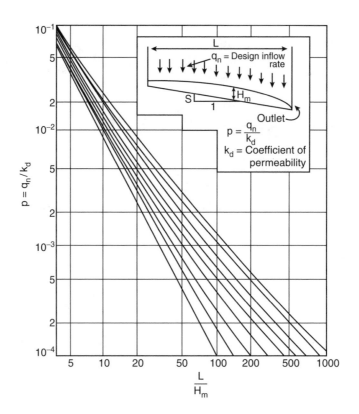

FIGURE 13.40   Chart for estimating maximum depth of flow caused by steady inflow (*Source*: From *Highway Sub-Drainage Design*, Report No. FHWA-TS-80-224, US Department of Transportation, Washington, DC, August 1980).

Assume:

Pavement unit weight $= 150$ lbs/ft^3
Subbase unit weight $= 125$ lbs/ft^3

$$\sigma_p = 150(9/12) + 125(6/12) = 175 \text{ lbs/ft}^2$$

Heave rate from Table 13.17 $= 20$ mm/day

Answer:

Using Figure 13.40 enter the graph at heave rate of 20 mm/day and go to the $\sigma_p = 175$ lbs/ft^2 find that

$$\frac{q_m}{\sqrt{k}} = \frac{1.32}{\sqrt{0.05}} = 0.295 = 0.3 \text{ ft}^3/\text{day}$$

13.5.10   Vertical Out-flow

Water that infiltrates the pavement may accumulate and flow vertically through the underlying drainage layers. Such flow will decrease the amount of water entering the drainage system and must be accounted for. For vertical flow to be deducted from the net in-flow, there must not be upward in-flow from any other source. There are three cases in which the vertical out-flow will take place, as defined in the FHWA

TABLE 13.17   Guidelines for Selection of Heave Rate or Frost Susceptibility Classification

		Unified Classification		
Soil Type	Symbol	Percent <0.02 mm	Heave Rate (mm/day)	Frost Susceptibility Classification
Gravels and sandy gravels	GP	0.4	3	Medium
	GW	0.7–1.0	0.3–1.0	Neg. to low
		1.0–1.5	1.0–3.5	
		1.5–4.0	3.5–2.0	
Silty and sandy gravels	GP–GM	2.0–3.0	1.0–3.0	Low to medium
	GW–GM	3.0–7.0	3.0–4.5	Medium to high
	GM			
Clayey and silty gravels	GW–GC	4.2	2.5	Medium
	GM–GC	15.0	5.0	High
	GC	15.0–30.0	2.5–5.0	Medium to high
Sands and gravely sands	SP	1.0–2.0	0.8	Very low
	SW	2.0	3.0	Medium
Silty and gravely sands	SP–SM	1.5–2.0	0.2–1.5	Neg. to low
	SW–SM	2.0–5.0	1.5–6.0	Low to high
	SM	5.0–9.0	6.0–9.0	High to very high
		9.0–22.0	9.0–5.5	
Clayey and silty sands	SM–SC	9.5–35.0	5.0–7.0	High
	SC			
Silts and organic silts	ML–OL	23.0–33.0	1.1–14.0	Low to very high
	ML	33.0–45.0	14.0–25.0	Very high
		45.0–65.0	25.0	Very high
Clayey silts	ML–CL	60.0–75.0	13.0	Very high
Gravely and sandy clays	CL	38.0–65.0	7.0–10.0	High to very high
Lean clays	CL	65.0	5.0	High
	CL–OL	30.0–70.0	4.0	High
Fat clays	CH	60.0	0.8	Very low

Source: From *Highway Sub-Drainage Design*, FHWA-TS-80-224, US Department of Transportation, Washington, DC, August 1980.

publication *Highway Sub-Drainage Design* (FHWA, 1980):

1. A water-table is located below the pavement section, and the flow is directed toward it in either a horizontal or sloping direction.
2. Vertical or nearly vertical flow is achieved as water flows through a highly permeable soil beneath the subgrade or along the embankment.
3. Flow passes through the embankment and foundation, vertically or laterally, and surfaces on the embankment slope or the foundation of the subgrade.

A more complete design analysis for vertical flow can be found in the FHWA publication *Highway Sub-Drainage Design* (FHWA, 1980).

13.5.11   Net In-flow

Net in-flow is the flow that the design must accommodate once all contributing sources have been taken into account. Net in-flow, q_n, should also take into account any deductions that are made due to vertical out-flow. Flow contributing to net in-flow should account for:

- infiltration, q_i
- ground water drainage, q_g
- artesian flow, q_a
- melt water, q_m
- vertical out-flow, q_v

TABLE 13.18   Guidelines for Using Equation 13.34 to Equation 13.38

Highway Cross-Section	Groundwater Inflow	Frost Action	Net Inflow Rate, q_n, Recommended For Design
Cut	Gravity	Yes	Maximum of Equations 17.35 and 17.37
		No	Equation 17.35
Cut	Artesian	Yes	Maximum of Equations 17.36 and 17.37
		No	Equation 17.36
Cut	None	Yes	Equation 17.37
		No	Equation 17.34
Cut	None	Yes	Equation 17.37
		No	Equation 17.38
Fill	None	Yes	Equation 17.37
		No	Equation 17.38

Source: From *Highway Sub-Drainage Design*, Report No. FHWA-TS-80-224, US Department of Transportation, Washington, DC, August 1980.

Due to the nature of in-flow, it is generally not necessary to account for all types of in-flow at all times. For example, if there is in-flow due to infiltration, it is most likely necessary to also account for in-flow due to ground water flow. Conversely, if there is in-flow due to melt water, there is likely no need to account for groundwater flow. Fine grained soils have little to no permeability when they are frozen, so it is not likely to have in-flow from ground water and melt water at the same time. The following equations are provided to determine net in-flow with different contribution combinations. The guidelines for using these equations are show in Table 13.18.

$$q_n = q_i \tag{13.34}$$

$$q_n = q_i + q_g \tag{13.35}$$

$$q_n = q_i + q_a \tag{13.36}$$

$$q_n = q_i + q_m \tag{13.37}$$

$$q_n = q_i - q_v \tag{13.38}$$

13.5.12   Design Thickness and Permeability

The thickness, H_d, and the permeability, k_d, can be computed and used to determine the required drainage layers. The drainage layers will then convey the flow to the design outlet. When this drainage layer is conveying water, of a constant depth, flow is directly related to the coefficient of transmissibility which is simply the product of H_d and k_d. The maximum depth of flow, H_m, can be found using Figure 13.41. However, if the value of H_m is known, the chart can be used to determine the permeability that is required. Values for the slope, S, and the length, L, of the drainage path must be known in all cases. Figure 13.41 is based on conditions where the in-flow is steady and uniformly distributed across the pavement surface. This phenomenon does not occur regularly in practice and the results should be considered conservative, as long as the recommendations for determining q_n have been followed.

Example 13.15

Drainage Depth Design
   Find the required depth for a drainage layer given the following:

$$q_n = 0.32 \frac{ft^3/day}{ft^2}$$

$$k_d = 2500 \ ft/day$$

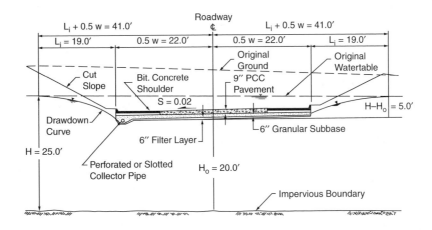

FIGURE 13.41 Rigid pavement section for Example 17.12.

$$p = \frac{q_n}{k_d} = \frac{0.32}{2500} = 1.28 \times 10^{-4}$$

$$S = 0.01$$

Enter Figure 13.41 at $p = 1.28 \times 10^{-4}$ and follow vertically to a slope of 0.01, $L/H_m = 155$ then $H_m = L/155 = 45/155 = 0.29$ ft $= 3.5$ in.

13.5.13   Filter Design

Drainage layers are designed so that they allow water to flow from fine grained material into coarser grained material for removal from the roadway. This often results in the migration of a portion of the fines into the coarser material. If this process continues for any amount of time, the coarse grained material will become filled or clogged with fines and the drainage systems will no longer function properly. Pumping due to traffic loads increases the probability of such an occurrence. This introduces the need for a filter layer to prevent such migration. The filter layer is generally composed of a particular gradation of particles that reduce this migration. The filter criteria are listed below; however it may be more cost effective to introduce a drainage fabric than a graded filter. A more complete cost analysis should be preformed when this is the case. Equation 13.39 to Equation 13.43 are normally used to check for the compatibility of the filter layer and the subgrade, while Equation 13.44 to Equation 13.46 check the compatibility of the filter and the base layers.

$$(D_{15})\text{filter} \leq 5(D_{85})\text{protected soil} \qquad (13.39)$$

$$(D_{15})\text{filter} \geq 5(D_{15})\text{protected soil} \qquad (13.40)$$

$$(D_{50})\text{filter} \leq 25(D_{50})\text{protected soil} \qquad (13.41)$$

$$(D_{5})\text{filter} \geq 0.074 \text{ mm} \qquad (13.42)$$

$$(\text{CU})\text{filter} = \frac{(D_{60})\text{filter}}{(D_{10})\text{filter}} \leq 20 \qquad (13.43)$$

$$(D_{15})\text{base} \leq 5(D_{85})\text{filter} \qquad (13.44)$$

$$(D_{15})\text{base} \geq 5(D_{15})\text{filter} \qquad (13.45)$$

$$(D_{50})\text{base} \leq 25(D_{50})\text{filter} \qquad (13.46)$$

Equation 13.41 is not necessary when the soil being protected is clay of medium or high plasticity. When using coarse materials, the design should be based on the portion of the material that is less than 1 in. in size.

13.5.14 Collector Pipe Design

The collection system should be designed so that the water from the drainage area flows into perforated pipes and travels to suitable outlet areas away from the roadway. When designing a collector system, the following must be taken into account:

- type and size of collector pipe
- location and depth of collector pipe
- outlet location
- slope of collector pipe
- adequate filter material to protect collector pipe

Longitudinal collector pipes are commonly used and their design depends on various factors. Some of these include the drainage that will be necessary along the shoulders of the roadway and the amount and depth of frost action.

13.5.14.1 Collector Placement

Longitudinal collector pipe design should take into account frost depth, economic considerations and the surrounding water-table. It is common to place the pipes in shallow ditches along the roadway when frost action is not present or the frost depth is shallow. However, as mentioned before, if it is necessary for the collector drain to remove water from a high water-table or to use it to eliminate saturation of the surrounding subbase, it may be necessary to place the collector in a deeper location. An example of collector pipe placement where there is little frost action is shown in Figure 13.42 and the deeper location in Figure 13.43. The benefits of locating the collector pipe in as shallow a ditch as possible are mostly economic. The cost of placement will increase as the depth of the trench increases due to an increase

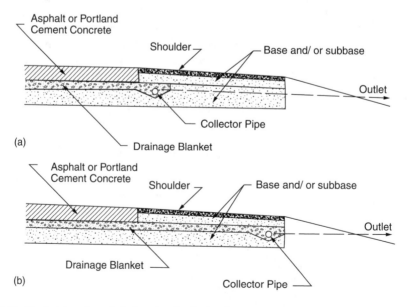

FIGURE 13.42 Typical location of shallow longitudinal collector pipes (*Source*: From *Highway Sub-Drainage Design*, Report No. FHWA-TS-80-224, US Department of Transportation, Washington, DC, August 1980).

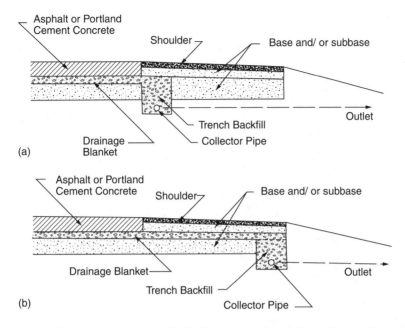

FIGURE 13.43 Typical location of deep longitudinal collector pipes (*Source*: From *Highway Sub-Drainage Design*, Report No. FHWA-TS-80-224, US Department of Transportation, Washington, DC, August 1980).

in the earthwork. The distance of the pipe for the roadway is often dependent on the necessity of drainage in the shoulder area. If the shoulder needs drainage, the design should be similar to that shown in Figure 13.42(b) and Figure 13.43(b). If the shoulder area of the roadway requires no drainage, the pipe design should resemble that of Figure 13.42(a) and Figure 13.43(a).

13.5.14.2   Collector Pipe Size

The diameter of a collector pipe is based on the amount of water it is required to carry to the outlet. Design of the diameter will be influenced by the pipe gradient, g, the quantity of water being carried by the pipe, q_n, and the distance between the outlets, L_o, as well as the hydraulic characteristics of the pipe. Figure 13.44 displays a nomograph which takes into account each of these parameters. The nomograph uses the flow rate in the drain, q_d, which can be computed using Equation 13.47. Conversely, if the gradient and the diameter are known, the nomograph will give the maximum spacing between the outlets.

$$q_d = q_n L \tag{13.47}$$

where:

q_d = flow rate in the drain (ft^3/day/ft).
q_n = net in-flow (ft^3/day/ft^2).
L = the length of the flow path (ft).

The length of the flow path, L, may vary within the section being considered. Caution should be used when calculating L in such cases and it is common to use an average of all the lengths.

13.5.14.3   Filter Material for Collector Pipes

Filter material surrounding the pipes should be large enough to allow the flow of water into the collector pipes, however this material should be small enough to prevent surrounding particles from entering and

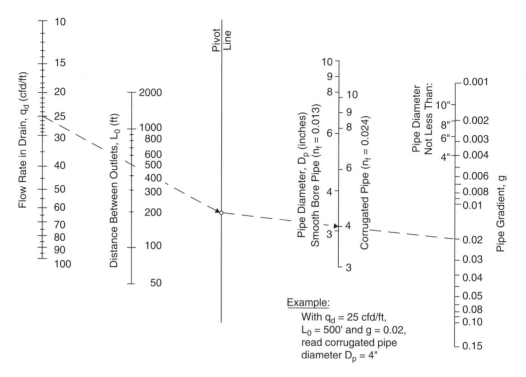

FIGURE 13.44 Nomogram Relating Collector Pipe Size with Flow Rate, Outlet Spacing and Pipe Gradient Source: *Highway subdrainage Design*, Report No. FHWA-TS-80-224, U.S. Department of Transportation, Washington D.C., August 1980.

clogging the drainage pipe. It is recommended that Equation 13.48 and Equation 13.49 be used as criteria for the filter material. Equation 13.48 is used in cases where the pipe is slotted and Equation 13.49 is used in cases where the pipe has circular holes. When filter material is not necessary, the surrounding fill material should have a high permeability.

$$(D_{85}) \text{ filter} > 1/2 \text{ slot width} \tag{13.48}$$

$$(D_{85}) \text{ filter} > 1.0 \text{ hole diameter} \tag{13.49}$$

13.6 Summary

Surface and subdrainage systems design are highly complex. Sound engineering judgment should be used in all cases. Past experience should be drawn upon, and conservative values should be used when economic constraints permit. The combination of many different methods and devices will ensure proper drainage of the roadway and improve its performance and durability. Many of the methods mentioned in this chapter are empirical and therefore are not exact. All charts and equations should be used with care to ensure a well designed roadway drainage system.

Acknowledgments

We would like to acknowledge George Huntingdon for his editorial comments.

References

Ayres Associates, 2001. *Culvert Design*, National Highway Institute (NHI). Course 13056.

Corps of Engineers, 1952. *Unified Soil Classification System*, Corps of Engineers and Bureau of Reclamation.

FHWA, 1980. *Highway Sub-Drainage Design*, Report No. FHWA-TS-80-224. US Department of Transportation, Washington, DC.

FHWA, 1984. *Drainage of Highway Pavements*, Highway Engineering Circular No. 12. Federal Highway Administration, Washington, DC.

FHWA, 1985. *Hydraulic Design of Highway Culverts Hydraulic Design Series No. 5*, Report No. FHWA-IP-85-15, September 1985.

FHWA, 1988. *Design of Roadside Channels with Flexible Linings*, Hydraulic Engineering Circular No. 15. Federal Highway Administration, Washington, DC, FHWA-IP-87-7.

FHWA, 1990. *FHWA Highways in the River Environment*, Training and Design Manual. Federal Highway Administration.

Garber, N.J. and Hoel, L.A. 2002. *Traffic and Highway Engineering, 3rd Ed.* Brooks/Cole, Belmont, CA.

Soil Conservation Service, 1986. *Engineering Field Manual for Conservation Practices*. US Soil Conservation Service, Washington, DC.

Soil Conservation Service, 1973. *A Method for Estimating Volume and Rate of Runoff in Small Watersheds*, Report No. SCS-TP-149. US Department of Agriculture.

Soil Conservation Service, 1968. *Hydrology Supplement A to Section 4*, Engineering Handbook. US Department of Agriculture, Soil Conservation Service.

US DOT, 1965. *Hydraulic Design Series No. 19*. US Department of Transportation, Washington, DC.

US DOT, 1979. *Design Charts for Open Channels*. US Department of Transportation, Washington, DC.

Further Reading

AASHTO, 1991. *Model Drainage Manual*, American Association of State Highway and Transportation Officials.

AASHTO, 2001. *A Policy on Geometric Design of Highways and Streets, 4th Ed.* American Association of State Highway and Transportation Officials.

FHWA, 1965. *Design of Roadside Drainage Channels*, Hydraulic Design Series No. 4. Federal Highway Administration, Washington, DC.

FHWA, 1973. *Design Charts for Open Channel Flow*, Hydraulic Design Series No. 3. Federal Highway Administration, Washington, DC.

Lewis, G.L. and Viessman, W. 1995. *Introduction to Hydrology Fourth Edition*. Addison-Wesley Educational Publishers, Inc., Reading, MA.

Appendix 13A

TABLE 13.A1 Conversion Factors From U.S. Customary Units to SI Units

U.S. Customary Units and Their SI Equivalents		
Unit	U.S. Customary	SI Equivalent
Length	in.	25.4 mm
	ft	0.3048 m
	mi	1.609 km
Area	ft^2	0.0929 m^2

(Continued)

TABLE 13.A1 Continued

	U.S. Customary Units and Their SI Equivalents	
Unit	U.S. Customary	SI Equivalent
Volume	$in.^2$	$645.2\ mm^2$
	ft^3	$0.02832\ m^3$
	$in.^3$	$16.39\ cm^3$
	gal	3.785 l
Mass	oz mass	28.35 g
	lb mass	0.4536 kg
Stress	lb/ft^2	47.88 Pa
	$lb/in.^2$	6.895 kPa
Work	ft·lb	1.356 J
Force	kip	4.448 kN
	lb	4.448 N
	oz	0.2780 N
Power	ft·lb/sec	1.356 W
	hp	745.7 W

Part C

Construction,
Maintenance and
Management of
Highways

14
Highway Construction

K.N. Gunalan
*Parsons Brinckerhoff Quade
and Douglas, Inc.
Murray, UT, U.S.A.*

14.1 Introduction

Highway construction has evolved over the years and as with any industry today, has become fairly sophisticated. Expectations of the tax-paying citizen with respect to quality, schedule and budget have also risen over time. Technology, and modern machinery have raised the level of skill sets required to understand and manage highway construction projects. The skill set required varies based on the nature and complexity of the project; differences between new construction and reconstruction projects; and differences for projects in urban settings and those in rural regions.

The quality of a project depends on material quality, skill and commitment of the work force. The majority of materials used in highway construction, such as fill materials for embankment construction, aggregates for base and surface courses, are naturally occurring materials. These materials are usually available within certain proximity to the project and meeting certain physical and chemical characteristics. The behavior of these materials under the local environment of traffic loads, ambient temperatures, annual rainfall and local practice of keeping highway serviceable during adverse weather conditions will have to be understood. Based on this understanding, the methods and means of constructing a highway will have to be adopted accordingly.

14.2 Highway Elements and Layout

Construction of a highway today requires broad knowledge of a number of elements which could be divided into technical and commercial elements of the system (Figure 14.1) and process, respectively. The technical elements include materials, material quality, installation techniques, and traffic control systems, etc. The commercial elements include understanding of the contract; environmental, political and legal aspects of a project; and public concerns.

The contractor's project organization for highway construction is structured around trades namely, project management, grading or earthwork, structures, specialty elements including mechanically

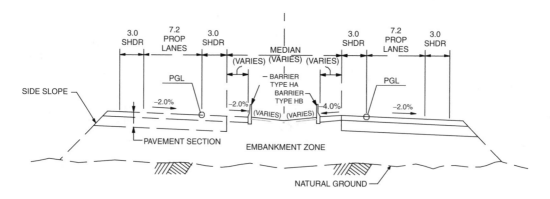

FIGURE 14.1 Typical roadway section.

stabilized earth walls, traffic control or systems, landscaping and paving (Figure 14.2). Project management covers scheduling, cost estimating, procurement, quality control, safety and environmental control. A good project management approach will include a well-defined structure outlining roles and responsibilities and ensure good communication among the various groups. A number of agencies are beginning to require the builder provide independent quality control and quality assurance. In addition to establishing a quality control or quality assurance program, management is expected to provide stop-work authority to this group.

The environmental aspects of a project are not limited to the elements of the project and to the project limits. Permit requirements and commitments will have to be understood to avoid expensive fines and shutdowns. Impacts to adjacent properties, surface and sub-surface drainage, noise and light pollution during and after construction will have to be understood and implemented.

Political aspects of a project can be complex, depending on the location, size of the project, funding for the project, etc. Though the contracting agency is responsible to keep the wheels turning, the constructor has a lot at stake if the funding for the project is not in place or not assured for the future years. Additionally, the constructor will need to be aware of other political pressures such as use of local work force, special needs of a community or group, etc.

Legal aspects of the project could include meeting commitments of the contract, subcontracting, claims, changed conditions, safety of traveling public, etc. These include the required level of participation of disadvantaged businesses, prescribed minimum wages, betterments established to placate local municipalities, and environmental requirements, etc.

14.3 Construction Methods and Means

The cornerstone of a successful project includes understanding of the process and requirements, a well thoughtout plan and effective management. Understanding the process includes studying the contract, aspects and requirements of the project. The proposed alignment of the highway to be constructed along with the drainage, utility, and geotechnical information provided should be reviewed and a site visit should be considered essential. These are the three areas of a typical project that pose a high risk, while some projects with unique characteristics may have other elements of risk that will need to be taken into account. Typical elements of risk include right-of-way (ROW) acquisition and uncertain unskilled labor market. In any case, it is prudent to conduct a risk-analysis so that proper resources may be allocated.

Study of the topography of the project site will provide some insight into the drainage patterns and existing facilities. This in turn will assist in construction sequencing and establishing starting points for construction of drainage facilities. Traditionally, construction starts at the lowest elevation whether it be

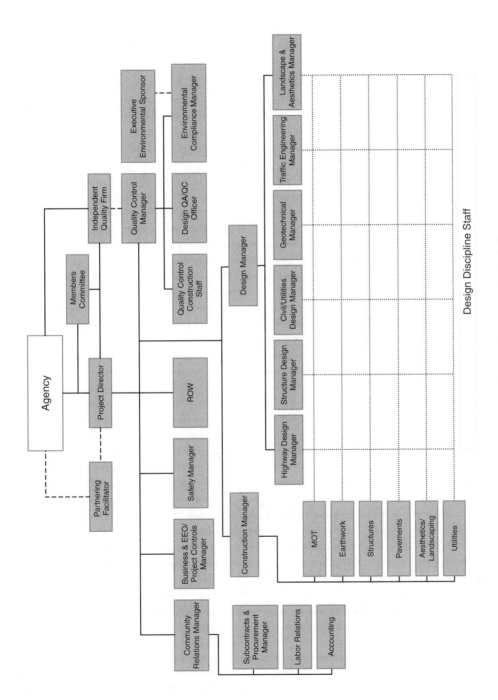

FIGURE 14.2   Typical construction organizational chart.

constructing a utility or a drainage facility and moves upward. Review of project specific geotechnical data is imperative, as it will provide insight into:

a) Existing ground conditions
b) Properties of materials to be excavated
c) Equipment that will be needed to excavate, grade, transport material to and from site
d) If dewatering will be necessary for below grade work
e) If shoring will be required to protect excavations and
f) Need for and quantity of water for dust control and compaction

Review of existing ground conditions will reveal the need for stripping, depth of required stripping, quantity of material to be stripped, equipment to be used and distance to disposal area of any unsuitable materials. If the stripped material is deemed suitable, it may be used as top-soil with appropriate amendments. The need for any stabilization of existing ground can be determined, as explained in the next section.

Plans are studied to develop what is called "block diagrams" or "mass diagrams" to identify cuts and fills. This provides information on the overall quantity of material that may be needed to be imported for constructing the embankment and also provides information (together with geotechnical data) on equipment to be used. The term grading is generally used to include excavating, hauling, spreading and compacting embankment material in place. Excavation is the process of loosening and removing rock or earth from its original location and transporting to a fill section or disposal site. Depending on the nature of the rock, it may be ripped or may need to be blasted. This material can then be bulldozed to nearby location or hauled off using loaders and trucks. On the other hand, earth material may be removed and hauled over short distance using self-loading scrapers. For hauling over a considerable distance or through existing highways and streets where axle loads are limited, the material may be loaded using front-end loaders on to rear or bottom dump trucks. Traffic control, dust control, and maintenance of affected public facilities will need to considered and planned for.

If the existing ground conditions are such that operation of construction equipment will be next to impossible, tractor mounted equipment may be needed to build a working platform. The platform may have to be built by pushing large-sized granular material in front of the equipment and working them into the unstable ground. Compaction is another element that requires understanding of not only the material but also knowing the type of equipment that will provide the necessary results in an efficient manner. The types of equipment available today include static compactors such as tamping or sheepsfoot rollers, pneumatic-tired rollers and smooth steel drum rollers. There are vibratory or dynamic compactors that may be needed for granular materials.

For reconstruction projects one needs to plan on extensive coordination with utility (wet and dry) companies, cities or municipalities, property owners and the traveling public. The time, sequence and equipment needed to demolish existing structures; vibration, noise, dust, traffic control, will all need to be considered and evaluated. Cost-effective means of recycling materials is becoming very important. One must also be aware of any potential for encountering hazardous and harmful materials and exposing the workers to any health hazards.

Another important element of the projects is to have a good survey crew. The crew should be able to review and interpret the information contained in the design drawings, break this information down and layout the center line; control points including edge of roadway alignment, fill limits including angle points and ROW limits.

Technology is available to feed information on elevations of the various courses to an onboard computer that assists the operator of grading equipment to finish within millimeters of the planned elevation. The added skills required for grading operations today include ability to download information from design drawings and to use commercially available software to generate information needed for field operation. Additionally the operator needs the skill to be able to understand and adjust his grading operations accordingly. Computer assisted grading operation takes away the human judgment of finish grades.

14.4   Subgrade Preparation and Stabilization

The term subgrade as used in this section refers to the natural or existing ground. For an elevated highway this will be below the embankment fill and for at-grade sections this would be immediately below the pavement section. It is customary to clear and grub within the staked fill limits of the highway and other designated areas. This task could mean a simple effort or an extensive one especially if the alignment is along an existing forest, swamp, or wooded area. On reconstruction projects it could include demolishing and disposal of existing facilities such as sidewalks and structures, and complications associated with existing utilities. As a general rule for embankments over 2 ft (0.6 m) in height (not including the pavement section), it is not necessary to strip and remove every blade of grass. It would be adequate to mow the vegetation down so as not to leave more than an inch or two of it above grade. Stumps, roots and nonperishable solid objects may be left in place.

Both the short-term and long-term performance (such as settlement and stability) of the newly constructed highway is highly dependent on the stability of the underlying soils. Care should be exercised to this aspect of building the highway. If unsuitable materials such as soft plastic silts or clays are encountered at shallow depths (within the top 3 to 5 ft or 0.9 to 1.5 m) and extending over large areas (10 ft × 10 ft or 3 m × 3 m), it will be prudent to remove and replace these materials with suitable granular materials (Figure 14.3). The backfill material should be compacted in lifts to ensure minimal settlement due additional load of the highway. If these materials extend to shallow depths but are over a small area (less than 10 ft × 10 ft, or 3 m × 3 m), these materials could be bridged over through the use of geo-grids, cobbles or gravels, cobbles or gravels inter-locked with geo-grids, or use of high strength

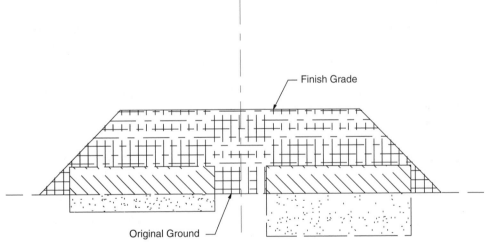

NOTE: If compaction of the such excavation working platform material is not achieved contact the Geotechnical Engineer before proceeding with embankment construction.

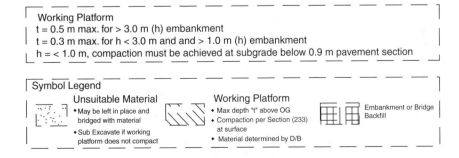

FIGURE 14.3   Unsuitable materials — subgrade preparation guidelines.

geo-textiles. The type of geo-grid or geo-textile will need to be evaluated with the assistance of a competent geotechnical engineer. If these materials extend to greater depths and area, other means of stabilization should be explored. These include the following displacement techniques:

Techniques to work large granular materials into the subgrade, for example through the use of stone columns;

Preloading with or without wick drains; wick drains have been found to be very effective if the soft soils are encountered at depths;

Chemical stabilization using cement, fly ash or lime.

The analysis and design of subgrade stabilization is generally left to the constructor. This is accomplished with the assistance of a geotechnical engineer or a specialty contractor.

Wick drains spaced closer than 4.5 ft (1.5 m) in a triangular pattern are found to be not very effective. A sand blanket is generally used at the base of the embankment to allow for the excess moisture coming up through the wick drains to dissipate; strip drains tied to a row or rows of wicks have also been used successfully in the past. The amount of preload to consolidate the underlying soils is dependent on the height of embankment and depth of these soils. As a rule of thumb, 15% of embankment height is considered to be the minimum height of surcharge above the embankment for fills greater than 10 ft (approximately 3 m) high. For fills less than or equal to 10 ft (approximately 3 m) high, it may be necessary to go up to 50% of embankment height to have the same influence.

14.5   Embankment Materials and Construction

Embankments are constructed over prepared and stabilized subgrade. Typical requirement for elevated embankment includes the use of good drainable material spread in thin lifts and compacted to the required density. Where good material sources are easily available, embankment materials falling between the classification of A-1-a and A-4 per AASHTO M145 are preferred. Materials meeting A-1-a classification are preferred around and behind structures, while A-4 material may be acceptable within the roadway prism but away from structures (Figure 14.4). Where a good source is not readily available, stabilization of existing materials should be considered.

Embankment construction should not be performed on frozen subgrade. Special care is needed during freezing weather to prevent frost from developing in an active fill. Broken up pieces of concrete may be used within the embankment fill provided the largest dimension is no greater than 1/3 of the fill height or 30 in. (0.76 m), whichever is smaller. Care should be taken to ensure that these large size materials do not nest to create large voids that could become detrimental to the performance of the embankment due to excessive settlement. Recycled asphalt can also be used within the embankment provided they are placed at least 3 ft (0.9 m) above the high ground water elevation. Recycled materials should not be extended to the face of the slope, but remain within the roadway prism.

The embankment surface should be finished to within 1 in. (25.4 mm) of the final grade. The finished surface should be protected from the elements of weather, construction traffic, etc. If settlement of the embankment is anticipated, elevation of fill should be monitored prior to stripping. If portions of the surcharge will settle into the final embankment zone, surcharge should also be placed in lifts and compacted as embankment fill. Monitoring of embankment during construction, for stability and settlement, should be planned for and implemented. The monitoring information generated has been found to be very useful not only for addressing to the short-term concerns but also long-term performance.

The optimum moisture and compaction effort required will depend on the materials used. Also, the equipment needed for compaction and moisture control will depend on the type of material and effort required. General practice has been to use construction traffic where and whenever possible. If needed, vibratory type pneumatic-tired equipment could be used to compact good granular material. An accelerated construction schedule sometimes does not allow for the use of traditional surcharge of

Bridge Backfill Zone

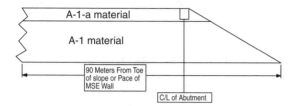

Material Spec	Max Size	Passing 200 (%)	Lift Thickness	Compaction Req's
A-1	100 mm	0-25	300 mm	T-180
				96 Average
				No Text < 92

Common Embankment Zone

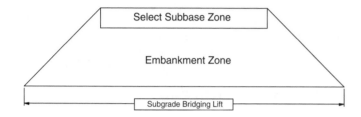

Material Spec	Max Size	Passing 200 (%)	Lift Thickness	Compaction Req's
A-1		0-25	300 mm	T-99
A-2		0-35		96 Average
A-3		0-10		No Text < 92
A-4		36-100		

Bridging Lift

Material Spec	Max Size	Passing 200 (%)	Lift Thickness	Compaction Req's
None	0.9 M	No Spec	No Spec	None

Wall Backfill

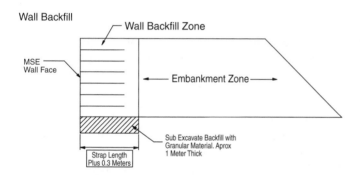

Material Spec	Max Size	Passing 200 (%)	Lift Thickness	Compaction Req's
A-1	102 mm	0-15	200 mm	95% Below Optimum
				T-99

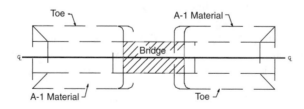

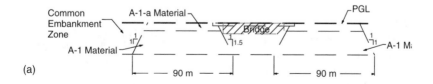

(a)

FIGURE 14.4 Embankment zones (a) bridge embankment zone.

soft subgrade soils and sufficient time for consolidation settlement to be completed. Embankments are constructed over wicked subgrades to accelerate settlement. Excess water coming up to the surface through the wicks will need to be handled, and embankment settlement and additional materials required to compensate for the settlement should be considered and addressed also.

Bridge abutments, specifically the vertical abutments, are now being built using mechanically stabilized earth (known as MSE walls). Care should be exercised in understanding and building these system walls. There are specialty wall suppliers who provide the design and hardware required to build these walls.

14.6   Subbase Course Materials and Construction

Subbase course with thickness typically between 4 and 16 in. (100 and 405 mm) is a layer of select material between the subgrade and the base course. Subbase course provides uniform support and adds to the required structural capacity of the pavement section. The material can be gravel, crushed stone or subgrade soil stabilized with cement, fly ash or lime. The use of permeable subbase course is becoming more common to accommodate drainage of water infiltrating from the surface or to keep subsurface water from reaching the surface. In certain regions, the course is also used to impede frost penetration into the subgrade and thereby minimizing frost heave damage to the pavement surface. Compaction and moisture control is generally achieved with specialized equipment as opposed to solely by construction traffic.

14.7   Base Course Materials and Construction

Base course is the portion of the pavement section that is immediately below the surface course. It is constructed on top of the subbase course or if there is no subbase course, directly on top of the subgrade. The base course is subjected to heavy loads and is an important component of the pavement structure. The requirements of the base course is more stringent than the subbase or subgrade material. Base course is generally constructed of untreated crushed aggregate (such as crushed stone, slag, or gravel) meeting the requirements of AASHTO Specification M147 and M75. The material should exhibit good drainage characteristics and stability under construction traffic.

Base course thickness is governed by the properties of the underlying layers. Its thickness is generally in the range of 4 to 6 in. (100 to 150 mm). Equipment and procedures for compacting base courses are similar to those of embankment and are the specialized equipment identified in Section 18.5. Surface tolerance of this course is very important as the cross slope requirements of the highway section is determined by how well the grades of the base course is set. Sacrificing base course thickness to meet cross slope requirement will be at the expense of drainage requirements. Base course material section is generally carried across underneath the shoulders or curb and gutters to not only provide a stable base but also to ensure continuity for drainage, away from the surface course of the pavement section. When these courses cannot be daylighted, it is traditional to install an under-drain system (Figure 14.5) to aid in draining excess moisture away. It is good practice to use good granular material around the underdrain system with filter fabric to keep fine-grained soils from clogging the drains. Care should be taken to keep construction traffic off these underdrains and from getting crushed. Adequate spacing of cleanouts should be provided for proper maintenance of the under-drain system.

Base course materials are sometimes treated with bitumen, cement, calcium chloride, sodium chloride, fly ash in lieu of the expensive cement and lime in the form of a powder or slurry. Use of bulk quicklime should be avoided to prevent burn injuries to workmen. The need for treatment is to cater for heavy loads, frost susceptibility, and to serve as moisture barrier. Cement treated base courses need to be protected from moisture loss until hydration of all cementitious materials. Tight control of grade is especially important when dealing with treated base courses. Compaction of treated bases can be accomplished using

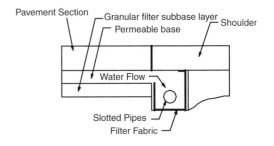

FIGURE 14.5   Under-drain system.

sheepsfoot roller or vibratory compactors. The top 1 to 2 in. (25 to 50 mm) is generally compacted using pneumatic-tired rollers preceded by shaping with a blade grader. It is common in construction today to use a laser-guided or (Global Positioning System) GPS controlled blade grader to achieve the desired tolerances of this course. Finished base course layer should be protected if necessary as outlined in the next section.

14.8 Surface Course Construction

This section will cover the two most common types of pavement surface courses used in highway construction today namely, hot-mix asphalt concrete (HMAC) and Portland cement concrete (PCC). HMAC sections are referred to as the flexible pavement and PCC sections are referred to as the rigid pavement. The two different pavement surface courses serve to provide a smooth and safe riding surface and to effectively transfer the traffic loads to the underlying subgrade soils through the various base courses discussed earlier. The properties of the two materials, PCC and HMAC, and their structural performance and load transfer mechanisms are very different. The process involved in the construction, the skill sets required to construct, and the problems associated with the two pavement types are not the same. This section is structured to address the two surface course types separately. Though the designed life and certain service conditions are taken care of through adequate design, the performance of the pavement section heavily depends on the experience and skill of the crew, quality of materials used, types and conditions of the equipment and tools used in the construction.

14.8.1 Portland Cement Concrete

Concrete pavements are generally classified as jointed plain concrete, jointed reinforced and continuously reinforced pavements (Figure 14.6). Jointed plain concrete pavements (JPCP) consist of a uniform concrete section with joints strategically placed to control cracking. The load transfer between sections take place through inter-locking of large aggregates in the mix or through load transfer devices in the transverse joints. Dowel bars are load-transfer devices used to provide an efficient load transfer across

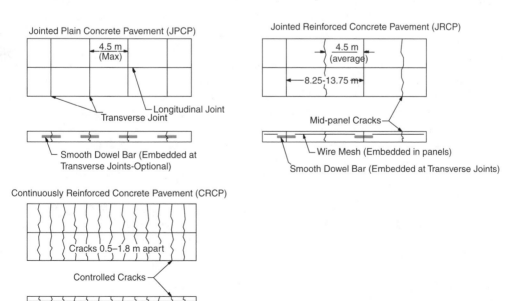

FIGURE 14.6 Concrete pavements types.

transverse joints; maintain horizontal and vertical alignments across the joint. Deformed steel bars called "tie-bars" are provided along longitudinal joints of adjacent pavement sections to hold them together. Dowel bars and tie-bars are required to perform over the entire service life of the pavement. These elements need to be protected from corrosion. Depending on the environment to which these elements are exposed, it is now common to epoxy coat the reinforcements used in pavement sections. There are other means to address to the same concern. Care should be taken to ensure that any kind of coating used to prevent onset of corrosion is not damaged during construction. Planning should include measures to mitigate this prior to placement.

Jointed reinforced concrete pavements (JRCP) consist of a uniform concrete section with reinforcement mesh. Continuously Reinforced Concrete Pavements (CRCP) consists of a concrete section that is continuously reinforced, usually about 0.6% by cross sectional area. These pavements have no contraction joints. Thermal cracks are anticipated to occur in this type of pavement every 3 to 5 ft (1 to 1.6 m).

14.8.1.1 Phases in Construction

Construction of concrete pavement sections involves varying levels of complexities and details. These can be broadly broken down into preparation, transportation, laydown and finishing operations. The preparation portion of the work starts with a good plan including preparation of the surface upon which the material is to be laid, placement of block-outs for drainage structures, defining the line and finished grade of the section, placement of reinforcement, selection of good quality materials, development of a durable mix, methods and means to ensure consistency of the mix, etc.

Transportation of the mix includes determination of haul distance, routes, time to transport, types of conveyance such as trucks, type of trucks, conveyor belts, etc., ambient conditions, and production rates. Laydown operations include determination of means to form the section, given the volume, location and other existing conditions such as block-outs, adjacent structures, etc. Concrete pavements are laid down in fixed forms or in long continuous sections. Fixed form paving uses wooden or steel forms that are set up along the edges of the section before placement. The placement in fixed forms involves the use of vibratory screeds or rolling tubes or larger form-riding self-propelled machines. For the continuous sections, a slip-form paving machine extrudes the concrete placed in front of the machine using a corkscrew mechanism, followed by spreading, consolidating and finishing in one pass. As the name implies, these pavers generally do not require a side form but form as they go and it is necessary to determine and maintain the consistency of the mix so as to prevent edge slumps. Hand placements using fixed forms is more labor intensive, requiring not only setting of the forms but also setting of the dowel and tie-bars on chairs, finishing, and curing. Currently, slip-form paving is a more common method of paving on large highway construction projects. Slip-form pavers come in varying widths (for example, two or three lane pavers), capacities and capabilities. Pavers are equipped to finish, float and cure in addition to be able to insert dowel bars and tie-bars.

Finishing operation includes ensuring that the surface of the section is smooth, with no exposed aggregates and other defects such as edge slumps. Finishing operation is an art requiring experienced personnel who understand when over-finishing could actually be detrimental to the performance of the concrete surface. In either hand finishing or machine finishing care should be taken to ensure that excess water is not used to finish, as it would be detrimental to the surface and long-term performance of the pavement. Section laydown, with the more sophisticated paving machines, also will require to keep a crew of finishers in readiness just in case problems arise. The machines are becoming more sophisticated and this need is becoming minimal.

14.8.1.2 Joints

Concrete sections will crack due to shrinkage, thermal contraction and moisture and thermal gradients within the concrete mass. Joints have been introduced in concrete pavements to address the above and are classified into contraction, construction and isolation joints. Typical details of the different types of joints are shown in Figure 14.7. The transverse and longitudinal joints are contraction joints that are

Contraction Joints

Note: T = Thickness of Concrete

Note: T = Thickness of Concrete

Construction Joints

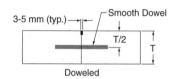

Doweled

Note: T = Thickness of Concrete

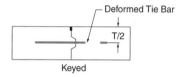

Keyed

Note: T = Thickness of Concrete

Other Joints

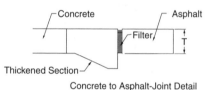

Concrete to Asphalt-Joint Detail

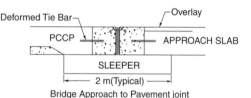

Bridge Approach to Pavement joint

Note: T = Thickness of Concrete

FIGURE 14.7 Joints.

sawed into the concrete. Sawing of joints should commence as soon as the concrete has hardened sufficiently to prevent raveling but before uncontrolled cracking develops. If random cracking has occurred prior to sawing of the joints, the design engineer should be consulted to make a decision on the appropriate action.

Construction joints are installed to join concrete sections paved at different times. Typically these joints are formed using a bulk-head and care should be taken to ensure that the concrete adjacent to the bulk-head is vibrated adequately so as not to form honeycombs. Isolation joints separate the pavement from objects or structures. They facilitate movements of the pavement section independent of the object to prevent damage. Bond breakers are typically used at isolation joints and construction joints. Bond breakers can consist simply of lubricants or pressboard or other materials.

Having introduced joints in the pavement section to prevent cracking, it is imperative to seal these joints to minimize infiltration of surface water and to keep out incompressible materials that could cause joint spalling and blow-outs. There are two major types of sealants available today, viz., liquid sealants and compression sealants. Among the liquid sealants, the hot pour sealant appears to be more commonly used due to the ease and flexibility in its application. Joints should be sealed after completion of the curing period and prior to opening to traffic including construction traffic. Joints should be thoroughly cleaned and dried prior to sealing. Cleaning of joints is usually accomplished through combinations of the following: flushing with water, high pressure water cleaning, air blowing and dry cutting. If dry cutting is used, care needs to be taken to maintain the specified cut widths. Care should be taken to ensure that the seal extends to prescribed depth within the joint and that no excess material is left on top of the surface. The excess material will cause the seal to be ripped out of the joint under traffic.

Compression seals should be of cross sectional dimension shown in the plans, installed with manufacturer-recommended tools, procedures and materials. The seals should be without elongation and of one piece.

14.8.1.3 Surface Texture

Performance of the section and safety of the traveling public are of utmost importance. Surface texture of concrete sections play a great role in the safety and agencies are specifying the requirements of surface textures. Textures are applied to the wet concrete surface using appropriate tools. Tine textures are more common on highway construction and are achieved by a mechanical device behind the paver, equipped with a tinning head or metal rake that moves across the width of the paving surface generally in a transverse direction with traffic. Care should be exercised to ensure randomness, spacing and depth of the texture. The randomness, spacing and width have an influence on the tire and road noise generated. Attention during installation can save costly alternatives such as grinding or saw cutting to meet skid resistance requirements.

14.8.1.4 Curing

Curing is the treatment applied to the concrete surface during the hydration period. The treatment preserves the moisture within the mix for proper hydration of the concrete. Where water is easily available, it is best to pond water on the surface during the hydration period. Other forms of curing include maintaining wet burlap sheets; covering with plastic sheets, insulating blankets and application of liquid membrane forming compounds to the surface. The type of protection by curing is dependent on the time of the year and conditions. Liquid membrane forming compounds are commonly used to cure the concrete sections. There are a number of types of curing compounds available on the market today. One has to carefully select the most appropriate one considering not only the impact on the product but also its impact to the environment.

14.8.1.5 Smoothness or Ride

Smoothness of the pavement is not only important to the long-term performance of the section, it is an expectation of the traveling public. Concrete pavement smoothness can be ensured by paying attention to the following: establishing and maintaining proper stringlines including the material from which it is made; building a stable and sound base to build on; maintaining a consistent speed of the paver; use of proper size and control of the paver head; maintaining a consistent mix; and using minimal hand finish. One needs to understand the process, methods, measurement parameters and equipment to be used to measure smoothness of pavement sections. Lightweight noncontact profilers are commonly used for quality control purposes. Differences among the various profilers and correlations among different machines need to be understood and established ahead of time. When smoothness criteria are not met, expensive grinding has to be undertaken.

14.8.2   Hot-mix Asphalt Concrete

HMAC has been in extensive use for a very long time. The Superpave System (SUperior PERforming Asphalt PAVEments) developed in the late 1980s has changed the approach to design, mix design, specification and testing of the material for quality. It has not changed much in terms of the construction of asphalt pavements. The selection of binder grade depends on the application.

Construction of long-lasting asphalt pavement sections requires experienced construction crew, committed to quality and functioning equipment in good condition. Areas of concern in accomplishing the objectives would include mix segregation, laydown, compaction and joint construction. Segregation is defined as nonuniform distribution of various aggregate sizes throughout the mass creating performance issues on top of texturing and smoothness concerns. Segregation can occur at the following stages: mix design, stockpiles; cold bin loading, cold bin feeding, hot bins, drum mixer operations, discharge systems, truck loading and unloading and paver operations. Attention should be paid to material source, quality, transportation, storage, mixing and discharge, and the equipment used.

Mixes must be placed and compacted before they cool to 185°F (85°C). The minimum mixing temperature will depend where it is placed as well as the ambient conditions. When asphalt pavements are placed in cold temperatures, the asphalt stiffens too rapidly and, therefore, a number of agencies place temperature or calendar restrictions on when asphalt paving can take place.

Prime coat is low viscosity asphalt that is applied to a granular base course prior to placement of the asphalt surface course. The purpose of the prime coat is to bond loose material providing a cohesive platform for paving operations and to provide adhesion between the base course and surface course. The prime coat serves to protect the surface of the base course from infiltration of moisture such as from rainfall. If there is a threat of rain prior to placement of surface course, it will be money well spent to protect the base course with a prime coat. However, if the surface course is to be placed immediately following trimming of the base course, omitting the prime coat will not be detrimental to the performance of the pavement section. Prime coat is generally applied at the rate of 0.1 to 0.3 gallons per square yard or 0.5 to 1.5 l/m^2.

Tack coat is required to ensure bond between an existing pavement surface and new asphalt overlay. Uniformity of application and use of proper application rate is important. Tack coat should be applied uniformly at a rate of 0.03 to 0.05 gal/yd^2 or 0.15 to 0.25 l/m^2 for cutback asphalt and undiluted asphalt and 0.02 gal/yd^2 or 0.1 l/m^2 for diluted asphalt emulsion. Additional time for emulsion to break prior to placing hot mix is critical. Tack coat is also generally sprayed on adjacent pavements-runs placed at different times, curbs, parapets and barriers to assist in the bond between the HMAC and the concrete. It also helps reduce water infiltration at these joint locations.

Paver speed should be geared to mix production, delivery and compaction. Every effort should be made to keep a constant paver speed. Reference should be made to manufacturers recommended paver speeds and other operational requirements.

The number of rollers required for proper compaction is based on the square yardage placed. Roller speed should be limited to 3 ml/h or 5 km/h. Many projects are compacted with three rollers: a breakdown roller, a compaction roller and a finish roller.

Readers are referred to Table 14.1 developed by the Federal Aviation Administration (FAA), for a brief summary on causes associated with various problems that can be encountered in a HMAC construction project.

14.9 Summary

This chapter highlights and draws the attention of the reader to aspects of a typical highway construction project. Some literatures and references are given in the reading list for further information. It is strongly recommended that the reader reviews these additional literatures on specific topics of interest and refer to local practices and experience.

TABLE 14.1 Causes Associated with Problems Encountered in HMAC Construction Project

Problem	Fluctuating head of material	Feeder screws overloaded	Finisher stopped too fast	Too much lead crown in screed	Too little lead crown in screed	Overcorrecting thickness control screws	Excessive play in screed mechanical connection	Screed riding on lift cylinders	Screed plate worn out or warped	Screed plates not tight	Cold screed	Moldboard on strike off too low	Running hopper empty between loads	Feeder gates set incorrectly	Kicker screws worn out or mounted incorrectly	Incorrect nulling of screed	Screed starting blocks too short	Screed extensions installed incorrectly	Vibrators running too slow	Grade control mounted incorrectly	Grade control hunting (sensitivity too high)	Grade control want bouncing on reference	Grade reference inadequate	Sitting long period between loads	Improper joint overlap	Improper mat thickness for max. agg. size	Trucks bumping finisher	Truck holding brakes	Improper base preparation	Improper rolling operation
Wavy surface–short waves (ripples)	✓	✓	✓				✓	✓		✓										✓	✓	✓	✓							×
Wavy surface–long waves	✓	✓				✓	✓	✓					✓	✓						✓			✓	✓					×	
Mat tearing–full width			✓						✓		✓															×				
Mat tearing–center streaks					✓				✓		✓	✓			✓															
Mat tearing–outside streak				✓					✓		✓			✓																
Mat texture non-uniform	✓		✓					✓	✓	✓	✓		✓						✓					✓		×			×	
Screed marks							✓											✓												
Screed not responding to correction			✓				✓	✓		✓								✓		✓							×	×		
Auger shadows		✓																	✓							×				
Poor Pre-compaction			✓			✓		✓																		×			×	×
Poor longitudinal joint	✓						✓	✓		✓															✓					×
Poor transverse joint	✓							✓								✓	✓			✓	✓	✓								×
Transverse cracking																													×	×
Mat shoving under roller																													×	
Bleeding or fat spots in mat																														
Roller marks																													×	×
Poor mix compaction																													×	×

Checks indicate causes related to the paver. X's indicate other problems to be investigated.

Note: Many times a problem can be caused by more than one item. Therefore, it is important that each cause listed is eliminated to assure solving the problem.

Source: FAA AC 150/5370-14 Appendix 1.

15

Project Management in Highway Construction

Derek Walker
RMIT University
Melbourne, Australia

Arun Kumar
RMIT University
Melbourne, Australia

15.1   Project Management Overview

Any activity undertaken involves time and resources, and has an intended outcome. An activity could be construction of a physical infrastructure such as, a highway, a bridge, a building, other forms of infrastructure; and major and minor maintenance works related to such physical infrastructure. In order to achieve the desired outcome it is essential that the entire activity is undertaken in a professional manner and that it is completed within the allocated time and budget. It is also essential that the entire process and the outcome have minimal social and environmental impact, and the stakeholders impacted by it are consult.

This entire activity can be broadly categorized as a project. Cleland and Ireland (2002) have defined a project as "a combination of organizational resources pulled together to create something that did not previously exist and that will provide a performance capability in the design and execution of organizational strategies".

The management of a project right from its inception to completion can be called *Project Management*.

In every project there are deliverables. Davidson (2000) defines deliverables as "Something of value generated by a project management team as scheduled, to be offered to an authorizing party, a reviewing committee, client constituent, or other concerned party, often taking the form of a plan, report, prescribed procedure, product, or service".

The one who coordinates all activities and is accountable for it is called the *Project Manager*. Davidson (2000) defines Project manager as, "An individual who has the responsibility for overseeing all aspects of the day-to-day activities in pursuit of a project goal, including coordinating staff, allocating resources, managing the budget, and coordinating overall efforts to achieve a specific desired result".

15.1.1   Need for Project Management

Every project has a specific goal or objective to achieve. In order to achieve that objective there have to be several input parameters and processes to realize the end result. All these activities will require coordination and optimization to produce the desired outcome within time and budget with least (and agreed) social and environmental impact. Hence there is a need to manage all activities of a project.

The Life cycle of a typical highway construction and maintenance activity is shown in Figure 15.1.

15.1.2   Components of Project Management

The broad components of project management include:

Project Planning
Organizational Framework
Monitoring and Control
Risk Management
Human Resources management

These are described in detail in this chapter.

15.2   Project Planning

15.2.1   Planning in General

Project management texts often focus on general application of tools and techniques to be used in project planning presenting detailed descriptions of project planning techniques such as, work breakdown structures (WBSs), network techniques, modeling, simulation, and line of balance scheduling. Useful further references can be found in (Cleland, 1999; Turner, 1999; PMI, 2000; Walker, 2002; Winch, 2003).

In this section an explanation of the context and rationale behind planning is presented. This is followed by a brief outline of how to plan so that project managers can create their own plans. To learn more about the techniques and their nuances the project management and construction planning texts should be consulted. Planning is a thoughtful exercise and software packages primarily assist in the administration of data processing and presentation or communication of planning data. They can be used as valuable templates providing checklists to develop plans.

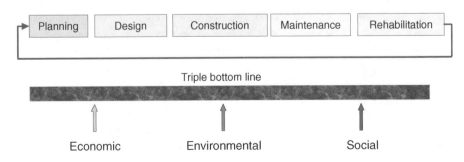

FIGURE 15.1   A typical life cycle of a highway project (Kumar, 2003).

In this section the broader context and fundamentals of planning are presented. This will also provide an appreciation of the various levels of planning.

15.2.2   Project Planning and the PMBOK

The project management body of knowledge (PMBOK) identifies nine areas of study: project integration management; project scope management; project time management; project cost management; project quality management; project human resource management; project communication management; project risk management; and project procurement management (PMI, 2000, p. 8). Many of these involve the fundamental concepts of planning.

Project integration management is a coordinating process. The inputs to this process are knowledge, historical and current information, assumptions and beliefs, and considerations of constraints and opportunities. Various techniques can be applied to these inputs, and the output of that exercise is, well-developed plans. These plans can be actioned and monitored to facilitate better control and direction and used as a means of communicating what needs to be done, what is and was being done, has already happened. So planning is the essential input to coordinating action.

Project scope management is the first step in planning projects. Project objectives need to be established, i.e., to construct a road/highway for vehicle movement (trucks, cars, bicycles, etc.) from location A to location B. This begins to shape the project brief and forms the basis for the project design. As the design becomes developed there is an opportunity to define the scope of the project. As more detailed information becomes available it provides more understanding to the scale and the complexity to the level of work. Various techniques and tools are used in managing this process. The use of two of these tools, the objectives breakdown structure and WBS are briefly outlined in this section. Scope management also involves a process of managing changes to the project design and scope of works. These are presented in detail in Turner (1999), Cleland (1999).

Project time management takes the knowledge inputs about the scope and complexity of the works and tasks and then defines activities to be undertaken and their characteristics, as well as methods to undertake these tasks. This part of the time planning work also involves using techniques to schedule activities.

Project cost management as defined in the PMBOK is somewhat misleading because it separately concentrates upon cost, though the PMBOK does also identify other resources as being encompassed within this section of project management. Planning should be viewed from a more integrated position to regard tasks as entities with characteristics or attributes that include resources (including cost) required to complete as well as their expected time duration and dependencies between activities. Further, tasks could be visualized in terms of a data file with many data fields that describe these attributes including work methods, constraints, risk profiles, and even less tangible characteristics such as impact upon stakeholders. For example, in road construction there may be an activity of trucking away excavation materials. Associated with this activity is an environmental impact — is the material contaminated? What is the community perception of this activity if materials spill onto pathways, roads, and nature strips of grass, etc.? In planning material removal with these characteristics in mind a totally different set of assumptions of the work methods, resources required, time and cost, etc. will emerge as part of the plan. Project cost and resource management, in addition to setting budgets and cost plans, is also about optimization of expenditure, monitoring costs as they become payable, minimizing waste, use of cost and other resource management information to be more effective, and a range of other resource management related aspects.

Project quality management is also a planning process. Objectives and requirements need to be established, a quality assurance plan developed, enacted and monitored to ensure that quality issues are being addressed (quality control). The PMBOK is also somewhat limited in its guidance on how to develop a quality culture and readers are advised that the many texts on quality management and organizational culture provide useful further reading. Quality management encompasses occupational health, safety, and environmental impact issues. Readers might gain further insights from the book chapter on addressing the quality culture issue by Keniger and Walker (2003).

The PMBOK identifies a section on project communication management that is closely linked to project planning. Plans need to be appropriately communicated for people to understand what is required of them and how they fit in with others in delivering a project. Communication planning not only concerns what needs to be communicated but how it is to be communicated, the channels and media. Additionally the target audience of the communication needs to influence the plan and the nature of the communication designed to suit the needs of that target audience. We are seeing a plethora of communications media being used on construction sites these days with mobile phones and text messaging, portable computer devices and web cameras. Hardcopy forms of plan communications can be charts, graphs, flow diagrams, lists, descriptive text as well as symbols.

Project risk management is very closely linked to project planning. Risk need to be identified, their potential impact assessed and mitigation strategies developed to minimize negative impact. There is also a branch of risk management that involves calculation of expected impact as well as how to deal with risks. A good text that can be referred to for this is Ward (1997).

Finally, the PMBOK recognizes procurement management as its ninth project management process. This involves developing strategies and plans to actually source required resources and obtaining them. It also concerns contractual issues and contracts administration. There is a wide body of literature that deals with procurement and there are many options that can be chosen to procure projects or for a main contractor to subcontract out (outsource) parts of the project. A chapter in a book on procurement provides a useful reference that readers may wish to review to obtain a better understanding of this part of project management (Walker and Hampson, 2003).

The key issue that this section addresses is that project management planning involves more than just scheduling activities or developing a bar chart to indicate a time line. It involves deep understanding of the nature of the work to be undertaken that integrates scope management to identify activities to be planned for. Cost and resource management aspects then come into play for planning what is required (resources) and how much to budget for these (cost management) as well as how to procure these resources. In developing a plan of how to deliver tasks, work methods need to be decided upon that achieve a quality result (quality management) addressing health, safety, and environment issues as well as identifying, preparing for and planning to mitigate identified risks. This requires coordination so that plans are enacted — requiring plans to be monitored and recalibrated to ensure that objectives are achieved. These plans need to be effectively communicated and coordinated to integrate the various parts to be orchestrated into the whole. Construction work, while it uses a high amount of machinery and equipment, is ultimately driven and controlled by teams of people.

15.2.3 Project Planning Effectiveness and Flexibility of Action

In three major Australian studies of the reasons why some construction projects are completed more quickly than others that (Walker, 1996; Walker and Sidwell, 1996; Walker and Vines, 1997) the issue of planning flexibility through sound planning was found to be a core factor influencing construction time performance. Further qualitative studies into this issue d upon an in depth case study approach of three construction projects confirmed the importance of integrated planning, particularly the use of method statements to describe tasks in detail — what was to be done, when, with what resources and equipment and the nature of the work methods/technologies to be used (Walker et al., 1998; Shen and Walker, 2001; Walker and Shen, 2002).

To be effective, people need to be flexible in their application of plans because plans rarely work out exactly as planned. To be flexible people need to fully understand the plans they are trying to comply to so that they cannot only change direction when required but that they can change in the right direction to maintain their project goals. Thus there is a strong link between understanding and flexibility as shown in Figure 15.2. Level of project understanding is dependent upon:

- Understanding the project scope through a thorough design definition that allows a useful and relevant WBS;
- The degree of design definition — being fixed as far as possible;

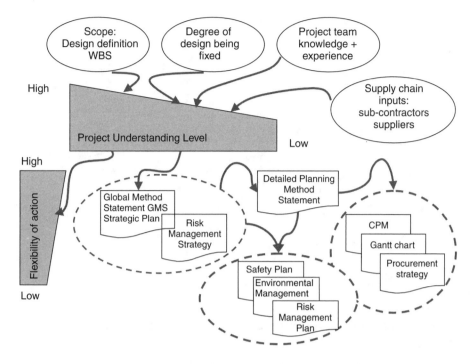

FIGURE 15.2   The link between project understanding and flexibility.

- The level of project team knowledge and experience;
- The level of supply chain knowledge input — suppliers and subcontractors that have expert and detailed tacit coalface knowledge that they can apply to solve the inevitable problems that arise.

Figure 15.2 indicates that higher the level of project understanding, the greater the ability to be flexible. Project understanding also affects the quality and depth of consideration that shapes the overall project strategy (the global method statement (GMS) that indicates how the project phases and major components are generally to be delivered) and the quality of consideration and planning that governs the risk management strategy. In many projects the GMS is also known as the business plan or sometimes the project initiation plan. The quality of the project strategy affects the quality of the detailed planning strategy that is expressed by a detailed method statement (DMS). These plans are used to develop safety, environmental management plans, and risk management plans. Detailed planning delivers a variety of detailed plans such as, a critical path method network (CPM) chart and Gantt (bar) charts. These are often produced as communication output from planning and scheduling software packages such as, Primavera or MS Project.

In a research study of a major freeway project in Australia to investigate the impact of planning on construction time performance the link between individual and group action d upon sound planning and deep understanding of the project plan became apparent (Walker and Shen, 2002). This reinforces the emphases placed in this chapter's section upon planning as understanding and preparation for flexible action.

Figure 15.3 illustrates that team/individual ability is very strongly influenced by understanding the complexity issues and options for working around problems and manifested risks as well as to have the core support and ability to exercise flexibility. Additionally, it also indicates that team and individual commitment is also very important. This motivation was driven by the team's internal support and encouragement to be flexible coupled with the organization's work-ethic culture and clear goals for construction time performance. This finding strengthens the emphasis on teams to be well-developed and coherent project leadership with high levels of engendered commitment.

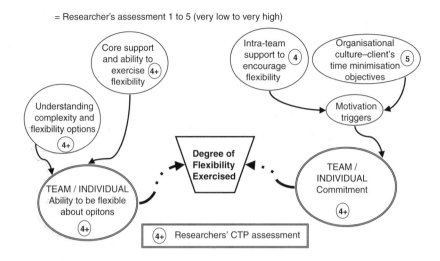

FIGURE 15.3   Flexibility and planning on an Australian freeway project.

15.2.4   Tools for Developing Project Understanding Through Project Time Planning

15.2.4.1   Tool 1 — Scoping Using Breakdown Structures

The research study example by Walker and Shen (2002) and Figure 15.3 clearly indicated the importance of commitment and understanding the project objectives. In the PMBOK, they describe a Scope Statement as:

"The scope statement provides a documented basis for making future project decisions and for confirming or developing common understanding of project scope among the stakeholders. As the project progresses, the scope statement may need to be revised or refined to reflect approved changes to the scope of the project. The scope statement should include, either directly or by reference to other documents: the project justification; the project product; project deliverables; and project objectives. It should provide documented supporting detail and a scope management plan" (PMI, 2000, p. 56). The principle of breakdown structure format is rational and simple. The idea is merely to take a complex arrangement, split it into simpler components and then reassemble these back into a coherent whole. Engineers feel comfortable with this as it appeals to their rational sense of order. Two examples of organizational structure are presented below: objectives breakdown structure and the WBS.

Figure 15.4 illustrates an objectives breakdown structure that indicates the multiple and complex nature of projects such as highways. The purpose of these projects is often not well understood by project participants. If the full scale of their potential impact for improving the life of others were to be fully appreciated then all parties may feel more motivated, engaged, and committed to project success. This example may be for a highway in a relatively sparsely populated region. The road could be providing commercial links or as an important transit way. It could be opening up an area to tourism while also being a transit route, or the highway could have strategic defense implications. Each of the subobjectives can be identified as indicated above and each sublevel then further split up into more discrete easily measured objectives. For example, ensuring economic value has a supply and demand side. There may be acceptable benchmark cost ranges for this type of project for example, $/kilometer of road at its coarsest level. Finer-grained benchmarks at the element level can be designed to be specific for roadway, bridges/tunnels, Mechanical and Electrical (M & E)[A1] services, etc. There would be a cost benefit objective with the roadway, either saving costs to some stakeholders and/or creating further business. In Hong Kong and some other places in the world the development of a train line and station complex provides development opportunities for industry and commerce that generate significant taxes and license

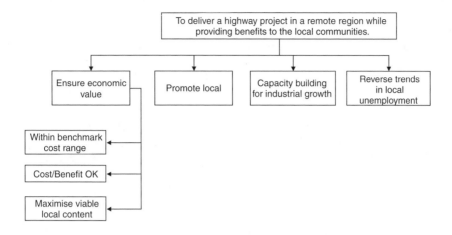

FIGURE 15.4 Objectives breakdown structure example.

revenues. Additionally, Figure 15.4 shows that a further consideration could concern maximizing viable economical and deliverable local content to stimulate the local economy without creating a painful boom-bust situation. You can see how a whole range of objectives can be defined, developed, and communicated and that this can have a significant impact on team motivation.

WBSs derive from the concept of a product breakdown structure (PBS) used in heavy engineering and manufacturing engineering industry sectors. They originally were used to define the elements of a product such as, a car, ship, airplane, or other engineered products. This was done by defining assemblies and subassemblies and then finally product components. Thus, each part of a product could be identified uniquely and data about the part recorded and used for planning, inventory control, and a host of other uses. The PBS idea is very attractive for developing into a WBS that can include work items or tasks rather than things stored in a warehouse. The PMBOK defines a WBS as "a deliverable-oriented grouping of project components that organizes and defines the total scope of the project; work not in the WBS is outside the scope of the project" (PMI, 2000, p. 59). Each level of the WBS describes the work in greater detail. One main issue that the project managers confront is to what level of details they should go down to?

Figure 15.5 provides an example of a WBS. The topmost level provides a general picture of what has to be done. Each successive level can only be defined and expanded upon as more detailed information

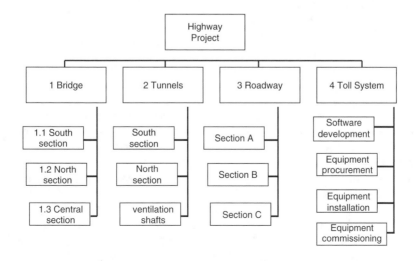

FIGURE 15.5 Work breakdown structure (WBS) highway project example.

emerges. It is common for large and multi-year projects to have a very sketchy WBS level of details for activities or tasks that are a long way ahead because of insufficient information about them or to avoid over-planning by wasting valuable planning time and energy on tasks that are likely to be subject to subsequent change. This is particularly true of components with subassembly configurations that have a number of options, but decisions on how to configure these are best left until some future time. WBS for top level and secondary levels make useful templates for planning, however, one must guard against avoiding innovation because a template exists that pushes one in a particular direction. In deciding upon a level of planning detail, it may be wise to stop when an activity can be described sufficiently for it to be measured in terms of: time to complete; quantities of resources required; and methods used to deliver that work package being described in the WBS. Note that each element can be uniquely identified through its WBS reference to avoid confusion with similar activities to be undertaken throughout the project.

15.2.4.2   Tool 2 — Method Statements

High-level WBS elements help us to define broad planning strategies. For example, WBS item 1.1 Bridge South Section may have a series of general elements flowing from it — abutments, foundations, etc. From these high-level views of the project scope one can start to construct a mental model of the project taking place and in developing this one can communicate it through a variety of communication channels to produce a GMS. This is very useful for joint problem solving where teams of experts in their own fields, can share knowledge and develop a feasible overall plan for the project so that it can be tested and validated and modified as needed to develop more detailed level plans. The types of communication devises used are often visual at this stage though written assumptions are frequently provided to supplement visual representations. Over past decades 3D-CAD modeling and simulation in the construction industry is increasingly being used, though this has been the norm for advanced manufacturing industries developing automobiles, aircraft, ships, weapon systems, etc. since the latter decades of the 20th century. The computer games industry has spurred animation and simulation development so that increasingly a GMS can substantially comprise an animation of the project being delivered as a kind of movie. This provides opportunities to fast-forward through the project, stop or look at slow-motion sequences to fully understand the more detailed technical problems involved in delivering these as indicated in the GMS.

The key purpose of a GMS is to provide an understanding of how the general logistical aspects of a project can be handled — such as, site layout; location of temporary, semi-permanent, or mobile equipment for resource handling; and to show how major components of the project can be constructed. With these assembly options being analyzed and tested beforehand, there is opportunity to better identify risks and likely problems and develop strategies to handle these before they occur so that, should they be needed, a range of preferred solutions can be advanced ready for application. A well-developed GMS can save considerable wasted time, effort, and resources. An example of part of a GMS is provided in Figure 15.6 that indicates the site layout for a commercial project. It indicates the safe crane radius reach to plan materials movement. Also it shows temporary material stores and access points as well as materials hoists.

This is a somewhat a simple example, however, a GMS plan could comprise many visual representations of this kind. At a more high-level component assembly level the GMS may illustrate how the work could proceed. Figure 15.7 illustrates a piling and excavation sequence.

The scope of this chapter is limited and many Engineers would have undertaken detailed study units in engineering equipment and methods and so the aim of introducing the above examples of a GMS was to re-acquaint readers with their existing extensive knowledge in this area to stress its application for construction planning.

Once a WBS work package task item has been defined, one can start to develop a DMS. A DMS defines a work package item in such sufficient detail that it can be used as an authoritative guide to be followed or improved upon. Again the automotive industry provides some useful examples of the DMS. For example, two articles in the Harvard Business Review (Sobek et al., 1998; Spear and Bowen, 1999), describe

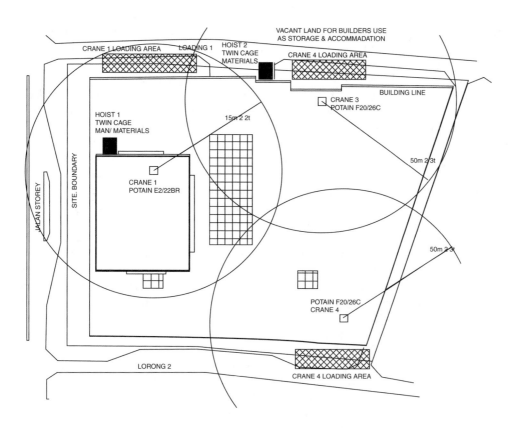

FIGURE 15.6 Site layout plan as part of a GMS.

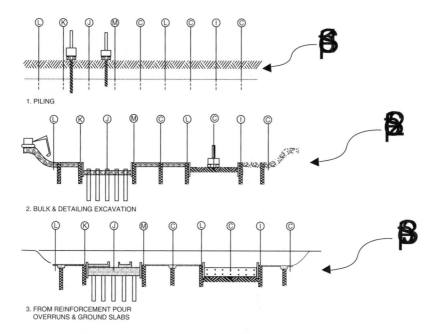

FIGURE 15.7 Piling and excavation sequence as part of a GMS.

Toyota's explicit process whereby work task definitions that carried knowledge about best practice are embedded in a database of work activities to be shared by all workers linked to that system. The first rule that Toyota worked by was that all work should be highly specified as to content, sequence, timing, and outcome. The "how-to" knowledge that in the construction industry is so tacitly held by workers is made explicit under the Toyota approach through the systematic use of a DMS. These centralized work report descriptions allow shared knowledge and conversion of tacit knowledge into valuable explicit knowledge that not only can be re-used, being a template for future use for what is probably a highly repeatable activity on many construction projects, but also enhances understanding of the nature of the activity.

In general, what might a DMS look like? First, it would link to the WBS via the WBS identifier. This would give it a unique identifier. There would need to be a brief activity (or work package) description. Also there needs to be a broad description of scope in measurable terms so that this data can be used to estimate likely time duration for the activity. The assumed production rate should be suggested to justify the time estimation for the activity. The estimated duration should be stated because this is a critical piece of data for scheduling. In the example indicated in Table 15.1 it can be noticed that a production rate of 2 cm/h is assumed and the quantity as defined from the scope is 15 cubic meters. Therefore, the time would be 15/2 or 7 h and that could be reasonably rounded up to one full day for that task. Work methods need to be specified to answer the important "how?" question of the way that the work will be undertaken and to be checked against risk, safety, environmental impact, and other QM issues. Resources need to be specified if the DMS is to be linked into a resource planning system to optimize workflow and resource management. Additionally, specifying the resources required is necessary for good time planning. The logical dependency of the activity should also be specified, i.e., this cannot happen before X and its completion links to Y, etc. This will help when developing a network of interlinked activities such as a CPM network. This provides added benefit from a tacit to explicit knowledge conversion that is all too rare in the construction industry and undoubtedly causes much rework. Many people get defensive about time or cost estimates and justifiably so, however, if the basis of their estimates can be clearly documented and understood then the process is transparent. With a transparent process it is easier to analyze why assumptions may not have eventuated or assess the impact of anticipated/unanticipated risks. A sample DMS is provided as an illustration in Table 15.1 below.

The above template may seem an onerous imposition on those involved in the planning process. This does represent a paperwork burden but the effort put in, if properly valued, can yield many benefits. First, a lot of time throughout a project is wasted in trying to re-plan activities because knowledge is lost. Often this is because different team members may be involved in the planning and the production phases and hence the knowledge never gets transferred. Alternatively, originators of plans simply forget how they originally planned to do the task in detail. So a system that captures this valuable knowledge is in fact a knowledge management initiative. Second, as intimated earlier, these DMS can form the basis of a template or default setting for similar activities either on the same or a different project. The approach we suggest accords with that adopted by Toyota (Sobek et al., 1998; Spear and Bowen, 1999). The greatest

TABLE 15.1 Detailed Method Statement Template Example

Activity	1-01-100-50
Description	Trench excavation for foundation works at South Bridge Section utility building K50
Scope	Excavate trenches and strip topsoil for foundations, dig trench for services (approx. 15 cubic meters).
Production Rate	Approx 2 cm/h
Duration	1 day
Work Methods	Machine dig with a laborer for sundry hand digging
Resources	Bobcat and truck (hired machine + driver), plus one laborer
Logic Sequence	Start after site establish + set out, finish before activity 1-01-100-090 concrete foundations, finish before activity 1-01-100-080 install in-ground services pipe work
Critical Assumptions	Dry soil, no rock, fine weather on ground access

benefit that we argue can accrue is that the GMS and DMS enhance project understanding and through that facilitates flexibility of action.

For full benefits to be realized the template as presented in Table 15.1 should be adopted. Currently the management culture of many construction industry organizations is one of focusing on the narrow task at hand and remaining "lean and mean" so spending time on recording DMS data such as the above is discouraged. This is unfortunate and short sighted; however, those that do commit themselves to this approach can reap competitive advantage. Competitive advantage can be gained through a cost advantage or developing a product/service perceived by its customers to offer unique differentiated benefits (Porter, 1985). This can occur from a cost point of view (saving wasted time, money, etc.) and/or from a competitive differentiation point of view. These organizations can become more efficient (delivering a cost advantage) and/or companies can become attuned to their knowledge assets and use them effectively to attract more favorable clients, more effective supply chain partners and more creative and innovative employees.

Now that work methods have been outlined to enable readers to understand the fundamental context of planning construction work as an intellectual exercise, Section 15.2.4.3 shows how this data can be used to produce project schedules and how these can be monitored to facilitate project control.

15.2.4.3 Tool 3 — Scheduling Techniques

The simplest schedule is the humble list. It would be useful to have it planned to minimize distance traveled or weight–distance measures of energy or time used. Lists can be logically ordered and they can be prioritized in some strategized manner. Lists can also evolve into flow charts. When flow charts contain or embed additional knowledge such as the logic of why certain activities are linked in a particular way, they begin to resemble network diagrams.

Figure 15.8 describes a project of seven activities. Activities are identified (A to G could be identified as numbers or alphanumeric symbols). A network indicates what is being planned (the activities) and their relationship to each other. In the DMS example in Table 15.1 the relationship of dependencies has been illustrated. In Figure 15.8 a default arrow indicates the dependency of one activity being totally complete before the following one may start. This finish-to-start dependency in Figure 15.8 assumes no time lag. There are other relationships depicted. The relationship A to E indicates that E cannot start until at least 3 time units after A finishes — note this *at least* means not a definitive 3 time units. There is no reason why other dependencies relationships may

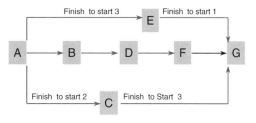

FIGURE 15.8 Network example.

not be described such a finish-to-finish or start-to-start or start-to-finish. Any logical relationship can be expressed in a network.

Figure 15.9 illustrates a similar network to indicate in a simplistic way in which network analysis can provide useful project scheduling data. For example, project managers and project teams would find it useful to ascertain not only the earliest possible times when activities can start and finish to maintain a project target completion date but also the latest start and finish times that can achieve this. It is often not wise to start all activities at the earliest opportunity because delaying some of them may result in a better resource demand optimization, this may be more convenient to various stakeholder and/or team members. Also by knowing the earliest and latest time it is possible to start or finish an activity you can have better control over your overall project plan.

Figure 15.9 shows a series of activities with more information represented on each activity of Figure 15.8, though there is little real difference. Activity E is shown as needing to start at least 3 time units after B starts. Activity F cannot be completed until at least 3 time units after activity C. Reasons are not evident and one would hope that these could be clear in a DMS if there is a record of one. The legend indicates the information presented in each activity box. The earliest time is derived from a forward pass

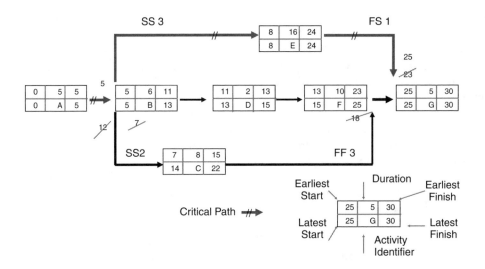

FIGURE 15.9 Network example indicating activity times.

calculation and by simply adding the duration of its successor plus any lag times. Activity B has an early start time of 5 because its predecessor has an earliest finish time of 5 and there is a zero time lag between its possible start. The latest start and finish times are derived from a backward pass chain that subtracts the duration from the latest times to arrive at the latest activity times and the similarly lags and relationship types between its successors' affect the value of the latest start and finish times. Scheduling software package manuals have an adequate description of the mathematical rules applied to the forward and backward pass calculations. The PMBOK also offers a brief summary on this (PMI, 2000, Chapter 6).

The type of network most widely known to civil engineers is most probably the CPM. It has a number of forms. In the 1950s the method was refined from a range of techniques developed by operation research scientist working on the Second World War (WW2) during the 1940s. Peter Morris has a book chapter that provides a useful history of project management from the WW2 years onwards (Morris, 1994). The refined version developed in the 1950s became known as Program Evaluation and Review Technique (PERT) and that name has remained even though the technique used then has modified several times and in fact PERT diagrams that are available through Microsoft Project is not a PERT diagram but a form of CPM called a Precedence Diagram Method (PDM). The old PERT approach was derived from a matrix algebra approach and calculated earliest and latest start and finish times on activity duration based on a probabilistic algorithm that used an estimate of the optimistic and pessimistic times. Because it was based on matrix algebra, it considered activities as events occurring between activity nodes and so it viewed activities as relationships. This was later refined to view activities as the nodes themselves and this solved some messy problems in the way that activity times were calculated. This approach became known as the PDM and by the early 1970s this predominated and also people tended to drop the activity duration probability profile and just allocate a single value time duration. This dropped the stochastic nature of the calculations and it did lose some added value information that is beyond the scope of this chapter to detail.

There are several features of this figure worth noting. First, there is an indication of a critical path. This is an important concept in scheduling. The critical path is the chain of activities that cannot be delayed without delaying the overall planned project duration. This chain has no leeway and it can be said to have zero "float" or "slack" time. When reading a network plan one generally looks for the critical path and focus on the activities in the chain indicated as being on that critical path. There are usually a number of these critical paths in any program. The useful part of knowing this is that one can be assured that focusing on these activities to ensure that they are conducted to plan is a wise way of managing the project because any delays in planned time for that chain, as a whole, will result in a time overrun.

Another aspect that is important is the concept of total float (or slack). For example, looking at Figure 15.9 for the chain of activities B, D, and F. B has an earliest start of 5 units and a latest finish of 25 units — that is a 20 time unit time to complete this chain of activities. The durations of these activities are 6, 2, and 10 units, respectively, and there are no lag times between finish to start of two activities, so that the budgeted time is 18 time units. This means that $25 - 18 = 7$ time units are floating or slack in the chain of those activities. This kind of information is very useful because if any of those activities were to slip in duration for some reason, one would not be unduly concerned provided that the total slippage was not greater that 7 time units.

Activity data represented in Figures 15.8 and 15.9 are limited. The DMS for example could pass data on resource requirements and cost. This data can be used in scheduling in a way similar to the time planning networks. If the time blocks for activities are known and the profile of cost expenditure or resource usage can be estimated then a network that uses this information can produce cash flow estimates/plans and resource histograms. These can be effectively used for management of the project. When this information becomes available project managers can plan to obviate problems associated with cash flow deficits that could adversely affect the organization. Similarly, knowing likely resource demand profiles, as illustrated by histograms for example, can point to potential bottlenecks in resource demands and supply deficits so that network information can be used to pre-empt problems. This is another example of how understanding the project and its demands more fully can enable teams to adjust their plans to flexibly cope with foreseen likely problems (risks).

There are a number of useful techniques that can be applied to network planning data that project managers can use to optimize and fine-tune their plans. One is resource leveling or resource smoothing. This entails modeling resource-use profiles using network analysis data. Total float can be used as a resource to know how to delay various activities to either smooth demand or to level activities and revisit assumptions of activity DMS to overcome potential foreseen problems before they occur. In a similar way when it becomes obvious from the network plan that a project target time cannot be achieved then the program network can be "crashed," that is revisited to seek out ways to flexibly change constraints, apply changed level of resources, re-sequence activities, and use a range of other measures to overcome foreseen problems before they occur. That is the purpose of planning. One cannot predict the future but one can model a range of potential futures that are credible and close to what is likely to happen given the assumptions and circumstances that describe the activities identified. By constructing such a model one can test different scenarios and undertake "what if" analysis to get ideas about how identified risks can be overcome. In doing this one gains a deeper understanding of the project and can prepare with a range of possible actions that one can choose to take.

Critical path analysis is not the only planning tool available. The return of stochastic approaches to planning has spawned a range of simulation tools. The most common is the Monte Carlo Simulation add on to most CPM software like Primavera and MS Project. These actually return that approach back to its original PERT form. Other simulation approaches model strings of activities, each with a probability profile of time duration, and then provide the users with an estimate of a range of project duration as well as its likelihood for that time estimate. The greatest value in this type of approach is that they often come with graphical interfaces that provide animation of the processes. This is useful because one can actually see any bottlenecks developing and the way that they subside. Such insights allow planner to try out numerous options of virtually building a project and, through varying the assumptions, to far better understand the outcomes and therefore the project. This means that they can pre-empt potential problems such as, material handling bottlenecks, resource shortage, etc. and can develop more reliable ways to obviate these problems than trying to do this "on the run" when faced with a crisis. The military uses highly sophisticated simulation tools for operational planning that the construction industry could make very profitable use of. Tools on the market can offer many similar features that the ones the military use. A good summary of project scheduling and monitoring current research status can be found on Ahuja and Thiruvengadam (2004).

The research into construction time performance indicated earlier supports the hypothesis that teams that spend sufficient energies simulating and testing their project plan, using models such as those described in this chapter, achieve far better construction time performance than those that do not. All this management focus depends upon data and projections based upon defining project scope at the broad level right down to the detailed level.

15.3 Organizational Framework

For any project to be effective, to deliver its intended outcomes efficiently and effectively it is desirable that individuals have clear responsibilities and accountabilities. All have to operate within an efficient organizational framework. Various activities such as, technical, financial, human resource management, procurement, contract management, monitoring and control, reporting etc. must operate within this framework.

The organizational structure for the project is the framework under which people work to achieve the most efficient operation for the project. The primary kinds of organizational structures are: Functional, Matrix, Pure Projects, and a Combination of the above (Kumar, 1998).

The *functional (traditional) organizational structure* has clear lines of operation based on the tasks or functions the groups perform. It has its advantages and constraints. Such structures are generally practiced in large organizations. Kerzner (2003) (pp. 92–93) has summarized the advantages and disadvantages of the traditional organization as below:

Advantages:

- Easier budgeting and cost control
- Better technical control
- Flexibility in the use of human resource and a broad base to work with
- Continuity in the functional disciplines: policies, procedures, and lines of responsibility are easily defined and understandable
- Control over personnel as each employee has only one person to report to
- Communication channels vertical and well established
- Quick reaction capacity.

Disadvantages:

- No one individual directly responsible for the total project
- Does not provide the project-oriented emphasis necessary to accomplish the project tasks
- Coordination becomes complex and additional lead lime is required for approvals and decisions
- Decisions normally favor the strongest functional group
- No customer focal point
- Response to customer needs is slow
- Difficulty in pinpointing responsibility
- Motivation and innovation are decreased
- Ideas tend to be functionally oriented with little regard for ongoing projects.

The *matrix organization structure* is more flexible in its operation, where individuals are drawn from line operations to work on a given project, which is managed by the project manager. Such a structure provides considerable flexibility and optimizes human resource.

Kerzner (2003) (pp. 105–106) has summarized the advantages and disadvantages of a pure matrix organizational form. Some of these are:

Advantages:

- The project manager maintains maximum project control (through the line managers) over all resources, including cost and personnel
- Policies and procedures can be set-up independently for each project, provided that they do not contradict company policies and procedures

- Project manager has the authority to commit company resources, provided that scheduling does not cause conflicts with other projects
- Rapid responses are possible to change, conflict resolution and project needs (as technology or schedule)
- The functional organizations exist primarily as support for the project
- Each person returns after project completion
- By sharing key personnel, project cost can be minimized
- Strong technical base and complex problems can be addressed
- Conflicts are minimal
- Better balance among time, cost, and performance.

Disadvantages:

- Multidimensional information and workflow
- Dual reporting
- Continuously changing priorities
- Management goals different from project goals
- Potential for conflict
- Difficulty in monitoring and control
- Functional managers may be biased according to their own set of priorities
- Balance of powers between functional and project organizations must be watched
- Balance of time, cost, and performance must be monitored.

A pure project structure is generally used for large projects where individuals are drawn from large organizations and are assigned to the projects for the duration of the project, say, 2 to 3 years.

Therefore, during the processes of setting up the plan the Project Manager has to make decisions as to what kind of type of Project structure, which would be best, suited to achieve the outcomes of the project.

For further reading books by Cleland and Ireland (2002) and Kerzner (2003) are recommended.

15.4   Monitoring and Control

15.4.1   Project Time Management — Monitoring

The need for monitoring is clearly necessary. An action that follows a plan should be evaluated during its operation to check if it is indeed going according to plan and what the consequences of not going to plan may be. Further, plans are constantly passing their use-by date because circumstances change. Assumptions made are found to be inappropriate or not matching the reality of a situation. Initial plans may be of poor quality and need to be fine-tuned as better information becomes available. Plans are not predictions of the future; nobody knows exactly what the future holds. So monitoring and recalibration of plans are essential to good project management. Monitoring also provides the necessary feedback to improve learning and experience and so it is not wasted effort but a value-adding activity.

In this section five tools used for monitoring using networks, bar charts, trend graphs, "S" curves, and Earned Value are illustrated. These show how they can be used to communicate progress and illustrate progress implications.

Monitoring a program first requires undertaking a survey of completed work and assessing progress on work under way using the program as a point of reference. The program could be derived using any of the previously mentioned techniques. One can compare progress achieved against the anticipated progress by the progress check (monitoring) date on the charts; this date is often referred to as the project status date or "time-now" date. One can then see from monitoring information, which activities are on schedule, ahead of schedule, and behind schedule. One can also infer from monitoring information what trends are developing and from one's own knowledge of the logic of the plan (particularly where a clear method statement exists or when those undertaking direct supervision have been well briefed regarding the planning assumptions) if these trends are important or not.

Recording progress, checking progress achieved against progress anticipated, i.e., *monitoring*, allows one to concentrate on important problems and let minor problems take a lower priority for action. Monitoring allows one to manage by exception. Steps to be taken when monitoring follow:

1. Refer to the current plan and list all items that one expect to be either complete or under way;
2. Make a physical check of progress by walking around the site or other work place and record which of those activities are complete and which are partially complete;
3. Make an assessment of the remaining duration of those activities in progress to determine how much time they require to be completed (estimate the time duration remaining or assess the percentage complete if the original times estimate of the activity duration is still valid);
4. Make an assessment of the program logic for those incomplete activities, have their relationships to other activities changed;
5. Mark up progress on the current program and redraw if necessary. One may have to substantially recalibrate the plan based on current information from progress check; and
6. Inform those in a position to take action on the plan of the results so that a decision can be made on what action is appropriate.

Monitoring sets the stage for management action based on current rather than outdated information. Progress check intervals should be kept appropriate to the currency and criticality of the planning level. If one is planning activities in terms of shifts then the shifts should be monitored. If planned in terms of workdays then one should monitor on daily or weekly basis, or monitor critical activities on a daily basis and the remaining activities on a weekly basis. These days many project sites have web cameras installed that make it easier for one to gauge the rate of progress and whether rapid progress is happening that demands more frequent monitoring. Increasingly, items of equipment and materials have barcodes embedded in or on them. This opens up possibilities to scan them enabling real-time updates of plans to be undertaken. In the late 1990s at least one of the major Japanese contractors in Singapore was undertaking trails of using barcode technology for planning and control.

15.4.1.1   Tool 1 — Monitoring with Bar Charts

Bar charts have a clear and understandable format for presenting planning information because planning information is often graphically conveyed. Most computerized planning tools use the bar chart as the output form generated to suit most users' needs. With some planning techniques (such as bar charts), one records progress by coloring in the hollow bars so that an immediate picture is given about progress (see Figure 15.10 below). Monitored programs indicate which activities are on schedule, ahead, or behind schedule.

When one wishes to undertake a progress check, one can use the bar chart by connecting colored thread between two pins located at the top and bottom of the bar chart along the monitoring date or

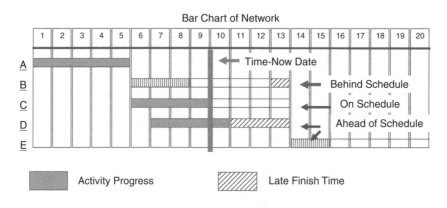

FIGURE 15.10   Bar chart indicating work complete and in progress.

(time-now) workday number. One can then look to the left of that line at what activities have not been colored in since the last monitoring check. These are the activities one needs to list for checking progress. One then undertakes the progress check, return to the bar chart and color or shade on the bars for those activities complete and partially color or shade in the bars of those activities under way.

One may need to extend the bars on some activities where the original estimated duration requires amendment. A trench may have been excavated on schedule, for example, subsequent heavy rain may have flooded the trench so that it needs to be pumped dry and the trench sides and base trimmed to remove mud and other debris. In this example one does not imply that the original time estimate was wrong, one merely records reality.

The progressed bar chart may need to be redrawn if circumstances have changed significantly. Perhaps there was an opportunity to undertake work out of planned sequence, or as in the case illustrated in Figure 15.10 planned progress was not achieved and activity durations were extended. The effect of such changes as these can alter the logic of the plan. It is important that the plan (bar chart or network) be current from both the progress and logic points of view. Network analysis techniques can produce output data that can then be used for producing monitoring information: bar charts, trend diagrams, "S" curves, and Earned Value Diagrams.

15.4.1.2   Tool 2 — Monitoring with Network Diagrams

One performs the same steps as that outlined above for monitoring bar charts except that the "time-now" or monitoring date line may not be a vertical line (unless the network was drawn to a time scale). With networks one lists all incomplete activities with an early or late finish less than the monitoring date. The check of work is undertaken in the same way, as when monitoring bar charts, however, updating the logic and redrawing the network may be easier than bar charts. Figures 15.11 and 15.12 illustrate a simple update of a network, each activity box has the same data representation as in Figure 15.9.

Figures 15.11 and 15.12 illustrate a network that has been monitored. In step 1, completed activities are crossed diagonally left to right and right to left while activities in progress are crossed only left to right. The remaining duration for activities is recorded to indicate the status of the project at the time-now date. In step 2 the network has been re-calculated from the time-now date using remaining duration times. It is to be noted how the forward pass calculations were undertaken; for activity "C" the EF is the time-now date + the remaining duration, i.e., $10 + 2 = 12$. Backward pass calculations are undertaken in the same manner as in the preparation of the initial program but using the remaining duration, e.g., $LS = LF − $ Remaining Duration. It should be remembered that some activities will *have* actually started and therefore there will be an Actual Start rather than Earliest Start date.

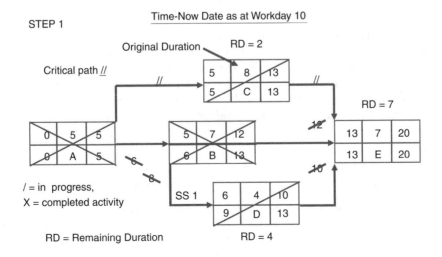

FIGURE 15.11   Step 1 of A PDM network.

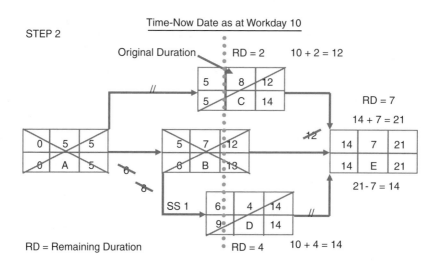

FIGURE 15.12   Step 2 of A PDM network.

The critical path may change as a result of the actual progress of the project (as illustrated in Figure 15.12 Step 2). The whole point of undertaking monitoring and a review of the plan is to establish what the current position is, and what impact (potential or real) it has on the plan. This monitoring system is a decision support tool to assist in deciding what control action (if any) is required.

It is fortunate that there are many PC based planning software packages allowing data selection and sorting, also allowing bar charts and/or check lists to be produced. Thus, for example, all incomplete activities can be selected with an early start date of less than X (where X is the monitoring date) sorted in any manner one requires (e.g., by project section and subsection). The technology allows networks to be prepared and amended quickly and also used to produce checklists of activities for monitoring. For example, one can generate a list or bar chart that has activities ordered in the physical direction that one would use to walk around to check progress, making it quicker and easier for one to monitor activities. Using this kind of software for network analysis has had a marked effect on the ability to closely monitor projects. Updated bar charts can also be readily produced enabling timely and accurate information to be available for decision making.

15.4.1.3   Tool 3 — Monitoring with Trend Diagrams

Key trend analysis can be used to monitor progress and provide information that can lead to the rectification of adverse trends. It can help in the planning process by alerting management to adverse trends and providing project status information in the form of trend diagrams that can then be used to help indicate how an adverse trend can be reversed. In any large network of activities it is possible to highlight critical activities. These activities must be closely watched to ensure that they do not "slip" and cause delay on the project. Monitoring allows to answer four questions, namely:

1. How far has one got?
2. How far did one anticipate getting?
3. How fast or slow is one currently going?
4. How fast or slow does one need to go to achieve desired objective?

The key trend method is an extension of normal monitoring procedures. It is used to highlight and closely monitor specific activities or groups of activities and map out planned progress for those tasks, for example, a group of critical activities for road construction. A picture of this map is produced on a time scale, and then progress is charted at regular and appropriate intervals. One can then project from the observed current progress a trend line to indicate required progress to maintain desired time objectives.

This method helps to inform management what progress has been made, what progress should have been made and to what extent a program must be "crashed" to bring the project back in line with the program. This helps management focus attention on key activities that might otherwise not receive sufficient attention. Critical activities that can be grouped together for monitoring purposes in this manner could include a concrete lift shaft core or typical floor slabs in a high-rise building or sections on a roadway or bridge.

Figure 15.13 illustrates a key trend where lack of progress is evident. Figure 15.14 illustrates a trend where the work package being trend graphed appears to be likely to finish ahead of program. By trend graphing a number of significant work packages or elements of the work, a broad picture of progress emerges. The number of "key trends" to be charted on a project will depend upon the number of activities on the critical path, the complexity of the project and the appropriateness of close monitoring of this kind. The key trend method provides an adjunct to planning and monitoring tools.

Trend graphs may be prepared in a number of forms. Often, they incorporate an image of the element of work being monitored as illustrated in Figure 15.14.

A key trend diagram is meant to be used to track and monitor key areas that realistically provide an indicator to the overall project or part of a project. The first step is, therefore, to identify key packages of work. From an analysis of the network, one will be able to identify several critical paths or perhaps

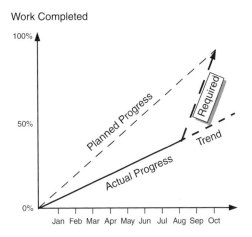

FIGURE 15.13 Key trend analysis diagram (behind schedule).

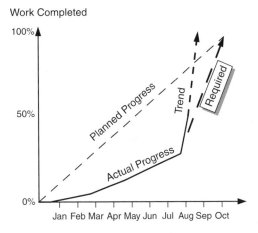

FIGURE 15.14 Key trend analysis diagram (ahead of schedule).

one critical path and several near-critical paths. Generally with high rise construction, the structural frame, services completion, facade treatment, and finishes will form key indicators. In housing projects, frame, weatherproofing to lock-up stage, and finishes may be appropriate packages of work to identify.

The next step is to identify milestone dates for completion of subwork packages. For roadway construction, this may be sections of roadway or on a percentage complete basis, i.e., dates at which 25, 50, 75, 90, and 100% of the construction is expected to be complete.

A graph can be drawn to illustrate the anticipated achievement of planned construction. In Figures 15.13 and 15.14, where the plan indicates that it is falling behind program, the start of the trend work package (0% complete) is shown as January, by May half the work is planned to be complete and in October it is planned to be 100% complete.

Monitoring is undertaken on a regular basis to assess actual progress. At each progress review a trend line can be sketched to indicate anticipated progress. The graph can then be used to map out required progress to meet the target set and the slope of this line can be measured to provide an indicator of how actual performance must be modified to achieve the target established.

The means to achieve the modification to performance can then be investigated. This will require a critical appraisal of the reasons for the performance variance and an assessment of how existing conditions can be changed or whether it is desirable to modify currently committed resources.

The main advantage of this technique is that it encourages management by exception and critical appraisal of current resource commitments and working methods. The following is a list of what an appraisal of Figure 15.13 may reveal:

1. That resources are being focused on other areas to the detriment of the area studied (perhaps other key areas are well ahead of anticipated progress);
2. That the delay was client induced through design changes and that a "hold" may be in place on areas affecting current progress;
3. That a lack of documentation by design consultants is hindering progress and that they should take responsibility for the delay and increase their output to allow a satisfactory level of output to be achieved;
4. That recent industrial relations problems (perhaps industry-wide) has caused the trend and that all disputes are resolved and that lost time can be readily made up; and
5. That the plan needs to be "crashed" to totally review assumptions originally made in order that the plan can be amended to achieve the target.

A network analysis technique used in conjunction with a bar chart and/or key trend diagram forms a powerful combination with which to monitor a project and provide excellent information to support decision-making.

15.4.1.4   Tool 4 — Monitoring with Cost Information

Monitoring with cash flow information or "S" curve graphing is similar in concept to key trend analysis except that it is generally used to monitor the expenditure of costs. In measuring and monitoring costs one is only indirectly measuring and monitoring units of production. "S" curves are also used for monitoring progress and providing information that can be used to prompt rectification of adverse trends in the progress of a project. It is not, therefore, a time planning technique although it often uses planning information to assist in the planning of cash flow. They can help in the planning process by alerting management to adverse trends and indicate how they may be reversed.

Many studies on the rate of construction of projects have been made by measuring the rate in expenditure in dollars (which can be updated by use of the building cost index data). It is believed that given sufficient data samples, an average rate of expenditure for various categories of projects can be derived. These expenditure patterns have been fitted to a cumulative distribution curve ("S" curve). "S" curves tend to be used to either monitor the whole project or large parts of the project. Figure 15.15 illustrates an example of an "S" curve cumulative cost diagram.

The main problem associated with using this tool for monitoring is the danger of information users confusing the concept of productive progress with that of funds expenditure. Funds expenditure is merely a historical record of what was deemed to be expended; it is not an accurate statement of productive output and indeed may contain misallocation of expenditure through cost coding errors. Monitoring funds expended has a backward looking focus (though trend lines may be projected from the information produced). Productive progress monitoring has as its focus a measure of the production of work undertaken; it has a much more direct link between the measurement criteria and output actually achieved.

It is possible, and often happens in industry, that resources are poured into what seems to be

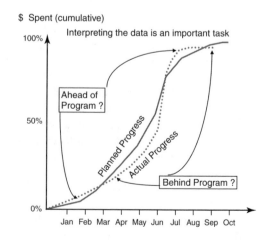

FIGURE 15.15 Typical "S" curve for project expenditure.

a bottomless pit. Let us look at the excavation of a trench as an example. Assume that you have estimated that trench excavation should consume $X or Y hours of labor. You may have started that excavation, and after 20% of the cost of the activity was expended, it was flooded. You may then have spent the remaining 80% of the cost budgeted for the activity, pumping it dry and trimming the sides and base of the trench. In the above example you may have expended 100% of your budget while completing less than 20% of the value of that work. The "S" curve would show that you are progressing on schedule but you would end up spending more resources than budgeted. It is only after that point that you will record an adverse trend. The "S" curve will give no indication on what other activities will be subsequently affected.

One disadvantage with "S" curves is that it indicates historical information in a way that may mask underlying problems. "S" curves need to reflect *actual* work complete as well as *budgeted* work complete so that the scheduled and budgeted variances become apparent. "S" curves can be used as a summary monitoring tool to give overview information to senior management.

The "S" curve for significant parts of the project or the entire project can be developed from an assessment of the construction costs to be expended over time. Segregating costs into labor, material, subcontract work, profit, and overhead can pose problems in constructing an accurate cash flow for monitoring because the invoices for materials, services, wages, and subcontractors are generally issued and paid at different lag times from the production of work, so this distorts the real picture of progress. Monitoring using "S" curve is a crude tool and less useful than that of key trend analysis (which is focused on productive units of expenditure and not cost expenditure).

Operational levels of management require more sophisticated tools allowing immediate focusing of attention on activities or logical groups of activities. There is also a further weakness in the use of this tool; one can argue that technological advances and market forces can distort the validity of stock standard "S" curve shapes over time. The theory of deriving a standard "S" curve shape is statistically sound but its practical application is debatable. "S" curves can be produced from computer packages allowing time and cost data to be entered and manipulated. These programs allow the program data to be graphically displayed, often showing expenditure and progress.

Figures 15.16 and 15.17 illustrate an "S" curve being constructed from a bar chart (one could just as easily draw one directly from a network diagram). It can be noted that a uniform expenditure of costs is

Barchart With Activity Costs $000's For Months 1 to 12

	1	2	3	4	5	6	7	8	9	10	11	12
A	2.0	2.0	1.5									
B		0.5	0.5									
C			10.5	11.0	12.0	12.0	9.0	6.0				
D			50.0	25.0	10.0							
E						8.0	8.0	8.0	8.0	8.0	8.0	
F					25.0	25.0	30.0	35.0	42.0	20.0	12.0	10.0
Ttl	2.0	2.5	62.5	36.0	47.0	45.0	47.0	49.0	50.0	28.0	20.0	10.0
Cum	2	4.5	67	103	150	195	242	291	341	369	389	399

FIGURE 15.16 Bar chart costed to produce an "S" curve.

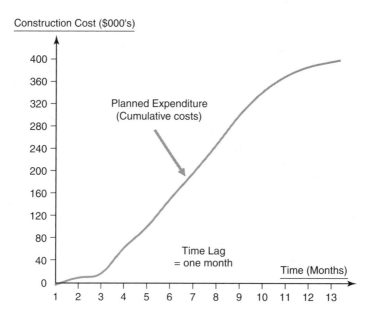

FIGURE 15.17   "S" curve generated from a bar chart.

not assumed on the bars in Figure 15.16. Assumed time lags may be built into the bar chart and cumulative rather than period-by-period expenditure drawn.

A second approach to preparing "S" curves is to cost out packages of work or activities (taking into account any time lags for payment). This information can then be used to draw a cumulative resource (cash) histogram. Figures 15.15 and 15.16 illustrate an "S" curve lagged by 1 month.

"S" curves are crude or indicative tools for monitoring and because of the time lag encountered in gaining access to funding commitments for work completed, the tool has limited use at the activity or macro-activity level. It is frequently used by the client as a coarse check on progress and is supplemented by other project status and monitoring systems.

15.4.1.5   Tool 5 — Monitoring with Earned Value

The U.S.A. materiel procurement establishment have sponsored and instigated a refinement of the "S" curve for monitoring projects; the concept that evolved was that of "Earned Value" analysis. With "Earned Value" analysis, planned expenditure is represented, as before, by a cumulative planned expenditure curve ("S" curve) and actual expenditure is also recorded and plotted. The extra ingredient is the inclusion of another measure, *the value of work authorized and undertaken*. Figure 15.18 illustrates an Earned Value diagram. The evolution from the "S" curve to this curve is based on the inclusion of a cumulative value of work done curve. Point (a) on the time-now line represents the cost of work actually expended at that time. Point (b) represents the planned expenditure at time-now. Point (c) represents the value of work performed at the time-now date line. Point (d) represents the planned expenditure for the actual value of work performed at the time-now date line. Point (e) represents the cost of the actual value of work performed at the time-now date line. This information is useful in deducing the likely amount of a budget over-run (or under-run) that represents a cost variance (a to e) and the part of that cost (d to e) that results from a time over-run (or under-run).

In Figure 15.18 the over budget amount is calculated from the difference between the value of work done and the actual expenditure of work authorized to be undertaken (a to c). Without information on Earned Value, the over-budget amount would have been assumed to be the difference between the actual

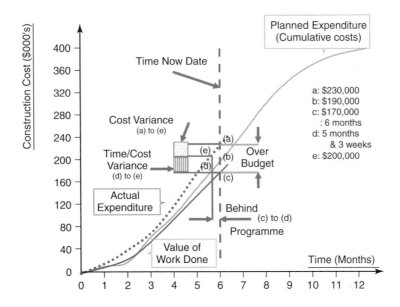

FIGURE 15.18 "Earned Value" curve example.

expenditure and the planned expenditure (a to b). This graph also indicates the likely time over-run or under-run (c to d). The value of work done (c) at the time-now date can be compared with the planned elapse time for that same value (d), this indicates the time over-run or under-run.

At the time-now date the following are illustrated in Figure 15.18:

1. Value of work completed (c) is $170,000;
2. Now the planned time for completing $170,000 worth of work is (d) 5 months 3 weeks;
3. The actual expenditure for planned expenditure at 5 months 3 weeks is shown at point (e) at $200,000;
4. The cost of the time delay is therefore (e) to (d) or $30,000;
5. The cost variance is therefore (a)−(e) or $30,000;
6. Total budget variance is a loss of (a) to (c) or $60,000.

Proponents of this approach maintain that this is a sound cost and time monitoring method giving a closer picture of reality than other methods. Critics of this system claim that the value of the information is less than the cost of obtaining that information. To gather the Earned Value information, one must not only obtain current valuation data of work in progress, but also a value that excludes unapproved variations to contracts let (i.e., unapproved extras or deductions from the contract sum). Good valuation procedures should, in any event, provide this information; this should not, therefore, represent extra effort being required of site management staff to gather.

The second cost curve (actual expenditure) should only reflect the costs of authorized work undertaken and should exclude unauthorized work. Actual expenditure often includes additional costs incurred. Elements of these additional costs include: greater waste than anticipated, re-work of faulty or poor quality work, and other items leading to a cost over-run.

The planned expenditure curve should represent a reasonably accurate forecast of cumulative expenditure costs over time.

The concept of "Earned Value" analysis appears to follow good practice, however, one should look at how this data is gathered to appreciate its weakness. It has been stated earlier that time lags exist between commitment of funds and the recording of the expenditure of those costs. Earned Value recording systems also have built-in time lags when gathering information and processing data.

Critics of the technique maintain that in practice one has two sets of data, both differently time-lagged, reflecting costs earned or expended at different times and therefore any comparison of the two sets of data is flawed.

Proponents of "Earned Value" analysis maintain that activities should have not only time but also resources and costs ascribed to them. When activities are monitored, cost incurred and value of work should also be ascribed to each activity to enable this data to be processed providing "Earned Value" information.

It is widely held by many proponents of "Earned Value" analysis that computer systems can use accounting information to integrate both cost and time status information into an integrated management information system (MIS). The planning and monitoring of time, resource, cost, and value for each activity, however, involves a great deal of extra work. Critics of this approach argue that network activities may not readily lend themselves to integration with cost coding systems used for cost accounting systems.

Take for example, the network activity "build wall". It may have been appropriate for the planner to consider a wall as an activity which includes wall framing, an outside skin of brickwork, and an inside skin of drywall plasterboard. From the accounting system's point of view it would be appropriate to split this activity into constituent component parts, coded in the accounting system by creditor or work package cost code. If this procedure is followed, it may be found that instead of dealing with a manageable network of say 200 activities, there could be 2000 activities generated — each time, resource, and cost loaded, and each requiring monitoring.

It can be seen that the result of being forced to split activities into inappropriately smaller activities (some of which may be completed in minutes), is probably going to result in the degeneration of the planning and monitoring system. All advantages of using a planning and monitoring system to provide control and sound management decision support information become a bureaucratic nightmare (due to inappropriate control systems being established) with predictable results.

15.5   Risk Management

The role of and need for risk management in project management and project planning has been briefly outlined earlier in this chapter. Robert Chapman is one of many leaders who have been studying this area in construction management risk for decades. He describes the overall process of risk management as being composed of risk analysis and risk management (Chapman, 2001). Chapman and Ward identify problems with the term *risk* as being focused on things that can go wrong and essentially having a perception of being confined to bad (potential) events (Ward and Chapman, 2003, p. 98). They prefer to focus on the term *uncertainly* as being more encompassing because it acknowledges a lack of knowledge and certainly regarding the feared events. This makes sense because once the word risk is raised in many quarters there is an expectation that there must be a plan to counter the risk despite the value of intelligence or knowledge that underpins its identification.

Every one is aware that, there is a constant risk that one could die from a range of events every second of our lives. It would be undesirable to have a risk management plan to attempt to meet all conceivable eventualities because then one would spend entire lives in fear and unproductive planning. Similarly, in project management there is a limit to the value of worrying about identifiable risks because often it is the unexpected that can have the most devastating effect. In contractual terms these are referred to *force majeure* or act of God — occuring once in a 100 year: storm, earthquake, terrorist attack, etc. Recognizing that *uncertainly* is a problem for planning is a positive step because it encourages one to remain flexible instead of being fixed upon risk management plans as providing an antidote.

Flexibility of action to counter problems is the best antidote to uncertainty as was stressed earlier in this chapter. This is also reliant upon efforts to reduce ignorance (or lack of knowledge) through a process of a combination of synthesis of existing knowledge and projecting and manipulating that knowledge to test hypotheses of how it may apply to unknown or uncertain situations. Indeed, that it was simulation is all about. It models a situation based upon firm assumptions and then introduces a range of possible variants some more likely to happen than others. The value of this exercise is not so much the mechanical exploration of case and effects of what may be billions of potential combinations of risks but the ability to discern a pattern. This pattern making and matching developed through investigating the sensitivity of one factor to an outcome develops a level of tacit knowledge that can later be deployed that is difficult to appreciate at the time. However, simulation exercises are time consuming and high-level intellect expensive and so, any way of narrowing the possible range of variables and characteristics to model, the easier the task becomes. Elements of the risk management approach and its methodology does help to identify key potential problems that can be investigated using simulation and other similar tools.

While one can and should be prepared to respect uncertainty one can take steps to reduce the levels of uncertainty through what is broadly understood as risk management processes. In this section the risk management processes are discussed and several tools as examples of how one can practically cope with identified risks are presented.

15.5.1   The Risk Management Process

The PMBOK describes six phases of the risk management process:

1. Risk management planning — deciding how to approach and plan the risk management activities for a project;
2. Risk Identification — determining which risks might affect the project and documenting their characteristics;
3. Qualitative Risk Analysis — performing a qualitative analysis of risks and conditions to prioritize their effects on project objectives;
4. Quantitative Risk Analysis — measuring the probability and consequences of risks and estimating their implications for project objectives;
5. Risk Response Planning — developing procedures and techniques to enhance opportunities and reduce threats to the project's objectives; and
6. Risk Monitoring and Control — monitoring residual risks, identifying new risks, executing risk reduction plans, and evaluating their effectiveness throughout the project life cycle (PMI, 2000, p. 127).

It is interesting to see the above in light of a time planning methodology; there are many common elements, specifically, the effort undertaken to understand the context of the project as a firm starting point for all action.

The British Standards version of the risk management process starts with five questions:

- What is at risk? — the context;
- What (and where) are the risks? — risk identification;
- What is known about them? — risk analysis;
- How important are they? — risk evaluation; and
- What should be done about them? — risk treatment (British Standards Institute, 2000, p. 2).

15.5.1.1   Understanding the Risk Context — What is the risk?

The risk context should be first developed in terms of the project objectives because these contain the assumptions that may be questioned. Indeed in the global and DMSs as well as the project's business plan, lie embedded assumptions that may be assumed to risk minimal when this may not be the case. An important risk that is often difficult to unearth is political risk be that national, local, or organizational. A valuable tool for understanding this level of context is to undertake stakeholder analysis. Stakeholders are those people who literally feel that they are entitled to have their stake in a project respected and considered in project plans. A useful easy-to-read summary from a project management perspective of this can be found in books by Cleland (1995), Briner et al. (1996), Pinto et al. (1998). Understanding the strength of potential impact of these stakeholders is also important in risk management. Understanding stakeholders and their value proposition for a project helps to develop an understanding of the political context of the project and its reasons. These are some of the many issues affecting the risk context.

The context relates to all of the aspects that were earlier discussed as part of project planning. It is for this reason that project planning and risk analysis were stated to be so inextricably linked.

15.5.1.2   Risk Identification — What (and/or Where) are the Risks?

A project risk pro-forma is a useful tool for identifying risks and starting to understand their potential impact. The risk management literature has many of these. A useful sources is the British Standards Institute, 2000 (Appendix E, page 21). Readers should acquaint themselves with the local standards because their location may present important risks not mentioned in other sources. The BS 6079-3:2000 for example identifies a range of common examples of general business risks including: human factors; environmental factors; legal; economic/financial; commercial; and technical and operational. Having now being exposed to tools such as the OBS and WBS one can see that it is possible to develop a company or project template of risks as a risk breakdown structure (RBS). If organizations make the investment in developing these along with associated knowledge about these risks then a fruitful foundation will be in place for effective risk management.

A useful tool that has been developed from the quality management literature that can be applied to risk management is the "fishbone diagram" often referred to as a "cause-and-effect" diagram. In this diagram an apparent cause of a problem is identified usually during a crisis or a simulation of a severe problem that warrants a lot of investigation. This problem is then challenged for likely causes if this were in fact not a cause but an effect, a symptom, the process can be repeated until a sound appreciation of the problem is gained and the sense is that saturation has occurred where each subsequent breakdown yields little new insights.

Figure 15.19 provides an example drawn from a construction project that relates to a lack of progress on plastering work, perhaps in utility offices and other facilities as part of highway engineering project. Note how the RBS comes into play here and how this tool unearths underlying assumptions and rigorously tests them. Also it is worth noting that in order to conduct this type of analysis one needs people with practical

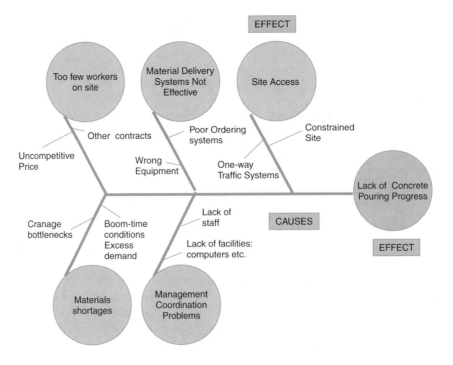

FIGURE 15.19   Fishbone or cause-and-effect diagram.

hands-on skills and tacit knowledge of the problem's context as well as strategic thinkers. These types of analyses can best be undertaken using cross-team and cross-hierarchy groups. Again, this may be viewed as a planning exercise, a risk management exercise, or even a knowledge management initiative.

A range of techniques can be used to identify risks. Generally through trying to make sense of the situation, a need for these will emerge. Figure 15.18 indicates that some levels of brainstorming is required as well as more formal research approaches such as finding out what current demand and supply levels may be or expert projects of these for the critical phase of the project lifecycle when this risk may have impact. The PMBOK suggests other tools and techniques such as, a Delphi technique, interviewing, checklists (similar to templates), assumptions analysis, and diagrams such as the cause and effect diagram (PMI, 2000, p. 132–133). It should be stressed that these are sense-making techniques to facilitate people to understand the project's context.

15.5.1.3 Risk Analysis — What is Known About the Risks?

The field of quantitative techniques for analyzing risk is a wide area of academic study well beyond the scope of this chapter. We will present only a sample of the most simple concepts, so readers are encouraged to seek further knowledge from risk management texts such as Ward (1997). The PMBOK also presents some techniques (PMI, 2000, Chapter 11). One approach or tool, the risk event tree, helps us to make a decision about which of several options to take, given a set of assumptions and judgments of the likely impact of the risk events and the probability of these occurring.

Risk impact should be assessed on the basis of the likelihood of the risk's occurrence and its potential consequences as indicated in Table 15.2 below (British Standards Institute, 2000, p. 12). Much of this assessment is judged using documented historical evidence from past histories of the risk events and of course direct (or indirect) experience. Modeling and experimentation may also be used and we have all seen examples of car crash testing, for example, where replicas of humans with all kinds of embedded measuring devices are placed in a car and the car driven into a concrete wall at various speeds to gauge the risk of various design practices of the car and road surface characteristics.

Scales of the likelihood of risk can be measured in descriptive terms such as, unlikely, possible, and likely or as an estimated percentage change of occurring, e.g., unlikely being the range 0–33%, possible being in the range 34 to 67% and likely being over 67% or other percentages may be deemed more appropriate. Similarly, degree of risk impact can be assessed in terms of low, medium, and high. This can lead to risk importance being judged as minimal, moderate, or significant.

Table 15.2, for example, shows a minimal risk impact being assessed for a risk being unlikely or possibly likely to occur with a low degree of assessed adverse consequences or a medium level of adverse consequences. These low, medium, or high descriptors could be linked to vignette descriptions if that were useful for types of risk, e.g., for a cost overrun risk this may be "costs exceed budget by up to 5%". There are a number of ways that you can populate the cells of the matrix described in Table 15.2, but the core message is that it should convey the severity of impact based upon likelihood and degree of adverse consequences. Naturally, because these are subjective and qualitative measures, they can hide bias and manipulation. However, one should not be lulled into believing that a quantitative approach is less likely to have inbuilt bias or capacity to be manipulated.

Another way of analyzing risk is in trying to understand its cause in numerical terms. This quantitative approach can be applied to decision tree approaches that allow estimates of decision point to be made. This is illustrated below with a simple example.

TABLE 15.2 Risk Analysis Assessment

Likelihood of Risk	Degree of Risk Impact		
	Minimal	Moderate	Significant
Likely	*Medium*	High	High
Possible	Low	*Medium*	High
Unlikely	Low	Low	*Medium*

For a solution to the site access risk or problem identified earlier, three approaches may be considered. One is to change the existing single site entrance to multiple points of entry but this will incur additional security. To simplify this example one need to only consider additional entry gate guards and that is estimated to cost an extra $300,000 within plus or minus 10%, if camera-only security surveillance and confusion about getting delivery authorization for material drop offs, etc. mean that any other option is not viable then the probability of the extra cost is 100%. Despite the additional guards the additional entry/exit access is assessed at having a problem of increased theft and pilfering taking place at an extra cost of $200,000 within plus or minus 5% and the chances of this are assessed at 10%. It is also accepted that to have dual entry/exit points will lead to confusion of deliveries and that this would result in double handling of materials with subsequent breakages and losses. This is estimated at an extra $60,000 plus or minus 5%, and the estimated probability of this is assessed at 30%. The probably cost of this action is thus assessed at most likely to be $340,000 with a range of between $371,000 and $318,000.

One can notice that each of the three options can be likewise assessed and the most suitable option adopted. Illustrating this example serves several purposes. First, it shows how risk assessment and planning are intertwined so one should only consider the chapter subsections as being convenient presentation devises rather than implying that risk management and planning are in any way discrete. Second, one may be thinking on how these dollar figures are estimated and also how the range of percentages are derived both for plus or minus ranges to the estimates and to the probability values. These are estimates only. An estimate is an intelligent guess based on assumptions and therefore is subject to bias and manipulation. So using a quantitative approach can be as subjective as using a qualitative approach. Both rely on judgment, experience and knowledge of past events, or good guesses. Risk assessment is a risky business. Finally it should be remembered that some risk can be controlled and others not.

15.5.1.4   Risk Evaluation — How Important are the Risks?

Risk evaluation is a process used to decide risk management priorities by evaluating and comparing the level of risk against predetermined standards, target risk levels or other criteria (British Standards Institute, 2000, p. 3).

Risk impact as being assessed as minimal, moderate, or significant has been discussed. Now what is to be done with these assessments? Clearly one needs to evaluate these in terms of the scale of potential damage to the expected project outcomes. Project teams do not have unlimited time or resources and so they must prioritize their attention and discard evaluation of some identified and assessed risks. Table 15.2 provides some guidance. Risks flagged as "high" warrant close scrutiny and priority 1 attention. Risks flagged as "medium" impact are worthy of consideration and would get a priority 2 attention level, and risks with a low risk impact assessment would either be set aside from immediate evaluation or logged and ignored. Maintaining a risk issues log of identified risks and their assessment is wise because circumstances change and what was considered insignificant can be affected by changed context to become moderate or significant.

Risk evaluation generally involves deeper analysis of the kind illustrated in Figure 15.20, but also includes the reasoning of embedded assumptions and both exploring and challenging these. This is where sensitivity analysis is a useful tool. Figure 15.20 only illustrated the first option being calculated but notice that it has been calculated as a range. There is likely to be overlap between offered solutions to obviate the risk for example, the low range of impact for providing additional security could overlap with the high range of increasing on-site stacking of materials. Thus assumptions need to be challenged to better understand the risk option's determining characteristics. Also, if risk evaluation is treated purely objectively as a numbers game, then opportunities for being innovative may be lost. In many cases even serious risks can be overcome by an innovative approach so we must be open to this regardless of the mathematical calculations and analysis being undertaken.

15.5.1.5   Risk Treatment — What Should Be Done About the Risks?

Once the previous parts of the risk management process have been undertaken it is possible to develop a risk event table to develop and communicate a recommended risk response (Table 15.3). Knowing

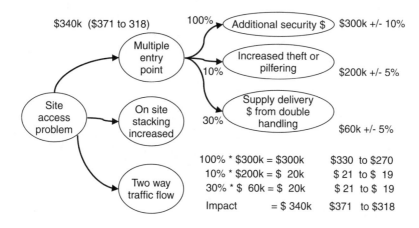

$340k ($371 to 318)   100% Additional security $ $300k +/- 10%

Multiple entry point

10%   Increased theft or pilfering $200k +/- 5%

Site access problem

On site stacking increased

30%   Supply delivery $ from double handling $60k +/- 5%

Two way traffic flow

100% * $300k = $300k $330 to $270
10% * $200k = $ 20k $ 21 to $ 19
30% * $ 60k = $ 20k $ 21 to $ 19
Impact   = $ 340k $371 to $318

FIGURE 15.20 Diagram of risk tree assessment for an identified risk event.

the level of controllability of the risk, the level of dependence on identified factors, consequences, probability of occurrence, and risk attitude of those dealing with the risk will help to shape responses that can be offered.

A useful tool in developing risk responses is to return to the fishbone or cause and effect diagram information and select through the risk assessment and evaluation candidates to focus upon (as for example illustrated in Figure 15.20). Having done that you can construct a "how–how" diagram that simply keeps posing the question "how can you do this" at each step. This leads to an action plan for recommendation. Figure 15.21 illustrates one of these.

Figure 15.21 steps us through the questions posed. The question how to increase concrete delivery performance is answered by three viable options. When asking the same question of each of these we can

TABLE 15.3 Risk Event Table

Risk Item	1	2	3	4
Identification	Water table problems	Poor soil for excavating	Poor soil for support	Underground services
Classification				
Level of control	high	very low	low	very low
Totally dependent		on geology	on geology	on past site and area's use
Partially dependent	on weather + on geology			
Independent				
Likely consequences	CRITICAL delays to ment excavation	Type of equipment required + $$ or unavailable	Review of support systems + time, + $$	Delays while locating + time + $$
Probability	High 80% due to planned start date	Low 10%	Medium 50%	Low 25%
Possible outcome	+30% to est $ + 50% to time	+25% to est $ + 25% to time	+100% to est $ + 50% to time	+10% to est $ + 100% to time
Attitude of those dealing with the risk	Transfer risk to client	Revise planned performance	Revise planned performance	Seek alternative options
Recommended response	Delay start of excavation by 3 weeks	Take additional soil bores	Take additional soil bores	Check all service details thoroughly — spend extra 1 week of research time

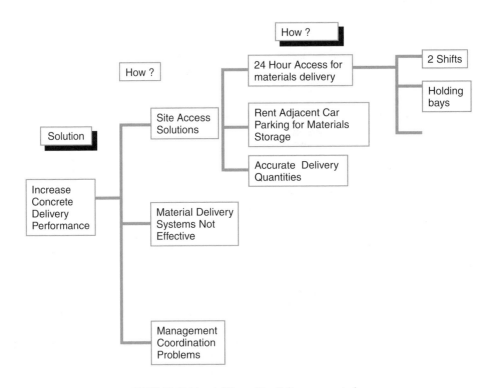

FIGURE 15.21   A "How–How" diagram example.

see that for site access solutions there are three viable options proposed and Figure 15.21 indicates that there are three solutions to that problem of which two are illustrated. This is a simple concept and a valuable tool.

Risks can be treated in a variety of ways. The key driver for the strategy adopted is that those best able to manage the risk should do so. That may, and should, result in a transfer of funds and/or resources to facilitate management of the risk. The options illustrated in Figure 15.22 are to retain the risk, reduce the risk, transfer the risk, or avoid the risk.

Retaining the risk involves absorbing the risk into the project plans and developing a contingency budget that is monitored and controlled just like any other budget (Figure 15.22).

Reducing the risk impact can be achieved by a combination of actions not limited to: spreading risks across groups of team members; minimizing the impact through developing risk contingency plans; seeking expert advice; investing in education and training about the risk (typically done for OHS and environmental risks); and mitigating the risk through design modification. This can involve developing innovation solutions that adhere to the project objectives and agreed outcomes.

Transferring risk includes: insuring against the risk; outsourcing or subcontracting to others that can better manage that risk; developing a strategy to be able to claim against those risks as part of the contractual arrangements or through other litigation; and finding a joint venture partner willing to bear these risks.

The final response is to simply avoid the risk through refusing it as a condition of contract or if that is not feasible, declining the contract.

One important aspect of dealing with risk is to recognize that there are uncertainties that are associated with any risk. Anything that will happen in the future is by definition not known in detail now. This is why it is important to document or ensure that assumptions made at the time of identifying, assessing, and evaluating risks. This is why it has been stressed that in the global and detail methods statements that assumptions should be made as explicit as is practicable. Also, that

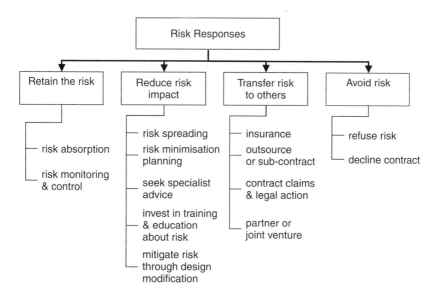

FIGURE 15.22   Risk response options.

uncertainty is about the unknown and that this unknown can be partially explored through analogy. Analogies or models can, as has been outlined earlier, use data from similar experiences in similar contexts, modeling can be used through simulation exercises.

Finally a risk management plan allocates responsibilities and outcomes. Table 15.4 illustrates a template for this.

Table 15.4 is only one format of many that can be adopted. The risk management system would most likely be web-enabled and linked to a database that stores data about identified risks. As each risk would be identified individually all data about it would be available to be presented in a report or on a computer screen. The main aim is to ensure that risk actions are sufficiently logged to allow them to be flagged for action when required. Table 15.4 indicates a brief description, the impact level, the action required by whom and when. There could be a check field for sign-off for the action by the supervisor of the persons taking the risk mitigation action; there could be a mechanism that allows all assumptions to be presented. Often, companies have their own internal systems and these vary in sophistication from a word document or spreadsheet file through to a fully integrated risk management system as part of project groupwide software.

15.5.1.6   Section Concluding Comments

Risk management is shown here to be integral to project planning. In this section some tools and techniques that can be applied have been outlined. Unlike time planning there are no dominant software tools such as, primavera or MS Project that construction companies can use for risk management.

The key to the process, as is the case with time and resource planning, is that plans are carefully analyzed and developed and that they are adequately monitored to ensure control.

TABLE 15.4   Risk Responsibility Schedule

Risk Identifier	Brief Description	Impact	Action Statement	By Whom by When
1-100-06	xxxxxxxxxxx	High	yyyyyyyyy	ZZ ABC

15.6   Human Resources Management

15.6.1   Introduction

Successful completion of any project depends on several factors. These include information technology, technical skills people possess and the management of human resources. People on the job play an important role to the success of the projects. It is essential to have good teamwork, leadership skills, staff motivation, delegation, conflict resolution, etc. for the success of the projects. Project managers of major civil engineering and other projects should have well-developed skills in human resources management as in the end a well motivated team with satisfied staff is a key ingredient in the success of projects.

Human behavior is one of the important factors to be considered by the project managers in the decision making process. Issues such as, teamwork and leadership style greatly influence the types and quality of decisions made by the project manager.

15.6.2   Leadership

Since one of the important elements to the success of the project is decision making the type of decision making largely depends on the style of leadership of the project manager. A leader is primarily one that best marries the goals of the project to the goals of individuals. There are several stages of decision making from autocratic where the manager takes all the decisions himself or herself to a pure democratic form where the manager gives the problem to the team and allows them to take the decision.

The project manager must possess the skills to use a style appropriate to the decision he or she is going to make.

As every individual is different the project manager should have well-developed skills to identify when to use the directive behavior and when to use the supportive behavior for individuals to achieve the best outcome. Therefore, the manager will have to choose when to use delegating, supporting, coaching, or directing style with various individuals.

A survey of the staff from different sectors of the industry (Public, consulting, and contracting) identified that the most important attributes they would like to see in their project managers are (Kumar, 1998):

- Leadership
- Communication

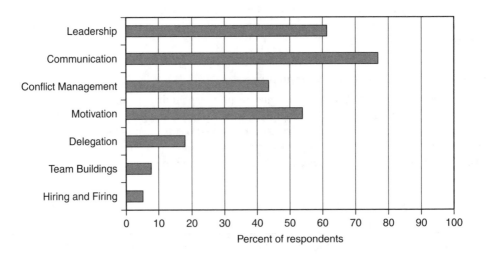

FIGURE 15.23   Most important skills you would like in your project manager — all sectors.

TABLE 15.5 Attributes staff *does not wish* to see in a Project Manager from a Human Resources Viewpoint (Kumar, 1998)

Public Sector	Contracting Industry	Consulting Industry
Not available at all times	Confronting	Lack of enthusiasm (vision)
Win/lose negotiation	Dictator	Aggressive or negative behavior
Aloofness	Tunnel vision	Inflexibility
Unwillingness to listen	Poor communication	Tunnel vision
Lacks communication	Poor team work	Impatient
Lack of commitment to team	Power seeker	Poor down communication
No people skills (negotiation etc.)	Arrogance	Soft — easy to influence
Directive	Unavailability	Poor listener
Task dictator	Rigidity/Stubbornness	Autocratic
Lacks decision making and is Aggressive	Unplanned activities	
Ego	Lack of appreciation	

- Conflict management
- Motivation.

These are presented in Figure 15.23.

A list of attributes that they did not wish to see in their project managers (Kumar, 1998) is presented in Table 15.5.

15.6.3   Working in Teams

It has been recognized that benefits of an effective team can provide much better outcomes than a cluster of individuals working for a project. For example, individuals working as a team will provide the following advantages to the outcome of the project.

- Working together more effective than alone
- Instrumental in management success
- Synergy as members change
- Collective experiences passed to new members
- Increased output
- Wide range of technical support.

Therefore, if the decisions are collective, individuals will own the outcome and therefore will be a lot more involved, enthusiastic, and energetic towards the successful outcome of the project. One of the key areas that the project manager has to address is to ensure that proper teams are formed and they operate effectively (Figure 15.23).

15.6.4   Conflict Resolution

When people are working together some conflict is bound to arise. This could be within the project, with clients or contractors. Moore (1996) (pp 60–61) has broadly described that conflicts could be broadly categorized under:

Relationship Conflicts
Data conflicts
Interest conflicts
Structural conflicts and Value Conflicts

Relationship conflicts, Moore (1996) explains, could be due to poor communication or misunderstanding, repetitive negative behavior, or strong emotions. The situation could be improved by better communication, positive problem solving attitude, clarification of perceptions and clear procedures, and ground rules.

Data conflicts (Moore, 1996) could be caused due to lack of information, different views on what is relevant, different interpretation, or different assessment procedures. Resolution can be reached by agreeing to what data is important, process of collection, its analysis, or by getting third party expert opinion.

Interest conflicts are due to actual or perceived competition over substantive interests, procedural, and psychological interests. To resolve one must focus on interests and not positions, and develop integrative solutions that address the needs of all parties, search for ways to expand options or resources, develop tradeoffs to satisfy interests of different interests (Moore, 1996).

Structural conflicts (Moore, 1996) could be due to unequal control, ownership, or distribution of resources, unequal power and authority, time constraints, destructive pattern of behavior or interaction. These could be addressed by defining and changing roles, relocating ownership or control of resources, development of a fair and acceptable decision making process, change time constraints, and replacing destructive behavior patterns.

Value conflicts, Moore (1996) explains, could be due to different criteria for evaluating ideas, intrinsically valuable goals or different ways of life. These could be addressed by defining problems in terms of value, allow parties to agree or disagree, search for goals that all parties share.

Overall, one must develop good listening skills, ability to appreciate other's viewpoint and most importantly address the problem and not people.

Recommended reading Baurch Bush and Folger (1994) and Moore (1996).

References

Ahuja, V. and Thiruvengadam, V., Project scheduling and monitoring: current research status, *Construct Innovation*, 4, 1, 19–31, 2004.

Baurch Bush, R.A. and Folger, J.A. 1994. *The Promise of Mediation*. Jossey-Bass Publishers, San Francisco.

Briner, W., Hastings, C., and Geddes, M. 1996. *Project Leadership*. Gower, Aldershot, UK.

British Standards Institute, 2000. *BS 6079-3:2000 Risk Management*. British Standards Institute, London.

Chapman, R.J., The controlling influences on effective risk identification and assessment for construction design management, *Int. J. Project Manage.*, 19, 2, 147–160, 2001.

Cleland, D.I., Leadership and the project management body of knowledge, *Int. J. Project Manage.*, 13, 2, 82–88, 1995.

Cleland, D.I. 1999. *Project Management Strategic Design and Implementation*. McGraw-Hill, Singapore.

Cleland, D.L. and Ireland, L.R. 2002. *Project Management: Strategic Design and Implementation, 4th Ed.* The McGraw Hill Companies, Inc.

Davidson, J. 2000. *10 Minute Guide to Project Management*. Pearson Education, Inc., USA.

Keniger, M. and Walker, D.H.T. 2003. *Developing a quality culture — project alliancing versus business as usual. Procurement Strategies: A Relationship Based Approach*, D.H.T. Walker and K.D. Hampson, eds., pp. 204–235. Blackwell Publishing, Oxford, chap. 8.

Kerzner, H. 2003. *Project Management: A Systems Approach to Planning, Scheduling, and Controlling, 8th Ed.* Wiley, New York.

Kumar, A. 1998. Management of human resources & success of projects, pp. 11–13. *First International Conference on New Information Technologies in Decision Making in Civil Engineering*, Montreal, Canada.

Kumar, A. 2003. Reflecting the triple bottom line in project appraisal and selection, *National Infrastructure Summit*, Sydney, Australia 5th and 6th August 2003 (organized by the Australian Financial Review).

Moore, C.W. 1996. *The Mediation Process: Practical Strategies for Resolving Conflict*. Jossey-Bass Publishers, San Francisco, pp. 60–61.

Morris, P.W.G. 1994. *The Management of Projects A New Model*. Thomas Telford, London.

Pinto, J.K., Thoms, P., Trailer, J.W., Palmer, T., and Govekar, M. 1998. *Project Leadership from Theory to Practice*. Project Management Institute, Sylva, NC.

PMI, 2000. *A Guide to the Project Management Body of Knowledge*. Project Management Institute, Sylva, NC, USA.

Porter, M.E. 1985. *Competitive Advantage: Creating and Sustaining Superior Performance*. The Free Press, New York.

Shen, Y.J. and Walker, D.H.T., Integrating OHS, EMS and QM with constructability principles when construction planning — a design & construct project case study, *TQM*, 13, 4, 247–259, 2001, MCB University Press, UK.

Sobek, D.K., Liker, J.K., and Ward, A.C., Another look at how Toyota integrates product development, *Harv. Bus. Rev.*, 76, 4, 36–49, 1998.

Spear, S. and Bowen, H.K., Decoding the DNA of the Toyota production system, *Harv. Bus. Rev.*, 77, 5, 97–106, 1999.

Turner, J.R. 1999. *The Handbook of Project-Based Management: Improving the Processes for Achieving Strategic Objectives*. McGraw-Hill, London, UK.

Walker, D.H.T., The contribution of the construction management team to good construction time performance — an Australian experience, *J. Construct. Procurement*, 2, 2, 4–18, 1996.

Walker, A. 2002. *Project Management in Construction*. Blackwell Publishing, Oxford.

Walker, D.H.T. and Hampson, K.D. 2003. *Procurement choices. Procurement Strategies: A Relationship Based Approach*, D.H.T.Walker and K.D.Hampson, eds., pp. 13–29. Blackwell Publishing, Oxford, chap. 2.

Walker, D.H.T., Lingard, H., and Shen, Y.J. 1998. *The Nature and Use of Global Method Statements In Planning*, Research Report. The Department of Building and Construction Economics, RMIT University, Melbourne.

Walker, D.H.T. and Shen, Y.J., Project understanding planning, flexibility of management action and construction time performance — two Australian case studies, *Construct. Manage. Econ.*, 20, 1, 31–44, 2002, E & F Spon. UK.

Walker, D.H.T. and Sidwell, A.C. 1996. *Benchmarking Engineering and Construction: A Manual For Benchmarking Construction Time Performance*. Construction Industry Institute Australia, Adelaide.

Walker, D.H.T. and Vines, M.W. 1997. Construction time performance in multi-unit residential construction — insights into the role of procurement methods, *XIII Annual ARCOM Conference*, Kings College, Cambridge, UK, 15–17 September, ARCOM: 93–101.

Ward, S.C. 1997. *Project Risk Management: Processes Techniques, and Insights*. Wiley, New York.

Ward, S.C. and Chapman, C., Transforming project risk management into project uncertainty management, *Int. J. Project Manage.*, 21, 2, 97–105, 2003.

Winch, G.M. 2003. *Managing Construction Projects*. Blackwell Publishing, Oxford.

16

Highway Maintenance

Ian van Wijk
Africon Engineering International
Pretoria, South Africa

16.1   General

The previous chapters discussed in depth how highways are planned, financed design and constructed. Once the highway with all the different elements has been constructed it must be managed and maintained. Chapter 17 (Highway Asset Management) deals with the management of the highway as an asset, while Chapters 18 (Pavement Management System) and 19 (Highway Condition Surveys and Serviceability Evaluation) cover the management and assessment of the pavement.

This chapter deals with highway maintenance. Highway maintenance can broadly be defined as actions taken to retain all the highway elements in a safe and usable condition. The definition of this "condition" will depend on the purpose of the highway, the traffic volumes, and other technical, social and political considerations. It can have a flexible, rigid or unpaved surface. A highway is defined as an engineered inter-urban road. Highway maintenance normally excludes upgrading or strengthening of the highway elements, but may be done if these appear to be the most cost-effective actions in the long-term. Maintenance could be of emergency, remedial (also called routine or recurrent), and preventative (also referred to as periodic) types.

This chapter addresses on the maintenance of all the elements of the highway, i.e.,

- Pavement (paved as well as unpaved)
- Road reserve (including fences)
- Road furniture (road-signs, guard-rails)
- Road markings and reflectors
- Road-side drainage
- Embankments and cuttings
- Bridges and culverts

Highway maintenance must be planned, managed, designed and executed. Planning and management are done by means of maintenance management systems and procedures. These systems and procedures are normally different from a pavement and bridge management systems since the latter do not focus on long-term and strategic repair and upgrading issues. Relevent pavement and bridges can be identified for protective and preventative actions as part of a highway maintenance program.

On time maintenance is extremely important. Studies have proven that it is more cost-effective to implement preventative maintenance actions on the pavement on a regular basis than neglecting the same and later trying to rehabilitate the pavement. Typically, premature periodic maintenance is 20 times costlier than proper routine and recurrent maintenance, while it is 3 times costlier to strengthen the pavement rather than properly maintain it (Robinson, 1988). Caltrans (Massey and Pool, 2003) found that for every US$1 spent on preventative maintenance, US$6 could be saved on rehabilitation and US$20 on reconstruction if applied timely. Timely maintenance can extend the pavement's service life by 5 to 10 years. Timely interventions are one third to one fifth the cost of a major repair (OECD, 1990a). Similar principles are true for other elements such as bridges, culverts and road signs. Maintenance produces very high benefit–cost ratios (Faiz and Horak, 1987). For the highway user it is equally important to avoid accidents due to unsafe conditions rather than to rectify the defect after an accident has occurred. Poor maintenance (in terms of quality and timing) can increase the vehicle operating costs by 15% and no maintenance by 50% (Robinson, 1988).

A further consequence of poor maintenance is the possible lack of accessibility. Poor maintenance of bridges, culverts, drainage structures and the pavement can lead to closure of roads for periods of time with the related detrimental social and economic consequences.

Highway authorities normally budget for remedial maintenance based on historical trends and preventative maintenance based on the outputs from pavement management systems (PMS). The amounts vary depending on the age of the road, the importance, the traffic levels, the location, etc. Typical annual remedial costs can range from US$1000 to US$5000/lane-kilometer and annual preventative maintenance costs from US$4000 to US$8000/lane-kilometer. The sum of the annual remedial and preventative cost is in the order of 1 to 4% of the initial construction cost of the highway (i.e. the asset

TABLE 16.1 Comparative Highway Maintenance Costs

Road Type	Activity	Cost Range (units)
Paved and unpaved	Routine and recurrent	1 to 5
Paved and unpaved	Periodic	4 to 5
Paved	Strengthening overlay	25 to 40
Paved	Rehabilitation	60 to 100

value). Table 16.1 shows typical comparative costs estimated by Robinson (1988) for highway maintenance.

16.2 Description and Definition of Highway Maintenance

Highway maintenance has the following objectives:

- the repair of the functional pavement defects
- prolonging the functional and structural life of the pavement
- maintaining road safety and signage
- keeping the road reserve in an acceptable condition

Road maintenance has been defined by the Permanent International Association of Road Congresses, (PIARC, 1982a) as "Suitable routine, periodic and urgent activities to keep pavement, shoulders, slopes, drainage facilities and all other structures and property within the road margins as near as possible to their as-constructed or renewed condition. Maintenance includes minor repairs and improvements to eliminate the cause of defects and avoid excessive repetition of maintenance efforts."

PIARC further defined periodic maintenance as "Operations that are occasionally required on a section of road after a period of a number of years. They are normally large scale and require specialist equipment and skilled resources to implement, and usually necessitate the temporary deployment of those resources on the road section. These operations are costly and require specific identification and planning for implementation, and often require design."

Periodic maintenance is akin to preventive maintenance defined by the American Association of State Highway and Transportation Officials (AASHTO) as follows (Galehouse et al., 2003): "...*preventive maintenance is a planned strategy of cost-effective treatments that preserves and maintains or improves a roadway system and its appurtenances and retards deterioration, but without substantially increasing structural capacity*." The U.S. Federal Highway Administration, FHWA (1999a) defines pavement preservation as: "*all activities undertaken to provide and maintain serviceable roadways; this includes corrective maintenance and preventive maintenance, as well as minor rehabilitation projects*."

Periodic or preventive maintenance is non-structural and is applied to extend the life of the pavement, to enhance the performance and to reduce user delays (Galehouse et al., 2003). Examples of preventative (or periodic) maintenance are regraveling of unpaved roads, resealing (with surface dressing, ultra thin asphalt, etc.) of paved roads, and regraveling of shoulders. Routine maintenance is defined by PIARC (PIARC, 1982a) as "*Operations required to be carried out once or more/year on a section of road. These operations are typically small-scale or simple, but widely dispersed, and require skilled or unskilled manpower. The need for some of these can be estimated and planned on a regular basis e.g., vegetation control*."

Non-preventive maintenance can also be described as routine, recurrent and urgent (TRRL, 1981), where routine refers to the "fixed-cost" activities such as grass cutting, drainage maintenance and road sign maintenance; recurrent to activities required throughout the year such as pothole patching, crack sealing, and grading, and urgent to the repair of defects caused by disasters (e.g., floods) or accidents. Routine maintenance activities are not influenced by the traffic, while the recurrent maintenance activities are. Both these sets of activities can also be defined as reactive maintenance, where activities are unscheduled and immediate response sometimes required.

The urgent maintenance is required to keep the highways open, protect property and road users. This has also been referred to as emergency maintenance (Galehouse et al., 2003) and includes repair of washouts, rigid pavement blow-ups, and earth slides.

AUSTROADS (1991) divides road maintenance into preventive and remedial maintenance, with preventative maintenance involving actions to prevent the roads from deteriorating and remedial maintenance involving the repair of defects.

The following classification of highway maintenance activities will be used in this chapter:

- preventive maintenance refers to actions associated with restoring the condition of the highway, reducing the rate of deterioration and increasing the life of the pavement. The restoration of the condition of the pavement is primarily related to the functional, i.e., skid resistance and riding quality, properties of the pavement. These activities are normally planned based on an assessment and processing of information in a pavement or bridge management system. All maintenance should include attention to drainage as water is the single most important factor affecting pavement performance.
- Remedial maintenance refers to actions associated with the rectification of defects on the carriageway or the road reserve.
- Emergency maintenances refers to activities associated with the urgent repair of defects caused by natural disasters or accidents.

The highway maintenance types and activities are summarized in Table 16.2. Some of the remedial activities can also be classified as preventive measures, since they prolong the life of the pavement, e.g., crack sealing, undersealing, and drainage repair, but they are normally considered remedial maintenance. Each of these activities will be discussed in detail in the proceeding sections.

16.3   Description of Defects and Appropriate Maintenance Actions

The purpose of highway maintenance is to rectify defects and preserve the pavement. It is necessary to define and record the defects, as well as to understand the mechanism of failure in order to select the most appropriate action. The commonly occurring defects are defined in this section and related to appropriate maintenance activities. The activities are discussed in details in Sections 16.6, 16.7 and 16.9. It is important to emphasize that drainage should always be addressed as part of all maintenance actions.

16.3.1   Flexible Pavements

Flexible pavements comprise all pavements with a bituminous surfacing even if the base or subbase layers have been stabilized. The defects and appropriate maintenance actions are summarized in Table 16.3.

16.3.1.1   Defects

(i) Cracking occurring in flexible pavements can be classified in one of three types, i.e. surfacing, fatigue and others.

Surfacing cracks are associated with the aging and deterioration of the surface bituminous layer due to shrinking and hardening of the bituminous binder with a loss of volatiles. These cracks are in general not load-related. One type of surfacing cracks is cracking in an irregular pattern. These cracks are also referred to as map cracks, star cracks and amorphous cracks. Another type is block cracking with cracks in well-defined rectangular blocks. Surfacing cracks occur across the full-width of the pavement. The surface cracks are more prominent on thin surfacings such as slurries, sand seals, and the block cracking more pronounced on asphalt concrete surfacings. In the latter the cracks start from the top and progress to the bottom asphalt layer.

Fatigue cracks (commonly called alligator or crocodile cracks) are a series of interconnected cracks in a chicken-wire pattern. The cracks are caused by traffic loading, occur only in wheel-paths and are often associated with deformation. Early signs of fatigue cracks are fine parallel longitudinal cracks in the wheel-path.

TABLE 16.2 Highway Maintenance Types and Activities

Type of Maintenance	Pavement Type	Activities
Preventive (periodic)	Flexible	Rejuvenation
		Mill and replacement of surfacings
	Rigid	Reinstatement of load transfer
		Undersealing
		Pressure relief joints
		Provision of edge support
		Retro fit drains
	Unpaved	Regraveling
		Regraveling of shoulders
		Road marking
Remedial (routine, recurrent, reactive)		Road sign repair and replacement
		Safety repair and replacement
		Drainage (subsurface, chutes, channels, batter drains, lined catch-drains) clearing and repair[a]
		Grass-cutting and pruning of vegetation
		Mending of fences
		Removal of litter
		Winter maintenance (preventative salting, salting and grouting of packed snow, snow clearing)
	Flexible	Resurfacing (with surface dressings, slurry, thin asphalt, micro thin asphalt, etc.)[a]
		Patching (flexible)
		Crack sealing (flexible)[a]
	Rigid	Full depth patching (rigid)[a]
		Partial depth patching (spalling repair)[a]
		Joint and crack sealing (rigid)[a]
		Slab jacking[a]
		Undersealing[a]
		Retexturing (grinding, grooving, cold milling)
		Recementation of cracks[a]
	Unpaved	Gravel and repair
		Blading
		Spot regraveling
		Bridge and culvert repairs[a]
Emergency (urgent)		Cleaning of spillages
		Replacement of damaged guard-rails and road signs at critical positions
		Repair of washouts, rock or earth slides
		Removal of dead animals, trees, etc.
		Clearing of accident sites

[a] These activities will also retard pavement deterioration and can be considered preservation activities.

Other types of cracks in flexible pavements are longitudinal, edge, transverse, reflection and stabilization cracks. These are well-defined fairly continuous cracks. Longitudinal and edge cracks are normally associated with the differential movement of the subgrade due to moisture variations (i.e. at embankments or on clays), and discontinuities in the pavement.

Transverse cracks occurring in isolation are caused by movements under the surfacing such as at a culvert or bridge joint. When these cracks appear at regular intervals they are classified as reflection cracks where joints of an overlaid concrete pavement reflect through the asphaltic concrete surfacing. Cracking on longitudinal joints would also be present in such cases. Reflection cracks follow the dimensions of the concrete slabs beneath the asphalt concrete surfacing.

A further type of cracking is stabilization cracks, which are rectangular blocks from 0.5 to 3 m in size and formed by the reflection of the cracked stabilized layer through the surfacing. Pumping, i.e., the movement

TABLE 16.3   Flexible Pavements: Defects and Appropriate Maintenance Actions (PIARC, 1982c; CSRA, 1985; TRRL, 1985; OECD, 1990a; SANRAL, 2000)

Defect	Appropriate Maintenance Action	
	Preventative	Remedial
Surface cracking	Rejuvenation, resurfacing	Mill and replace (if severe)
Fatigue cracking	Resurfacing	Mill and replace
Longitudinal or transverse cracks		Crack sealing
Block or stabilization cracks	Resurfacing	Mill and replace crack sealing (if widely spaced)
Potholes (patch deterioration)		Patching
Rutting and shoving	Resurfacing	Rut filling mill and replace
Polished aggregate raveling and weathering	Resurfacing	Mill and replace
Poor binder condition	Rejuvenation, resurfacing	
Edge breaks		Patching of the edges
Lane-to-shoulder drop-off (unpaved shoulders)	Shoulder blading	Shoulder reconstruction
Roughness	Resurfacing	
Skid resistance	Resurfacing	Sand blasting
Macro texture	Resurfacing	

Note: 1. Resurfacing entails thin asphalt overlays, ultra thin asphalt overlays, double and single surface dressings, and slurries. Resurfacing should be accompanied by the repair of the underlying defects to the extent that the defects remaining will not significantly affect the performance of the resurfacing. Timeous resurfacing can prevent many of the defects occurring.
2. HIPR is an alternative to milling and replacing.

of fine material from the stabilized layer through the cracks can occur as stabilization cracks. Closely spaced stabilization cracking (i.e., small blocks) is an indication that the stabilized layer has broken up into small pieces and is close to the end of its functional life. Stabilization cracks should not be confused with block cracks. Stabilization cracks develop from the bottom of the surfacing and block cracks from the top. Stabilization cracks tend to be more prominent in the wheel tracks, than between the wheel-tracks, especially when the block sizes are small.

Isolated and well-defined cracks, such as transverse, longitudinal and reflection cracks can be sealed with a penetration grade or polymer modified bitumen (PMB). Where cracks are active the application of a thin (150 to 300 mm) strip of geo-textile on the crack is appropriate. Extensive cracking cannot be sealed individually cost-effectively and the application of a membrane of bitumen, or PMB or geo-textile is more feasible (AUSTROADS, 1998a).

Low severity surface cracking can be addressed with the application of a rejuvenator to the surface. Extensive and severe surface cracking can be rectified by removal and replacement of the defective portion of the layer or by hot in-place recycling (HIPR) depending on the thickness which is affected.

Where fatigue cracking is extensive and associated with deformation, the appropriate action is to remove the defective material and replace it with new material.

(ii) Potholes are bowl-shaped holes of various sizes on the pavement surface. A pothole is normally only considered significant if the diameter of the hole is more than 150 mm and the depth more than 25 mm. Potholes are secondary forms of distress that develop from cracks. The ingress of moisture into the pavement layers reduces the structural capacity of the layers and thereby accelerates the progression of the pothole. The condition of patches must be recorded and repaired.

Repair of the potholes entails patching which is the removal of the defective layers and the replacement with, normally, a bituminous mixture. Patch deterioration refers to the condition of the patch and requires the same remedial actions as potholes.

(iii) Rutting is the longitudinal surface depression in the wheel-path. Rutting is caused by compaction or shear deformation of the pavement layers through traffic loading. Wide-shaped rutting is normally an indication of deformation of the lower pavement layers or subgrade and narrower and more sharply defined rutting, of deformation in the upper layers (i.e., surfacing or base). Rutting can be repaired by the filling of the ruts with a slurry or thin asphalt or milling and replacement. A resurfacing will rectify rutting, and if properly designed, reduce further rutting. If rutting is severe more than one application of the resurfacing layer may be required.

(iv) Shoving refers to the longitudinal displacement of localized areas of the pavement caused by shear forces induced by traffic loading. Shoving is most evident where vehicles stop and start. The repair is similar to that of rutting.

(v) Other defects associated with the surfacing are bleeding, polished aggregate, raveling (aggregate loss), weathering and the binder condition.

Bleeding is the condition where a film of bituminous binder is present on the surface which creates a shiny, reflective surface which may be tacky in hot weather. Polished aggregate refers to the smoothness of the exposed aggregate. Bleeding and polished aggregate reduce the skid resistance of the pavement with resulting safety consequences.

Raveling (aggregate loss) describes the process where the aggregate particles are dislodged and weathering where the asphalt binder is removed. Raveling is caused by the abrasive action of traffic. This can be extensive on surface dressings when the binder content is too low, chippings contaminated or bituminous binder too cold to effectively adhere to the chippings during construction.

Binder condition refers to the freshness and the elasticity of the binder. The binder loses its elasticity with time, resulting in raveling and weathering.

Repair of these defects comprises milling and replacement where the defects are isolated and resurfacing where the defects are extensive. Poor binder condition can be rectified with the application of a rejuvenator.

(vi) Other miscellaneous distresses are edge breaks and lane-to-shoulder drop-off or separation. Edge breaks are the breaking away of the surfacing at the edges of the pavement and often caused by poor unpaved shoulder maintenance. Lane-to-shoulder drop-off or separation describes the difference in elevation or width of joint between the pavement and the shoulder.

Edge breaks are repaired by patching and the lane-to-shoulder drop-off by the reinstatement of the shoulder. The defects can be prevented by regular blading of the unpaved shoulder.

(vii) Roughness is defined as deviations of the surface from a true plain. It is caused by one or a combination of the preceding defects, e.g., rutting, cracking, and potholes. The roughness can be improved slightly by patching, crack sealing and elimination of rutting, but only completely be rectified with a resurfacing of slurry (normally coarse slurry) or asphalt.

(viii) Skid resistance. Skid resistance is affected by the polishing stone value (PSV) of the chippings and the micro-texture. The skid resistance can effectively be improved by resurfacing, while sandblasting techniques have also been successful in some cases.

(ix) The macro-texture reflects the drainage capacity of the pavement under wet conditions and is an important safety issue. Resurfacing with the appropriate type will improve the skid resistance.

16.3.1.2 Recording of Defects

Although progress is made in the automated detection of pavement defects, most road authorities still use visual methods to record the defects. The only defects which are measured by means of mechanical devices on a regular basis are riding quality, rutting and skid resistance. Skid resistance can be seen as an indirect measurement of bleeding and polished aggregate.

The magnitude of defects is normally recorded in terms of severity (or degree) and extent. The severity is commonly expressed as low, moderate, or high or on a scale of one to five with one indicating a very low severity and five a very high severity. The presence or extent of the defect is expressed as length, area or percentage, or in terms of occurrence i.e. as isolated, intermittent or extensive. Defects are recorded/preidentified section. The section length can range from 50 m to 1 km. The shorter sections are normally used for detailed level assessments and the longer sections for network level assessments.

The length of the road segments, the defects recorded and the definition of the degree and extent vary from road agency to road agency. Two examples of how defects are recorded are given in Table 16.4 and Table 16.5, namely the Strategic Highway Research Program (SHRP) distress identification procedures for the monitoring of long-term pavement performance (SHRP, 1993) and the Highway Design and Maintenance Model (HDM) 4 procedures (Odoki and Karali, 1999).

TABLE 16.4   SHRP-Flexible Pavement Surfaced Pavement Defect Types and Severity Levels (SHRP, 1993)

Defect Type and Description	Unit of Measure	Defined Severity Level		
		Low	Moderate	High
Fatigue cracking. A series of interconnected cracks in a chicken-wire or alligator pattern in areas subject to traffic loading	Square meters	An area of cracks with no or only a few connecting cracks; cracks are not spalled or sealed; pumping is not evident	An area of interconnected cracks forming a complete pattern; cracks may be slightly spalled; cracks may be sealed; pumping is not evident	An area of moderately or severely spalled interconnected cracks forming a complete pattern; pieces may move when subjected to traffic; cracks may be sealed; pumping may be evident
Block cracking. A pattern of approximately rectangular cracks from 0.1 to 10 m² in size	Square meters	Cracks with a mean width ≤ 6 mm or sealed cracks with sealant material in good condition and with a width that cannot be determined	Cracks with a mean width > 6 mm and ≤ 19 mm or any crack with a mean width ≤ 19 mm and adjacent low severity random cracking	Cracks with a mean width > 19 mm or any crack with a mean width ≤ 19 mm and adjacent moderate to high severity random cracking
Edge cracking. Crescent shaped or fairly continuous cracks which encroach on the pavement edge	Meters	Cracks with no breakup or loss of material	Cracks with some breakup and loss of material for up to 10% of the length of the affected portion of the pavement	Cracks with considerable breakup and loss of material for more than 10% of the length of the affected portion of the pavement
Longitudinal cracking. Cracks predominantly parallel to the pavement centerline, in or outside the wheel-path	Meters, and position (wheel-path, non-wheel-path)	A crack with a mean width ≤ 6 mm or a sealed crack with sealant material in good condition and with a width that cannot be determined	Any crack with a mean width > 6 mm and ≤19 mm or any crack with a mean width ≤ 19 mm and adjacent low severity random cracking	Any crack with a mean width > 19 mm or any crack with a mean width ≤ 19 mm and adjacent moderate to high severity random cracking
Reflection cracking at joints. Cracks in asphalt concrete overlays that occur over joints in concrete pavements	Number and meters (distinguish between transverse and longitudinal)	An unsealed crack with a mean width ≤ 6 mm or a sealed crack with sealant material in good condition and with a width than cannot be determined	Any crack with a mean width > 6 mm and ≤19 mm or any crack with a mean width ≤ 19 mm and adjacent low severity random cracking	Any crack with a mean width > 19 mm or any crack with a mean width ≤ 19 mm and adjacent moderate to high severity random cracking
Transverse cracking. Cracks predominantly perpendicular to the pavement centerline	Number and meters	An unsealed crack with a mean width ≤ 6 mm or a sealed crack with sealant material in good condition and with a width than cannot be determined	Any crack with a mean width > 6 mm and ≤19 mm or any crack with a mean width ≤ 19 mm; and adjacent low severity random cracking	Any crack with a mean width > 19 mm or any crack with a mean width ≤ 19 mm and adjacent moderate to high severity random cracking

Distress	Unit of measurement	Low severity	Moderate severity	High severity
Patch by patch deterioration. Portion of the pavement surface of > 0.1 m^2 that has been removed or replaced	Number and square meters	Patch has a most low severity distress of any type	Patch has moderate severity distress of any type	Patch has high severity distress of any type
Potholes. Bowl-shaped holes $>$ 150 mm	Number and square meters	Less than 25 mm deep	25 mm to 50 mm deep	More than 50 mm deep
Rutting. Longitudinal surface depression in the wheel path	Millimeters	Not applicable. Severity levels could be defined by categorizing the measurements taken. A record of the measurements taken is much more desirable, because it is more accurate and repeatable than are severity levels		
Shoving. Longitudinal displacement of a localized area	Number and square meters	Not applicable. Severity levels can be defined by the relative effect of shoving on riding quality		
Bleeding. Excess bituminous binder occurring on the pavement surface	Square meters	An area of pavement surface discolored relative to the remainder of the pavement by excess asphalt	An area of pavement surface that is losing surface texture due to excess asphalt	Excess asphalt gives the pavement surface a shiny appearance; the aggregate may be obscured by excess asphalt; tyremarks may be evident in warm weather
Polished aggregate. Surface binder worn away to expose coarse aggregate	Square meters	Not applicable. The degree of polishing may be reflected in a reduction of surface friction		
Raveling. Wearing away of the pavement surface in hot-mix asphalt concrete	Square meters	The aggregate or binder has begun to wear away but has not progressed significantly. Some loss of fine aggregate	Aggregate and binder has worn away and the surface texture is becoming rough and pitted; loose particles exist; loss of fine aggregate and some loss of coarse aggregate	Aggregate and binder has worn away and the surface texture is very rough and pitted; loss of coarse aggregate
Lane-to-shoulder drop-off. Difference in elevation between the traveled surface and the outside shoulder	Millimeters	Not applicable. Severity levels could be defined by categorizing the measurements taken. A record of the measurements taken is much more desirable, because it is more accurate and repeatable than are severity levels		
Water bleeding and pumping. Seeping or ejection of water from beneath the pavement through cracks	Number and meters	Not applicable. Severity levels are not used because the amount and degree of water bleeding and pumping changes with varying moisture conditions		

TABLE 16.5   HDM4 — Definitions of Types of Defect Types for Flexible Pavements (Odoki and Karali, 1999)

Defect	Definitions
All cracking	Narrow and wide structural cracking inclusive
Narrow cracking	Interconnected or line cracks of 1–3 mm crack width
Wide cracking	Interconnected or line cracks of 3 mm crack width or greater, with spalling
Indexed cracking	Normalized sum of narrow and wide cracking weighted by class
Transverse thermal cracking	Unconnected crack running across the pavement
Raveling	Loss of material from wearing surface
Pothole	Open cavity in the road surface with at least 150 mm diameter and at least 25 mm depth
Edge-break	Loss of bituminous surface material (and possibly base materials) from the edge of the pavement
Rutting	Permanent or unrecoverable traffic-associated deformation within pavement layers which, if channeled into wheel-paths, accumulates over time and becomes manifested as a rut
Rut depth	Maximum depth under 2 m straightedge placed transversely across a wheel-path
Roughness, IRI (International Roughness Index)	Deviations of a surface from a true planar surface with characteristic dimensions that affect vehicle dynamics, ride quality, dynamic loads and drainage — typically in the ranges of 0.1 to 100 m wavelength and 1 to 100 mm amplitude
Mean texture depth	The average depth of the surface of a road surfacing expressed as the quotient of a given volume of standardized material [sand (sand patch test), glass spheres] and the area of that material spread in a circular patch on the surface tested
Skid resistance	Resistance to skidding expressed by the sideways force coefficient (SFC) measured using the Sideways Force Coefficient Routine Investigation Machine (SCRIM)

Area (of distress): sum of rectangular areas circumscribing manifest distress (line cracks are assigned a width of 0.5 m), expressed as a percentage of the carriageway area.

16.3.2   Rigid Pavements

Three types of rigid (concrete) pavement are commonly in use i.e. jointed concrete pavement (JCP), jointed reinforced concrete pavements (JRCP), and continuously reinforced concrete pavements (CRCP). A fourth type, prestressed concrete pavements, have been used on airport pavements. Most of the defects apply to all three types of concrete pavements. Table 16.6 summarizes the defects which can occur on rigid pavements and the appropriate maintenance actions.

16.3.2.1   Defects

(i) Surface defects are defects at the surface which do not protrude deeply into the concrete slab i.e. map cracking, scaling, polished aggregate and pop-outs. Map cracking is a series of random cracks that extend only into the upper surface of the slab. Scaling defines the process when thin layers of the surfacing get dislodged. Polished aggregate refers to the surface mortar and texturing worn away to expose coarse aggregate. Pop-outs are small pieces of pavement broken loose from the surface, normally ranging in diameter from 25 to 100 mm and depth from 13 to 50 mm.

These defects are normally not of major concern, but if action is needed milling, grooving, or resurfacing can be considered.

(ii) Durability cracking (D-cracking) forms closely spaced crescent-shaped hairline cracking patterns and occurs adjacent to joints, cracks, or free edges initiated at the intersection of cracks and a free edge. Dark coloring of the cracking pattern and surrounding area are common. D-cracking is a progressive condition in that the first cracks that appear are at junctions of joints, cracks and pavement edges and progress outward to involve greater slab areas. The development of D-cracking is strongly dependent on the physical and chemical character of the ingredients of the concrete mixture, especially the coarse aggregate. No maintenance procedures are known that will slow or stop the progression of deterioration

TABLE 16.6 Rigid Pavements — Defects and Appropriate Maintenance Actions (Faiz and Yoder, 1974; NCHRP, 1979; ERES Consultants Inc., 1982; FHWA, 1984; Van Wijk, 1985; CSIR, 1986; Chou and McCullough, 1987)

Defect	Appropriate Maintenance Action	
	Preventative	Remedial
Map cracking and scaling Polished aggregate	Resurfacing	Grooving, milling
Pop-outs		No action
Durability (D) cracking		Patching (full depth)
Longitudinal cracking	Undersealing (in specific cases)	Crack sealing
Diagonal cracking	Undersealing (in specific cases)	Crack sealing
		Patching (full depth) if shattered
Blow-ups	Relief joints Joint maintenance	Patching (full depth)
Corner breaks		Patching (full depth)
Transverse cracking (jointed)	Underseal (in specific cases)	Crack sealing
Spalling of joints or cracks		Patching (partial depth)
Joint seal damage		Joint seal replacement
Patch deterioration		Patching
Pumping or water bleeding		Underseal
Punch-outs (CRCP)		Patch (full depth)
Steel rupture (CRCP)		Patch (full depth)
Corner breaks (JCP or JRCP)		Patch (full depth)
Faulting (JCP or JRCP)	Underseal	Grinding, slab jacking
Lane-shoulder drop-off or separation		Shoulder institution

CRCP = continuously reinforced concrete pavement; JCP = jointed reinforced concrete pavement; JRCP = jointed reinforced concrete pavement.

Note: The distinction between preventative and remedial is not always clear and actions can often fall into both categories.

in susceptible concrete. D-cracking is considered to be a concrete material problem attributable to freeze–thaw cycles and a peculiar aggregate pore structure.

Repair of D-cracking is by means of full-depth patching.

(iii) Longitudinal cracking describes cracks that are predominantly parallel to the pavement centerline. Most longitudinal cracking, and especially that which results when fracture does not take place at weakened-plane longitudinal joints as planned (for instance due to late sawing or shallow cut-depths), does not become a performance problem nor is it usually a precursor of problem development elsewhere. Occasionally, and especially where progressive widening or spalling of the crack becomes evident, a longitudinal crack can signify the beginning of foundation settlement or base erosion that prompt treatment may control. If unattended, the open crack can provide easy entry for surface water to aggravate an already damaging condition. Longitudinal cracking, if not already present, may take place during the development of punch-outs in CRCPs.

Timeous crack sealing is the most appropriate maintenance action, unless the slab support is affected in which case undersealing may have to be considered.

(iv) Cluster cracking is a grouping of three or more closely spaced transverse cracks where spacings may be in the order of 150 to 600 mm. This can occur in CRCP and often develops into punch-outs. Cluster cracking is believed to be associated with such factors as localized areas of change in subgrade support, lack of concrete consolidation, inadequate pavement thickness, high base friction, poor drainage, and high temperatures at the time of construction. Concrete mixture variability might be added to the foregoing list. Because cluster cracking is so often seen early in the life of a pavement, traffic loads probably are not a basic cause.

The appropriate repair is full depth patching.

(v) Diagonal cracking is cracking in a direction oblique to the pavement centerline and can be viewed as an indicator of the existence of a foundation problem (settlement, expansion or erosion) that may be controllable through early treatment.

These cracks should be sealed to prevent water from entering and affecting the support. Undersealing to stabilize the slab and fill the voids should be considered if appropriate.

Where more than one diagonal crack develops on a slab, a shattered slab situation will develop which requires the replacement of the slab.

(vi) Blow-ups are localized upward movements of the pavement surface at transverse joints or cracks, often accompanied by shattering of the concrete in that area. The rise can be sufficient magnitude to endanger the safe passage of traffic. The dominant feature of a shattering blow-up is an accumulation of debris from the shattering process. Blow-ups usually occur during periods of high temperature and precipitation. Blow-ups occur in concrete pavements when available expansion space is not sufficient to accommodate expansion. They almost never occur in new pavements. The longitudinal pressure build-up that produces them is initiated by the entry of incompressibles into previously available expansion space in cracks and joints during periods of pavement contraction.

Relief joints cut early in the blow-up process may avoid later major repair or at least allow some latitude in selecting the time for repair. When badly deteriorated full depth patching will be required to replace the damaged portions of the slab and construct proper expansion joints.

(vii) Corner breaks are recorded if a portion of the slab is separated by a crack which intersects the adjacent transverse and longitudinal joints, describing approximately a 45° angle with the direction of traffic. The length of the sides is from 0.3 m to one-half the width of the slab, on each side of the corner. These cracks are often caused by the loss of slab support due to erosion and pumping at the corners. Full-depth patching is required to correct the integrity of the slab and load transfer.

(viii) Transverse cracking is cracking that is predominantly perpendicular to the pavement centerline and can also take the shape of a "Y." This cracking is expected in a properly functioning CRCP and ideally 1 to 1.5 m apart. Tight transverse cracks showing little spalling in CRCP are, like transverse joints in a well-designed and well-constructed jointed concrete pavement, deficiencies only to the extent that they are interruptions in what might be considered the ideal but so far unattainable concrete pavement — a single, unbroken ribbon of concrete. Transverse cracks in groups at close intervals, i.e. cluster cracking, can be the sign of a potential problem area (e.g., punch-outs). Any abnormal increase in the amount of transverse cracking, especially when accompanied by an increase in spalling, also can be a sign of problem development.

Y-cracking appears to develop mostly when nominally transverse cracks meander sufficiently to either branch or join. Like closely spaced transverse cracking, it can be a problem location if the underlying support weakens.

Transverse cracking in jointed concrete pavements is often caused by late sawing of joints, constrained movements at the joints or the lack of slab support. Crack sealing is required as a maintenance option. Transverse cracking in CRC pavements is expected and only requires attention if it has developed into closely spaced cluster) cracking and punch-outs in which case full-depth patching is required.

(ix) A punch-out is the area enclosed by two closely spaced (usually less than 0.6 m) transverse cracks, a short longitudinal crack, and the edge of the pavement or a longitudinal joint present only on CRCP. It also includes "Y"-cracks that exhibit spalling, break-up and faulting and occurs when two transverse cracks at a critical distance apart are connected by a longitudinal crack. Once a transverse cracks has begun to open, repeated heavy loads continue the breakdown of aggregate interlock. The process may be aided by corrosion of the steel, which begins when the crack width becomes large enough to admit water and de-icing salts.

A punch-out is a structural failure in which a small segment of pavement is loosened from the main body and displaced downward under traffic. The punch-out is usually bounded by two closely spaced transverse cracks, a longitudinal crack, and the pavement edge and sometimes by the branches of a Y-crack and the pavement edge. More rarely, a punch-out may occur in the interior of a pavement away from the edge, but these differ from edge punch-outs in severity and cause. Edge punch-outs are the major structural distress experienced in CRCPs.

The repair of a punch-out entails deep patching of the portion of the slab which had failed.

(x) Water bleeding and pumping are the seeping or ejection of water from beneath the pavement through cracks or joints. Most of the pumping in CRCPs is at the edges, while it can be at the joints or the edges in jointed pavements. In some cases it is detectable by deposits of fine material left on the pavement surface, which were eroded (pumped) from the support layers and have stained the surface. Inadequate pavement thickness and erodible underlying material type are the main causes of pumping. Pumping can be rectified by undersealing (i.e., filling of the voids) and prevented by keeping water out of the pavement and proper load transfer.

(xi) Faulting of transverse joints and cracks is defined as the difference in elevation across a joint or crack. This is caused by the loss of slab support due to erosion or settlement. Faulting is the most important single defect influencing the riding quality of rigid pavements. Further faulting can be prevented by undersealing and repaired by grinding or slab jacking.

(xii) Patch deterioration describes the condition of a patch (asphalt or concrete) present on the rigid pavement. Deteriorated patches should be replaced.

(xiii) Steel rupture in CRCP is almost invariably associated with open cracks or open construction joints (>3 mm). Steel rupture in CRCP is breakage and full loss of continuity in the steel reinforcement.

Loss of tightness at any crack or joint in CRCP is evidence that the steel reinforcement is not functioning properly and is either in a weakened condition or has already ruptured. Excessive spalling is likely to be present. Rust stains on the pavement surface suggest loss of cross-sectional area and the possibility of rupture. Faulting or slab settlement often indicates that rupture has taken place. Although the actual fracture that takes place may be a rather sudden occurrence, the weakening process may be gradual, taking place over a period of time.

Asphalt sealing or application of cement grout at an early stage can postpone or avert failure. Failure usually requires repair by removing the damaged material and patching the area.

(xiv) Spalling of joints or cracks is the cracking, breaking, chipping, breakdown, disintegration or fraying of slab edges within 0.6 m of the joint or crack. Slight spalling, in which the flaking is confined mostly to the mortar in the concrete matrix, is sometimes referred to as raveling. Spalling alone, without accompanying deterioration, rarely becomes so severe that some form of maintenance is required. Progressive spalling often is a good indicator of future distress. Progressive spalling that deepens more rapidly than it expands outward is primarily related to structural weakness. Spalling that widens more quickly than it deepens (often widening to over 50 mm while deepening less than 12 mm) usually is related more to weakness of the surface concrete.

Slight spalling needs no maintenance but more severe spalling may produce holes that require partial depth patching for riding comfort.

(xv) Joint seal damage is any condition which enables incompressible materials or a significant amount of water to infiltrate into the joint from the surface. Typical types of joint seal damage are:

- Extrusion, hardening, adhesive failure (bonding), cohesive failure (splitting), or complete loss of sealant.
- Intrusion of foreign material in the joint.
- Weed growth in the joint.

Damage of construction joints due to the ingress of incompressibles can manifest as a series of closely spaced transverse cracks or a large number of interconnecting cracks occurring near the construction joint.

(xvi) Defects on shoulders can be:

- Lane-to-shoulder drop-off which is the difference in elevation between the edge of slab and outside shoulder; and typically occurs when the outside shoulder settles.
- Lane-to-shoulder separation which is widening of the joint between the edge of the slab and the shoulder.
- Deterioration of the flexible shoulder

16.3.2.2 Recording of Defects

Table 16.7 gives the defects and severity definition proposed for the SHRP distress identification (SHRP, 1993) and Table 16.8 those used on HDM-4 (Odoki and Karali, 1999).

TABLE 16.7 Rigid Pavement Defect Types and Severity Levels (SHRP, 1993)

Defect Type and Description	Unit of Measure	Defined Severity Level		
		Low	Moderate	High
D-cracking. Closely spaced crescent-shaped hairline cracks adjacent to joints, cracks, and free edges	Number of slabs and square meters	"D" cracks are tight, with no loose or missing pieces, and no patching is in the affected area	"D" cracks are well-defined, and some small pieces are loose or have been displaced	"D" cracking has a well-developed pattern, with a significant amount of loose or missing material. Displaced pieces, up to 0.1 m^2, may have been patched
Longitudinal cracking. Cracks that are predominantly parallel to the pavement centerline	Meters	Crack widths <3 mm, no spalling, and no measurable faulting; or well sealed and with a width that cannot be determined	Crack width ≧ 3 mm and <13 mm; or with spalling <75 mm; or faulting up to 13 mm	Crack widths ≧ 13 mm; or with spalling ≧ 75 mm; or faulting ≧ 13 mm
Transverse cracking. Cracks those are predominantly perpendicular to the centerline	Number and meters	Crack widths <3 mm, no spalling, and not measurable faulting; or well-sealed and the width cannot be determined	Crack widths ≧ 3 mm and <6 mm; or with spalling <75 mm; or faulting up to 6 mm	Crack widths ≧ 6 mm; or with spalling ≧ 75 mm; or faulting ≧ 6 mm
Map cracking. Series of cracks extending only into the upper portion of the slab	Number and square meters	Not applicable	Not applicable	Not applicable
Scaling. Deterioration of the upper 3 to 12 mm of the concrete surface, resulting in loss of surface mortar	Number and Square meters	Not applicable	Not applicable	Not applicable
Surface mortar and texturing worn away	Square meters	Not applicable. The degree of polishing may be reflected in a reduction of surface friction		
Pop-outs. Small pieces of pavement broken loose from the surface	Number or square meters	Not applicable. Severity levels can be defined in relation to the intensity of pop-outs as measured below		
Blowups. Localized upward movement of the pavement surface at transverse joints or cracks	Number	Not applicable. Severity levels can be defined by the relative effect of a blowup on ride quality and safety		
Lane-to-shoulder drop-off. Difference in elevation between the slab edge and the outside shoulder	Millimeters	Not applicable. Severity levels could be defined by categorizing the measurements taken. A complete record of the measurements taken is much more desirable, because it is more accurate and repeatable than are severity levels		

Distress	Units	Low severity	Moderate severity	High severity
Lane-to-shoulder separation. Widening of the joint between the slab edge and the shoulder	Millimeters	Not applicable. Severity levels could be defined by categorizing the measurements taken. A complete record of the measurements taken is much more desirable, because it is more accurate and repeatable than are severity levels		
Patch by patch deterioration. A portion of >0.1 m² of original surface removed	Number and square meters	Patch has a most low severity distress of any type; and no measurable faulting or settlement at the perimeter of the patch	Patch has moderate severity distress of any type; or faulting or settlement up to 6 mm at the perimeter of the patch	Patch has a high severity distress of any type; or faulting or settlement ≧ 6 mm at the perimeter of the patch
Water bleeding and pumping. Seeping or ejection from beneath the slabs through cracks	Number and meters	Not applicable. Severity levels are not used because the amount and degree of water bleeding and pumping changes with varying moisture conditions		
Corner breaks. A portion of slab separated by a crack which intersects the adjacent joints at a 45° angle with the direction of traffic	Number	Crack is not spalled for more than 10% of the length of the crack; there is not measurable faulting; and the corner piece is not broken into two or more pieces	Crack is spalled at low severity for more than 10% of its total length; or faulting of crack or joint is < 13 mm; and the corner piece is not broken into two or more pieces	Crack is spalled at moderate to high severity for more than 10% of its total length; or faulting of the crack or joint is ≧ 13 mm; or the corner piece is broken into two or more pieces
Transverse joint seal damage. Extension, hardening, loss, or loss of adhesion	Number	Joint seal damage as described above exists over less than 10% of the joint	Joint seal damage as described above exists over 10–50% of the joint	Joint seal damage as described above exists over more than 50% of the joint
Longitudinal joint seal damage repeat	Number, meters	None	None	None
Spalling of longitudinal joints. Cracking, breaking, chipping, or fraying of slab edges within 0.6 m repeat of the joint	Meters	Spalls less than 75 mm wide, measured to the centre of the joint, with loss of material, or spalls with no loss of material and no patching	Spalls 75 mm to 150 mm wide, measured to the centre of the joint, with loss of material	Spalls greater than 150 mm wide, measured to the centre of the joint, with loss of material
Spalling of transverse joints	Number, meters	Spalls less than 75 mm wide, measured to the centre of the joint, with loss of material, or spalls with no loss of material and no patching	Spalls 75 mm to 150 mm wide, measured to the centre of the joint, with loss of material	Spalls greater than 150 mm wide, measured to the centre of the joint, with loss of material
Faulting of transverse joints and cracks. Difference in elevation across joint or crack	Millimeters	Not applicable. Severity levels could be defined by categorizing the measurements taken. A complete record of the measurements taken is much more desirable, because it is more accurate and repeatable than are severity levels		
Punch-outs. The area enclosed by 2 closely spaced transverse cracks a short longitudinal cracks, and the edge or joint	Number	Longitudinal and transverse cracks are tight; and may have spalling < 75 mm or faulting <6 mm. Does not include "Y" cracks	Spalling ≧ 75 mm and <150 mm or faulting ≧ 6 mm and <13 mm exists	Spalling ≧ 150 mm or concrete within the punch-out is punched down by ≧ 13 mm or is loose and moves under traffic
Longitudinal joint seal damage	Number, meters	Not applicable	Not applicable	Not applicable

TABLE 16.8   Distress Modes Modelled in HDM-4 for Rigid Pavements

No.	Defect	Units of Measurement	Pavement Surface Type
1	Cracking[a] (transverse, longitudinal and durability)	Percent of slabs cracked	Jointed plain
	Cracking[a] (transverse, longitudinal and durability)	Number of miles	Jointed reinforced
2	Faulting	Inches	Jointed plain and reinforced
3	Spalling[b]	Percent of spalled joints	Jointed plain and reinforced
4	Failures	Number/mile	Continuously reinforced
5	Serviceability loss	Dimensionless	Jointed and continuously reinforced
6	Roughness	Inches/mile (or m/km)	All

[a] Severity level for cracking:

low = cracks with a width of less than 3 mm, without visible spalling or faulting; or well-sealed, with a non-determinable width;

medium = cracks with a width between 3 and 6 mm, or with spalling less than 75 mm, or faulting less than 6 mm;

high = cracks with a width greater than 6 mm, or spalling greater than 75 mm, or faulting greater than 6 mm.

[b] Severity level for spalling:

low = spalling of <75 mm of distress width, measured from the centre of the joint, with or without loss of material;

medium = spalling of between 75 and 150 mm of distress width, measured from the centre of the joint, with or without loss of material;

high = spalling of >150 mm of distress width, measured from the centre of the joint, with or without loss of material.

16.3.3   Unpaved Highways

This section covers defects on highways which do not have a bituminous or concrete surface. These highways would normally have a designed vertical and horizontal alignment and cross-section, and a wearing course of sound quality material. The relevant defects and appropriate maintenance actions are summarized in Table 16.9.

(i) Potholes in unpaved roads are normally caused by traffic action on the surface by the replacement of loose material or weak (soft) material. Potholes are bowl shaped depressions and usually considered significant if the diameter is more than 200 mm and the depth more than 25 mm.

(ii) Rutting is the presence of longitudinal depressions in the wheel-paths and caused by the shear (plastic) deformation or compaction under traffic loading. Wearing courses with high clay contents are more prone to rutting, especially in wet conditions.

(iii) Loose material describes the accumulation of loose material on the road surface and is also caused by the traffic action. The phenomenon is more prevalent on roads with sandy or non-cohesive fine-grained wearing courses.

TABLE 16.9   Unpaved Highways: Defects and Appropriate Maintenance Actions
(PIARC, 1982b; TRRL, 1985; OECD, 1990a; AUSTROADS, 1991; ARRB, 1993)

Defect	Appropriate Maintenance Action	
	Preventative	Remedial
Potholes	Regraveling	Spot regraveling, blading
Rutting	Regraveling	Blading
Corrugations	Regraveling	Blading
Gravel loss	Regraveling	–
Stoniness	Regraveling	–
Slipperiness	Regraveling	–
Erosion	Regraveling	Spot regraveling, blading
Cross-section	Regraveling (with camber correction)	–
Roughness	Regraveling	Blading

(iv) Corrugations (washboardings) are fairly regular evenly spaced transverse ridges and caused by traffic actions in conjunction with loose aggregate. The intervals of the ridges are normally less than 1 m. Corrugations most often occur on hills, on curves, and in areas of acceleration or deceleration.

(v) Gravel loss is an important defect, since it is indicative of the amount of wearing course left on the subgrade. The gravel loss is caused by the sweeping action of traffic. A high loss of gravel can result in exposed rock on the surface. It is measured as the loss of surface aggregate (wearing course aggregate) and degree of exposure of the subgrade. A high degree of gravel loss points to the need for regraveling.

(vi) Slipperiness indicates the presence of a material with a high clay content which becomes slippery when wet (similar to embedded stone).

(vii) Stoniness refers to the oversized material present in the wearing course. Oversized material is normally defined as material in excess of 40 mm and adversely affects riding quality and blading actions and may cause damage to vehicle tires.

(viii) Erosion is an effect peculiar to unpaved roads and is caused by free-flowing water across the surface of the road.

(ix) Erosion, gravel loss and blading practices can alter the cross-section of the road detrimentally with the associated safety risks and riding discomfort.

(x) Roughness refers to the riding comfort. The rougher the road, the poorer the riding comfort. Potholes, rutting, corrugations erosion and stoniness are major contributors to roughness.

Rutting and loose material can create extremely unsafe driving conditions, while potholes and corrugations adversely affect riding comfort and vehicle operating costs.

Defects on unpaved roads can be corrected by either preventative (i.e. regraveling) or remedial maintenance (i.e. spot regraveling or blading).

Regraveling should be used to correct stoniness and gravel loss, while regraveling or blading can be used to rectify potholes, rutting, loose aggregate, corrugations, erosions, and roughness. Spot regraveling can be used to fill potholes and erosion, depending on the extent of the latter. The cross-section can be improved by regraveling (and the necessary camber correction).

As in the case of flexible and rigid pavements, the defects in unpaved roads are normally assessed manually. Table 16.10 gives the defects and typical definition of severity levels.

16.3.4 Roadside Drainage

Roadside drainage refers to water standing on the road surface or road shoulder, as well as the facilities built to drain water away from the pavements. These facilities include side-drain or ditch (long narrow excavations designed or intended to collect and drain off surface water), miter or turn-out drains to (lead water away from the side drains), chutes (inclined pipes, drains or channels constructed in or on a slope) and subsoil drains (pipes or permeable material provided to collect and dispose of groundwater). Drainage structures are not included under roadside drainage

A defect occurs when the water is not effectively removed from the road surface or the pavement layers after rain. The reasons for this can be, *inter alia*, improper cross-falls, slacks, flat or high gravel shoulders, vegetation on the shoulder, lack of capacity of outlets, damaged or silted chutes or kerbs, and blocked subsurface drains.

The drainage problems can best be observed after heavy rain. The condition and effectiveness of roadside drainage features should be assessed visually on a regular basis to determine the extent and severity of the defect. The frequency of the inspections depends on the importance of the highway and can range from weekly to 6 monthly inspections. The extent is normally recorded as percentage of the highway section on which the defects are present. The severity is commonly reported as:

- Low: Drainage effective and features on a satisfactory condition. No maintenance action required.
- Moderate: Concern about the effectiveness and condition. Maintenance probably required.
- High: Drainage ineffective. Immediate maintenance required.

TABLE 16.10 Unpaved Highways — Pavement Defect Types and Severity Details

Defect Type	Unit of Measure of Extent	Defect Severity Levels		
		Low	Moderate	High
Potholes	Number/100 m	<20 mm in depth (under straightedge)	20–40 mm in depth	>40 mm in depth
Rutting	Percentage of subsection length	<20 mm (deformation under 2 m straightedge)	20–50 mm (deformation under 2 m straightedge)	≥50 mm (deformation under 2 m straightedge)
Corrugations	Percentage of subsection length	≤20 mm (deformation under 2 m straightedge)	20–50 mm (deformation under 2 m straightedge)	≥50 mm (deformation under 2 m straightedge)
Gravel loss (gravel thickness)	Percentage of subsection length	>100 mm thickness of gravel subgrade		
Stoniness	Percentage of subsection length	<50 mm maximum size of aggregate	50–100 mm maximum size of aggregate	>100 mm maximum size of aggregate
Slipperiness (clay)	Percentage of subsection length	–	–	–
Erosion	Percentage of subsection length	≤20 mm (deformation under 2 m straightedge)	20–50 mm (deformation under 2 m straight edge)	≥50 mm (deformation under 2 m straightedge)
Cross section (camber and cross fall)	Percentage of subsection length	≥50 mm difference between centerline and edge of gravel	20–50 mm difference between centerline and edge of gravel	≤20 mm difference between centerline and edge of gravel

16.3.5   Road Markings, Reserve and Furniture

Road signs, markings reflectors, and guard-rails are features directly influencing the safety of the road, while the condition of the fences and the shoulders, and the length of grass and vegetation, can adversely affect road safety. The clearing of litter from the road reserve is important, not only from an aesthetic point of view, but also to prevent environmental contamination and clogging of drainage features.

The maintenance of fencing is of particular importance in residential areas and in areas where livestock or game is present, mainly due to safety concerns. Fences are normally mended when damaged, but general replacement of old and deteriorated fences from time to time may be cost-effective. Removal of litter should take place on a regular basis. The frequency depends on the traffic volumes and location of the section of road (for instance, more litter is present close to towns or settlements). Frequencies vary from once a month to every 6 months.

Damaged guard-rails are themselves a hazard and should be repaired promptly. The overall guard-rail system (i.e. the guard-rail and posts) should be inspected on a regular (often weekly) basis with special attention given to the condition of guard-rails in high risk or accident areas.

Grass cutting is required for reasons of visibility (especially at curves and intersections), drainage and fire hazard. The frequency and width (e.g., from fence to fence or to a width of 4 m from the shoulder) vary, depending on the growth of the grass (influenced by the climate), the traffic volume and importance of the road. Performance criteria such as a limit on the height of the grass can be used to dictate the frequency.

Traffic signs, road markings and reflectors are important from a road user safety point of view. Damaged signs should be replaced or repaired timely. Signs and road markings must be inspected on a regular (annual) basis and be replaced before they become ineffective. The life of a road sign could be 7 to 8 years and that of road markings 2 to 4 years. The life of the latter depends on the traffic volume, climate and type of surfacing.

16.3.6   Geotechnical Features

The maintenance of earth and rock slopes, cuttings and fills also falls under the domain of road maintenance. The failure of road embankments (fills), cuttings and lateral support can have severe consequences such as closure of the road, damage to property and injury or loss of life.

Stable slopes can become unstable for a variety of reasons. Failure can take place within hours or over a long period of months or even years. Little can be done to prevent failure triggered by events such as abnormal downpours, floods, natural disasters or bust water pipes. Many failures can be prevented or the consequences reduced by monitoring typical indicators of instability.

Regular inspections should be done to assess and monitor these features. Table 16.11 and Table 16.12 (SANRAL, 2000) give typical indicators which could be monitored for embankments and cuttings to preempt catastrophic failures. These procedures rely on visual inspection of signs of cracking, offset fences, unexpected water flows, etc. More sophisticated measures, such as surface surveys, extensometric measurements across tension cracks, and the installation of piezometers, can also be considered.

Routine maintenance of cuttings and embankments essentially entails controlling the flow of water and vegetation.

16.3.7   Drainage Structures

Drainage structures refer to structures constructed to allow water to flow from the one side of the road to the other, i.e., bridges and culverts.

Defects occurring on bridge and drainage structures are associated with the waterway, the substructure (abutment and piers), superstructure, roadway (including expansion joints) and approaches.

TABLE 16.11   Typical Instability Indicators — Embankments

Description	Action
Longitudinal or semicircular cracks with level difference. Particularly where level differences are large immediate action is necessary	Monitor the position, extent and direction of the cracks and displacement at suitable intervals. Should further movement or failure appear imminent particularly in wet weather such measurements could be at hourly intervals and would help to decide whether further or total road closure is necessary
Lateral and vertical displacement of guard-rails, kerbs or road edge markings	Note position, extent and direction of movement and monitor at suitable intervals. Check for cracking and displacement in the road surface
Circular depressions in the road surface or on the fill slopes indicating sinkhole development (chimneying) caused by void formation within or below the embankment	Note position and size of depression, check for cracks, inspect toe of fill and any nearby drainage structures for collapse or open joints. Monitor daily if the embankment is high and inform road authority immediately
Bulging of slope or displacement at the toe of the fill	Note position and extent and monitor to see if any enlargement occurs with time
Road surface heave or settlement indicating possible deep-seated foundation problems	Note position and extent and monitor
Seepage out of the slope or at the toe (often shown by greener areas or reeds) especially where there is a regular (perennial) flow	Note position, check for signs of movement such as cracks and bulging in the slope and also check the road pavement for signs of structural distress
Trees and shrubs that are not vertical can indicate movement of the slope	Note position and check slope for other signs of movement such as bulging, cracks and seepage
Culvert deformation or collapse and, in prefabricated culverts, open joints	Note position and monitor. A collapsed culverts must be repaired

TABLE 16.12   Typical Instability Indicators — Cuttings

Description	Action
Boulders, rocks and soil in the side-drain, at the base of protective nets and on the road surface	Note the position and monitor the frequency and quantities of debris removed to determine the extent and severity of the problem. Where rocks and boulders fall directly onto the road surface the road authority should be immediately informed. Warning signs should be erected and interim protective measures taken such as the erection of guard-rails or gabion baskets, the deepening of side-drains where they are unlined and the construction of a catchwater drain above the cut
Weathering debris blocking the side drain. Where harder rocks like sandstone overlie degradable rocks such as mudstones and shales there is a possibility of toppling failure	Note the position, evaluate the possibility of toppling and inform the road authority if there is a likelihood of serious failure. Monitor the cutting
Tension cracks above or in the cut slope	Note the position, length and monitor daily if movement is taking place. Close the cracks with impermeable material to prevent water getting in. Inform the road authority
Open joints in rock slopes	Note the position and width of joints. Monitor to determine whether the joints are opening further and check for signs of seepage and fines washed out. Assess whether there is any danger of sliding or falling rock as a result of the cracks and if this is likely inform the road authority
Trees and shrubs that are not vertical can indicate movement of the slope	Note position and check for other signs of slope movement including tension cracks and bulging
Bulging or toe kick often accompanied by seepage	Note position, extent and check for other signs of movement such as tension cracks. Monitor and if movement continues or accelerates inform and the road authority immediately
Seepage, especially if high flows are noted at specific locations (as opposed to widespread seepage along the toe)	Note the position and describe flow. Check for open tension cracks, depressions where ponding can take place (e.g., blocked catchwater drain) or where piping is evident. Monitor the position for any deterioration
Erosion can concentrate water flow and allow wetting up of the slope	As soon as possible take preventative action such as constructing a catchwater drain and reinstating the eroded area with vegetation or gabions

16.3.7.1   Defects

(i) *Waterway.* The assessment of the waterway covers the free flow of water under the bridge, the stability of the waterway and the probable effect on the bridge. Specific defects are the build-up of debris, scouring, erosion damage of gabion stone pitching or paving blocks.

Maintenance actions removal of debris, sand and vegetation; cutting of trees and brushes; erosion repair; and the repair of gabions, stone pitching and concrete slabs.

(ii) *Substructure* (*abutment and piers*). Elements to be assessed include abutment and pier foundations, concrete or masonry abutments, piers, concrete or masonry wing or retaining walls, and abutments bearings. Typical defects are excessive movement or scouring at abutments; foundations; cracking, spalling, honeycombing and staining of concrete abutments and walls; and damaged or contaminated bearings. Maintenance actions range from backfilling or underpinning, to repair of spalling, to cleaning, and to cleaning and replacement of bearings.

(iii) *Superstructure.* The superstructure comprises soffits, deck slabs, concrete beams, steel girders and bracings, and steel trusses. All of these elements need to be assessed in terms of cracking, spalling, corrosion, or damage. Appropriate maintenance activities are repair of spalling, crack sealing, application of protective coating, repair of honeycombed concrete and strengthening or replacement of steel girder or truss elements.

(iv) *Roadway (on the bridge)*. The relevant elements include the asphalt or concrete surface, as well as guard-rails, parapets, drainage, kerbs and sidewalks. Defects on the road surface are the same as for pavements. While others are related to water ponding, cracking and spalling of concrete parapets, and corrosion and damage of guard-rails and metal parapets. Maintenance actions include those of pavements, as well as for concrete and metal elements.

(v) *Expansion joints*. Expansion joints can become dysfunctional due to debris in the joints, vertical or horizontal movement, spalling or cracking of nosings, and missing or displaced fillers. Maintenance entails cleaning of joints, replacement of sealants, and repair of concrete nosings.

16.3.7.2 Recording of Defects

The assessment of bridge and drainage structures should be done by experienced staff and often requires special equipment to be able to inspect all elements. The urgency of the repair can range from critical (to be done as soon as possible), to urgent (to be done within 12 months), to priority (to be done within 2 years), to future (to be done within 5 years).

16.4 Planning and Management

A highway asset management system is universally seen as an essential design support tool (AASHTO, 1990; Markow et al., 1994). An asset management system comprises a number of subsystems. Four of these subsystems are used in the identification, planning and execution of highway maintenance activities, i.e. PMS, bridge management system, unpaved road management system and management system for remedial maintenance. The last is often called a maintenance management system.

Preventive maintenance activities should be managed, planned and programed by means of a pavement (paved and unpaved) management system as discussed in Chapter 18. A formal management system can be utilized to plan, program and monitor remedial maintenance, while emergency maintenance has to be managed on an ad hoc basis (Robinson, 1986; Robinson et al., 1998; Zimmerman and Peshkin, 2003).

A highway agency's budget for highway maintenance can come from:

- A PMS for the preventative maintenance requirements
- A bridge management system
- A remedial maintenance management system
- An allowance for the repair of unsafe and unacceptable conditions caused by accidents and natural disasters

While pavement and bridge management system are widely used, specific management systems for remedial activities have not been developed universally. There are many reasons for this, amongst others,

- The definition of remedial maintenance is not uniform.
- Remedial work can be done either on contract or by in-house staff. In both cases the requirement for such a system varies considerably. For example, some authorities want the system to incorporate the management of material, staff input and cost. Such a system has often to be linked to a financial and payment system.

Examples of two types of systems to manage the remedial maintenance activities will be given here. The first is a system to identify the remedial maintenance needs, as well as to plan and program the execution. The second is to monitor the expenses and costing.

16.4.1 Remedial Maintenance Management System

This system has the same elements as a PMS, i.e.:

Referencing system and inventory. All highway elements must be identified and referenced. The reference system can either be a kilometer position or an X, Y and Z coordinate.

Inspection or assessment of the element. The condition of each element is described in terms of a predetermined rating scheme. The rating is normally done in terms of type, degree and extent of the defect in the element. Degree refers to the severity of the defect (e.g., poor or good riding quality, width of crack, depth of rut, degree of reflectivity of a road sign, etc.), while the extent defines the frequency of occurrence (e.g., isolated or general) or the affected area (percentage of area with cracks). Table 16.13 to Table 16.15 depict typical inspection procedures for flexible highways roads, unpaved roads, and roadside features, respectively (OECD, 1990b). The examples indicate criteria for three levels of extent. The severity of the defect is classified as low, moderate, or severe. Information in Tables 16.4 and 16.10 can be used to devise criteria for the three severity levels. The frequency of the assessments will depend on the use of the information. If it is only for budgeting purposes it can be once a year. If it is used to plan and monitor the execution of the maintenance, the assessment can be done on a monthly basis. The length of a section to be assessed range typically from 100 m to 1 km if the information is to be used for more than a network level assessment. The survey is typically done by walking or slowly driving the section.

Determination of the maintenance needs. The assessed condition (or rating) is compared with predetermined minimum or maximum allowable condition or intervention criteria. The acceptable condition is normally based on economic and safety evaluations and criteria and will vary for different highway categories and regions. Where the assessed condition falls outside the acceptable condition, a certain type of maintenance is required to rectify the condition. For example, if the accepted condition is that no pothole must be open for more than 12 H, pothole repair teams must be mobilized to fix the pothole. Another example is improvement of riding quality. If the maximum accepted average riding quality on a sector of road is 4 mm/m and the measured (or assessed) riding quality is 6 mm/m remedial action such as the application of a thin asphalt overlay is required.

Costing. Once the required maintenance action is defined, unit costs for each identified maintenance activity are used to determine the required budget.

Prioritization and optimization. If the required budget exceeds the available budget a process of prioritization and optimization must be followed to select the maintenance activities which should be executed.

Scheduling of the execution. The identified maintenance activities must be scheduled for execution taking cognizance of the urgency, the available resources, traffic conditions (e.g., avoid work during peak traffic flows), etc. This is followed by the execution and the monitoring of the quality of work and the costs, either in-house or on contract. This information should ideally be fed back into the management system to adjust the unit costs and review the acceptable criteria.

16.4.2   Expense and Costing System

The objectives of this system are to capture requests for remedial or emergency maintenance actions, to measure the work done, and to keep track of expenses. This system is often called a maintenance management system, which can be defined as: "the process of organizing, scheduling and controlling maintenance activities in order to make the best possible use of resources available." (OECD, 1990a).

Requests for maintenance work can come from needs identified through regular inspections, customer complaints, and accidents or disasters. These requests are then reported on a job instruction form, which is given to the executing party for implementation within a predetermined time. The job instruction form (see example in Table 16.16) contains information on the type of maintenance work required (e.g., replacement of a damaged guard-rail), the position and an estimate of the quantity involved (number of guard-rails and posts).

A proper system would then allocate material and staff resources to the identified maintenance needs, and based on the urgency for repair and available resources devise an optimized implementation plan.

After the work has been completed, the work is inspected to confirm compliance with the standards and to verify the quantities. A certificate for payment can then be issued based on the quantities and an agreed schedule of rates.

TABLE 16.13 Example of Visual Inspection Form for Flexible Highways

Name of inspector:

Date:

District:

km from:_____ to _____

Weather: clear or rainy

Carriageway surface: ☐ wet/☐ drying/☐ dry

Topography: ☐ flat/☐ rolling/☐ hilly

Pavement type:

Carriageway width:

CARRIAGEWAY		SEVERITY					
		LEFT			RIGHT		
DAMAGE TYPE	EXTENT	LOW	MODERATE	HIGH	LOW	MODERATE	HIGH
Rutting	< 10%						
	10–50%						
	> 50%						
Corrugations	< 10%						
	10–50%						
	> 50%						
Depressions	< 10%						
	10–50%						
	> 50%						
Transverse cracking	< 2(no./100 m)						
	2–15						
	> 15						
Longitudinal cracking	< 10%						
	10–50%						
	> 50%						
Alligator cracking	< 10%						
	10–50%						
	> 50%						
Holes	< 5(no./100 m)						
	5–15						
	> 15						

(*Continued*)

TABLE 16.13   Continued

Edge distress	< 10%						
	10–50%						
	> 50%						
Stripping, fretting or raveling	< 10%						
	10–50%						
	> 50%						
Stripping, fretting or raveling of surface	< 15%						
	15–30%						
	> 30%						
Bleeding	< 5%						
	5–50%						
	> 50%						
Remarks:	_____						
	_____						
	_____						

Note: The information in Table 16.10 can be used as guide for the definition of low, moderate and high severity.

16.5 Maintenance Criteria

Road authorities use different criteria to trigger highway maintenance actions (also called intervention levels) and to determine response times or urgency. The criteria depend on the traffic volumes, importance of the highway, available manpower and equipment, and budgets.

Defects can be classified according to the required response:

(a) *immediate response* (i.e., within hours). These are normally the ones which cause major traffic disruptions and impede the safety of the highway. Examples are debris on the road, broken down vehicles, severely damaged or missing important road signs (e.g., stop signs), large potholes or punch-outs, rock or earth slides, wash-aways and collapsed culverts. These maintenance actions would fall under emergency action.

(b) *intermediate response* (e.g., within a day or week). Defects such as all severely damaged or missing road signs, damaged guard-rails, ineffective fences, severe lane-shoulder drop-offs, sections of highway with very low skid-resistance in critical areas (e.g., sharp curves, intersections), and missing bridge balustrades need to be repaired within a short time. These defects affect the safety of the road and cannot be left unattended for a long period of time. The appropriate maintenance actions will fall under emergency or remedial maintenance.

(c) *delayed response* (i.e. within a period of months). Remedial maintenance actions such as grass cutting, cleaning of drainage structures, cleaning of roadside drains, sealing of severe cracks, cleaning or replacement of defective road signs, spot regraveling of unpaved roads, etc., can be executed according to a schedule based on the amount of work and resources available, as long as it is done at certain critical times, e.g., cleaning of culverts before the rainy season.

(d) *preventive maintenance*. Preventive maintenance actions are planned, programed and executed on an annual basis.

Typical maintenance criteria for paved and unpaved highways are given in Table 16.17 and Table 16.18, while Table 16.19 contains criteria for other highway elements (TRRL, 1981; AUSTROADS, 1991).

TABLE 16.14 Example of Visual Inspection Form for Unpaved Highways

Name of inspector:

Date:

District:

km from:_____ to _____

Weather: clear or rainy

Carriageway surface: ☐ wet/☐ drying/☐ dry

Topography: ☐ flat/☐ rolling/☐ hilly

Pavement type:

Carriageway width:

| CARRIAGEWAY | | SEVERITY | | | | | |
| | | LEFT | | | RIGHT | | |
DAMAGE TYPE	EXTENT	LOW	MODERATE	HIGH	LOW	MODERATE	HIGH
Rutting	< 10%						
	10–50%						
	> 50%						
Corrugations	< 10%						
	10–50%						
	> 50%						
Loss of camber	<10%						
	10–50%						
	>50%						
Gravel thickness	<10%						
	10–50%						
	>50%						
Erosion gullies	<10%						
	10–50%						
	>50%						
Potholes	<5 (no./100 m)						
	5–15%						
	>15						
Clay	<5%						
	5–50%						
	>50%						

Remarks: _____

_____

Note: The information in Table 16.10 can be used as guide for the definition of low, moderate and high severity.

TABLE 16.15　Example of Roadside Feature Condition Survey Form						
Name of inspector:						
Date:						
District:						
km from:_____ to _____						
Weather: clear or rainy						
Carriageway surface: ☐ wet/☐ drying/☐ dry						
Topography: ☐ flat/☐ rolling/☐ hilly						
Pavement type:						
Carriageway width:						
Roadside				Left		Right
Damage type				Left		Right
Roadside elements	Shoulder	Deformation	1	2 3	1	2 3
		Scour	1	2 3	1	2 3
	Side-drains	Siltation	1	2 3	1	2 3
		Scour	1	2 3	1	2 3
	Debris or vegetation encroachment		1	2 3	1	2 3
	Obstacles or obstruction		1	2 3	1	2 3
Roadside/ Furniture	Chainage					
	Dirty		1	2 3	1	2 3
	Damaged		1	2 3	1	2 3
	Missing		1	2 3	1	2 3
			1	2 3	1	2 3
Remarks: _____ _____						
Note: 1, 2 and 3 are used to define the condition of the element.						

TABLE 16.16 Example of Job Instruction Form for Highway Maintenance

Job Category:	Routine	Special	Accident	Other

Position of Work:	Carriageway	Interchange	General

Route: ....   Section: ....

Start km: ....   End km: ....

Direction:	Positive (+)	Negative (−)	Both

Description of Work: ....

....

....

....

....

Completion due by

or   or   Date: ....   Time: ....

Respond by

Engineer's Representative: ....   Date: ....   Time: ....

Commenced on: Date: ....   Time: ....

Completed on: Date: ....   Time: ....

Allocated Project:	Item No	Rate	Quantity	Cost
Remarks				

(Continued)

TABLE 16.16   Continued

Example of Job Instruction Form for Highway Maintenance				
		Total estimated costs		

Instruction received

on behalf of Contractor: ....   Date: ....   Time: ....

16.6   Remedial Maintenance Activities

(a) *Road sign repair and replacement.* Road signs include all road signs, guide posts, and delineators. Road signs must be clearly visible at all times, including at night. Road signs can lose their effectiveness from deterioration due to environmental weathering or traffic accidents or vandalism. Missing or severely damaged road signs must be immediately replaced. Others should be replaced or repaired before they reach a critical level of effectiveness, e.g., cannot be seen at night-time under dimmed vehicle head lights. The life of a road sign is typically 7 to 8 years.

Road signs are normally assessed annually by means of visual inspection or the measurement of the reflectivity. Repair and replacement can then be programed. Repair methods entail cleaning, repair of indentations, filling of holes, repainting and the replacement of adhesive coatings. Cleaning agents that can be used include kerosene or mineral spirits (for tar and bituminous products on the sign), a solution of sodium hypochlorite (for pollen and fungi), petrol, heptane or napthane (for certain types of graffiti).

Repairs of road signs and delineators include maintenance of the posts which can either be timber, concrete or metal.

TABLE 16.17   Typical Maintenance Criteria — Paved Highways

Defect	Criteria	Action	Response
Cracking but not rutting	Small local areas	Local sealing	Delayed
	Large areas	Surface sealing	Delayed
Stripping or fretting	Small local areas	Local sealing	Delayed
	Large areas	Surface sealing	Delayed
Fatting-up	Small local areas	Sanding	Delayed
	Large areas	Surface dressing	Intermediate
Rutting	<10 mm	No action	Delayed
	10–25 mm	Overlay	Intermediate
	>25 mm	Reconstruction	Intermediate
Cracking and rutting	<1 m/m^2	No action	Delayed
	1–2 m/m^2	Surface dressing	Delayed
	2–5 m/m^2	Patch and overlay	Delayed
	>5 m/m^2	Reconstruction	Delayed
Potholes	Fewer than 10 holes/100 m	Patch	Intermediate[a]
	10–20 holes/100 m	Patch and overlay	Intermediate[a]
	More than 20 holes/100 m	Reconstruction	Intermediate[a]
Reflection cracking	Widely spaced cracks	Crack sealing	Delayed
	Closely spaced cracks	Surface dressing	Delayed
Edge breaks	Visual assessment	Patch road edge and repair shoulder	Intermediate[a]
Deformation (shoulders)	Visual assessment	Fill	Intermediate[a]

[a] Urgent if a safety hazard.

TABLE 16.18 Typical Unpaved Highways

Defect	Criteria	Action	Response
Corrugations	Less than 15 mm mean depth	Dragging or brushing	Delayed
	15–40 mm mean depth	Grading	Delayed
	Greater than 40 mm mean depth	Heavy grading	Intermediate
Rutting (unpaved)	Less than 15 mm mean depth	Dragging or brushing	Delayed
	15–40 mm mean depth	Grading	Delayed
	Greater than 40 mm mean depth	Heavy grading	Intermediate
Potholes	Fewer than 15 holes/100 m	Fill	Intermediate
	15–40 holes/100 m	Grading	Intermediate
	More than 40 holes/100 m	Heavy grading	Intermediate
Defective camber or crossfall	Less than 1 in 25 (4%)	Grading	Delayed

(b) *Safety barrier repair and replacement.* Safety barriers, guard-rails and New Jersey concrete barriers are placed in areas where extra safety precautions are required such as at high fills, at sharp curves and as direction dividers. Any ineffective or missing guard-rails must immediately be repaired, replaced or installed. This is normally the case after a traffic accident. Where accidents often occur, the road geometry should be checked and improved, if necessary. In general guard-rails should be inspected on a regular basis (monthly). The inspections for guard-rails include observation of the overall alignment, the condition of the reflectors, the poles and the spacer blocks, the overlapping, the height of the guard-rails and missing bolts. Guard-rails, which are effective but damaged, or deteriorated can be repaired and replaced as part of a maintenance program.

(c) *Mending of fences.* Fences are primarily provided to prevent animals and livestock from getting onto the road. Broken or missing fences should be repaired and replaced immediately. Fences can be damaged or lost as a result of ageing, accidents, theft or cutting to provide access to the road. The condition of fences is normally inspected on a monthly basis.

(d) *Grass-cutting and pruning of vegetation.* Grass-cutting and pruning of vegetation are required for reasons of visibility (at curves and intersections), drainage (shoulders and side-drains), plant invader control and fire hazard. The width of cutting grass can vary from 4 m from the shoulder to the full road reserve depending on the road function, layout and climatic conditions. The frequency of grass-cutting varies considerably due mainly to climatic conditions, but is normally done at regular intervals ranging form 1 month to 6 months. Alternatively performance criteria are used which specifies a maximum height of grass in the road reserve.

(e) *Removal of litter.* Litter needs to be removed from the road reserve on a regular basis for aesthetic reasons and to prevent clogging up of the drainage system. The frequency of the removal of litter can

TABLE 16.19 Typical Roadside Features and Signs

Defect	Criteria	Action	Response
Debris on road	Any	Remove	Immediate
Broken down or damaged vehicles	Any	Inform traffic police/remove	Immediate
Silt in drains and ditches	Ditch depth reduces to less than 1 m	Clean out	Delayed
Scour in side-drains	Erosion channels in ditch	Build check-dams fill	Delayed
Drains too shallow	Ditch depth less than 1 m	Deepen	Delayed
Silted or blocked bridges or culverts	Visual assessment	Clean out	Delayed
Outfalls scoured at bridges or culverts	Any	Build scour control works fill	Delayed
Structural damage	Any	Repair	Delayed
Dirty or corroded road signs	Any	Clean or repaint	Intermediate
Damaged or unreadable road signs	Any	Repair and replace	Intermediate
Missing road signs	Any	Replace	Immediate
Scour	Visual assessment	Fill	Delayed
High vegetation growth (shoulders)	Interferes with line of sight	Cut	Intermediate

range from 1 month in residential and rest areas, to 6 months in low volume traffic areas where drainage systems cannot easily be effected.

(f) *Drainage.* Drains are constructed to remove or keep water from the pavement surface and groundwater road formation. Subsurface drains are used to intercept water flowing towards the road formation while a variety of drain configurations are used to intercept or remove water from the road surface and shoulders. The latter are:

   (i) Side-drains or table-drains are channels parallel to the road with the function of collecting water that has fallen on the carriageway, the shoulder, or the batters of a cutting.
   (ii) Berms are shallow embankments or mounds usually placed transversely in the side-drain to deflect the flow of water.
   (iii) Miter-drains (or turn-outs) provide a means for the water in the side-drain to be discharged away from the road. The spacing depends on the material and gradient. On erodible soils it can vary from 50 m spacings for gradients of <1% to every 10 m for gradients of >10%.
   (iv) Batter-drains (chutes) are drains constructed to channel water down either a fill or a cut batter.
   (v) Catch water drains are positioned on the upslope side of a cut face parallel to the road to intercept sheet flow and to prevent erosion of the cut face.

These drains become defective due to silting (or blockages) and erosion. The silting and blockages can be removed and the erosion be repaired. More often than not additional measures such as reshaping and realignment, reduction of ditch length and special erosion protection measures will be required to reduce the amount of silting and erosion.

Defects to the surface erosion system can easily be picked up by visual inspections. Regular inspections (often monthly) are recommended and remedial work should be done before rainy seasons.

The effectiveness of subsurface drains is more difficult to assess and use has been made of stand-pipes, flow notches and piezometers to monitor the water-table levels and outflow. Most of the problems with the ineffective performance of subsurface drains can be related to the incorrect selection of materials, poor installation and inadequate outlets.

Preventative and remedial maintenance may also warrant the installation of drainage during the repair of pavement defects such as potholes, and crocodile cracking.

(g) Patching (flexible pavement) entails the removal and repair of edge breaks, potholes, showing severely cracked and deformed areas and surface failures.

The patching can either be confined to the surface or involve replacement of a number of pavement layers. With all patching it is important to remove the failed area entirely and preferably cut 50 to 100 mm into the sound material as well. The side and bottom of the patch should be squared-off and all loose material removed. Shallow patches, up to 100 m can be done with cold and hot asphalts, while deeper patches can be filled with a high quality gravel, crushed stone, cement or emulsion-stabilized material, or large aggregate asphalt before the bituminous surfacing is placed. The selection of the material will depend on the strength required (taking cognizance of the existing bearing capacity of the surrounding layers and the required design life), the traffic volumes, available materials, and required performance. For example cold-mix asphalt does not perform as well as hot-mix asphalt in patches. A surface dressing can be used instead of asphalt if the support layers in the patch are of high quality, e.g., stabilized with cement or bitumen.

(h) *Crack sealing (flexible pavements).* Crack sealing comprises the cleaning of cracks, the application of a weed killer and primer (if necessary), and the application of a bituminous product as sealant. A bitumen emulsion is normally used to fill cracks less then 3 mm wide. Application is preferably by means of a lance fitted with a small nozzle, but the product can also be sprayed and spread with a broom or squeegee, in which case if must be covered with grit or clean sand. A primer should be used for cracks of more than 3 mm and applied to cover the sides of the top 20 mm of the crack. A bitumen emulsion modified with a polymer or hot poured PMB is the most effective for cracks more than 5 mm in width. Very large cracks can also be effectively filled with rubber crumb slurry or covered with a geo-textile strip of 100 to 150 mm. In such case a tack coat is sprayed and the geo-textile strip placed centrally over the crack and rolled. The geo-textile strip must be covered with clean sand or grit if not overlayed by a surface dressing or asphalt. Large and well-defined cracks can effectively be sealed.

When cracks are fine and closely spaced it is often more cost-effective to apply a surface dressing or thin asphalt overlay directly.

(i) Partial depth patching (rigid pavement) is used to correct severe spalling at joints comers, and cracks, or damage to the surface of the slab which does not extend more than 50 to 100 mm into the slab. The latter can be caused by mechanical damage, spillage of certain hazardous materials or burning of material on the slab during accidents. The patching entails removal of the spilled (or damaged) section to below the damage, cleaning the excavated area, the application of a bond breaker and the placement of fresh concrete. A normal or a quick setting concrete can be used depending on the time the patch can be kept closed to traffic. For normal concrete this time is about 3 days.

No provision has to be made for the replacement of reinforcing, tie or dowel bars since partial depth patching does not extend to the depth of the reinforcement. Provision must be made for joints and joint seals.

(j) Full depth patching (rigid pavement) entails the replacement of sections of slabs to their full depth. This includes the replacement of reinforcement, dowels, tie bars and joint seals. Full depth patching is used to correct blow-ups, corner breaks, durability D-cracking, punch-outs, joint load transfer associated deterioration and spalling which extends to more than half the depth of the slab.

Different patch layouts are used depending on the position of the defective area i.e., partial-lane (where the distress is confined to only a portion of the lane), full-lane (where the distress occurs across more than one half of a lane), and mid-slab (where the distress is more than 0.8 m from a joint or edge). In the first two cases joints and pavement edge define the outside edges of the patch in jointed pavements. Special care must be taken to reinstate dowel bars and joints. At CRC pavements the patch may extend to the outside pavement edge and longitudinal joint.

As in the case of partial depth patching, the concrete mixture selected depends on the available curing time before the patch must be opened to traffic.

(k) Joint and crack sealing (rigid pavement) is one of the major maintenance activities for rigid pavements — jointed pavement in particular. All joints and cracks, i.e., transverse as well as longitudinal, should be sealed to prevent water from infiltrating into the pavement effectively. A number of sealants can be used, e.g., preformed compression seals, hot-powered sealants and cold-applied sealants. Hot-poured sealants are injected into the prepared sealant reservoir through nozzles shaped to penetrate into the joint and fill the reservoir from the bottom. The sealant should be applied when the ambient and pavement temperature is at least 10 °C and filled to 4 to 8 mm from the top of the pavement surface.

Cold-applied sealants comprise two components combined in a special mixing head before applied.

All existing sealants should be removed from the joints and cracks; the openings widened (or routed), if necessary, to obtain the required opening/closing conditions and reservoir, and openings be blown clear before the sealant is applied. All sealants must comply with specifications. All sealant types can perform satisfactory, but it is advisable to obtain information on local experiences before a certain type of sealant is recommended.

The main factors affecting joint and crack sealant performance include; the joint and crack, the sealant reservoir shape and width; bond between the sealant and the concrete; and the properties of the sealant (extensibility and durability).

(m) Blading (or grading) is an important maintenance activity for unsealed highways and intended to keep the road well drained and the riding quality in a satisfactory condition. The blading process consists bringing in material from the sides and cutting down corrugations and filling low spots. The effectiveness of grading is increased if a roller is utilized and the material is moist.

(n) Spot regraveling or patching of the wearing courses of unpaved roads is done where the material performs satisfactory but small potholes are present in isolated cases. A material with at least the same properties as that of a well-performing wearing course should be used.

(o) *Mill and replacement.* When the surfacing is in a poor condition the top 40 to 50 mm can be removed and replaced with asphalt cement having the same thickness as the removed portion. This is an effective technique where the surfacing has aged and is cracked, but no structural strengthening is required. This method is particularly cost-effective where only portions of the road, e.g., only the slow lane, needs attention or where clearance requirement limits the addition of layers of the road.

An alternative to mill and replace is HIPR also called remixing. The process entails the heating, scarifying, rejuvenation and remixing of up to 50 mm in depth of aged asphalt (FHWA, 2001).

HIPR is an alternative to the milling and placement of a distressed asphalt layer. The major advantages are the potential savings in transport and processing costs, and the reutilization of existing materials. HIPR does not address any structural deficiencies since only the top 50 mm, at most, is treated and possible changes to the composition of the existing mix are limited. The structural equivalency should be the same as that of conventional asphalt.

Pavements exhibiting base failures, irregular packing and poor drainage, are not suitable drainage candidates for HIPR.

(p) Slab jacking is the process of pumping cement grout under pressure beneath the slab to slowly raise the slab until it reaches a smooth profile. Slabs can be lifted up to 9 mm. This procedure can be used to correct depressions and faulting but will not eliminate the causes of these defects. The areas to be treated can fairly easily be identified as opposed to the void to be undersealed.

A general guideline for the position of the holes for slab jacking is that the hole should be placed in about the some location as hydraulic jacks would be placed if it were possible to get them under the slabs. Holes (60 mm in diameter) should not be spaced less than 500 mm from a transverse joint or slab edge, and 1 to 2 m centre-to-centre.

The slab is lifted by pumping a cement grout mixture (similar to that used for undersealing) into the holes at a pressure of about 1.5 MPa. Typical gradings of the grout are given in Table 16.20 (AUSTROADS, 1991).

The slab is gradually lifted in increments of less than 6 mm lift/hole. A string line method or survey can be used to control the levels. Care must be taken not to contaminate the drainage system.

(q) *Retexturing (grinding, grooving and cold milling)*. Diamond grinding, grooving and cold milling are three forms of retexturing and surface restoration. These maintenance activities address poor skid resistance, surface water (transverse) drainage, faulting, rutting from studded tires, map cracking, sealing and pop outs.

   (i) *Diamond grinding*. Patterns cut into the hardened concrete with closely spaced diamond blades. The spacing of the blades will depend on the hardness of the aggregate and normally varies between 2 and 3 mm. The major purpose of grinding is to remove material, to provide a smooth surface and to improve skid resistance. Grinding is mainly used to remove faulting and ruts, improve riding quality and skid resistance. Grinding can be performed only at faulted joints but must be feathered back. A general guideline is 10 mm for every 1 mm of faulting removed. Grinding will correct the levels, but not prevent further faulting. Grinding has been found not to be cost-effective if the faulting is more than 6 mm. In such cases milling is more effective. This must be done by more extensive maintenance such as undersealing and slab replacement.

   (ii) *Grooving*. Patterns cut into the hardened concrete with a center-to-center blade spacing of 20 mm or greater. The major purpose of grooving is to reduce hydroplaning accidents i.e. to increase the texture depth and improve the drainage.

TABLE 16.20   Typical Cement Grout Gradings

Seive Size (mm)	Percent Passing	
	Fine	Coarse
2.36	100	100
1.18	85–100	95–100
0.15	14–45	15–40
0.075	10–25	5–25

(iii) *Cold milling*. Use of carbide teeth cutting bits to chip off the surface of the pavement primarily to remove material and provide a textured pattern. This is an effective may of removing surface defects such as map cracking and scaling, and to produce of roughened clean surface for bonding a concrete overlay.

(s) Recementation of cracks has been used to restore the structural integrity of cracked slabs. Recementation is accomplished by injecting a liquid epoxy under pressure into the cracks. The pavement condition will be changed since the cracks are removed, but pumping and faulting rates may not be greatly changed. The ingress of water into the cracks should be greatly reduced. The performance of recemented cracks has been very poor to good.

(t) *Bridge and culvert repairs*. Preventative maintenance of bridges and culverts primarily comprises cleaning and clearing of the water pathways, controlling of scour and erosion and the repair of minor structural damage.

(i) The waterways should be cleared of wind blown debris, water-borne soils, gravel and vegetation. The area in the vicinity of the structure should also be kept clear of undergrowth and rubbish. Regular usual inspection programs should be used to identify defects. The assessed severity can range from free flow, where no action is required to total obstruction where action is required before rain. It is important to get the waterways cleared before rainy seasons and after flooding.

In same cases silting is caused by design deficiencies such as incorrect waterway openings, and incorrect culvert invert levels and longitudinal slopes. Some of these can be alleviated by the construction of protective structures upstream, but will in most cases require major reconstruction work.

(ii) *Control of scour and erosion*. This type of damage consists of the loss of materials from both ends of the culvert, or the erosion of slopes and bedding at bridges. The erosion can affect the stability of the structure. The damage can be prepared by backfilling of the eroded areas and the installation of gabion protection, rock beds, and concrete protection structures.

(iii) Damage can comprise degradation of parapets, guard-rails, kerbs, sidewalks or roadway surface. If any of these defects are significant enough to endanger the structure or the safety of the traffic immediate action must be taken. Normal maintenance actions include tightening of guard-rails, replacement of joint filler, replacement of concrete nose and seals, crack sealing and repainting.

(iv) *Minor damage to the sub- and superstructure*. These include concrete deterioration, cracking, and scouring or displacement of foundations. Repair of these defects are often of a very specialized nature requiring expertise, special materials and equipment.

16.7 Preventive Maintenance Activities

Highway agencies are increasingly applying low-cost pavement preservation treatments to extend the service life of the pavement instead of rehabilitating them when they are in a poor condition. Emphasis is placed on designing and constructing long-lasting structural pavements where very little strengthening is required over a long design period with timely, relatively low-cost preventive maintenance at regular intervals. In France and South Africa (Batac and Ray, 2001; FHWA, 2002), for example, the philosophy is to build a strong base so that the wearing surface needs repairing only every 10 to 15 years and structural overlay is only required every 20 years. Surface dressing, thin (40 to 60 mm) and ultra-thin asphalt overlays, as well as the use of proven polymer-modified binders (PMB) and high quality aggregates (chippings) have been found to be cost-effective preventive maintenance treatments, warranties, usually 4 years in duration, are often used in contracts when applying preventative maintenance.

Preventive maintenance actions have as objectives the restoration of the condition of the highway and reducing the rate of deterioration. The actions are identified and scheduled through some form of management system.

(a) Rejuvenation is an application of a diluted emulsion or commercial product with the aim of replenishing the bituminous binder. The rejuvenation makes the binder more elastic and pliable, and retards the aging (hardening) process. Small (hairline) cracks are normally also sealed in the process. Single and double surface dressings greatly benefit from the application of a rejuvenator every 3 to 4 years. The life of slurries can also be increased, but care must be taken that the rejuvenator penetrates sufficiently. A typical diluted emulsion application is 0.8 l/m^2 of an emulsion with 30% bitumen (i.e., a net bitumen application of 0.24 l/m^2).

The normal practice is to determine the appropriate application through a trial process.

(b) Resurfacing can be one of the following:

   (i) Single surface dressing (single seal) where a single form of binder (the tack coat) is followed by a single layer of chippings. The size of the chippings (or aggregate) can range from 6 to 20 mm. The larger the size of aggregate the larger the application of bituminous binder that can be accommodated. The chipping must have a certain hardness, polishing value, and flakiness.

  (ii) Double surface dressing (double seal) has two layers of chippings and two applications of binder, the second (the penetration coat) placed between the layers of chippings. The size of the second layer of chippings is normally half or less the size of the first layer of chippings, and can be sand used to lock in the first layer of chippings. A slurry can also be used to lock in the first layer of chippings (this is also called a Cape seal). The use of sand or crusher dust to lock the larger chippings is very effective at intersections where the turning movement of vehicles can easily dislodge single chippings. Double surface dressings produce lower texture depths than single surface dressings.

       Typical double surface dressings comprise 20 and 10 or 6 mm chippings or 14 and 6 mm chippings.

 (iii) Inverted double surface dressing comprises the application of a single dressing of small chippings followed by an application of larger chippings. This surface dressing is used on a non-uniform surface to produce a uniform surfacing before the larger chippings are applied.

  (iv) Slurry seal, which is a mixture of crusher sand, bitumen emulsion, water and often cement or lime combined to form a fluid mixture. The mixture can be spread by machine or by hand.

       A slurry seal can be applied as a fine slurry (1 to 3 mm in thickness), a medium slurry (4 to 6 mm in thickness), or a coarse slurry (7 to 10 mm in thickness). Quick-set polymer bitumen can be used instead of the emulsion to increase the setting time.

       Thin slurry is often applied on a non-uniform surface before a single surface dressing, while thick slurry, especially when modified bitumen is used, is effective in filling ruts.

   (v) Otta seal is similar to a single chip seal except that graded chippings (instead of single size) are placed on a relatively thick film of comparatively soft binder which on rolling and trafficking, makes its way upwards through the aggregate (Botswana, 1999).

  (vi) Surface enrichment is the application of a bituminous product such as a diluted emulsion or rejuvenator to add volatiles to an aged bituminous binder. This is a very cost-effective maintenance treatment of brittle surfacing.

 (vii) Thin ($<$50 mm) asphalt concrete wearing courses are used extensively and successfully as preventative maintenance action.

(viii) Ultra thin ($<$25 mm) asphalt concrete wearing courses have also successfully been used as preventative maintenance action. Some of the major advantages of an ultra thin-surfacing are that it can be opened to traffic soon after construction, is normally very rut resistant and can easily be tied-in to the surrounding surfacing.

  (ix) Porous asphalt is an open graded ($>$20% voids) mixture, which is very effective in draining rain water from the surface of the pavement, has a good skid resistance and reduces road noise. It has been used successfully as overlay on CRCP and on roads where water drainage and road noise are issues.

(c) *Undersealing.* The loss of support from underneath concrete slabs is a major factor in the deterioration of the pavement. The support must be reestablished to reduce the rate of further deterioration and before an overlay can be applied.

Undersealing is defined as the insertion of a cement grout mixture or bitumen under pressure beneath the slab through holes in the slab. Only the voids are filled and the support reinstated. Undersealing is a very specialized operation and care must be taken not to lift the slabs and create additional voids.

Various types of cement grout mixtures have been used, e.g., cement–loam topsoil, cement–lime stone dust, cement–pozzolan, and cement–fine sand slurry. The bitumen used is specially prepared penetration grade bitumen heated to more than 200 °C to have the required viscosity to be pumped into the pavement.

Holes of about 50 mm in diameter are drilled in areas where the voids have to be filled. This is typically at joints, medium to high severity cracks, sags and depressions in the slow lanes. Deflections and visual pumping are the primary criteria used in the location of voids beneath the slabs. Provision must be made for the effect of the curling of the slabs when deflections are used. Examples of deflection criteria for the identification of voids are: corner deflection of more than 0.6 mm under an 80 kN single axle measured during early morning or defections in the upper 20 to 40 percentile on CRC pavements.

Holes are drilled through the slab into the sub-base (often to a depth of 100 mm). Holes are typically about 0.6 m from the joints or pavement edge. The grout is then pumped into the holes to fill the voids without raising the slab 1 m above grade. Vertical movement of the slab of less then 3 mm is normally acceptable. Pumping should also be ceased if grout appears at the other holes, joints or cracks or the upward movement is more than 1 to 3 mm.

Undersealing will be successful only if care is taken that the void is not overfilled and new voids thereby created. Undersealing alone does not stop pumping. It should be supplemented with sealing or retrofit drains (Thornton and Gulden, 1980). The filling of voids will restore the slab support conditions, but may not reduce subsequent pumping or erosion. Subsealing does not increase the initial design structural capacity or eliminate faulting. The riding quality is not changed by undersealing, but the rate of cracking will be changed due to the restoration of the slab support conditions. It is conceivable that undersealing will effect the initially, since the slab deflections are reduced.

The effectiveness of undersealing can be assessed by the repeat measurement of deflections or the monitoring of the long-term performance.

(d) *Load transfer improvement.* Perfect load transfer can reduce stresses and deflections to half that of a pavement with no load transfer. All faulty joints and cracks with load transfer of less than 50 to 60%, when measured in the early morning, should be provided with load transfer devices (NCHRP, 1983). A number of different load transfer devices are available, e.g., Vee, Double Vee, Figure Eight, Georgia split pipe device, and dowels.

The short-term experience with load transfer restoration has been satisfactory. The long-term performance has not been established. Load transfer devices will not change, riding quality, but should reduce slab deflections and reduce the pavement deterioration rate.

(e) Pressure relief joints relieve the stresses in the slab due to restriction of the slab movement. A relief joint usually consists of replacing a piece of the slab with a strip of asphalt material. Neither the pavement serviceability (such as the Pavement Serviceability Index, PSI) nor the rate of pumping, faulting or cracking will be changed significantly.

(f) *Provision edge support.* Slab deflections can be reduced by providing edge support. Edge support can be in the form of a tied PCC-shoulder or the installation of an edge beam. Tied PCC-shoulders with 100% load transfer reduce the deflection and stresses of the slab corners by 50%. A 600 mm-wide edge beam can also reduce slab deflections by at least 50%. A wider beam will reduce the deflections even more (NCHRP, 1983).

The provision of edge support does not change the condition of pavement, but it does change the rate of pumping, faulting, cracking and pavement deterioration. The general performance of this type of rehabilitation has been fair to good.

Retrofit (edge, longitudinal or trench) drains have been used to reduce pumping and faulting. Retrofit drains increase the drainage of water from the pavement and can improve the shoulder stability. These drains should be used in conjunction with other rehabilitation activities, e.g., undersealing, grinding, resurfacing. The installation of retrofit drains is a preventive measure, since the condition of the pavement is not changed. The drains can reduce the rate of deterioration and prolong the life of pavement. The effects of retrofit drains on the pavement performance vary.

Drains extended the life of the pavement from 20–25 years to 30–35 years. The rate of faulting after the installation of the drains was found to be about an eighth of the rate before installation. A typical example would be a rate of 0.05 mm/yr compared to an initial rate of 0.43 mm/yr (Ames, 1985). The best results were obtained with the top of the drainage pipe at least 125 mm below the bottom of the slab (Marks, 1981; Ridgeway, 1982).

Retrofit drains can increase the removal of fines, due to the loosening of material during construction of the drains. Retrofit drains can reduce faulting and cracking if the subbase is not highly erodible. Drains should not be used on pavements with erodible subbases, in high rainfall areas under heavy traffic, or when the subbase is broken and rests on an unstabilized subgrade under heavy traffic. The particle removal rate from the subbases can be increased by drains along pavements with subbases with high pumping and erosion rates. These particles will clog filters and drains. Slab deflections can be increased soon after the placement of retrofit drains due to the removal of blocked fines.

The difference in performance of retrofit drains seems to depend on the condition of the pavement and the properties of the subgrade. Retrofit drains used in pavements with pump- and erosion-susceptible subbases will probably not be effective. Drains will increase the rate of water flow over or through the subbase which can increase the erosion. This can be even more pronounced in high rainfall areas. Eroded material clogs the drains. This can cause an accumulation of water under the slab. The same criteria used in the design of nonerodible stabilized layers should be used to determine if retrofit drains will perform properly. Retrofit drains will have little influence if the permeability of the existing sub-base is less than 0.009 cm/sec., but will have appreciable influence on the pore water pressure and pumping when the permeability is more than 0.09 cm/sec (Dempsey et al., 1982).

Retrofit drains are not recommended in pavements with high pumping and faulting without other remedial measures, since the drains will increase the removal of fines. Retrofit drains are recommended for pavements in good condition in areas with high rainfall and heavy traffic. Badly deteriorated pavements with high deflections and broken slabs will probably not benefit from sub-drains.

The design, construction and maintenance of the drains are important. Care should be taken during construction not to disturb the slab and weaken the support conditions, especially along the sides.

(g) *Regraveling*. Material from the wearing course on a gravel road gets removed by the traffic, rain and wind actions and needs to be replenished. Action, i.e., regraveling, is normally taken when the wearing course is 25 mm thick. The properties of a good quality wearing course are:

   (i) Plasticity (PI) between 4 and 15. The lower PI is appropriate for wetter climates and higher traffic volumes.
  (ii) Maximum size < 26.5 mm for ease of compaction.
 (iii) Percentage retained on the 2.36 mm sieve should be between 20 and 60% for raveling resistance.
 (iv) The fines (< 0.075 mm) to sand (< 2.36 mm) ratio should be between 0.2 and 0.6 for the best stability and lowest permeability.

Regraveling gives the best performance if the layer is compacted at optimum moisture content to 95% Modified AASHTO density.

The control of dust can be enhanced by good construction practices, and the use of well-graded materials. The properties of available materials can be enhanced with the addition of better quality natural materials or dust suppressants (or dust palliatives). Table 16.21 contains descriptions of commonly available dust suppressants and selection criteria.

(h) Road markings have to be reapplied on a regular basis. The frequency depends on the environment (faster deterioration in region with high percentage of sunshine days); traffic (faster under high traffic volumes), type of surfacing (faster deterioration on coarser surfaces), and the type of paint used.

TABLE 16.21 Guidelines on the Selection of a Dust Suppressant

	Traffic			Fines					Climate		
	Light <100	Medium 100–250	Heavy >250	<5%	5–10%	10–20%	20–30%	>30%	Rainy	Normal	Dry
Electro chemical	●	○	○	○	●	●	●	○	○	●	○
Salts	✔	✔	●	○	●	✔	●	○	○	✔	●
Organic, non-bituminous	✔	●	○	○	●	✔	✔	●	●	✔	✔
Petroleum products	●	○	○	○	●	●	○	○	○	●	●

✔ = Good, ● = Fair, ○ = Poor.

(i) *Repair of shoulders.* Material on unpaved shoulders, like wearing courses, needs to be graded and replenished with time. The types of appropriate maintenance action are smoothing and reshaping by means of a grinder or by hand; the addition of new material to replace material lost from the action of traffic, water erosion and grading operations; and watering and rolling.

Properties of a good quality shoulder material are: for ease of grading and compaction, the maximum size of aggregate should be 15 mm; for stability, linear shrinkage (LS) should be between four and ten on material passing 425 μm; and for low permeability a uniform grading curve with fines to sand ratio between 0.20 and 0.40 is desirable, along with a Plasticity Index (PI) × % less than 75 μm of 300 maximum, to limit shrinkage cracking.

Care must be taken during construction not to damage the edge of the surfacing during the grading operation. Watering and rolling in conjunction with grading or the placement of additional material will produce a stronger more stable and less permeable surface on the shoulder. The best results are obtained if the shoulder material is compacted at optimum moisture content. Larger material spilt on the road surface should be broomed off. The shoulder should not be allowed to become higher than the edge of the road surface as this will retard run-off.

Where shoulders have been paved, their maintenance should be similar to that of a carriageway with a similar pavement composition. If the shoulder has been stabilized with cement, lime or bitumen, patching with a similar material can be used to address potholes, while large cracks can be sealed.

16.8 Dust Suppressants

A number of products have been developed to eliminate or reduce the loss of fine material, and thereby reduce the development of dust (AUSTROADS, 1991; Foley et al., 1998). The suppressants can be classified into one of four types (also see Table 20.21), i.e.:

(i) *Chlorides or salts (calcium, magnesium or sodium chloride).* These products are most effective for materials with fines (i.e., material <0.075 mm) between 10 and 20%, moderate or low PI and a moderate or higher CBR. The material leaches into the pavement, but can only leach out. The surface can become slippery during wet weather but the products can lose effectiveness during long dry periods.

(ii) Organic, non-bituminous (calcium, lignosulphonate) products are most effective for materials with high percentage of fines (i.e., 10 to 30%) in a densely graded material. The construction process is critical. The product remains effective during long dry periods with low humidity.

(iii) Petroleum-based products (bitumen emulsion, waste oils), typically performs the best with materials containing low percentages of fines (<10%). Maintenance is more cumbersome than with the previous two types where surface defects can be corrected by blading. With petroleum-based products some type of patching is normally required to rectify defects.

(iv) *Electro-chemical stabilizers (sulphonated petroleum, ionic products, and enzymes).* These stabilizers can be effective in all climatic conditions and work best with a material with PIs in excess of 8 and more than 15% fines. A curing period is required.

None of these dust suppressants produces maintenance-free surfaces and allowance must be made for reapplications, regular bladings and patching.

Other measures that have been used successfully to control dust are:

(a) Good construction and maintenance practices such as the provision of a crowned cross-section, well-graded materials, and adequate drainage.

(b) Mechanical stabilization where natural materials are used to enhance the properties of the material to meet the requirements of a high quality wearing surface. (i.e., proper grading and plasticity content.)

The selection of the most appropriate dust control measure should include the consideration of the traffic volumes, the importance of the road, the cost (initial and maintenance), the materials available and the maintenance resources available.

General guidelines are:

(a) for roads carrying less than 50 vehicles/day (vpd) the use of dust suppressant will probably not be justifiable. Mechanical stabilization will be an option.

(b) for roads carrying between 50 and 250 vpd dust suppressants could be viable.

(c) for roads carrying more than 250 vpd sealing of the road will probably be more cost-effective than regular spending on maintenance even with a dust suppressant.

16.9   Design and Construction of Surface Dressings

16.9.1   Design Considerations

The design of a single or double surface dressing entails the determination of the size and amount of chippings to be applied and of the amount of bitumen to be added to partially cover the voids between the chippings.

For a single surface dressing one layer of bitumen (the tack coat) is applied and for a double surface dressing a tack coat and a penetration coat. In both cases an additional application of bitumen in the form of a diluted emulsion (fog spray) can be applied on top of the constructed surface dressing. The chippings can be precoated with products such as diluted bitumen emulsions, cut-back bitumen further diluted with aromatic paraffin, diesel fuel, tar cut back with a suitable tar oil or other commercially available products to enhance the adhesion between the chippings and the bitumen.

The following factors must be considered on the selection and design of a surface dressing (PIARC, 1989; TPA, 1994; Nicholls, 1996; COLTO, 1998).

- *Traffic.* Traffic loading (mainly by commercial vehicles) causes the embedment of the chippings resulting in flushing and bleeding. Larger size chippings are used for higher traffic volumes allowing for more embedment to take place before flushing occurs. There is a limit on the size of chipping that can be used before it affects the ride and becomes too noisy. Surface dressings are not recommended for very high traffic volumes. The upper limit for traffic volumes for which surface dressings can be used varies among road authorities, but is in the order of 4000 heavy (commercial) vehicles/day/lane.

  Although heavy vehicles cause the most damage, cognizance must also be taken of the volumes of light vehicles by increasing the size of the chippings or reducing the application rate of the binder. Another related factor is the traveling speed of the vehicles. Special care must be taken to prevent loose chippings on roads with relatively high vehicle speeds (85th percentile > 100 km/h).

- *Highway layout.* On steep gradients, at traffic circles or in places where frequent stopping and starting occurs, traffic imposes such great stresses on the surfacing that a surface dressing is liable to be damaged, particularly in its early life. For these special conditions the choice of surfacing type and the design are most critical. A normal single surface dressing is likely to prove inadequate even for light traffic conditions. Under these conditions, surface dressings in which binders with higher viscosity than normal (modified binders) are used, together with the use of precoated stone, or surface dressings with a relatively fine texture finished off with a fog spray, or surface dressings in which the stone particles are firmly held in place by slurry, will perform better.

  Normal binder can be used for surface dressings on highways on gradients of less than 6%, while care must be taken in the selection of the appropriate chipping and binder type for surface dressings on gradients of more than 6%. At gradients of more than 10% a surface dressing may not be the appropriate type of surfacing.

  The design of the surface dressing at junctions and crossings is of particular importance since a high skid resistance is required. The texture depth and good polishing characteristics of the chippings must be maintained. This is also a requirement at sharp curves.

- *Temperature and climate conditions.* The ambient temperature influences the embedment characteristics. More embedment takes place at higher road temperatures and the binder application has to be reduced accordingly.

  Aging of the binder is different in different climatic conditions and must be considered in the selection of the binder and, to some extent, the application rate. In areas where aging is severe, modified binders will prolong the life of the surface dressing, as well as a higher application of the binder.

- *Road surface.* The road hardness dictates the amount of embedment that will take place. The hardness is normally measured by some type of penetration test where a steel rod is forced into the road surface under a constant load (Nicholls, 1996) or a load is dropped on a small steel ball (COLTO, 1998). The penetration and temperature are recorded and used to quantify the hardness of the road surface for a specific climate region. The temperature profile of the region affects the long-term embedment characteristics.

- *The surface texture.* The texture of the existing surface has a great influence on the final quantity of binder required. The coarser the surface (greater texture depth), the more binder is required. If the chippings in the new surface are large relative to the nature of the existing texture, the chippings will lie on top of the existing stone and may easily be dislodged. A much greater quantity of binder will be required because of the binder collecting in the voids. If the chipping is small relative to the nature of the existing texture, the chippings could be pushed into the depressions which will lead to bleeding.

  If the existing texture is very coarse or varying, it should first be treated by means of a texture treatment consisting of a slurry seal, or coarse sand, applied to the existing surface in a very thin layer by sweeping or brushing it into the voids.

  The surface texture can be measured by means of laser profiling or the sand patch method. A texture treatment is normally required if the mean texture depth is more than 0.6 mm.

  Dry and porous surfaces tend to absorb binder, resulting in the loss of chippings from the overlaying surface dressing. This can be prevented by the application of a diluted emulsion prior to the surface dressing.

- *Shaded areas.* Areas where the road surface is shaded by trees, buildings, tunnels or bridges tend to be cooler and more resistant to chipping embedment than non-shaded areas. Allowance must be made in the determination of the binder application of lower embedment in these areas.

- *Skid resistance.* The skid resistance is influenced by the micro and macro texture. Skid resistance is of particular importance at intersections, curves and for high volume roads. A chipping with a PSV will improve the micro texture and larger chipping the texture depth (or macro texture).

  Typical criteria for the PSV of chippings are higher than 55 for normal conditions, more than 60 for high traffic volumes and more than 70 in high accident risk areas should as approaches, high gradients, bends and roundabouts.

- *Noise.* Surfacing dressings can be noisy, especially if larger aggregates are used and may not be appropriate in areas where noise is a concern.
- Condition and structural strength of the existing pavement. A surface dressing has no structural strength and cannot be used to improve the bearing capacity of the pavement. It will improve the surface characteristics and prolong the life of the pavement. A surface dressing will not be effective if placed on a pavement in a very poor condition.

16.9.2   Relevant Bituminous Products

(SABITA, 1987; Department of Transport, 1994; AUSTROADS, 1998b): There are five types of bituminous product

  (i) Road grade bitumen is the least expensive of the bituminous products and is used for surface and hot mix asphalt. It must be heated for application and requires sophisticated and expensive plant for spraying or mixing and laying.
 (ii) Cutback bitumen can be used for prime and tack costs, surface dressings and cold-mix asphalts. Less heating of the bitumen is required than for road grade bitumen, but curing is necessary for the solvents to evaporate. One of the main advantages is that cutback bitumens can be used in cold weather. The major disadvantages are high energy consumption and fire risk due to the flammable solvents.
(iii) Bitumen emulsion has a similar application to cutback bitumen, but does not require heating and does not contain the flammable solvents. Emulsions can be applied at low application rates and chemical handling improves the adhesion between the borders and chips. Time is required for the emulsion to "break" (i.e., for the water to evaporate). This time depends on the composition of the emulsion, the type of chipping and the temperature.
 (iv) Tar can be used as prime or in hot-mix asphalt, but is not environmentally friendly and should be used with the utmost care so as not to burn the workers or the environment.
  (v) PMB refer to bituminous products where a polymer has been added to penetration grade bitumen to enhance the properties. The main benefits of this modification are improved consistency, stiffness, cohesion, flexibility, toughness, binder aggregate adhesion and resistance to in-service aging. The modified products also are less temperature susceptible. PMB's are more expensive than conventional binders and the higher cost will only be justifiable where the properties of the concentred binders are adequate to meet the requirements.

The polymers can be divided into three types, i.e.:

- Elastomeric polymer, including SBS (stirene–butadiene–stirene), (stirene–butadiene rubber) and PBD (polybutadiene). These are elastic materials which increase flexibility and toughness in spray seal binders or improve flexibility and deformation resistance in asphalt.
- Plastomeric polymer, including EVA (ethylene vinyl acetate) and EMA (ethylene methacrylate). They form a tougher, more rigid binder compared to elastomeric types, and are used in asphalt to improve deformation resistance as well as for increased durability in open graded asphalt.
- Crumb rubber, usually from old tires. Crumb rubber PMBs provide properties which are similar to elastomeric polymer types.
- Elastomeric and plastomeric polymers can also be clarified as homogenous binder since there are two distinct detectable phases.

Typical conditions where the PMBs can be advantageous are for:

- surface seals where improved resistance to cracking as strain alleviating membrane or strain alleviating membrane inter layer. This type of membrane is required if a surface dressing is placed on active cracks of less than 3 mm (cracks more than 3 to 5 mm wide should preferably be sealed with a crack sealant). Polymer modification will also improve the properties of slurry and can be used where the slurry is used for heavy duty applications (e.g., rut filling).

The reduced temperature susceptibility makes the PMBs feasible in conditions where large daily or seasonal temperature fluctuations are present.

Lastly, surface dressings constructed with PMBs has improved durability making their use attractive in remote areas.

- Hot-mix asphalt where the use of a PMB will enhance the rut-resistance, and the flexible strength of the mix. The former is relevant for asphalts under heavy traffic of steep inclines, intersections and sharp curves, and the latter on relatively highly flexible pavements. Different types of modifiers are used to get the required properties, e.g., electromeric types give a lower stiffness, but higher flexibility, and the plastometric types a high stiffness and deformation resistance. Crumb rubber has been used to increase flexibility and resistance to reflection cracking. Another advantage of the use of a PMB is that a higher film thickness can be achieved which is beneficial in opening mixes.
- The sealing of wide (3 to 5 mm) and highly active cracks.

16.9.3 Design

Although the surface dressing designs are based on the same principles, road authorities over the world have developed their own design procedures. These design procedures take cognizance of local climatic conditions, properties of locally available chippings, available binders, preferred construction techniques, user requirements (e.g., on skid levels, noise), etc. Detailed descriptions of the design and construction procedures can be found in manuals and guidelines developed by these road authorities, e.g., British (Nicholls, 1996), Southern Africa (COLTO, 1998), and AUSTROADS (AUSTROADS, 1998c)

The basic procedures followed in each of these are:

- *Selection of the most appropriate type of surfacing, size of chipping and type of binder.* Table 16.22 contains some broad guidelines. A decision must also be made on the use of precoated chips and a fog spray. The decision depends mainly on the chemical and mineralogical composition of chips. Precoated chips have successfully been used in South Africa and Australia (FHWA, 2002). Commonly used chipping sizes for single surface dressings are 9.5 (or 10 mm), 13.2 (or 14 mm) and 16 mm; 19 (or 20 mm) chippings tend to create a very noisy surface, while 6.7 (or 6 mm) chippings are only applicable for very low volume roads. The larger chippings are used for higher traffic volumes since the higher the traffic, the lower the binder application and larger chippings require more binder. If small chippings are used for high traffic volumes the binder may not be enough to keep the chippings down.

  Double surface dressings perform best if the second layer of chipping is half the size of the first layer e.g., 13.2 and 6.7 mm or 19 and 9.5 mm.

TABLE 16.22 Guidelines on the Selection of a Surfacing Type

| | Surface Dressing | | | | |
	Coarse Slurry	Single (with Conventional Binder)	Double (with Conventional Binder)	PMB	Hot-Mix Asphalt
Traffic	Low	Low	Medium	High to very high	High to very high
		Medium (larger chippings)	High (larger chipping)		
Gradient	Low	Low	Medium	All	All
Maintenance capacity	High	High	Medium	Same as conventional	Low
Surface condition	Fair to good	Fair	Fair to poor	Poor	Poor
Expected performance	4–7 years	7 to 10 years	9 to 12 years	10 to 15 years	10 to 15 years

- *Determination of the application rate of the chippings.* The application rates depend on the size of the chipping and the required density of the chipping matrix, i.e., lean to dense.
  (i) Different countries use different sieve sizes, e.g., 13.2 or 14 mm chippings are basically the same.
- *Calculation of the traffic volumes to be used on the design.* The traffic volumes used are the daily volumes, as opposed to the cumulative volumes over the structural life of the highway used in pavement and overlay design. The traffic volumes used are the daily total traffic volume, the number of commercial (heavy) vehicles/day (Nicholls, 1996) or a combination of heavy and light vehicles, e.g., an adjusted volume which is equal to the light vehicles plus 20 to 40 times the heavy vehicles (COLTO, 1998)/lane/day.
- *Determination of the application rate of the binder.* The application rate of the binder is a function of the traffic volume and the size of the chippings [often expressed in average least dimension (ALD)]. The binder application rate is normally calculated as residual (or cold) bitumen and adjustments must be made to convert the amount to the equivalent amount for the binder to be used. Most design methods contain graphs from which the application rates can be determined).
- Adjustment of the determined binder application rate to cater for embedment, high gradients, shaded areas, climatic conditions, and condition of the pavement surface.
- Conversion of the calculated residual binder to that of the binder to be used (i.e., equivalent binder).
- A further reapportionment of the binder is necessary if the surfacing is a double surface dressing and if a fog spray is used.
- *Specification of the chippings and binder.* Specifications vary among road authorities, but most contain requirements grading, shape, hardness, polishing, and absorption for the chippings. Typical specifications for the chippings are given in Table 16.23. These standard bitumen specifications are used.
- The design procedure is outlined in Table 16.24, while Table 16.25 and Table 16.26 contain examples.

16.9.4 Preparation of a Pavement for Resealing

All existing pavement defects should be corrected in advance of the resealing to ensure effectiveness of the resealing. This entails the following:

- Repair of potholes, wide cracks and other structural deficiencies.
- *Sealing of cracks.* Where the extent of cracking is not excessive the individual cracks can be sealed. In cases where the cracking is extensive the treatment of the entire surface should be considered as it may be more cost-effective. Where cracks are small (less than 2 mm) the tack coat applied as part of the surface dressing is normally sufficient to seal the cracks. If the cracks are larger (2 to 5 mm)

TABLE 16.23 Typical Grading Requirements for Chippings

Sieve Size (mm)	19.0 mm Nominal Size	13.2 mm Nominal Size	9.5 mm Nominal Size	6.7 mm Nominal Size
37.50	–	–	–	–
26.50	100	–	–	–
19.00	85–100	100	–	–
13.20	0–30	85–100	100	–
9.50	0–5	0–30	85–100	100
6.70	–	0–5	0–30	85–100
4.75	–	–	0–5	0–30
3.35	–	–	–	–
2.36	–	–	–	–
Fines content: Material passing a 0.425 mm sieve (max)	1.5	1.5	1.5	2.0
Dust content: Material passing a 0.075 mm sieve (max)	0.5	0.5	0.5	1.0
Flakiness index	30	30	35	35

TABLE 16.24	Surface Dressing Design

1.	Calculate the residualbinder application rate $(\ell/m^2) = 0.16* (ALD_1 + ALD_2)$
	where ALD_1 = average leastdimensionlayer 1 (in mm)
	ALD_2 = average leastdimensionlayer 2 (in mm)

(a)   Adjust for Traffic

Category	Vehicles/day (vpd)	Commercial vehicles / day (cv)	Adjustment (both directions)
Very light	≤ 2500	≤ 400	Subtract 0.10 ℓ/m^2
Light	$> 2500 - 5000$	$> 400 - 500$	Subtract 0.05 ℓ/m^2
Medium	$> 5000 - 7500$	$> 800 - 1000$	0
Heavy	$> 7500 - 10,000$	$> 1000 - 1200$	Add 0.05 ℓ/m^2
Very heavy	$> 10,000$	>1200	Add 0.10 ℓ/m^2

(b)   Adjust for road hardness

Category	Ball penetration*	Adjustment
Hard	≤ 2 mm	Reduce by 5%
Medium	$> 2 - 3$ mm	0
Soft	> 3 mm	Add 10%

* Ball penetration = The test entails the measurement of the penetration of a 19 mm steel ball into the pavement surface when a standard Marshall hammer is dropped once on it. (TMH6 1984).

(c)   Adjust for climate

Category	Adjust
Wet	Reduce by 15%
Moderate	0
Dry	Add 15%

(d)   Adjust for gradients

   • Gradient: $\leq 4\%$: no adjustment

   • Gradient: $> 4 - 5\%$: reduceby 3%

   • Gradient: $> 5 - 6\%$: reduceby 7%

   • Gradient: $> 6 - 7\%$: reduceby 9%

   • Gradient: $> 7\%$: reduce by 10%

(Continued)

TABLE 16.24   Continued

(e)	Adjust for existing texture	

- Texture depth = $< 0.4 - 0.6$ mm: add $0.20 \; \ell/m^2$
- Texture depth = $< 0.3 - 0.4$ mm: add $0.15 \; \ell/m^2$
- Texture depth = $< 0.2 - 0.3$ mm: add $0.10 \; \ell/m^2$
- Texture depth ≤ 0.2 no adjustment

(f)	Adjust for shade

- Add 2.5% for partlyshaded areas
- Add 5% for fully shaded areas

2· Calculate the hot application rate by multiplying the adjusted residual binder application with a factor to adjust to equivalent binder application rate:

- Penetration grade bitumen = 1.1
- Cutback bitumen (MC 3000) = 1.25
- Cutback bitumen (MC 30) =1.9
- 60% Emulsion = 1.7
- 70% Emulsion = 1.45

3. Distribute binder between the different layers, i.e.,

60% to the tack coat and a 40% to the penetration coat after the fog-spray, if any,

has been subtracted.

4· Calculated the application rate of the chippings

	Application rate (m³/m²)	
ALD (mm)	First Layer	Second Layer
4	0.006	0.005
6	0.008	0.006
8	0.010	0.008
10	0.012	0.010
12	0.014	0.012

a PMB can be used as tack coat. This is also called a stress alleviating membrane (SAM) or stress alleviating membrane interlayer (SAMI) if placed beneath an asphalt overlay.

- Where cracks are very wide (5 to 15 mm) and active (movements of more than 0.5 to 1 mm), consideration should be given to the utilization of a geo-textile fabric beneath the surface dressing. Good performances of up to 15 years have been reported in Australia for geo-textile reinforced sprayed seals (FHWA, 2002). Double surface dressings are more effective on highly cracked areas than single surface dressings since the application of the binder is higher.

TABLE 16.25   Example of Single Surface Dressing

Situation:	Highway to be resealed with a 13.2 mm single surface dressing using penetration grade bitumen and no fog-spray
Terrain:	Rolling with maximum gradients of 5%, moderate climate
Current traffic:	Heavy vehicles = 150/day and light vehicles = 1000/day (both directions)
Existing surfacing:	9.5 mm single seal treated 3 years ago with a fine slurry texture treatment
Texture depth:	Texture depth/uniform section is 0.3 mm
Embedment potential:	The average of ball penetration values is less than 1 mm
Aggregate:	ALD of 7.9 mm

1. Design
Residual binder required $\qquad$ $0.16 \times ALD_1 = 0.16 \times 7.9 = 1.26 \; \ell/m^2$
Adjustments
(a) Traffic: $\qquad$ Reduce by 0.1 ℓ/m^2
(b) Road hardness: $\qquad$ Add 5%
(c) Climate: $\qquad$ No adjustment
(d) Gradients: $\qquad$ Max. = 5%, Reduce binder by 3% on uphills
(e) Existing texture: $\qquad$ Depth = 0.3 mm, add 0.1 ℓ/m^2
Adjusted residual binder content

$$= (1.26 - 0.1) \times 1.1 + 0.1$$
$$= 1.32 \; \ell/m^2 \text{ on flat and down hill areas}$$
$$= 1.32 \times (1 - 0.3) = 1.28 \; \ell/m^2 \text{ on uphills}$$

2. Equivalent binder (penetration grade bitumen) $= (1.1 \times 1.32) = 1.45 \; \ell/m^2$ (on flat and down hill areas) and $(1.1 \times 1.28) = 1.4 \; \ell/m^2$ on the uphills

3. Application rate of the chippings
$= 0.008 + ((0.010 - 0.008)/(8 - 6)) \times (7.9 - 6)$
$= 0.0099 = 0.010 \; m^3/m^2$

TABLE 16.26   Example of Double Surfacing Design

Situation:	Highway to be resealed with a 13.2 and 6 mm double surface dressing using penetration grade bitumen and no fog-spray
Terrain:	Rolling with maximum gradients of 4%, dry climate
Current traffic:	Heavy vehicles = 100/day and light vehicles = 1000/day (both directions)
Existing surfacing:	Thick slurry
Texture depth:	Texture depth/uniform section is 0.5 mm
Embedment potential:	The average of ball penetration values is less than 2 mm
Aggregate:	ALD of 7.7 mm for the 13.2 mm and 4.0 mm for the 6.7 mms

1. Design
Residual binder required $\qquad$ $= 0.16 \times (7.8 + 4.0) = 1.87 \; \ell/m^2$
Adjustments
(a) Traffic: $\qquad$ Reduce by 0.1 ℓ/m^2
(b) Road hardness: $\qquad$ Add 5%
(c) Climate: $\qquad$ Add 15%.
(d) Gradients: $\qquad$ Max. = 4%, no adjustment
(e) Existing texture: $\qquad$ Depth = 0.5 mm, add 0.20 ℓ/m^2
Adjusted residual binder content

$$= (1.87 - 0.1) \times 0.15 \times 1.15 + 0.2 = 2.38 \; \ell/m^2$$

2. Equivalent binder (penetration grade bitumen) $= 1.1 \times 2.38 = 2.62 \; \ell/m^2$

3. The application of the tack coat is $60\% \times 2.62 = 1.57 \; \ell/m^2$
The application of the penetration coat is $40\% \times 2.62 = 1.05 \; \ell/m^2$

4. Application rate of the chippings
First layer $\qquad$ $= 0.008 + ((0.010 - 0.008)/(8 - 6)) \times (7.7 - 6)$
$= 0.0097 = 0.010 \; m^3/m^2$
Second layer $\qquad$ $= 0.005 \; \ell/m^2$

- *Ruts and depressions.* These must be repaired. The depth of rut to be filled will dictate the type of action required. Typical repair action ranges from a fine-graded hot-mix asphalt surfacing coat for deep ruts, to a modified emulsion slurry or a normal slurry for shallow ruts.
- Flushing or bleeding of the underlying pavement surface will quickly cause bleeding of the applied surface dressing as well and need to be treated. The most effective treatment entails the light application of a binder, such as an emulsion (typically diluted by 50% water and applied of 0.7 to 1 l/m^2) and covered with a small chippings (6.7 or 9.5 mm) which are rolled in with a pneumatic roller.
- Dry and porous surfaces should be enriched with the application of a diluted emulsion before the application of a surface dressing to prevent absorption of the binder into the old surface.
- *Uneven or coarse surface texture.* Surfaces with highly uneven or coarse surface textures should be treated with slurry to obtain a uniform texture before the surface dressing is placed.

The appropriate application placement of the rate of the diluted emulsion will depend of the dryness and porosity. Typical application rates are 0.5 to 1.0 l/m^2 of 30% emulsion.

Best results are obtained of the surface texture treatment is left open to traffic for a period of 1 to 12 months (depending on the traffic, and climatic conditions) before the surface dressing is placed. Alternatively the slurry can be rolled with a pneumatic roller to shorten the time.

16.9.5   Construction

The construction process can be divided into the following actions.

- *Implementation of traffic control procedures.* Cognizance should also be taken from the fact that surface dressings cannot immediately be opened to traffic after compaction. A period of curing of about 12 H is often required before the highway can be opened to traffic of lower speeds for 2 or 3 days.
- *Repair and clearing of the surface prior to the spraying of the binder.* The details of the repair are given in preceding sections.
- *Spraying of the binder.* Spraying should be done with properly calibrated sprayers, and at the binder temperature that would ensure an even application. Appropriate spraying viscosities for slotted types of nozzles are in the order of 43×10^{-6} to 107×10^{-6} m^2/second. For bitumen emulsions with 40% water the spraying temperatures are between 45 and 60 °C. When stored, the storage temperatures must not be excessive so as to cause aging of the binder. For normal bitumens emulsions and cutback the road surface temperature should preferable be above 10 °C and for penetration grade bitumen above 20 °C.

  Furthermore, care must be taken that the road to be sprayed is of reasonable length and that transverse and longitudinal overlaps are done properly.
- *Spreading of chippings.* The spreading of chippings should preferably be done with a self-propelled chip spreader to ensure a continuous uniform application of chips and follows directly after the binder application. It is essential that the chippings are clean and dust free.

  Precoating of chips will enhance the adhesion characteristics and can be carried out with a low viscosity bituminous binder such as bitumen emulsion, cut-back bitumen blended with paraffin, tar oil or petroleum oil. The amount of binder added is typically between 8 and 13 l/m^3 (higher for bitumen emulsions). Mixing is by means of a front-end loader or concrete mixer. After precoating, the chips should present a slightly oily damp surface. The precoated chips are applied as soon as they are dry (normally only after 2 days).

  A variation of this precoating procedure is the coating of the chips with penetration-grade bitumen at a coating plant. Recommended binder contents are 1% bitumen for 6 mm chippings, 0.8% for 10 mm chippings, 0.6% for 14 mm chippings and 0.5% bitumen for 20 mm chippings (Nicholls, 1996).

Commercially available adhesion agents (proprietary chemical products) can also be used to enhance the adhesion between the chippings and the binder.

Chippings can be heated to 60 to 120 °C to improve adhesion, but may cause problems with bitumen emulsion.

- Rolling of the chippings

  To ensure that the surface dressing performs well, the spraying of binder, spreading of chippings and initial rolling should follow directly on each other.

  Initial rolling with a pneumatic roller should be carried out immediately after the chippings are spread, except when the chippings are damp, in which case rolling should be delayed until they are almost dry. Rolling should commence at the road edges and proceed towards the crown. Each roller pass should overlap the previous one by half the width of the roll. After completion of the initial rolling, light broom dragging may be carried out, if necessary.

  Rolling should be continued with pneumatic-tired rollers until the stone has become well-embedded and a uniform surface has been obtained. With a double seal, before a further coat of binder is applied, the layer should be given one or two roller passes.

  Steel-wheeled rollers tend to crush the stone. On double seals, provided the crushing is not excessive, this can be beneficial since it increases the mechanical interlock of the particles and reduces the void content of the layer.

- Application of a fog spray, if required.

The construction of a double surface dressing is similar to the process described above except that some of the activities are repeated. The construction of surface dressings with PMBs is also similar, but the application and rolling temperature of the binder is more critical than for conventional binders.

Normal quality and process control procedures are applicable to the construction of surface dressings.

16.10   Design and Construction of a Slurry Seal

A slurry seal is a surfacing seal comprising a mixture of suitably graded fine chippings, cement or hydrated lime, bitumen emulsion and water. A polymer modified emulsion can also be used as bituminous binder. Chippings with three types of gradings, namely fine, medium and coarse are typically used. The gradings and required chipping properties are given in Table 16.27.

The slurries can be designed using the Marshall method, but most road authorities have standard designs based on past experience.

Typical bitumen emulsion contents prescribed by one road authority are 200 to 300 l/60% stable-grade bitumen emulsion/m^3 of dry sand for the fine graded chippings. The emulsion content is reduced by

TABLE 16.27   Typical Grading for Slurry Chippings

Sieve Size (mm)	Fine	Percent Passing	
		Medium	Coarse
6.7	–	100	100
4.75	100	82–100	70–90
2.36	90–100	56–95	45–70
1.18	65–95	37–75	28–50
0.600	42–72	22–50	19–34
0.300	23–48	15–37	12–25
0.150	10–27	7–20	7–18
0.075	4–12	4–8	2–8

The aggregate must further comply with:
Fineness modulus for coarse slurry sand = 2.0–3.8.
Sand equivalence (min) = 35.
ACV of the parent rock (max) = 30.

6.25% if the medium-graded chippings are used and by 12.5% if the coarse graded chippings are used (Gautrans). Another recommends the use of 8% residual bitumen for more than 5000 equivalent light vehicles (e/v)/lane/day, 10% for between 500 and 5000 e/v/lane/day and 12% residual bitumen for less than 500 e/v/lane/day.

The addition of cement or lime as a filler is advantageous if the percentage of fines is less than 6% to improve the workability and the dissipation of the particles. The filler should be less than 2% by mass of dry aggregate. Typical percentages are 1.0 to 1.5%.

Slurry seals can be mixed and distributed by means of a slurry machine and slurry box or by means of a concrete mixer and squeegees. The minimum thickness is normally 1.25 times the maximum aggregate size. No rolling is required for thin slurries, but pneumatic-tired rollers should be used if the layer is thicker than 0.0085 m^3/m^2. The roller compaction can only take place once the breaking of the emulsion has occurred. Aspects such as traffic control, preparation of the road surface and quality control are the same as for surface dressings.

16.11   Winter Maintenance

Winter maintenance consumes a significant portion of the maintenance budget of some countries. In the United States, for instance, more than US$2 billion is spent on winter maintenance each year (Hyman and Vary, 1999). This is as high as 50% of the maintenance budget in some Canadian provinces. In Switzerland the cost of winter maintenance varies between 16 and 36% of the total road maintenance cost (Schlup, 1993) and in the U.K. it is about a third of the cost of routine maintenance (Atkinson, 1990). On the other hand, winter maintenance has significant benefits such as:

- Reduction of accident costs (by more than 80%).
- Savings in road user cost, mainly time. For example the average road user savings in road user costs can be US$6.50 for each $1 spent on winter maintenance during the first 4 hours of salt spreading (Hanbali, 2001). Certain types of winter maintenance have detrimental effects on the environment, vehicles, the pavement and bridges. Damages attributable to road de-icing have been estimated to be between US$5 and 8 billion annually in the United States (Hyman and Vary, 1999).

Winter highway maintenance comprises antiicing and de-icing. Antiicing is the practice of preventing the development of a bond between precipitation and road surfaces through the application of chemicals before the precipitation or soon after the start of a snowfall. The success relies heavily on accurate information on the roadway and weather conditions. De-icing entails the removal of the snow and ice after it has accumulated on the roadway surface. The normal practice is ploughing followed by the application of chemicals to weaken the bond between the ice or snow and the pavement and abrasives, such as sand, to increase the friction. The term de-icing is also used to describe anticipatory measures to prevent the formation.

Although salt is still widely used as de-icing substance, detrimental environmental effects have lead to the increased use of alternatives. These alternatives are mainly additives, such as calcium magnesium acetate, magnesium chloride and calcium chloride. Another effective practice is the prewetting of the salt with an agent such as brine, calcium chloride, sodium chloride and water. The use of wet salt has the advantages of a reduced (up to 20%) dosage compared to dry salt and the quicker initiation of the de-icing process (Schlup, 1993). The practice on the Netherlands is to apply 5.5 G sodium chloride/m^2 to prevent the formation of ice during the dropping of temperature (Noort, 1997). On porous asphalt the rate is 11 g/m^2. Snow is prevented from attaching to the asphalt by the application of 15 G of sodium chloride/m^2, while 15 to 20 g/m^2 of the same product is applied during snow fall.

The de-icing material is dispersed on the road by means of spreaders. New generation spreaders can produce the de-icing material at a uniform rate. The quality of the equipment is important. Another method used for road surfaces on bridge space is a sprinkler system which will be activated if the conditions become hazardous. The system sprays a 20% sodium chloride solution onto the road.

The effectiveness of a winter maintenance system depends on the information received, the processing of the information and the reaction. A number of systems are in place which uses and monitors weather information to trigger winter maintenance actions (Noort, 1997; Axelsor, 1995; Matsuzawa and Takeuchi, 2001; Schlup, 1993). Temperature sensors on the highway can be used to monitor road temperatures (AUSTROADS, 1991).

16.12 Execution of Highway Maintenance

Different models can be used in the execution of highway maintenance, ranging from the traditional approach using in-house teams to the complete outsourcing (privatization) of all work for a period of time. The appropriate model will vary among road authorities (owners) and depend on a number of factors such as resources available, national strategic initiatives, legislation, etc.

(a) Traditional approach where all planning, programming and execution are done by the road authority (or road owner). The traditional approach can be efficient. The main advantages of this approach are that the road administration controls the complete process and has the skills and resources to handle increases in maintenance requirements and emergencies. The main disadvantages include low flexibility, high fixed costs, irregular utilization of resources and few incentives to improve effectiveness and efficiency. The maintenance can be controlled by the resources available and not the needs. Many of these deficiencies can be addressed by improved management systems, and training.

(b) *Partial outsourcing*. The planning and programming are done by the road authority, but some aspects of the execution or the provision of equipment are contracted out. The main reason for contracting out the work is a lack of resources. A common split is the execution of remedial maintenance (e.g., crack sealing, patching, shoulder repair, regraveling, etc.) by in-house (departmental) teams and preventative maintenance by contractors. Consultants are often engaged to assist in the design, documentation and supervision. With partial outsourcing the highway authority retains control of the management of the works, while peaks in work and lack of resources or skills can be handled by outsourcing. This normally leads to the high utilization of plant and staff and low fixed costs. The disadvantages are that it requires a high level of management and supervision, and there are few incentives for innovation. Care must be taken to maintain an optimal staff and plant complement. If not, work may have to be outsourced without adequate procurement and supervision mechanisms, or in-house maintenance teams can become involved in unproductive work just to keep busy, i.e., the maintenance can become resource and not needs driven. Typical examples will be appointment of consultants to design and supervise the resurfacing or regraveling of a section of highway. The consultants will do an assessment, propose a design, prepare tender documents (including a bill of quantities), invite tenders, adjudicate, the tenders and supervise on behalf of the road authority. Preventative maintenance contracts are easier to handle than remedial or emergency maintenance contracts with this approach since the scope of work is well-defined, the quantities easy to establish and to monitor. Performance based specifications can effectively be used in these contracts.

(c) *Complete outsourcing (also referred to a privatization)*. All aspects of the road maintenance are outsourced. Various models exist on how this can be done, e.g.:

- Where the road owner appoints a consultant to manage part or the whole of a network. A contractor is then engaged to do the work. The consultant is paid on a time and cost basis or a fixed fee, while the contractor is appointed on a schedule of rates or bill of quantities. The contractor proposed a maintenance plan, which is assessed by the consultant and approved by the client. The manager and service provider are independent. This model has been used for 10 years in the U.K. where consulting firms have been appointed as agencies to plan and administer all aspects of highway maintenance.

- Where the road owner appoints an entity, normally a contractor to manage the road network and to execute the road maintenance. There is no independence between the manager and the service provider.

Many countries encourage private contractors to undertake road maintenance. In Latin America for instance the concession services to the private sector is a major emphasis. The Road Maintenance Initiative (RMI) launched in 1987 by the World Bank also has the commercialization of road maintenance as one its key concepts for the more effective use of funds and for better roads in the sub-Saharan African countries. In some countries the outsourcing or privatization of road maintenance has be implemented years ago (for instance the U.K., Spain, Canada, Australia and New Zealand). In others, such as the Netherlands, Zambia and Ghana pilot projects are embarked upon. Almost 40% of all cities in the United States outsource road maintenance services with cost savings of 25 to 50%.

The main reasons given for outsourcing by state road authorities in the U.S.A. are staff constraints, the need for specialized skills or equipment and the focus on establishing public–private partnership (Warne, 2001).

Other reasons for the privatization of road maintenance encompass; solicitation of additional sources of funding (i.e., from the private sector), the anticipated increase in the effectiveness of maintenance with the resultant cost-savings and greater road user satisfaction and empowerment and the building of a local road maintenance industry.

Although the benefits of privatization of maintenance are difficult to quantify, savings of 10% and 50% in maintenance costs have been reported in California, British Colombia and Florida. The Department of Transport of the U.K. has saved 15% in maintenance costs over the last 10 years by privatization of road maintenance. States in the U.S.A. have reported satisfaction with outsourcing efforts of 7.2 on a scale from 1 to 10, with 10 being very satisfied (Warne, 2001).

The wide variety of reasons for the cost savings exists. They include:

- Better management techniques
- Better and more productive equipment
- Greater incentives to innovate
- Incentive pay structures
- More efficient deployment of workers
- Greater use of part-time and temporary employees
- Utilization of comparative cost information and
- More work schedules for off-peak hours

Although most privatization efforts in other countries have been successful, some concern still exists on whether the reported benefits are real. For instance, some argue that increased effectiveness could have been achieved by better management and training of departmental resources or the maintenance can become resource instead of needs driven. Privatization of maintenance can also lead to retrenchments of departmental work forces. It can place undue risk on the concessionaires, which lessens the expectance of the private sector.

In the U.K., consulting firms are appointed as agencies to plan and administer all aspects of routine and cyclic road maintenance. Initially a schedule of rates was used to specify maintenance work, but a bill of quantities is currently used. The first contracts were let in 1998 for a period of 3 years. A total of 24 such contracts are envisaged naturally.

In British Colombia (Canada), the privatization of maintenance of road (4500 km) and bridges (2500) started in 1987. A maintenance standard is developed for each activity. Contractors are paid a lup-sum for general maintenance and on a Unit Rate Basis for emergency and additional maintenance services. A maintenance plan is proposed by the contractor for approval by the client. Later refinements of the procedures include the implementation of a compulsory maintenance management system by the contractor. The owner employs a large number of inspectors to monitor the work. Consultants are not involved. Contracts in British Columbia are 2 to 3 years long and vary in size between US$5 and

US$35 million. Payments are made monthly. Preferential treatment is given to employee-aimed corporations. Most of these comprise former government employees.

The same system is used in Spain where maintenance contracts were underway on 3852 km highways in 1994. The duration of these contracts is 4 years and they include ordinary maintenance, emergency services and data collection.

All Australian states are engaged in contract maintenance projects. Arrangements range from the contracting out of preventative maintenance only as tendered contracts (e.g., Victoria) to 3-year schedule of rates contracts (Northern Territory), to 10-year performance based contracts (Western Australia) (Cottman and Chamala, 2002). In the latter roughness, cracking, rutting and texture are used as performance criteria. Vicroads has reported a reduction in routine road maintenance costs from US$920/lane-km in 1993 to US$730/lane-km in 2000 by benchmarking and monitoring of costs (FHWA, 2002). Ten-year lump-sum performance based contracts are also used in New Zealand, where a 20% saving has been reported compared to schedule of rates contracts (Cottman and Chamala, 2002).

Quantities for remedial activities are invariably difficult to define in advance, resulting in a high level of road authority and consultant involvement during execution. Only a few of the activities lend themselves to performance-based specifications.

The main advantages in the complete outsourcing of highway maintenance are the increase in innovation, reduced input from the road authority and increased cost-effectiveness. The disadvantages are that highly sophisticated agreements are required and road authorities can convert to fixed long-term funding. Another consequence, which can be very controversial in some countries, is that government employees can become redundant. On the other hand this does create self-employment opportunities for staff.

A number of approaches can be used in the complete outsourcing (or privatization) of highway maintenance. The approach will depend on, *inter alia*, the main objective of the outsourcing, the level of contracting expertise available, and the resources available in the road authority. Many developing countries (e.g., Malawi, Zambia and Tanzania) view the outsourcing of road maintenance as a means of employment creation and the development of a local contracting industry. The main emphasis in British Colombia was cost savings, while Mexico focused on access to private funding. Others, like Western Australia, used short-term maintenance contract to gain experience before letting a performance-based Team Network contracts.

Two types of contracts have mainly been used in the past. The first is a schedule of rates. The supplier of the services (the contractor) bids and is awarded the contract based on a schedule of rate for the maintenance activities. The maintenance activities to be executed is determined by the highway manager (which can be a consultant or the contractor) based on regular inspections and a management problem. Transit NZ, the controlling authority for state highways in New Zealand, uses such an approach. The second is performance-based contracts, where the appointed contractor has to maintain the road network to a certain predescribed condition (i.e., certain criteria must be complied with) for a tendered lump sum.

16.13   Performance-Based Specifications

Performance-based specifications refer to the allowable condition in which each highway element must be. This is often used in highway maintenance contracting. Their use reduces detailed method descriptions and extensive supervision. The contractor is required to maintain the highway elements to a specified standard and the road agency only monitors compliance with these standards. As in the case of defect assessment procedures, and maintenance criteria, each highway authority devises its own performance-based criteria. Table 16.28 and Table 16.29 give examples of the performance-based specifications used by highway authorities in New Zealand (OPUS, 1998).

TABLE 16.28 Performance-Based Maintenance Specifications: Non-Pavement Related Activities

Activity	Standard (Allowed)	Frequency of Inspection	Response Time
Vegetation control			
Shoulder and traffic islands	Height of grass for AADT > 2000 vpd: 20–150 mm / AADT <2000 vpd: 25–200 mm	Grass: once in 2 months / Vegetation: quarterly	Two weeks
Line of sight	Visibility at all times	As for shoulders	Two weeks
Around roadside furniture	Visibility at all times	Once/month	Two weeks
Side-drains	Not to impede water flow and <150 mm where growth is permissible	Once every 2 months	One month
Removal of loose material on sealed road surface	No loose material > 1 m^2 as a group or >2 linear m and no obscured pavement markings	AADT > 2000 vpd: monthly / AADT <2000 vpd: 3 monthly	24 hours / 48 hours
Litter removal	Removal of all litter within the road reserve	Yearly or after reported accidents	AADT > 2000 vpd: monthly / AADT <2000 vpd: 3 monthly
Sign maintenance	In place and visible at all times	AADT > 2000 vpd: monthly / AADT <2000 vpd: 3 monthly	AADT > 2000 vpd: 1 week / AADT < 2000 vpd: 2 weeks / In safety issue: 48 hours
Safety beams (guard-rails)	No misalignment, no loose connections and intact	Yearly or after reported accidents	One week
Pavement marking	All pavement markings easily visible under dipped headlights (distance of 80 m) during a night time survey	AADT > 2000 vpd: quarterly / AADT <2000 vpd: half yearly	
Skid resistance	Side force coefficient (SFC) above specified investigatory levels (i.e., normally above 0.35 to 0.55 depending on the site)	Every 3 years	1 month

TABLE 16.29 Performance-Based Maintenance Specifications: Pavement-Related Activities

Activity	Standard (Allowed)
Roughness	AADT > 10,000 vpd: 100 NAASRA counts/km
	AADT: 2000–5000 vpd: 130 NAASRA counts/km
	AADT < 2000 vpd: 150 NAASRA counts/km
	Unpaved: 180 NAASRA counts/km
Rutting	% of ruts > 30 mm (measured under 2 m straight edge)
	AADT > 10,000 vpd: 5% max
	AADT: 4000–10,000 vpd: 10% max
	AADT: 2000–4000 vpd: 15% max
	AADT < 2000 vpd: 20% max
Shoving	<1% of shoving in the wheel path/rating section
Crocodile cracking	Percentage of cracking in wheel path
	AADT > 10,000 vpd 1% max
	AADT: 2000–10,000 vpd 5% max
	AADT < 2000 vpd 10% max
Flushing	Percentage of the wheel path in which flushing is a problem
	AADT > 10,000 vpd 5% max
	AADT: 4000–10,000 vpd 10% max
	AADT: 2000–4000 vpd: 15% max
	AADT < 2000 vdp 20% max
Edge breaks	Percentage of seal edge where the sealed width is reduced by > 100 mm
	AADT > 10,000 vpd 5% max
	AADT: 4000–10,000 vpd 10% max
	AADT < 4000 vpd 15% max
Adequate drainage	Water drains freely from the pavement and moisture does not enter the pavement layers
Grading of unsealed highways	AADT < 50 vpd blade 3 times a year
	AADT: 50–100 vpd blade 4 times a year
	AADT: >100 vpd blade 6 time a year
Pothole repairs	Number of potholes > 50 mm deep and > 200 mm wide
	AADT > 2000 vpd 5 max
	AADT < 2000 vpd 10 max
Routine bridge repairs	
i. Deck joints	Free movement and adequate seal
ii. Piers	Free of debris
iii. Handrails	No misalignment, no loose connection

16.14 Labor-Based Techniques

Labor-intensive is a phrase used in economics to describe an operation in which proportionately more labor is used than other factors of production (McCutcheon and Parkins, 2003). Labor-intensive (and maintenance) may be defined as the economically efficient employment of as great a proportion of labor as is technically feasible — ideally throughout the construction or maintenance process, including the production of materials — to produce as high a standard of construction as demanded by the specification and allowed by the funding available. Labor-intensive construction results in the generation of a significant increase in employment opportunities/unit of expenditure by comparison with conventionally capital-intensive methods. By significant is meant at least 300% to 600% increases in employment generated/unit of expenditure.

As mentioned, labor-based techniques are defined as the cost-effective substitution of equipment by labor. This does not exclude the use of equipment but comprises the optimum mix of plant and labor for a given set of circumstances, i.e., national strategies to create employment, the cost of employment and plant, the type of maintenance activity and the time available for the execution of the activity. Highway maintenance, especially on lower volume highways lends to the utilization of labor-based techniques in the execution of the work resulting in the creation of employment which is a major focus in many developing countries. Highway maintenance also provides an opportunity to establish and utilize small, local or citizen contractors. This is an equally important focus in developing countries where the local contracting industry needs to be strengthened.

Labor-based techniques have proven to be cost-effective compared to equipment-based techniques and can be done to acceptable standards (i.e., no compromise in quality). Labor-based construction can only be cost-effective and sustainable if accompanied by relevant designs, appropriate equipment, training and a program rather than a project approach. All the remedial maintenance activities listed in Table 16.2 can be executed by means of labor-based technologies. A constraint may be the duration of maintenance if labor-based techniques are used on high volume roads can effectively be done by means of labor-based techniques.

As far as preventative maintenance is concerned regraveling and road marking can be done competitively through labor-based techniques.

The three most common labor-based maintenance type contracts are (Anderson et al., 1996):

- Single length-person contract where a defined section of road (1 to 2 km) is given to a single person to maintain.
- Petty contract (or labor group) where a small-scale contractor employs five to ten laborers to maintain a defined section of road (5 to 20 km).
- Small-scale contract for a particular road or network. A small to medium size contractor, who employs laborers, is appointed to maintain a particular road section (20 to 200 km) or a specific road network (100 to 300 km).

16.15   Other Considerations

A number of other aspects need to be considered in the planning and execution of highway maintenance. The two most important are traffic control and safety, and environmental issues.

16.15.1   Traffic Control and Safety

Traffic control during the highway maintenance operations and the safety of the workers are of the utmost importance. Highway maintenance work can be divided into four categories in terms of safety

TABLE 16.30   Safety Guidelines

Always be aware that it is a potentially dangerous situation
Always wear a safety jacket when working on or close to the road
Do not step onto the traveled surface without first checking for oncoming traffic
When not in a delineated work area as far as possible face oncoming traffic and always be watching for traffic
When driving on the road do not make sudden stops and when stopping try to park off the road surface
All vehicles should be equipped with yellow flashing lights and "Maintenance Vehicle" signs
Making a U-turn on the road is dangerous and increases the risk of an accident
Remove all loose or foreign objects such as tools, material, broken exhausts, rubber from burst tires and signs not in use, from the road surface
Maintain appropriate temporary signage — when not required remove or cover signs
Do not hold discussions on the carriageway or shoulder — inspect problem area and move to a safe place in the road reserve for any discussions
Do not work simultaneously on both sides of the road at one location
Clear accident scenes as soon as possible

TABLE 16.31 Selected Environmentally Friendly Maintenance Practices

Maintenance Area	Practice
Roadway and shoulder surfaces	Recycling roadway millings for shoulder backup and roadway surfaces
	Restricted use of cutback asphalts to prime and winter coat
	Use of liquid emulsions only in surface treatment
	Reuse existing shoulder material
Roadside vegetation	Integrated roadside vegetation management plans
	Reduction in or limited mowing
	Reduction in herbicide application
	Use of native plans
	Use of salt-tolerant grasses along roadside
	Right-of-way reestablishment of native vegetation
Drainage	Grass- or stone-lined treatment swales
	Annual drainage cleaning, regular cleaning, inspection of drainage, inlets and lines
	Watercourse maintenance guidelines
Cleaning and litter pickup	Annual neighborhood cleanup
	Cleaning services contracted to handicapped and vending concessions to blind
Bridge maintenance	Use of calcium sulfonate paint
	Use of settling basins and cofferdams (sheet pile, sandbags) or dewatering work areas
Snow and ice control	Increased use of magnesium chloride
	Use of salt alternatives, accurate application equipment
	Snow disposal sites selected based on the environmental impact of contaminants
	Rare use of sand or abrasives
Signs, stripping and pavement markers	Water-based paints
	Use of lead-free, solvent-free pavement marking materials
	Testing of spray guns on paved surfaces only; paint residue and overspray recycled
	Plastic sheeting under illuminated arrow boards to prevent fuel seepage
Equipment	Conversion of some fleet to alternate fuels
	Energy-efficient air conditioning units
Material and handling	Major effort to comply with legal requirements
Stormwater run-off	Erosion and sediment control program
Water quality	Timely inspection of storm drains
	Erosion protection at ends of roadway or itch conduits
	Stormwater pollution prevention plans for all facilities
	Stormwater management plan for maintenance activities
Air quality	Elimination of sandblasting and vehicle spray painting
	No burning allowed on construction or maintenance projects
Noise	New equipment must meet specified noise levels
Soil contamination	Temporary fuel storage spill program
Worker health and safety	Elimination of solvents in sign fabrication, water-based road paints
	No sandblasting, vacuum attachments on tools
	Annual blood lead levels
	Material data safety sheets acquisition contracted out
	Formal safety policies and procedures for trenching or excavation or lead in construction or shoring and use of explosives and asbestos management
	Training in right-to-know, work in confined places, workplace safety and hazardous waste disposal

and traffic control; each with a different traffic control guidance scheme:

- Work not affecting the carriageway (e.g., cleaning of side drains, mending of fences, cutting of grass)
- Work requiring partial closure of the carriageway (e.g., small-scale patching, shoulder maintenance)
- Work on the centerline (e.g., centerline road marking)
- Work requiring total closure of lanes (e.g., resealing, reconstruction of damaged culvert, regraveling)

The layout of the road signs and the road signs used must comply with the standards applicable in the specific country and region. All traffic control and guidance schemes must be properly planned, designed, installed, operated and removed. The following broad principles apply:

- The layout must give the road users time to understand and respond to the information which the signs convey.
- Standard signs must be used in a standard layout to reduce the possibility of confusion.
- All signs must be clear and in a good condition, and be visible at night (if appropriate).
- All schemes must take cognizance of governing legislations, and standards.

The safety of the workers is of equal importance. Table 16.30 contains useful safety guidelines.

16.15.2 Environment

The term "environment" includes the natural environment (plants and animals) as well as the social environment (surrounding communities) and consideration should be given to in the execution of the maintenance. Aspects of particular importance are: disposal of solid waste (e.g., construction debris, chemical waste, bitumens, domestic waste, etc.) as soon as possible to an approved site; noise levels; and safety of public.

As in the case of traffic control guidance schemes, pertaining environmental requirements and guidelines should be adhered to. Table 16.31 contains practices which will enhance the environmental sensitivity of highway maintenance work.

16.16 Summary

Highway maintenance is an important, but integral part, of the design, construction and management of a highway system. Timely and effective maintenance will produce significant benefits to the road authority and the user.

Although the principles of highway maintenance are uniform, many of the procedures, strategies, specifications, and implementations vary among road authorities. It is prudent to obtain information on governing procedures and specifications in the area where the maintenance is to be executed. The general principles and guidelines, as well as some typical criteria, procedures and specifications have been given in this chapter. The references listed provide sources for very useful further detailed information and specifications. This chapter should not be read in isolation as the issues relate to many other chapters in this Handbook.

Acknowledgments

The author would like to express his appreciation of the contributions of his colleagues at AFRICON Engineering International, Alan Campbell and Prof. Robert McCutcheon (labor-based techniques). The efforts of Leonie Pretorius in the preparation of the text are greatly appreciated.

References

AASHTO, 1990. *AASHTO Guidelines for Pavement Management Systems*. American Association of State Highway and Transportation Officials, Washington, DC.

Ames, W.H. 1985. Concrete pavement design and rehabilitation in California. In *Proceedings of the Third International Conference on Concrete Pavement Design and Rehabilitation*, West Lafayette, Indiana, April 1985.

Anderson, C., Beusch, A., and Miles, D. 1996. *Road Maintenance and Regraveling (ROMAR) Using Labor-Based Methods: Handbook*. Stylus Publishing, London.

ARRB, 1993. *Unsealed Roads Manual: Guidelines to Good Practice*. ARRB (Australian Road Research Board) Transport Research, Vermont South.

Atkinson, K. 1990. *Highway Maintenance Handbook*. Telford Thomas, London.

AUSTROADS, 1991. *Road Maintenance Practice*. AUSTROADS, Sydney.

AUSTROADS, 1998a. Treatment of cracks in flexible pavements. Pavement Work Tips No. 8. AUSTROADS, Sydney. Available on the Internet: http://www.aapa.asn.au/docs/worktip8.pdf

AUSTROADS, 1998b. Polymer modified binders. Pavement Work Tips No. 6. AUSTROADS, Sydney. Available on the Internet: http://www.aapa.asn.au/docs/worktip6.pdf

AUSTROADS, 1998c. Preparing pavements for resealing. Pavement Work Tips No. 9. AUSTROADS, Sydney. Available on the Internet: http://www.aapa.asn.au/docs/worktip9.pdf

Axelsor, L.B. 1995. Winter road maintenance system. *Maintenance Management: Proceedings of the Seventh Maintenance Management Conference*, Orlando, FL, July 18–21, 1994.

Batac, G., Ray, M. 2001. *French Strategy for Preventive Road Maintenance: Why and How?* TRR 1183. Transportation Research Board, Washington, DC.

Botswana Ministry of Works, Transport and Communications, 1999. *The Design, Construction and Maintenance of Otta Seals*, Guideline No. 1. Botswana Roads Department, Gabarone.

Chou, C.-P., McCullough, B.F. 1987. *Development of a Distress Index and Rehabilitation Criteria for CRCP Using Discriminant Analysis*, TRR 1117. Transportation Research Board, Washington, DC.

COLTO, 1998. *Surfacing Seals for Urban and Rural Roads*, TRH3 (Draft). Department of Transport, Committee of Land Transport Officials, Pretoria, South Africa.

Cottman, E., Chamala, R. 2002. *Investigation of Alternative Road Maintenance Delivery Methods*. Cottman Consulting, Brisbane, Australia.

CSIR, 1986. *Standard Nomenclature and Methods for Describing the Condition of Jointed Concrete Pavements*, TRH9 (Draft). Council for Scientific and Industrial Research, Pretoria.

CSRA, 1985. *Nomenclature and Methods for Describing the Condition of Asphalt Pavements*, TRH6. Committee of State Road Authorities, Pretoria, South Africa.

Dempsey, B.J., Carpenter, S.H., Darter, M.I. 1982. *Improving Sub-drainage and Shoulders of Existing Pavements*, FHWA/RD-81/078. University of Illinois, Urbana-Champaign.

Department of Transport, 1994. *The use of Bitumen Emulsion in the Construction and Maintenance of Roads*, TRH7. Department of Transport, Pretoria, South Africa.

ERES Consultants Inc., 1982. *Techniques for Pavement Rehabilitation: A Training Course*. U.S. Department of Transportation, FHWA, Washington, DC.

Faiz, A. and Horak, C. 1987. The road deterioration problem in developing countries: the magnitude and typology of the problem. *66th Annual TRB Meeting*. Transportation Research Board, Washington, DC.

Faiz, A., Yoder, E.J. 1974. *Factors influencing the performance of continuously reinforced concrete pavements*, TRR 485. Transportation Research Board, Washington, DC.

FHWA, 1984. *Construction Handbook on PCC Pavement Rehabilitation*. U.S. Department of Transportation, Federal Highway Administration, Washington, DC.

FHWA, 1999. *Pavement Preservation: A Strategic Plan for the Future*, FHWA-SA-99-015. U.S. Department of Transportation, Federal Highway Administration, Washington, DC.

FHWA. 2001. Reclaimed asphalt pavement user guideline: asphalt concrete (hot recycling). Available: http://www.tfhrc.gov/hnr20/recycle/waste/rap132.htm

FHWA, 2002. *FHWA: Pavement Preservation Technology in France, South Africa and Australia*. U.S. Department of Transportation, Federal Highway Administration, Washington, DC.

Foley, G., Cropley, S., and Giummarra, G. 1998. Road dust control techniques. ARRB Special Report No. 54. ARRB Transport Research, Melbourne.

Galehouse, L., Moulthrop, G.S., and Hicks, R.G. 2003. *Principles of Pavement Presentation: Definitions, Benefits, Issues and Barriers*, TR News No. 228. Transportation Research Board, Washington, DC.

Hanbali, R.M. 2001. *Economic impact of winter road maintenance on road users*, TRR 1442. Transportation Research Board, Washington, DC.

Hyman, W.A. and Vary, D. 1999. *Best management practices for environmental issues related to highway and street maintenance.* NCHRP Synthesis of Highway Practice 272. Transportation Research Board, Washington, DC.

Markow, M.J., Harrison, F.D., and Thompson, P.D. 1994. Role of highway maintenance in integrated management systems. NCHRP Report 363. Transportation Research Board, Washington, DC.

Marks, V.J. 1981. *Use of Longitudinal Sub-drains in the 3R Program.* Iowa Department of Transportation, Ames, IA.

Massey, S. and Pool, P. 2003. *Preserving Pavements and Bridges: California's Strategies Leverage Limited Funds,* TR News No. 228. Transportation Research Board, Washington, DC.

Matsuzawa, M. and Takeuchi, M. 2001. *Field Test of Road Weather Information Systems and Improvement of Winter Maintenance in Hokkaido.* Hokkaido Development Bureau, Civil Engineering Research Institute, Japan.

McCutcheon, R.T. and Parkins, T., eds., 2003. FLM. *Employment and High Standard Infrastructure,* p. 640, University of the Witwatersrand, Research Centre for Employment Creation in Construction, Johannesburg.

NCHRP. 1979. Failure and repair of continuously reinforced concrete pavements. NCHRP Synthesis of Highway Practice 60. Transportation Research Board, Washington, DC.

NCHRP. 1983. *Design and construction guidelines and guide specifications for repair of jointed concrete pavements.* Draft Report NCHRP No. 1-21. University of Illinois, Urbana-Champaign

Nicholls, J.C. 1996. *Design guide for road surface dressing,* TRL Road Note 39. Transport and Road Research Laboratory, Crowthorne, Berklure, U.K.

Noort, M. 1997. Winter maintenance in the Netherlands. Snow removal and ice control technology. *TRB Conference Proceedings,* No. 16. Transportation Research Board, Washington, DC.

Odoki, J.B. and Karali, G.R. 1999. *Highway Development and Management, HDM-4,* Technical Reference Manual, Vol. 4. PIARC, Paris.

OECD, 1990a. *Road Monitoring for Maintenance Management,* Damage Catalogue for Developing Countries, Vol. 2. Organisation for Economic Co-operation and Development, Paris.

OECD, 1990b. *Road Monitoring for Maintenance Management,* Manual for Developing Countries, Vol. 1. Organisation for Economic Co-operation and Development, Paris.

OPUS International Consultants. 1998. Review of proposed funding allocation model, international levels and maintenance standards for Transfund New Zealand. Napier, New Zealand.

PIARC, 1982a. *International Road Maintenance Handbook,* Maintenance of Roadside Areas and Drainage, Vol. I. World Road Association, Paris.

PIARC, 1982b. *International Road Maintenance Handbook,* Maintenance of Unpaved Roads, Vol. II. World Road Association, Paris.

PIARC, 1982c. *International Road Maintenance Handbook,* Maintenance of Paved Roads, Vol. III. World Road Association, Paris.

PIARC Technical Committee on Roads in Developing Regions, 1989. *Surface Dressings: Synthesis of International Experiences.* Permanent International Association of Road Congresses, Paris.

Ridgeway, H.H. 1982. Pavement subsurface drainage systems. NCHRP Syntheses of Highway Practice 96. Transportation Research Board, Washington, DC.

Robinson, R. 1986. Road maintenance planning and management for developing countries. Highways and Transportation, June–July 1986.

Robinson, R. 1988. A view of road maintenance economics, policy and management in developing countries. Research Report 145. Transport and Road Research Laboratory, Crowthorne, England.

Robinson, R., Danielson, U., and Snaith, M. 1998. *Road Maintenance Management: Concepts and Systems.* Macmillan, London.

SABITA, 1987. *Bituminous Products for Road Construction and Maintenance,* SABITA Manual 2. Southern African Bitumen and Tar Association, Roggelbaai, South Africa.

SANRAL, 2000. *Routine Road Maintenance Guidance Manual.* South African National Road Agency, Pretoria, South Africa.

Schlup, U. 1993. *Snow Removal and Ice Control Technology on Swiss Highways*, TRR 1387. Transportation Research Board, Washington, DC.

SHRP, 1993. *Distress Identification Manual for the Long-Term Pavement Performance Project*, SHRP-P-338. Transportation Research Board, Strategic Highway Research Program, Washington, DC.

Thornton, J.B. and Gulden, W. 1980. Pavement restoration measures to proceed joint sealing. Transportation Research Record No. 752.

TPA, 1994. *Surface Treatment Manual.* Transvaal Provincial Administration, Roads Department, Pretoria, South Africa.

TRRL, 1981. *Maintenance Management for District Engineers*, Overseas Road Note 1. Transport and Road Research Laboratory, Crowthorne, England.

TRRL, 1985. *Maintenance Techniques for District Engineers*, Overseas Road Note 2. Transport and Road Research Laboratory, Crowthorne, England.

Van Wijk, A.J. 1985. *Rigid Pavement Pumping*, JHRP-85-10. Indiana Department of Highways, West Lafayette, IN.

Warne, T.R. 2001. State DOT outsourcing and private sector utilization. NCHRP Synthesis of Highway Practice 313. Transportation Research Board, Washington, DC.

Zimmerman, K.A. and Peshkin, D.G. 2003. *Pavement Management Perspective on Integrating Preventative Maintenance into a Pavement Management System*, TRR 1822. Transportation Research Board, Washington, DC.

Further Reading

Appropriate standards for bituminous surfacings for low volume roads, 1992. SABITA, Manual 10. Southern African Bitumen and Tar Association, Cape Town.

AASHTO, 1995a. *Standard Specifications for Transportation Materials and Methods of Sampling and Testing, Part II: Tests.* American Association of State Highway and Transportation Officials, Washington, DC.

AASHTO, 1995b. *Standard Specifications for Transportation Materials and Methods of Sampling and Testing, Part I: Specification.* American Association of State Highway and Transportation Officials, Washington, DC.

ASTM, 2000. *Annual Book of ASTM standards, Road and Paving Materials; Vehicle-Pavement Systems*, Vol. 04.03. American Society for Testing and Materials, West Conschocken, PA.

Binder on coated chippings and for measurement of the rate of spread of coated chippings, 1990. British Standard 598: Part 108. British Standards Institution, London.

British Standards Institution, 1984. *Bitumen road emulsion (anionic and cationic): Part 1. Specification for bitumen road emulsion.* British Standard 434: Part 1: 19.

British Standards Institution, 1987. *Road aggregates: Part 1, specification for single-sized aggregate for general purposes.* British Standard 63: Part 1. British Standards Institution, London.

British Standards Institution, 1987. *Road aggregates: Part 2, specification for single-sized aggregate for surface dressing.* British Standard 63: Part 2. British Standards Institution, London.

British Standards Institution, 1989. *Bitumens for building and civil engineering: Part 1. Specification for bitumens for roads and other paved areas.* British Standard 3690: Part 1. British Standards Institution, London.

British Standards Institution, 1989. *Specification for hot binder distributors for road surface dressing.* British Standard 1707. British Standards Institution, London.

British Standards Institution, 1990. *Sampling and examination of bituminous mixtures for roads and other paved areas: methods for determination of the condition of the binder on coated chippings and for measurement of the rate of spread of coated chippings.* BS 598, Part 108.

British Standards Institution, 1996. *Surface dressing — test methods — Part 1: rate of spread and accuracy of spread of binders and chippings.* Draft for Public Comment prEN12272-1, Document 96/100355. British Standards Institution, London.

Highway materials, Part 2: standard methods of sampling and testing highway materials, 9th Ed. 1966. American Association of State Highway Officials, Washington, DC.

Interim specifications for bitumen rubber road binder for spray applications, 1992. SABITA, Manual 6. Southern African Bitumen and Tar Association, Cape Town.

Interim specifications for rubber crumb for use in bitumen rubber binders, 1988. SABITA, Manual 4. Southern African Bitumen and Tar Association, Cape Town.

International Slurry Seal Association, Technical Bulletin No. 106 (Revised 1990). ISSA, Washington, DC.

International Slurry Seal Association, Technical Bulletin No. 115 (Revised 1990). ISSA, Washington, DC.

Liantand, G. and Faiz, A. 1994. *Factors Influencing the Transferability of Maintenance Standards for Low-Volume Roads*. TRR 1434. Washington, DC.

Mildenhall, H.S. and Northcott, G.D.S. 1986. *A Manual for the Maintenance and Repair of Concrete Roads*, London.

Road Surface Dressing Association, 1995. *Advice Note on Surface Dressing Binders*. Road Surface Dressing Association, Matlock.

Road Surface Dressing Association, 1995. *Code of Practice for Surface Dressing*. Road Surface Dressing Association, Matlock.

Road Surface Dressing Association, 1995. *Guidance Note on Preparing Roads for Surface Dressing*. Road Surface Dressing Association, Matlock.

Road Surface Dressing Association, 1995. *Guidance Note on Surface Dressing Aggregates*. Road Surface Dressing Association, Matlock.

Standard method for determining the aggregate crushing value of coarse aggregates, 1994. Standard Method, SM 841. South African Bureau of Standards, Pretoria.

Standard method for determining the FACT value (10% fines aggregate crushing value) of coarse aggregates, 1994. Standard Method, SM 842. South African Bureau of Standards, Pretoria.

Standard method for determining the flakiness index of coarse aggregates, 1994. Standard Method, SM 847. South African Bureau of Standards, Pretoria.

Standard method for determining the polished stone value of aggregates, 1994. Standard Method, SM 848. South African Bureau of Standards, Pretoria.

Standard method for determining the sieve analysis, fines content and dust content of aggregates, 1994. Standard Method, SM 829. South African Bureau of Standards, Pretoria.

Standard method of test for penetration of bituminous materials, 1976. Standard Method, IP-49. Institute of Petroleum, Barking, Essex, UK.

Standard specification for penetration grade bitumens (amended), 1972. Standard Specification, 307. South African Bureau of Standards, Pretoria.

Standard specifications for anionic bitumen road emulsions (amended), 1972. Standard Specification, 309. South African Bureau of Standards, Pretoria.

Standard specifications for cationic bitumen road emulsion (amended), 1972. Standard Specification, 548. South African Bureau of Standards, Pretoria.

Standard specifications for cut-back bitumen (amended), 1973. Standard Specification, 308. South African Bureau of Standards, Pretoria.

Standard test method ASTM D 36, 1993. Part 4, Vol. 04-04. American Society for Testing and Materials, Philadelphia, PA.

Standard test method ASTM D 3910, 1993. Part 4, Vol. 04-03. American Society for Testing and Materials, Philadelphia, PA.

Technical guidelines for seals using homogenous modified binders, 1994. SABITA, Manual 15. Southern African Bitumen and Tar Association, Cape Town.

Test methods for bitumen rubber, 1992. SABITA, Manual 3. Southern African Bitumen and Tar Association, Cape Town.

TG1, 2001. *Technical Guideline: The Use of Modified Bituminous Binders in Road Construction*. Asphalt Academy, Pretoria, South Africa.

TMH1, 1986. Standard methods of testing road construction materials. *Technical Methods for Highways*, TMH1. Department of Transport, Pretoria.

TMH6, 1984. Special methods of testing roads. *Technical Methods for Highways*. Department of Transport, Pretoria.

TRH15, 1984. *Subsurface drainage for roads (draft)*. Committee of State Road Authorities, Pretoria, South Africa.

TRH18, 1987. The investigation, design, construction and maintenance of Road Cutting Committee of State Road Authorities (draft), South Africa.

Appendix 16A
Repair of Defects in Flexible Pavements

16A.1 Crack Sealing

16A.1.1 Scope

Work in connection with the sealing of cracks in the road surface. Cracks shall be classified as either active or passive cracks. The active cracks originate from levels below the surfacing while the passive cracks apply to the surfacing.

16A.1.2 Methods of Repair of Active Cracks

(i) Stabilization cracks

Active cracks with a very distinctive block form, which with time, deteriorates to secondary cracking at closer spacings and eventually if untreated forms large open, closely-spaced cracks. These cracks are associated with cemented pavement layers.

The repair method entails:

- Blowing out all loose material and grit from the cracks.
- Priming cracks with an inverted emulsion primer as specified and filling the cracks with a stable-grade anionic emulsion modified with an anionic latex (8% net rubber on net bitumen), or a cationic spray-grade emulsion modified with a neutralized latex. Materials are to be injected with the equipment under pressure as specified.

Where the block cracking degenerates to secondary cracking, initially these are hairline cracks with pumping of fines. These cracks shall be treated with a geo-textile bandage to reduce or stop the pumping of fines. Where the secondary cracks are open they shall be treated as open block cracks as specified above.

Where isolated areas of small blocks are rocking under traffic these should be removed and repaired.

(ii) Volcano cracks are active cracks, which usually occur along with stabilization cracks in the base in areas where there is little or no traffic such as on paved shoulders. The cracks are open and up to 10 mm wide with a raised edge like the rim of a volcano.

The crack shall be prepared by blowing out all loose material under pressure. The surface for a width of 300 mm on either side of the crack should be treated with a rejuvenator or a solution of one part cut-back bitumen (or similar) and two parts diesoline. The crack should be primed with an inverted emulsion prime as specified, tacked with a modified emulsion and filled with rubber crumb slurry as specified. The raised areas around the crack shall be compacted with a vibratory roller until the area is level with the surrounding surface.

(iii) Expansive soil cracks are active cracks with cyclical movements related to the wet and dry seasons of the year. These cracks are open, wide and deep, extending down through the pavement to the subgrade. The cracks are often parallel to the centerline and occur mainly towards the edge of the road along the extent of shallow fills, fields and marshy areas.

Longitudinal cracks close to the edge of the road which are open, wide and deep, caused by settlement often occur where there are newly constructed high fills or widened sections and are normally parallel to the road centerline. Usually there is little vertical displacement across the settlement cracks. Conversely where slip failures occur in the pavement fill there are often noticeable vertical steps across the crack (lower towards the outside of the pavement) and the cracks form an arc towards the shoulder edge rather than a straight crack.

The crack should be cleaned of loose material and filled with fine slurry of clean fine sand and lime (in equal parts) to the underside of the base layer. Alternatively wider cracks could be filled with fine dry sand. More than one filling may be necessary to fill the crack. The crack must be primed with an emulsion prime then tacked with a modified emulsion and filled with rubber crumb slurry. A geo-fabric or a prefabricated bitumen rubber can be applied after the emulsion has broken the seal patch. Where cracks occur in the fill slope, fill the crack with bentonite (2%) and sand slurry to prevent ingress of water.

16A.1.3   Methods of Repair of Passive Cracks

(i) Surfacing cracks occur randomly over the road surface in a map format, (diamond shape). These cracks are often referred to as map cracks. In extreme cases the surface deteriorates to a pattern, which resembles and can be mistaken for crocodile cracks. This distress mode is not accompanied by any marked deformation or pumping of fines.

The repair method entails blowing out the cracks to remove all loose material, the application of bitumen emulsion to the surface, brooming of emulsion into the cracks and application of fine slurry.

(ii) Crocodile cracking is a series of small inter-linked near circular cracks often associated with pumping of fines in or after wet weather. It is accompanied by rutting of the pavement in the wheel-tracks and precedes pavement failure.

Limited areas of crocodile cracking can be treated as a holding measure by applying a geo-fabric prefabricated road patch. The geo-fabric should be protected or armored by treating it with a further application of latex modified emulsion and a nominal 4.75 mm grit.

(iii) Longitudinal cracks are fairly straight single cracks which often occur along construction joints, in joints in the surfacing or base. Also quite common where the surfacing meets concrete channels and kerbs. The cracks are open but not wide (say less than 5 mm). They tend to catch and hold water. Other random single passive cracks can be grouped under this description.

Narrow cracks (<3 mm) should be repaired by blowing out all loose material and grit from the cracks. Priming of cracks with an inverted emulsion primer as specified and filling the cracks with a stable-grade, anionic emulsion modified with an anionic latex (8% net rubber on net bitumen), or a cationic spray-grade emulsion modified with a cationic latex. Materials are to be injected with the equipment under pressure as specified.

Cracks could also be filled with fine dry sand. More than one filling may be necessary to fill the crack. Prime the crack with an inverted emulsion prime (MSP1), then tack the crack with a modified emulsion and fill the crack with rubber crumb slurry as specified. Allow the emulsion to break and then apply a geo-fabric or a prefabricated bitumen rubber seal patch. Where cracks occur in the fill slope fill the crack with bentonite (2%) and sand slurry to prevent ingress of water.

16A.2   Repair of Potholes, Edge Breaks and Surface Failures

16A.2.1   Scope

The repair of potholes, edge breaks and surface failures on an *ad hoc* basis. A pothole is defined as a surface failure, which has extended into the base layer forming a hole with an area smaller that 0.5 m^2. Potholes are isolated and are not associated with displacement. Potholes with areas larger than 0.5 m^2 must be treated as pavement failures.

Edge break is defined as the failure of the edge of the surfacing up to a maximum width of 300 mm from the continued edge of the surfacing usually accompanied by a loss of gravel on the shoulder. Edge breaks wider than 300 mm will be treated as pavement failures.

Surfacing failures often proceeded by map or diamond-like cracking is the breaking up of only the surfacing layer (surface dressing, slurry, or asphalt) exposing but not affecting the underlying layer. The resulting depression is usually of uniform thickness. Surfacing failures with areas larger than 2 m^2 must be treated as pavement failures.

Distinction shall be made in terms of temporary and permanent repairs of potholes, edge breaks and surface failures.

16A.2.2   Method of Work

Potholes, edge breaks and surface failures shall consist of trimming away raveled edges and loose material to the full depth and then backfilling as specified.

(i) Potholes and edge breaks

Potholes: The existing material shall be removed in a neat rectangle to sound base, with a minimum dimension of 200 mm × 200 mm. All sides shall be perpendicular or parallel to the direction of traffic.

Edge breaks: Loose and cracked edges shall be trimmed back to a neat rectangular shape, parallel and perpendicular to the centerline of the road to sound surrounding surfacing or base layer. All edges shall be saw-cut to a minimum depth of 30 mm below the road surface and the maximum thickness of each layer shall be 50 mm.

The floor of the excavation shall be cleaned of all undulations to ensure a firm flat base and sides and shall be tacked with 60% cationic stable-grade bitumen emulsion at a rate of 0.6 l/m². Continuously-graded, medium asphalt shall be placed and compacted to the level of the existing adjacent surfacing.

The asphalt shall be placed and compacted in layers not exceeding 40 mm in thickness after compaction.

A cold premixed bituminous mixture can be used for temporary pothole or edge break repair. Within a period of two months after placement of the cold mix, the cold mix must be replaced by hot asphalt mix as specified.

The mixture shall either be obtained from approved commercial sources or prepared and mixed in a suitable concrete or other approved type of mixer in the following proportions:

- 9.5 mm nominal sized aggregate: one part
- 6.7 mm nominal sized aggregate: one part
- Crusher sand (fine grade): one part
- 60% Stable mix-grade emulsion: Between 125 and 150 l/m³ (prepared from 80/100 penetration aggregate mix bitumen grade bitumen)
- 1% cement to promote breaking

Before spreading the mixture, the surface shall be prepared by painting it with one layer of bituminous emulsion at a rate of 0.6 l/m², which shall be allowed to dry. The mixture shall then be placed on the areas to be sealed and screeded off in a layer of uniform thickness. After the emulsion has broken and the layer has attained sufficient stability, it shall be compacted with a steel-wheeled roller or whacker. The thickness of the layer shall be the same as that of the adjacent seal.

(ii) Surfacing failures

The surface repair shall have a neat rectangular shape, at right angles to the direction of traffic. Before starting any repair work the areas adjacent to the holes should be checked for debonding by tapping the surface with a hammer. A dull sound indicates lack of bond. Debonded material must be removed and can be lifted off with a flat spade.

Surfacing failures should be well cleaned (if contaminated with fumes by washing) and a tack coat of 60% cationic emulsion applied at a rate of 0.40 l/m² (road penetration and distribution of the tack coat must be achieved). This can be done by scrubbing the floor and sides with a bristle broom. The hole must be backfilled with either coarse slurry or fine asphalt to specifications as specified.

16A.3   Pavement Layer Repair

16A.3.1   Scope

Pavement layer repair refers to work in connection with the repair of localized pavement failures other than potholes, edge breaks and surface failures. It involves the excavation of the deformed areas and reconstructing the pavement and surfacing layers, including treatment of the floor of the excavation prior to backfilling.

Pavement failures consist of a combination of rutting, cracking and displacement of the road surface and base layer or the surfacing layer only usually accompanied by disintegration of the surfacing.

16A.3.2 Method of Work

(i) Removal of distressed areas

Failed areas to be repaired shall be demarcated. The repaired area shall have a neat rectangular shape, at right angles to the direction of traffic. The existing material shall be excavated and removed to the specified depth.

Excavation for pavement failures shall be cut or trimmed with side slopes perpendicular to the horizontal; for each excavated layer, a step shall be created with the horizontal distance equal to the vertical distance to a maximum of 150 mm.

Excavation can be done with approved milling equipment. The equipment to be used for the conventional breaking-up and excavation of existing pavement layers will be determined by the size and depth of the pavement section to be processed or excavated, taking into consideration the fact that work may have to be carried out in small areas.

Where all or a part of the existing surfacing material is to be reprocessed together with the underlying layer, the surfacing shall be properly broken down and mixed up to the full depth of the existing base material. Fragments of bituminous material shall be broken down to sizes not exceeding 37.5 mm.

The existing bituminous material shall first be removed before the underlying layers are broken up.

Bituminous material may be milled out or otherwise broken up and removed to approved stockpile sites for recycling or to spoil sites, whichever is required. Where the underlying material is to be reprocessed as base, the exposed surface shall be cleaned after removal of the bituminous material. All remaining fragments of bituminous material shall be removed, and not more than 5% of the surface may be covered with bituminous material.

The existing pavement material shall be broken down to the specified depth and removed, or reprocessed *in situ*, whichever may be required. The underlying layers may not be damaged, and material from one layer may not be mixed with that of another layer.

(ii) Backfilling of excavation for pavement failures

Prior to backfilling, the base and sides of the excavation shall be cleaned of all loose material. The excavation shall be backfilled with approved gravel, crushed stone or asphalt and compacted to a density as specified below:

> Base (0 to 150 mm below final base level): 98%
> Subbase (150 to 300 mm below final base level): 95%
> Selected (300 to 600 mm below final base level): 93%
> Fill (Below 600 mm of final base level): 90%

Backfilling of the excavation shall be done as follows:

- Stabilized material excavated from the existing pavement, may be used as backfilling, either for subbase layers only or for subbase and base layers.

  Material shall be broken down and 60 kg/m^3 of cement (OPC) shall be added. Water shall be uniformly mixed into the material. The material shall then be returned to the road and compacted as specified above.

- Backfilling for the base layer shall be done with imported crushed stone material treated with bitumen emulsion. Cement (OPC) shall be added at a rate of 1% (mass/mass) and mixed off the road by means of a concrete mixer or hand labor. All mixing shall result in a homogenous mixture of additives and parent material.

Thereafter, the material shall be treated with a stable-grade bitumen emulsion at a rate of 3% (mass/mass). Before the emulsion is added to the stone, it shall be diluted with water so that the moisture content of the mix will be the optimum moisture content for compaction.

The mixed material shall then be transported to the excavated area, placed and compacted, all within 5 hours of the commencement of the mixing process. Thereafter, 0.6 I/m^2 of the diluted 60% bitumen emulsion shall be applied to the base or layer to ensure a sealed surface.

The density of the backfilling of the base layer shall be at least 98% of modified AASHTO density.

- Where required the backfilling of the base layer shall be done with continuously graded asphalt. The asphalt shall be tested, accepted and placed according to the normal governing specifications.

Appendix 16B
Repair of Defects in Rigid Pavements

16B.1 Full Depth Repair of Continuously Reinforced and Jointed Reinforced Concrete Slabs (Full Depth Patching)

16B.1.1 Scope

The full depth breaking up, excavation, removal to spoil sites, repair of the underlying layer, selection of material and patching of existing continuously reinforced and jointed reinforced concrete slabs. This type of treatment can be used to repair concrete pavements exhibiting severe cracking, punch-outs or failures.

16B.1.2 Method of Repair

For local repair work (areas < 10 m^2, but a minimum length should preferably be 2 m), the perimeter of the area has to be saw-cut approximately 40 mm deep and less than the depth of the reinforcement, whereafter the concrete has to be broken out without damaging the sides of the existing concrete which is left in place, the longitudinal reinforcement or the underlying layers. The concrete cut surfaces must be vertical, without any loose material. Should it become necessary to cut the reinforcement in order to remove the concrete, the cut must be made at least one lap length away from the edge. Bars made of high yield steel shall not be bent and subsequently straightened.

For restricted area repair work (areas > 10 m^2), the perimeter of the demarcated area has to be saw-cut full-depth and the concrete removed by crane or by any other quick and cost-effective method. Explosives may not be used to break up the concrete. The longitudinal reinforcement must be replaced and tied into the existing concrete by means of tie-bars and the edges of the surrounding concrete must be scabbled to provide aggregate interlock. The concrete cut surfaces must be vertical, without any loose material.

After the concrete has been removed, the subbase should be examined to determine its condition. Where required, existing unbound pavement layers shall be removed and be replaced with fresh material. Where the stabilization of the underlying pavement layers is required, the stabilizing agent shall be mixed with the pavement material in a plant or concrete mixer suitable for this purpose. Where the layer consists of cemented material or cemented crushed stone, its surface shall be swept clean and all loose, soft or poorly cemented material shall be removed and the surface treated for irregularities. A primer shall then be applied.

Tie-bars are to be installed in either wet-to-wet concrete joints or in wet-to-dry concrete joints or in dry-to-dry joints if required. The type of bars should match the original in grade, quality and number. The length of tied laps shall be at least 35 bar diameter or 450 mm, whichever is the greater for longitudinal bars and 300 mm for transverse bars. The minimum length of any welded lap shall be 150 mm. The specified method, dimensions and positions of drill holes for installation of tie-bars into the existing concrete shall be in accordance with the project drawings.

Where concrete repair work is done with the longitudinal sides of the new concrete coinciding with existing longitudinal joints, the existing concrete shall be removed, the existing tie-bars shall be cleaned and straightened and new tie-bars installed, if necessary. A bond breaking agent shall be applied to the existing concrete along the vertical face of the longitudinal joint, new concrete placed and a new joint sawn in one operation to a specified width and depth and the back-up material and liquid sealant installed.

Except where concrete is placed in continuous lengths exceeding 50 m, placing, compacting and finishing with hand equipment shall be carried out. Where new concrete is placed next to existing concrete, the edge of the existing pavement shall be properly cleaned and all bituminous and other jointing material shall be removed.

Holes should be drilled a minimum of 150 mm and not more than 200 mm long in the vertical exposed faces of the slab, parallel to the surface and sides of the slabs. The diameter of the holes shall be the minimum that is necessary to accommodate the size of dowel bar that is to be used. Drilled holes must be cleaned out using oil-free compressed air, primed and plugged with resin mortar before dowel bars accurately inserted and aligned parallel to the surface and sides of the slab. The dowel bars should be debonded with thin, tight-fitting plastic sheaths.

If the edge of the repair is at a joint groove forming strips should be stuck along the top edge of the adjacent slab.

All the longitudinal joints shall be resealed.

Clean out the area with compressed air and thoroughly dampen the sub-base and edges of the repair before placement of the concrete. Place and evenly spread pavement quality concrete to the appropriate surcharge, compact using internal and surface vibration and finish flush with the surface of the adjacent slab and to within a tolerance of ± 3 mm and with a difference of not more than 4 mm between the surface of the repair and a 3 m straightedge. Particular care shall be taken to ensure thorough compaction around the reinforcement and the edges of the repair.

No concrete shall be placed during rainy weather. The contractor shall at all times have available frame-mounted waterproof covers for protecting the surface of the unhardened concrete. When slipforms are used, the contractor shall also provide acceptable emergency protection for the slab edges.

When paving is done during hot weather and when the temperature of the fresh concrete can be expected to exceed 24°C, the contractor shall implement appropriate precautionary measures to place the concrete at the coolest temperature practicable. Paving operations shall cease when the concrete temperature as discharged at the paver exceeds 32°C.

Where concrete operations are carried out at ambient temperatures in excess of 32°C the contractor shall carry out relevant recommended actions for hot weather concreting, such as windbreaks, and applying a recommended and approved liquid curing compound which forms a moisture retaining membrane on the surface. No concrete shall be placed if the temperature is less than 5°C unless the patch is adequately insulated.

Curing time of ordinary cement is 7 days or when cube strength is >30 Mpa, and of high early strength when the 7 day cube strength is >30 MPa.

The contractor shall produce a surface texture on the new concrete slab which shall match, as far as is practical, that of the existing surface in terms of color and texture.

16B.2 Thin Bonded Surface Repair of Spalled Joints and Cracks (Partial Depth Patching)

16B.2.1 Scope

The repair of joints and cracks which have been spalled by partial depth breaking up, excavation and patching or saw widening and routing.

16B.2.2 Method

The area to be repaved should be demarcated to cover the full extent of the damage as a square or rectangular area. The minimum size should be 150 mm × 150 mm and extend a minimum of 50 mm beyond all unsound concrete. Delineating grooves of at least 20 mm in depth are then cut along the marked lines around the perimeter of the repair intersecting at corners. The delineating groove should be chased out to at least 10 mm below the required depth using either a small single-headed scabbling tool or a router and template to form a vertical edge to the repair.

The concrete is removed from within the repair area to a reasonably even surface at a depth that ensures that all unsound concrete is eliminated and the repair area cleaned out using oil-free compressed air to remove all dust and loosened concrete. Any partially loosened concrete that remains should be removed by wire brushing after which the area shall be cleaned out with compressed air again.

The repair and surrounding area should be kept thoroughly damp and remain so until the repair material is placed but no excess water should be present where the prime or repair material is placed.

A groove forming strip should be placed where a joint is to be formed so that the top is flush with the required finished surface.

A prime in the form of a neat cement grout or approved bonding agent can be applied to the surfaces if required.

The repair material immediately loosely spread to a 20% surcharge and thoroughly compacted by vibration to work the fine material into the prepared surface, paying particular attention to compaction in the corners and around the edges of the repair to ensure good bond to the old concrete. Compaction shall be by means of a vibrating hammer.

Finishing should be flush with the level of the surrounding slab to within a tolerance of ± 3 mm. Lightly brush the repair material against the hardened edges around the perimeter with a soft brush and apply a surface texture to match the existing texture.

Cementitious repairs shall be cured immediately after texturing by the application of a resin-based, aluminized curing compound to the surface. During hot weather, wet sand or hessian covered by polythene sheeting shall be placed over the area as soon as it can be put on without damaging the applied surface texture and maintained for at least the minimum period. During cold weather, thermal protection should be required. Curing time should be 5 days for ordinary cement and 4 hours for resin mortar. The manufacturer's recommendations should be used for special cements.

The groove former is removed, the joint groove prepared and sealed.

The use of epoxy resin mortar to repair spalling along transverse or longitudinal joints is not recommended.

Where very high severity spalling or breakaways occur, the treatment comprises the removal of a specified width and depth of existing concrete along the spalled joint. A new joint is then formed making use of fresh concrete tied down onto the existing concrete by means of rawl bolts. The concrete to be used shall be manufactured using rapid-hardening cement. A bond-breaking agent shall be applied to the vertical surface of the old concrete along the joint before casting of the infill concrete. Sawing of the joint and installation of a new sealant shall be performed as specified. Fresh concrete (along joints) shall be rounded and existing concrete (along joints) shall be mechanically beveled.

16B.3   Resealing of Joints

16B.3.1   Scope

The removal of damaged seals at joints, cleaning of the joints and the replacement of new seals.

16B.3.2   Method

All existing sealant and caulking material from the joint groove must be raked out and removed.

If necessary, to obtain the minimum dimensions and width to depth ratios required, or to eliminate minor spalling or to completely remove the old sealant, widen and deepen the sealing groove by sawing up to a maximum width of 30 mm. (The minimum depth of sealing groove shall be obtained by adding together the depth of the seal below the surface, the minimum depth of the seal and the thickness of the caulking strip).

The sides of the groove must be scoured by abrasive blasting and thoroughly cleaned out the sealing groove using oil-free compressed air at a minimum pressure of 0.5 N/mm^2. Alternatively, when compression seals are to be used, the sides of the groove may be prepared by grinding or wire brushing.

The sealing groove is then primed and sealed. When either hot or cold applied sealant are used, a caulking strip of compressible material at least of 5 mm thick should be inserted into the bottom of the groove completely filling the bottom of the sealing groove.

Appendix 16C
Repairs of Other Defects

16C.1   Cleaning of Prefabricated Culverts

16C.1.1   Scope

The cleaning of prefabricated culverts. The work involved under this section is the removal of silt and debris from the prefabricated culverts and the removal and clearing of vegetation, silt and debris from the inlet and outlet areas as shown on the drawing.

16C.1.2   Method of Work

All prefabricated culverts, inlet and outlet areas up to the end of the road reserve shall be cleaned. Material removed shall be disposed of in close proximity of the culverts within the road reserve or loaded and transported to spoil at dumping areas. Material spoiled near the culverts shall be spread neatly well clear of the top of drainage trenches where it cannot wash back.

16C.2   Cleaning of Concrete Drains and Channels

16C.2.1   Scope

Work in connection with the removal of silt, debris and vegetation causing obstruction to flow in concrete drainage channels, *inter alia*, side-drains, median drains, kerb-channeling combinations, down chutes and any other concrete drains or concrete channels and armco or plastic chutes.

16C.2.2   Method of Work

Silt and debris at grid inlets and concrete channel outlets shall be removed at specified intervals during the rainy season and thereafter. Material removed from channels shall either be loaded and transported to designated spoil sites or disposed of adjacent to channels where it cannot wash back into the channel within the road reserve.

Where material is spoiled adjacent to channels, the material should be spread neatly and well clear of the top of the channels where it cannot wash back. Material removed from kerb and channel combinations, side-drains in cuts and median drains or from other channels shall be transported to spoil.

Vegetation growing in channel joints and cracks shall be removed with roots to prevent regrowth. Vegetation growing over channels from the edges shall be slashed at the concrete edges and disposed of. Undesirable vegetation shall be removed with roots and spoiled.

Silt debris and vegetation removed shall not be thrown up against cut or down fill slopes.

16C.3   Cleaning and Maintenance of Existing Earth Channels

16C.3.1   Scope

The work involved in cleaning of all earth drains and channels, and repairs to damaged earth drains and channels.

16C.3.2   Method of Repair

Earth side-drains and channels shall be cleaned of all debris, silt and vegetation. Silt and debris excavated from the drains shall be deposited and spread neatly in close proximity of the drains where it will not wash back. Materials removed from drains shall not be deposited against cut and fill slopes.

Scoured and eroded sections of drains shall be backfilled with suitable material obtained from the side of the road or from suitable sources. The backfill material shall be compacted at the optimum uniform moisture content in layers not exceeding 100 mm after compaction. The material shall be compacted to an acceptable standard.

Protective covering against scouring and erosion shall be applied as necessary.

17

Asset Management

Sue McNeil
University of Illinois
Chicago, IL, U.S.A.

Pannapa Herabat
Asian Institute of Technology
Pathumthani, Thailand

17.1 Introduction

Highway agencies in the United States moved from a focus on expansion to preservation in the period 1960 to 1985. In the years 1985 to 2000 the agencies began to focus on reinventing government. In recent years they have shifted to an emphasis on good business practices. The focus on good business means embracing quality, an emphasis on strategic rather than tactical issues, the integration of economics and engineering, and taking advantage of the information age.

In the first phase, highway expansion, (particularly the Interstate System), was completed, and preservation became the bulk of the program with the major focus being on preventive maintenance, rehabilitation, upgrading, and reconstruction. In the second phase, the emphasis was on "right sizing" including outsourcing and streamlining procedures. Operations were performance based, and there were several public/private ventures and opportunities for privatization. During this period, government retained fiduciary responsibility, and responsibility for stewardship, overall management, and programming.

All of the changes resulted in an investment of $1.75 trillion in highway systems throughout the United States. (Peters, 2003). As the physical infrastructure ages, the focus remains on preservation; and while the role of government changes, there are opportunities to manage change. Together with emphases on performance-based management and increased accountability, asset management provides the tools for doing this (Kane, 2000).

Over the past two decades, highway and road agencies have adopted the term "asset management" from the private sector to refer to the development, maintenance, operation, improvement, and upgrading of highway assets in a systematic manner. Concerns with aging infrastructure, increased public demands for more accountability, and tighter budgets have stimulated interest in asset management both in the U.S. and internationally.

This chapter provides an introduction to the concepts of asset management and presents a framework for developing an asset management system. The chapter develops the concepts in more detail and delves into issues related to maintenance and upgrading of systems, the role of asset management in financial management, accountability and lastly, the decision making process. The chapter also provides examples from the public and private sectors in the United States and Asia to illustrate the application of the concepts to real-world problems. This section provides an introduction to highway asset management.

This section begins with several definitions of asset management (AM) and its historical background, and continues with a comparison of private and public assets. The factors motivating interest in implementation of asset management are addressed, and finally the benefits of asset management are presented.

17.1.1 Defining Highway Asset Management

Highway assets are economic resources that provide services to the public (McNeil et al., 2000). Asset management is the process for managing these assets. Asset management can be used as a tool to communicate between system users, stakeholders, state governmental officials, and managers (US DOT, 1999).

The Federal Highway Administration (FHWA) has promoted asset management since the mid 1990s to assist highway agencies in managing different types of highway assets. In 1999 the FHWA formed the Office of Asset Management. Also in the U.S., professional organizations have formed taskforces and committees to promote the concept of transportation asset management. These include the AASHTO Task Force on Asset Management formed in 1997, the American Public Works Association (APWA) task force on asset management formed in 1998, and the Transportation Research Board (TRB) task force on asset management established in 2000 (McNeil et al., 2000).

Asset management is described as a strategic enterprise that combines engineering practices and analysis with sound business practices and economic theory (US DOT, 1999). The transportation community is continually refining the definition of asset management to meet the needs of specific organizations. The following definitions illustrate this point:

- "A systematic process of maintaining, upgrading, and operating physical assets cost-effectively" (US DOT, 1999).
- "A methodology needed by those who are responsible for efficiently allocating generally insufficient funds amongst valid and competing needs" (Danylo and Lemer, 1998).
- "A comprehensive and structured approach to the long-termed management of assets as tools for the efficient and effective delivery of community benefits" (Austroads, 1997).
- "A comprehensive business strategy employing people, information, and technology to improve the allocation of available funds amongst valid and competing asset needs" (TAC, 1996).

- "Asset Management ... goes beyond the traditional management practice of examining singular systems within the road networks, i.e., pavements, bridges, etc., and looks at the universal system of a network of roads and all of its components to allow comprehensive management of limited resources" (OECD, 1999).
- "Asset management ... a set of concepts, principles, and techniques leading to a strategic approach to managing transportation infrastructure. Transportation asset management enables more effective resource allocation and utilization, based upon quality information and analyses, to address facility preservation, operation, and improvement." (Cambridge Systematics, 2002c).

An asset management system (AMS) can be defined as a dynamic management system that combines and integrates engineering, business, and technological aspects to optimize infrastructure management under budget constraints. The system collects asset inventory and condition, analyzes the impacts and the asset deterioration process to further establish alternative maintenance strategies, estimates costs and benefits received from maintenance actions, determines the tradeoffs analysis for different infrastructure assets, and establishes the prioritization and optimization programs for asset maintenance planning. Optimal maintenance planning helps decision makers properly allocate maintenance budget and establish assets' short-term and long-term planning.

Figure 17.1 illustrates the interaction between various elements of a generic asset management system guided by an agency's objectives and goals; policy; system monitoring; system planning and evaluation; and system implementation. These elements support the concept of an asset management system in the sense that the agency has a clear policy, goals, and objectives to achieve. System monitoring, planning, evaluation, and implementation support the establishment of optimal or cost effective maintenance management.

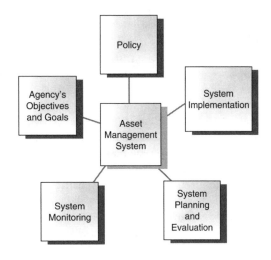

FIGURE 17.1 Elements of an asset management system.

Implementing AM is a challenge for highway agencies. Agencies can choose to either design and develop a new AMS or integrate different platforms of existing management systems. Agencies also need to select the appropriate assets, actions, business processes, and asset management concepts (Cambridge Systematics, 2002c). A good AMS should improve the assets cost effectively, while enhancing the credibility and accountability of the agency.

17.1.2 Historical Background

The concept of asset management is adopted as a way to help agencies simultaneously manage their various assets while staying within their limited budget. Infrastructure management received significant attention from agencies as early as the late 1960s. Negligence, in the form of not maintaining infrastructure, caused many catastrophic infrastructure failures such as the collapse of the Silver Bridge in Ohio, in 1967 (US DOT, 1986). The failure resulted in loss of life and property in addition to several major lawsuits. This legal action prompted Congress to pass new legislation. The resulting laws require the inspection of all U.S. bridges at least every two years.

Pavement management system concepts were introduced in the 1960s as a systematic way to manage pavements. Pavement management systems (PMS) are designed to manage, plan, allocate budget, and schedule all the pavement maintenance activities for agencies. Similarly, bridge management systems (BMS) were introduced in the 1970s. AASHTO, FWHA, and other organizations have developed guidelines and published papers to support and encourage agencies to adopt pavement and bridge management systems (TRB, 1994, 2000; AASHTO, 2001). At the same time maintenance management systems have evolved to support the day-to-day and tactical issues related to asset preservation (Markow, 1995). The detailed development and implementation of each type of management system depends on the agencies. This is due to the fact that physical assets by their nature are both complex and unique.

From the 1960s to the 1980s, most agencies focused more on infrastructure construction. Shifts in federal, state, and local policies relative to infrastructure management and expansion, budgeting decisions, and staff resource allocations over the last 20 to 30 years have impacted transportation investment decisions and are likely to play a key role in the future. First, federal transportation policy and legislation have seen significant shifts in response to constrained budgets and shifting priorities at all levels. Specifically, the Intermodal Surface Transportation Efficiency Act (ISTEA) adopted in 1991 implemented a national intermodal transportation system approach intended to "link highway, rail, air, and marine transportation. Prior to ISTEA, the Federal-Aid Highway Program had been directed primarily toward the construction and improvement of four federal-aid systems: the Interstate, primary, secondary, and urban highways" (Schweppe, 2002). ISTEA required public sector agencies to shift their focus away from capacity expansion and emphasize the preservation and operation of the country's $1 trillion investment in its highways and bridges (USDOT, 1999). Among other things, it established an Interstate Maintenance Program for resurfacing, restoring, and rehabilitating the Interstate Highway System (Schweppe, 2002). The Transportation Equity Act of the 21st Century continued to build on the policies of ISTEA and offered greater spending flexibility to fund highway safety and transit programs.

Prior to 1995 the view of asset management in the United States was that asset management was something private sector companies did, transportation agencies in Australia and New Zealand said they practiced (Norwell and Youdale, 1997), and state departments of transportation (DOTs) thought they should be practicing. In September 1996, the American Association of State Highway and Transportation Officials (AASHTO) and the FHWA held the first asset management workshop focused on sharing experiences in both the public and private sectors (FHWA, 1996). Since 1996, a series of activities has helped to advance the state of the art and state of the practice of asset management. These activities include (Oberman et al., 2002):

- *21st Century Asset Management*, a second workshop (US DOT, 1997)
- AASHTO Transportation Asset Management Task Force formed (AASHTO, 1998)
- FHWA Office of Asset Management formed to provide technical support
- Survey of state agencies conducted (McNeil et al., 2000)
- Asset Management Peer Exchange: Using Past Experiences to Shape Future Practice, a third workshop, held (AASHTO, 2000)
- Transportation Research Board Asset Management Task Force formed
- Community of Practice website Transportation Asset Management Today (TAMT) launched by FHWA and AASHTO (Winsor et al., 2004)
- Multiple National Cooperative Highway Research Program (NCHRP) studies
- Project 20-24(11), Asset Management Guidance for Transportation Agencies — completed in 2002. (http://www4.trb.org/trb/crp.nsf/All+Projects/NCHRP+20-24(11))
- Project 20-57, Analytic Tools to Support Transportation Asset Management — scheduled for completion in 2003. (http://www4.trb.org/trb/crp.nsf/All+Projects/NCHRP+20-57)
- Project 20-60, Performance Measures and Targets for Transportation Asset Management — initiated in 2002. (http://www4.trb.org/trb/crp.nsf/All+Projects/NCHRP+20-60)
- Project 19-04, A Review of DOT Compliance with GASB 34 Requirements — scheduled for completion in 2003. (http://www4.trb.org/trb/crp.nsf/All+Projects/NCHRP+19-04)

- Taking the Next Step, a fourth workshop (Wittwer et al., 2002)
- Moving from Theory to Practice, a fifth workshop (Wittwer et al., 2004)

Today, the emphasis has shifted from new construction to maintenance and management of existing infrastructures. Realizing how infrastructure contributes to the nation and its significant initial investment, relevant agencies have a commitment and responsibility to the public to maintain current infrastructure and provide a safe environment to the community and the nation. Most highway agencies manage more than one type of asset such as, bridges, pavement, culverts, guardrails, intersections, and sign structures. However, the highway preservation and improvement funds are received from the same pool (usually a combination of state and federal money). Analyzing the costs and benefits of different types of infrastructure is beyond the scope of an individual infrastructure management system. The new trend in infrastructure management systems has thus shifted to the concept of AMS to improve overall short-term and long-term planning for infrastructure investment and preservation across various types of assets.

17.1.3 Public Assets versus Private Assets

Physical assets are economic resources that generate benefit streams for an entity or society. Land, buildings, machinery, pavements, bridges, water and wastewater, waste management, and parkland are examples of physical assets (Anthony and Reece, 1989). Groups of infrastructure assets may be owned or managed by public and private sector entities. Infrastructure assets provide essential services to individuals, communities and nations.

Highway assets can be divided into two types: physical highway assets and other operational types (US DOT, 1999). Physical highway assets are composed of pavement, structures, tunnels, and hardware, i.e., guardrail, signs, lighting, barriers, impact attenuators, electronic surveillance and monitoring equipment, and operating facilities. Other operational highway assets include construction and maintenance equipment, vehicles, real estate, materials, human resources, and corporate data.

Public highway assets are under the responsibility of public agencies or government-owned agencies, such as, municipalities, and departments of transportation. These agencies receive insufficient capital funds from the central government to build new and enhance existing public assets. Ongoing significant investments are required to maintain the physical and operational quality of the public highway assets to ensure safety to the public and to maintain an overall condition above the minimally acceptable level. In the United States, maintenance has traditionally been the responsibility of state and local agencies. Maintenance budgets, whether allocated from any central government or based on local revenue sources, are usually inadequate compared with the overall maintenance needs.

Private highway assets are managed by the private sector. The private sector is more business-oriented since it generally has to be self-sustaining. The management activities of the private sector are similar to those of the public sector: management and maintenance. Due to a limited central maintenance budget on a national level, some public highway assets are privatized, in other words, the responsibility for their maintenance is transferred to an agency in the private sector. National government may partially support the private sector or may not. Regardless, private asset owners usually have to find their own resources to maintain their assets. Highway toll is an example of finding the resources to pay for needed and anticipated repairs. In the case that privatized infrastructures are partially supported by public agencies, the concept of a shadow toll is sometime used to negotiate the budget allocation from the central government to the private sector. In addition, the private sector has the advantage of greater flexibility to develop systematic procedures and data uniformity, and to generate a more complex financial analysis (US DOT, 1996).

17.1.4 Motivation for Asset Management

Traditional infrastructure practices are managed based on the collection of infrastructure inventory data, condition, maintenance, cost and other relevant information. Data analysis procedures are used to probe the information to support maintenance planning under limited budgets. Traditional highway preservation and improvement is a dynamic process that combines the condition of infrastructures with engineering economic principles to determine the viability or cost effectiveness of different maintenance plans.

Asset management has long been an important component of the management practices in the private sector. Recently, asset management has been receiving significant interest in the public sector around the globe. Many agencies are implementing asset management concepts as a way to expand their infrastructure management practices. Examples can be found in the U.S., Australia, and Canada. Several factors motivated the different agencies to include asset management strategies in their agency's objectives. Several are listed below (US DOT, 1999, 2002):

- To improve the highway management efficiency and capability,
- To support the shifted paradigm from new construction to maintenance management,
- To reinforce budget demands by providing rational justification for investment in infrastructure when competing with other publicly supported programs,
- To increase public acceptance and accountability,
- To support tradeoff decisions as demand continues to grow causing increased congestion and wear and tear on the system,
- To overcome personnel constraints from downsizing problems and the competition of the employment market,
- To improve communication with customers, owners, and elected officials.

17.1.5   Benefits of Highway Asset Management System

Infrastructure development and management affect operators, owners, and users. Maintaining assets in the most cost effective manner for short-term and long-term planning under budget constraints indirectly improves the quality of life for the citizens, communities, and the nation. Tangible benefits received from applying a highway AMS are the optimal benefit-cost analysis and long-term cost reduction. Intangible benefits received from using a highway AMS include a safe, convenient, and comfortable highway system. Additional benefits obtained from using an asset management system described by (US DOT, 1996) are presented in Table 17.1. These benefits require a commitment to asset management. This commitment involves (Cambridge Systematics, 2002c):

- Strong executive leadership.
- Buy-in at all levels of the organizations.

TABLE 17.1   Summary of AMS Benefits (US DOT, 1996)

Benefits to Owners/Operators	Benefits to Users
Improvements to program quality	Improved convenience
Improved information and its accessibility	Improved service (e.g., comfort, reliability, safety in transportation context)
Improved economic assessment of various scenarios	Savings passing on from the owner/operator to the customer
Improved documentation of decisions	Increased accessibility of facilities and services due to more efficient operations
Enhanced communication	
Improved information on return on investment and value of investments	
Reduction of both short- and long-term costs	
Improved facility performance	
Streamlining of the decision making process	
Inclusion of public opinion on facility and network performance	
Justification for resources in competitive funding situations	

- A multi-disciplinary perspective.
- Sustained commitment to the process of asset management.
- Meeting the needs of the organization providing the user-friendly approach to accessing data and information, and the appropriate tools to support the process that link technical, management, and budgetary processes.

17.2 Highway Asset Management Framework

This section establishes the framework for asset management. The section begins with a discussion of the essential elements of highway asset management systems. The section then develops the concepts, performance measures, processes, model development, decision making, and optimization to integrate these elements. This section serves as a foundation for the tools presented in the following section.

17.2.1 Essential Elements of Highway Asset Management System

Asset management systems involve the development and implementation of structured processes to improve the decision-making capability of an agency. As shown in (Figure 17.2), an overall framework of asset management systems suggested by (US DOT, 1999) consists of asset inventory, condition assessment and performance modeling, maintenance alternative selection and evaluation, methods of evaluating the effectiveness of each strategy, project implementation, and performance monitoring. System complexity substantially depends on the types of assets that are being managed and the available budgets or resources.

In order to illustrate the AMS concept, (Figure 17.3) maps the detailed dataflow and data analysis of asset management systems. Asset management can be examined on two levels: (1) network-level management and (2) project-level management. Network-level management provides decision makers with the optimal maintenance planning and strategy selections. It considers all asset inventories, determines overall network deficiency, evaluates the maintenance alternatives, prioritizes maintenance actions, performs economic analysis for the entire network, and optimizes the maintenance planning within the constraints of a limited budget. The decision makers could view the overall improvement in terms of an improved asset performance, total maintenance costs, overall benefits received or total user savings based on selected maintenance actions. Project-level management concentrates more on the detailed information of the asset and its deficiency, detailed maintenance actions, and the implementation process of each project.

In the process shown in Figure 17.2 each highway management system (e.g., PMS, BMS, guardrail management system, and sign management system) collects its inventory information in a database management system (DBMS). An asset-shared DBMS is the first essential element of a highway asset management system, which stores all the relevant asset information. Large amounts of asset information are stored and retrieved for further data analysis as shown in Figure 17.3. Some examples of the asset inventory data are asset names, locations, age, condition, construction dates, and maintenance history.

To avoid data duplication and inconsistencies, it is essential to identify systematically the location of the assets. In many highway agencies, different management systems such as, PMS and BMS are

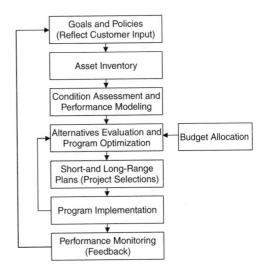

FIGURE 17.2 Generic asset management system components (modified from US DOT, 1999).

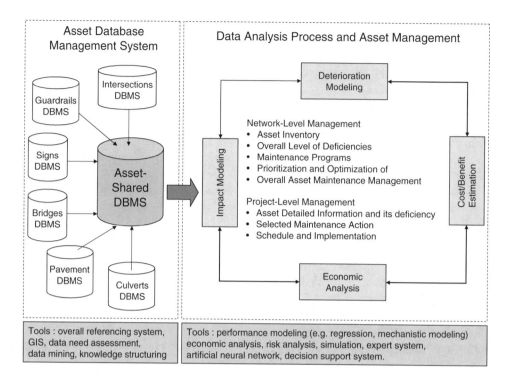

FIGURE 17.3   Mapping of asset management system.

often incompatible in terms of location referencing systems, condition evaluation, analytical procedure, and data transformation (Lee and Deighton, 1995; Gharaibeh et al., 1999). This is why data sharing and communication among these systems may be impractical and expensive. Sinha and Fwa (1987) listed the limitations of the existing management systems as lacking comprehensiveness, system coordination interdependence of functional activities, quality of data management, organizational structure, and linkage to pricing and taxation.

Individual management systems should be developed with proper coordination from other systems and with a clear understanding of the requirements of the entire system (Sinha and Fwa, 1987). For example, if an aggregate bridge-pavement maintenance schedule is anticipated through the use of an integrated bridge and pavement management system, then all the resources will be effectively used (Tonias, 1994). There have been many efforts to combine several management systems into a unique integrated management system in order to improve the effectiveness of the system and to provide better utilization of an overall infrastructure both at the project level and the network level (Ravirala and Grivas, 1995; Gharaibeh et al., 1999; Sadek et al., 2003). These efforts have met with varying degrees of success. However, data sharing and linking appropriate functions have also been proved effective.

A consistent location referencing system should be established in the asset-shared DBMS to record the geographical location of the assets in a unique standard format. The state-of-the-art application that provides this capability is a Geographic Information System (GIS). GIS provides not only spatial reference, but it also has spatial analysis capabilities. In addition, several studies have been conducted that incorporate other tools and technologies to enhance the DBMS's capability and integrity. For example, knowledge structuring can be used in conjunction with data mining and artificial neural networks.

In general these computer tools can be used to catalogue, synchronize, and archive asset information. Agencies often collect a large amount of asset information such as, asset basic inventory, usage, and maintenance history. This information is cumulatively collected through time and can be separately collected in different units' databases. However, data dispersion results in difficulties for agencies to acquire or access the complete information for different assets. One of the core elements of any traditional management systems is the database, which is designed to store the basic information of each asset type. Asset management systems recognize the multi-disciplinary characteristics of different asset types and allow various units of a single agency to catalogue all necessary information in one system.

Data integration is one technique that can be used to help agencies effectively manage their large amounts of asset information. The process of data integration can be defined as "the process of combining or linking two or more data sets from different sources to facilitate data sharing, promote effective data gathering and analysis, and support overall information management activities in an organization" (US DOT, 2001). Benefits received from data integration include easier accessibility and data retrieval, and improved data accuracy and integrity, more comprehensive data, reduction of data duplications, and a better decision-making process (US DOT, 2001). Figure 17.4 presents the data integration process.

According to Figure 17.4, requirement analysis identifies the need for data integration, which reflect the business process, data characteristics, user requirements, agency's goals and constraints, and the information system requirements (US DOT, 2001). Data and process flow modeling support the needs and characteristics identified in the requirements component and further map the data flow and its interconnections. For example, an agency collects pavement data and its inventory data. Condition data is then collected from the field to select and plan maintenance activities. Basic inventory data can aid

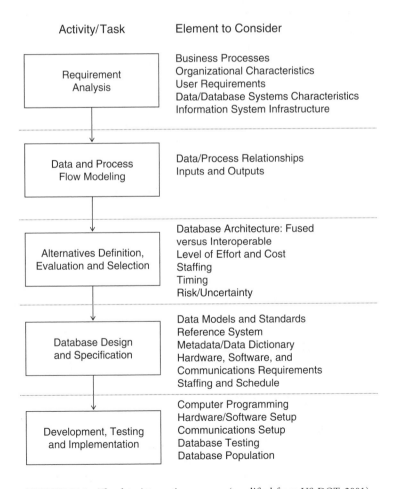

FIGURE 17.4 The data integration process (modified from US DOT, 2001).

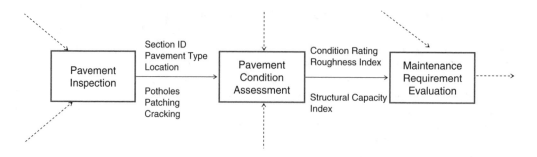

FIGURE 17.5   Section of data and process flow diagram for pavement management (adapted from US DOT, 2001).

the maintenance process in terms of identifying the pavement type and location, and provide an identification code (ID) to link the different types of data. Figure 17.5 presents a data and process flow diagram for pavement management.

Two other approaches to the data integration process are suggested in (US DOT, 2001): the first is a fused database or data warehouse, and the second is interoperable data (federated data) as illustrated in Figure 17.6. Data fusion or warehousing extracts data from different data sources and exports the data into the centralized database. The fused database allows users to access a vast collection of data in a centralized database. Interoperable databases or distributed database systems allow "collection of separate and possibly diverse interoperating database systems over multiple sites that are connected through a computer communications network" (US DOT, 2001). This type of database allows users to directly access different databases rather than the specific data item selected for inclusion in the data warehouse.

Database system specification should be in accordance with the requirements analysis and data and process flow modeling components. The concept of data models can be used to structure data and set/network their relations. Issues of hardware, software, and communications requirements need to be addressed in this process. Once the database system is implemented, verification and validation should be conducted to ensure the accuracy of the database.

Data integration requires a large coordination of database servers and clients. The most common technical and organizational issues are (1) heterogeneous data, (2) bad data, (3) lack of storage capacity, (4) unanticipated costs, (5) inadequate cooperation from staff, and (6) lack of data management expertise.

Data analysis is composed of four main processes: (1) deterioration modeling; (2) impact modeling; (3) cost-benefit estimation; and (4) economic analysis. Deterioration modeling captures the deterioration process and attempts to forecast the future deficiency of an asset. Deterioration modeling can be characterized in many different ways, empirical/mechanistic, probabilistic/deterministic, and static/dynamic to name a few. The data collection and facility monitoring component, as illustrated in the asset-shared DBMS, provides basic data on the current condition. This information is then input into the impact modeling component for evaluation and deterioration prediction. Two types of data are obtained from the asset-shared DBMS. The first type is related to the asset and includes causal data and exogenous variables that affect condition. Causal variables include usage or demand, as-built condition, environment and age. These variables contribute to facility deterioration. Data on facility condition are also used for facility monitoring. These data may be collected using automated methods, subjective judgment or special-purpose testing machines or vehicles. The data collection and monitoring may be modified to reflect changes in the objectives. The second type of data are cost related and include both costs to the agency for performing maintenance and improvements, and costs to the user related to the condition of the facility.

Impact modeling develops and maintains three types of models for the evaluation of the impact of maintenance and recording of the deterioration of current and future facility performance, users of

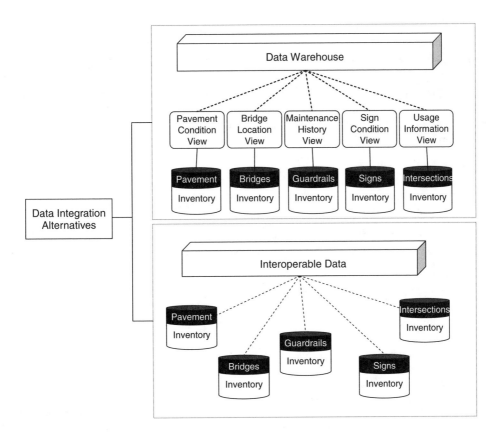

FIGURE 17.6   Data integration alternatives.

the facility, and the agency. The models are:

- Facility deterioration models for evaluation of facility performance.
- User impact models that include the effects of causal variables, maintenance strategies, and facility performance on vehicle operating costs, ride comfort, safety factors and other relevant variables.
- Strategy identification and cost prediction models that account for agency resource constraints recognize new technology and estimate costs using engineering or statistical cost estimates.

The impact models may be reformulated and estimated or updated, as additional data become available. The impact models are used to predict the future performance, possible strategies and current and future impacts and costs. The output from the impact modeling components becomes the input for the strategy selection component. There are several approaches to strategy selection that may be used independently or in combination, such as:

- Preplanned decision rules or policies,
- Engineering judgment or heuristics,
- Optimization techniques.

The strategies selected may vary with the objectives selected for the management process as reflected by the interaction indicated in Figure 17.3. Related costs and benefits need to be estimated. Economic analysis can then further evaluate the viability of different strategy selections and thus optimize the overall maintenance planning. The results of this stage give an "optimal" strategy as defined by the objective.

The data analysis module provides effective tools or procedures for users to address their needs and improve the decision-making process of the AMS. The information retrieved from the asset-shared database, and the results from the data analysis, need to be presented in a user-friendly manner so that the users can make use of the information. The elements described above are similar to the components of traditional infrastructure management systems such as, BMS and PMS. However, the detailed implementation varies among the different types of assets being managed.

Customizing the asset management framework to a particular organization is essential to the successful implementation and application of asset management, there are many ways to look at the key components or elements of an asset management system. For comparison purposes, the components of an asset management system are briefly reviewed as identified by Austroads (2002). The components of the comprehensive planning framework are:

- Policy development,
- Planning,
- Planed execution,
- Verification.

These components embrace the eight elements of asset management for road networks as described in the *Road Asset Management Guidelines* (Austroads, 1994):

- Community benefits,
- Road system performance,
- Asset features,
- Asset condition,
- Asset usage,
- Physical treatments,
- Management of use,
- Asset management strategy.

This structure accomplishes the same end results but puts slightly different emphasis on different elements. This is clearly demonstrated by a comparison of Figures 17.7 and 17.8. Figure 17.7 is an example

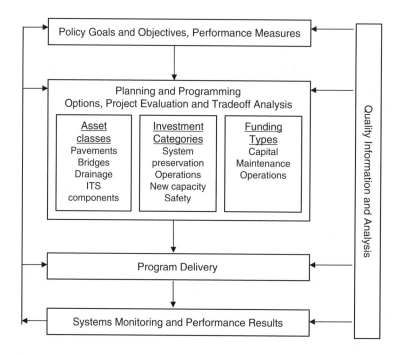

FIGURE 17.7 Resource allocation and utilization process in asset management (modified from Cambridge Systematics, 2002c).

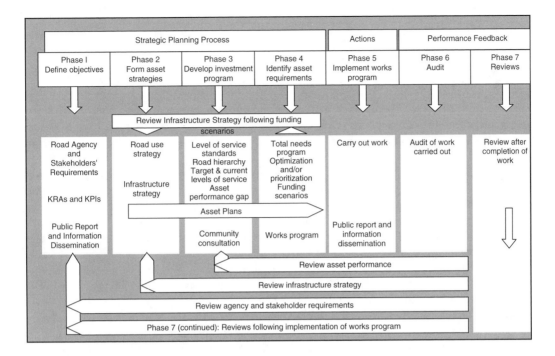

FIGURE 17.8 Integrated asset management process flow diagram (modified from Austroads, 2002).

of resources allocation and utilization in asset management modified from the AASHTO *Transportation Asset Management Guide* (Cambridge Systematics, 2002c), and Figure 17.8 is an asset management process flow diagram modified from the Austroads *Integrated Asset Management Guidelines for Roadwork* (Austroads, 2002). Every element, (albeit with different labels), and interconnection in Figure 17.7 can be found in Figure 17.8 and vice versa. Both processes are assumed to occur within the context of a life cycle analysis.

17.2.2 Highway Asset Performance Measures

In a narrow sense, performance measures are developed to measure the condition of assets in order to determine appropriate actions to prolong the asset's life (Hudson et al., 1997; Danylo and Lemer, 1998). In the broader context of asset management as a strategic approach to managing transportation infrastructure, performance measures are developed to promote the more effective allocation and use of resources to address preservation, operation, and improvement of transportation infrastructure. Technical measures are used for tradeoff analyses and to support investment decisions. Non-technical measures, such as, security, social welfare, environment, and economic development are also important to asset management.

Many transportation agencies have developed system-level performance measures to help track the impact of program investment, maintenance, and operational improvements. These performance measures are usually technical in nature, capturing an engineering or operational attribute of the transportation system. For example, Austroads uses ten different categories of measures in the national performance database. These categories are (Austroads, 2003):

- Road safety,
- Registration and licensing,
- Road construction and maintenance,
- Environmental,
- Program/project assessment,

- Travel time,
- Lane occupancy rate,
- User cost distance,
- User satisfaction index,
- Consumption of road transport, freight and fuel indicators.

Alternatively, performance measures can be grouped into four broad categories: service and user perception, safety and sufficiency, physical condition, and structural integrity/load-carrying capacity (Hudson et al., 1997). Systematic and uniform performance measures are essential in infrastructure maintenance planning.

Figure 17.9 shows examples of physical condition performance measures for sign structures and pavements. The sign condition rating is a numerical scale from 1 to 5 in which 1 represents the failure condition while 5 refers to an excellent condition. The sign structures are divided into five components, which are sign face, sign sheet, substrate, post, and foundation. Each component has its detailed condition rating (Cunard, 1992). For example, retro-reflectivity needs to be measured for each sign face. Five condition ratings of retro-reflectivity for sign face are classified as (1) non retro-reflectivity, (2) shade between one and three, (3) partial retro-reflectivity, (4) shade between three and five, and (5) high retro-reflectivity. Once the condition rating of each sign component is assessed, weighting factors can be determined to calculate the overall sign performance index (SPI).

Pavement performance is a broad term that describes how the pavement condition changes with time or how pavement performs (George et al., 1989). Pavement performance can be defined as the ability of pavement to serve traffic demand within the acceptable level. As shown in Figure 17.4, major groups of pavement performance measures are categorized as roughness measure (e.g., present serviceability index [PSI] and international roughness index [IRI]), surface distress (e.g., pavement condition index [PCI]), deflection, surface friction (e.g., skid number [ASTM]), combined index (e.g., overall pavement condition index [OPI]), vehicle operating costs and traffic delays. These performance indices are developed to represent the actual condition of the pavement. The details of each pavement performance measure can be found in (Haas et al., 1994).

Performance information provides the basic data in the data analysis process. Traditional infrastructure preservation collects this information and stores it in an individual management system for data analysis. The concept of asset-shared DBMS provides a greater flexibility to users and decision makers as they manage the collection of highway assets in any one place. The challenge arises when different assets have different performance measure platforms as seen in the example of signs and pavement. These performance measures are customized to reflect the actual condition of each asset. The overall asset performance could be viewed as the combined performance measures from all highway assets as illustrated in the performance curve of Figure 17.9. There are several techniques, which can be used to combine different performance platforms, e.g., analytical hierarchy process (AHP), weighting factors, and statistical approaches. (Kim and Bernadin, 2002; Cafiso et al., 2002)

The combination of performance measures needs to be developed with caution since the aggregate nature of the overall asset performance may override the importance and critical nature of each asset. However, this process may not represent the true degradation process of an individual asset thus affecting the reliability, accuracy, and superiority of maintenance strategy selection.

17.2.3 Activity and Cost Model Development

Defining highway asset classes, and activities or actions included in the asset management system depends on the agency's objectives, policy, and goals as illustrated in Figure 17.1 as well as on who owns and controls the assets and the particular geographical area of interest. Ideally investments in the preservation, capital improvement, and operations of all highway assets (right of way, structures, operations equipment, pavements, and drainage) must be considered for the full benefits of asset management. Once the scope, extent, and nature of the assets and activities are identified, models and data are needed to support the decision making process.

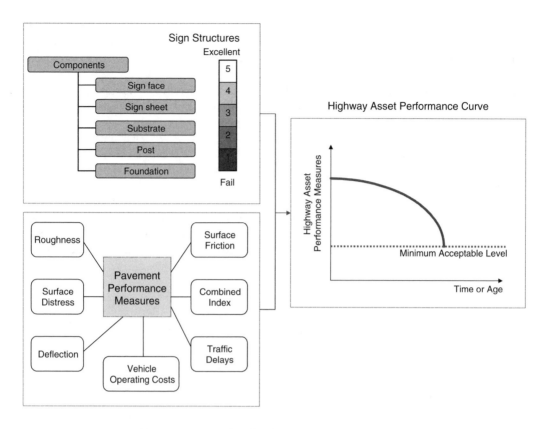

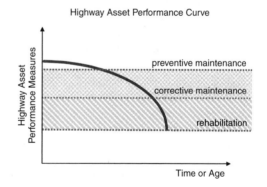

FIGURE 17.9 Examples of highway asset performance measures.

For example, a highway agency needs to establish several criteria for different maintenance actions in terms of minimally acceptable levels of condition for preventive maintenance, corrective maintenance, and rehabilitation action. Figure 17.10 illustrates how an agency could establish different criteria for various maintenance actions.

Definitions of different maintenance actions are described as follows (Hudson et al., 1997):

- *Preventive maintenance.* The maintenance type that is performed to prolong service life. Preventive maintenance should be performed on a regular basis. Examples of preventive maintenance actions are surface cleaning, joint sealing, and painting.

FIGURE 17.10 Maintenance actions and their established criteria.

- *Corrective maintenance.* The maintenance type that is performed to restore the degradation process of an asset to satisfy the functionality, operation, and safety of an asset.
- *Rehabilitation.* The maintenance action that is performed to restore the asset and to prevent structural failure. This may include major repair activities or asset reconstruction.

Highway asset management covers a wider range of actions, including pavement maintenance, bridge maintenance, roadway and bridge widening, intersection improvements, signal upgrading, and guardrail maintenance. The sophistication of activity classification and development significantly affects the data analysis process. Similarly the associated cost models are an important component of the asset management process. The cost models and data should be developed to support the activities defined by an agency.

The cost functions should be calibrated to reflect the agencies' objectives and local environment by taking into account historical maintenance information. Cost functions can be expressed as linear and non-linear relationships between maintenance expenditures and other relevant factors such as, traffic level (in terms of ESAL), and asset age (the number of years since the asset was last maintained). Zaghloul et al. (1989) stated that a successful maintenance cost model could reduce a wide gap between the estimates and actual project cost. A cost model could be used to quantify the consequence of different highway maintenance activities.

In addition to obtaining the direct relationship between costs and other variables, some agencies use breakdown items and their unit costs for each maintenance activity to estimate construction and maintenance costs. Each maintenance activity may consist of many sub-cost functions For example, the pavement overlay cost function is composed of the transportation cost function, the material cost function, and the operational cost function.

17.2.4 Decision Making

Highway asset management systems are designed to support and enhance different decision-making processes ranging from the detailed technical aspect of each maintenance project to the administrative level. The technical and engineering aspects focus on the detailed deficiency of each project and detailed maintenance strategy selection. This aspect is the project-level management in the asset management system.

Higher management level requires extensive calculation based on the data analysis module to determine the optimal maintenance planning. Network-level maintenance planning is prepared for administration and politicians for budget allocation as illustrated in Figure 17.11. Highway maintenance budget has to compete with other publicly supported programs such as, education and healthcare. Reliable reports for the needs of asset maintenance are essential. Received benefits from maintenance actions and the improvement

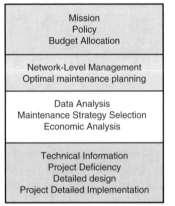

Administration & Political

Mission
Policy
Budget Allocation

Network-Level Management
Optimal maintenance planning

Data Analysis
Maintenance Strategy Selection
Economic Analysis

Technical Information
Project Deficiency
Detailed design
Project Detailed Implementation

Technical Level & Engineering

FIGURE 17.11 Levels of decision making (adapted from Haas et al., 1994).

of the overall asset performance could help convince administrators and politicians to allocate the maintenance budget to highway agencies. Hatry and Steinthal (1984) classifies decision making into eight categories as follows:

- Only do crisis maintenance — "if it ain't broke don't fix it," leads to low initial costs, high life cycle costs, reactive approach.
- Worst first — does not preserve prior investments, based on condition.
- Opportunistic scheduling — encourages deferral of maintenance.
- Prespecified maintenance cycle standards — (e.g., prespecified time intervals) — set repair cycle, rarely enforced, based on averages, good for long range planning.
- Repair components at risk — identify & repair components likely to fail, ignores condition, attempts to reduce service interruptions.

- Preventative maintenance — regular inspection and maintenance where frequency is a function of condition and hours of service or usage.
- Reduce the demand for and/or wear and tear on a facility — reduce the need for maintenance, e.g., fees based on usage.
- Economic comparison of alternatives — system economic analysis of tradeoffs of the costs of implementation and impacts.

17.2.5   Optimal Highway Asset Management System

The optimization module of highway asset management is a decision support tool that balances the objective function, (i.e., minimizes maintenance costs or maximizes the overall benefits), and selects the optimum set of Maintenance and Rehabilitation (MR&R) actions for the entire highway network. Its determination is subject to user-defined budget constraints and other considerations. Optimization provides a bundle of projects that satisfies a set of criteria including budget and other constraints over an analysis period. Many optimization techniques, such as, integer programming, integer linear programming, and dynamic programming, can be used to determine the optimal management planning.

Linear programming (LP) techniques can maximize or minimize an aggregate consequence of individual actions given a set of limitations or constraints. (Hudson et al., 1997) suggests that LP is a useful formulation for multi-year optimization because it can be set up to model the tradeoffs between project timing and benefit losses.

A concept of a highway asset management system could be envisioned as a three-dimensional highway management system (Sinha and Fwa, 1987). The three dimensions are highway facility, operational function, and system objective. The three-dimensional matrix structure indicates that highway agencies have a number of facilities to manage within a given highway system. Using this framework, the highway management process can certainly be viewed as a multiple-objective system. The optimization module of the traditional infrastructure management system is not designed to handle a multiple-objective system.

Two issues are then raised when optimizing the maintenance strategy selection across various types of assets which are (1) measuring and comparing benefits; and (2) allocating component budget allocation (Gharaibeh et al., 1999). Quantifying the benefits received from maintenance actions across various types of assets in a standardized way may be unattainable, such as in the case of comparing road user savings, reduction of traffic delays, cost effectiveness of bridge improvement, and sign replacement. The "efficiency" of a particular type of infrastructure is suggested as a comparative performance measure to help measure and compare benefits across competing asset components (Gharaibeh et al., 1999). The efficiency of linear features, (e.g. pavement), is "defined as the ratio of the total vehicle miles traveled over adequate infrastructure (VMT-A) over the infrastructure life to the total VMT-A computed with an unlimited budget" (Gharaibeh et al., 1999). The efficiency of point features (e.g. intersections) is defined as "a ratio of the total number of vehicles using an adequate infrastructure (VU-A) to the total VU-A computed with an unlimited budget" (Gharaibeh et al., 1999). Overall performance based on these benefit measures was found to be highly correlated with investment levels and simple models can be used to optimize budget allocation across components (Gharaibeh et al., 1999). This technique is one approach suggested to provide a uniform measure of received benefits from maintenance actions, which would result in an enhanced optimization module of the highway asset management.

17.3   Tools for Highway Asset Management

Tools for highway asset management are intended to support decision-making. As discussed in the previous sections, asset management goes beyond the individual pavement, bridge, sign, and maintenance management systems. As stand alone tools, the individual management systems do not allow for tradeoffs between investments in different types of assets, such as, maintenance and rehabilitation; or between competing policy objectives, such as, mobility, safety, preservation and economic development;

and among modes (Cambridge Systematics, 2002a). However, these management systems provide a foundation from which other tools can be developed to support asset management. The asset management system components shown in Figure 17.2 form the fundamental building blocks for the tools for asset management.

In 1996 a GAO report (Government Accounting Office, 1996) found that based on survey of the 50 states, the District of Columbia and Puerto Rico, management systems were being widely explored for transportation infrastructure. However, very few states were addressing the difficult issues related to integrating their systems although several states indicated that this was an objective. In 1999 a survey was sent to each of the 50 states to determine "who is doing what" (McNeil et al., 2001). Thirty-three states responded. Not surprisingly, the 1999 survey found that fewer states were using management systems than indicated in the 1996 GAO survey as shown in Table 17.2. The results of the 1999 survey also provided considerable insight into the diversity of approaches and the different ways in which states are implementing and applying the tools, particularly as it relates to integrating management systems. According to Table 17.2, the results can be summarized as follows:

- The majority of states (80%) are able to assess the impacts of investments or expenditures using one or more of their management systems. Of these states, 84% do this for pavements and 68% for bridges.
- A variety of tools are used by states for making decisions. Only one state did not use any tools and 82% of states using tools used more than one tool. In fact, three states used four or more methods. The most popular tools were cost-benefit analysis, used by 83% of states using tools; and life cycle cost analysis, used by 80% of states using tools.
- The questions focusing on how these tools were used indicated that a significant number, but not the majority, of states used tools such as benefit-cost analysis across modes to analyze maintenance expenditures, operational improvements, and impacts on system performance.
- Respondents indicating the use of feedback mechanisms, approximately 0% usually cited bridge management systems as the application.

Providing easy to use cost effective tools for asset management is challenging. Such tools can be "data intensive." This means that the tools would require a lot of data in order to provide meaningful results. One strategy is to use commonly available data such as, the National Bridge Inventory (NBI) data,

TABLE 17.2   States Using Management Systems [Modified from (McNeil et al., 2000)]

Type of Management System	% of Respondents with Management System [a]	% Agencies Developing and Implementing ISTEA Management Systems [b] (Government Accounting Office, 1996)
Pavement	96.7	100
Bridge	96.7	96
Safety	70.0	96
Congestion	56.7	90
Public Transportation	43.3	65
Financial	56.7	NA
Maintenance	70.0	NA
Program Management	53.3	NA
Intermodal Transportation	20.0	60
Traffic Monitoring	66.7	NA
Operations Management	13.3	NA
Airport (Pavement)	33.3	NA
Port & Waterway	3.3	NA
Other	10.0	NA

[a] 30 respondents.
[b] 50 states, Puerto Rico, District of Columbia.

(US DOT, 1995), or the Highway Performance Monitoring System (HPMS) data, (US DOT, 2000), that is routinely collected for reporting purposes. There are also issues related to compatibility with the existing legacy systems and access to commonly available software such as, spreadsheets, geographic information systems (GIS), and databases.

Key data to support asset management includes geography (location), inventory, condition, traffic, crash statistics, work history, and programmed work. These data are used to support decision-making related to preservation versus mobility, maintenance versus capital, project selection on the basis of cost benefit or cost effectiveness, identification of combinations of projects, assessment of standards for needs assessment, and multi-objective analysis. Several different types of tools can provide this functionality. They are classified as follows (Cambridge Systematics, 2002a):

- Simulation models — These models attempt to capture the behavior of the assets and the impact of specific policies over time. They are capable of capturing the interactions among influential variables but require well-developed models and comprehensive data sets.
- Sketch planning tools — These tools use approximations to capture changes in performance and cost over time. They are less accurate than simulation models but faster to use.
- "What if" tools — These tools use sketch planning tools or simulation models to assemble a series of responses to specific strategies. These responses can be use to explore the impact of specific policies.
- Databases — Databases provide a method for organizing data and information. The data can be manipulated or summarized to identify trends or average conditions.
- Optimization — Optimization tools attempt to find the best solution to a specific problem that is formulated in terms of an objective and constraints.

Figure 17.12 shows how these different types of tools relate to the key data, the analysis parameters, the business rules of the organization and the decision support elements. Examples of tools that address some specific decision support issues include (Cambridge Systematics, 2002a):

- HDM-4 — the Highway Development and Management model evaluates roadway management and investment alternatives (http://hdm4.piarc.org/main/home-e.htm).
- MicroBenCost analyzes the benefits and costs of proposed highway improvements (Texas Transportation Institute, 1999).
- NBIAS — the National Bridge Investment Analysis System predicts national bridge maintenance, improvement, and replacement needs (Cambridge Systematics Inc, 2002b).
- STEAM analyzes the benefits, costs and impacts of multimodal tradeoffs using a traditional four step transportation model (www.fhwa.dot.gov/steam).
- StratBENCOST estimates the impact of proposed highway improvements (Hickling Lewis Brod, Inc., 1999).

The following sections explore three other tools in a little more detail. They are:

- The Performance Planning Process (P^3) developed by Montana Department of Transportation,
- The Highway Economic Requirements System (HERS-ST) developed by the FHWA to support the budgeting process. HERS-ST has been adapted for use by states for asset management,
- The pavement life cycle cost analysis tools developed by the FHWA. Each tool illustrates a different strategy for supporting decision-making. These three tools illustrate the integrative and strategic nature of asset management.

17.3.1 Performance-based planning: An example — Montana DOT

Montana Department of Transportation (MDT) does not use the term "asset management," but asset management does occur through its Performance Planning Process (P^3) system (Montana Department of Transportation, 2003). MDT has well developed pavement, bridge, congestion, and safety management

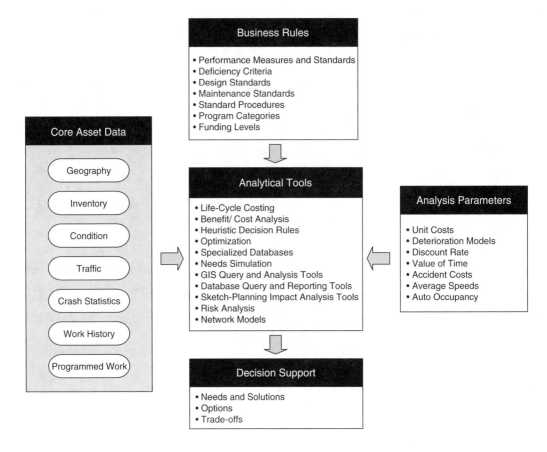

FIGURE 17.12   Tools for asset management (modified from Cambridge Systematics, 2002a).

systems in place. These systems include inventory, condition assessment, performance measures and evaluation. The Performance Planning Process (P^3) integrates the component systems of asset management in the following manner:

- P^3 links ongoing, annual and multi-year activities,
- P^3 serves as a project nomination process,
- P^3 ensures consistent goals and goal measurement.

The initial data entry for the P^3 system is processed by the Statewide Long Range Transportation Plan (Montana Department of Transportation, 2002), the Funding Distribution Plan, the Construction Program Delivery and System Monitoring, and the Statewide Transportation Improvement Program. The dynamics of this system are illustrated in Figure 17.13.

MDT uses PONTIS as their bridge management system. MDT has taken a leadership role in creating a web-accessible version of PONTIS. The general public can access data and use basic queries through the web (http://webdb2.mdt.state.mt.us/pls/bms_pub/pontis40_site.htm). The web-based system also provided access to route clearance and posting information.

MDT's pavement management system was developed by Texas Research and Development Institute (TRDI) (http://www.trdi.com/content/montana.htm). This network level system includes:

- Inventory,
- History — construction, maintenance, condition,
- Condition survey data,

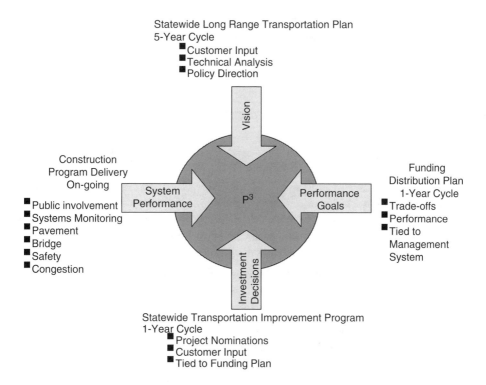

FIGURE 17.13   Montana's performance planning process (*Source*: Montana Department of Transportation, 2003).

- Traffic data,
- Database system,
- Data analysis capability,
- Report generation.

The congestion management system and the safety management system have been developed in-house. The safety management system is evolving to focus on localized spot improvements. There are also an intermodal management system and a public transportation management system.

Maintenance is integrated throughout all the systems. For example, performance goals include reactive maintenance dollar. Decisions related to these goals are made on basis of the pavement management system output. Similar efforts will be developed for signs, guardrails, and other hardware.

The "Performance Programming Process" is the system that integrates the various components of asset management. The performance planning process (P^3) links ongoing annual and multi-year activities to plan, program, and deliver highway improvements. P^3 is a project nomination process that is closely related to the evaluation of performance measures. The inputs are:

- Statewide Long Range Transportation Plan — This is updated on a five-year cycle and includes customer input, technical analysis and policy direction. This provides the *vision*.
- Funding Distribution Plan — On a one-year cycle this plan involves trade-off analysis and performance measures that are derived from the management systems. This provides the *performance goals*.
- Construction Program Delivery and System Monitoring — These are ongoing efforts that provide system *performance measures* through existing systems and public involvement.
- Statewide Transportation Improvement Program — The project nominations and customer input are updated annually. These are the *investment decisions*.

Ultimately, the process will include the REMI model, a regional socioeconomic forecasting and policy model, to quantify the impacts (REMI, 2003). Other modules include:

- Tracking Program delivery
- Tracking Economic Development Impacts

17.3.2   Highway Economic Requirements System — State Version (HERS-ST)

The Highway Economic Requirements System (HERS) is a simulation tool developed by FHWA to provide an estimated highway budget or understanding the impact on performance of a specific budget level. HERS-ST is a modified version of HERS for use by state departments of transportation. The tool includes capacity analysis, condition assessment and safety analyses to assess the cost of expanding, upgrading and maintaining the highway system. On a pilot basis, FHWA is working with states to explore the applicability of this tool to asset management (US DOT, 2003a).

The questions that can be addressed using the HERS-ST model are:

- What is the impact of a change in budget over a specified time frame?
- What investment is required to maintain current conditions?
- What is the change in budget to improve the network to a specified level?

The model functions by simulating network segments in terms of the cost to implement a specific strategy and its impact in travel time, safety, vehicle operating costs, emissions and improvement costs. Using an incremental benefit costs analysis in HERS-ST can assemble a series of projects that address the overall policy goals.

Amekudzi (1999) used HERS and Monte Carlo simulation to explore the impacts of uncertainty in data inputs on model outputs. The simulation explored the impact of variations in traffic on performance variables such as, delay and pavement roughness. As HERS has evolved over several decades the relative effects of model and data changes over time were also explored. Such analyses provide a deeper understanding of the models, their robustness to variation in input variables and their relative importance.

17.3.3   Pavement Life Cycle Cost Analysis (LCCA) Tool

Life cycle cost analysis (LCCA) is a key concept in asset management (Lee, 2002). LCCA uses the same economic analysis concepts as benefit-cost analysis, but assumes that benefits remain constant (US DOT, 2002). LCAA works best for the comparison of projects because it ensures consideration of all relevant costs. The basic process is as follows (US DOT, 2001):

1. Establish design alternatives
2. Determine activity timing
3. Estimate costs including both agency costs and user costs
4. Compute life cycle costs accounting for the time value of money
5. Analyze the results

The software tool Pavement LCCA (http://fhwa.dot.gov/infrastructure/asstmgmt) is a spreadsheet tool that implements the life cycle cost principles described in FHWA's guide (US DOT, 1998). The focus is on alternative pavement designs. An added feature is a risk analysis based on Monte Carlo simulation. Inputs include traffic, value of time and disruption. Impacts are based on construction cost and user impacts. Designs and timing are evaluated based on the impact.

17.3.4   Future Directions

An alternative strategy is to explore integration of existing management systems. Given the diversity of state responsibilities, organizational structures, and state politics, the details of how and when state DOTs approach management system integration will vary. Specifically, there are following common issues and factors that are critical to the success of management system integration (Cambridge systematics, 2002a):

- A plan for management system integration that is tied to an agency's strategic plan is required. The plan must recognize two conditions: (1) the politics in that specific state, both within the state department of transportation and the state legislature, and with respect to the constituents; and (2) the organizational structure of the agency.
- A departmental champion must take ownership of the integration process.
- Internal cultural changes within the department must occur and buy-in must occur at all levels with continuing acceptance of new paradigms.
- The time required to implement integrated systems must be clearly communicated and the length of time to develop a functional system must be recognized. Where possible, the implementation should be incremental.
- Performance measures must be developed and used as the process is implemented.
- Effective technical and human communication networks must be developed and maintained, as they are keys to the continuing success of the integration process.
- Organizational commitment to the integration process is essential. The resources for successful and continued integration have to be identified by management and then these resources have to be dedicated by management.
- Initial and continued education and training opportunities for all stakeholders must be available and utilized.
- Continuing and effective internal and external involvement are required.

There are several strategies for management system integration (Cambridge Systematics, 2002c). Historically, asset management has been practiced based on a "silo" or "stovepipe" organizational structure in which decisions, approvals and evaluation have occurred by type of asset. Like the trend to matrix type organizational structures in the private sector, these systems have moved to embrace multi-disciplinary teams in their design and implementation. Fully integrated management systems and decision-making may be achieved through one or a combination of three different methods:

- Massive integration — Advances in databases and GIS present opportunities for massive system integration with any decision maker being able to see the status of all assets.
- Modular development — Following principles of software engineering, modular systems are intended to be more efficient as not all users need to access all elements.
- Evolution — Realistically management systems integration will evolve over time to meet the needs of the agency given the resource constraints. Most state DOTs approach integration by building on and modifying existing systems.

17.4   An Application: GASB 34 and Asset Valuation

In June 1999, the Government Accounting Standards Bureau (GASB) (GASB, 1999) issued a reporting guideline (Statement 34) that state and local governments in the United States show the value of the infrastructure assets that they own. The use of these guidelines is very controversial. However, it is mandated in most state statutes for state and local agencies to follow the guidelines. Similar accounting practices are in place in Canada, Australia, and New Zealand and agencies privatizing their assets also need to understand the value of those assets.

In the United States, the GASB guidelines provided an option for not depreciating physical assets if the agency demonstrates responsible stewardship of those assets using an asset management system. While it is clear that the application of GASB Statement 34 is consistent with principles of good asset management, the reverse is not true: the application of the GASB guidelines does not constitute asset management. Accountability is just one aspect of asset management.

This section describes different approaches to asset valuation, reviews the GASB guidelines and explores the role of asset value in decision making.

17.4.1 Asset Valuation Approaches

Like asset management, valuing assets can be interpreted in many different ways. The value of an asset depends on whether you are interested in the financial or the economic value. There are also many different methods for determining the value of an asset including (Lemer and McCarthy, 1997):

- Book value — current value based on historical cost adjusted for depreciation,
- Written down replacement cost — current value based on replacement cost depreciated to current condition,
- Equivalent present worth in place — historic cost adjusted for inflation and wear,
- Market value — price buyer is willing to pay,
- Productivity realized value — net present value of benefit stream for remaining service life.

Both the book value and the written down replacement cost have no real meaning. They are accounting representations of the value of infrastructure and should not be used to support decision making. The equivalent present worth in place is also an artifact. It represents the sunk economic cost and again is not appropriate for supporting infrastructure investment decisions. The market value and the productivity realized value, if they could actually be computed or observed provide a sense of the economic value of the assets. Lemer and McCarthy (1997) and a report for the Transportation Association of Canada (Stantech, 2000) provide examples of the application of various computations of asset value.

17.4.2 GASB Statement 34 Guidelines

Historically, public sector agencies have used revenue and expense reports and have not reported the value of their investments or assets. However, consistent with other business principles, there is considerable interest in moving to a balance sheet that includes assets and enhances public accountability. Also, asset valuation is a key element for evaluating success within organizations (GASB, 1999).

Public agencies are required to report the value of infrastructure assets such as, roads, bridges, and tunnels. Although the requirements are effective June 1999, a transition period has been defined and the earliest implementation is June 2002. The value may be reported as an historical cost minus depreciation, or using a modified approach. Using the modified approach (GASB, 1999):

Infrastructure assets are not required to be depreciated if (1) the government manages those assets using an asset management system that has certain characteristics and (2) the government can document that the assets are being preserved approximately at (or above) a condition level established and disclosed by the government. Qualifying governments will make disclosures about infrastructure assets in required supplementary information (RSI), including the physical condition of the assets and the amounts spent to maintain and preserve them over time.

The asset management system must have an up-to-date inventory, include condition assessments and estimate the annual amount required each year to preserve these assets at some level of performance specified by the reporting agency.

Statement No. 34 provides an example of asset value based on book value using an estimated historical cost and straight line depreciation as follows (GASB, 1999):

In 1998, a government has sixty-five lane miles of roads in a secondary road subsystem, and the current construction cost of similar roads is $1 million per lane-mile. The estimated total current replacement cost of the secondary road subsystem of a highway network, therefore, is $65 million. The roads have an estimated weighted average age of fifteen years. Therefore 1983 is considered to be the acquisition year. Based on U.S. Department of Transportation, Federal Highway Administration's "Price Trend Information for Federal Aid highway Construction for 1983 and 1998", 1983 constructions costs were 69.03 percent of 1998 costs. The estimated historical cost of the subsystem, therefore, is $44,869,500. In 1998, the government would have report the subsystem in its financial statements to have an estimated cost of $44,869,500 less accumulated depreciation for fifteen years based on that deflated amount. ...... assume that the road system had a total useful life of twenty-five years. Assuming no residual value at the end of the time the straight-line depreciation expense would be $1,794,780 per year, and accumulated depreciation in 1998 would be $26,921,700.

In deciding on a method, the availability of data, and what the results will be used for are critical factors. The value of the asset can be used for establishing accountability, decision making and decision support. It is important to recognize that the value of an asset should also include the question "to whom?" Answering this question requires knowledge of the users of the asset and consideration of time in the sense of whether or not the value of the asset should reflect its value for future generations. For example, an underutilized section of roadway may be in the same condition as a heavily traveled section. To the user they have very different values, but their value based on condition may be the same (Kadlec and McNeil, 2001).

17.4.3 Incorporating Asset Valuation into Highway Asset Management Systems

Incorporating asset valuation into highway asset management systems provides additional data to support the management process. Selecting one or more of the asset valuation techniques presented in the previous section depends on many factors such as, public disclosure requirements, agency's objectives and policy, and the nature of highway assets. In addition, the value of an asset depends on whether the analysts are interested in financial or economic value (Kadlec and McNeil, 2001).

Figure 17.14 illustrates a simplified concept of a highway asset management system. An asset-shared DBMS stores basic information for different assets and their condition. An example of asset-shared data is asset location referencing, historical maintenance activities of assets, asset characteristics, and the performance of the asset.

Other components that further analyze the asset data are deterioration modeling, maintenance selection, tradeoff analysis, prioritization and optimization. These components are the same as for individual/traditional management systems. However, the detailed derivation of each component in the asset management system is much more complex, since the system has to deal with several assets at the same time. For example, the tradeoff analysis performed in the asset management system has to take

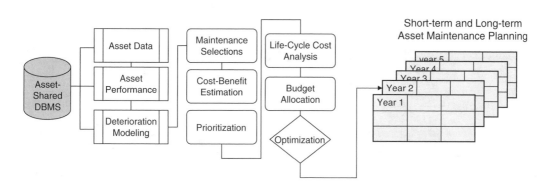

FIGURE 17.14 A simplified framework of highway asset management system.

into account benefits/savings from across disciplines such as, the benefit comparison between sign replacement, bridge expansion, and highway routine maintenance.

Budget allocation controls the amount of available budget in the optimization module. Optimal maintenance planning could be performed for both short-term and long-term planning depending on an agency's goals and objectives as shown in Figure 17.14. The challenge is when asset valuation becomes a requirement for states and agencies to report their financial status and values of their assets and how an agency could make use of the available asset data in the asset valuation module.

The following section presents an example of applying the cost approach to a highway asset management system. This approach may not be suitable for all highway asset types. However, the discussion here could provide a basic concept on how one could move from their traditional engineering-based management system to a value-based asset management system. Figure 17.15 illustrates a value-based asset management system that selects the cost approach to be used in the asset valuation component.

Cost approach could be considered as one valuation technique in which its calculated values relate to highway asset performance and time. This relation is the key element in highway asset management since further data analysis builds on this relation. A traditional engineering-based management system requires the basic information of asset inventory and the condition of the assets. Other components in the management system draw on this basic information to further analyze its future condition, prioritize, and optimize maintenance activities.

Asset values derived from the cost approach can be calculated from the summation of land value, replacement cost of the asset and accrued depreciation at time t. Land value data may be taken as the current sale price or governmental land value data. The replacement cost is estimated based on

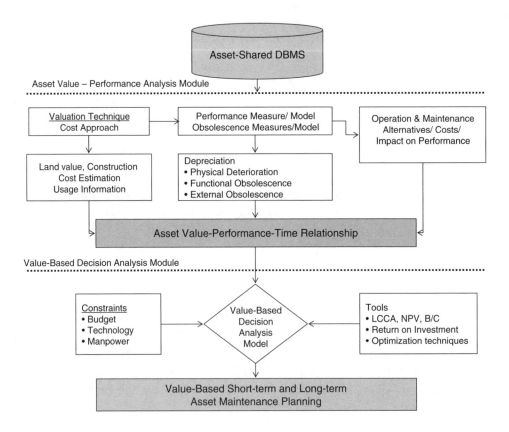

FIGURE 17.15 Value-based asset management system using cost approach (modified from Sirirangsi and Herabat, 2001).

materials used, required labor, overhead expenses and contractor's profits (Sirirangsi and Herabat, 2001). Accrued depreciation is a loss of value from replacement cost (AIREA, 1987). Depreciation involves three components, which are: physical deterioration, functional obsolescence, and external obsolescence.

Physical deterioration is the loss in serviceability, which is caused by damage or decay in the components of a particular asset (Nutt et al., 1976). The change in performance indices of an asset reflects the loss in serviceability. When a maintenance or rehabilitation is applied to an asset, its condition improves and the corresponding asset value should increase. Functional obsolescence is a loss of value caused by a change in design standards or regulations (AIREA, 1987). Functional obsolescence is calculated as the difference between the costs of any modifications or adjustments required for an asset to meet new design standards or updated regulations and the cost of the asset installed at the time of original construction (Sirirangsi and Herabat, 2001). External obsolescence is a loss of value affected by a change in external factors such as, technology, environment, society and economy (AIREA, 1987). External obsolescence is determined by the repair costs due to external factors.

The value-based decision analysis module is suggested in this framework in order to utilize the calculated asset values from the asset value-performance-time module. The output of the asset management system is the optimal short-term and long-term planning.

Figure 17.15 further extends the concept of value-based pavement maintenance management to further plan as a value-based management system for short-term and long-term planning. In the engineering-based management system, optimal planning is determined by the LCAA that considers the tradeoffs among total maintenance costs and received benefits of highway improvements under different funding scenarios.

The concept of return on investment (ROI) is selected to be applied to capture the returns or benefits received when an improvement is taken in the value-based pavement maintenance management. However, the concept of ROI is traditionally used to measure a corporation's profitability or how effective the firm uses its capital to generate profit in the financial management aspect. Applying ROI in the pavement maintenance planning may diversify the definition of ROI used in financial management. The ROI can be computed as the change of asset value due to its maintenance divided by the sum of the maintenance and opportunity costs (Sirirangsi et al., 2003). Maintenance costs are derived from the cost functions used by the relevant agency. Opportunity costs are the additional costs that result from the delay of a maintenance treatment. Alternatives are ranked based on the basis of ROI estimates. The alternative with the maximum ROI is considered to be the best alternative. The ROI concept is comparable to the tradeoff analysis in the traditional engineering-based management system.

17.5 Implementation and System Sustainability Issues

The importance of leadership commitment, and buy-in at all levels of the organization to the success of asset management has been well documented (US DOT, 1996; McNeil et al., 2001; Wittwer et al., 2002). Training and research has also seen some attention (Wittwer, 2001; Sanford and McNeil, 2001). However, the implementation process and the ongoing maintenance improvement of asset management systems warrant some further discussion. This section discusses implementation issues and strategies, and provides some guidance on the maintainability and sustainability of both the asset management process and the systems that support that process.

17.5.1 Implementation Issues and Strategies

The AASHTO *Asset Management Guide* (AASHTO, 2002) serves as a "how to" manual for implementing asset management, particularly at the level of a state department of transportation. The guide recognized that all agencies have some elements of asset management in place and serves as a guide to improvement rather than implementation of asset management in the absence of existing data, business practices, and institutional structure. A similar strategy is also presented in the Austroads guide (Austroads, 1997, 2002). The process to develop an asset management strategy involves three steps:

- Defining the scope of asset management for a specific agency.
- Conducting a self-assessment process that serves as a foundation for the improvement strategy that is customized to a specific organization.
- Developing an asset management implementation strategy and plan.

17.5.1.1   Defining the Scope of Asset Management

Defining the scope of asset management involves determining which assets are included, what actions are covered, and which business processes are included and identifying particular concepts to be emphasized. Which assets are covered will often be determined by the nature of the organization. In terms of highway assets this should include right of way, drainage, pavements, structures, and assets that support operations and safety functions such as, traffic signals and guardrail. Some organizations include construction and maintenance equipment and buildings. Austroads (Austroads, 2002) recommended that the assets of interest should be identified on the basis of the most expensive in terms of life cycle costs, the key contributors to performance in terms of addressing using needs, and components most prone to deterioration and in need of ongoing maintenance. On this basis, the components of interest are: road formations, drainage, pavements, bridges, traffic control equipment and roadside ITS installations. Which assets to include are also influenced by who owns the asset and is responsible for its ongoing preservation and improvement. In some cases, notably bridges, complex ownership agreements are in place as a result of historical accidents or local politics.

Asset management should embrace all activities that occur over the life cycle of the asset. However, in reality budget and time constraints may limit the types of activities to be included in the asset management process. Determining which actions to include is based on constraints in terms of activities that the agency has control over and resources available, and which activities are likely to realize the most benefit to the system when thought of in an holistic way consistent with the principles and concepts of asset management. A similar argument can also be made for determining which business principles to include in the process.

Finally, determining which asset management concepts to include also involves consideration of resource constraints. This includes the presence of buy-in to the concept, the availability of data and analytical tools, institutional and legislative constraints (AASHTO, 2002).

17.5.1.2   Conducting a Self-Assessment

The self-assessment provides an opportunity for an organization to understand what they already have in place, what they would like to have in place and what the constraints are. A self-assessment can be undertaken using a variety of tools include strategic planning tools such as, SWOT (Strengths, Weaknesses, Opportunities and Threats) or a tool such as the self-assessment presented in the AASHTO guide (AASHTO, 2002). The self-assessment tool is intended to build consensus, identify strengths, weaknesses and constraints, develop priorities, and form the formation for asset management improvement. It is a diagnostic tool rather than an analysis organized around the four areas of asset management: policy goals and objectives, planning and programming, program delivery, and information and analysis. The assessment asks a number of questions with statements that require responses on a scale ranging from strongly disagree to strongly agree. The management team involved in setting the direction of asset management and implementing the process should respond to the questions individually but discussion should occur in a group setting. For example, the question "Do Resource Allocation decisions reflect good practice in Asset Management?" has a statement "Our agency's long-range plan is consistent with currently established goals and objectives." This statement is one of a group of statements relating policy, planning and programming. The response to this type of statement would indicate the need to focus on these links as part of asset management improvement (AASHTO, 2002).

17.5.1.3 Developing an Asset Management Implementation Strategy and Plan

The structure of the implementation strategy and plan is strongly influenced by the how asset management is positioned within the organizations. Some organizations have established task forces, some have developed a special office with responsibility for coordination with other functions, and others have integrated the process into their existing organizational structure. Establishing roles and responsibilities is a critical first step. This includes determining who has the lead role, what the responsibilities are, and then who provides support to the leadership function. It is also important to distinguish among the need for guidance (for example, this may be through an executive steering committee), technical support, and stakeholder involvement.

The next step is to build an action plan. The action plan must identify the areas needing improvement, establish a timeframe, and develop an implementation plan. The AASHTO guide suggest five types of improvement that are relevant to any of the four areas covered in the self-assessment (policy goals and objectives, planning and programming, program delivery, and information and analysis) and provides specific ideas in each area (AASHTO, 2002):

- Desired improvement in technique
- Consistency and alignment among practices
- Gaps in practice
- Constraints on the practice
- Desired improvement in principles of asset management.

The time frame for the improvements takes into consideration the importance of the areas, the logical relationships among the proposed improvements and with the funding, budgeting and programming cycles in the organization, and the resources available. The plan, like most other strategic documents, addresses objectives, activities, benefits, time frame and other critical information.

17.5.2 System Maintainability and Sustainability

Sustaining a successful asset management system requires attention to the institutional structure that supports the system including processes, data, hardware and software. The interconnections between different components of an asset management system are dynamic process as shown in Figure 17.8. Policy, goals and objectives addressed by the institutional structure need to match the program delivery and system monitoring as shown in Figure 17.7. Matching the needs and outputs of the upper-, middle- and lower-management levels and program delivery of an asset management in a timely manner would help sustain the system and provide an effective asset management system.

Like other systems an asset management system requires continuous upgrading of hardware and software; and ongoing training and development of support systems. Hardware and software needs to be upgraded as technology advances. Computer capabilities continue to evolve to help users perform more complex analyses, provide faster access to more data, and provide more informative visual displays of data and information.

Personnel training should be conducted on a regular basis to ensure the continuity of an asset management system. Training can be a continuous process in response to changes to the system such as system upgrades and modifications. It can help users better understand the system modifications. Timing is also critical. Having systems and processes in place without training is not a sustainable approach. Asset management is not a one time effort but a commitment to the process and involves re-evaluation and response as illustrated by the feedback in Figure 17.7 and Figure 17.8.

17.6 An Application: Asset Financial Management

Limited funding is a major constraint in asset management projects. These projects include construction, different types of maintenance activities; and could further extend to reconstruction and rehabilitation projects. A well-defined budget and its detailed cost estimation are essential to manage these activities. Available funding has a major influence on the decision-making process and the optimal highway management planning as presented in the highway asset management framework. Understanding how available funds are allocated to different management activities and how funds could be obtained from different sources helps agencies enhance their decision-making process. This section presents different funding usage for different types of highway assets and different sources of funding.

17.6.1 Funding Usage versus Sources of Funding

Funding usage is an important issue in highway asset management since it indicates the deteriorating conditions of national highway stock and the financial needs to construct, upgrade, and maintain highway assets to cope with higher road user demand for highway services (National Council on Public Works Improvement, 1986). The funding usage is determined from the required construction costs of new highways and the maintenance or rehabilitation costs of existing highway stocks.

Determination of the capital costs of the highway stock needs assessment for the requirement of new highway construction and the optimal maintenance treatments of existing highway routes should be studied. Total highway construction costs are usually estimated based on construction unit cost and different sub-construction projects. The optimal maintenance treatments are determined from relevant highway management systems such as the pavement management system, and the bridge management system. Maintenance cost can then be calculated based on the maintenance unit cost or the predetermined cost functions of recommended maintenance treatments.

Once funding usage is determined, sources of funding should be studied to verify the availability of funds in order to finance the construction and maintenance of highway projects. There are various sources of funding that can be applied in the highway industry. Governmental budgeting is a traditional financing method to fund these types of projects. Developing countries usually rely on central budgeting since other available funding sources are scarce. Debt financing and loan syndication are typically used in highway construction. However, it is difficult for lenders to finance routine maintenance projects. Initial public offering (IPO) and public offering (PO) are two types of privatization techniques. These financial instruments can be applied for highway organizations that would like to raise external equity in domestic or international capital markets. The funding objective to issue shares to the public can be used for operational purposes, especially for future construction and maintenance plans for toll-road projects. Bond financing is another financing scheme in the capital market that highway agencies may implement to construct, operate and maintain their highways, especially toll roads.

17.6.2 Funding Usage for Different Types of Highway Assets

The formulation of funding usage begins with an estimate of total external funds requirement from the construction or maintenance costs of a particular highway (Finnerty, 1996). For highway investment projects, the amount of required funding usage is calculated from the sum of total highway construction costs, financial-related charges incurred during the construction period and other initial operating expenses prior to the completion of the project construction. In the case of toll roads, the funding usage can be reduced when the project generates its own revenues during the construction period. This realization of toll revenues can be found when there are many phases in the toll road project and toll operation is begun before the completion of final phase.

The highway asset construction cost is estimated from the unit cost of material, labor, equipment, subcontract work, overhead, and profit (Peurifoy and Oberlender, 2002). The cost should consider factors of location and construction difficulties. The financial-related charges are: interest during construction, commitment fees and the cost of financial guarantees or other credit supports (Finnerty, 1996). The interest during construction must be paid during the construction period and it is calculated from the amount of funds that are drawn from the financial institutions multiplied by the interest rate during the time of construction. The commitment fee is charged based on the unused loan balance, which is committed to the lenders. Other initial operating expenses include the engineering, design or supervision fee in addition to internal highway agencies' labor and other expenses.

The maintenance cost for a particular highway section or route is calculated based on the type of maintenance treatment and its relevant unit cost. The process in selecting appropriate maintenance and rehabilitation alternatives is usually done by engineering judgment, experience, engineering economics, or a predetermined maintenance standard, which is based on a range of various factors such as, highway condition, user cost, usage, or budget constraints.

Three main types of maintenance activities are maintenance, rehabilitation, and reconstruction (Hudson et al., 1997). Details of cost functions and their components vary depending on different maintenance activities. The cost functions can further be decomposed into different line items of each activity and their associated costs.

Based on the actual practice of a highway agency, maintenance costs can be determined from an internally developed cost function or treatment unit cost and its breakdown items for each maintenance activity. There are many factors involved to calculate the total maintenance costs, which are, for example, material costs, transportation costs, labor costs and operation costs. Figure 17.16 illustrates a simplified pavement overlay cost function that can be subdivided into the cost estimation of each layer, maintenance factors, and traffic factors. These factors are calibrated to assign weights for different contributing factors such as, the location of the resources and level of usage.

17.6.3 Sources of Funding

The development of funding usage requires a thorough analysis of potential sources of funding in relation to the project's year-to-year funding requirements, available cash flows, and the availability of credit support mechanisms to support project debt. There are various types of sources of funding which are: governmental budgeting; debt financing and loan syndication; IPO and PO; and bond financing.

17.6.3.1 Governmental Budgeting

Government budgeting in the public asset domain usually finances highway activities, which are new construction, maintenance, rehabilitation and reconstruction. Budget allocation is the key element of the decision-making process when different requirements from different agencies are compared in the budget allocation process (Hudson et al., 1997). The funding needs in highway sectors compete with the requirements from other industries such as, education, military services, and healthcare. Political and policy issues may lead to poor decision-making solutions. One example showing a poor decision making process is that a politician may approve to build more unnecessary roads, even though the highway construction costs are too expensive, and overlook the importance of maintenance projects

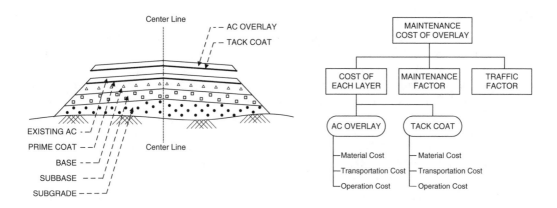

FIGURE 17.16 Maintenance cost function for pavement overlay.

(ADB, 2001). The importance of maintenance budget allocation is often overlooked since the benefits received from maintenance work are intangible. This could cause an insufficient budget allocation for maintenance work, which results in significant backlogs and creates a burden to highway agencies.

17.6.3.2   Debt Financing and Loan Syndication

Commercial and development banks are two types of financial institutions for debt financing of projects. The commercial bank is involved in debt financing for construction financing, a short-term loan that is needed to be repaid at the time of construction completion in addition with the long-term financing (Finnerty, 1996). The criterion for construction financing approval for lender's perspective is the commitment of long-term financing from other lenders. The construction financing normally occurs for large projects, which is time consuming for long-term loan approval and involves many prospective lenders.

The criteria for long-term loan approval concerns the profitability of the projects, project leverage and project risks. The project should be profitable and have suitable capital structure, i.e., debt and equity. The objective of profitability of the projects and project leverage study is to confirm that a project has the ability to repay its debt. In addition, lenders will assess relevant project risks in order to plan for risk mitigation, interest, and other fees charged to a project. Examples of risks involved in the project are completion risk, economic risk, financial risk, currency risk and political risk. Development banks such as Asian Development Bank or World Bank have been involved in long-term financing for public and toll highways in many developing countries.

For sources of funding in highway maintenance activities, development banks such as Asian Development Bank have been involved in many road maintenance projects (ADB, 2001). The bank has financed road rehabilitation projects, especially roads that have suffered from accelerated deterioration, and provided loans for technical assistance to developing countries for a large sum of money. However, from the lender's viewpoint it is worthless to fund routine maintenance (ADB, 2001).

17.6.3.3   Initial Public Offering (IPO) and Public Offering (PO)

IPO and PO are types of privatization techniques (Miller, 2000). The main difference between the IPO and the PO is that the IPO issues shares at the first time that the organization decides to sell their shares to the public. Otherwise, it is considered to be the public offering (PO). The success of the IPO and PO depends on the public confidence that an organization would be profitable. These types of financial instruments are suitable for financing the construction, operation and maintenance of profitable toll roads. There are several advantages of IPO and PO, which include stockholder diversification, increased liquidity and greater ease in raising new capital (Weston and Brigham, 1990). There are various disadvantages of using these instruments such as, high cost of reporting and disclosure requirements from public.

17.6.3.4   Bond Financing

A bond is one type of debt instrument (Fabozzi, 1994). A bond requires a borrower to repay amount borrowed on the predetermined day (maturity date) that the amount is due plus the interest incurred. A typical bond pays interest to lenders or bondholders every six months. There are four important features of bond, which are issuer, term to maturity, principal and coupon rate. There are three types of bond issuers. Federal government and its agencies, municipal government and corporations are players in the bond market.

The term to maturity of a bond is the number of years over which a borrower redeems the bond by paying principal back to bondholders. Generally, short-term bond refers to a bond with a maturity of between one to five years. Bonds with a maturity between five to twelve years are called intermediate-term bonds while long-term bonds have a maturity of more than 12 years.

The principal of a bond is the amount that the borrower agrees to repay the bondholder at its maturity date. The coupon rate is the interest rate that the borrower agrees to pay to the bondholder annually and the coupon is the interest payment that is paid annually to bondholders.

Federal government and its agencies, municipal government and corporations which construct, operate and maintain highways can issue their bonds to the capital market. The objectives of this funding are to generate cash flows to construct, operate and maintain their highway assets. Bond repayment sources should be studied before the bond is issued in order to verify the repayment sources and timing. Repayment of public highway debt should come from future government budgets, shadow tolls, or actual toll revenues (Dornan, 2001).

Dornan (2001) recommended that bond financing could be considered as one alternative type of maintenance funding source in the future. It is further suggested that reduced costs due to appropriated asset management can generate significant long-term cost savings in highway maintenance. These cost savings should provide a sufficient source of income to be considered as a source of repayment in the debt service of bond instruments.

17.7 Efforts and Challenges

Asset management involves significant organizational change. To illustrate both successes and challenges, this section highlights efforts in the United States and Thailand. It also devotes a section to show examples of how assets are managed in the private sector. These examples are not intended to be either comprehensive or representative but illustrate both the concepts and the issues.

17.7.1 Current Development of Highway Asset Management Systems in Different DOTs in the U.S.

State Departments of Transportation in the United States have made significant progress in implementing various aspects and components of asset management. This section describes four of these efforts: Michigan, New York, Oregon, and Pennsylvania. The FHWA has developed case studies describing each of these efforts. The FHWA has developed case studies presenting detailed descriptions including the context for the development effort, the strategy, lessons learned and future directions.

17.7.1.1 The Michigan Experience (US DOT, 2003b)

Since the early 1990s, Michigan Department of Transportation (MDOT) has been developing an integrated management system that serves as a foundation for several asset management applications. The system is built on the principle "collect data once, store it once, use it many times" and has reduced some 20,000 files to five major databases. Central to this activity is a single statewide linear referencing system that provides location information as part of a geographic information system.

The initial effort was influenced by the requirement of the 1991 ISTEA that required state departments of transportation to develop and implement management systems. While this requirement was later rescinded MDOT seized the opportunity to re-engineer its business processes and computer systems. The effort began in 1993 with an assessment of existing systems and data, followed by prototypes of the new systems. In 1997, a business plan formalized the structure of the Transportation Management System (TMS) and the subsequent development process.

The end result is a system that recognizes data as a corporate asset and has demonstrated benefits in terms of (US DOT, 2003b):

- Better data management and utilization
- New approach to system management
- Stabilized program development and project delivery.

MDOT identifies the following factors as key to a successful data integration process (US DOT, 2003b):

- Establish cooperation between information technology staff and business process owners.
- Maintain buy-in using a business plan that supports technical efforts.
- Commit to a statewide referencing system.

- Decide whether to buy or build.
- Ensure effective project management.
- Recognize that it is easier and cheaper to stay current than it is to catch us
- Satisfy system users.

The Michigan Legislature has now mandated asset management. TMS serves as a foundation for establishing performance targets, developing strategies and reporting to the legislature.

17.7.1.2   The New York Experience (US DOT, 2003c)

New York State Department of Transportation (NYSDOT) is one of the first states in the U.S. that applied information system and economic analysis concepts for highway planning and investments. NYSDOT places a high priority to their customer satisfaction in regards to "a safe, balanced, and environmentally sound transportation system" (US DOT, 2003c).

In 1960s, NYSDOT used an automated system to collect their infrastructure information. Economic analysis has been taken into account in NYSDOT highway planning for more than 40 years. Infrastructure deficiencies become a significant problem in 1980s. The need to improve their highway planning made the upper management reform their overall decision-making process. A process that was based on the goal-driven program and provided a quality assurance was created to enhance to their highway planning. In 1997, an internal task force of NYSDOT was formed to prepare a blueprint for advancing the implementation of transportation asset management (TAM).

NYSDOT defines transportation asset management as "a strategic approach to maximize the benefits from resources used to operate, expand, and preserve the transportation infrastructure" (US DOT, 2003c). Tradeoff analysis used to determine optimal lists of improvement activities is a crucial element in the highway planning since budget constraint influences the amount of highway improvements.

NYSDOT had developed "a prototype TAM Tradeoff Model that employs economic tradeoff analysis to compare the dollar value of customer benefits to investment costs among competing investment candidates" (US DOT, 2003c). The prototype is designed to assess program-level tradeoffs among its pavements, bridges, safety, and mobile goal areas.

Continuous efforts are made to enhance the TAM prototype "to incorporate additional LCCA and benefit data into the model equations and taking steps to ensure that economic comparisons among projects use consistent values for benefit and cost elements" (US DOT, 2003c). The common measure of benefits in TAM Tradeoff Model is based on the concept of "excess user cost". Excess user cost is "the cost to travelers that exceeds a level (or threshold) deemed to be reasonable by NYSDOT" (US DOT, 2003c). The TAM system combines the concepts of engineering, mathematics, and economic analysis to constitute the core of the program.

17.7.1.3   The Oregon Experience (USDOT, 2003c)

Oregon Department of Transportation (ODOT) had implemented a computer model to support its investment decision-making process since 1991, called Highway Performance Monitoring System Analytical Process (HPMS-AP). The HPMS-AP is an investment/performance model that simulates the required investment and forecasts the future highway deficiencies, assesses the physical condition, safety, service and efficiency of highway systems. This approach primarily relies on the engineering judgment, which may not be realistic when the impacts of the highway users are ignored. In the late 1990s, ODOT was in process of updating its Oregon Highway Plan.

In 1999, HERS model was introduced to ODOT by the U.S. Department of Transportation. ODOT is one of the first state departments that incorporates HERS model for their planning and policy analysis at the state level. HERS was developed to enhance the economic analysis simulation to incorporate the performance prediction, analyze the tradeoffs among different projects for limited available investment, and take into account of the impacts of highway users.

In addition, ODOT had decided to tailor the HERS approach for use by the Oregon's Planning staff, called HERS-OR. "The HERS-OR is revised to capture a more *real world* conditions, for example, the changes in the traffic volume based on known land use changes. The output of HERS-OR combines the model-identified improvements as well as the ODOT-identified improvements" (FHWA, 2003c). HERS-OR performs a "what-if" analysis to capture the changes in physical condition, relevant costs and benefits and further optimize the lists of improvements for different funding scenarios. The recommended results of HERS-OR include relevant costs, received benefits, changes in physical condition if the improvement is made. The results of HERS-OR provide an excellent basis for budgetary process and legislative program evaluation and development. However, HERS-OR model is insensitive to the Oregon's land use law, and the travel demand model.

A continuous effort is essential in order to enhance the statewide planning program. ODOT is recently trying to synchronize their available data with other agency tools and moving towards adopting the FHWA-sponsored State version of HERS and HERS-ST. HER-ST is a direct extension of the national-level HERS model, but provides additional features that are not available in HERS-OR. Additional features are, for example, user-friendly interfaces and GIS capabilities. HERS-ST is viewed as a significant upgrade of HERS-OR and ODOT is working on revising and updating their internal functions, calculations, and formulas found in HERS-ST.

17.7.1.4 The Pennsylvania Experience (US DOT, 2003e)

Pennsylvania has used LCAA as part of its pavement selection process since the mid 1980s. This engineering economic analysis tool accounts for both agency costs and the impact on users of different pavement designs. The end results are: improved overall pavement performance, lower costs for new pavement and rehabilitation work, and improved credibility. The process includes design and engineering costs, construction costs, reconstruction costs, preservation costs and costs of maintenance of traffic over a 40-year period. It also recognizes user costs associated with work zones in the form of delays and vehicle operating costs.

LCAA is a data intensive process as it uses both cost and performance data. Over time the process has been streamline building on existing databases for contract management and roadway data, and standardizing the design processes.

The end results have been better pavement condition and pavements that last longer. It has also resulted in secondary benefits in terms of professional development, improved products as suppliers recognize the challenge of demonstrating a better product, improved credibility, and reduced user costs.

17.7.2   Asset Management in the Private Sector

Having coined the term asset management from the private sector, public sector agencies are interested in how the private sector approaches asset management. In September 1996, the American Association of State Highway and Transportation Officials (AASHTO) and the FHWA held the first asset management workshop focused on sharing experiences in both the public and private sectors (US DOT, 1996). The workshop brought together 65 participants from 23 states and included representatives from retailing, trucking, manufacturing, electric utilities, railroads, telecommunications, toll roads and port authorities. The private sector participants described near state-of-the-art Asset Management Systems that vary in complexity and sophistication depending on the nature of the business, need for data, layout and breakdown of the network, and role of the customer. All the systems described were highly streamlined and cost effective, and were demonstrated to be responsive to top management, user and customer needs.

A recent study (McNeil et al., 2002) documented asset management practices at seven near-transportation, private sector companies. The companies included: two railroads, two airlines, two energy companies, and one railcar leasing company. The study found that in the private sector transportation industry, railroads have aggressively integrated formalized asset management systems in their management processes. They have focused primarily on the use of advanced software systems to improve both customer service and asset utilization. In the private sector customer-oriented approach and proactive maintenance are the two attributes that have been at the heart of all asset management practices. In the public sector, asset management has been similar to that in a private industry in the fact that the emphasis is still on maintenance and operating physical assets within budgetary limitations.

Table 17.3 provides a summary of the various attributes discussed in the case studies. The case studies reveal that the main objective of the private industry is to provide improved profitability while not sacrificing customer satisfaction. However, performance measurement and proactive maintenance is an aspect, which is at the heart of most of the asset management programs discussed in this paper. While these concepts of customer satisfaction and proactive maintenance need to be the focus of any asset management program, a few other important considerations identified below will help or aid the public sector or state departments of transportation.

1. With increasing personnel constraints within state DOTs, the need for use of advanced technology and information systems becomes critical.
2. Effective communication of information is needed between the various stakeholders, and coordinating mechanisms need to be established between various asset classes of the organization. Although many tools are already available to sustain asset practices, including inventory and condition assessment, pavement and bridge management systems, and optimization methods; most of these are at grass roots level of an agency. Integration is required at the level of inventory and condition assessment.
3. Methods for assigning asset value, measuring return on investment, and evaluating investment tradeoffs need to be developed.
4. Educating and training legislators, decision-makers, and public officials about asset management and the need for maintenance of assets is essential. One way to promote awareness is by continuing to hold workshops and conferences on asset management and other related areas.

Organizations/agencies that strive to have a successful asset management process need to make sure that these areas of focus are addressed and defined in industry specific terms. The analyses of case studies have shown that an established asset management system has been critical to corporate profitability and has contributed to effective maintenance and management of assets. They also emphasize the use of information systems to oversee their maintenance schedules. An important inference that can be drawn from this analysis is that most of the systems had proactive maintenance systems in place with emphasis on performance measurement. Perhaps, in the future, these two features should form the essence of any asset management system to be employed by the transportation agencies.

17.7.3 Current development of Highway Asset Management Systems in Different Organizations in Thailand

In the past several years, the government of Thailand has undergone a major reform and restructuring process. The reform process of the budget system is aimed at increasing the government's efficiency in budget allocation and improving the current budget system by focusing on increased reporting and accountability of the government entity's performance (Jatusripitak, 2002). The performance-based budgeting concept is a systematic budgeting method that defines an organization's mission, goals, and objectives, and a procedure for regularly evaluating its performance in terms of inputs, outputs and outcomes. There are various recommended key performance indices such as, profit margin, returns on asset, returns on investment and asset value. The concept of performance-based budgeting is based on business principles.

Highway agencies in Thailand are responsible for constructing, maintaining and managing all highway assets such as, bridges, pavements, and culverts under their own jurisdiction. In the past several decades, these agencies view their construction and maintenance management of their assets in terms of individual basis. The reformed performance-based budgeting concept imposes these agencies to view all their assets as a whole. The key performance indices of these agencies are evaluated based on their asset values and returns on investment (i.e., the returns on investment of construction and maintenance activities) for further use in the central budget allocation of the government and to monitor the overall performance of different agencies. This motivates the agencies to endorse the concepts of asset management system and asset valuation. The challenge lies in how the agencies could move from individual management systems to operate in terms of an asset management system. Continuous efforts

TABLE 17.3 Private Sector Case Studies (Modified from Sriraj et al., 2003)

Inventory	Case Study	Major Assets	Formalized Asset Management System	Prime Objective	Performance Measurement	Proactive Maintenance	Integrated Information System
Railroads	Wisconsin Central	Tracks, Locomotives, Railroad R.O.W.	No	Monitor economic performance	Yes	No	No
Railroads	Union Pacific	Tracks, Locomotives, Railroad R.O.W.	Yes	Meet the demand in lieu of decreasing supply	Yes	Yes	Yes
Railcar Leasing Company	TTX	Railcars	Yes	Maximize revenue through managed capacity	Yes	Yes	No
Airlines-ISD	United Airlines	Network computers, Data feeds and Software	Yes	Provide reliable assets for efficient network planning and scheduling	Yes	Yes	Yes
Airlines	Midwest Express	Jets, Turboprops	Yes	Improving system and economic performance	Yes	No	No
Energy	Xcel Energy	Coal plants, Nuclear plants, Hydro-electric plants, and Other distribution and transmission lines	Yes	To fix assets before they fail	Yes	Yes	Yes
Energy	Great Lakes	Pipeline, Gas utility plants	Yes	Improve network performance and profitability	Yes	Yes	No

are being contributed from agencies and researchers in Thailand to integrate individual system into one highway asset management system. However, the most frequent difficulty that agencies in Thailand encountered at this time is how to transform their traditional management approach to comply with a performance-based budgeting approach and enable reporting of the agency's accountability.

Asset valuation is one technique that can be used to express the performance of any asset, as well as its deterioration, in monetary terms (Herabat et al., 2002). This common valuation in monetary terms can be used to avoid any discrepancies that may arise due to different asset performance measures. However, other management tasks are still required (i.e., prioritization, optimization, budgeting and scheduling). Integrating the asset valuation concept with existing management systems raises many issues for researchers and practitioners. Objectives and directions need to be clarified in order to manage assets in the most cost effective ways, while maintaining or enhancing the value of these assets. There are several different asset valuation techniques (Herabat et al., 2002). Selecting the most appropriate technique involves various considerations that are linked to an agency's asset management objectives.

Meaningful representation of asset values and its accuracy is very *critical* since it directly reflects the overall key performance indices of an agency. In addition, there is also the practical issue of how to actually integrate asset valuation concepts and techniques with existing and historical data, performance measures, prediction models, LCAA and other data and decision making tools that have been used in practice for several years, in meaningful and non-redundant ways.

In response to the reformed performance-based budgeting concept, Thailand DOH has recently established the new asset management standard in 2003. Thailand DOH is the main public agency that holds a major share of the highway network in the country. DOH had developed Thailand Pavement Management System (TPMS) since 1984, which uses the basic concept of a pavement management system. Thailand DOH estimates the useful lives of their highway assets as 20 years for concrete roads, 7 years for asphalt concrete roads, 1 year for gravel roads and 40 years for bridges (Teeraganont, 2003). Percentage depreciation can be derived based on the estimated useful lives of highway assets by applying a straight-line depreciation technique. Therefore, the percentage depreciations are, for example, 5% for concrete roads and 14.25% for asphalt concrete road. The residual values of pavement are defined as 50% of initial asset values (Teeraganont, 2003). The book value of an asset using the straight-line depreciation technique may not reflect the actual condition of the assets or the return on investment received from reconstruction or maintenance activities.

Other examples of contributing research for the incorporation of asset valuation into the TPMS are presented in (Sirirangsi and Herabat, 2001). This research selected to apply the cost approach in their study to estimate pavement value based on the collected data from TPMS. The cost approach is the only approach that could relate asset value, performance and depreciation. This relation could be viewed as a medium between a traditional engineering-based and a financial-based management. The detailed discussion of the cost approach is presented in Section 17.4. This development enhances the capability of the TPMS to be able to value their pavement sections.

Sirirangsi et al. (2003) further attempt to apply the calculated pavement values based on the cost approach to transform the traditional engineering-based management system into value-based prioritization and optimization models using the TPMS data and the calculated pavement values. The concepts of the normalized pavement value and ROI are proposed for use in the prioritization and optimization models as discussed in (Sirirangsi et al., 2003).

The optimal strategy for the network level can be classified into two groups: the maximization of network benefits subject to budget constraints and the minimization of total expected costs subject to certain network performance levels (Golabi, 2002). One suggestion for a value-based optimization model could be formulated to maximize the total pavement values under a budget constraint and under an acceptable level of pavement condition (Sirirangsi et al., 2003).

These developments enhance the capability of TPMS to enable valuing pavement sections, to apply the calculated values to prioritize and to further optimize the maintenance planning. In addition, the calculated pavement values based on the cost approach could be treated as supplemental information on book values derived from DOH standards.

17.8   Conclusion and Future Directions

Asset management is viable, useable and practical framework to make informed transportation investment decisions necessary to operate and manage transportation facilities systematically and cost effectively. Maintaining the quality and reliability of our highway networks system, with growing constraints on funding and other regulation, requires maximum use of all the resources. Currently, asset management approaches to address infrastructure maintenance are critical but are underutilized resources.

Objective, data-based, reproducible, systematic approaches that can enhance the dialogue among decision-makers regarding the type and timing of capital investments to address preservation, capacity expansion, mobility, accessibility and operations are needed. However, there are still significant research gaps as well as issues related to the implementation of research results (US DOT, 2002; Switzer and McNeil, 2004). Key areas include: quantifying the benefits of asset management, modeling costs and decisions, improving LCAA and benchmarking with other organizations.

Despite these gaps, the last decade has seen significant progress in the application of asset management to highway networks. Most of these applications build on earlier versions of pavement management and bridge management, but they demonstrate the willingness of agencies to break down institutional issues to deliver transportation services that are better and more cost effective.

References

American Association of State Highway and Transportation Officials (AASHTO), 1998. *AASHTO Asset Management Task Force*, Strategic Plan.

American Association of State Highway and Transportation Officials, 2000. Washington, DC.

American Association of State Highway and Transportation Officials (AASHTO), 2001. *Pavement Management Guide*, Washington, DC.

American Institute of Real Estate Appraisers (AIREA), 1987. *The Appraisal of Real Estate*, 9th Ed., Appraisal Institute, Chicago, IL.

Amekudzi, A. 1999. *Evaluating Uncertainty, Analyzing Risk and Managing Information Quality in National Highway Investment and Performance Analysis*, unpublished PhD Thesis, Department of Civil and Environmental Engineering, Carnegie Mellon University, Pittsburgh, PA, May.

Anthony, R.N. and Reece, J.M. 1989. *Accounting: Text and Case, 8th Ed.*, Irwin, IL.

Asian Development Bank (ADB), 2001. An Approach to Sustainable Funding of Road Maintenance: ADB Publication No. 120401.

Association of Australian and New Zealand Road Authorities (AUSTROADS), 1994. Road Asset Management Guidelines (AP-109/94), New South Wales, Australia.

Association of Australian and New Zealand Road Authorities (AUSTROADS), 1997. Strategy for Improving Asset Management Practices, (AP-53/97), New South Wales, Australia.

Association of Australian and New Zealand Road Authorities (AUSTROADS), 2002. Integrated Asset Management Guidelines for Road Networks (AP-R202/02), New South Wales, Australia.

Association of Australian and New Zealand Road Authorities (AUSTROADS), 2003. National Performance Database, http://www.algin.net/austroads/, Accessed December 20.

Cafiso, S., Di Graziano, A., Kerali, H.R., and Odoki, J.B. 2002. *Multicriteria Analysis Method for Pavement Maintenance Management*, Transportation Research Record, No. 1816, pp. 73–84.

Cambridge Systematics Inc., 2002a. *Analytical Tools for Asset Management*, Interim Report NCHRP Project No. 20-57, December.

Cambridge Systematics Inc., 2002b. NBIAS 3.0 User Manual, October.

Cambridge Systematics Inc., 2002c. *Transportation Asset Management Guide*, Prepared for the National Cooperative Highway Research Program (NCHRP) Project 20-24(11), November.

Cunard, R.A., Sign Maintenance Basics: Needs and Cycles, *Better Roads*, 62, 8, 32–33, 1992.

Danylo, N.H. and Lemer, A. 1998. *Asset Management for the Public Work Manager Challenges and Strategies*, Findings of the APWA Task Force on Asset Management.

Dornan, D.L. 2001. *Asset Management and Innovative Finance*, Paper prepared for TRB 2nd National Conference on Transportation Finance, Scottsdale, AZ

Fabozzi, F.J. 1994. *Bond Markets: Analysis and Strategies, 2nd Ed.*, Prentice Hall, New Jersey.

Finnerty, J.D. 1996. *Project Financing: Asset-based Financial Engineering*. Wiley, New York.

George, K.P., Rajaopal, A.S., and Lim, L.K. 1989. *Model Predicting Pavement Deterioration*, Transportation Research Record, No. 1215, pp. 1–7.

Gharaibeh, N.G., Dater, M.I., and Uzarski, D.R., Development of prototype highway asset management system, *Journal of Infrastructure Systems, ASCE*, 5, 2, 61–68, 1999.

Golabi, K. 2002. A Pavement Management System for Portugal. *Proceeding of the 7th International Conference on Application of Advanced Technologies in Transportation*, MA, USA.

Government Accounting Office (GAO), 1996. Transportation Infrastructure: States' Implementation of Transportation Management Systems, GAO/RCED-97-32, Washington, DC.

Government Accounting Standard Board (GASB), 1999. Statement No. 34, Basic Financial Statements and Management's Discussion and Analysis for State and Local Government, June.

Haas, R., Hudson, W.R., and Zaniewsk, J., 1994. *Modern Pavement Management*. Krieger, Malabar.

Hatry, H.P. and Steinthal, B. 1984. *Guide to Selecting Maintenance Strategies for Capital Facilities*, Guides to Managing Urban Capital, Vol. 4. The Urban Institute Press, Washington, DC.

Herabat, P., Amekudzi, A., and Sirirangsi, P. 2002. *Application of Cost Approach for Pavement Valuation and Asset Management*, Transportation Research Record, No. 1812, pp. 219–227.

Hickling Lewis Brod, Inc., 1990. *User's Manual for STRATBENCOST32*, NCHRP 2-18(4), August 20.

Hudson, W.R., Haas, R., and Uddin, W. 1997. *Infrastructure Management: Integrating Design, Construction, Maintenance, Rehabilitation, and Renovation*. McGraw Hill, New York.

Jatusripitak, S. 2000. *Ministry of Finance's Mission and 2002 Annual Plan*, http://www.mof.go.th/mof_speech_03Jan45.htm, December.

Kadlec, A. and McNeil, S. 2001. *Applying the Government Accounting Standards Board Statement 34: Lessons from the Field*, Transportation Research Record, No. 1747.

Kane, A., Why asset management is more critically important than ever before, *Public Roads*, 63, 5, 22–24, 2000, March–April.

Kim, K. and Bernadin, V. 2002. Application of an Analytical Hierarchy Process at the Indiana Department of Transportation for Transportation for Prioritizing Major Highway Capital Investments, Seventh TRB Conference on the Application of Transportation Planning Methods, Transportation Research Board, Boston, MA, pp. 266–278.

Lee, D. 2002. *Fundamentals of Life-Cycle Cost Analysis*, Transportation Research Record, No. 1818, pp. 203–210.

Lee, H. and Deighton, R., Developing infrastructure management systems for small public agency, *Journal of Infrastructure Systems, ASCE*, 1, 4, 230–235, 1995.

Lemer, A. and McCarthy, P. 1997. The Infrastructure Balance Sheet & Income Statement (IBS/IIS): Design and Illustrative Applications of a Full-Cost Management Accounting Report for Metropolitan Infrastructure, I^2 Partnership, Purdue University, West Lafayette, IN, July.

Markow, M., Highway management systems: state of the art, *Journal of Infrastructure Systems*, 1, 3, 186–191, 1995, September.

McNeil, S. Tischer, M.L., and DeBlasio, A. 2000. *Asset Management: What is the Fuss?* Transportation Research Record, No. 1729, pp. 21–25.

McNeil, S., Sriraj, P.S., Shaumik, P., and Ogard, L. 2002. *Evaluation of Near-Transportation Private Sector Asset Management Practices*, Final Report to Midwest Regional University Transportation Center, University of Wisconsin, Madison, Project 01–02.

Miller, R.R. 2000. *Doing Business in Newly Privatized Markets: Global Opportunities and Challenges*. Quorum Books, Westport.

Montana Department of Transportation, 2000. Performance Planning Process, A Tool for Making Transportation Investment Decisions (ftp://ftp.mdt.state.mt.us/planning/tranplanp3.pdf, Accessed 5/20/03).

Montana Department of Transportation, 2002. TRANPLAN 21 2002, November.

National Council on Public Works Improvement, 1986. *The Nation's Public Works: Defining the Issues*, Report to the President and Congress, National Council on Public Works Improvement, Washington, DC.

Norwell, G. and Youdale, G. 1997. *Managing the Road Asset*, PIARC Special Report on Concessions.

Nutt, B., Walker, B., Holiday, S., and Sears, D. 1976. *Obsolescence in Housing*. Saxon House, Hampshire, UK.

Oberman, W., Bittner, J., and Wittwer, E. 2002. *Synthesis of National Efforts in Transportation Asset Management* Midwest Regional University Transportation Center, May. Available: http://www.mrutc.org/research/0101/report0101.pdf, Accessed December 20 2003.

OECD (Organization for European Cooperation and Development), 1999. *Asset Management System*, Project Description, Paris, France.

Peters, M. 2003. Planning for Choice, Mobility and Livability: The Reauthorization of TEA-21, *National Planning Conference Proceedings*, Denver. Available: http://www.caed.asu.edu/apa/proceedings03/PETERS/peters.htm, Accessed December 20 2003.

Peurifoy, R.L. and Oberlender, G.D. 2002. *Estimating Construction Costs*. McGraw Hill, New York.

Ravirala, V. and Grivas, D.A., Goal programming methodology for integrating pavement and bridges program, *Journal of Transportation Engineering*, ASCE, 121, 4, 345–351, 1995.

Regional Economic Models Inc (REMI), 2003. Overview, http://www.remi.com/index.html, Accessed December 20.

Sadek, A.W., Kvasnak, A., and Segale, J., Integrated infrastructure management system for a small urban area, *Journal of Infrastructure Systems*, 9, 3, 98–106, 2003, September.

Sanford, K. and McNeil, S., Infrastructure and public works education: one size does not fit all, *Journal of Public Works Management and Policy*, 5, 4, 318–328, 2001.

Schweppe, E., Do better roads mean more jobs?, *Public Roads*, 65, 6, 19–22, 2002, May/June.

Sinha, K.C. and Fwa, T.F. 1987. *On the Concept of Total Highway Management*, Transportation Research Record, No. 1229, pp. 79–88.

Sriraj, P.S., Shaumik, P., Ogard, L., and McNeil, S. 2003. *Evaluation of Private Sector Asset Management Practices*, Transportation Research Record, No. 1848, pp. 29–36.

Sirirangsi, P. and Herabat, P., Pavement performance and its valuation, *Journal of the Eastern Asia Society for Transportation Studies*, 1, 1, 147–160, 2001.

Sirirangsi, P., Amekudzi, A., and Herabat, P. 2003. *Capturing the Effects of Maintenance Practices in Highway Asset Valuation: the Replacement Cost Approach versus the Book Value Method*, Transportation Research Record, No. 1824, pp. 57–65.

Stantech Consulting Ltd and University of Waterloo, 2000. *Measuring and Reporting Highway Asset Value, Condition and Performance*, Report prepared for the Transportation Association of Canada.

Switzer, A. and McNeil, S. A road map for transportation asset management research, *Public Works Management and Policy*, 8, 3, 162–175, 2004.

Teeraganont, P. 2003. *Asset Management Standards of Department of Highways*, Financial Division, Department of Highways.

Texas Transportation Institute, 1999. *MicroBENCOST Version 2.0 User's Manual*, prepared for National Cooperative Highway Research Program Project 7-12(2), June.

Transportation Association of Canada (TAC), 1996. *Primer on Highway Asset Management Systems*, Transportation Association of Canada, Ottawa.

Tonias, D.E. 1994. *Bridge Engineering: Design Rehabilitation, and Maintenance of Modern Highway Bridges*. McGraw Hill, New York.

TRB (Transportation Research Board), 1994. *Characteristics of Bridge Management Systems*, Transportation Research Circular No. 423, April.

TRB (Transportation Research Board), 2000. *International Bridge Management Conference*, Transportation Research Circular No. 498, June.

US DOT (U.S. Department of Transportation), Federal Highway, 1986. *Inspection of Fracture Critical Bridge Members: Supplement to the Bridge Inspector's Training*, FHWA-IP-86-26, September.

US DOT (U.S. Department of Transportation), Federal Highway, 1995. Bridge Inspector's Training Manual/90, Revised March.

US DOT (U.S. Department of Transportation), Federal Highway, 1996. *Asset Management, Advancing the State of the Art into the 21st Century through Public-Private Dialogue*, FHWA-RD-97-046, Washington, DC, September.

US DOT (U.S. Department of Transportation), Federal Highway, 1997. *21st Century Asset Management, Executive Summary*, Prepared by the Center for Infrastructure and Transportation Studies at Rensselaer Polytechnic Institute, Troy, New York, October.

US DOT (U.S. Department of Transportation), Federal Highway, 1998. *Life Cycle Cost Analysis in Pavement Design*, Pavement Division Interim Technical Bulletin, Report No. FHWA-SA-98-079, September.

US DOT (U.S. Department of Transportation), Federal Highway, 1999. Asset Management Primer, Publication No: FHWA-IF-00-010, Washington, DC.

US DOT (U.S. Department of Transportation), Federal Highway, 2001. Data Integration Primer, U.S. Department of Transportation, Office of Asset Management, August.

US DOT (U.S. Department of Transportation), Federal Highway, 2002. Life-Cycle Cost Analysis Primer, U.S. Department of Transportation, August.

US DOT (U.S. Department of Transportation), Federal Highway, 2003b. Data Integration: The Michigan Experience, Transportation Asset Management Case Studies, FHWA-IF-03-027, Washington, DC.

US DOT (U.S. Department of Transportation), Federal Highway, 2003c. Economics in Asset Management: The New York Experience, Transportation Asset Management Case Studies, FHWA-IF-03-039, Washington, DC.

US DOT (U.S. Department of Transportation), Federal Highway, 2003d. Highway Economic Requirements System: The Oregon Experience, Transportation Asset Management Case Studies, FHWA-IF-03-037, Washington, DC.

US DOT (U.S. Department of Transportation), Federal Highway, 2003e. Life Cycle Cost Analysis: The Pennsylvania Experience, Transportation Asset Management Case Studies, FHWA-IF-03-038, Washington, DC.

US DOT (U.S. Department of Transportation) 2000. Federal Highway Administration (FHWA), *Highway Performance Monitoring System Field Manual*, December, OMB No. 21250028.

US DOT (U.S. Department of Transportation), 2003. Federal Highway Administration (FHWA), *Highway Economic Requirements System — State Version, HERS-ST*, http://www.fhwa.dot.gov/infrastructure/asstmgmt/hersindex.htm, Accessed December 20 2003.

Weston, J.F. and Brigham, E.F. 1990. *Essentials of Managerial Finance, 9th Edition*, Dryden, Chicago.

Winsor, J., Adams, L., Ramasubramanian, L., and McNeil, S. 2004. *Transportation Asset Management Today: An Application of Community of Practice Collaboration in the Transportation Industry*, Transportation Research Record, Journal of the Transportation Research Board, No. 1885, pp. 88–95.

Wittwer, E., Bittner, J., and Switzer, A., The fourth national transportation asset management workshop, *International Journal of Transport Management*, 1, 2, 87–99, 2002, October.

Wittwer, E., Zimmerman, K., McNeil, S., Bittner, J., and The Workshop Planning Committee. 2004. Key Findings from the Fifth National Workshop on Transportation Asset Management, Midwest Regional University Transportation Center, July 2004, http://www.mrutc.org/outreach/Final Report 5th NTAM.pdf (Accessed March 9, 2005).

Zaghloul, S.M., Sharaf, E.A., and Gadallah, A.A. 1989. *A Simplified Pavement Maintenance Cost Model*, Transportation Research Record, No. 1216, pp. 29–38.

18

Pavement Management Systems

Waheed Uddin
University of Mississippi
University, MS, U.S.A.

18.1   Historical Overview of PMS

18.1.1   Societal Perspectives

The highway network of a country forms the backbone of its economy and represents a huge investment in millions of dollars that allows for the safe and efficient movement of people and goods. Economically feasible and technically sound decisions on design, construction, maintenance, rehabilitation, and reconstruction of highway pavements are crucial for preserving the highway network in an acceptable condition. The economic prosperity of a country is strongly associated with the relative size and physical condition of its road network, which is the most important component of its transportation infrastructure (Hudson et al., 1997). High paved road density values in Km per million inhabitants (Queiroz et al., 1994; Mobility, 2001; Uddin, 2002a) are reported for industrialized countries (examples: 12,517 for the United States; 9,330 for the European Union of Western European countries; 9,200 for Japan; and 7,880 for Central and Eastern European countries). Comparatively low values of this important economic development indicator are observed in developing countries (examples: 104 for Afghanistan; under 200 for China and India; 310 for Nigeria; and 763 for Brazil). These trends show that new construction and upgrade road projects consume most road funds in many developing countries, and more funds are allocated in industrialized nations to maintenance, rehabilitation, and preservation of the existing pavement assets. As competition for funding among different sectors of the economy and society has grown, there is a critical need to implement modern management and systems engineering tools to assist decision makers for cost-effective and longer lasting pavement construction and effective use of funds for timely maintenance to preserve the condition and prolong the life of pavement assets.

Good pavement construction and maintenance practices in the PreChristian era were established first by Romans who built an impressive road network throughout Europe and the Middle East primarily for military and commercial use. For many hundreds of years the road network and pavement technology did not improve until the pioneering pavements constructed in the late 1700s and early 1800s for the French and British kings. The invention of the motor vehicle led to modern highway pavement design and construction in the late 19th century. Historically, design and construction of pavements have been based on methodical specifications, and the maintenance functions were treated mostly on an *ad hoc* basis reflecting local experience. In modern times the primary economic development is focused on the general public and commercial users who benefit from a well maintained highway network. For that reason and to provide valuable assistance to decision makers, the concept of pavement management system (PMS) evolved in the late 1960s and 1970s. Since that time PMS has been recognized as the most effective way to select cost-effective strategies for constructing and maintaining pavements in serviceable condition within the constraints of available funds.

The first major conference on pavement management was hosted in 1985 in Toronto by the Ministry of Transport in Ontario, Canada, followed by the second conference in 1987 in the same city (Kher, 1985; MOT, 1987). Since then it became a major international event with the third conference in 1994 in San Antonio, Tx, U.S.A. organized by the Transportation Research Board (TRB), the fourth in South Africa in 1998 (Visser, 1998), the fifth in Seattle, Washington, U.S.A. in 2001, and the sixth in Brisbane, Australia in 2004. These conferences have become excellent venues to exchange technology innovations and PMS development and implementation experience for worldwide participants from all sectors of the society. Many other international and regional conferences have also contributed with sessions on pavement management (Haas, 2003). This is a remarkable achievement as just 20 years ago, the PMS concept was not widely known or accepted by most highway agencies. The hard work of many people throughout the world has made PMS a reliable system for quantifying pavement maintenance needs and evaluating pavement performance as it relates to the cost-effective management of road and highway networks.

18.1.2   Evolution of PMS Concept

The following list of key milestones is recognized for their contributions in the process of successful PMS development and implementation.

(a) *AASHO Road Test — 1959 to 1961*: The American Association of State Highway Officials (AASHO) Road Test gave the first comprehensive concept of relating pavement serviceability to performance and methods of their measurements (AASHO, 1962). This was followed by an unprecedented freeway and highway construction program throughout the United States (U.S.) of America.

(b) *NCHRP Project 1-10, 1968*: National Cooperative Highway Research Program (NCHRP) Application of systems approach for pavement design and PMS concept (Hudson et al., 1968).

(c) *Systems Approach to Pavement Design*: Texas State Highway Department Project 123, "Application of PMS concept for the development of pavement design systems" (Hudson et al., 1970), working system models for flexible and rigid pavement designs (Hudson et al., 1972, 1973).

(d) *International In-Service Road Performance Studies, early 1070s and 1980s*: The Brazilian study on road performance and vehicle operating cost (VOC), sponsored by The World Bank and United Nations Development Program (UNDP), made significant contributions in pavement engineering as it included inadequately maintained paved and unsurfaced roads (Watanatada et al., 1987).

(e) *NCHRP Report 215, 1979*: Application of PMS concept for highway network (Hudson et al., 1979).

(f) *PMS Workshop and Conferences in North America*: FHWA Pavement Management Workshop Proceedings, 1978 (FHWA, 1978); Pavement Management Conferences, 1985, 1987 (Kher, 1985; MOT, 1987).

(g) *FHWA — PMS Policy, 1989*: The U.S. Department of Transportation Federal Highway Administration (FHWA) required all state highway agencies in the U.S. to implement basic PMS by 1993 (FHWA, 1989).

(h) *AASHTO — PMS Guidelines, 1990*: The American Association of State Highway and Transportation Officials (AASHTO) issued the PMS guidelines (AASHTO, 1990) to support state highway agencies in the U.S. in their efforts to implement PMS.

The serviceability–performance concept developed at the AASHO Road Test (AASHO, 1962) provided the first comprehensive approach to assess pavement physical condition by periodic monitoring, and quantify the effect of different design strategies on pavement life. Pavement performance is the history of serviceability or the area under the curve in Figure 18.1 (Haas and Hudson, 1978; Fwa and Sinha, 1991). This figure shows this concept in terms of pavement life on the abscissa. The ordinate shows serviceability or Present Serviceability Rating (PSR) on a subjective rating scale from five (excellent) to zero (failed or impassable) when the pavement is extensively broken and unsafe for use by vehicles. The Present Serviceability Index (PSI) is an objective assessment of serviceability on the same scale from five to zero by measuring one or more appropriate pavement condition attributes, such as longitudinal roughness measurement, on a subjective rating scale from five (excellent) to zero (failed). The minimum acceptable serviceability or PSR or PSI is taken as 2.5 for freeway and superhighways with high traffic volume or 2.0 for secondary highways with less traffic volume. The minimum PSR or PSI is, therefore, a decision

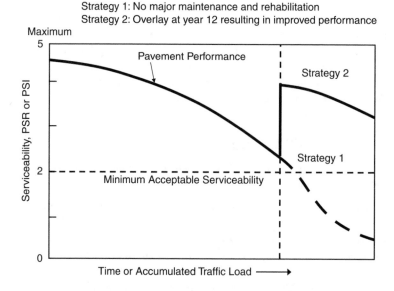

FIGURE 18.1 Effect of maintenance and rehabilitation strategy on pavement performance.

criterion set by the owner agency to consider functional "pavement failure" and schedule an appropriate maintenance, rehabilitation & reconstruction (M,R&R) strategy to improve the condition at a lower cost and prolong the pavement life. An alternative is to permit it to deteriorate until failure and reconstruct at relatively higher cost and inconvenience to traveling public.

The traditional practice of 20 years, used arbitrarily for pavement design life, has been a short-sighted approach as the highway is built forever until it fails to serve the traveling public. For example, it may have reached its capacity and become unsafe because of high traffic volume and congestion, abandoned because of no traffic, or disintegrated because of catastrophic failure from earthquake and war damage. The next 20 years after the AASHO Road Test witnessed a period of unprecedented road construction in the U.S. But many pavements built during that period approached the end of their design lives and needed reconstruction or rehabilitation, which was never considered a part of pavement design methods. Other pavements showed signs of serious distress much earlier because of unexpectedly high traffic volume, which was a warning sign to highway managers requiring immediate M,R&R actions. The major outcomes of the AASHO Road Test research studies and these observations of premature failure had profound effects on future pavement design and related technologies, which led to life-cycle analysis of pavement assets and systematic approach to pavement management.

The PMS concept was first used in late 1960s to develop the systems approach for pavement design (Hudson et al., 1968), which integrated systems engineering, management principles, engineering analysis, and economic evaluation. It was followed by advances in related technologies in the 1970s (Terrel and LeClerc, 1978; Finn, 1994) and the knowledge was documented in a book on PMS (Haas and Hudson, 1978). In the late 1970s PMS evolved as a process to address the maintenance needs for the entire road network in the jurisdiction of an agency (Hudson et al., 1979). A review of the early development and implementation of PMS for network maintenance and rehabilitation planning shows the pioneering work in pavement performance modeling, optimization of project selection with budget and overall network condition constraints, and communication with agency staff and decision makers (Finn, 1994). One of the early working PMS for highway network maintenance and rehabilitation planning was developed for Washington State (LeClerc and Nelson, 1982).

The World Bank's Highway Design and Maintenance model (HDM III) was originally developed from the results of the Brazilian study on road performance and VOC (Watanatada et al., 1987), which included inadequately maintained paved and unsurfaced roads. The VOC models were also incorporated in the first major VOC study conducted in the U.S. on well maintained paved roads (Zaniewski et al., 1982). These important studies provided quantifiable benefits for life-cycle analysis in terms of user cost savings from reduced VOC by bringing pavements to good condition in timely manner.

As stated earlier the pavement design of Interstate freeway and highway network built mostly during the 1960s and 1970s in the U.S. was based on 20 years design life. During the early 1980s many parts of these superhighways reached their design lives prematurely because of unexpectedly higher truck traffic volumes. Pavement maintenance, rehabilitation, resurfacing, restoration, and reconstruction became priorities for state highway agencies that required monitoring of the pavement condition and application of timely M,R&R strategies. These concerns helped the implementation of PMS concepts for cost-effective planning and management of highway network programs. Since the 1980s the development and implementation of high-speed multifunction equipment have been helping in efficient monitoring of large highway network. The institutional support of the decision-making administrators at federal level in the U.S. (FHWA, 1989) by requiring the state highway agencies to implement the major components of PMS in the 1990s and the development of AASHTO guidelines on PMS were major impetus in the successful implementation of PMS in the United States. The modern PMS concepts and related technologies are well documented in a book by Haas et al. (1994).

In summary, the modern PMS concept of pavement management has now progressed to a widely accepted approach worldwide for cost-effective planning of road investment and maintenance management of road and highway networks. The FHWA and AASHTO have encouraged states in the U.S. to implement PMS. Pavement management has also become a required component of international funding agencies, such as the World Bank, for most road and highway projects in developing countries for improved maintenance planning.

18.2 Network Level vs. Project-Level PMS Functions

18.2.1 Key Components of PMS

The following definition of PMS is provided by the FHWA and AASHTO (FHWA, 1989; AASHTO, 1990). "A set of tools or methods that (can) assist decision markers in finding cost-effective strategies for providing, evaluating, and maintaining pavements in a serviceable condition." In other words, a PMS is a systematic approach that provides quantifiable engineering information to help administrators and engineers manage highway pavements. The decision-making process is based on information from a working PMS and involves engineering experience, budget constraints, scheduling needs, management priorities, public input, and political considerations.

The pavement management process is developed to respond to managing the most substantial area of investment in transportation infrastructure. It incorporates all activities required to provide and maintain pavements. Figure 18.2 illustrates key components and major activities of a generic PMS, which may need adjustments based on the purpose of PMS process and administrative setup of a highway agency. The central database is the heart of a working PMS which supports all related activities. Overall these activities are grouped into six major components or categories: (1) administration and planning, (2) design, (3) construction, (4) maintenance, and (5) evaluation and research. Figure 18.3 shows the influence level concept of these major PMS components over the life-cycle of a highway

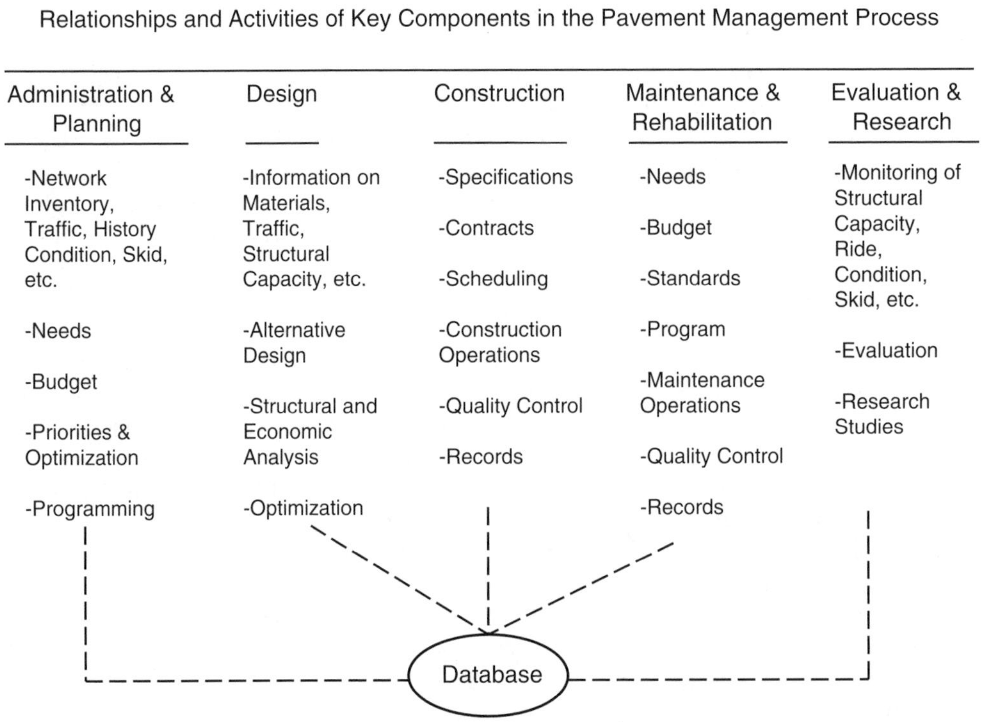

Relationships and Activities of Key Components in the Pavement Management Process

Administration & Planning	Design	Construction	Maintenance & Rehabilitation	Evaluation & Research
-Network Inventory, Traffic, History Condition, Skid, etc.	-Information on Materials, Traffic, Structural Capacity, etc.	-Specifications -Contracts -Scheduling	-Needs -Budget -Standards	-Monitoring of Structural Capacity, Ride, Condition, Skid, etc.
-Needs	-Alternative Design	-Construction Operations	-Program	-Evaluation
-Budget -Priorities & Optimization	-Structural and Economic Analysis	-Quality Control -Records	-Maintenance Operations -Quality Control	-Research Studies
-Programming	-Optimization		-Records	

FIGURE 18.2 Relationships and activities of key PMS components, after (Haas et al., 1994).

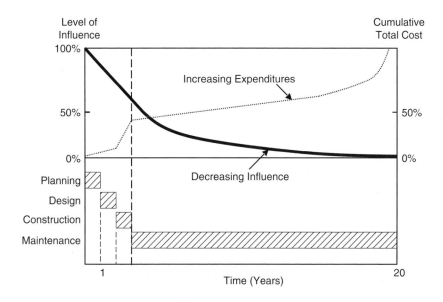

FIGURE 18.3 Influence levels of PMS subsystems on the total costs, after (Haas et al., 1994).

project (Haas et al., 1994), which clearly demonstrates that decision taken during construction phase can greatly impact the cost of pavement maintenance and rehabilitation.

Haas et al. have grouped the PMS related issues and expected outputs at legislative level, administrative level, and technical level.

18.2.2   Legislative Level Issues

The legislative level issues as given below are fairly broad in scope and must be recognized by the administrative and technical levels.

1. Justification of budget requests.
2. Effects of less capital and maintenance funding.
3. Effects of deferring work or lowering standards.
4. Effects of budget request on future status of the network.
5. Effects of increased load limits.

18.2.3   Administrative Level Issues

The administrative level issues are related to the decision-making process including budget and programing priorities.

1. An objectively based priority program to provide justification for budget requests.
2. A summary assessment of the current status of the network, in graphical and tabular format, based on inventory measurements.
3. The means for quantitatively determining the effects of lower budget levels, and the budget level required to keep the network in its present state.
4. The means for quantitatively demonstrating the effects of deferring maintenance or rehabilitation.
5. Estimates of the future status of the network (in terms of average serviceability, condition, safety, etc.) for the expected funding.
6. Benefits of a PMS, its major features or "deliverables," etc.

7. Costs of pavement management implementation, including inventory; assignment of responsibility and manpower requirements; implementation staging and schedule, etc.
8. Implementation experience of others; documentation of their experience.
9. Relationship between pavement management and existing maintenance management system.
10. Interfacing of a PMS with highway management in general.

18.2.4 Technical Level Issues

From a technical perspective, pavement management involves a large number of issues and questions. In addition, the questions and issues faced at the administrative level must be appreciated if technical activities are to be meaningful. The following is an example listing of some of the key questions for this level, involving both network and project considerations:

1. Inventory database design and operation.
2. Methods and adequacy of inventory database.
3. Models for predicting traffic, performance, distress, skid, etc., — their reliability, consistency, reasonableness, deficiencies.
4. Criteria for minimum serviceability, minimum skid, maximum distress, minimum structural adequacy — reasonableness, effects of changes in criteria, etc.
5. Models for priority analysis and network optimization.
6. Verification of models.
7. Relating project (sub) optimization to network optimization.
8. Methods for characterizing materials and using results.
9. Sensitivity of model analysis results to variations in factors.
10. Relationships between VOCs and pavement characteristics.
11. Construction quality control.
12. Effects of construction and maintenance on pavement performance.
13. Communication among design, construction, and maintenance, within existing administrative structure.
14. Guidelines of pavement management implementation.
15. Relating pavement management to maintenance management.
16. Improving the technology of pavement management and making use of implementation projects for this purpose.

18.2.5 Network Level vs. Project-Level Pavement Management

The pavement management process has two basic working levels to serve different sets of decisions through different levels of analysis of the collected data using a shared database. There is strong interaction of activities and feedback between these two levels through the common database.

Network level: At the network level the primary goal of PMS is to assess the overall condition of the network, produce a prioritized work program and schedule by examining several different possible scenarios of time and budget constraints, and allocate funds. Decisions made at this level affect the entire network and involve tradeoffs between projects and activities.

Project-level: At the project-level the actual implementation of a scheduled project is carried out in the most cost-effective way so that the allocated funds are used to construct, rehabilitate, or maintain longer lasting pavements. Decisions are made about specific projects or pavement sections.

Figure 18.4 shows an example of these two working levels of PMS and related functions and activities developed for Dubai PMS in the United Arab Emirates (Uddin and Al-Tayer, 1993). A successful pavement management process must consider the consequences of deferred maintenance and rehabilitation for one pavement section at the project-level or for all pavement sections in the network, which are below the minimum acceptable condition (for example 2.0 using PSI scale). After a certain drop in pavement serviceability, the rate of deterioration accelerates, and it may cost significantly more to

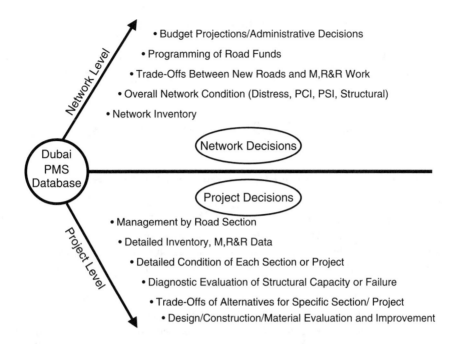

FIGURE 18.4 PMS Functions at project-level and network level for Dubai PMS.

implement corrective action of a pavement, as shown in Figure 18.5 by an example of a typical asphalt pavement performance curve.

Extent of Data and Level of Details: It must be realized that data requirements and levels of details are different at these network- and project-levels of the pavement management process and depend on the available funding and resources for PMS development. In the past many agencies in the U.S. and other countries embarked upon extensive data collection for 100% network at a frequency of one to three years. In many cases the useful products of the PMS development could not be produced in a reasonable period, and the results suffered from development and operational cost overruns. Therefore, it is important at the outset of the initial PMS development to: agree upon overall objective and goals, estimate the available funding and resources, select PMS analysis software option and associated database system, list data categories in priority ranking, plan acquisition of essential pavement condition monitoring equipment,

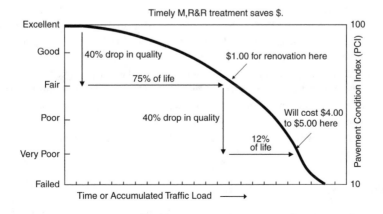

FIGURE 18.5 Pavement condition deterioration effects on maintenance costs, after (Shahin and Walther, 1990).

develop data collection frequency, and develop decision criteria and priority ranking method for network level M,R&R programs. A good feasible plan for PMS development can not be made without studying the existing agency practices related to these aspects and having inputs of agency staff involved in planning, design, materials, construction, maintenance, research and evaluation, safety audit, and project management.

18.3   Data Needs, Decision-Support Systems, GIS, Database Design

18.3.1   Data Needs and Related Information Technologies

Pavement management data needs require that data be stored in an organized way so that it can be efficiently expanded, updated, and retrieved rapidly for various different users. Computerized database and information technologies are now integral to all levels of management because of rapid developments in computer hardware and software in the last two decades, and electronic database storage and retrieval capabilities using desktop computer stations. There has been a computing revolution from the mainframe and expensive computers and associated database management system software in the 1970s to the relational and object-oriented databases on powerful personal computers and workstations in the late 1990s. Other information technologies available for enhanced pavement management database applications include knowledge-based expert systems, artificial neural networks, video logging and digital photos, satellite imagery and spatial data visualization using a GIS, and three dimensional virtual reality visualization. The advancement in networking and remote data communications commonly used for intelligent transportation systems (ITS) are also available for pavement management database applications.

All highway agencies have a computer based system for their data, and the PMS database design must consider the existing computer equipment and database management software at the outset of system design. The database may be organized as a single file or multiple files. Information technology professionals in the agency must be consulted to know the existing computer systems in the agency and assess the need for an upgrade based on an estimate of PMS data needs. Typically, the computer operating system software and database management software are commercially available off-the-shelf products. Examples of operating software are Windows for PCs, Unix for minicomputers and workstation, and MVS-XA for mainframe computers. Currently, most of large databases are established on mainframe and workstations which can be accessed by PCs through client–server network applications. Database management software is commercially available products designed for specific category of computer platforms and operating systems. The database software enables application programers to write programs that can take advantage of data structure facilities and perform functions of data access, retrieve, manipulation, and queries. The relational database management system (RDBMS) software is most commonly used on mainframe and PCs, such as Oracle. The relational database software facilitates the use of high-level query languages using plain English-like commands to perform complex retrievals form the database. A popular example is the structures query language (SQL) used by most relational databases.

Application programs are specifically written software for data access and manipulation, analysis functions, and user input and query interface for end-users. These user friendly programs are nowadays written using high level programing languages and object-oriented programing tools. On the other hand, application programs, such as PMS analysis software, are usually unique to the application. These can become very costly component software to develop and need careful consideration at the time of system design and development.

18.3.2   Decision-Support Systems

A decision-support system (DSS) term refers to the computerized data storage, processing, queries, analysis, and display of results and information used to support decision-making. Therefore, DSS

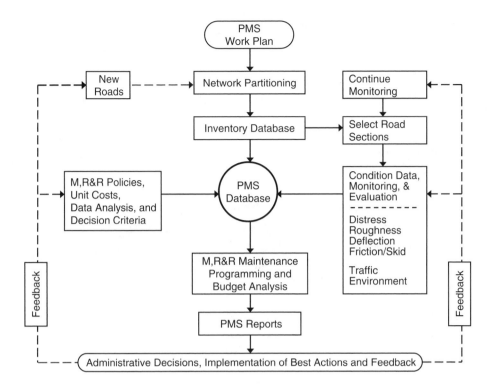

FIGURE 18.6   A DSS framework for network level PMS applications, after (Uddin, 1992).

is an integral part of a PMS, which replaces the traditional paper based subjective judgment practice and provides rational decision making and better service to the public. Once a DSS is established, the preparation of network level M,R&R work programs is streamlined as an automatic PMS output, which can be conveniently updated based on changed constraints and feedback to the database. Figure 18.6 shows a DSS framework used for Dubai PMS (Uddin, 1995). A successful DSS application program requires the flow of support requests from the decision maker to the technical staff and relies on efficient data management and evaluation of alternatives.

In summary, the DSS uses data tables and necessary analytical models to generate decision-support rationales and reports. A centralized and maintained database is the heart of a DSS. The size of the highway network and complexity of data collection activities dictate the computer memory and hard-disk storage requirements. Graphical databases demand several times more memory and storage requirements than the traditional nongraphic databases. Temporal and spatial data attributes are important for good database management and DSS applications. This requires a spatial location referencing system to identify the boundaries and physical characteristics of homogeneous pavement sections. Traditionally, this was based on Km or milepost system along the highway length. The development and implementation of GIS can greatly enhance the spatial location referencing for database management and field data collection by using geographical coordinates to designate the limits of a pavement section.

18.3.3   GIS Applications

A graphical display of the highway network on a map and queries to any feature on the map using a GIS application makes it easy for end-users to work with complex databases and DSS applications. Therefore, GIS applications have also become popular for PMS development as shown in a detailed example by Hudson et al. (1997). A GIS database consists of geographic and non-geographic data. The geographic

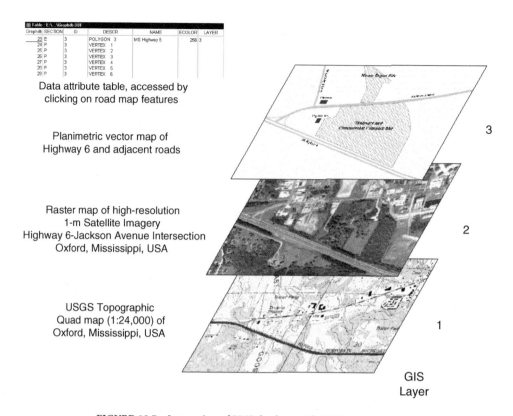

Data attribute table, accessed by
clicking on road map features

Planimetric vector map of
Highway 6 and adjacent roads

Raster map of high-resolution
1-m Satellite Imagery
Highway 6-Jackson Avenue Intersection
Oxford, Mississippi, USA

USGS Topographic
Quad map (1:24,000) of
Oxford, Mississippi, USA

GIS
Layer

FIGURE 18.7   Integration of PMS database with GIS layers and maps.

data are map features represented by geo-coordinates of vector maps (points, lines, polygon areas) and raster maps (pixel, cell, grid). Therefore, complex database tables describing the locations and attributes of roads meeting at an interchange can be simply displayed by an easy-to-read map. A geographic data element can be associated with non-geographic data attribute tables. A series of layers can be used to show different map features. Each layer of homogenous features is separated through logical relationships and linked to other layers through the common geo-coordinate data.

Figure 18.7 shows an example of a GIS data attribute table and layers of the United States Geologic Survey (USGS) map, high-resolution IKONOS satellite imagery, and highway and road vector map developed for MS Highway six and adjacent roads in Oxford, MS, U.S. Most GIS software vendors offer SQL functions for data queries and options for integration with third party relational databases. The DSS results can be simply displayed as GIS layers, such as separate layer of all roads in different colors showing (a) different condition in terms of ranges of PSR values, (b) different categories of traffic volumes, and (c) the roads which were treated with major M,R&R during the last five years.

18.3.4   PMS Data Needs and Database Design

RDBMS software serves PMS needs better because it allows a large number of different database tables, which can be linked through one or more common key field. A key field is simply a unique alphanumeric code assigned to each homogenous pavement section in the PMS database. The PMS database should have some common data for use at network level and project-level. The network level data summaries are often available on-line through computer networking. However, detailed condition data for both network and project-levels are made available off-line on as-needed basis. Research & development data needs are more detailed, precise, and study specific than the data needs at the project-level. The following list gives general PMS data needs and data collection frequency at network- and project-levels.

- Inventory data (section-specific road classification, historical data describing construction history and costs, physical dimensions and boundaries, and pavement material data) — *one time* (collected once) and updated after every major rehabilitation and reconstruction.
- Traffic history data (highway-specific traffic volume, directional split, vehicle mix, truck load data, equivalent single axle load (ESAL) data used for design and accumulated history, and speed data) — *every other year*; initially based on design information and periodically updated from traffic monitoring sites and truck weigh station records.
- Environmental data (general climatic data in different region, detailed environmental data history such as air temperature needed for pavement and hot mix asphalt design, special pavement-specific data collected on site) — *annually* acquired from publicly available weather station data and appropriate historical data summaries stored in the database; annual or periodic updates (section-specific or regional).
- Pavement condition monitoring and evaluation data (section-specific real-time pavement quality measurement of longitudinal roughness, manual and video-based visual distress survey including cracking and rutting, friction or skid resistance, deflection testing for structural evaluation, safety and noise audits) — *every 3–5 years* periodically collected in field on regularly basis. Data collection frequency depends on the network decision criteria for M,R&R intervention and annual monitoring budget because condition data are the most costly components of PMS database.
- M,R&R history data (section-specific date and type of treatment including routine maintenance, major intervention, reconstruction, cost data) — *annually* and as soon as available.
- Decision criteria and constraints (M,R&R intervention policies, decision criteria related to pavement quality and performance measure, unit cost and budget data established for the entire network) — *one time* and updated as needed.
- Annual and multiyear M,R&R work programs (prioritized list of roads and allocation of funds for the entire network) — *annually* and as required by executive management and decision makers.
- Feedback from as-built records and M,R&R work program implementation (section-specific) — *annually* and as soon as available for new construction and major M,R&R.

Exact levels of details and data collection frequency will depend on overall PMS work plan and database design (Figure 18.6), as well as committed funding sources. The detail and complexity of analysis models limit the amount and details of data at the network level. However, detailed section-specific data of material properties and deflection data for structural evaluation are needed for performance modeling and M,R&R analysis at project-level. Some lack of details can be compensated by having higher number samples in order to increase the reliability and accuracy of each data point. The agency should compile a comprehensive data file of terminology (data dictionary) and appropriate units of measurement. Data standards, such as those developed by American Society of Testing and Materials (ASTM), should be followed as much as possible (ASTM, 1999). Database quality, integrity, and security are important issues when working with high volume PMS databases because considerable time and money is involved in the database development and the data are often located in several departments of a highway agency. For example, the condition data are collected in the field at significantly high cost and other databases are developed after lots of effort. Usually one person in the central office is given the authority to maintain tight control on the functioning aspects of the PMS database to ensure database integrity and security through database manager software and specially developed client–server application programs. In addition to periodic backups of the PMS database, it is important to keep an off-site permanent backup copy with annual updates against the data loss resulting from an unfortunate catastrophic disaster.

18.4   Network Partitioning, Inventory and Monitoring Databases

18.4.1   Network Partitioning and Location Referencing System

The partitioning of a highway or road network in homogeneous sections is the most important first step for the development of PMS database. The purpose is to create relatively uniform and homogeneous pavement sections. A homogenous PMS section should have the same functional classification and similarity with respect to pavement width, surface type, layer structure, and other criteria along its length. This step is important for efficient condition data analysis and assigning uniform M,R&R and routine maintenance treatment on the candidate sections. Network partitioning of highways or roads in homogenous PMS sections is conducted by identifying key data and selected criteria related to:

(a) Planning data (same administrative boundaries, zones, functional class, highway or road number, adjacent road references).
(b) Construction data (same construction project number, construction year or year of last major M,R&R, pavement surface type).
(c) Geometric data (same number of lanes, width, shoulder type and width)
(d) Traffic data (similar daily traffic volume range, % truck).
(e) Sections should be easily identifiable on plans and in the field using adjacent landmarks and hand-held global positioning system (GPS) receivers.
(f) Homogeneous section length limits should be based on maximum length that can be surveyed for condition assessment in a day (for example 5 Km). There should be a practical minimum length limit to avoid unnecessary large number of sections in the database. For example, 0.5 Km minimum section length can be assigned on a national highway network.

In order to be effective, inputs must be sought from members of the PMS steering committee representing all key departments and file maintenance staff. A prioritized list of selected partitioning factors from the above should be established. The level of complexity of network partitioning depends on the use of data and available resources. The partitioning procedure should be computerized as a part of the DSS framework (Figure 18.6) and programs be developed to assign appropriate inventory data automatically to all PMS sections on a highway within one jurisdiction. These common inventory data may include jurisdiction and administrative zone, highway number and name, functional class and surface type, date of original construction, design traffic volume and design ESAL applications. The homogenous pavement sections are necessary to facilitate automatic calculations of age, surface area, volume of M,R&R work, and prediction of future condition. However, more sophisticated dynamic segmentation procedures can be used to merge several sections into one large segment and establish new boundaries based on behavioral attributes (distress index, pavement condition index [PCI], serviceability, deflection, skid, roughness, etc.) without losing other physical inventory and condition data in the original sections. Dynamic segmentation can be easily displayed on a GIS database using its query options.

RDBMS software uses one or more common key field to link the section-specific data stored in a large number of different PMS database tables. The key field is a unique alphanumeric code assigned to each homogenous pavement section in the PMS database. The key field is often a common location referencing system or a cross-correlated referencing system. Most highway agencies use kilometer or milepost markers as the most common referencing points to identify the beginning or end of the section. In addition, GIS-based geographic coordinates (longitude, latitude, altitude) at both ends of a section are also being specified or provided in inventory data. Longitudinal chainage and transverse location referencing systems are also needed to define positions of standalone features like a bridge or overpass, culverts and ditches, traffic control devices and large overhead traffic information features (directional sign or ITS variable message sign), and roadside appurtenance items (guardrail, crash protection devices, signs). These are additional benefits of a well planned PMS inventory database. Some of them may need to be reinstalled during major M,R&R operations. In summary, location referencing codes defining unique section numbers should not be confused with record serial numbers generated by the database software or created by users. Adequately defined unique location referencing codes define the section, used as key fields by the database management and application programs to track the section attributes in different database tables, and enables the user to identify the section on PMS reports and in the field.

18.4.2 Inventory of Pavement Sections and Related Assets

The inventory data of pavement for PMS sections and related assets can be numerous. These must be listed and prioritized based on their importance for both network and project-levels before devoting too much effort to collect detailed project-level data. Some of these details can be left blank in the initial stages of PMS database development. A typical highway and road inventory database file contains the following data (description and codes).

- Planning and identification data
  PMS section ID, location referencing codes, geographic coordinates, name
  Administrative zone, functional class (freeway, expressway, arterial, toll, etc.)
- Carriageway geometry data
  Single or divided, number of lanes, lane width, pavement width, length
  Inside and outside shoulder width, sidewalk, fill and cut data
- Construction and design data
  Construction number and date, project ID, pavement surface type (asphalt or concrete), construction costs, annual maintenance cost, design traffic mix, design ESALs and typical annual traffic volume, etc.
- Pavement structure and material data
  Layers, layer thickness and material data (codes), layer material properties
  Shoulder material type, pavement marking materials, etc.
- Other structure data
  Location of interchange, bridge, culvert and drainage structures, ditch, etc.
  Crossings over or under roads, railroad, utility pipeline, dirt, river or stream, etc.
- Appurtenance features (identified by either, a point, line, area, or a reference file)
  Traffic control devices, overhead signs, lighting, guardrail, barriers, etc.
  ITS video surveillance equipment, etc.

Construction and maintenance history data should start from the original construction (construction number 1) and other major M,R&R intervention construction identified as construction number two and so on. The construction number is used as a key identifier to represent any major changes in the inventory data items that can significantly affect subsequent pavement performance. The record of all historical construction data is useful for assessing condition deterioration and performance models needed for life-cycle analysis (Uddin, 1995). Inventory data collection and processing needs are accomplished by locating and using any existing electronic data files and computer databases (the most preferred way with least errors), extracting data from plans and "as-built" project documents, and conducting windshield surveys using hand-held dataloggers or through multifunction PMS data collection vans along with some verification by field visits and coring records. In any event, a paper and similar electronic inventory data collection form must be adapted to collect data manually and enter on computer using similar data interface screens. Other innovative methods include the use of video logs, high resolution aerial photo, and satellite imagery. It is always wise to design a comprehensive database and collection forms, which does not take too much time and cost. Nondestructive testing, such as ground-penetrating radar equipment, is often used to establish pavement layer thickness on pavements with no reliable construction record.

The cost of inventory data collection and processing can vary depending upon the existing resources, level of details, and scope of any field inventory data collection. The extent of geometry, structural, and material data can vary from a few mandatory items to hundreds of items, depending on the facility and intended use of data. Therefore, only essential items should be prioritized and collected in the initial PMS development phase and more details be collected in later stages. Most of the inventory data do not change from year to year, and any changes resulting from a major M,R&R work are updated upon the completion of that work. The major data collection and development efforts for inventory database are the initial one, taken at the time of establishing the PMS database for the network. The reader is referred to the literature for a detailed example of detailed inventory and historical database developed for highway and road network of Dubai Emirate (Uddin, 1993a). The database was used to get summary statistics of pavement assets by functional class, which are: 2.5% Freeway, 10.2% expressway, 8.1% arterial, 10.7% collector, and 68.5% local roads. A summary of the road assets available to the agency and useful for public information is an immediate output and benefit of the PMS inventory database.

18.4.3   Monitoring Database

In-service monitoring and evaluation of pavement condition is an essential component of PMS. Good monitoring and evaluation information is required to (a) to establish condition database for pavement maintenance management functions, (b) measure the effectiveness of different M,R&R strategies in different environmental regions and considering local design and construction practices, and (c) develop and calibrate performance models to improve design, construction and maintenance policies. The evaluation phase of pavement management involves the determination and continuous monitoring of the condition of the roadways within the agency's purview. The evaluation provides the primary source of information for use at all levels and in all activity areas of a PMS. Monitoring involves the routine-collection of field data and recording such data in a useful form. Evaluation encompasses monitoring, but involves a judgment or determination of the meaning of the information collected. The first basic requirement for condition data collection in the field is some reference system for identifying locations. It is desirable to have a common location referencing scheme across an agency so that the databases for inventory, condition monitoring, and maintenance, etc., can be linked, such as the use of GPS and geographical coordinates.

18.5   Traffic Data History and Environmental Data Needs

Historically, pavement design methods have considered traffic mix repetitions and material properties as the primary input variables for structural designs without considering the effect of environment, material degradation over time, and maintenance scheduling. The detrimental effects of environment and its interaction with traffic loads on pavement performance is reflected in "aging" as condition deterioration accelerates with time (Figure 18.5). An early benefit of condition monitoring and performance evaluation of highways is the identification of important environmental factors, construction quality, and interaction effects. These must be identified and considered in improved project-level design and network level maintenance programs, in addition to traffic usage and aging that affect the in-service condition deterioration. These condition deterioration mechanisms cause with surface defects, deformation, cracking and failures. The interaction of traffic loads with one or more environmental mechanisms is more critical for pavement condition deterioration than the effect of the environmental alone. Heavy traffic load repetitions will in all cases accelerate the damage caused by environmental factors and material degradation. Thermal cracking in concrete slabs can be caused by three to four times larger tensile stress due to high temperature differential from the top surface to the bottom, compared to the load induced tensile stress (Uddin and Chung, 1997). Low temperature thermal cracking of asphalt pavements (Haas, 1973) is initiated by very low ambient temperatures. Pothole formation during the spring–thaw season represents a good example of interaction of three mechanisms; load repetitions, weakening of sublayers and roadbed soil, and thaw (environmental) conditions. The earlier chapter on traffic presents detailed discussion of traffic data collection and its use for designing highway pavements. Traffic history, loading, axle configuration, and volume data for roads and bridges are sometimes collected and stored in the central transportation agency database. Traffic monitoring and environmental data collection are briefly discussed in this section.

18.5.1 Traffic Monitoring Equipment and Data Collection

Traffic volume and traffic weight distribution data for each highway section are necessary inputs to the PMS, performance modeling, design of pavement thickness and rehabilitation, traffic control studies, and statewide transportation studies. Traffic volume and truck weight data are often used as weighting factors in priority ranking calculations and optimization approaches used for M,R&R work programs at the network level. Therefore, the measurement of traffic is an important part of in-service monitoring data. For highways the traffic data are collected on a sampling basis using traffic counters or other sensors at specific locations for several days per year. These data are then expanded to estimate the annual average daily traffic for each highway location. Load information is estimated by taking a sample weighing of axle loads by various classes of trucks and by doing automatic classifications of vehicles in various classes of passenger cars, light trucks, 3-axle trucks, etc., through 5-axle tandem trucks. There are several types of automatic data collection devices used for monitoring traffic and collecting traffic volume and traffic weight data. These include portable counters, fixed counters, weigh-in-motion (WIM) devices, portable scales, and permanent weigh stations. Portable and fixed counters are capable of determining the number and types of vehicles. Vehicle weights and axle load distributions are obtained from WIM devices, portable scales, and permanent weigh stations. There are several sources of information on equipment for collecting traffic volume and weight data (Cunagin et al., 1986; French and Soloman, 1986; Epps and Monismith, 1996; FHWA, 2001). The WIM equipment is installed directly in pavements in the highway lane with no interfere with the traffic. However, the WIM sensors should be located away from areas of vehicle acceleration or deceleration and should not be located where the pavement is rough and in poor condition. The WIM data accuracy is affected by the vehicle speed and the roughness in the approach area. If adequate weigh station facilities are not available throughout the network, it may be useful to acquire and install additional cheaper WIM equipment in coordination with the planning and traffic departments of the agency. Many long-term pavement performance (LTPP) sections established by the Strategic Highway Research Program (SHRP) in the U.S. for traffic and pavement performance monitoring have produced good information on the cost and reliability of the WIM devices. Traffic data are collected in separate traffic monitoring databases. The data are analyzed and the following section-specific summaries are entered into the PMS traffic database.

- *Traffic volume data*: average annual daily (ADT) traffic volume, % of truck volume, directional split and lane distribution, growth factors, design and cumulative past traffic
- *Vehicle classification*: traffic mix including different truck axle groups, wheel and axle configuration, tire pressure, vehicle speed
- *Truck weight data*: gross vehicle load, wheel load and axle load
- *Truck Load Equivalency data*: calculation of ESAL applications (using fourth power law or other acceptable load equivalency factors) for past cumulative and future design applications, growth factors

As shown in Figures 18.1 and 18.5, the serviceability history plotted against accumulated traffic applications provide the performance curve because highway pavements are designed for a certain design number of ESAL applications to last the design period. For example, 1 million ESAL applications on a highway section may be over in 10 years or 20 years of the performance period of service life depending on the volume of truck traffic and yearly growth factor. A plot of the serviceability history against accumulated years shows the combined effect of aging on serviceability history in terms of traffic load, environment, and their interaction. This concept is discussed in detail in the chapters on pavement design.

18.5.2   Environmental Data Collection

Collection of climatological data is the least costly of all environmental monitoring activities. For example, the weather and climate data recorded on airports, national weather stations, and local weather news channels are available from the National Oceanic and Atmospheric Administration (NOAA) on the Internet for most of the United States (NOAA, 1994). In most countries the environmental data are available in electronic and published documents from meteorological departments for all of their climatic regions. Specific environmental history data for pavement performance analysis applications are: ambient air temperature (daily maximum, minimum, and average), average wind direction and wind speed (daily maximum, minimum, and average), daily total precipitation, typical monthly solar radiation data, and seasonal subsurface water content changes.

In addition to the meteorological data, section-specific data contributing to environmental related deterioration are equally important. These include surface drainage, subsurface water content changes, frost depth, freeze–thaw information, and pavement temperature. Pavement temperature affects the properties of the asphalt layer and the movement of concrete slabs. Pavement temperatures can be calculated from the meteorological data and thermal properties of the surface pavement layer (Uddin and Chung, 1997). Initial water content and seasonal water content changes in granular layers and subgrade soils adversely influence resilient modulus properties and pavement performance. Most of these data items can be provided in the environmental database and priority should be assigned to most items for the project-level analysis only. For the network level application only general environmental region categories can suffice in terms of a regional factor because pavements perform differently in arid, temperate, dry-freezing, or wet-freezing climates. The continental U.S. has been divided in several environmental regions, which helps to group pavement performance data in separate pavement and environmental families for reliable performance modeling and thickness design (AASHTO, 1986; Datapave, 2001).

18.6   Pavement Condition Monitoring and Evaluation Technologies

18.6.1   In-Service Monitoring and Evaluation Approaches

Monitoring of the physical condition of pavement assets is conducted by the collection of field inspection data, in-service testing data, and traffic history data. Evaluation is the process of analyzing, interpreting, and assessing the collected data to determine its current condition state and rate of deterioration. It is the function of pavement monitoring and evaluation in a PMS to measure pavement condition periodically for:

   (a) updating performance predictions and overall network condition,
   (b) assessing structural integrity and predicting risk of failures,
   (c) checking and improving the performance prediction models, developing new models
   (d) evaluating and improving construction and maintenance techniques, and
   (e) improving and scheduling M,R&R programs as indicated by these updated predictions.

Table 18.1 describes several pavement monitoring data items and evaluation of pertinent attributes, which influence pavement condition and overall performance (Hudson et al., 1997). The description of condition monitoring data collection technologies and their evolution have been documented extensively in literature (Epps and Monismith, 1986; Hudson et al., 1987a, 1987b, 1987c, 1997; Janoff, 1988; FHWA, 1990; Haas et al., 1994), and are discussed in detail in the following chapters.

The approach for pavement condition monitoring and evaluation depends upon the "definition" of pavement failure with respect to the user perception, the functions of safety and sufficiency, and the structural integrity and physical condition. The performance indicators based on these approaches of pavement evaluation dictate the calculations of pavement life estimates and M,R&R needs.

TABLE 18.1 Concepts for In-Service Monitoring and Evaluation of Pavements (After Hudson et al., 1997)

Monitoring	Evaluation
1. Longitudinal roughness	Serviceability
2. Surface distress and defects (cracking, deformation, patches, disintegration, surface defects)	Deterioration, overall composite condition index, and maintenance needs
3. Deflection testing	Material properties and structural capacity
4. Skid resistance or surface friction	Safety against skidding
5. Ride quality	User evaluation of overall pavement quality
6. Appearance and noise	Aesthetics, noise reduction control strategy
7. Traffic	Performance and remaining life
8. Costs (construction, maintenance, user)	Unit cost summaries for economic evaluation
9. Location reference, geometric and structure data, longitudinal and cross fall deficiency, coring for layer thickness, video log	Verification of inventory database, inputs for structural evaluation, safety against hydroplaning potential
10. Environment (climate, pavement temperature, drainage, water below surface, freeze or thaw)	Material degradation, distress and defects progression, structural integrity, performance

The evaluation of pavements can involve one or more of the following: structural capacity, physical deterioration or distress, user-related factors (such as riding comfort), safety and appearance, and user-related costs and benefits associated with varying serviceability and with various rehabilitation measures. Since the 1980s many advanced pavement evaluation technologies have been introduced for increased productivity and reliability (Hudson and Uddin, 1987). Several evaluation methods with examples of application to evaluate various types of infrastructure including pavements are described by Hudson et al. (1997). The common pavement evaluation approaches are further discussed here.

18.6.2   User Evaluation

Common factors in the user evaluation are: riding quality and comfort, safety, aesthetics aspects, and overall satisfaction with the quality of service. These are subjective in nature and often depend on visual observations and opinions formed during a normal ride on the pavement. A numeric subjective rating scale is used, ranging from zero (the worst or failed condition) to a maximum of 5, 10, or 100 (the excellent condition).

Serviceability, PSR: The primary operating characteristic of a pavement is the quality of service it provides to the user, both today and in the future. It is important to (a) measure or evaluate this level of service to establish the current status of a pavement, and (b) predict the change of level of service in the future, for either an existing pavement or for pavements to be constructed. Pavement design systems involve determination of the pavement thickness required to hold certain computed stresses or strains below some specified levels. It is known that cracks will occur if the pavement is overstressed, but not much information was available prior to the time of the AASHO Road Test (AASHO, 1962) to relate such cracks to functional behavior. Serviceability is defined relative to the purpose for which the pavement is constructed, that is, to give a smooth, comfortable, and safe ride. In other words, the measurement should relate explicitly to the user, who is influenced by several attributes of the pavement. For the case of a highway pavement, the users are the traveling public. The PSR scale (Figure 18.1) has been correlated with the objective measurement of longitudinal roughness, cracking, and rutting distress. The resulting serviceability indicator is termed PSI.

Riding Comfort: The evaluation of riding quality is a complex problem (Janoff, 1988), depending on: (a) the pavement, (b) the vehicle and the pavement roughness, and (c) interactions among the first two. The PSR scale (0 to 5) of serviceability, defined in Figure 18.1, represents a subjective rating of the driver. If a roughness equipment is not available then raters can be trained and riding comfort criteria used to estimate PSR, as adapted for Dubai PMS (Uddin, 1992; Uddin and Al-Tayer, 1993).

18.6.3 Functional Evaluation

The functional evaluation of a pavement is the effectiveness in fulfilling its intended function of good serviceability shown by "smooth ride" without bumps and high noise level. The functional evaluation is often similar to the user evaluation. Major concerns are safety, sufficiency based on capacity and demand, serviceability or quality of service, noise pollution, and physical appearance. The use of objective measurements are preferred because of their better repeatability and higher productivity. Examples are: PSI based on pavement roughness measurements, insufficient number lanes indicated by poor level-of-service and congestion, hazards and safety problems due to poor drainage on roads, reduced skid resistance, and hydroplaning (aquaplaning). Where possible, the condition or performance indices should be based on objective measurements of selected performance indicators and an analysis of maintenance records.

Roughness or Unevenness: The longitudinal roughness or unevenness of a pavement is strongly correlated with PSR, as shown by the development of the PSI scale (0 to 5) at the AASHO Road Test. Therefore, longitudinal roughness provides an objective measurement of the serviceability. Roughness is defined as "irregularities of the pavement surface which impact undesirable vehicle accelerations and forces to the vehicle and to its riders, contributing to an undesirable, uneconomical, unsafe, or uncomfortable ride." Equation 18.1 shows the functional form.

$$\text{Roughness} = f(\text{Road profile, speed, vehicle parameters}) \tag{18.1}$$

Response-type roughness measurement devices operate at 80 Km/h and need regular calibration on selected pavement sections with known roughness values. The measurement of longitudinal profiles in the two wheel paths using a road profiling device provides the best sampling of roadway surface roughness. The IRI in m/Km (Paterson, 1986; Sayers et al., 1986) is currently the most acceptable and used unit of roughness measurements. The following equation (Equation 18.2) relates IRI to PSR (Paterson and Attoh-Okine, 1992).

$$\text{IRI} = 5.5 \log_n(5/\text{PSR}) \tag{18.2}$$

Serviceability and Roughness Measurement Equipment: The serviceability of a pavement is largely a function of its roughness. There are several methods for measuring roughness and serviceability (Hudson et al., 1987c), which fall into three broad categories.

- Measuring riding comfort index by subjective panel rating procedure, such as PSR.
- Measuring pavement roughness by response-type road roughness meter (RTRRM).
  (1) U.S. Bureau of Public Road type of roughometer (BPR)
  (2) Maysmeter or car road meter that uses accelerometer response
  (3) Rolling Straightedge (RSE), Profilographs
- Measuring pavement roughness by more precise or sophisticated profiling methods.
  (1) CHLOE type profilometer used at the AASHO Road Test
  (2) Precise leveling method for profile determination (LEVEL)
  (3) Surface Dynamic Profilometer (SDP)
  (4) Noncontact type profilometers (laser)
  (5) Law 8300 Roughness Surveyor (ultrasonic sensor)
  (6) Dip-stick Profile
  (7) Fench APL profiler
- Measuring roughness with high-speed multifunction vans using built-in noncontact laser profiling equipment.

Safety: Safety may be evaluated by the skid resistance measurement or surface friction measurement. Empirically, it can be assessed through correlations with texture parameters or determination of the locations with high accident rates. However, this may not be due to only pavement-related factors (such as aggregate wear resistance or very slippery surface asphalt mix) but could, for example, indicate adverse weather conditions, an alignment problems, or human factors reflecting driver's mistakes. Such factors may be included in the PMS, at the discretion of the agency involved. The typical current practice in the U.S. is to use skid resistance as the primary measure of safety related to pavements. Geometric deficiencies coupled with adverse environmental conditions can lead to unsafe pavement conditions, such as hydroplaning. The safety evaluation of pavements based on the skid resistance, friction, and texture measurement is a major concern in European countries (Ergun et al., 2002; Woodward et al., 2002; Benedetto and Angio, 2003).

Skid resistance and Friction Measurement Equipment: The measurement is reported as skid number (SN) at the designated speed or as International friction index (IFI) for friction testing devices. The most widely used method by state highway agencies in the U.S. for measuring skid resistance is the locked-wheel trailer method in accordance with the American Society for Testing and Materials (ASTM) Method E-274 (ASTM, 2003). The result of the test is reported as SN40, if the speed of the test trailer is 40 ml/h (64 Km/h). The test is conducted by measuring the forces obtained with a towed trailer riding on wet pavement, equipped with a standardized tire (ribbed or smooth). The apparatus is brought to the test speed. The surface is wetted and the braking system is actuated to lock the test tire. The resulting friction force acting between the test tire and the pavement surface with the exact speed of the test vehicle are recorded (Murad, 2004). Equation 18.3 can be used to calculate SN, as follows:

$$SN_{40} = \frac{\text{Force to slide the locked tire at 40 mph}}{\text{Effective wheel load}} \times 100 \qquad (18.3)$$

Table 18.2 gives a summary of devices, which are used to measure pavement friction in the U.S.A. and other countries worldwide (Murad, 2004). Texture measurements from noncontact laser road profile surveys have been correlated with the skid resistance. More details are discussed in a later chapter on skid resistance.

Noise: The noise pollution from highway pavements and urban roads may become a discomfort and nuisance if the sound from the tire-pavement interaction exceeds the designated A range in decibel (dBA). Noise or loudness doubles with each increase of 10 dBA. Jointed concrete pavement

TABLE 18.2 Friction and Skid Measuring Devices in the U.S. and Other Countries (Murad, 2004)

Method	Friction/Skid Measuring Device	Test Tire	Country
Locked-wheel	Skid Trailer (ASTM E274)	Ribbed and smooth	Most U.S.A. states, Ontario, Taiwan
	LCPC Skid Trailer	Smooth	France
	Stuttgrater Reibungmesser	Ribbed	Germany, Switzerland
	Skiddometer	Ribbed	Switzerland
	Diagonal Braked Vehicle (DBV)	Various	U.S.A. (NASA)
Fixed slip	Skiddometer BV 11	Ribbed	Sweden, Slovakia
	Griptester	Smooth	United Kingdom, Australia
	Stuttgrater Reibungmesser	Ribbed	Switzerland
	Norsemeter OSCAR	Smooth	Norway
	Dienst Weg-en Waterbouwkunde Friction Tester (DWW)	Smooth	Netherlands
	Komatsu Skid Trailer	Smooth	Japan
Side force	Sideway-Force Coefficient Routine Investigation Machine (SCRIM)	Smooth	Quebec, U.K., France, Hungry, Italy, Spain, Ireland, Portugal
	Stradograph	Smooth	Denmark
	MuMeter	Special	U.K., U.S.A., Norway
	Odoliograph	Smooth	Belgium
Slider	British Pendulum Tester (BPT)	N/A	United Kingdom, U.S.A.
	Dynamic Friction Tester (DFTester)	N/A	Japan

produce more noise than asphalt pavements. Certain asphalt surfaces like open graded friction course and asphalt–rubber surfaces have been successfully used to reduce noise level. Alternatively at project-level analysis, noise barriers can be justified to protect areas neighboring highway corridors. Noise measurement and control strategies are discussed in detail on the earlier chapter on road noise.

18.6.4   Structural Integrity Evaluation

The structural evaluation is conducted to assess structural integrity and load carrying capacity. The structural evaluation can be assessed using (a) direct visual inspection and distress measurement of physical structure, (b) measurement of layer thickness and material strength by physical destructive field coring or sampling and laboratory testing, and (c) nondestructive test methods, which are desirable in most cases. Structural condition indices, based on objective measurements, indicate structural integrity and load carrying capacity.

18.6.5   Distress

Distress data manifest condition deterioration of the pavement, provide clues to the predominant mechanisms associated with the distress, and lead to appropriate corrective maintenance strategies so that the problems do not appear again soon. The identification of various distress types for measurement in a routine pavement condition survey is made on the basis of the experience of the individual agency regarding which distress types are most important. The specific variable recorded and the units in which they are measured will vary from agency to agency. A more manageable list of distress parameters can be prioritized for less time consuming condition surveys at the network level. Comprehensive distress identification manuals and standards have been developed for detailed distress identification and measurements (Hudson and Uddin, 1987; Shahin and Walther, 1990; SHRP, 1993; ASTM, 1999). Detailed procedures of condition surveys and serviceability evaluation are described in the following chapter. The distress data can be divided into three groups:

Cracking: Examples of load induced cracking are: longitudinal, transverses, and alligator (or fatigue cracking in asphalt pavements associated with repeated loads). The environmentally associated block cracking is causes by thermal cycles. The reflection cracking appears regularly spaced transverse cracks on the asphalt layer paved over an old jointed concrete pavement or a strong base stabilized with cement or lime. The cracking in concrete shows as linear (mostly load associated), durability D, map, and shrinkage cracks.

Deformation: Examples of deformation distresses in asphalt pavements are: rutting (associated with poor asphalt mixes and interaction with repeated traffic loads and high ambient temperatures), shoving and bumps (related to mix problems), and depression (caused by the settlement of subgrade soils). The faulting of transverse joints and depression is a common distress in concrete pavements. The lane or shoulder drop-off distress is found in both pavement types, and it can be hazardous on gravelly or unpaved shoulders.

Surface Defects: Examples of surface defects in asphalt pavements are: flushing or bleeding (indicating excessive asphalt content), raveling and weathering (interaction of mix problems and environment), potholes (appearing more in spring–thaw interacting with traffic loads), and patches (indicating the repair of localized areas of distresses or utility cuts and trenches in the pavement). Examples of surface defects in concrete pavements are: joint deterioration, punch-out, corner break, spalling, aggregate pop-outs, and pumping (subsurface water damage and interaction with traffic load repetitions). Other shoulder distresses can be grouped in this category.

Distress Data and Rutting Measurement Equipment: Almost all distresses can be identified at three severity level (low, medium, and high) and the extent of each distress can be classified at three extent levels (low, medium, and high). Distress data analysis to calculate a summary statistics, such as PCI based on distress severity and extent measurements (Shahin and Walther, 1990), is discussed in the next section on pavement condition deterioration and performance modeling. Distress data can be monitored by manual visual inspections or with the help of less intrusive photographic and video recording in the field followed by data interpretation in the office.

Visual Inspection: Visual inspection and observation is the most common method of condition monitoring. However, manual visual inspections are labor intensive, expensive and subject to inspector's judgments even if the survey is conducted from the windshield of a slow moving van. However, in some emergency cases, the manual inspection is necessary, such as catastrophes (hurricane, floods) and postearthquake evaluation. The visual procedures of pavement distress identification has been standardized in the U.S. by the Army and by SHRP for LTPP monitoring (Shahin and Walther, 1990; SHRP, 1993). The cost of data collection can be reduced through sampling for a small network size.

Photographic and Optical Methods: Photographic and optical methods include video logging digital photography, 35 mm photologging, borescope inspections, and other optical methods with permanent records (Hudson et al., 1987a, 1987b). The records may be obtained from selected positions, slow-walk speed or high-speed dedicated vehicles. Photo records must be read or interpreted to produce data. Automated pattern recognition techniques can identify and quantify data but are still in the development stage. Video based methods are being used successfully for traffic studies. Pavement distress data collection costs and time can be saved for larger highway networks by high-speed video logging vans followed by the data interpretation in the office on sampled sections (for example one 100 m section each 1 Km centerline length). This technology has been adapted for network level PMS by most highway agencies in the U.S. including the Mississippi Department of Transportation (George et al., 1994). Video records of the right-of-way at the same time provides a useful source of information for verifying or making inventory of signs and other appurtenances, managing highway asset, and identifying corridor problems.

18.6.6 Nondestructive Testing and Evaluation

Nondestructive testing provides response measurements, which are used to infer the information about the physical structure of the pavement from behavioral evaluations. It should be remembered that the nondestructive testing techniques evaluate only the response of the pavement and not the physical properties directly.

Deflection Measurement Equipment: Load-deflection testing of all types, including plate load tests for static deflection measurements, slow moving wheel load using the Benkelman Beam and deflectograph, and dynamic deflection measurements using the Dynaflect and falling weight deflectometer (FWD) fall into this category. Most of these devices stop at the test location, make measurements, and then driven to the next test location. The continuously driven deflection testing equipment, such as the Curviameter, has been used extensively in France for the network level application on asphalt pavements (Hudson et al., 1987b).

Deflection data can be used to establish homogeneous pavement design sections from a plot of the last sensor along the length of the pavement. The peak deflection bowl and thickness information can be analyzed to back-calculate Young's modulus of different pavement layers and subgrade using the static layered linear elastic theory. This is the project-level analysis. Deflection data are not used for structural evaluation at the network level (Uddin and Torres-Verdin, 1998). In some cases more simplified indicators from deflection data are used for the network level, such as a structurally adequate (not needing structural maintenance or rehabilitation) or inadequate (needing detailed deflection testing and analysis) pavement section based on a limiting maximum deflection criterion. The analysis of nondestructive deflection data for load carrying capacity calculations at the project-level is discussed in detail in the chapter on structural evaluation.

Other NDE Methods: The structural integrity of pavements can also be checked using other nondestructive evaluation (NDE) methods which do not subject the pavement to actual loading (such as heavy load deflection testing) or destructive testing. These NDE methods include seismic evaluation (such as the wave propagation), vibration methods (such as the modal analysis), acoustic methods, electromagnetic method, and electrical resistivity methods (Hudson et al., 1997). Other noncontact and nondestructive testing technologies for the structural evaluation include the ground penetrating radar (GPR), infrared thermography, high-speed video and related optical methods, Moire technique, and ultrasonic sensors. Noncontact GPR data is used for determining surface layer thickness. A van-mounted GPR equipment has been used successfully to evaluate pavement surface layer thickness nondestructively for several kilometers a day. The infrared thermography and GPR have been used for the detection of voids under concrete pavements (Uddin et al., 1996).

18.6.7 Destructive Testing and Evaluation

For the diagnostic evaluation of distressed or failed pavement sections at the project-level, it occasionally becomes necessary to conduct *destructive testing* by removing portions of the pavement to identify the layers where problems are occurring and why. This also gives an opportunity to collect material samples from different pavement layers and subgrade. Subsequently, the pavement is repaired or replaced. For example, destructive testing has been used on the AASHO Road Test (AASHO, 1962) and the SHRP-LTPP study (SHRP, 1989a, 1989b). Destructive testing techniques include coring in bound layers, boring in soft layers, and Dynamic Cone Penetrometer (DCP) testing in subgrade soils (Uddin, 2002b). The process also involves removing samples of the various layers, examining the samples in the field, and then testing them in the laboratory. Only a limited number of samples are tested in the laboratory (for example, resilient modulus test of subgrade soils) because of the time and cost involved in the destructive testing. Then the results are inferred to the remaining units or used as independent tests to verify and validate the results of deflection testing.

18.6.8 Other Monitoring Data and Related Evaluation

Other monitoring data used for the performance prediction include traffic history and environmental parameters, which are discussed in an earlier section. General automated records for weather and climate data in most cities and regions are available from NOAA on the Internet for most of the U.S. Similarly, in most countries the environmental data are available in electronic and published documents from meteorological departments. Traffic history, loading, axle configuration, and volume data for roads and bridges are sometimes collected and stored in the central database of a highway agency.

18.6.9 Automated Geometric Data Collection

Automated measurements of geometrical characteristics can be cost-effective for the network level evaluation, for example, longitudinal grade and cross fall deficiencies on paved surfaces and slabs (Hudson et al., 1987b). Such data collection can be especially useful and efficient on highways, interchanges, and intersections. The cross fall deficiency coupled with frequent rain and poor drainage may lead to hydroplaning, which is a serious safety hazard as discussed earlier.

18.6.10 Cost Data Collection

Automated Cost models for the PMS budget analysis should be made using the local cost data, therefore, their collection from past projects becomes important so that appropriate cost models or cost tables can be developed for more precise M,R&R analysis and work programs. This is also an important feedback activity in the overall DSS framework for a PMS, as shown in Figure 18.6.

Agency Costs: Construction and M,R&R costs are not measured as a part of the pavement monitoring and evaluation program. Each activity area is charged with recording costs incurred in carrying out its own specific functions. Routine maintenance costs, for example, are reported by the maintenance division.

User Costs: User costs can be very high on rough and deteriorated pavements. It is calculated from special VOC models as a function of pavement condition and other vehicle parameters (Zaniewski et al., 1982; Uddin, 2002a). Accident costs, travel time costs, and discomfort costs are also used to compare alternatives.

Societal Costs: Additionally, societal costs can be enormous resulting from the damage to environment by polluted surface run off and public health problems of respiratory and lung diseases caused by vehicular emissions and associated ground-level Ozone pollution (Uddin, 2003a; Boriboonsomsin, 2004). These costs need to be considered for the evaluation of future transportation projects, as the number of mega cities and urban areas is increasing throughout the developing world.

18.6.11 Selection of Monitoring Sections and Data Collection Frequency

The selection of monitoring sections for periodic pavement evaluations and the frequency of measurements involve a number of considerations.

(a) type of facility (i.e., highway type like freeway, local road, etc.),
(b) type of measurement (i.e., roughness, skid resistance, deflection, etc.),
(c) purpose of measurement (i.e., for detailed project-level evaluation, for mass inventory at the network level, etc.),
(d) users of evaluation information (design, maintenance or construction people, administrators, researchers),
(e) resources of the agency,
(f) age and condition of the section,
(g) ecologic and topographic features of the area,
(h) traffic and geometric conditions, and
(i) maintenance and rehabilitation history.

A monitoring section should be relatively homogeneous over its length and with respect to traffic, pavement type, and geometrics. The following considerations provide the basis for selecting boundaries of a section.

(a) beginning and end of original construction contract,
(b) intersection with another major facility, for a major change in traffic volume, or pavement types (for example, high quality pavement and pavement surface type),
(c) beginning or end of maintenance district or county for a highway facility
(d) section beginning or end,
(e) major change in subgrade soil type or drainage characteristics, and
(f) change in pavement structure (thickness and surface type).

The selection of the frequency of measurements depends upon the size of the network and agency resources. It is not possible to develop absolute standards for the frequency with which evaluation measurements should be taken. Nevertheless, it is possible to develop some general guidelines related to mass inventory evaluation based on available resources and data processing speed. For example, ideally for the network level purpose, major roads should be evaluated at least once in three years. The initial first time monitoring program should be implemented on a small pilot sample of the network to gain experience and estimate of productivity before proceeding to the full network survey. The cost of data collection can be reduced through sampling for a small network size, such as Dubai, where the manual distress data collection frequency was 35% each year enabling a complete survey of all highways and major roads within 3 years, and 20% each year for local roads with one cycle of distress data targeted within 5 years (Uddin, 1992).

A benefit–cost study is helpful to determine the possible use of service providers for some parts of the monitoring program, especially the distress survey. A multifunction high-speed video based distress survey equipment may cost about one-half million dollars, but the annual service contract may be a cost-effective alternative for most agencies. It is imperative that evaluation measurements be properly indexed by section and subsection for efficient data management. Sections should be indexed by geographical coordinates using GPS receivers, contract number, facility number and station, and offset from a landmark. Measurements should be indexed by date, or subsection as a whole, or by graphic coordinates, or station within subsection (where precise location is desired). It is highly desirable that monitoring and evaluation sections and subsections be properly located and indexed so they are completely compatible with design, construction, and maintenance.

Condition assessment immediately after construction should be included in the evaluation database because the constructed quality affects the pavement life. This consideration is important for developing countries where the construction of new highways is a major part of transportation infrastructure as conceived in the PMS work plan for Dubai in Figure 18.6. The rate of physical deterioration is often a critical element of in-service evaluation, and it is important to keep records that are consistent from year to year. Database entries on usage should be updated regularly as new data becomes available and in-service monitoring and evaluation database files should be maintained by survey date so that all historical evaluation data files are readily accessible for M,R&R needs assessment at the network level, improvement of performance prediction models, and other project-level applications.

18.6.12   Selection of Pavement Evaluation Equipment

The selection of a specific technology and equipment for each monitoring data category depends on the relationship with the PMS network level analysis software specifications, desired data quality, details, collection speed and productivity, network size, and available budget. The equipment can be grouped into:

- Distress and rutting evaluation using manual visual survey, wind-shield surveys, high-speed photographic and video recording
- Stand-alone equipment for serviceability and roughness, skid resistance and friction, noise, and deflection testing
- Multipurpose automated high-speed vans to collect distress, rutting, roughness, geometrics within a single pass at highway speed

Several examples and capabilities of the above equipment have been discussed in earlier sections. A detailed review of equipment is presented in the final report of COST 325 study in Europe (COST, 1997). Most multifunction equipments offer the productivity, cost-effectiveness, and fast data processing for network level pavement evaluation applications (Hudson et al., 1997).

18.7   Pavement Condition Deterioration and Performance Models

Highway and road pavements do not last 20 to 25 years without heavy maintenance and overlays because of the adverse effect of traffic loads, environmental degradation, and interaction of load and environment. Therefore, periodic condition monitoring and evaluation conducted during the PMS implementation provides the basis to select the candidate projects in need of maintenance and rehabilitation. Figure 18.8 illustrates the evaluation history trend plots of several pavement performance indictors and the associated decisions criteria for reaching a minimum acceptable limit or "failure" status, which can then be used as a trigger for maintenance or rehabilitation interventions (Haas et al., 1994). The condition monitoring and evaluation data are summarized into simple statistics and used in conformance to an appropriate network level PMS analysis software to support the objectives of a PMS. For example the PAVER procedure is based only on summarizing the distress measurements for the PCI calculation, which is used for the network level analysis of single year M,R&R program

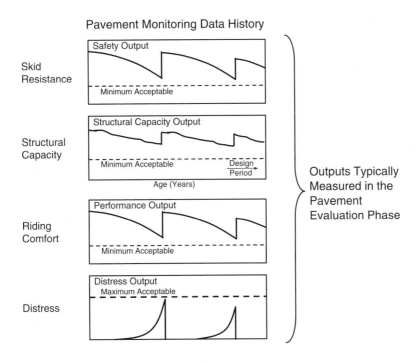

FIGURE 18.8 Typical pavement monitoring data and evaluation histories, after (Haas et al., 1994).

and multiyear budget. Most highway agencies in the U.S. compute a composite condition rating based on roughness or serviceability and distress data for M,R&R analysis at the network level. Each agency formulates specific decision criteria and the weighing factors used in the analysis based on its own preferences to reflect the goals defined by the agency. Pavement safety evaluation based on skid resistance and friction is a major concern in U.K. because of predominantly wet weather and slippery wet pavement (Woodward et al., 2002). Additionally, a complete life-cycle analysis needs performance models to estimate the effectiveness of M,R&R strategies in the immediate improvement of pavement condition and prediction of deterioration due to the time and traffic.

18.7.1 Calculation of Condition Index

Distress surveys are carried out to assess the degree of physical pavement deterioration, which is a function of;

- Type of distress
- Severity of distress
- Extent of distress (amount or density of distress)

Each of the above three characteristics of pavement distress has a significant influence on the determination of the overall pavement deterioration. Because there are many types of distresses and a variety of ways to define severity levels and extent measurement, it is important to use or adapt standard procedures for distress identification and measurements of extent at each severity level. For the practical and meaningful performance evaluation of a network, most distress data are combined into an overall condition index. Washington state PMS first used the concept of deduct values for each measured distress from a perfect score of 100 for an excellent pavement with no distress, the riding quality was calculated in a similar way, and a combined index of Pavement Condition Rating (PCR) was established (LeClerc and Nelson, 1982). The condition index data from the PMS database can be used to develop performance models for different environmental regions and different functional class of pavements, which are more reliable for a specific region and reflect the effect of pavement construction and maintenance practices.

The PAVER distress survey procedure combines the effect of various distress types and measurements of distress severity and extent into a single index, PCI to evaluate overall pavement condition of the surveyed section. PCI varies between 0 to 100. A value of 100 implies the pavement is in excellent condition and zero means a failed pavement (Shahin and Walther, 1990). The detailed calculation of PCI using distress data measured by the PAVER procedure is described here because the network level PAVER analysis software has been implemented successfully in the U.S. and abroad for highway networks, urban and rural road networks, local road network, and airport pavements. A new ASTM standard has evolved from this method (ASTM, 1999). The visual distress survey methodology and PCI calculation have been adapted by many agencies and service providers for distress data measured by wind-shield surveys and interpreted in the office using video records. Following steps are used to determine the PCI of a pavement section using the PAVER procedure and the deduct value curves developed for asphalt and concrete pavements (Shahin and Walther, 1990).

1. Divide pavement section into sample units.
2. Inspect sample units. Determine distress types and severity level; record extent and compute density.
3. Determine deduct value for each distress type and severity, for example a (for alligator cracking) and b (for longitudinal and transverse cracking). Distress density must first be computed to use the deduct curves. In general, the density is the amount of distress (extent) divided by the sample unit area (for asphalt pavement) or the number of slabs in the sample unit (for jointed concrete pavement). A deduct value is a number from 0 to 100 with a zero indicating the distress has no impact on pavement condition and 100 indicating an extremely serious distress which causes the pavement to fail.
4. Sum all individual deduct values to compute total deduct value (TDV) = $a + b$
5. Use correction curves to determine the corrected deduct value (CDV) from the TDV. The correction curves are for different q (number of entries with deduct value over five points), which allow for the proper summation of individual distress deduct values when more than one distress type is observed.
6. Adjust deduct value by using correction chart. The result is CDV. This is needed to ensure the total deduct value not to exceed 100 if the pavement is badly deteriorated.
7. Compute PCI using Equation 18.4 for each sample unit inspected.

$$PCI = 100 - CDV \qquad (18.4)$$

8. Compute PCI of entire section (average of all PCI values).
9. Determine pavement section condition rating using the following PCI and rating scale.

PCI:	0–10	11–25	26–40	41–55	56–70	71–85	86–100
Rating:	Failed	Very poor	Poor	Fair	Good	Very good	Excellent

18.7.2 Methodologies for Performance Modeling

The slope and form of the condition deterioration history determines the performance. Figure 18.5 shows the PCI performance curve and rating plot and the effect of delays in M,R&R work on increased costs. Traditional pavement design practices based on a preselected service life of 20 years did not consider the effect of environment and interaction with traffic loads on performance as shown in Figures 18.1 and 18.5. The PMS database provides good opportunities to use the monitoring historical data to check and calibrate the existing models, improve performance prediction models, and develop new models that apply to local conditions of traffic, subgrade soils, pavement materials, and environment. Performance data and models are essential to predict future conditions and future M,R&R

needs for life-cycle cost (LCC) analysis, and enhance decision-making both at network- and project-levels. For network level purposes, simplified time-series models may be sufficient, whereas on a project-level basis a more comprehensive set of data on materials properties, traffic loads, and environment factors may be needed to develop a sufficiently accurate model. Performance modeling can be accomplished using the following modeling methodologies.

18.7.3 Regression Analysis Techniques

Multiple linear regression analysis techniques are very well established (Draper and Smith, 1981), which use several statistically significant explanatory variables to model the historical variations of the dependent variable (PSI, PCI, condition index or other performance measures). A scatter-plot of the historical data should be reviewed first to estimate the model shape. The analysis of variance (ANOVA) should be used to identify statistically significant independent variables. The statistical regression analysis develops empirical performance models by estimating parameters and coefficients of independent or explanatory variables in deterministic mathematical equations that can explain most of the variations in the dependent variable. The multiple regression model may include linear and nonlinear terms of transformed variables. The goodness of the fitted model is evaluated with a desired high value of Pearson correlation coefficient R ranging from $(-)$ one to $(+)$ one between observed and predicted values of the dependent variable (y), and a low value of root mean square error (RMSE) that measures the spread of the residuals (or errors between observed y and predicted y) about the fitted regression line. Some examples of deterministic empirical models are:

- The AASHO Road Test data were analyzed by multiple regression analysis and produced the first performance models to predict number of accumulated ESAL applications as a function of, drop in serviceability, layer material properties and thicknesses, subgrade strength, and environmental factors (AASHO, 1962). These models are more applicable for project-level design.
- The HDM III performance models predict incremental progression of roughness and several distresses on asphalt pavements such as, cracking, and rutting, etc. (Watanatada et al., 1987). The independent variables include present distress level, subgrade strength, environmental factor, traffic load, and time. These models have been used for project-level design and have been adapted by many investigators for network level analysis (Ferreira et al., 2003).
- Nonlinear power function regression models were developed for different pavement types from the State of Washington PMS database in the U.S. for network level applications (LeClerc and Nelson, 1982; Jackson and Mahoney, 1990). Some of these regression equations are summarized by Haas et al. (1994). The model for predicting PCR is defined in Equation 18.5 and examples are shown in Figure 18.9 (FHWA, 1990).

$$PCR = C - mA^P \qquad (18.5)$$

where:

$C = 100$.
$A =$ pavement age in years.
$m =$ slope coefficient.
$P =$ constant controlling the shape of the curve.

- Nonlinear sigmoidal (S-shaped) power function used for regression models for different pavement types, used in Canada (Hajek et al., 1985) and several other studies, is defined as:

$$PCI = PCI_0 - c\, e^{-[\alpha/A]\beta} \qquad (18.6)$$

where:

PCI = Pavement Condition Index at any time defined as a function of Riding Comfort Index (RCI) at a scale of 0–10 and a composite distress index.

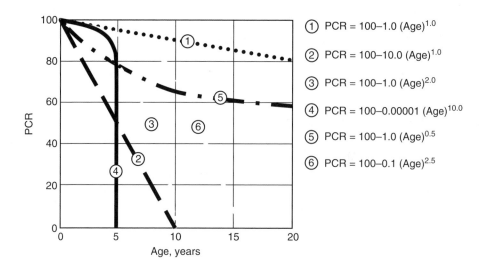

FIGURE 18.9   Examples of performance model curve shapes, after (FHWA, 1990).

PCI_0 = PCI at time zero or initial PCI.
   A = pavement age in years.
   c = nonlinear regression coefficient.
   α, β = nonlinear regression coefficients.
- Pavement damage function equations for the FHWA cost allocation study were developed using the regression technique and nonlinear sigmoidal model (Rauhut et al., 1982).

Other examples are mechanistic-empirical models, where a calculated or measured pavement response is included in the deterministic equations derived from the regression analysis. These models are applicable to project-level design because of the complexity of response calculations. Some examples are: (a) roughness model involving surface deflection, tensile strain in asphalt layer, and cumulative ESALs (Queiroz, 1983) and (b) OPAC design model involving roughness and surface deflection (Jung and Phang, 1974; Kher and Phang, 1977).

   To ensure accuracy and reliability of predictions, the empirical performance models developed from regional (for example Washington State in the U.S.), national (the AASHO Road test and LTPP in the U.S.), and international (The World Bank's HDM) studies, must be verified and the coefficients calibrated using local performance data. Regression modeling has its limitations, especially if the scatter-plot does not show a known model shape, if many different independent variables influence the dependent variable, or if the goodness of fit is low (low R values). Many new analysis and modeling techniques have emerged during the last decade to develop empirical models in these situations, which are reviewed in the literature (Hudson and Haas, 1994; Flintsch, 2003). A knowledge based expert system is an artificial intelligence technology. This technology is suited better for maintenance management decision-making and is discussed later. Other approaches are briefly discussed here.

18.7.4   Artificial Neural Networks Modeling

Artificial Neural Network (ANN) technology has gained considerable popularity for empirical modeling using parallel computations for knowledge representation and information processing. Because of their fundamental processing unit similarity to that of the human brain, ANNs have some unique, human-like capabilities in information processing. ANNs computing systems are made up of several simple and highly interconnected elements, which process information by dynamic response to external inputs.

The ANN model does not execute a series of fixed instructions like a traditional computer program or statistical analysis; rather it responds, in parallel, to the inputs presented to it during a training period. The major variables of a neural network are: network topology (the number of nodes and their connectivity), rules of computation of the activations of the processing units, rules of propagation, and the rules of self-organization and learning (Ghaboussi, 1992). One of the common types of rules involves back-propagation networks. The processing units in back-propagation networks are arranged in layers. Each neural network has an input layer, an output layer and a number of hidden layers. A neural network performs "computations" by propagating changes in activation amongst the processors. Usually a binary threshold or a sigmoid function is used as the activation or transfer function. The neural network gains its knowledge through training. Once a neural network processes all of its training data and achieves a state of equilibrium, new input data can be processed for evaluation. ANN models are capable of learning complex, highly nonlinear relationships and associations from a large body of data, such as PMS databases. Detailed discussions on the ANN modeling are provided in literature (Ghaboussi, 1992; Hudson et al., 1997; Attoh-Okine, 1998, 2000).

18.7.5   Probabilistic Performance Modeling

Deterministic regression equations predict only one value and do not provide probabilistic distributions of the expected value. Bayesian and Markov probabilistic modeling approaches are other alternatives, which have been used successfully for network level performance modeling applications.

The Bayesian statistical decision theory allows both subjective data from prior knowledge and experience and objectively obtained actual monitoring data to be combined for developing regression equations. These are used to predict posterior estimates of condition deterioration. In the Bayesian regression approach the model parameters are assumed to be random variables associated with probability distributions (Smith et al., 1979). The Bayesian approach has been used in long-term pavement performance modeling in the Canadian SHRP studies (Haas et al., 1994). The main advantage of using the Bayesian approach, compared to the classical regression analysis, is that a comprehensive historical database is not needed.

Markov transition probability models are particularly useful for network level applications where a historical database and reliable regression equations are not available for performance predictions. They capture experience of engineers or experts in a structured way by using different combinations of pavement classes or situations and condition states. For example, one pavement model class for developing a road deterioration prediction model might be high traffic combined with a minimum thickness structure combined with a strong foundation. The combined boundaries for the variables comprising the classes must be provided (for example, high traffic is greater than 10,000 ADT). The basic assumptions are that *current* pavement condition is only dependent on its preceding *prior* condition state, *future* pavement condition is dependent on the preceding *current* condition state, and the *future* year state is independent of how the pavement acquired a *current* year condition state. A transition probability matrix defines the probabilities that a pavement in its current state will be "transitioning" in some future states. Referring to the vertical PCI scale of Figure 18.5, PCI measures could be in 10 condition states with each state representing 10 points, for example, $PCI_1 = 100$ to 90, $PCI_2 = 89$ to 80, $PCI_3 = 79$ to 70, $PCI_4 = 69$ to 60, $PCI_5 = 59$ to 50, etc. An illustration of transitional probability is that a pavement section in current state 3 (PCI of 70) may have next year a probability of 0.1 or one in 10 chances that it will be in state three (PCI of 70, no change in the condition) or in state four (PCI of 66), and a probability of 0.3 or three in 10 chances that it will be in state five (PCI of 58). The probability state can be based on a preselected interval of one, two, or more years. Many references are available on Markov performance modeling for network level applications (Finn, 1974; Cook and Lytton, 1987; FHWA, 1990; Haas et al., 1994).

Some advantages of the Markov models are: use of the judgment of experienced engineers to develop transitional probabilities for the modeling process, a probability distribution of the expected value of the dependent variable (future condition) indicating sections with different future performance, consideration of performance trends from field observations regardless of nonlinear trends with time, and an easy means to incorporate feedback of field measurements into prediction models. The disadvantages include no guidance to the physical causes for the condition deterioration and no consideration of pavement aging on transitional probabilities (Finn et al., 1974; FHWA, 1990). The primary application of Markov approach is the M,R&R decision-making process at the network level. Depending on the PMS sections and network size, a large-scale linear programing software and computer system may be required to reach an optimal solution for M,R&R priorities. This may be a limitation on the resources.

18.7.6 Network level Performance Models

Markov transitional probability performance models have been developed and implemented for network level applications in the states of Arizona, KS, and Washington in the U.S. (Finn et al., 1974; FHWA, 1990). The condition index and roughness data from the PMS database have been used by numerous agencies to develop empirical performance models, which are more reliable for a specific region and pavement practices. These models are necessary to predict future conditions for use in the LCC analysis and M,R&R programing. Figure 18.9 illustrates some examples of the rate of deterioration for PCR using different values of regression parameters for the nonlinear power function model. Several examples are presented by Molenaar for network level applications of the performance models based on the AASHO Road Test serviceability data, HDM performance models for roughness and cracking or rutting, and skid resistance analysis in the Netherlands (Molenaar, 2003).

   The PAVER and MicroPAVER programs allow the prediction of future pavement conditions at any point in time based on a "pavement family relationship" developed from the measured performance of local pavements. The following description is taken from the MicroPAVER reference (Shahin and Walther, 1990). Initial predictions are based on default relationships until a local PCI database can be established. A pavement family is defined as a group of pavement sections with similar deterioration characteristics, for example all thin asphalt pavement sections with similar traffic volumes. It allows the user to define a family based on a number of factors, such as pavement type, usage, rank, construction dates, PCI, etc. As a database is developed from the PCI measurements, the family relationship becomes more unique to the roadway network, environment, pavement types, subgrade, traffic and other factors that define the deterioration of the pavements. The family prediction function is a fourth-degree polynomial constrained least squared error relationship. Provisions are provided to remove the data that are considered as errors or outliers. The best-fit family curve for the family analysis extends only as far as the available data. Predictions of future conditions are made by extrapolation of the tangent of the slope of the curve for the past few years. The prediction function for a pavement family represents the average behavior of all sections of that family. Comparing the section to the family deterioration provides invaluable feedback on the effects of maintenance, traffic, drainage, and other factors on the pavement behavior. The family curve method allows for continuously updating the deterioration model as more data is incorporated into the database (Shahin and Walther, 1990). Figure 18.10 shows an example of an asphalt pavement family performance curve. Equation 18.7 presents the regression equation developed from the time series PCI data where X is pavement age in years.

$$PCI = 100 - 6.391411X + 1.368907X^2 - 0.1516525X^3 + 0.004647153X^4 \qquad (18.7)$$

18.7.7 Project-Level Performance Models

Traditional pavement design practices have relied on a preselected service life using safety factors in material properties and design traffic volume to take into account uncertainties in design and demand inputs, without the concept of field performance as shown in Figures 18.1, 18.5, and 18.7. This old practice is inadequate because of recent developments in pavement materials and construction methods; for example, the use of polymer-modified asphalt has shown to outperform virgin asphalt pavements

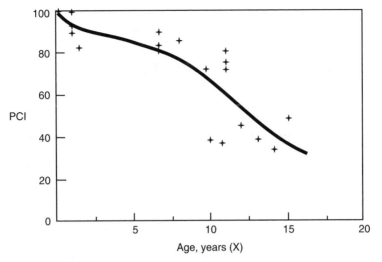

$$PCI = 100 - 6.391411\ X + 1.368907\ X^2 - 0.1516525\ X^3 + 0.004647153\ X^4$$

FIGURE 18.10 A typical pavement family performance curve, after (Shahin and Walther, 1990).

(Uddin and Nanagiri, 2002). Most of the cracking and rutting models being used or developed for mechanistic-based pavement design methods and project-level evaluations have been derived from extensive laboratory testing and accelerated pavement testing (Roberts et al., 2002; Molenaar, 2003). They need to be calibrated using the field performance data because they do not adequately consider material variability, real world environment and traffic loads, and their interaction. Therefore, appropriate performance measures should be identified and monitored so that the historical data can be used to develop and improve performance prediction models. The LTPP database provides this ability for data being collected over 1000 in-service pavement sections of all types including existing and newly constructed pavements in North America (Datapave, 2001). However, in the absence of the LTPP data because of a lack of resources, the PMS database provides the unique opportunity to collect extensive project-level performance monitoring data on selected pavements and use for validating and calibrating new models derived from the laboratory and accelerated pavement testing data.

Some examples using the performance models developed from the study of in-service pavements in Brazil and the U.S. for project-level applications are presented here. A simplified IRI progression model for high quality asphalt paved road developed from the analysis of a database generated from the HDM computations (Paterson and Attoh-Okine, 1992) is:

$$(IRI)_t = 1.04e^{mt}[(IRI)_0 + 263(1 + SNC)^{-5}(CESAL)_t] \qquad (18.8)$$

where:

$(IRI)_t$ = roughness IRI at time t, m/Km,
$(IRI)_0$ = initial roughness IRI at time t (equal to 1.5 for new pavements), m/Km,
t = time since last construction or major maintenance and rehabilitation, years,
m = environmental coefficient which varies between 0.01 to 0.70 (from dry, non-freeze to wet, freeze environment),
SNC = structural number modified by subgrade strength, as used in the HDM III program, and
$(CESAL)_t$ = cumulative equivalent standard axle load (ESAL) applications at time t, in millions.

The model is applicable to the full roughness range of up to 12 m/Km IRI, particularly for maintained pavements with the area of cracking not exceeding 30%. Figure 18.11 compares the condition history on a Mississippi highway in the U.S. using the detailed HDMIII model and the implementation of the

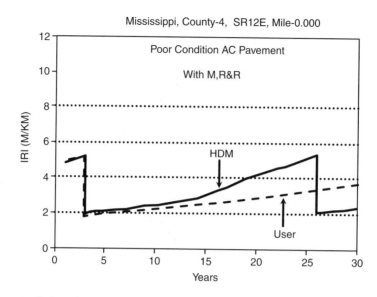

FIGURE 18.11   Prediction of pavement condition deterioration for an asphalt pavement, after (Uddin, 1993).

simplified IRI model version in the USER life-cycle analysis program (Uddin, 1993). The IRI roughness prediction for a poor section (2 Km long, structural number of 3.2) of an asphalt pavement on this highway shows that a lower rate of deterioration is predicted by the USER program, particularly after 15 years. On the other hand, a second overlay is predicted by HDM III in year 26 indicating a slower rate of pavement condition deterioration.

The COPES models developed for jointed concrete pavements in a national study in the U.S. (NCHRP, 1985) have been incorporated in the USER program. These models predict PSR at any time t as a function of cumulative ESALs at time t, pavement strength, and environmental parameters. These models predict PSR at time t, (PSR_t) as a function of initial PSR or (PSR_0), cumulative equivalent standard axle load applications at time t in millions or $(CESAL)_t$, environment parameters and pavement strength. These models predict changes in PSR every year of the analysis period without maintenance and consideration of the effect of a condition-responsive major M,R&R action. The COPES model for jointed reinforced concrete pavements (JRCP) is shown in the generalized form by Equation 18.9 and Equation 18.10 describes the COPES performance model for jointed plain concrete pavements (JPCP).

$(PSR)_t = f[(PSR)_0, (CESAL)_t,$ edge stress and concrete modulus of rupture, dummy variables (for transverse joint spacing, reactive aggregates and base type), freezing index, average annual precipitation, average monthly temperature]   (18.9)

$(PSR)_t = f[(PSR)_0, (CESAL)_t,$ edge stress and concrete modulus of rupture, average annual precipitation, average monthly temperature]   (18.10)

The USER program uses a stepwise linear model (linear rate of deterioration vs. years) for continuously reinforced concrete (CRC) pavements. The default user-defined model is based on Texas performance data (Uddin et al., 1987). These performance models for concrete pavements and the simplified IRI prediction model for asphalt pavements, incorporated in the USER program, have been implemented for network level analysis in Mexico (Uddin and Torres-Verdin, 1998). The input data for these models are easily extracted from the PMS database.

18.8   Pavement LCC Analysis and User Cost Models

18.8.1   LCC Analysis Methodology

The prediction of service life is an important and fundamental aspect of performance modeling. This may or may not coincide with the original estimated design life. Traditionally, 20 years of design life is associated with highway pavements. However, the public expects the highway to serve forever, unless a catastrophic failure occurs or the area is abandoned. In the real world, the pavement may reach the design life in less than 20 years because of accelerated heavy truck traffic volume, which happened in the late 1970s and 1980s on many Interstate highways in the U.S. Decision makers in a highway agency recognize that the highway can not provide adequate service after certain number of years without an appropriate M,R&R intervention because it may become:

(1) structurally unsafe,
(2) functionally obsolete,
(3) congested causing long delays and inconvenience to the public, and
(4) costly to preserve doing only routine or emergency maintain.

This leads to the definition of physical service life within a life-cycle. The same type of pavement (for example, a standard highway asphalt pavement) may have variation in its initial and total service life because of the varying influence of traffic history, environmental inputs, and maintenance practices. The concept of the LCC analysis was not applied in the traditional pavement design procedures until the development of early project-level pavement design systems developed in the 1970s (Haas and Hudson, 1978). Its significance to decision-making process has been recognized in the 1980s for pavement surface type selection (Peterson, 1985) and pavement thickness design guide (AASHTO, 1986). The LCC application for pavement design problems is discussed in an earlier chapter on engineering economy. Only a discussion of LCC topics related to the pavement management process is presented here.

An adequately and timely maintained pavement will have a better chance of extended service life, as compared to a pavement with no or poor maintenance (Figure 18.1). The timing or schedule of these future or long-term M,R,&R interventions is an important aspect of a rational LCC analysis. As shown in Figure 18.1, appropriate timing and implementation of the M,R,&R treatment in strategy 2 leads to better serviceability, performance, and longer lasting pavements. Additionally, it will cost more if maintenance is deferred because the rate of deterioration accelerates (Figure 18.5) when the pavement is not timely maintained, neglected, or lacks agency l funding. Figure 18.3 shows the influence of maintenance on life-cycle costs. Maintenance, rehabilitation, and reconstruction costs may be a large sum and even more than the initial construction over the entire pavement service life. A good PMS recognizes the importance of whole service life analysis including life-cycle agency costs (for construction, maintenance, rehabilitation, and replacement), as well as user costs and benefits. To preserve the pavement at an acceptable level of serviceability, timely condition-responsive M,R&R and preventive maintenance strategies are essential. Preventive maintenance treatments also result in improved performance. This requires appropriate decision criteria as shown in Figure 18.8.

A comprehensive LCC methodology considers several alternatives for resurfacing of existing pavements, pavement restoration, maintenance and rehabilitation cycles, and peripheral activities (Fwa and Sinha, 1991; FHWA, 1998; Uddin, 2002a). The scheduling of competing M,R&R alternatives may be (a) user-inputs or (b) the results of condition-responsive M,R&R policies. For example, the LCC1 microcomputer program (Uddin et al., 1987) incorporates the user-input methodology for life-cycle M,R&R actions and focuses on a comprehensive economic evaluation of competing design strategies. The condition-responsive M,R&R policies require pavement performance models to predict the time(s) and type(s) of M,R,&R treatment during the pavement service life. This is addressed by specifying a desired minimum level of serviceability at which an appropriate M,R&R action is triggered. This is a component of the M,R&R policy set by the agency that includes appropriate M,R&R treatments and their unit cost data., as shown by examples in the next section. The criteria of serviceability threshold will vary from agency to agency and with the functional class of roadway. For example, on urban highways and expressways one will probably use a PSR or PSI value of 2.5 to determine the time to initiate rehabilitation. On rural highways and other low-volume roads, a threshold value of 1.5 will suffice. These minimum acceptable values will vary with what the user public is ready to accept and what the agency can afford. The following recent examples of LCC methodologies incorporate the condition responsive M,R&R interventions, and calculate VOCs and other user costs.

- The World Bank's HDM III (asphalt and gravel roads) and HDM4 (concrete, asphalt, and gravel roads) programs are mostly used for project-level (Watanatada et al., 1987; HDM, 2002). The HDM pavement deterioration models and VOC models have been implemented in many network level applications.
- The USER LCC analysis program (Uddin, 1993, 2002a) incorporates condition deterioration models for asphalt and concrete pavements, as described in the previous section, and the FHWA VOC models (Zaniewski et al., 1982). The methodology has been implemented in the PAM (Planning and Budget Analysis of Maintenance) program for network level analysis of pavement management projects in Mexico (Uddin and Torres-Verdin, 1998).
- The probabilistic LCC analysis methodology, developed in West Virginia in the U.S., considers user costs including VOCs, accident costs during normal operations and work zone hours, and uncertainty in input parameters (Reigle and Zaniewski, 2002).
- The LCCA program, developed in Washington in the U.S., is implemented in the project-level pavement investment decision (PID) program, which accesses condition data from the PMS database and uses deterministic dynamic programing to select M,R&R treatments with maximum user benefits in terms of VOC savings from improved condition (Papagiannakis, 2003).
- The LCC analysis procedure of calculating work zone user costs (FHWA, 1998) is adapted in Japan, which shows that queue delay cost due to congestion is very large compared to VOC costs for an interchange of a national highway in a densely populated district of metropolitan Tokyo (Taniguchi and Yoshida, 2003).

18.8.2   Example of LCC Analysis of Costs and Benefits

18.8.2.1   Agency and User cost Models

Agency costs are those costs directly represented by the agency budget. External or exogenous costs, such as those associated with user delays during M,R&R work and VOC, may be combined with agency costs in a total economic analysis as shown in the FHWA procedure (FHWA, 1998). They may or may not affect agency decision making, but they do not appear in an agency's budget. All types of unit costs and fixed cost components are user inputs. However, default values for these costs components are provided in the LCC analysis computer programs. The major initial and recurring costs over the life-cycle that a public agency may consider in the LCC analysis of project alternatives include the following:

- Initial capital costs of construction
- Future costs of maintenance, rehabilitation, and reconstruction
- Cost of maintenance and protection of traffic
- Salvage return or residual value at the end of the period ("negative cost")
- Engineering and administration costs
- Costs of borrowing (for projects not financed from allocated public funds or toll revenues)

Nonagency costs can involve the road user, or they can be incurred by nonusers. The nonuser costs are associated with the environmental pollution (ground-level ozone pollution, vehicle emissions, noise, visual, etc.) and neighborhood disruptions. Some of the above costs are difficult to quantify. The nonuser costs are societal costs, which are not considered in a routine LCC analysis except when these are mostly critical in urbanized and metropolitan areas with high traffic volume and congestion. The user costs are quantifiable and include the following:

TABLE 18.3 Vehicle Operating Costs, in Dollars Per Vehicle-Mile for all Vehicles Combined Data from FHWA-VOC Report (Zaniewski et al., 1982)

PSI	2-Lane	4-Lane Undivided	4 or more Lanes, Divided	2-Lane	4-Lane, Undivided	4 or more Lanes, Divided	Average VOC	% Increase in VOC[a]
1.5	0.258	0.238	0.227	0.230	0.231	0.233	0.236	32.58
2.0	0.227	0.216	0.213	0.218	0.219	0.219	0.219	23.03
2.5	0.210	0.199	0.199	0.208	0.210	0.211	0.206	15.73
3.0	0.199	0.189	0.189	0.198	0.198	0.198	0.195	9.55
3.5	0.191	0.181	0.181	0.189	0.189	0.189	0.187	5.06
4.0	0.185	0.176	0.177	0.183	0.183	0.183	0.181	1.69
4.5	0.182	0.173	0.173	0.180	0.180	0.180	0.178	0

[a] % increase in VOC from PSI 4.5 to 1.5 PSI (divide by 1.6 to calculate cost per vehicle-Km).

- User cost associated with VOCs
- User costs due to traffic delays in work zones
- User costs for occupancy time on the road and time delays
- Accident costs

18.8.2.2 Vehicle Operating Cost Models

Most agencies do not consider nonagency costs because (a) these do not affect the road work funding and (b) these costs are difficult to quantify objectively. However, reasonably good VOC (VOC) models are available for LCC analysis, as reviewed in the preceding section (Watanatada et al., 1987; Uddin, 1993b, 2002a; FHWA, 1998; Papagiannakis, 2003). Detailed information can be found in the cited references. The USER software incorporates pavement deterioration and performance models described above and the VOC model developed in a study by the FHWA in the U.S. (Zaniewski et al., 1982). The FHWA study produced detailed consumption rate tables for individual VOC components, as well as total VOC tables as functions of

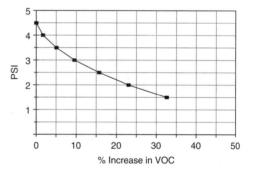

FIGURE 18.12 Increase in VOC as a function of pavement condition deterioration.

the pavement condition. The USER program incorporates these VOC consumption tables which enable the use of site-specific VOC unit cost components. The VOC associated with fuel consumption, oil consumption, tire wear, repair and maintenance of vehicles, depreciation, and pavement condition history over the analysis period are calculated by the built-in VOC consumption rate tables. The consumption rate for each VOC attribute is a function of the vehicle type and yearly volume, constant running speed, speed change cycles, grade and curvature of the road section. The average running speed and consumption rates are adjusted for the prevailing pavement condition. Table 18.3 shows the summary VOC data for different classes of roads at different PSI levels. Figure 18.12 shows % increase in VOC as a function of drop in PSI.

These VOCs are calculated for each analysis year of service life and discounted to present worth. The USER results of VOC show US$ 0.23 per vehicle Km, which is about 20% less than compare reasonably with the HDM III results of US$ 0.29 per vehicle Km for the poor asphalt pavement section described earlier in Figure 18.11 with M,R&R intervention (Uddin, 1993b), and for a good asphalt pavement the USER shows US$ 0.18 per vehicle Km, which is more reasonable than the HDM III results similar to the poor pavement section. In this case the USER program demonstrates that early maintenance will preserve the pavement condition and will lead to lower life-cycle VOC results by 24%. Figure 18.13 shows

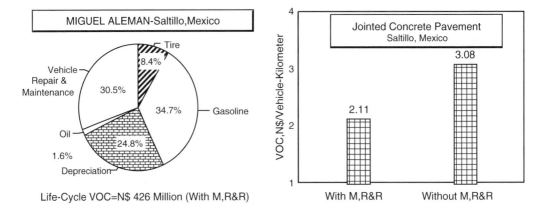

FIGURE 18.13   VOC in Mexican $ per vehicle-Km and distribution of VOC components predicted for jointed concrete pavement sections in Saltillo, Mexico, after (and Torres-Verdin, 1998).

VOC in Mexican $ per vehicle-Km and distribution of VOC components predicted for a jointed concrete pavement section in Saltillo, Mexico as a part of the network level PMS analysis (Uddin and Torres-Verdin, 1998).

18.8.2.3   Benefits

The benefits of a pavement design or M,R&R alternative accrue primarily from direct reductions in user costs, mainly in life-cycle VOC. It is also possible to consider benefits in terms of additional taxes or total income generated by a project, if a road user fee or toll is planned for the project. Other examples of benefits are user savings because of a reduction in travel time cost and a reduction in accident costs. The benefit in the USER analysis is calculated as a sum of reductions in life-cycle VOC, travel time saving, and reduction in accident costs due to the improved condition of road and traffic congestion, compared to a base alternative (Uddin and Torres-Verdin, 1998). The calculation of life-cycle benefit–cost ratio is calculated and used in priority ranking of candidate projects for M,R&R programing.

18.8.2.4   Economic Evaluation

The economic evaluation of competing M,R&R alternatives is based on discounting the life-cycle costs and benefits to the present value of dollars. The economic analysis has no relationship to the financing of a project. Such financing considerations can either limit the number of feasible projects (on a network level basis) or limit the amount available for a particular project. The period of analysis should be chosen so that it is possible to forecast the factors involved in the analysis with a reasonable degree of reliability. A general guideline for selecting the length of an analysis period is that it should not extend beyond the period of reliable forecasts (Hudson et al., 1997). For short-term or high-uncertainty M,R&R projects, 10 years could be an upper limit. For other long-term M,R&R projects a range of 20 to 30 years is reasonable. Forty years or more may not be unreasonable; the present worth (PW) of costs or benefits at future times beyond 40 years may not be significant, depending on the discount rate used.

PW analysis is the best and most widely accepted method for LCC evaluations. The PW and equivalent uniform annual cost (EUAC) or simply annual cost methods are directly comparable. Using the PW method, the analysis period should be the same for all alternative design strategies. The PW method involves the discounting of all future sums to the present, using a selected discount rate and ignoring inflation. The PW factor for discounting either a future single cost or benefit is calculated by Equation 18.11. It answers the following question. How much future sum F, "invested" (n) years from now, is worth if "invested" today at a discount rate (interest rate) of $i\%$?

$$P = F\left[\frac{1}{(1+i)^n}\right]$$ (18.11)

Equation 18.12 is used to calculate the PW of a series of equal pavements. It answers: How much a series of payments of A, "invested" at the end of each of (n) periods at a discount rate (i), is worth today?

$$P = A\left[\frac{(1+i)^n - 1}{i(1 = i)^n}\right] \tag{18.12}$$

Using the time value of money, the different costs occurring through the lifetime or analysis period of a project can be transferred to an equivalent cost at some reference year. For example, the present value of a pavement project will last (n) years, with an initial cost C, a yearly maintenance cost M, and a salvage value S is equal to:

$$P = C + M\left[\frac{(1+i)^n - 1}{i(1+i)^n}\right] - S\left[\frac{1}{(1+i)^n}\right] \tag{18.13}$$

Example of LCC Analysis: Calculate the PW of a pavement section using Equation 18.13 where the following investments need to be made:

(a) Initial construction cost	$1,000,000
(b) Maintenance costs	$3,000/per year
(c) Salvage value	$200,000

Assume an analysis period and service life of 10 years and a discount rate of 5%.

$$P = 1,000,000 + 3,000(7.72217) - 200,000(0.6139)$$
$$P = 1,000,000 + 23,165 - 122,780$$
$$P = 900,385 \text{ (PW of all life-cycle agency costs)}$$

Similarly, PW of user costs and associate benefits for an alternative strategy are calculated. By comparing the life-cycle costs and benefits of alternatives with the base "do nothing" case, the alternatives can be ranked and the most economical alternative selected. Many agencies in the U.S. prefer a comparison of benefit–cost ratios. The "rate of return" method also considers both costs and benefits and determines the discount rate at which the discounted costs and benefits for the alternative are equal. Projects funded by the World Bank and other international institutions often require economic evaluations and comparisons of alternatives based on the "rate of return" method.

18.9 Maintenance, Rehabilitation & Reconstruction (M,R&R) Policies and Analysis

After the analysis of condition data for all pavement sections in the network with the required inventory and condition data in the database, the next step is the selection of maintenance, rehabilitation and reconstruction alternatives and the analysis of related costs for all candidate sections in the need of M,R&R. An appropriate M,R&R analysis methodology is incorporated in the PMS network analysis software, which is the most critical component of a framework for the PMS development and implementation at the network level. The results of the M,R&R analysis and priority ranking of the candidate sections are then used by the PMS analysis software to generate reports of annual and multiyear maintenance work programs. Therefore, this section discusses the following key database tables and steps involved in the M,R&R analysis and ends up with specific guidelines to select an appropriate PMS analysis software for the network level pavement management.

- Decision criteria to trigger minor or routine maintenance, short-term major M,R&R, and long-term M,R&R (requiring condition data summaries)

- Policies and criteria for selection of minor or routine maintenance, short-term major M,R&R, and long-term M,R&R alternatives
- Maintenance treatment catalogues (type of maintenance, unit costs, effect on pavement condition improvement after M,R&R)
- Pavement condition deterioration or performance models
- Global decision criteria (analysis period, discount rate, inflation rate, M,R&R analysis year, budget constraints)
- Input data for user cost analysis (VOC input data on vehicle fleet, vehicle speed, traffic delay cost, travel time and accident cost data, and other related input data)

The last two types of PMS database tables in the above list provide inputs for analysis at both the network level and project-level.

18.9.1   M,R&R Decision Criteria, Policies, and Catalogue Tables

The M,R&R analysis methodology depends on the following criteria and information.

- Decision Criteria
  - Pavement condition parameters for the network level analysis (severity and extent of distresses, rutting, roughness, structural condition, loss in skid resistance or friction, noise, geometric deficiencies, etc.)
  - Minimum acceptable condition for selected attributes (may vary with functional class, traffic level, etc.)
  - Capital improvement needs (more lanes, upgrade to a higher functional class, etc.)
  - Other executive priority criteria (urgency, safety, emergency based on accidental or natural disasters, overall M,R&R budget constraints, etc.)
- M,R&R Policies for minor maintenance, short-term or single year major M,R&R, and long-term major M,R&R (related condition data, scope of application, section area, and executive priority criteria)
- M,R&R Catalogue tables (M,R&R treatments by pavement types including life estimate or deterioration rate model, resulting improvement in pavement condition, unit cost, and productivity as required in the agency work performance standard)
- PMS analysis software, which is linked to the requirements of monitoring data collection and evaluation, performance prediction models, M,R&R analysis, and prioritized work programs.

As the pavement condition deteriorates, user cost and frequency of public complaints increase, therefore, the identification of a highway section for timely M,R&R is a great benefit of PMS to the agency, as well as the public. Decision criteria and M,R&R policies should be established as soon as possible as a part of the overall PMS development with the assistance of folks involved in planning, design, construction, research and pavement evaluation, and maintenance. The decision for PMS analysis software should be made at the outset. Finally, these critical decisions should be approved by the agency executive management. Because any major change in the criteria and the selected analysis software option may have drastic effects on the M,R&R reports and agency resources. These are important elements of the DSS framework involving a common database and applied for PMS (Figure 18.6). For the owners of large networks (for example, state highway departments and urban public works agencies), the PMS objective must be integrated with the objectives of other highway elements of the network. Sinha and Fwa (1989) present a three-dimensional matrix of highway facility elements, operational features, and system objectives to develop an integrated highway management system. The elements of a highway facility include pavement, bridge, roadside and related appurtenance assets, signs and other traffic control devices. This concept of integrated highway management needs to be tailored to the specific agency needs so that there is no duplication and wastage of resources in the central database management, GIS implementation, and implementation of M,R&R work programs.

18.9.1.1 Decision Criteria for M,R&R Intervention

Decision criteria are needed for identifying and scheduling candidate pavement sections at the network level, for making a rational selection of M,R&R strategies, and for long-term LCC analysis. Also called maintenance intervention criteria, these criteria are used to establish intervention levels of selected pavement condition attributes at which appropriate M,R&R treatments are triggered and a strategy is selected based upon the built-in M,R&R policy tables. These or similar decision criteria and M,R&R policy table are needed in the PMS database to predict long-term M,R&R needs and evaluate life-cycle costs and benefits. The condition data may be collected for all performance indictors discussed in earlier sections, only some prioritized condition attributes may be desirable for M,R&R analysis at the network level. Some examples are shown in Figure 18.8. Appropriate database tables are required to access the decision criteria, which include the following:

- Minimum acceptable serviceability or minimum PSR
- Maximum acceptable roughness level or IRI
- Minimum acceptable PCI or maximum distressed area based on distress survey
- Minimum acceptable composite index such as PCR (based on both distress data and roughness)
- Maximum sensor 1 deflection normalized to a standard peak load and temperature, and related deflection basin parameters
- Minimum threshold for structural adequacy or remaining life
- Minimum threshold for skid resistance, friction, or other safety rating
- Minimum threshold for geometric deficiencies (grade and cross fall, drainage, etc.)
- Minimum threshold for pavement noise (critical for heavily trafficked urban areas).

The M,R&R decision criteria are established for identifying and scheduling candidate pavement sections for a rational selection of M,R&R strategies. At least one of the above criteria is essential for the M,R&R analysis at the network level. Combinations of two or more criteria are often used considering whichever can first trigger maintenance intervention. Of course, this will add to the complexity of the M,R&R policy and analysis. The usefulness and versatility of a good PMS analysis software is reflected by the number of condition related decision criteria and the rational approach in their use with M,R&R policies. Examples of some decision criteria with simple M,R&R treatment selection for asphalt surfaced flexible pavements at the network level are:

(a) Localized maintenance (asphalt patching) is triggered to correct rutting, pothole distress.
(b) IRI 50 m/Km (or PSI of 2.0) triggers asphalt overlay on the entire section.
(c) 30% or more section area with cracking distresses at medium and high severity level triggers asphalt overlay on the entire section.

Table 18.4 lists some examples of typical maintenance intervention criteria for JRCP and JPCP.

Some condition data elements (like deflection data) are collected on selected or flagged candidate sections only, are stored off-line because of the large data file size, and only some summary results are entered into the network level database. The summary data may include average of sensor 1 maximum

TABLE 18.4 Maintenance Intervention Levels of Concrete Pavement Deterioration

Pavement Condition	JPCP	JRCP
Pumping	Three (high severity)	Three (high severity)
Joint deterioration	140 joints/mile	53 joints/mile
Cracking	818 feet/mile	15,000 feet/mile
Depressions and swells	10/mile	10/mile
Faulting	0.25 in.	0.40 in.
Skid number	30	30
PSI	2.5	2.5

Source: The 1984 Report on Cost Allocation Study conducted by the FHWA in the U.S.

deflection (normalized to a standard load) and standard deviation, remaining (residual) life, and required overlay thickness. The structural evaluation is conducted by stand-alone analysis programs for remaining life and overlay design at the project-level.

18.9.1.2 M,R&R Policies and Catalogues

The establishment of appropriate M,R&R policies is important for the development of a PMS analysis software. The policy should recognize the level of needs in terms of *localized* routine or minor maintenance and *global* major M,R&R applied on 100% of the pavement area. The M,R&R policy (or sometime called standard) identifies a specific M,R&R treatment associated with intervention levels of pavement condition attributes for candidate sections. The policy table describes a specific M,R&R treatment, unit cost, productivity, extent of application on the pavement section, and effect on condition improvement. In the network level PMS applications the most widely used policies are based on intervention levels of distress data and roughness data. Typical policies and methods for maintenance treatment selection currently in use include:

- "ad-hoc" based on past experience, subjective judgment, or personal preference,
- based upon distress type and severity, such as the PAVER procedure (Shahin and Walther, 1990),
- use of a composite index; for example, PCR as a function of distress attributes and IRI,
- calculation of the life-cycle costs and benefits considering roughness and distress attributes for interventional levels and selection of the most economical one
- a decision tree approach considering all distress type and other condition attributes, and
- application of modern artificial intelligence technologies (expert system, fuzzy set, ANN).

Many limitations can be identified with some of the above approaches. The traditional *ad hoc* policy was the norm before the PMS concept evolved. The use of one or more composite index in some current PMS programs is oversimplification; it will miss the mechanism leading to condition deterioration and, therefore, may result in an inappropriate maintenance treatment. The decision tree approach is the most popular but it can be very complex if other condition attributes (such as IRI and deflection) are considered besides distress data. Some examples of M,R&R policies follow.

18.9.1.3 Example — HDM III and HDM-4 Programs

The World Bank's HDM program is the recommended choice of the World Bank and other international banks on the highway construction and maintenance management projects funded by these institutions. These programs perform a comprehensive life-cycle analysis of agency costs, user costs, and benefits using condition deterioration models for roughness, cracking, and rutting. The roughness and distress attributes are used for interventional levels. The M,R&R treatments (called maintenance standards in HDM) are arbitrary. The most economical M,R&R strategy is then selected for the project-level design (Watanatada et al., 1987). The HDM-4 version has extensive features for creating the network level database and M,R&R work programs (HDM, 2002). The HDM performance prediction models are empirical regression models and needs calibration (specially, the distress progression models), if used in different geographical areas. Therefore, it requires experienced HDM users to calibrate the models and establish rational M,R&R treatments and policies for a new geographical application region. These models were mostly developed using performance data on relatively thinner and inadequately maintained pavements in Brazil during the 1970s, which do not present the recent and current practice of thicker and high quality highway pavements. The HDM standard condition deterioration models predict a higher rate of deterioration for well maintained asphalt highway pavements (Uddin, 2002a). The HDM methodology does not consider deflection and associated rational overlay design procedures in its maintenance standards.

18.9.1.4 Example — PAVER and MicroPAVER Procedure

The PAVER procedure for the M,R&R analysis, applied to both asphalt and composite (asphalt over concrete) pavements, uses distress data and rutting data to trigger the maintenance treatment selection, and the PCI scale for rating pavement condition. It allows maintenance and rehabilitation policies for single-year work program based only on the severity levels of these distresses. The use of rational M,R&R treatments is critical for a meaningful M,R&R analysis. Separation of *localized* maintenance treatments from *global* M,R&R treatments (on 100% area) is important in such policies. For a multiyear M,R&R program, the PAVER procedure allows a maintenance policy of *different* maintenance intervention levels and gross M,R&R unit costs based on the PCI ranges, without consideration of any rational M,R&R treatment. The PAVER procedure primarily considers distress type and severity and does not consider extent, therefore, it may result in entirely different treatments within the section (seal cost, local repair, and overlay, etc.), which may not be practical and rational for real world implementation. It also does not consider deflection data, roughness, and other condition data for maintenance analysis. The PAVER procedure provides only gross estimates for long-term maintenance costs and does not estimate VOCs and benefits. It does not use optimization and only applies a simple priority ranking procedure to rank order the candidate sections by high maintenance and rehabilitation costs and low PCI.

18.9.1.5 Example — PAM Methodology

The Program for PAM software, integrated with the GIS and implemented for the network level pavement management in several cities of Mexico, is a comprehensive PMS analysis program (Uddin and Torres-Verdin, 1998). It incorporates the use of several condition data attributes for M,R&R policies, prediction of long-term M,R&R costs, and LCC and benefit analysis. It uses a heuristic procedure of incremental benefit–cost ratio to prioritize candidate sections. The PAM methodology recognizes the following maintenance and rehabilitation categories for each pavement type (asphalt, concrete, composite, and block pavements):

(1) annual routine or emergency maintenance policy, as defined by the agency,
(2) minor maintenance treatments on localized basis depending on the extent and severity of visual distress types (for example, crack and joint sealing, patching),
(3) major maintenance and rehabilitation used on 100% area in conjunction with an appropriate short-term major M,R&R policy for a specific pavement type, and
(4) long-term maintenance and rehabilitation intervention programed on 100% area using the long-term M,R&R policy for the specific pavement type.

Vehicle operating cost calculations are made for both short-term major maintenance and long-term maintenance policies.

Annual Routine and Emergency Maintenance Policy: Total annual routine and emergency cost category is used by some agencies. This policy includes routine peripheral activities (for example, scheduled maintenance of pavement marking, slope or drainage repair) and emergency repair (for example, pothole, surface damage caused by accidents, and utility cuts). In this case annual cost is calculated every year by multiplying the section lane-length with the $/lane-Km/year input for the specific pavement type. This cost item is also used to calculate cost of deferred maintenance if the maintenance year is delayed by more than one year. The user may give a zero value if this cost is not desirable.

Minor Maintenance Policy: Policy guidelines, established for the localized minor maintenance, are based on the visual distress type, extent and severity level. For example, the following local minor maintenance treatments are used to calculate total local minor maintenance cost for asphalt pavements: (1) asphalt full depth patching, (2) asphalt partial depth patching, (3) asphalt crack sealing, and (4) miscellaneous surface repair. Minor maintenance work quantity and cost is based on distress quantity with appropriate conversion to the units used in the catalogue. In the case of immediate major maintenance action of milling and overlay, the agency may be able to ignore all local minor maintenance treatments except potholes which are repaired as emergency maintenance.

Short-term Major M,R&R Policy: Short-term maintenance and rehabilitation policies, developed for major M,R&R strategies, are used to identify candidate sections for the current analysis year. A major maintenance alternative (with the exception of concrete pavement restoration and partial repaving of brick and block pavements) implies its application over 100% of the pavement area. A typical policy table for asphalt pavement contains asphalt overlay, milling and inlay, recycling, and reconstruction options. Overlay and concrete pavement restoration options are provided for concrete pavements. The strategy selection is based on a comprehensive analysis of the condition attributes and related decision criteria for maintenance intervention. The policy tables can be updated and modified, and new policy tables can be created by the user. Many different policies can be established for each type of pavement. Two major M,R&R treatments can be assigned for a given condition scenario. Two or more treatment alternatives are provided in the M,R&R catalogue tables and additional alternatives can be added by the user.

Long-term Major M,R&R Policy: A long-term M,R&R strategy is selected as a part of the agency cost calculation for the total service life analysis. The long-term maintenance and rehabilitation policy considers two types of major M,R&R for each pavement type on 100% of the pavement area, similar to the short-term major M,R&R. Pavement condition deterioration predictions and intervention criteria are used for each type of pavement to determine the IRI and PSR in future years during the analysis period. If the PSR of any section falls below the maintenance intervention threshold (typically two for paved roads), then a global major maintenance M,R&R for 100% area is selected according to the long-term policy. For example, asphalt overlay or reconstruction can be selected by the user for asphalt pavements. The long-term maintenance cost of each section is calculated in conjunction with the VOC and other user costs and resulting benefits for priority ranking of candidate sections using a heuristic approach, which is described in an earlier section.

There are numerous other examples of M,R&R policies, needs analysis, and priority methods in published papers and in the proceedings of past international conferences on pavement management. The reader is encouraged to learn from this extensive literature on the state-of-the-practice before embarking upon the development a new methodology. The following sections describe decision trees and artificial intelligence technologies.

18.9.2   Decision Tree

A decision tree approach is often used to develop M,R&R policies based on distress data attributes as illustrated in Figure 18.14 (FHWA, 1990). This is the most simple but a subjective approach. The development of maintenance policy standards becomes very complicated if other condition attributes are also considered, e.g., rutting, deflection, skid resistance, and cross fall deficiency. A common problem is that the decision tree may not lead to a unique set of M,R&R alternatives because it depends on the distress type a the start and becomes complex if one tries to consider adequately the influence of both severity and extent.

18.9.3   Knowledge-Based Expert Systems, Fuzzy Logic, and ANN

These are included in the group of artificial intelligence technologies. Expert systems have been extensively applied fro the selection of M,R&R treatments. They develop a knowledge base and solve engineering problems through simple decision rules, as discussed in detail in the section on performance models. They can take advantage of the judgment of experienced pavement engineers and analyze problems, which do not have well established programing solutions. A more comprehensive discussion of knowledge based expert systems technology for maintenance decision-making applications and description of several software packages have been provided in an entire chapter by Haas et al. (1994).

Fuzzy logic systems are an extension of the traditional rule-based reasoning (knowledge-based expert systems), and especially useful for the decision-making process combined with descriptive rules through fuzzy logic. Fuzzy logic was developed to make inferences about imprecise and qualitative data and use the data. It enables the variables to partially belong to a particular set and, at the same time, makes use of the generalizations of conventional Boolean logic operators in data processing. The main advantages of this approach include the possibility of introducing and using rules from expert opinions, past experience, intuition, and heuristics, providing functional transparency, and not requiring an empirical regression model of the process. More information on these artificial intelligence topics is available in

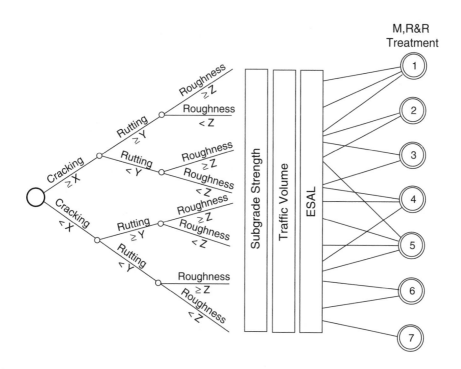

FIGURE 18.14   Illustration of a decision tree for M,R&R selection, after (FHWA, 1990).

the literature (Flintsch, 2003). Detailed discussion of the ANN methodology is included in an earlier section on performance models. ANN technologies and fuzzy systems have been used as alternatives to the traditional tools, such as decision trees for M,R&R selection and needs analysis (Fwa and Chan, 1993; Fwa and Shanmugam, 1998).

18.9.4   M,R&R Catalogues

The detailed data for each pavement-specific M,R&R treatment are needed in separate database tables including the unit cost and expected condition immediately after the treatment. Information on production rate per day is useful to determine the time required to apply this treatment to a section. A catalogue table for major M,R&R alternatives for each pavement type also includes the estimate of expected life and years of service before the next maintenance. The attribute data from these tables are accessed by the PMS analysis program for the M,R&R analysis and work program. Examples of routine and minor maintenance treatments for concrete pavements are cleaning and sealing of joints, spall repair, subsealing, slab jacking, patching, and crack sealing. Bituminous patching, crack sealing, seal coat, and surface treatments can be considered for flexible pavements. Base repair can be specified for both concrete and asphalt pavements. Gravel regarding and upgrading to surface treated pavements are examples of shoulder maintenance.

Table 18.5 lists many examples of rehabilitation alternatives in the group of major M,R&R strategies. These include overlays, concrete pavement restoration techniques, and reconstruction. The alternatives for asphalt (flexible) pavements are applicable to pavements with either flexible or rigid (concrete or cement stabilized) base (Uddin et al., 1987). The M,R&R policy and LCC methodology should be designed to analyze a combination of rehabilitation techniques under a single strategy. For example, joint rehabilitation, subsealing, grinding, spall repair, and full-depth concrete patching can be combined as a single rehabilitation scheme in a concrete pavement restoration strategy. Resurfacing or overlay is a major rehabilitation option. Most agencies use standard overlay alternatives for the network level analysis according to traffic and roadbed soil categories. Determining overlay thickness requires a structural

TABLE 18.5   Examples of Rehabilitation Alternatives

Items	Applicable to Pavement Type
1. Milling	Asphalt or bituminous pavements
2. Leveling course	Asphalt or bituminous pavements
3. Recycling	Asphalt or bituminous pavements
4. Scratch course	Asphalt or bituminous pavements
5. Thin white-topping	Asphalt or bituminous pavements
6. Joint rehabilitation	PCC (Portland Cement Concrete) pavements
7. Spall repair	PCC pavements
8. Subsealing	PCC pavements
9. Slab jacking	PCC pavements
10. Slab replacement	PCC pavements
11. Diamond grinding	PCC pavements
12. Recycling	PCC pavements
13. "Do-nothing"	All pavement types
14. Asphalt or bituminous overlay [a]	All pavement types
15. Continuously reinforced concrete overlay [a]	All pavement types
16. Plain jointed concrete overlay [a]	All pavement types
17. Jointed reinforced concrete overlay [a]	All pavement types
18. Reconstruction	All pavement types

[a] Overlays can be considered with or without bond breakers layers.

evaluation of the existing pavement by deflection testing and the material properties of new overlay or recycled materials, which is handled at the project-level for a selected candidate section from the M,R&R work program list. Some design procedures can generate an array of rehabilitation designs for both concrete and flexible pavements (Haas et al., 1994).

The effectiveness of various maintenance and rehabilitation alternatives on pavement condition after treatment and expected future "life" in years must be known in order to perform a LCC analysis of alternative strategies. The scheduling of M,R&R activities depends on the effectiveness (expected lives) of these treatments. This information must be developed from local experience practice and stored in the M,R&R catalogue tables. If not available, the default data can be enhanced later after getting feedback from the implementation of M,R&R strategies in the field. For the LCC1 program (Uddin et al., 1987) a survey of pavement experts of the Pennsylvania Department of Transportation (Penn DOT) in the U.S. working in various technical areas (maintenance, construction, design, and administration, for example) was performed to estimate the expected life of selected M,R&R alternatives. Table 18.6 presents some recommended ranges of various alternatives based on these results for asphalt pavements. Caution should be exercised in using these results because these gross estimates are regional and appropriate to specific

TABLE 18.6   Recommended Ranges for Estimating Expected Lives of Various Rehabilitation Alternatives (et al., 1987)

Rehabilitation Alternative	Asphalt (Flexible) Pavements Expected Life (years)	
	Low Traffic	High Traffic
1. Crack sealing	3–5	2–3
2. Bituminous patching	4–6	3–4
3. Seal coat	4–5	2–3
4. Level and seal coat	5–7	2–4
5. Milling and recycling	7–9	5–7
6. Thin overlay	5–8	3–6
7. Thick overlay	9–12	7–10

TABLE 18.7 Typical Data Included in an M,R&R Catalogue Table

M,R&R Code: A01 Description: Asphalt Structural Overly (100 mm Default Thickness)								
Pavement Type	Expected Life, Years	Cost		Production Rate/Day (m^2)	Improved Condition			Change in Surface Elevation[a]
		US$	Unit		PSR	IRI	PCI	
Flexible (Asphalt)	20	10.00	m^2	500	4.5	2.0	100	*Yes*
Composite[b]	15	10.00	m^2	500	4.5	2.0	100	*Yes*
Jointed Plain Concrete	10	10.00	m^2	500	4.5	2.0	100	*Yes*
Jointed Reinforced Concrete	12	10.00	m^2	500	4.5	2.0	100	*Yes*
Continuously Reinforced Concrete	12	10.00	m^2	500	4.5	2.0	100	*Yes*

[a] Change in surface layer elevation can be used for lane and section rationalization to develop an automated methodology of maintenance contract packaging. The entry *Yes* implies shoulder and peripheral upgrading because of 100 mm rise in the pavement elevation; *No* means no change in the surface elevation, such as milling and inlaying, and there is no additional shoulder and peripheral costs involved (used in the PAM methodology).

[b] Composite: Asphalt over old concrete pavement, IRI in m/Km.

highway traffic levels; significantly longer life may be possible for low volume roads. Survivor curve techniques have been used to develop overlay life estimates for asphalt pavements in Arizona in the U.S. (Flintsch et al., 1994). An example of M,R&R catalogue data items for an asphalt overlay alternative is shown in Table 18.7.

If noise is a concern on a pavement section and the section needs to be rehabilitated because of extensive cracking then the M,R&R policy should be designed to select an appropriate rehabilitation alternative from the M,R&R catalogue table. For example, porous or open graded asphalt surface is a preferred choice as it has better acoustic absorption properties (Losa et al., 2003). The M,R&R treatment strategies are implemented along with the following peripheral items:

- Construction, maintenance, and rehabilitation of shoulders (unpaved, asphalt surfaced, surface treated, or PCC)
- Maintenance, rehabilitation, and replacement of guardrails, drainage structures, and miscellaneous appurtenance items
- Other peripheral items such as drainage and surface improvements

18.9.5 LCC Analysis and Deterioration Models for Rehabilitated Pavements

The role of a comprehensive LCC analysis has been discussed in detail in an earlier section. It requires the performance prediction of the rehabilitated pavement. Two types of models are required for a complete life-cycle analysis and maintenance effectiveness calculations. These models are: (a) for predicting the immediate improvement of pavement condition after rehabilitation, and (b) for estimating deterioration of the rehabilitated pavement due to time and traffic. The maintenance effects on pavement condition can be modeled by assuming that the maintenance effectiveness can be represented by either a change in rate of deterioration or an improvement in pavement condition. Routine and minor maintenance (whether scheduled or emergency, routine, periodic, or corrective) comprises those activities which can be represented mathematically by corrections to the deterioration function. For example, 10% improvement in the PSR will bring PSR from 2.0 to 2.2. This is a general and flexible approach that allows the selection and interpretation of maintenance activities with different mixes. It also distinguishes routine and minor localized maintenance from global major rehabilitation that brings a greater improvement in pavement condition.

Other models are used to quantify the effect of a major M,R&R. One approach is to account for major rehabilitation actions (for example, overlay) by measurable changes or to apply mostly subjective adjustment factors (AF) in the current condition of the pavement. For example, the PSR can improve from two to four. Another example is shown in Table 18.7. The routine maintenance can be represented as an adjustment in the slope of the deterioration curve. The most effective deterioration models for rehabilitated pavements are regression models developed from careful monitoring a devaluation of in-service pavements. The in-service evaluation and review of performance trends using the PMS database can lead to an effective M,R&R intervention policy based on a preselected level of deterioration. A major restoration and rehabilitation intervention can extend the service life of a pavement significantly, as shown in Figures 18.1 and 18.5, and reduce VOCs.

18.9.6   PMS Analysis Software

The PMS analysis software is a very critical component of the DSS framework for the PMS development and implementation, as illustrated in Figure 18.6. The decision for PMS analysis software should be made at the outset of the PMS development and implementation plan. A RDBMS software and a compatible PMS analysis software are required for database management and M,R&R planning and budgeting analysis. Pavement inventory and condition reports could be easily designed and generated by in-house effort using the database software. However, annual or single-year M,R&R work programs reports, prioritized listing of roads, and multiyear M,R&R program and budget reports require the use of an appropriate PMS analysis software. The analysis software is also linked with the decisions for type and extent of condition data collection and evaluation. Basically there are three optional approaches to develop the PMS analysis component of the DSS framework.

1. *Adopt Existing Shelf PMS Software*: This is the least expensive option if the decision is made early and the data is collected and processed accordingly. Most of the shelf PMS software packages developed by public agencies are relatively in-expensive. The PMS analysis in the MicroPAVER software selects the M,R&R alternatives, based only on distress data only; and M,R&R policies are based on distress severity only. It will limit the inventory and condition data collection and management as per the required formats only. This implies that only visual distress data will be the basis of maintenance planning and budgeting analysis if the analysis software uses only distress data. The shelf packages developed by commercial vendors are relatively more expensive. This option may be infeasible because of the high cost of commercial software packages, and may have serious limitations because the resulting reports may not satisfy the goals and objectives of the agency.

2. *Modify and Adapt Existing Shelf PMS Software*: This is a better option but it will have additional costs depending on the extent of enhancements, required adaptation, and implementation efforts. This option is available with all commercial software packages. It will provide improved PMS analysis reports and improved database management. However, it is less likely to upgrade and expand the scope of maintenance planning and budgeting analysis unless a significant amount of effort is developed to accommodate the specific agency's needs. As expected, there will be sacrifice in the scope of the generated PMS reports and system incompatibility problems may pose difficulties in satisfactory implementation.

3. *Develop Customized PMS Analysis Software*: This is an ideal but most expensive option. A PMS analysis software developed for the exclusive use of an agency permits the agency to select the most efficient method of collecting distress and other pavement condition data, processing customized M,R&R policies, and work program reports. A customized analysis application can be developed using a preferably dedicated or an existing database management software, which can accommodate the analysis of all types of condition data that the agency likes to collect, implement desired M,R&R analysis and optimization methodologies, and generate custom designed management reports. The customized analysis software and database can also be integrated easily with a preferred GIS program for enhanced graphical displays.

The third option of a customized PMS analysis software is the most expensive to implement but it will provide the most cost-effective maintenance and rehabilitation programs. At the state level and national level, this should be the preferred choice, and every effort should be made to secure funding sources to implement this option. The agency will also own the software and, therefore, it can upgrade the software according to future enhancement needs. The second option of modifying and adapting existing shelf software should be the next preference if the agency has limited funding in the initial phase. The first option of adopting a shelf package is suitable for small to medium size cities and local governments which can not afford the staff and funding resources to implement any of the last two options (Uddin, 1992).

18.10 Priority Ranking and Prioritization Methodologies

The effective annual or single year M,R&R work program should contain prioritized lists of roads for executive managers. The priority ranking can be as simple as "by decreasing total PW of M,R&R cost" and as complex as "nonlinear mathematical optimization." The following factors should always be considered in any priority ranking option; agency, cost, pavement serviceability and condition index, functional classification, and traffic level. The total M,R&R costs for the selected candidate sections should not exceed the preselected budget levels. The state highway agencies in the U.S. commonly set priorities for network needs based on condition, load, skid resistance, and, in some cases, deflection. Quantitative pavement performance data is a major input in estimating the network level needs and setting priorities. Each highway agency formulates its own analytical methods fro priorities, such as a condition index, priority index, decision tree, and mathematical optimization models based on the nature and performance characteristics of the network. Agency engineers formulate the engineering decision criteria and the weighing factors used in the analytical procedures based on the goals set by the agency and the network conditions.

18.10.1 Priority Ranking Procedures

Priority ranking and optimization procedures are built in the PMS maintenance and rehabilitation analysis software. For the case of a shelf type software, such as MicroPAVER, there is not much flexibility to make any changes in its simple priority ranking procedure. Common prioritization approaches evaluate inter-project tradeoffs in selecting M,R&R strategies, and select strategies which are guaranteed to adhere to budget limits. On the other hand, mathematical optimization is more complex and effective than prioritization. It allows (a) not only tradeoffs among candidate sections (as done in priority ranking), but also evaluates any number of strategies for each section in the course of making these tradeoffs and (b) multiyear M,R&R planning and programing at the network level aimed at moving the overall system towards a defined level of performance (FHWA, 1990). Table 18.8 is based on a comparison by Haas et al. (1994) of priority ranking methods commonly used in PMS.

18.10.2 PAM Prioritization of M,R&R Work Programs Based on Incremental B/C Ratio

An example of the use of priority ranking type 3 is Dubai PMS, which was developed in the United Nations institutional support project (Uddin and Al-Tayer, 1993). This prioritization methodology has been enhanced by incorporating LCC and benefit analysis for type 5 (near optimization heuristic approach), adapted in the PAM program, and implemented for several PMS studies in Mexico (Uddin and Torres-Verdin, 1998). The PW of all major M,R&R treatments predicted over the service life is calculated, which constitutes the life-cycle M,R&R cost. Similarly, the PW of the resulting life-cycle benefits are calculated from savings in VOCs due to improved pavement condition for a typical analysis period of 30 years. The candidate sections are then prioritized by the incremental benefit–cost (B/F) ratio until the allocated budget limit reaches in the first year. As discussed later, the heuristic methods, such

TABLE 18.8 Comparison of Prioritization Methods After (Haas et al., 1994)

Class of Method	Advantages and Disadvantages
1. Simple subjective ranking of projects based on judgment, overall condition index, or decreasing first cost (single year or multiyear)	Quick, simple; subject to bias and inconsistency; may be far from optimal
2. Ranking based on condition parameters, such as serviceability or distress; can be weighted by traffic (single year or multiyear)	Simple and easy to use; may be far from optimal, particularly if traffic weighting not used
3. Ranking based on condition parameters and traffic, with economic analysis including decreasing PW cost or benefit–cost ratio (single year or multiyear)	Reasonably simple: may be closer to optimal
4. Annual optimization by mathematical programing model for year-by-year basis over analysis period	Less simple; may be close to optimal; effects of timing not considered
5. Near-optimization using heuristics approaches including incremental benefit–cost ratio and marginal cost-effectiveness (M,R&R timing taken into account)	Reasonably simple; suitable for microcomputer environment; close to optimal results
6. Comprehensive optimization by mathematical programing models taking into account the effects of M,R&R timing	Most complex and computationally demanding; can give optimal program (maximization of benefits or cost-effectiveness)

as the incremental B/C ratio method used in the PAM methodology, are computationally efficient and provide reasonably optimal prioritization.

The heuristic approach of prioritization, used in the PAM methodology at the network level, is based on functional class, traffic volume, maintenance needs and incremental life-cycle B/C ratio. First, a priority index (PI) is calculated to sort the surveyed sections based upon the urgency and type of maintenance needs (routine and emergency maintenance, minor maintenance, major M,R&R). The sections with executive and safety priority are always given top priority commitment by the program. The priority methodology computes the traffic AF as a function of traffic volume and road classification, groups sections by maintenance type needs, and then calculates an adjusted priority index (API). The sections are then prioritized in the order of decreasing API (maintenance group priority) and incremental B/C ratio. The prioritized list also contains the cost of maintenance and rehabilitation for each section and the cumulative cost of maintenance for all those sections requiring maintenance. This list becomes useful either to estimate the budget required for the "need analysis" or to select a list of candidate projects within the available budget. The PAM software generates a single year M,R&R work program report and a projected five year M,R&R work program report for the entire road network. The sections which can not be included in the first year short-term M,R&R prioritized program because of the budget constraints, are kicked into the following year program. For the five-year work program report, the program calculates the deferred maintenance cost for the candidates sections in the following next year, recalculates VOC and B/C ratio for each deferred section, and prioritizes the sections in the second year M,R&R program, and so on for five consecutive years. This incremental B/C approach, based upon the service life analysis of agency costs and benefits (reduction in VOC), is more practical and near-optimal. It is convincing to decision makers as compared to the marginal cost-effectiveness analysis or more complex mathematical optimization techniques.

18.10.3   Mathematical Optimization for Prioritization of M,R&R Work Programs

As with priority ranking methods, optimization methodologies need the current condition data on all pavement sections at the network level, M,R&R alternatives, associated costs, and improvement in pavement condition from M,R&R. The annual optimization procedure defines an objective function with the goal to minimize the overall cost or maximize the B/C ratio subject to certain constraints, and find cost-effective solutions. For example, the minimum cost function requires the constraint that the overall network average condition never reaches below a minimum acceptable value. The focus of this type of optimization is to choose the best set of candidate sections based on their current performance. After the first year budget is exhausted then simply predict the future condition in the next year, select the best set of the remaining projects, and so on. An example of this optimization for a multiyear program is shown in Figure 18.15, where the total required funding for different M,R&R programs and resulting overall network conditions are plotted for several years. This plot is meaningful to the executive management who can clearly see the advantage of the multiyear M,R&R programs and the payoff of increased budgets in early years on the improved condition of the network. This type of optimization may not provide true optimal solution because it does not consider the best M,R&R timing for the candidate sections. A "true" optimization also addresses the dimension of timing and selects the best sequence of M,R&R strategies over several years (FHWA, 1990). The "true" optimization is computationally complex depending on the number of pavement sections and size of the network, number of condition states, and constraints.

The mathematical optimization (or near optimization) functions are used to generate PMS work programs and budgets consistent with their performance goals and financial constraints. The mathematical optimization problems can be solved for true and exact optimal solutions by several approaches: linear and nonlinear programing, integer programing, and dynamic programing (FHWA, 1990; Zimmerman, 1995). These mathematical programing methods produce exact solutions. Markov decision theory is a probabilistic optimization approach based on transitional probabilities. Heuristic approaches are computationally more efficient for near-optimal solutions. These are discussed here.

18.10.4   Markov Probabilistic Optimization for Prioritization of M,R&R Work Programs

Some of the early PMS developments at the network level, such as for the state highway agencies in Kansas and Arizona in the U.S. (Golabi et al., 1982; FHWA, 1990), incorporated Markov decision process

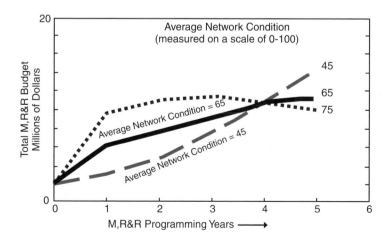

FIGURE 18.15   Network prioritization and multiyear M,R&R work program for different budget levels.

to produce prioritized work programs. This probabilistic approach considers uncertainty in prediction of pavement performance and considers a large number of alternative M,R&R strategies. One basic assumption of the Markov process is that it only has a one-step memory. This means that the transition probability of a pavement section (i.e., the probability that the section makes a transition to a particular condition state in a unit time following a particular action) depends only on the present condition state of the section and the action selected at that time, and not on how the section reached that condition state. The Markov process of transitional probabilities to predict future condition states from the current condition state has been discussed in more detail in the earlier section on condition deterioration and performance models. By including such factors as age and expected life of the last M,R&R action in the definition of a condition state, the one-step memory can be made to take into account the effect of type and time of the last action.

As the section continues to make its annual transitions under a given M,R&R policy, various costs are incurred. These include: the one-time cost of a major M,R&R treatment, and routine maintenance and user costs during the year following the M,R&R treatment. For a given policy, all the expected annual costs can be accumulated to find the total expected PW cost. The objective of the analysis is to find the policy that minimizes the total expected cost subject to selected performance-related constraints. For example, keeping the overall average network condition to the acceptable threshold of a pavement condition attribute, such as PSR = 2.0, IRI = 5 m/Km, or PCI = 40. An alternative approach is to estimate the expected annual benefit (in terms of reduced user costs or a subjective value scale) of operating the section in a given condition state. The objective then would be to maximize the total expected benefit subject to budgetary constraints. The following discussion provides useful insight of the issues related to the implementation of Markov decision process for network level optimization (FHWA, 1990).

- Since an extremely large number of alternative M,R&R treatments can be generated, it is complex and impractical to find the exact optimal policy solution. Techniques of mathematical programing can be used to find the optimal policy efficiently. A linear programing (LP) algorithm is particularly effective to analyze a Markov decision process which involves a large number of condition states, transitional probabilities, and a large number of M,R&R treatment alternatives.
- The computational limits of the LP software and computer memory put constraints on the size of Markov process. Therefore, the number of condition states needs to be limited to a small number. This translates into a maximum three pavement distress types for a given pavement type and into four levels defined for each distress type.
- The mathematical concepts of a Markov decision process are derived from the discipline of operations research and more difficult to explain to the executive management and decision makers than those of a simple priority ranking procedure.

Because of these reasons, computationally efficient heuristic approaches have been traditionally implemented more widely for pavement management applications.

18.10.5   Heuristic Approaches for Prioritization of M,R&R Work Programs

Heuristic methods producing near-optimal solutions are well suited for large-scale problems, such as large PMS networks. Heuristic methods include marginal cost-effectiveness and incremental B/C ratio methods (Markow et al., 1994). Other advanced heuristic methods are based on artificial intelligence techniques. Effectiveness can be calculated as the additional area under the performance curve due to the M,R&R application that can be further multiplied by weighting functions, such as traffic. The marginal cost-effectiveness method has been used in several states in the U.S. including Minnesota and North Carolina and the Canadian Province of Alberta (FHWA, 1990). The method has been used with the PAVER results where the benefit, considered as the area under the PCI curve, is calculated as a non-monetary benefit and used to compute incremental B/C ratio (Shahin et al., 1987).

The PAM methodology, described earlier for Mexico PMS studies (Uddin and Torres-Verdin, 1998), also uses incremental B/C ratio method where the benefits are calculated from the savings in VOC and other user costs resulting from timely M,R&R treatments, compared to the "do-nothing" base policy.

18.10.6 Artificial Intelligence Techniques for Prioritization of M,R&R Work Programs

Artificial intelligence techniques include fuzzy mathematical programing, ANNs, and evolutionary computing (including genetic algorithms). These techniques are particularly appropriate for pavement management because the information may be uncertain and incomplete. The data may involve combinations of objective measurements, subjective rating, and expert inputs, such as those data used to create M,R&R decision criteria and policy tables. Artificial neural networks and fuzzy systems have been used for needs analysis as alternatives to the traditional priority ranking tools, such as decision trees (Fwa and Chan, 1993; Fwa and Shanmugam, 1998). These have been discussed in earlier sections. More information on the PMS applications of these artificial intelligence topics is available in the literature (Flintsch, 2003). Genetic Algorithms are another artificial intelligence technique which has been pioneered by Fwa in pavement management application at the network level (Fwa et al., 1994, 1998, 2000; Chan et al., 2003).

18.10.6.1 Genetic Algorithms

The genetic algorithm (GA) search technique is based on the mechanics of natural selection and evolutionary computing used in solving complex optimization problems. These algorithms were developed through an analogy based on Darwin's theory of evolution and the basic principle of "survival of the fittest." The search runs in parallel from a population of solutions. New generations of solutions are produced through reproduction, crossover, and mutation until a prespecified stopping condition is satisfied. The main advantage of this approach is that it provides an efficient heuristic to find "good" solutions of difficult large-scale optimization problems. The GA technique has been uses in pavement management for network level programing (Chan et al., 1994; Fwa et al., 1994, 1998). The GA optimization technique proposed by Fwa has been recently implemented for the network level programing in the development of Lisbon PMS in Portugal (Ferreira et al., 2003).

18.11 PMS Program Management Reports and Contract Packaging

18.11.1 Frequency of PMS Reports and Implementation

An early benefit of PMS implementation is the production of summary inventory and condition reports, such as a summary of road length by functional class and by average pavement condition. Detailed PMS program management reports on inventory, condition, and maintenance programing and budgeting should be generated after the first cycle and each following cycle of field data collection. Table 18.9 shows a typical list of the network level PMS reports generated by the MicroPAVER software for Dubai PMS (Uddin and Al-Tayer, 1993). These include: detailed section reports on inventory, visual inspection, and condition; network summary reports on inspection schedule, PCI frequency, family curve and PCI predictions, and budget condition forecasts; and M,R&R program reports for needs analysis and funding requests.

Graphical outputs and GIS maps of network summary condition and M,R&R work program are important for presentation to the executive management. For example, for Dubai PMS, the reports from the MicroPAVER and customized inventory database were used to produce customized network inventory and PCI pie chart, PCI histograms to compare the current conditions with the previous year data, and histograms of requested annual budgets for different maintenance activity groups

TABLE 18.9   PMS Reports Generated by the MicoPAVER 2.1 Software

Report No.	Report Title
1	Branch Listing Report
2	Inventory Report
3	PCI Report
4	Inspection Report
5	PCI Frequency Report
6	Budget Condition Forecasting Report
7	Inspection Schedule Report
8	Condition History
9	Family Analysis Report
10	Section Prediction Report
11	Maintenance & Rehabilitation Report
12	Network Maintenance Report
13	Preventive Maintenance Report

(Uddin and Al-Tayer, 1993). Numerous other examples can be found in the published literature and proceedings of the international conferences on pavement management.

18.11.2   Maintenance Contract Packaging

The M,R&R work program report generated from most PMS analysis programs lists section-by-section treatment types and costs. However, it may not be practical to implement these because of no consideration of the rationalization of maintenance alternatives in the field with respect to across-the-lane and across-the-adjacent sections incompatibilities (Uddin, 1993b). These aspects need to be implemented in future PMS enhancements. The single-year prioritized maintenance work program should be used for preparing contract packaging or the annual budget through a central PMS implementation committee, which requires proper communication among the users of the PMS reports.

18.12   PMS Development, Implementation, and Institutional Issues

18.12.1   Overview of System Development Issues

There is no single ideal PMS for universal applications. The development of a PMS for a highway network and road network and its efficient operation require cost-effective decisions for the selection of equipment and software and system development based on the objectives of the agency. Most of the critical development and operation issues are related to:

- staff resources, equipment, and training,
- database management and analysis software,
- inventory and condition data collection and database development,
- generation of maintenance program reports and implementation, and
- system maintenance, security, and feed back.

These issues and corresponding decision are interrelated and should be considered at the outset of the system design and work plan. Selected case studies of effective PMS development and implementation are presented in a later section, which required different levels of system complexity.

18.12.2   PMS Objectives

The executive management should be fully involved in estimating the equipment, staff training and system development costs, and long-term operation costs; and committing the required sources.

All components of the PMS process should be involved in the PMS task force for the purpose of handling tactical and strategic decisions and acceptance by all involved from planning, design, materials, construction, research and pavement evaluation, safety audit, and maintenance. Two primary objectives are:

- System development, implementation and staff planning
- Maintenance work program and budget reports.

For large networks (for example, state highways and urban public works agencies) the PMS objective must be integrated with the objectives of other highway elements of the network (Sinha and Fwa, 1989). This concept of integrated highway management needs to be tailored to the agency needs so that there is no duplication and wastage of resources in the database management and M,R&R planning and implementation.

As illustrated in Figure 18.6, the basic PMS components are (a) inventory, historical, and traffic database development, (b) pavement condition monitoring data collection and database development, (c) decision criteria and maintenance and rehabilitation policies, (d) implementation of M,R&R planning and budget analysis software (or simply stated as PMS analysis software), (e) generation of PMS management reports, and (f) implementation of M,R&R programs in the field, feedback, and continuation of condition monitoring. The agency resource requirements and impacts of system complexity will be greatly influenced by the size of road and highway network and short-term and long-term goals related to data collection and PMS analysis. The following sections discuss various issues related to the system development, complexity, operation, and feedback from implementation.

18.12.3 System Specifications

It is important to develop specifications for the PMS analysis software keeping in view of the objectives, network size, and available funds. The condition monitoring and evaluation data are summarized into simple statistics and used in conformance to an appropriately PMS network level analysis software to support the objectives of a PMS. For example the MicroPAVER procedure is based only on summarizing the distress measurements, which is used for PCI calculations, the network level analysis of a single year M,R&R program, and multiyear budget. Most highway agencies in the U.S. compute a composite condition rating based on roughness or serviceability and distress data. The States commonly set priorities for network needs based on condition, load, and skid resistance. Deflection data are used for the project-level analysis of structural capacity and strengthening design. Quantitative pavement performance data is now a major input in estimating network level needs and setting priorities.

18.12.4 Agency Resource Requirements; Costs and Benefits

Dedicated staff, office space and vehicles are required to begin the PMS development plan and continue operations. In order to fully understand the current PMS technology, capabilities and limitations of related equipment, the designated PMS Engineer should have prior experience in this area or should attend a short course in pavement management to learn the state-of-the-practice. Visits to and interaction with other agencies of comparable size of road network can provide useful insight to make rational recommendations to the executive management for the system development while satisfying the objectives. Hands-on training of the dedicated PMS staff should be an integral part of the agency's PMS work plan. This should be an on-going process so that a back-up team is always available to ensure continuity of routine PMS operations. A good feasible plan for the PMS development can not be made without studying the existing agency practices and having inputs of agency staff involved in planning, design, materials, construction, maintenance, research and evaluation, safety audit, and project management.

Most costs are related to development and implementation efforts, which include: staff, office facility and transport, pavement monitoring equipment and specialized pavement evaluation services, PMS analysis software and database system development and operation, and overall PMS operation and data collection (inventory, historical, environmental, traffic, construction and maintenance costs). Benefits of PMS development and operation are: improved pavement condition, cost-effective alternatives, effective use of funds, better M,R&R budgeting and planning, and benefits to the executive management with respect well informed decision-making information.

18.12.5 Implementation and Operational Issues

18.12.5.1 Equipment and Database

Computer Equipment: Current generation of microcomputers provides adequate desktop computing power, speed, and memory for PMS system development and operation. System integration will be required to access the existing data files on a mainframe or workstation environment for large and existing databases. In that case client–server applications can provide the easy-to-use user friendly windows interfaces. Computer networking and remote data entry facilitates the decentralized data processing, if needed for routine operations. Suitable back-up medium or equipment are also required. Back-up options vary from traditional back-up floppy diskettes to optical disks (such as CD and DVD). Good estimates of yearly data processing and storage requirements are useful to make the right selection of computer equipment. The optimization program embedded in the PMS analysis software should be considered in making computer acquisition decisions.

Pavement Monitoring and Evaluation Equipment: Depending on the choice of PMS analysis and database software, the pavement condition evaluation equipment is selected from the following condition data categories:

- Distress and rutting data (the most desirable and "minimum" monitoring data)
- Longitudinal roughness (to evaluate riding comfort and serviceability)
- Deflection data (to evaluate structural capacity and residual life)
- Skid resistance or pavement friction data (to evaluate safety)
- Geometric and cross fall deficiency (to evaluate areas of poor surface drainage and potential hydroplaning hazard)
- Other equipment (noise and texture measurements)

For example, if the shelf package of MicroPAVER (Shahin and Walther, 1990) analysis software is selected then all network maintenance program reports are based on distress and rutting data only. A customized software can be used to consider additional condition data categories for PMS analysis. The selection of a specific type of technology for each of the above condition data category depends on the data quality, data collection speed and productivity, and available budget (Uddin, 1992). The equipment and methods can be grouped in the following three classes.

1. *Visual Distress and Rutting* (detailed slow walk survey or high speed photographic survey, faster wind-shield survey)
2. *Stand-Alone Automated Equipment* (for example, dedicated roughness equipment like Maysmeter or Bump Integrator, deflection device like Dynaflect or falling weight deflectometer, and skid resistance measuring equipment)
3. *Multi-purpose Automated Vehicles* (high speed equipment used to collect distress, roughness, rutting and other condition or video data during a single pass).

A tradeoff analysis between purchasing major equipment (for example road proofing, high-speed video recording van, or a multi-function equipment) costing several hundred thousand dollars and leasing the equipment, or contracting with a service provider for annual monitoring and evaluation is extremely beneficial considering benefits and costs of each option over 3 to 5 years. Useful technical data, accuracy evaluation, and cost comparisons for most devices are available in the literature (Hudson et al., 1987a, 1987b, 1987c; Benson et al., 1988; Humplick, 1992; COST, 1997). The following major decision attributes and related criteria are recommended for selecting and purchasing appropriate pavement evaluation equipment:

- Cost (capital, annual maintenance and operating)
- Operational Characteristic (speed, training requirements, calibration, data recording, operating constraints)
- Quality (accuracy, repeatability, compatibility with existing procedures)
- Versatility (number of different tests, additional project-level applications)
- Reliability (downtime, maintainability)
- Degree of Development (time in service, user recommendations).

It may be beneficial to get assistance from an expert in pavement evaluation equipments to review the agency needs and establish specifications for pavement condition monitoring equipment, which can be helpful in making a cost-effective decision for purchase, lease, or contract with service providers.

- Relationship with analysis software
- Network size (number of roads, staff)
- Project-level considerations vs. network level considerations
- Reduced operating constraints
- Ease of operation and productivity
- Equipment reliability and repeatability
- Calibration and accuracy and precision
- Computerized data collection and processing
- Automation
- High-speed operation
- Nondestructive and noncontact measurement
- Cost-effectiveness.

Data Collection and Processing: The first step of the PMS database development is network partitioning into uniform and homogeneous PMS road sections with respect to road name and number, functional class (freeway, arterial, etc.), road or carriageway type (divided or undivided), pavement surface type (asphalt, or Portland cement concrete or PCC, unpaved), section width (one lane, 2 lanes, etc.), and administrative zone. All data is referenced with a unique PMS identification code, also called reference location code including geographical coordinates for easy identification in the filed using a portable GPS receiver. The inventory database should have as a minimum the above identification data, construction date, last major maintenance, available traffic history, and environmental data. Pavement condition monitoring data are collected on maximum possible percentage of the network using the selected condition evaluation equipment. Condition database interface programs are required to process the raw data and transfer appropriate summaries into the PMS database. Similarly, M,R&R history and all construction related cost data should be entered into the database.

18.12.6 System Complexity

The PMS development and implementation time table and required level of effort are functions of the road network size. Further, the system complexity and time requirement increase significantly depending upon the type and extent of inventory data elements, condition attributes, analysis requirements, and research needs. Some of these issues are discussed here (Uddin, 1992).

- *Segmentation Complexity*: The reference location code used to uniquely designate uniform PMS road sections needs further extension if traffic pattern or pavement thickness is also required for homogeneous segmentation. System complexity will further increase if dynamic segmentation is also warranted using one or more condition attribute (like roughness and deflection) for detailed project-level reconstruction and rehabilitation design. However, these issues can be handled by good system analysis and computer programing.

- *Extent of Condition Data*: Distress and other attributes are collected but not all of them used by the maintenance planning and budgeting analysis software. Many agencies routinely collect deflection data for project-level applications, but lack a rational network level PMS analysis considering deflection data. System complexity will also increase if condition evaluation of each lane is required for comprehensive maintenance analysis. Necessary extrapolations will be required for the network budget analysis if the condition data are collected only on a small fraction of selected sample sections.
- *Single-year or Annual M,R&R Program and Budget*: More complexity and computational effort needed if mathematical optimization is used. Heuristic methods, such as the incremental B/C ratio, are reasonable options in the beginning.
- *Uncertainty in Condition Deterioration and Performance Models*: These are important issues to consider in the implementation of the generated M,R&R work program reports and LCC and benefit calculations. However, the PMS feedback from the M,R&R implementation and historical condition data should be used to enhance the prediction models.
- *Multiyear M,R&R Program and Budget*: Multiyear M,R&R program and budget analysis require the use of appropriate pavement deterioration prediction models, maintenance intervention policies and user cost models. This will increase the complexity of the analysis software and input data requirements significantly.
- *Maintenance Contract Packaging*: Reasonable maintenance contract packaging is probably the most useful output from the single-year maintenance program report that the PMS users would like to get. Most working PMS software packages lack an automated computerized output of a rational list of candidate M,R&R contract packages for larger segments of highways and roads. This will require complex analysis including lane-by-lane condition data analysis, across-the-lane rationalization, and across-the-adjacent sections rationalization of M,R&R alternatives.

18.12.7   Feedback and Other Institutional Issues

The historical inventory data, segmentation and location reference codes should be updated whenever a major maintenance is performed, changes in uniform pavement section occur, or when new roads are constructed. Traffic database entries should be updated as the new data becomes available. The monitoring database files should be maintained by survey dates so that all historical condition data files are accessible for the development of improved pavement deterioration prediction models and other project-level applications. Reasonable guidelines should be developed for data collection frequency based on the results of a pilot study. Maintenance and construction performance and cost data should be monitored and considered as a critical feedback to improve condition prediction models and M,R&R analysis.

Quality control (QC) should be exercised by following manuals and protocols implemented for specific data collection and processing activities. Check lists, random checking and correction protocols should be used to ensure data quality assurance (QA). The use of built-in permissible ranges is the most acceptable practice to ensure database integrity.

Computer system maintenance and database Security are equally important concerns. Computer system maintenance for hardware and software is an important activity for smooth PMS operation. An appropriate back-up protocol is important for database security. Both on-site and off-site permanent back-up mediums should be saved on regular basis.

All institutional issues related to the PMS development and operation in a highway agency must be recognized and resolved through a designated PMS champion in the agency. Pavement management activities and information generated by the PMS reports assist technical staff and decision makers at all levels and it may require education process through appropriate training and periodic participation in short courses, workshops, and conferences. Proper communication among the users of the PMS reports is the key to PMS success.

18.13 PMS Case Studies

18.13.1 Network Level Applications

Key issues related to the PMS system development, implementation and routine operation have been discussed. Examples of various case studies, presented in the pavement management conferences and literature, show that there is no ideal single system that is suitable for universal applications. Based on the expected level of system complexity, the requirements must be tailored to fit the road network environment and the agency objective. The following case studies of PMS development and implementation at different levels of system complexity demonstrate the significant impact of an agency's specific needs and resources on the final PMS outputs.

18.13.2 The United States — Kansas PMS for Network Level Applications

The Kansa Department of Transportation (KDOT) developed the network level PMS in the 1980s for a highway network of 10,017 miles (16,027 Km) maintained by the state. The distribution of this mileage by pavement type is as follows: 7.3% Portland cement concrete pavement (PCCP), 10.8% Composite (asphalt over concrete), 28.1% asphalt — full design bituminous (FDBIT), and 53.8% asphalt — partial design bituminous (PDBIT). The location reference system is the county, route or milepost system. The pavement evaluation equipment used by KDOT includes the following; KJ Law Surface Dynamics pavement surface friction testers (2), Dynaflect deflection equipment (2), Photolog equipment (1), trailer-mounted Mays Meters (3), network level survey vans with computerized equipment for data collection (3), and South Dakota profilometer (1). The description of the PMS is extracted from the FHWA advanced PMS course notebook (FHWA, 1990).

The network optimization system (NOS) is designed to identify pavement maintenance and rehabilitation policies which would minimize total costs subject to meeting desired network performance standards, or maximize performance standards for a fixed budget. For the PMS, the highway system is divided into one-mile long segments to facilitate modeling using the Markov technique. For computational convenience, the statewide highway network is divided into a total of 23 road categories which are defined using the following factors: (a) functional classification, (b) pavement type, (b) roadway width, and (c) traffic loading. At the network level, distresses of individual, mile-long segments are monitored yearly. The distresses selected for the different pavement types are: PCCP (roughness, joint distress, faulting), Composite (roughness, transverse cracking, block cracking), FDBIT (roughness, transverse cracking, block cracking), and PDBIT (roughness, transverse cracking, fatigue cracking). In addition to these distress types, rutting is monitored on all pavement types and used in safety evaluations.

The Markov optimization model in NOS uses "condition states" to evaluate cost and performance. Condition states are defined as specific combinations of distress levels and levels of variables that influence the rate of pavement deterioration. Two influence variables are used — index to first distress and rate of change in the distress. A total of 216 possible condition-states are defined for each pavement types. The network level monitoring program determines the current condition state of each individual section. "Distress States" are subsets of the sets of condition states and are defined as combinations of the three levels of each distress type. The total 27 distress states are used to simplify assignment of feasible rehabilitation actions, costs, and prediction models. The NOS identifies optimal M,R&R actions for individual roadway sections. The major M,R&R outputs of NOS include: (a) annual "minimum" rehabilitation budgets over a selected planning horizon (such as 5 years), (b) locations of candidate rehabilitation projects, (c) minimum performance requirements for a fixed budget, and (d) optimal rehabilitation actions. For practical purposes, several contiguous sections are combined to form a single project. The system also assists in the development of project pavement design. However, the system uses of a complex optimization methodology, which is computer intensive and requires an annual network survey.

18.13.3   The United States — Mississippi PMS for Network Level Applications

The Mississippi Department of Transportation (MDOT) has implemented a network level PMS for its 12,000 ml (19,200 Km) of state-maintained highway network, using a customized inventory database, an exclusive PCR scale, and pavement performance models developed for asphalt, asphalt overlaid PCC, jointed PCC, and continuously reinforced concrete pavements (George et al., 1994). The information management system was developed starting in the late 1980s, followed by the implementation of condition data collection and analysis software in the early 1990s. High speed automated survey equipment for video imaging, rutting measurement, and longitudinal profiling for roughness has been deployed on the outside travel lane of all mainline pavements for annual surveys.

Specially created interfaces programs enhance the video condition database developed from the office sampling and distress data processing of video footage. The M,R&R treatments are selected using a decision tree that considers roughness, rutting and video distress data. The customized PMS analysis software generates an annual prioritized maintenance work programs and budget analysis reports for the network based on agency costs and the life-cycle VOC (George et al., 1994). The PMS has been integrated with the MDOT GIS database and it can be accessed from the MDOT district offices to allow partially decentralized operations.

18.13.4   The United Arab Emirates — Dubai PMS for Network Level Applications

As a part of the United Nations technical assistance program to Dubai Municipality in UAE, a comprehensive PMS project was initiated in 1989. This unique United Nations project resulted in the development and implementation of Dubai Road Pavement Management System (DRPMS) within two years. The project was completed with in-house training, visual distress data collection and processing efforts, and the adoption of the MicroPAVER 2.1 analysis software. Customized interface programs were developed to enhance the MicroPAVER maintenance program reports for local terminology, priority ranking, and extrapolation of annual and multiyear budgets for the entire network by functional class (Uddin, 1992, 1995; Uddin and Al-Tayer, 1993).

The customized inventory database contains over 100 data elements related to all classes of highways and urban roads. The enhanced maintenance reports were generated using the distress data collected on 7% of the entire asphalt road network of 3,400 lane-Km. The overall pavement condition of the network, based on the 1991 survey data analysis, was very good with an average PCI of 81. The analysis produced a maintenance and rehabilitation program report for major arterial roads. A new PMS Unit was established with locally trained staff and customized user manuals, which is fully operational within the Roads Department since 1991.

18.13.5   Portugal — Lisbon PMS for Network Level Applications

The following description is based on a paper by Ferreira et al. (2003). In the beginning of 1999 the City Council of Lisbon, in Portugal, decided to build a GIS-based PMS for the vast road network under its administration. The ArcView GIS was implemented to visualize the information stored and treated within the road network database. Nodes and segments that constitute the road network have a unique system of number identification. The "VIZIROAD" equipment for visual survey was installed on a commercial vehicle (Ford Transit). The "VIZIROAD" data acquisition system is based on a notebook computer two keyboards to enter field observations.

The Quality Evaluation Tool was mainly used to analyze the present pavement state. This tool calculates a modified version of the PSI for each road segment, which ranges from zero (poor condition) to five (new pavement). The PSI calculation was applied for the Lisbon road network by assigning specific weights to condition parameters. These are: longitudinal roughness, rutting depth, cracking, and superficial degradation (combined areas of zones with pot-holes and lost of surface material and areas of zones with patching) influence. Minimum acceptable condition quality level criteria for the M,R&R intervention are: Cracking (area = 10% maximum), Rutting (average depth = 10 mm maximum), Superficial degradations (area = 10% maximum), Longitudinal roughness (IRI = 3500 mm/Km maximum), and Global quality index (PSI = 2.5 minimum). PSI equal to 2.5 is the level that indicates the necessity for the M,R&R intervention on the pavement.

The pavement performance models describe pavement condition with regard to the above five degradations. Vehicle operating costs are calculated and budget constraints are specified. The Decision-Aid Tool proposed for the Lisbon PMS is an optimization model aimed at minimizing the expected costs of M&R actions over a given planning period, keeping some specified road network standards. The optimization model uses deterministic pavement performance models and allows the definition of the M&R actions that must be applied in road segments in each year of the planning period. It can be solved through heuristic methods. In this study, the genetic algorithm, called GENETIPAV-D, was developed to solve the model. The method follows the GA research initiated by Fwa in the 1990s (Chan et al., 1994; Fwa et al., 1998). More detailed information on the model and the results of a case study, involving a subset network of 161 Km partitioned into 1244 segments, are provided by Ferreira et al. (2003).

18.13.6 Project-Level PMS Applications

The evaluation of *in situ* structural capacity and design of overlay thickness are good examples of the project-level application of the PMS database. Nondestructive deflection data, collected on selected or flagged candidate sections only, are stored off-line because of the large data file size, and are analyzed for structural capacity using stand-alone computer programs at the project-level. Results from deflection analysis and modulus backcalculation programs are used to compute remaining life and to design overlay thickness, with some summary results entered into the network level database. On the other hand, some agencies use standard design alternatives based on traffic and roadbed soil categories. The thickness design for resurfacing and overlays can be computed using standard AASHTO design guide procedure (AASHTO, 1986, 1993) using pavement information from the PMS database for the project-level applications. Some design procedures can generate an array of rehabilitation designs for both concrete and flexible pavements. The pavement design procedure can be computerized for applications at the project-level and network level. Several examples are presented by Molenaar for the project-level applications of deflection data and performance models (Molenaar, 2003). More discussion is presented in a later chapter on structural evaluation of pavements.

18.14 Innovative Technologies and Future Directions

As the pavement condition deteriorates, the user cost and frequency of public complaints increase. Pavement deterioration and congestion also result in increased vehicular emission, greater air quality degradation, and more pollution of stream and subsurface water due to pavement surface water runoff. In the future, the adverse effects of all these factors on public health risks (and associated medical costs) as quantifiable societal costs should be investigated further. These effects have been ignored in the PMS until now. Preservation and improvement in the pavement condition through timely implementation of PMS maintenance and rehabilitation programs will result in tremendous payoffs including increased public satisfaction, reduced vehicle operating and other user costs, as well as other related societal costs of lost time and worker productivity. Some related key areas of technology innovations and future directions to improve the pavement management process are discussed in the following sections.

18.14.1   Pavement Evaluation Technologies

An excellent overview of pavement evaluation technologies, projected improvements in methods for pavement condition data collection, technology innovations, and potential pay off was made at the second North American Pavement Management conference (Hudson et al., 1987a). Automation, productivity and speed, cost-effectiveness, and the use of objective measurement approaches were identified as the primary objectives for applying innovative technologies to pavement monitoring and evaluation with emphasis on using nondestructive and noncontact methods, real-time computerized data handling, and reducing operating constraints (like traffic interference and environmental conditions). Data interpretation for some of these relatively new technologies needs further refinement with the help of independent tests and field data.

- GPR is being used to nondestructively determine the surface layer thickness, delamination of concrete overlays on rigid pavements, and detection of voids beneath concrete pavements. The interpretation of the response outputs is still the most critical element in correct and effective use of radar.
- Multipurpose pavement evaluation vehicles offer potential benefits because they are capable of measuring several components of pavement condition at a greater savings of time and costs. Such pavement evaluation technologies are also useful for heavily trafficked urban pavements where traffic interference is a major operating constraint for pavement monitoring and evaluation.
- Inadequate automatic pattern recognition techniques are being used currently to extract distress data from the video image recorded by the multipurpose vehicles. Commercially available or public domain software products are essential to get the full benefit of the video images and digital photos.
- Inadequate testing and quality control exists on newly purchased equipment. Believing in claims made by sales people for new equipment purchase without adequate evaluation, acceptance testing, and trial runs is a major concern. A practical code of ethics is needed to solve this problem. Often the new equipment is compared to the routinely used traditional equipment and the new measurements are claimed to have greater accuracy. Therefore, standards must be developed to assure quality of data. ASTM is involved in setting quality control standards for accuracy and repeatability of several pavement evaluation technologies.

18.14.2   Traffic Monitoring and ITS Technologies

Traffic volume and axle data are the most important inputs for designing reliable pavements and M,R&R strategies. Being a moving target and due to its nature of having a high variability in fleet mix, traffic volume and axle loads are the least reliable data available for pavement management applications at both network level and project-level. Traditional traffic monitoring technologies are being improved with new sensors. New opportunities are available from the ITS video surveillance equipment and real-time data acquisition, which are being installed in many urban corridors and congested areas.

- Several studies were carried out before and during the SHRP-LTPP project in the U.S. For example, one study evaluated several types of portable equipment for monitoring traffic data including traditional surface mounted sensors such as pneumatic tubes and loop detectors, as well as a wide variety of sensors for vehicle detection from above or at the side of roadway. Radar is the most common of these devices and not only senses the vehicle but measures speed. Other sensors used include microwave, sonic, optical, infrared, video, laser, magnetic, seismic or radio frequency devices. The study identified piezoelectric cable, infrared, and laser sensors as having strong potential for use at temporary locations for traffic data collection (Cunagin et al., 1986).
- Reasonable estimates of actual vehicle axle loads are essential for pavement management activities. Significant strides have been made in this area in the last two decades. The equipment

manufacturer's claims about instrument accuracy and quality need to be checked. Caution is needed in evaluating low-cost alternatives such as piezoelectric cable and rubber mat systems. Claims of accuracy of $\pm 5\%$ or less are unjustified. In general, the accuracy of such systems is inversely related to cost and is nonlinear (Hudson et al., 1987a).

- The ITS deployment projects have seen tremendous growth in commercial technologies and applications throughout the world. The successful payoff of enormous investments in ITS technologies and their deployment is well quantified (ITS, 2003) because they address and resolve many problems related to the safety and comfort of the traveling public on roads. The related ITS technologies include driver's information systems, video surveillance of traffic, GPS based real-time tracking of public transit, evaluation of WIM technologies to improve commercial vehicle traffic flow, and airborne and spaceborne remote sensing technologies for enhanced GIS base maps and other uses (Uddin, 2002d). The PMS can get direct benefits from the ITS technologies deployed by the agency to improve traffic data and safety assessment of heavily trafficked and congested segments of highways, especially in urban areas.

18.14.3 New Materials, Acceptance Testing, and Analysis Methods

Many new materials have been developed in recent years and applied on pavements, which show superior performance compared to traditional materials. Diminishing material resources and environmental concerns require the use of more and more recycled and waste materials. Advanced analysis methods for improved pavement design and smoothness acceptance testing should be the goal of future enhancements in the PMS activities, as discussed here.

- New and improved performance models for the recycled pavement materials, waste materials, and new materials and construction methods are needed to realize their benefits in terms of the extended life even at a higher initial cost. With reliable performance models, these new alternative strategies can be justified based on the LCC analysis and higher B/C ratio. Some examples are: rutting resistant polymer-modified and rubber–asphalt pavements, high-performance concrete pavements, preventive maintenance and preservation strategies to prolong the pavement life, and the recent concept of perpetual pavements for heavily trafficked highways.
- Smoothness acceptance testing of the paved surface of a newly constructed pavement or rehabilitated pavement is the first condition or serviceability data-point on the performance curve, which should be entered into the PMS database as a monitoring feedback. This initial condition should be established by using reliable measuring equipment, such as that used for roughness. However, the smoothness measurement must be very precise, accurate, and repeatable with a high resolution to be legally defensible (Hudson et al., 1987c; Woodstrom, 1990).
- Advanced analysis and computer simulation methods for the correct analysis of the deflection history data should be required for the project-level pavement design on large highway projects. The deflection- and load-time history data have been collected since the early 1990s on the LTPP sections in the U.S., and are currently collected by all FWD models. The traditional myth that the there dimensional-finite element (3D-FE) dynamic analysis requires costly computing sources and time is not valid in today's and future computing environments. Several examples are available in the literature where the 3D-FE modeling and simulation conducted in the early 1990s on a supercomputer is now being performed on desktop computers (Uddin et al., 1994, Garza, 2003; Uddin, 2003b).

18.14.4 Innovative Remote Sensing and Spatial Technologies

Environmental concerns should be addressed while planning and implementing the construction or expansion of transportation corridors for sustainable community and transportation development. New airborne and spaceborne remote sensing technologies can be used for cost-effective environmental assessment of transportation projects.

- The societal cost, associated with the public health and related medical expenses from the adverse effects of degradations of water and air quality, is enormous. Therefore, the evaluation of road upgrading and expansion should consider innovative M,R&R alternatives in the PMS tradeoff analysis. An example is the use of pervious concrete on shoulders, and peripheral areas to reduce runoff pollution in watersheds, streams, and subsurface water sources. Good pavement condition and improved traffic flows, combined with good horticulture and tree plantation practices in highway corridors, can improve air quality by reducing vehicle emissions and ground-level Ozone pollution. These considerations can be used to quantify additional benefits in the LCC analysis at the project-level and network level.
- Modern airborne and spaceborne remote sensing and spatial technologies provide a unique opportunity to expedite and streamline environmental assessment of transportation projects. These technologies include airborne laser or Light Detection And Ranging (LIDAR) and high-resolution commercial spaceborne satellite imagery, which are available at affordable costs. The LIDAR topographic mapping is a relatively new airborne survey technology that can be used to survey large areas and produce digital terrain maps by reducing months of manual traditional surveying to weeks. The reader is encouraged to read the related literature for more information (Al-Turk and Uddin, 1999; Uddin et al., 2001; Uddin, 2002a, 2002b, 2002c, 2002d; USDOT, 2002).
- The use of the GIS for mapping and PMS applications just started in the early to mid 1990s (Uddin, 1992; Hudson et al., 1997) that provided two dimensional (2D) displays of the transportation network on topographic maps. The accuracy of these maps can be tremendously increased by developing road planimetrics on the high-resolution satellite imagery layer by using the GIS tools, as shown in Figure 18.7. With the availability of 15-cm accurate LIDAR data, three dimensional (3D) virtual reality models of highway corridors can be created by draping the geo-referenced satellite imagery or aerial photo over the LIDAR data. The 3D visualization in the virtual world can help to analyze traffic safety aspects (due to presence of commercial truck traffic), geometric safety aspects (line of sight and obstructions), and the use of vector maps for evaluation of traffic parameters and related air pollution mapping from mobile sources and congestion scenarios.

Ultimately, in the world of diminishing natural resources all those involved in the pavement management process must embrace new technology innovations for enhancing PMS capabilities to improve the highway corridor management, serve the public, support the sustainable socioeconomic growth, and preserve the environment.

References

AASHO, 1962. *The AASHO Road Test Report 5: Pavement Research*, Special Report 61 E. Highway Research Board, National Research Council, Washington, DC.

AASHTO, 1986. *AASHTO Design Guide for Pavement Structures*. American Association of State Highway and Transportation Officials, Washington, DC.

AASHTO, 1990. *AASHTO Guidelines of Pavement Management System*. American Association of State Highway and Transportation Officials, Washington, DC.

AASHTO, 1993. *AASHTO Design Guide for Pavement Structures*. American Association of State Highway and Transportation Officials, Washington, DC.

Al-Turk, E. and Uddin, W. 1999. *Infrastructure Inventory and Condition Assessment using Airborne Laser Terrain Mapping and Digital Photography*, Transportation Research Record 1690. Transportation Research Board, National Research Council, Washington, DC, pp. 121–125.

ASTM, 1999. American Society for Testing and Materials, *Standard Practice for Roads and Parking Lots Pavement Condition Index Survey*, Designation: D6433. Book of ASTM Standards.

ASTM, 2003. American Society for Testing and Materials. Skid Resistance of Paved Surfaces using a Full-Scale Tire. Designation: E 274-97. Book of ASTM Standards, Vol. 04.03, pp. 902–907.

Attoh-Okine, N.O. 1998. Artificial Immune Systems and Networks for Pavement Deterioration Process. *Proceedings, International Workshop on Artificial Intelligence and Mathematical Methods in Pavement and Geomechanical Systems*, Miami, Florida, November, pp. 73–81.

Attoh-Okine, N.O. 2000. Computational Intelligence Applications in Pavement and Geomechanical Systems. *Proceedings, Second International Workshop on Artificial Intelligence and Mathematical Methods in Pavement and Geomechanical Systems*, Newark, Delaware, August 11–12, 219 p.

Benedetto, A. and Angio, C. 2003. Macrotexture Effectiveness Maintenance for Aquaplaning Reduction and Road Safety. *Proceedings, MAIREPAV — Third International Symposium on Maintenance and Rehabilitation of Pavements and Technological Control*, University of Minho, Guimarães, Portugal, July 7–11, pp. 545–554

Benson, K.R., Elkins, G.E., Uddin, W., and Hudson, W.R. 1988. *Comparison of Methods and Equipment to Conduct Pavement Distress Surveys*, Transportation Research Record 1196. Transportation Research Board, Washington, DC, pp. 40–50.

Boriboonsomsin, K., Transportation-Related Air Quality Modeling and Analysis Based on Remote Sensing and Geospatial Data, Ph.D. Dissertation, Department of Civil Engineering, The University of Mississippi, May 2004.

Chan, W.T., Fwa, T.F., and Tan, C.Y., Road maintenance planning using genetic algorithms: formulation, *J. Transport. Eng., ASCE*, 120, 5, 693–709, 1994.

Chan, W.T., Fwa, T.F., and Tan, J.Y., Optimal fund-allocation analysis for multi-district highway agencies, *ASCE J. Infrastructure Syst.*, 9, 4, 167–175, 2003.

Cook, W.D. and Lytton, R.L. 1987. Recent Developments and Potential Future Directions in Ranking and Optimization Procedures for Pavement Management. *Proceedings, Vol. 2, 2nd North American Pavement Management Conference*, Toronto, November.

COST, 1997. *New Rpad Monitoring Equipment and Methods*, COST 325 Final Report of the Action. European Commission, Directorate General Transport, Brussels, Luxembourg.

Cunagin, W.D., Grubbs, A.B. Jr., and Ayoub, N.A. 1986. *Portable Sensors and Equipment for Traffic Data Collection*, Transportation Research Record 1060. Transportation Research Board, Washington, DC, pp. 127–139.

Datapave, 2001. Explore and Retrieve LTPP Data, Datapave CD. Federal Highway Administration, U.S. Department of Transportation. http://www.datapave.com\

Draper, N. and Smith, H. 1981. *Applied Regression Analysis*, 2nd Ed. Wiley, New York.

Epps, J.A. and Monismith, C.L. 1986. *Equipment for Obtaining Pavement Condition and Traffic Loading Data*, NCHRP Synthesis 126. National Cooperative Highway Research Program, National Research Council, Washington, DC.

Ergun, M., Iyinam, S., and Iyinam, A.F. 2002. Does Surface Texture of Asphalt Pavement Affect Skid Resistance of Road? *CD Proceedings, 3rd International Conference on Bituminous Mixtures and Pavements*, Thessaloniki, Greece, November 21–22.

Ferreira, A., Picado-Santos, L., Antunes, A., and Pereira, P. 2003. A Deterministic Optimization Model Proposed for the Lisbon's PMS. *Proceedings, MAIREPAV — Third International Symposium on Maintenance and Rehabilitation of Pavements and Technological Control*, University of Minho, Guimarães, Portugal, July 7–11, pp. 793–804.

FHWA, 1989. Pavement Management Systems, a National Perspective, FHWA-PAVEMENT Newsletter, PAVEMENT, Issue 14, Federal Highway Administration Newsletter, Washington, DC, Spring.

FHWA, 1990. Advanced Course in Pavement Management Systems, Course Notebook, Federal Highway Administration Newsletter, Washington, DC.

FHWA, 1998. Life-Cycle Cost Analysis in Pavement Design, Report FHWA-SA-98-079, Federal Highway Administration, Washington, DC, September.

FHWA, 2001. Traffic Monitoring Guide, Office of Highway Policy Information, Federal Highway Administration, Washington, DC.

Finn, F.N. 1994. Keynote Address. *Proceedings, Third International Conference on Managing Pavements*, San Antonio, Vol. 3, pp. 9–15.

Finn, F.N., Kulkarni, R., and Nair, K. 1974. *Pavement Management System Feasibility Study*, Final Report. Prepared for Washington Highway Commission, Department of Highways, Olympia, Washington.

Flintsch, G.W. 2003. Pavement Management Enhancement using Soft Computing. *Proceedings, MAIREPAV — Third International Symposium on Maintenance and Rehabilitation of Pavements and Technological Control*, University of Minho, Guimarães, Portugal, July 7–11, pp. 783–792.

Flintsch, G.W., Scofield, L.A., and Zaniewski, J.P. 1994. *Network Level Performance Evaluation of Asphalt–Rubber Pavement Treatments in Arizona*, Transportation Research Record 1435. Transportation Research Board, National Research Council, Washington, DC.

French, A. and Soloman, D. 1986. *Traffic Data Collection and Analysis: Methods and Procedures*, NCHRP Synthesis 130. National Cooperative Highway Research Program, National Research Council, Washington, DC.

Fwa, T.F. and Sinha, K.C., Pavement performance ands life-cycle cost analysis, *J. Transport. Eng., ASCE*, 117, 1, 33–46, 1991.

Fwa, T.F. and Chan, W.T., Priority rating of highway needs by neural networks, *J. Transportation Eng., ASCE*, 118, 3, 419-432, 1993.

Fwa, T.F. and Shanmugam, R. 1998. Fuzzy Logic Technique for Pavement Condition Rating and Maintenance-Needs Assessment. *Proceedings, Fourth International Conference on Managing Pavements*, Durban.

Fwa, T.F., Tan, C.Y., and Chan, W.T., Road maintenance planning using genetic algorithms: analysis, *J. Transport. Eng., ASCE*, 120, 5, 710–722, 1994.

Fwa, T.F., Chan, W.T., and Hoque, K.Z. 1998. *Analysis of Pavement Management Activities Programming by Genetic Algorithms*, Transportation Research Record 1643. Transportation Research Board, National Research Council, Washington, DC, pp. 1–7.

Fwa, T.F., Chan, W.T., and Hoque, K.Z., Multi-objective optimization for pavement management programming, *J. Transport. Eng., ASCE*, 126, 5, 367–374, 2000.

Garza, S.G., Integration of Pavement Nondestructive Evaluation, Finite Element Simulation, and Air Quality Modeling for Enhanced Transportation Corridor Assessment and Design, Ph.D. Dissertation, Department of Civil Engineering, The University of Mississippi, May 2003.

George, K.P., Uddin, W., Ferguson, J.P., Crawley, A.B., and Shekharan, A.R. 1994. *Maintenance Planning Methodology for Statewide Pavement Management*, Transportation Research Record 1455. Transportation Research Board, National Research Council, Washington, DC, pp. 123–131.

Ghaboussi, J., 1992. Potential Application of Neuro-biological Computational Models in Geotechnical Engineering, *Numerical Models in Geomechanics*, In Pand, G.N. & Pietruszczak, S.P. eds., Rotterdam, pp. 543-555.

Golabi, K., Kulkarni, R.B., and Way, G.B., A statewide pavement management system, *Interfaces*, 12, 6, 5–21, 1982.

Haas, R.C.G. 1973. *A Method for Designing Asphalt Pavements to Minimize Low Temperature Shrinkage Cracking*, Research Report 73-1. Asphalt Institute, Lexington, Kentucky.

Haas, R. 2003. Good Technical Foundations are Essential for Successful Pavement Management, Invited Lecture, Proceedings, MAIREPAV — Third International Symposium on Maintenance and Rehabilitation of Pavements and Technological Control, University of Minho, Guimarães, Portugal, July 7–11, pp. 3–28.

Haas, R. and Hudson, W.R. 1978. *Pavement Management Systems*. McGraw-Hill, New York.

Haas, R., Hudson, W.R. and Zaniewski, J.P. 1994. *Modern Pavement Management*. Krieger Publishing Company, Malabar, Florida.

Hajek, J.J., Phang, W.A., Prakash, A., and Wrong, G.A. 1985. Performance Prediction for Pavement Management. *Proceedings, North American Conference on Pavement Management*, Toronto, Vol. 1, pp. 4.122–4.134.

HDM, HDM-4: http://hdm4.piatc.org/info/HDMfeatures-e.htm, accessed February 2002.Hudson, W.R. and McCullough, B.F. 1973. *Flexible Pavement Design and Management Systems Formulation,*

NCHRP Report 139. National Cooperative Highway Research Program, National Research Council, Washington, D.C.

Hudson, W.R. and Uddin, W. 1987. Future Evaluation Technologies: Prospective and Opportunities. *Proceedings, 2nd North American Pavement Management Conference*, Toronto, Vol. 3, November, pp. 3.233–3.258.

Hudson, W.R. and Haas, R. 1994. What are the True Costs and Benefits of Pavement Management. *Proceedings, Third International Conference on Managing Pavements*, San Antonio, Vol. 3, pp. 29–36.

Hudson, W.R., Kher, R.K., and McCullough, B.F. 1972. *A Working System Model for Rigid Pavement Design*, Highway Research Record 407, Transportation Research Board, Washington, D.C.

Hudson, W.R., Haas, R., and Pedigo, R.D. 1979. *Pavement Management System Development*, NCHRP Report 215. National Cooperative Highway Research Program, National Research Council, Washington, DC.

Hudson, W.R., Uddin, W., and Elkins, G.E., 1987c. Smoothness Acceptance Testing and Specifications for Flexible Pavements. *Proceedings, Second International Conference on Pavement Management*, Toronto, Vol. 3, November 1987, pp. 3.331–3.362.

Hudson, W.R., Haas, R., and Uddin, W. 1997. *Infrastructure Management*. McGraw-Hill, New York.

Hudson, W.R., McCullough, B.F., Scrivner, F.H., and Brown, J.L. 1970. *A Systems Approach Applied to Pavement Design and Research*, Research Report 123-1. Texas Highway Department, Texas Transportation Institute, Texas A&M University, and the Center for Highway Research, The University of Texas at Austin, March.

Hudson, W.R., Elkins, G.E., Uddin, W., and Reilley, K. 1987a. *Improved Methods and Equipment to Conduct Pavement Distress Surveys*, Report No. FHWA-TS-87-213. ARE, Inc. Report for Federal Highway Administration.

Hudson, W.R., Elkins, G.E., Uddin, W., and Reilley, K. 1987b. *Evaluation of Deflection Measuring Equipment*, Report No. FHWA-TS-87-208. ARE, Inc., Report for Federal Highway Administration.

Hudson, W.R., Finn, F.N., McCullough, B.F., Nair, K., and Vallerga, B.A. 1968. *Systems Approach to Pavement Design, System Formulation, Performance Definition, and Materials Characterization*, Final Report, NCHRP Project 1-10. Materials Research and Development, Inc., Oakland, California.

Humplick, F. 1992. *Identifying Error-Generating Factors in Infrastructure Condition Evaluations*, Transportation Research Record 1344. Transportation Research Board, National Research Council, Washington, DC, pp. 106–115.

ITS, 2003. *Intelligent Transport Systems Benefits and Costs: 2003 Update*, Report FHWA-OP-03-075. Federal Highway Administration, U.S. Department of Transportation, Washington, DC.

Jackson, N. and Mahoney, J.P. 1990. *Washington State Pavement Management System*, Advanced Course in Pavement Management Notebook. Federal Highway Administration, Washington, DC.

Janoff, M.S. 1988. *Pavement Roughness and Rideability Field Evaluation*, NCHRP Report 308. National Cooperative Highway Research Program, National Research Council, Washington, DC.

Jung, F.W. and Phang, W.A. 1974. *Elastic Layer Analysis Related to Performance in Flexible Pavement Design*, Transportation Research Record 521. Transportation Research Board, National Research Council, Washington, DC, pp. 14–29.

Kher, R. 1985. Forward. *Proceedings, North American Conference on Pavement Management*, Toronto, Vol. 1.

Kher, R. and Phang, W. A. 1977. OPAC Design System (Ontario Pavement Analysis of Costs). *Proceeding, 4th International Conference on Structural Design of Asphalt Pavements*, Michigan, and MTC Report IR61, August.

LeClerc, R.V. and Nelson, T.L. 1982. Washington States Pavement Management Systems. *Proceedings, 5th International Conference on the Structural Design of Asphalt Pavements*, Vol. 1, August, pp. 534–552.

Losa, M., Bonomo, G., Licitra, G., and Cerchiai, M. 2003. Performance Degradation of Porous Asphalt Pavements. *Proceedings, MAIREPAV — Third International Symposium on Maintenance and*

Rehabilitation of Pavements and Technological Control, University of Minho, Guimarães, Portugal, July 7–11, pp. 475–484.

Markow, J.J., Harrison, F.D., Thompson, P.D., Harper, F.A., Hyman, W.A., Alfelor, R.M., Mortenson, W.G., and Alexander, T.M. 1994. *Role of Highway Maintenance in Integrated Management Systems*, NCHRP Report 363. National Cooperative Highway Research Program, National Research Council, Washington, DC.

Mobility, 2001. *Mobility 2001 — World Mobility at the End of the Twentieth Century and its Sustainability*. World Business Council for Sustainable Development, Geneva, Switzerland.

Molenaar, A.A.A. 2003. Pavement Performance Evaluation and Rehabilitation Design, Invited Lecture. *Proceedings, MAIREPAV — Third International Symposium on Maintenance and Rehabilitation of Pavements and Technological Control*, University of Minho, Guimarães, Portugal, July 7–11, pp. 29–70.

MOT, 1987. *Proceedings, Second International Conference on Pavement Management*, Ministry of Transportation of Ontario, Toronto, November.

Murad, M.M., Asphalt concrete pavement friction: variables related, accident-models, and the role in design and maintenance practices, *Int. J. Pavements, IJP 2004*, 3, 1–2, 2004, 63–75.

NCHRP, 1985. *Portland Cement Concrete Pavement Evaluation System*, NCHRP Report 277. National Cooperative Highway Research Program, Transportation Research Board, National Research Council, Washington, DC.

NOAA, 1994. *Climatological Data*. National Oceanic and Atmospheric Administration, National Climatic Center, Asheville, North Carolina.

Papagiannakis, A.T. 2003. Accounting for Agency and User Costs in Pavement Life-Cycle Cost Analysis. *Proceedings, MAIREPAV — Third International Symposium on Maintenance and Rehabilitation of Pavements and Technological Control*, University of Minho, Guimarães, Portugal, July 7–11, pp. 805–814.

Paterson, W.D.O. 1986. *International Roughness Index: Relation to Other Measures of Roughness and Riding Quality*, Transportation Research Record 1084, Pavement Roughness and Skid Resistance. Transportation Research Board, National Research Council, Washington, DC, pp. 49–59.

Paterson, W.D.O. and Attoh-Okine, B. 1992. *Simplified Models of Paved Road Deterioration Based on HDM-III*, Transportation Research Record 1344. Transportation Research Board, National Research Council, Washington, DC, pp. 99–105.

Peterson, D. 1985. *Life-Cycle Cost Analysis of Pavements*, Synthesis 122. National Cooperative Highway Research Program, Transportation Research Board, National Research Council, Washington, DC.

Queiroz, C.A.V., A Mechanistic Analysis of Asphalt Pavement Performance in Brazil, Proceeding, Association of Asphalt Paving Technology, Vol. 52, 1983, pp. 474–488.

Queiroz, C., Haas, R., and Cai, Y. 1994. *National Economic Development and Prosperity Related to Paved Road Infrastructure*, Transportation Research Record 1455. Transportation Research Board, National Research Council, Washington, DC, pp. 147–152.

Rauhut, J.B., Lytton, R.L., and Darter, M.I. 1982. *Pavement Damage Functions for Cost Allocation*, Report No. FHWA/RD-82/126. Federal Highway Administration, U.S. Department of Transportation, Washington, DC.

Reigle, J. and Zaniewski, J.P., Probablistic life-cycle cost analysis for pavement management, *Int. J. Pavements, IJP 2002*, 1, 2, 71–83, 2002.

Roberts, F.L., Qin, H., and Mohammad, L.N., Modeling permanent deformation of the Louisiana ALF crumb rubber modified test sections, *Int. J. Pavements, IJP 2003*, 2, 1–2, 100–111, 2003.

Sayers, M.W., Gillespie, T.D., and Queiroz, C.A.V. 1986. *The International Roughness Index Experiment: a Basis for Establishing a Standard Scale for Road Roughness Measurements*, Transportation Research Record 1084, Pavement Roughness and Skid Resistance. Transportation Research Board, National Research Council, Washington, DC, pp. 76–85.

Shahin, M.Y. and Walther, J.A. 1990. Pavement Maintenance Management for Roads and Streets using the PAVER System, U.S.A. CERL Technical Report M-90/05, Champaign, IL.

Shahin, M.Y., Kohn, S.D., Lytton, R.L., and McFarland, W.F. 1987. Pavement Budget Optimization using the incremental Benefit–Cost Technique. *Proceedings, Second International Conference on Pavement Management*, Toronto, Vol. 3, November.

SHRP, 1989a. *SHRP-LTPP Guide to Field Sampling and Handling*, SHRP 5021. Strategic Highway Research Program, National Research Council, Washington, DC, revised 1992.

SHRP, 1989b. *SHRP-LTPP Laboratory Guide for Testing Pavement Samples*, SHRP 5025. Strategic Highway Research Program, National Research Council, Washington, DC.

SHRP, 1993. *Distress Identification Manual for the Long-Term Pavement Performance Project*, Report SHRP-P-338. Strategic Highway Research Program, National Research Council, Washington, DC.

Sinha, K.C. and Fwa, T.F. 1989. *On the Concept of Total Highway Management*, Transportation Research Record 1229. Transportation Research Board, National Research Council, Washington, DC, pp. 79–88.

Smith, W., Finn, F., Kulkarni, R., Saraf, C., and Nair, K. 1979. *Bayesian Methodology for Verifying Recommendations to Minimize Asphalt Pavement Distress*, NCHRP Report 213. National Cooperative Highway Research Program, Transportation Research Board, National Research Council, Washington, DC.

Taniguchi, S. and Yoshida, T. 2003. Estimating Work Zone User Cost using Integration Curve, Proceedings, MAIREPAV — Third International Symposium on Maintenance and Rehabilitation of Pavements and Technological Control, University of Minho, Guimarães, Portugal, July 7–11, pp. 835–844.

Terrel, R.L. and LeClerc, R.V. 1978. Pavement Management Workshop, Tumwater, Washington, Report No. FHWA-TS-79-206, September.

Uddin, W. 1992. Highways and Urban Road Maintenance Management: Development and Operations Issues, Compendium. *Proceedings, 4R Conference & Road Show*, Atlanta, pp. 28–32.

Uddin, W. 1993b. Application of User Cost and Benefit Analysis for Pavement Management and Transportation Planning, Compendium. *Proceedings, 4R Conference & Road Show*, Philadelphia, December, pp. 24–27.

Uddin, W. 1995. *Pavement Material Property Databases for Pavement Management Applications*, Computerization and Networking of Materials Databases: Fourth Volume, ASTM STP 1257. American Society for Testing and Materials, Philadelphia, pp. 96–109.

Uddin, W. 2002a Life-cycle analysis for investment decision-making to revive ancient silk road and enhance the economies of central Asian countries, *Int. J. Pavements, IJP 2002*, 1, 2, 84–97.

Uddin, W. 2002b. In Situ Evaluation of Seasonal Variability of Subgrade Modulus using DCP and FWD Tests. *Proceedings, International Conference on the Bearing Capacity of Roads and Airfields*, Lisbon, Portugal, Vol. 1, June, pp. 741–750.

Uddin, W. 2002c. Application of Remote Sensing and Spatial Technologies for Oxford ITS Integration Project, CD *Proceedings, 9th World Congress on Intelligent Transportation Systems*, Session SP5: Traffic Surveillance, Chicago, Illinois, October 14–18.

Uddin, W. 2002d. Application of Remote Sensing Tunable Laser Technology for Measuring Transportation Related Air Pollution. *Proceedings, Seventh International Conference on Applications of Advanced Technology in Transportation*, Massachusetts, August 5–7, 2002, pp. 257–265.

Uddin, W. 2003a. Air Quality Analysis Considering Mobile and Aviation Sources and Monitoring using Remote Sensing Tunable Laser Technology, *CD Proceedings, 2003 International Conference on Airports*, Rio de Janeiro, Brazil, June 8–11.

Uddin, W. and Al-Tayer, M. 1993. Implementation of Pavement Management Technology for Dubai Emirate Road Network. *Proceedings, XII International Road Federation Meeting*, Madrid, Spain, Vol. IV, May 16–22, pp. 305–314.

Uddin, W. and Chung, T. 1997. Effects of Thermal Stresses on Jointed Concrete Pavement Performance. *Proceedings, Second International Symposium on Thermal Stresses*, Rochester, New York, June 8–11, pp. 621–624.

Uddin, W. and Torres-Verdin, V. 1998. Service Life Analysis for Managing Road Pavements in Mexico. *Proceedings, Fourth International Conference on Managing Pavements*, Durban, Vol. 2, May 17–21, pp. 882–898.

Uddin, W. and Nanagiri, Y., Performance of Polymer-Modified Asphalt Field Trials in Mississippi Based on Mechanistic Analysis and Field Evaluation, *Int. J. Pavements, IJP 2002*, 1, 1, 84–97, 2002.

Uddin, W., Carmichael III, R.F., and Hudson, W.R. 1987. A Methodology for Life-Cycle Cost Analysis of Pavements using Microcomputer. *Proceedings, Vol. I, 6th International Conference, Structural Design of Asphalt Pavements*, Ann Arbor, pp. 773–794.

Uddin, W., Zhang, D., and Fernandez, F. 1994. *Finite Element Simulation of Pavement Discontinuities and Dynamic Load Response*, Transportation Research Record 1448. Transportation Research Board, National Research Council, Washington, DC, pp. 100–106.

Uddin, W., Yiqin, L., and Phillips, L.D. 2001. Integration of Remote Sensing and Geospatial Technologies for Managing Transportation Infrastructure Assets. *CD Proceedings, Second International Symposium on Maintenance and Rehabilitation of Pavements and Technological Control*, Auburn, Alabama, July 29–August 1.

Uddin, W., Garza, S., and Boriboonsomsin, K. 2003b. A 3D-FE Simulation Study of the Effects of Nonlinear Material Properties on Pavement Structural Response Analysis and Design. *Proceedings, MAIREPAV — Third International Symposium on Maintenance and Rehabilitation of Pavements and Technological Control*, University of Minho, Guimarães, Portugal, July 7–11, pp. 805–814.

Uddin, W., Hackett, R.M., Noppakunwijai, P., and Chung, T. 1996. Nondestructive Evaluation and *In Situ* Material Characterization of Jointed Concrete Pavement Systems. *Proceedings, 2nd International Conference on Nondestructive Testing of Concrete in the Infrastructure*, Nashville, pp. 242–249.

USDOT, U.S. Department of Transportation, Achievements of the DOT-NASA Joint Program on Remote Sensing and Spatial Information Technologies Application to Multimodal Transportation, 2000–2002, April 2002.

Visser, A.T., ed. 1998. *Proceedings, Fourth International Conference on Managing Pavements*, Durban, May 17–21.

Watanatada, T., Harral, C.G., Patterson, W.D.O., Dreshwar, A.M., Bhandari, A., and Tsunokawa, K. 1987. The Highway Design and Maintenance Standards Model, Vol. 2. John Hopkins University Press, Baltimore, Maryland.

Woodstrom, J.H. 1990. *Measurements, Specifications, and Achievement of Smoothness for Pavement Construction*, Synthesis of the Highway Practice 167. National Cooperative Highway Research Program, National Research Council, Washington, DC.

Woodward, W.D.H., Woodside, A.R., and Jellie, J.K. 2002. Development of Early Life Skid Resistance for High Stone Content Asphalt Mixes. *CD Proceedings, 3rd International Conference on Bituminous Mixtures and Pavements*, Thessaloniki, Greece, November 21–22.

Zaniewski, J.P., Butler, B.C., Cunningham, G., Elkins, G.E., Paggi, M.S., and Machemehl, R. 1982. *Vehicle Operating Costs, Field Consumption and Pavement Type and Condition Factors*, Final Report. Federal Highway Administration, Washington, DC.

Zimmerman, K. 1995. *Pavement Management Methodologies to Select Projects and Recommend Preservation Treatments*, Synthesis of the Highway Practice 222. National Cooperative Highway Research Program, National Research Council, Washington, DC.

19

Highway Condition Surveys and Serviceability Evaluation

Zahidul Hoque
Pavement Specialist
Melbourne, Victoria, Australia

19.1 Introduction

19.1.1 Purpose and Need of Condition Surveys

Highway pavement condition deteriorates with time due to one or more of the following factors: design inadequacies, traffic loading, material ageing, construction deficiencies, and environmental forces, etc. A typical deterioration trend of pavement condition with time may be seen in Figure 19.1, assuming no maintenance treatment is applied. As can be seen from Figure 19.1, it is important for a pavement manager or engineer to know the condition of its pavements at various stages of their lives to plan for maintenance and rehabilitation strategies. Pavement condition surveys, also known as distress surveys, are conducted as a part of Pavement Management System (PMS) in order to assess current pavement condition and the need for maintenance and rehabilitation treatments. The focus of these surveys is to determine pavement surface condition at a given point of time, not including the evaluation of pavement structural condition. These surveys usually help to address the causes of the distresses present.

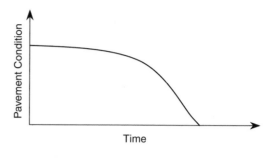

FIGURE 19.1   Pavement condition versus time.

Though distress surveys generally provide detailed distress information including distress type, severity, extent and location, the level of information collected in a condition survey depends on the intended use of the data. For a network level study, for example, some highway agencies consider it adequate to conduct windshield survey. On the other hand, for project level survey, detailed distress information is often required. It is desirable that a distress survey must also identify, classify, and quantify the causes of all distresses and factors that may influence pavement performance.

Traditionally, condition surveys are conducted periodically through visual inspection by traveling along the pavement, and identification and classification of the distresses present. Though these ratings may be subjective in nature, they are very useful for project level surveys or where technology or resource constraints exist. Many authorities and organizations have devised methods and equipment to objectively measure pavement surface conditions. The latest developments in this area include the use of laser sensors and video imaging technology to classify and quantify pavement distresses objectively.

19.1.2   Concept of Rideability and Serviceability

Pavement serviceability refers to the ability of a pavement to provide the desired level of service to the user. The ability of a pavement to perform at its desired level of service is affected by pavement condition. Figure 19.2 shows a general trend of loss of serviceability due to pavement condition affected by time or traffic loading. The "serviceability-performance-concept" was developed by Carey and Irick (1960) during the AASHO Road Test (HRB, 1962) to evaluate pavement performance in terms of pavement ride quality. The Present Serviceability Index (PSI) is the subjective assessment of "service-

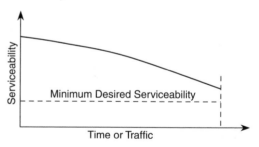

FIGURE 19.2   Pavement serviceability versus time or traffic.

ability" by a panel of raters and is related to objective measures of surface condition by response-type road roughness measuring systems (RTRRMS). PSI is expressed on a scale of zero to five where scale five refers to an excellent ride conditions and scale zero indicates a very poor ride quality, i.e., the pavement has failed to provide any level of service. The serviceability of a pavement section is obtained from a group of raters who drive on the pavement and assign ratings based on their subjective judgment of the ride condition.

Janoff et al. (1985) questioned if the public's perception of serviceability was still the same as it was during the AASHO Road Test (HRB, 1962) due to the fact that many travel parameters such as vehicle and road characteristics had changed. He argued that serviceability is not exclusively a measurement of rideability but is confounded by surface defects and rideability, and surface defects should have separate measures.

19.1.3   Relationship between Serviceability and Roughness

Pavement serviceability is sensitive to many parameters. Studies have concluded that serviceability is highly correlated with roughness and can be adequately predicted using pavement roughness measurements (Haas et al., 1994). Pavement roughness is defined as the longitudinal deviations of a pavement surface from a true surface. The acceptable level of service or pavement roughness varies with the road type or its importance. For instance, high roughness level may be acceptable to the users of low-speed roads, while the same level of roughness may not be acceptable for high-speed roads.

Apart from pavement roughness measurements, evaluation of pavement distresses is important and these terms were included in the original serviceability regression equation used in the AASHO Road Test. However, as these parameters had little influence on the value of pavement serviceability, the serviceability can be reliably predicted using roughness only (Haas et al., 1994). Haas et al. pointed out that serviceability-roughness relationships are empirical in nature and serviceability-roughness experiments should be based on proper statistical design to accommodate the range of perceptions of the highway users.

19.2   Types of Pavement Distresses and Their Identification

Pavement distresses are generally related to the deficiencies in construction, materials and maintenance strategies, but not directly to pavement design (Huang, 1993). Identification of pavement distresses and their causes are important in establishing pavement distress survey procedures. There are excellent references available for identifying distresses in flexible and rigid pavements (Chong et al., 1982; Austroads, 1987; Chong et al., 1989a; 1989b; 1989c; Miller and Bellinger, 2003). The definitions of pavement distresses are generally the descriptions of their physical appearance. While descriptions of various distresses are somewhat similar in most references, the definition of level of severity of distresses and measuring techniques vary somewhat among the road authorities around the world. The definitions and photographs presented here are reproduced from the *Austroads Guide to the Visual Assessment of Pavement Condition (Austroads, 1987)*.

19.2.1   Distresses in Flexible Pavements and Causes

Distresses in flexible pavements can be grouped into four broad categories:

- Deformation;
- Cracking;
- Surface Defects;
- Edge Defects.

19.2.1.1   Deformation

Deformation in pavement is defined as the change in the pavement surface profile and it can affect roughness condition and skid resistance when water ponding occurs. Deformations may also accelerate crack initiation. The common pavement deformations are corrugations, depressions, rutting, and shoving, as shown in Figure 19.3.

19.2.1.1.1   Corrugations

Corrugations in flexible pavements are identified as closely and regularly spaced transverse undulations, as shown in Figure 19.4. Corrugations are plastic movements of the pavement surface and usually occur where acceleration and deceleration take place such as, bus stop, bends, and intersections. Lack of stability in the surface or base course is the possible cause for corrugations.

19.2.1.1.2   Depressions

Depressions are localized bowl shaped settlement in the pavement (see Figure 19.5) caused by one or combination of the following: (i) settlement of service trenches, (ii) consolidation in the poorly

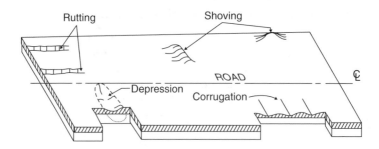

FIGURE 19.3   Sketch of deformations in flexible pavements (*Source*: Austroads, 1987).

FIGURE 19.4   Corrugations in flexible pavements (*Source*: Austroads, 1987).

compacted subgrade, and (iii) change of moisture in the subgrade. Water ponding in the depressions causes cracking and other pavement damages such as roughness and skidding problem.

19.2.1.1.3 Rutting

Rutting is the longitudinal depression (see Figure 19.6) that occurs in the wheel path due to inadequate surface thickness, and lack of compaction or stability in the surface or base course. Rutting in early pavement life may be due to poor compaction, high moisture content, or lack of lateral resistant. On the other hand, under-strength material and over loading are among the causes for rutting in the later part of pavement life. Untreated significant rutting may lead to further damage such as cracking and hydroplaning especially when water ponding occurs.

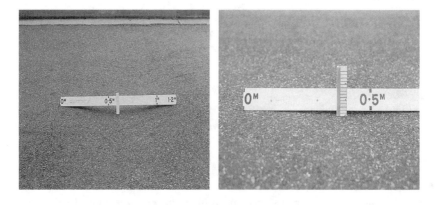

FIGURE 19.5   Depressions in flexible pavements (*Source*: Austroads, 1987).

FIGURE 19.6 Rutting in flexible pavements (*Source*: Austroads, 1987).

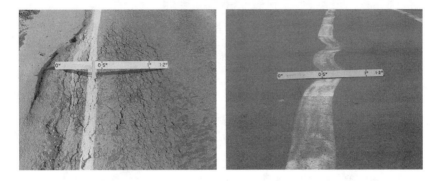

FIGURE 19.7 Shoving in the flexible pavements (*Source*: Austroads, 1987).

19.2.1.1.4 Shoving

Shoving or creep is the horizontal displacement of surfacing materials occurring mainly in the direction of traffic where braking or acceleration actions take place, as shown in Figure 19.7. Permanent displacements can cause single ridges or depressions similar to corrugations. Possible causes for shoving are weak bond between pavement layers, lack of edge support, and insufficient pavement thickness or stability.

19.2.1.2   Cracks

Cracks are fractured pavement condition resulting from a number of causes and are found in a variety of patterns ranging from single to an interconnected pattern. Possible causes of cracks are deformations, fatigue life exceeded, reflection of cracks from underlying layers, shrinkage, and poor construction joints. Presence of cracks causes numerous problems in the pavement including limiting load spreading capability through loss of structural strength, loss of waterproofing, and loss of roughness level. As seen in Figure 19.8, the following types of cracks are commonly found in the flexible pavements:

- Crocodile cracks;
- Longitudinal cracks;
- Transverse cracks;
- Block cracks;
- Diagonal cracks;
- Meandering cracks;
- Crescent shaped cracks.

19.2.1.2.1   Crocodile Cracks

As the term implies, crocodile cracks are interconnected cracks forming a series of small polygons resembling a crocodile skin (see Figure 19.9). Crocodile cracks are usually fatigue cracks and are the early

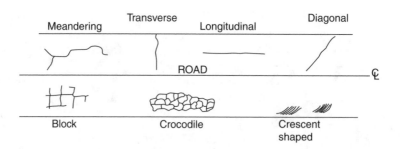

FIGURE 19.8   Sketch of common types of cracks in flexible pavements (*Source*: Austroads, 1987).

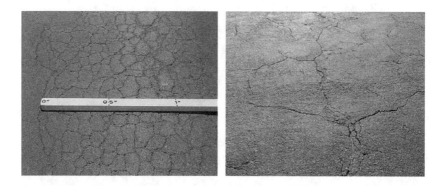

FIGURE 19.9   Crocodile cracks in flexible pavements (*Source*: Austroads, 1987).

signs of pavement distress. They indicate a pavement condition where excessive movements occur in or more underlying layers. These cracks are caused due to inadequate pavement thickness, low modulus base course and aged wearing course, shrinkage of aged bitumen, very low temperature, and over-loaded vehicles. Crocodile cracks usually appear only in the traffic loading areas and allow water entering in the pavement.

19.2.1.2.2 *Longitudinal Cracks*

Single or a number of parallel cracks appear longitudinally along the pavement (see Figure 19.10) due to reflection of a shrinkage crack or joint in the underlying layer, weak bond between pavement lanes, asphalt hardening, volume change in the subgrade, and differential settlement between cut and fill. Poorly constructed lane joint may also cause longitudinal cracks. These cracks do not usually result from load but wide cracks will permit water to enter the pavement structure.

FIGURE 19.10   Longitudinal cracks in flexible pavements (*Source*: Austroads, 1987).

FIGURE 19.11   Transverse cracks in flexible pavements (*Source*: Austroads, 1987).

19.2.1.2.3 *Transverse Cracks*

Transverse cracks run transversely across the pavement (see Figure 19.11) and are common in aged asphalt surfaces. These cracks are caused due to primarily reflection of a shrinkage crack or joint in the underlying layer, construction joint, and structural failure in the cement base. Similar to longitudinal cracks, wide cracks permits water in the pavement.

19.2.1.2.4 *Block Cracks*

Interconnected cracks that form a series of rectangular blocks over the entire pavement are known as block cracks (see Figure 19.12). Block cracks are generally non-load assisted, but repeated loads increase severity. These cracks indicate that the asphalt has hardened significantly and the possible causes are (i) shrinkage and fatigue cracking in the underlying cemented materials, (ii) joints in the underlying layer, and (iii) fatigue and shrinkage cracking in the asphalt surfacing layers.

19.2.1.2.5 *Crescent Shape Cracks*

As shown in Figure 19.13, crescent shaped cracks commonly appear on pavements as a set of closely spaced parallel cracks along with shoving, usually in the direction of traffic. These cracks appear when part of the asphalt surface moves away from the rest of the surface caused by the lateral and shear stresses developed in the pavement due to traffic loading. Weak bonding between surface layers and underlying layer, inadequate surface thickness, and high stresses due to braking and acceleration are primary causes for these types of cracks.

19.2.1.2.6 *Diagonal Cracks*

These are single cracks that form diagonally across the pavement, as shown in Figure 19.14. Differential settlement, service trenches, and shrinkage crack or joint in the underlying layers are possible causes for diagonal cracks.

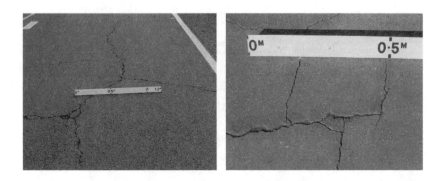

FIGURE 19.12   Block cracks in flexible pavements (*Source*: Austroads, 1987).

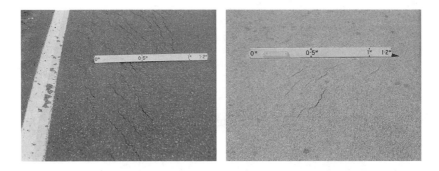

FIGURE 19.13   Crescent shaped cracks in flexible pavements (*Source*: Austroads, 1987).

FIGURE 19.14   Diagonal cracks in flexible pavements (*Source*: Austroads, 1987).

19.2.1.2.7 *Meandering Cracks*

These are unconnected irregular single cracks that run in any direction (see Figure 19.15). Possible causes are reflection of a shrinkage crack in the underlying layer, weak pavement edge, differential settlement between cut and fill, and tree roots.

19.2.1.3   Surface Defects

As the term implies, surface distresses are associated with the pavement surface and usually do not indicate structural problem in the pavement layers. However, they cause significant effect on pavement serviceability and skid resistance, and, if not treated, may lead to structural problems. Common types of

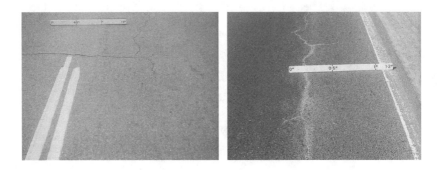

FIGURE 19.15   Meandering cracks in flexible pavements (*Source*: Austroads, 1987).

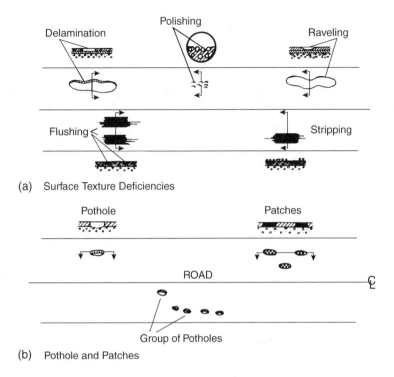

FIGURE 19.16 Sketch of surface defects in flexible pavements (*Source*: Austroads, 1987) (a) Surface texture deficiencies (b) Pothole and patches.

surface defects are:

- Delamination;
- Flushing;
- Polishing;
- Raveling;
- Stripping;
- Pothole; and
- Patching.

Figure 19.16 shows sketch of common surface defect types.

19.2.1.3.1 *Delamination*

Delamination is defined as the distress condition when wearing course debonds from the underlying layers and as a result part of pavement wearing course of uniform thickness is lost (see Figure 19.17).

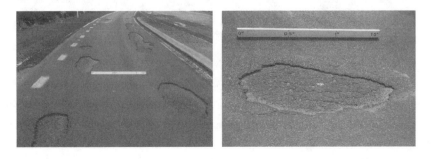

FIGURE 19.17 Delamination in flexible pavements (*Source*: Austroads, 1987).

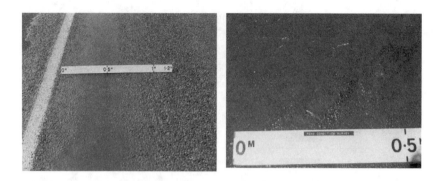

FIGURE 19.18   Flushing in flexible pavements (*Source*: Austroads, 1987).

Possible causes of delamination are lack of bonding due to poor construction practice, inadequate surface thickness, seepage of water through surface affecting bond between layers, weak underlying layer, and adhesion of binder to tires.

19.2.1.3.2   Flushing or Bleeding

Flushing occurs when aggregate is immersed in the bitumen resulting excess binder on the surface with very little texture, as shown in Figure 19.18. Flushing causes soft pavement surface in hot weather and slippery in wet or cold weather. It is normally associated with sprayed seal pavements, though it may appear in asphalt pavements. Possible causes for flushing are (i) application of excess binder, (ii) use of excess prime coat, (iii) excess binder in the underlying surface, (iv) flaky aggregate, and (v) low strength base causing aggregate penetration into the base.

19.2.1.3.3   Polishing

Polishing is referred to the effect of smoothing or rounding of the upper part of the aggregates by the action of heavy vehicular traffic (see Figure 19.19). Polishing usually occurs at the wheel paths and can be identified by the relative appearance of the trafficked and untrafficked areas. Use of aggregate of low polishing resistance is the primary cause of polishing.

19.2.1.3.4   Raveling

Raveling is used to define a distress condition where both aggregate and binder are progressively lost from pavement surface, as shown in Figure 19.20. Raveling is caused due to inadequate binder, deterioration of binder condition, or inferior asphalt mix design. Inadequate compaction or construction in the wet or cold periods may also cause raveling. Delamination and raveling in sprayed seal pavements may not be easily differentiated.

FIGURE 19.19   Polishing in flexible pavements (*Source*: Austroads, 1987).

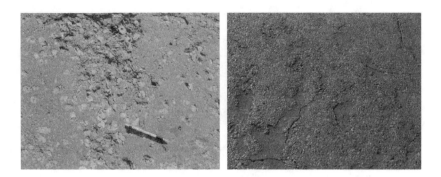

FIGURE 19.20   Raveling in flexible pavements (*Source*: Austroads, 1987).

19.2.1.3.5 *Stripping*

Stripping is normally associated with sprayed seal pavements and is a condition where only coarse aggregate is lost leaving the binder in good condition (see Figure 19.21). Possible causes of stripping are (i) poor mix design, (ii) weak bonding between binder and stone, (iii) aging or absorption of binder, (iv) use of weak aggregate, (v) presence of water, and (vi) poor compaction.

19.2.1.3.6 *Pothole*

Potholes, bowl-shaped cavity in the pavements (see Figure 19.22), are developed due to the loss of surface and base course material that become weak and loose when water enters into the pavement though cracking. Localized disintegration and freeze–thaw cycles also cause potholes in the pavement. Traffic and water in the cavity accelerate the development of pothole.

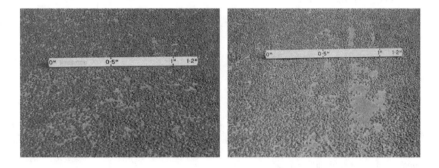

FIGURE 19.21   Stripping in flexible pavements (*Source*: Austroads, 1987).

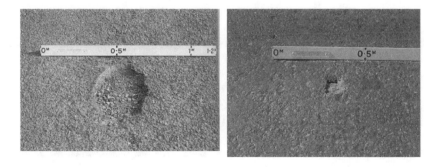

FIGURE 19.22   Pothole in flexible pavements (*Source*: Austroads, 1987).

FIGURE 19.23   Patches in flexible pavements (*Source*: Austroads, 1987).

19.2.1.3.7 *Patches*

Pavement sections repaired to fix loss of serviceability or structural capacity are termed as patches (see Figure 19.23). Patches are considered as the distress when it is raised or depressed below the surface level. Other distresses may also occur in the patches. Patches include surface repair without digging out pavement layers, reconstruction of surface or other layers, and excavation for services.

19.2.1.4   Edge Defects

Defects that occur along the joint of the pavement edge and shoulder are known as edge defects. These defects may appear in isolated areas or continuously along the joint and they are particularly significant for unsealed shoulders. Edge defects reduce pavement width and affect ride quality. They also allow water entry into the pavement. Two types of edge defects, such as, edge break and edge drop-off, as shown in Figure 19.24 are common in flexible pavements.

19.2.1.4.1 *Edge break*

As shown in Figure 19.25, irregular or broken pavement edge is caused due to lack of edge support, inadequate pavement width, poor bonding between seal and base course, and improperly designed pavement alignment.

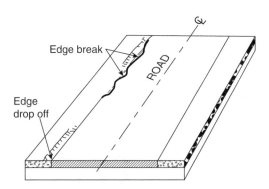

FIGURE 19.24   Sketch of edge defects in flexible pavements (*Source*: Austroads, 1987).

FIGURE 19.25   Edge break in flexible pavements (*Source*: Austroads, 1987).

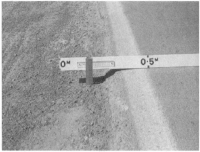

FIGURE 19.26   Edge drop-off in flexible pavements (*Source*: Austroads, 1987).

19.2.1.4.2 *Edge drop-off*
The difference between the elevation of pavement surface and shoulder is termed as edge drop-off (see Figure 19.26). Inadequate pavement width, lack of support to protect shoulder erosion, and pavement resurfacing are common causes for the drop-off.

19.2.2   Distresses in Rigid Pavements and Causes

Distresses in rigid pavements can be classified into five groups, such as:

- Deformations;
- Cracking;
- Joint seal and Spalling;
- Surface defects; and
- Edge defects.

19.2.2.1   Deformations

Deformation in rigid pavements results from cracking of slabs or from relative movement between slabs due to load assisted and non-load assisted factors. Deformation affects pavement ride quality and helps water entering in the pavement structures. The common types of deformations found in the rigid pavement are depression, faulting, pumping, and rocking.

19.2.2.1.1 *Depression*
Depressions in rigid pavement occur across a crack or joint, as shown in Figure 19.27, and are generally associated with significant cracking. Poor compaction, weak subgrade support, and differential settlement of subgrade are possible causes for depressions. It allows water ponding and increases the chance of water entering through joints and cracks.

FIGURE 19.27   Depressions in rigid pavements (*Source*: Austroads, 1987).

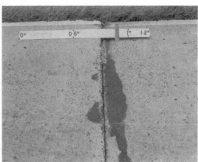

FIGURE 19.28 Faulting in rigid pavements (*Source*: Austroads, 1987).

19.2.2.1.2 Faulting

Faulting is a condition where concrete slab breaks into two pieces along joints or cracks and the vertical displacements between the broken slabs create a step, as shown in Figure 19.28. Possible causes for faulting are (i) volume change in the subgrade, (ii) poor support in the subbase or subgrade, (iii) rocking, (iv) pumping as result of loss of fine material under one slab, and (v) warping effects due to temperature and moisture gradients. Faulting is a major distress in rigid pavements as it allows water entry in the pavement and affects ride quality.

19.2.2.1.3 Pumping

Pumping in the concrete slab occurs when excessive moisture enters into the pavement through crack or poorly constructed joint and the excess water is ejected through cracks and joints, due to the upward wrap and curl of the slab near the joint or crack, and movements of traffic. Pumping causes erosion of fine particles in the basecourse and thus causes structural deterioration such as, rocking, faulting and cracking (see Figure 19.29).

19.2.2.1.4 Rocking

Rocking is felt at a joint or crack with the passage of a vehicle where pumping has caused loss of support (see Figure 19.30). Inadequate subbase or subgrade support or differential settlement in the subgrade may also cause rocking.

19.2.2.2 Cracking

Shrinkage cracks are common in rigid pavements but it does not affect pavement performance if the pavement is properly designed and constructed. Many patterns of cracks ranging from single isolated cracks to interconnected multiple cracks are found in rigid pavements. The possible causes of cracking in rigid pavements include: insufficient slab thickness, shrinkage, weak subbase or subgrade, and differential subgrade movement. Cracking causes a number of problems such as loss of load spreading capability, loss

FIGURE 19.29 Pumping in rigid pavements (*Source*: Austroads, 1987).

FIGURE 19.30　Rocking in rigid pavements (*Source*: Austroads, 1987).

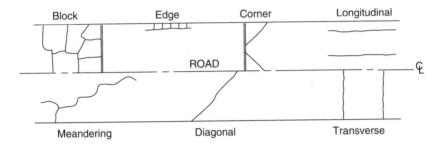

FIGURE 19.31　Sketch of cracks in rigid pavements (*Source*: Austroads, 1987).

of appearance, deteriorate ride quality, and entry of water into the pavement structures. As shown in Figure 19.31, the common types of cracks in the rigid pavements are block cracks, longitudinal cracks, transverse cracks, diagonal cracks, corner cracks, and meandering cracks.

19.2.2.2.1 *Block Cracks*
Interconnected cracks that form a series of rectangular blocks over the entire pavement are known as block cracks (see Figure 19.32). Possible causes for block cracking are (i) inadequate slab thickness, (ii) loss of subbase or subgrade support, and (iii) subgrade settlement.

19.2.2.2.2 *Longitudinal Cracks*
Single or a number of parallel cracks appear longitudinally along the pavement, as shown in Figure 19.33. Possible causes for longitudinal cracks are (i) lateral shrinkage, (ii) insufficient slab thickness, (iii) differential settlement, (iv) lateral shrinkage, and (v) longitudinal joint too shallow or too close to traffic lane. Poorly constructed joints can also cause longitudinal cracks.

FIGURE 19.32　Block cracks in rigid pavements (*Source*: Austroads, 1987).

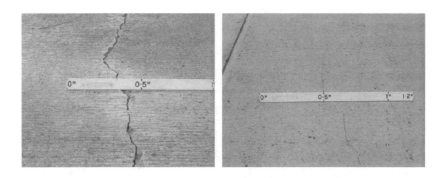

FIGURE 19.33   Longitudinal cracks in rigid pavements (*Source*: Austroads, 1987).

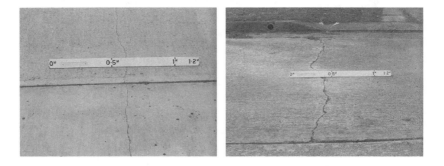

FIGURE 19.34   Transverse cracks in rigid pavements (*Source*: Austroads, 1987).

19.2.2.2.3   Transverse Cracks

As shown in Figure 19.34, transverse cracks run transversely across the slab. Shrinkage, rocking action, and insufficient slab thickness are possible causes for these cracks. Late saw cutting of construction joint can also cause these cracks. Transverse cracks generally start near construction joints.

19.2.2.2.4   Diagonal Cracks

Diagonal cracks are single cracks run diagonally across the slab, as may be seen in Figure 19.35. Settlement, shrinkage in the slab, inadequate slab thickness, and rocking actions are possible causes for diagonal cracks in the rigid pavement.

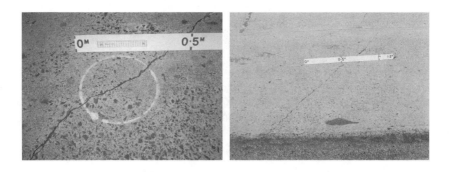

FIGURE 19.35   Diagonal cracks in rigid pavements (*Source*: Austroads, 1987).

 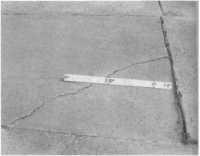

FIGURE 19.36   Corner cracks in rigid pavements (*Source*: Austroads, 1987).

19.2.2.2.5   Corner Cracks

As seen in Figure 19.36, corner cracks appear at the corners of slabs diagonally from a longitudinal edge to a transverse joint. The causes for corner cracks are inadequate slab thickness, loss of support of subbase or subgrade, overloading, curling of corners, and poor joint design.

19.2.2.2.6   Meandering Cracks

Similar to flexible pavements, meandering cracks in rigid pavements are irregular single cracks that run in any direction (see Figure 19.37). Rocking action, settlement, shrinkage, and inadequate slab thickness are possible causes for meandering cracks.

19.2.2.3   Joint Seal and Spalling

19.2.2.3.1   Joint Seal Defects

Joint seal defects are the most common defects in jointed concrete pavements. They can be easily identified when seal is lost leaving a gap between the joints or when sealant extrudes from the joint, as shown in Figure 19.38. Defective seal joints permit water to enter in the pavement structures and allow incompressible rubbish to lodge in the joint. Incompressible material keeps the joint permanently open and limits the horizontal expansion of slabs. As a result, high stresses are developed in the pavement. Sealant construction quality, ageing of sealant, too much or too little sealant in the joint, and poor sealant performance are primary causes for joint seal defects, though rocking and pumping actions may also cause these defects.

19.2.2.3.2   Spalling

Spalling is found when small pieces of concrete, usually in angular shape, are separated from the pavement surface. As seen in Figure 19.39, spalling generally occurs at joints, edges, corners or cracks, or directly over reinforcing bars. Infiltration of incompressible particles into joints or cracks, corrosion

 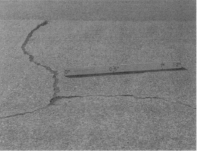

FIGURE 19.37   Meandering cracks in rigid pavements (*Source*: Austroads, 1987).

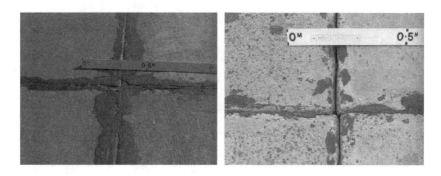

FIGURE 19.38   Joint seal defects in rigid pavements (*Source*: Austroads, 1987).

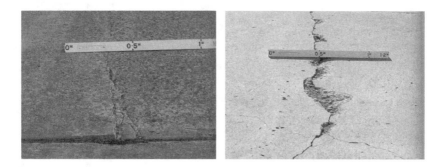

FIGURE 19.39   Spalling in rigid pavements (*Source*: Austroads, 1987).

of rebars or dowels, misalignment of dowels, subbase movement, and poor quality concrete aggregate are possible causes for spalling.

19.2.2.4   Surface Defects

Surface defects in rigid pavements include scaling, skid resistance, pothole, and patching.

19.2.2.4.1   *Scaling or Raveling*

As shown in Figure 19.40, raveling or scaling is a condition when mortar and aggregate are lost through progressive breakdown of slab surface due to poor aggregate quality, inadequate curing, local cement deficiency, or overworking of surface during construction.

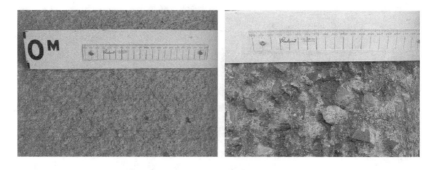

FIGURE 19.40   Scaling or raveling in rigid pavements (*Source*: Austroads, 1987).

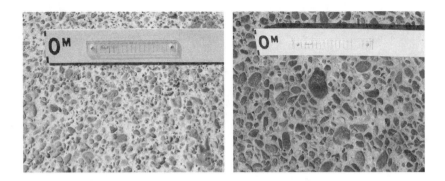

FIGURE 19.41   Low skid resistant concrete surface (*Source*: Austroads, 1987).

19.2.2.4.2 Skid Resistance

Smooth, polished or glassy appearances indicate lack of skid resistance (see Figure 19.41). Loss of skid resistance can occur from low microtexture and macrotexture of the aggregates, as well as due to spillages, poor construction finishing, and poor quality mortar that worn by traffic.

19.2.2.4.3 Pothole

Potholes in rigid pavement are bowl-shaped cavity in the surface (see Figure 19.42). Potholes are developed due to localized cracks, freeze–thaw action, and rebars too close to the surface. Traffic and water in the cavity accelerate the development of pothole.

19.2.2.4.4 Patch

Original material from localized area of the concrete surface is removed and replaced with asphalt or concrete material to repair localized distresses (see Figure 19.43). Variety of distresses may occur at the patches including cracking, spalling, and distortion.

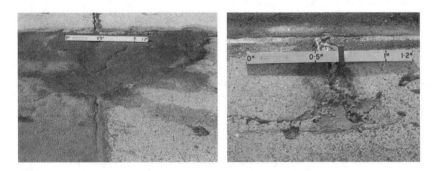

FIGURE 19.42   Potholes in rigid pavements (*Source*: Austroads, 1987).

FIGURE 19.43   Patches in rigid pavements (*Source*: Austroads, 1987).

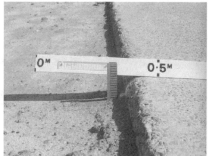

FIGURE 19.44   Edge defects in rigid pavements (*Source*: Austroads, 1987).

19.2.2.5   Edge Defects

As seen in Figure 19.44, shoulder drop-off is the most common pavement edge defect found in concrete pavements. Shoulder drop-off is characterized as the differential settlement between the shoulder and the edge of the slab that may cause due to incorrect geometry, poor shoulder drainage, and loose shoulder materials.

19.3   Measurement and Reporting of Pavement Distress Data

19.3.1   Characterization and Measurement of Distresses

In measuring pavement distresses, detailed information related to the distresses are usually collected and stored for maintenance planning and research purposes. Such details are commonly known as distress parameters and they include location of the distress, extent or density, and severity. Distress measurement methods are ought to be simple and easy-to-follow. Complex measurement procedures should be avoided as they are costly and may cause confusion to survey crews.

The details of data requirements are usually driven by the need of the data. For example, project level decisions such as estimation of needed routine maintenance (e.g., crack sealing or patching works) require detail information in terms of location, extent, and severity of the identified distresses. Network decisions, on the other hand, require much less information as these decisions are based on the overall network conditions, not on the worst or excellent condition of any particular section in the network.

19.3.2   Recording Forms and Reporting Distress Data

There is no single standard method or procedure for distress measurements and recording. In fact these methods vary widely among the road authorities. Most road authorities have developed their own system of distress survey procedures depending on their available resources, need, and network characteristics.

In the United States, the PAVER system (Shahin and Khon, 1984; Shahin and Walther, 1990) is widely used, though many states have their own survey systems. The U.S. Federal Highway Administration (FHWA) developed a procedure for distress evaluation surveys of the Long Term Pavement Performance (US-LTPP) Study (Miller and Bellinger, 2003) sites. The Ministry of Transportation and Communications, Ontario, Canada has four manuals for condition rating of four pavement types such as, rigid pavements (Chong et al., 1982), flexible pavements (Chong et al., 1989a), surface treated pavements (Chong et al., 1989b), and gravel pavements (Chong et al., 1989c). In Australia, Austroads (the national association of road transport and traffic authorities in Australasia) has a guide for visual assessment of pavement condition (Austroads, 1987). However, member authorities of this organization have their own

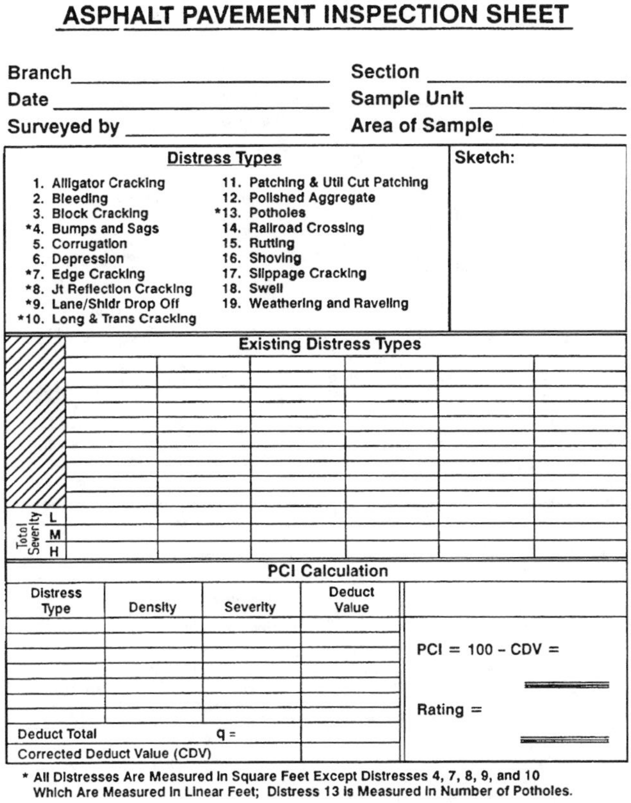

ASPHALT PAVEMENT INSPECTION SHEET

Branch_____ Section _____

Date _____ Sample Unit _____

Surveyed by _____ Area of Sample _____

Distress Types

1. Alligator Cracking	11. Patching & Util Cut Patching
2. Bleeding	12. Polished Aggregate
3. Block Cracking	*13. Potholes
*4. Bumps and Sags	14. Railroad Crossing
5. Corrugation	15. Rutting
6. Depression	16. Shoving
*7. Edge Cracking	17. Slippage Cracking
*8. Jt Reflection Cracking	18. Swell
*9. Lane/Shldr Drop Off	19. Weathering and Raveling
*10. Long & Trans Cracking	

Sketch:

Existing Distress Types

Total Severity L / M / H

PCI Calculation

Distress Type	Density	Severity	Deduct Value	
				PCI = 100 – CDV =
				Rating =

Deduct Total   q =

Corrected Deduct Value (CDV)

* All Distresses Are Measured In Square Feet Except Distresses 4, 7, 8, 9, and 10
Which Are Measured In Linear Feet; Distress 13 Is Measured In Number of Potholes.

FIGURE 19.45   PAVER pavement condition form for flexible pavements (*Source*: Shahin and Walther, 1990).

survey systems and procedures. LTPP Study (LTPP-Australia, 2005) also implemented a detailed data collection procedure similar to the US-LTPP Study.

Pavement Condition Rating forms used in the PAVER system is shown in Figure 19.45 and Figure 19.46. Flexible Pavement Condition Evaluation Form used in the Ministry of Transportation and Communication, Ontario, Canada can be found in (Chong et al., 1989a). A number of survey forms and data sheet used in the US-LTPP study can be found in the LTPP distress identification manual (Miller and Bellinger, 2003).

19.4   Methods and Equipment for Pavement Condition Surveys

19.4.1   Detailed Manual Inspection

Detailed manual inspection surveys provide the most precise data about the pavement surface condition and are performed by walking along the pavement section. These surveys are time consuming and expensive and thus only a number of sample sites of the road network is selected for manual inspection. Selection of sample sites is critical in that the sections must represent the overall network condition. Among the methods of sampling, random selection produces a good assessment of the overall highway network condition provided the sample size is sufficient (Haas et al., 1994). In the random selection method, some sections will have less distress and other sections will have more distresses than the typical condition. Other methods of sampling include sampling at fixed distance intervals and picking a representative sample by field personnel or survey gang. While picking a sample by field personnel may

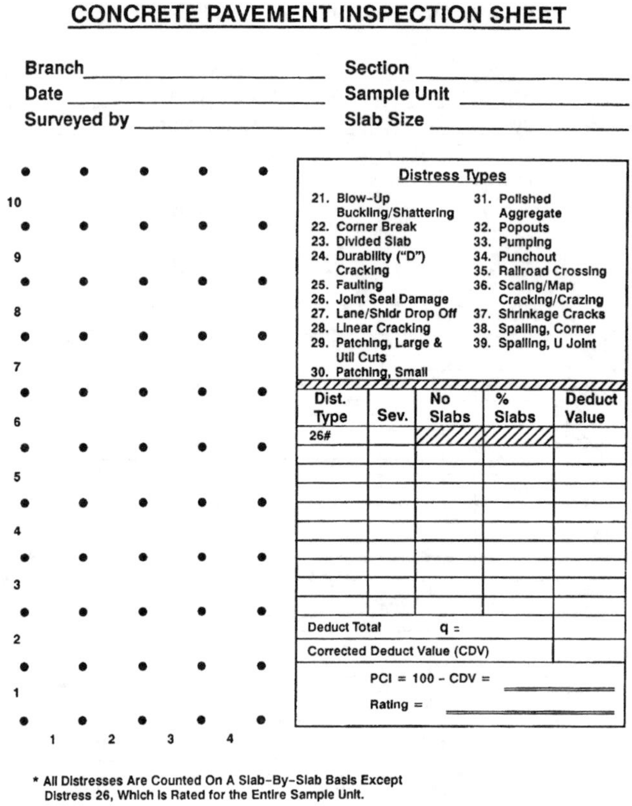

CONCRETE PAVEMENT INSPECTION SHEET

Branch _____   Section _____
Date _____   Sample Unit _____
Surveyed by _____   Slab Size _____

Distress Types	
21. Blow-Up Buckling/Shattering	31. Polished Aggregate
22. Corner Break	32. Popouts
23. Divided Slab	33. Pumping
24. Durability ("D") Cracking	34. Punchout
25. Faulting	35. Railroad Crossing
26. Joint Seal Damage	36. Scaling/Map Cracking/Crazing
27. Lane/Shldr Drop Off	37. Shrinkage Cracks
28. Linear Cracking	38. Spalling, Corner
29. Patching, Large & Util Cuts	39. Spalling, U Joint
30. Patching, Small	

Dist. Type	Sev.	No Slabs	% Slabs	Deduct Value
26#				

Deduct Total	q =	
Corrected Deduct Value (CDV)		
PCI = 100 – CDV =		_____
Rating =		_____

* All Distresses Are Counted On A Slab-By-Slab Basis Except Distress 26, Which Is Rated for the Entire Sample Unit.

FIGURE 19.46   PAVER Pavement Condition Form for rigid pavements (*Source*: Shahin and Walther, 1990).

be a biased selection that may misrepresent the network condition, sample selection by distance is also questionable in terms of network condition as appearance of distresses is not related to distance. Statistical sampling methods are employed to determine the sample size considering the distress variability and level of survey accuracy required. There are a number of detailed inspection methods available including the following:

- PAVER Method, developed by U.S. Army Construction Engineering Research Laboratory (APWARF 1983; Shahin and Kohn 1984; Shahin and Walther, 1990).
- COPES Method, developed in an National Cooperative Highway Research Program Study to survey rigid pavements (Darter et al., 1985).
- US-LTPP Distress survey method, developed to collect pavement distress data from LTPP sites (Miller and Bellinger, 2003).
- The Ministry of Transportation Ontario Method (Chong et al., 1982; 1989a; 1989b; 1989c) covers four different types of pavements.
- Austroads Method, provides a guide to the visual assessment of pavement condition (Austroads, 1987).

19.4.2   Windscreen Survey

Windscreen surveys are conducted by driving along the road at low speed, typically at about 7 to 15 km/h. These types of surveys are usually preferred for network level survey where the requirement of quality of the data is not as critical as for the project level survey.

19.4.3 Automated Distress Survey Methods

Automated distress survey methods have been evolved to minimize the difficulties of manual data collection. The difficulties of manual data collection include costs of data collection, safety during surveys, and consistency of data collection among the raters. There are numerous methods and equipment available in the market ranging from high-speed contact-less laser sensors to devices recording video images of the pavement surfaces. While it is possible to instantly analyze distress data collected using laser sensors, data collected using video images are usually post-processed in the office. A few of the methods and equipment are listed here. Details of these equipment and survey procedure can be found in the respective references.

19.4.3.1 Automated Data Logger

Carmichael et al. (1985) described the use of an Automated Data Logger System as a part of the pavement evaluation and management system for the Rhode Island Department of Transportation. The process involves a rater who walks or slowly drives on a pavement section and identifies the distress types, extent, and severity present on the section, and records these distress data into a field data recorder. These recorded data are then analyzed using the above pavement evaluation and management system.

19.4.3.2 PASCO ROADRECON System

The PASCO ROADRECON System (PASCO, 1985; 1987) developed by PASCO Corporation in Japan. The system has undergone a number of developments since its original form in 1965. It is widely used in Japan (Hudson et al., 1987) and has been used in the US-LTPP study. In its current form, pavement surface condition, roughness, and rutting data are recorded on a continuous filmstrip. The system is fitted with laser sensors and camera that automatically takes photographs at regular intervals while traveling at a speed up to 80 km/h. The system uses artificial lights to operate at night. Photographs recorded during the survey are manually interpreted to evaluate pavement distresses.

19.4.3.3 GERPHO System

The principle of recoding continuous images of pavement surface as used in the PSACO ROADRECON system was utilized in the development of the GERPHO System by the Laboratory Central Des Ponts et Chausses (LCPC) in France. The GERPHO system also has artificial lighting facility to operate at night but was designed only for pavement surface distress surveys, not rutting nor faulting. The system is widely used in a number of European countries (DeWilder, 1985; MAP, 1986).

19.4.3.4 Automatic Road Analyzer

The Automated Road Analyzer (ARN) records where pavement distresses observed through the windshield and data are entered using two rating keyboards (HPI, 1985). Rut depths are measured by ultrasonic sensors installed on a front bumper bar, and roughness is measured by an accelerometer installed on the rear axle. Images of pavement surface and right-of-way are captured using two video cameras. The equipment is also capable of measuring grade and crossfall with additional gyroscopes.

Though pavement distress survey can be conducted through windshield at a speed of 48 km/h and roughness survey at a speed between 40 and 88 km/h, the ARN system is highly labor dependent as it needs three operators to run the system. The ARN system can only be operated in daytime and in dry weather.

19.4.3.5 Laser Road Surface Tester (LRST)

As the terminology implies, the Swedish Road and Traffic Research Institute utilized laser technology in developing the LRST (Novak, 1985). The system has eleven sensors mounted on the front of the vehicle. An on-board computer simulates transverse profile using the data supplied by all the sensors to record maximum rut depths. Laser sensors are also used to measure cracking parameters such as, depth and width, longitudinal profile, macrotexture, and cross profile. In addition to the sensors, there are eight manual switches to identify and record other information including patching works, longitudinal, and crocodile cracks. Similar to the ARN system, the LRST system needs three operators to run the system. The system can be operated at a speed of 80 km/h both at daytime and night.

19.4.3.6   RoadCrack

RoadCrack (Ferguson and Pratt, 2000; 2002) is an automated device for collecting cracking information developed by the Commonwealth Scientific and Industrial Research Organization (CSIRO), Australia in association with the Roads and Traffic Authority (RTA) of New South Wales, Australia. The RoadCrack development process started since early 1990s and it is believed that the latest model has achieved world class performance in terms of measuring the width of cracking (Ferguson and Pratt, 2000; 2002). The system is capable of detecting and reporting cracking type, extent, and severity of cracking width greater than one millimeter.

The system is equipped with an array of sensors in the transverse direction and a high-speed linear array image acquisition system, and is mounted on a truck that can travel at road speed. In addition to the driver, it needs an operator to control the system. RTA has been using extensively the system for its network level cracking surveys.

19.4.3.7   Hawkeye

ARRB Group Ltd. has developed and integrated a number of its automated survey equipment in a series of scaleable survey products know as HAWKEYE (ARRB, 2005). As seen in Figure 19.47, HAWKEYE can be customized for a number of individual packages such as, video package, digital profiler package, and Global Positioning System (GPS) packages to conduct network and project level surveys, road-side inventory and asset management surveys, road geometry and mapping surveys, and contractor quality control. The video package is a sophisticated video acquisition system that is used to identify and locate pavement distresses (e.g., cracking) and roadside features accurately at highway speeds (up to 110 km/h). It produces crisp high-resolution frames using latest digital camera technology (Camera format IEEE-1394, 400 Mb/sec). The digital profiler package comes with a variety of sensor systems to capture longitudinal and transverse profile, roughness, rutting, slab faulting, and pavement macrotexture. Output from the digital profiler meets a number of international standards including ASTM, AASHTO, World Bank, ISO, and Austroads.

19.4.4   Other methods

Other methods of post-processing distress data are also widely used. Such methods include photologging method, videotaping method, and image processing method.

FIGURE 19.47   HAWKEYE series (*Source*: ARRB, 2005) (Copyright ARRB Group Ltd., 2005).

19.5 Pavement Condition Indices

19.5.1 Condition Assessment of Pavement Distresses

An index derived from the combined effects of all distresses is found to be useful in preparing a priority list of maintenance need sections. Many indices are being used to suit the purpose and resources of the road authorities. The indices and procedures include simple subjective condition evaluation to highly mathematical formulations.

Subjective condition evaluation include assigning a rating by an experienced engineer who visually inspects a site and assigns a rating based on engineering experience and judgment, and photo-matching method where raters assign rating by comparing the photographs of a pavement section with the photographs of a rated section.

The mathematical formulations that have been used to compute pavement condition indices include Fuzzy Logic Technique and Factor Analysis of Distresses. Fwa and Shanmugam (1998) used Fuzzy Logic Technique as a tool to characterize the subjective judgment of different raters and uncertainty involved in pavement condition rating and assessments. In assessing pavement condition, there exists an uncertainty about the extent of severity of distresses present and also it is likely that subjective judgment by one rater will differ from another rater. Because of the fuzziness in the pavement condition rating and assessment, fuzzy logic is found to be an excellent choice in accommodating the above uncertainties.

Often distress manifestations are interrelated and some are statistically highly correlated. Factor analysis can be employed in order to assess if all the distresses are caused by the same mechanism. This helps to identify the actual cause of the distress and hence assist in deciding appropriate maintenance treatments. In factor analysis, distress types are grouped into fundamental uncorrelated categories and the problem is solved to identify the contribution of various factors causing the distresses.

While the subjective condition evaluation techniques are constrained with a number of limitations and mathematical formulations are primarily in the research stage, aggregated condition indices are widely used by many road authorities. Some of the indices are deduct value based condition index, visual condition index (VCI), and statistically derived condition index, as explained in the next section.

19.5.2 Aggregated Pavement Condition Indices

19.5.2.1 Deduct Value Based Condition Index

LeClerc and Marshall (1971) and LeClerc et al. (1973) developed this procedure to determine a composite index of pavement distresses for Washington State, U.S.A. In this approach, a perfect pavement is assigned with an index of 100. Pavement condition at any point of time is assessed by deducting the cumulative value based on the level of severity and extent of the distresses present. The deduct value scores for individual distress types were developed empirically or subjectively.

19.5.2.2 Sum of Weighted Distress Contributions

In this procedure, pavement condition is expressed as the sum of weights assigned to each distress type. An example of this approach is the Ministry of Transportation and Communications, Ontario, Canada's Distress Manifestation Index (DMI). DMI is expressed as:

$$\text{DMI} = \sum_{i=1}^{N} W_i(S_i + D_i) \tag{19.1}$$

where:

W_i = weight for distress type.
S_i = weight for distress severity.
D_i = weight for distress extent (severity).

TABLE 19.1   VCI Score and Descriptions

VCI	Description
90 ~ 100	Very good
70 ~ 89	Good
50 ~ 69	Fair
30 ~ 49	Poor
0 ~ 29	Very poor

19.5.2.3   Nonlinear Aggregate Condition Index

VCI is an example of nonlinear aggregate index that is used in South Africa. VCI of a pavement is determined in a two-step process where an initial index VCI_p is calculated in the first step using the Equation 19.2 from observed distresses. The range of VCI scores and their descriptions are given in Table 19.1.

$$VCI_p = 100 \left\{ 1 - C \left(\sum_{n=1}^{N} F_n \right) \right\} \tag{19.2}$$

Where,

$$C = \frac{1}{\sum_{n=1}^{N} F_n \max}, \tag{19.3}$$

$$F_n = D_n \times E_n \times W_n \tag{19.4}$$

where:

D_n = degree rating of distress n (0 to 5).
E_n = extent rating of distress n (0 to 5).
W_n = weighting factor for distress n.

In Step 2, VCI is calculated from Equation 19.5.

$$VCI = (0.02509\ VCI_p + 0.0007568\ VCI_p^2)^2 \qquad 0 \le VCI \le 100 \tag{19.5}$$

19.5.2.4   PCI of PAVER System

The deduct value based condition index developed by LeClerc and Marshall (1971), and LeClerc et al. (1973) was further developed for the PAVER (Shahin and Kohn, 1984; Shahin and Walther, 1990) system, which is one of the most detailed and commonly used distress survey method in the U.S.A. In the PAVER system, a pavement section is divided into sample units for inspection. The size of the sample units for flexible pavement is 1500 ft² ~ 3500 ft² (recommended: 2500 ft²) and for rigid pavement is 12 ~ 28 slabs (recommended: 20 slabs). The quality of pavement is expressed in terms of Pavement Condition Index (PCI) that has a scale of 0 to 100, where index 0 refers that a failed condition and index 100 is the excellent condition. The meaning of various PCI scale is shown in Figure 19.48.

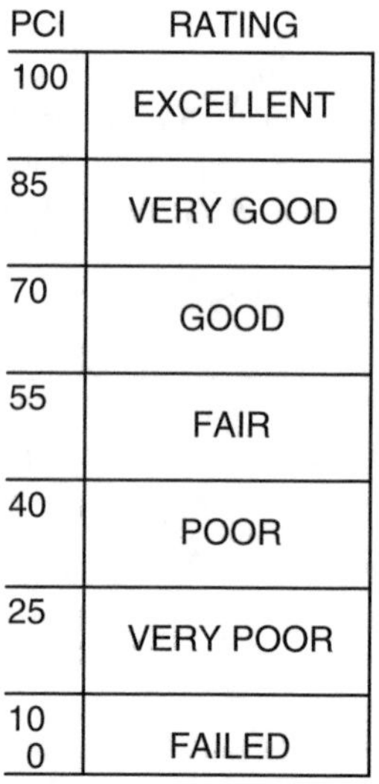

PCI	RATING
100	EXCELLENT
85	VERY GOOD
70	GOOD
55	FAIR
40	POOR
25	VERY POOR
10 / 0	FAILED

FIGURE 19.48   Interpretation of pavement condition index.

In order to ensure that the estimated PCI is within $+5$ points of true mean PCI about 95% of the time, a minimum number of sample units should be surveyed. The minimum number of sample units is calculated using the Equation 19.6.

$$n = \frac{N\sigma^2}{\dfrac{e^2}{4}(N-1) + \sigma^2} \geq 5 \tag{19.6}$$

where:

n = minimum number of samples to be surveyed ($n = N$, if $N \leq 5$).
N = total number of sample units.
σ = standard deviation of PCI sample units. For initial survey, use $\sigma = 10$ for flexible pavements and $\sigma = 14$ for concrete pavements.
$e = 5$.

In estimating the pavement condition, an index of 100 is assigned assuming the pavement was in excellent condition initially and deduct values are obtained for each distress type based on the current distress extent and severity. A series of deduct curves is available in Shahin and Walther (1990), which were developed for a specific distress types and severity levels. Appropriate modifications to these curves may be warranted in situations where the distress type definitions and severity levels differ from those that were considered in developing these curves. Figure 19.49 demonstrates the PCI calculation procedure. The processes are listed below:

Step 1
For flexible pavement, determine distress type, severity, and extent of each distress type for the sample unit surveyed. Distress severity is expressed in low, medium, and high scale, while the extent is expressed in linear feet, square feet, or number depending on the distress type.
For rigid pavement, determine distress type and severity only.

Step 2
For flexible pavement, compute distress density by:

$$\text{Density} = \frac{\text{Distress in square feet}}{\text{Sample unit area in square feet}} \times 100$$

$$\text{Density} = \frac{\text{Distress in linear feet}}{\text{Sample unit area in square feet}} \times 100$$

$$\text{Density} = \frac{\text{Number of pothole}}{\text{Sample unit area in square feet}} \times 100$$

For rigid pavement, compute distress density by:

$$\text{Density}_i = \frac{\text{number of slabs with distress type, } i}{\text{number of slabs in sample unit}} \times 100$$

Step 3
Obtain deduct points (Dp) from "Deduct Value Curves" for each distress type.

Step 4
Calculate total deduct points (TDV) for the sample unit:

$$\text{TDV} = \sum_{i=1}^{Y} (\text{Dp})_i \qquad \text{where, } Y = \text{number of distress types.} \tag{19.7}$$

Step 5
Determine corrected deduct point (CDV) for the sample unit.

Step 6
Calculate PCI_i of the sample unit by deducting CDV from 100.

STEP 1. DIVIDE PAVEMENT SECTION INTO SAMPLE UNITS.

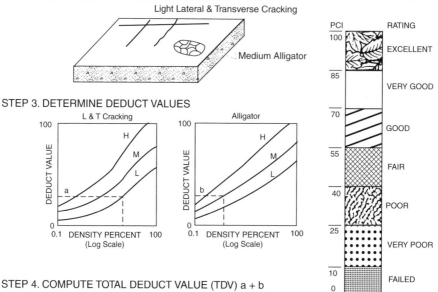

STEP 2. INSPECT SAMPLE UNITS DETERMINE DISTRESS TYPES AND SEVERITY LEVELS AND MEASURE DENSITY

STEP 8. DETERMINE PAVEMENT CONDITION RATING OF SECTION

STEP 3. DETERMINE DEDUCT VALUES

STEP 4. COMPUTE TOTAL DEDUCT VALUE (TDV) a + b

STEP 5. ADJUST TOTAL DEDUCT VALUE

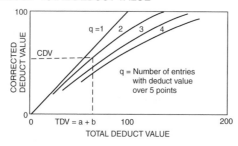

STEP 6. COMPUTE PAVEMENT CONDITION INDEX (PCI) 100-CDV FOR EACH SAMPLE UNIT INSPECTED

STEP 7. COMPUTE PCI OF ENTIRE SECTION (AVERAGE PCI'S OF SAMPLE UNITS)

FIGURE 19.49   PCI calculation procedure for PAVER system (*Source*: Shahin and Walther, 1990).

Step 7

Calculate PCI of the pavement section by

$$PCI = \frac{1}{n} \sum_{i=1}^{n} (PCI)_i \tag{19.8}$$

Example Problem 19.1

A condition survey on a sample unit of a flexible and rigid pavement produced the following distress summary. The area of the flexible pavement sample unit was 1800 m² and for rigid pavement the sample unit was 16 slabs. Calculate the pavement condition in PCI using PAVER system. The deduct value curves shown in Figure 19.50 to Figure 19.57 may be used.

Pavement Type	Distress Type	Extent	Severity
Flexible	Crocodile cracking	90 m^2	Medium
	Rutting	65 m^2	High
	Pothole	5 numbers	Low
Rigid	Faulting	3 slabs	High
	Joint Spalling	2 slabs	Low
	Linear crack	5 slabs	Medium

Solution

Flexible Pavement

Distress	Severity	Density (extent/sample area) × 100	Deduct Value
Crocodile Cracking	Medium	90/1800 × 100 = 5.0	38
Rutting	High	65/1800 × 100 = 3.61	43
Pothole	Low	5/1800 × 100 = 0.28	8

As shown above, the deduct values for crocodile cracking, rutting, and potholes are obtained from Figure 19.50, Figure 19.51 and Figure 19.52.

Total deduct value = 38 + 43 + 8 = 89 and q (number of deduct value > 5) = 3

Corrected deduct value = 57 (from Figure 19.53).

PCI for the sample flexible pavement = 100 − 57 = 43, which represents a fair condition (see Figure 19.48).

Rigid Pavement

Distress	Severity	Density (no. of slabs with distress/no. of sample slabs) × 100	Deduct Value
Faulting	High	3/16 × 100 = 18.75	29
Joint Spalling	Low	2/16 × 100 = 12.5	3
Linear crack	Medium	5/16 × 100 = 31.25	20

As shown above, the deduct values for crocodile cracking, rutting, and potholes are obtained from Figure 19.54, Figure 19.55 and Figure 19.56.

Total deduct value = 29 + 3 + 20 = 52 and q (number of deduct value > 5) = 2

Corrected deduct value = 40 (from Figure 19.57).

PCI for the sample flexible pavement = 100 − 40 = 60, which means the sample rigid pavement slabs are in good condition (see Figure 19.48).

19.6 Serviceability and Roughness Measurements

19.6.1 Serviceability Evaluation

PSI was the first and most commonly used method to relate the objective measures of surface condition to the public's perception of serviceability. The original PSI equations were:

For Flexible pavement

$$PSI = 5.03 - 1.91 \log(1 + S_v) - 0.01\sqrt{(C_l + P_a)} - 1.38(R_d)^2 \tag{19.9}$$

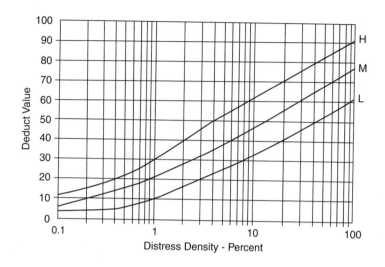

FIGURE 19.50   Deduct value curves for crocodile cracks in flexible pavements (*Source*: Shahin and Walther, 1990).

For Rigid Pavement

$$PSI = 5.41 - 1.8 \log(1 + S_v) - 0.09\sqrt{(C_l + P_a)} \tag{19.10}$$

where:

S_v = slope variance [$\log(1 + S_v)$ = function of profile roughness]:
C_l = crack length in inch (1 in. = 25.4 mm).
P_a = patching area in ft^2.
R_d = Rut depth in inch.
E_r = Statistical error.

Carey and Irick (1960) pointed that PSI was developed by multiple regression techniques and was only intended to predict the Pavement Serviceability Rating (PSR) to a satisfactory approximation. In this process, a number of objective measurements (i.e., PSI) of physical parameters were related to the

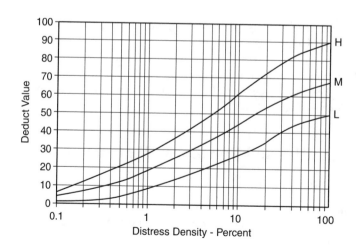

FIGURE 19.51   Deduct value curves for rutting in flexible pavements (*Source*: Shahin and Walther, 1990).

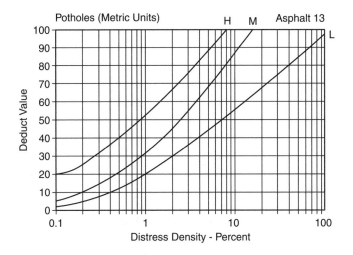

FIGURE 19.52 Deduct value curves for pothole in flexible pavements (*Source*: Shahin and Walther, 1990).

subjective panel ratings (i.e., PSR).

$$PSR = PSI + E_r \qquad (19.11)$$

where:

E_r = statistical error term.

19.6.2 Interpretation of Road Roughness Profile

Different profiling equipment measure and record pavement profiles in different way. Interpretation of these profiles provides a consistent and reliable way to compare pavement ride quality regardless

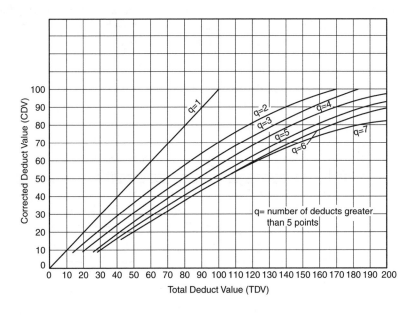

FIGURE 19.53 Corrected deduct value curves for flexible pavements (*Source*: Shahin and Walther, 1990).

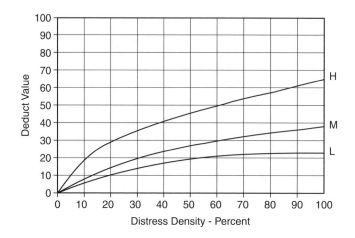

FIGURE 19.54   Deduct value curves for linear cracks in rigid pavements (*Source*: Shahin and Walther, 1990).

of the equipment used. Roughness indices are derived from these profiles through a systematic procedure in order to evaluate different roads with different equipment.

19.6.2.1   Deviations in Elevations

19.6.2.1.1   *Accumulation of Vertical Derivations*
Cumulative deviations or displacements (see Figure 19.58) of an axle relative to a fixed reference such as vehicle body are referred to as roughness. An example is the Bump Integrator (Jordan and Young, 1980) that records counts of displacements (in either downward direction only or both upward and downward directions) in inches.

Roughness $= \sum |d|$, where, $d =$ deviation from a fixed reference.

19.6.2.1.2   *Slope Variance*
Carey and Huckins (1962) used Slope Variance as a measure for roughness statistic during the AASHO Road Test from 1958 to 1960. The Slope Variance is a profile-based roughness statistic obtained from the slope profilometer that was used in the AASHO Test. It is calculated from the statistical variance of

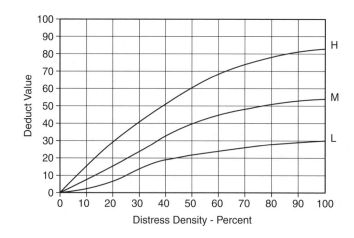

FIGURE 19.55   Deduct value curves for faulting in rigid pavements (*Source*: Shahin and Walther, 1990).

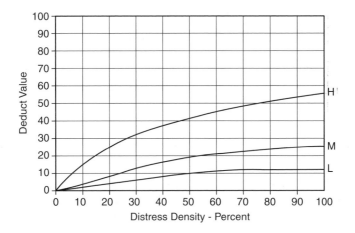

FIGURE 19.56 Deduct value curves for joint spalling in rigid pavements (*Source*: Shahin and Walther, 1990).

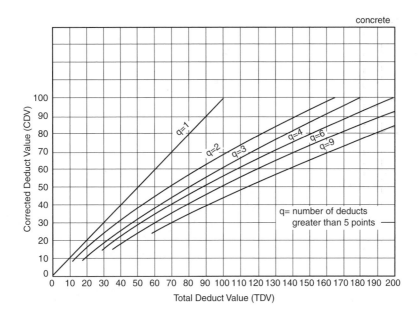

FIGURE 19.57 Corrected deduct value curves for rigid pavements (*Source*: Shahin and Walther, 1990).

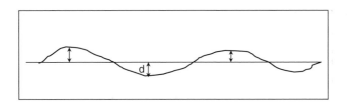

Roughness = Σ|d|, where, d = deviation from a fixed reference.

FIGURE 19.58 Sketch of a longitudinal profile of a road.

surface slope (see Equation 19.12) defined for a constant distance of 1 ft (305 mm).

$$SV = \frac{\sum_{i=1}^{n} Y_i^2 - \frac{1}{n}\left(\sum_{i=1}^{n} Y_i\right)^2}{n-1} \tag{19.12}$$

where:

    SV = Slope Variance.
    Y_i = difference in elevation between two successive points at a constant distance of 1 ft (305 mm).
    n = number of intervals.

19.6.2.1.3 Squares of Deviations

Square of deviations is used in the PCA Road Meter (Brokaw, 1967) that measures deviations in terms of 1/8 inch increments at idle vehicle positions. Correlations have been established between the sums of squares of deviations and the slope variance. Equation 19.13 is used to calculate squares of deviations and Equation 19.14 shows the above correlation.

$$\sum d^2 = \frac{1}{64}(A + 4B + 9C + 16D + 25E + \cdots) \tag{19.13}$$

where:

    d = deviations.
    a = number of 1/8 in. (1 in. = 25.4 mm) deviations.
    B = number of 2/8 in. deviations.
    C = number of 3/8 in. deviations.
    D = number of 4/8 in. deviations.
    E = number of 5/8 in. deviations, etc.

$$SV = 0.68 \sum (d^2) + 0.8 \tag{19.14}$$

Example Problem 19.2

A road profile survey results are shown below. Values in the table are profile elevations in mm with respect to a reference plane and are taken at 300 mm intervals. Calculate accumulation of vertical deviations, slope variance, and squares of deviations.

Lane	Profile Elevations (mm)									
Lane 1	−1.05	2.19	0.80	−1.50	1.26	1.09	−2.01	−1.00	2.30	−0.79
Lane 2	2.11	2.12	−1.56	−1.96	1.34	0.53	−1.67	−1.25	2.10	−0.92

Solution

    (a) Sum of Vertical Deviations
        Lane 1: $\sum |d|$ = 13.99 mm
        Lane 2: $\sum |d|$ = 15.56 mm
    (b) Slope Variance

$$SV = \frac{\sum_{i=1}^{n} Y_i^2 - \frac{1}{n}\left(\sum_{i=1}^{n} Y_i\right)^2}{n-1}$$

    where Y_i = difference between two points and n = number of intervals = 9

	Lane 1				Lane 2		
Readings	Y_i	Y_i^2	d^2	Readings	Y_i	Y_i^2	d^2
-1.05	—	—	1.10	2.11	—	—	4.45
2.19	-3.24	10.50	4.80	2.12	-0.01	0.00	4.49
0.8	1.39	1.93	0.64	-1.56	3.68	13.54	2.43
-1.5	2.3	5.29	2.25	-1.96	0.4	0.16	3.84
1.26	-2.76	7.62	1.59	1.34	-3.3	10.89	1.80
1.09	0.17	0.03	1.19	0.53	0.81	0.66	0.28
-2.01	3.1	9.61	4.04	-1.67	2.2	4.84	2.79
-1	-1.01	1.02	1.00	-1.25	-0.42	0.18	1.56
2.3	-3.3	10.89	5.29	2.1	-3.35	11.22	4.41
-0.79	3.09	9.55	0.62	-0.92	3.02	9.12	0.85
Sum	-0.26	56.43	22.52	Sum	3.03	50.61	26.91

$$SV_{lane1} = 56.43 - \{1/9(-0.26)^2\}/(9-1) = 7.05$$

$$SV_{lane2} = 50.61 - \{1/9(3.03)^2\}/(9-1) = 6.20$$

(c) Sum of Squares of Deviations
Lane 1 $= \sum d^2 = 22.52$
Lane 2 $= \sum d^2 = 26.91$

19.6.2.2 Power Spectral Density (PSD)

Pavement profile data can be expressed in terms of amplitudes and frequencies of superimposed waveforms consisting of many wavelengths. PSD sorts the profile data into its component frequencies and determines their relative amplitudes. The basis of PSD functions to interpret profile data are explained below.

(a) A combination of sine waves when added together constitutes the original profile, and roughness can be expressed as a continuous spectrum of wavelengths that have different significance on vehicle vibration and pavement condition.
(b) Though the term "PSD" has been derived from its earlier applications in electronics and communication field, a more appropriate term could be Variance Spectral Density.
(c) PSD function is continuous at all wavelengths and the magnitude of individual wavelength can be observed.
(d) The area under the PSD function curve of a variable is equal to the overall mean square value of a variable.

There are three types of PSD functions generally used to interpret profiles. They are: PSD function of elevation profile, PSD function of surface slope, and PSD function of spatial acceleration. Examples of PSD functions derived from a single measured profile can be found in Sayers et al. (1986a).

The PSD Function of Elevations Profile has the following characteristics:

• Lower frequency features (i.e., lowest wavenumbers or longest wavelengths) have larger amplitudes.
• Amplitude decreases with increasing frequency.
• PSD curve tends to move downward for smoother roads and upward for rougher roads.
• Different construction techniques result in slightly different characteristic PSD slopes.
• Each road has a unique PSD slope. Thus, roads constructed by the same construction technique would have similar PSD curves.

Roughness expressed as PSD function of surface slope is derived from PSD function of elevation profile by taking spatial derivative of elevation. Roughness obtained in this process tends to be close to constant slope across the spectrum, which is much narrower in amplitude than the elevation, especially at higher wavenumbers.

PSD function of spatial acceleration is also used to express roughness which is derived from the spatial derivative of slope. Because of its close relationship with vehicle vibration, this method is most commonly used.

Aggregated PSD functions can be plotted for different pavement types in order to show that different surface types have characteristically different "signatures", which is reflected by the distribution of roughness over wavenumbers. Examples of aggregated PSD "signatures" for asphalt surface, surface treated, gravel, and earth pavements can be found in (Sayers et al., 1986a). It is obvious from those example "signatures" that the relationship between roughness and wavenumbers are quite different among the surface types presented in the example.

19.6.2.3 Quarter Car Simulation (QCS)

A mathematical model was developed based on the concept of a single wheel to obtain roughness numeric from direct profile measurement. The conceptual quarter-car suspension model (Sayers et al., 1986a) is shown Figure 19.59. Parameters that influence the response from the model include two masses, two spring rates, a damping rate, and a simulation speed.

The behavior of quarter-car is described by the following two second-order differential equations:

$$M_s \ddot{Z}_s + C_s(\dot{Z}_s - \dot{Z}_u) + K_s(Z_s - Z_u) = 0 \tag{19.15}$$

$$M_s \ddot{Z}_s + M_u \ddot{Z}_u + K_t(Z_u - Z) = 0 \tag{19.16}$$

where:

Z = road profile elevation points.
Z_u = elevation of unsprung mass (axle).
Z_s = elevation of sprung mass (body).
K_t = tire spring constant.
K_s = suspension spring constant.
C_s = shock absorber constant.
M_u = unsprung mass (axle).
M_s = sprung mass.

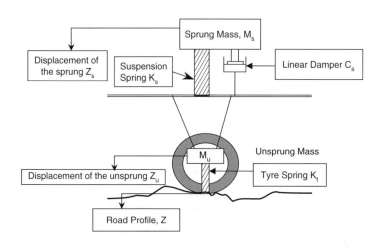

FIGURE 19.59   Quarter-car representation of response to uneven road surface.

Two common forms of QCS statistics are:

$$QCS_1 = \frac{1}{T} \int_0^T |\dot{Z}_s - \dot{Z}_u| dt \quad \text{(m/sec)} \tag{19.17}$$

$$QCS_2 = \frac{1}{L} \int_0^T |\dot{Z}_s - \dot{Z}_u| dt \quad \text{(mm/m)} \tag{19.18}$$

where, T = total time and L = total distance.

Note that Velocity = QCS_2/QCS_1.

The Roughness parameter, called Average Rectified Slope (ARS) and expressed in mm/m, is calculated from QCS that measures the axle motion relative to the vehicle body normalized by distance. The simulation speed is noted as a subscript of the ARS numeric which is usually expressed in m/km, mm/km, or in./mile. The ARS numerics can be assumed as the average slope of the profile where a zero ARS means an ideal smooth profile. Another parameter known as Average Rectified Velocity (ARV) is also used. ARV measures axle motion relative to car body normalized by time.

Example Problem 19.3

The following displacements were produced when a quarter-car analysis of a road surface profile was performed. The traveling speed was 80 km/h and the constant distance interval was 0.25 m. Calculate the roughness of the road surface. The velocity can be approximated by $(U_2 - U_1)/dt$.

Sprung/unsprung				Readings					
U_s (m)	0.871	0.869	0.868	0.865	0.866	0.864	0.862	0.860	0.859	0.857
U_u (m)	0.871	0.870	0.869	0.868	0.870	0.867	0.869	0.865	0.862	0.863

Solution

Consider Equation 19.18,

$$\text{Roughness} = \frac{1}{L} \int_0^T |\dot{Z}_s - \dot{Z}_u| dt, \text{ where } z = \text{velocity}(U_2 - U_1/dt).$$

$$\text{Roughness} = \frac{1}{L} \sum_{i=1}^{n} |\dot{Z}_1 - \dot{Z}_2|_i dt = \frac{1}{L} \sum_{i=2}^{n} |(U_{2s} - U_{1s}) - (U_{2u} - U_{1u})|_i$$

U_s	$U_{2s} - U_{1s}$	U_u	$U_{2u} - U_{1u}$	$|(U_{2s} - U_{1s}) - (U_{2u} - U_{1u})|$
0.871	—	0.871	—	—
0.869	−0.002	0.87	−0.001	0.001
0.868	−0.001	0.869	−0.001	0
0.865	−0.003	0.868	−0.001	0.002
0.866	0.001	0.87	0.002	0.001
0.864	−0.002	0.867	−0.003	0.001
0.862	−0.002	0.869	0.002	0.004
0.86	−0.002	0.865	−0.004	0.002
0.859	−0.001	0.862	−0.003	0.002
0.857	−0.002	0.863	0.001	0.003
Sum	—	—	—	0.016

Total distance, $L = 0.25 \times 9 = 2.25$ m

Roughness $= -1/(0.25 - 9) - 0.016 = 7.11 \times 10^{-3}$ m/m $= 7.11$ m/km.

19.6.3 Roughness Indices

Pavement roughness indices are generally estimated from applying mathematical models or from physical response of a vehicle.

19.6.3.1 Root Mean Square Vertical Acceleration (RMSVA)

The positive and negative elevations in a profile tend to average out with traversing distance. RMSVA calculates deviations without canceling the positive and negative elevations. Though RMSVA is not used as a roughness index, it is a building block for defining an index, such as quarter-car index (Sayers et al., 1986a).

$$\text{RMSVAs} = \sqrt{\sum_{i=2}^{n-1}\left(\frac{l_{i+2} - 2l_i - l_{i-1}}{s^2}\right)^2 /(n-2)} \tag{19.19}$$

where:

l = level measurement.

s = interval of level measurement.

Mean Absolute Vertical Acceleration (MAVA) is used to express the mean of the absolute change in grade with distance and it can be calculated by the following equation.

$$\text{MAVAs} = \frac{1}{n-2}\sum_{i=2}^{n-1}\left|\frac{l_{i+2} - 2l_i + l_{i-1}}{d^2}\right| \tag{19.20}$$

Example Problem 19.4

Consider the problem explained in Example Problem 19.2 and calculate RMSVA and MAVA.

Solution

For RMSVA calculation use Equation 19.19 where $s = 300$ mm and $n = 10$. Calculations are shown below:

	Lane 1				Lane 2		
l_i	$l_{i+2} - 2l_i + l_{i-1}$	$\|l_{i+2} - 2l_i + l_{i-1}\|$	$(l_{i+2} - 2l_i + l_{i-1})^2$	l_i	$l_{i+2} - 2l_i + l_{i-1}$	$\|l_{i+2} - 2l_i + l_{i-1}\|$	$(l_{i+2} - 2l_i + l_{i-1})^2$
−1.05	—	—	—	2.11	—	—	—
2.19	−4.63	4.63	21.44	2.12	−3.69	3.69	13.62
0.8	−0.91	0.91	0.83	−1.56	3.28	3.28	10.76
−1.5	5.06	5.06	25.60	−1.96	3.7	3.7	13.69
1.26	−2.93	2.93	8.58	1.34	−4.11	4.11	16.89
1.09	−2.93	2.93	8.58	0.53	−1.39	1.39	1.93
−2.01	4.11	4.11	16.89	−1.67	2.62	2.62	6.86
−1	2.29	2.29	5.24	−1.25	2.93	2.93	8.58
2.3	−6.39	6.39	40.83	2.1	−6.37	6.37	40.58
−0.79	—	—	—	−0.92	—	—	—
Sum	—	29.25	128.01	Sum	—	28.09	112.91

$$\text{RMSVA}_{(\text{lane 1})} = \{(1/(10-2))\times(128.01/300^4)\}^{0.5} = 4.44\times10^{-5}$$

$$\text{RMSVA}_{(\text{lane 2})} = \{(1/(10-2))\times(112.91/300^4)\}^{0.5} = 4.17\times10^{-5}$$

For MAVA calculation use Equation 19.20.

$$\text{MAVA}_{(\text{lane 1})} = \{(1/(10-2))\times(29.25/300^2)\} = 4.06\times10^{-5}$$

$$\text{MAVA}_{(\text{lane 2})} = \{(1/(10-2))\times(28.09/300^2)\} = 3.9\times10^{-5}$$

19.6.3.2 International Roughness Index (IRI)

IRI is a statistic calculated from a mathematical function of the pavement longitudinal profile. It was developed from the World Bank sponsored International Roughness Experiment in Brazil in 1982 (Sayers

TABLE 19.2 Description of IRI Categories and Their Applications

IRI Class	Description
Class 1	Index obtained from highest quality measurements such as Rod and Level Measurements where measurement interval is less than 250 mm. This class of IRI is suitable for all applications. However, Class 1 category IRI is desirable when incremental deterioration of special test sections is to be tracked.
Class 2	This group of index is obtained with less accurate instruments such as lower precision measurements or longer sampling interval.
Class 3	Response type road meters provide Class 3 type IRI which is satisfactory for network level study.

et al., 1986a), which was aimed at establishing correlation and calibration procedures for roughness measurements. IRI is expressed as the average longitudinal road profile that represents the vertical response of a hypothetical quarter-car traveling at 80 km/h to the measured longitudinal road profile. Based on the quality and use, IRI values can be grouped into three categories as given in Table 19.2.

19.6.3.3 NAASRA counts

In Australia, a response type road roughness measuring system known as National Association of Australian Road Authorities (NAASRA) roughness meter has been in use since 1970s. The measured roughness is reported by NAASRA Roughness Counts per kilometer (NRM). The NRM is the cumulative total relative vertical displacement between axle and body of a standard vehicle where one NRM equals to 15.2 mm of axle-to-body separation. NAASRA Roughness meter can be operated at two principal standard speeds, 80 or 50 km/h. PREM (1989) showed that there is a very good correlation between IRI and NRM ($r^2 > 0.99$) and the relationship is shown by the Equation 19.21. Despite such an excellent correlation and wide use, there are concerns about the NAASRA Roughness Meters calibration, repeatability, and reproducibility of the results (Austroads, 2001).

$$\text{NAASRA (counts/km)} = 26.49 \times \text{Lane IRI}_{qc}\text{(m/km)} - 1.27 \qquad (19.21)$$

where, Lane IRI_{qc} = average lane roughness measured in IRI using the quarter car method.

19.6.4 Types of Roughness Measurements and Devices

The requirements for roughness measuring devices include the capability of measuring pavement surface distortions that may consist long and short wavelength of varying frequencies. Yoder and Witczak (1975) noted that an all purpose roughness measuring device should have (i) capability of measuring large number of samples in short time, (ii) flexibility to construction control, and (iii) ability to measure long wavelengths and sharp changes in surface profile.

Two generic types of roughness measuring devices are commonly used. One of them uses the profilometric methods in which longitudinal elevations are measured to calculate roughness indices, while the other is a response type road roughness measuring system (RTRRMS) that produces roughness indices from motion of a vehicle that traversing on the road. As roughness varies with vehicle type and condition, time, and weather during testing, RTRRMS equipment produce comparatively less accurate results compared to the profilometric methods (Sayers et al., 1986b). All RTRRMS devices are sensitive to the vehicle characteristics and operating conditions regardless of their specially designed sensors and transmitters (Haas et al., 1994). Thus these devices are usually calibrated to produce standard roughness results.

Roughness measuring devices can also be grouped into three categories: static devices, inertial profilers and profilographs. Rod and level, Walking profiler, walking dipstick and rolling dipstick fall into the static devices group which measures the relative elevation of the pavement profile. Non-contact distance measuring transducers and accelerometers are used in the inertial profilers where vertical movements of the vehicle are translated into pavement profile. Profilograph, on the other hand, produces pavement profile based on the vertical displacement of the measuring wheel.

A variety of roughness measuring devices are being used in many countries around the world. Recognizing the fact that it is not possible to list and describe all the devices, attempt is made to name the basic devices that were used during the AASHO Road Test, as well as a few other common devices that are being used in a number of countries.

19.6.4.1   Slope Profilometer and CHLOE Profilometer

There were no mechanical devices available to measure roughness during the AASHO Road Test. An instrument known as Slope Profilometer (Yoder and Witczak, 1975) was build and used in the AASHO Road Test. The instrument was cumbersome and operated on a complex principle of imaginary plane established by pendulum or gyroscope. A modified version of this instrument, known as CHLOE Profilometer (Haas et al., 1994), was developed at the later stage. The CHLOE Profilometer was simple and did not have the fixed horizontal reference system. Though the instrument was widely used in the past, due to its slow operating speed and limitations in measuring roughness for wavelengths longer than 12 ft, there has been limited use of this equipment.

19.6.4.2   Rod and Level Survey

ASTM E 1364-95 (ASTM, 1995) outlines detailed procedure of this method. In this method, true ground profile is measured by comparing the elevation of a point with respect to a reference point. Korchi (2000) described how the elevations are measured. Advanced forms of survey equipment are available today that allow highly accurate ground profile to be measured to meet the Class 1 requirement of the IRI (see Section 19.6.3.2).

19.6.4.3   Walking Dipstick and Rolling Dipstick

Similar to Rod and Level device the Walking Dipstick measures true profile, but at a much faster rate. It comes with an on-board computer and software that records and analyses data and plots longitudinal profile of the surface measured. The Rolling Dipstick is a variant of the Walking Dipstick with the exception that it has rolling wheels. The maximum operating speed of the Rolling Dipstick is 3 mi/h (about 5.0 km/h). IRI and pavement profile are automatically calculated using the data it captured during the survey. Photographs of the Walking Dipstick and the Rolling Dipstick are shown in Figure 19.60.

FIGURE 19.60   Walking and rolling dipstick (*Source*: Korchi, 2000) (used by permission).

FIGURE 19.61 TRRL high speed road monitor (*Source*: Croney and Croney, 1998) (Copyright McGraww-Hill, New York, 1998, used by permission).

19.6.4.4 The Road Meter

It is commonly known as PCA Road Meter and was developed by Brokaw (1967). A schematic diagram of the equipment can be found in Yoder and Witczak (1975). The instrument can be attached to a passenger car where it measures the deviations between the body frame and the rear axle of the car. The car is driven over the pavement about a speed of 80 km/h and it measures the number of deviations in 1/8 inch increments by comparing vehicle idle position. Squares of deviations (see Equation 19.13) are used to estimate roughness condition.

19.6.4.5 TRRL High-Speed Road Monitor

Transport Road Research Laboratory (TRRL) developed a high-speed towed profilometer based on the CHLOE profilometer (Croney and Croney, 1998). The equipment (see Figure 19.61) uses laser sensors to measure longitudinal profile, rut depth, and surface macrotexture. The laser sensors are attached to a 5-meter beam which is mounted on a two-wheel unsprung axle. The beam is attached to a towing vehicle and the vehicle can be driven at a speed of 80 km/h.

19.6.4.6 Walking Profiler

Commonly known as ARRB Walking Profiler (see Figure 19.62), ARRB Group Ltd. has recently launched an upgraded and improved version, Walking Profiler G2 (ARRB, 2005). It is primarily designed for project level survey to precisely and accurately measure true profile, grade, and level of pavement surface. Output parameters include IRI, California Profilograph emulation, and Straight Edge simulation. The equipment meets World Bank Class 1 Profilometry requirements and it has an excellent correlation ($R^2 = 0.999$) with Rod and Level derived IRI. The Walking Profiler is accepted as a reference tool for the calibration of roughness measuring equipment by Permanent International Association of Road Congress (PIARC) and Federal Highway Authority (FHWA) in U.S.A.

FIGURE 19.62 ARRB walking profiler (*Source*: ARRB, 2005) (Copyright ARRB Group Ltd., 2005).

19.6.4.7   Roughometer

Figure 19.63 shows a Roughometer II, developed by ARRB Group Ltd., as in its latest version (ARRB, 2005). It is a cost effective alternative to laser profiling equipment to objectively measure road roughness. It is highly suitable for unsealed roads as there is no loss of data integrity due to dusty or wet conditions. However, for sealed roads it is suitable for indicative assessments. The equipment can be easily fitted to most vehicles and survey can be conducted at a speed between 40 and 60 km/h. Road roughness is reported in IRI or NAASRA counts and the accuracy of the roughness results is about ± 1 m/km.

19.6.4.8   BPR Roughometer & Bump Integrator

The Bureau of Public Roads (BPR) developed one of the earliest roughometers which was initially mounted in an automobile. This device was later converted to a trailer type device which

FIGURE 19.63 ARRB roughometer (*Source*: ARRB, 2005) (Copyright ARRB Group Ltd., 2005).

represents the response of one quarter of a passenger car. It measures the difference in movements between the trailer body and the wheel axle produces roughness which is dependent on the trailer characteristics and travel speed.

The Bump Integrator (see Figure 19.64), developed by TRRL, is a form of BPR roughometer. TRRL made significant improvements to the device by using special shock absorbers that have damping properties that are reasonably insensitive to time and operating conditions.

19.6.4.9   The California Profilograph

This profilograph machine consists of a center-profiling wheel that moves freely in the vertical direction and records the profile on a paper. One of the uses of this profilograph is to determine pavement

FIGURE 19.64 Bump integrator (*Source*: Croney and Croney, 1998) (Copyright McGraww-Hill, New York, 1998, used by permission).

smoothness by pushing it over the pavement at walking speed. Thus measuring speed is its significant limitations. These profilographs also respond poorly to long wavelengths such as wavelengths greater than 7.6 m.

19.6.4.10 The Rainhart Profilograph

The concept of the Rainhart Profilograph (Haas et al., 1994) is similar to the California profilograph, though they have a different form and application. It measures the true profile by covering a wide section of roadway where all the 12 of its wheels are arranged on a series of triangular trusses and each travel at a different path.

19.7 Summary and Future Developments

The development of the data collection procedures and equipment is an ongoing process. For instance, Wang et al. (2004) have experimented on the use of stereovision technology for comprehensive pavement distress data collection. In this research, digital highway data vehicle was used and a three-dimensional space of the pavement surface is established. ARRB Group Ltd. (ARRB, 2005) is developing the Hawkeye scaleable series of survey equipment that uses the latest digital video recording technology and the sophisticated laser sensor technology to collect a variety of pavement distresses at highway speed. Research is also being undertaken elsewhere to remove noises from the data collected in the filed, particularly from images of pavement cracks, and to automate the distress detection algorithms (Wang et al., 2004).

Evaluation of the distress data and quantitative representation of the distressed pavement are important requirements of a practical pavement management system. Subjective condition evaluation techniques have the limitation that the evaluation may be affected by the subjective nature of the rater. In the assessment of the severity and maintenance needs of individual distresses, as well as the aggregated distressed state (i.e., overall condition) of a pavement section, expert opinion and subjective judgment are necessary. Mathematical techniques such as Fuzzy Logic are effective in handling the subjective nature of human opinions and are preferred over subjective evaluation procedures (Fwa and Shanmugam, 1998). Other artificial intelligence techniques such as expert systems and neural network (Fwa and Chan, 1993) have been explored and found to be useful tools in automating the distress evaluation process.

Pavement serviceability, which is a measure of pavement ride quality, is mainly determined by the roughness condition. Since the AASHO Road Test there have been enormous developments in this area both in the theoretical side and in equipment range. Among the number of roughness indices commonly used to express roughness condition, the use of IRI is fast gaining acceptance worldwide. Correlations among the various indices have been developed.

The representation of pavement surface profiles in terms of pavement roughness has been an area of active research in pavement engineering. A recent study at the National University of Singapore (Wei and Fwa, 2004; Wei et al., 2004) uses wavelet transform to characterize road roughness. In this method pavement roughness is quantitatively represented by wavelet energy in different wave bands. The wavelet transform provides additional information of a roughness profile compared to the common roughness indices such as IRI and RMSVA. The common roughness indices represent the overall pavement roughness of a pavement section, while the wavelet transform represents roughness characteristics in terms of frequency and position. Pavements with different roughness profiles may have very similar roughness index values when expressed by the common indices, but may have different wavelet transform characteristics if analyzed by the wavelet transform. Wavelet analysis can be used to identify features such as road humps that should not be taken into account in assessing pavement roughness condition. It can also determine the contributing distresses that cause increases in the roughness level of a pavement section. This methodology was successfully tested using roughness data from 400 US-LTPP test sections.

A number of sophisticated response type road roughness measuring systems (RTRRMS) are available in the market that uses the latest laser sensor technology and are capable of measuring roughness at the highway speed. The accuracy of the measured roughness is generally very high. A great deal of improvement has also been achieved in the accuracy and operation of the static devices. For instance, the ARRB Walking profilometer and the Walking Dipstick are excellent examples that meet World Bank Class 1 profilometry requirements. They are accepted reference tools for the calibration of roughness measuring equipment by authorities such as PIARC and FHWA.

References

ARRB, 2005. ARRB Group Ltd. URL: http://www.arrb.com.au. Vermont South, Melbourne, Australia.

LTPP - Australia, 2005. Long Term Pavement Performance Study, Australia. URL: http://www.arrb.com.au/projects/ltpp.

ASTM, 1995. Standard Test Method for Road Roughness by Static Level Method, E 1364-95 (2000), American Society for Testing and Materials.

Austroads, 1987. *A Guide to the Visual Assessment of Pavement Condition.* Austroads, Sydney, Australia.

Austroads, 2001. *Guidelines for Road Condition Monitoring: Part 1 — Pavement Roughness.* Austroads, Sydney, Australia.

APWARF, 1983. *Draft APWA PAVER Implementation Manual.* American Public Works Association Research Foundation, Chicago, Illinois.

Brokaw, M.P., 1967. Development of the PCA Road Meter: A Rapid Method for Measuring Slope Variance, Highway Research Board 189.

Carmichael, R.F. III, Halbach, D.S., and Moser, L.O., 1985. Pavement Condition Survey Manual for Rhode Island Pavement Evaluation and Management System. ARE Inc Report No. RI - 2/2, Final Report.

Carey, W.N. and Huckins, H.C. 1962. *Slope Variance as a Measure of Roughness and the Chloe Profilometer,* Highway Research Board Special Report No. 73. Highway Research Board, Washington, DC.

Carey, W.N. and Irick, P.E., 1960. The Pavement Serviceability — Performance Concept. Highway Research Bulletin 250.

Chong, G.J., Phang, W.A., and Wrong, G.A. 1982. *Manual for Condition Rating of Rigid Pavements — Distress Identifications.* Ministry of Transportation and Communications, Ontario, Canada.

Chong, G.J., Phang, W.A., and Wrong, G.A. 1989a. *Manual for Condition Rating of Flexible Pavements — Distress Manifestations,* Report SP - 024. Ministry of Transportation and Communications, Ontario, Canada.

Chong, G.J., Phang, W.A., and Wrong, G.A. 1989b. *Manual for Condition Rating of Surface-Treated Pavements — Distress Manifestations.* Ministry of Transportation and Communications, Ontario, Canada.

Chong, G.J., Phang, W.A., and Wrong, G.A. 1989c. *Manual for Condition Rating of Gravel Surface Roads,* Report SP - 025. Ministry of Transportation and Communications, Ontario, Canada.

Croney, D. and Croney, P. 1998. *The Design and Performance of Road Pavements,* 3rd Ed., McGrawHill, New York.

Darter, M.I., Becker, J.M., Snyder, M.B., and Smith, R.E. 1985. Portland Cement Concrete Pavement Evaluation System, COPES, NCHRP Report 277.

DeWilder, M. 1985. GERPHO and APL. Road Evaluation Workshop, Proceedings, Report No. FHWA-TS-85-210.

Ferguson, R.A. and Pratt, D.N. 2000. Automated Detection and Classification of Cracking in Road Pavements. 1st European Pavement Management Systems Conference, 24–27 September, Budapest, Hungary.

Ferguson, R.A. and Pratt, D.N. 2002. Automated Detection and Classification of Cracking in Road Pavements (RoadCrack™). IRF and ARF Asia Pacific Roads Conference and Exhibition. 1–5 September, Sydney, Australia.

Fwa, T.F. and Chan, W.T. 1993. Priority rating of highway maintenance needs by neural networks, *J. Transport. Eng.*, 119, 3, 419–432.

Fwa, T.F. and Shanmugam, R. 1998. Fuzzy Logic Technique for Pavement Condition Rating and Maintenance-Needs Assessment. Fourth International Conference on Managing Pavements, South Africa.

Haas, R., Hudson, Q.R., and Zaniewski, J. 1994. *Modern Pavement Management.* Krieger Publishing Company, Malabar, Florida.

Highway Products International Inc., 1985. Automatic Road Analyzer - Mobile Data Acquisition Vehicle. Product Bulletin, Paris, Ontario, Canada.

Highway Research Board, 1962. The AASHO Road Test: Report 5 — Pavement Research. Highway Research Board Special Report 61-E.

Huang, Y.H. 1993. *Pavement Analysis and Design.* Prentice Hall, Englewood Cliffs, New Jersey, 07632.

Hudson, W.R., Elkins, G.E., Uddin, W., and Reilley, K.T. 1987. *Improved Methods and Equipment to Conduct Pavement Distress Surveys*, Report No. FHWA-TS-87-213. Federal Highway Administration, Virginia.

Janoff, M.S., Nick, J.B., Davit, P.F., and Hayhoe, G.F., 1985. Pavement Roughness and Rideability. NCHRP Report No. 275. Washington, DC.

Jordon, P.G. and Young, J.C. 1980. Developments in the Calibration and Use of the Bump Integrator for Ride Assessment. Transport and Road Research Laboratory, Supplementary Report 604, Crowthorne, UK.

Korchi, El.T. 2000. Correlation Study of Ride Quality Assessment Using Pavement Profiling Devices. Research Report CEE00-0122. Worcester Polytechnique Institute. Worcester, MA 01609.

LeClerc, R.V. and Marshall, T.R. 1971. Washington Pavement Rating System: Procedures and Applications. Highway Research Board Special Report 116.

LeClerc, R.V., Marshall, T.R., and Anderson, K.W. 1973. Use of the PCA Road Meter in the Washington Pavement Condition Survey System. Highway Research Board Special Report 133.

MAP Inc., 1986. Pavement Condition Monitoring Methods and Equipment Final Report on GERPHO Survey for ARE Inc - Austin. Federal Highway Administration Strategic Research Program, Washington, DC.

Miller, J.S. and Bellinger, W.Y. 2003. *Distress Identification Manual for the Long-Term Pavement Performance Program*, Fourth Ed., Report no. FHWA-RD-030031. Federal Highway Administration, Virginia.

Novak, R.L. 1985. Swedish Laser Road Surface Tester. Roadway Evaluation Equipment Workshop; Proceedings, Report No. FHWA-TS-85-210.

PASCO US Inc., 1985. Report on Pavement Monitoring Methods and Equipment. Lincoln Park, NJ.

PASCO Corp. Mitsubishi International Co., 1987. 1 for 3 PASCO Road Survey System (PRS System) from Theory to Implementation.

PREM, H., 1989. NAASRA Roughness Meter Calibration via the Road Profile-based International Roughness Index (IRI). ARRB Transport Research Ltd. ARR 164. Vermont South, Melbourne, Australia.

Sayers, M.W., Gillespie, T.D., and Queiroz, C.A.V. 1986a. *The International Road Roughness Experiment: Establishing Correlation and a Calibration Standard for Measurements*, World Bank Technical Paper No. 45. The World Bank, Washington, DC.

Sayers, M.W., Gillespie, T.D., and Queiroz, C.A.V. 1986. The International Road Roughness Experiment: A Basis for Establishing a Standard Scale for Road Roughness Measurements. Transportation Research Record 1084. pp. 76–85.

Shahin, M.Y. and Kohn, S.D. 1984. *APWA-PAVER Pavement Condition Index Field Manual.* American Public Works Association Research Foundation, Chicago, Illinois.

Shahin, M.Y. and Walther, J.A. 1990. Pavement Maintenance Management for Roads and Streets Using the PAVER System. U.S. Army Construction Engineering Research Laboratory, Report No. USACERL TR M-90/05. Champaign, IL.

Yoder, Y.J. and Witczak, M.W. 1975. *Principle of Pavement Design*, 2nd edn, John Wiley and Sons, Inc., New York, USA.

Wang, K.C.P., Gong, W., Elliott, R.P., and Zaniewski, J.P. 2004. A Feasibility Study on Data Automation for Comprehensive Pavement Condition Survey. 6th International Conference on Managing Pavements. 19–24 October, Brisbane, Australia.

Wei, L. and Fwa, T.F. 2005. Characterizing Road Roughness by Wavelet Transform. *J. Transp. Res. Rec*, 1869, 152–158.

Wei, L., Fwa, T.F., and Zhe, Z. 2005. Wavelet Analysis and Interpretation of Road Roughness. *J. Transp. Engin.*, 131(2), 120–130.

20

Structural Evaluation of Highway Pavements

T.F. Fwa
National University of Singapore
Republic of Singapore

20.1 Approaches of Structural Evaluation of Pavements

20.1.1 Evolution of Structural Evaluation Techniques

Structural evaluation of the material properties of in-service pavements is a key activity for both the project and network level pavement management systems. It provides the necessary information for pavement remaining-life analysis and overlay design, network level monitoring of pavement performance, and for decision making by highway agencies concerning pavement maintenance programming and rehabilitation planning.

The approaches for structural evaluation of pavements can be broadly classified into destructive and nondestructive testing. The most common methods of destructive testing are (a) extraction of samples of various pavement layers by means of coring or other techniques, followed by laboratory testing; and (b) test pits for *in situ* testing of various pavement layers. There are instances where a full-scale vehicular load test or an accelerated simulated load test may be conducted to cause failure of a pavement section for determining its load bearing capacity. Destructive test methods are rarely used today for normal structural evaluation of in-service pavement sections because they are costly, tend to cause traffic obstruction and are time-consuming. There is also the problem of sample disturbance in coring or open-pit testing.

Nondestructive structural evaluation of pavements has become the norm for either project or network pavement management systems. Several approaches of nondestructive testing have been adopted for structural evaluation of in-service pavement sections. Among them the deflection-based approach and the wave propagation technique are the two that have received the most attention by the highway community.

The nondestructive testing of pavement using wave propagation involves the measurement of velocities and wavelengths of surface waves transmitted from a vibration source on the pavement surface. Four different types of waves are generated from a vibration on the surface of an ideal linear elastic material: (a) Primary waves, also known as compression, dilatational, or P-waves, (b) Shear waves, also referred to as secondary, distortional, or S-waves, (c) Rayleigh waves, also termed as surface waves, or R-waves, and (d) Loves waves. The velocity of each type of the waves is related to the properties of the pavement materials. An array of wave propagation analysis methods, such as seismic analysis of surface wave (SASW) and inversion of dispersion curve, have been applied to extract pavement properties. Since the interpretation of test results is not straight forward, and due to the difficulties in relating the computed parameters with those under the different stress- and strain-states of pavement materials generated by moving traffic loads, this approach has remained largely a research tool. Readers are referred to technical reports by Douglas and Elle (1986), Nazarian and Stokoe (1989), Gartin (1991), Tawfig et al. (2000), and Ryden et al. (2004) for further information.

Since the early 1970s, the surface deflection-based approach has been widely adopted as the most practical approach for nondestructive evaluation of pavement properties. Surface deflections represent an overall response of the entire depth of the pavement system tested. Back-analysis is performed on the measured deflections to determine the structural properties of different layers of the pavement, and these properties are in turn applied to assess the service condition of the pavement, to predict pavement performance and remaining service life, to formulate maintenance and rehabilitation strategies, to select types of treatment and repair, and to perform overlay design if required.

Deflection test methods employing static or slow-moving loads were used initially in the 1960s and 1970s. With the increasing acceptance of pavement management systems by highway agencies since the late 1970s, the need for nondestructive deflection testing over large lengths of pavement sections has made the speed of testing a major operational requirement. To meet the higher test-speed requirements for network level testing, higher speed deflection test devices such as steady state vibratory test equipment were introduced and become popular in the late 1970s and early 1980s. This form of test equipment, however, has given way to the Falling Weight Deflectometer (FWD) that produces an impulse load more closely resembling the loading pattern of moving wheel loads on the pavement.

In conjunction with the development of deflection test devices, a large number of empirical and analytical procedures have been developed to estimate the material properties of different pavement layers using the measured deflections. Initially, semimechanistic methods based on the magnitudes of surface deflections and shapes of deflection bowls have been used (Wang et al., 1978). Curvature indices were computed as indicators of the relative quality of different pavement layers. The information gathered is inadequate to serve the needs of modern day pavement management systems. More sophisticated analytical procedures, collectively known as backcalculation analysis, have since been developed to back-derive material properties of individual pavement layers from surface deflection measurements.

20.1.2   Issues in Backcalculation of Pavement Structural Properties

20.1.2.1   Backcalculation Analysis as an Inverse Problem

Backcalculation is an inverse problem that determines *in situ* material properties of pavement layers from measured surface deflections under the action of a known applied load. The problem is typically poorly conditioned and the solution is sensitive to small changes in the measured surface deflections. Furthermore, there is the nonuniqueness of the backcalculated solution that makes the problem a complicated one (Uddin et al., 1986; Stolle and Hein, 1989). Fortunately, when the likely ranges of moduli of different pavement layers are specified, most backcalculation procedures can often converge to a proper solution. As there exist more than one possible combination of elastic moduli of different pavement layers that would give deflections identical or close to the measured deflections, the selection of appropriate initial seed values of the unknown pavement parameters is an important aspect of the backcalculation analysis.

While most backcalculation algorithms require the user to select the seed values of the unknown pavement properties, some researchers have proposed the use of predictive equations to address the nonuniqueness and user dependence problem (Uddin, 1984). Expert systems have also been developed to incorporate the knowledge of experienced pavement engineers in the evaluation of backcalculated results (Greenstein and Berger, 1989). To ensure that only realistic results are obtained, some backcalculation algorithms require the range of each unknown material property to be specified.

As indicated in Figure 20.1, the pavement backcalculation problem can be viewed as a numerical analysis involving the following components: (a) a surface loading model, (b) pavement and material models for various pavement layers and the subgrade, (c) a pavement response model, and (d) a back-analysis model. Each of these is described in the following subsections.

20.1.2.2   Surface Loading Models

Depending on the type of deflection test adopted, the applied load may be represented by a static load, a moving load, a vibratory load, or an impulse load. Unfortunately, due to the relative complexity of modeling a moving or dynamic load, it is common practice to adopt the assumption of static applied load in the backcalculation analysis. It has also been reasoned that since most of the current pavement design procedures are based on static analysis, the pseudo-properties obtained from static load backcalculation are applicable.

Dynamic loading generates additional effects such as inertia and resonance. The main objection against using static analysis in backcalculation appears to be directed at cases of pavements with a shallow depth to the bedrock. Chang et al. (1992) reported that when dynamic effects occurred in the deflection

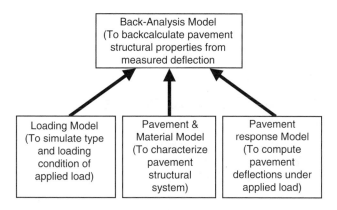

FIGURE 20.1   Main components of analytical models in backcalculation analysis of pavement properties based on deflection test.

measurements but were not taken into account in the backcalculation analysis, the modulus of the subgrade could be underestimated by 50% or more, and the base and subbase moduli were overestimated by about the same margin.

Roesset and Shao (1985) analyzed simulated Dynaflect tests and concluded that when the subgrade thickness was more than 35 ft (11.48 m), there was no significant difference between dynamic deflections and the static deflections computed by the elastic layered theory. Uzan (1994) suggested that static analysis could be used for analysis with FWD data when no bedrock existed within the top 6 m.

20.1.2.3 Pavement and Material Models

Pavements are generally represented by either rigid or flexible model, although there exists other models that analyze the behavior of composite pavements and segmental pavements. Descriptions of pavement models for flexible pavements are found in Chapters 8 and 11, and those for rigid pavements in Chapters 9 and 12. The applications of these models for backcalculation analysis of rigid and flexible pavements are presented in Sections 20.3 and 20.4.

Granular materials are commonly used for pavement base and subbase construction. Their properties are stress-dependent and are nonlinear in their response under loads. The behavior of subgrade materials, either sandy or clayey, is also known to be stress-dependent and nonlinear. The modulus of a sandy subgrade will increase with depth under the action of the overburden, while that of a clayey subgrade will increase as the horizontal distance from the load increases or as the deviator stress decreases.

The resilient modulus models in use today for granular materials are empirical relationships derived from experimental studies. It is usually expressed as a function of mean effective normal compressive stress and deviator stress. The most widely used model in pavement backcalculation studies appears to be the following that relates the resilient modulus positively with the bulk stress (Hicks, 1970):

$$E = k_1 \theta^{k_2} \tag{20.1}$$

where θ is the bulk stress or first invariant of stress, and k_1 and k_2 are material constants. Tam and Brown (1988) have proposed a modification to the model by incorporating the effect of deviator stress as follows:

$$E = c_1 \left(\frac{p}{q} \right)^{c_2} \tag{20.2}$$

where p is the mean effective normal compressive stress, q is the deviator stress, and c_1 and c_2 are material constants.

The nonlinear behavior of fine-grained subgrade soil has been characterized by a bilinear model (Thompson and Robnett, 1979) given by

$$E = A_2 - A_3(A_1 - q) \qquad \text{for } A_1 > q \tag{20.3a}$$

or

$$E = A_2 + A_4(q - A_1) \qquad \text{for } A_1 < q \tag{20.3b}$$

where A_1, A_2, A_3, and A_4 are material constants. Alternatively, Moosazadeh and Witczak (1981) have suggested a nonlinear relation of the form

$$E = b^1 q^{b_2} \tag{20.4}$$

where b_1 and b_2 are material constants. Another nonlinear model proposed by Brown et al. (1987) is given by

$$E = A \left(\frac{p_o}{q_r} \right)^B \tag{20.5}$$

where p_o is the mean normal effective stress due to self weight of the pavement, q_r the deviator stress due to wheel loading, and A and B are material constants.

As the response of asphaltic mixtures under loads is time and temperature dependent, Stubbs et al. (1994) have made an attempt to model an asphalt pavement as a layered visco-elastic half-space. The asphalt concrete surface layer was represented as a visco-elastic medium while the granular layers and the subgrade were represented as damped elastic solid. The following three-parameter model has been adopted for asphalt concrete:

$$D(t) = D_0 + D_1 t^m \tag{20.6}$$

where $D(t)$ is the creep compliance, t is time, and D_0, D_1, and m are material constants. The disadvantages of the visco-elastic models include the need for creep compliance data at very small time intervals and difficulty in considering the nonlinear behavior of pavement materials.

20.1.2.4 Pavement Response Models

Regardless of the type of pavement and material model and surface loading model employed, a backcalculation procedure relying on matching of computed and measured pavement responses consists of the following three major steps: (a) selection of a trial set of values for the unknown pavement parameters, (b) forward computation of pavement response based on the parameter values selected, and comparison of the computed response with the measured, and (c) changing of the selected parameter values by means of an appropriate search algorithm to achieve improved matching of the computed and measured responses.

The common pavement response models used for forward computation of pavement response can be classified into four general classes, depending on the type of material model and surface loading model employed. These four classes are: (a) linear static analysis, (b) nonlinear static analysis, (c) linear dynamic analysis, and (d) nonlinear dynamic analysis.

Linear static analysis is undoubtedly the most widely used procedure primarily because of its simplicity in application and wide availability of such pavement response programs based on linear static analysis. Plate theory is usually adopted for analysis of rigid-pavement responses under loads (Westergaard, 1948; Ioannides, 1986; Fwa et al., 1996). For flexible pavements, layered elastic programs such as BISAR (Shell Koninklijke, 1972), CHEVRON (Michelow, 1963), WESLEA (Van Cauwelaert et al., 1989), ELSYM (Kopperman et al., 1986), and KENLAYER (Huang, 1993) have been used. Since pavement layer thicknesses and Poisson's ratios are usually assumed to be known, there is only one unknown (i.e., elastic modulus) for each layer. The measured deflection basin is defined by the peak deflections at all the sensor points, and the peak load level is considered in the analysis. The backcalculated moduli in this case represent the respective "average" values for different pavement layers. As explained in the preceding sections, corrections may have to be incorporated to account for nonlinear behavior of pavement materials and the effect of dynamic loading, especially when the depth to bedrock is shallow.

In nonlinear static analysis, the deflection basin and applied load representations are the same as the linear static analysis. The main difference lies with the material models with which the nonlinear behavior of pavement materials can now be considered. This means that there are more than one unknown model parameters for each pavement layer. The reliability of backcalculated values of these parameters is an important issue to consider. The need to model stress-dependence of pavement materials makes layered elastic programs unsuitable for this type of analysis. Finite element methods are necessary to represent the variations of material moduli at different points within the pavement consistent with their stress levels. ILLI–PAVE (Raad and Figueroa, 1980) and MICH–PAVE (Harichandran et al., 1989) are examples of finite element computer programs. The additional computation requirements due to larger number of unknown parameters and stress dependence of the unknown moduli, as well as the use of finite element method have led to considerable longer computing time as compared to linear static analysis. This is a major obstacle to the adoption of nonlinear static analysis for routine nondestructive pavement evaluation.

Linear and nonlinear dynamic analyses require the time history data of applied load and the deflection basin defined by the peak deflections. Alternatively, the time history of deflections may be used instead of the peak-value deflection basin. The use of deflection time histories at different sensor points provides additional information that describes the response of the pavement tested. Either frequency domain or time domain fitting may be used for matching the computed and measured responses. Dynamic backcalculation analysis with pavement layers modeled as linear visco-elastic materials has been demonstrated by Uzan (1994). Nonlinear dynamic backcalculation analysis that considers dynamic loading function as well as nonlinear behavior of pavement materials has so far remained as a topic of research interest due to the complexity of problem modeling and the long computation time required.

Each of the four pavement response models describes a different idealized physical problem. It would not be appropriate to compare their backcalculated parameters for a problem directly as none of the models could be considered to give a perfect representation of the test procedure and pavement response under the test load. It can be imagined that when identical deflection basins are used for backcalculation analyses using the four pavement response models, respectively, the backcalculated parameters in each case are pseudo-material properties which are meaningful only as far as the specific model is concerned.

Besides examining how closely a model could simulate a deflection test and the corresponding pavement response, a useful basis for comparing the suitability of the different pavement response models is whether these results can be applied to achieve the following objectives: (a) accurate identification of structurally deficient pavement sections and the structurally weak layers of these pavement sections, (b) reliable estimation of the length of remaining service life, and (c) adequate overlay design to serve the design traffic specified.

20.1.2.5   Back-Analysis Models

Back-analysis is performed to identify the values of the unknown parameters of the selected material models which, when introduced as input to the adopted pavement response model, would produce a computed response sufficiently close to the corresponding *in situ* measured response. Two general approaches of determining pavement parameters to match the computed with measured pavement deflections may be identified. They are the so-called inverse back-analysis and direct back-analysis (Gioda, 1985).

The inverse back-analysis approach adopts an "equation error criterion" to minimize the errors of pavement response equations in estimating the actual *in situ* surface deflections. In an ideal case where the pavement response models describe the actual response exactly, there are no errors involved in all the response equations. In a real-life situation, there is an error associated with each response equation, known as an "equation error." The solution is one that minimizes a combined measure of all "equation errors" of the system by means of a proper choice of the values of the unknown material parameters. Different solution techniques, such as linear programming, quadratic programming, and matrix inversion may be employed to arrive at the "best" choice of material parameter values.

The direct back-analysis approach is based on the minimization of the "output error," i.e., the discrepancy between the measured and computed surface deflections. In general, the error is a complicated nonlinear function of the unknown parameters and its analytical expression often cannot be determined. Different iterative search techniques have been employed to obtain values of material parameters leading to computed deflections sufficiently close to the corresponding measured values.

In the pavement back-analysis problem, the direct approach based on the "output error" criterion has been used in virtually all the backcalculation algorithms reported in the literature so far. The three common measures of output errors used by researchers are: (a) the sum of the absolute differences, (b) the sum of the squared differences, and (c) the sum of the squared relative errors. While many researchers assign identical weighting factors to all deflection error terms, different weighting factors have also been used to scale the magnitudes of different terms so that the smaller components would not be ignored in the error minimization process. Besides the deflection error criterion, some researchers have also employed additional termination criteria such as the percent change in predicted modulus values in successive iterations and the maximum number of iteration cycles.

Numerous techniques have been used to arrive at a solution that gives an acceptable match between the computed and measured deflection basin. The most common approach is one that employs an iterative gradient search algorithm, such as the Gauss–Newton method. Compared with the so-called database methods (Anderson, 1989; Tia et al., 1989) and the regression equation-based procedures (Zaghloul et al., 1994; Fwa and Chandrasegaram, 2001), this approach usually takes much longer time due to the need to execute the forward structural response model repeatedly. A database method precalculates solutions based on measured deflections for a large number of pavement sections, and stores them in organized database for efficient matching of deflection basins. Regression equation-based methods relate surface deflections to pavement layer moduli using statistical regression techniques, allowing almost instantaneous computation of the moduli once the measured deflections are known. The transferability of the regression equations and database are restricted to pavements of the same design with materials of similar characteristics under similar environmental conditions.

20.1.2.6   Backcalculated vs. Laboratory Measured Material Properties

Laboratory testing of pavement materials, either on field samples or laboratory prepared specimens, has been the traditional practice to determine material properties for pavement design and analysis. It is of interest to examine how backcalculated pavement properties compare with the corresponding laboratory measured values. Studies by most researchers have found that the agreement between the two forms of measured values has not been always satisfactory (Houston et al., 1992; Akram et al., 1994; Van Deusen et al., 1994; Nazarian et al., 1998). The following contributing factors have been cited by researchers: (a) a nondestructive field test covers a large volume of pavement materials as compared to the much smaller sizes of laboratory test specimens; (b) the backcalculated values are averaged layer properties as opposed to the "point estimates" given by individual laboratory tests; (c) the modes of load applications are not the same, (d) the states of stress are different; and (e) cored samples used for laboratory testing have been disturbed.

20.1.2.7   Sources of Errors in Backcalculation

20.1.2.7.1   *Measurement Errors*
Deflection readings as obtained from the recording device contain three general forms of measurement errors, viz., setting errors, system errors, and random errors. Setting errors refer to errors caused by improper seating of the sensors, or imperfect setting up of the test device. System errors are related to the inherent features of the measuring system, such as the sensitivity or precision of a device. For example, a typical FWD commercially available today is able to achieve a measurement precision of better than $(\pm)2\%$. A poorly calibrated measuring device would produce biased measurements with systematic errors, which cannot be eliminated by making multiple repeat deflection measurements. Random errors are unavoidable errors associated with individual measurements, but their effects can be reduced by taking the average values of repeated measurements.

Irwin et al. (1989) conducted an analysis of the effects of input deflection measurements containing random measurement errors with a standard deviation of 1.95 μm. Table 20.1 summarizes the findings of the study. Shown in the table are the statistics of 30 sets of backcalculated layer moduli for a four-layer pavement. It shows that even a seemingly inconsequential measurement error can have a significant effect on the backcalculation results. The same table also demonstrates the benefit of having three repeat measurements per point. The results also serve to illustrate the general findings that the backcalculation moduli of the subgrade have the best reliability, while that of the surface is comparatively the least reliable. Similar findings and order of magnitude of effects were found for rigid pavements (Li et al., 1996).

20.1.2.7.2   *Input Errors*
Among the various input parameters to a backcalculation problem, layer thicknesses, and bedrock depth are the two most often cited due to their important impacts on the backcalculated results. The effects of bedrock depth have been explained earlier. The effect of errors in layer thicknesses on backcalculated moduli is illustrated in Table 20.2 based on a study by Irwin et al. (1989). The following variations in layer

TABLE 20.1 Effects of Random Deflection Measurement Error with Standard Deviation of 1.95 μm Based on Study by Irwin et al. (1989)

Test Scheme	Statistics of Backcalculated Modulus (ksi) (1 ksi = 6.9 Mpa)			
	Surface Layer (True modulus = 300 ksi)	Base Layer (True modulus = 45 ksi)	Subbase Layer (True modulus = 21 ksi)	Subgrade (True modulus = 7.5 ksi)
Single FWD test per point for 30 cases	$R = (196, 426)$ $\mu = 306$ COV = 16.3%	$R = (32.3, 59.9)$ $\mu = 44.6$ COV = 13.2%	$R = (18.7, 25.0)$ $\mu = 21.3$ COV = 6.6%	$R = (7.39, 7.67)$ $\mu = 7.49$ COV = 1.2%
Three FWD tests per case for 10 case	$R = (273, 391)$ $\mu = 311$ COV = 16.3%	$R = (34.1, 48.6)$ $\mu = 43.7$ COV = 9.8%	$R = (20.2, 24.9)$ $\mu = 21.5$ COV = 6.5%	$R = (7.31, 7.55)$ $\mu = 7.48$ COV = 0.9%

Note: R = range, μ = mean, and COV = coefficient of variation.

TABLE 20.2 Effects of Variable Layer Thicknesses on Backcalculated Moduli Based on Study by Irwin et al. (1989)

Test Scheme	Statistics of Backcalculated Modulus (ksi) (1 ksi = 6.9 MPa)			
	Surface Layer (True modulus = 300 ksi)	Base Layer (True modulus = 45 ksi)	Subbase Layer (True modulus = 21 ksi)	Subgrade (True modulus = 7.5 ksi)
Single FWD test per point for 30 cases	$R = (236, 410)$ $\mu = 311$ COV = 12.2%	$R = (39.8, 53.7)$ $\mu = 46.0$ COV = 8.0%	$R = (15.8, 26.4)$ $\mu = 20.9$ COV = 11.0%	$R = (7.35, 7.67)$ $\mu = 7.50$ COV = 0.9%

Note: R = range, μ = mean, and COV = coefficient of variation.

thickness were assumed by Irwin et al.: (a) the standard deviation in the asphalt concrete thickness was ±6.4 mm (±0.25 in.), (b) the standard deviation in the base layer thickness was (±)25 mm ((±)1 in.), (c) the standard deviation in the subbase layer thickness was ±38 mm ((±)1.5 in.), and (d) the layer thicknesses were rounded to the nearest 6.4 mm (0.25 in.), simulating typical field measurement accuracy. The order of magnitude of the effect is similar to that caused by random measurement errors.

20.1.2.7.3 Modeling Related Errors

Systematic errors are introduced by employing models which do not give exact representation of the actual pavement. Since none of the material models and the pavement response models described earlier in this chapter could represent perfectly the real pavement structure, errors are involved in the backcalculated moduli. The contributing factors include nonhomogeneity and anisotropy of pavement materials, moisture and thermal conditions, and development of failure conditions in the unbound materials. Errors are also introduced by loading characteristics of the deflection test. For example, Akram et al. (1994) reported that the deflection signal durations of a moving truck wheel were longer than those of FWD by as much as three to five times. They also noted the FWD deflection pulse duration was more or less constant with depth, whereas the deflection pulse under truck loading changed with depth and vehicle speed.

20.2 Equipment for Nondestructive Deflection Test

As mentioned in Section 20.1.1, devices for deflection test can be classified in accordance with the following modes of load application: static, moving, vibratory, impact, and rolling. The sequence of load application modes cited is also approximately the chronological sequence in which the various modes of devices were introduced. It reflects the attempt by pavement engineers and researchers to simulate the effects of actual traffic loading more closely as more advanced theoretical representations are possible, and sophisticated sensors are made available.

20.2.1 Static-Load Test Equipment

Static-load deflection tests are seldom performed nowadays as they are clumsy to perform due to the magnitude of load required, and they do not represent realistically the loading mode of actual traffic. The concept is similar to pile tests in building construction and has been used for pavement construction and testing in the early days of road building. For instance, Westergaard (1926) and Stratton (1944) conducted plate-loading tests to determine properties of subgrade soils and base materials, and Anderson et al. (1972) applied a repetitive static plate load to measure load–deformation response and evaluate the capacity of pavements in a test road study.

A plate load test in the field can be conducted with either repetitive loading or nonrepetitive loading. ASTM Standard Test Methods D1195 (ASTM, 2003a) and D1196 (ASTM, 2003b) describes the repetitive and nonrepetitive static plate load tests, respectively, of soils and flexible pavements for use in the evaluation of airport and highway pavements. These tests are cumbersome to perform. They require elaborate field set-up of the test apparatus for each point. The slow speed of operation severely limits its application to modern day high traffic-volume roads. It is rarely used in routine structural evaluation of pavements at the network level, although at the project level, this form of test could still be used when other deflection devices are unavailable.

20.2.2 Moving-Load Test Equipment

The most well known moving load deflection test is the Benkelman Beam test. The test procedure is covered in ASTM Test Method D4695 (ASTM, 2003c). As shown in Figure 20.2, the device consists of a pivoted beam with a 8 ft (2.440 m) long measurement probe on one end and a dial gauge positioned at the other end of the beam 24 in. (610 mm) from the pivot point. The deflection test makes use of a loaded truck that has a dual-wheel single axle with an axle load of 80 kN (18,000 lb) and tire pressure of between 480 and 550 kPa (70 and 80 lb/in^2). The test point (usually in the outer wheel path of a pavement) is first selected and the loaded truck is driven to position the dual-wheel of its single axle approximately on the test point. The tip of the measurement probe is next placed on the pavement surface between the dual wheels. The rebound deflection of the pavement surface is measured by the dial gauge by means of level arm action as the truck moves away slowly at about 1.6 to 3.2 km/h (1 to 2 miles/h). The rebound deflection must be recorded to within 0.001 in. (0.025 mm). Benkelman Beam is a low cost, easy-to-operate and easy-to-maintain equipment. Its slow speed of operation and reliance on manual setting up has made it unsuitable for pavement management applications at the road network level.

To increase the speed of measurement for application at the road network level, several automated Benkelman Beams have been produced. They include the traveling deflectometer developed

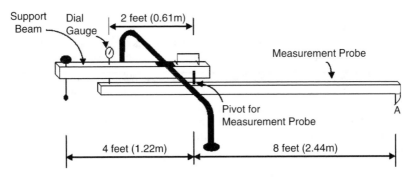

Note: Point A is the tip of measurement probe resting on the test surface

FIGURE 20.2   Schematic diagram of Benkelman Beam.

by the California Division of Highways, the LaCroix deflectometer developed in France, and the Deflectograph manufactured in the United Kingdom. All of these improvements were introduced in 1960s, consisting essentially of a mechanized beam assembly that is carried forward by the truck between the points of measurement. During measurement, the truck travels at a more or less constant speed of about 2 to 4 km/h (1.25 to 2.5 miles/h). The LaCroix deflectometer is popular in Europe, while the Deflectograph is used in the United Kingdom, some British Commonwealth countries, and China. An often mentioned drawback of these equipment is that the support of the beam, which is about 9 ft (2.75 m) from the tip of the measurement probe, may be within the deflection basin and thus does not provide a good reference for deflection measurements, especially for stiff flexible pavements and rigid pavements. There is also the potential problem that other wheels away from the test point might influence the readings made on thick heavy duty pavements.

20.2.3   Steady-State Vibratory-Load Test Equipment

A steady state vibratory-load test equipment measures pavement deflections under a sinusoidal loading generated by the equipment. The applied loading is normally generated by means of either counter rotating masses or electro-hydraulic systems. To prevent the measuring device from bouncing off the pavement surface, a sufficiently heavy static load is required. The dynamic force amplitude of the generated steady state sinusoidal vibration must not be greater than the static load provided. Many steady state vibratory loading devices offer a range of test frequencies and magnitudes of the applied load. The total time taken at each test point is about 30 sec although the loading test may lasts for less than 15 sec. The testing requirements and procedure are standardized by ASTM Standard Test Method D4602 (ASTM, 2003d).

The Dynaflect and the Road Rater are two common vibratory-load deflection test devices. The Dynaflect is a trailer unit that uses a vibratory force with a peak-to-peak magnitude of 1000 lb (454 kg) applied at a frequency of 8 Hz to a pair of steel wheels which carry a static weight of about 2100 lb (952 kg). It is brought to the test point and the deflection sensors lowered onto the pavement surface before the dynamic load is applied. Deflections are measured by five velocity transducers (also known as geophones), with the first located midway between the steel wheels, and the remaining four at 1 ft (305 mm) spacing along the axis of the trailer unit. The mode of operation of the Road Rater is similar to the Dynaflect, except that it offers a choice of a range of load magnitudes and frequencies. The static preload ranges from 2800 to 8000 lb (1270 to 3268 kg), while the load frequency can vary at 0.1 Hz increments from 5 to 70 Hz, while the dynamic load amplitude range is from 250 to 4000 lb (113 to 1814 kg). Other similar devices include the Federal highway Administration (FHWA) Thumper developed by FHWA and the WES vibrator developed by the U.S. Army Engineer Waterways Experiment Station (Bush, 1980). The former has a static load of 15,000 lb (6800 kg) and a dynamic amplitude of up to 10,000 lb (4535 kg), while the corresponding loads of the latter are 16,000 lb (7255 kg) and 15,000 lb (6800 kg).

A considerable amount of measurements have been made using steady state vibratory equipment by various states of the United States in the 1980s. The main drawback of such equipment is the use of heavy static preload which could significantly alter the state of stress in the pavement before the dynamic test loads are applied. Other drawbacks are related to the vibratory loads generated by the equipment. They could cause resonance in certain pavements. The harmonic loads without rest periods do not offer a close simulation of the actual traffic loadings. It may also be added that, since each test measurement is conducted at a fixed point, it is unable to simulate the rotation of principal planes of stresses within the pavement structure under actual moving traffic loads.

20.2.4   Impact-Load Test Equipment

Impact-load deflection test devices deliver to the pavement surface a transient impulse load that approximates a half-sine impulse with a duration typically of the order of 0.02 to 0.06 sec. The most common form of impact-load deflection test equipment is the FWD that generates a load impulse by falling a weight from a predetermined height. The main components of an FWD include a falling mass, a load platen, and a spring damping system, as shown in Figure 20.3. The height of fall and the spring stiffness are selected to produce load pulses to simulate the loading conditions of a moving wheel. The idea of producing a load impulse by a falling weight on a set of springs was first tried out in France in 1963, and subsequently further developed in Denmark (Bohn et al., 1972). One of the first FWD was built in the late 1960s in Denmark on the basis of a falling-weight device developed in France (Thomsen, 1982).

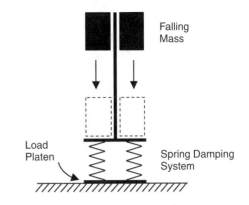

FIGURE 20.3 Schematic diagram of mechanism falling-weight deflection test.

An FWD is typically equipped with seven or nine deflection sensors (geophones) to define the deflection basin. The so-called deflection basin is represented by measured peak displacements at different sensor locations. It is noted that the peak displacements at different locations occur at different time instants, and the FWD reported deflection basin does not represent any of the actual transient deflection basins. In backcalculation analyses using FWD recorded deflection basins, a "static" status is generally assumed and the time lags among the peak displacements at different locations are ignored. Testing procedures with impact-load deflection test devices are documented in ASTM Standard Test Method D4694 (ASTM, 2003e).

Many different FWDs have been developed by commercial manufacturers and research institutions. Dynatest, Phoenix, and Kaub are the most common FWDs available commercially. The Dynatest FWD is a trailer unit that provides a load range of 7 to 105 kN (1500 to 23,600 lb) by dropping a selected mass from a height of 20 to 380 mm (0.75 to 15 in.) on to spring system and a loading plate of 300 mm diameter. The mass of the falling weight varies from 50 to 300 kg (110 to 660 lb). Deflections at seven locations are measured with velocity transducers. With the first deflection sensor positioned at the center of the loading plate, the positions of the remaining six sensors can be fixed according to the thickness of the pavement structure tested. Shorter distance coverage is suitable for thinner pavements, and wider spacing (with the furthest sensor at up to 2.25 m or 7.38 ft from the center of the loading plate) may be used on thick pavements.

The Phoenix FWD offers an impulse load within the range of 10 to 50 kN (2250 to 11,250 lb) using a weight between 30 and 150 kg (66 to 110 lb) and a falling height between 50 and 400 mm (2 to 16 in.). The KAUB FWD manufactured in Sweden has a unique feature in that the impulse load is created by providing two weights from different heights. The two-mass loading system is meant to provide a smooth rise of the load impulse exerted on the pavement. The impulse force can be varied from 12 to 150 kN (2700 to 33,750 lb).

FWDs are the most widely used form of pavement deflection test device today. Compared with other forms of loading mode, these impulse type loading forms offer the best simulation in terms of load effect and stress levels which are comparable with normal truck traffic. Ullidtz (1973) reported that, in addition to providing a close estimate of the surface deflections under moving wheel loads, FWD also measured vertical subgrade stresses and strains that matched well with those under moving wheel loads. The FWD impulse load simulation of moving wheel loads is still not exact. Through field experimental comparison of the effects of FWD impulse loads and moving wheel loads, Bohn et al. (1972) found that the durations of deflection were quite different in the two loading cases, and the stresses due to moving wheel loads had appreciable greater durations in the deeper layers. The FWD impulse load also introduces inertia or dynamic effects not present under a moving wheel load. Operation wise, like all other fixed point tests, the FWD test causes disruption to traffic and presents a potential safety hazard in busy high-speed roads.

20.2.5 Rolling-Load Test Equipment

All the test equipment described in the preceding sections are meant for fixed point tests that provide pavement properties at discrete points. Attempts have been made by researchers to conduct deflection test from a moving loaded device so as to achieve a higher test speed and provide a continuous record of pavement condition data. The idea of having a continuous record of deflection profile is particularly attractive as it effectively eliminates the risk of missing critical pavement sections. Three such defelctometers, viz., the Rolling Dynamic Deflectometer (RDD), the Rolling Weight Deflectometer, and the Rolling Wheel Deflectometer (RWD), that are able to produce continuous deflection measurements have recently been reported although all have not been commercially marketed for deflection testing.

20.2.5.1 RDD

The RDD is a device recently developed at the University of Texas at Austin (Bay and Stokoe, 1998). The RDD is a steady state vibratory-load device that applies large sinusoidal dynamic forces to the pavement. The dynamic forces applied to the pavement can be varied from 4.45 to 310 kN (1000 to 70,000 lb) peak-to-peak at frequencies from 5 to 100 Hz. The static hold-down force applied by the RDD can be varied from 4.45 to 195 kN (1000 to 44,000 lb). The testing speed of about 2.4 km/h (1.5 miles/h) is still rather low as compared with normal traffic speeds. It also suffers from the need to rely on a large static preload that could significantly alter the state of stresses within the pavement.

20.2.5.2 Rolling Weight Deflectometer

The Rolling Weight Deflectometer is a trailer-mounted device developed by Quest Integrated, Inc. (2004). The static force provided by removable weights can be varied from 76 to 89 kN (17 to 20 kip), and is applied through a single 76 cm (30 in.) diameter pneumatic tire-mounted on the rear of the trailer. The device uses the "spatially coincident" methodology for measuring pavement deflections (Johnson and Rish, 1996). Testing is performed by pulling the deflectometer at a velocity of up to 20 miles/h (33 km/h), and an average deflection is determined over every 0.3-m (1-ft) interval. Optical sensors are installed to measure the distance from the beam to the surface without a load and then as the trailer moves forward, the sensors measure the distance from the beam to the surface under load. The difference between the loaded and unloaded surface deflections provides a measure of the maximum pavement surface deflection under the load. This device only measures the maximum deflection directly under the load and it does not record the entire surface deflection basin. Though a maximum towing speed of 20 miles/h (33 km/h) is permitted, the average operating speed is about 10 miles/h (16 km/h) which is still too slow for the traffic in most roads.

20.2.5.3 RWD

The RWD is a semitrailer truck that travels at 88 km/h (55 ml/h) while collecting deflection response data under the truck impact in real time. The semitrailer carries 8172 kg (18,000 lb) of dead weight over its rear axle, and has a 7.8-m (25.5-ft) aluminum beam that houses four lasers spaced 2.6 m (8.5 ft) apart. Rapid processing time allows for deflection measurements every 12.2 mm (0.48 in.) at 88 km/h (55 miles/h) (Grogg and Hall, 2004). Since the deflection basin under rolling wheels is not symmetrical, the RWD measures the basin ahead of (leading) the rolling wheel for comparison with the undeflected pavement surface. It is also noted that the leading part of the basin may be less influenced by hysteresis effects. Hall (2004) reported an attempt to use a scanning laser system (SLS) that is able to scan a 14-ft (4.25-m) length of pavement in 1 msec from a single reference point. This offers a significant advantage over a spot sensor that only measures the point on the pavement directly beneath the sensor. It records the full deflection basin by collecting both deflected and undeflected surface data in one sweep.

20.3   Backcalculation of Rigid-Pavement Properties

20.3.1   Layered Elastic Theory vs. Plate Theory

Solutions based on plate theory, those by Westergaard (1926, 1948) in particular, have been widely used in the design of rigid pavements. In backcalculation analysis of pavement properties, plate theories as well as layered elastic theory have both been used. Proponents of the latter contend that by restricting the deflection tests to cases of interior loading only, layered elastic theory can be applied without much loss of accuracy in the backcalculated results (Uddin et al., 1986). This restriction does not appear to be a serious practical constraint. Other researchers have shown that the use of layered elastic theory could lead to significant errors in some practical cases (Ioannides, 1986). One source of error arises from the finite dimensions of jointed pavement slabs. Using a theoretical solution for rectangular thick plate, Fwa et al. (1996) have shown that the slab length A must be at least 4 times the radius of relative stiffness ℓ (i.e., $A/\ell > 4.0$) for the infinite slab assumption to be valid for deflection measurements. Another source of error is related to the effect of relative plate stiffness expressed in terms of the ratio (a/ℓ), where a is the radius of loading area. Analyzing a two-layer system of a slab resting on a Boussinesq half-space, Ioannides (1986) found that good agreement (less than 5% difference) in pavement response computed by the plate and layered elastic theories was obtained when (a/ℓ) is between 0.2 and 1.0, but large discrepancies (exceeding 20% difference) occurred outside this range. The main concern is with cases where (a/ℓ) was less than 0.2, which are not uncommon in nondestructive testing of rigid pavements.

20.3.2   Closed-Form Solutions for Two-Layer Rigid Pavement

For a two-layer rigid pavement with unknowns E_c (elastic modulus of concrete slab) and E_s (elastic modulus of subgrade) or k (modulus of subgrade reaction), only two surface deflection measurements are required theoretically to solve for the unknowns. Mathematically,

$$D_{mi} = f_1(E_c, E_s \text{ or } k) \tag{20.7a}$$

$$D_{mj} = f_2(E_c, E_s \text{ or } k) \qquad i \neq j \tag{20.7b}$$

from which E_c and E_s (or k) can be determined. When more than two measurements are available, any two deflection measurements can be used to solve for the two unknowns. In mathematical terms,

$$\sum_{i=1}^{N_i} W_i D_{mi} = \sum_{i=1}^{N_i} W_i f_j(E_c, E_s \text{ or } k) \tag{20.8a}$$

$$\sum_{i=1}^{N_j} W_j D_{mj} = \sum_{i=1}^{N_j} W_j f_j(E_c, E_s \text{ or } k) \tag{20.8b}$$

where D_{mi} and D_{mj} should have at least one dissimilar value, N_i and N_j are, respectively, the number of deflection measurements chosen to be included, and W_i and W_j are numerical weighting factors.

20.3.2.1   ILLI-BACK

A closed-form backcalculation procedure that gives unique material properties of a two-layer rigid pavement was first developed by Ioannides et al. (1989). Named as ILLI-BACK, the procedure computes the elastic properties of a concrete pavement slab and the underlying foundation based on a unique relationship between these properties and the shape of pavement surface deflection basin. It defines a parameter AREA to represent the cross-sectional area of the deflection basin:

$$\text{AREA} = 6\left(1 + 2\left[\frac{D_1}{D_0}\right] + 2\left[\frac{D_2}{D_0}\right] + \left[\frac{D_3}{D_0}\right]\right) \qquad \text{(inches)} \tag{20.9}$$

in which D_0, D_1, D_2, and D_3 are measured deflections at the subscripted sensor locations. Sensors 0, 1, 2, and 3 are located at 0, 30.5, 61.0, and 91.4 cm, respectively, from the center of loading plate. The use of Equation (20.9) is equivalent to solving Equation (20.8a) and Equation (20.8b) with the following choice of deflection measurements and weighting factors:

$$\sum_{i=1}^{N_i} W_i D_{mi} = 6D_0 + 12D_1 + 12D_2 + 6D_3 \qquad (20.10a)$$

$$\sum_{i=1}^{N_j} W_j D_{mj} = D_0 \qquad (20.10b)$$

ILLI-BACK is thus only one of an infinite number of possible ways of using the surface deflection measurements for backcalculation. The fact that AREA represents an approximate cross-sectional area of deflection basin (normalized by D_0) does not have any bearing on the method of analysis. One could replace Equation (20.10a) by any linear combination of the four deflection measurements and achieve the same results, on the assumption that the deflection inputs are exact with no measurement errors. When there are measurement errors, it can be shown that the built-in choice of weighting factors in AREA as used by ILLI-BACK may not be the best choice (Li et al., 1997).

The backcalculation concept based on AREA has been adopted by the pavement design guide of AASHTO (1993) in the overlay design of concrete pavement. There are certain limitations associated with the use of built-in weighting factors and rigid solution scheme of ILLI-BACK. The adoption of AREA, which is not a nondimensional parameter, has made the procedure rather inflexible in handling measurement errors. In an evaluation study of backcalculation program for the Long-Term Pavement Performance (LTPP) monitoring program by SHRP (1993), it was found that ILLI-BACK, being a backcalculation program for two-layer rigid pavements, is not particularly suitable for backcalculation of the slab modulus when a base layer is present. The study reported that the backcalculated slab moduli were considerably higher than what would be normally expected (i.e., 3 to 7 million lb/in.², or 20 to 48 million kPa) for a number of sections.

Example 20.1

A deflection test on a concrete pavement produces the following results:

Radial distance	0.00 m (0.00 in.)	0.305 m (12 in.)	0.610 m (24 in.)	0.915 m (36 in.)
Deflection	0.077 mm (0.0030 in.)	0.072 mm (0.0028 in.)	0.063 mm (0.0025 in.)	0.053 mm (0.0021 in.)

The pavement slab has a thickness of 254 mm (10 in.), and the load of 34.7 kN (7792 lb) is applied on a loading plate of 150 mm (6 in.) in radius. The Poisson's ratio of the concrete slab is 0.15, and that of the subgrade is 0.45. Assuming a liquid foundation model and adopting ILLI-BACK backcalculation procedure, determine the elastic modulus of the slab and the modulus of subgrade reaction.

(a) *Liquid foundation model.* AREA $= 6[1 + 2(0.072/0.077) + 2(0.063/0.077) + (0.053/0.077)] = 31.17$ (inches). Enter Figure 12.1 of Chapter 12 with $D_0 = 0.00349$ in. $= 3.49$ mils under 9000 lb, and AREA $= 31.17$ to obtain modulus of subgrade reaction $k = 215$ pci (58.4 MN/m³). Enter Figure 12.2 of Chapter 12 with AREA and k to obtain $E_c D^3 = 6 \times 10^9$ psi-in.³. For $D = 10$ in., $E_c = 6.0 \times 10^6$ psi (41.4 GPa). $< \ell = 28.5$ in. $E_c = 12(1 - \mu^2) k \ell^4 / h^3 = 5.50 \times 10^6$ $psi = 38.0$ GPa)??

20.3.2.2 NUS-BACK

NUS-BACK is a PC-based closed-form backcalculation program which presents a more general solution scheme as compared to ILLI-BACK. It can be used with deflections measured by any type of nondestructive measuring device directly. The backcalculation analysis can be performed when surface deflections at any two points are available. The two unknowns are determined by setting up two independent pavement response equations as shown in Equation (20.11) and Equation (20.12), or Equation (20.13) and Equation (20.14).

For liquid foundation:

$$D_{m1} = \frac{P}{k\pi a^2} F_k(\ell_k, r_1) \tag{20.11}$$

$$D_{m2} = \frac{P}{k\pi a^2} F_k(\ell_k, r_2) \tag{20.12}$$

For solid foundation:

$$D_{m1} = \frac{P(1 - \mu_s^2)}{E_s \ell_E} F_E(\ell_E, r_1) \tag{20.13}$$

$$D_{m2} = \frac{P(1 - \mu_s^2)}{E_s \ell_E} F_E(\ell_E, r_2) \tag{20.14}$$

where D_{m1} and D_{m2} are known deflection measurements, a is the radius of loaded area, P the magnitude of applied load, k the modulus of subgrade reaction, E_s the elastic modulus of the subgrade, μ_s the Poisson's ratio of the subgrade, E_c the elastic modulus of the pavement slab, h_c the slab thickness, r_1 and r_2 are, respectively, the horizontal distances of points 1 and 2 from the center of loaded area, ℓ_k and ℓ_E are radii of relative stiffness defined as

$$\ell_k = \left(\frac{E_c h_c}{12(1 - \mu_c^2)k} \right)^{(1/4)} \tag{20.15}$$

$$\ell_E = \left(\frac{E_c h_c^3 (1 - \mu_s^2)}{6(1 - \mu_c^2)E_s} \right)^{(1/3)} \tag{20.16}$$

and $F_k(\ell_k, r)$ and $F_E(\ell_E, r)$ are deflection factors computed by subroutines for the given values of r_1 and r_2. The expressions for $F_k(\ell_k, r_1)$ and $F_E(\ell_E, r_2)$ are given by Panc (1975).

Practically all surface deflection test devices provide deflection measurements at more than two points. For example, most FWDs give seven deflection readings per test. With seven deflection points, 21 sets of solutions can be computed. The normal routine of the NUS-BACK reports the mean values of all possible solutions. Backcalculation solutions are displayed almost instantly when run on a PC with a 100 MHz Pentium IV processor. The reliability of the backcalculated material properties can be determined if the measurement precision of individual sensors are known. Li et al. (1996) have demonstrated that NUS-BACK provides more reliable solutions than ILLI-BACK when measurement errors are involved. The NUS-BACK algorithm can also be used as an effective tool to identify faulty sensors or erroneous deflection readings (Fwa et al., 1997d).

Graphical solutions for the NUS-BACK algorithm, as presented in Figures 20.4, 20.5, and 20.6, have been prepared by Fwa et al. (1997d). The following steps are applicable:

1. Select a deflection ratio (D_{mi}/D_{mj}), and use Figures 20.4 or 20.5 as appropriate to obtain ℓ_k or ℓ_E.
2. Using Figure 20.6(a) or (b) as appropriate, and the value of ℓ_k or ℓ_E determined from Step 1, obtain F_k or F_E for either r_i (corresponding to D_{mi}) or r_j (corresponding to D_{mj}).
3. Compute k from either Equation (20.11) or Equation (20.12), or E_s from either Equation (20.13) or Equation (20.14).
4. Compute E_c from Equation (20.15) or Equation (20.16) as appropriate.

In addition to the graphical solutions described in this section, the microcomputer program NUS-BACK is also provided with the Handbook. The computer software is accessible on the CRC Press website. The users' manual and input instructions are found in Appendix 20A. The required input parameters are concrete slab thickness h_c, Poisson's ratios μ_c and μ_s of concrete slab and subgrade soil, respectively,

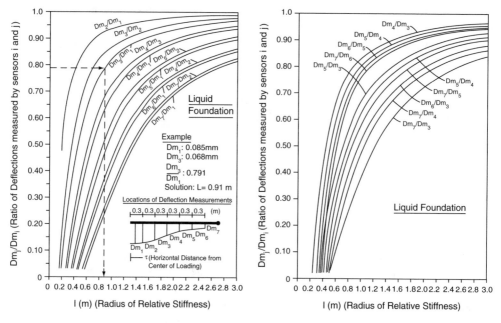

FIGURE 20.4 Graphical solution for backcalculating radius of relative stiffness of rigid pavement on liquid foundation (after Fwa et al., 1997, with permission).

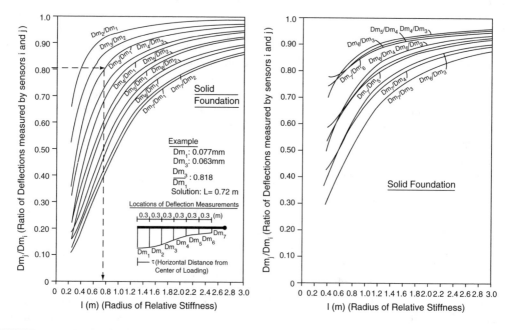

FIGURE 20.5 Graphical solution for backcalculating radius of relative stiffness of rigid pavement on solid foundation (after Fwa et al., 1997d, with permission).

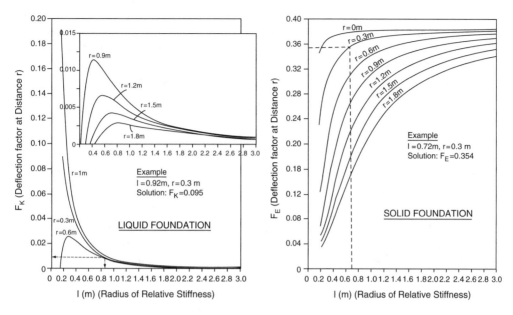

FIGURE 20.6 Graphical solution for determining deflection factor (after Fwa et al., 1997d, with permission).

magnitude of applied load P, radius of loading area a, deflection measurements of the sensors, and the distances of the corresponding sensors form the center of loading area.

Fwa et al. (1999) studied the reliability of the backcalculated properties of two-layer rigid pavement against input errors. Based on the typical measurement errors of 2% ± 2 μm stated in the supplier specifications of most commercial FWDs, 100 sets of surface deflections were generated with random deflection measurement errors. The study covered the following ranges of pavement properties: 0.5 to 3.0 m for radius of relative stiffness (ℓ), 10 to 220 MN/m^3 for modulus of subgrade reaction (k), and 25 to 600 MPa for elastic modulus of subgrade (E_s). For the liquid foundation model, the order of magnitude of the coefficient of variation caused by the random measurement errors was about 3% for ℓ, 8% for E_c, and 11% for k. The corresponding values for ℓ, E_c, and E_s for the case of solid foundation model were 5, 4, and 5%. The effect of errors in pavement slab thickness measurements was also examined. It was found that the input errors in the slab thickness only affects the backcalculated slab modulus, but has no effects on the radius of relative stiffness (ℓ), nor the modulus of subgrade reaction (k) or the elastic modulus of subgrade (E_s). For -105%, -5%, $+5\%$, and $+10\%$ errors in slab thickness input, the corresponding deviations in the backcalculated slab modulus were 14.5, 7.5, -5.8, and -11.1% for the liquid foundation model, and 8.4, 4.0, -3.3, and -6.9% for the solid foundation model.

TABLE 20.3 Comparison of Backcalculated and Measured Concrete Moduli by NUS-BACK and ILLI-BACK

Measured Concrete Modulus (GPa)	Backcalculation Results Based on Deflection Tests				
	Applied Load	Concrete Modulus Based on Liquid Foundation (GPa)		Concrete Modulus Based on Solid Foundation (GPa)	
		NUS-BACK	ILLI-BACK	NUS-BACK	ILLI-BACK
42.4; 45.8; 54.8	6 kips (26.7 kN)	50.7	40.1	37.7	35.3
	8 kips (35.6 kN)	51.7	36.1	38.2	32.0
	11 kips (48.9 kN)	49.5	35.7	36.2	31.8
	15 kips (66.7 kN)	48.9	33.3	36.3	29.9
Average = 47.7	—	Average = 50.2	Average = 36.3	Average = 37.1	Average = 32.25

Fwa et al. (1999) also applied both NUS-BACK and ILLI-BACK to backcalculate concrete pavement properties and compared with the experimentally measured values of a concrete pavement reported by Ioannides et al. (1989). The direct measurements of concrete moduli by Ioannides et al. and the backcalculated values are summarized in Table 20.3. The results suggest that in this case, the liquid foundation model gives moduli values that are much closer to the measured ones. The moduli values computed by the solid foundation model are too low. Comparing the backcalculated moduli by NUS-BACK and ILLI-BACK, the results show that NUS-BACK produced backcalculated results in better agreement with the measured values.

Example 20.2

A deflection test on a concrete pavement produces the following results:

Radial distance	0.00 m (0.00 in.)	0.305 m (12 in.)	0.610 m (24 in.)	0.915 m (36 in.)
Deflection	0.086 mm (0.0034 in.)	0.080 mm (0.0032 in.)	0.069 mm (0.0027 in.)	0.056 mm (0.0022 in.)

The pavement slab has a thickness of 254 mm, and the load of 34.7 kN is applied on a loading plate of 150 mm in radius. The Poisson's ratio of the concrete slab is 0.15, and that of the subgrade is 0.45. (a) Assuming a liquid foundation model and adopting NUS-BACK backcalculation procedure, determine the elastic modulus of the slab and the modulus of subgrade reaction, and (b) Assuming a solid foundation model and adopting NUS-BACK backcalculation procedure, determine the elastic moduli of the slab and the subgrade.

The problem can be solved either graphically using Figures 20.4, 20.5, and 20.6, or by means of the microcomputer software NUS-BACK. The solutions are given in the following tables.

(a) *Liquid foundation model*

Pavement Property	Backcalculated Value						
	D_{m2}/D_{m1}	D_{m3}/D_{m1}	D_{m4}/D_{m1}	D_{m3}/D_{m2}	D_{m4}/D_{m2}	D_{m4}/D_{m3}	Mean
ℓ_k (m)	0.931	0.931	0.931	0.931	0.931	0.931	0.931
k (MN/m^3)	57.8	57.8	57.7	57.8	57.7	57.7	57.75
E_c (GPa)	31.0	31.0	31.5	31.0	31.5	31.0	31.17

(b) *Solid foundation model*

Pavement Property	Backcalculated Value						
	D_{m2}/D_{m1}	D_{m3}/D_{m1}	D_{m4}/D_{m1}	D_{m3}/D_{m2}	D_{m4}/D_{m2}	D_{m4}/D_{m3}	Mean
ℓ_k (m)	0.712	0.712	0.712	0.711	0.712	0.712	0.712
k (MN/m^3)	190.9	190.7	190.5	190.6	190.3	190.1	190.5
E_c (GPa)	31.0	30.9	30.5	30.9	30.9	30.8	30.83

20.3.3   Closed-Form Solutions for Three-Layer Rigid Pavement

Today virtually all concrete pavements are constructed with a subbase layer. While the NUS-BACK algorithm described in the preceding section can be used to compute an effective material property for the subbase and the subgrade combined, it would be preferable to determine the actual properties of the two layers. This calls for a backcalculation procedure for three-layer concrete pavements. Presented in this section is a closed-form algorithm developed by Li et al. (1997) for backcalculating the layer properties of a pavement slab resting on a two-layer elastic solid foundation. The unknown problem parameters are the elastic moduli of the three pavement layers, viz., E_c of the concrete slab, E_b of the subbase and E_s of the subgrade.

20.3.3.1   Theoretical Basis and Solution Procedure

The theoretical approach is similar to that adopted for the backcalculation analysis of two-layer pavements. That is, the unknowns are determined by solving simultaneous equations that describe pavement surface response under load. As there are three unknown material parameters from the three pavement layers, three pavement response equations are required for the solution. The three unknown parameters are the layer moduli E_c, E_b, and E_s. Based on the work of Burmister (1945) and Panc (1975), the following pavement response equations for surface deflections are used in the present analysis:

$$D_{m1} = \frac{2(1 - \mu_b^2)P}{\pi a E_b} F(\ell, c, h_b, r_1) \tag{20.17}$$

$$D_{m2} = \frac{2(1 - \mu_b^2)P}{\pi a E_b} F(\ell, c, h_b, r_2) \tag{20.18}$$

$$D_{m3} = \frac{2(1 - \mu_b^2)P}{\pi a E_b} F(\ell, c, h_b, r_3) \tag{20.19}$$

where D_{m1}, D_{m2}, and D_{m3} are known deflection measurements sensor points 1, 2, and 3, respectively, a is the radius of loaded area, P the magnitude of applied load, E_b the elastic modulus of the subbase, μ_s the Poisson's ratio of the subgrade, E_c the elastic modulus of the pavement slab, r_1, r_2, and r_3 are, respectively, the horizontal distances of points 1, 2, and 3 from the center of loaded area, μ_b is the Poisson's ratio of the subbase, h_b the thickness of subbase layer, and c and ℓ are given by

$$c = \frac{(1 + \mu_b)E_s}{(1 + \mu_s)E_b} \tag{20.20}$$

$$\ell = \left(\frac{E_c h_c^3 (1 - \mu_b^2)}{6(1 - \mu_c^2)E_b} \right)^{(1/3)} \tag{20.21}$$

where E_s is the elastic modulus of the subgrade, μ_s the Poisson's ratio of the subgrade, and h_c the slab thickness.

For any three measured deflections d_{m1}, d_{m2}, and d_{m3}, unique values of ℓ, c, and E_b can be determined by solving Equation (20.17), Equation (20.18), and Equation (20.19). Computationally, ℓ and c are first solved by combining Equation (20.17), Equation (20.18), and Equation (20.19) into two equations as follows,

$$d_{m1}F(\ell, c, r_2) - d_{m2}F(\ell, c, r_1) = 0 \tag{20.22}$$

$$d_{m2}F(\ell, c, r_3) - d_{m3}F(\ell, c, r_2) = 0 \tag{20.23}$$

Unique values of ℓ and c can be solved from Equation (20.22) and Equation (20.23). The expression for the deflection factor $F(\ell, c, r)$ has been derived by Panc (1975), assuming a no-slip interface between the subbase layer and the subgrade. A trial-and-error iterative approach is adopted to compute the values of $F(\ell, c, r)$. The next step is to determine E_b from Equation (20.17) or Equation (20.18) or Equation (20.19). Having determined ℓ, c, and E_b, the remaining parameters E_s and E_c are computed from Equation (20.20) and Equation (20.21), respectively.

20.3.3.2 Backcalculation Program NUS-BACK3

A microcomputer program NUS-BACK3 has been written to determine the unknown parameters of a concrete pavement resting on two-layer foundation (Fwa et al., 1997b). Although initial ranges of ℓ and c values are required for the computation, this is not a problem as the solutions are unique and are not dependent on the initial values assumed. The program assumes a default range of ℓ values between 0.05 and 3.0 m, and a range of c values from zero to one. These ranges of ℓ and c values practically cover all three-layer concrete pavement structures in road construction. A backcalculation analysis by the program takes about 20 sec of execution time on a PC with a 100 MHz Pentium IV processor.

The microcomputer program NUS-BACK3 is provided with the Handbook. The computer software NUS-BACK2 and NUS-BACK4 are accessible on the CRC Press website. The users' manual and input instructions are found in Appendix 20B. The required input parameters are concrete slab thickness h_c, thickness of subbase layer h_b, Poisson's ratios μ_c, μ_b, and μ_s of concrete slab, subbase, and subgrade soil, respectively, magnitude of applied load P, radius of loading area a, deflection measurements of the sensors, and the distances of the corresponding sensors form the center of loading area.

The reliability of the NUS-BACK3 backcalculated results against errors in input parameters was investigated by Fwa et al. (1999). Hundred sets of deflection measurements with random measurement errors of 2% $\pm$ 2 μm in the deflection readings of the seven sensors were considered. The analysis showed that coefficients of variation for the backcalculated concrete slab modulus, subbase modulus, and subgrade modulus were of the order of 8, 12, and 3%, respectively.

Example 20.3

The measured deflections of a deflection test are as follows: 0.141 mm at $r = 0$ m, 0.125 mm at $r = 0.305$ m, and 0.123 mm at $r = 0.610$ m. The load of 40 kN is applied on a loading plate of 150 mm radius. The thickness of the pavement slab and subbase are 254 and 203 mm, respectively. The Poisson's ratios of the slab, subbase, and subgrade are 0.15, 0.15, and 0.35, respectively. Apply NUS-BACK3 to determine the layer properties of the pavement.

Applying the microcomputer program NUS-BACK3, the elastic moduli of the pavement slab, subbase, and subgrade are, respectively, 28.2 GPa, 1510, and 50.4 MPa.

20.3.4   Semiclosed-Form Solutions for Rigid Pavement with Two Slab Layers

Another form of rigid pavements found in practice is one that has two rigid concrete slabs. For example, it can be a concrete pavement with a lean concrete base. There are also concrete pavements with a concrete overlay. Two semiclosed-form backcalculation algorithms developed by Fwa et al. (1997b) are described in this section, one for two-slab concrete pavements resting on a single-layer foundation, and another for two-slab concrete pavements supported on a two-layer foundation. These algorithms are termed as semiclosed-form because they provide unique closed-form solutions for layer properties of the foundation materials and the equivalent combined elastic modulus of the two concrete slabs, but do not provide unique solutions for the individual elastic moduli of the concrete slabs.

20.3.4.1   Concept of Equivalent Slab Layer

In order to apply either NUS-BACK or NUS-BACK3, it is necessary to replace the two slab layers by an equivalent single slab of the same total thickness. The equivalent slab is derived based on the assumption of equal flexural rigidity. Depending on the bonding condition at the interface of the two slab layers, the following two possible cases can be considered: (a) bonded slabs, and (b) unbonded slabs. The expressions that relate the elastic modulus E_{eq} of the equivalent slab and the actual modulus values E_{c1} and E_{c2} of the two slabs can be shown to be as follows:

(a) *Bonded slabs*

$$E_{eq} = \frac{4E_{c1}}{(h_{eq})^3}[h_0^3 - (h_c - h_{c1})^3] + \frac{4E_{c1}}{(h_{eq})^3}[(h_{c1} + h_{c2} - h_0)^3 - (h_0 - h_{c1})^3] \qquad (20.24)$$

with

$$h_0 = \frac{E_{c1}(h_{c1})^2 + 2E_{c2}h_{c1}h_{c2} + E_{c2}(h_{c2})^2}{2(E_{c1}h_{c1} + E_{c2}h_{c2})} \qquad (20.25)$$

(b) *Unbonded slabs*

$$E_{eq} = E_{c1}\left(\frac{h_{c1}}{h_{eq}}\right)^3 + E_{c2}\left(\frac{h_{c2}}{h_{eq}}\right)^3 \qquad (20.26)$$

where μ_c is the Poisson's ratio of concrete, h_{eq} is the thickness of the equivalent slab layer, h_{c1} and h_{c2} are the original upper and lower slab thicknesses, respectively, and E_{c1} and E_{c2} are the elastic moduli of the upper and lower slab, respectively.

The backcalculation analysis consists of the following steps:

Step 1

Closed-Form Backcalculation Analysis of Equivalent Pavement — Having replaced the two slab layers by an equivalent slab, the equivalent pavement structure becomes either (i) one of a concrete slab supported on a single-layer foundation, or (ii) one of a concrete slab supported on a two-layer foundation. For (i) NUS-BACK is used for the backcalculation analysis and the following unique material properties are obtained: ℓ_{eq} and E_{eq} of the equivalent slab, k for liquid foundation, or E_s for rigid foundation. For (ii) NUS-BACK3 is used and the following unique material properties are obtained: ℓ_{eq} and E_{eq} of the equivalent slab, the elastic modulus of subbase E_b, and the elastic modulus of subgrade E_s.

Step 2

Estimation of Individual Slab Properties — This step computes the elastic moduli E_{c1} and E_{c2} of the two original slabs from E_{eq} The knowledge of E_{eq} alone is inadequate. To determine the elastic moduli of the two slabs, another equation relating E_{c1} and E_{c2} is needed. In practice, the two slab moduli may be determined by providing additional information in one of the following ways: (a) Based on the pavement design or construction records, an estimate of the ratio (E_{c1}/E_{c2}) can be made based on the mix design crushing strength or other related properties of the two concrete materials; or (b) obtain another E_{eq} equation based on FWD test on a separate pavement section of the same materials but with different slab thicknesses; or (c) determine the elastic modulus of the top slab layer by laboratory testing of cored samples; or (d) determine the elastic modulus of the top slab layer by other nondestructive probing techniques.

20.3.4.2 Backcalculation Programs NUS-BACK2 and NUS-BACK4

Two microcomputer backcalculation programs, NUS-BACK2 and NUS-BACK4, have been developed for rigid pavement with two slab layers (Fwa et al., 1997b). NUS-BACK2 is for a two-slab-layer pavement resting on a single layer foundation, while NUS-BACK4 is for a two-slab-layer pavement supported on a two-layer foundation. NUS-BACK2 and NUS-BACK4 have NUS-BACK and NUS-BACK3, respectively, as the backcalculation subroutine. As such, the reliability characteristics of the backcalculated parameters by NUS-BACK2 and NUS-BACK4 are similar to those of NUS-BACK and NUS-BACK3, respectively. Both NUS-BACK2 and NUS-BACK4 cover unbonded as well as fully bonded slab interface conditions. The NUS-BACK2 program, as in the NUS-BACK program, can solve for both liquid and solid foundations. The NUS-BACK4 program, like the NUS-BACK3 program, considers only the two-layer elastic solid foundation.

In the process of the backcalculation analysis, the two programs also compute the upper and lower limits for the modulus values of the two slabs. The upper and lower limits of the elastic modulus of the top slab are obtained as follows: (a) The upper bound for E_{c1} can be obtained by assuming that the lower slab does not exist, and performing backcalculation analysis using the measured deflections for the actual pavement as if there is only one slab layer of thickness h_{c1} and elastic modulus E_{eq}. The value of backcalculated E_{eq} thus obtained is the upper bound value for E_{c1}. (b) The lower bound for E_{c1} is obtained by assuming that $E_{c1} = E_{c2} = E_{eq}$. The corresponding h_{eq} can be obtained from Equation (20.24) for bonded slabs, and Equation (20.26) for unbonded slabs. The backcalculation analysis may then be performed using NUS-BACK or NUS-BACK3 with the assumed thickness of h_{eq} for the equivalent slab. The value of backcalculated E_{eq} thus obtained is the lower bound value for E_{c1}. Next, for the lower slab, it can be reasoned that the value of E_{eq} computed in step (b) is in fact the upper bound value for its elastic modulus.

The microcomputer programs NUS-BACK2 and NUS-BACK4 are provided with the Handbook. The computer software NUS-BACK2 and NUS-BACK4 are accessible on the CRC Press website. The users' manual and input instructions are found in Appendix 2OC and 2OD, respectively. The required input parameters for NUS-BACK2 are the upper and lower concrete slab thicknesses h_{c1} and h_{c2}, respectively, interface bonding condition of the two slabs (i.e., bonded or unbonded), their corresponding Poisson's ratios μ_{c1} and μ_{c2}, Poisson's ratio μ_s of subgrade soil, magnitude of applied load P, radius of loading area a, deflection measurements of the sensors, and the distances of the corresponding sensors form the center of loading area. For NUS-BACK4, the additional input parameters are the thickness of subbase layer h_b, and its Poisson's ratio μ_b.

20.4 Backcalculation of Flexible-Pavement Properties

20.4.1 Backcalculation for Elastic Multi-Layer System

A typical flexible pavement is a multi-layer system consisting of three basic structural layers overlying the subgrade soil. These basic structural layers are surface course (which could be constructed in two layers, a wearing course and a binder course), base course and the subbase course. Backcalculation analysis is performed to estimate the structural properties of all these layers, as well as those of the subgrade soil, based on deflection measurements.

Practically all the backcalculation methods in common use today assume the flexible pavement as a linear elastic multi-layer system, and employ iterative approach with trial layer moduli to identify a set of layer moduli that would produce a deflection basin that closely matches the measured deflection basin. When the FWD is used, the measured deflection basin is represented by the peak deflections recorded by the spot sensors. The computed deflection basin is typically obtained using multi-layer elastic theory by assuming static loading. The objective function of the approach is thus to minimize the output errors, i.e., the differences between the computed and the measured deflections. As explained in Section 20.1.2, the common forms of objective functions are:

(a) Minimize the sum of the absolute differences of deflections:

$$\text{Minimize} \sum_{i=1}^{N} w_i(D_i - d_i)$$

(b) Minimize the sum of squared differences of deflections:

$$\text{Minimize} \sum_{i=1}^{N} w_i(D_i - d_i)^2$$

(c) Minimize the sum of the squared relative errors of deflections:

$$\text{Minimize} \sum_{i=1}^{N} w_i \left(\frac{D_i - d_i}{D_i} \right)^2$$

(d) Minimize the maximum percent error of deflection:

$$\text{Minimize} \left\langle \text{Maximum} \left(\frac{D_i - d_i}{D_i} \right) \times 100\%, \quad i = 1, 2, \dots N \right\rangle$$

(e) Minimize the maximum percent error of computed layer modulus in successive iterations:

$$\text{Minimize}\left(\text{Maximum}\left(\frac{E_j^{(k)} - E_j^{(k-1)}}{E_j}\right) \times 100\%, \qquad j = 1, 2, \ldots m\right)$$

where w_i the weighting factor for sensor i, D_i and d_i are, respectively, the measured and computed deflections at sensor point i, N is the total number of deflection sensors that are available for deflection measurement, $E_j^{(k)}$ the computed elastic modulus of layer j in the kth iteration, and m the total number of pavement layers. It must be mentioned that the same weight is often assigned to all the N number of deflection measurements. This, however, can give rise to the problem of having the computed value governed to a larger extent by the larger deflections recorded nearer to the center of loading area. To overcome this problem, some backcalculation software applies unequal weights to different deflection measurements, while others consider percent deviations of the computed deflection basin from the measured deflection basin.

Figure 20.7 shows the flow diagram of the steps involved in a typical backcalculation analysis. To safeguard against nonconvergence caused by erroneous input, poor choices of seed moduli, or other unforeseen circumstances, virtually all backcalculation programs include the setting of a maximum number iterations as one of the termination criteria. Many backcalculation computer programs developed by either research institutions or commercial organizations are available for backcalculation based on this general procedure. The end results of backcalculated layer moduli vary depending on the pavement response model and the forward calculation software employed, method of generating seed moduli, nature of moduli search algorithm, the criterion selected for matching of computed and measured deflection basin, and the sensitivity of the backcalculation algorithm with respect to errors in deflection measurements and input layer thicknesses, etc. Some of the backcalculation computer programs are described in the following paragraphs.

MODULUS is a microcomputer linear elastic backcalculation program developed at the Texas Transportation Institute (Uzan et al., 1988; Scullion et al., 1990). It utilizes the five-layer elastic solution WESLEA (Van Cauwelaert et al., 1989) as the forward calculation scheme. MODULUS searches for the best fit between the measured and computed deflection basins. The program expresses all layer modulus as ratios with respect to the subgrade soil modulus. A database of computed deflection basins is developed for different ratios of layer moduli. In the process of minimizing the error function which is the sum of the squared relative errors of deflections, the search algorithm repeatedly picks trial sets of pavement layer moduli and the associated computed deflection basins from the database through assuming a series of different ratios of layer moduli. The search algorithm is driven by taking the derivative of the error function with respect to layer moduli, and setting it to zero. For each backcalculation analysis involving a fixed set of modulus ratios, a seed value of subgrade modulus is first assumed. With the assumed subgrade modulus and the chosen layer modulus ratios, the seed moduli of all pavement layers are defined. The program can be applied to a flexible pavement of up to four-layers, with or without a rigid bedrock layer.

WESDEF was developed by the U.S. Army Corps of Engineers Waterways Experiment Station (WES) (Van Cauwelaert et al., 1989) to backcalculate layer moduli of flexible pavements. It is an improved version of the BISDEF backcalculation program (Bush and Alexander, 1985) by replacing the BISAR n-layer forward calculation scheme with WELSEA. With given seed moduli, the program adopts a gradient search technique and uses the WESLEA computer code within an iterative process. The terminating criteria for the iterative process are governed by either the absolute sum of the percent differences between the computed and measured deflections, or the maximum changes of the predicted layer moduli in successive iterations. The program incorporates into the analysis a stiff layer with a modulus of elasticity of 1,000,000 lb/in.^{2} (6900 MPa) and infinite thickness below the subgrade. The maximum number of layers with unknown modulus values in WESDEF should be limited to three in the backcalculation process.

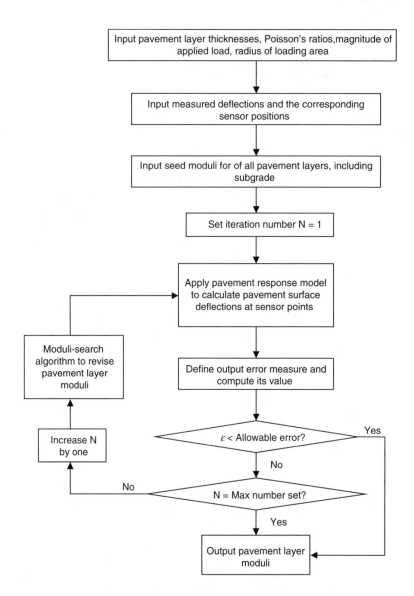

FIGURE 20.7   Flow diagram of typical backcalculation analysis.

EVERCALC is a microcomputer backcalculation analysis program based on the CHEVRON n-layer elastic solution scheme. The program was developed at the University of Washington for the WSDOT (Lee, 1988; WSDOT, 1999). It begins with assumed seed layer moduli and uses an iterative procedure to match the computed and the measured deflections. For pavements with up to three layers, users can either input their own seed moduli or apply the seed moduli generated by the program based on a set of regression equations. Seed moduli have to be user-defined when the number of layers exceeds three. The iterative backcalculation process ceases when the summation of the absolute values of the discrepancies between the computed and measured deflections falls within a preset allowable tolerance (usually 10%). Alternatively, the iteration can be terminated when the maximum change in the computed moduli is less than a prescribed level, say 0.1%. An efficient minimization scheme, known as the Levenberg–Marquardt algorithm, was implemented in EVERCALC to facilitate an efficient convergence with only a modest number of calls to the forward solution code CHEVRON (Sivaneswaran et al., 1991).

MICHBACK uses CHEVRONX, an extended precision version of CHEVRON as the forward calculation scheme for better accuracy in deflection computation (Harichandran et al., 2000). It employs a modified Newton iterative algorithm to obtain the least-squares solution of an over-determined set of equations. The iteration is terminated when the changes in the computed layer moduli between successive iterations are sufficiently small or when the root-mean-square error of the computed deflections with respect to the measured deflections as given in Equation (20.27) and Equation (20.28) is less than a preset tolerable limit. That is:

$$\varepsilon_i = \left(\frac{E_k^i - E_k^{i-1}}{E_k^{i-1}} \right), \qquad k = 1, 2, \ldots, n \tag{20.27}$$

$$\text{RMS error in deflections} = \sqrt{\frac{1}{N} \sum_{j=1}^{N} \left(\frac{D_j - d_j}{D_j} \right)^2} \tag{20.28}$$

where ε_i is the change in the modulus of layer k from the $(i-1)$th iteration to the ith iteration, E_k^i the modulus of layer k from the ith iteration, n is the total number of layers, N the total number of deflection sensors, D_j and d_j the measured and computed deflections, respectively. Using internal regression equations, MICHBACK is able to provide default seed values either with a representative deflection basin, the first deflection basin in the pavement segment of interest, or the average deflection from multiple load drops.

20.4.2 Closed-Form Solutions for Two-Layer Asphalt Pavement

A closed-form backcalculation algorithm for a two-layer asphalt pavement was first proposed by Scrivner et al. (1973) based on the deflection equation developed by Burmister (1945). Burmister's solution permits one to compute the deflection, w_i, of a point i in the pavement surface at the radial distance, r_i, from the center of the loaded area by the following expression:

$$w_i r_i E_1 = \frac{3P}{4\pi} F_i \tag{20.29}$$

where E_1 is the elastic modulus of the surface layer of the pavement, P the total applied load, and F_i a deflection factor which is a function of the thickness of the surface layer, the radial distance r_i, and the ratio of the elastic moduli of the pavement surface layer and the subgrade, as given by the equation below,

$$F_i = \int_{x=0}^{\infty} \left(\frac{1 + 4Nme^{-2m} - N^2 e^{-4m}}{1 - 2N(1 + 2m^2)e^{-2m} + N^2 e^{-4m}} \right) J_0(x) dx, \tag{20.30}$$

$$\text{with } x = \frac{mr_i}{h} \text{ and } N = \frac{1 - \theta}{1 + \theta}$$

where $J_0(x)$ is the Bessel function of the 0th order, m is a continuous variable of integration, h the thickness of the surface layer, and θ is the ratio given by (E_2/E_1), E_2 and E_1 are, respectively, the elastic moduli of the subgrade and the surface layer of the pavement.

Consider the surface deflections at any two points i and j, it can be shown that the following equation holds (Li et al., 1996):

$$w_i F_j - w_j F_i = 0 \tag{20.31}$$

It is noted that in the above equation, θ is the only unknown which can be solved by trial-and-error. Fwa and Jia (2004) developed a computer backcalculation program 2L-BACK using the bisection method to solve for the unknown θ. Once θ is known, E_1 can be computed from Equation

(20.1)), and E_2 is given by θ times E_1. The execution time of the backcalculation analysis on a Pentium 4 personal computer with a clock speed of 2.4 GHz takes less than 1 sec.

The microcomputer program 2L-BACK is provided with the Handbook. The computer software is accessible on the CRC Press website. The users' manual and input instructions are found in Appendix 20E. The required input parameters are asphalt surface layer thickness h, Poisson's ratios μ and μ_s of the asphalt surface layer and subgrade soil, respectively, magnitude of applied load P, radius of loading area a, deflection measurements of the sensors, and the distances of the corresponding sensors form the center of loading area.

Example 20.4

A Dynaflect test applying a load 1000 lb (4.45 kN) recorded deflections of 0.00216 in. (0.0549 mm) and 0.00096 in. (0.0244 mm) at sensor points 1 ($r = 0$ in. or 0 mm) and 3 ($r = 24.0$ in. or 610 mm). The diameter of the loading area is 2.82 in. (71.6 mm). The thickness of the asphalt layer is 7.5 in. (191 mm). The Poisson's ratios of the asphalt layer and the subgrade are 0.30 and 0.40, respectively. Apply the closed-form backcalculation software 2L-BACK to determine the elastic moduli of the asphalt surface layer and the subgrade.

Applying the microcomputer program 2L-BACK, the elastic moduli of the asphalt surface layer and subgrade are, respectively, 86,234 lb/in.^{2} (595 MPa) and 10,849 lb/in.^{2} (74.9 MPa).

20.4.3 Generation of Seed Elastic Moduli for Backcalculation Analysis of Multi-Layer Flexible Pavement

The choice of seed moduli in backcalculation analysis of multi-layer flexible pavements can have significant impacts on the performance of backcalculation software, and sometimes the final solutions of the backcalculated moduli. (Harichandran et al., 1993; May and Von Quintus, 1994). Ullidtz (1987) noted that a small error in the estimation of the subgrade elastic modulus would lead to large errors in the backcalculated elastic moduli of other pavement layers. Some backcalculation programs provide internally generated seed moduli for backcalculation analysis. As the internally generated seed moduli do not always produce satisfactory results, user-input seed moduli are generally encouraged.

Fwa and Rani (2005) developed a seed-generation algorithm 2L-BACK for multi-layer flexible pavements based on the closed-form moduli backcalculation solution for two-layer structures, 2L-BACK. The proposed algorithm does not require any subjective judgment by the user. Based on an evaluation analysis of the effectiveness of the proposed procedure by testing with two backcalculation software, MICHBACK and EVERCALC, and using LTPP measured and computed data of flexible pavement segments, it was found that the seed-generation algorithm led to enhanced program performance with respect to convergence characteristics and accuracy of backcalculated solutions. The 2L-BACK software provided with this Handbook (see Section 20.4.2), with second-generation capability, takes less than 1 sec to generate the seed moduli on a Pentium 4 personal computer with a clock speed of 2.4 GHz. The proposed seed-generation algorithm can be effectively incorporated into a backcalculation software for multi-layer flexible pavements, and it does not suffer from location and pavement type transferability constraints of most regression-based seed-generation methods.

20.4.3.1 Generating Seed Moduli Using 2L-BACK

This section describes the procedure that generates pavement layers seed moduli using the closed-form two-layer backcalculation program 2L-BACK. Using the four-layer pavement section in Figure 20.8(a) as example, the steps involved in the procedure that applies 2L-BACK in the computation are as follows:

(1) Estimate surface layer modulus E_1
   (i) Consider a two-layer transformed pavement section as shown in Figure 20.8(b) that has the same surface layer as the original pavement section of Figure 20.8(a), but with an equivalent

subgrade that provides the same supporting capacity as the combined effect of the base, subbase, and subgrade of the original pavement section.

(ii) Obtain E_1 by performing backcalculation analysis using 2L-BACK with deflection measurements at radial distances of $r_1 = 203$ mm and $r_3 = 305$ mm as input.

(2) Estimate subgrade modulus E_4

(i) Consider a two-layer transformed pavement section as shown in Figure 20.8(c) that has a surface layer having the same thickness as the combined thickness of the surface, base, and subbase layers of the original pavement section of Figure 20.8(a), and an identical subgrade to that of the original pavement section.

(ii) Obtain E_4 by performing backcalculation analysis using 2L-BACK with deflection measurements at radial distances of $r_1 = 203$ mm and $r_6 = 610$ mm as input.

(3) Estimate seed moduli for elastic moduli E_3 and E_4 of the intermediate layers

The seed moduli of the intermediate layers of a flexible pavement were estimated from known values of E_1 and E_4 by means of the Corps of Engineers method (Smith and Witczak, 1981). This method estimates base modulus E_2 and subbase modulus E_3 from the thickness of the layer concerned and the subgrade elastic modulus as follows,

$$E_2 = (1 + 0.17h_2)E_4 \tag{20.32}$$

$$E_3 = (1 + 0.075h_2)E_4 \tag{20.33}$$

The program also provides a range control of the moduli of the various layers. Adopting the recommendations of Chou et al. (1989), the following ranges of moduli were specified.

(a) For the asphalt surface layer and subgrade, the lower and upper limits were set as -60% and $+60\%$ of the respective seed moduli; and

(b) For the base and subbase layers, the lower and upper limits were one-third and three times the respective seed moduli.

The microcomputer software 2L-BACK provided with this Handbook permits the user a choice of three different analyses, viz., a closed-form backcalculation analysis for two-layer flexible pavement systems, a seed-generation analysis for three-layer flexible pavement to provide seed moduli for the asphalt surface layer, base layer and the subgrade, and a seed-generation analysis for four-layer flexible pavement to provide seed moduli for the asphalt surface layer, base layer, subbase layer, and the subgrade.

Surface Layer (E_1, μ_1, h_1)	Surface Layer (E_1, μ_1, h_1)	Equivalent Surface Layer ($E'_1, \mu'_1, h_1+h_2+h_3$)
Base (E_2, μ_2, h_2)	Equivalent Subgrade (E'_2, μ'_2)	
Subbase (E_3, μ_3, h_3)		
Subgrade (E_4, μ_4)		Subgrade (E_4, μ_4)
(a) Original pavement section	(b) Model for computing surface modulus	(c) Model for computing subgrade modulus

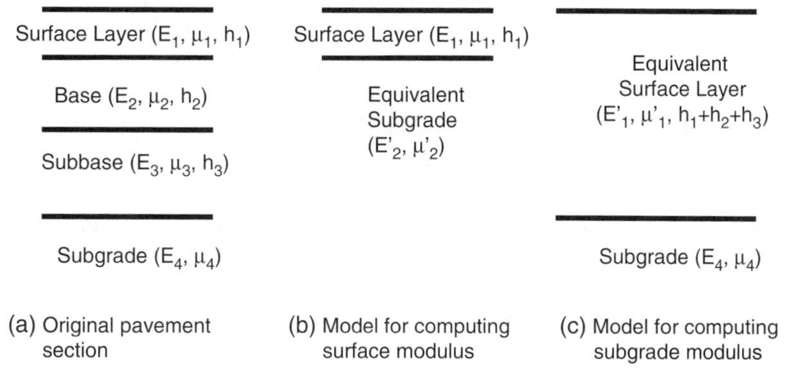

FIGURE 20.8 Pavement models for backcalculating surface layer modulus and subgrade modulus.

Example 20.5

The deflection test results of two pavement sections under the LTPP program are as follows:

Pavement Section	Layer Thickness (mm)	Sensor Location (mm)	0	203	305	457	610	914	1524
A	AC surface = 71.1 base = 157.5	Deflection (μm)	1290	896	685	427	300	165	88
B	AC surface = 53 base = 168 subbase = 152	Deflection (μm)	652	456	297	180	126	79	43

The test load of 71.1 kN was applied on a loading plate of 150 mm radius. The Poisson ratio for the AC surface layer was 0.30, and that of all other layers were 0.35. Select the seed moduli for all layers of the two pavement sections using software 2L-BACK.

Applying the software 2L-BACK, the seed values in MPa for the various layers of the two sections are as follows:

Pavement Section	E_1 (MPa) (Surface)	E_4 (MPa) (Subgrade)	E_2 (MPa) (Base)	E_3 (MPa) (Subbase)
A	7208.099	99.975	205.3487	—
B	17,301.77	196.233	416.4064	284.5378

20.5 Artificial Intelligence Techniques for Backcalculation Analysis

20.5.1 Application of Neural Network Technique

A major obstacle to the use of more elaborate backcalculation methods, such as static nonlinear or dynamic nonlinear back-analysis procedures, has been their relatively long-computation time. The problem of computation speed can be overcome by using either the regression equation-based approach (Zaghloul et al., 1994; Fwa and Chandrasegaram, 2001) or the database backcalculation method (Anderson, 1989; Tia et al., 1989). A common problem with these methods is that they are only applicable to the specific pavement structure and test loading conditions for which they have been derived. Another problem is the nonuniqueness of solutions, which cannot be handled effectively by either of these methods. This problem is further exacerbated by the inevitable presence of noise in deflection measurements in the field.

Recently, Meier et al. (1997) proposed the use of artificial neural networks as a forward computation tool to overcome some of the problems identified in the preceding paragraph. Neural networks are computational models with the ability to learn from training data. In a neural network, the unit analogous to the biological neuron of the human brain is referred to as a processing unit. A processing element has many input paths coming from other processing elements. The input is processed and modified by a transfer function in the processing element. The resultant output is passed to one or more output elements that hold the response of the network to the given input. In the learning process, connecting weights between processing elements are adjusted in response to stimuli from training data presented at the input buffer, and also at the output buffer. After a learning process phase, a trained neural network is ready to give virtually instant response to a fresh set of input data of similar nature to the training data.

As most of the computation time of a conventional backcalculation analysis is spent in the forward calculation of pavement response, a neural-network forward computation model would be able to cut down the computation time considerably. Meier et al. have demonstrated that the proposed approach could reduce the computation time by more than 40 times for their example problems. Another advantage of the approach is that the forward pavement response analysis has a unique solution for a given set of pavement layer properties. The neural-network training phase thus can be performed based on an exact one-to-one mapping between layer properties and surface deflections. Meier et al. proposed that a very broad range of pavement types and test conditions could be handled by breaking the training data into subsets and training a separate neural network on each subset.

20.5.2 Application of Genetic Algorithms

From the discussion of nonuniqueness earlier in this chapter, it is known that the solution surface of the backcalculation problem of pavement layer moduli contains many local minima, and that the backcalculated modulus values and their accuracy are procedure dependent. This implies that a potentially good backcalculation procedure would be one that has a strong global search ability to overcome the problem of local minima. Genetic algorithms (GA) are a technique that satisfies this requirement. A computer code NUS-GABACK that applies genetic algorithms to backcalculation analysis has been developed by Fwa et al. (1997c). It should be mentioned that a separate effort that employed the genetic algorithm technique for the same purpose was reported later in the same year by Kameyama et al. (1997).

20.5.2.1 Theoretical Basis of Genetic Algorithms

Genetic algorithms are robust search techniques formulated on the mechanics of natural selection and natural genetics (Holland, 1975). It adopts the Darwinian approach by borrowing the concept of natural adaptation in the evolution where species adapt themselves to the environment for survival. Genetic algorithms generate new feasible solutions from an existing pool of feasible solutions (known as a parent pool), and select "fitter" solutions from the new solutions to form the next parent pool. This repeated process of generating new solutions (known as offspring) and selecting fitter (i.e., better) solutions to form the next parent pool has been found to produce better and better solutions when appropriate procedures of offspring generation are followed. In the backcalculation analysis of pavement layer moduli, the objective is to identify a set of pavement layer moduli that would produce a deflection basin matching the measured deflection basin. A fitter solution is thus one that produces a closer match with the measured deflections.

20.5.2.2 Merits of Genetic Algorithm Approach

Genetic algorithms present an attractive approach to overcome the local optima problem because of the following characteristics:

(a) Genetic algorithm operations such as mutation of genes offer repeated perturbations to move out of local optima;
(b) The stochastic generation of offspring in the search for new solutions offers an effective global search of the solution space;
(c) A large number of similar solutions can be processed through the coded string structures; and
(d) Genetic algorithms are able to efficiently exploit past information to explore new regions of the decision space with a high probability of finding solutions with improved fitness.

Most of the iterative backcalculation procedures are sensitive to seed modulus values set by the users. Seed modulus values are not required by NUS-GABACK. Only the range of possible modulus values for each pavement layer is needed as user input. This should make the approach easier to use for most users, and effectively eliminate the dependence of solutions on input seed values.

20.6 Future Developments

The practices of nondestructive structural evaluation of pavements since the 1970s have been dominated by the surface deflection-based backcalculation approach. Because of the stress and strain dependence of the structural properties of pavement materials, it is likely that deflection-based procedures will continue to be the main mode of nondestructive pavement evaluation test in the years to come. Several limitations of the current practices of nondestructive pavement evaluation have been identified in this chapter.

Among these, there are two main limitations that will probably dictate the directions of future research and development efforts:

(a) *Method of deflection testing.* From the standpoint of field operation, the single most unsatisfactory aspect of the current deflection testing using FWDs is the need to conduct fixed point stationary-vehicle falling weight tests at selected intervals along the test pavement section. This presents a safety hazard to moving traffic and creates traffic delays in busy streets and highways. From the consideration of deflection testing, the fixed point falling-weight test does not produce a perfect simulation of the stress and strain fields caused by an actual moving truck. The RWD (see Section 20.2.5) moving at normal traffic speed presents definite improvements in terms of operation productivity, safety, and more realistic simulation of the deflection basin, stresses and strains caused by trucks. A key challenge is to accurately pick up the "total" surface deflection under the loading wheel, and at points that define the complete deflection basin. The determination of the "total" surface deflection at each point within the deflection basin requires the knowledge of the "unloaded" and "loaded" surface elevations of the same point. The measurement of "unloaded" surface elevation of a loaded point could present a practical problem on strong and stiff pavements with a large moving deflection basin.

(b) *Method of backcalculation analysis.* Practically all the backcalculation algorithms in common applications today are based on static load analysis using multi-layered elastic theory. Since the structural response of pavement in the FWD test or the RWD test is inherently a dynamic process, more sophisticated backcalculation procedures based on dynamic analysis are required. Either linear or nonlinear dynamic analyses that make use of the time history data of deflection tests lead to additional information for backcalculation analysis.

Depending on the choice of test load and material models adopted, different analysis approaches can be used for backcalculation of pavement layer properties. None of these approaches gives an exact representation of the actual test conditions. The backcalculated pavement layer properties are pseudo material properties meaningful only for the material models used. Nevertheless, regardless of the models adopted, it should be possible that appropriate pavement remaining-life analysis and overlay design procedures be developed for each approach. This calls for research in the areas of material characterization, pavement design, and performance prediction using information derived from nondestructive *in situ* deflection testing based on applied loading that closely simulates actual moving vehicular wheel loads.

References

AASHTO, 1993. *AASHTO Guide for Design of Pavement Structures.* American Association of State Highway and Transportation Officials, Washington, DC.

Akram, T., Scullion, T., and Smith, R.E. 1994. *Comparing Laboratory and Backcalculated Layer Moduli on Instrumented Pavement Sections*, ASTM STP No. 1198. American Society for Testing and Materials, West Conshohocken, PA, pp. 170–202.

Anderson, M. 1989. *A Data Base Method for Backcalculation of Composite Pavement Layer Moduli*, ASTM STP No. 1026. American Society for Testing and Materials, West Conshohocken, PA, pp. 201–216.

Anderson, O., Orborn, B., and Ringstrom, G. 1972. The bearing capacity of pavements with frost retarding layers — a test road study, pp.3–16. In *Proc. 3rd Int. Conf. Structural Design of Asphalt Pavements.* London, UK.

ASTM, 2003a. *ASTM D-1195 Standard Test Method for Repetitive Static Plate Load Tests of Soils and Flexible Pavement Components, for Use in Evaluation and Design of Airport and Highway Pavements*, Annual Book of ASTM Standards. American Society for Testing and Materials, West Conshohocken, PA.

ASTM, 2003b. *ASTM D-1196 Standard Test Method for Non-Repetitive Static Plate Load Tests of Soils and Flexible Pavement Components, for Use in Evaluation and Design of Airport and Highway Pavements*, Annual Book of ASTM Standards. American Society for Testing and Materials, West Conshohocken, PA.

ASTM, 2003c. *ASTM D4695 Standard Guide for Pavement Deflection Measurements*, Annual Books of ASTM Standards. American Society for Testing and Materials, West Conshohocken, PA.

ASTM, 2003d. *ASTM D4602 Nondestructive Testing of Pavements Using Cyclic-Loading Dynamic Deflection Equipment*, Annual Books of ASTM Standards. American Society for Testing and Materials, West Conshohocken, PA.

ASTM, 2003e. *ASTM D4694 Standard Test Method for Deflections with a Falling-Weight-Type Impulse Load Device*, Annual Books of ASTM Standards. American Society for Testing and Materials, West Conshohocken, PA.

Bay, J.A., Stokoe, K.H. 1998. *Development of a Rolling Dynamic Deflectometer for Continuous Deflection Testing of Pavements*, Research Report No. FHWA/TX-99/1422-3F. Federal Highway Administration, Washington, DC.

Bohn, A., Ullidtz, P., Stubstad, R., and Sirenson, A. 1972. Danish experience with the French FWD, pp. 1119–1128. In *Proc. 3rd Int. Conf. Structural Design of Asphalt Pavements*, London, UK.

Brown, S.F., Tam, W.S., and Brunton, J.M. 1987. Structural evaluation and overlay design: analysis and implementation, pp. 1013–1028. In *Proc. 6th Int. Conf. on the Structural Design of Asphalt Pavements*, 13–17 July, University of Michigan, Ann Arbor, MI.

Burmister, D.M., The general theory of stresses and displacements in layered system, *J. Appl. Phys.*, 16, 89–94, 1945, see also 126–127.

Bush, A.J. 1980. *Dynamic Surface Deflection Measurements on Rigid Pavements Compared with the Model of an Infinite Plate on an Elastic Foundation*, Transportation Research Record, No. 756. Transportation Research Board, Washington, DC.

Bush, A.J. and Alexander, D.R. 1985. *Pavement Evaluation Using Deflection Basin Measurements and Layered Theory*, Transportation Research Record, No. 1022, pp. 16–28.

Chang, D.W., Kang, Y.V., Roesset, J.M., and Stokoe, K.H. II 1992. *Effect of Depth to Bedrock on Deflection Basins obtained with Dynaflect and FWD Tests*, Transportation Research Record 1355, pp. 8–16.

Chou, Y.J., Uzan, J., and Lytton, R.L. 1989. *Backcalculation of Layer Moduli from Nondestructive Pavement Deflection Data using the Expert System Approach. Nondestructive Testing of Pavement and Backcalculation of Moduli*, ASTM STP 1026, A.J. Bush, III and G.Y. Baladi, eds., pp. 341–354. American Society for Testing and Materials, West Conshohocken, PA.

Douglas, R.A. and Elle, G.L. 1986. *Nondestructive Pavement Testing by Wave Propagation: Advanced Method of Analysis and Parameter Analysis*, Transportation Research Record No. 1070, pp. 53–62.

Fwa, T.F. and Chandrasegaram, S., Regression models for backcalculation of concrete pavement properties, *J. Transport. Eng.*, ASCE, 127, 4, 353–356, 2001.

Fwa, T.F. and Jia, X.M. 2004. *Backcalculation of Asphalt Pavement Material Properties*, Project Report CTR-04–21. Center for Transportation Research, National University of Singapore.

Fwa, T.F., Li, S., and Tan, K.H. 1997a. Backcalculation of material properties for concrete pavement, Vol. 2, pp. 101–114. In *Proc. 6th Int. Conf. on Concrete Pavement Design and Materials for High Performance*. 18–21 Nov., Indianapolis, IN, USA.

Fwa, T.F., Li, S., and Tan, K.H. 1997b. *Closed-Form Backcalculation of Concrete Pavement Parameters*, Project Report CTR-97 No.14. Center for Transportation Research, National University of Singapore.

Fwa, T.F. and Rani, T.S. 2005. Seeds-generation algorithm for moduli backcalculation of flexible pavements. In *Paper presented at 84th Annual Meeting of Transportation Research Board*, Jan. 9–13, Washington, DC, Transport Research Record, in press.

Fwa, T.F., Shi, X.P., and Tan, S.A., 1996. Analysis of concrete pavements by rectangular thick-plate model, *J. Transport. Eng., Am Soc. Civ. Eng.*, 122, 2, 146–154.

Fwa, T.F., Tan, C.Y., and Chan, W.T. 1997c. Backcalculation analysis of pavement layer moduli using genetic algorithms, pp. 134–142. In *Presented at 76th Annual Meeting of Transportation Research Board*, Transportation Research Record No. 1570. Jan 12–16, Washington, DC.

Fwa, T.F., Tan, K.H., and Li, S. 1997d. *Graphical Backcalculation Solutions for Two-Layer Rigid Pavements*, Research Report CTR-97-15. Center for Transportation Research, Department of Civil Engineering, National University of Singapore, Singapore.

Fwa, T.F., Tan, K.H., and Li, S. 1999. *Closed-Form and SemiClosed-Form Algorithms for Backcalculation of Concrete Pavement Properties*, ASTM STP No. 1375. American Society for Testing and Materials, West Conshohocken, PA, pp. 267–280.

Gartin, R.S. 1991. *An Introduction to Wave Propagation in Pavements and Soils: Theory and Practice*, Report No. AK-Rd-91-02. Alaska Department of Transportation and Public Facilities, Fairbank, AK.

Gioda, G. 1985. Some remarks on back-analysis and characterization problems in geomechanics, Vol. I, pp. 47–61. In *Proc. 5th Int. Conf. on Numerical Methods in Geomechanics*. Balkema, AA.

Greenstein, J., and Berger, L. 1989. *Using NDT Aided by an Expert System to Evaluate Airport and Highway Systems*, ASTM STP No. 1026. American Society for Testing and Materials, West Conshohocken, PA, pp. 525–539.

Grogg, M.G. and Hall, J.W. 2004. *Measuring Pavement Deflections at 55 MPH, Public Roads*, Jan–Feb 2004, Federal Highway Administration, Washington, DC.

Hall, J.W. 2004. *Rolling Wheel Deflectometer*, ERES Consultants. Champaign, Illinois.

Harichandran, R.S., Baladi, G.Y., and Yeh, M. 1989. *Development of a Computer Program for Design of Pavement Systems Consisting of Bonded and Unbonded Materials*. Department of Civil and Environmental Engineering, Michigan State University, East Lansing, MI.

Harichandran, R.S., Mahamood, T., Raab, A.R., and Baladi, G.Y. 1993. *Modified Newton Algorithm for Backcalculation of Pavement Layer Properties*, Transportation Research Record, No.1384. Transportation Research Board, Washington, DC, pp. 15–22.

Harichandran, R.S., Ramon, C.M., and Baladi, G.Y. 2000. *MICHIBACK User's Manual*. Department of Civil and Environmental Engineering, Michigan State University, East Lansing, MI.

Hicks, RG. 1970. *Factors Influencing the Resilient Response of Granular Materials*, Ph.D. Dissertation, University of California, Berkeley.

Holland, J.H. 1975. *Adaptation in Natural and Artificial Systems*. The University of Michigan Press, Ann Arbor.

Houston, W.N., Mamlouk, M.S., and Perera, W.S., Laboratory vs. Nondestructive Testing for Pavement Design, *J. Transport. Eng.*, 118, 2, 1992.

Huang, Y.H. 1993. *Pavement Analysis and Design*. Prentice-Hall, Englewood Cliffs, NJ, USA.

Ioannides, A.M. 1986. *Discussion on Rigid Pavement Analysis*, Transportation Research Record No. 1070, pp. 27–29.

Ioannides, A.M., Barenberg, E.J., and Lary, J.A. 1989. Interpretation of FWD results using principles dimensional analysis, pp. 231–247. In *Proc. 4th Int. Conf. on Concrete Pavement Design and Rehabilitation*. April 18–20, Purdue University, Lafayette, IN.

Irwin, L.H., Yang, W.S., and Stubstad, R.N. 1989. *Deflection Reading Accuracy and Layer Thickness Accuracy in Backcalculation of Pavement Layer Moduli*, ASTM STP No. 1026. American Society for Testing and Materials, West Conshohocken, PA, pp. 229–244.

Johnson, R.F. and Rish, J.W., III. 1996. Rolling weight deflectometer with thermal and vibrational bending compensation. In *Presented at TRB Annual Meeting*. January 1996, Washington, DC.

Kameyama, S., Himeno, K., Kasahara, A., and Maruyama, T. 1997. Backcalculation of pavement layer moduli using genetic algorithms, pp. 1375–1386. In *Proc. 8th Int. Conf. on Asphalt Pavements*. 10–14 August, Seattle.

Kopperman, S., Tiller, G., and Tseng, M. 1986. *ELSYM5 User's Manual*, Report No. FHWA-TS-87-206. Federal Highway Administration, U.S. Department of Highways.

Lee, S.W. 1988. *Backcalculation of Pavement NDT Moduli by Use of Pavement Surface Deflections*, Ph.D. Dissertation, University of Washington, Seattle, WA.

Li, S., Fwa, T.F., and Tan, K.H., Closed-Form Backcalculation of Rigid Pavement Parameters, *J. Transport. Eng.*, 122, 1, 1996.

Li, S., Fwa, T.F., and Tan, K.H., Backcalculation of parameters for slabs on two-layer foundation system, *J. Transport. Eng.*, 123, 6, 84–489, 1997.

May, R.W. and Von Quintus, H.L. 1994. *The quest for a standard guide to NDT backcalculation. Non-Destructive Testing of Pavements and Backcalculation of Moduli*, ASTM STP No. 1198, H.L. Von Quintus, A.J. Bush, III and G.Y. Baladi, eds., pp. 505–520. American Society for Testing and Materials, West Conshohocken, PA.

Meier, R.W., Alexander, D.R., and Freeman, R.B. 1997. *Using Artificial Neural Network as a Forward Approach to Backcalculation*, Transportation Research Record No. 1570, pp. 126–133.

Michelow, J. 1963. *Analysis of Stresses and Displacements in an N-Layered Elastic System under a Load Uniformly Distributed in a Circular Area.* California Research Corporation, Richmond, CA.

Moosazadeh, J. and Witczak, M.W. 1981. *Prediction of Subgrade Modulus for Soil that Exhibits Nonlinear Behavior*, Transportation Research Record No. 810, pp. 9–17.

Nazarian, S., Rojas, R., and Pezo, R. 1998. Relating laboratory and field moduli of Texas base materials. In *Paper presented at 77th Transportation Research Board Annual Meeting*, January 11–15, Washington, DC.

Nazarian, S. and Stokoe, K.H. 1989. *Nondestructive Evaluation of Pavements by surface Wave Method*, ASTM Special Technical Publication STP No. 1026, pp. 119–154.

Panc, V. 1975. *Theories of Elastic Plates.* Noordhoff International Publishing, Leyden.

Quest Integrated, Inc., 2004. *Rolling Weight Deflectometer.* Quest Integrated, Inc., Kent, Seattle, WA.

Raad, L., Figueroa, J.L. 1980. Resilient response of transportation support systems. *J. Transport. Eng.*, 106, 1, 111–128.

Roesset, J.M. and Shao, K.Y. 1985. *Dynamic Interpretation of Dynaflect and FWD Tests*, Transportation Research Record No. 1022, pp. 7–16.

Ryden, N., Park, C.B., Ulriksen, P., and Miller, R.D., Multimodal Approach to Seismic Pavement Testing, *J. Geotech. Geoenviron. Eng*, ASCE, 130, 6, 636–645, 2004.

Scrivner, F.H., Michalak, C.H., and Moore, W.M. 1973. *Calculation of the Elastic Moduli of a Two-Layer Pavement System from Measured Surface Deflection*, Highway Research Record No. 431. Highway Research Board.

Scullion, T., Uzan, J., and Paredes, M. 1990. *MODULUS: A Microcomputer Based Backcalculation System*, Transportation Research Record, No. 1260, pp. 180–191.

Shell Koninklijke, 1972. *BISAR Users Manual — Layered System under Normal and Tangential Loads.* Shell Laboratorium, Amsterdam, The Netherlands.

SHRP, 1993. *Layer Moduli Backcalculation Procedure: Software Selection Strategic Highway Research Program*, Report SHRP-P-651. National Research Council, Washington, DC.

Sivaneswaran, N., Krammer, S.L., and Mahoney, J.P. 1991. Advanced backcalculation using a nonlinear least squares optimization technique. In *Presented at the 70th Annual Meeting of the Transportation Research Board*, January 13–17, Washington, DC.

Smith, B.E. and Witczak, M.W., Equivalent granular base modulus: prediction, *J. Transport. Eng.*, ASCE, 107, 6, 1981.

Stolle, D. and Hein, D. 1989. *Parameter Estimates of Pavement Structure Layers and Uniqueness of the Solution*, ASTM STP No. 1026. American Society for Testing and Materials, West Conshohocken, PA, pp. 313–322.

Stratton, JH. 1994. Military airfields, pp. 27-89. In *Proceedings of the American Society of Civil Engineers*, 70.

Stubbs, N., Torpunuri, V., Lytton, R., and Magnuson, A. 1994. *Methodology for Material Properties in Pavements Modeled as Layered Visco-elastic Half Spaces*, ASTM STP No. 1198. American Society for Testing and Materials, West Conshohocken, PA, pp. 53–67.

Tam, W.S. and Brown, S.F. 1988. Use of FWD for insitu evaluation of granular materials in pavements, pp. 155–163. In *Proc. 14th ARRB Conf., Part 5*, Australian Road Research Board.

Tawfig, K., Armaghani, J., and Sobanjo, J. 2000. *Seismic Pavement Analysis vs. FWD for Pavement Evaluation: Comparative Study*, ASTM STP No. 1375. American Society for Testing and Materials, West Conchohocken, PA, pp. 327-345.

Thompson, M.R., and Robnett, Q.L., Resilient Properties of Subgrade Soils, J. of Transportation Engineering, *American Society of Civil Engineers*, 105, 1, 71–89, 1979.

Thomsen, T.M. 1982. Phoenix FWD and Registration Equipment, Vol. 1, pp. 457–463. In *Proc. Int. Symposium on Bearing Capacity of Roads and Airfields*, Trondheim, Norway.

Tia, M., Eom, K.S., and Ruth, B.E. 1989. *Development of the DBCONPAS Computer Program for Estimation of Concrete Pavement Parameters from FWD Data*, ASTM STP No. 1026. American Society for Testing and Materials, West Conshohocken, PA, pp. 291–312.

Uddin, W. 1984. *A Structural Evaluation Methodology for Pavements based on Dynamic Deflections*, Ph.D. Dissertation, University of Texas at Austin.

Uddin, W., Meyer, A.H., and Hudson, W.R. 1986. *Rigid Bottom Consideration for Nondestructive Evaluation of Pavements*, Transportation Research Record No. 1070, pp. 21–29.

Ullidtz, P. 1973. The Use of Dynamic Plate Loading Tests in Design of Overlays. In *Proc. Conf. on Road Engineering in Asia and Australasia*, Kuala Lumpur, Malaysia.

Ullidtz, P. 1987. *Pavement Analysis*. Elsevier, Amsterdam.

Uzan, J. 1994. *Advanced Backcalculation Techniques*, ASTM STP No. 1198, pp. 3–37.

Uzan, J., Scullion, T., Michalak, C.H., Paredes, M., and Lyttom, R.L. 1988. *A Microcomputer Based Procedure for Backcalculating Layer Moduli from FWD Data*, Research Report 1123-1. Texas Transportation Institute, College Station.

Van Cauwelaert, F.J., Alexander, D.R., White, T.D., and Barker, W.R. 1989. *Multilayer Elastic Program for Backcalculating Layer Moduli in Pavement Evaluation*, ASTM STP No. 1026. American Society for Testing and Materials, West Conshohocken, PA, pp. 171–188.

Van Deusen, D.A., Lenngren, C.A., and Newcomb, D.E. 1994. *A Comparison of Laboratory and Field Subgrade Moduli at the Minnesota Road Research Project*, ASTM STP No. 1198. American Society for Testing and Materials, West Conshohocken, PA, pp. 361–379.

Wang, M.C., Larson, T.D., Bhajandas, A.C., and Cumberledge, G. 1978. *Use of Road Rater Deflections in Pavement Evaluation*, Transportation Research Record No. 666. Transportation Research Board, pp. 2–39.

Westergaard, H.M., Stresses in Concrete Pavements Computed by Theoretical Analysis, *Public Roads*, 7, 2, 25–35, 1926.

Westergaard, H.M. 1948. *New Formulas for Stresses in Concrete Pavements of Airfield*, ASCE Transaction, Vol. 113. American Society of Civil Engineers.

WSDOT. 1999. *WSDOT Pavement Guide. Volume 3 — Pavement Analysis Computer Software and Case Studies*. Washington State Department of Transportation, Seattle, WA.

Zaghloul, S.M., White, T.D., Drnevich, V.P., and Coree, B. 1994. *Dynamic Analysis of FWD Loading and Pavement Response using a Three-Dimensional Dynamic Finite Element Program*, ASTM STP No. 1198. American Society for Testing and Materials, West Conshohocken, PA, pp. 125–138.

Appendix 20A
User's Manual for NUS-BACK

20A.1   Overview of NUS-BACK

NUS-BACK is a computer software to backcalculate the layer moduli of a two-layer concrete pavements (i.e., a slab on an elastic foundation), based on a closed-form backcalculation algorithm developed at the National University of Singapore (NUS).

The NUS-BACK accepts deflection measurements from any surface deflection measuring device without the restriction on the number of sensors and their spacing, and without the need to assume initial values of pavement layer elastic properties. In NUS-BACK, there are two concrete pavement structure models available for users to select based on their requirements. The first is a slab-on-liquid-foundation model and the second is a slab-on-solid-foundation model. For the first model, the output values are the elastic modulus of concrete slab and the modulus of subgrade reaction. For the second model, the output values are the elastic modulus of concrete slab and the elastic modulus of subgrade.

20A.2   System Requirement

The NUS-BACK program runs on IBM compatible computer with Windows 2000 or Windows XP professional Version Operating System.

20A.3   Input

This required input data for the NUS-BACK are listed in the following table

Input Parameter	Description	Example
Number of sensors, N	Number of sensors used in deflection measurement	4
Force (kN)	Load applied in deflection test	34.7
Slab thickness (m)	Thickness of concrete pavement slab	0.254
Radius of loading plate (m)	Radius of loading plate	0.15
Slab Poisson's ratio	Poisson's ratio of concrete pavement slab	0.15
Subgrade Poisson's ratio	Poisson's ratio of subgrade soil	0.45
Sensor location (m) $(L_i, i = 1, ..., N)$	Locations of deflection sensors that measured from the center of loading plate	0, 0.3, 0.6, 0.9
Deflection (mm) $(D_i, i = 1, ..., N)$	Measured surface deflections	0.086, 0.080, 0.060, 0.056.

Screen before keying in input

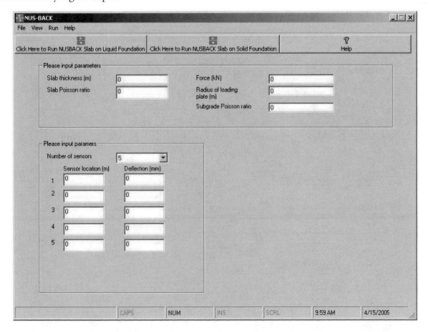

Screen after keying in input

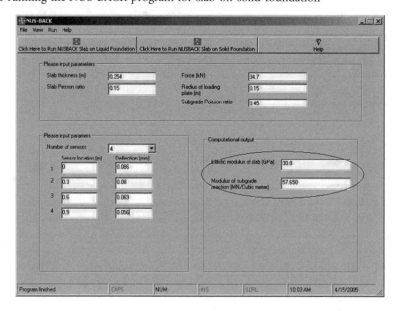

20A.4   Output

The output of NUSBACK depends on the model one chooses to perform backcalculation analysis.
Output for the slab-on-liquid-foundation model:

- Elastic modulus of slab (GPa)
- Modulus of subgrade Reaction (MN/m^3).

Output for the slab-on-solid-foundation model:

- Elastic modulus of upper slab (GPa)
- Elastic modulus of subgrade (GPa).

Screen after running the NUS-BACK program for slab-on-solid-foundation

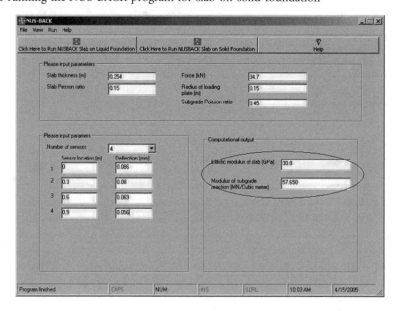

Screen after running slab-on-solid-foundation model

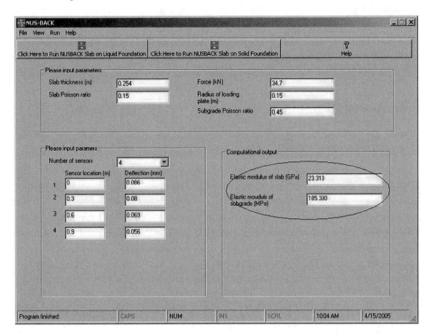

20A.5   Example

Input

Applied load = 34.7 kN
Radius of loading plate = 0.15 m
Slab thickness = 0.254 m
Slab Poisson's ratio = 0.15
Subgrade Poisson's ratio = 0.45

Deflection data:

0.086 mm at 0 m radial distance
0.080 mm at 0.3 m radial distance
0.069 mm at 0.6 m radial distance
0.056 mm at 0.9 m radial distance

Output If the liquid foundation model is chosen:

Elastic modulus of upper slab = 30.800 GPa
Modulus of subgrade Reaction = 56.650 MN/m^3

If the solid foundation model is chosen:

Elastic modulus of upper slab = 23.313 GPa
Elastic modulus of subgrade = 185.380 GPa

Appendix 20B
User's Manual for NUS-BACK3

20B.1   Overview of NUS-BACK3

NUS-BACK3 is a computer software to backcalculate the layer moduli of a concrete pavement slab with two support layers, i.e., a slab with a base or subbase layer supported on the subgrade soil. The closed-form backcalculation algorithm reduces a system of three equations with three unknowns into one single nonlinear equation and gives unique closed-form solutions without the need to select initial elastic moduli of pavement layers to start with. NUS-BACK3 accepts deflection measurements from any surface deflection measuring device without restriction on the number of sensors and their spacing. The output values are the elastic moduli of concrete slab, subbase and subgrade soil, respectively.

20B.2   System Requirement

The NUS-BACK3 program runs on IBM compatible computer with Windows 2000 or Windows XP professional Version Operating System.

20B.3   Input

This required input data for the NUS-BACK3 are listed in the following table

Input Parameter	Description	Example
Force (kN)	Load applied in deflection test	35
Slab thickness (m)	Thickness of concrete pavement slab	0.254
Subbase thickness (m)	Thickness of subbase	0.203
Radius of loading plate (m)	Radius of loading plate	0.15
Slab Poisson's ratio	Poisson's ratio of concrete pavement slab	0.15
Subbase Poisson's ratio	Poisson's ratio of subgrade soil	0.15
Subgrade Poisson's ratio	Poisson's ratio of subgrade soil	0.35
IKEY	Continuity factor between concrete slab and subbase, which ranges between 0 and 1	1
Number of sensors, N	Number of sensors used in deflection measurements	3
Sensor location (m) (L_i, $i = 1, ..., N$)	Location of each sensor, measured from the center of loading plate	0, 0.305, 0.610
Deflection (mm) (D_i, $i = 1, ..., N$)	Measured surface deflection	0.141, 0.135, 0.124

Screening display before keying in input

Screening display after keying in input

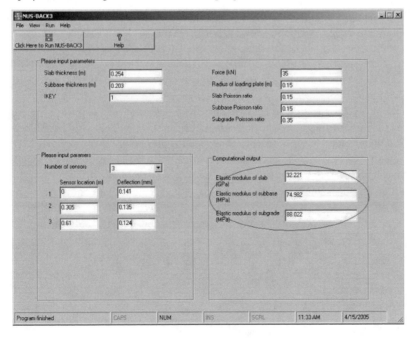

20B.4 Output

Elastic modulus of slab (GPa)
Elastic modulus of subbase (MPa)
Elastic modulus of subgrade (MPa).

Screening display after running NUS-BACK3 model program

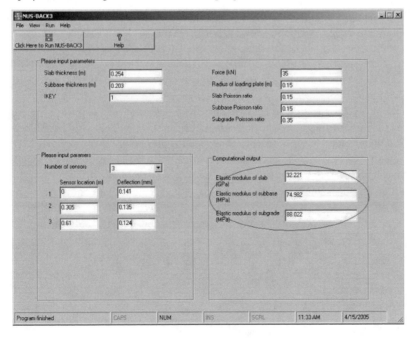

20B.5   Example

Input

> Applied load = 35 kN
> Radius of loading plate = 0.15 m
> Slab thickness = 0.254 m
> Subbase thickness = 0.203 m
> Slab Poisson's ratio = 0.15
> Subbase Poisson's ratio = 0.15
> Subgrade Poisson's ratio = 0.35

Deflection data:

> 0.141 mm at 0 m radial distance
> 0.135 mm at 0.305 m radial distance
> 0.124 mm at 0.610 m radial distance

Output

> Elastic modulus of slab = 32.221 GPa
> Elastic modulus of subbase = 74.982 MPa
> Elastic modulus of subgrade = 88.022 MPa

Appendix 20C
User's Manual for NUS-BACK2

20C.1   Overview of NUS-BACK2

NUS-BACK2 is a computer software to backcalculate the layer moduli of a concrete pavement with two slab layers, i.e., an upper slab and a lower slab supported on an elastic foundation system. The evaluation of the subgrade modulus is carried out based on the closed-form backcalculation method slab-on-elastic-foundation concrete pavement, the NUS-BACK1 (see Appendix). The evaluation of the elastic moduli of the two slab layers are carried out by a trial-and-error approach to mach the flexural rigidity of an equivalent single slab.

   NUS-BACK2 accepts deflection measurements from any surface deflection measuring device without the restriction on the number of sensors and their spacing, and without the need to assume initial seed modulus values of the pavement layer elastic properties. In NUS-BACK2, there are two concrete pavement structure models available for users to select based on their own requirements. The first is a slab-on-liquid-foundation model and the second is a slab-on-solid-foundation model. For the first model, the output values are the elastic moduli of the upper and lower concrete slabs, respectively, and the modulus of subgrade reaction. For the second model, the output values are the elastic moduli of the upper and the lower concrete slabs, respectively, and the elastic modulus of subgrade.

20C.2   System Requirement

The NUS-BACK2 program runs on IBM compatible computer with Windows 2000 or Windows XP professional Version Operating System.

20C.3   Input

The required input data for the NUS-BACK2 are listed in the following table

Input Parameter	Description	Example
Force (kN)	Load applied in deflection test	40
Upper slab thickness (m)	Thickness of the upper pavement slab	0.18
Lower slab thickness (m)	Thickness of the lower pavement slab	0.23
Radius of loading plate (m)	Radius of loading plate	0.15
Slab Poisson's ratio	Poisson's ratio of concrete pavement slab	0.15
Subgrade Poisson's ratio	Poisson's ratio of subgrade soil	0.40
Continuity factor between the upper and lower slabs	1—fully continuous, 0—fully discontinuous or total separated	0
Number of sensors, N	Number of deflection measurements	4
Sensor location (m) (L_i, $i = 1, \ldots, N$)	The locations of sensors	0, 0.3, 0.6, 0.9
Deflection (mm) (D_i, $i = 1, \ldots, N$)	Measured surface deflection	0.0791, 0.0734, 0.0621, 0.0498.

Screening display before keying in input

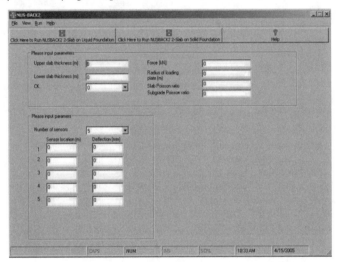

Screening display after keying in input

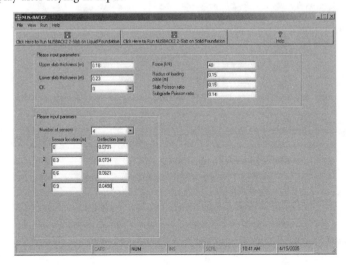

20C.4   Output

The output of NUSBACK2 depends on the model one chooses to perform the backcalculation analysis.
Output for the slab-on-liquid-foundation model:

- Elastic modulus of upper slab (GPa)
- Elastic modulus of lower slab (GPa)
- Modulus of subgrade Reaction (MN/m^3)

Output for the slab-on-solid-foundation model:

- Elastic modulus of upper slab (GPa)
- Elastic modulus of lower slab (GPa)
- Elastic modulus of subgrade (GPa)

Screening display after running NUS-BACK2 for slab-on-solid-foundation model

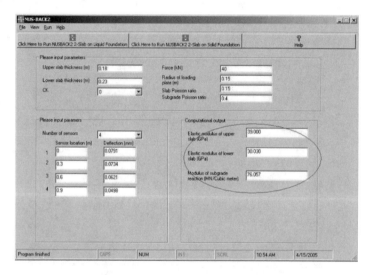

Screening display after running NUS-BACK2 for slab-on-solid-foundation model

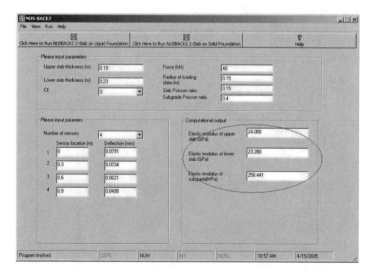

20C.5   Example

Input

Applied load = 40 kN
Radius of loading plate = 0.15 m
Upper Slab thickness = 0.18 m
Lower Slab thickness = 0.23 m
Slab Poisson's ratio = 0.15
Subgrade Poisson's ratio = 0.40

Deflection data:

0.0791 mm at 0 m radial distance
0.0734 mm at 0.300 m radial distance
0.0621 mm at 0.600 m radial distance
0.0498 mm at 0.900 m radial distance

Output If the liquid foundation model is chosen:

Elastic modulus of upper slab = 39.000 GPa
Elastic modulus of lower slab = 30.030 GPa
Modulus of subgrade Reaction = 76.057 MN/m^3

If the solid foundation model is chosen:

Elastic modulus of upper slab = 24.000 GPa
Elastic modulus of lower slab = 23.280 GPa
Elastic modulus of subgrade = 258.441 GPa

Appendix 20D
User's Manual for NUS-BACK4

20D.1   Overview of NUS-BACK4

NUS-BACK4 is a computer software to backcalculate the layer moduli of a concrete pavement with two slab layers resting on two supporting layers. It accepts deflection measurements from any surface deflection measuring device without the restriction on the number of sensors and their spacing, and without the need to assume initial values. The output values are the elastic moduli of the two concrete slab layers, subbase and subgrade soil, respectively.

20D.2   System Requirement

The NUS-BACK4 program runs on IBM compatible computer with Windows 2000 or Windows XP professional Version Operating System.

20D.3   Input

This required input data for the NUS-BACK4 are listed in the following table

Input Parameter	Description	Example
Force (kN)	Load applied in deflection test	35
Upper slab thickness (m)	Thickness of concrete pavement slab	0.15
Lower slab thickness		0.1

(Continued)

Input Parameter	Description	Example
Radius of loading plate (m)	Radius of loading plate	0.15
Slab Poisson's ratio	Poisson's ratio of concrete pavement slab	0.15
Subbase Poisson's ratio	Poisson's ratio of subbase soil	0.15
Subgrade Poisson's ratio	Poisson's ratio of subgrade soil	0.35
CK	Continuity factor between the upper slab and the lower slab, 1—fully continuous, 0—fully discontinuous or total separated	1
IKEY	Continuity factor between concrete slab and subbase, which ranges between 0 and 1	1
Number of sensors, N	Number of sensors used in deflection measurements	5
Sensor location (m) (L_i, $i = 1, ..., N$)	Location of each sensor, measured from the center of loading plate	0, 0.305, 0.610, 0.915 1.220
Deflection (mm) (D_i, $i = 1, ..., N$)	Measured surface deflection	0.128, 0.122, 0.114, 0.105, 0.094.

Screen display before keying in input

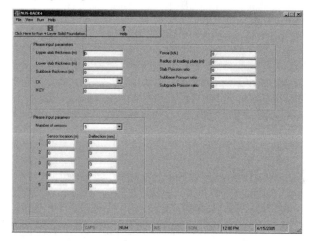

Screen display after keying in input

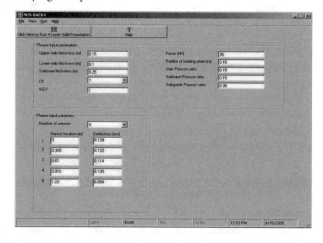

20D.4   Output

The following are the output of NUSBACK4 backcalculation analysis:

Elastic modulus of upper slab (GPa)
Elastic modulus of lower slab (GPa)
Elastic modulus of subbase (MPa)
Elastic modulus of subgrade (MPa).

Screen display after running NUS-BACK4 model program

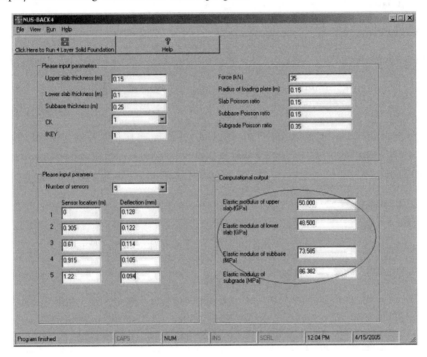

20D.5   Example

Input

Applied load = 35 kN
Radius of loading plate = 0.15 m
Upper slab thickness = 0.15 m
Lower slab thickness = 0.1 m
Subbase thickness = 0.25 m
Slab Poisson's ratio = 0.15
Subbase Poisson's ratio = 0.15
Subgrade Poisson's ratio = 0.35

Deflection data:

0.128 mm at 0 m radial distance
0.122 mm at 0.305 m radial distance
0.114 mm at 0.610 m radial distance
0.105 mm at 0.915 m radial distance
0.094 mm at 1.220 m radial distance

Output

Elastic modulus of upper slab = 50.000 GPa
Elastic modulus of lower slab = 48.500 GPa
Elastic modulus of subbase = 73.585 MPa
Elastic modulus of subgrade = 86.382 MPa

Appendix 20E
User Manual for 2L-BACK

20E.1 Overview

2L-BACK is a backcalculation software that generates seed modulus values for multi-layer flexible pavements. Based on a closed-form moduli backcalculation solution for 2-layer structures, the 2L-BACK seed-generation algorithm provides seed modulus values of all layers for 2-Layer, 3-Layer and 4-Layer flexible pavements. It does not suffer from the location and pavement type transferability constraints of most regression-based seed-generation methods. The algorithm is valid for flexible pavement in which the elastic modulus of an upper layer is lager than that of a lower layer. The 2L-BACK program runs on IBM compatible computer with Windows 2000 or Windows XP professional Version Operating System.

20E.2 Execution

1. Download the 2L-BACK folder from CRC website to desktop computer.
2. Double click the 2L-BACK folder to run the program.

20E.3 Input

The required input data for the 2L-BACK are listed in the following table

Input Parameter	Description	Example
Number of layers	Number of layers in the flexible pavement structure analyzed	4
Surface layer thickness (m)	Thickness of pavement surface obtained by field measurement or from design data	0.2
Surface layer Poisson's ratio	Poisson's ratio of pavement surface layer	0.25
Base thickness (m)	Thickness of pavement base course obtained by field measurement or from design data	0.3
Base Poisson's ratio	Poisson's ratio of base course	0.3
Subbase thickness (m)	Thickness of subbase obtained by field measurement or from design data	0.25
Subbase Poisson's ratio	Poisson's ratio of subbase layer	0.4
Subgrade Poisson's ratio	Poisson's ratio of subgrade soil	0.4
Force (kN)	Load applied in deflection test	70
Radius of loading plate (m)	Radius of loading plate	0.15
Number of sensors, N	Number of sensors used in deflection measurement	4
Sensor location (m) (L_i, $i = 1, ..., N$)	Locations of deflection sensors, measured from the center of loading plate	0, 0.3, 0.6, 0.9
Deflection (mm) (D_i, $i = 1, ..., N$)	Measured surface deflections	0.532, 0.411, 0.325, 0.235

Screen display before keying in input

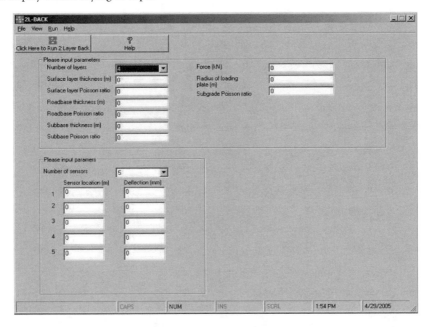

Screen display after input

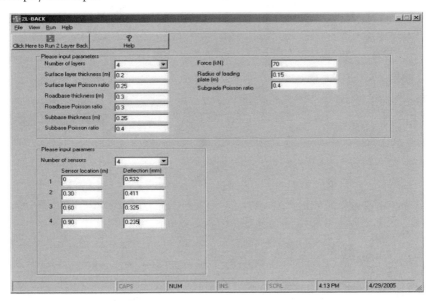

20E.4   Output

- Elastic modulus of surface (GPa)
- Elastic modulus of base (MPa)
- Elastic modulus of subbase (MPa)
- Elastic modulus of subgrade (MPa)

Screen display after running 2L-BACK

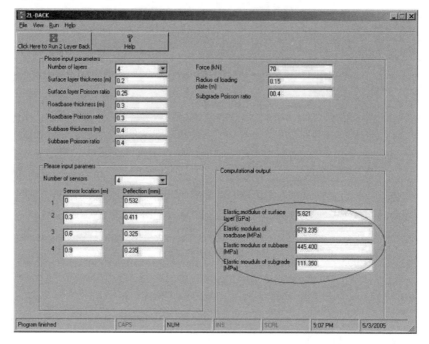

20E.5   Example

Input

> Applied load = 70 kN
> Radius of loading plate = 0.15 m
> Surface layer thickness = 0.2 m
> Surface layer Poisson's ratio = 0.25
> Base thickness = 0.3 m
> Base Poisson's ratio = 0.3
> Subbase thickness = 0.25 m
> Subbase Poisson's ratio = 0.4
> Subgrade Poisson's ratio = 0.4

Deflection data:

> 0.532 mm at 0 m radial distance
> 0.411 mm at 0.3 m radial distance
> 0.325 mm at 0.6 m radial distance
> 0.235 mm at 0.9 m radial distance

Output

> Elastic modulus of surface layer = 5.821 GPa
> Elastic modulus of base = 679.235 MPa
> Elastic modulus of subbase = 320.131 MPa
> Elastic modulus of subgrade = 111.350 MPa

21

Pavement Skid Resistance Management

Kieran Feighan
PMS Pavement Management Services Ltd.
Dublin, Ireland

21.1 Introduction

The measurement and management of pavement skid resistance is of critical importance to highway agencies worldwide. The key point of interest is the frictional resistance generated between the vehicle tires and the pavement surface that allow a moving vehicle to be brought to a stop. The greater the frictional resistance mobilized, the quicker the vehicle can be slowed or stopped.

There have been many improvements in tire technology over the past 30 years, including advances in tire compounds, tread patterns, and tread depths. All of the improvements are aimed at ensuring maximum contact between the tire and the pavement surface at all speeds, with generation of high frictional resistance under braking conditions.

From a road agency standpoint, the aim must be to supply and maintain pavement surfaces that provide maximum frictional resistance to braking tires. In frost, snow, and ice conditions, this is achieved by clearing the road of the temporary low friction surfacing. Road gradients and crossfall are designed to reduce or eliminate the accumulation of water ponding on the pavement surface; if the water thickness is too great for the tire to make contact with the pavement surface, hydroplaning can occur where the tire is effectively sliding along a water surface with almost no ability to generate frictional resistance. In this chapter, we are dealing with situations where there is contact between the tire and the pavement surface. Issues of winter maintenance and geometric design affect skid resistance but are outside the scope of this chapter.

21.1.1   Factors Influencing Skid Resistance

There are many factors that affect the skidding resistance provided by any particular pavement surface. A TRL report (Hosking and Woodford, 1976a) quantifies many of the pertinent factors. The most critical factor is whether the pavement surface is wet or dry. Under completely dry conditions, resistance to skidding is very high, and the frictional resistance as measured by the SCRIM machine (discussed in detail in Section 21.2) is almost identical for a wide range of different surfaces. There is a very significant drop in skidding resistance when the surface becomes even slightly wet. There are further smaller drops as the water film thickness on the surface increases. Surfaces showing identical skidding resistance in dry conditions exhibit a wide range of skidding resistance under wet conditions. Accordingly, most field skid resistance measurement programs are carried out under self-wetting conditions to test the lower range of skid resistance provided by a wet road surface.

Frictional resistance (as measured by testing equipment) has been found to fall as the speed of testing increases. Figure 21.1 illustrates the nature of the relationship, with relatively steep falls in frictional resistance at lower speeds, and much shallower gradients at higher speeds.

Seasonal effects also can have a very significant role on the frictional resistance provided by a given pavement surface. The frictional resistance of a wet road surface is greater in winter than in summer. The proportion of the time that the road is wet seems to influence the variation in frictional resistance more than the actual temperature. Thus wet winters produce higher frictional resistance on a wet road surface than cold dry winters.

The main difference between summer and winter conditions relates to changes in microtexture. It is believed that in dry summer conditions, heavy traffic polishes the aggregate faces using dust and other fine materials on the road. In wet conditions, the dust is washed away, and in wet winter conditions, the presence of salt and grit works to improve the microtexture. As this polishing or gritting action involves traffic, and in particular heavy commercial vehicles, greater volumes of traffic on a particular road section will lead to greater variation in skid resistance over an annual cycle. Higher volumes of heavy vehicles also generally lead to lower average skid resistance values compared to similar surfaces carrying lower volumes of traffic.

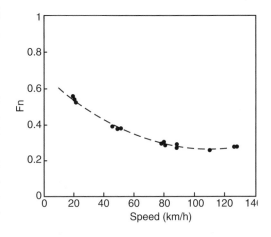

FIGURE 21.1   Frictional resistance versus speed.

Temperature also exerts a considerable influence on the variability in skidding resistance results. Resistance to skidding tends to decrease as the temperature increases, due mainly to temperature effects on the material properties of the vehicle tires. This further increases the pattern of skidding resistance being lower in the summer than in the winter.

The age of the pavement surface can also be a significant factor. New pavement surfaces relying on exposed aggregate faces for skid resistance provision tend to polish under traffic action most rapidly in the first year, leading to a noticeable decrease in skidding resistance. However, the rate of decrease in skidding resistance levels off rapidly after the first year. This pattern is reversed for some of the newer polymer-modified bitumen surfacing. Initial skid resistance values can be quite low as the binder rather than the aggregate is providing the surface at the tire/surface interface. Under traffic action over the first 6 to 12 months, the binder skin is abraded, exposing the aggregate surface, and leading to a noticeable increase in skidding resistance.

Finally, contamination of the pavement surface with oil or other lubricants, grit, clay, or other organic materials; and the deposition of small particles of tire rubber over time can all lead to very significant drops in pavement skid resistance on any given pavement.

21.1.2   Relationship between Skid Resistance and Accidents

The horrendous social and economic costs associated with road accidents must be minimized by road agencies and maintenance of adequate skid resistance throughout the agency's network is a key requirement in this regard. There is also the reality for road agencies of the growth in litigation in relation to accidents worldwide. The ability to demonstrate standards, policies, and monitoring of road networks to ensure adequate skid resistance is becoming essential in this context as well as in the broader public interest.

It has been estimated that over one third of all wet road accidents are related to road surface characteristics (Rado, 2000). A U.K. study showed over 25 percent of all wet road accidents are related to skidding conditions (Kennedy et al., 1990).

A classic piece of research in the U.K. attempted to establish a link between frictional resistance and accident frequency by using the seasonal variation in frictional resistance and examining if accident frequency varied in line with the seasonal variation (Hosking, 1986). The main findings of the research were:

a very highly significant correlation was found between month-to-month changes in wet-road skidding resistance and wet-road skidding rates for all categories of vehicles examined (cycles, motorcycles, cars, light and heavy trucks),

the estimated reduction in skidding frequency per unit change in skidding resistance was similar for most categories,

the frictional resistance was measured with the SCRIM machine. It was estimated that an average increase of 0.1 in the SCRIM coefficient can be expected to lead to a 13 percent reduction in wet-road skidding rate, seasonal changes in traffic volumes and light conditions have a relatively small effect on skidding rates,

higher-speed roads have a higher rate of skidding than lower-speed roads,

heavy goods vehicles have a higher rate of skidding than cars and light goods vehicles.

Another excellent research study examined the relationship between the surface texture of roads and accident rates (Roe et al., 1991). Surface texture depth was measured on a large number of in-service roads using a laser-based texture meter. Frictional resistance was measured on the in-service roads using the SCRIM. The study was concerned primarily on relatively high-speed roads with speed limits in excess of 64 km/h. The roads tested were primarily surfaced with bituminous materials.

The study found that the texture depth measurements (representing surface macrotexture) and SCRIM results (representing surface microtexture) were not correlated — the two parameters were effectively independent of one another. It was concluded overall that accidents of all types tended to occur on sites with low textures, with relatively few at the higher end of the distribution. It further found that all categories of accident, i.e., dry and wet, skidding and non-skidding accidents, were similarly affected by low texture depth. The texture depth level below which accident risk begins to increase significantly was estimated to be approximately 0.7 mm sensor-measured texture depth (SMTD).

When examining links between microtexture (as measured by SCRIM) and accident rates, it was concluded in this study that microtexture was not having as dominant an effect on accidents as was macrotexture. This can be seen as contradicting the findings of the 1986 TRL report. However, the in-service roads included in the study had a relatively narrow spread of SCRIM results, with very few low SCRIM values and relatively few very high SCRIM values. This in turn resulted from the adoption of minimum SCRIM and Polished Stone Values (PSV) in the U.K. from 1988.

21.1.3   Microtexture and Macrotexture

In the provision of pavement surfaces with adequate skid resistance, there are two primary parameters that need to be provided and measured — microtexture and macrotexture. Figure 21.2 shows a diagrammatic representation of microtexture and macrotexture.

Microtexture is primarily a property of the individual pieces of aggregate making up the road surface. The microtexture of the aggregate controls the contact between the tire rubber and the pavement surface. The microtexture characteristics of an aggregate are governed by the crystalline nature of the aggregate itself. From a skid resistance viewpoint, these characteristics are primarily measured by some type of polishing test to determine the ability of the aggregate to retain relatively rough microtexture under the abrasive action of tire forces.

In the provision of sufficient frictional resistance to bring a moving vehicle to a stop, microtexture provides the primary source of frictional resistance at low speeds (less than 50 km/h) due to the direct interaction between the tire rubber and the aggregate surfaces. Accordingly, in urban areas and other areas within speed limits, the microtexture properties of the surface are of primary concern.

The macrotexture of the pavement surface refers to the coarser texture defined by the shape of the individual aggregate chips and by the spaces between the individual aggregate chips. Macrotexture is particularly important in relation to wet conditions. In traditional surfacing materials, the individual aggregate chips provide a capability to penetrate a water film on the pavement surface, while the spaces between the aggregate chips provide drainage paths for the water to be dispersed.

It has also been suggested (Roe et al., 1991) that macrotexture plays a critical role in tire hysteresis. Hysteresis or distortion of the tire surface occurs as the surface of the tire passes over projections and depressions in the pavement surface. The primary means of reducing the kinetic energy of a moving vehicle is through the vehicle braking system, where the kinetic energy is converted into heat and dissipated. Hysteresis effects, while much smaller than the braking/heat conversion mechanism, also convert kinetic energy into heat, with the tire surface heating up. Roe et al., claim that on a surface with coarse macrotexture, hysteresis losses make a significant contribution to energy dissipation before skidding occurs, and this contribution in turn reduces the likelihood of an accident and the severity of the accident collision if one occurs.

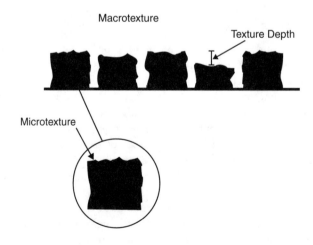

FIGURE 21.2   Microtexture and macrotexture.

The particles protruding above the plane of the surface provide positive macrotexture. This type of positive texture is provided in the various forms of surface dressing, specialist surfacing such as Hot Rolled Asphalt, and in the texture provided by brushing a concrete surface for example.

Newer materials such as porous asphalt and other thin-surfacing materials provide negative texture. Under this model, the texture is composed of voids below the plane of the road surface. The voids may or may not be interconnecting. In a porous asphalt material, the voids are generally interconnecting, so that water can drain laterally below the pavement surface. Some of the thin-surfacing materials provide voids that act as a storage compartment for surface water, but the degree of interconnectedness is substantially lower than porous asphalt. Grooving of concrete is also a form of negative macrotexture provision.

In all cases, the aim is to provide sufficient macrotexture to allow a safe and rapid dispersion of water away from the interface of tire surface and pavement surface. Macrotexture is the dominant factor in providing adequate frictional resistance at higher vehicle speeds (>50 km/h). However, it must be emphasized that microtexture does have some influence on frictional resistance at higher speeds, and macrotexture also has an influence at lower speeds. A recent study in the U.K. (Roe et al., 1998) emphasizes this, pointing out that the level of high-speed friction of surfacing depends on the microtexture, which provides low-speed friction and also that texture depth (a measure of macrotexture) has an impact on friction loss at low speeds. This is an important point as traditionally it has been assumed that measurement of macrotexture in urban and other low-speed locations is not important.

Ultimately it is the provision of proper microtexture and macrotexture that is required for adequate surface frictional resistance over the full spectrum of vehicle speeds. A vehicle traveling at high speed in wet conditions relies in the first instance on the tire tread depth and pattern to disperse the water film continuously, ensuring consistent contact with the pavement surface. When the vehicle brakes, the driver is relying on the pavement surface to provide adequate frictional resistance to the tires. At high speeds, the macrotexture of the pavement surface provides drainage passages for the water, and provides particle to tire rubber contact. As the vehicle slows due to the frictional resistance, the interaction between the tires and the individual aggregate faces (the microtexture) becomes more and more important in mobilizing the required frictional resistance to stop the vehicle.

Microtexture by definition exists on a microscopic scale, dealing as it does with the crystalline structure of the aggregate. The PIARC International Experiment to compare and harmonize texture and skid resistance measurements (Wambold et al., 1995), defines microtexture as existing in the range of peak-to-peak amplitudes of 0.001 to 0.5 mm. Under the same definitions, macrotexture exists in the range of peak-to-peak amplitudes of 0.5 to 50 mm. For completeness, megatexture relates to pavement longitudinal profile (roughness) and exists in the range of peak-to-peak amplitudes of greater than 50 mm. Measurement and assessment of megatexture is outside the scope of this chapter.

21.1.4 Scope of Chapter

In Section 21.2, the various types of equipment used to measure frictional resistance will be described. Section 21.3 covers the measurement of texture, both microtexture and macrotexture. Section 21.4 covers the International Friction Index (IFI), an attempt to standardize a single approach to frictional resistance measurement over many equipment types. In Section 21.5, development of skid resistance policies for individual sites is outlined, covering definitions of investigatory and intervention levels, and drawing on experience in road agencies worldwide. Section 21.6 covers the development of network management strategies to implement skid resistance policies. Finally, Section 21.7 examines maintenance options available to road agencies to improve skid resistance on bituminous and concrete road pavements.

21.2 Measurement of Frictional Resistance

In the case of a braking, moving tire on a pavement surface, the measurement of the resistance to movement at the interface between the tire and pavement surfaces is the friction coefficient. The friction coefficient is defined as the force resisting motion (the tire/pavement interface frictional force) divided by the vertical load. The higher the friction coefficient, the greater the force mobilized at the interface to resist the vehicle movement, and in turn, the vehicle can be slowed or stopped more rapidly.

However, it is known that the friction coefficient between a rubber tire and a pavement surface is not a constant. The friction coefficient is dependent on many factors. These include the tire compound properties, the tire tread pattern and tread depth, the pavement surface microtexture and macrotexture, and the presence of surface water or other lubricants or contaminants. In particular, for a given tire and pavement surface, the presence of a thin film of water significantly reduces the coefficient of friction.

Slippage occurring between the tire and pavement surface also affects the frictional resistance generated. The percent slip is defined by the following equation (ASTM E1859-97)

$$S = \frac{v - v_r}{v} \times 100\% \tag{21.1}$$

where:

    S = the percent slip,
    v = velocity of the test vehicle,
    v_r = rotational velocity of the test tire.

The slip speed in turn is defined by

$$s = v \frac{S}{100} \tag{21.2}$$

where:

    s = slip speed,
    S = percent slip (from Equation 21.1 above).

Thus, in a fully locked wheel sliding along the surface, the rotational velocity of the tire = 0 as the tire is not spinning at all, the percent slip is 100% and the slip speed is equal to the velocity of the test vehicle. On low-slip devices, where v_r is significantly greater than 0, the percent slip is much lower than 100% and the slip speed is much lower than the velocity of the test vehicle.

The friction coefficient varies with speed, increasing very rapidly from zero at 0 km/h to a maximum value at a relatively low slip speed, termed the critical slip speed value, and then decreasing gradually depending on the slip speed of the wheel. Figure 21.3 illustrates this general relationship.

Between zero and the critical slip speed, the tire is the dominant influence in the friction slip speed curve. Beyond the critical slip speed (where the friction coefficient is maximum), the pavement surface is the dominant influence in the friction slip speed curve.

Some friction measurement devices operate at low slip speeds, and for these devices, the tire characteristics have a large influence on the results obtained. Accordingly, variations in tire character-

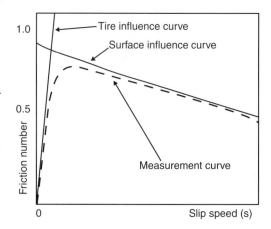

FIGURE 21.3   Friction number versus slip speed.

istics can influence the results very significantly, and the tires used in testing must be specified and controlled to very tight specifications. Other types of friction measurement equipment operate at relatively high slip speeds, and for these devices, the results obtained are primarily surface-influenced rather than tire-influenced. Variations in tire properties will affect the results less than on the low slip speed devices (Rado, 2000).

Given that the influence of the surface on frictional resistance is of paramount importance to the road agency, it is reasonable to ask why low slip speed devices are used. The discussion on microtexture and macrotexture in the introduction to this chapter is relevant here. The results from low slip speed devices in general are capturing a greater part of the microtexture influence of the pavement surface, while the results from high slip or locked wheel devices are capturing the combined effects of microtexture and macrotexture. If a macrotexture-measuring device is used in conjunction with a low slip speed device, separate parameters on microtexture and macrotexture performance may be obtained, giving greater scope for targeted use of the results.

There are four categories of friction-measuring devices that can be used to measure skid resistance on pavement surfaces. They are:

Locked wheel devices
Fixed slip devices
Variable slip devices
Sideway-force devices

21.2.1   Locked Wheel Testers

Equipment using this method measures the force on a wheel that is sliding but not rotating. This is expressed as a 100% slip condition, and the slip speed is the same as that of the test vehicle. A brake is applied to the test wheel, causing the wheel to become fully locked. Once the wheel is fully locked, the force measurement system measures and averages the vertical and longitudinal forces over a 1 sec period.

Equipment based on this principle is most commonly used in the United States where all 50 states use locked wheel data in the measurement and management of skid resistance. A number of European and Asian countries also use locked wheel devices. The equipment specification is outlined in an ASTM standard (ASTM E274-97).

The test wheel may be built into the testing vehicle, or may be part of a trailer towed by the testing vehicle. The vast majority of locked wheel testers are trailers with two testing wheels built into the trailer. The test wheel is equipped with a standardized test tire (ASTM E501 for a standard rib or treaded tire, E524 for a standard smooth [untreaded] tire). The locked wheel tester can be operated in wet or dry conditions, but for typical monitoring purposes, it is operated in wet conditions, with water supplied from a storage tank incorporated into the testing vehicle at a specified rate. Water storage requirements are typically 200 to 500 gallons. The testing can be carried out at any speed, with the ASTM standard requiring an ability to maintain any given test speed between 65 and 100 km/h to the specified test speed ± 1 km/h while the test wheel is locked. The ASTM specified standard test speed is 65 km/h.

The procedure may be summarized as follows. The test vehicle/trailer is brought up to the desired test speed. Water is delivered ahead of the test tire, and the braking system locks the test tire. The resulting frictional force is measured by some type of force transducer. The speed of the testing vehicle/trailer is also simultaneously measured. The reported output under the ASTM standard is the skid number (SN), defined as the force required to slide the locked test wheel at the stated speed, divided by the effective vertical load applied to the test wheel and multiplied by 100.

There are a number of key parameters in the design and development of locked wheel devices. The specified vertical static load on the test wheel should be maintained during testing. The braking system on the test wheel must be capable of keeping the wheel in locked condition throughout the test duration. The water supply system must be capable of delivering a volume of water that is directly proportional to the test speed. The force-measuring transducer must be capable of accurate and repeated measuring the drag force on the wheel. It is desirable to also accurately measure the vertical force being applied at the time of measurement given that the SN is defined by the ratio of horizontal (drag) force to vertical force. The vehicle speed at the time of testing must also be very accurately measured so that measurements can be corrected to a standard speed.

ASTM E274 lays out tolerances on all of the above parameters. The vertical load under the ASTM standard is 4800 N, ± 65 N. The water supply system must deliver 0.6 l/min per mm of wetted width, and the total wetted width must be at least 25 mm wider than the test tire width. A nominal water film thickness of 0.5 mm is usually used. There is a tolerance of $\pm 10\%$ on the water supply quantity. The force measuring transducers should have a tolerance of $\pm 1\%$ of the measured force. The speed measuring system should have an accuracy of $\pm 1.5\%$ of the measured speed. Clearly, the tolerances are very tight, reflecting the need for the equipment to generate repeatable and reproducible results in the field.

The tires used in testing are also mentioned with exacting specifications. On ribbed or grooved tires, a total of six grooves are specified, each 5.08 mm wide and 9.8 mm deep. The tire tread width for both ribbed and smooth tires is 148.6 mm. There are six wear indicators (lines) spaced around the tire circumference. If the tire is damaged in testing, or if the tire is worn down to the wear line, the tire can no longer be used as a standard test tire. The test tire inflation pressure is 165 kPa. New tires must be conditioned by running them at normal traffic speeds on the test vehicle for at least 300 km before they can be used for standardized testing purposes.

In carrying out the locked wheel test, the apparatus is brought up to the desired speed, and water is delivered at the specified rate ahead of the wheel. The brake is then applied to the test wheel to lock the wheel completely. The test measurement of horizontal and vertical force is carried out over an interval of between 1 and 3 sec, (usually close to 1 sec) after full wheel lock-up is achieved, and the average values of horizontal and vertical force are used to calculate the skid number.

A five-stage test cycle is carried out in practice by most automated locked wheel testers. Once the test vehicle is at the desired test speed, the stages are (Roe et al., 1998).

Start water pump before testing (0.5 sec)
Brake test wheel from rolling to locked (lock-up must occur in less than 2 sec)
Allow locked test wheel to settle (0.5 sec)
Calculate locked wheel friction (over at least 1 sec)
Allow test-wheel to spin back up to vehicle speed (0.5 sec)

The load and drag forces are measured at 0.01 sec intervals and averaged over the 1 sec interval in stage 4.

The main advantage of a locked wheel tester is that the slip speed of the locked wheel is the same as the testing vehicle speed. In this way, it is measuring a frictional resistance similar to that experienced by a member of the traveling public braking on the same pavement surface. The main disadvantage is that locking the wheel puts a very high stress on the testing tire, and it is not practical or realistic to carry out continuous measurements of skid resistance over a road network. Other types of equipment were developed at lower slip speeds to allow this type of continuous measurement to be carried out over the entire length of a road section or road network.

21.2.2   Fixed Slip Devices

Fixed slip devices can carry out continuous measurement of frictional resistance provided a small slip ratio is selected. Fixed slip devices usually operate between 10 and 20% slip. Thus, for a fixed slip device operating at a slip ratio of 20%, if the test vehicle is moving at 65 km/h, the device is measuring at a tire speed of $65 \times 0.2 = 13$ km/h. Clearly, these devices are measuring low speed frictional resistance, and in turn are reflecting to a large extent the microtexture of the pavement surface.

The Griptester device is representative of fixed slip devices. The operation of this device is covered by a British Standard specification (BS 7941-2). The Griptester is a small trailer-towed device, approx. 1 m in length and 0.8 m in width. It has been used on roads, airports and helipads, and because of the small size, it can also be operated in manual (pushed) mode in areas inaccessible to vehicles such as, footpaths, platforms, industrial floors and road markings. The Griptester is shown in Figure 21.4.

In the roadway context, the Griptester can be operated at vehicle speeds between 5 km/h and 130 km/h, subject to the road being sufficiently smooth that the trailer is in constant contact with the road surface.

FIGURE 21.4 Griptester in operation.

Frictional results are measured continuously along the road section length. Results can be averaged and summarized over specified intervals, usually over either 5 or 10 m intervals for road applications.

The trailer has three wheels in total, arranged in a tricycle formation. All three wheels have a 250 mm diameter, and are in constant contact with the roadway surface. The trailer has two drive wheels with standard wheels and patterned tires. The drive wheels support 75% of the trailer weight. There is a single test wheel mounted on a stub axle. The test wheel has a smooth tread. The test wheel is mechanically braked by a gearing system linked to the drive wheels. Through this gearing system, a fixed slip ratio between the drive wheels and the test wheel is maintained. On the Griptester, the percent slip is maintained at just over 15%.

On roadways, testing is carried out in wet conditions. The water supply is carried in the load bay of the towing vehicle. Water supply flow rate is controlled by pumps and flowmeters to deliver the specified water film thickness. As in the case of the locked wheel tester, flow rate is varied with speed to achieve a constant film thickness.

Vehicle speed and distance measurement are carried out by sensors attached to the trailer drive wheels. Vertical and horizontal (drag) forces are measured by strain gauge sensors attached to the stub axle mounting of the test wheel. All sensor readings are sent to and stored on a data capture device in the towing vehicle.

The vertical force on the Griptester test wheel is much lower than in the locked wheel device at approximately 250 N under normal operating conditions. Testing speed can be varied. In the United Kingdom it is usually carried out at 50 km/h. Test tires are sourced through the manufacturer of the Griptester. They have wear guides incorporated, and the test tire must be replaced when the tire has worn down to the wear line, or if it is damaged in any way.

Before carrying out standardized testing, the Griptester must be operated over at least 500 m with the water deposition system in operation. In routine testing operation, the operator maintains the towing vehicle at the standard speed, and the operating system continuously records the sensor outputs and produces summary GripNumber outputs averaged over specified intervals. The GripNumber calculated is similar to the Skid Number for locked wheel devices, being the ratio of horizontal drag force to vertical load force.

21.2.3 Variable Slip Devices

Variable slip devices are designed to measure at any specified slip. They can typically also carry out automated measurements through a range of specified slips, and can also be programmed to seek a maximum friction level over the range of slip values.

The Roar device, manufactured by Norsemeter in Norway, is the most commonly used variable slip device. It is a trailer-towed device, relatively similar in scale to the Griptester. More generally, it falls under the ASTM specification, E1859 (ASTM E1859-97). Equipment covered under this specification should be able to operate in fixed slip or variable slip modes.

Essentially, the rotational velocity of the test wheel can be controlled and maintained at a fixed slip ratio gradient, similar to the fixed slip devices. However, variable slip devices can maintain a wide range of slip ratio gradients through feedback sensors measuring the rotational velocity of the test wheel and interacting with the braking mechanism on the test wheel. The brake system is specified to be capable of controlling test wheel rotational velocities to within $7 \pm 2\%$ for any specified slip ratio.

When the test trailer is moving at test speed, commonly 65 km/h, and the test tire is unbraked, the rotational speed of the tire is equal to the test speed. When the brake is applied, the rotational speed of the tire decreases. In the extreme case when the test wheel has zero rotational speed, the test wheel is skidding along the road. This is basically a locked wheel situation. In reality, the slip speed is controlled to a value close to but greater than zero to avoid excess and uneven wear on the test tire.

The slip friction number at any slip speed is the ratio of longitudinal friction force to the vertical load force, as in all friction measurement devices. When, the slip friction number is plotted against slip speed over a range of slip speeds, the peak slip friction number is the maximum value recorded. This is usually at a slip speed close to zero (i.e., close to a fully locked wheel condition). The great value of the variable slip device is the ability to produce this relationship between slip friction and slip speed over a wide range of slip speeds — all other devices are measuring at a single slip speed.

With the feedback system in place, any given slip ratio can be specified. For example, in New Zealand, the Roarmeter is used within a specified 34% fixed slip configuration to give good correlation with Griptester and SCRIM results. Comparisons with locked wheel trailers can be made using calculations based on the peak slip friction number. The variable slip devices are also very useful if the IFI is to be calculated, as the output generated from the variable slip devices is directly used in IFI calculations, and the underlying approach to the IFI derivation is closely related to the variable slip device approach.

21.2.4   Sideway Force Devices

Sideway force devices apply a different principle to the fixed slip devices with the same aim in mind   to allow continuous measurement of the frictional resistance of the pavement. Sideway force systems maintain the test wheel with its vertical plane at an angle to the longitudinal plane of the test vehicle. The test wheel is allowed to rotate freely, and the test wheel is maintained in a controlled slipping condition by virtue of the angular difference between the plane of the test wheel and the plane of the test vehicle.

The relative velocity between the test tire and the pavement surface is quite low, as the relative velocity is calculated as ($V \, sin\theta$) where V is the velocity of the vehicle, and θ is the angle between the plane of the test wheel and the longitudinal plane of the test vehicle. The SCRIM used in the United Kingdom, and described below, has a vehicle test speed of 50 km/h, and a 20 degree offset angle on the test wheel, yielding a sliding speed of the test wheel of 17 km/h. A schematic is shown in Figure 21.5.

21.2.4.1   SCRIM

The most common type of equipment using sideway force as a measure of frictional resistance is the Sideway force Coefficient Routine Investigation Machine (SCRIM). The Mu-Meter, used in frictional resistance measurement on airports, is also based on the sideway force principle. As the SCRIM is by far the most commonly used equipment for road measurement, it is described in detail here. The SCRIM is pictured in Figure 21.6.

The SCRIM is covered by a British Standard (BS 7941-1). The equipment was produced originally in the early 1970 s in Britain for measuring the skid resistance of the road network. The equipment is mounted on a truck chassis, with a large capacity water tank (approximately 4000 l). It is specifically designed to allow large lengths of road network to be continuously tested. Daily outputs of approximately 200 km can be expected using the SCRIM.

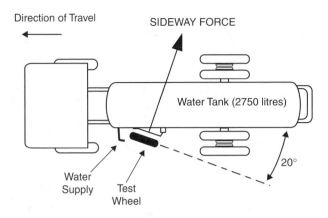

Direction of Travel

SIDEWAY FORCE

Water Tank (2750 litres)

20°

Water
Supply

Test
Wheel

FIGURE 21.5 SCRIM diagram.

The SCRIM can be either single-sided or double-sided. Double-sided machines allow testing in the inner and outer wheel paths simultaneously. The test wheel is contained within a loading frame that is mounted mid-machine, between the front and rear axles of the truck. Two vertical shafts are mounted within the frame, and the test wheel assembly moves up and down on the vertical shafts.

The test wheel assembly, shown in Figure 21.7, consists of the test wheel, a single damper/spring suspension unit, connections to the vertical shafts and load-cells to measure the horizontal and vertical force. The load-cell to measure vertical force is a very recent addition — previously the mass of the test assembly was assumed to be providing a constant vertical force of 200 N, and only the horizontal drag force was being measured. The vertical plane of the test wheel is fixed at 20 degrees, with a tolerance of $+1$ and -0.5 degrees, to the line of the truck chassis. The test tire is allowed to rotate freely in its own plane, which ensures even wear on the test tire. Distance measurement to a high degree of accuracy is carried out using a distance measuring instrument (DMI) attached to the truck axle. As the SCRIM is used for network testing, accurate positioning of the data is essential. Geo-referencing of the data in real-time using GPS equipment is also common with the SCRIM.

When the test vehicle moves forward along the road section, the test wheel is rotating, but slides in the forward direction due to the angular difference. The standard test speed for SCRIM is 50 km/h in

FIGURE 21.6 SCRIM in operation.

FIGURE 21.7   SCRIM test wheel apparatus.

the United Kingdom, with a permitted range of operation of 30 to 67 km/h. Readings are speed-corrected back to the standard test speed. On bends with low radii of curvature (less than 100 m) and on roundabouts, testing is carried out at 20 km/h.

BS 7941 Part 1 calls for a theoretical water film thickness of 2 mm for the SCRIM. Using the prescribed water delivery system and nozzle, this equates to a water discharge rate of 0.95 l per sec based on the standard test speed of 50 km/h. At higher testing speeds (55 to 65 km/h), a water discharge rate of 22.2 l per sec is called for. Typically, approximately 50 km of roadway can be tested before the water tank must be refilled. The water supply valve is connected to the raise/lower mechanism of the test wheel to ensure that water is flowing before the test wheel is in contact with the ground and to ensure that water flow stops after the test wheel is raised off the ground.

The test tire is a pneumatic, tubed tire with a smooth tread. The tire is inflated to 3.5 kg/cm^2 at ambient temperature. New tires must be conditioned by operating over a length of at least 2 km before use in standardized testing. There are wear indicators built into the tire, and the tire must be discarded if the tire wear exceeds 3 mm, or if the tire is damaged in any way.

The horizontal and vertical forces are measured continuously when the test wheel is in contact with the ground. The ratio of sideway force to vertical force is referred to as the sideway-force coefficient (SFC). A SCRIM reading (SR) is output for each subsection of road tested; subsections can be 5, 10, or 20 m in length. Typically, SR values are calculated at 10 m intervals on rural roads and 5 m intervals on urban roads. The SR value is the average SFC value over the subsection length, multiplied by 100. The SR is expressed as an integer value. The SR values coming directly from the machine are raw values, uncorrected for speed, temperature, seasonal adjustment, etc. In the United Kingdom, the value used for assessment of frictional resistance and comparison with skid resistance standards in operation is the SCRIM coefficient (SC). Effectively, this is an SFC value, corrected for speed, temperature and machine variability. It is a decimal fraction, being a corrected ratio of sideway force to vertical force, and is expressed as a decimal fraction to two decimal places. Care must be taken when comparing values derived from SCRIM testing to determine if the values are expressed as SFC, SR, or SC.

Calibration of the SCRIM is required on a routine basis. Static calibration of the horizontal load sensor is carried out using a rolling trolley calibration rig that can apply a range of horizontal loads in the plane of the test wheel. Distance calibration over a known length of at least 1 kilometer is carried out on a regular basis.

Dynamic calibration of the equipment output to determine consistency of results over time is carried out at least once per week during the testing season. BS 7941 recommends identification of a check site with separate sections of low, medium and high levels of skidding resistance, or alternatively separate sites with these characteristics. Sections should be at least 100 m long, reasonably straight and flat and with a reasonably smooth pavement profile. The calibration section is tested at 50 km/h using a 10 m subsection length. The average value for each calibration run should not differ by more than 5 units (in terms of SR) from the previous average on the same section.

Countries currently using side force systems for measurement of network skidding resistance include the United Kingdom, Ireland, Belgium, France, Italy, Spain, Germany and Denmark in Europe, Australia and New Zealand, and Singapore and Malaysia in Asia. The Belgian and French devices use a testing setup with the test tire at a 15 degree angle to the longitudinal plane of the truck chassis. The Mu-Meter, used primarily for airport testing, has two test wheels mounted at an angle of 7.5 degrees to the longitudinal plane of the trailer chassis.

As previously indicated, the sliding speed of the test wheel is quite low, approximately 17 km/h on the United Kingdom devices. Clearly this assists greatly in enabling continuous assessment of the frictional resistance of the road network without unsustainable rates of tire consumption. However, at such low tire/pavement interface speeds, the friction measuring system is primarily susceptible to the influence of microtexture. As a result, road agencies using SCRIM as a monitoring tool frequently also require the measurement of macrotexture on a routine basis. The details of macrotexture measurement will be outlined in Section 21.3. At this point, it is sufficient to point out that many SCRIM machines have been fitted with a laser macrotexture measurement system to allow simultaneous measurement of SCRIM values and macrotexture values.

21.2.5 Summary

There are basically four types of friction-measuring devices. Fixed slip devices and sideway force measurement devices produce relatively comparable outputs as they are measuring at relatively low tire/pavement surface interface velocities. At these relatively low interface velocities, pavement surface microtexture has a large influence on the values recorded. These two device types induce relatively low tire wear, and are suitable for assessment of long, continuous stretches of pavement.

Locked wheel systems carry out measurements at much higher tire/pavement surface interface velocities, with typical values at 65 km/h. In these systems, the test wheel is traveling at the same velocity as the test vehicle, yielding a 100% slip condition. At these speeds, the friction results are heavily influenced by the pavement macrotexture. These devices are very hard-wearing on the test tires by the nature of the testing regime, and continuous testing of long stretches of pavement is not feasible using this approach.

Variable slip devices have the capability of measuring at any specified slip. They can be compared with fixed slip and sideway force devices at low slip rates, and locked wheel systems at high slip rates.

The most suitable device to be used for friction measurement is primarily dependent on the use to which the friction measurement will be put, in conjunction with data and standards accumulated over time by the road agency. Further discussion of these factors is included in Sections 21.5 and 21.6. At this stage, it is sufficient to stress that the output from one device is not necessarily correlated with the output from another device type.

21.3 Measurement of Texture

The frictional resistance generated at the tire/pavement surface interface is dependent on both the microtexture and macrotexture of the pavement. Accordingly, it is desirable to be able to measure both properties, and to incorporate suitable minimum values into specifications for pavement surfacing. PIARC (Wambold et al., 1995) have defined microtexture, macrotexture and megatexture in global terms based on measurements of amplitude and wavelength. These definitions are independent of any particular test method.

Macrotexture is defined by PIARC as the deviation of a pavement surface from a true planar surface in the texture wavelength range of 0.5 to 50 mm. The peak-to-peak amplitudes of the wavelengths are normally in the range of 0.01 to 20 mm. This type of texture is visible to the naked eye, and is primarily representative of the patterns observable on the pavement surface and on the tire surface.

Microtexture is defined by PIARC as the deviation of pavement surface from a true planar surface in the texture wavelength range of less than 0.5 mm. Peak-to-peak amplitudes of these wavelengths are normally in the range of 0.001 to 0.5 mm. There is some overlap in the amplitudes characteristic of macrotexture at the lower range, so it is the combination of wavelength and amplitude that distinguishes the two types of texture. Under this definition, microtexture is generally not observable with the unassisted eye. In relation to the aggregates making up the pavement surface, variations in microtexture can be observed under a microscope, and can also be distinguished by touch, feeling relatively harsh or smooth under fingertip touch.

For completeness, the PIARC definitions of megatexture and unevenness are also included here to cover the entire range of wavelength/amplitude characteristics of pavement surfaces. It must again be stressed that, in relation to frictional resistance characterization and measurement, only microtexture and macrotexture are relevant.

Megatexture is defined by PIARC as the deviation of the pavement surface from a true planar surface in the texture wavelength range of approximately 50 to 500 mm. Peak-to-peak amplitudes of these wavelengths are normally in the range of 0.1 to 50 mm. This texture is characteristic of pavement defects such as potholes and localized bumps.

Unevenness is defined by PIARC as the deviation of the pavement surface from a true planar surface in the texture wavelength range of greater than 500 mm. Wavelengths in this range are completely outside the texture definitions and are in the area of transverse and longitudinal profile measurement. Further information on unevenness (i.e., roughness) measurement is contained in Chapter 19 of this book.

Returning to microtexture and macrotexture, it is desirable to have clear definitions of the mathematical properties so that appropriate test equipment can be developed to measure in the specified ranges. In the profile-measuring equipment, the characteristic wavelengths can be derived from the raw profile. In the measurement of macrotexture and microtexture, the test methods used at the moment are surrogate measurements rather than actual measurements. Test methods to characterize macrotexture primarily measure some form of texture depth. Test methods to measure microtexture measure frictional resistance in the microtexture range.

21.3.1 Texture Depth Measurement

Traditionally, texture depth measurements have been carried out as a measure of pavement macrotexture. The sand patch or volumetric patch test is by far the most commonly prescribed test worldwide. It is covered in numerous national and international specifications including ASTM E965 (ASTM E965-96). The test basically consists of a known volume of standardized sand or small glass spheres being spread out over the pavement surface using a flat disk. The sand or glass spheres are distributed to form a circular patch. A small circle diameter indicates a high average texture depth; a large circle diameter indicates a low average texture depth. The volume of sand (known) divided by the surface area covered by the sand (measured on site) yields the average texture depth. Figure 21.8 shows the volumetric patch test being carried out on site.

ASTM E965 points out that in spreading the material (sand or glass spheres), the surface voids are completely filled to the tips of the surrounding aggregate particles. The plane from which the texture depth is measured is effectively defined by the bottom surface of the spreading disk passing over the tips of the aggregate particles. The volumetric patch method is not suitable for pavements with large surface voids or for use on grooved surfaces.

The main advantage of the volumetric patch test is its simplicity. The equipment is relatively low-cost, the calculations of volume divided by area are straightforward, and results are immediately attainable. The main disadvantages are that the test is slow and labor-intensive, and the results are dependent on the skill of the technician to produce a circular surface area. Traditionally, Ottawa sand was used for spreading, hence the "sand patch" designation. The use of glass spheres rather than sand is now advocated, as the uniformity

FIGURE 21.8 Volumetric patch test.

in particle size and shape is easier to achieve. In addition, the glass spheres are less susceptible to the influence of trace amounts of moisture in the sub-surface voids.

Volumetric patch tests are still used as the standard in specifications worldwide. ASTM E965 recommends that at least four random-spaced measurements of average texture depth be measured, and the arithmetic average value be used to characterize the pavement surface texture depth. Under the relevant British Standard (BS 598-105, 1990), five volumetric patch tests are carried out over a 50 m length, and the average value is used to represent the stretch.

ASTM E965 contains estimates of the precision and bias that can be expected from texture depth measurement using the volumetric patch test. It is estimated that the standard deviation of repeated measurements by the same operator on the same surface can be as low as 1% of the average texture depth, while the standard deviation of the repeated measurements by different operators on the surface can be as low as 2% of the average texture depth. In this author's experience, much higher deviations in practice can occur, particularly with relatively inexperienced operators. There is a misconception that because the test is relatively straightforward, it is also simple and repeatable. There is in fact a lot of operator skill required to produce meaningful and representative results, and this should always be borne in mind in examining the results from volumetric patch tests.

Finally, ASTM E965 notes that the standard deviation of individual measurements within a nominally homogeneous pavement section may be as high as 27% of the average texture depth, even with very high standards of operator repeatability. Thus, a large number of tests may be necessary to estimate the average texture depth reliably where there is large variation in texture. Clearly this has time, cost and potential contractual implications, particularly where acceptance/rejection of pavement surfacing are dependent on volumetric patch test results.

21.3.1.1 Outflow Volumetric Test

There is a widespread recognition that volumetric patch tests (and laser-based texture measurement devices described in the following section) do not fully capture the beneficial texture offered by negative-textured surfacing materials such as porous asphalt and some of the thin surfacing materials. Undoubtedly such materials offer greater drainage capabilities than the volumetric patch test results on such materials would indicate. There is also a significant amount of variability in results on such materials using the traditional tests for texture depth.

The outflow volumetric tests attempt to capture the texture and drainage capabilities of these negative-textured materials more realistically. A clear cylinder with a known volume of water is placed on the pavement surface. A rubber seal between the edge of the cylinder and the pavement surface ensures that all water draining from the cylinder is draining vertically through the pavement surface — it is then free to drain laterally below the pavement surface through interconnected voids. The cylinder is filled with water, and the time for a known volume to drain out (usually measured between two lines marked on the cylinder) is measured using a stopwatch. The faster the water drains away, the greater the negative texture and effective texture depth in the material, as it is the texture that determines the outflow rate. This type of test is not practical on relatively impervious surfaces, and is intended for pavement surfaces with high drainage potential.

21.3.1.2 Texture Depth by Laser Measurement

The drawbacks of the volumetric patch and outflow volumetric tests are that they are slow and essentially very localized tests. In addition, while they are relatively easy to carry out on closed roads (new construction or traffic-controlled overlay sites), they are difficult to carry out on *in situ* roads in a routine way. Accordingly, a variety of mechanized equipment to carry out texture depth or wavelength/amplitude measurement using laser-based devices has been developed.

In the United Kingdom, research into the use of high-speed texture meters has been underway since the late 1970s. The High-Speed Road Monitor (HRM) used for multi-parameter road evaluation at highway speeds incorporated laser devices for profile measurement and texture measurement. The HRM has been superceded by the more recently developed equipment TRACS that also uses lasers for texture depth measurement. Many other equipment manufacturers have incorporated laser measurement of texture depth as a standard module in multifunction high-speed pavement monitoring equipment.

There are a number of variations in the application of lasers to texture depth measurement. The most common variation is the projection of light from the laser onto the pavement surface, and measurement of the intensity of the reflected light by an array of light sensors after the reflected light is first focused onto the array by a receiving lens. The sensors are placed at different heights, and there is a correspondence between the height of the sensor measuring the highest intensity of reflected light and the height of the pavement surface at the point reflecting the light. The lasers used produce pulsed light, with the pulse rate varying from 4 to 64 kHz depending on the equipment.

The variance of the individual displacement measurements over a given length can be calculated, and representative values are stored for specified intervals. Common results reported from laser texture measurement include: (a) SMTD which is calculated from the root-mean-square (RMS) of the individual displacements, and (b) Mean Profile Depth (MPD) is the average difference between the profile and a line through the top of the highest particle within the profile sample.

Additionally, the PIARC experiment (Wambold et al., 1995) calculated an Estimated Texture Depth (ETD), which is based on an MPD calculation, but with a linear equation transformation to produce closer correlation with volumetric patch test texture depth. In addition, some post-processing of profile results involves calculation of the area under the profile curve rather than the average or RMS depth. The PIARC analysis indicated that calculations using the area under the profile curve improved correlations between laser-measured texture depths and patch-measured texture depths on positive texture surfaces compared to RMS values, but produced poorer correlations on negative texture surfaces.

The MPD is now more commonly reported worldwide (ISO 13473-1). There are correlations between MPD, SMTD and texture depth measured and reported under the Volumetric Patch test, but the correlations are surface-specific and cannot be generalized. The sample length over which the texture depth parameters are estimated varies widely depending on the equipment used. It must be stressed that the results reported in terms of MPD can vary between equipment types for the same pavement surface due to differences in implementation, measurement and processing of results.

Finally, it is also possible to process the raw data by filtering the profile using Fourier transforms to break the profile down into separate wavelength components. In this way, the megatexture and unevenness components can be removed from the measured profile, leaving the components in the macrotexture wavelength/amplitude spectrum for further analysis.

There is no doubt that correlations between laser texture depth and patch texture depths are significantly better on positive texture surfaces than on negative texture surfaces. Overall, the PIARC experiment which compared results from Volumetric Patch tests and Outflow tests with 11 laser profile measuring devices concluded that "it is possible to very accurately predict MTD (Mean Texture Depth from Volumetric Patch test) from profile measurements. The best devices gave an unexplained variance of only 8 to 10% when porous surfaces are disregarded, and around 15% when porous surfaces are included."

There are some practical considerations that limit the applicability of laser profile measurement. If the pavement surface is very dark, it may not reflect sufficient light to be detected by the sensors. This is potentially a problem on materials with relatively high bitumen contents, but also on surfaces that are damp. If there is any surface water on the pavement surface, the macrotexture is obscured. In addition, the laser beam can be refracted by the surface of the water, leading to inaccurate measurement.

21.3.2 Measurement of Microtexture

As already discussed, direct measurement of microtexture is not feasible due to the microscopic nature of the factors influencing differences between aggregate sources. On pavement surfaces, the polishing susceptibility of a particular stone under the scrubbing action of vehicle tires at the tire/surface interface is of primary interest. The PSV test (BS EN 1097-8) or the Small-Wheel Circular Track Wear and Polishing Machine test (ASTM E660) using the skip pendulum or British Pendulum Tester device are the principal measures of the relative resistance of different aggregate sources to polishing under traffic.

The British Pendulum Tester is described in ASTM E303 (ASTM E303-93). Basically, it consists of a swinging arm with a rubber slider base. The arm is released from a specified height, and allowed to swing down, acting as a pendulum. At the lowest point of the swing arc, the rubber slider comes in contact with a prepared surface of aggregate to be tested. The greater the friction developed between the rubber slider and the test surface, the more the swing of the pendulum is retarded. A drag pointer attached to the pendulum arm indicates the height reached by the pendulum arm after it has dragged over the test surface. Lower height measurements indicate greater frictional resistance. The height measurements are read off as British Pendulum Number (BPN) values.

Effectively, the pendulum tester is measuring the energy loss when the rubber slider is propelled by a standard force (achieved by dropping a specified weight at the bottom of the swing arm from a specified height) across the test surface. Due to the slow speed of the dropping arm in conjunction with the properties of the rubber slider and the prepared test surface, the energy loss is directly related to the microtexture of the aggregate surface. Accordingly, relative rankings of aggregate microtexture from different aggregate sources in relation to their frictional resistance are achievable.

A standard method of preparing and conditioning aggregate samples for measurement of frictional properties using the pendulum tester is also necessary. The accelerated polishing machine (Figure 21.9) is used widely for this purpose (ASTM D3319-00; BS EN 1097-8). Aggregate chippings are prepared in a test slide which in turn is attached to a test wheel (Figure 21.10). Up to 14 slides can be attached to the wheel, with different aggregate sources in each slide if required. The slides are prepared by hand, placing a single

FIGURE 21.9 PSV polishing machine.

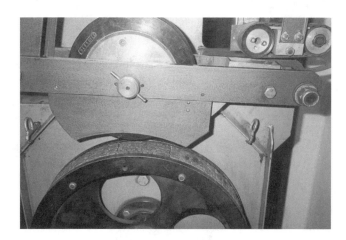

FIGURE 21.10   PSV test wheel.

layer of aggregate chippings as densely as possible. The chippings are typically held in place with a bonding resin.

The wheel then rotates for a number of hours (six or nine depending on the specification followed)at 320 rpm, and the test samples are subjected to constant polishing by the introduction of a gritty sand abrasive and water between the rotating test slides and a rubber-tired wheel. The conditioning of the samples over the period is designed to produce samples similar to pavement surfaces that have been subjected to traffic over a considerable period. The prepared sample slides can then be tested using the pendulum tester. The resulting pendulum number is expressed as a PSV or Polished Value (PV).

ASTM E660 (ASTM E660-90) describes another conditioning machine, the Small-Wheel Circular Track Wear and Polishing Machine, that is used to prepare polished aggregate samples for friction testing. In this case, 12 sample slides are prepared and joined together to form a continuous circular test track, 914 mm in diameter. Four small wheels with smooth pneumatic rubber tires are attached to a rotating shaft that drives the wheels over the test track continuously. Usually the samples are conditioned for eight hours. No abrasives or water are introduced between the rubber tires and the sample slides. The conditioned aggregate samples can then be tested using the pendulum tester. It is recommended that a standard aggregate be used in one or more of the test slides for comparison and quality control purposes when using both the Accelerated Polishing Machine and the Small-Wheel Machine.

The Small-Wheel machine can also be used to condition bituminous and concrete samples for testing of polished properties. In the case of concrete samples, an additional period of accelerated wear using studded steel wheels to expose the coarse aggregate is incorporated before final polishing with the rubber wheels. The specification advises that the British Pendulum Tester has been found to be ineffective in obtaining friction measurements on concrete surfaces and bituminous surface treatment surfaces, and recommends use of the North Carolina State University Variable-Speed Friction Tester (ASTM E707-90) for assessment of these surface types.

The results from the polishing tests are used to rank aggregates in terms of their frictional resistance. Standards can be set for minimum acceptable levels of PSV for general use and for limited use in high-risk areas such as, approaches to traffic lights, pedestrian crossings, roundabouts, etc. This approach is outlined in greater detail in Section 21.5.

21.4   International Friction Index

It is clear by now that there is a variety of different measures of frictional resistance at the interface between the pavement surface and the vehicle tire. Frictional resistance at the interface varies with speed, and both microtexture and macrotexture play significant roles in the frictional resistance generated depending upon the speed. Surrogate measures of microtexture can be carried out in the laboratory (PSV-type testing) and in the field using devices with low slip speeds such as SCRIM. Measures of macrotexture in the field include labor-intensive tests such as, Volumetric Patch and Volumetric Outflow tests, and measurement of texture depth using laser equipment at speeds up to 70 km/h. High speed frictional resistance in the field can be measured with locked wheel devices, or devices with high slip speed capabilities.

A similar situation existed in the measurement of the unevenness or roughness of road surfaces. An international road roughness experiment was carried out in the mid 1970s, leading to the development of the International Roughness Index (IRI). This measure could be computed from the data gathered by many different types of profiling and response-type equipment, and provided a common basis of comparison.

In the early 1990s, PIARC established an experiment to harmonize the different pavement friction measurement equipment and methods used throughout the world (Wambold et al., 1995). A primary goal of the experiment was to develop an international friction scale that all equipment could generate from the raw data collected. As in the case of the IRI, the perceived benefits of such a scale from a road agency viewpoint include international exchange of data, research, experience and developed standards. From an equipment manufacturer view, an international scale increases the potential use of equipment in other countries, with similar benefits accruing to contractors operating friction-measuring equipment.

The index resulting from the PIARC experiment is the IFI. It involves estimation of two parameters from the friction/texture measurements, $F60$ and S_p. $F60$ is termed the Friction Number, and is the frictional resistance value generated at a slip speed of 60 km/h. S_p is termed the Speed Number and is a measure of variation in the relationship between frictional resistance and speed at speeds other than 60 km/h. The principle behind the use of both parameters is shown in Figure 21.11.

Effectively, the belief is that certain combinations of Friction Number and Speed Number will yield acceptable performance over the full range of vehicle speeds. These combinations fall within the acceptable region shown as "good" in the figure. All other combinations of Friction Number and Speed Number will give unsatisfactory performance, due to either poor microtexture or macrotexture or both.

The Speed Number S_p is derived from texture depth measurements. It is usually a linear transform of a texture measure such as MPD derived from laser measurement.

$$S_p = a + b \times \text{Texture(mm)} \qquad (21.3)$$

where Texture is a defined measure such as MTD or MPD and a and b are texture measure-specific constants.

To estimate the Friction Number, it is clear that locked wheel devices or devices with high rates of slip can calculate the $F60$ parameter directly, as testing can be carried out at slip speeds of 60 km/h. This is not

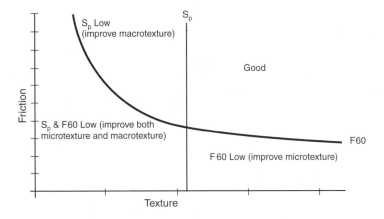

FIGURE 21.11 IFI matrix of values.

the case for low slip speed devices. Values of frictional resistance measured at lower speeds must be adjusted to an equivalent value at 60 km/h. This is achieved through a calculation as follows:

$$FR60 = FRS \times \exp\left(\frac{60 - S}{S_p}\right) \tag{21.4}$$

where S is the slip speed in km/h, FR60 is the derived frictional resistance at 60 km/h, FRS is the measured skid resistance value at speed S, and S_p is the Speed Number derived from texture measurement.

FR60 is device-specific, and is transformed into the general Friction Number, $F60$, through a linear transformation

$$F60 = A + B \times FR60 \tag{21.5}$$

where A and B are device-specific constants.

It is then proposed that the IFI is reported in a two parameter format, i.e., IFI($F60$, S_p). Transformation constant values for a range of different equipment types used during the friction experiment are included in the PIARC report, as well as procedures for deriving the transformation constant values for equipment that has not previously calibrated. Briefly, a minimum of ten sites with wide ranges of texture and friction are tested at a number of different speeds, and linear regression is used to derive speed adjustment curves, and ultimately device-specific transformation constants.

To date, there is not widespread use of IFI as a standard reporting procedure. There are some difficulties in relation to the universal application of the transformation constants derived in the original experiment. A New Zealand study (Cenek et al., 2000) attempted to apply the IFI concepts to coarse textured surfaces. The Speed Number was to be derived from Volumetric Patch test texture measurements. The Friction Number was to be derived from Griptester measurements. The Griptester is a fixed low-slip device, and the Friction Number $F60$ has to be transformed from slip speeds lower than 60 km/h (typically at approximately 10 km/h). As shown earlier, the Speed Number is a critical parameter in the derivation of Friction Number.

The New Zealand experiment derived Speed Number based on the Volumetric Patch results and PIARC transformation constants. It also independently derived a speed number relationship with locally derived transformation constants, based on carrying out Griptester results over a wide range of speeds from 40 to 100 km/h and deriving Speed Numbers from these results in conjunction with the measured texture depths. The conclusion was that the variability in FR60 (and in turn the $F60$ Friction Number) results was much reduced if actual measurements of speed sensitivity were used to derive the Speed Number rather than using texture depth and PIARC transformation constants to derive the Speed Number for the purposes of calculating the Friction Number based on low slip speed results.

Undoubtedly as more research is carried out worldwide, transformation equations and methodology will be improved to reduce the variability in IFI calculation. When such measures are sufficiently repeatable, the advantages of a single, simple system of frictional resistance measurement and reporting outlined in the introduction to this section will accrue. At the moment, many countries continue to use other types of two parameter reporting systems. For example, road agencies using low slip speed devices such as SCRIM also typically record a measurement of macrotexture using laser devices, and establish parameters of minimum acceptability for both the measured SCRIM coefficient and MPD. In all cases, the aim is to capture the reality of different frictional resistance developed over the full spectrum of vehicle speeds.

21.5   Skid Resistance Policy

The previous sections have outlined the importance of adequate microtexture and macrotexture in the provision of acceptable levels of frictional resistance to the road user. Laboratory and field tests are available to characterize the performance of a pavement surface over the full range of expected vehicle speeds. While the particular test methods and equipment used may vary across the world, the fundamental underlying aim to ensure adequate microtexture and macrotexture remains.

A road agency charged with provision of pavements with adequate frictional resistance can use the information and available equipment in a variety of ways. Essentially however, these reduce to one of the following three main areas of interest:

- Controlling the frictional resistance of the road agency's pavements through specification of microtexture and macrotexture;
- Using field results of frictional resistance to investigate accidents and to identify high-risk accident locations; and
- Managing the *in situ* frictional resistance of a road agency's network of pavements through systematic network surveys.

The first two areas are addressed in this section. Section 21.6 deals with network management of skid resistance. While some of the same tests and equipment are used under each of the areas, the approach and underlying philosophy is fundamentally different and needs separate treatment.

21.5.1 Control by Specification

21.5.1.1 Control of Microtexture

Control of the aggregate types used in pavement construction and surface treatment is the primary means of ensuring that sufficiently high microtexture will be available over the long-term under the polishing action of vehicle tires. At the most basic level, general aggregate classification procedures can be adopted based on historical records of satisfactory/unsatisfactory performance. In this way, aggregate sources that are known to have produced unsatisfactory performance in the field can be restricted from use in the surfacing layers, while still being allowed to be used in sub-surface layers. Similarly, aggregate sources with satisfactory performance in the field may be specified for use in surfacing layers.

This approach has the great advantage of simplicity, but has a number of practical disadvantages. First, aggregate properties from a given source can vary from batch to batch in the short-term, and vary significantly in the longer term as different sections of the rock layer are quarried. In addition, aggregates yielding satisfactory performance in the past may not give satisfactory performance in the future under different traffic volumes, different traffic speeds, or changes in tire technology. Furthermore, a restrictive listing of satisfactory aggregate sources can lead to premium pricing policies that may not be to the economic advantage of the road agency. It may also be difficult for new entrants into the aggregate supply market to establish satisfactory performance of their aggregate sources in the absence of prior historical performance on the road agency's network.

Specification of materials based on some testing measure related to polishing value as obtained from the accelerated polishing machine/British Pendulum tests is more commonly used. This type of specification overcomes most of the disadvantages listed for general aggregate classification procedures while still allowing any road agency to tailor the required properties to prior experience of satisfactory performance.

Practice in the United Kingdom is well advanced in the specification of acceptable aggregates based on PSV. In addition, aggregate durability, as measured currently by the Aggregate Abrasion Value (AAV) test (BS EN 1097-8), is specified. The principle of the AAV test is similar to that in the Los Angeles Abrasion Test and other abrasion tests, i.e., to determine the resistance to abrasion of an aggregate under traffic action. This is not the same property as a resistance to polishing — an aggregate may have a high resistance to polishing but may not be durable, leading to frequent replacement of the surfacing layer.

The U.K. Design Manual for Roads and Bridges (DMRB Vol.7, HD36/99) lays out a comprehensive matrix of minimum PSV to be applied to aggregates used in new surface courses. The minimum values vary significantly depending upon traffic volume and site category type. For example, on the site category covering multi-lane separated roads such as motorways and dual-carriageways where there are no junctions or ramps (termed non-event sections), minimum PSV values range from 50 to 65 depending upon traffic volume. Traffic volume is based on commercial vehicle volumes, as this traffic has been found to have the greatest polishing effect on aggregates. The minimum PSV of 50 is specified up to 2000 commercial vehicles per lane per day, while a minimum PSV of 65 is specified for volumes in excess of 5000 commercial vehicles per lane per day.

There are a total of 14 different site categories specified, and the minimum PSV varies significantly, reflecting variable likelihood in the frequency and severity of braking being required by the driver. The high-risk category covering approaches to roundabouts, traffic signals, pedestrian crossings, railway crossings, etc., has much higher minimum PSV requirements than the motorway/dual-carriageway category discussed earlier. On this high-risk category, minimum PSV from 0 to 750 commercial vehicles per lane per day is specified at 68 + , rising to 70 + at volumes in excess of 750 commercial vehicles per lane per day.

The AAV specifications vary based on traffic volume, but do not vary by site category. The maximum AAV varies from 14 at sites with less than 250 commercial vehicles per lane per day down to 10 on sites with greater than 1750 commercial vehicles per lane per day.

The site categories used in the PSV specification are the same as those used in the designation of minimum SCRIM values — this is to ensure consistency of approach between aggregate specification and *in situ* frictional resistance testing. It is stated that using the appropriate PSV for a particular site category/traffic loading combination should yield a surfacing with acceptable SCRIM values.

In reality it is difficult to source aggregates with PSV values over 68 and AAV values of 10 or less. Artificial aggregate such as calcined bauxite (a clay fired at high temperature to produce a material with very high polish resistance) is used in these circumstances in conjunction with a resin, effectively glueing the aggregate to the road. As a result, it is a very expensive material to lay on a square-meter basis due to the specialized materials and attention to detail required. While this material provides excellent microtexture, it does not provide very good macrotexture and its use on high-speed roads should be very tightly controlled. Clearly, it is ideal for low-speed approaches to traffic signals, pedestrian crossings, etc., where the extra cost is justified in the light of the increased accident risk to vehicles and pedestrians.

New Zealand practice has developed a simple equation to specify the PSV required for the aggregate (Transit New Zealand, 2002):

$$PSV = 100 \times SR + 0.0063 \times CVD + 2.6 \qquad (21.6)$$

where SR = Investigatory Level for the site, and CVD = Commercial vehicles per lane per day.

The investigatory level is a minimum SCRIM value which is related to the site category. There are five site categories in total, reflecting groupings of the U.K. site categories. For hot mix asphalt, the New Zealand specification requires that 85% or more of the coarse aggregate shall have a PSV as determined by the equation above.

21.5.1.2   Control of Macrotexture

Provision of adequate macrotexture is equally important in the performance of pavement surfaces. The most straightforward way of ensuring adequate macrotexture in a new pavement surface is through a texture depth specification. Traditionally this was achieved through a minimum texture depth based on the Volumetric Patch (Sand Patch) test. This test is still the reference test method under many specifications. It is very suitable for positive texture surfaces, but not as suitable for negative texture surfaces. It is a slow test to carry out with a high skilled labor input. However, in new road construction, this is not always a key problem as the road is closed to traffic.

Laser-based measurements of texture depth have been used increasingly often in recent years. The Mini-Texture Meter (BS 598:105) is a laser-based device installed on a handcart. It can be pushed by hand over newly-laid surfaces and gives very rapid results over a wide area compared to the Volumetric Patch test. The Mini-Texture Meter shows good correlation with the Volumetric Patch test on positive texture — however, for routine quality control on any particular site, specific correlation between the Volumetric Patch texture depth and the Mini-Texture meter texture depth should be developed for that site. High-speed laser measurement equipment can also be used to check on texture depth, but this is usually on long stretches of completed new surfacing or overlays rather than on a daily quality control basis.

Traditionally the specification of minimum texture depth for a particular surfacing material was laid out in conjunction with recipe-type specifications for the surfacing material. Material gradations, binder contents, maximum chipping size, etc., were all specified with a view to designing a material that would provide sufficient macrotexture to meet the texture depth specification and the subsequent provision of acceptable frictional resistance at high speed. Thus, in the U.K., materials such as Hot Rolled Asphalt were developed to specifically provide high levels of macrotexture by rolling in single size chips into a binder-rich mastic. Surface dressings also by their design provide high levels of macrotexture with cubic-shaped chippings held by a relatively thin layer of binder.

More recently, there has been an increasing trend towards performance specifications in relation to surfacing materials. The road agency specifies a minimum (and sometimes maximum) texture depth to be maintained over a specified period. It is up to the contractor and material supplier to produce a material that meets the specification. This type of specification has produced innovative materials including some of the new thin surfacing. Difficulties can arise in achieving and retaining adequate texture depth with some thin surfacing unless the design and construction is very carefully controlled. However, the concept of the road agency specifying a performance parameter such as texture depth as a surrogate for macrotexture is attractive and encourages the development of innovative and imaginative solutions rather than a "one-size fits all" approach.

21.5.2 Accident Reduction Management

The primary purpose of the road agency is to provide safe roads for the traveling public. Realistically, not all sections of the road network will have the same level of risk in relation to accidents. Higher traffic volumes, poorer geometric features, junctions, areas of greater interaction with pedestrians all increase the risk of accidents. Provision of higher levels of frictional resistance at these sites can compensate for the increased risk of accidents to some extent. On the other hand, locations with poor levels of frictional resistance will show a higher accident rate than the norm.

There are two different triggers that should launch a study of a particular location to determine if increasing the frictional resistance at the site will help to reduce the accident rate. First, a reactive accident investigation at a particular site often brought about by a serious or fatal traffic accident. Second, a proactive investigation of a site identified through monitoring of accident rates and locations on a systematic, continuous basis.

The second approach, while being more costly and time-consuming in terms of initial data collection, data collation and analysis, is clearly superior to the first approach in that a number of minor accidents may be sufficient to highlight the higher accident risk without any death or serious injury arising. In addition, when the full economic, legal and social costs of accidents which could have been avoided through proper monitoring are included, the reactive approach is much more costly overall. A road agency should be carrying out continuous monitoring of accident rates and locations with a view to identifying high risk sites, while also carrying reactive studies to determine if lack of frictional resistance was a contributory factor in the accident recorded.

The U.S. Federal Highway Administration has issued a technical advisory on the elements to be included in skid accident reduction program (FHWA, 1980). It is recommended that wet weather accident location studies be carried out on an ongoing basis. The purpose is to identify locations with high incidence of wet weather accidents so that corrective measures, if needed, can be carried out in a timely and systematic manner.

Accident records compiled under U.S. highway safety standards have the weather conditions at the time of the accident recorded as a required item. Total accident rates, wet weather accident rates and wet/dry accident rate ratios at specific sites, locally and regionally are all easily retrieved from a well-design accident database. This information should then be supplemented by rainfall data over the same period as the accident data.

Wet Pavement Time (WPT) can be estimated from annual rainfall records. A relationship is put forward in the technical advisory,

$$WPT = 3.45 \ln(AR) - 5.07 \tag{21.7}$$

where AR is the annual rainfall in inches.

Dry pavement time is estimated by subtracting the WPT along with ice and snow cover times from 100, if ice and snow cover is significant over the analysis period.

A measure to evaluate the relative safety of the location, termed the Wet Safety Factor (WSF), is also put forward. The factor is defined as

$$WSF = (DA)(PWT)/(WA)(PDT) \tag{21.8}$$

where DA = number of dry weather accidents, WA = number of wet weather accidents, PWT = percent of WPT, and PDT = percent of dry pavement time.

Each of the variables for a particular accident site must be estimated over the same time period. The lower the WSF, the higher the relative risk of a wet accident compared to a dry accident. A WSF of less than 0.67 is indicative of a wet weather problem. This cut-off assumes that the wet weather problem is completely related to poor frictional resistance.

Practice in the U.K. in relation to accident analysis and frictional resistance is broadly similar to the U.S. practice (DMRB Vol. 7, HD28/94). National accident rates are not provided or recommended for use as variation in rainfall rates, snow and ice cover, etc. render comparison with an overall national rate misleading and meaningless. Local information on accident rates taking into account regional climate etc. is much preferred for identifying outlier sites with significantly higher rates.

In the U.K., conditions at the time of the accident are recorded, and in addition, whether or not skidding occurred in the course of the accident is recorded. The recording of skidding is intended to further aid in the analysis and identification of accident sites that would benefit from an improvement in frictional resistance. The following two parameters are put forward that can be used in this type of comparison:

Wet Skidding Rate(%) = (Wet skid accidents/Wet accidents) × 100

Wet Accident Rate(%) = (Wet Accidents/Total Accidents) × 100

The Wet Skidding Rate parameter is the most logical parameter to use, as it does give a good correlation with changes in frictional resistance. It is very useful as a monitoring tool for assessing changes over time and impacts of an accident reduction program over a network of roads where a relatively large data sample can be gathered.

However, there are practical difficulties in using the Wet Skidding Rate for a single site. The sample size may be too small to draw meaningful conclusions. In addition, skidding is often not reported on the accident report and recording forms even though it may have occurred. Also, when the pavement surface is damp but without visible water, it is frequently recorded as dry even though it may in fact have a low frictional resistance due to the damp conditions. Thus, the number of wet skid accidents may be significantly lower than the real number.

The Wet Accident Rate is a coarser measure, susceptible to weather conditions as periods of dry weather will reduce the value obtained. The WSF described earlier is more satisfactory in this regard. However, in the U.K., the Wet Accident Rate is frequently used as it gives a larger accident sample for a given site compared to the Wet Skidding Rate and more robust conclusions for a given site can be derived as a result.

Finally, the increased use of Geographic Information Systems (GIS) to link accident databases with the road agency network graphically should be commented on. This tool is very effective for identifying high-risk sites, particularly when linked with other road agency information on frictional resistance. It also allows identification of accidents that have incorrectly recorded grid co-ordinates but are clearly related to the accident site, as well as nearby accident locations on approach roads etc. that would not be obvious without the map-based display of data.

21.6   Skid Resistance Management for Networks

There are two fundamentally different approaches to the management of road networks in relation to provision of frictional resistance. First, an approach that is governed by the belief that testing of the entire road network on an ongoing basis is impractical, unnecessary or unjustified on a cost-benefit basis. The second approach is governed by a belief that monitoring of the entire road network is necessary and justified on a cost-benefit basis. Clearly the local climate, traffic volumes, accident rates, percentage of WPT, etc., influence the road agency in the approach adopted. In addition, the development of these two very different approaches has been influenced greatly by the type of equipment routinely used for frictional resistance measurement by the road agency.

It would be very unusual for any road agency to survey all roads in the network every year to determine the frictional resistance. The FHWA advisory (FHWA, 1980) advises that roads and streets with low speed limits do not need to be considered in the testing program at all. Thus, in urban areas, arterial routes would be tested but residential streets and roads would not. Rural roads with low traffic volumes, approximately 1000 vehicles per day, are also excluded from routine testing on a cost-effectiveness basis. On the remainder of the candidate sections, the FHWA publication advises that few roads vary significantly in frictional resistance within a 2 to 3 year period, and accordingly a 2 to 3 year gap between repeat testing can be allowed for.

Guidelines similar to these are used in most countries, with the emphasis being on identifying the conditions on the most critical elements of the network. However, there are many countries that have separated the management of the high-speed, high-volume network of roads (usually designated national roads) from the management of the remaining road network, with a national road agency taking responsibility for national roads and regional/local agencies managing the remainder of the road network. The national road agency, dealing with a smaller road network composed of high-speed, high-volume traffic, will commonly measure the frictional resistance of the entire road network on an ongoing basis.

21.6.1   Sampling Approach to Network Management

The general approach in the United States is to monitor the frictional resistance performance of the network through selective sampling of pavement sites throughout the network. The FHWA advisory on a Skid Accident Reduction Program (FHWA, 1980) states that a total skid inventory of all roads and streets in a highway system has proven to be impractical.

This perceived impracticality is a reflection of the equipment used for frictional resistance measurement in the United States, the locked-wheel skid tester. This equipment, fully described in Section 21.2, carries out measurements with a fully locked tire at speeds of approximately 60 km/h. The testing methodology is very severe on the test tires. Realistically, only very short stretches of pavement can be evaluated using this equipment. Accordingly, a sampling system where the network condition is derived statistically from many short samples of locked-wheel results is necessary. Testing under this regime at low-speed, high accident-risk sites such as curves and junctions requires use of the locked-wheel tester at lower speeds, supplemented with macrotexture measurements to allow adjustment to frictional resistance at normal test speeds for statistical purposes.

21.6.2   Continuous Skid Resistance Surveys

The alternative to the sampling approach is continuous measurement of the frictional resistance over very large stretches of the road network on an annual basis. This approach was pioneered by the U.K. well over 30 years, and has been adopted by many other countries and road agencies since then. In large part, this reflects the use of the SCRIM, a truck-mounted machine designed to carry out continuous measurement of frictional resistance. The SCRIM uses a low slip-speed tire that can rotate freely which testing, and gives a long test life as a result. Typically, approximately 50 km of road can be measured using a tank load of water, with results available for each 5 or 10 m stretch. Daily outputs of approximately 200 lane-kilometers would be typical.

Access to high-output equipment providing continuous measurement fundamentally changes the approach that can be taken to management of frictional resistance, and brings this area in line with other pavement condition data such as profile and visual condition data. As a result, typical pavement management tools for reporting and managing this type of data become applicable.

When frictional data is used directly in the management and allocation of resources, the consistency and comparability of the data is of utmost importance. The U.K. approach (DMRB Vol. 7, HD28/94) to this problem was to develop the Mean Summer SCRIM Coefficient (MSSC). It is known that frictional resistance varies throughout the year, being lowest in the summer season (Figure 21.12). A standard testing season is specified, between May 1 and September 30 in the U.K., and all SCRIM measurements to be used for network management purposes should be taken within this testing season.

However, it was still believed that the true seasonal variability in results within this period was too high for management purposes. Accordingly, it has been standard practice in the U.K. to carry out three SCRIM surveys, reasonably evenly-spaced over the May to September time period, and use the mean value from the three surveys to calculate the MSSC. Typically, one-third of the road network is surveyed every year, with three surveys being carried out on each road segment in the year of testing. This approach is time-consuming and costly, but it does result in a value that can be used to underpin the entire skid resistance and accident reduction program in the U.K.

Road management in New Zealand also operates the SCRIM on a routine basis for management of the frictional resistance of the road network (Transit New Zealand, 2002). The Equilibrium SCRIM Coefficient (ESC) is used as the base value for management of road sections. This value is obtained by averaging the SCRIM coefficient measured within the testing season over a period of three consecutive years. In this way, for example, the entire road network could be surveyed every year within the testing season. The timing of the network survey for individual sections is varied from year to year. For a particular section, it may be tested early in the season in year 1, in the middle of the test season in year 2 and towards the end of the test season in year 3. Thus, over a 3 year period, the same amount of SCRIM testing is carried out compared with the MSSC approach, but the entire network is tested every year, giving a basis for comparison of within-year results using the SC as well as a stable, repeatable parameter

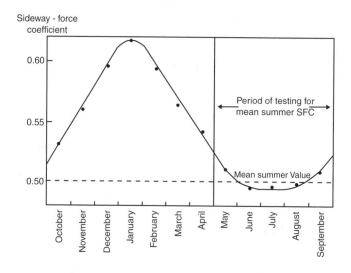

FIGURE 21.12   Seasonal variation in friction — Northern Hemisphere.

(ESC) for multi-year comparisons and other road management purposes. The U.K. Highways Agency is currently moving its approach to SCRIM measurement towards the ESC.

In addition, it is essential in the interests of having a consistent basis for comparison of results across sections of the network, and over time, that the equipment used to collect the network-level frictional resistance data is all measuring in the same way. Each SCRIM is calibrated statically at the beginning and end of each day's testing under the U.K. Highway Agency requirements. Dynamic calibration on test sites with known levels of frictional resistance is carried out at least weekly, and preferably daily during periods of operation. The dynamic calibration sites should be relatively straight and flat, not newly surfaced, in reasonably good condition and with a reasonably uniform frictional resistance over the calibration site length (at least 100 m). Finally, all SCRIM equipment to be used on the U.K. trunk road network are simultaneously tested and approved annually at a group correlation in London. This level of monitoring and calibration is necessary if the data generated is to be used as a basis for decision-making and allocation of resources.

21.6.3 Approach to Policy Definition — Full Network Testing

An explicit linkage between accident risk and wet skidding resistance values has been determined in the U.K. based on extensive research. These categories were previously referred to in the discussion on control of microtexture by specification, where minimum levels of PSV were established for different site categories. A total of 13 categories are defined, with Investigatory Levels of MSSC defined for each category. If the MSSC at a particular site is measured to be less than the defined Investigatory Level for that site category, site investigation must be initiated to determine the source of the low frictional resistance, and appropriate remedial measures suggested.

The Investigatory Levels range from 0.35 (for mainline Motorway sections) up to 0.55 (for approaches to roundabouts, traffic signals, etc.) when the MSSC is measured at 50 km/h. For measurements on roundabouts and very tight (small radius) bends, the MSSC is measured at 20 km/h and different Investigatory Levels are specified. Each of the Investigatory Levels has an associated Risk Rating, with values ranging from one to eight.

The Risk Rating and Investigatory Level for a particular site category do not usually vary, e.g., single carriageway with minor junctions usually has a Risk Rating of 4 with an associated Investigatory Level of 0.45. However, the risk rating system allows for cases where particular sites are at increased risk. For example, a particular site in site category E (single carriageway with minor junctions) may warrant an increase in Risk Rating from 4 to 5, based on the site's geometry, traffic volumes, proportion of turning movements, past history of accidents, etc. If the increase in Risk Rating is justified, the site will be assigned an Investigatory Level of 0.5, corresponding with the Risk Rating of 5. In this way, flexibility is built into the system to cater for particular cases, but generally the standard Investigatory Levels for site categories are used.

It is suggested that in cases where the site fits into multiple categories, e.g., a single carriageway with a minor junction (category E) on a bend (category H1) with a steep gradient (category G2), the risk rating of the site may increase by more than 1 above the highest risk rated category. Variables put forward as influencing the risk rating include geometric features such as, narrow lane widths, sub-standard horizontal and vertical alignment and turning areas, particularly where traffic is turning across oncoming traffic and there is not sufficient room to provide a turning lane. Roadside developments also influence the assignment of a risk rating, particularly high-density developments generating significant pedestrian traffic, including shops, schools, office developments, etc.

In any case, when an appropriate risk rating and associated Intervention Level have been assigned to a site, and the MSSC measured at the site is below the Intervention Level, a site investigation must be carried out. In the U.K., it is a requirement that signs warning that a road is slippery must be erected at any such site until either remedial action is taken to increase the frictional resistance at the site to an acceptable level or until the site investigation indicates that surface treatment is not required.

The basic aim of the site investigation is to determine what appropriate remedial action, if any, is needed at the site. In particular, the appropriate action may involve improving the frictional resistance and/or improving the site in other ways (geometric improvements, traffic calming, better separation of vehicles from pedestrians etc.) It is recommended when assessing the site that other factors affecting the choice of appropriate remedial action should be explicitly documented. This includes assessing the structural condition of the site using visual assessment or other structural testing, assessing the macrotexture provided at the site, and assessing the current and projected future traffic levels, both total traffic volumes and commercial traffic volumes. In this way, a more appropriate choice of maintenance action combining strengthening of the pavement in conjunction with improvement of frictional resistance may be appropriate.

A modified approach to network management of frictional resistance is used in New Zealand (Transit New Zealand, 2002). The SCRIM is again used as the basic equipment for network testing. A total of five site categories are defined, with Investigatory Levels ranging from 0.35 to 0.55. A further modification in New Zealand is the definition of a Threshold Level, set at 0.1 below the Investigatory Level for each category. The ESC measured for the site can thus fall into one of three ranges:

Low: Value of ESC below the Threshold Level
Medium: Value of ESC between Threshold Level and Investigatory Level
High: Value of ESC above Investigatory Level

In practice, the ranges are used to prioritize treatment. Sites in the High range have no required action. Sites in the Medium range are inspected and prioritized for future treatment. Sites in the Low range must be investigated and programmed for treatment as soon as possible within the maintenance program.

It is emphasized in the New Zealand specification that a sharp reduction in SCRIM value between tests may be due to a temporary reduction in frictional resistance. The presence of contaminants on the road surface is an obvious source of reduction. Bleeding of bitumen to the surface in high temperatures is a common cause of sudden decrease, and usually produces a much greater decrease than increased polishing of the aggregate under traffic loading. Seasonal variation in values does also occur as previously discussed in this section, and the ESC calculation is aimed at reducing this variation over time.

The IFI provides another potential network tool for management of frictional resistance of an agency's road network. The IFI explicitly incorporates two measures of frictional resistance, the Friction Number and Speed Number, within the index. A two-dimensional plot of acceptable combinations of Friction Number and Speed Number can be produced by a road agency to reflect their requirements and practices.

The IFI has the great advantage that it is device-independent. If it gains in acceptance, there will be much greater transferability of standards and experiences between countries using different equipment but having similar traffic, climate and road surface materials. Currently, countries such as the U.K. and New Zealand that are carrying out network monitoring on an annual basis lead the way in terms of development of standards for frictional resistance and relating appropriate treatments to the levels of frictional resistance measured. However, the experience developed is realistically only transferable to other road agencies operating the SCRIM. More widespread adoption of the IFI may allow this experience to be applied in other jurisdictions.

21.6.4   Pavement Management and Frictional Resistance

When network monitoring of the pavement frictional resistance is carried out on a continuous widespread basis, the results can be used to generate statistics for use at network level as well as for project-specific decisions discussed in the previous section. Frictional resistance parameters can then be used through pavement management and asset management reports, in conjunction with network assessment of other condition parameters such as, roughness, rutting, cracking extents, etc., to monitor changes in network condition over time, to assess the impact of changes in funding (both real and projected) on network condition and to provide road agency management and staff, political representatives and the general public with an informed picture of the state of the network.

In New Zealand, performance measures related to texture depth and skid resistance are reported on annually through the State Highway National Pavement Condition Report (Transit New Zealand, 2003). Results are reported for the entire road network, and also by Network Management Area. Typically, results are also further broken down by road class, allowing a wide variety of comparisons across road class and management area within each year and over multiple years.

There are two parameters currently reported for texture depth. These are average texture depth, and the percentage of texture depth below 0.5 mm (MPD measurement), where 0.5 mm is the minimum acceptable texture depth.

There are a number of network summary calculations based on the SCRIM results. The use of Investigation Levels and Threshold Levels to identify sites with unsatisfactory levels of frictional resistance was discussed in the previous section. The percentage of SCRIM readings below Threshold Level for different road classes and nationally is reported. A second measure, Good Skid Exposure, is also calculated to reflect the importance of provision of good frictional resistance where traffic is heaviest. The total vehicle kilometers traveled nationally, and by road class is calculated. The total vehicle kilometers on all sections with values above the Threshold Levels is also calculated nationally, and by road class. The total vehicle kilometers traveled above the Threshold Levels expressed as a percentage of the total vehicle kilometers traveled is the definition of the Good Skid Exposure parameter.

21.7 Maintenance Treatments to Restore Frictional Resistance

The appropriate maintenance treatment to restore frictional resistance to an acceptable level is dependent on a number of factors such as:

- Source of loss of friction (Microtexture or Macrotexture)
- Traffic Volumes
- Extent of loss of friction (localized, wheel paths, general)
- Length of section
- Traffic Volumes
- Heavy Commercial Vehicle Volumes
- Accident Rates
- Time of Year (it may not be possible to lay certain materials in very cold or wet conditions)
- Existing pavement surface condition
- Other pavement defects (ride quality, structural etc.)

A summary of the range of treatments available is included here for completeness. There is no reference in this chapter to the design of new surface layer materials such as, Asphaltic Concrete, Stone Mastic Asphalt, Porous Asphalt, Hot Rolled Asphalt, Open-graded Macadams, etc. Reference should be made to the relevant chapters in this handbook for more information on these materials. The design and gradation of these materials reflects the need to provide acceptable long-life frictional resistance over the full range of traffic speeds.

21.7.1 Localized Treatments

Decontamination of the Surface
    Remove Decontamination by oil, mud, etc., by washing the surface
Bleeding/Fatting Up of the surface with excess binder
    Treatment with Hot Sand to "soak up" the excess bitumen
    Burning off the excess bitumen with burning equipment
    Waterblasting of the surface to remove excess binder
Polished Aggregates (loss of Microtexture)
Restoration of a coarse microtexture by machine abrasion using mechanized hammers

21.7.2 Surface Treatments

Surface Dressing with binder and chippings. There are a wide variety of surface dressings available, including single and double surface dressings, racked-in systems and sandwich dressings for surfaces that have a binder-rich surface. Chip size can vary from 6 to 14 mm depending on the system used, volume and speed of traffic, existing surface condition, etc. Surface dressing treatments generally provide very good macrotexture; the microtexture achieved is primarily dependent on the aggregate used in the treatment.

Slurry Seals. A mix of bitumen emulsion and small size aggregate (sand-size and smaller) laid by machine in a single pass, usually giving a thickness of approximately 5 to 6 mm. Not capable of providing significant macrotexture — suitable for lower speeds only, usually urban situation.

Thin/Ultra-Thin/Micro-Thin Systems. Usually ranging from approximately 10 to 25 mm in thickness, a graded mix of aggregates and binder, often polymer-modified. Usually a proprietary material, can have a range of fillers including cement, fibers, rubber, etc. The layer is machine-laid in a single pass. These system types have increased market share very rapidly over the past 10 years. They can be designed to improve macrotexture, microtexture, or both. Increasingly popular in urban areas compared with surface dressing treatments as quicker and less messy.

High Friction (Anti-Skid) Surfacing. It is an expensive proprietary treatment using very high PSV aggregate, often an "artificial" aggregate such as calcined bauxite. The aggregate is effectively glued to the existing surface using resin binders. Due to the expense, usually used in high-accident risk sites such as, approaches to junctions, traffic lights, roundabouts, etc., particularly where there is high pedestrian/traffic interaction. More suited to urban, lower-speed roads — macrotexture is not high.

Retexturing. This treatment covers a range of abrasive treatments to restore microtexture, macrotexture, or both. Retexturing can be applied to bituminous or concrete surfaces. The existing surface must be in good condition as significant forces are applied to the surface to restore the properties. Retexturing can be carried out using hammering or shot blasting techniques, grinding of the pavement using combinations of blades, or other mechanical methods.

Grooving. Transverse grooves can be inserted into concrete (usually) or bituminous surfacing to improve the macrotexture and provide drainage channels to allow water to get away quickly from the pavement/tire interface. Again, the surface needs to be structurally sound to withstand the large mechanical forces applied. Grooving can significantly increase the noise generated at the pavement/tire interface and may not be appropriate in urban areas.

References

American Society for Testing and Materials, D3319-00. Practice for the Accelerated Polishing of Aggregates Using the British Wheel, Philadelphia, U.S.A.

American Society for Testing and Materials, E274-97. Test Method for Skid Resistance of Paved Surfaces Using a Full-Scale Tire, Philadelphia, U.S.A.

American Society for Testing and Materials, E303-93. Test Method for Measuring Surface Frictional Properties Using the British Pendulum Tester, Philadelphia, U.S.A.

American Society for Testing and Materials, E501-94. Specification for Standard Rib Tire for Pavement Skid-Resistance Tests, Philadelphia, U.S.A.

American Society for Testing and Materials, E524-88. Specification for Standard Smooth Tire for Pavement Skid-Resistance Tests, Philadelphia, U.S.A.

American Society for Testing and Materials, E660-90. Practice for Accelerated Polishing of Aggregates or Pavement Surfaces Using a Small-Wheel, Circular Track Polishing Machine, Philadelphia, U.S.A.

American Society for Testing and Materials, E707-90. Test Method for Skid Resistance Measurements Using the North Carolina State University Variable-Speed Friction Tester, Philadelphia, U.S.A.

American Society for Testing and Materials, E965-96. Test Method for Measuring Pavement Macrotexture Depth Using a Volumetric Technique, Philadelphia, U.S.A.

American Society for Testing and Materials, E1859-97. Test Method for Friction Coefficient Measurements between Tire and Pavement Using a Variable Slip Technique, Philadelphia, U.S.A.

British Standards Institution BS EN 1097-8: 2000. Method for determination of aggregate abrasion value (AAV), London, England.

British Standards Institution BS EN 1097-8: 2000. Method for determination of the polished-stone value, London, England.

British Standards Institution BS 7941-1:1999. Methods for measuring the skid resistance of pavement surfaces. Sideway force coefficient routine investigation machine, London, England.

British Standards Institution BS 7941-2:1999. Methods for measuring the skid resistance of pavement surfaces. Test method for measurement of surface skid resistance using the GripTester braked wheel fixed slip device, London, England.

British Standards Institution BS 598: Part 105 1990. Methods of test for the determination of texture depth, London, England.

Cenek, P.D., Fong, S., and Donbavand, D. 2000. *New Zealand's Experience in Applying the IFI to Coarse Textured Road Surfaces.* Transit New Zealand Research, Wellington, New Zealand.

Federal Highway Administration. 1980. T5040.17, Skid Accident Reduction Program, Washington, DC, U.S.A.

Hosking, J.R. and Woodford, G.C. 1976a. Measurement of resistance Part 2: Factors affecting the slipperiness of a road surface, TRRL Laboratory Report 738, Transport Research Laboratory, Crowthorne, United Kingdom.

Hosking, J.R. 1986. Relationship between skidding resistance and accident frequency — estimates based on seasonal variation, TRRL Research Report 76, Transport Research Laboratory, Crowthorne, United Kingdom.

International Standards Organization, ISO 13473-1. 1997. Characterization of pavement texture utilizing surface profiles — Part 1, Determination of mean profile depth. Geneva, Switzerland.

Kennedy, C.K., Young, A.E., and Butler, I.C. 1990. *Measurement of Skidding Resistance and Surface Texture and the Use of Results in the United Kingdom.* ASTM STP 1031, Philadelphia, U.S.A.

Rado, Z. 2000. Analysis of road surface friction in relation to vehicle braking performance and its application to PMS, *Proceedings of the First European Pavement Management Systems Conference,* Budapest, Hungary.

Roe, P.G., Webster, D.C., and West, G. 1991. The relation between the surface texture of roads and accidents, Research Report RR296, Transport Research Laboratory, Crowthorne, United Kingdom.

Roe, P.G., Parry, A.R., and Viner, H.E. 1998. High and low speed skidding resistance: the influence of texture depth, TRL Report 367, Transport Research Laboratory, Crowthorne, United Kingdom.

Stationery Office, Design Manual for Roads and Bridges, Volume 7 Pavement Design and Maintenance, HD28/94, London, England.

Stationery Office, Design Manual for Roads and Bridges, Volume 7 Pavement Design and Maintenance, HD36/99, London, England.

Transit New Zealand. 2002. TNZ T10: 2002, Specification for Skid Resistance Investigation and Treatment Selection, Wellington, New Zealand.

Transit New Zealand. 2003. State Highway National Pavement Condition Report, Wellington New Zealand.

Wambold, J., Antle, C.E., Henry, J.J., and Rado, Z. 1995. International PIARC experiment to compare and harmonize texture and skid resistance measurements, *Association Internationale Permanente des Congres de la Route,* Paris, France.

22

Bridge Management Systems

Yi Jiang
Purdue University
West Lafayette, IN, U.S.A.

22.1   Introduction

The Silver Bridge over the Ohio River between West Virginia and Ohio collapsed during rush hour and it resulted in the loss of 46 lives in 1967. This disaster caused public outcry and indicated the need for a formalized national inspection program on the condition of the nation's bridges. The 1968 U.S. Federal-Aid Highway Act required the establishing of national bridge inspection standards (NBIS) for the Federal-aid highway system. In April 1971, NBIS (FHWA, 1971) were issued for the states to inventory, inspect, and report on the condition of the 274,000 bridges on the Federal-aid highway system. By the end of 1973, the states had inventoried most of the bridges on the Federal-aid highway system. In 1978, the Surface Transportation Assistance Act extended the inventory and inspection program to include bridges on all public roads. The NBIS were revised in December 1979 to comply with the new regulation to include a total of 577,000 bridges on all public roads in the United States. Additional revisions to the NBIS were made in September 1988 to modify some of the bridge inspection requirements, including frequency of routine inspection, special inspection requirements, inspector certification, and reporting requirements. The NBIS include five provisions: inspection procedures, frequency of inspections, qualifications of personnel, inspection reports, and inventories. The primary purpose of the NBIS is to locate, evaluate, and act on existing bridge deficiencies to assure that the bridges are safe for the traveling public.

In early 1980s, some individual states started to study how to effectively utilize the increasing database from the NBIS mandated bridge condition inventory. Their studies focused on system-wide analysis of bridges. This concept of reliance on systematized procedures for making bridge programming decisions was a sharp departure from previous practice of applying engineering expertise on a case-by-case basis. A North Carolina study (Johnston and Zia, 1984; Chen and Johnston, 1987) established a level of service to evaluate the adequacy of the state's bridges in serving public needs and developed procedures for optimizing system-wide level of service. Pennsylvania (Bridge Management Work Group, 1987), Virginia (McKeel and Andrews, 1985), Nebraska (Committee on Bridge, 1986), and Kansas (Kansas Department of Transportation, 1984) all developed priority systems for bridge project selections. Several studies were performed to develop methods to predict bridge condition deterioration rates (Fitzpatrick et al., 1981; Hymon et al., 1983; Busa et al., 1985a, 1985b). Efforts were also made to develop methods for bridge life-cycle cost and cost-benefit analysis (Lemmerman 1983; McFarland et al., 1983). Optimization theories and techniques have been explored for their applications in maximizing system-wide benefit with respect to highway and bridge systems (McFarland et al., 1983; Subramanian et al., 1983).

Based on the results of these individual studies, the Federal Highway Administration (FHWA) conducted a demonstration project DP-71 in 1987 to promote research and development of comprehensive bridge management system (BMS). Since then, many BMSs have been developed and implemented in many countries as well as in the United States. Although no two BMSs are exactly the same, the basic characteristics and functions of different BMSs are essentially same or similar. Because of the complexity of bridge management, BMSs are usually computerized so that the complicated mathematical and statistical calculations are handled by computers.

Among the many BMSs in operation, Pontis (Cambridge Systematics 1996) and BRIDGIT (National Engineering Technology, 1994) are the two most commonly applied BMSs in the United States. Both of these systems provide users with a powerful computer package that is capable of managing bridge systems with thousands of bridges.

In addition to Pontis and BRIDGIT, there are several BMSs that have been developed and implemented by various individual states. The Indiana Bridge Management System (IBMS) was first developed in 1989 through a joint effort between the Indiana Department of Transportation and Purdue University (Sinha et al., 1989) and has been continuously enhanced since then. The IBMS includes four modules: decision tree, economic analysis, ranking, and optimization (Woods 1994). The Alabama DOT has been working on developing and improving the Alabama Bridge Management System (ALBMS) since 1989 (Green and Richardson, 1994). The ALBMS contains incidental module, data capturing module, and data analysis/manipulation module. The North Carolina DOT started to develop its BMS in 1982 (Johnston and Lee, 1994). The North Carolina BMS has the capability to assess the optimum timing and selection of bridge projects and to determine the optimum use of constrained budgets. The Pennsylvania DOT began development of its BMS in 1984 and has been implementing the BMS since 1986 (Oraves, 1994). The Pennsylvania's BMS has a data storage and analysis subsystem, a bridge rehabilitation and replacement subsystem, a bridge maintenance subsystem, a bridge performance modeling system, an automated permit rating and routing system, and a report subsystem.

22.2   Components and Functions of Bridge Management System

A comprehensive BMS is defined as a systematic process of making maintenance, rehabilitation, and replacement decisions on a given population of bridges based on comprehensive bridge condition inspection and bridge data analysis and subject to available funds for preserving the given bridge system. A BMS typically contains defined procedures or models for bridge condition inspection and recording, bridge condition prediction, bridge cost analysis, and bridge project selection. These procedures or models require applications of many theories and techniques of mathematics, statistics, and computer programming. Typically, a BMS has an established bridge inspection program, is capable of assessing the current condition and predicting the future condition of the bridges, contains the models for bridge cost analysis, and provides the functions to yield alternative bridge project plans. Although BMS software packages enable users to obtain mathematical and statistical results with minimal effort or knowledge of the underlying theories, it is essential for bridge engineers and managers to understand the concepts, principles, and characteristics of bridge management so that a BMS can be effectively implemented.

The objective of a BMS is to assist bridge engineers and managers to make decisions on selection and planning of bridge maintenance, rehabilitation and replacement projects with limited and constrained budgets so that the total system benefit can be maximized. A BMS involves a system level of decision making process, which is different from a project level of decision making process that treats each bridge in isolation. When managing a bridge system, a great number of combinations and tradeoffs between costs and benefits come into play. System analysis techniques must be utilized to choose the projects with an optimal combination among the hundreds or thousands of possible tradeoffs. That is, the impacts of individual bridge projects on the bridge system must be determined from a system-wide perspective in order to select the bridge projects that will maximize the total or system benefit. This project selection process is an optimization process that considers many variables and funding constraints. It is a complex process and requires the use of techniques from different disciplines. A comprehensive BMS performs system analysis that is different from evaluating bridges one at a time (FHWA, 1987). The BMS's system analysis approach has the following characteristics:

1. It compares the benefits and costs of project alternatives for individual bridges and combination of a number of bridges to select the optimal combination of bridge projects for limited funding. In contrast, evaluation of a single bridge is compartmentalized. The treatments to a bridge are not compared to those of other bridges in terms of system benefits and costs.
2. The benefits and costs of all types of treatments to bridges are quantified with respect to their impacts on the bridge system, so that system-wide benefits can be maximized with the optimal combination of bridge projects. Isolated bridge project evaluation fails to consider bridges from a system-wide perspective.
3. It is inter-temporal and dynamic, but project analysis is static. A BMS examines the tradeoffs and implications of bridge treatments over time and then determines a set of actions that is systematically and dynamically optimal.
4. It is concerned with how different variables affect the behavior of the system. The variables include type and timing of bridge treatments and budget constraints. A BMS has the capability of analyzing how varying any or combination of these variables affects the output of the system.

The ultimate goal of a BMS is to provide bridge managers with scheduled bridge projects that will maximize the system benefit for a given budget. Bridge managers can use the BMS selected list of bridge projects as a basis for making decisions on maintenance, rehabilitation and replacement of bridges. To optimize project selections, it is necessary for a BMS, as a minimum, to have the following major components:

1. bridge condition data collection and management;
2. bridge performance analysis and prediction;
3. bridge cost analysis and need assessment;
4. bridge project selection and system optimization.

Each of these components is an integral part of BMS and performs unique functions, as described below.

22.2.1 Data Collection and Management

This component of a BMS is the basis of the whole system. A BMS applies scientific principles and techniques from many disciplines to generate the final results. All of these complicated operations, however, are performed on the historical and current bridge conditions. Therefore, only a sound data collection and management component can provide quality input for a BMS to perform as desired. No matter how sophisticated a BMS is, it can not produce meaningful or useful results without a bridge condition database of high quality. As the phrase "garbage in, garbage out" implies, a bridge condition database of low quality will certainly cause a BMS to generate undesirable or even harmful results. The reason for the BMS results being harmful is that the results would mislead bridge managers to take actions that are supposed to maximize system benefit, but in fact would utilize the limited resources in a wasteful manner. A database for a BMS generally contains bridge design data (bridge type, dimensions, and materials), bridge inspection data (condition ratings of bridge components), construction and maintenance records, and traffic volume and composite (vehicle types). The functions of data collection and management include: 1). inspecting bridge conditions at regular time intervals and following standardized procedures, 2). screening bridge data to minimize errors, 3). storing historical bridge data, including design, construction, and condition data, in the format required by BMS, and 4). updating bridge data after each inspection.

22.2.2 Bridge Performance Analysis and Prediction

A BMS is to assist bridge managers in making consistent and cost-effective decisions related to maintenance, rehabilitation, and replacement of bridges. Bridge structural condition is the most important variable considered in the process of bridge project selection. The decision making, either at the system level or at the project level, is based on bridge conditions at present and in the future. The accuracy of the future condition prediction directly affects the effectiveness of BMS's optimization results. This component of a BMS is to analyze the trends and patterns of bridge condition deterioration and to predict the future structural conditions of individual bridges based on the historical and present conditions. That is, bridge performance in BMS means bridge conditions at different points in time. Bridge performance prediction is also called as bridge condition deterioration prediction. After the future conditions of a bridge are predicted, the maintenance and rehabilitation needs of the bridge at different times can be planned according to the future conditions. For each bridge, various maintenance and rehabilitation options are usually proposed and then compared to determine the optimal option.

22.2.3 Bridge Cost Analysis and Need Assessment

A bridge requires periodic maintenance and rehabilitation until the end of its service life when the bridge cannot be used safely and a new bridge must be built to replace it. In addition to maintenance, rehabilitation and replacement costs, user costs may also be incurred for bridges with functional deficiency. User costs are the extra vehicle operating costs related to detour, lost travel time, and higher accident rates as a result of deteriorated deck, load posting or clearance restriction. Costs for a bridge include a series of costs for maintenance, rehabilitation and replacement, and the user costs throughout the bridge's service life. A bridge's life-cycle cost can be calculated once the maintenance and rehabilitation activities are planned for the bridge's entire service life. A life-cycle cost is usually expressed as the present worth of all the costs during the service life. These activities are planned according to the predicted bridge conditions from a BMS's performance prediction component. In addition to life-cycle cost analysis, benefit and cost ratios are also commonly utilized to compare bridge maintenance and repair options. Typically, several sets of maintenance and rehabilitation activities are proposed for each bridge. The different bridge activity profiles can be compared through cost analysis in terms of benefits and costs to determine the most cost-effective activity profile.

A large percentage of bridges in the United States are either functionally obsolete or structurally deficient. A functionally obsolete bridge is a bridge with substandard width or clearance that would restrict some types of vehicles from using the bridge. A structurally deficient bridge is a bridge with condition ratings lower than a critical value that a remedy action is needed to make it safe for use. A need is defined as the necessary action and its associated cost to upgrade a bridge condition to meet the functional and structural requirements of the bridge. With respect to BMS, a need is often defined as the most net beneficial improvement to each bridge on the given highway system (FHWA, 1987). To obtain a need estimate, it is necessary to compare a range of alternative improvement options for each bridge and determine which one maximizes net benefits. A net benefit of a bridge improvement option is defined as the difference between the life-cycle cost of the improvement alternative and the life-cycle cost of a selected less expensive improvement alternative. The less expensive improvement alternative is called the base alternative. A do-nothing or maintenance only alternative can also serve as a base alternative. For example, a bridge will need to be replaced in 5 years if no rehabilitation is done, i.e., the do-nothing option, and the life-cycle cost will be $700,000. If limited deck rehabilitation is performed now to the bridge, the bridge will need to be replaced in 15 years instead of in 5 years and the life-cycle cost will be $500,000. Then the net benefit of the deck rehabilitation option is $700,000 $-$ $500,000 $=$ $200,000. For a BMS, the need of each bridge on the highway system must be determined in order to maximize the system benefit. With the needs of all bridges, the present and future system-wide needs can be shown along the time line. The system-wide needs are then utilized as the basis for optimizing the total system benefit.

22.2.4   Project Selection and System Optimization

The results from the above components provide the estimated future conditions, proposed bridge maintenance and rehabilitation activities and costs, and expected bridge condition improvement and benefit from these activities. These results plus the amount of expected or available budgets are used in BMS as the input for project selection and system optimization. Bridge project selection is a typical multi-objective decision making process that involves many interrelated factors such as, bridge condition, safety, and cost. Selecting bridge projects through optimization can assure optimal use of resources. The optimal solution is obtained either by maximizing the system benefit or by minimizing the total negative effect on the system by undertaking the selected bridge projects. Optimization techniques select bridge projects that contribute the most benefit to the bridge system while all of the constraints are satisfied simultaneously. In the process of optimization, all possible tradeoffs among benefit, cost, and constraints are considered systematically and mathematically so that the limited budget can be spent efficiently to achieve the maximum system benefit. Therefore, this component of a BMS is to produce an optimal selection of scheduled bridge maintenance, rehabilitation, and replacement projects through optimization techniques. For each selected bridge, the results of this component will specify when the bridge project should be undertaken, and what types of maintenance or rehabilitation activities should be performed. The list of scheduled bridge projects produced by the BMS can then be utilized by bridge managers as a basis for making final decisions on bridge projects.

   Literature shows that no two existing BMSs contain exactly the same components and many BMSs contain more components than the four components discussed above. However, these four components are the basic and essential elements of bridge management systems. The typical process of a BMS with the basic components is exhibited in Figure 22.1.

22.3   Bridge Condition Data Collection and Management

22.3.1   Provisions of the NBIS

There are five major provisions of the NBIS:

1. *Inspection Procedures*: Bridge inspections are to be performed in accordance with the Manual for Maintenance Inspection of Bridges by the American Association of State Highway and Transportation Official (AASHTO, 1983).
2. *Frequency of Inspections*: Typically, each bridge should be inspected once every two years. Each state may determine certain types or groups of bridges to be inspected at intervals less than two

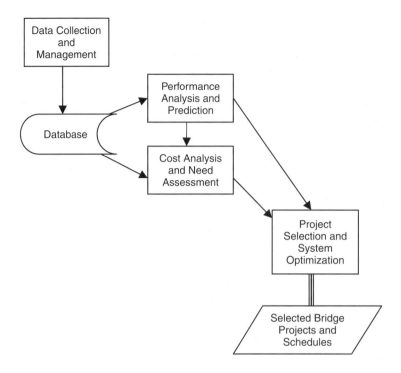

FIGURE 22.1 Typical process of bridge management system.

years and establish appropriate frequency and depth of inspections. A state may request approval from FHWA to inspect certain types or groups of bridges for inspections at intervals longer than two-year intervals. Once approved by FHWA, these bridges on the approval list may be inspected at the specified intervals exceeding two years.

3. *Qualifications of Inspection Personnel*: Qualifications for two types of bridge inspection personnel are specified: 1). the individual in charge of the organizational unit that is responsible for bridge inspection, reporting, and inventory, and 2). the bridge inspection team leader.

   To be qualified as an individual in charge of the organizational unit, one must satisfy at least one of the following qualifications:
   1) Registration as a professional engineer;
   2) Eligibility for registration as a professional engineer;
   3) Completion of a comprehensive course in bridge inspection and a minimum 10 years of bridge inspection experience.

   To be qualified as a bridge inspection team leader, one must satisfy at least one of the following qualifications:
   1) Registration as a professional engineer;
   2) Eligibility for registration as a professional engineer;
   3) Completion of a comprehensive course in bridge inspection and a minimum five years of bridge inspection experience;
   4) Current certification as a Level III or IV Bridge Safety Inspector under the National Society of Professional Engineer's Program for National Certification in Engineering Technologies.

4. *Inspection Reports*: The inspection findings and results of each bridge are recorded on standard forms. The information recorded for each bridge includes identification, classification, structure data, condition of bridge components, structural condition appraisal, and proposed improvements and associated costs.

5. *Inventories*: It is required for each state to prepare and maintain an inventory of all bridge structures subject to the Standards. The structure inventory and appraisal data must be collected and entered in the inspection reports and the computer inventory files as promptly as practical. Information on newly completed structures and modification of existing structures must be included in the inventory within 90 days after the change in the status for the state owned bridges and within 180 days for all other bridges on public roads within the state.

22.3.2 Structure Inventory and Appraisal of Bridges

To assess bridge conditions, bridges must be inspected regularly following specified procedures. The FHWA specifies the required data items and criteria in its publication, *Recording and Coding Guide for the Structure Inventory and Appraisal of the Nation's Bridges* (FHWA, 1988). Since the first version of this guide was published in 1971, it has been revised several times. The 1988 version of the guide requires 116 items for each bridge to be included in the bridge inventory database. In addition to the recording and coding guide, FHWA has also prepared inspection guidance documentation for a series of special emphasis items such as culverts, moveable bridges, scour, and fracture-critical bridges members. These documents have provided uniform guidelines for states to inspect and record bridge data. The bridge database created by the states will be used for the FHWA to make reports to the Congress for future legislation. The bridge database is also a major source for states' BMSs.

The inspection items listed in the recording and coding guide include several categories such as, bridge identification, classification, structure, and proposed improvements. Among the 116 required inspection items, the most important ones for BMS are those related to the condition data of the bridge and its components. The bridge components to be inspected include deck, superstructure, substructure, channel and channel protection, culvert and retaining walls, and approach roadway alignment (in relation to the effect on the use of the bridge). Based on the specified criteria, a condition rating ranging from 0 to 9, with 9 being the condition of a new bridge, is assigned to each of these components. In addition, an appraisal rating, also ranging from 0 to 9, is assigned to the bridge's functional characteristics. The items for appraisal rating include structural condition, deck geometry, clearance (vertical and horizontal), safe load capacity, waterway adequacy, and approach roadway alignment (in relation to the adequacy of approach roadway). An appraisal rating is used as an indication of the effect of the rated item on the bridge as a unit. Table 22.1 shows the descriptions of the condition and appraisal ratings as defined by FHWA (1988).

Besides the 116 required inspection items, many states collect additional data items to satisfy their own needs for bridge management. A survey by Turner and Richardson (1994) shows that the median number of data items lies between 270 items in California and 280 items in North Carolina, and the maximum number of data items is 700 data items in New York. The survey also shows that it takes 2 to 16 hours to inspect a bridge, depending on the number of data items and the number of persons per inspection team. It has been a common practice to collect bridge data by filling in paper forms in the field and then entering the recorded data into computer in the office. However, more and more bridge inspectors enter field data directly into portable computers. Small size computers, such as, palm pads, are easy for inspectors to carry around on a bridge during inspection. As palm-held computers become more powerful, it is increasingly common for bridge inspectors to directly enter inspection data into portable computers in the field.

Some of the bridge data items, such as traffic counts, are not directly measured by inspectors. Instead, they are obtained from other data sources. These data values may be entered manually or transferred electronically from other database. Bridge inspectors often take photos of bridges during inspections. A bridge database with photos, sketches, and narrative information is very helpful for bridge managers to plan bridge maintenance and rehabilitation activities. The availability of digital cameras has made it easy to include bridge photos into bridge database. Many bridge database programs allow bridge managers to view bridge photos on computer screen and thus to be able to connect condition ratings with actual images of bridge defects.

TABLE 22.1   Descriptions of Bridge Rating

Rating	Bridge Components	Appraisal
9	New condition	Conditions superior to present desirable criteria
8	Good condition — no repair is needed	Condition equal to present desirable criteria
7	General good condition — potential exists for minor maintenance	Condition better than present minimum criteria
6	Fair condition — potential exists for major maintenance	Condition equal to present minimum criteria
5	General fair condition — potential exists for minor rehabilitation	Condition somewhat better than minimum adequacy to tolerate being left in place as is
4	Marginal condition — potential exists for major rehabilitation	Condition meeting minimum tolerate limits to be left in place as is
3	Poor condition — repair or rehabilitation required immediately	Basically intolerable condition requiring high priority of repair
2	Critical condition — the need for repair or rehabilitation is urgent. Facility should be closed until the indicated repair is complete	Basically intolerable condition requiring high priority of replacement
1	Critical condition — facility is closed. Study should determine the feasibility for repair	Immediate repair necessary to put back in service
0	Critical condition — facility is closed and is beyond repair	Immediate replacement necessary to put back in service

22.4   Bridge Performance Analysis and Prediction

Bridge condition rating is the most important variable considered in the process of bridge project selection. The decision making, either at the system level or at the project level, is based on bridge conditions at present and in the future. The accuracy of the future condition prediction directly affects the outcome of optimization in selecting bridge projects. Therefore, it is essential for a BMS to have the capacity of accurately predicting future conditions.

The objective of bridge performance prediction, also often referred as bridge condition deterioration prediction, is to estimate the future condition of individual bridges based on the historical bridge condition data. Regression analysis and Markov process are the two most commonly utilized methods for bridge performance prediction. Regression analysis is applied to develop equations to express the condition ratings of bridge elements as a function of such variables as bridge age, bridge type, and traffic volume. A Markov process in bridge performance prediction is a probability-based model to estimate condition rating changes at different points in time. The two methods of bridge performance prediction are discussed in the following sections.

22.4.1   Regression Models for Bridge Performance Prediction

The objective of a bridge performance prediction model is to predict the future conditions of bridge components. In regression terms, the dependent variable is the condition rating of a bridge component as a function of one or more independent variables. It is essential to predict the conditions of bridge deck, superstructure, and substructure for a BMS. Since condition deterioration rates are different for deck, superstructure and substructure, a separate bridge performance prediction model should be developed for each of the components. There are various types of bridge performance models based on the statistical regression theory. Presented below are some examples of regression models for bridge performance prediction.

22.4.1.1 Linear Regression with Single Independent Variable

Bridge condition is affected by many factors, including design, age, type, traffic volume, weather, and construction quality. Among these factors, bridge age is most highly correlated with condition rating of bridge components. Because of its simplicity, the linear regression model with one independent variable has been widely applied in bridge performance prediction. Such a simple linear regression model uses bridge age as the independent variable and bridge condition rating as the dependent variable. For each type of bridges, three linear regression equations can be developed for bridge components as follows:

$$\text{Deck}_i = \beta_0 + \beta_1 T_i + \varepsilon_i \quad (22.1)$$

$$\text{Sup}_i = \beta_0 + \beta_1 T_i + \varepsilon_i \quad (22.2)$$

$$\text{Sub}_i = \beta_0 + \beta_1 T_i + \varepsilon_i \quad (22.3)$$

where Deck_i = Condition rating of deck, from 0 to 9; Sup_i = Condition rating of superstructure, from 0 to 9; Sub_i = Condition rating of substructure, from 0 to 9; T_i = bridge age, in years; β_0, β_1 = the parameters to be estimated; and ε_i = random error term with a standard normal distribution.

Figures 22.2 to 22.4 show some examples of regression equations and their corresponding curves. The regression equations are obtained by the method of least squares. The bridge data for this simple linear regression model includes the condition ratings of bridge components and their corresponding bridge ages. To predict the future condition rating of a bridge component, one simply inserts the given bridge age (T) into the appropriate regression equation. For example, the condition rating of a bridge deck at 40 years of age can be predicted using the regression equation in Figure 22.2 as: $\text{Deck} = 9 - 0.117(40) = 4.32 \approx 4$. That is, the condition rating of a 40-year old bridge deck is expected to be 4.

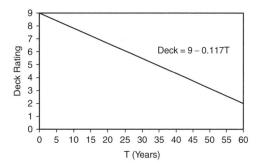

FIGURE 22.2 Example of deck condition regression equation.

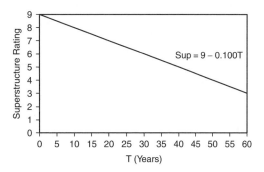

FIGURE 22.3 Example of superstructure condition regression equation.

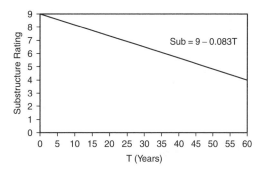

FIGURE 22.4 Example of substructure condition regression equation.

Apparently, these regression equations exclude all other factors, such as, traffic volume, bridge type, and weather condition. The significance of these factors should be determined through statistical multifactor studies (Neter et al., 1985). If a factor is determined to be statistically significant, the linear regression model with single independent variable may be used by dividing bridges into groups according to the factor. For example, if traffic volume is found to be a significant factor of bridge condition, the bridges should be divided into groups according to the levels of traffic volumes, such as bridge groups with high, medium, or low traffic volumes. Thus, regression equations can be derived for each of the groups. The bridge groups can be further divided into subgroups based on additional factors if necessary. Such factors may include type of bridges (e.g., steel bridges and concrete bridges), type of highways (e.g., interstate and non-interstate), and location (e.g., north and south). Within each subgroup, the single independent variable regression equations can be developed for each of the bridge components. By grouping bridges, appropriate factors can be taken into account, so that the accuracy of bridge condition predictions can be improved.

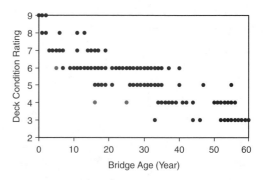

22.4.1.2 Piecewise Linear Regression Model

FIGURE 22.5 Distribution of deck condition rating.

The linear regression with single independent variable has been commonly applied in bridge condition predictions because of its simplicity. However, this model may not fit actual bridge condition well. Figure 22.5 is a plot of bridge deck condition ratings versus bridge ages from a sample data of the concrete bridges on Indiana's non-interstate highways. The plot exhibits that deck condition ratings decreased at a fast pace in the early years of bridge life, and then became relatively more stable for a period of time before a quick drop in the late years. The condition rating distribution indicates that the condition deterioration rate changes with bridge age. That is, the regression of condition rating on bridge age follows a particular

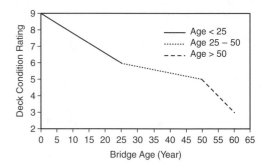

FIGURE 22.6 Piecewise regression of deck condition rating.

linear relation in a range of bridge age, and follows a different linear relation in another range of bridge age. The piecewise linear regression method can be applied to fit different linear regressions for different ranges of independent variable values.

To illustrate piecewise regression, the deck condition data shown in Figure 22.5 is used to develop a three-piecewise regression curve. The three-piecewise regression curve should have different slopes at the following three ranges of bridge ages: from age 0 to 25, from age 25 to 50, and age older than 50. The piecewise regression equation can be expressed as follows:

$$Y_i = \beta_0 + \beta_1 X_{i1} + \beta_2 (X_{i1} - 25) X_{i2} + \beta_3 (X_{i1} - 50) X_{i3} + \varepsilon_i \tag{22.4}$$

where Y_i = deck condition; β_0, β_1, β_3 = parameters to be estimated; X_{i1} = bridge or deck age; X_{i2} and X_{i3} = indicator variables; $X_{i2} = 1$ if $25 \leq X_{i1} \leq 50$, $X_{i2} = 0$ otherwise; $X_{i3} = 1$ if $X_{i1} > 50$; $X_{i3} = 0$ otherwise.

Figure 22.6 shows the three-piecewise regression curves for the deck condition rating data. As can be seen from the figure, the deterioration rates of deck condition are different at the three age periods. The connected segments of straight lines fit bridge condition data better than a single straight line of linear regression (Fitzpatrick et al., 1981; Hymon et al., 1983).

22.4.1.3 Polynomial Regression Model

The second-order or third-order polynomial regression model can be utilized when the relation between the dependent variable and independent variable is not a straight line. For example, a third-order polynomial regression model with one independent variable is expressed as follows:

$$Y_i = \beta_0 + \beta_1 X_i + \beta_2 X_i^2 + \beta_3 X_i^3 + \varepsilon_i \tag{22.5}$$

Although in Equation 22.5 the independent variable appears in the second and third power, it can be treated as a special case of the general linear regression model. The conversion of the polynomial equation to a general linear model can be achieved as follows. Let $X_{i1} = X_i$, $X_{i2} = X_i^2$ and $X_{i3} = X_i^3$, then Equation 22.5 can be written as:

$$Y_i = \beta_0 + \beta_1 X_{i1} + \beta_2 X_{i2} + \beta_3 X_{i3} + \varepsilon_i \quad (22.6)$$

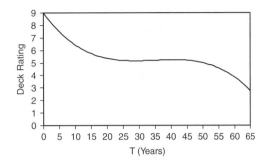

FIGURE 22.7 Polynomial regression of deck condition.

Equation 22.6 is a general linear model with three independent variables. That is, the polynomial regression model is a particular case of the general linear model. Therefore, the least square method for linear regression can be applied to fit the polynomial regression model. An example of polynomial regression equation for deck condition rating is given below (Jiang and Sinha 1989):

$$\text{Deck} = 9 - 0.3498T + 0.0104T^2 - 0.0001T^3 \quad (22.7)$$

The corresponding polynomial curve of this regression equation is plotted in Figure 22.7. As depicted by the polynomial curve, the deck condition ratings deteriorated rapidly at the beginning of a bridge's life, then became more stable as the bridge age increased and fell rapidly again after the condition rating reached five or less. The trend of the polynomial curve would match well with the bridge rating data shown in Figure 22.5.

Polynomial regression method was successfully applied to predict bridge conditions in Indiana (Jiang and Sinha, 1989). As can be seen from Figures 22.6 and 22.7, the piecewise regression model and the polynomial regression model show similar trends of condition deterioration rates. However, the polynomial regression model is easier to use because it uses a single equation throughout a bridge's service life, while the piecewise regression uses different equations for different periods of a bridge life.

22.4.1.4 Linear Regression with More Than One Independent Variable

Theoretically, the general linear regression model with more than one variable can be utilized to establish the relationship of condition rating and two or more variables. However, because the complexity of regression will be significantly increased with each additional independent variable, linear regressions with more than two independent variables are rarely used in BMS. Linear regression with two independent variables, bridge age and average daily traffic (ADT), was applied to predict the conditions of bridge deck, superstructure, and substructure (Busa et al., 1985a, 1985b). The regression equations are in the following general form:

$$Y_i = \beta_0 + \beta_1 X_{i1} + \beta_2 X_{i2} + \varepsilon_i \quad (22.8)$$

where Y_i = condition rating of deck, superstructure, or substructure; X_{i1} = bridge age; X_{i2} = ADT; β_0, β_1 and β_2 = parameters to be estimated; and ε_i = error term.

Because ADT is included in the regression equation, traffic volume will not be used to divide bridge into sub-groups. However, bridges should be divided into groups according to other factors that are significantly correlated with bridge conditions. Such factors may include bridge type and climate region. The statistical methods for determining the significances of factor effects are beyond the scope of this chapter, but they can be found in many statistics books.

22.4.2 Applications of Regression Models

Linear regressions with single independent variable, including polynomial regression, have been widely used in bridge management because of their simplicity in model development and in interpretation of regression curves. A curve of linear regression with single independent variable represents the mean values of bridge condition ratings at different bridge ages. For example, in Figure 22.7, the value on the curve corresponding to age 13 is rating 6, indicating that the average or mean condition rating of bridge decks at age 13 is 6. That is, even though individual bridge decks at age 13 may have condition ratings above or below 6, the average value of the condition ratings of these bridge decks is expected to be 6.

The first application of regression models is to predict the average condition ratings of bridge components at different ages. In other words, the average condition ratings are used in regression models to represent the general trend of bridge condition deterioration.

The second application of regression models is to find the deterioration rates at different ages. From a performance curve, such as the polynomial regression curve shown in Figure 22.8, one can see that the deterioration rate of bridge condition is different at different ages. This figure shows a tangent line on the polynomial regression curve. The absolute value of $\tan(\alpha)$ is the deterioration rate at age t. It is evident that when the value of

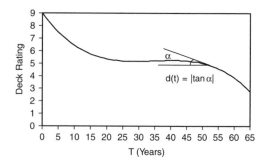

FIGURE 22.8 Determination of deterioration rate on regression curve.

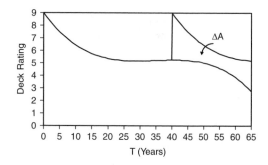

FIGURE 22.9 Area of benefit gained by rehabilitation.

deterioration rate is small, rehabilitation may be delayed for a period of time. Otherwise, an immediate repair should be applied because a rapid rehabilitation is expected due to the large deterioration rate. It should be noted that since a bridge with a condition rating greater than 6 does not need rehabilitation, the tangent values on the curve section between condition rating of 6 and 9 are not considered in selecting bridge rehabilitation activities, even though the deterioration rate is relatively large at the curve segment.

The third application of regression models is to define the improvement benefit gained by rehabilitation. When a rehabilitation activity is performed on a bridge, the condition rating of various bridge components increases depending on the type of improvement. As shown in Figure 22.9 a particular rehabilitation activity causes a jump in the deck condition rating. As the bridge age increases, the condition rating gradually decreases from the new condition rating. The area between the deterioration curves of the bridge with and without rehabilitation, ΔA, represents a benefit gained in terms of condition rating and service life of the bridge. This benefit can be used as an important input for bridge project selection.

22.4.3   Markov Chain Approach

Markov chain is a special case of stochastic processes. The theory of stochastic processes has been applied in many areas of engineering and other applied science. In 1980s the theory was applied in pavement management system (Butt et al., 1987) and in BMSs (Jiang and Sinha, 1989). Essentially, a stochastic process is a probability-based process describing the changes of random variables in time.

22.4.3.1   Markov Process

A stochastic process is said to be Markovian if given the value of bridge condition $X(t)$ at time t, the value of $X(s)$ for $s > t$ does not depend on the value of $X(\mu)$ fro $\mu < t$. In other words, the future behavior

TABLE 22.2 Correspondence of Condition Ratings, States, and Transition Probabilities

		$R=9$ $S=1$	$R=8$ $S=2$	$R=7$ $S=3$	$R=6$ $S=4$	$R=5$ $S=5$	$R=4$ $S=6$	$R=3$ $S=7$	$R=2$ $S=8$	$R=1$ $S=9$	$R=0$ $S=10$
$R=9$	$S=1$	$p_{1,1}$	$p_{1,2}$	$p_{1,3}$	$p_{1,4}$	$p_{1,4}$	$p_{1,5}$	$p_{1,6}$	$p_{1,7}$	$p_{1,8}$	$p_{1,9}$
$R=8$	$S=2$	$p_{2,1}$	$p_{2,2}$	$p_{2,3}$	$p_{2,4}$	$p_{1,4}$	$p_{2,5}$	$p_{2,6}$	$p_{2,7}$	$p_{2,8}$	$p_{2,9}$
$R=7$	$S=3$	$p_{3,1}$	$p_{3,2}$	$p_{3,3}$	$p_{3,4}$	$p_{3,4}$	$p_{3,5}$	$p_{3,6}$	$p_{3,7}$	$p_{3,8}$	$p_{3,9}$
$R=6$	$S=4$	$p_{4,1}$	$p_{4,2}$	$p_{4,3}$	$p_{4,4}$	$p_{4,4}$	$p_{4,5}$	$p_{4,6}$	$p_{4,7}$	$p_{4,8}$	$p_{4,9}$
$R=5$	$S=5$	$p_{5,1}$	$p_{5,2}$	$p_{5,3}$	$p_{5,4}$	$p_{5,4}$	$p_{5,5}$	$p_{5,6}$	$p_{5,7}$	$p_{5,8}$	$p_{5,9}$
$R=4$	$S=6$	$p_{6,1}$	$p_{6,2}$	$p_{6,3}$	$p_{6,4}$	$p_{6,4}$	$p_{6,5}$	$p_{6,6}$	$p_{6,7}$	$p_{6,8}$	$p_{6,9}$
$R=3$	$S=7$	$p_{7,1}$	$p_{7,2}$	$p_{7,3}$	$p_{7,4}$	$p_{7,4}$	$p_{7,5}$	$p_{7,6}$	$p_{7,7}$	$p_{7,8}$	$p_{7,9}$
$R=2$	$S=8$	$p_{8,1}$	$p_{8,2}$	$p_{8,3}$	$p_{8,4}$	$p_{8,4}$	$p_{8,5}$	$p_{8,6}$	$p_{8,7}$	$p_{8,8}$	$p_{8,9}$
$R=1$	$S=9$	$p_{9,1}$	$p_{9,2}$	$p_{9,3}$	$p_{9,4}$	$p_{9,4}$	$p_{9,5}$	$p_{9,6}$	$p_{9,7}$	$p_{9,8}$	$p_{9,9}$
$R=0$	$S=10$	$p_{10,1}$	$p_{10,2}$	$p_{10,3}$	$p_{10,4}$	$p_{10,4}$	$p_{10,5}$	$p_{10,6}$	$p_{10,7}$	$p_{10,8}$	$p_{10,9}$

Note: R, condition rating; S, state; $p_{i,j}$, transition probability from state i to state j.

of the process depends only on the present state but not on the past. In formal terms, a process is said to be Markovian if

$$P[a < X(t) \le b | X(t_0) = x_0, \ldots, X(t_n) = x_n] = P[a < X(t) \le b | X(t_n) = x_n] \quad (22.9)$$

where $t_0 < t_1 < \cdots < t_n < t$.

Whenever the parameter set T is discrete, the Markovian processes are called Markov chains. In a Markov process, therefore, if the state is known for any specific value of the time parameter t, that information is sufficient to predict the behavior of the process beyond that point.

22.4.3.2 Markov Chain Approach to Performance Prediction

The Markov chain as applied to bridge performance prediction is based on the concept of defining states in terms of bridge condition ratings and obtaining the probabilities of bridge condition changing from one state to another. These probabilities are represented in a matrix form that is called the transition probability matrix or simply, transition matrix, of the Markov chain. Knowing the present state of bridges, or the initial state, the future conditions can be predicted through multiplications of initial state vector and the transition matrix.

To model the bridge condition deterioration process as a Markov chain, the ten condition ratings, from 9 to 0, are defined as ten states with each condition rating corresponding to one of the states. For example, condition rating 9 is defined as state 1, rating 8 as state 2, and so on. Without repair or rehabilitation, the bridge condition rating decreases as the bridge age increases. Therefore, there is a probability of condition changing from one state, say i, to another state, j, during a given period of time. This probability is denoted by $p_{i,j}$ as shown in Table 22.2 with the correspondence of condition ratings, states and transition probabilities.

Let the transition probability matrix of the Markov chain be **P**, given by

$$\mathbf{P} = \begin{bmatrix} p_{1,1} & p_{1,2} & \cdots & p_{1,10} \\ p_{2,1} & p_{2,2} & \cdots & p_{2,10} \\ \vdots & \vdots & \ddots & \vdots \\ p_{10,1} & p_{10,2} & \cdots & p_{10,10} \end{bmatrix} \quad (22.10)$$

Then the state vector for any given time T, $\mathbf{Q}_{(T)}$, can be obtained by the multiplication of initial state vector $\mathbf{Q}_{(0)}$ and the Tth power of the transition probability matrix **P**:

$$\mathbf{Q}_{(T)} = \mathbf{Q}_{(0)}\mathbf{PP}\cdots\mathbf{P} = \mathbf{Q}_{(0)}\mathbf{P}^T \quad (22.11)$$

Thus, a Markov chain is completely specified when its transition matrix **P** and the initial state vector are known. Since the initial state vector $\mathbf{Q}_{(0)}$ is the current condition of bridges and is usually known, the main problem of the Markov chain approach as applied in bridge management is to determine the transition probability matrix, which is discussed in the following section.

22.4.3.3   Transition Probability Matrix

The inspection of bridges includes ratings of individual components such as, deck, superstructure and substructure. Unless rehabilitation or repair is applied, bridge structures would be gradually deteriorating so that the bridge condition ratings are either unchanged or changed to a lower value during one year period. That is, a bridge condition rating should decrease as the bridge age increases. Therefore, the probability $p_{i,j}$ is null for $i > j$, where i and j represent the states in the Markov chain.

A Markov process requires a presumption of homogeneity (Bhat and Miller, 2002). Since the rate of deterioration of bridge condition is different at different bridge ages, the transition process of bridge condition is not homogeneous with respect to bridge age. Thus, if only one transition matrix were used throughout a bridge's life span, the inaccuracy of condition estimation would occur as a result of non-homogeneity of the condition transition process. To avoid overestimating or underestimating the bridge condition, the bridge's service life should be divided into several age groups and a transition matrix should be developed for each of the age group. The Indiana study (Jiang and Sinha 1989) found that six-year age groups were appropriate for Indiana's bridge data base. That is, a transition matrix must be determined for each six-year period of the bridge age. Therefore, in order to study bridge performance up to sixty years, ten transition matrices must be developed for the ten six-year age periods.

To obtain the transition probabilities, the number of transitions during one year period from condition state i to condition state j for all the condition states defined in Table 22.2 must be known. Let $n_{i,j}$ denote the number of transitions from state i to state j in one year, then the number of bridges in state i before the transition can be defined as:

$$n_i = \sum_j n_{i,j} \tag{22.12}$$

It can be proved (Bhat and Miller 2002) that the estimated transition probability is

$$\hat{p}_{i,j} = \frac{n_{i,j}}{n_i} \tag{22.13}$$

This method of estimating transition probabilities is simple and straightforward. In order for the estimations to be accurate, it requires a sufficient number of transitions from one condition state to every other condition state for each year of every age group. However, the condition state transitions may not always exist for some time period or the number of transitions may not always be sufficient for probability estimation. It is because there are more old bridges than new bridges, so that there may not be sufficient bridges in all age periods to develop transition probabilities. In addition, there may not be sufficient number of bridges in certain age ranges to cover all condition ratings for estimating transition probabilities. For example, very few bridges younger than six year old would have condition ratings less than 6. Similarly, not many bridges older than 50 years would have condition ratings greater than 7. Although there were always some bridges with exceptionally high or low condition ratings with respect to their ages, the number of such bridges would usually not be sufficient for probability estimations.

Consequently, Equation 22.13 might not be always applicable in bridge management because of data limitations.

Jiang and Sinha (1989) derived an alternative method to obtain the transition probabilities. To simplify the transition matrix expressed in Equation 22.10, some realistic assumptions can be made according to actual bridge condition data. First, it is assumed that the bridge condition rating would not drop by more than one in a single year. This is reasonable because, in reality, bridge condition rating seldom drops more than one in a single year. Second, it is assumed that the lowest bridge condition rating is 3, because it is a FHWA requirement that a bridge be repaired or replaced when its condition rating reaches 3. Any bridge with condition rating below 3 must be closed due to safety concerns. With the two assumptions, the transition matrix of condition ratings has the form:

$$\mathbf{P} = \begin{bmatrix} p(1) & q(1) & 0 & 0 & 0 & 0 & 0 \\ 0 & p(2) & q(2) & 0 & 0 & 0 & 0 \\ 0 & 0 & p(3) & q(3) & 0 & 0 & 0 \\ 0 & 0 & 0 & p(4) & q(4) & 0 & 0 \\ 0 & 0 & 0 & 0 & p(5) & q(5) & 0 \\ 0 & 0 & 0 & 0 & 0 & p(6) & q(6) \\ 0 & 0 & 0 & 0 & 0 & 0 & 1 \end{bmatrix} \tag{22.14}$$

where $q(i) = 1 - p(i)$. In Equation 22.14, $p(i)$ corresponds to $p_{i,i}$ and $q(i)$ to $p_{i,i+1}$ in Table 22.2 or Equation 22.10. Therefore, $p(1)$ is the transition probability from rating 9 (state 1) to rating 9, and $q(1)$, from rating 9 to rating 8 (state 2), and son on. It should be noted that the lowest rating number before a bridge is repaired or replaced is 3. Consequently, the corresponding transition probability $p(7)$, or $p_{7,7}$, equals to 1.

To estimate the transition matrix probabilities, for each age group the following non-linear programming objective function can be used (Jiang and Sinha, 1989):

$$\min \sum_{t=1}^{N} |Y(t) - E(t, \mathbf{P})| \tag{22.15}$$

Subject to: $0 \le p(i) \le 1, i = 1, 2, 3, \ldots, I$ where $N =$ the number of years in one age group. Jiang and Sinha (1989) used $N = 6$, which is also the number of unknown transition probabilities in Equation 22.14, so that the number of unknowns is equal to the number of equations in the objective function; $I =$ the number of unknown transition probabilities; $\mathbf{P} = [p(1), p(2), \ldots, p(I)]$, a vector of transition probabilities of length I; $Y(t) =$ the condition rating at time t, estimated by a regression function, e.g., Equation 22.5; $E(t, \mathbf{P}) =$ estimated condition rating by Markov chain at time t.

The objective function is to minimize the absolute distance between the average condition rating estimated by regression and the predicted condition rating generated by the Markov chain with the transition probabilities. The solution to the non-linear programming provides the values of the six transition probabilities, $p(1)$ through $p(6)$, in Equation 22.14. Since $q(i) = 1 - p(i)$, the values of $q(1)$ through $q(6)$ can also be obtained. The non-linear programming (Equation 22.15) can be solved using mathematic computer software.

The maximum rating of bridge condition is 9 and it represents a near-perfect condition of a bridge component. It is almost always true that a new bridge has condition rating 9 for all of its deck, superstructure and substructure. In other words, a bridge at age 0 has condition rating 9 for its components with unit probability. Thus, the initial condition state vector $\mathbf{Q}_{(0)}$ for deck, superstructure or substructure is always [1, 0, 0, 0, 0, 0, 0], where the numbers are the probabilities of having condition rating 9, 8, 7, 6, 5, 4, and 3 at age 0, respectively. Condition ratings below 3 are not included in the state vector because 3 is the minimum condition rating for a bridge to be usable. Thus, the initial state vector of the first age group for the Markov chain is known as $Q_{(0)} = [1, 0, 0, 0, 0, 0, 0]$ and can be used obtain $E(t, P)$ in Equation 22.15. With the initial state vector $Q_{(0)}$ and the transition matrix P, the state vectors for the first age group can be estimated as:

$$Q_{(1)} = Q_{(0)}P$$
$$Q_{(2)} = Q_{(0)}P^2 \tag{22.16}$$
$$Q_{(t)} = Q_{(0)}P^t$$

where $Q_{(t)}$ represents the condition state rating at age t. Once all of the state vectors are obtained for the first age group, age group 2 takes the last state vector of age group 1 as its initial state vector. Similarly, age group n takes the last state vector of age group $n-1$ as its initial state vector.

Let R be a vector of condition ratings:

$$R = \begin{bmatrix} 9 \\ 8 \\ 7 \\ 6 \\ 5 \\ 4 \\ 3 \end{bmatrix} \tag{22.17}$$

Then the estimated condition rating at age t by Markov chain is:

$$E(t, P) = Q_{(t)}R \tag{22.18}$$

or

$$E(t, P) = Q_{(0)}P^tR \tag{22.19}$$

For example, the polynomial regression function for Indiana's substructures of steel bridges on interstate highways gives the following values of predicted condition ratings for the first six years:

$$Y(1) = 8.67$$
$$Y(2) = 8.37$$
$$Y(3) = 8.10$$
$$Y(4) = 7.86$$
$$Y(5) = 7.64$$
$$Y(6) = 7.44$$

The predictions of the condition ratings by the Markov chain method can be expressed by the following functions:

$$E(1, P) = Q_{(0)}PR$$
$$E(2, P) = Q_{(0)}P^2R$$
$$E(3, P) = Q_{(0)}P^3R$$

TABLE 22.3 Transition Probabilities for Deck Condition of Concrete Bridges in Indiana

Age	$p(1)$	$p(2)$	$p(3)$	$p(4)$	$p(5)$	$p(6)$
0−6	0.700	0.780	0.874	0.600	0.500	0.400
7−12	0.690	0.770	0.870	0.720	0.610	0.540
13−18	0.690	0.780	0.950	0.850	0.760	0.660
19−24	0.616	0.720	0.980	0.970	0.930	0.850
25−30	0.560	0.700	0.980	0.980	0.950	0.940
31−36	0.520	0.680	0.980	0.980	0.970	0.960
37−42	0.480	0.620	0.980	0.980	0.970	0.960
43−48	0.460	0.600	0.980	0.980	0.930	0.900
49−54	0.440	0.570	0.970	0.960	0.900	0.880
55−60	0.400	0.500	0.800	0.820	0.750	0.600

$$E(4, \mathbf{P}) = \mathbf{Q}_{(0)}\mathbf{P}^4\mathbf{R}$$

$$E(5, \mathbf{P}) = \mathbf{Q}_{(0)}\mathbf{P}^5\mathbf{R}$$

$$E(6, \mathbf{P}) = \mathbf{Q}_{(0)}\mathbf{P}^6\mathbf{R}$$

These $E(t, \mathbf{P})$s are functions of transition probabilities, $p(1)$ through $p(6)$. Insert all polynomial regression values, $Y(1)$ through $Y(6)$ and $E(t, \mathbf{P})$s into the objective function (Equation 22.15), the non-linear programming can be solved to find the values of these transition probabilities.

22.4.3.4 Applications of the Markov Chain Model

Once the transition matrix is obtained, the prediction of the future condition by Markov chain becomes a matter of simple multiplication of matrices. Let us use the deck condition ratings of the concrete bridges on Indiana's non-interstate highways as an example. The transition probabilities for the deck conditions of concrete bridges are presented in Table 22.3 (Jiang and Sinha 1989). For illustration, $p(1) = 0.700$ for bridge age group 1 (age 6 or less) indicates that the probability of deck condition changing from state 1 (condition rating 9) to state 1, or remaining in state 1, in one year period is 0.700, and the probability of changing from state 1 to state 2 (condition rating 8) is $q(1) = 0.300$. Similarly, $p(2) = 0.780$ for age group 1 indicates that the probability of deck condition transiting from state 2 to state 2 (remaining in state 2) in one year period is 0.780, and the probability of transiting from state 2 to state 3 (condition rating 7) is $q(2) = 0.220$.

Using the values for age 0 to 6 in Table 22.3, the transition matrix of Equation 22.14 for age group 1 is:

$$\mathbf{P} = \begin{bmatrix} 0.700 & 0.300 & 0.000 & 0.000 & 0.000 & 0.000 & 0.000 \\ 0.000 & 0.780 & 0.220 & 0.000 & 0.000 & 0.000 & 0.000 \\ 0.000 & 0.000 & 0.874 & 0.126 & 0.000 & 0.000 & 0.000 \\ 0.000 & 0.000 & 0.000 & 0.600 & 0.400 & 0.000 & 0.000 \\ 0.000 & 0.000 & 0.000 & 0.000 & 0.500 & 0.500 & 0.000 \\ 0.000 & 0.000 & 0.000 & 0.000 & 0.000 & 0.400 & 0.600 \\ 0.000 & 0.000 & 0.000 & 0.000 & 0.000 & 0.000 & 1.000 \end{bmatrix} \quad (22.20)$$

As mentioned earlier, the initial state vector of the first age group (age 0 to age 6) for the components of a new bridge is:

$$\mathbf{Q}_{(0)} = \begin{bmatrix} 1 & 0 & 0 & 0 & 0 & 0 & 0 \end{bmatrix} \quad (22.21)$$

Therefore, the condition state vector and condition rating of age group 1 for year *t* can be obtained by as follows:

$$
\mathbf{R} = \begin{bmatrix} 9 \\ 8 \\ 7 \\ 6 \\ 5 \\ 4 \\ 3 \end{bmatrix}
$$

$\mathbf{Q}_{(0)} = [\,1 \quad 0 \quad 0 \quad 0 \quad 0 \quad 0 \quad 0\,]$ $\qquad\qquad$ $E(0, \mathbf{P}) = \mathbf{Q}_{(0)}\mathbf{R} = 9.0$

$\mathbf{Q}_{(1)} = \mathbf{Q}_{(0)}\mathbf{P} = [\,0.70 \quad 0.30 \quad 0.00 \quad 0.00 \quad 0.00 \quad 0.00 \quad 0.00\,]$ $\qquad$ $E(1, \mathbf{P}) = \mathbf{Q}_{(1)}\mathbf{R} = 8.70$

$\mathbf{Q}_{(2)} = \mathbf{Q}_{(0)}\mathbf{P}^2 = [\,0.49 \quad 0.44 \quad 0.07 \quad 0.00 \quad 0.00 \quad 0.00 \quad 0.00\,]$ $\qquad$ $E(2, \mathbf{P}) = \mathbf{Q}_{(2)}\mathbf{R} = 8.42$

$\mathbf{Q}_{(3)} = \mathbf{Q}_{(0)}\mathbf{P}^3 = [\,0.34 \quad 0.49 \quad 0.16 \quad 0.01 \quad 0.00 \quad 0.00 \quad 0.00\,]$ $\qquad$ $E(3, \mathbf{P}) = \mathbf{Q}_{(3)}\mathbf{R} = 8.17$

$\mathbf{Q}_{(4)} = \mathbf{Q}_{(0)}\mathbf{P}^4 = [\,0.24 \quad 0.49 \quad 0.24 \quad 0.03 \quad 0.00 \quad 0.00 \quad 0.00\,]$ $\qquad$ $E(4, \mathbf{P}) = \mathbf{Q}_{(4)}\mathbf{R} = 7.94$

$\mathbf{Q}_{(5)} = \mathbf{Q}_{(0)}\mathbf{P}^5 = [\,0.17 \quad 0.45 \quad 0.32 \quad 0.05 \quad 0.01 \quad 0.00 \quad 0.00\,]$ $\qquad$ $E(5, \mathbf{P}) = \mathbf{Q}_{(5)}\mathbf{R} = 7.72$

$\mathbf{Q}_{(6)} = \mathbf{Q}_{(0)}\mathbf{P}^6 = [\,0.12 \quad 0.40 \quad 0.38 \quad 0.07 \quad 0.02 \quad 0.01 \quad 0.00\,]$ $\qquad$ $E(6, \mathbf{P}) = \mathbf{Q}_{(6)}\mathbf{R} = 7.50$

Then, $\mathbf{Q}_{(6)}$ obtained above for age group can be taken as the initial condition state vector of age group 2 and the corresponding transition matrix for age group 2 can be used to continue the procedure. By this procedure, taking the last vector of age group $n - 1$ as the initial vector of age group n, the bridge condition at any bridge age can be predicted in terms of appropriate initial state vector and transition matrix. In this example, the Markov chain predictions are obtained by using the new bridge condition as the initial condition vector. The predicted condition ratings should either fall on or be close to the polynomial regression curve because, as indicated in the non-linear programming (Equation 22.15), the transition probabilities are obtained on the basis of regression values. Figure 22.10 illustrates that the condition rating predictions by the Markov method (the dots) and the regression method (the polynomial curve) are very close. Therefore, both Markov chain method and regression method can be used to predict the mean or average condition ratings of bridges. It should be emphasized that the predictions by the two methods are the same only if the initial condition rating falls on the regression curve. As in the above example, the initial condition rating is the condition rating of a new bridge condition, 9, which is on the regression curve at age 0.

The following example shows that the Markov chain method has great advantage over the regression method in predicting conditions of individual bridges when the initial condition rating

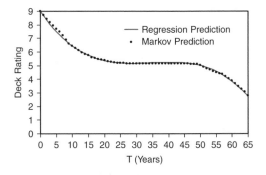

FIGURE 22.10 Comparison of Markov and regression predictions.

is not on the regression curve. Figure 22.11 presents the regression curve of concrete deck conditions, a bridge is presently 19 years old with a deck condition rating 5, which is denoted by r_{19}. It is desired to predict the deck condition rating at age 24, i.e., to predict the deck condition rating in five years. Because the probability of current condition rating being 5 is 1, the initial condition state vector is:

$$Q_{(0)} = [0 \quad 0 \quad 0 \quad 0 \quad 1 \quad 0 \quad 0].$$

From Table 22.3, the corresponding transition matrix for the given bridge age is:

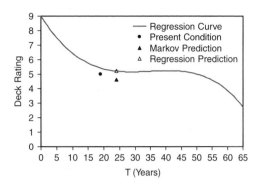

FIGURE 22.11   Condition predictions by Markov and regression models.

$$P = \begin{bmatrix} 0.616 & 0.384 & 0.000 & 0.000 & 0.000 & 0.000 & 0.000 \\ 0.000 & 0.720 & 0.280 & 0.000 & 0.000 & 0.000 & 0.000 \\ 0.000 & 0.000 & 0.980 & 0.020 & 0.000 & 0.000 & 0.000 \\ 0.000 & 0.000 & 0.000 & 0.970 & 0.030 & 0.000 & 0.000 \\ 0.000 & 0.000 & 0.000 & 0.000 & 0.930 & 0.070 & 0.000 \\ 0.000 & 0.000 & 0.000 & 0.000 & 0.000 & 0.850 & 0.150 \\ 0.000 & 0.000 & 0.000 & 0.000 & 0.000 & 0.000 & 1.000 \end{bmatrix}$$

The vector of condition ratings is:

$$R = \begin{bmatrix} 9 \\ 8 \\ 7 \\ 6 \\ 5 \\ 4 \\ 3 \end{bmatrix}$$

The deck condition is predicted using the Markov chain:

$$Q_{(5)} = Q_{(0)}P^5 = [0.000 \quad 0.000 \quad 0.000 \quad 0.0000 \quad 0.696 \quad 0.220 \quad 0.084]$$

$$E(5, P) = Q_{(5)}R = 4.612$$

That is, using the Markov chain method, the deck condition is predicted to change from condition rating 5 to 4.612 in five years.

The regression method, on the other hand, gives a prediction of condition rating at age 24 of 5.213, which is even greater than the current condition rating of 5 and therefore is apparently inaccurate. As can be seen, the regression method is appropriate only in estimating the mean or average condition rating of a group of bridges. However, the Markov chain method is useful in estimating both the average condition rating and the condition rating of a particular bridge.

The reason for the difference between the predictions from the two methods is that the Markov chain model uses the most recent available information to predict the future condition. As in the above example, the present condition rating, i.e., condition rating 5 at age 24, is used in the initial condition state vector, so that the condition rating in five years (or at age 24) is estimated based on the actual current condition of the given bridge instead of on the mean condition rating of a group of bridges. This dynamic feature of Markov chain well reflects the actual bridge condition deterioration process that the future condition is directly related to the current condition. In addition, the dynamic feature makes it possible to update bridge condition predictions at each stage of bridge project selections.

22.5   Cost Analysis and Needs Assessment

In addition to the design and construction cost, a bridge requires a series of expenditures for maintenance, repair, and rehabilitation during its service life. The bridge will be eventually replaced by a new bridge at the end of its service life. These expenditures of various amounts occur at different times during the bridge's service life. The quantities of these expenditures or costs are often depicted along a time line of the bridge age. Figure 22.12 shows an example of costs incurred during a bridge's service life.

The drawing of bridge costs along bridge service life in Figure 22.12 is called the bridge cost profile. Obviously, a bridge can have alternative cost profiles that would have different type, amount, and timing of maintenance and rehabilitation actions and costs. To determine a cost-effective plan of actions, bridge managers would propose several alternative cost profiles and then choose the best alternative based on their equivalent costs. The best alternative for preserving and improving the condition of a bridge is the need for the bridge. There are three main methods for bridge cost analysis, including life-cycle cost analysis, benefit-cost ratio analysis, and incremental benefit-cost analysis. These methods are applied to analyze bridge project alternatives at project level to determine bridge needs based on costs and benefits. The three cost analysis methods are introduced in the following sections.

22.5.1   Life-Cycle Cost Analysis

Life-cycle cost analysis is a method used to bring all the costs during a bridge's service life to an equal basis so that the costs from various alternatives can be compared. A bridge can be taken care of in different forms, including various types of routine maintenance, repair, and rehabilitation. The amount of costs and timing of these activities may vary considerably, depending on the desired level of service, available resources, importance of the bridge, and bridge conditions. For the purpose of bridge management, it is necessary to compare a number of proposed cost profiles so that the most cost-effective alternative can be selected for each bridge. The comparison of different cost distributions during a bridge's service life has to be conducted on the same basis. To convert the costs at different times to costs on the same reference basis, it is necessary to recognize the time of money, interest rate and the concept of cost equivalence.

22.5.1.1   Interest Rate

Interest rate is often called discount rate in many cost analysis applications. Interest may be defined as money paid for the use of borrowed money. Alternatively, interest may be thought of as the return obtainable by the productive investment of capital (Grant et al., 1982). The rate of interest is the ratio of the interest payable at the end of a period of time to the money owed at the beginning of that period. Even though interest can be payable more often than once a year, an interest rate usually means the interest rate per annum. For example, an interest rate of 8% implies that $8 interest is payable annually on a debt of $100, which

FIGURE 22.12   Example of bridge cost profile.

is calculated as $8/$100 = 0.08$ or 8%. The interest of each year is based on the total amount owed at the end of the previous year, which is the principal plus the accumulated interest that had not been paid when due. Therefore, interest is compounded annually. With an interest rate of 8%, the interest for a debt of $100 is $8 at the end of the first year. If the principal and the interest is not paid, the interest at the end of second year will be $0.08($100 + $8) = 8.64. Similarly, the interest at the end of the third year will be $0.08($108 + 8.64) = 9.33.

22.5.1.2 Equivalence

The above example indicates that, with an interest rate of 8%, a debt of $100 now will become $108, $116.64, and $125.97, at the end of first year, second year, and third year, respectively. In other words, the worth of $100 at present is equivalent to $108 one year later, $116.64 two years later, and $125.97 three years later. Therefore, the values of money at different points in time are comparable for a given interest rate. This concept of equivalence of money values is essential for cost analysis in many engineering applications. Based on this concept, the present worth of money can be converted to the equivalent worth at a future time, as long as the interest rate is known. Conversely, the worth of money at a future time can be converted to the equivalent present worth.

In engineering cost analysis, the commonly utilized conversions involve the present worth of money, the future worth of money, and the uniform money series of end-of-period payments or receipts. The formulas for the conversions can be found in many books of engineering economy. The symbols and formulas for cost analysis are presented as follows (Grant et al., 1982).

- Symbols
  i = annual interest rate.
  n = number of interest years.
  P = present worth of money.
  F = future worth of money at the end of n years from the present date that is equivalent to P with interest rate i.
  A = uniform annual series of money, or end-of-year payment or receipt in a uniform series continuing for the coming n years, the entire series equivalent to P at interest rate i.
- Formulas
  Given P, to find F

$$F = P(1 + i)^n \qquad (22.22)$$

Given F, to find P

$$P = F\left[\frac{1}{(1 + i)^n}\right] \qquad (22.23)$$

Given F, to find A

$$A = F\left[\frac{i}{(1 + i)^n - 1}\right] \qquad (22.24)$$

Given P, to find A

$$A = P\left[\frac{i(1 + i)^n}{(1+i)^n - 1}\right] \qquad (22.25)$$

or

$$A = P\left[\frac{i}{(1+i)^n - 1} + i\right] \tag{22.26}$$

Given A, to find F

$$F = A\left[\frac{(1+i)^n - 1}{i}\right] \tag{22.27}$$

Given A, to find P

$$P = A\left[\frac{(1+i)^n - 1}{i(1+i)^n}\right] \tag{22.28}$$

or

$$P = A\left[\frac{1}{\dfrac{i}{(1+i)^n - 1} + i}\right] \tag{22.29}$$

As shown in these formulas, the conversions of money values are achieved through multiplying one money value by a factor containing i and n. The conversion factors can be expressed by the following functional symbols:

$(F/P, i, n)$ is the conversion factor from the present worth (P) to the future worth (F), i.e., the conversion for finding F when P is given. $(F/P, i, n) = (1 + i)^n$. It is called the single payment compound factor. To convert P to F, the formula is $F = P(F/P, i, n)$.

Similarly,

$(P/F, i, n)$ is the single payment present worth factor: $1/[(1 + i)^n]$
$(A/F, i, n)$ is the sinking fund factor: $i/[(1 + i)^n - 1]$
$(A/P, i, n)$ is the capital recovery factor: $i(1 + i)^n/[(1 + i)^n - 1]$
$(F/A, i, n)$ is the uniform series compound amount factor: $[(1 + i)^n - 1]/i$
$(P/A, i, n)$ is the uniform series present worth factor: $(1 + i)^n - 1/[i(1 + i)^n]$

22.5.1.3 Life-Cycle Cost

Applying the concept of equivalence, various types of costs during a bridge service life can be converted to a single equivalent cost at a selected point in time. Present worth is commonly used as the single equivalent cost in life-cycle cost analysis, even though other equivalent values, such as annual uniform series of cost, can also be used. In terms of bridge cost, the life-cycle cost is the total equivalent cost of various costs incurred during a bridge's service life. For a given bridge, there could be several proposed cost profiles. These profiles can be compared through life-cycle cost analysis.

To illustrate the method of life-cost analysis, a simple example is given for a bridge with two proposed cost profiles. It is a 100-m-long concrete bridge with an estimated service life of 60 years. The bridge's design and construction cost is $2,000,000. There are two possible alternative cost profiles as described in the following.

Alternative A:

- Annual maintenance and inspection cost will be $1,500 from year 1 to year 30 and $2,500 from year 31 to year 60. The increased cost after year 30 is due to the aged bridge decks and superstructure components.

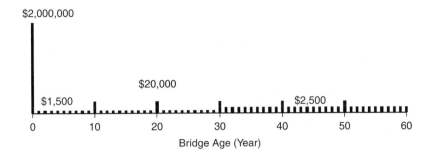

FIGURE 22.13   Cost profile of Alternative A.

- The asphalt deck overlay will be replaced every 10 years and the cost of each replacement will be $20,000.
- The bridge will be replaced at the end of 60 years with on salvage value.

Alternative B:

- Annual maintenance and inspection cost will be $1,500 throughout the bridge's service life. This is because in year 30 a deck and superstructure rehabilitation will be performed at a cost of $100,000.
- The asphalt overlay will be replaced every 10 years and the cost of reach replacement will be $20,000. The overlay replacement will not be needed in year 30 because of the deck and superstructure rehabilitation.
- The bridge will be replaced at the end of 60 years with on salvage value.

The cost profiles of the two alternatives are plotted in Figures 22.13 and 22.14. The life-cycle costs of the two alternatives are computed to determine which one is the more cost-effective, as shown below. An annual interest rate of 6% is assumed in the life-cycle cost analysis.

With an interest rate of 6%, the life-cycle cost of Alternative A in terms of present worth is:

$$P_A = \$2,000,000 + \$1,500(P/A, 6\%, 30) + \$2,500(P/A, 6\%, 30)(P/F, 6\%, 30)$$
$$+ \$20,000[(P/F, 6\%, 10) + (P/F, 6\%, 20) + (P/F, 6\%, 30)$$
$$+ (P/F, 6\%, 40) + (P/F, 6\%, 50)] = \$2,000,000 + \$1,500(13.765)$$
$$+ \$2,500(13.765)(0.1741) + \$20,000[0.5584 + 0.3118 + 0.1741 + 0.0972 + 0.0543]$$
$$= \$2,050,555$$

In the calculation of the life-cycle cost of Alternative A, year 0 is used as the base year so that all costs are converted to the equivalent present worth in year 0. No conversion is needed for the design and construction cost, $2,000,000, because it is a cost in year 0 or the base year. The annual maintenance

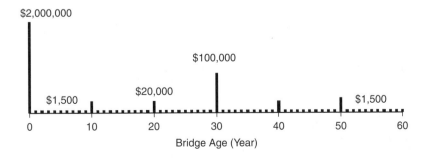

FIGURE 22.14   Cost profile of Alternative B.

and inspection costs from year 1 to year 30 are converted to the base year using the uniform series present worth factor $(P/A, 6\%, 30) = 13.765$. The annual maintenance and inspection costs from year 31 to year 60 are converted in two steps. First, the annual costs are converted to the present worth in year 30 through the factor $(P/A, 6\%, 30)$. Second, this present worth in year 30 is converted to the present worth in year 0 through the factor $(P/F, 6\%, 30)$, because the present worth in year 30 is a future worth with respect to year 0. The asphalt deck overlay costs are converted to the present worth values in year 0 using the appropriate single payment present worth factors, $(P/F, 6\%, n)$, where n is 10, 20, 30, 40, or 50.

The life-cycle cost of Alternative B in terms of present worth is:

$$P_B = \$2,000,000 + \$1,500(P/A, 6\%, 60) + \$100,000(P/F, 6\%, 30) + \$20,000[(P/F, 6\%, 10)$$
$$+ (P/F, 6\%, 20) + (P/F, 6\%, 40) + (P/F, 6\%, 50)]$$
$$= \$2,000,000 + \$1,500(16.161) + \$100,000(0.1741) + \$20,000[0.5584 + 0.3118$$
$$+ 0.0972 + 0.0543] = \$2,062,086$$

In Alternative B, the annual maintenance and inspection cost is \$1,500 for 60 years. Therefore, the conversion factor $(P/A, 6\%, 60)$ is used. The cost of the deck and superstructure rehabilitation in year 30 is converted to the equivalent present worth in the base year using $(P/F, 6\%, 30)$. It should be noted that the conversion factor $(P/F, 6\%, 30)$ for the asphalt deck overlay is not used in calculation of P_B. This is because an asphalt deck overlay will not be performed in year 30 when the deck and superstructure rehabilitation is applied.

Thus, the present worth of Alternative A is $P_A = \$2,050,555$ and the present worth of Alternative B is $P_B = \$2,062,086$. Since $P_A < P_B$, Alternative A is more cost-effective and should be selected as the need of the bridge.

22.5.1.4 Life-Cycle Cost in Perpetuity

In the above example of life-cycle cost analysis, the service life of the bridge is 60 years for both of the bridge cost profiles. In fact, it is a requirement for life-cycle cost comparisons that the service lives of different alternatives be the same. If two alternatives with different service lives are to be compared, the least common multiple of the two estimated service lives of the two alternatives must be used (Grant et al., 1982). For bridge management purpose, it is reasonable to assume that a bridge must be replaced at the end of its service life in order for the bridge to continue its function. In reality, a bridge at the end of its service life is often replaced with a new bridge at the same location. The new bridge may be built even at a different location near the old bridge, but the need for the bridge in the vicinity of the area would almost never vanish. Therefore, it can be assumed that the life-cycles of a bridge repeat indefinitely or in perpetuity in the same pattern of the first life-cycle. With perpetual life-cycle costs of alternatives, the alternatives of unequal service lives can be compared. If a bridge's service life is SL years and the present worth of the first life-cycle cost is P, then P would be repeated at intervals of SL years. The present worth of the perpetual life-cycle cost, P_P, can be computed based on the present worth of one life-cycle cost, P, in the following manner:

$$P_P = \frac{P}{1 - (1 + i)^{-SL}} \tag{22.30}$$

In the last example, the present worth of Alternative A $(P_A = \$2,050,555)$ and the present worth of Alternative B $(P_B = \$2,062,086)$ are both the present values of the costs in one 60-year life-cycle. If the cost profiles are repeated every 60 years, the perpetual life-cycle costs can be computed with an interest rate of 6% as follows:

For Alternative A,

$$P_{\rm P} = \frac{P_{\rm A}}{1 - (1+i)^{-\rm SL}} = \frac{2{,}050{,}555}{1 - (1+0.06)^{-60}} = 2{,}114{,}660$$

For Alternative B,

$$P_{\rm P} = \frac{P_{\rm B}}{1 - (1+i)^{-\rm SL}} = \frac{2{,}062{,}086}{1 - (1+0.06)^{-60}} = 2{,}126{,}551$$

22.5.2 Benefit-Cost Ratio Analysis

Benefit-cost analysis is another method for comparing alternative bridge improvement actions. This method considers both agency costs and user costs, while life-cycle cost analysis involves only agency costs. Agency costs are the costs incurred by a government agency responsible for bridges. User costs are the extra costs incurred by bridge users (motorists) because the actual bridge condition is worse than the desired or ideal bridge condition. Bridge deficiencies that cause user costs include rough deck surface, narrow deck width, low vertical clearance, poor alignment, and low road capacity. These extra costs incurred by bridge users due to bridge deficiencies are the monetary values of increased travel time, vehicle wear, fuel usage, and accidents caused by the less-than-perfect condition. Major user costs include additional user costs due to detours, additional user cost during bridge maintenance, rehabilitation, or replacement, and additional accident costs resulted from high accident probabilities due to bridge deficiencies.

22.5.2.1 Agency Costs

In BMSs, major agency costs include maintenance costs, element rehabilitation costs, element replacement costs, and bridge replacement costs. Generally, agency costs are observable with measurable monetary values, as described in the following:

- *Routine Maintenance Costs*: These costs include periodic activities required to maintain a bridge's condition above acceptable levels and to alleviate bridge condition deterioration due to minor defects. Routine maintenance activities are conducted to correct minor defects as well as to apply preventive measures so that future major repairs can be avoided or delayed. Bridge routine maintenance includes joint sealing, maintaining proper deck drainage, restoring or replacing bridge bearings, repairing bridge beam ends and beam bearing areas, and bridge painting.
- *Element Rehabilitation Costs*: These costs are associated with substantial actions to repair major bridge elements, such as, deck, superstructure, and substructure.
- *Element Replacement Costs*: These costs are related to the costs to replace damaged bridge elements. The most common element replacement activity is bridge deck replacement.
- *Bridge Replacement Costs*: These are the costs for building a new bridge to replace the existing bridge that has deteriorated below the acceptable level for the bridge to be used. The bridge replacement costs include the costs for design, engineering, acquisition, and construction.

22.5.2.2 User Costs

User costs include all extra costs incurred by bridge users due to bridge deficiencies. These costs are mainly caused by increased fuel consumption, time lost and accident rate.

- *Traffic Delay Costs*: A bridge is closed or some of its lanes are closed during bridge maintenance and rehabilitation work. A bridge closure will require vehicles to take detour and lane closures will cause traffic congestion because of reduced traffic capacity. Insufficient vertical clearance or load capacity may also require some vehicles to take detour. The delays imposed on bridge users result in time lost and increased vehicle-operating costs primarily due to increased fuel consumption.

Traffic delay costs increase as a function of traffic volume and traffic composition (percent of various vehicle types), duration of bridge closure or lane closure, and assumed monetary values of lost time.

- *Accident Costs*: Accident rate increases as a function of bridge deficiencies. Traffic congestion due to bridge lane closures may also induce accidents.
- *Vehicle Wear/Damage Costs*: Bridge deck deficiencies and work zone conditions may increase vehicle wear or damage by rough surface and obstructions.

User costs are directly proportional to traffic volume, traffic condition, and traffic composition. The total user cost on a bridge is equal to the unit user cost (user cost of a vehicle) times the total number of vehicles of that given type. As traffic volume increases, the total user cost increases. Bridge deficiencies and bridge repair work disturb traffic flow and cause traffic congestion. The disturbed traffic condition results in traffic delays and frequent vehicle speed changes. The frequent vehicle accelerations and decelerations increase travel time, vehicle wear, fuel consumption and air pollution, and likelihood of accidents.

22.5.2.3 Benefit-Cost Ratio

Benefit-cost ratio analysis compares the benefits and costs of proposed alternatives in terms of the benefit (B) to cost (C) ratio, or B/C. The benefit to cost ratio of an alternative must be greater than one, or $B/C > 1$, for the alternative to be cost-effective. In order to evaluate the benefit and cost of a bridge improvement alternative, the alternative must be compared to a base alternative. A base alternative could be a bridge replacement alternative (building a new bridge) or a maintenance only alternative. The cost of the bridge improvement alternative is the total extra cost over the base alternative. A benefit to cost ratio greater than one implies $B > C$ or the gained benefit from the alternative exceeds its extra cost.

The benefit of a bridge improvement alternative includes agency benefit and user benefit (FHWA, 1987; Farid et al., 1994). Agency benefit is defined as the difference between the life-cycle costs in terms of present worth of a base alternative and the proposed alternative. User benefit is equal to the reduced user costs resulted from the proposed alternative. The total benefit is the agency benefit plus the user benefit. The cost in the benefit-cost ratio is defined as the initial agency cost of the proposed alternative.

The following example is given to illustrate the benefit-cost ratio method. To upgrade a bridge, three alternatives are proposed. The first alternative is to perform a minor repair to the bridge at a cost of $100,000. The second alternative is to do a major rehabilitation at a cost of $250,000. The third alternative is to build a new bridge to replace the old one at a cost of $600,000. To conduct a benefit-cost ratio analysis, the total benefit of each alternative must be determined against a base alternative. As a common practice of bridge management, a do-nothing alternative can be assumed as the base alternative. The do-noting alternative means no repair or rehabilitation work will be performed. The benefit and cost of a do-nothing alternative will both be zero. If the total benefit for the three proposed alternatives is calculated against the base alternative as $400,000, $1,600,000, and $2,100,000, respectively, then the benefit-cost ratios of the three alternatives can be computed as shown in Table 22.4.

As indicated by the benefit-cost ratios, the alternative of major rehabilitation has the highest ratio value. Therefore, the alternative of major rehabilitation should be selected because it will yield the highest benefit with each unit of agency cost. Of course, the final decision on alternative selection also depends on available funds.

TABLE 22.4 Benefit-Cost Ratio Example

Alternative	Benefit (B; in $)	Cost (C; in $)	Benefit-Cost Ratio (B/C)
Do-nothing	0	0	—
Minor repair	400,000	100,000	4
Major rehabilitation	1,600,000	250,000	6.4
Bridge replacement	2,100,000	600,000	3.5

TABLE 22.5 Incremental Benefit-Cost Ratio Example

Alternative	Benefit (B; in $)	Cost (C; in $)	B/C	ΔBenefit (ΔB; in $)	ΔCost (ΔC; in $)	$\Delta B/\Delta C$
Do-nothing	0	0	—	—	—	—
Minor repair	400,000	100,000	4	400,000	100,000	4
Major rehabilitation	1,600,000	250,000	6.4	1,200,000	150,000	8
Bridge replacement	2,100,000	600,000	3.5	500,000	350,000	1.4

22.5.3 Incremental Benefit-Cost Ratio Analysis

The incremental benefit-cost ratio is defined as the ratio of the extra benefits of advancing from one improvement level to the next, divided by the corresponding extra cost (FHWA, 1987). When incremental benefits exceed incremental costs, the incremental benefit-cost ratio is greater than 1. Otherwise, the incremental benefit-cost ratio is less than 1. Therefore, an incremental benefit-ratio is an indication whether a higher level of bridge improvement alternative will generate sufficient extra benefits to compensate the extra costs. The steps of incremental benefit-cost analysis for selecting an alternative for a bridge project under no budget constrains are as follows (Farid et al., 1994):

1. Sort all proposed alternatives in increasing order of their initial costs.
2. Set the least-cost alternative as the base alternative.
3. Calculate the incremental benefit-cost ratio for the second least-cost alternative against the base alternative. If the ratio equals or exceeds 1, replace the base alternative with the current alternative and the current alternative becomes the base alternative. If the ratio is less than 1, the current alternative is removed from the alternative list.
4. Repeat Step 3 for all alternatives.
5. Select the last alternative remained on the list, which is the highest-cost alternative with an incremental benefit-cost ratio of at least 1.

To illustrate the incremental benefit-cost ratio method, the data in Table 22.4 is used to calculate the values of incremental benefit-cost ratios. The incremental benefit-cost ratios are listed in Table 22.5 along with the simple benefit-cost ratios. The extra benefit and extra cost are denoted as ΔBenefit (or ΔB) and ΔCost (or ΔC). The incremental benefit-cost ratio is expressed as $\Delta B/\Delta C$.

In Table 22.5, the $\Delta B/\Delta C$ value of each alternative is computed based on the next lower level of bridge improvement alternative. That is, the $\Delta B/\Delta C$ of minor repair alternative is computed using do-nothing as the base alternative, the $\Delta B/\Delta C$ of major rehabilitation alternative is computed using minor repair as the base alternative, and the $\Delta B/\Delta C$ of bridge replacement alternative is computed using major rehabilitation as the base alternative. For instance, the extra benefit of bridge replacement against major rehabilitation is $\Delta B = \$2,100,000 - \$1,600,000 = \$500,000$, and the corresponding extra cost is $\Delta C = \$600,000 - \$250,000 = \$350,000$. The incremental benefit-cost ratio is thus $\Delta B/\Delta C = \$500,000/\$350,000 = 1.43$. The $\Delta B/\Delta C$ values in Table 22.5 are all greater than 1, implying that the extra benefits of advancing from any improvement level to the next are sufficient to compensate for the corresponding extra costs. As indicated in the analysis steps, the bridge replacement alternative should be selected under no budget constraints.

22.6 Bridge Project Selection and System Optimization

The ultimate goal of a BMS is to provide bridge managers with scheduled bridge projects that will maximize the system benefit for a given budget. Bridge managers can use the BMS selected list of bridge projects as a basis for making decisions on maintenance, rehabilitation and replacement of bridges. Through the modules introduced above, i.e., bridge data collection and management; performance analysis and prediction; and cost analysis and need assessment, a BMS has obtained necessary information for systematically selecting bridge projects. The previously discussed modules deal with individual bridges, which is often called project level analysis. Unlike these modules, the bridge project selection module is a system-wide or network level analysis. A network level analysis utilizes condition and cost data of individual bridges as its input and generates results based on the overall benefit of the whole bridge system. There are several methods for bridge project selections, including incremental benefit-cost ratio, ranking, and optimization.

22.6.1 Incremental Benefit-Cost Ratio Analysis at System Level

As described early, at project level, the incremental benefit-cost ratio is to select bridge improvement alternatives with incremental benefit-cost ratio greater than 1. The incremental benefit-cost ratio method can also be used at system level to select bridge projects for given level of available funds. McFarland et al., (1979) developed project selection procedures through incremental benefit-cost ratio analysis. Farid et al. (1994) described applications of incremental benefit-cost ratio analysis in bridge project priority ranking and budget allocation.

Incremental benefit-cost ratio analysis at system level requires the following input data for project selection and budget allocation:

1. Identification of bridges that need maintenance, rehabilitation, or replacement;
2. Determination of possible improvement alternatives of each of the bridges;
3. Initial cost of every improvement alternative;
4. Total benefit expected from every improvement alternative;
5. Available budget.

With the input data, project selection for a given available budget can be processed. In the algorithm proposed by Farid et al., (1994), it is required to list alternatives for a bridge in descending order of the incremental benefit-cost ratios. If an alternative's ratio exceeds that of the previous, the incremental benefits and costs for the two alternatives are combined. The overall ratio of the two alternatives is then used to represent the more expensive alternative of the two. The logic behind this requirement is not clear and using a combined ratio may not be reasonable or necessary. Farid's algorithm can be modified with the following steps for project selection and budget allocation using incremental benefit-cost ratios:

1. Alternatives for each bridge are sorted in the increasing order of their initial costs.
2. If two or more alternatives for a bridge have the same initial costs, only the alternative with the highest total benefit is retained.
3. The incremental benefit-cost ratio are calculated for all bridges.
4. Discard alternatives with incremental benefit-cost ratio of 1 or less.
5. All improvement alternatives for all bridges are ranked and listed in descending order of their incremental benefit-cost ratios.
6. Select bridge improvement alternatives from the highest ranking and proceed downward until the available budget is exhausted. No more than one alternative can be selected for each bridge. Therefore, when the selection process proceeds to a bridge alternative, if no alternative has been selected for this bridge, then select this alternative if the cumulative cost is within the available budget. If an alternative has been selected for this bridge and the previously selected alternative is less-expensive, then select the current alternative to replace the less-expensive alternative previously selected for the same bridge. If the previously selected alternative is a more-expensive alternative, then skip the current alternative and the previous selection remain in the list.
7. When the selection of an alternative causes the cumulative cost to exceed the budget, this selection should be replaced with the next less-costly alternative until the budget is exhausted.

22.6.1.1 Example

An application of the above method is presented through the following example. Table 22.6 contains the information on improvement alternatives for five bridges and their calculated incremental benefit-cost ratios. It is assumed that there are three improvement alternatives for each bridge, i.e., repair (minor repair), rehabilitation (major deck and superstructure rehabilitation), and replacement (replacement of the old bridge with a new one). Alternatives for each bridge are listed in Table 22.6 in the increasing order of their initial costs. For each bridge, the repair alternative is used as the base alternative for calculating the incremental benefit-cost ratios. Therefore, the initial cost of the repair alternative is subtracted from the initial cost of the rehabilitation or replacement alternative to obtain the incremental cost. Similarly, the total benefit of the repair alternative is subtracted from that of other alternative to yield the incremental benefit. The incremental benefit-cost ratio for the repair alternative is calculated using a do-nothing alternative as the base alternative. A do-nothing alternative has an initial cost of 0 and a total benefit of 0.

For Bridge 3, the repair alternative is eliminated because it has the same initial cost but lower total benefit than the rehabilitation alternative. For bridge 4, the replacement alternative is eliminated because its incremental benefit-cost ratio is less than 1. The retained alternatives are then ranked in decreasing order of their incremental benefit-cost ratios, as shown in Table 22.7. The obtained ranking in Table 22.7 can then be used to select bridge improvement projects for a given amount of budget. The project selection and budget allocation process is illustrated in Table 22.8. This table shows that if sufficient budget is available, the selected bridge improvement alternatives include Replacement for Bridge 2, Replacement for Bridge 3, Replacement for Bridge 5, Rehabilitation for Bridge 4, and Replacement for Bridge 1. The project selection and budget allocation process proceeds as follows:

- The top three alternatives are first selected with cumulative budget values of $250,000, $430,000, and $510,000, respectively.
- The Replacement alternative for Bridge 2 is now ready for selection. Because the Repair alternative for this bridge was previously selected, the Replacement alternative is added to the list of selected projects and the Repair alternative is deleted from the list. This is done to satisfy two conditions: 1). No more than one alternative can be selected for each bridge; and 2). Choose

TABLE 22.6 Incremental Benefit-Cost Ratios of Bridge Improvement Alternatives

Bridge	Alternative	Initial Cost ($1,000)	Total Benefit ($1,000)	ΔC ($1,000)	ΔB ($1,000)	$\Delta B/\Delta C$	Comments
1	Repair	100	400	100	400	4.00	
	Rehabilitation	250	1600	150	1200	8.00	
	Replacement	600	2100	350	500	1.43	
2	Repair	80	380	80	380	4.75	
	Rehabilitation	160	500	80	120	1.50	
	Replacement	450	1800	290	1300	4.48	
3	Repair	120	380	–	–	–	"Repair" is deleted because it has same cost but less benefit than "Rehabilitation"
	Rehabilitation	120	420	120	420	3.50	
	Replacement	490	1600	370	1180	3.19	
4	Repair	60	250	60	250	4.17	"Replacement" is deleted because $\Delta B/\Delta C < 1.0$
	Rehabilitation	85	300	25	50	2.00	
	Replacement	315	500	230	200	0.87	
5	Repair	90	350	90	350	3.89	
	Rehabilitation	180	900	90	550	6.11	
	Replacement	700	2300	520	1400	2.69	

TABLE 22.7   Ranking of Bridge Improvement Alternatives

Bridge	Alternative	Initial Cost ($1,000)	Total Benefit ($1,000)	ΔC ($1,000)	ΔB ($1,000)	$\Delta B/\Delta C$
1	Rehabilitation	250	1600	150	1200	8.00
5	Rehabilitation	180	900	90	550	6.11
2	Repair	80	380	80	380	4.75
2	Replacement	450	1800	290	1300	4.48
4	Repair	60	250	60	250	4.17
1	Repair	100	400	100	400	4.00
5	Repair	90	350	90	350	3.89
3	Rehabilitation	120	420	120	420	3.50
3	Replacement	490	1600	370	1180	3.19
5	Replacement	700	2300	520	1400	2.69
4	Rehabilitation	85	300	25	50	2.00
2	Rehabilitation	160	500	80	120	1.50
1	Replacement	600	2100	350	500	1.43

the more-expensive alternative for a bridge if the available budget is sufficient for the new selection. As shown in the table, the new cumulative budget is $880,000, which is obtained as the previous cumulative budget ($510,000) minus the initial cost of the deleted alternative ($80,000) and plus the initial cost of the newly selected alternative ($450,000).
  • The Repair alternative for Bridge 4 is selected.

TABLE 22.8   Project Selection and Budget Allocation

Bridge	Alternative	Initial Cost ($1,000)	Cumulative Budget Allocation ($1,000)	Remarks
1	Rehabilitation	250	250	
5	Rehabilitation	180	430	
2	Repair	80	510	
2	Replacement	450	880	This alternative replaces "Repair" for Bridge 2, the less-expensive alternative. Cumulative Cost = 510 − 80 + 450 = 880
4	Repair	60	940	
1	Repair	100	—	Skip this alternative because the more-expensive "Replacement" alternative was selected for Bridge 1
5	Repair	90	—	Skip this alternative because the more-expensive "Rehabilitation" alternative was selected for Bridge 5
3	Rehabilitation	120	1060	
3	Replacement	490	1430	This alternative replaces "Rehabilitation" for Bridge 3, the less-expensive alternative. Cumulative Cost = 1060 − 120 + 490 = 1430
5	Replacement	700	1950	This alternative replaces "Rehabilitation" for Bridge 5, the less-expensive alternative. Cumulative Cost = 1430 − 180 + 700 = 1950
4	Rehabilitation	85	1975	This alternative replaces "Repair" for Bridge 4, the less-expensive alternative. Cumulative Cost = 1950 − 60 + 85 = 1975
2	Rehabilitation	160	—	Skip this alternative because the " Replacement" alternative was selected for Bridge 2
1	Replacement	600	2325	This alternative replaces "Rehabilitation" for Bridge 1, the less-expensive alternative. Cumulative Cost = 1975 − 250 + 600 = 2325

- The Repair alternative for Bridge 1 and the Repair alternative for Bridge 5 are not selected because the more-expensive alternatives for these two bridges were already selected.
- The process is continued until the last bridge improvement alternative in the list is covered.

The project selection shown in Table 22.8 is obtained with an assumption of sufficient available budget. In reality, however, BMS almost always deals with insufficient available budget. The output of project selection depends on the available budget. The results in Table 22.8 can be used for project selection and budget allocation at different levels of available budget.

If the available budget is $510,000, the selected bridge improvement alternatives should be Rehabilitation for Bridge 1, Rehabilitation for Bridge 5, and Repair for Bridge 2. In this case, Bridge 3 and Bridge 4 will not receive any improvement treatment.

If the available budget is $880,000, then the selected alternatives will include Rehabilitation for Bridge 1, Rehabilitation for Bridge 5, and Replacement for Bridge 2. Thus, when the budget is increased from $510,000 to $880,000, the alternative for Bridge 2 is changed from Repair to Replacement while the alternatives for Bridge 1 and Bridge 5 remain the same.

If the available budget is $1,435,000, then the selected alternatives will be Rehabilitation for Bridge 1, Rehabilitation for Bridge 5, Replacement for Bridge 2, Repair for Bridge 4, and Replacement for Bridge 3. Thus, all five bridges will receive some improvement treatments. As shown in Table 22.8, the total cost for these selected bridge improvement alternatives is $1,430,000. That is, there will be a balance of $5,000 after the selected bridge improvements.

22.6.2 Ranking Methods

Ranking techniques evaluate several related factors of a project simultaneously and yield a quantitative ranking value based on the evaluation on these factors. Thus, all the considered projects are ranked according to their corresponding ranking values. The ranking methods do not necessarily give an optimal solution that maximizes the system benefit or minimizes the total system cost. Nevertheless, a ranking approach is simple to use and provides the relative order of importance of different projects. Such an ordered list can be used for decision-makers to make final decisions on the basis of project ranking values.

22.6.2.1 Sufficiency Rating

Setting priorities on bridge related projects is usually a multi-attribute decision making problem, requiring decision-makers to evaluate simultaneously several related factors. Ranking techniques have been applied in many BMSs. Although they are in different forms, ranking models in BMS are generally based on similar concepts and principles, that is, setting priorities on bridge projects based on their composite indices. A composite index for a bridge is a numerical value used to indicate the bridge's overall condition. When bridges are sorted in the order of their composite indices, the composite index for each bridge represents the bridge's relative importance and condition status. The sufficiency rating (FHWA, 1988) is such an index representing the overall condition of a bridge. Although it is not directly used for ranking, the sufficiency rating has been used for ranking either in a modified form or in combination with additional bridge attributes. Therefore, it is necessary to briefly discuss the sufficiency rating system of bridges as defined by FHWA (1988). The sufficiency rating of a bridge is calculated based on the bridge's structural adequacy and safety, serviceability and functional obsolescence, and essentiality for public use. Sufficient rating factors include the following:

- Structural Adequacy and Safety:
  1. Superstructure
  2. Substructure
  3. Culvert
  4. Inventory Rating

- Serviceability and Functional Obsolescence:
    1. Defense Highway
    2. Lanes on Structure
    3. ADT
    4. Roadway Width
    5. Structure Type
    6. Bridge Roadway Width
    7. Vertical Clearance over Deck
    8. Deck Condition
    9. Structural Condition
    10. Deck Geometry
    11. Under Clearance
    12. Waterway Adequacy
    13. Roadway Alignment
- Essentiality for Public Use:
    1. Defense Highway
    2. Detour Length
    3. ADT

Sufficiency ratings range from 0 to 100, with 100 being the rating for a perfect bridge condition. The general criteria for federal funding based on sufficiency rating are: sufficiency rating from 0 to 50 for replacement or rehabilitation eligibility, and 50 to 80 for rehabilitation only. This general rule can be used by individual states to select bridge projects according to their ranking results.

Sufficiency rating is a good indicator of bridge conditions. However, it has shortcomings if it is used as a sole priority ranking index. One shortcoming is that sufficiency rating places heavy weights to load capacity and deck width for all classes of roadways. As a result, some bridges with low load capacity and narrow deck will be assigned low sufficiency ratings while they actually are in good condition and adequate for the traffic they serve. Another shortcoming is that it lacks sensitivity to some key ranking factors, such as, traffic volume and detour length (FHWA, 1987). Therefore, several alternative ranking methods were developed for identifying and ranking improvement needs. These methods are based on the concept of level of service. The level of service is represented by a composite index that takes into consideration of not only bridge structural conditions but also such factors as traffic volume, detour length, and cost.

22.6.2.2   Ranking by Level of Service

22.6.2.2.1 The North Carolina Ranking Method
Johnston and Zia (1984) developed a ranking system for North Carolina DOT based on deficiency points. A value of deficiency point represents a bridge's overall condition with respect to four factors: load capacity, clear deck width, vertical over-clearance or under-clearance, and remaining service life. Deficiency points are calculated from the four factors with weights of 0.7, 0.12, 0.12, and 0.06, respectively. The first three factors are expressed as functions of traffic volume. The load capacity also includes detour length as a variable. The total deficiency points range from 0 to 100 with 0 being no deficiency and 100 being highly deficient. Bridges with higher deficiency points will have higher priorities. In addition, this ranking method also considers the cost factor in priority ranking. A cost factor of a bridge is the value of the replacement cost divided by the total deficiency points of the bridge. The cost factor measures the cost per deficiency point eliminated by bridge improvement activities.

22.6.2.2.2 The Virginia Ranking Model
McKeel and Andrews (1985) developed a ranking model for the state of Virginia. The Virginia model combines the sufficiency rating with the level of service concept. Similar to the North Carolina ranking model, the Virginia model also uses deficiency points to rank bridge projects. However, the Virginia model calculates deficiency points based on the factors that are converted and modified from the

sufficiency rating criteria. The factors included in the Virginia ranking model are:

- Capacity (Inventory Rating) with weight of 0.3
- Clearance bridge deck width with weight of 0.12
- Vertical roadway under/over clearance with weight of 0.12
- Sufficiency (condition) with weight of 0.46

22.6.2.2.3 *The Nebraska Ranking Model*
The Nebraska ranking model (Committee on Bridge, 1986; FHWA, 1987) is also based on the level of service in terms of deficiency points. The ranking method considers four bridge attributes:

- Single vehicle load capacity with weight of 0.50
- Clear bridge deck width with weight of 0.12
- Vertical roadway over/under clearance with weight of 0.33
- Estimated remaining life with weight of 0.10

22.6.2.2.4 *The Pennsylvania Ranking Model*
Similar to the Virginia ranking system, the Pennsylvania ranking system (Bridge Management Work Group, 1987; FHWA, 1987) also combines the features of the sufficiency rating and level of service system. Bridges are ranked according to the total deficiency rating on a scale of 0 to 100. Deficiencies in this system are measured by eight weighted factors, including load capacity, clear deck width, under clearance, over clearance, bridge condition, remaining life, approach roadway alignment, and waterway adequacy.

22.6.2.3 Ranking by the Delphi Process

The Kansas Department of Transportation (1984) applied the Delphi process in its ranking system. The Delphi process (Linstone and Turoff, 1975) is a procedure designed to obtain the most reliable consensus amongst a group of experts by a series of questionnaires interspersed with controlled feedback to minimize bias of opinions. Using the Delphi technique, the need functions of bridge attributes and their weights and adjustment factors were obtained. The attributes include user safety measured by horizontal clearance and bridge roadway restriction, preservation of investment measured by deck and structural conditions, and travel time and operating costs measured by detour length. The adjustment factors include functional classification, traffic volume, bridge importance, accident rate, and posted speed limit.

22.6.2.4 Ranking by Analytical Hierarch Process and Utility Values

Saito and Sinha (1989) developed a ranking model for Indiana BMS. This model applies the analytical hierarch process (AHP) (Saaty, 1980) in combination with utility functions of bridge condition attributes. The AHP method is a technique that compares variables of a problem in a pairwise manner in a hierarchy structure. A hierarchy structure is used to analyze a problem that involves judgment by breaking the problem into sub-problems.

22.6.2.4.1 *AHP Scaling System*
In the process of pairwise comparison, judgments are translated into absolute numbers by using a scale value which would reflect the human judgment on the considered variables. Saaty (1980) proposed the use of a one-to-nine scale to express the decision-maker's preference and intensity of preference of one element over the other. The levels of this scaling system consist of the following:

- 1 = equal importance
- 3 = weak importance of one over another
- 5 = essential or strong importance
- 7 = very strong or demonstrated importance
- 9 = absolute importance
- 2, 4, 6, 8 = intermediate values between adjacent scale values

This scaling system uses the five attributes: equal, weak, strong, very strong and absolute important, to make qualitative distinctions of judgments.

22.6.2.4.2 Determination of Criterion Weights

The Indiana ranking model specifies several bridge evaluation criteria with respect to bridge cost, condition, safety, and community impact. The bridge evaluation criteria may not be equally important. Therefore, their weights should be determined based on the experts' judgments. When several criteria are being considered for ranking by a group, the group's goals would be: (1) to provide judgments on the relative importance of these criteria; and (2) to insure that the judgments are qualified to the extent that would permit a quantitative interpretation of the group's judgment on all subjects. The first goal can be achieved by performing pairwise comparisons and by assuming the transitivity of comparison results. The transitivity assumption states that if criterion A is preferred to criterion B, and at the same time, criterion B is preferred to criterion C, then criterion A must be preferred to criterion C. The second goal, which is of special interest to bridge management, can be achieved by applying the procedure proposed by Saaty (1980). The procedure is called eigenvector approach, which is based on the mathematical matrix theory. The reciprocal matrix consisting of resulting values of pairwise comparisons is used to obtain the values of factor weights. The main steps to determine the weight values are:

1. Construct a reciprocal matrix by the results of pairwise comparisons of the factors included using the one-to-nine scale;
2. Obtain the eigenvalues and eigenvector of the reciprocal matrix.

22.6.2.4.3 Development of Utility Functions

Utility functions for each evaluation criterion were developed in order to simplify the AHP ranking process. Utility is an indication of the level of overall effectiveness that can be achieved by undertaking a project. This is because a direct application of the AHP method is not practical if the number of projects is large. For example, even when there are 22 bridge projects to compare and rank, there will be $22(22 - 1)/2 = 231$ pairwise comparisons for each evaluation criterion. If there are four bridge evaluation criteria, the number of pairwise comparisons will be 924. In reality, the number of bridge projects range between hundreds and thousands, and the direct use of the AHP is thus impractical. Therefore, the AHP method was modified by including utility functions to the ranking model so that a large number of bridge projects can be prioritized efficiently. Utility is often expressed by values ranging from 0 to 100, with 100 corresponding to the worst condition. For instance, if a bridge is in a very poor condition, the decision-maker would place a high expected utility because the urgency for improvement is high and resulting benefits to the highway user and the surrounding community are expected to be high. However, if a bridge is still in fair condition, the improvement work may not be an urgent matter. Then, the top priority project is the alternative with the highest expected utility value.

There are several procedures to develop utility functions. However, it was found that the eigenvalue approach to determining the factor weights could also be used to develop utility functions (Saito and Sinha, 1989). One can consider the levels of each evaluation criterion as the elements of pairwise comparisons. In other words, one can place different expected utilities to the levels of each criterion. For instance, a bridge with condition rating 3 is far more eligible for replacement than a bridge with condition rating 5. Also, the difference between the condition ratings of 2 and 3 is far more significant to the decision-maker than the difference between the condition ratings of 5 and 6. These differences are exactly the topic of pairwise comparisons. The inclusion of the utility concept reduces the number of pairwise comparisons, while still maintains the advantage of such comparisons. A utility function can be developed by first defining the boundaries of interest within each criterion, and then dividing the region between the boundaries into appropriate intervals, upon consensus of the participating decision-makers. Then, reciprocal matrices of pairwise comparisons would be constructed for these intervals. Using mathematical software for eigenvalues and eigenvector of square matrices, the maximum eigenvalue and the corresponding eigenvector can be obtained. The values in the eigenvector are translated into utility points. Figure 22.15 shows an example of a utility curve.

22.6.2.5 An Example of Ranking Model

As discussed above, ranking methods are based on similar concepts and principles. That is, they all use some kind of composite indices to represent the relative importance of individual projects. To illustrate the structure and use of ranking techniques, the Indiana's BMS ranking model is selected as an example and is briefly introduced in the following.

The Indiana's BMS ranking model (Saito and Sinha, 1989) was developed following the AHP principles. The ranking model utilizes four ranking criteria (Woods 1994):

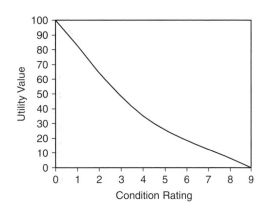

FIGURE 22.15 Utility curve of structural condition.

- Cost-effectiveness;
- Bridge condition;
- Bridge safety;
- Community impact.

The structure of the Indiana ranking model is depicted in Figure 22.16. The numbers on the figure are the weights of the criteria. As shown in Figure 22.16, the total weight of the four criteria is 1.0. Each criterion is measured in terms of one or more factors. The criterion of cost-effectiveness is measured by a cost factor, which is resulted from life-cycle analysis. The criterion of bridge condition is evaluated through structural condition and remaining service life. Similarly, the criterion of bridge safety is evaluated by three factors: clear deck with, vertical clearance, and inventory rating. Community impact is measured by the detour length. A utility function was developed for each of the factors. Since two of the four criteria, i.e., bridge condition and bridge safety, are evaluated through multi-factors, the utility values of these factors are combined according to their weights. The criterion weights, factor weights, and utility functions were all developed using the AHP method, as previously discussed.

With the weights and utility functions determined, the total weighted utility value for each bridge can be obtained. The total weighted utility value for a bridge is calculated using the following equation:

$$U = \sum_i W_i \left(\sum_j w_{i\cdot j} u_{i,j} \right) \tag{22.31}$$

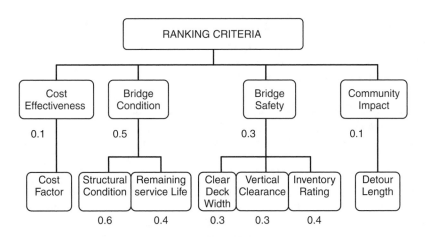

FIGURE 22.16 Ranking model structure.

where U = total weighted utility value of a bridge; W_i = weight of criterion i; $w_{i,j}$ = weight of factor j of criterion i; $u_{i,j}$ = utility value of factor j of criterion i.

Both the individual utility $u_{i,j}$ and the total weighted utility U range from 0 to 100. The following shows a simple example of the use of Equation 22.31 to compute the total weighted utility value of a bridge project. Assume a bridge improvement alternative has the following utility values:

- 60 for cost factor;
- 65 for structural condition;
- 68 for remaining service life;
- 20 for clear deck width;
- 30 for vertical clearance;
- 70 for inventory rating;
- 80 for detour length.

Based on weights shown in Figure 22.16, the total weighted utility value for this bridge project can be computed as:

$$U = 0.1 \times 60 + 0.5(0.6 \times 65 + 0.4 \times 68) + 0.3(0.3 \times 20 + 0.3 \times 30 + 0.4 \times 70) + 0.1 \times 80 = 93.1$$

That is, the total weighted utility value for this project is 93.1. With the total weighted utility value calculated for each bridge project, bridge projects can then ranked in the descending order of their total weighted utility values.

22.6.3   Optimization Methods

Ranking techniques sort projects in priority order through evaluation of several factors for each project in the system. The projects are usually selected from the top of the priority list until the available budget is used up. This approach to selecting projects is virtually based on the rule of "choosing the project with the worst conditions". Although this rule is considered rational by many decision-makers and is widely adopted in the project selection practice, it may not maximize benefit or minimize negative effects of a system. On the other hand, optimization techniques manipulate the tradeoffs between the objective and constraints systematically or mathematically, so that an optimal solution to the problem among many possible solutions can be obtained. In managing a bridge system, optimization techniques can be applied to produce optimal strategies in project selection by maximizing the system benefit subject to its constraints, such as available resources.

The concept of optimization is now applied as a principle underlying the analysis of many complex decision or allocation problems. Using optimization techniques, one approaches a complex decision problem, involving the solution of values for a number of interrelated variables, by focusing attention on a single objective designed to quantify performance and measure the quality of the decision (Luenberger, 1965). This objective is maximized (or minimized depending on the formulation) subject to constraints that may limit the selection of decision variable values. If a suitable single aspect of a problem can be isolated and characterized by an objective, optimization may provide a suitable framework for analysis and produce the best solution to the problem from a set of alternatives.

In managing a bridge system, problems such as selecting projects to maximize system benefit with a limited budget are difficult because many related factors are involved. It is virtually impossible to fully represent all the complexities of variable interactions, constraints, and appropriate objectives when a statewide bridge system is concerned. Thus, modeling a problem and formulating it quantitatively are actually processes of approximation. An optimal solution, then, should be regarded as the best solution corresponding to the specific formulation rather than the absolutely correct solution to the real system. Many different optimization techniques, such as, dynamic programming, linear programming, integer linear programming, and goal programming, have been applied to roadway and bridge management problems. The use of one technique instead of another depends on the nature of the given problem and various considerations of the model to be developed.

22.6.3.1 Linear Programming and Integer Linear Programming

Programming techniques are used to determine the efficient use or allocation of limited resources to meet desired objectives. Linear programming is a tool for solving such a problem that has a set of simultaneous linear equations, which represent the conditions of the problem, and a linear function, which expresses the objective of the problem. A linear programming problem can be expressed mathematically as: Maximize (or minimize):

$$Z = \sum_{j=1}^{p} c_j X_j \qquad (22.32)$$

Subject to:

$$\sum_{j=1}^{p} a_{ij} X_j \leq (\text{or} =, \text{ or} \geq) b_i \qquad i = 1, 2, \ldots, m \qquad (22.33)$$

where c_j, a_{ij}, and b_i are known constants for all i and j, and X_j are variables. Computer packages are commonly available today to solve linear programming problems with hundreds of constraints and variables (Winston, 1994).

If the variables of a linear programming problem are required to be integers, the linear programming becomes an integer linear programming. A particular useful type of the integer linear programming is an integer linear programming in which all the variable must equal 0 or 1. Such an integer linear programming is called a zero–one integer linear programming (Winston 1994). The mathematically equations of a zero-one integer liner programming problem are as follows:
Maximize (or minimize):

$$Z = \sum_{j=1}^{p} c_j X_j \qquad (22.34)$$

Subject to:

$$\sum_{j=1}^{p} a_{ij} X_j \leq (\text{or} =, \text{ or} \geq) b_i \qquad i = 1, 2, \ldots, m \qquad (22.35)$$

$$X_j = 0 \text{ or } 1 \qquad (22.36)$$

Compared to the linear programming equations, the zero–one integer linear programming has an added constraint on the variables in Equation 22.36. In bridge management, each activity of a bridge is usually defined as a zero–one decision variable. When the value of one of the decision variables is one, the corresponding activity is selected; otherwise, routine maintenance is assumed for the bridge. The objective function of an integer linear programming in bridge management can be either to maximize the system effectiveness or to minimize the negative impact of the system, such as total cost. The branch-and-bound procedure is often used to solve zero–one integer linear programming.

22.6.3.2 Dynamic Programming

Dynamic programming is a way of looking at a problem which may contain a large number of interrelated decision variables so that the problem is regarded as if it consisted of a sequence of problems, each of which required the determination of only one (or a few) variables (Cooper and Cooper, 1981).

Dynamic programming approach substitutes n single variable problems for solving one n-variable problem, so that it usually requires much less computational effort. The principle that makes the transformation of an n-variable problem to n single variable problems possible is known as the principle of optimality, which is stated as: an optimal policy has the property that whatever the initial state and the initial decision are, the remaining decisions must constitute an optimal policy with respect to the state which results from the initial decision. A simpler expression of this principle consists of the following statement: every optimal policy consists only of optimal sub policies (Cooper and Cooper, 1981).

An important advantage of dynamic programming is that it determines absolute (global) maxima or minima rather than relative (local) optima. Also, dynamic programming can easily handle integrality and non-negativity of decision variables. Furthermore, the principle of optimality assures that dynamic programming results in not only the optimal solution of a problem, but also the optimal solutions of sub-problems. For example, for a 10 year program period, dynamic programming gives the optimal project selections for the entire 10 year period as well as the optimal project selections for any period less than 10 years. These optimal solutions of the sub-periods are often of interest to bridge programmers. In virtually all other optimization techniques, certain kinds of constraints can cause significant problems. For example, the imposition of integrality on the variables of a problem will destroy the utility of these computational methods; however, in dynamic programming the requirement that some or all of the variables be integers greatly simplifies the computation process. Similar considerations apply to such restriction as non-negativity of the decision variables (Cooper and Cooper, 1981).

The key elements of a dynamic programming are: stages, states, decision and return (Cooper and Cooper, 1981). A bridge system can be considered to progress through a series of consecutive stages, each year is viewed as a stage. At each stage, the system is described by states, such as, bridge condition and available budget. Decisions (project selections) are made at each stage by optimizing the returns (system benefit). The bridge conditions are predicted and updated and the system undergoes the next stage. A dynamic programming problem can be solved by proceeding forwards from stage 0 to stage N, as well as by proceeding backwards from stage N to stage 0.

22.6.3.3   An Optimization Model for Bridge Project Selection

Two major BMSs, Pontis and the Indiana BMS, utilize optimization techniques for bridge project selections. Pontis (Golabi et al., 1990) applies zero-one integer linear programming to optimize bridge project selections. The Indiana BMS (Jiang and Sinha 1989) utilizes zero–one integer linear programming in combination with dynamic programming and Markov chain to maximize the system benefit of statewide bridge activities. The Indiana's model is discussed in the following to illustrate the optimization techniques applied in BMS.

The optimization model for the IBMS requires that it handles about 1000 bridges with about 3000 decision variables, if three improvement alternatives are considered (deck reconstruction, deck replacement and bridge replacement). Because the stochastic nature of bridge systems and the large number of variables involved in bridge project selection, the dynamic programming, integer linear programming, and Markov chain in combination were chosen for the optimization model. The dynamic programming divides the federal and state budgets of each year into several possible spending portions and the integer linear programming selects projects by maximizing yearly system effectiveness subject to different given budgets. The dynamic programming chooses the optimal spending policy, which maximizes the system effectiveness over a program period, by comparing the values of effectiveness of these given budgets resulted by the integer linear programming for each year. For example, suppose the program period T equals to 2 years, and the possible spendings for year 1 are 50, 60, 70, 80, 90 and 100 millions, and the possible spendings for year 2 are 150, 140, 130, 120, 110 and 100 millions, respectively. Any combination of spendings for the individual years can be considered. The task of the dynamic programming is to determine the optimal policy among possible combinations of spendings, i.e., (50, 150), (60, 140), (70, 130), (80, 120), (90, 110) and (100, 100), and to obtain the corresponding optimal project selections. Similarly, if T is larger than 2, say 10, the model can determine the optimal policy from year 1 to year 10 and give the corresponding project selections.

In terms of dynamic programming, each year of the program period is a stage. The federal and state budgets are state variables. Each activity of a bridge is a decision variable of the dynamic programming as well as of the integer linear programming. The effectiveness of the entire system is used as the return of the dynamic system.

At each stage, decision must be made as to the optimal solution from stage 1 to the current stage. When a decision is made, a return (or reward) is obtained and the system undergoes a transformation to the next stage. The bridge conditions are updated for the next stage by the Markov transition probabilities. Figure 22.17 is a flow chart of the optimization model which illustrates the optimization process. For a given program period, the objective of the model is to maximize the effectiveness of the entire system. The definition of system effectiveness and some other assumptions are discussed as follows.

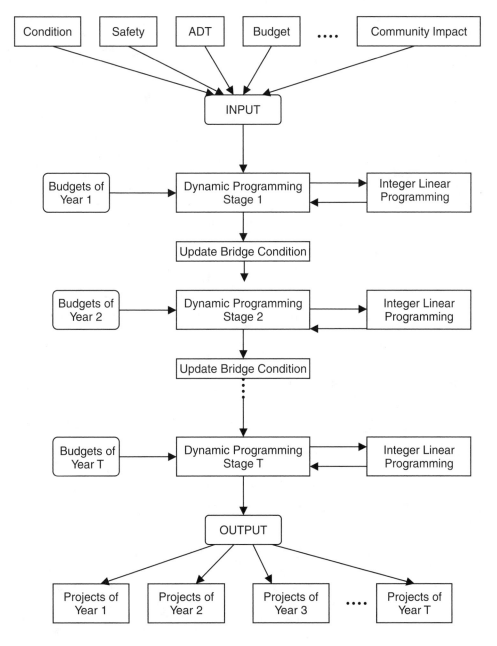

FIGURE 22.17 Flow chart of the optimization model.

22.6.3.3.1 Assumptions and Definitions

The purpose of the dynamic optimization model is to select bridge projects that would provide the maximum system-wide benefit to the bridge system with a given budget. Different bridge deficiency problems call for different treatments. To develop the optimization model, the bridge activities must be clearly identified and defined. In Indiana, rehabilitation activities mainly include deck reconstruction and deck replacement. Deck reconstruction work includes shallow and/or full-depth patching of deteriorated deck spots and an overlay of the deck after scarifying the wearing surface. Along with this reconstruction, curbs, railing, and expansion joints are replaced in most cases. Other related works include guardrails, approach slab reconstruction, approach shoulder reconstruction, and small amounts of substructure repairs. The deck replacement alternative is a more extensive rehabilitation work than deck reconstruction. Deck replacement consists of a replacement of the entire deck, including rehabilitation of parts of the superstructure and the top portion of the substructure. The replacement of the entire bridge is considered when reconstruction and rehabilitation cannot adequately correct the existing deficiencies. Thus, bridge rehabilitation and replacement activities were grouped into three options:

1. Deck reconstruction;
2. Deck replacement; and
3. Bridge replacement.

When a rehabilitation activity is applied on a bridge, the condition rating of various bridge components increases depending on the type of improvement. As shown in Figure 22.18, a particular rehabilitation activity causes a jump in the deck condition rating. As the bridge age increases, the condition rating gradually decreases from the new condition rating. The area between the deterioration curves of bridge i with and without rehabilitation, $A_i(a)$, represents an improvement in terms of condition rating and service life of the bridge.

From a performance curve, one can see that the deterioration rate of bridge condition is different at different ages. Figure 22.19 shows a tangent line on a performance curve, the absolute value of $\tan\alpha$, d_i, is the deterioration rate at the corresponding time. It is evident that when the value of deterioration rate is small, a rehabilitation activity may be delayed for a period of time. Otherwise, an immediate repair should be applied because a rapid deterioration is expected. It should be noted that since a bridge with a condition rating greater than 6 does not need rehabilitation, the tangent values on the curve section between condition rating of 6 and 9 are not considered in selecting bridge activities in the optimization model.

In order to develop an optimization program, the first task is to define the objective function of the program. As discussed above, the area $A_i(a)$ shown in Figure 22.18 represents the condition improvement that can be expected from under-taking a rehabilitation or replacement activity. In addition, other factors, such as, average daily traffic (ADT), traffic safety condition, and community impact of a bridge, should also be considered in the optimization program.

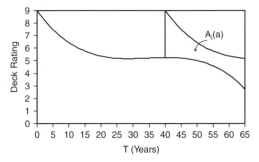

FIGURE 22.18 Condition improvement by rehabilitation.

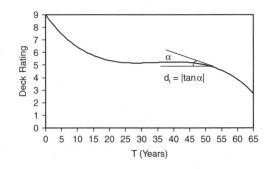

FIGURE 22.19 Slope of performance curve.

There are several ways the effectiveness of a bridge activity can be defined. Because ADT_i is the number of vehicles served by bridge i, the multiplication of ADT_i and $A_i(a)$, i.e., $ADT_i\, A_i(a)$, can be interpreted as the measure of the improvement that can be experienced by the users or vehicles on bridge i. Traffic safety condition and community impact of a bridge are two other factors affecting decisions on bridge rehabilitation or replacement activities in addition to structural condition. As discussed early, utility functions for these two factors were developed for ranking purpose. The utility values of these two factors can be converted to coefficients ranging from 0 to 1 by dividing the utility values by 100. The effectiveness of a bridge improvement activity is defined as follows:

$$E_i = ADT_i\, \Delta A_i(a)(1 + Csafe_i)(1 + Cimpc_i) \qquad (22.37)$$

where E_i = effectiveness gained by bridge i if activity a is chosen; a = improvement activity with values of 1, 2 and 3 representing deck reconstruction, and bridge replacement, respectively; ADT_i = average daily traffic on bridge i; $\Delta A_i(a)$ = area under performance curves of components of bridge i obtained by activity a; $Csafe_i$ = coefficient of safety condition of bridge i, converted from bridge safety utility value; and $Cimpc_i$ = coefficient of community impact of bridge i, converted from community impact utility value.

22.6.3.3.2 *Formulation*

Considering that budgets can be carried over from year to year, the mathematical model for maximizing the overall effectiveness of various activities over a program period T is formulated as follows:Objective function:

$$\max \sum_{t=1}^{T}\left[\sum_i \sum_a X_{i,t}(a)\, E_i\, d_i(t)\right] \qquad (22.38)$$

Subject to the following constraints:

(a) available federal budget:

$$\sum_{t=1}^{T}\left[\sum_i \sum_a X_{i,t}(a)\, c_i(a)\, F_i\right] \le C_{BF} \qquad (22.39)$$

(b) available state budget:

$$\sum_{t=1}^{T}\left[\sum_i \sum_a X_{i,t}(a)c_i(a)(1 - F_i)\right] \le C_{BS} \qquad (22.40)$$

(c) one activity cannot be undertaken more than once on one bridge in T years:

$$\sum_{t=1}^{T} X_{i,t}(a) \le 1 \qquad (22.41)$$

Constraints (d) to (h) correspond to the integer linear programming problem:
(d) maximize system effectiveness of year t:

$$\max \sum_i \sum_a [X_{i,t}(a)E_i d_i(t)] \qquad (22.42)$$

(e) spending constraint of year t for federal budget:

$$\sum_i \sum_a [X_{i,t}(a)c_i(a)F_i] \leq \eta_{tF} \qquad (22.43)$$

(f) spending constraint of year t for state budget:

$$\sum_i \sum_a [X_{i,t}(a)c_i(a)(1 - F_i)] \leq \eta_{tS} \qquad (22.44)$$

(g) no more than one activity can be chosen for one bridge in year t:

$$\sum_{a=1}^{3} X_{i,t}(a) \leq 1 \qquad (22.45)$$

(h) zero–one integer decision variable:

$$X_{i,t}(a) = 0 \text{ or } 1 \qquad (22.46)$$

The Markov chain model is incorporated into the optimization model to update bridge conditions: If bridge i is not selected in year t, its condition rating in year $t + 1$ is estimated as:

$$R_{i,t+1} = R_{i,t}p_i(R, t) + (R_{i,t} - 1)(1 - p_i(R, t)) \qquad (22.47)$$

If bridge i is selected in year t for activity a, its condition will be improved:

$$R_{i,t+1} = R_{i,t} + \Delta R_i(a) \qquad (22.48)$$

where $X_{i,t}(a) = 1$, if bridge i is chosen for activity a; $X_{i,t}(a) = 0$, otherwise; $d_i(t) =$ the absolute tangent value on performance curve of bridge i at time t, as shown in Figure 22.19, $d_i(t)$ reflects the deterioration rate of bridge condition at time t; $C_{BF} =$ total available federal budget for the program period; $C_{BS} =$ total available state budget for the program period; $F_i =$ federal budget share of bridge i; $1 - F_i =$ state budget share of bridge i; $c_i(a) =$ estimated cost of activity a on bridge i; $\eta_{tF} =$ spending limit of federal budget in year t; $\eta_{tS} =$ spending limit of state budget in year t; $R_{i,t} =$ condition rating of bridge i in year t; $p_i(R, t) =$ Markov condition transition probability of bridge i with condition rating R in year t; and $\Delta R_i(a) =$ condition rating gained by bridge i for activity a.

22.6.3.3.3 Solution Techniques

Equation 22.38 through Equation 22.48 constitute a dynamic programming which includes an integer linear program (Equation 22.42 to Equation 22.46) as a part of the constraints. The objective of the model is to obtain optimal budget allocations and corresponding project selections over T years so that the system effectiveness can be maximized. Let us denote the number of spending combinations by N, the number of possible spendings of each year by s, and the program period by T, then N can be expressed by s and T, $N = sT - 1$. When T is large, the number of possible spending combinations becomes so large that the search for the optimal path of spendings from year 1 to year T needs great effort and computation time.

Dynamic programming is an efficient technique to search for the optimal path among the combinations of spendings. Rather than examining all the paths, dynamic programming looks at only a small part of these paths. According to the principle of optimality, at each stage the programming finds the optimal subpath up to the current stage, and only this subpath is used to search for the optimal subpath up to the next stage. The paths that do not belong to the optimal subpath are abandoned as the search goes on, which makes the search efficient and saves a great deal of time. The search for the optimal path can be performed by expressing the problem as recurrence relations (Cooper and Cooper, 1981). In doing so, Equation 22.38 to Equation 22.40 are rewritten as follows,

$$\max \sum_{t=1}^{T} \Phi_t(Y(t)) \qquad (22.49)$$

subject to:

$$\sum_{t=1}^{T} Y_F(t) \le C_{BF} \qquad (22.50)$$

$$\sum_{t=1}^{T} Y_S(t) \le C_{BS} \qquad (22.51)$$

where:

$$\Phi_t(Y(t)) = \sum_i \sum_a [X_{i,t}(a)E_i d_i(t)]$$

$$Y_F(t) = \sum_i \sum_a [X_{i,t}(a)c_i(a)F_i] \le C_{BF}$$

$$Y_S(t) = \sum_i \sum_a [X_{i,t}(a)c_i(a)(1 - F_i)] \le C_{BS}$$

$$Y(t) = Y_F(t) + Y_S(t)$$

We define state variable as:

$$\lambda_t = \lambda_{t+1} - Y(t+1) \qquad (22.52)$$

We also define the optimal return function as:

$$g_1(\lambda_1) = \max \Phi_1(Y(1)), \qquad 0 \le Y(1) \le \lambda_1 \qquad (22.53)$$

$$g_2(\lambda_2) = \max[\Phi_2(Y(2)) + g_1(\lambda_2 - Y(2))], \qquad 0 \le Y(2) \le \lambda_2 \qquad (22.54)$$

$$g_t(\lambda_t) = \max[\Phi_t(Y(t)) + g_{t-1}(\lambda_t - Y(t))], \qquad 0 \le Y(t) \le \lambda_t \qquad (22.55)$$

By the recurrence relations of Equations 22.53, 22.54 and Equation 22.55, the dynamic programming process starts at year 1, or stage 1, and $g_1(\lambda_1)$ can be obtained for all the possible spendings of year 1. Then the bridge conditions are updated by Equation 22.47 or Equation 22.48 according to the project selections corresponding to $g_1(\lambda_1)$, and $g_2(\lambda_2)$ can be solved based on the information of $g_1(\lambda_1)$ as well as the updated bridge conditions. This forward recursion is performed for every successive year of the program period until $g_T(\lambda_T)$ is obtained, and therefore the optimal spending policy and project selection from year 1 to year T are obtained.

The value of $\Phi_t(Y(t))$ can be obtained by solving the integer linear program (Equation 22.42 to Equation 22.46). The value of the objective function (Equation 22.42) of the linear program equals $\Phi_t(Y(t))$ if η_{tF} and η_{tS} of Equation 22.43 and Equation 22.44 are substituted by possible spending limitations of year t.

A computer program of the optimization model was coded and incorporated into the computer package for the IBMS (Sinha et al., 2002). The output of the program is a list of selected bridges, activities and the corresponding costs for each year of the program period. The output of this model depends on the available budgets. As budget changes, the program gives different project selections so that the system effectiveness could be maximized by efficiently spending available budget in the program period.

The use of dynamic programming in combination with integer programming and the Markov chain provides bridge managers an optimization tool for managing bridge systems. The model selects projects by maximizing the effectiveness of entire system over a given program period subject to budget constraints. Therefore, for any available budget, the model always gives a mix of projects that maximizes the system effectiveness for the given budget. That is, the model always offers optimal solutions. The priority ranking methods as used in BMSs, however, usually do not guarantee optimal solutions because they are based solely on the comparison of rankings. In a ranking procedure the following two important ingredients may be missing (Cook and Lytton, 1987):

1. evaluation of inter-project tradeoffs in selecting projects,
2. selection of optimal strategies which are guaranteed to adhere to existing budget limitations.

The principle of optimality assures that dynamic programming results in not only the optimal solution for the program period T, but also for any period less than T. These optimal solutions for the sub-periods are of importance to bridge managers in scheduling bridge activities. Furthermore, these solutions are also guaranteed by the principle of optimality to be absolute optima rather than relative optima. The optimization model has a powerful capability of handling a system with hundreds of bridges. It can be used by highway programmers to gain maximum return by effectively allocating the limited bridge budgets in both short-term and long-term planning horizons.

22.6.3.4   An Application Example of the Optimization Model

To show the application of the model, an example problem is presented as follows. Fifty bridges are given, 25 of them are recommended for rehabilitation, and another 25 bridges are recommended for bridge replacement. A five year program period is used (i.e., $T = 5$). Suppose the bridges being considered are eligible for a 90% federal budget share (F_i) on interstates and an 80% federal budget share on non-interstate highways. The program is run for different budget inputs and the outputs corresponding to these different budget scenarios are used to compare the project selection and the values of system effectiveness. The output of the program is a list of selected bridges, activities and the corresponding costs for each year of the program period. The output of this model depends on the available budgets. As budget changes, the program gives different project selections so that the system effectiveness could be maximized by efficiently spending available budget in the program period. Tables 22.9 and 22.10 present results obtained by available budgets equal to 100% and 40% of the needed budgets, respectively. Figure 22.20 shows the comparison of project selections and system benefits obtained with respect to the size of different available budgets. The results indicate that at a lower level of budget most of the projects are rehabilitations. This trend continues until 60% level of budget. Then, replacement projects increase at a higher rate, while rehabilitation projects start to decrease. The reason for a higher number of rehabilitation projects at lower budget levels is that these projects are less expensive and more projects can be accommodated to maximize system effectiveness. It can also be seen that the benefit does not decrease as quickly as budget goes down. This phenomenon indicates that the optimization model always attempts to select projects so that the system benefit is as large as possible.

22.6.4   Combination of Optimization and Ranking Methods

Ranking and optimization are two of the most widely used techniques applied in highway project selections. However, these two approaches are very different in concepts. Ranking techniques evaluate several related factors of a project simultaneously and yield a quantitative ranking value based on the evaluation on these factors. Thus, all the considered projects are ranked according to their corresponding ranking values. The ranking methods do not necessarily give an optimal solution. Nevertheless, a ranking approach is simple to use and provides the relative order of importance of different projects. Such an ordered list can be used for decision-makers to make final decisions on the basis of project ranking values. On the other hand, an optimization technique produces an "optimal" solution of a highway system while the projects are selected subject to a set of constraints. The optimal solution is obtained

TABLE 22.9 Output of Optimization Program (100% Needed Budget)

Year	Bridge Number	Activity	Federal Cost ($1,000)	State Cost ($1,000)	Total ($1,000)
1	7	DRC	189	21	8502
	11	DRC	242	27	
	14	DRC	428	48	
	16	DRC	139	15	
	21	DRC	198	49	
	22	DRC	148	16	
	32	BRP	2727	682	
	34	BRP	1037	259	
	35	BRP	508	127	
	40	BRP	154	39	
	43	BRP	823	206	
	48	BRP	336	84	
2	24	DRC	79	20	9124
	25	DRC	1594	399	
	26	BRP	400	100	
	29	BRP	1754	438	
	30	BRP	968	242	
	33	BRP	536	109	
	37	BRP	672	168	
	39	BRP	236	59	
	41	BRP	308	77	
	46	BRP	772	193	
3	2	DRC	221	55	8369
	8	DRC	1040	260	
	9	DRC	205	51	
	18	DRC	53	13	
	19	DRC	99	25	
	20	DRC	95	24	
	27	BRP	4000	1000	
	36	BRP	672	168	
	44	BRP	310	78	
4	6	DRC	109	12	7749
	10	DRC	297	33	
	15	DRC	67	7	
	23	DRC	237	59	
	31	BRP	1727	432	
	38	BRP	2522	631	
	47	BRP	1239	310	
	49	BRP	52	13	
5	1	DRC	188	47	10969
	3	DRC	310	77	
	4	DRC	224	56	
	5	DRC	86	21	
	12	DRC	161	40	
	13	DRC	161	40	
	17	DRC	972	218	
	28	BRP	4983	1246	
	42	BRP	1257	314	
	45	BRP	230	58	
	50	BRP	224	56	

DRC, deck reconstruction; BRP, bridge replacement.

TABLE 22.10 Output of Optimization Program (40% Needed Budget)

Year	Bridge Number	Activity	Federal Cost ($1,000)	State Cost ($1,000)	Total ($1,000)
1	1	DRC	188	47	3295
	2	DRC	221	51	
	6	DRC	109	12	
	11	DRC	242	27	
	15	DRC	67	7	
	20	DRC	95	24	
	21	DRC	198	49	
	22	DRC	148	16	
	26	BRP	400	100	
	35	BRP	508	508	
	40	BRP	154	154	
	48	BRP	336	336	
	49	BRP	52	52	
2	16	DRC	139	139	3528
	18	DRC	53	13	
	24	DRC	79	20	
	30	BRP	968	242	
	39	BRP	236	59	
	43	BRP	823	206	
	44	BRP	310	78	
	45	BRP	230	58	
3	14	DRC	428	48	3543
	33	BRP	436	109	
	34	BRP	1037	259	
	37	BRP	672	168	
	41	BRP	308	77	
4	19	DRC	99	25	3557
	38	BRP	2522	631	
	50	BRP	224	56	
5	3	DRC	310	77	3568
	4	DRC	224	56	
	5	DRC	86	21	
	7	DRC	189	21	
	8	DRC	1040	260	
	9	DRC	205	51	
	10	DRC	297	33	
	12	DRC	161	40	
	13	DRC	161	40	
	23	DRC	237	59	

DRC, deck reconstruction; BRP, bridge replacement.

either by maximizing the system benefit or by minimizing the total negative effect on the system that is caused by undertaking the selected projects. Different from ranking methods, optimization techniques do not follow the rule of "choosing projects with the worst conditions"; instead, the optimization techniques select projects that contribute the most benefit to the highway system while all of the constraints are satisfied simultaneously.

The IBMS provides two separate procedures: ranking and optimization models for selecting bridge rehabilitation and replacement projects. Thus, decision-makers have two alternative methods for bridge project selection. However, because of the different concepts of the two techniques, the two models would produce two different sets of results. It would, however, be desirable to combine the techniques so that the ranking and optimization models would have a direct connection and the results could

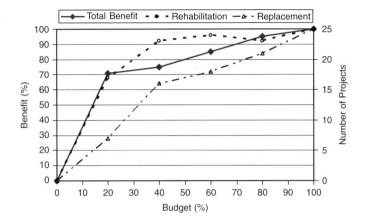

FIGURE 22.20   Optimization results at different available budgets.

be compared and analyzed on the basis of common criteria. This section presents an approach of combining ranking and optimization techniques.

With utility values as common measures for the ranking and optimization models, the proposed approach can be developed easily by modifying the existing models. Because the utility values range from 0 to 100, with 0 being the utility value of a "perfect" bridge and 100 the value of the "worst" bridge, the utility value of a bridge will decrease after undertaking a rehabilitation activity. Thus, the difference between utility values of before and after undertaking a bridge activity indicates the improvement in overall utility. This difference, therefore, can be defined as the effectiveness or benefit of the bridge activity. Incorporating this definition into the dynamic optimization model, the objective function is to maximize the total decrease of utility values of the bridge system subject to the budget constraints.

In order to combine the two models, the only modification of the dynamic programming formulation was to change Equation 22.37 to the following:

$$E_i = U_{ib} - U_{ia} \qquad (22.56)$$

where E_i = effectiveness gained by bridge i if an activity is undertaken; U_{ib} = utility value of bridge i before the activity is undertaken; and U_{ia} = utility value of bridge i after the activity is undertaken.

The formulation of the new approach is obtained by substituting Equation 22.56 for Equation 22.37, while Equation 22.38 to Equation 22.48 would remain unchanged. The value of E_i would be available from the ranking model. This value is the weighted summation of individual utility differentials for cost efficiency, bridge condition, bridge safety, and community impact.

The change of Equation 22.37 to Equation 22.56 combines the ranking model and the optimization model. Thus the result obtained from the optimization model can be directly compared with that of the ranking model in terms of the total gain in utility values. The change in the computation will be to have Equation 22.56 as a subroutine of the dynamic optimization program. This subroutine is, in effect, the ranking program. At each stage of the dynamic optimization process, the ranking program, as a subprogram, computes the system benefit, or the total gain of utility values, and the dynamic programming as the main program makes optimal project selection according to the system benefit.

22.6.4.1   An Application Example of the Combination Model

To illustrate the combination model, the same 50 bridges used in the last application example are utilized to run the combination model. Table 22.11 presents the results from the ranking portion of the model. The bridges in Table 22.11 are listed in the order of priorities so that one can select rehabilitation projects from the top of the list. The symbols used in the table are defined as follows: U_i = Utility Value of Bridge i; E_i = Effectiveness of Bridge i; C_i = Cost of the Activity of Bridge i, in \$1000; BRP = Bridge Replacement; and DRC = Deck Reconstruction.

TABLE 22.11 Output of the Ranking Model

Bridge Number	Priority	U_i	E_i	ΣE_i	C_i	ΣC_i	Activity
31	1	72.9	52	52	2159	2159	BRP
30	2	72.5	50	102	1210	3369	BRP
47	3	72.3	46	148	1549	4918	BRP
27	4	70.4	53	201	5000	9918	BRP
49	5	69.9	51	252	65	9983	DRC
24	6	69.0	21	273	98	10081	DRC
25	7	68.9	18	291	1993	12074	DRC
26	8	68.0	52	343	500	12574	BRP
46	9	67.6	50	393	965	13539	BRP
33	10	65.0	50	443	545	14084	BRP
50	11	65.0	50	483	280	14364	BRP
28	12	63.2	48	541	6228	20593	BRP
37	13	61.2	50	592	840	21433	BRP
48	14	60.5	50	642	420	21853	BRP
32	15	60.1	51	693	3409	25262	BRP
42	16	60.1	50	743	1571	26833	BRP
40	17	59.4	50	793	193	27026	BRP
17	18	59.0	46	839	1090	28116	DRC
43	19	59.0	50	889	1029	29145	BRP
44	20	59.0	50	939	388	29533	BRP
45	21	59.0	50	989	288	29821	BRP
23	22	55.7	40	1029	296	30117	DRC
35	23	52.7	42	1071	635	30759	BRP
34	24	51.7	42	1113	1297	32049	BRP
10	25	51.7	29	1142	330	32379	DRC
38	26	50.0	41	1183	3153	35532	BRP
36	27	49.2	38	1221	840	36372	BRP
39	28	46.0	37	1258	295	36667	BRP
11	29	42.0	15	1273	269	36936	DRC
29	30	42.0	33	1306	2192	39128	BRP
13	31	36.0	23	1329	201	39329	DRC
22	32	36.0	13	1342	164	39493	DRC
41	33	35.9	29	1372	385	39878	BRP
9	34	32.6	15	1387	257	40135	DRC
12	35	32.0	19	1406	201	40336	DRC
21	36	31.9	9	1415	247	40583	DRC
8	37	30.3	15	1430	1300	41883	DRC
18	38	28.6	8	1438	66	41949	DRC
7	39	28.4	12	1450	210	42159	DRC
3	40	28.2	12	1462	387	42546	DRC
6	41	27.0	10	1472	121	42668	DRC
20	42	26.9	8	1480	119	42787	DRC
14	43	26.8	4	1484	476	43262	DRC
16	44	26.6	10	1494	154	43416	DRC
15	45	23.0	4	1498	74	43491	DRC
4	46	22.3	8	1506	281	43771	DRC
5	47	21.9	8	1513	107	43878	DRC
1	48	20.4	8	1521	235	44113	DRC
2	49	20.0	8	1529	276	44389	DRC
19	50	19.9	4	1533	124	44512	DRC

Since the project selection in the combination approach depends on available budgets, the optimization program was run several times using different given budgets. The results of one of the runs are shown in Table 22.12. It should be noted that the bridges in Table 22.12 are not presented in a priority list as those in Table 22.11. This run was made with a given budget of $11,128,000, or about 25% of the total budget needed for repairing and replacing all of the 50 bridges. The total gain in utility, or the system-wide benefit,

TABLE 22.12 Output of the Combination Model

Bridge Number	E_i	C_i	Activity
1	8	235	DRC
4	8	281	DRC
6	10	121	DRC
7	12	210	DRC
9	15	257	DRC
10	29	330	DRC
11	15	269	DRC
12	19	201	DRC
13	23	201	DRC
15	4	74	DRC
16	10	154	DRC
17	46	1090	DRC
18	8	66	DRC
19	4	124	DRC
20	8	119	DRC
21	9	247	DRC
22	13	164	DRC
23	40	296	DRC
24	21	98	DRC
26	52	500	BRP
33	50	545	BRP
35	42	635	BRP
36	38	840	BRP
37	50	840	BRP
39	37	295	BRP
40	50	193	BRP
41	29	385	BRP
44	50	388	BRP
45	50	288	BRP
46	50	965	BRP
48	50	420	BRP
50	50	280	BRP

Available budget $= \$11,128,000$; total objective value (benefit) $= 900.0$; E_i, effectiveness, or change of utility value, of Bridge i; C_i, cost of the activity of Bridge i, in $1000; BRP, bridge replacement; DRC, deck reconstruction.

was 900.0. With the same amount of budget, one can also select bridge projects from the ranking list in Table 22.11. Selecting the bridges from top of the list, the first six bridges in Table 22.11 could be chosen with the given budget. Thus, with this selection the total cost is $10,081,180, and the total gain of the utility is 272.6.

By dividing the total gain of utility by its corresponding total cost, the gain of utility per million dollars for the combination approach is:

$$\frac{9000.0 \text{ (utility)}}{11128000 \text{ (\$)}} = 81 \text{ units per million dollars}$$

and that for the ranking method is:

$$\frac{272.6 \text{ (utility)}}{10081180 \text{ (\$)}} = 27 \text{ units per million dollars}$$

Therefore, the value of the combination optimization method is three times as large as the value of the ranking method in this example.

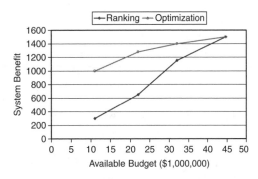

FIGURE 22.21 Comparison of project selections by ranking and optimization approaches.

Figure 22.21 is a comparison of the results from the two approaches in terms of system benefits and available budget. It can be seen that the optimization approach always gives better solution than the ranking approach when the available funds are less than 100% of the need. That is, optimization techniques are better in respect of maximizing system benefit.

Through this example, it is shown that by defining the system benefit as the total gain in utility value changes, the ranking and optimization models can be combined into one model. In BMSs, both ranking and optimization techniques are used for project selection. However, an optimization approach would insure an optimal use of resources.

22.7   Summary

A comprehensive BMS is a systematic process of making maintenance, rehabilitation, and replacement decisions on a given population of bridges based on comprehensive bridge condition inspection and bridge data analysis and subject to available funds for preserving the given bridge system. Typically, a BMS has an established bridge inspection program, is capable of assessing the current condition and predicting the future condition of the bridges, contains the models for bridge cost analysis, and provides the functions to yield alternative bridge project plans. It is essential for bridge engineers and managers to understand the concepts, principles, and characteristics of bridge management so that a BMS can be effectively implemented.

The ultimate goal of a BMS is to provide bridge managers with scheduled bridge projects that will maximize the system benefit for a given budget. To optimize project selections, it is necessary for a BMS, as a minimum, to have the following major components:

1. bridge condition data collection and management;
2. bridge performance analysis and prediction;
3. bridge cost analysis and need assessment;
4. bridge project selection and system optimization.

Each of these components is an integral part of a BMS and performs unique functions. No two existing BMSs contain exactly the same components and many BMSs contain more components than the four components discussed above. However, these four components are the basic and essential elements of BMSs. This chapter introduced the functions of BMS with discussions of BCM's typical structure, data requirements and collection, techniques and theories applied in each component, and application examples.

The functions of data collection and management include: 1). inspecting bridge conditions at regular time intervals and following standardized procedures, 2). screening bridge data to minimize errors, 3). storing historical bridge data, including design, construction, and condition data, in the format required by BMS, and 4). updating bridge data after each inspection.

The purpose of bridge performance analysis and prediction is to analyze the trends and patterns of bridge condition deterioration and to predict the future structural conditions of individual bridges based on the historical and present conditions. Bridge structural condition is the most important variable considered in the process of bridge project selection. The decision making is based on bridge conditions at present and in the future. The accuracy of the future condition prediction directly affects the effectiveness of BMS's optimization results. This chapter presented the commonly utilized methods for bridge performance analysis and prediction. Several statistical regression methods were introduced, including linear regression with one independent variable, piecewise linear regression, polynomial regression, and linear regression with multiple independent variables. A stochastic bridge performance model was also presented. The Markov chain based stochastic method utilizes the most recently obtained bridge condition data to predict the future condition. Its dynamical feature reflects the nature of bridge condition changes and thus often provides good bridge condition predictions.

Through bridge cost analysis and need assessment, the cost-effectiveness of bridge rehabilitation and replacement alternatives can be determined. A bridge requires periodic maintenance and rehabilitation until the end of its service life when the bridge cannot be used safely and a new bridge must be built to replace it. In addition to maintenance, rehabilitation and replacement costs, user costs may also be incurred for bridges with functional deficiency. The concepts and methods for bridge cost analysis were illustrated, including life-cycle cost analysis, benefit-cost ratio, and incremental benefit-cost ratio. A need is defined as the necessary action and its associated cost to upgrade a bridge condition to meet the functional and structural requirements of the bridge. For a BMS, the need of each bridge on the highway system must be determined in order to maximize the system benefit. This can be achieved through bridge performance prediction and cost analysis to select the most cost-effective action to a bridge. With the needs of all bridges, the present and future system-wide needs can be shown along the time line. The system-wide needs are then utilized as the basis for optimizing the total system benefit.

The results from the above components provide the estimated future conditions, proposed bridge maintenance and rehabilitation activities, and expected bridge condition improvement and benefit from these activities. These results plus the amount of expected or available budgets are used in BMS as the input for project selection and system optimization. Incremental benefit-cost ratio analysis, ranking and optimization techniques are widely used to select projects while the available budgets are limited. The procedure using incremental benefit-cost ratio to select bridge projects was presented through an application example. The concepts of ranking were introduced and several ranking models were briefly discussed. The ranking method based on the AHP was illustrated in a more detailed manner. The AHP method is a technique that compares variables of a problem in a pairwise manner in a hierarchy structure. The AHP based ranking method applies AHP in combination with utility functions of bridge condition attributes.

Optimization techniques applied in BMS include linear programming, integer linear programming, and dynamic programming. These techniques and their applications were described. A dynamic optimization model for bridge project selection was described. The model applied dynamic programming in combination with integer linear programming and Markov process techniques. The use of these techniques makes it possible to manage bridge systems with hundreds of bridges dynamically. The principle of optimality assures that dynamic programming results in not only the optimal solution for the program period T, but also for any sub-period less than T. The optimal solutions for the sub-periods are of importance to bridge managers in scheduling bridge activities. Furthermore, these solutions are also guaranteed by the principle of optimality to be absolute optima rather than relative optima.

An alternative optimization method that combines the AHP ranking model and the dynamic optimization model was also illustrated. By defining the system benefit as the total decrease in utility values, the ranking and optimization models can be combined with the ranking model as a subprogram of the optimization model. The combination model makes it possible that an optimal use of resource is insured and at the mean time the quantitative comparison between bridge projects is provided.

It should be emphasized that BMSs vary from agency to agency. It is impossible or impractical to cover all aspects of existing BMSs. The objective of this chapter is to provide a basis for readers to understand the concepts, functions, and essences of BMSs. Bridge management has been continuously enhanced and will continue to be in an improving process. Therefore, new concepts and techniques of bridge management will certainly emerge and be incorporated into future systems.

References

American Association of State Highway and Transportation Official (AASHTO), 1983. *Manual for Maintenance Inspection of Bridges*. AASHTO, Washington, DC.

Bhat, U.N. and Miller, G.R. 2002. *Elements of Applied Stochastic Process*. Wiley-Interscience, Hoboken, NJ.

Bridge Management Work Group 1987. The Pennsylvania Bridge Management System, Draft Final Report. Bureau of Bridge and Roadway Technology, Pennsylvania Department of Transportation, Harrisburg, PA.

Busa, G., Ben-Akiva, M., and Buyukozturk, O. 1985a. *Modeling Concrete Deck Deterioration*. Department of Civil Engineering, Massachusetts Institute of Technology, Cambridge, MA.

Busa, G., Cassella, M., Gazda, W., and Horn, R. 1985b. *A National Bridge Deterioration Model*. Transportation Systems Center (SS-42-US-26), U.S. Department of Transportation, Cambridge, MA.

Butt, A.A., Feighan, K.J., and Shahin, M.Y. 1987. Pavement Performance Prediction Model Using the Markov Process, Transportation Research Record 1123. Transportation Research Board, Washington, DC.

Cambridge Systematics, 1996. *PONTIS Release 3.1 — User's Manual*. AASHTO, Washington, DC.

Chen, C.-J. and Johnston, D.W. 1987. *Bridge Management under a Level of Service Concept Providing Optimum Improvement Action, Time, and Bridge Prediction*, FHWA/NC/88-004. Center for Transportation Engineering Studies, North Carolina State University, Raleigh, NC.

Committee on Bridge Maintenance, Rehabilitation and Replacement Procedures. 1986. Interim Report on Bridge Maintenance, Rehabilitation and Replacement Procedures. Nebraska Department of Roads, Lincoln, Nebraska.

Cook, W.D. and Lytton, R.L. 1987. Recent Development and Potential Future Directions in Ranking and Optimization Procedures for Pavement Management, *Proceedings, Second North American Conference on Managing Pavements*, Vol. 2, Toronto, Canada.

Cooper, L. and Copper, M.W. 1981. *Introduction to Dynamic Programming*. Pergamon Press, New York.

Farid, F., Johnston, D.W., Rihani, B.S., and Chen, C.J. 1994. Feasibility of Incremental Benefit-Cost Analysis for Optimal Budget Allocation in Bridge Management Systems, Transportation Research Record 1442. Transportation Research Board, Washington, DC.

Federal Highway Administration (FHWA), 1971. *National Bridge Inspection Standards, Section 23, Code of Federal Regulations, Part 650.3*. U.S. Department of Transportation, Washington, DC.

Federal Highway Administration (FHWA), 1987. *Bridge Management Systems*, Demonstration Project No. 71. U.S. Department of Transportation, Washington, DC.

Federal Highway Administration (FHWA), 1988. *Recording and Coding Guide for the Structure Inventory and Appraisal of the Nation's Bridges*. U.S. Department of Transportation, Washington, DC.

Fitzpatrick, M., Law, D., and Dixon, W. 1981. Deterioration of New York State Highway Structures, Transportation Research Record 800. Transportation Research Board, National Research Council, Washington, DC.

Golabi, K., Thompson, P.D., and Jun, C.H. 1990. Network Optimization System for Bridge Improvements and Maintenance. Report to California Department of Transportation and FHWA, Cambridge Systematics/Optima.

Grant, E.L., Ireson, W.G., and Leavenworth, R.S. 1982. *Principles of Engineering Economy, 7th Ed.* Wiley, New York.

Green, S.G. and Richardson, J.A. 1994. Development of a Bridge Management System in Alabama, Characteristics of Bridge Management Systems, Transportation Research Circular 423. Transportation Research Board, National Research Council.

Hymon, W., Hughes, D., and Dobson, T. 1983. The Least Cost Mix of Bridge Replacement and Repair Work on Wisconsin's State Highways over Time — a Computer Simulation, Draft Technical Report. Wisconsin Department of Transportation, Madison, Wisconsin.

Jiang, Y. and Sinha, K.C. 1989. The Development of Optimal Strategies for Maintenance, Rehabilitation and Replacement of Highway Bridges, Final Report Vol. 6: Performance Analysis and Optimization, FHWA/IN/JHRP-89/13.

Johnston, D.W. and Lee, J.D. 1994. Analysis of Bridge Management Data in North Carolina, Characteristics of Bridge Management Systems, Transportation Research Circular 423. Transportation Research Board, National Research Council.

Johnston, D.W. and Zia, P. 1984. A Level of Service System for Bridge Evaluation. Center for Transportation Engineering Studies, North Carolina State University, Raleigh, NC.

Kansas Department of Transportation. 1984. Development of a Highway Improvement Priority System for Kansas. Division of Planning and Development, Office of Analysis and Evaluation.

Lemmerman, J.H. 1983. A Quick Benefit-Cost Procedure for Evaluating Proposed Highway Projects, Transportation Analysis Report 38. Planning Division, New York Department of Transportation, Albany, New York.

Linstone, H.A. and Turoff, M. 1975. *The Delphi Method: Techniques and Applications.* Addison-Wesley, Boston, MA.

Luenberger, D.G. 1965. *Introduction to Linear and Nonlinear Programming.* Addison-Wesley Publishing Company, Reading, MA.

McFarland, W.F., Griffin, L.I., Rollins, J.B., Stockton, W.R., Phillips, D.T., and Dudek, C.L. 1979. Assessment of Techniques for Cost-Effectiveness of Highway Accident Countermeasures, Final Report, FHWA-RD-53-79. Federal Highway Administration, U.S. Department of Transportation, Washington, DC.

McFarland, W.F., Rollins, J.B., and Harris, F. 1983. *Documentation for Integer Programming Technique.* Texas Transportation Institute, Texas A&M University System, College Station, TX.

McKeel, W.T., Jr. and Andrews, J.E. 1985. *Establishing a Priority of Funding for Deficient Bridges.* Virginia Department of Highways and Transportation, Richmond, VA.

National Engineering Technology Cooperation, 1994. *BRIDGIT Bridge Management System: Technical Manual — Version 1.00*, NCHRP Project 12-28(2)A. TRB, National Research Council, Washington, DC.

Neter, J., Wasserman, W., and Kutner, M.H. 1985. *Applied Linear Statistical Models, 2nd Ed.* Richard D. Irwin, Inc., Homewood, IL.

Oraves, J.D. 1994. PennDOT's Bridge Management Decision Support Process, Characteristics of Bridge Management Systems, Transportation Research Circular 423. Transportation Research Board, National Research Council.

Saaty, T. 1980. *The Analytic Hierarchy Process: Planning, Priority Setting, and Resource Allocation.* McGraw-Hill, Inc., New York.

Saito, M. and Sinha, K.C. 1989. The Development of Optimal Strategies for Maintenance, Rehabilitation and Replacement of Highway Bridges, Final Report Vol. 5: Priority Ranking Method, FHWA/IN/JHRP-89/12.

Sinha, K.C., Saito, M., Jiang, Y., Murthy, S., Tee, A., and Bowman, M.D. 1989. The Development of Optimal Strategies for Maintenance, Rehabilitation and Replacement of Highway Bridges, Final Report Vol. 1: The Elements of Indiana Bridge Management System (IBMS), FHWA/IN/JHRP-88/15.

Sinha, K.C., Zhang, Y., Singh, M., Kepaptsoglou, K., Haryopratomo, A., and Woods, R. 2002. *Instruction Guide for the Operation of the Indiana Bridge Management System.* Purdue University and Indiana Department of Transportation, West Lafayette, IN.

Subramanian, B., Schafer, D., and Tyler, J. 1983. *Documentation for Dynamic Programming Technique.* Texas Transportation Institute, Texas A&M University System, College Station, TX.

Turner, D.S. and Richardson, J.A. 1994. Bridge Management System Data Needs and Data Collection, Characteristics of Bridge Management Systems, Transportation Research Circular 423. Transportation Research Board, National Research Council, Washington, DC.

Winston, W.L. 1994. *Operations Research: Applications and Algorithms, 3rd Ed.* Duxbury Press, Belmont, CA.

Woods, R.E. 1994. Indiana's Approach to a Bridge Management System, Characteristics of Bridge Management Systems, Transportation Research Circular 423. Transportation Research Board, National Research Council.

Conversion Table for Metric and Imperial Units

Imperial to Metric	Metric to Imperial
1 in. = 2.540 cm = 25.40 mm	1 mm = 0.03937 in.
1 ft = 0.3048 m = 30.48 cm	1 m = 3.281 ft = 1.094 yd
1 yd = 3 ft = 0.9144 m	
1 mile = 5,280 ft = 1.609 km	1 km = 1,000 m = 0.6214 mile
1 in.^{2} = 645.2 mm^2	1 mm^2 = 0.00155 in.^{2}
1 ft^2 = 0.0929 m^2	1 m^2 = 10.76 ft^2
1 mile^2 = 640 acres = 2.589 km^2	1 km^2 = 100 hectare = 0.3861 mile^2
1 acre = 43,560 ft^2 = 0.4047 hectare	1 hectare = 10,000 m^2 = 0.01 km^2 = 2.471 acres
1 lb = 0.4536 kg = 4.448 N	1 kg = 9.807 N = 2.205 lb
	1 N = 0.1020 kg = 0.2248 lb
1 ton = 2,000 lb = 907.2 kg = 0.9072 tonne[a]	1 tonne = 1,000 kg = 9.807 kN = 1.102 ton[a]
1 psi = 1 lb/in^2 = 6.895 kPa = 0.006895 Mpa	1 MPa = 1 N/mm^2 = 1 MN/m^2 = 145.0 psi
1 psf = 1 lb/ft^2 = 0.006944 psi = 0.04788 kPa	1 kPa = 1 kN/m^2 = 0.1450 psi = 20.88 psf
1 pci = 1 lb/in^3 = 271.4 kN/m^3	1 kN/m^3 = 0.003684 pci = 6.366 pcf
1 pcf = 1 lb/ft^3 = 0.0005787 pci = 0.157 kN/m^3	1 tonne/m^3 = 1,000 kg/m^3 = 1 Mg/m^3 = 1 g/cm^3 = 62.4 lb/ft^{3b}
1 USA gallon = 1 imperial gallon = 3.785 litres	1 litre = 1,000 cm^3 = 0.2642 USA gallons = 0.2200 imperial gallon
Temperature n°F equals to {5(n − 32)/9}°C	Temperature n°C equals to (1.8n + 32)°F
	Acceleration of gravity g = 9.80665 m/s^2 = 32.17405 ft/s^2
	1 Hz = 1 cycle per second
	Base of natural algorithms e = 2.71828
	π = 3.141593
	1 radian = (180/π) degree = 57.296 degree

[a] The ton in this table refers to the imperial unit used in North America. Formerly in Britain, there are 2,240 lb (1,016 kg) in the British ton. To avoid confusion, the British ton is also called the **long ton** and the American ton is the **short ton**.

[b] At 4°C pure water has a specific gravity of 1, and a density of 1 g/cm^3, 1 g/mℓ, 1 kg/litre, 1000 kg/m^3, 1 tonne/m^3 or 62.4 lb/ft^3.

Index